XXXth International Colloquium on Group Theoretical Methods in Physics (ICGTMP) (Group30)

Journal of Physics: Conference Series Volume 597

Ghent, Belgium
14 – 18 July 2014

ISBN: 978-1-5108-0254-4
ISSN: 1742-6588

XXXth International Colloquium on Group Theoretical Methods in Physics (ICGTMP) (Group30)

Journal of Physics: Conference Series Volume 597

Ghent, Belgium
14 – 18 July 2014

ISBN: 978-1-5108-0254-4
ISSN: 1742-6588

IOP Institute of Physics

Printed from e-media with permission by:

Curran Associates, Inc.
57 Morehouse Lane
Red Hook, NY 12571

Some format issues inherent in the e-media version may also appear in this print version.

Printed by Curran Associates, Inc. (2015)

For permission requests, please contact the Institute of Physics at the address below.

Institute of Physics
Dirac House, Temple Back
Bristol BS1 6BE UK

Phone: 44 1 17 929 7481
Fax: 44 1 17 920 0979

techtracking@iop.org

Additional copies of this publication are available from:

Curran Associates, Inc.
57 Morehouse Lane
Red Hook, NY 12571 USA
Phone: 845-758-0400
Fax: 845-758-2634
Email: curran@proceedings.com
Web: www.proceedings.com

TABLE OF CONTENTS

PLENARY TALKS

CONTRIBUTED TALKS

Journal of Physics: Conference Series **597** (2015) 011001 doi:10.1088/1742-6596/597/1/011001

Preface

The XXXth International Colloquium on Group Theoretical Methods in Physics (ICGTMP), also known as the Group30 conference, took place in Ghent (Belgium) from Monday 14 to Friday 18 July 2014. The conference was organised by Ghent University (Department of Applied Mathematics, Computer Science and Statistics, and Department of Mathematical Analysis). The website http://www.group30.ugent.be is still available.

The ICGTMP is one of the traditional conference series covering the most important topics of symmetry which are relevant to the interplay of present-day mathematics and physics.

More than 40 years ago a group of enthusiasts, headed by H. Bacry of Marseille and A. Janner of Nijmegen, initiated a series of annual meetings with the aim to provide a common forum for scientists interested in group theoretical methods. At that time most of the participants belonged to two important communities: on the one hand solid state specialists, elementary particle theorists and phenomenologists, and on the other mathematicians eager to apply newly-discovered group and algebraic structures. The conference series has become a meeting point for scientists working at modelling physical phenomena through mathematical and numerical methods based on geometry and symmetry. It is considered as the oldest one among the conference series devoted to geometry and physics. It has been further broadened and diversified due to the successful applications of geometric and algebraic methods in life sciences and other areas.

The first four meetings took place alternatively in Marseille and Nijmegen. Soon after, the conference acquired an international standing, especially following the 1975 colloquium in Nijmegen and the 1976 colloquium in Montreal. Since then it has been organized in many places around the world. It has become a bi-annual colloquium since 1990, the year it was organized in Moscow. This was the first time the colloquium took place in Belgium.

There were 246 registered participants at the Group30 colloquium from 46 different countries. This high number of participants makes the 30th edition of ICGTMP one of the most successful meetings of the series. We were particularly happy with the attendance of 32 students and many young postdocs: this is promising for the future of the research field and of the colloquium series. The attraction of young scientists was promoted by a reduced participation fee, the availability of cheap accommodation in the University Student Guest House, grants from a US-NSF project and grants from the Organising Committee supporting participants from countries with limited financial resources.

The colloquium brought together international top researchers, with a variety of different backgrounds but with "group theoretical or algebraic methods" as common ground. The selection of plenary speakers and topics was made by the Organizing Committee, following the suggestions of the International Advisory Committee (consisting of prominent scientists in their domain), and ensuring the diversity of the subjects treated. There were 11 plenary talks, given as well by distinguished world experts as by young rising stars: Matthias Christandl, Alberto De Sole, Rui Loja Fernandes, François Gay-Balmaz, Gitta Kutyniok, Amiram Leviatan, Karl-Hermann Neeb, Christoph Schweigert, Yuji Tachikawa, Luc Vinet and Joshua Zak. Next to the plenary talks, there were 170 talks in six parallel sessions and 10 poster presentations.

All scientific activities took place in the Joseph Plateau Building of the Faculty of Engineering and Architecture of Ghent University, a historical building conveniently located in the city centre of Ghent. In this building we could make use of the main lecture hall (400 seats) for the plenary lectures, six lecture rooms for parallel sessions, two offices for organizational activities and registrations, two PC-rooms and a large foyer for coffee breaks, book exhibitions and the poster session.

As part of the colloquium, a panel discussion on recent advances and future challenges in group theoretical methods in physics was held on Wednesday morning. The panel members were: Maia Angelova, Heinz-Dietrich Doebner, Jean-Pierre Gazeau, Norbert Hounkonnou, Peter Olver and Luc Vinet, with Willard Miller Jr. as moderator. In this discussion, the importance of "symmetry" in physics (and other sciences) was emphasized, and some challenges were made explicit.

On Wednesday evening there was a separate ceremony in the Aula Academica (the "University Palace") of Ghent University, for the presentation of the 2014 Wigner Medal and Hermann Weyl Prize. The Hermann Weyl Prize was established in 2000 to provide recognition for young scientists having performed original work of significant scientific quality in the area of the understanding of physics through symmetries. The 2014 Hermann Weyl Prize was awarded to Yuji Tachikawa (Tokyo University). He is praised for his outstanding contributions to the understanding of supersymmetric quantum field theories and in particular, to the discovery of the Alday-Gaiotto-Tachikawa correspondence that has led to spectacular advances in both mathematics and quantum physics. The Wigner Medal, established in 1978, is an award designed "to recognize outstanding contributions to the understanding of physics through Group Theory", and can be considered as a "lifetime achievement award". It was awarded to Joshua Zak (Technion, Haifa, Israel) for his discovery of the magnetic translation group and the phase and transform that bear his name. In his work, the relationship between mathematics and physics could hardly be closer, with mathematicians being familiar with the "Zak transform" and physicists with the "Zak phase", both concepts originating in group theory. The ceremony was graced by a saxophone quintet, "Sax Obsession", and followed by a reception in the adjacent Peristylium.

The social programme included a variety of events. On Monday evening, the welcome address by Christophe Peeters, Deputy Mayor of the City of Ghent, was followed by a reception in the historical "Pacificatiezaal" of the City Hall. On Wednesday afternoon the participants had the opportunity to take a guided tour through medieval Ghent, admiring its wide range of monuments. The tour was followed by a much-appreciated boat trip exploring the canals and rivers of Ghent, all in sunny weather. On Thursday evening the conference banquet was held in the "Brasserie HA". Located in the Handelsbeurs Ghent. In the majestic banquet hall, the participants were not only treated to an exclusive dinner, but also to some fine piano music by Dmitry Gal'tsov and Richard Kerner.

The conference was sponsored by: American Institute of Physics (AIP Publishing), Clay Mathematics Institute, Clifford Research Group - Faculty of Engineering and Architecture (Ghent Unviversity), Elsevier, Faculty of Sciences (Ghent University), Fonds de la Recherche Scientifique (FNRS), Foundation Compositio Mathematica, FWO Research Foundation - Flanders, International Association of Mathematical Physics, International Solvay Institutes, International Union of Pure and Applied Physics, National Science Foundation (USA) and Springer Birkhäuser. We would like to thank all our sponsors for their generous support.

It took more than two years to organise a conference of such a size and importance. We express our gratitude to the International Advisory Committee for its help in selecting the plenary speakers and to the external members of the Organising Committee for their helpful suggestions and advice. We thank Wouter Dewolf for his devoted secretarial, administrative and organisational work and Vera Vanden Driessche for arranging the "Accompanying persons' programme". Finally, a big 'thank you' to all the local people (administrative and technical staff, research assistants and research students) for their devoted and tireless work in preparing and running the conference.

Fred Brackx, Hennie De Schepper and Joris Van der Jeugt

Journal of Physics: Conference Series **597** (2015) 011002 doi:10.1088/1742-6596/597/1/011002

Committees and List of participants

1. Sponsors

- International Union of Pure and Applied Physics (IUPAP)
- International Solvay Institutes
- International Association of Mathematical Physics (IAMP)
- Clay Mathematics Institute (CMI)
- Elsevier
- Springer Birkhäuser
- American Institute of Physics (AIP Publishing)
- Foundation Compositio Mathematica
- National Science Foundation US
- The Research Foundation - Flanders (FWO)
- Le Fonds de la Recherche Scientifique (FNRS)
- Ghent University, Faculty of Sciences
- Ghent University, Clifford Research Group, Faculty of Engineering and Architecture

2. Standing Committee for the ICGTMP-series (2014)

J.P. Gazeau (Chairman, Paris Diderot University)
S.T. Ali (Concordia University, Montreal)
L. Boya (Salamanca University)

C. Burdik (Czech Technical University Prague)
E. Celeghini (Firenze University, INFN)
M. del Olmo (Valladolid University)
V. Dobrev (Bulgarian Academy of Sciences)
M. Ge (Nankai University)
G. Goldin (Rutgers University)
F. Iachello (Yale University)
R. Kerner (P&M Curie University, Paris)
P. Kielanowski (Cinvestav, Mexico)
A. Kostelecky (Indiana University)
G. Pogosyan (UNAM, Mexico; JINR, Dubna)
C. Schweigert (Hamburg University)
F. Toppan (CBPF, Rio de Janeiro)
R. Twarock (York University)
J. Van der Jeugt (Ghent University)
P. Van Isacker (Ganil, Caen)
L. Vinet (Montreal University)
A. Vourdas (Bradford University)
K. Wendland (Augsburg University)
K.B. Wolf (UNAM, Cuernavaca)
H.D. Doebner (Honorary Member, Clausthal University)

3. International Advisory Committee

E. Corrigan	York, UK
M. de León	Madrid, Spain
H.-D. Doebner	Clausthal, Germany
J. Draayer	Lousiana State University
L. Faddeev	St. Petersburg, Russia
J. Fuchs	Karlstad, Sweden
J.N. Ginocchio	Los Alamos National Laboratory, USA
M.E.H. Ismail	Florida, USA
M. Jimbo	Kyoto, Japan
V.G. Kac	MIT, Cambridge, USA
T. Koornwinder	Amsterdam, Netherlands
Y.I. Manin	MPIM, Bonn, Germany
G. Marmo	Napels, Italy
W. Miller	Minnesota, USA
P.J. Olver	Minnesota, USA
T. Ratiu	EPFL, Lausanne, Switzerland
I.T. Todorov	Sofia, Bulgaria
C. Vafa	Harvard, Cambridge, USA

4. Organising Committee

Jean-Pierre Antoine	UCL, Louvain-La-Neuve University, Belgium
Francoise Bastin	Mathematics Department, Liège University, Belgium
Pierre Bieliavsky	Mathematics Department, UCL, Louvain-La-Neuve, Belgium
Fred Brackx	Ghent University, Belgium
Stefaan Caenepeel	VUB Brussels, Belgium
Frans Cantrijn	Ghent University, Belgium

Journal of Physics: Conference Series **597** (2015) 011002 doi:10.1088/1742-6596/597/1/011002

Hennie De Schepper	Ghent University, Belgium
David Eelbode	Antwerp University, Belgium
Simone Gutt	Mathematics Department, ULB Brussels, Belgium
Marc Henneaux	Physics Department, ULB Brussels, Belgium
Erik Koelink	Mathematics Department, Nijmegen, Netherlands
Joris Van der Jeugt	Ghent University, Belgium
Piet Van Isacker	GANIL, Caen, France
Pierre Van Moerbeke	Mathematics Department, UCL, Louvain-La-Neuve, Belgium

5. List of participants

Firstname LASTNAME	Affiliation	Country
Ana AGORE	Vrije Universiteit Brussel	Belgium
Naruhiko AIZAWA	Osaka Prefecture University	Japan
Noud ALDENHOVEN	Radboud Universiteit Nijmegen	Netherlands
Alexander ALLDRIDGE	University of Cologne	Germany
Amedeo ALTAVILLA	University of Trento	Italy
Maia ANGELOVA	Northumbria University	United Kingdom
Jean-Pierre ANTOINE	Université Catholique de Louvain	Belgium
Natig ATAKISHIYEV	Instituto de Matematicas, UNAM, Unidad Cuernavaca	Mexico
Max ATKIN	Université Catholique de Louvain	Belgium
Chengming BAI	Chern Institute of Mathematics	China
Sigiswald BARBIER	Ghent University	Belgium
Glenn BARNICH	Université Libre de Bruxelles & International Solvay Institutes	Belgium
Ehud Moshe BARUCH	Technion	Israel
Tathagata BASAK	Iowa State University	United States
Françoise BASTIN	University of Liège	Belgium
Bruno BAUWENS	Ghent University	Belgium
Joe BENSON	University of Minnesota - Twin Citites	United States
Stefan BERCEANU	"Horia Hulubei" National Institute of Physics and Nuclear Engineering	Romania
Swanhild BERNSTEIN	TU Bergakademie Freiberg, Institute of Applied Analysis	Germany
Cinzia BISI	Department of Mathematics and Computer Science, Ferrara University	Italy
Alfonso BLASCO-SANZ	Universidad de Burgos (Burgos University)	Spain
Arno BOHM	University of Texas at Austin	United States
Laurence BOYLE	Vine Cottage	United Kingdom
Fred BRACKX	Ghent University	Belgium
Hai BUI	University of Texas at Austin	United States
Stefaan CAENEPEEL	Vrije Universiteit Brussel	Belgium
Frans CANTRIJN	Ghent University	Belgium
Giovanni CASSINELLI	Physics Department - University of Genoa	Italy
Enrico CELEGHINI	University of Florence	Italy
Paula CEREJEIRAS	CIDMA - Universidade de Aveiro	Portugal
Oleg CHALYKH	University of Leeds	United Kingdom
Claudia CHANU	University of Turin	Italy
Matthias CHRISTANDL	University of Copenhagen	Denmark

Journal of Physics: Conference Series **597** (2015) 011002 doi:10.1088/1742-6596/597/1/011002

Firstname LASTNAME	Affiliation	Country
Tom CLAEYS	Université Catholique de Louvain	Belgium
Lander CNUDDE	Ghent University	Belgium
Fabrizio COLOMBO	Dipartimento di Matematica, Politecnico di Milano	Italy
Kevin COULEMBIER	Ghent University	Belgium
Rolf DAHM	Beratung für IS, Mainz	Germany
Wen-Rong DAI	Shanghai Normal University	China
Stijn DE BAERDEMACKER	Ghent University	Belgium
Hendrik DE BIE	Ghent University	Belgium
Kenny DE COMMER	Vrije Universiteit Brussel	Belgium
Axel DE GOURSAC	FNRS, Université Catholique de Louvain	Belgium
Hubert DE GUISE	Lakehead University	Canada
Manuel DE LEÓN	Instituto de Ciencias Matemáticas	Spain
Hilde DE RIDDER	Ghent University	Belgium
Nele DE SCHEPPER	Antwerp University	Belgium
Hennie DE SCHEPPER	Ghent University	Belgium
Alberto DE SOLE	University of Rome La Sapienza	Italy
Alexis DE VOS	Ghent University	Belgium
Pierre-Philippe DECHANT	Durham University	United Kingdom
Mariano A. DEL OLMO	Universidad de Valladolid	Spain
Laurent DELISLE	Université de Montréal	Canada
Maksym DEREVYAGIN	KU Leuven	Belgium
Wouter DEWOLF	Ghent University	Belgium
Erik DIAZ-BAUTISTA	Centro de Investigacion y de Estudios Avanzados del IPN	Mexico
Yassir DINAR	Sultan Qaboos University	Oman
Vladimir DOBREV	Bulgarian Academy of Sciences, Institute for Nuclear Research and Nuclear Energy, Sofia	Bulgaria
Heinz-Dietrich DOEBNER	Technical University of Clausthal	Germany
Andrew DOUGLAS	City University of New York	United States
David EELBODE	Antwerp University	Belgium
Alexander ELASHVILI	Razmadze Mathematical Institute	Georgia
Sirkka-Liisa ERIKSSON	Tampere University of Technology, Department of Mathematics	Finland
Thomas ERNST	Department of Mathematics, Uppsala University	Sweden
Charalampos EVRIPIDOU	University of Cyprus	Cyprus
Nelson FAUSTINO	University of Campinas (UNICAMP)	Brazil
Laszlo FEHER	Department of Theoretical Physics , University of Szeged and Wigner RCP, Budapest	Hungary
Bojana FEMIC	Universidad de la República	Uruguay
Thomas FERGUSON	Southwestern Assemblies of God University	United States
Rui Loja FERNANDES	University of Illinois at Urbana-Champaign	United States
David J. FERNÁNDEZ	Physics Department, Cinvestav	Mexico
Galina FILIPUK	University of Warsaw	Poland
Edward FRENKEL	University of California, Berkeley	United States
Rodrigo FRESNEDA	UFABC - Universidade do ABC	Brazil

Journal of Physics: Conference Series **597** (2015) 011002 doi:10.1088/1742-6596/597/1/011002

Firstname LASTNAME	Affiliation	Country
Tamar FRIEDMANN	University of Rochester	United States
Bahruz GADJIEV	International University for Nature, Society and Man	Russian Federation
Giuseppe GAETA	Università di Milano	Italy
Dmitry GAL'TSOV	Moscow State university	Russian Federation
Eduardo GARCÍA-TORAÑO ANDRÉS	Ghent University	Belgium
François GAY-BALMAZ	Ecole Normale Superieure de Paris	France
Jean Pierre GAZEAU	University Paris Diderot	France
Vincent GENEST	Centre de Recherches Mathematiques, Universite de Montreal	Canada
Graziano GENTILI	Università di Firenze	Italy
Vladimir GERDT	Joint Institute for Nuclear Research and Dubna International University	Russian Federation
Gerald A. GOLDIN	Rutgers University	United States
Jose Francisco GOMES	Instituto de Fisica Teorica IFT-Unesp	Brazil
Karan GOVIL	The Pennsylvania State University	United States
Nikolai GROMOV	Komi Science Centre RAS	Russian Federation
Hendrik GRUNDLING	School of Mathematics, University of New South Wales	Australia
Simone GUTT	Université Libre de Bruxelles	Belgium
Christian HAGENDORF	Université Catholique de Louvain	Belgium
Lenka HÁKOVÁ	Czech Technical University in Prague	Czech Republic
Masashi HAMANAKA	Nagoya University	Japan
John HARNAD	Centre de recherches mathématiques and Concordia University	Canada
Jérémy HAUT	Université Catholique de Louvain	Belgium
Inge Svein HELLAND	Department of Mathematics, University of Oslo	Norway
Thomas HEMPFLING	Springer Basel AG / Birkhäuser	Switzerland
Willy HEREMAN	Colorado School of Mines	United States
Alberto HERNÁNDEZ	University of Porto	Portugal
Francisco J. HERRANZ	Universidad de Burgos	Spain
Eckhard HITZER	International Christian University	Japan
Ladislav HLAVATY	Czech Technical University in Prague	Czech Republic
Mahouton Norbert HOUNKONNOU	University of Abomey-Calavi	Benin
Jiri HRIVNAK	Czech Technical University in Prague	Czech Republic
Véronique HUSSIN	Université de Montréal	Canada
Ivan IP	Kavli IPMU, University of Tokyo	Japan
Phillip ISAAC	The University of Queensland	Australia
Yakov ITIN	Hebrew University and JCT	Israel
Elchin JAFAROV	Institute of Physics, Azerbaijan National Academy of Sciences	Azerbaijan
Alban JAGO	Université Catholique de Louvain-la-Neuve	Belgium
Jonas JOOS	Ghent University	Belgium
Madeleine JOTZ LEAN	University of Sheffield	United Kingdom
Uwe KAEHLER	CIDMA - Universidade de Aveiro	Portugal
Dalibor KARÁSEK	Czech Technical University in Prague	Czech Republic

Journal of Physics: Conference Series **597** (2015) 011002 doi:10.1088/1742-6596/597/1/011002

Firstname LASTNAME	Affiliation	Country
Birgit KAUFMANN	Purdue University	United States
Richard KERNER	Université Pierre et Marie Curie (Paris-VI University)	France
Erik KOELINK	IMAPP, Radboud Universiteit Nijmegen	Netherlands
Tom KOORNWINDER	University of Amsterdam	Netherlands
Venkata Krishna Brahmam KOTA	Physical Research Laboratory, Ahmedabad	India
Arno KUIJLAARS	KU Leuven	Belgium
Atsuo KUNIBA	University of Tokyo	Japan
Seckin KURKCUOGLU	Middle East Technical University (METU)	Turkey
Sengul KURU	Ankara University	Turkey
Gitta KUTYNIOK	Technische Universität Berlin	Germany
Seiichi KUWATA	Hiroshima City University	Japan
Zhanna KUZNETSOVA	Federal University of ABC	Brazil
Pekka LAHTI	University of Turku, Department of Physics and Astronomy	Finland
Kristina LAUNEY	Louisiana State University	United States
Roman LAVICKA	Charles University in Prague, Faculty of Mathematics and Physics	Czech Republic
Ci LEI	University of Bradford	United Kingdom
Amiram LEVIATAN	The Hebrew University	Israel
Eugenio LEY KOO	Instituto de Física, UNAM	Mexico
Libin LI	Yangzhou University	China
Matvei LIBINE	Indiana University, Bloomington	United States
Karl LIECHTY	DePaul University	United States
Jon LINKS	The University of Queensland	Australia
Stephen LOW	Austin, Texas	United States
Inna LUKYANENKO	School of Mathematics and Physics, The University of Queensland	Australia
Ady MANN	Technion - Israel Institute of Technology	Israel
Ian MARQUETTE	The University of Queensland	Australia
Eduardo MARTÍNEZ	University of Zaragoza	Spain
Michal MARVAN	Mathematical Institute in Opava	Czech Republic
Tom MESTDAG	Ghent University	Belgium
Jouko MICKELSSON	University of Helsinki	Finland
Willard MILLER JR	University of Minnesota, School of Mathematics	United States
Michael MONASTYRSKY	Institute of Theoretical and Experimental Physics	Russian Federation
Vicente Said MORALES SALGADO	Centro de Investigación y de Estudios Avanzados del Instituto Politécnico Nacional (Cinvestav-IPN)	Mexico
Alexi MORIN-DUCHESNE	Université Catholique de Louvain	Belgium
Ali MOSTAFAZADEH	Koc University	Turkey
Zouhair MOUAYN	Sultan Moulay Slimane University, Faculty of Sciences & Technics	Morocco
Patrick MOYLAN	The Pennsylvania State University, Abington College	United States
Jun MURAKAMI	Department of Mathematics, Waseda University	Japan

Firstname LASTNAME	Affiliation	Country
Romain MURENZI	World Academy of Sciences (TWAS) for the advancement of science in developing countries	Italy
Ian MUSSON	UW-Milwaukee	United States
Laura NASTASESCU	University of Bucharest	Romania
Karl-Hermann NEEB	Friedrich-Alexander University Erlangen-Nuremberg	Germany
Javier NEGRO	Universidad de Valladolid	Spain
Giuseppe NISTICO	Universita della Calabria	Italy
Yousuke OHYAMA	Department of Mathematics, Osaka University	Japan
Peter OLVER	University of Minnesota	United States
Roy OSTE	Ghent University	Belgium
Dominik OSTERMAYR	University of Cologne	Germany
Georgios PAPAMIKOS	University of Kent	United Kingdom
Kishor Chandra PATI	N.I.T. Rourkela,Odisha,India	India
Andrei PATRASCU	University College London	United Kingdom
Melissa PATTERSON	AIP Publishing	United States
Yufeng PEI	Shanghai Normal University	China
Hilde PELEMAN	Ghent University	Belgium
Karol PENSON	Université Paris VI, France	France
Alessandro PEROTTI	University of Trento	Italy
George POGOSYAN	CUCEI, Universidad de Guadalajara and Physics Department, Yerevan State University	Armenia
Elena POLETAEVA	University of Texas-Pan American	United States
Sarah POST	U. Hawaii	United States
Christiane QUESNE	Université Libre de Bruxelles	Belgium
Fabian RADOUX	University of Liège	Belgium
Tim RAEYMAEKERS	Ghent University	Belgium
Julien RANDON-FURLING	SAMM - Université Paris-1 Panthéon-Sorbonne	France
Jorgen RASMUSSEN	University of Queensland	Australia
Giovanni RASTELLI	University of Turin	Italy
Tudor RATIU	Ecole Polytechnique Federale de Lausanne	Switzerland
Michel RAUSCH DE TRAUBENBERG	Institut Pluridisciplinaire Hubert Curien/Strasbourg University	France
Christopher RAYMOND	University of Queensland	Australia
Felix REHREN	University of Birmingham	United Kingdom
Trevor REMPEL	Perimeter Institute for Theoretical Physics	Canada
Regina RICOTTA	Faculdade de Tecnologia de São Paulo, FATEC-SP	Brazil
David RIDOUT	Australian National University	Australia
Matthias ROELS	University of Antwerp	Belgium
Irene SABADINI	Dipartimento di Matematica, Politecnico di Milano	Italy
Yvan SAINT-AUBIN	Université de Montréal	Canada
Masoumeh SAJEDI	Université de Montréal, Département de mathématiques et de statistique	Canada
Norisuke SAKAI	Keio University	Japan
Pavle SAKSIDA	University of Ljubljana	Slovenia
Giulia SARFATTI	University of Bologna	Italy

Firstname LASTNAME	Affiliation	Country
Lawrence S SCHULMAN	Clarkson University	United States
Christoph SCHWEIGERT	Algebra und Zahlentheorie, Fachbereich Mathematik, Universität Hamburg	Germany
Ulrich SEYFARTH	Max Planck Institute for the Science of Light	Germany
Guilherme SILVA	KU Leuven	Belgium
Dalibor SMID	Charles University in Prague	Czech Republic
Walter SMILGA	No institution (retired)	Germany
Libor SNOBL	Czech Technical University in Prague, Faculty of Nuclear Sciences and Physical Engineering	Czech Republic
Frank SOMMEN	Ghent University	Belgium
Stephen SONTZ	Centro de Investigacion en Matematicas, CIMAT	Mexico
Vladimir SOUCEK	Mathematical Institute, Charles University	Czech Republic
Irina STEPANOVA	Institute of Computational Modeling of Sibirian Branch of Russian Academy of Science	Russian Federation
Daniel STERNHEIMER	Rikkyo University & Université de Bourgogne	France
Dries STIVIGNY	KU Leuven	Belgium
Caterina STOPPATO	Istituto Nazionale di Alta Matematica - Unità di Ricerca di Firenze	Italy
Aleksander STRASBURGER	Warsaw University of Life Sciences - SGGW	Poland
Yossi STRAUSS	Ben-Gurion University of the Negev	Israel
Fumihiko SUGINO	Okayama Institute for Quantum Physics	Japan
Osamu SUZUKI	Department of Information Science, Nihon University	Japan
Franciszek SZAFRANIEC	Uniwersytet Jagiellonski	Poland
Yuji TACHIKAWA	University of Tokyo	Japan
El Hassan TAHRI	Mohamed First University	Morocco
Daisuke TARAMA	Department of Mathematics, Kyoto University	Japan
Piergiulio TEMPESTA	Universidad Complutense de Madrid and Instituto de Ciencias Matemáticas	Spain
Valerie TENG-BROUG	Elsevier	Netherlands
Lee Peng TEO	University of Nottingham Malaysia Campus	Malaysia
Jiri TOLAR	Czech Technical University in Prague	Czech Republic
Francesco TOPPAN	CBPF, Rio de Janeiro	Brazil
Mihaela B VAJIAC	Chapman University	United States
Daniele VALERI	SISSA, Trieste	Italy
Francis VALIQUETTE	State University of New York at New Paltz	United States
Joris VAN DER JEUGT	Ghent University	Belgium
Jan-Willem VAN HOLTEN	Nikhef and Leiden University	Netherlands
Piet VAN ISACKER	GANIL, Caen	France
Peter VAN LANCKER	Clifford research group, Ghent University	Belgium
Mario VAN RAEMDONCK	Ghent University	Belgium
Vera VANDEN DRIESSCHE	Ghent University	Belgium
Luc VINET	CRM, Université de Montréal	Canada
Bart VLAAR	University of Amsterdam	Netherlands

Journal of Physics: Conference Series **597** (2015) 011002 doi:10.1088/1742-6596/597/1/011002

Firstname LASTNAME	Affiliation	Country
Apostolos VOURDAS	University of Bradford	United Kingdom
Goedele WAEYAERT	Ghent University	Belgium
Dong WANG	National University of Singapore	Singapore
Jing Ping WANG	University of Kent	United Kingdom
Wei WANG	Southeast University/Vrije Universiteit Brussel	Belgium
Katrin WENDLAND	Freiburg University	Germany
Kurt Bernardo WOLF	Instituto de Ciencias Físicas, Universidad Nacional Autónoma de México	Mexico
Michael WUTZIG	Ghent University	Belgium
Lode WYLLEMAN	Ghent University	Belgium
Joshua ZAK	Technion	Israel
Anton ZEITLIN	Columbia University / IHES	France
Yinhuo ZHANG	Universiteit Hasselt	Belgium
J. ZHENG-JOHANSSON	Inst. Fundamental Physics Research	Sweden
Yongchang ZHU	Hong Kong University of Science and Techology	Hong Kong
Federico ZULLO	Roma Tre University	Italy

Journal of Physics: Conference Series **597** (2015) 011003 doi:10.1088/1742-6596/597/1/011003

Alexander Isaac Solomon – In Memoriam

R Kerner

Université Pierre et Marie Curie (Paris-VI University), 4 Place Jussieu, 75005 Paris, France

E-mail: richard.kerner@upmc.fr

Allan Solomon: Glasgow, 11.12.1936 – Watford, 3.4.2013

Our colleague and friend Allan Solomon passed away in his home in Watford, England, UK, on April 3, 2013, after a long and brave fight with a terrible cancer.

Allan was one of the most brilliant and active members of the Standing Committee of the International Colloquia on Group-Theoretical Methods in Physics (ICGTMP) until 2009, and Honorary Member until his last days. He was also a permanent member of the Conference Board of the International Symposia on Quantum Theory and Symmetries (QTS).

Allan's life was successful and exemplary in many ways:

- He had a beautiful family and was a beloved husband, father and grandfather;
- As a scientist, he was successful as an outstanding researcher in both applied and theoretical physics;
- He displayed his pedagogical skills in a novel form of popularising science on radio and television;
- He served as President of the Watford & District Synagogue and was a great friend and defender of Israel;
- He was a staunch promoter of academic freedom, international collaboration and tolerance.

Allan's scientific activity, rich and original, can be characterized as a deep exploration of the physical consequences of various symmetries in physics, especially in quantum phenomena. Besides the ICGTMP colloquia, Allan participated many times in the conference series "Symmetries in Science" and in other workshops and colloquia.

Alexander Isaac Solomon was born in Glasgow on December 11th, 1936. Soon after, his family moved to Belfast where his father worked with Allan's uncle in the family music business. Allan attended primary school in Belfast. Later on, Allan's family moved to Dublin where Allan continued his education at Wesley College and won a Foundation Scholarship in 1956 to fund his studies at university. He studied science at Trinity College, Dublin, and later on in Cambridge, England.

For his postgraduate studies Allan went to Peterhouse College, Cambridge, where he worked on elementary particles and then to Orsay, France, where he obtained a doctorate in Theoretical Physics in 1963. The same year he married Paulette Baroukh, who he had met in Paris while studying at the Sorbonne.

It was in Paris that Allan began his lifetime love of all things French and especially a certain appreciation of *joie de vivre*.

Paulette was always a great support for Allan in both good times and bad and was fascinated by her husband's academic work. She accompanied him all over the world on his scientific trips, while continuing her own career with people affected by learning difficulties. She passed away in June 2014, less than two years after Allan's death.

Allan and his wife Paulette with their daughters Annabel (centre) and Nathalie (right) in 1997

In 1963 when Americans were coming to Europe and causing a true brain-drain by hiring young graduates in scientific and medical fields, Allan was at the Dublin Institute for Advanced Studies and was offered a job in the United States to work in the "Republic Aviation" industry. He also worked for N.A.S.A. studying cosmic radiation and its effects on supersonic travel.

Although Allan very much enjoyed being in New York and for his area of research he pined for academia and accepted a considerably lower-paid job teaching at New York Polytechnic University in Brooklyn. Academic research was confirmed as Allan's passion and when Yuval Ne'eman invited him to work at the recently-established Tel Aviv University, he gladly accepted despite a move to the Levant and a further drop in salary. He and Paulette then returned to the Dublin Institute for Advanced Studies with their baby daughter, Nathalie. There he spent two fruitful years working alongside and forging life-long friendships with Professor J.L. Synge and Professor C. Lanczos.

Journal of Physics: Conference Series **597** (2015) 011003 doi:10.1088/1742-6596/597/1/011003

Allan and Paulette, 48 years after their first meeting in Paris

His next job was in England, at the newly-established Open University and it was there where Allan found his *raison d'être.*

He was a joyful character and liked to report some of his travels and conferences with humour when he described them in *The Times Higher Education Supplement* under the column known as "Don's Diary".

Allan's position at the Open University gave him plenty of free time and during his life he was able to pursue research in collaboration with colleagues in Italy, Israel, China, Singapore and France as well as giving lectures in many other places. His scientific interests concentrated around the influence of symmetries on various phenomena in quantum physics, including magnetism, superconductivity, ferroelectricity and, later on, coherent and entangled states, quantum information and cryptography. In the eighties he collaborated with Jacob Katriel and Mario Rasetti, and above all with Joseph Birman, with whom he co-authored 15 publications, including their *magnum opus* on an SU(8) grand unification of superconductivity and various charge and spin density waves.

After writing numerous textbooks for the Open University, Allan went on to make over 150 radio and television broadcasts with the BBC, lecturing to Open University students around the world.

Allan thoroughly enjoyed making television programmes even if his long hair, flared trousers and horn-rimmed square glasses inspired much laughter in Irish and British comic circles.

In the late seventies Allan joined an International group of scientists, headed by Joe Birman and several other eminent Western scholars, and travelled to the former USSR to visit Soviet scientists who were *"refuseniks"* who had been kicked out of their universities. Allan and other visiting scientists gave them lectures in their homes, often huddled in small kitchens, to keep their Soviet counterparts up-to-date with the latest research in their fields.

Starting from the mid-nineties, after his retirement from the Open University, Allan developed an intense collaboration with the Parisian group headed by Karol Penson, working on coherent states and combinatorics. His co-authors of this period included Sonia Schirmer, Apostol Vourdas, Hongchen Fu, Gérard Duchamp, Pawel Blasiak and Kasia Górska.

During all this time, Allan was greatly contributing to the activities of the ICGTMP Standing

Committee, bringing his wise and considered views and opinions, combining scientific honesty with both tolerance and humanity.

Allan and Paulette came to cherish Allan's colleagues as friends, sharing many good times with all of us. He always loved meeting us and loved visiting our homes and meeting our families. He bonded with all of us in different ways and accompanied us through different stages over 40 years.

In early 2011 he underwent a major operation which appeared to have been successful and he attended the 7th QTS Symposium in Prague in August 2011. Shortly afterwards, however, his health deteriorated rapidly. He fought the terrible disease very bravely and Paulette tried all possible avenues to find an efficient treatment. Unfortunately, however, this was to no avail. Allan passed away on April 3, 2013.

He will be missed and remembered by his family and friends and by all colleagues who had the privilege to meet him and to appreciate his deep knowledge, his openness, his inimitable affability and his wonderful sense of humour.

Acknowledgments
The author expresses his gratitude to Nathalie and Annabel Solomon for their help during the preparation of the manuscipt. Thanks are also due to Laurence Boyle and Joris Van der Jeugt for the careful reading and improving the quality of this text.

The Wigner Medal 2014

J Avron[1], A Bohm[2] and L S Schulman[3]

[1] Department of Physics, Technion – Israel Institute of Technology, Haifa 32000, Israel
[2] Department of Physics, The University of Texas at Austin, Austin, TX 78712-1081, USA
[3] Physics Department, Clarkson University, Potsdam, NY 13699-5820, USA

E-mail: avron@physics.technion.ac.il, bohm@physics.utexas.edu, schulman@clarkson.edu

1. Introduction

The Wigner Medal was established in 1977/78 and is administered by the Group Theory and Fundamental Physics Foundation. The first Wigner Medal was awarded to Eugene P. Wigner and Valentine Bargmann at the Integrative Conference on Group Theory and Mathematical Physics / VIIth International Group Theory Colloquium in 1978 in Austin, Texas, U.S.A.

The purpose of the Wigner Medal is to recognize outstanding contributions to the understanding of physics through group theory.

The Medal is in principle awarded every two years and the awarding ceremony is held at the colloquium on Group Theoretical Methods in Physics. The previous recipients of the Wigner Medal are:

1978	E.P. Wigner	1996	V. Kac and R.V. Moody
1978	V. Bargmann	1998	M. Moshinsky
1980	I.M. Gel'fand	2000	L. O'Raifeartaigh
1982	L. Michel	2002	H.J. Lipkin
1984	Y. Ne'eman	2004	E. Inönü
1986	F. Gürsey	2006	S. Okubo
1988	I.M. Singer	2010	M. Jimbo
1990	F. Iachello	2012	C.A. Mead
1992	J. Wess and B. Zumino		

The winner of the Wigner Medal is chosen by the Selection Committee which consists of five members. The chairman of the Group Theory and Fundamental Physics Foundation is the member ex officio. The remaining members are assigned for a four year term: two by the Standing Committee of the Group Theory Colloquia and two by the Board of the Group Theory and Fundamental Physics Foundation.

The 2014 International Selection Committee consisted of: Alden Mead (Chairman), Arno R. Bohm, Mo-Lin Ge, Michio Jimbo and Lawrence S. Schulman.

The 2014 Wigner Medal was awarded to:

Joshua Zak

Technion, Haifa, Israel

FOR HIS DISCOVERY OF THE MAGNETIC TRANSLATION GROUP AND THE PHASE AND TRANSFORM THAT BEAR HIS NAME

Journal of Physics: Conference Series **597** (2015) 011004 doi:10.1088/1742-6596/597/1/011004

The ceremony for the Wigner Medal took place on July 16, 2014 in the Aula Academica of Ghent University, a historical building dating from 1826. The ceremony was graced by the saxophone quintet, "Sax Obsession", playing music from Gershwin, Shostakovich and more. The participants were welcomed by Debora Van Durme, master of ceremonies, who welcomed the awardees, committee members and the invitees. She gave a brief presentation on the history of the university and the Aula building, and on Adolphe Sax and the instruments that bear his name. The Hermann Weyl Prize was also presented during this ceremony.

The Wigner Medal was presented by Arno Bohm (University of Texas at Austin) after a short picture show of the history of the Medal. The laudatio for Professor Joshua Zak was delivered by Lawrence Schulman (Clarkson University). The full text of the laudatio is given in the following pages.

The ceremony was concluded by a reception with drinks and cocktail snacks in the adjacent Peristylium.

In figure 1 the front and back face of the 2014 Wigner Medal are shown. Figure 2 shows some pictures taken at the ceremony.

Figure 1. The 2014 Wigner Medal

2. Laudatio for Professor Joshua Zak

It is a pleasure and an honor to deliver the laudatio for Professor Joshua Zak, the 2014 Wigner medal Laureate.

As I will recount below, besides Joshua's scientific achievements, there has been a thread through much of his life connecting him to the person implicitly also honored at this ceremony, Eugene Wigner. But before I get serious let me mention one tiny strand of that thread. For many years, when I was on the faculty of the Technion, Joshua's office was down the hall from mine and we'd often chat in front of the blackboard. He would ask me some "simple" question, as if I were the real authority. So I'd give him the benefit of my wisdom, but then he'd have one "little" question, and pretty soon my glib answer was completely unraveled and I'd appreciate that there was a really deep issue involved. Now I ask those of you who knew Wigner: is this man his disciple or not?

Let me begin then with Joshua's personal history.

Joshua Zak was born in 1929 in Wilno, Poland, now Vilnius, capital of Lithuania, where he went to a Hebrew grammar school. The Molotov-Ribbentrop pact of 1939 partitioned Poland between Germany and the USSR. Teaching Hebrew was forbidden once the Red army took over

Journal of Physics: Conference Series **597** (2015) 011004 doi:10.1088/1742-6596/597/1/011004

Figure 2. Arno Bohm presenting the Medal to Professor Joshua Zak (left). Lawrence Schulman delivering the Laudatio (right). Professor Joshua Zak at the reception, together with friends and relatives (bottom).

in Wilno, and the 10-year old Joshua switched to Yiddish. He was 12 when the German army invaded Vilnius and for the next two years Joshua lived in Vilnius Ghetto (1941-1943). At age 14 he was taken to a labor camp in Estonia and at 15 to the Stutthof concentration camp (1944-1945). At 16 he participated in the Death Marches from Stutthof to the West. While still 16 he was liberated by the Red Army, only to be drafted into that same army for the next three years (until April 1948).

At age 19 Joshua returned to Vilnius and resumed his schooling. Two years later he graduated from high school with a silver medal. He then went on to study physics in the University of Vilnius.

In 1950 the Korean war started and Joshua was again drafted into the Red Army. He was on the train to go to war, when he heard his name on the loudspeaker. He was taken off the train and allowed to resume his studies. Whence this miracle? It was the successful intervention of his brother, Benzion, who later became a well-known physician in Tel Aviv.

Journal of Physics: Conference Series **597** (2015) 011004 doi:10.1088/1742-6596/597/1/011004

The interest of Joshua in group theory started early. As an undergraduate he studied Wigner's book on symmetries in German, since the English or Russian translations of the book were yet to be made. You can now count how many languages Joshua speaks.

In 1953 Joshua won the championship of Lithuania in 1000m singles in kayaking.

In 1955 Joshua got his Master's degree from the University of Vilnius and applied to the Graduate School of Herzen University in Leningrad. In the Soviet Union after graduating one had to perform three years of national service, and Joshua was assigned to a military school in the Latvian Republic. He was relieved from this obligation by being accepted as a graduate student in the group of Professor Lev Emanuilovich Gurevich. A second miracle.

The third miracle was the political upheaval in Poland and the rise of Gomulka in 1956. Among the reforms Gomulka introduced was allowing Jews born in Poland to emigrate to Israel. This was done partly to relieve anti-Semitic pressures in Poland. Joshua and his two brothers seized the opportunity and in March 1957 moved to Wałbrzych, where Joshua worked in a book store for nine months. He and his brothers arrived in Haifa, Israel in December 1957 and Joshua became a graduate student of Professor Nathan Rosen, the R of EPR, at the Technion.

Nathan Rosen was mostly interested in general relativity while Joshua wanted to apply group theory to solid state physics. So Joshua was on his own and as a topic for his thesis he chose *Irreducible Representations of non-Symmorphic Space Groups*. There was nobody in the Technion who was expert in this area, but Joshua did have Giulio Raccah, at the Hebrew University in Jerusalem with whom he would consult on a weekly basis. Joshua finished his Ph.D. thesis in 1960 with George Koster of MIT as referee.

When Wigner visited Israel for the first time, he also visited the Physics Department of the Technion. It was a small department and Rosen introduced all its members to Wigner. When Joshua's turn came Wigner asked him what was the subject of his thesis. Joshua told him. Wigner's reaction was, "Oh really, I had a student, Frederick Seitz, who also did some work on representations of Space Groups." It was well known that after Seitz there was very little left to do in this field.

In 1961 Benjamin Lax, Director of the National Magnet Laboratory (NML) at MIT, invited Joshua for the academic year 1963-64. It was at there that Joshua discovered the Magnetic Translation Group. He then returned to the Technion as a faculty member, where he has remained ever since. In 1967 Joshua was invited again to MIT as visiting professor. During that year he discovered the kq representation, also known as the Zak Transform. The Zak phase was invented at the Technion in 1989.

Let me now turn to the scientific side, with a short description of Zak's main discoveries.

First the kq representation. Weyl recognized that the canonical commutation relations between position and momentum cannot be implemented by finite dimensional matrices: P and Q are necessarily unbounded operators. As we all know, Q can be implemented as multiplication by the coordinate on the line and P as a first order differential operator. By a theorem of Stone and von Neumann all other representations of the canonical commutation relations are unitarily related to this one. For example, the momentum representation, where P is a multiplication and Q a first order differential operator is unitarily equivalent by the Fourier transform. The kq representation is a representation of the canonical commutation relations which maintains the PQ symmetry. Actually, kq is a one parameter family of representations parametrized by a period. The q of the kq is associated with the position Q modulo the period and the k is associated with P modulo the dual period in momentum space. The two periods define a lattice in phase space whose area is one Planck unit. In the kq representation both P and Q are first order differential operators acting on functions on a compact two-dimensional cell in phase space. The boundary conditions need to be twisted so that P and Q are self-adjoint and have the correct spectrum.

The kq representations are universal, and like the position representation, can be used to study any quantum problem. However, they are naturally tailored to situations with discrete translational symmetry where the Schrödinger equation retains its form as a differential operator, rather than a pseudo-differential operator. The kq representation has the Bloch theorem built in and gives elegant proofs of fundamental results in the dynamics of Bloch electrons. It also has applications to condensed matter physics and to signal processing. Google gives more than 640,000 entries under the entry "Zak transform." It is recognized by Mathematica, Wolfram alpha, and MATLAB. A 2012 US patent makes practical use of the Zak transform and books have been written about it.

The magnetic translation group: Emmy Noether taught us that each continuous symmetry in the Lagrangian implies a conservation law. Bloch then found that in quantum mechanics the discrete symmetry of translations also leads to a conserved momentum. An interesting situation arises when translations in different directions fail to commute. This happens for translations in phase space, but, surprisingly, it can also happen for translations in coordinate space in the presence of a constant magnetic field. Since the vector potential associated with a homogenous magnetic field is a linear function, it is translation invariant up to a gauge transformation. The magnetic translations introduced by Zak are translations accompanied by a gauge transformation. They do not commute in general but rather satisfy the same Weyl algebra as translation in phase space. This observation is fundamental to the modern theory of the quantum Hall effect and is at the root of Chern-Simons theory. Another important application is to periodic Schrödinger operators with a magnetic field and in particular the Hofstadter model where the symmetry and spectral properties are sensitive to the rationality of flux per unit cell. When Joshua studied this problem in the 1960's it looked like a mathematical delicacy, with little relation to observable physical phenomena. Fifty years later, experimental work on the quantum Hall effect demonstrates this sensitivity to the flux. This work now has the fetching name, "moiré butterflies" (see [1]).

The Zak phase: In 2005 Kane and Mele discovered that there are distinct topological classes of insulators labeled by integers. These phases are characterized by non-dissipative chiral edge currents. The subject of topological insulators is arguably one of the hottest topics in condensed matter physics today. The Zak phase, which Joshua found in 1989, is the one-dimensional granddaddy of topological insulators. It associates a parity index with insulators in one dimension. (There are no edge currents in one dimension, unfortunately). The Zak phase is a close relative of the Mead-Truhlar phase in the adiabatic pseudo-rotations of molecules. The Zak phase has been measured in recent experiments by Immanuel Bloch (Max Planck Institute) and Eugene Demler (Harvard) on cold atoms.

This brings my laudatio to its end. Eugene Wigner was the scientific hero of Joshua Zak. If Wigner could be watching this ceremony, I am sure he would be pleased to see his disciple, Joshua Zak, honored with the medal that bears his name.

References

[1] Dean C R *et al* 2013 *Nature* **497** 598

Journal of Physics: Conference Series **597** (2015) 011005 doi:10.1088/1742-6596/597/1/011005

The Hermann Weyl Prize 2014

E Frenkel[1] and J P Gazeau[2]

[1] Department of Mathematics, University of California at Berkeley, Berkeley, CA 94720-3840, USA

[2] University Paris Diderot, Rue A. Dumon and L. Duquet 10, 75013 Paris, France

E-mail: frenkel@math.berkeley.edu, gazeau@apc.univ-paris7.fr

1. Introduction by Jean-Pierre Gazeau

The Hermann Weyl Prize was created in the year 2000 by the Standing Committee of the International Colloquia on Group Theoretical Methods in Physics (ICGTMP), under the initiative of its chairman H.D. Doebner. The purpose of the Weyl Prize is to provide recognition for young scientists (in principle younger than 35) who have performed original work of significant scientific quality in the area of understanding physics through symmetries.

The Hermann Weyl prize consists of a certificate citing the accomplishments of the recipient, prize money of US$ 500 and an allowance towards attendance of the bi-annual International Colloquium on Group Theoretical Methods in Physics at which the award is presented.

The previous winners of the award are: Edward Frenkel (2002), Nikita A. Nekrasov (2004), Boyko Bakalov (2006), Mohammad M. Sheikh-Jabbari (2008), Giulio Chiribella (2010) and Razvan Gurau (2012).

The Selection Committee of the Weyl Prize 2014 consisted of Edward Frenkel (Chair, UC Berkeley, USA), Gitta Kutyniok (TU Berlin, Germany), Joris Van der Jeugt (Ghent University, Belgium), Katrin Wendland (Freiburg University, Germany) and Pavel Winternitz (University of Montreal, Canada).

The Committee received the nominations of 7 excellent candidates by members of the Standing Committee of ICGTMP. The committee members came to the unanimous conclusion to award the 2014 Hermann Weyl Prize to

Yuji Tachikawa

Department of Physics, University of Tokyo

> for outstanding contributions to our understanding of supersymmetric quantum field theories; in particular, to the discovery of the Alday-Gaiotto-Tachikawa correspondence that has led to spectacular advances in both mathematics and quantum physics.

According to Edward Frenkel, Chairman of the Selection Committee, Tachikawa and his collaborators found unexpected connections between models of quantum physics in two and four dimensions. "This work shed new light on many important phenomena in physics, and it also led to some astonishing predictions in geometry that have been rigorously proved by mathematicians" he said. "Without a doubt, this is one of the most important developments in mathematical physics of the past decade."

The ceremony for the Hermann Weyl Prize took place on July 16, 2014 in the Aula Academica of Ghent University, a historical building dating from 1826. The ceremony was graced by a

Journal of Physics: Conference Series **597** (2015) 011005 doi:10.1088/1742-6596/597/1/011005

saxophone quintet, "Sax Obsession", playing music from Gershwin and Shostakovich, amongst others. The participants were welcomed by Debora Van Durme, master of ceremonies, who welcomed the awardees, committee members and the invitees. She presented an introduction to some history of the university and the Aula building, and to Adolphe Sax and his instruments. The laudatio for Professor Tachikawa was delivered by Edward Frenkel. The full text of his laudatio is given in the next section. Then followed the actual presentation of the Certificate and the Prize by Jean-Pierre Gazeau and Joris Van der Jeugt. The ceremony was concluded by a reception with drinks and cocktail snacks in the adjacent Peristylium.

In figure 1 the certificate of the 2014 Hermann Weyl Prize is are shown. Figure 2 shows some pictures taken at the ceremony.

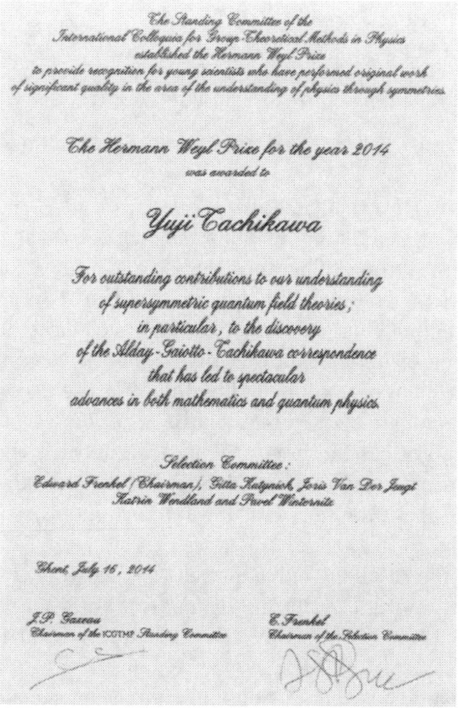
The Standing Committee of the
International Colloquia for Group-Theoretical Methods in Physics
established the Hermann Weyl Prize
to provide recognition for young scientists who have performed original work
of significant quality in the area of the understanding of physics through symmetries

The Hermann Weyl Prize for the year 2014
was awarded to

Yuji Tachikawa

For outstanding contributions to our understanding
of supersymmetric quantum field theories;
in particular, to the discovery
of the Alday-Gaiotto-Tachikawa correspondence
that has led to spectacular
advances in both mathematics and quantum physics

Selection Committee:
Edward Frenkel (Chairman), Gitta Kutyniok, Joris Van Der Jeugt
Katrin Wendland and Pavel Winternitz

Ghent, July 16, 2014

J.P. Gazeau
Chairman of the ICGTMP Standing Committee

E. Frenkel
Chairman of the Selection Committee

Figure 1. The 2014 Hermann Weyl Prize certificate

2. Laudatio for Professor Yuji Tachikawa by Edward Frenkel

It gives me a great pleasure to present, on behalf of the Selection Committee, the 2014 Hermann Weyl Prize to Yuji Tachikawa of University of Tokyo.

Hermann Weyl was a pioneer of applications of symmetries in physics. He once said: "I tried to unite the true and the beautiful. And when I had to choose one or the other, I usually chose beautiful."

True and beautiful are the attributes that indeed may be used to describe the work of Yuji Tachikawa for which he is receiving the Prize today. The main focus of this work is on supersymmetric gauge theories, or Yang–Mills theories, in four dimensions. These are very important models of quantum field theory, because three out of four known forces of nature (electromagnetic, weak, and strong) are described by gauge theories. The latter are non-supersymmetric, but physicists have been studying in great detail the supersymmetric cousins

Figure 2. Jean-Pierre Gazeau presenting the certificate to Professor Yuji Tachikawa (left). Edward Frenkel delivering the Laudatio (right). Professor Tachikawa at the reception, with Edward Frenkel and Joris Van der Jeugt (Chairman of the 30th ICGTMP) (bottom).

of these models in recent decades. Even if supersymmetry is not found in nature, these models are very useful because they are simpler and allow a more detailed analysis. The hope is that they will shed light on various important effects in gauge theories, such as the appearance and behavior of instantons.

The origin of the work of Tachikawa and his collaborators that we honor today is in the observation made by Edward Witten and others in the mid-1990s that for each simply-laced simple Lie algebra $\mathfrak{g}$ one can construct a remarkable six-dimensional quantum field theory (known as the "(0,2) superconformal field theory" or "theory X"). While these theories are still quite mysterious, their very existence provides us with a host of useful information. Indeed, we can consider such a theory on a product $M_d \times M_{6-d}$ manifolds of dimensions d and $6-d$ and consider the limit when the size of M_d goes to zero. The limiting theory is effectively a theory on M_{6-d} which we can analyze.

For example, it has been known that if we take as M_d the elliptic curve $E_\tau = \mathbb{C}/(\mathbb{Z}+\mathbb{Z}\tau)$, we obtain the maximally supersymmetric four-dimensional gauge theory with the gauge group G (simply-connected compact Lie group associated to $\mathfrak{g}$) and the complexified coupling constant τ. This has far-reaching consequences, such as the celebrated electromagnetic duality of gauge theories which is explained by the fact that E_τ is isomorphic to $E_{-1/\tau}$ (it may be generalized to non-simply connected and non-simply laced Lie groups). The electromagnetic duality plays a paramount role both in quantum physics (possibility to relate strongly coupled and weakly coupled quantum field theories) and in mathematics (connection to the Langlands Program).

The work of Yuji Tachikawa and his collaborators, Luis Alday and Davide Gaiotto, focused on what happens if we consider instead of E_τ an arbitrary Riemann surface X. It turns out that it is also possible to introduce certain defects corresponding to a collection of n marked points (x_i) on X. For any such $(X,(x_i))$ and any simply-laced simple Lie algebra $\mathfrak{g}$ we then obtain a four-dimensional supersymmetric quantum field theory and these theories are expected to exhibit a plethora of dualities generalizing the electromagnetic duality.

In the simplest case that $\mathfrak{g} = sl_2$, it has been argued that these four-dimensional theories have several Lagrangian descriptions: one for each decomposition of the marked curve $(X,(x_i))$ into pairs of pants. Note that by shrinking the cycles appearing in such a decomposition, we obtain a degenerate curve. The gauge group of this theory turns out to be $SU(2)$ to the power equal to the number of shrinking cycles, and hence the theory has coupling constants corresponding to the lengths of these shrinking cycles.

This is a very powerful statement because it immediately tells us that there should be a non-perturbatively defined four-dimensional quantum field theory parametrized by the *entire* moduli space of n-pointed curves $(X,(x_i))$, which has many different perturbative descriptions at different "corners" corresponding to the degenerate curves.

And there is more. We can study various quantities in these four-dimensional theories and ask whether they may be interpreted in terms of two-dimensional field theories on the Riemann surface X. One such quantity that has recently been studied extensively is the "Nekrasov partition function" $Z_{\rm Nek}$, which is the generating function of the equivariant volumes of the moduli spaces of instantons of this theory (note that Nikita Nekrasov was the recepient of the 2004 Hermann Weyl Prize). This function depends on various parameters, including two parameters ϵ_1, ϵ_2 corresponding to the action of two circles on two orthogonal planes in $\mathbb{R}^4$.

Since $Z_{\rm Nek}$ is produced by compactifying a six-dimensional theory on a punctured curve $(X,(x_i))$, we ask how exactly $Z_{\rm Nek}$ depends on $(X,(x_i))$. A startling *conjecture of Alday–Gaiotto–Tachikawa* is that $Z_{\rm Nek}$ is a conformal block in a two-dimensional conformal field theory on X, with operator insertions at the punctured points. This theory has been identified with the Liouville theory with the Virasoro central charge $c = 1 + 6(\epsilon_1 + \epsilon_2)^2$.

Thus, Alday, Gaiotto and Tachikawa discovered an unexpected link between four-dimensional supersymmetric gauge theories and two-dimensional conformal field theories. This has already led to an avalanche of papers by both physicists and mathematicians in which the conjecture has been proved in many cases and new conjectures have been formulated.

For example, for compactifications of the six-dimensional theories of other types one expects the Nekrasov partition functions to reproduce conformal blocks of conformal field theories with W-algebra or an affine Kac–Moody algebra symmetry. Also, in the limit $\epsilon_1 \to 0$, the leading term of $\log Z_{\rm Nek}$ is intimately connected to quantum integrable systems, closely related to the quantization of the Hitchin system.

The work of Tachikawa and his collaborators has illuminated an exciting area of research. This work has brought together symmetries of different kinds in quantum field theory: gauge symmetries and conformal symmetries (such as Virasoro and Kac–Moody). It represents an important new step in our quest to understand the mysteries of the physical world using its symmetries.

Journal of Physics: Conference Series **597** (2015) 011006 doi:10.1088/1742-6596/597/1/011006

Peer review statement

All papers published in this volume of *Journal of Physics: Conference Series* have been peer reviewed through processes administered by the proceedings Editors. Reviews were conducted by expert referees to the professional and scientific standards expected of a proceedings journal published by IOP Publishing.

The Bannai-Ito algebra and some applications

Hendrik De Bie[1], **Vincent X Genest**[2], **Satoshi Tsujimoto**[3], **Luc Vinet**[2] **and Alexei Zhedanov**[4]

[1]Department of Mathematical Analysis, Faculty of Engineering and Architecture, Ghent University, Galglaan 2 Galglaan 2, 9000 Ghent, Belgium
[2]Centre de recherches mathématiques, Université de Montréal, P.O. Box 6128, Centre-ville Station, Montréal (QC) Canada, H3C 3J7
[3]Department of applied mathematics and physics, Kyoto University, Kyoto 6068501, Japan
[4]Donetsk Institute for Physics and Technology, Donetsk 340114, Ukraine

E-mail: `hendrik.debie@ugent.be`, `vincent.genest@umontreal.ca`, `tujimoto@i.kyoto-u.ac.jp`, `luc.vinet@umontreal.ca` and `zhedanov@yahoo.com`

Abstract. The Bannai-Ito algebra is presented together with some of its applications. Its relations with the Bannai-Ito polynomials, the Racah problem for the $sl_{-1}(2)$ algebra, a superintegrable model with reflections and a Dirac-Dunkl equation on the 2-sphere are surveyed.

1. Introduction

Exploration through the exact solution of models has a secular tradition in mathematical physics. Empirically, exact solvability is possible in the presence of symmetries, which come in various guises and which are described by a variety of mathematical structures. In many cases, exact solutions are expressed in terms of special functions, whose properties encode the symmetries of the systems in which they arise. This can be represented by the following virtuous circle:

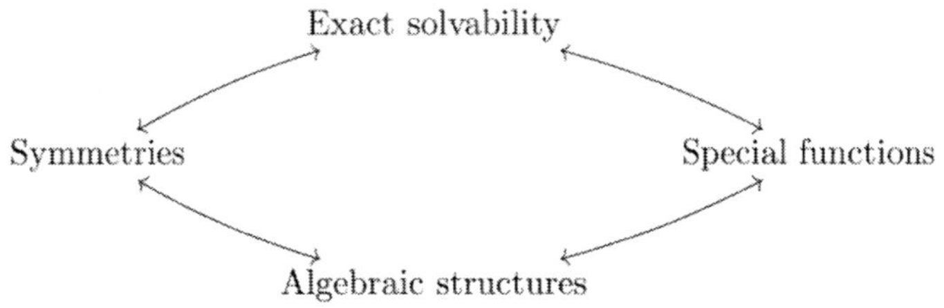

The classical path is the following: start with a model, find its symmetries, determine how these symmetries are mathematically described, work out the representations of that mathematical structure and obtain its relation to special functions to arrive at the solution of the model. However, one can profitably start from any node on this circle. For instance, one can identify and characterize new special functions, determine the algebraic structure they encode, look for models that have this structure as symmetry algebra and proceed to the solution. In this paper, the following path will be taken:

$$\text{Algebra} \longrightarrow \text{Orthogonal polynomials} \longrightarrow \text{Symmetries} \longrightarrow \text{Exact solutions}$$

The outline of the paper is as follows. In section 2, the Bannai-Ito algebra is introduced and some of its special cases are presented. In section 3, a realization of the Bannai-Ito algebra in terms of discrete shift and reflection operators is exhibited. The Bannai-Ito polynomials and their properties are discussed in section 4. In section 5, the Bannai-Ito algebra is used to derive the recurrence relation satisfied by the Bannai-Ito polynomials. In section 6, the paraboson algebra and the $sl_{-1}(2)$ algebra are introduced. In section 7, the Racah problem for $sl_{-1}(2)$ and its relation with the Bannai-Ito algebra is examined. A superintegrable model on the 2-sphere with Bannai-Ito symmetry is studied in section 8. In section 9, a Dunkl-Dirac equation on the 2-sphere with Bannai-Ito symmetry is discussed. A list of open questions is provided in lieu of a conclusion.

2. The Bannai-Ito algebra

Throughout the paper, the notation $[A, B] = AB-BA$ and $\{A, B\} = AB+BA$ will be used. Let ω_1, ω_2 and ω_3 be real parameters. The Bannai-Ito algebra is the associative algebra generated by K_1, K_2 and K_3 together with the three relations

$$\{K_1, K_2\} = K_3 + \omega_3, \quad \{K_2, K_3\} = K_1 + \omega_1, \quad \{K_3, K_1\} = K_2 + \omega_2, \tag{1}$$

or $\{K_i, K_j\} = K_k + \omega_k$, with (ijk) a cyclic permutation of $(1, 2, 3)$. The Casimir operator

$$Q = K_1^2 + K_2^2 + K_3^2,$$

commutes with every generator; this property is easily verified with the commutator identity $[AB, C] = A\{B, C\} - \{A, C\}B$. Let us point out two special cases of (1) that have been considered previously in the literature.

(i) $\omega_1 = \omega_2 = \omega_3 = 0$

The special case with defining relations

$$\{K_1, K_2\} = K_3, \quad \{K_2, K_3\} = K_1, \quad \{K_3, K_1\} = K_2,$$

is sometimes referred to as the *anticommutator spin algebra* [1, 2]; representations of this algebra were examined in [1, 2, 3, 4].

(ii) $\omega_1 = \omega_2 = 0 \neq \omega_3$

In recent work on the construction of novel finite oscillator models [5, 6], E. Jafarov, N. Stoilova and J. Van der Jeugt introduced the following extension of $\mathfrak{u}(2)$ by an involution R ($R^2 = 1$):

$$[I_3, R] = 0, \quad \{I_1, R\} = 0, \quad \{I_2, R\} = 0,$$
$$[I_3, I_1] = iI_2, \quad [I_2, I_3] = iI_1, \quad [I_1, I_2] = i(I_3 + \omega_3 R).$$

It is easy to check that with

$$K_1 = iI_1R, \quad K_2 = I_2, \quad K_3 = I_3R,$$

the above relations are converted into

$$\{K_1, K_3\} = K_2, \quad \{K_2, K_3\} = K_1, \quad \{K_1, K_2\} = K_3 + \omega_3.$$

3. A realization of the Bannai-Ito algebra with shift and reflections operators

Let T^+ and R be defined as follows:

$$T^+ f(x) = f(x+1), \quad Rf(x) = f(-x).$$

Consider the operator

$$\widehat{K}_1 = F(x)(1-R) + G(x)(T^+R - 1) + h, \qquad h = \rho_1 + \rho_2 - r_1 - r_2 + 1/2, \tag{2}$$

with $F(x)$ and $G(x)$ given by

$$F(x) = \frac{(x-\rho_1)(x-\rho_2)}{x}, \quad G(x) = \frac{(x - r_1 + 1/2)(x - r_2 + 1/2)}{x + 1/2},$$

where ρ_1, ρ_2, r_1, r_2 are four real parameters. It can be shown that $\widehat{K}_1$ is the most general operator of first order in T^+ and R that stabilizes the space of polynomials of a given degree [7]. That is, for any polynomial $Q_n(x)$ of degree n, $[\widehat{K}_1 Q_n(x)]$ is also a polynomial of degree n. Introduce

$$\widehat{K}_2 = 2x + 1/2, \tag{3}$$

which is essentially the "multiplication by x" operator and

$$\widehat{K}_3 \equiv \{\widehat{K}_1, \widehat{K}_2\} - 4(\rho_1\rho_2 - r_1 r_2). \tag{4}$$

It is directly verified that $\widehat{K}_1$, $\widehat{K}_2$ and $\widehat{K}_3$ satisfy the commutation relations

$$\{\widehat{K}_1, \widehat{K}_2\} = \widehat{K}_3 + \widehat{\omega}_3, \quad \{\widehat{K}_2, \widehat{K}_3\} = \widehat{K}_1 + \widehat{\omega}_1, \quad \{\widehat{K}_3, \widehat{K}_1\} = \widehat{K}_2 + \widehat{\omega}_2, \tag{5}$$

where the structure constants $\widehat{\omega}_1$, $\widehat{\omega}_2$ and $\widehat{\omega}_3$ read

$$\widehat{\omega}_1 = 4(\rho_1\rho_2 + r_1 r_2), \quad \widehat{\omega}_2 = 2(\rho_1^2 + \rho_2^2 - r_1^2 - r_2^2), \quad \widehat{\omega}_3 = 4(\rho_1\rho_2 - r_1 r_2). \tag{6}$$

The operators $\widehat{K}_1$, $\widehat{K}_2$ and $\widehat{K}_3$ thus realize the Bannai-Ito algebra. In this realization, the Casimir operator acts as a multiple of the identity; one has indeed

$$\widehat{Q} = \widehat{K}_1^2 + \widehat{K}_2^2 + \widehat{K}_3^2 = 2(\rho_1^2 + \rho_2^2 + r_1^2 + r_2^2) - 1/4.$$

4. The Bannai-Ito polynomials

Since the operator (2) preserves the space of polynomials of a given degree, it is natural to look for its eigenpolynomials, denoted by $B_n(x)$, and their corresponding eigenvalues λ_n. We use the following notation for the generalized hypergeometric series [8]

$${}_rF_s\left(\begin{matrix} a_1, \ldots, a_r \\ b_1, \ldots, b_s \end{matrix} \,\middle|\, z\right) = \sum_{k=0}^{\infty} \frac{(a_1)_k \cdots (a_r)_k}{(b_1)_k \cdots (b_s)_k} \frac{z^k}{k!},$$

where $(c)_k = c(c+1)\cdots(c+k-1)$, $(c)_0 \equiv 1$ stands for the Pochhammer symbol; note that the above series terminates if one of the a_i is a negative integer. Solving the eigenvalue equation

$$\widehat{K}_1 B_n(x) = \lambda_n B_n(x), \qquad n = 0, 1, 2, \ldots \tag{7}$$

it is found that the eigenvalues λ_n are given by [7]

$$\lambda_n = (-1)^n (n + h), \tag{8}$$

and that the polynomials have the expression

$$\frac{B_n(x)}{c_n} = \begin{cases} {}_4F_3\left(\begin{matrix} -\frac{n}{2}, \frac{n+1}{2}+h, x-r_1+1/2, -x-r_1+1/2 \\ 1-r_1-r_2, \rho_1-r_1+\frac{1}{2}, \rho_2-r_1+\frac{1}{2} \end{matrix} \middle| 1 \right) & \\ \quad + \frac{(\frac{n}{2})(x-r_1+\frac{1}{2})}{(\rho_1-r_1+\frac{1}{2})(\rho_2-r_1+\frac{1}{2})}\, {}_4F_3\left(\begin{matrix} 1-\frac{n}{2}, \frac{n+1}{2}+h, x-r_1+3/2, -x-r_1+1/2 \\ 1-r_1-r_2, \rho_1-r_1+\frac{3}{2}, \rho_2-r_1+\frac{3}{2} \end{matrix} \middle| 1 \right) & n \text{ even}, \\ {}_4F_3\left(\begin{matrix} -\frac{n-1}{2}, \frac{n}{2}+h, x-r_1+\frac{1}{2}, -x-r_1+\frac{1}{2} \\ 1-r_1-r_2, \rho_1-r_1+\frac{1}{2}, \rho_2-r_1+\frac{1}{2} \end{matrix} \middle| 1 \right) & \\ \quad - \frac{(\frac{n}{2}+h)(x-r_1+\frac{1}{2})}{(\rho_1-r_1+\frac{1}{2})(\rho_2-r_1+\frac{1}{2})}\, {}_4F_3\left(\begin{matrix} -\frac{n-1}{2}, \frac{n+2}{2}+h, x-r_1+\frac{3}{2}, -x-r_1+\frac{1}{2} \\ 1-r_1-r_2, \rho_1-r_1+\frac{3}{2}, \rho_2-r_1+\frac{3}{2} \end{matrix} \middle| 1 \right) & n \text{ odd}, \end{cases} \tag{9}$$

where the coefficient

$$c_{2n+p} = (-1)^p \frac{(1-r_1-r_2)_n(\rho_1-r_1+1/2, \rho_2-r_1+1/2)_{n+p}}{(n+h+1/2)_{n+p}}, \qquad p \in \{0, 1\},$$

ensures that the polynomials $B_n(x)$ are monic, i.e. $B_n(x) = x^n + \mathcal{O}(x^{n-1})$. The polynomials (9) were first written down by Bannai and Ito in their classification of the orthogonal polynomials satisfying the *Leonard duality* property [9, 10], i.e. polynomials $p_n(x)$ satisfying both

- A 3-term recurrence relation with respect to the degree n,
- A 3-term difference equation with respect to a variable index s.

The identification of the defining eigenvalue equation (7) of the Bannai-Ito polynomials in [7] has allowed to develop their theory. That they obey a three-term difference equation stems from the fact that there are grids such as

$$x_s = (-1)^s(s/2 + a + 1/4) - 1/4,$$

for which operators of the form

$$H = A(x)R + B(x)T^+R + C(x),$$

are tridiagonal in the basis $f(x_s)$

$$Hf(x_s) = \begin{cases} B(x_s)f(x_{s+1}) + A(x_s)f(x_{s-1}) + C(x_s)f(x_s) & s \text{ even}, \\ A(x_s)f(x_{s+1}) + B(x_s)f(x_{s-1}) + C(x_s)f(x_s) & s \text{ odd}. \end{cases}$$

It was observed by Bannai and Ito that the polynomials (9) correspond to a $q \to -1$ limit of the q-Racah polynomials (see [11] for the definition of q-Racah polynomials). In this connection, it is worth mentioning that the Bannai-Ito algebra (5) generated by the defining operator $\widehat{K}_1$ and the recurrence operator $\widehat{K}_2$ of the Bannai-Ito polynomials can be obtained as a $q \to -1$ limit of the Zhedanov algebra [12], which encodes the bispectral property of the q-Racah polynomials. The Bannai-Ito polynomials $B_n(x)$ have companions

$$I_n(x) = \frac{B_{n+1}(x) - \frac{B_{n+1}(\rho_1)}{B_n(\rho_1)} B_n(x)}{x - \rho_1},$$

called the *complementary* Bannai-Ito polynomials [13]. It has now been understood that the polynomials $B_n(x)$ and $I_n(x)$ are the ancestors of a rich ensemble of polynomials referred to as "-1 orthogonal polynomials" [7, 13, 14, 15, 16, 17, 18]. All polynomials of this scheme are eigenfunctions of first or second order operators of Dunkl type, i.e. which involve reflections.

5. The recurrence relation of the BI polynomials from the BI algebra

Let us now show how the Bannai-Ito algebra can be employed to derive the recurrence relation satisfied by the Bannai-Ito polynomials. In order to obtain this relation, one needs to find the action of the operator $\widehat{K}_2$ on the BI polynomials $B_n(x)$. Introduce the operators

$$\widehat{K}_+ = (\widehat{K}_2 + \widehat{K}_3)(\widehat{K}_1 - 1/2) - \frac{\widehat{\omega}_2 + \widehat{\omega}_3}{2}, \quad \widehat{K}_- = (\widehat{K}_2 - \widehat{K}_3)(\widehat{K}_1 + 1/2) + \frac{\widehat{\omega}_2 - \widehat{\omega}_3}{2}, \tag{10}$$

where $\widehat{K}_i$ and $\widehat{\omega}_i$ are given by (2), (3), (4) and (6). It is readily checked using (5) that

$$\{\widehat{K}_1, \widehat{K}_\pm\} = \pm K_\pm.$$

One can directly verify that $\widehat{K}_\pm$ maps polynomials to polynomials. In view of the above, one has

$$\widehat{K}_1\widehat{K}_+B_n(x) = (-\widehat{K}_+\widehat{K}_1 + \widehat{K}_+)B_n(x) = (1-\lambda_n)\widehat{K}_+B_n(x),$$

where λ_n is given by (8). It is also seen from (8) that

$$1-\lambda_n = \begin{cases} \lambda_{n-1} & n \text{ even},\\ \lambda_{n+1} & n \text{ odd}.\end{cases}$$

It follows that

$$\widehat{K}_+B_n(x) = \begin{cases} \alpha_n^{(0)}B_{n-1}(x) & n \text{ even},\\ \alpha_n^{(1)}B_{n+1}(x) & n \text{ odd}.\end{cases}$$

Similarly, one finds

$$\widehat{K}_-B_n(x) = \begin{cases} \beta_n^{(0)}B_{n+1}(x) & n \text{ even},\\ \beta_n^{(1)}B_{n-1}(x) & n \text{ odd}.\end{cases}$$

The coefficients

$$\alpha_n^{(0)} = \frac{2n(\frac{n}{2}+\rho_1+\rho_2)(r_1+r_2-\frac{n}{2})(\frac{n-1}{2}+h)}{n+h-\frac{1}{2}}, \quad \alpha_n^{(1)} = -4(n+h+1/2),$$
$$\beta_n^{(0)} = 4(n+h+1/2), \quad \beta_n^{(1)} = \frac{4(\rho_1-r_1+\frac{n}{2})(\rho_2-r_1+\frac{n}{2})(\rho_1-r_2+\frac{n}{2})(\rho_2-r_2+\frac{n}{2})}{n+h-1/2},$$

can be obtained from the comparison of the highest order term. Introduce the operator

$$V = \widehat{K}_+(\widehat{K}_1+1/2) + \widehat{K}_-(\widehat{K}_1-1/2). \tag{11}$$

From the definition (10) of $\widehat{K}_\pm$, it follows that

$$V = 2\widehat{K}_2(\widehat{K}_1^2 - 1/4) - \widehat{\omega}_3\widehat{K}_1 - \widehat{\omega}_2/2. \tag{12}$$

From (7), (11) and the actions of the operators $\widehat{K}_\pm$, we find that V is two-diagonal

$$VB_n(x) = \begin{cases} (\lambda_n+1/2)\alpha_n^{(0)}B_{n-1}(x) + (\lambda_n-1/2)\beta_n^{(0)}B_{n+1}(x) & n \text{ even},\\ (\lambda_n-1/2)\beta_n^{(1)}B_{n-1}(x) + (\lambda_n+1/2)\alpha_n^{(1)}B_{n+1}(x) & n \text{ odd}.\end{cases} \tag{13}$$

From (12) and recalling the definition (3) of $\widehat{K}_2$, we have also

$$V B_n(x) = \left[(\lambda_n^2 - 1/4)(4x+1) - \widehat{\omega}_3\lambda_n - \widehat{\omega}_2/2\right] B_n(x). \tag{14}$$

Upon combining (13) and (14), one finds that the Bannai-Ito polynomials satisfy the three-term recurrence relation

$$x\, B_n(x) = B_{n+1}(x) + (\rho_1 - A_n - C_n)B_n(x) + A_{n-1}C_nB_{n-1}(x),$$

where

$$A_n = \begin{cases} \frac{(n+1+2\rho_1-2r_1)(n+1+2\rho_1-2r_2)}{4(n+\rho_1+\rho_2-r_1-r_2+1)} & n \text{ even}, \\ \frac{(n+1+2\rho_1+2\rho_2-2r_1-2r_2)(n+1+2\rho_1+2\rho_2)}{4(n+\rho_1+\rho_2-r_1-r_2+1)} & n \text{ odd}, \end{cases}$$
$$C_n = \begin{cases} -\frac{n(n-2r_1-2r_2)}{4(n+\rho_1+\rho_2-r_1-r_2)} & n \text{ even}, \\ -\frac{(n+2\rho_2-2r_2)(n+2\rho_2-2r_1)}{4(n+\rho_1+\rho_2-r_1-r_2)} & n \text{ odd}. \end{cases} \tag{15}$$

The positivity of the coefficient $A_{n-1}C_n$ restricts the polynomials $B_n(x)$ to being orthogonal on a finite set of points [19].

6. The paraboson algebra and $sl_{-1}(2)$

The next realization of the Bannai-Ito algebra will involve $sl_{-1}(2)$; this algebra, introduced in [20], is closely related to the parabosonic oscillator.

6.1. *The paraboson algebra*

Let a and $a^\dagger$ be the generators of the paraboson algebra. These generators satisfy [21]

$$[\{a, a^\dagger\}, a] = -2a, \quad [\{a, a^\dagger\}, a^\dagger] = 2a^\dagger.$$

Setting $H = \frac{1}{2}\{a, a^\dagger\}$, the above relations amount to

$$[H, a] = -a, \quad [H, a^\dagger] = a^\dagger,$$

which correspond to the quantum mechanical equations of an oscillator.

6.2. *Relation with* $\mathfrak{osp}(1|2)$

The paraboson algebra is related to the Lie superalgebra $\mathfrak{osp}(1|2)$ [22]. Indeed, upon setting

$$F_- = a, \quad F_+ = a^\dagger, \quad E_0 = H = \frac{1}{2}\{F_+, F_-\}, \quad E_+ = \frac{1}{2}F_+^2, \quad E_- = \frac{1}{2}F_-^2,$$

and interpreting $F_\pm$ as odd generators, it is directly verified that the generators $F_\pm$, $E_\pm$ and E_0 satisfy the defining relations of $\mathfrak{osp}(1|2)$ [23]:

$$[E_0, F_\pm] = \pm F_\pm, \quad \{F_+, F_-\} = 2E_0, \quad [E_0, E_\pm] = \pm 2E_\pm, \quad [E_-, E_+] = E_0,$$
$$[F_\pm, E_\pm] = 0, \quad [F_\pm, E_\mp] = \mp F_\mp.$$

The $\mathfrak{osp}(1|2)$ Casimir operator reads

$$C_{\mathfrak{osp}(1|2)} = (E_0 - 1/2)^2 - 4E_+E_- - F_+F_-.$$

6.3. $sl_q(2)$

Consider now the quantum algebra $sl_q(2)$. It can be presented in terms of the generators A_0 and $A_\pm$ satisfying the commutation relations [24]

$$[A_0, A_\pm] = \pm A_\pm, \quad [A_-, A_+] = 2\frac{q^{A_0} - q^{-A_0}}{q - q^{-1}}.$$

Upon setting

$$B_+ = A_+ q^{(A_0-1)/2}, \quad B_- = q^{(A_0-1)/2} A_-, \quad B_0 = A_0,$$

these relations become

$$[B_0, B_\pm] = \pm B_\pm, \quad B_- B_+ - q B_+ B_- = 2\frac{q^{2B_0} - 1}{q^2 - 1}.$$

The $sl_q(2)$ Casimir operator is of the form

$$C_{sl_q(2)} = B_+ B_- q^{-B_0} - \frac{2}{(q^2-1)(q-1)}(q^{B_0-1} + q^{-B_0}).$$

Let j be a non-negative integer. The algebra $sl_q(2)$ admits a discrete series representation on the basis $|j, n\rangle$ with the actions

$$q^{B_0}|j,n\rangle = q^{j+n}|j,n\rangle\,, \qquad n = 0, 1, 2, \ldots.$$

The algebra has a non-trivial coproduct $\Delta : sl_q(2) \to sl_q(2) \otimes sl_q(2)$ which reads

$$\Delta(B_0) = B_0 \otimes 1 + 1 \otimes B_0, \quad \Delta(B_\pm) = B_\pm \otimes q^{B_0} + 1 \otimes B_\pm.$$

6.4. *The $sl_{-1}(2)$ algebra as a $q \to -1$ limit of $sl_q(2)$*

The $sl_{-1}(2)$ algebra can be obtained as a $q \to -1$ limit of $sl_q(2)$. Let us first introduce the operator R defined as

$$R = \lim_{q \to -1} q^{B_0}.$$

It is easily seen that

$$R|j,n\rangle = (-1)^{j+n}|j,n\rangle = \epsilon(-1)^n|j,n\rangle\,,$$

where $\epsilon = \pm 1$ depending on the parity of j, thus $R^2 = 1$. When $q \to -1$, one finds that

$$\begin{array}{l} q^{B_0} B_+ = q B_+ q^{B_0} \\ B_- q^{B_0} = q q^{B_0} B_- \end{array} \longrightarrow \{R, B_\pm\} = 0,$$

$$B_- B_+ - q B_+ B_- = 2\frac{q^{2B_0}-1}{q^2-1} \longrightarrow \{B_+, B_-\} = 2B_0,$$

$$C_{sl_q(2)} \longrightarrow B_+ B_- R - B_0 R + R/2,$$

$$\Delta(B_\pm) = B_\pm \otimes q^{B_0} + 1 \otimes B_\pm \longrightarrow \Delta(B_\pm) = B_\pm \otimes R + 1 \otimes B_\pm.$$

In summary, $sl_{-1}(2)$ is the algebra generated by J_0, $J_\pm$ and R with the relations [20]

$$[J_0, J_\pm] = \pm J_\pm, \quad [J_0, R] = 0, \quad \{J_\pm, R\} = 0, \quad \{J_+, J_-\} = 2J_0, \quad R^2 = 1. \tag{16}$$

The Casimir operator has the expression

$$Q = J_+J_-R - J_0R + R/2, \tag{17}$$

and the coproduct is of the form [25]

$$\Delta(J_0) = J_0 \otimes 1 + 1 \otimes J_0, \quad \Delta(J_\pm) = J_\pm \otimes R + 1 \otimes J_\pm, \quad \Delta(R) = R \otimes R. \tag{18}$$

The $sl_{-1}(2)$ algebra (16) has irreducible and unitary discrete series representations with basis $|\epsilon, \mu; n\rangle$, where n is a non-negative integer, $\epsilon = \pm 1$ and μ is a real number such that $\mu > -1/2$. These representations are defined by the following actions:

$$J_0|\epsilon,\mu;n\rangle = (n+\mu+\frac{1}{2})|\epsilon,\mu;n\rangle, \quad R|\epsilon,\mu;n\rangle = \epsilon(-1)^n|\epsilon,\mu;n\rangle,$$
$$J_+|\epsilon,\mu;n\rangle = \rho_{n+1}|\epsilon,\mu;n+1\rangle, \quad J_-|\epsilon,\mu;n\rangle = \rho_n|\epsilon,\mu;n-1\rangle,$$

where $\rho_n = \sqrt{n + \mu(1-(-1)^n)}$. In these representations, the Casimir operator takes the value

$$Q|\epsilon,\mu;n\rangle = -\epsilon\mu|\epsilon,\mu;n\rangle.$$

These modules will be denoted by $V^{(\epsilon,\mu)}$. Let us offer the following remarks.

- The $sl_{-1}(2)$ algebra corresponds to the parabose algebra supplemented by R.
- The $sl_{-1}(2)$ algebra consists of the Cartan generator J_0 and the two odd elements of $\mathfrak{osp}(1|2)$ supplemented by the involution R.
- One has $C_{\mathfrak{osp}(1|2)} = Q^2$, where Q is given by (17). Thus the introduction of R allows to take the square-root of $C_{\mathfrak{osp}(1|2)}$.
- In $sl_{-1}(2)$, one has $[J_-, J_+] = 1 - 2QR$. On the module $V^{(\epsilon,\mu)}$, this leads to

$$[J_-, J_+] = 1 + 2\epsilon\mu R.$$

7. Dunkl operators

The irreducible modules $V^{(\epsilon,\mu)}$ of $sl_{-1}(2)$ can be realized by Dunkl operators on the real line. Let R_x be the reflection operator

$$R_x f(x) = f(-x).$$

The $\mathbb{Z}_2$-Dunkl operator on $\mathbb{R}$ is defined by [26]

$$D_x = \frac{\partial}{\partial x} + \frac{\nu}{x}(1 - R_x),$$

where ν is a real number such that $\nu > -1/2$. Upon introducing the operators

$$\widehat{J}_\pm = \frac{1}{\sqrt{2}}(x \mp D_x),$$

and defining $\widehat{J}_0 = \frac{1}{2}\{\widehat{J}_-, \widehat{J}_+\}$, it is readily verified that a realization of the $sl_{-1}(2)$-module $V^{(\epsilon,\mu)}$ with $\epsilon = 1$ and $\mu = \nu$ is obtained. In particular, one has

$$[\widehat{J}_-, \widehat{J}_+] = 1 + 2\nu R_x.$$

It can be seen that $\widehat{J}_\pm^\dagger = \widehat{J}_\mp$ with respect to the measure $|x|^{2\nu}\,dx$ on the real line [27].

8. The Racah problem for $sl_{-1}(2)$ and the Bannai-Ito algebra

The Racah problem for $sl_{-1}(2)$ presents itself when the direct product of three irreducible representations is examined. We consider the three-fold tensor product

$$V = V^{(\epsilon_1,\mu_1)} \otimes V^{(\epsilon_2,\mu_2)} \otimes V^{(\epsilon_3,\mu_3)}.$$

It follows from the coproduct formula (18) that the generators of $sl_{-1}(2)$ on V are of the form

$$J^{(4)} = J_0^{(1)} + J_0^{(2)} + J_0^{(3)}, \quad J_{\pm}^{(4)} = J_{\pm}^{(1)}R^{(2)}R^{(3)} + J_{\pm}^{(2)}R^{(3)} + J_{\pm}^{(3)}, \quad R^{(4)} = R^{(1)}R^{(2)}R^{(3)},$$

where the superscripts indicate on which module the generators act. In forming the module V, two sequences are possible: one can first combine (1) and (2) to bring (3) after or one can combine (2) and (3) before adding (1). This is represented by

$$\left(V^{(\epsilon_1,\mu_1)} \otimes V^{(\epsilon_2,\mu_2)}\right) \otimes V^{(\epsilon_3,\mu_3)} \quad \text{or} \quad V^{(\epsilon_1,\mu_1)} \otimes \left(V^{(\epsilon_2,\mu_2)} \otimes V^{(\epsilon_3,\mu_3)}\right). \tag{19}$$

These two addition schemes are equivalent and the two corresponding bases are unitarily related. In the following, three types of Casimir operators will be distinguished.

- The initial Casimir operators

$$Q_i = J_+^{(i)}J_-^{(i)}R^{(i)} - (J_0^{(i)} - 1/2)R^{(i)} = -\epsilon_i\mu_i, \qquad i = 1,2,3.$$

- The intermediate Casimir operators

$$\begin{aligned} Q_{ij} &= (J_+^{(i)}R^{(j)} + J_+^{(j)})(J_-^{(i)}R^{(j)} + J_-^{(j)})R^{(i)}R^{(j)} - (J_0^{(i)} + J_0^{(j)} - 1/2)R^{(i)}R^{(j)} \\ &= (J_-^{(i)}J_+^{(j)} - J_+^{(i)}J_-^{(j)})R^{(i)} - R^{(i)}R^{(j)}/2 + Q_iR^{(j)} + Q_jR^{(i)}, \end{aligned}$$

  where $(ij) = (12), (23)$.

- The total Casimir operator

$$Q_4 = [J_+^{(4)}J_-^{(4)} - (J_0^{(4)} - 1/2)]R^{(4)}.$$

Let $|q_{12}, q_4; m\rangle$ and $|q_{23}, q_4; m\rangle$ be the orthonormal bases associated to the two coupling schemes presented in (19). These two bases are defined by the relations

$$Q_{12}|q_{12}, q_4; m\rangle = q_{12}|q_{12}, q_4; m\rangle, \quad Q_{23}|q_{23}, q_4; m\rangle = q_{23}|q_{23}, q_4; m\rangle,$$

and

$$Q_4|-, q_4; m\rangle = q_4|-, q_4; m\rangle, \quad J_0^{(4)}|-, q_4; m\rangle = (m + \mu_1 + \mu_2 + \mu_3 + 3/2)|-, q_4; m\rangle.$$

The Racah problem consists in finding the overlap coefficients

$$\langle q_{23}, q_4 \,|\, q_{12}, q_4\rangle,$$

between the eigenbases of Q_{12} and Q_{23} with a fixed value q_4 of the total Casimir operator Q_4; as these coefficients do not depend on m, we drop this label. For simplicity, let us now take

$$\epsilon_1 = \epsilon_2 = \epsilon_3 = 1.$$

Upon defining

$$K_1 = -Q_{23}, \qquad K_3 = -Q_{12},$$

one finds that the intermediate Casimir operators of $sl_{-1}(2)$ realize the Bannai-Ito algebra [28]

$$\{K_1, K_3\} = K_2 + \Omega_2, \quad \{K_1, K_2\} = K_3 + \Omega_3, \quad \{K_2, K_3\} = K_1 + \Omega_1, \tag{20}$$

with structure constants

$$\Omega_1 = 2(\mu_1\mu + \mu_2\mu_3), \quad \Omega_2 = 2(\mu_1\mu_3 + \mu_2\mu), \quad \Omega_3 = 2(\mu_1\mu_2 + \mu_3\mu), \tag{21}$$

where $\mu = \epsilon_4\mu_4 = -q_4$. The first relation in (20) can be taken to define K_2 which reads

$$K_2 = (J_+^{(1)}J_-^{(3)} - J_-^{(1)}J_+^{(3)})R^{(1)}R^{(2)} + R^{(1)}R^{(3)}/2 - Q_1R^{(3)} - Q_3R^{(1)}.$$

In the present realization the Casimir operator of the Bannai-Ito algebra becomes

$$Q_{\mathrm{BI}} = \mu_1^2 + \mu_2^2 + \mu_3^2 + \mu_4^2 - 1/4.$$

It has been shown in section 3 that the Bannai-Ito polynomials form a basis for a representation of the BI algebra. It is here relatively easy to construct the representation of the BI algebra on bases of the three-fold tensor product module V with basis vectors defined as eigenvectors of Q_{12} or of Q_{23}. The first step is to obtain the spectra of the intermediate Casimir operators. Simple considerations based on the nature of the $sl_{-1}(2)$ representation show that the eigenvalues q_{12} and q_{23} of Q_{12} and Q_{23} take the form [29, 30, 28, 20]:

$$q_{12} = (-1)^{s_{12}+1}(s_{12} + \mu_1 + \mu_2 + 1/2), \quad q_{23} = (-1)^{s_{23}}(s_{23} + \mu_2 + \mu_3 + 1/2),$$

where $s_{12}, s_{23} = 0, 1, \ldots, N$. The non-negative integer N is specified by

$$N + 1 = \mu_4 - \mu_1 - \mu_2 - \mu_3.$$

Denote the eigenstates of K_3 by $|k\rangle$ and those of K_1 by $|s\rangle$; one has

$$K_3|k\rangle = (-1)^k(k + \mu_1 + \mu_2 + 1/2)|k\rangle, \quad K_1|s\rangle = (-1)^s(s + \mu_2 + \mu_3 + 1/2)|s\rangle.$$

Given the expressions (21) for the structure constants Ω_k, one can proceed to determine the $(N+1) \times (N+1)$ matrices that verify the anticommutation relations (20). The action of K_1 on $|k\rangle$ is found to be [28]:

$$K_1|k\rangle = U_{k+1}|k+1\rangle + V_k|k\rangle + U_k|k-1\rangle,$$

with $V_k = \mu_2 + \mu_3 + 1/2 - B_k - D_k$ and $U_k = \sqrt{B_{k-1}D_k}$ where

$$B_k = \begin{cases} \frac{(k+2\mu_2+1)(k+\mu_1+\mu_2+\mu_3-\mu+1)}{2(k+\mu_1+\mu_2+1)} & k \text{ even}, \\ \frac{(k+2\mu_1+2\mu_2+1)(k+\mu_1+\mu_2+\mu_3+\mu+1)}{2(k+\mu_1+\mu_2+1)} & k \text{ odd}, \end{cases}$$

$$D_k = \begin{cases} -\frac{k(k+\mu_1+\mu_2-\mu_3-\mu)}{2(k+\mu_1+\mu)2)} & n \text{ even}, \\ -\frac{(k+2\mu_1)(k+\mu_1+\mu_2-\mu_3+\mu)}{2(k+\mu_1+\mu_2)} & n \text{ odd}. \end{cases}$$

Under the identifications

$$\rho_1 = \frac{1}{2}(\mu_2 + \mu_3), \quad \rho_2 = \frac{1}{2}(\mu_1 + \mu), \quad r_1 = \frac{1}{2}(\mu_3 - \mu_2), \quad r_2 = \frac{1}{2}(\mu - \mu_1),$$

one has $B_k = 2A_k$, $D_k = 2C_k$, where A_k and C_k are the recurrence coefficients (15) of the Bannai-Ito polynomials. Upon setting

$$\langle s \mid k \rangle = w(s) 2^k B_k(x_s), \qquad B_0(x_s) \equiv 1,$$

one has on the one hand

$$\langle s \mid K_1 \mid k \rangle = (-1)^s (s + 2\rho_1 + 1/2) \, \langle s \mid k \rangle ,$$

and on the other hand

$$\langle s \mid K_1 \mid k \rangle = U_{k+1} \langle s \mid k+1 \rangle + V_k \langle s \mid k \rangle + U_{k-1} \langle s \mid k-1 \rangle .$$

Comparing the two RHS yields

$$x_s B_k(x_s) = B_{k+1}(x_s) + (\rho_1 - A_k - C_k) B_k(x_s) + A_{k-1} C_k B_{k-1}(x_s),$$

where x_s are the points of the Bannai-Ito grid

$$x_s = (-1)^s \left(\frac{s}{2} + \rho_1 + 1/4 \right) - 1/4, \quad s = 0, \ldots, N.$$

Hence the Racah coefficients of $sl_{-1}(2)$ are proportional to the Bannai-Ito polynomials. The algebra (20) with structure constants (21) is invariant under the cyclic permutations of the pairs (K_i, μ_i). As a result, the representations in the basis where K_1 is diagonal can be obtained directly. In this basis, the operator K_3 is seen to be tridiagonal, which proves again that the Bannai-Ito polynomials possess the Leonard duality property.

9. A superintegrable model on S^2 with Bannai-Ito symmetry

We shall now use the analysis of the Racah problem for $sl_{-1}(2)$ and its realization in terms of Dunkl operators to obtain a superintegrable model on the two-sphere. Recall that a quantum system in n dimensions with Hamiltonian H is maximally superintegrable it it possesses $2n - 1$ algebraically independent constants of motion, where one of these constants is H [31]. Let $(s_1, s_2, s_3) \in \mathbb{R}$ and take $s_1^2 + s_2^2 + s_3^2 = 1$. The standard angular momentum operators are

$$L_1 = \frac{1}{i}\left(s_2 \frac{\partial}{\partial s_3} - s_3 \frac{\partial}{\partial s_2} \right), \quad L_2 = \frac{1}{i}\left(s_3 \frac{\partial}{\partial s_1} - s_1 \frac{\partial}{\partial s_3} \right), \quad L_3 = \frac{1}{i}\left(s_1 \frac{\partial}{\partial s_2} - s_2 \frac{\partial}{\partial s_1} \right).$$

The system governed by the Hamiltonian

$$H = L_1^2 + L_2^2 + L_3^2 + \frac{\mu_1}{s_1^2}(\mu_1 - R_1) + \frac{\mu_2}{s_2^2}(\mu_2 - R_2) + \frac{\mu_3}{s_3^2}(\mu_3 - R_3), \tag{22}$$

with μ_i, $i = 1, 2, 3$, real parameters such that $\mu_i > -1/2$ is superintegrable [32].

(i) The operators R_i reflect the variable s_i: $R_i f(s_i) = f(-s_i)$.

(ii) The operators R_i commute with the Hamiltonian: $[H, R_i] = 0$.

Journal of Physics: Conference Series **597** (2015) 012001 doi:10.1088/1742-6596/597/1/012001

(iii) If one is concerned with the presence of reflection operators in a Hamiltonian, one may replace R_i by $\kappa_i = \pm 1$. This then treats the 8 potential terms

$$\frac{\mu_1}{s_1^2}(\mu_1 - \kappa_1) + \frac{\mu_2}{s_2^2}(\mu_2 - \kappa_2) + \frac{\mu_3}{s_3^2}(\mu_3 - \kappa_3),$$

simultaneously much like supersymmetric partners.

(iv) Rescaling $s_i \to r s_i$ and taking the limit as $r \to \infty$ gives the Hamiltonian of the Dunkl oscillator [27, 33]

$$\widetilde{H} = -[D_{x_1}^2 + D_{x_2}^2] + \widehat{\mu}_3^2(x_1^2 + x_2^2),$$

after appropriate renormalization; see also [34, 35, 36].

It can be checked that the following three quantities commute with the Hamiltonian (22) [29, 32]:

$$C_1 = \left(iL_1 + \mu_2 \frac{s_3}{s_2} R_2 - \mu_3 \frac{s_2}{s_3} R_3\right) R_2 + \mu_2 R_3 + \mu_3 R_2 + R_2 R_3/2,$$
$$C_2 = \left(-iL_2 + \mu_1 \frac{s_3}{s_1} R_1 - \mu_3 \frac{s_1}{s_3} R_3\right) R_1 R_2 + \mu_1 R_3 + \mu_3 R_1 + R_1 R_3/2,$$
$$C_3 = \left(iL_3 + \mu_1 \frac{s_2}{s_1} R_1 - \mu_2 \frac{s_1}{s_2} R_2\right) R_1 + \mu_1 R_2 + \mu_2 R_1 + R_1 R_2/2,$$

that is, $[H, C_i] = 0$ for $i = 1, 2, 3$. To determine the symmetry algebra generated by the above constants of motion, let us return to the Racah problem for $sl_{-1}(2)$. Consider the following (gauge transformed) parabosonic realization of $sl_{-1}(2)$ in the three variables s_i:

$$J_{\pm}^{(i)} = \frac{1}{\sqrt{2}}\left[s_i \mp \partial_{s_i} \pm \frac{\mu_i}{s_i} R_i\right], \quad J_0^{(i)} = \frac{1}{2}\left[-\partial_{s_i}^2 + s_i^2 + \frac{\mu_i}{s_i^2}(\mu_i - R_i)\right], \quad R^{(i)} = R_i, \tag{23}$$

for $i = 1, 2, 3$. Consider also the addition of these three realizations so that

$$J_0 = J_0^{(1)} + J_0^{(2)} + J_0^{(3)}, \quad J_{\pm} = J_{\pm}^{(1)} R^{(2)} R^{(3)} + J_{\pm}^{(2)} R^{(3)} + J_{\pm}^{(3)}, \quad R = R^{(1)} R^{(2)} R^{(3)}. \tag{24}$$

It is observed that in the realization (24), the total Casimir operator can be expressed in terms of the constants of motion as follows:

$$Q = -C_1 R^{(1)} - C_2 R^{(2)} - C_3 R^{(3)} + \mu_1 R^{(2)} R^{(3)} + \mu_2 R^{(1)} R^{(3)} + \mu_3 R^{(1)} R^{(2)} + R/2.$$

Upon taking

$$\Omega = QR,$$

one finds

$$\Omega^2 + \Omega = L_1^2 + L_2^2 + L_3^2 + (s_1^2 + s_2^2 + s_3^2)\left(\frac{\mu_1}{s_1^2}(\mu_1 - R_1) + \frac{\mu_2}{s_2^2}(\mu_2 - R_2) + \frac{\mu_3}{s_3^2}(\mu_3 - R_3)\right), \tag{25}$$

so that $H = \Omega^2 + \Omega$ if $s_1^2 + s_2^2 + s_3^2 = 1$. Assuming this constraint can be imposed, H is a quadratic combination of QR. By construction, the intermediate Casimir operators Q_{ij} commute with the total Casimir operator Q and with R and hence with Ω; they thus commute with $H = \Omega^2 + \Omega$ and are the constants of motion. It is indeed found that

$$Q_{12} = -C_3, \qquad Q_{23} = -C_1,$$

in the parabosonic realization (23). Let us return to the constraint $s_1^2 + s_2^2 + s_3^2 = 1$. Observe that

$$\frac{1}{2}(J_+ + J_-)^2 = (s_1 R_2 R_3 + s_2 R_3 + s_3)^2 = s_1^2 + s_2^2 + s_3^2.$$

Because $(J_+ + J_-)^2$ commutes with $\Omega = QR$, Q_{12} and Q_{23}, one can impose $s_1^2 + s_2^2 + s_3^2 = 1$. Since it is already known that the intermediate Casimir operators in the addition of three $sl_{-1}(2)$ representations satisfy the Bannai-Ito structure relations, the constants of motion verify

$$\{C_1, C_2\} = C_3 - 2\mu_3 Q + 2\mu_1\mu_2,$$
$$\{C_2, C_3\} = C_1 - 2\mu_1 Q + 2\mu_2\mu_3,$$
$$\{C_3, C_1\} = C_2 - 2\mu_2 Q + 2\mu_3\mu_1,$$

and thus the symmetry algebra of the superintegrable system with Hamiltonian (22) is a central extension (with Q begin the central operator) of the Bannai-Ito algebra. Let us note that the relation $H = \Omega^2 + \Omega$ relates to chiral supersymmetry since with $S = \Omega + 1/2$ one has

$$\frac{1}{2}\{S, S\} = H + 1/4.$$

10. A Dunkl-Dirac equation on S^2

Consider the $\mathbb{Z}_2$-Dunkl operators

$$D_i = \frac{\partial}{\partial x_i} + \frac{\mu_i}{x_i}(1 - R_i), \qquad i = 1, 2, \dots, n,$$

with $\mu_i > -1/2$. The $\mathbb{Z}_2^n$-Dunkl-Laplace operator is

$$\vec{D}^2 = \sum_{i=1}^{n} D_i^2.$$

With γ_n the generators of the Euclidean Clifford algebra

$$\{\gamma_m, \gamma_n\} = 2\delta_{nm},$$

the Dunkl-Dirac operator is

$$\not{D} = \sum_{i=1}^{n} \gamma_i D_i.$$

Clearly, one has $\not{D}^2 = \vec{D}^2$. Let us consider the three-dimensional case. Introduce the Dunkl "angular momentum" operators

$$J_1 = \frac{1}{i}(x_2 D_3 - x_3 D_2), \quad J_2 = \frac{1}{i}(x_3 D_1 - x_1 D_3), \quad J_3 = \frac{1}{i}(x_1 D_2 - x_2 D_1).$$

Their commutation relations are found to be

$$[J_i, J_k] = i\epsilon_{jkl} J_l (1 + 2\mu_l R_l). \qquad (26)$$

The Dunkl-Laplace equation separates in spherical coordinates; i.e. one can write

$$\vec{D}^2 = D_1^2 + D_2^2 + D_3^2 = \mathcal{M}_r + \frac{1}{r^2}\Delta_{S^2},$$

where Δ_{S^2} is the Dunkl-Laplacian on the 2-sphere. It can be verified that [36]

$$\begin{aligned}\vec{J}^2 &= J_1^2 + J_2^2 + J_3^2 \\ &= -\Delta_{S^2} + 2\mu_1\mu_2(1 - R_1R_2) + 2\mu_2\mu_3(1 - R_2R_3) + 2\mu_1\mu_3(1 - R_1R_3) \\ &\qquad - \mu_1R_1 - \mu_2R_2 - \mu_3R_3 + \mu_1 + \mu_2 + \mu_3.\end{aligned} \tag{27}$$

In three dimensions the Euclidean Clifford algebra is realized by the Pauli matrices

$$\sigma_1 = \begin{pmatrix} 0 & 1 \\ 1 & 0 \end{pmatrix}, \quad \sigma_2 = \begin{pmatrix} 0 & -i \\ i & 0 \end{pmatrix}, \quad \sigma_3 = \begin{pmatrix} 1 & 0 \\ 0 & -1 \end{pmatrix},$$

which satisfy

$$\sigma_i\sigma_j = i\epsilon_{ijk}\sigma_k + \delta_{ij}.$$

Consider the following operator:

$$\Gamma = (\vec{\sigma} \cdot \vec{J}) + \vec{\mu} \cdot \vec{R},$$

with $\vec{\mu} \cdot \vec{R} = \mu_1R_1 + \mu_2R_2 + \mu_3R_3$. Using the commutation relations (26) and the expression (27) for $\vec{J}^2$, it follows that

$$\Gamma^2 + \Gamma = -\Delta_{S^2} + (\mu_1 + \mu_2 + \mu_3)(\mu_1 + \mu_2 + \mu_3 + 1).$$

This is reminiscent of the expression (25) for the superintegrable system with Hamiltonian (22) in terms of the $sl_{-1}(2)$ Casimir operator. This justifies calling Γ a Dunkl-Dirac operator on S^2 since a quadratic expression in Γ gives Δ_{S^2}. The symmetries of Γ can be constructed. They are found to have the expression [37]

$$M_i = J_i + \sigma_i(\mu_jR_j + \mu_kR_k + 1/2), \quad (ijk) \text{ cyclic},$$

and one has $[\Gamma, M_i] = 0$. It is seen that the operators

$$X_i = \sigma_iR_i \qquad i = 1, 2, 3$$

also commute with Γ. Furthermore, one has

$$[M_i, X_i] = 0, \qquad \{M_i, X_j\} = \{M_i, X_k\} = 0.$$

Note that $Y = -iX_1X_2X_3 = R_1R_2R_3$ is central (like Γ). The commutation relations satisfied by the operators M_i are

$$[M_i, M_j] = i\epsilon_{ijk}\left(M_k + 2\mu_k(\Gamma + 1)X_k\right) + 2\mu_i\mu_j[X_i, X_j].$$

This is again an extension of $\mathfrak{su}(2)$ with reflections and central elements. Let

$$K_i = M_iX_iY = M_i\sigma_iR_jR_k.$$

It is readily verified that the operators K_i satisfy

$$\begin{aligned}\{K_1, K_2\} &= K_3 + 2\mu_3(\Gamma + 1)Y + 2\mu_1\mu_2, \\ \{K_2, K_3\} &= K_1 + 2\mu_1(\Gamma + 1)Y + 2\mu_2\mu_3, \\ \{K_3, K_1\} &= K_2 + 2\mu_3(\Gamma + 1)Y + 2\mu_3\mu_1,\end{aligned}$$

showing that the Bannai-Ito algebra is a symmetry subalgebra of the Dunkl-Dirac equation on S^2. Therefore, the Bannai-Ito algebra is also a symmetry subalgebra of the Dunkl-Laplace equation.

11. Conclusion

In this paper, we have presented the Bannai-Ito algebra together with some of its applications. In concluding this overview, we identify some open questions.

(i) Representation theory of the Bannai-Ito algebra

Finite-dimensional representations of the Bannai-Ito algebra associated to certain models were presented. However, the complete characterization of all representations of the Bannai-Ito algebra is not known.

(ii) Supersymmetry

The parallel with supersymmetry has been underscored at various points. One may wonder if there is a deeper connection.

(iii) Dimensional reduction

It is well known that quantum superintegrable models can be obtained by dimensional reduction. It would be of interest to adapt this framework in the presence of reflections operators. Could the BI algebra can be interpreted as a W-algebra ?

(iv) Higher ranks

Of great interest is the extension of the Bannai-Ito algebra to higher ranks, in particular for many-body applications. In this connection, it can be expected that the symmetry analysis of higher dimensional superintegrable models or Dunkl-Dirac equations will be revealing.

Acknowledgments

V.X.G. holds an Alexander-Graham-Bell fellowship from the Natural Science and Engineering Research Council of Canada (NSERC). The research of L.V. is supported in part by NSERC. H. DB. and A.Z. have benefited from the hospitality of the Centre de recherches mathématiques (CRM).

References

[1] Arik M and Kayserilioglu U 2003 *Int. J. Mod. Phys. A* **18** 5039–5046
[2] Gorodnii M and Podkolzin G 1984 *Irreducible representations of a graded Lie algebra* (Inst. Math. Acad. Sci.) pp 66–76 Spectral theory of operators and infinite-dimensional analysis
[3] Brown G 2013 *Elec. J. Lin. Alg.* **26** 258–299
[4] Ostrovskii V and Sil'vestrov S 1992 *Ukr. Math. J.* **44** 1395–1401
[5] Jafarov E I, Stoilova N I and Van der Jeugt J 2011 *J. Phys. A: Math. Theor.* **44** 265203
[6] Jafarov E I, Stoilova N I and Van der Jeugt J 2011 *J. Phys. A: Math. Theor.* **44** 355205
[7] Tsujimoto S, Vinet L and Zhedanov A 2012 *Adv. Math.* **229** 2123–2158
[8] Andrews G, Askey R and Roy R 1999 *Special functions* (Cambridge: Cambridge University Press)
[9] Leonard D 1982 *SIAM J. Math. Anal.* **13** 656–663
[10] Bannai E and Ito T 1984 *Algebraic Combinatorics I: Association Schemes* (San Francisco: Benjamin/Cummings)
[11] Koekoek R, Lesky P and Swarttouw R 2010 *Hypergeometric orthogonal polynomials and their q-analogues* 1st ed (Springer) ISBN 978-3-642-05013-8
[12] Zhedanov A 1991 *Theor. Math. Phys.* **89** 1146–1157
[13] Genest V X, Vinet L and Zhedanov A 2013 *SIGMA* **9** 18–37
[14] Genest V X, Vinet L and Zhedanov A 2014 *SIGMA* **10** 38–55
[15] Vinet L and Zhedanov A 2011 *J. Phys. A: Math. Theor.* **44** 085201
[16] Vinet L and Zhedanov A 2012 *Trans. Amer. Math. Soc.* **364** 5491–5507
[17] Tsujimoto S, Vinet L and Zhedanov A 2013 *Proc. Amer. Math. Soc.* **141** 959–970
[18] Tsujimoto S, Vinet L and Zhedanov A 2011 *J. Math. Phys.* **52** 103512
[19] Chihara T 2011 *An Introduction to Orthogonal Polynomials* (Dover Publications)
[20] Tsujimoto S, Vinet L and Zhedanov A 2011 *SIGMA* **7** 93–105
[21] Green H S 1953 *Phys. Rev.* **90** 270–273

[22] Ganchev A C and Palev T D 1980 *J. Math. Phys.* **21** 797
[23] Kac V G 1977 *Adv. Math.* **26** 8–96
[24] Vilenkin N J and Klimyk A U 1991 *Representation of Lie Groups and Special Functions* (Amsterdam: Kluwer Academic Publishers)
[25] Daskaloyannis C and Kanakoglou K 2000 *J. Math. Phys.* **41** 652–660
[26] Dunkl C F 1989 *Trans. Amer. Math. Soc.* **311** 167–183
[27] Genest V X, Ismail M, Vinet L and Zhedanov A 2013 *J. Phys. A: Math. Theor.* **46** 145201
[28] Genest V X, Vinet L and Zhedanov A 2014 *Proc. Am. Soc.* **142** 1545–1560
[29] Genest V X, Vinet L and Zhedanov A 2013 *Comm. Math. Phys. (to appear)*
[30] Genest V X, Vinet L and Zhedanov A 2013 *J. Math. Phys.* **54** 023506
[31] Miller W, Post S and Winternitz P 2013 *J. Phys. A: Math. Theor.* **46** 423001
[32] Genest V X, Vinet L and Zhedanov A 2014 *J. Phys. A: Math. Theor.* **47** 205202
[33] Genest V X, Ismail M, Vinet L and Zhedanov A 2014 *Comm. Math. Phys.* **329** 999–1029
[34] Genest V X, Lemay J M, Vinet L and Zhedanov A 2013 *J. Phys. A: Math. Theor.* **46** 505204
[35] Genest V X, Vinet L and Zhedanov A 2013 *J. Phys. A: Math. Theor.* **46** 325201
[36] Genest V X, Vinet L and Zhedanov A 2014 *J. Phys.: Conf. Ser.* **512** 012010
[37] De Bie H, Genest V X and Vinet L 2014 *In preparation*

Journal of Physics: Conference Series **597** (2015) 012002 doi:10.1088/1742-6596/597/1/012002

Surface defects and symmetries

Jürgen Fuchs[1] **and Christoph Schweigert**[2]

[1] Teoretisk fysik, Karlstads universitet, Universitetsgatan 21, 651 88 Karlstad, Sweden
[2] Algebra und Zahlentheorie, Fachbereich Mathematik, Universität Hamburg, Bundesstraße 55, 20148 Hamburg, Germany

E-mail: `juergen.fuchs@kau.se`, `christoph.schweigert@uni-hamburg.de`

Abstract. In quantum field theory, defects of various codimensions are natural ingredients and carry a lot of interesting information. In this contribution we concentrate on topological quantum field theories in three dimensions, with a particular focus on Dijkgraaf-Witten theories with abelian gauge group. Surface defects in Dijkgraaf-Witten theories have applications in solid state physics, topological quantum computing and conformal field theory. We explain that symmetries in these topological field theories are naturally defined in terms of invertible topological surface defects and are thus Brauer-Picard groups.

1. Introduction

In this contribution we discuss the relation between two important structures in quantum field theory: symmetries on the one hand, and defects of various codimensions on the other. While it is hardly necessary to emphasize the importance of symmetries, defects in quantum field theories or, in a different formulation, quantum field theories on stratified spaces, and their relation to symmetries are a more recent topic of interest.

Our exposition is organized as follows. We first discuss a few general aspects of topological codimension-1 defects, both in two-dimensional and in three-dimensional quantum field theories. A special emphasis is on the relation between *invertible* topological defects and symmetries. Afterwards we consider defects for a particular class of three-dimensional topological field theories, Dijkgraaf-Witten theories. An advantage of these theories is that they admit a mathematically precise formulation as gauge theories, see [7, 10, 23] and references therein. Apart from providing insights in a language that is close to standard field theoretical formulations, this allows one to uncover interesting relations between categories of (relative) bundles and recent results in representation theory.

2. Topological defects in quantum field theories

It has been recently realized in several different contexts that boundary conditions and defects of various codimensions constitute important parts of the structure of a quantum field theory. (Alternatively, one might say that it is instructive to consider the quantum field theory not only on smooth manifolds, but also on stratified spaces.) Codimension-1 defects, also called interfaces, separate regions that can support different quantum field theories. Such defects arise naturally in applications, ranging from condensed matter systems, where they appear as interfaces between different phases of matter, to domain walls in cosmology. The aspect we wish to emphasize in this contribution is the fact that such defects provide a lot of structural insight into quantum field theories, including in particular their symmetry structures.

Similarly as in the case of boundary conditions, there are various different types of defects one can consider. In particular, one can impose various conservation conditions of physical quantities on defects. An important subclass of codimension-1 defects are *topological* defects. They are characterized by the property that the values of correlators for configurations with defects do not change when the location of a defect is only slightly changed, that is, changed by a homotopy without crossing any field insertions or other defects. Examples of such defects have been known for a long time (see e.g. [8]). Consider, for instance, the two-dimensional Ising model, defined by a $\mathbb{Z}_2$-valued variable on the vertices of a two-dimensional CW-complex. Select a line that crosses bonds transversally, and change the coupling on each bond that is crossed by the line from ferromagnetic to anti-ferromagnetic. In the continuum limit this provides a topological defect line in the critical Ising model.

2.1. Symmetries from invertible topological defects

A particularly important subclass of topological defects are the *invertible* topological defects. Owing to their topological nature, two topological defects can be brought to coincidence, leading to a fusion product of defects. The precise mathematical formulation of the relevant monoidal structure depends on the dimension in which the quantum field theory is defined. For the moment, let us consider two-dimensional quantum field theories. All statements made in the sequel have the status of theorems [12, 13] in the case of two-dimensional rational conformal field theories; the general picture should be much more widely applicable, though. For a two-dimensional rational conformal field theory, the topological line defects form a monoidal category, with morphisms provided by field insertions which can change the type of defect. In particular, there is an "invisible defect", which is a monoidal unit $\mathbf{1}$ in the category of defects.

An *invertible* defect D is characterized by the fact that there exists another defect $D^\vee$ with the property that the fusion of D and $D^\vee$ gives the invisible defect,

$$D \otimes D^\vee \cong \mathbf{1} \cong D^\vee \otimes D\,.$$

(Thus in particular for every theory the invisible defect is invertible.) This behaviour leads to the relation [1]

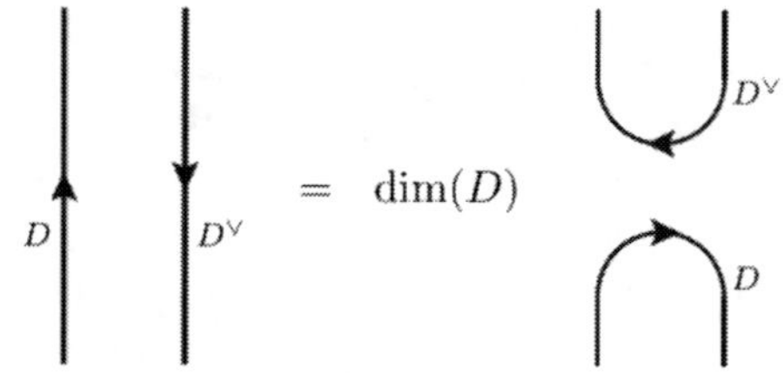

which is to be understood as an identity of correlators when applied locally in any configuration of fields and defect lines. With this relation, it is immediate to deduce a connection with symmetries: one has equalities of the form

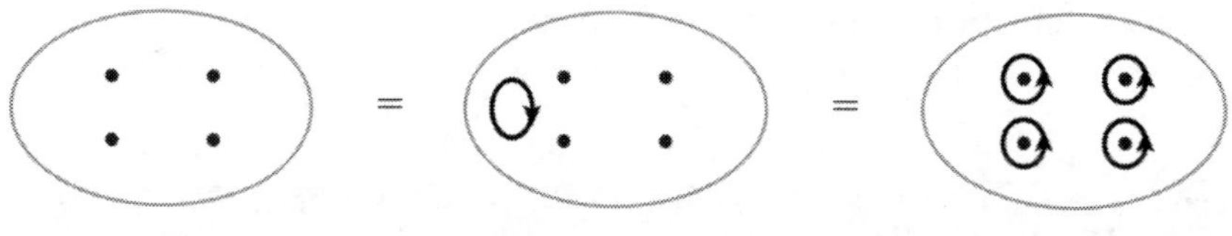

[1] Here $\dim(D)$ is the (quantum) dimension of the defect, which for an invertible defect can take only the values $+1$ or -1. This factor is suppressed in the sequel. In unitary theories one has $\dim(D) = 1$ for every invertible defect.

between correlators of different bulk fields.

Let us discuss this issue explicitly for the case of the two-dimensional critical Ising model. There are then three primary fields, commonly referred to as the identity field 1, the spin field σ and the energy field ϵ. Their scaling dimensions are 0, $\frac{1}{16}$ and $\frac{1}{2}$, respectively. Indecomposable topological line defects of the critical Ising model turn out to be in bijection with these fields: in particular, 1 corresponds to the invisible defect and ϵ to an invertible defect. (It is in fact the latter defect that amounts to changing the couplings from ferromagnetic to antiferromagnetic.) The action of this specific invertible defect on bulk fields is as follows: it leaves the bulk fields corresponding to 1 and to ϵ invariant and changes the sign of σ. Inside RCFT correlators for the critical Ising model we thus have the equalities

As can be seen from these pictures, there is a natural action of invertible topological line defects on field insertions: wrapping such a defect around a bulk field insertion yields another bulk field.

A further advantage of having a realization of symmetries in terms of invertible topological defects is seen by studying what happens when the defect is moved to a boundary with boundary condition M: this process yields another boundary condition M', according to

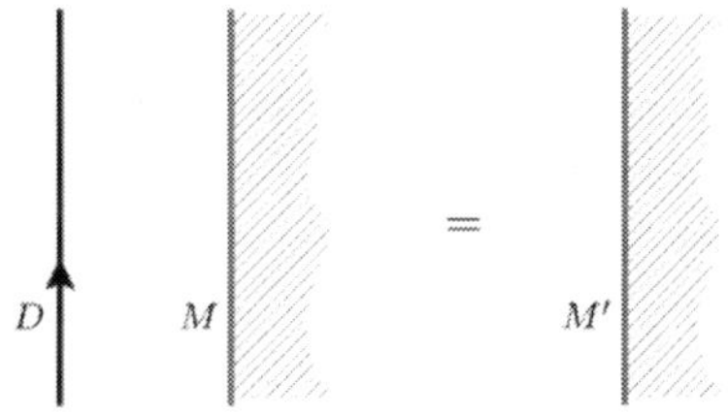

In a similar way, by wrapping around various structures in an obvious way, the group of invertible topological defects acts as a symmetry group on *all* data of the field theory, including field insertions of bulk, boundary and disorder fields, boundary conditions and types of defects. In the framework of the TFT construction of correlators of two-dimensional rational conformal field theories [25], isomorphism classes of invertible topological defects can be explicitly classified [12, 13]. In the example of the critical Ising model one finds a symmetry group $\mathbb{Z}_2$, consisting of the invisible defect and the ϵ-defect described above, while e.g. for the critical three-state Potts model one obtains a non-abelian symmetry group S_3.

For the present purposes, it is important to realize that topological codimension-1 defects can also exist in more general classes of quantum field theories. Moreover, in the general case geometric considerations suggest a natural action of such defects on field theoretic structures, like boundary conditions and defects of various codimensions, as well. In short, defects can wrap around field theoretic structures.

2.2. T-dualities and Kramers-Wannier dualities from topological line defects

Before turning to higher-dimensional field theories, we point out that topological line defects are indeed of much wider use: they can also implement Kramers-Wannier dualities and T-dualities.

If a general topological defect wraps around a bulk field, the following situation is created:

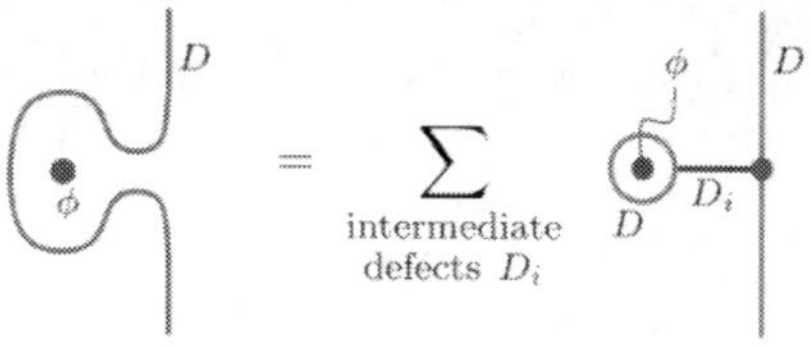

In this way a bulk field ϕ is turned by the defect D into a disorder field. To obtain an order-disorder duality, one also needs the opposite process, turning disorder fields into ordinary local bulk fields. It can be shown [12, 13] that to this end the dual defect $D^\vee$ must be used, and that in this case one turns the disorder field back into a bulk field if and only if the fused defect $D \otimes D^\vee$ is a direct sum of invertible defects. This condition can be examined in concrete models. It is in particular satisfied for the defect corresponding to the spin field σ in the critical Ising model, thanks to the well known fusion rule $\sigma \otimes \sigma \cong 1 \oplus \epsilon$; this defect indeed produces the action of the Kramers-Wannier duality in the critical model. Again, we have a natural action on correlators,

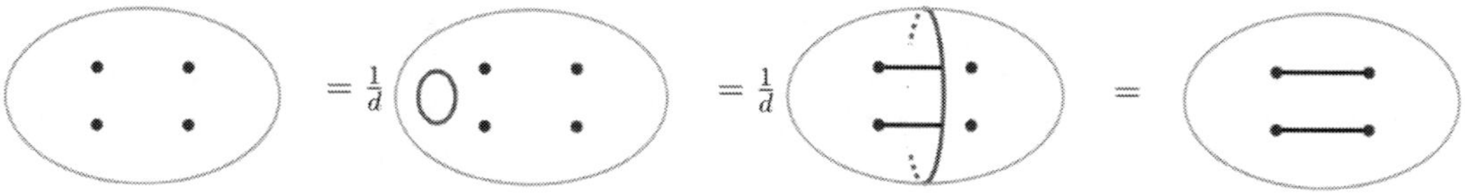

which shows how correlators involving only bulk fields are related to correlators involving disorder fields. Specifically, through the action of the σ-defect the correlator of four spin fields on a sphere can be seen to be equal to the correlator of four disorder fields, according to

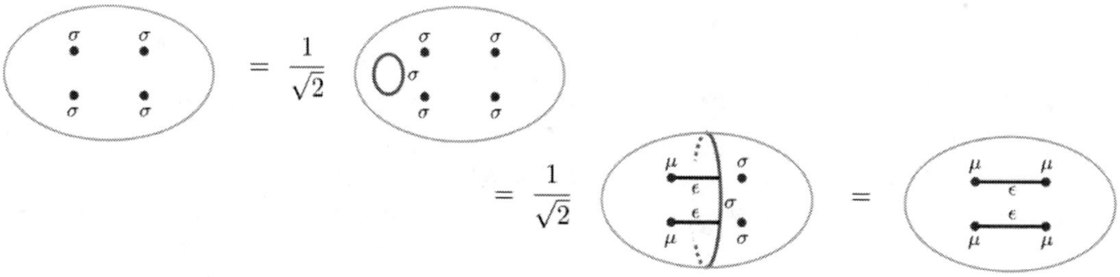

and the correlator of two spin fields on a torus can be expressed as

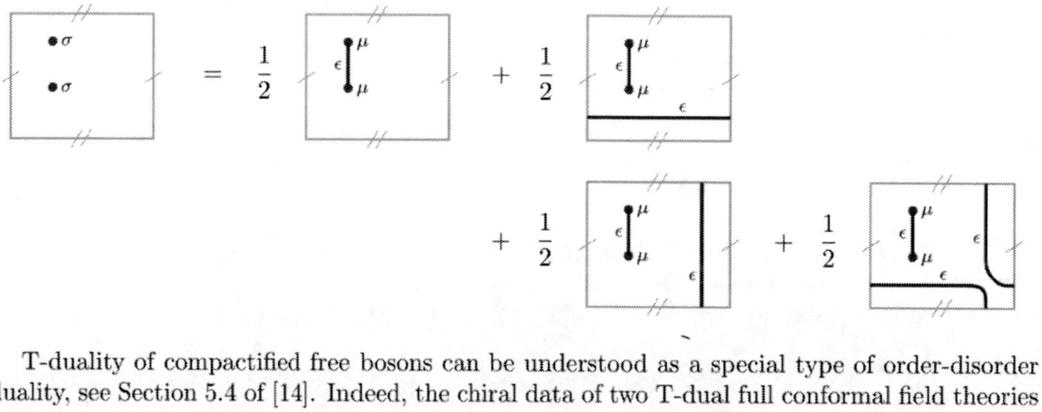

T-duality of compactified free bosons can be understood as a special type of order-disorder duality, see Section 5.4 of [14]. Indeed, the chiral data of two T-dual full conformal field theories

can be described by the same theory, namely the one describing a $\mathbb{Z}_2$-orbifold of the free boson theory. The line defect implementing the duality is then constructed from twist fields of this orbifold theory.

2.3. Relative field theories
Before turning to defects in the specific class of three-dimensional Dijkgraaf-Witten theories, we briefly explain two applications of defects. The first is in the context of *relative field theories*. By definition, a $(d-1)$-dimensional relative field theory is a field theory on the boundary of a d-dimensional field theory or, more generally, on a codimension-1 defect. There are interesting cases in which the latter is a topological field theory; we mention two particular situations:

- The case that the d-dimensional topological field theory is even an invertible theory. This provides a geometric setup for describing anomalous theories; see [11].
- The case $d=3$, with the three-dimensional topological field theory being of Reshetikhin-Turaev type and built from a modular tensor category that is the representation category of a vertex algebra. In this case the relative theory on a topological surface defect is a local rational conformal field theory whose underlying chiral theory is the one built from the vertex algebra. This provides a geometric interpretation [21] of the TFT construction [25] of RCFT correlators.

2.4. Quantum codes
Another application of surface defects in three-dimensional topological field theories are quantum codes. It is well known that the space of qubits of the toric code on a surface Σ can be described by a topological field theory of Turaev-Viro type. Indeed, the vector space assigned by the topological field theory to the surface Σ contains the information on a vector space together with an action of mapping class groups which is used to construct quantum gates. Two problems are apparent: First, the surfaces Σ of concrete samples typically have a low genus, and thus, according to the Verlinde formula, the TFT state space space $\mathrm{tft}(\Sigma)$ has low dimension. Moreover, the representation of the mapping class group might be too small to allow for universal gates. To circumvent these problems the idea has been put forward [3, 2, 1] to consider samples with a bilayer (or multi-layer) system and twist defects that create branch cuts. They effectively lead to systems defined on surfaces of higher genus and thus to higher-dimensional quantum codes with richer representations of mapping class groups. Such systems have been analyzed mathematically [17] by exploiting results on permutation equivariant categories.

3. Defects and boundary conditions in three-dimensional topological field theories

We now turn our attention to a particularly accessible subclass of three-dimensional topological field theories, Dijkgraaf-Witten theories. These are topological field theories of Turaev-Viro type, and they can be constructed explicitly in a gauge-theoretic setting with finite gauge group. We treat them as 3-2-1-extended topological field theories.

3.1. Construction of extended Dijkgraaf-Witten theories from G-bundles
As a first input datum, we select a finite group G. The space of field configurations on a closed oriented compact smooth three-manifold M is then taken to be the groupoid $\mathrm{Bun}_G(M)$ of G-bundles on M. By considering formally a path integral with vanishing action, we obtain a partition function that evaluates to

$$\mathrm{tft}_G(M) = \int_{\mathrm{Bun}_G(M)} \mathrm{D}A \; \mathrm{e}^0 = \big|\mathrm{Bun}_G(M)\big| .$$

Here we use the fact that the groupoid $\mathrm{Bun}_G(M)$ is essentially finite, so that there is a well-defined counting measure, its groupoid cardinality. The groupoid cardinality of an (essentially) finite groupoid Γ is

$$|\Gamma| := \sum_{\gamma\in\pi_0(\Gamma)} \frac{1}{|\mathrm{Aut}_\Gamma(\gamma)|}\,,$$

where the summation is over the set $\pi_0(\Gamma)$ of isomorphism classes of objects of Γ, each of which is counted with the inverse of the cardinality of its automorphism group.

Trivially, the number $\mathrm{tft}_G(M) = |\mathrm{Bun}_G(M)|$ defines an invariant of the three-manifold M. It is non-trivial, though, that this invariant is local in the following sense. Let us cut the closed three-manifold into two pieces, consisting of three-manifolds with boundaries which are closed oriented surfaces. Suppose that the topological field theory can associate meaningful quantities to such manifolds as well, which would allow us to reduce the problem of computing a three-manifold invariant to computing invariants of simpler manifolds. To see what the topological field theory tft_G should associate to a closed oriented two-manifold Σ, we take a three-manifold M with boundary Σ. For any G-bundle P on Σ, consider the groupoid $\mathrm{Bun}_G(M,P)$ of G-bundles on M restricting to the G-bundle P on the boundary ∂M. Then $\mathrm{Bun}_G(M,P)$ is an essentially finite groupoid: its groupoid cardinality provides a function

$$\begin{array}{rrcl} \Psi_M: & \mathrm{Bun}_G(\Sigma) & \longrightarrow & \\ & P & \longmapsto & |\mathrm{Bun}_G(M,P)| \end{array}$$

that depends only on the isomorphism class of the bundle P. To the surface Σ itself, we should associate the recipient for all these functions, the vector space $[\pi_0(\mathrm{Bun}_G(\Sigma))]$ of complex-valued functions on the set $\pi_0(\mathrm{Bun}_G(\Sigma))$ of isomorphism classes of G-bundles on Σ. We thus recover the well known feature that a three-dimensional topological field theory assigns vector spaces to surfaces. In the case of Dijkgraaf-Witten theories, these vector spaces are obtained by linearization of (isomorphism classes of objects of) categories of bundles.

Closed two-manifolds can, in turn, be decomposed, e.g. in a pair-of-pants decomposition, by cutting them along circles. Hence let S be a closed oriented one-manifold and ask what an extended topological field theory should associate to S. Following the same type of analysis as above, we choose a surface Σ with boundary S. Fixing a G-bundle P on the boundary S, we assign to it a vector space,

$$\Psi_\Sigma: \quad P \longmapsto \quad [\pi_0(\mathrm{Bun}_G(\Sigma,P))]\,.$$

The so obtained map $\Psi_\Sigma: \mathrm{Bun}_G(S) \to \mathrm{Vect}$ is a vector bundle on the space of field configurations on S. Thus the extended Dijkgraaf-Witten theory based on G should associate to S the -linear category of vector bundles over the space of field configurations. Denoting, for an essentially finite groupoid Γ, the functor category from Γ to a category $\mathcal{C}$ by $[\Gamma, \mathcal{C}]$, this is the functor category

$$\mathrm{tft}_G(S) = [\mathrm{Bun}_G(S), \mathrm{Vect}]\,.$$

We thus learn that a 3-2-1-extended topological field theory assigns to a one-manifold a -linear category. Moreover, the assignment consists of two steps: one first assigns to S the groupoid $\mathrm{Bun}_G(S)$, and in a second step to this groupoid its linearization $[\mathrm{Bun}_G(S), \mathrm{Vect}]$.

Hereby we naturally arrive at the general definition of 3-2-1-extended topological field theories as a symmetric monoidal 2-functor

$$\mathrm{tft}: \quad \mathrm{cobord}_{3,2,1} \to \text{2-Vect}(\quad)\,,$$

which we discuss in the rest of this subsection. This definition naturally generalizes Atiyah's classical definition of (non-extended) topological field theory. Here 2-Vect( ) is the symmetric monoidal bicategory of finitely semisimple -linear abelian categories, with the Deligne product as a monoidal structure. The bicategory $\mathrm{cobord}_{3,2,1}$ is an extended cobordism category whose objects are closed oriented one-manifolds S, whose 1-morphisms $S \to S'$ are two-manifolds Σ with boundary together with a decomposition $\partial\Sigma \xrightarrow{\simeq} S' \sqcup -S$, and whose 2-morphisms are three-manifolds with corners, up to diffeomorphism, again with an appropriate decomposition of the boundary.

According to the definition, the 2-functor tft assigns to the oriented circle $\mathbb{S}^1$ a finitely semisimple -linear category $\mathcal{C} := \mathrm{tft}(\mathbb{S}^1)$. The trinion (pair of pants), regarded as a cobordism $\mathbb{S}^1 \sqcup \mathbb{S}^1 \to \mathbb{S}^1$ gives a functor $\otimes\colon\ \mathcal{C} \boxtimes \mathcal{C} \to \mathcal{C}$. Finally, suitable three-manifolds with corners provide natural isomorphisms that endow this functor with an associativity constraint and the category $\mathcal{C}$ with the structure of a braiding. The constraints on these natural transformations, expressed by pentagon and hexagon diagrams, can be deduced from homotopies between three-manifolds with corners. In this way, the category $\mathcal{C} = \mathrm{tft}(\mathbb{S}^1)$ is expected to be endowed with the structure of a modular tensor category (see [4] for some recent progress). The Reshetikhin-Turaev construction can be seen as a converse, constructing a 3-2-1-extended topological field theory from a modular tensor category.

3.2. Extended three-dimensional topological field theories with boundaries and defects

Incorporating defects and boundaries into a topological field theory amounts to consider an enlarged bicategory $\mathrm{cobord}^{\partial}_{3,2,1}$ of cobordisms. Recently several definitions have been proposed for this enlarged category, see e.g. [22, Sect 4.3]; we do not present a formal definition, but rather formulate some of the physical requirements that such a category should satisfy. Again we start from oriented three-manifolds, but this time we also allow them to have oriented boundaries that are actual physical boundaries to which boundary conditions need to be assigned, rather than the cut-boundaries that were already present in Section 3.1, which are instead a device for incorporating some aspects of locality into the theory. We also admit codimension-1 submanifolds, called surface defects. Such a defect separates two three-dimensional regions each of which supports a topological field theory, and the theories associated to the two regions may be different. Labels must be assigned to such defects, too. In this way, we get in particular invariants of three-manifold with embedded surfaces. In the present setting of 3-2-1-extended topological field theories, it is natural to go one step further and allow for codimension-2 defects as well. In particular, there will be generalized Wilson lines separating two surface defects from each other, and likewise boundary Wilson lines that are confined to a boundary component.

In a next step one then imposes locality by cutting such a three-manifold to produce three-manifolds with cut-boundaries. It is natural to impose the condition that the cut-surface intersects all structure in the three-manifold, e.g. surface defects and boundaries, transversally. Boundary manifolds now carry a decoration consisting of lines labeled by surface defects and boundaries labeled by boundary conditions. Imposing further locality, as in Section 3.1, requires to cut such decorated surfaces transversally. This yields new types of labeled one-manifolds, which are to be taken as the objects in the bicategory $\mathrm{cobord}^{\partial}_{3,2,1}$. Let us describe these objects in detail: There are oriented intervals and circles, with finitely many marked points in the interior that divide them into subintervals. Each subinterval is to be marked with a topological field theory. Points adjacent to two subintervals are to be marked by a surface defect, boundary points by a boundary condition. The possible boundary conditions and types of surface defects have to be determined by a field-theoretic analysis. For the case of Dijkgraaf-Witten theories we present this analysis in the next subsection.

3.3. Defects and boundaries in Dijkgraaf-Witten theories

We now describe an explicit gauge-theoretic construction [19] of boundary conditions and surface defects for Dijkgraaf-Witten theories. A key idea is to keep the two-step procedure outlined in

Section 3.1: first we assign to manifolds of various dimension groupoids of (generalized) bundles, and then we linearize those categories.

Our ansatz is inspired by the notion of a *relative bundle*: Given a morphism $j\colon Y \to X$ of smooth manifolds and a group homomorphism $\iota\colon H \to G$, the category of relative bundles has as objects triples consisting of a G-bundle $P_G \to X$, an H-bundle $P_H \to Y$, and an isomorphism

$$\alpha: \quad \mathrm{Ind}_H^G(P_H) \xrightarrow{\cong} j^*P_G$$

of G-bundles on Y. A morphism is a pair consisting of a morphism $P_G \xrightarrow{\varphi_G} P'_G$ of G-bundles on X and a morphism $P_H \xrightarrow{\varphi_H} P'_H$ on Y of H-bundles, obeying the obvious compatibility constraint

$$\begin{array}{ccc} \mathrm{Ind}_H^G(P_H) & \xrightarrow{\alpha} & j^*P_G \\ {\scriptstyle \mathrm{Ind}_H^G(\varphi_H)}\downarrow & & \downarrow{\scriptstyle j^*\varphi_G} \\ \mathrm{Ind}_H^G(P'_H) & \xrightarrow{\alpha'} & j^*P'_G \end{array}$$

with the isomorphisms α and α'.

In our construction, relative bundles are the appropriate tool for describing boundaries. In the case of a surface defect $\Sigma \hookrightarrow M$ the situation is somewhat more subtle. Suppose that Σ divides the three-manifold M into two disjoint connected components $M_\pm$. We consider the situation that Dijkgraaf-Witten theories for two, possibly different, finite groups $G_\pm$ are associated to the components M_+ and M_-. Denote by $\overline{M}_\pm := M_\pm \cup \Sigma$ the closure of $M_\pm$ in M, and by $j_\pm\colon \Sigma \hookrightarrow \overline{M}_\pm$ the corresponding embeddings. We then consider, for a given group homomorphism $H \to G_+ \times G_-$ as field configurations the groupoid whose objects consist of two $G_\pm$-bundles $P_\pm \to M_\pm$, an H-bundle $P_H \to \Sigma$ and an isomorphism

$$\beta: \quad \mathrm{Ind}_H^G(P_H) \xrightarrow{\cong} j_+^*P_+ \times j_-^*P_-$$

of $G_+ \times G_-$-bundles on Σ.

We have to add yet one further datum to Dijkgraaf-Witten theories. Namely, typically, the linearization of groupoids is twisted by a 2-cocycle on the relevant groupoid of generalized bundles. This requires the choice of additional data on the finite groups involved. We first discuss the situation without boundaries and without surface defects, for a Dijkgraaf-Witten theory based on a finite group G. Then the additional datum is a three-cocycle $\omega \in Z^3(G,\ {}^\times)$ on the group. A more geometrically inclined reader is invited to think about this cocycle as a three-cocycle on the stack Bun_G of G-bundles, and thus as a (Chern-Simons) 2-gerbe on Bun_G. Such a three-cocycle leads to a holonomy on closed three-manifolds and thus furnishes a topological Lagrangian, yielding a three-dimensional topological field theory $\mathrm{tft}_{G,\omega}$. To determine the category $\mathrm{tft}_{G,\omega}(\mathbb{S}^1)$, we need a 2-cocycle $\tau(\omega)$ on the groupoid $\mathrm{Bun}_G(\mathbb{S}^1) \cong G//G$, the action groupoid for the adjoint action of G on itself. The transgressed 2-cocycle $\tau(\omega)$ can be obtained [26] from the group cocycle ω by evaluating ω on a suitable triangulated three-manifold. This indeed produces the well-known 2-cocycle [6] for Dijkgraaf-Witten theories.

We now illustrate the generalization of this prescription to boundaries and defects in an example. Let $\mathbb{I}$ be an interval with a marked point in its interior, corresponding to a surface defect. To each of the two subintervals $\mathbb{I}_{1,2}$ we a assign a finite group $G_{1,2}$ and three-cocycles $\omega_{1,2} \in Z^3(G_{1,2},\ {}^\times)$, of which we think of as topological bulk Lagrangians. To the end point adjacent to the first interval, we assign a group homomorphism $\iota_1\colon H_1 \to G_1$ and a boundary Lagrangian as follows. We think about the three-cocycle ω_1 on G_1 as a Chern-Simons 2-gerbe on Bun_{G_1}. The group homomorphism ι_1 induces a morphism $\mathrm{Bun}_{H_1} \to \mathrm{Bun}_{G_1}$ of 2-gerbes. A similar situation, one categorical dimension lower, is familiar from the study of D-branes, see e.g. [15]: for D-branes, one has a gerbe on the bulk manifold and a morphism of gerbes on the boundary, from the trivial gerbe to the restriction of the bulk gerbe. This is an instance of the general principle that in gauge theories, holonomies of manifolds with boundary, and thus topological actions on such manifolds, can only be defined if a trivialization on the boundary is chosen.

In the situation at hand, what is to be trivialized is the restriction of the Chern-Simons 2-gerbe to Bun_{H_1}. Thus in terms of group cochains, we are looking for a 2-cochain $\theta_1 \in C^2(H_1, \ ^\times)$ that represents a morphism from the trivial 2-gerbe on Bun_H to the 2-gerbe described by the restriction of the three-cocycle ω_1 on G_1 to Bun_{H_1}. Accordingly we impose the condition $\mathrm{d}\theta_1 = \iota_1^*(\omega_1)$. We think about θ_1 as a topological boundary Lagrangian. The situation at the other end point is, in complete analogy, described by a group homomorphism $\iota_2\colon H_2 \to G_2$ and a 2-cochain $\theta_2 \in C^2(H_2, \ ^\times)$ such that $\mathrm{d}\theta_2 = \iota_2^*(\omega_2)$.

In a similar vein, the situation for the surface defect is a higher-categorical analogue of bibranes [20], which describe the target space physics of topological defects: We now have a group homomorphism $\iota_{12}\colon H_{12} \to G_1 \times G_2$ and a 2-cochain $\theta_{12} \in C^2(H_{12}, \ ^\times)$ satisfying $\mathrm{d}\theta_{12} = \tilde\iota_2^*(\omega_2) \cdot \tilde\iota_1^*(\omega_1)^{-1}$, with the group homomorphisms $\tilde\iota_i\colon H_{12} \xrightarrow{\iota_{12}} G_1 \times G_2 \xrightarrow{p_i} G_i$ obtained from ι_{12} and the canonical projections. Indecomposable boundary conditions and defects correspond to group homomorphism $\iota_{1,2}$, respectively ι_{12}, that are injective, i.e. are subgroup embeddings.

In [19], an explicit prescription was given for transgressing these cochains to a 2-cocycle on the relevant category of (generalized) relative bundles. This prescription is a generalization of the one given in [26] for Dijkgraaf-Witten theories without boundaries or defects. It makes the categories associated to one-manifolds with defects and boundaries quite explicitly computable.

On the other hand, there are also general results for boundary conditions and defects in three-dimensional topological field theories of Reshetikhin-Turaev type. Suppose the theory is based on a modular tensor category $\mathcal{C}$. A model independent analysis [18] reveals that topological boundary conditions only exist if the category $\mathcal{C}$ is braided equivalent to the Drinfeld center of a fusion category $\mathcal{A}$. For Dijkgraaf-Witten theories based on a finite group G and a three-cocycle $\omega \in Z^3(G, \ ^\times)$, this condition is satisfied automatically: the category $\mathcal{C} = \mathcal{C}_{G,\omega}$ is the Drinfeld center of the category $\mathrm{Vect}(G)^\omega$ of G-graded vector spaces with associator twisted by the three-cocycle ω.

The general analysis [18] then shows that boundary conditions are in bijection with module categories over the fusion category $\mathcal{A}$. For the fusion category $\mathrm{Vect}(G)^\omega$ relevant for Dijkgraaf-Witten theories, indecomposable module categories are, in turn, known [24] to be in bijection with subgroups $\iota\colon H \hookrightarrow G$ and 2-cochains $\theta \in C^2(H, \ ^\times)$ such that $\mathrm{d}\theta = \iota^*\omega$. This matches exactly the gauge-theoretic description of types of boundary conditions. According to the general results [18], the topological field theory should assign to an interval with Dijkgraaf-Witten theory based on (G,ω) and boundaries with boundary conditions (H_1,θ_1) and (H_2,θ_2) the category

$$\mathrm{Fun}_{\mathrm{Vect}(G)^\omega}(\mathcal{M}(H_1,\theta_1), \mathcal{M}(H_2,\theta_2))$$

of module functors from the indecomposable module category $\mathcal{M}(H_1,\theta_1)$ over the fusion category $\mathrm{Vect}(G)^\omega$ to the indecomposable module category $\mathcal{M}(H_2,\theta_2)$. A non-trivial calculation [19] shows that this category coincides, as a finitely semisimple abelian category, with the category obtained by linearizing categories of generalized relative bundles.

3.4. Symmetries from invertible topological surface defects

We now turn to the relation between symmetries of three-dimensional topological field theories of Turaev-Viro type and invertible surface defects. For such a theory, the modular tensor category $\mathcal{C}$ of bulk Wilson lines is the Drinfeld center of a fusion category $\mathcal{A}$, i.e. $\mathcal{C} = \mathcal{Z}(\mathcal{A})$.

According the paradigm discussed in Section 1, symmetries should correspond to invertible topological surface defects, and their action should be described by wrapping those defects. By the analysis presented in the preceding subsection, invertible surface defects are described by invertible bimodule categories over the fusion category $\mathcal{A}$. These form a monoidal bicategory; restricting to only invertible 1-morphisms and 2-morphisms in this bicategory, one obtains a categorical 2-group, the Brauer-Picard 2-group [9]. Its isomorphism classes form a finite group $\mathrm{BrPic}(\mathcal{A})$; in the sequel work with this group, rather than the underlying 2-group.

Applying an extended topological field theory to a two-manifold with boundary yields a functor. Take this two-manifold to be a cylinder with a surface defect of type D wrapping the non-contractible cycle:

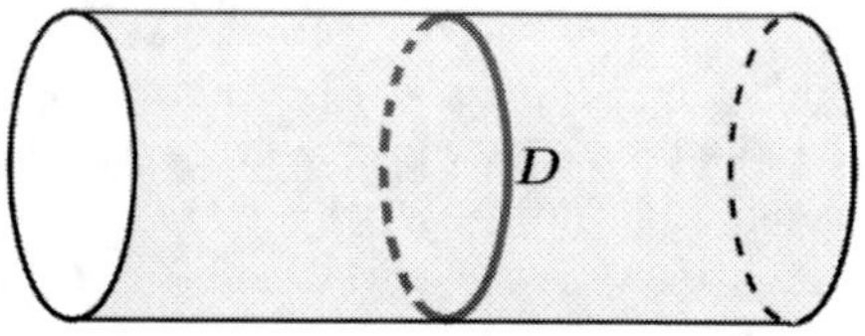

This yields an endofunctor $F_D\colon \mathcal{C} \to \mathcal{C}$. A detailed analysis [16] shows that if the defect D is invertible, then the functor F_D has the structure of a braided monoidal autoequivalence. In the case of Dijkgraaf-Witten theories, the modular tensor category of bulk Wilson lines is $\mathcal{C} = \mathcal{Z}(\mathrm{Vect}(G))$, and an invertible (and thus indecomposable) defect is given by a subgroup $\iota\colon H \hookrightarrow G \times G$ and a 2-cochain $\theta \in C^2(H, {}^\times)$. The functor $F_D \equiv F_{(H,\theta)}$ can then be explicitly obtained by linearizing the span of action groupoids

$$\begin{array}{ccc} & H//H & \\ {}^{\hat\pi_1}\swarrow & & \searrow^{\hat\pi_2} \\ G//G & & G//G \end{array}$$

for adjoint actions which describe the relevant categories of (relative) bundles. Here $\hat\pi_i$ is the functor on groupoids that is induced from the group homomorphism $H \xrightarrow{\iota} G \times G \xrightarrow{p_i} G$. One can then compute a pull-push functor on the linearizations to arrive at the functor

$$F_{(H,\theta)}: \quad \mathcal{Z}(\mathrm{Vect}(G)) = [G//G, \mathrm{Vect}] \longrightarrow [G//G, \mathrm{Vect}]\,,$$

which comes with a natural monoidal structure and which is braided. This construction provides an explicit group homomorphism

$$\mathrm{BrPic}(\mathcal{A}) \longrightarrow \mathrm{brdEq}(\mathcal{Z}(\mathrm{Vect}(G)))$$

from the Brauer-Picard group to the group $\mathrm{brdEq}(\mathcal{Z}(\mathrm{Vect}(G)))$ of braided autoequivalences of the Drinfeld center of $\mathrm{Vect}(G)$. A detailed comparison [16] shows that this group homomorphism is exactly the one considered in [9]; this embeds the construction of that paper naturally into the framework of topological field theories with defects. It is also shown in [9] that the group homomorphism is actually an *iso*morphism. According to the considerations above, this representation theoretic result can be reinterpreted as the statement that symmetries of topological field theories of Turaev-Viro type can already be detected from their action on the category of bulk Wilson lines.

Journal of Physics: Conference Series **597** (2015) 012002 doi:10.1088/1742-6596/597/1/012002

3.5. Symmetries for abelian Dijkgraaf-Witten theories

We now restrict our attention to the class of Dijkgraaf-Witten theories with vanishing three-cocycle and based on a finite abelian group A. In this case the Brauer-Picard group is explicitly known [9]:

$$\mathrm{BrPic}(\mathrm{Vect}(A)) \cong \mathrm{O}_q(A \oplus A^*)\,,$$

where A^* is the character group of A, $q\colon A \oplus A^* \to \ ^{\times}$ is the natural quadratic form determined by $q(g,\chi) = \chi(g)$ for $g \in A$ and $\chi \in A^*$, and $\mathrm{O}_q(A \oplus A^*)$ is the subgroup of those automorphisms of the finite group $A \oplus A^*$ that preserve the quadratic form q.

There are three obvious types of symmetries the Dijkgraaf-Witten theory should possess:

(i) Symmetries of Bun_A.
As explained in Section 3, we can think about the stack Bun_A as a target space for Dijkgraaf-Witten theories. Accordingly, symmetries of Bun_A, i.e. group automorphisms of A, can be expected to be kinematical symmetries of the Dijkgraaf-Witten theory.
Indeed, for $\varphi \in \mathrm{Aut}(A)$ the graph $graph(\varphi) \subset A \oplus A$ is a subgroup; together with the trivial 2-cochain $\theta = 1 \in C^2(graph(\varphi),\ ^{\times})$ it describes an invertible surface defect. Following the prescription given in Section 3.4, we compute the corresponding braided autoequivalence to be the element

$$\varphi \oplus (\varphi^*)^{-1}: \quad A \oplus A^* \to A \oplus A^*$$

of $\mathrm{O}_q(A \oplus A^*)$.

(ii) Automorphisms of the Chern-Simons 2-gerbe on Bun_A.
Recall that the three-cocycle $\omega \in C^3(A,\ ^{\times})$ has the interpretation of a Chern-Simons 2-gerbe on Bun_A. In the case at hand this 2-gerbe is trivial. Nevertheless its automorphisms are not trivial; rather, they are given by 1-gerbes on Bun_A. Their isomorphism classs are given by the cohomology group $H^2(A,\ ^{\times})$. By transgression [26], any cohomology class gives an alternating bihomomorphism $\beta\colon A \times A \to \ ^{\times}$, which in physical terms can be interpreted as a "B-field". (Indeed, for a finite abelian group, transgression provides a group isomorphism between the cohomology group $H^2(A,\ ^{\times})$ and the group of alternating bihomomorphisms.)
Again we can explicitly identify an invertible topological surface defect: the one given by the diagonal subgroup $A_{\mathrm{diag}} \subset A \oplus A$ together with a representative of a cohomology class in $H^2(A,\ ^{\times})$, which we identify with the associated alternating bihomomorphism β. The corresponding braided equivalence can be computed to act as

$$\begin{aligned} A \oplus A &\longrightarrow A \oplus A \\ (g,\chi) &\longmapsto (g, \chi + \beta(g,-))\,. \end{aligned}$$

Thus this type of automorphism shifts the character by the contraction of the B-field β with the group element.

(iii) Electric-magnetic dualities.
It is finally expected that electric-magnetic dualities form a class of symmetries of Dijkgraaf-Witten theories (see e.g. [5] for a discussion of such symmetries for Turaev-Viro theories based on finite-dimensional semisimple Hopf algebras).
To present a simple example, assume that A is a cyclic group and fix a group isomorphism $\delta\colon A \to A^*$. Then there is a natural braided equivalence

$$\begin{aligned} A \oplus A^* &\longrightarrow A \oplus A^* \\ (g,\chi) &\longmapsto (\delta^{-1}(\chi), \delta(g))\,. \end{aligned}$$

This can again be described by an invertible surface defect, namely the one given by the diagonal subgroup $A_{\text{diag}} \subset A \oplus A$ and the alternating bihomomorphism β defined by

$$\beta(g_1, g_2) := \frac{\delta(g_1)(g_2)}{\delta(g_2)(g_1)} \quad \text{for } g_1, g_2 \in A\,.$$

A careful investigation [16] of the structure of the finite group $O_q(A \oplus A^*)$ shows that in fact the three types of symmetries listed above generate this group, i.e. they already provide *all* symmetries of the theory.

4. Conclusions

We conclude by briefly summarizing the crucial insights presented in this contribution:

- Topological defects are important structures in quantum field theories.
- By general principles they describe symmetries and dualities of quantum field theories in such a way that the action of these symmetries and dualities on all different structures of a quantum field theory becomes apparent. This includes the action on field insertions as well as the action on boundary conditions and defects of various codimensions.
- Topological codimension-1 defects and boundaries provide a natural setting for studying relative field theories. This includes in particular anomalous field theories and the TFT construction of RCFT correlators.
- Applications of topological defects virtually concern all fields of physics to which quantum field theory is applied.

There are several classes of quantum field theories in which topological defects can be treated quite explicitly. Besides two-dimensional rational conformal field theories, these include three-dimensional topological field theories of Reshetikhin-Turaev type. We have demonstrated, for the subclass of Dijkgraaf-Witten theories, how defects and boundary conditions can be formulated in a gauge-theoretic setting using (generalizations of) relative bundles. This provides a natural embedding of structures of categorified representation theory, in particular of the theory of module and bimodule categories over monoidal categories, into the more comprehensive setting of (extended) topological field theory.

Acknowledgments

JF is supported by VR under project no. 621-2013-4207. CS is partially supported by the Collaborative Research Centre 676 "Particles, Strings and the Early Universe - the Structure of Matter and Space-Time", by the RTG 1670 "Mathematics inspired by String theory and Quantum Field Theory" and by the DFG Priority Programme 1388 "Representation Theory". We are grateful to the Erwin-Schrödinger-Institute (ESI) for the hospitality during the program "Modern trends in topological field theory" while part of this work was done.

References

[1] Barkeshli M, Jian CM and Qi XL 2013 Twist defects and projective non-Abelian braiding statistics *Phys. Rev. B* **87** 045130 (cond-mat/1208.4834)
[2] Barkeshli M and Qi XL 2012 Topological nematic states and non-abelian lattice dislocations *Phys. Rev. X* **2** 031013 (cond-mat/1112.3311)
[3] Barkeshli M and Wen XG 2010 $U(1)\times U(1)\rtimes\mathbb{Z}_2$ Chern-Simons theory and $\mathbb{Z}_4$ parafermion fractional quantum Hall states *Phys. Rev. B* **81** 045323 (cond-mat/0910.2483)
[4] Bartlett B, Douglas CL, Schommer-Pries CJ and Vicary J 2014 Extended 3-dimensional bordism as the theory of modular objects (math.GT/1411.0945)
[5] Buerschaper O, Christandl M, Kong L and Aguado M 2013 Electric-magnetic duality of lattice systems with topological order *Nucl. Phys. B* **876** 619-636 (cond-mat/1006.5823)

[6] Dijkgraaf R, Pasquier V and Roche P 1990 Quasi Hopf algebras, group cohomology and orbifold models *Nucl. Phys. B (Proc. Suppl.)* **18B** 60-72

[7] Dijkgraaf R and Witten E 1990 Topological gauge theories and group cohomology *Commun. Math. Phys.* **129** 393-429

[8] Drühl K and Wagner H 1982 Algebraic formulation of duality transformations for abelian lattice models *Ann. Phys.* **141** 225-253

[9] Etingof P I, Nikshych D, Ostrik V and Meir E 2010 Fusion categories and homotopy theory *Quantum Topology* **1** 209-273 (math.QA/0909.3140)

[10] Freed D S 1995 Classical Chern-Simons theory, Part 1 *Adv. Math.* **113** 237-303 (hep-th/9206021)

[11] Freed D S and Teleman C 2014 Relative quantum field theory *Commun. Math. Phys.* **326** 459-476 (hep-th/1212.1692)

[12] Fröhlich J, Fuchs J, Runkel I and Schweigert C 2004 Kramers-Wannier duality from conformal defects *Phys. Rev. Lett.* **93** 070601 (cond-mat/0404051)

[13] Fröhlich J, Fuchs J, Runkel I and Schweigert C 2007 Duality and defects in rational conformal field theory *Nucl. Phys. B* **763** 354-430 (hep-th/0607247)

[14] Fuchs J, Gaberdiel M R, Runkel I and Schweigert C 2007 Topological defects for the free boson CFT *J. Phys. A* **40** 11403-11440 (hep-th/0705.3129)

[15] Fuchs J, Nikolaus T, Schweigert C and Waldorf K 2012 Bundle gerbes and surface holonomy in: *European Congress of Mathematics, A.C.M. Ran, H. te Riele, and J. Wiegerinck, eds. European Math. Society, Zürich 167-195* (math.DG/0901.2085)

[16] Fuchs J, Priel J, Schweigert C and Valentino A 2014 On the Brauer groups of symmetries of abelian Dijkgraaf-Witten theories (hep-th/1404.6646)

[17] Fuchs J and Schweigert C 2014 A note on permutation twist defects in topological bilayer phases *Lett. Math. Phys.* **104** 1385-1405 (hep-th/1310.1329)

[18] Fuchs J, Schweigert C and Valentino A 2013 Bicategories for boundary conditions and for surface defects in 3-d TFT *Commun. Math. Phys.* **321** 543-575 (hep-th/1203.4568)

[19] Fuchs J, Schweigert C and Valentino A 2014 A geometric approach to boundaries and surface defects in Dijkgraaf-Witten theories *Commun. Math. Phys.* **332** 981-1015 (hep-th/1307.3632)

[20] Fuchs J, Schweigert C and Waldorf K 2008 Bi-branes: Target space geometry for world sheet topological defects *J. Geom. and Phys.* **58** 576-598 (hep-th/0703145)

[21] Kapustin A and Saulina N 2011 Surface operators in 3d topological field theory and 2d rational conformal field theory in: *Mathematical Foundations of Quantum Field and Perturbative String Theory H. Sati and U. Schreiber, eds. American Mathematical Society, Providence* 175-198 (hep-th/1012.0911)

[22] Lurie J 2009 On the classification of topological field theories *Current Developments in Mathematics* **2008** 129-280 (math.CT/0905.0465)

[23] Morton J C 2013 Cohomological twisting of 2-linearization and extended TQFT *J. Homotopy and Related Structures* **8** 1-61 (math.QA/1003.5603 math.QA/1003.5603)

[24] Ostrik V 2003 Module categories over the Drinfeld double of a finite group *Int. Math. Res. Notices* **27** 1507-1520 (math.QA/0202130)

[25] Schweigert C, Fuchs J and Runkel I 2006 Categorification and correlation functions in conformal field theory, in: *Proceedings of the ICM 2006*, M Sanz-Solé, J Soria, J L Varona and J Verdera, eds. European Math. Society, Zürich 443-458 (math.CT/0602079)

[26] Willerton S 2008 The twisted Drinfeld double of a finite group via gerbes and finite groupoids *Alg. & Geom. Topol.* **8** 1419-1457 (math.QA/0503266)

Partial and quasi dynamical symmetries in quantum many-body systems

A Leviatan

Racah Institute of Physics, The Hebrew University, Jerusalem 91904, Israel

E-mail: ami@phys.huji.ac.il

Abstract. We introduce the notions of partial dynamical symmetry (PDS) and quasi dynamical symmetry (QDS) and demonstrate their relevance to nuclear spectroscopy, to quantum phase transitions and to mixed systems with regularity and chaos. The analysis serves to highlight the potential role of PDS and QDS towards understanding the emergent "simplicity out of complexity" exhibited by complex many-body systems.

1. Introduction

The concept of dynamical symmetry (DS) is now widely recognized to be of central importance in our understanding of complex many-body systems. It had major impact on developments in diverse areas of physics, including, hadrons, nuclei and molecules [1–3]. Its basic paradigm is to write the Hamiltonian of the system under consideration in terms of Casimir operators of a chain of nested algebras, $G_0 \supset G_1 \supset \ldots \supset G_n$. The following properties are then observed. (i) All states are solvable and analytic expressions are available for energies and other observables. (ii) All states are classified by quantum numbers, $|\alpha_0, \alpha_1, \ldots, \alpha_n\rangle$, which are the labels of the irreducible representations (irreps) of the algebras in the chain. (iii) The structure of wave functions is completely dictated by symmetry and is independent of the Hamiltonian's parameters.

The merits of a DS are self-evident. However, in most applications to realistic systems, the predictions of an exact DS are rarely fulfilled and one is compelled to break it. More often one finds that, in a given system, the assumed symmetry is not obeyed uniformly, *i.e.*, is fulfilled by only some states but not by others. In describing a transition between different structural phases, the relevant Hamiltonian, in general, involves competing interactions with incompatible symmetries. The need to address such situations has led to the introduction of partial dynamical symmetry (PDS) [4–6] and quasi dynamical symmetry (QDS) [7–10]. These intermediate-symmetry notions and their implications for dynamical systems, are the subject matter of the present contribution.

1.1. The interacting boson model

In order to illustrate the various notions of symmetries and demonstrate their relevance, we employ the interacting boson model (IBM) [2], widely used in the description of low-lying quadrupole collective states in nuclei in terms of N monopole (s) and quadrupole (d) bosons representing valence nucleon pairs. The bilinear combinations $\{s^\dagger s,\, s^\dagger d_m,\, d^\dagger_m s,\, d^\dagger_m d_{m'}\}$ span a U(6) algebra, which serves as the spectrum generating algebra. The IBM Hamiltonian is expanded in terms of these generators and consists of Hermitian, rotational-scalar interactions which conserve the total number of s- and d- bosons, $\hat{N} = \hat{n}_s + \hat{n}_d = s^\dagger s + \sum_m d_m^\dagger d_m$. The three dynamical symmetries of the IBM are

$$\begin{array}{ccccccc} \mathrm{U}(6) & \supset & \mathrm{U}(5) & \supset & \mathrm{O}(5) & \supset & \mathrm{O}(3) \\ \downarrow & & \downarrow & & \downarrow & & \downarrow \\ [N] & & \langle n_d \rangle & & (\tau) & n_\Delta & L \end{array} , \tag{1a}$$

$$\begin{array}{ccccc} \mathrm{U}(6) & \supset & \mathrm{SU}(3) & \supset & \mathrm{O}(3) \\ \downarrow & & \downarrow & & \downarrow \\ [N] & & (\lambda,\mu) & K & L \end{array} , \tag{1b}$$

$$\begin{array}{ccccccc} \mathrm{U}(6) & \supset & \mathrm{O}(6) & \supset & \mathrm{O}(5) & \supset & \mathrm{O}(3) \\ \downarrow & & \downarrow & & \downarrow & & \downarrow \\ [N] & & \langle \sigma \rangle & & (\tau) & n_\Delta & L \end{array} , \tag{1c}$$

where, below each algebra, its associated labels of irreps are given. n_Δ and K are multiplicity labels needed in the O(5) $\supset$ O(3) and SU(3) $\supset$ O(3) reductions, respectively. These solvable limits correspond to known benchmarks of the geometric description of nuclei [11], involving vibrational [U(5)], rotational [SU(3)] and γ-soft [O(6)] types of dynamics.

A geometric visualization of the model is obtained by an energy surface

$$E_N(\beta,\gamma) = \langle \beta,\gamma;N|\hat{H}|\beta,\gamma;N\rangle , \tag{2}$$

defined by the expectation value of the Hamiltonian in the coherent (intrinsic) state [12, 13]

$$|\beta,\gamma;N\rangle = (N!)^{-1/2}(b_c^\dagger)^N\,|0\rangle , \tag{3a}$$

$$b_c^\dagger = (1+\beta^2)^{-1/2}[\beta\cos\gamma d_0^\dagger + \beta\sin\gamma(d_2^\dagger + d_{-2}^\dagger)/\sqrt{2} + s^\dagger] . \tag{3b}$$

Here (β,γ) are quadrupole shape parameters whose values, $(\beta_{\rm eq},\gamma_{\rm eq})$, at the global minimum of $E_N(\beta,\gamma)$ define the equilibrium shape for a given Hamiltonian. The shape can be spherical $(\beta = 0)$ or deformed $(\beta > 0)$ with $\gamma = 0$ (prolate), $\gamma = \pi/3$ (oblate), $0 < \gamma < \pi/3$ (triaxial), or γ-independent. The equilibrium deformations associated with the dynamical symmetry limits are $\beta_{\rm eq} = 0$ for U(5), $(\beta_{\rm eq} = \sqrt{2}, \gamma_{\rm eq} = 0)$ for SU(3) and $(\beta_{\rm eq} = 1, \gamma_{\rm eq}\,\text{arbitrary})$ for O(6).

One particularly successful approach within the IBM is the extended consistent-Q formalism (ECQF) [14, 15], which uses the following Hamiltonian

$$\hat{H}_{\rm ECQF} = \omega\left[(1-\xi)\,\hat{n}_d - \frac{\xi}{4N}\,\hat{Q}^\chi\cdot\hat{Q}^\chi\right] . \tag{4}$$

Here $\hat{Q}^\chi = d^\dagger s + s^\dagger\tilde{d} + \chi\,(d^\dagger\tilde{d})^{(2)}$ is the quadrupole operator, $\tilde{d}_m = (-)^m d_{-m}$ and the dot implies a scalar product. ξ and χ are the sole structural parameters of the model since ω is a scaling factor. The parameter ranges $0 \le \xi \le 1$ and $-\frac{\sqrt{7}}{2} \le \chi \le 0$ interpolate between the U(5), O(6) and SU(3) DS limits, which are reached for $(\xi,\chi) = (0,\chi)$, $(1,0)$, and $(1,-\frac{\sqrt{7}}{2})$, respectively. It is customary to represent the parameter space by a symmetry triangle [16], shown in Fig. 1, whose vertices correspond to these limits. The ECQF has been used extensively for the description of nuclear properties and it was found that the vast majority of nuclei are best described by ECQF parameters in the interior of the triangle, away from any DS limit. In this context, a key question, addressed in the present contribution, can be phrased in the form: are there any remaining "symmetries" inside the triangle?

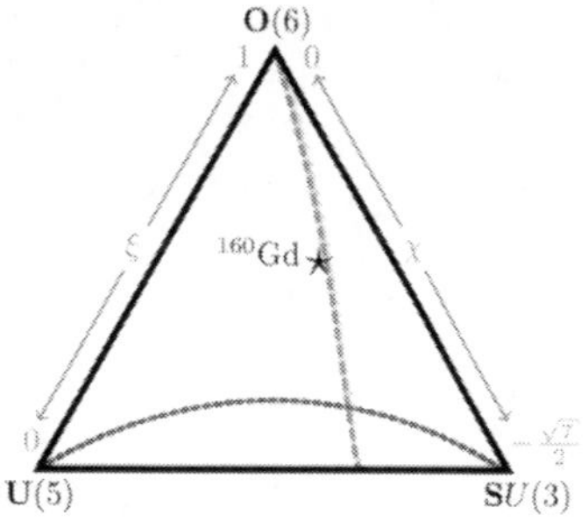

Figure 1. The ECQF symmetry triangle with the position of the nucleus ^{160}Gd indicated by a star. The calculated green curve and its approximation by a red dashed line, correspond to a region of enhanced purity in the ground state, exemplifying an O(6)-PDS discussed in Sections 5.1 and 7. The blue dotted line shows the "arc of regularity" mentioned in Section 8, exemplifying SU(3)-QDS. Adapted from [25]

2. Partial dynamical symmetry (PDS)

In algebraic models, such as the IBM, the required symmetry breaking is achieved by including in the Hamiltonian terms associated with (two or more) different sub-algebra chains of the parent spectrum generating algebra. In general, under such circumstances, solvability is lost, there are no remaining non-trivial conserved quantum numbers and all eigenstates are expected to be mixed. A partial dynamical symmetry (PDS) [4–6] corresponds to a particular symmetry breaking for which some (but not all) of the virtues of a dynamical symmetry are retained. The essential idea is to relax the stringent conditions of *complete* solvability so that the properties (i)–(iii) of a DS, mentioned above, are only partially satisfied. It is then possible to identify the following types of partial dynamical symmetries [6]

- *PDS type I:* **some** of the states have **all** the dynamical symmetry
- *PDS type II:* **all** the states have **part** of the dynamical symmetry
- *PDS type III:* **some** of the states have **part** of the dynamical symmetry.

In PDS of type I, only part of the eigenspectrum is analytically solvable and retains all the dynamical symmetry (DS) quantum numbers. In PDS of type II, the entire eigenspectrum retains some of the DS quantum numbers. PDS of type III has a hybrid character, in the sense that some (solvable) eigenstates keep some of the quantum numbers. In what follows we discuss algorithms for constructing Hamiltonians with partial dynamical symmetries of various types and demonstrate their relevance to quantum many-body systems.

3. PDS (type I)

PDS of type I corresponds to a situation for which the defining properties of a dynamical symmetry (DS), namely, solvability, good quantum numbers, and symmetry-dictated structure are fulfilled exactly, but by only a subset of states. An algorithm for constructing Hamiltonians with PDS has been developed in [4] and further elaborated in [5]. The analysis starts from the chain of nested algebras

$$\begin{array}{ccccccc} G_{\rm dyn} & \supset & G & \supset & \cdots & \supset & G_{\rm sym} \\ \downarrow & & \downarrow & & & & \downarrow \\ [h] & & \langle\Sigma\rangle & & & & \Lambda \end{array} \tag{5}$$

Eq. (5) implies that $G_{\rm dyn}$ is the dynamical (spectrum generating) algebra of the system such that operators of all physical observables can be written in terms of its generators; a single irrep of $G_{\rm dyn}$ contains all states of relevance in the problem. In contrast, $G_{\rm sym}$ is the symmetry algebra and a single of its irreps contains states that are degenerate in energy. Assuming, for simplicity, that particle number is conserved, then all states, and hence the representation $[h]$, can then be assigned a definite particle number N. For N identical particles the representation $[h]$ of the dynamical algebra $G_{\rm dyn}$ is either symmetric $[N]$ (bosons) or antisymmetric $[1^N]$ (fermions) and will be denoted, in both cases, as $[h_N]$. The occurrence of a DS of the type (5) signifies that its eigenstates can be labeled as $|[h_N]\langle\Sigma\rangle\ldots\Lambda\rangle$; additional labels (indicated by ...) are suppressed in the following. Likewise, operators can be classified according to their tensor character under (5) as $\hat{T}_{[h_n]\langle\sigma\rangle\lambda}$.

Of specific interest in the construction of a PDS associated with the reduction (5), are the n-particle annihilation operators $\hat{T}$ which satisfy the property

$$\hat{T}_{[h_n]\langle\sigma\rangle\lambda}|[h_N]\langle\Sigma_0\rangle\Lambda\rangle = 0 \ , \tag{6}$$

for all possible values of Λ contained in a given irrep $\langle\Sigma_0\rangle$ of G. Equivalently, this condition can be phrased in terms of the action on a lowest weight (LW) state of the G-irrep $\langle\Sigma_0\rangle$, $\hat{T}_{[h_n]\langle\sigma\rangle\lambda}|LW; [h_N]\langle\Sigma_0\rangle\rangle = 0$, from which states of good Λ can be obtained by projection. Any n-body, number-conserving normal-ordered interaction, $\hat{H} = \sum_{\alpha,\beta} A_{\alpha\beta}\hat{T}^\dagger_\alpha\hat{T}_\beta$, written in terms of these annihilation operators and their Hermitian conjugates (which transform as the corresponding conjugate irreps), can be added to the Hamiltonian with a DS (5), while still preserving the solvability of states with $\langle\Sigma\rangle = \langle\Sigma_0\rangle$. If the operators $\hat{T}_{[h_n]\langle\sigma\rangle\lambda}$ span the entire irrep $\langle\sigma\rangle$ of G, then the annihilation condition (6) is satisfied for all Λ-states in $\langle\Sigma_0\rangle$, if none of the G irreps $\langle\Sigma\rangle$ contained in the $G_{\rm dyn}$ irrep $[h_{N-n}]$ belongs to the G Kronecker product $\langle\sigma\rangle \times \langle\Sigma_0\rangle$. So the problem of finding interactions that preserve solvability for part of the states (5) is reduced to carrying out a Kronecker product. The arguments for choosing the special irrep $\langle\Sigma\rangle = \langle\Sigma_0\rangle$ in Eq. (6), which contains the solvable states, are based on physical grounds. A frequently encountered choice is the irrep which contains the ground state of the system.

3.1. SU(3) PDS (type I) in nuclei

The SU(3) DS chain of the IBM and related quantum numbers are given in Eq. (1b). The DS Hamiltonian involves the Casimir operators of SU(3) and O(3), with eigenvalues $\lambda^2+\mu^2+\lambda\mu+3(\lambda+\mu)$ and $L(L+1)$, respectively. The spectrum resembles that of an axially-deformed rotovibrator and the corresponding eigenstates are arranged in SU(3) multiplets. The label K corresponds geometrically to the projection of the angular momentum on the symmetry axis. In a given SU(3) irrep (λ,μ), each K-value is associated with a rotational band and states with the same angular momentum L, in different K-bands, are degenerate. The lowest SU(3) irrep is $(2N,0)$, which describes the ground band $g(K=0)$ of a prolate deformed nucleus. The first excited SU(3) irrep $(2N-4,2)$ contains degenerate $\beta(K=0)$ and $\gamma(K=2)$ bands. This β-γ degeneracy is a characteristic feature of the SU(3) limit which, however, is not commonly observed. In most deformed nuclei the β band lies above the γ band. In the IBM framework, with at most two-body interactions, one is therefore compelled to break SU(3) in order to conform with the experimental data.

The construction of Hamiltonians with SU(3)-PDS is based on identification of n-boson operators which annihilate all states in a given SU(3) irrep (λ,μ), chosen here to be the ground band irrep $(2N,0)$. For that purpose, we consider the following two-boson SU(3) tensors, $B^\dagger_{[n](\lambda,\mu)K;Lm}$, with $n=2$, $(\lambda,\mu)=(0,2)$ and angular momentum $L=0,\,2$

$$B^\dagger_{[2](0,2)0;00} \;\propto\; P^\dagger_0 = d^\dagger\cdot d^\dagger - 2(s^\dagger)^2 \ , \tag{7a}$$

$$B^\dagger_{[2](0,2)0;2m} \;\propto\; P^\dagger_{2m} = 2d^\dagger_m s^\dagger + \sqrt{7}\,(d^\dagger\, d^\dagger)^{(2)}_m \ . \tag{7b}$$

The corresponding Hermitian conjugate boson-pair annihilation operators, P_0 and P_{2m}, transform as $(2,0)$ under SU(3), and annihilate all L-states in the $(2N,0)$ irrep

$$P_0\,|[N](2N,0)K=0,L\rangle \;=\; 0\;, \tag{8a}$$

$$P_{2m}\,|[N](2N,0)K=0,L\rangle \;=\; 0\;. \tag{8b}$$

Equivalently, these operators annihilate the coherent state, $|\beta=\sqrt{2},\gamma=0;N\rangle$, of Eq. (3), which is the lowest-weight state of this irrep and serves as an intrinsic state for the SU(3) ground band. The relations in Eq. (8) follow from the fact that the action of the operators P_{Lm} leads to a state with $N-2$ bosons in the U(6) irrep $[N-2]$, which does not contain the SU(3) irreps obtained from the product $(2,0)\times(2N,0)=(2N+2,0)\oplus(2N,1)\oplus(2N-2,2)$. In addition, P_0 satisfies

$$P_0\,|[N](2N-4k,2k)K=2k,LM\rangle \;=\; 0\;, \tag{9}$$

where for $k>0$ the indicated L-states span only part of the SU(3) irreps $(\lambda,\mu)=(2N-4k,2k)$ and form the rotational members of excited $\gamma^k(K=2k)$ bands.

Following the general algorithm, a two-body Hamiltonian with partial SU(3) symmetry can now be constructed as [17]

$$\hat{H}(h_0,h_2) \;=\; h_0\,P_0^\dagger P_0 + h_2\,P_2^\dagger\cdot\tilde{P}_2\;, \tag{10}$$

where $\tilde{P}_{2m}=(-)^m P_{2,-m}$. For $h_2=h_0$, the Hamiltonian is an SU(3) scalar, related to the quadratic Casimir operator of SU(3): $h_2[-\hat{C}_{SU(3)}+2\hat{N}(2\hat{N}+3)]$. For $h_0=-5h_2$, it transforms as a $(2,2)$ SU(3) tensor component. Although in general $\hat{H}(h_0,h_2)$ is not invariant under SU(3), Eqs. (8)-(9) ensure that it retains selected solvable states with good SU(3) symmetry, which are members of the ground $g(K=0)$ and $\gamma^k(K=2k)$ bands with the following characteristics

$$\begin{array}{lll} |N,(2N,0)K=0,L\rangle & E=0 & L=0,2,4,\ldots,2N \\ |N,(2N-4k,2k)K=2k,L\rangle & E=h_2\,6k\,(2N-2k+1) & L=K,K+1,\ldots,(2N-2k)\;. \end{array} \tag{11}$$

The remaining eigenstates of $\hat{H}(h_0,h_2)$ do not preserve the SU(3) symmetry and therefore get mixed. One can add to $\hat{H}(h_0,h_2)$ the Casimir operator of O(3), $\hat{C}_{O(3)}$, which contributes an $L(L+1)$ splitting without affecting the wave functions. The resulting Hamiltonian has, by construction, SU(3)-PDS.

The empirical spectrum of ^{168}Er is shown in Fig. 2 and compared with SU(3)-DS, SU(3)-PDS and broken SU(3) calculations [17]. The SU(3)-PDS spectrum shows an improvement over the schematic, exact SU(3) dynamical symmetry description, since the β-γ degeneracy is lifted. The quality of the calculated PDS spectrum is similar to that obtained in the broken-SU(3) calculation, however, in the former the ground $g(K=0_1)$ and $\gamma(K=2_1)$ bands remain solvable with good SU(3) symmetry, $(\lambda,\mu)=(2N,0)$ and $(2N-4,2)$ respectively. At the same time, the excited $K=0_2^+$ band involves about 13% SU(3) admixtures into the dominant $(2N-4,2)$ irrep [18]. Since the wave functions of the solvable states (11) are known, one can obtain *analytic* expressions for matrix elements of observables between them. The SU(3) generator, $\hat{Q}$, is obtained from $\hat{Q}^\chi$, Eq. (4), for $\chi=-\frac{\sqrt{7}}{2}$, and hence one can write the general E2 operator as $\hat{Q}^\chi=\hat{Q}+\theta(d^\dagger s+s^\dagger\tilde{d})$. Since $\hat{Q}$ cannot connect different SU(3) irreps, only the second term contributes to $\gamma\to g$ transitions. Accordingly, the calculated B(E2) ratios for these interband transitions are parameter-free predictions of SU(3)-PDS. Based on this observation, an extensive test of SU(3)-PDS was conducted recently [19], showing evidence for its relevance not only for ^{168}Er but also for a wide range of deformed rare earth nuclei. Representative examples of the comparison are shown in Fig. 3. Their detailed analysis provides insights into the complementary role of finite-nucleon number and band-mixing in nuclei.

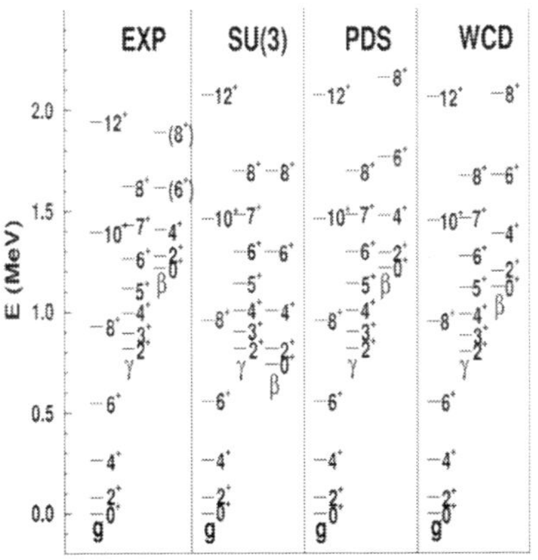

Figure 2. Spectra of ^{168}Er ($N = 16$). Experimental energies (EXP) are compared with IBM calculations in an exact SU(3) DS [SU(3)], in a broken SU(3) symmetry (WCD) and in SU(3) PDS. The latter employs the Hamiltonian of Eq. (10), $\hat{H}(h_0, h_2) + C\,\hat{C}_{O(3)}$ with $h_0 = 8$, $h_2 = 4$, $C = 13$ keV. Adapted from [17].

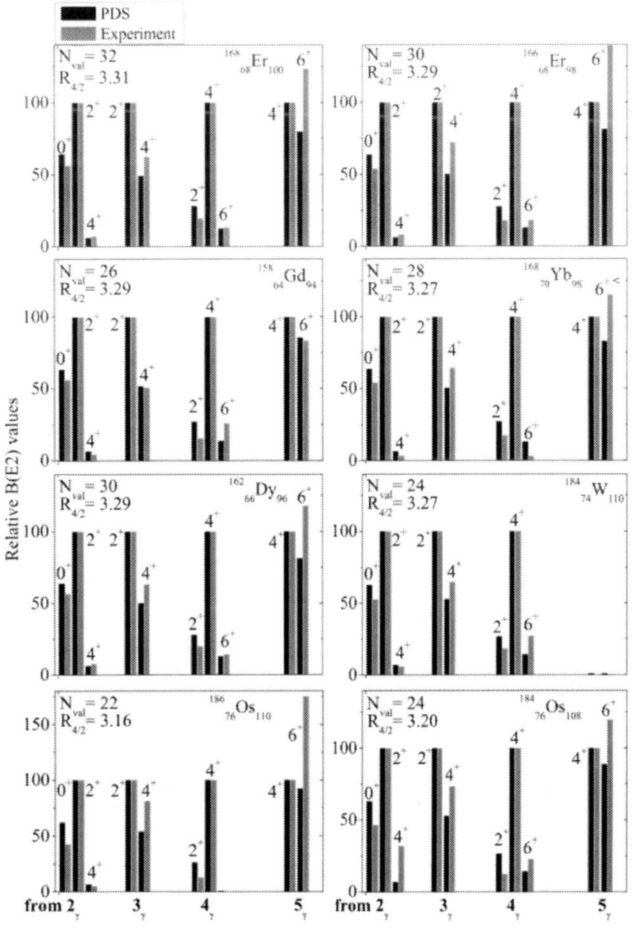

Figure 3. Comparison of SU(3)-PDS predictions (black bar) with the data (red bar) on the relative γ-band to ground-band E2 transitions in several rare earth nuclei. Adapted from [19].

4. PDS (type II and type III)

PDS of type II corresponds to a situation for which *all* the states of the system preserve *part* of the dynamical symmetry, $G_0 \supset G_1 \supset G_2 \supset \ldots \supset G_n$. In this case, there are no analytic solutions, yet selected quantum numbers (of the conserved symmetries) are retained. This occurs, for example, when the Hamiltonian contains interaction terms from two different chains with a common symmetry subalgebra, *e.g.*,

$$G_0 \supset \left\{ \begin{array}{c} G_1 \\ G_1' \end{array} \right\} \supset G_2 \supset \ldots \supset G_n \,. \tag{12}$$

If G_1 and G_1' are incompatible, *i.e.*, do not commute, then their irreps are mixed in the eigenstates of the Hamiltonian. On the other hand, since G_2 and its subalgebras are common to both chains, then the labels of their irreps remain as good quantum numbers.

In the IBM, such a situation arises in Hamiltonians combining terms from both the U(5) and O(6) chains, Eqs. (1a) and (1c) [20]. All eigenstates now mix U(5) irreps (n_d) and O(6) irreps $\langle\sigma\rangle$, but retain the (τ, L) labels of the O(5) $\supset$ O(3) segment, common to both chains, as good quantum numbers. Hamiltonians of this type have been used in the study of shape-phase transitions between spherical [U(5)] and γ-soft [O(6)] nuclei [21].

An alternative situation where PDS of type II occurs is when the Hamiltonian preserves only some of the symmetries G_i in the DS chain and only their irreps are unmixed. Let $G_1 \supset G_2 \supset G_3$ be a set of nested algebras which may occur anywhere in the chain, in-between the spectrum generating algebra G_0 and the invariant symmetry algebra G_n. A systematic procedure [22] for identifying interactions with PDS of type II, is based on writing the Hamiltonian in terms of generators, g_i, of G_1, which do not belong to its subalgebra G_2. By construction, such Hamiltonian preserves the G_1 symmetry but, in general, not the G_2 symmetry, and hence will have the G_1 labels as good quantum numbers but will mix different irreps of G_2. The Hamiltonians can still conserve the G_3 labels *e.g.*, by choosing it to be a scalar of G_3. The procedure involves the identification of the tensor character under G_2 and G_3 of the operators g_i and their products, $g_i g_j \ldots g_k$. The Hamiltonians obtained in this manner belong to the integrity basis of G_3-scalar operators in the enveloping algebra of G_1 and, hence, their existence is correlated with their order.

In the IBM, such a scenario can be realized by considering an interaction term of the form $((\Pi^{(2)} \times \Pi^{(2)})^{(2)} \cdot \Pi^{(2)}$, constructed from the O(6) generator, $\Pi^{(2)} = d^\dagger s + s^\dagger \tilde{d}$, which is not a generator of O(5) [22]. Such a term cannot connect states in different O(6) irreps but can induce O(5) mixing subject to $\Delta\tau = \pm 1, \pm 3$. Consequently, all eigenstates of the resulting Hamiltonian have good O(6) quantum number σ but do not possess O(5) symmetry τ. These are the necessary ingredients of an O(6) PDS of type II associated with the chain of Eq. (1c).

PDS of type III has a hybrid character, for which *some* of the states of the system under study preserve *part* of the dynamical symmetry [23]. In relation to the dynamical symmetry chain of Eq. (5), with associated basis, $|[h_N]\langle\Sigma\rangle\Lambda\rangle$, this can be accomplished by relaxing the condition of Eq. (6), so that it holds only for *selected* states Λ contained in a given irrep $\langle\Sigma_0\rangle$ of G and/or selected (combinations of) components λ of the tensor $\hat{T}_{[h_n]\langle\sigma\rangle\lambda}$. Under such circumstances, let $G' \neq G_{sym}$ be a subalgebra of G in the aforementioned chain, $G \supset G'$. In general, the Hamiltonians, constructed from these tensors, in the manner described in Section 3, are not invariant under G nor G'. Nevertheless, they do possess the subset of solvable states, $|[h_N]\langle\Sigma_0\rangle\Lambda\rangle$, with good G-symmetry $\langle\Sigma_0\rangle$ (which now span only part of the corresponding G-irrep), while other states are mixed. At the same time, the symmetry associated with the subalgebra G', is broken in all states (including the solvable ones). Thus, part of the eigenstates preserve part of the symmetry. These are precisely the requirements of PDS of type III.

In the IBM, such a generalized partial symmetry associated with the O(6) chain of Eq. (1c), can be realized by an Hamiltonian constructed of boson-pair operators which are not invariant under O(6) nor O(5), but annihilate the coherent state, $|\beta = 1, \gamma = 0; N\rangle$, of Eq. (3), which has $\sigma = N$ [23]. Such an Hamiltonian has a solvable ground band with good O(6) symmetry, which is not preserved by other states. All eigenstates, including the solvable ones, break the O(5) symmetry. An empirical manifestation of such type of O(6)-PDS is presented in Section 5.1.

5. Measures of PDS

The PDS notion reflects the purity of selected eigenstates with respect to a DS basis. The above algorithms provide a procedure for an explicit construction of Hamiltonians with such property. More general (and realistic) Hamiltonians often exhibit features of a PDS to a certain approximation. In such cases, one needs to assess the quality and applicability of the PDS notion. In what follows, we discuss two quantitative measures of PDS, based on wave-function entropy and quantum number fluctuations.

Consider an eigenfunction of the IBM Hamiltonian, $|L\rangle$, with angular momentum L. Its expansion in a DS basis reads $|L\rangle = \sum_i C_i \,|[N], \alpha_i, L\rangle$, where C_i stands for the expansion coefficients $C^{(L)}_{n_d,\tau,n_\Delta}$, $C^{(L)}_{(\lambda,\mu),K}$, $C^{(L)}_{\sigma,\tau,n_\Delta}$, in the U(5), SU(3), O(6) bases, Eq. (1), respectively.

The probability distributions of U(5): $P^{(L)}_{n_d}$, SU(3): $P^{(L)}_{(\lambda,\mu)}$, and O(6): $P^{(L)}_{\sigma}$, are calculated as

$$P^{(L)}_{n_d} = \sum_{\tau,n_\Delta} |C^{(L)}_{n_d,\tau,n_\Delta}|^2 \quad , \quad S_{\mathrm{U5}}(L) = -\sum_{n_d} P^{(L)}_{n_d} \ln P^{(L)}_{n_d} \; , \tag{13a}$$

$$P^{(L)}_{(\lambda,\mu)} = \sum_{K} |C^{(L)}_{(\lambda,\mu),K}|^2 \quad , \quad S_{\mathrm{SU3}}(L) = -\sum_{(\lambda,\mu)} P^{(L)}_{(\lambda,\mu)} \ln P^{(L)}_{(\lambda,\mu)} \; , \tag{13b}$$

$$P^{(L)}_{\sigma} = \sum_{\tau,n_\Delta} |C^{(L)}_{\sigma,\tau,n_\Delta}|^2 \quad , \quad S_{\mathrm{O6}}(L) = -\sum_{\sigma} P^{(L)}_{\sigma} \ln P^{(L)}_{\sigma} \; . \tag{13c}$$

The indicated Shannon state entropy $S_G(L)$, $[G = \mathrm{U}(5), \mathrm{SU}(3), \mathrm{O}(6)]$ is a convenient tool to evaluate the purity of eigenstates with respect to a DS basis [24]. It vanishes when the considered state is pure with good G-symmetry $[S_G(L)=0]$, and is positive for a mixed state. Normalized entropies are obtained by dividing by $\ln D_G$, where D_G counts the number of possible G-irreps for a given L. In this case, the maximal value $[S_G(L) = 1]$ is obtained when $|L\rangle$ is uniformly spread among the irreps of G, *i.e.*, for $P^{(L)}_G = 1/D_G$. Intermediate values, $0 < S_G(L) < 1$, indicate partial fragmentation of the state $|L\rangle$ in the respective DS basis.

The degree of a symmetry of a state $|L\rangle$ can also be inferred from the fluctuations of the corresponding quantum number. As an example, for the O(6) symmetry of the IBM, the fluctuations in σ can be calculated as [25]

$$\Delta\sigma_L = \sqrt{\sum_i C_i^2\,\sigma_i^2 - \left(\sum_i C_i^2\,\sigma_i\right)^2} \; , \tag{14}$$

where the sum is over all basis states in the chain, Eq. (1c). If $|L\rangle$ carries an exact O(6) quantum number, σ fluctuations are zero, $\Delta\sigma_L = 0$. If $|L\rangle$ contains basis states with different O(6) quantum numbers, then $\Delta\sigma_L > 0$, indicating that the O(6) symmetry is broken. Note that $\Delta\sigma_L$ also vanishes for a state with a mixture of components with the same σ but different O(5) quantum numbers τ, corresponding to a $|L\rangle$ with good O(6) but mixed O(5) character. $\Delta\sigma_L$ has the same physical content as $S_{O6}(L)$ (13c) and both can be used as measures of O(6)-PDS.

5.1. *O(6)-PDS (type III) in nuclei*

A recent study [25] has examined the fluctuations $\Delta\sigma_L$, Eq. (14), for the entire parameter space of the ECQF Hamiltonian (4). Results of this calculation for the ground state, $|L = 0^+_1\rangle$, with $N = 14$, are shown in Fig. 4. At the O(6) DS limit ($\xi = 1$, $\chi = 0$), $\Delta\sigma_0 \equiv \Delta\sigma_{L=0_1}$ vanishes per construction whereas it is greater than zero for all other parameter pairs. Towards the U(5) DS limit ($\xi = 0$), the fluctuations reach a saturation value of $\Delta\sigma_0 \approx 2.47$. At the SU(3) DS limit ($\xi = 1$, $\chi = -\frac{\sqrt{7}}{2}$) the fluctuations are $\Delta\sigma_0 \approx 1.25$. Surprisingly, one recognizes a valley of almost vanishing $\Delta\sigma_0$ values, two orders of magnitude lower than at saturation. This region (depicted by a green arc in the triangle of Fig. 1), represents a parameter range of the IBM, outside the O(6) DS limit, where the ground-state wave function exhibits an exceptionally high degree of purity with respect to the O(6) quantum number σ.

The ground-state wave functions in the valley of low $\Delta\sigma_0$, can be expanded in the O(6)-DS basis (1c). At the O(6) DS limit only one O(6) basis state, with $\sigma = N$ and $\tau = 0$ contributes, while outside this limit the wave function consists of multiple O(6) basis states. Investigation of the wave function for parameter combinations inside the valley reveals an overwhelming dominance of the O(6) basis states with $\sigma = N$. This is seen in Fig. 5 for the ground-state wave function of $\hat{H}_{\mathrm{ECQF}}$ (4), with parameter values that apply to the nucleus ^{160}Gd. The $\sigma = N$

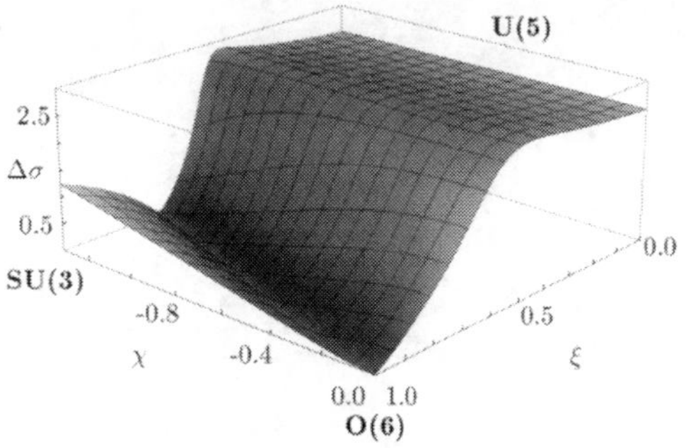

Figure 4. Ground-state ($L=0_1^+$) fluctuations $\Delta\sigma_0$ (14) for $\hat{H}_{\mathrm{ECQF}}$ (4) with $N = 14$. The fluctuations vanish at the O(6) DS limit, saturate towards the U(5) DS limit, and are of the order 10^{-2} in the valley. Adapted from [25].

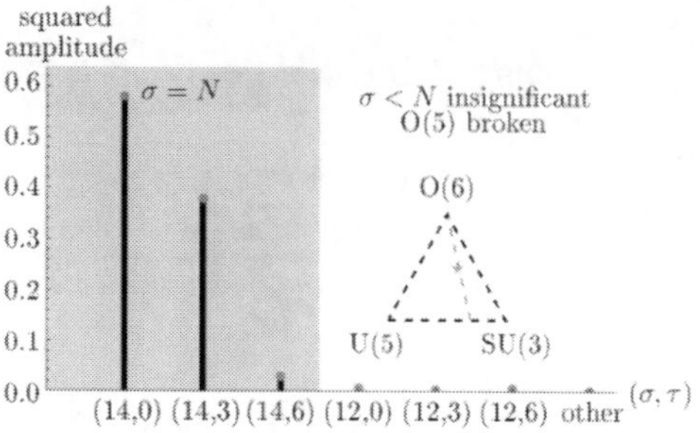

Figure 5. Squared amplitudes C_i^2 in the expansion in the O(6) basis (1c), of the $L=0_1^+$ ground state of the ECQF Hamiltonian (4) with parameters $\xi = 0.84$, $\chi = -0.53$, $N = 14$, appropriate for ^{160}Gd. Adapted from [25].

Table 1. Calculated σ fluctuations $\Delta\sigma_L$, Eq. (14), for rare earth nuclei in the vicinity of the identified region of approximate ground-state-O(6) symmetry [25]. Also shown are the fraction $f^{(L)}_{\sigma=\mathrm{N}}$ of O(6) basis states with $\sigma = N$ contained in the $L = 0, 2, 4$ states, members of the ground band. The structure parameters ξ and χ are taken from [26].

Nucleus	N	ξ	χ	$\Delta\sigma_0$	$f^{(0)}_{\sigma=\mathrm{N}}$	$\Delta\sigma_2$	$f^{(2)}_{\sigma=\mathrm{N}}$	$\Delta\sigma_4$	$f^{(4)}_{\sigma=\mathrm{N}}$
^{156}Gd	12	0.72	-0.86	0.46	95.3%	0.43	95.8%	0.38	96.6%
^{158}Gd	13	0.75	-0.80	0.35	97.2%	0.33	97.5%	0.30	97.9%
^{160}Gd	14	0.84	-0.53	0.19	99.1%	0.19	99.2%	0.17	99.3%
^{162}Gd	15	0.98	-0.53	0.41	96.0%	0.40	96.0%	0.40	96.1%
^{160}Dy	14	0.81	-0.49	0.44	96.2%	0.39	96.4%	0.36	96.8%
^{162}Dy	15	0.92	-0.31	0.07	99.9%	0.07	99.9%	0.06	99.9%
^{164}Dy	16	0.98	-0.26	0.13	99.6%	0.13	99.6%	0.13	99.6%
^{164}Er	14	0.84	-0.37	0.39	96.5%	0.37	96.7%	0.35	97.1%
^{166}Er	15	0.91	-0.31	0.12	99.7%	0.11	99.7%	0.10	99.7%

states comprise more than 99% of the ground-state wave function at the bottom of the valley and their dominance causes $\Delta\sigma_0$ to be small. At the same time, the O(5) symmetry is broken, as basis states with different quantum number τ contribute significantly to the wave function. Consequently, the valley can be identified as an entire region in the symmetry triangle with an approximate O(6)-PDS of type III. Outside this valley the ground state is a mixture of several σ values and $\Delta\sigma_0$ increases.

Detailed ECQF fits for energies and electromagnetic transitions of rare-earth nuclei, performed in [26], allow one to relate the structure of collective nuclei to the parameter space of the ECQF Hamiltonian (4). From the extracted (ξ, χ) parameters one can calculate the fluctuations $\Delta\sigma_L$ and the fractions $f_{\sigma=N}$ of squared $\sigma = N$ amplitude. Nuclei with $\Delta\sigma_0 < 0.5$ and $f_{\sigma=N} > 95\%$ in the ground-state ($L = 0_1^+$) are listed in Table 1. These quantities are also calculated for yrast states with $L > 0$ and exhibit similar values in each nucleus. It is evident that a large set of rotational rare earth nuclei, such as ^{160}Gd, are located in the valley of small σ fluctuations. They can be identified as candidate nuclei with an approximate O(6)-PDS of type III not only for the ground state, but also for the members of the band built on top of it.

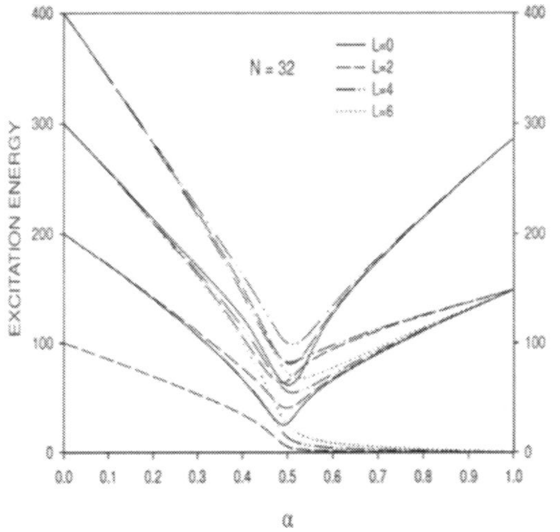

Figure 6. Spectrum of $\hat{H}(\alpha)$, Eq. (15), interpolating between the U(5) ($\alpha = 0$) and SU(3) ($\alpha = 1$) DS limits. From [29].

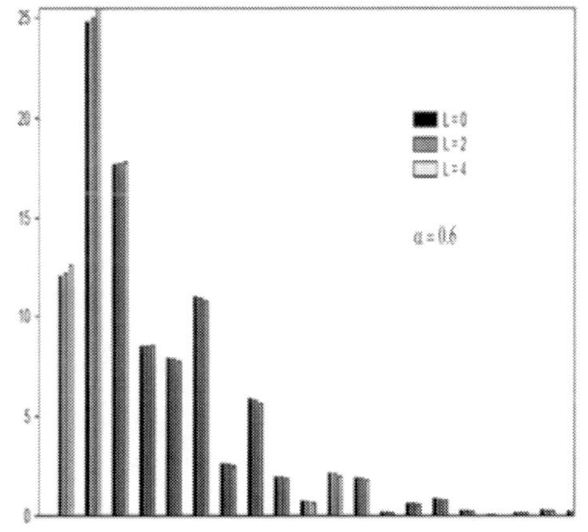

Figure 7. Squared amplitudes for angular momenta $L = 0, 2, 4$ yrast eigenstates of the U(5)-SU(3) Hamiltonian $\hat{H}(\alpha)$, Eq. (15), with $\alpha = 0.6$, in the SU(3) basis (1b). From [29]

6. Quasi dynamical symmetry (QDS)

A second kind of intermediate-symmetry occurring in algebraic modeling of dynamical systems, is that of quasi dynamical symmetry (QDS) [7–10]. While QDS can be defined mathematically in terms of embedded representations [27, 28], its physical meaning is that several observables associated with particular eigenstates, may be consistent with a certain symmetry which in fact is broken in the Hamiltonian. This typically occurs for a Hamiltonian transitional between two DS limits

$$\hat{H}(\alpha) \;=\; (1-\alpha)\,\hat{H}_{G_1} + \alpha\,\hat{H}_{G_2}\ . \tag{15}$$

$\hat{H}(\alpha)$ involves competing incompatible (non-commuting) symmetries. For $\alpha = 0$ or $\alpha = 1$, one recovers the limiting symmetries. For $0 < \alpha < 1$, both symmetries are broken and mixing occurs. A detailed study [8–10] of such Hamiltonians has found that for most values of α, selected states continue to exhibit characteristic properties (*e.g.*, energy and B(E2) ratios) of the *closest* DS limit. Such an "apparent" persistence of symmetry in the face of strong symmetry-breaking interactions, defines a QDS. The indicated persistence is clearly evident in the spectrum shown in Fig. 6, for an IBM Hamiltonian, $\hat{H}(\alpha)$, interpolating between the $G_1 = U(5)$ and $G_2 = SU(3)$ DS limits, relevant to shape-phase transitions between spherical and axially-deformed nuclei [21]. The "apparent" symmetry is due to the coherent nature of the mixing. As seen in Fig. 7, the mixing of SU(3) irreps is large, but is approximately independent of the angular momentum of the yrast states, *i.e.*, the SU(3) expansion coefficients $C^{(L)}_{(\lambda,\mu),K} \approx$ independent of L. The set of states exhibiting SU(3)-QDS thus have a common structure and form a band (the ground band) associated with a single intrinsic state.

The criterion for the validity of QDS is the similarity of the decomposition in the given DS basis. Thus, in the IBM, a quantitative measure of SU(3)-QDS can be defined as $\sqrt{1-\bar{\Theta}}$ [30], where $\bar{\Theta}$ is the average of $\Theta_{LL'} = \sum_i C_i^L C_i^{L'}$ for all pairs $L \neq L'$, and C_i^L are the amplitudes of the chosen set of eigenstates with angular momentum L in the SU(3) basis (1b).

The coherent decomposition signaling SU(3) QDS, implies strong correlations between the SU(3) components of different L-states in the same band. This can be used as an alternative criterion for the identification of rotational bands with SU(3)-QDS [31]. Focusing on the

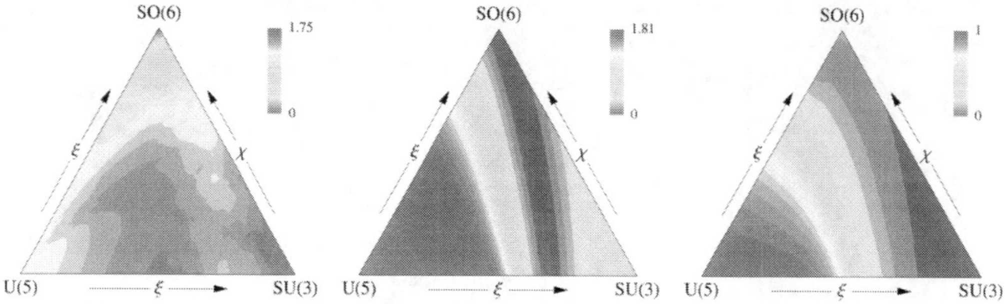

Figure 8. PDS and QDS for eigenstates of the ECQF Hamiltonian, Eq. (4), with $N=15$. Left panel: the O(6)-PDS measure [Shannon entropy $S_{O6}(L)$, Eq. (13c)], averaged for all $L=0$ eigenstates. Middle panel: $S_{O6}(L)$ for the ground state $L=0_1$. Right panel: the quantity $\sqrt{1-\Theta}$, which is the SU(3)-QDS measure (discussed in the text) for yrast (lowest energy) states with $L=0,2,\ldots,10$. From [32].

$L=0,2,4,6$, members of $K=0$ bands, given a $L=0_i^+$ state, among the ensemble of possible states, we associate with it those $L_j>0$ states which show the maximum correlation, $\max_j\{\pi(0_i,L_j)\}$. Here $\pi(0_i,L_j)$ is a Pearson coefficient whose values lie in the range $[-1,1]$. Specifically, $\pi(0_i,L_j)=1,-1,0$, indicate a perfect correlation, a perfect anti-correlation, and no linear correlation, respectively, among the SU(3) components of the 0_i and L_j states. To quantify the amount of coherence (hence of SU(3)-QDS) in the chosen set of states, one can adapt the procedure of [31] and employ the following product of the maximum correlation coefficients: $C_{SU3}(0_i\!-\!6)\equiv\max_j\{\pi(0_i,2_j)\}\,\max_k\{\pi(0_i,4_k)\}\,\max_\ell\{\pi(0_i,6_\ell)\}$. The set of states $\{0_i,\,2_j,\,4_k,\,6_\ell\}$ is considered as comprising a $K=0$ band with SU(3)-QDS, if $C_{SU3}(0_i\!-\!6)\approx 1$.

6.1. PDS and QDS in the symmetry triangle

Recently, a comprehensive analysis of the PDS and QDS properties of $\hat{H}_{ECQF}$, Eq. (4), was carried out [30], employing the symmetry measures discussed above. Representative results are shown in Fig. 8. The left panel of Fig. 8, shows the O(6) wave function entropy, $S_{O6}(L=0)$, Eq. (13c), averaged over *all* $L=0$ eigenstates. As seen, only a small region (marked in blue) near the O(6) vertex where an exact O(6) DS occurs, shows a high degree of purity with respect to the O(6) quantum number, σ. The middle panel displays $S_{O6}(L=0_1)$ of only the lowest $L=0_1$ state. Here, σ is conserved in the ground state throughout an entire region (marked in blue) of ECQF Hamiltonians, reflecting the O(6)-PDS of type III discussed in Section 5.1. The right panel displays the QDS measure $\sqrt{1-\Theta}$, with respect to the SU(3) basis. It is seen that large areas of the triangle are blue, *i.e.*, display SU(3)-QDS. These results illustrate the wider applicability in nuclei of the extended concepts, PDS and QDS, as compared to an exact DS.

7. Linking PDS and QDS

The concept of PDS reflects the purity of selected states, hence is different from the concept of QDS which reflects a coherent mixing. Nevertheless, a link between these two hitherto unrelated symmetry concepts can be established and shown to be empirically manifested in rotational nuclei [25].

The experimental spectrum of ^{160}Gd, along with its ECQF description (4), is shown in the left panel of Fig. 9. The middle and right panels show the decomposition into O(6) and SU(3) basis states, respectively, for yrast states with $L=0,2,4$. It is evident that the SU(3) symmetry is broken, as significant contributions of basis states with different SU(3) quantum numbers (λ,μ) occur. It is also clear from Fig. 9c that this mixing occurs in a coherent manner with similar patterns for the different members of the ground-state band. As explained in Section 6,

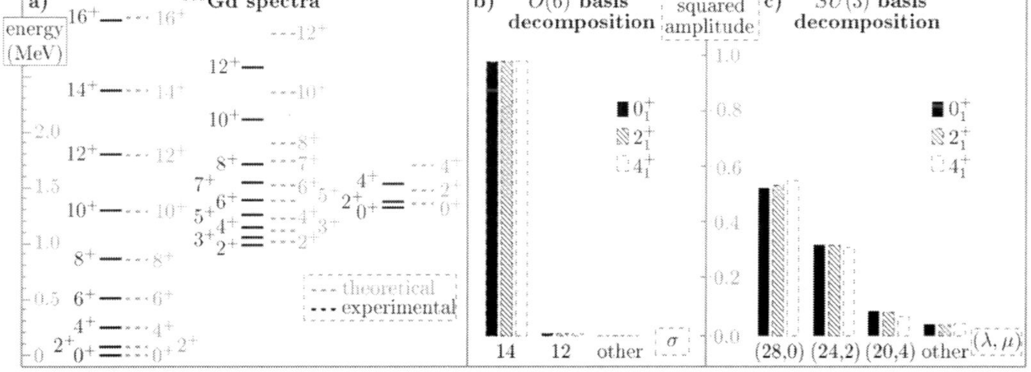

Figure 9. a) The experimental spectrum of ^{160}Gd compared with IBM calculation using $\hat{H}_{\rm ECQF}$ (4) with $\xi = 0.84$, $\chi = -0.5$, $N = 14$. b) The O(6) decomposition in σ components of yrast states with $L = 0, 2, 4$. c) The SU(3) decomposition in (λ, μ) components of the same states. Adapted from [25].

this is the hallmark of SU(3) QDS. On the other hand, as seen in Fig. 9b, the yrast states with $L = 0, 2, 4$ are almost entirely composed out of O(6) basis states with $\sigma = N = 14$ which implies small fluctuations $\Delta\sigma_L$ (14) and the preservation of O(6) symmetry in the ground-state band. At the same time, as shown in Fig. 5, the O(5) symmetry is broken in these states. Thus an empirically-manifested link is established between SU(3) QDS and O(6) PDS of type III.

The SU(3) QDS property for the members of the ground band results from the existence of a single intrinsic state, which is the coherent state, $|\beta = 1, \gamma = 0; N\rangle$ of Eq. (3) that has $\sigma = N$. Indeed, the (ξ, χ) parameter range of the ECQF Hamiltonian for which the equilibrium deformations are $(\beta = 1, \gamma = 0)$, shown by a red dashed line in Fig. 1, coincides with the region of an approximate ground-state O(6) symmetry for large N. These results demonstrate that coherent mixing of one symmetry (QDS) can result in the purity of a quantum number associated with partial conservation of a different, incompatible symmetry (PDS).

8. Impact of PDS and QDS on mixed regular and chaotic dynamics

Hamiltonians with a dynamical symmetry are always completely integrable. The Casimir invariants of the algebras in the chain provide a set of constants of the motion in involution. The classical motion is purely regular. A symmetry-breaking is connected to non-integrability and may give rise to chaotic motion. Hamiltonians with PDS and QDS are not completely integrable, hence can exhibit stochastic behavior, nor are they completely chaotic, since some eigenstates preserve the symmetry exactly in a PDS or mix in a coherent fashion in a QDS. Consequently, Hamiltonians with such intermediate symmetries are optimally suitable to the study of mixed systems with coexisting regularity and chaos.

The dynamics of a generic classical Hamiltonian system is mixed; KAM islands of regular motion and chaotic regions coexist in phase space. In the associated quantum system, the statistical properties of the spectrum are usually intermediate between the Poisson and the Gaussian orthogonal ensemble (GOE) statistics. In a PDS, the symmetry of the subset of solvable states is exact, yet does not arise from invariance properties of the Hamiltonian. Several works have shown that a PDS is strongly correlated with suppression of chaos [33, 34]. This enhancement of regularity was seen in both the classical measures of chaos, *e.g.*, the fraction of chaotic volume and the average largest Lyapunov exponent, and in quantum measures of chaos, *e.g.*, the nearest neighbors level spacing distribution, whose parameter interpolates between the

Poisson and GOE statistics. The reduction in chaos occurs even when the fraction of solvable states approaches zero in the classical limit, suggesting that the existence of a PDS increases the purity of other neighbouring states in the system.

The coherent mixing common to a set of states, characterizing a QDS, results from the existence of a single intrinsic state for each such band and imprints an adiabatic motion and increased regularity [35]. This was verified for low- [10] and high-lying [31] rotational bands using the ECQF Hamiltonian, Eq. (4). SU(3) QDS has been proposed [36] to underly the "arc of regularity" [37], a narrow zone of enhanced regularity in the parameter-space of $\hat{H}_{\rm ECQF}$. The arc, shown by a blue dotted line in Fig. 1, resides in the interior of the symmetry triangle and connects the U(5) and SU(3) vertices.

9. PDS and QDS in a first-order quantum phase transition

Quantum phase transitions (QPTs) are qualitative changes in the properties of a physical system induced by a variation of parameters in the quantum Hamiltonian. Such structural changes are currently of great interest in different branches of physics [38]. The competing interactions in the Hamiltonian that drive these ground-state phase transitions can affect dramatically the nature of the dynamics and, in some cases, lead to the emergence of quantum chaos [39–42]. Here we show that PDS and QDS can characterize the remaining regularity in a system undergoing a QPT, amidst a complicated environment of other states [42].

Focusing on the dynamics at the critical-point of a first-order QPT between spherical and deformed shapes, the relevant IBM Hamiltonian [43], upto a scale, can be taken to be the second term of Eq. (10), $\hat{H}_{\rm cri} = P_2^\dagger \cdot \tilde{P}_2$. The latter has the SU(3) basis states of Eq. (11) and the following U(5) basis states

$$|N, n_d = \tau = L = 0\rangle \qquad E = 0\ , \tag{16a}$$

$$|N, n_d = \tau = L = 3\rangle \qquad E = 6(2N-1)\ , \tag{16b}$$

as solvable eigenstates, while all other states are mixed with respect to both U(5) and SU(3). As such, $\hat{H}_{\rm cri}$ exhibits a coexistence of SU(3)-PDS and U(5)-PDS [44].

The classical limit of the Hamiltonian is obtained through the use of Glauber coherent states. This amounts to replacing $(s^\dagger,\, d_\mu^\dagger)$ by c-numbers $(\alpha_s^*,\, \alpha_\mu^*)$ rescaled by $\sqrt{N}$ and taking $N \to \infty$, with $1/N$ playing the role of $\hbar$ [45]. Setting all momenta to zero, yields the classical potential $V(x,y)$, which coincides with the surface of Eq. (2), with (β,γ) as polar coordinates. The classical dynamics constraint to $L = 0$, can be depicted conveniently via Poincaré surfaces of sections in the plane $y = 0$, plotting the values of (p_x, x) each time a trajectory intersects the plane. Regular trajectories are bound to toroidal manifolds within the phase space and their intersections with the plane of section lie on 1D curves (ovals). In contrast, chaotic trajectories randomly cover kinematically accessible areas of the section.

The Poincaré sections associated with the classical critical-point Hamiltonian are shown in Fig. 10 for representative energies. The bottom panel displays the classical potential which has two degenerate spherical and deformed minima. The dynamics in the region of the deformed minimum is robustly regular. The trajectories form a single island and remain regular even at energies far exceeding the barrier height $V_{\rm bar}$. In contrast, the dynamics in the region of the spherical minimum shows a change with energy from regularity to chaos, until complete chaoticity is reached near the barrier top. The clear separation between regular and chaotic dynamics, associated with the two minima, persists all the way to the barrier energy, $E = V_{\rm bar}$, where the two regions just touch. At $E > V_{\rm bar}$, a layer of chaos develops in the deformed region and gradually dominates the surviving regular island for $E \gg V_b$.

The quantum manifestations of such an inhomogeneous phase space structure, can be studied by the method of Peres lattices [46]. The latter are constructed by plotting the expectation

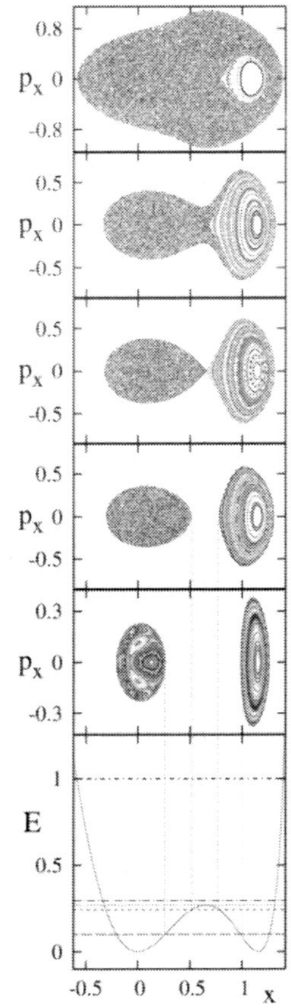

Figure 10. Poincaré sections at the critical point of the first-order QPT (upper five rows). The bottom row displays the classical potential. The five energies at which the sections were calculated consecutively, are indicated by horizontal lines. Adapted from [42].

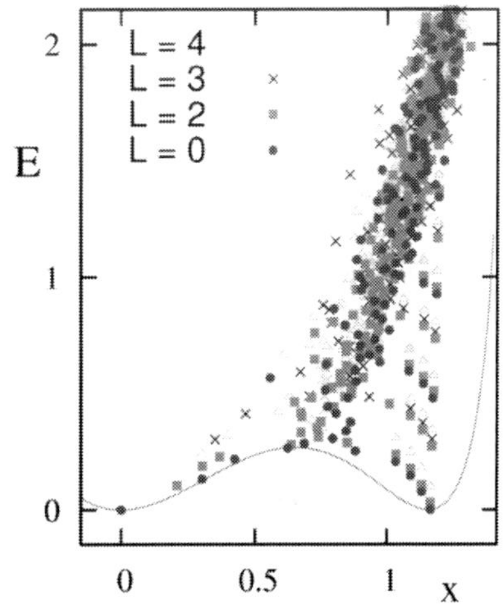

Figure 11. Peres lattices $\{x_i, E_i\}$ for $L = 0, 2, 3, 4$ and $N = 50$ eigenstates of the critical-point Hamiltonian, discussed in the text. The lattices are overlayed on the corresponding classical potential. Adapted from [42].

values $O_i = \langle i|\hat{O}|i\rangle$ of an arbitrary operator, $[\hat{O}, \hat{H}] \neq 0$, versus the energy $E_i = \langle i|\hat{H}|i\rangle$ of the Hamiltonian eigenstates $|i\rangle$. The lattices $\{O_i, E_i\}$ corresponding to regular dynamics display an ordered pattern, while chaotic dynamics leads to disordered meshes of points [46, 47]. In the present analysis, we choose the Peres operator to be $\hat{n}_d$, whose expectation value is related to the coordinate x in the classical potential. The lattices $\{x_i, E_i\}$, with $x_i \equiv \sqrt{2\langle i|\hat{n}_d|i\rangle/N}$, can then distinguish regular from irregular states and associate them with a given region in phase space.

The Peres lattices corresponding to ($N = 50$, $L = 0, 2, 3, 4$) eigenstates of $\hat{H}_{\rm cri}$ are shown in Fig. 11, overlayed on the classical potential. They disclose regular sequences of states localized within and above the deformed well. They are comprised of rotational states with $L = 0, 2, 4, \ldots$ forming regular $K = 0$ bands and sequences $L = 2, 3, 4, \ldots$ forming $K = 2$ bands. Additional K-bands (not shown in Fig. 11), corresponding to multiple β and γ vibrations about the deformed shape, can also be identified. The states in each regular band share a common structure, to be discussed below. Such ordered band-structures persist to energies above the barrier and are not present in the disordered (chaotic) portions of the Peres lattice. At low-energy, in the vicinity of the spherical well, one can also detect multiplets of states with $L = 0$, $L = 2$ and $L = 0, 2, 4$, typical of quadrupole excitations of a spherical shape.

An important clue on the nature of the surviving regular sequences of selected states, in the presence of more complicated type of eigenstates, comes from a symmetry analysis of their wave functions. The left column of Fig. 12 shows the U(5) n_d-probabilities, $P^{(L_i)}_{n_d}$ (13a), for eigenstates of $\hat{H}_{\rm cri}$, selected on the basis of having the largest components with $n_d = 0, 1, 2, 3, 4$, within the given L spectra. The states are arranged into panels labeled by 'n_d' to conform with

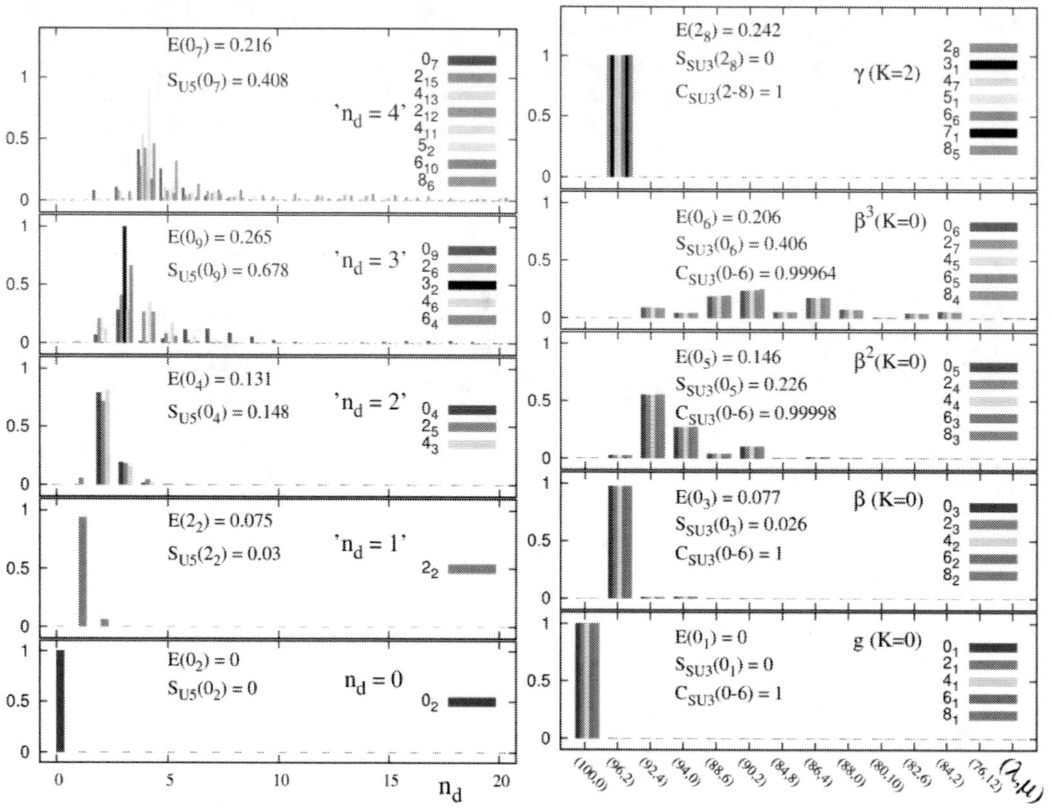

Figure 12. U(5) n_d-probability distribution, $P^{(L_i)}_{n_d}$ (13a) [left column], and SU(3) (λ, μ)-probability distribution, $P^{(L_i)}_{(\lambda,\mu)}$ (13b) [right column], for selected eigenstates of the critical-point Hamiltonian, with $N = 50$. The U(5) Shannon entropy, $S_{U5}(L_i)$ (13a), and SU(3) correlator, $C_{SU3}(0-6)$ are indicated for spherical and deformed type of states, respectively. Adapted from [42].

the structure of the n_d-multiplets of the U(5) DS limit. The normalized U(5) Shannon entropy $S_{U5}(L_i)$, Eq. (13a), is indicated for representative eigenstates. In particular, the zero-energy $L = 0^+_2$ state is seen to be a pure $n_d = 0$ state, with $S_{U5} = 0$, which is the solvable U(5)-PDS eigenstate of Eq. (16a). The state 2^+_2 has a pronounced $n_d = 1$ component (96%) and the states ($L = 0^+_4, 2^+_5, 4^+_3$) in the third panel, have a pronounced $n_d = 2$ component and a low value of $S_{U5} < 0.15$. All the above states with '$n_d \leq 2$' have a dominant single n_d component, and hence qualify as 'spherical' type of states. These are the lowest left-most states in the Peres lattices of Fig. 11, mentioned above. In contrast, the states in the panels '$n_d = 3$' and '$n_d = 4$' of Fig. 12, are significantly fragmented. A notable exception is the $L = 3^+_2$ state, which is the solvable U(5)-PDS state of Eq. (16b) with $n_d = 3$. The existence in the spectrum of specific spherical-type of states with either $P^{(L)}_{n_d} = 1$ [$S_{U5}(L) = 0$] or $P^{(L)}_{n_d} \approx 1$ [$S_{U5}(L) \approx 0$], exemplifies the presence of an exact or approximate U(5) PDS at the critical-point.

The states shown on the right column of Fig. 12 have a different character. They belong to the five lowest regular sequences seen in the Peres lattices of Fig. 11, in the region $x \geq 1$. They have a broad n_d-distribution, hence are qualified as 'deformed'-type of states, forming rotational bands: $g(K=0)$, $\beta(K=0)$, $\beta^2(K=0)$, $\beta^3(K=0)$ and $\gamma(K=2)$. Each panel depicts the SU(3) (λ, μ)-

distribution, $P^{(L_i)}_{(\lambda,\mu)}$ for the band members, the normalized SU(3) Shannon entropy $S_{\rm SU3}(L)$ (13b) for the bandhead state, and the Pearson correlator $C_{\rm SU3}(0_i{-}6)$ defined in Section 6. The ground $g(K=0)$ and $\gamma(K=2)$ bands are pure $[S_{\rm SU3}=0]$ with $(\lambda,\mu)=(2N,0)$ and $(2N-4,2)$ SU3) character, respectively. These are the solvable bands of Eq. (11) with SU(3) PDS. The non-solvable K-bands are mixed with respect to SU(3) in a coherent, L-independent, manner, hence exemplify SU(3)-QDS. As expected, we find $C_{\rm SU3}(0_i{-}6)\approx 1$ for these K-bands. The persisting regular U(5)-like [SU(3)-like] multiplets reflect the geometry of the classical Landau potential, as they are associated with the different spherical (deformed) minimum. One can use the corresponding measures of PDS and QDS as fingerprints of the QPT, not only at the critical point, but also throughout the coexistence region, where the two minima interchange [42].

10. Concluding remarks

The many examples of PDS and QDS, discussed in the present contribution, demonstrate that these intermediate-symmetries are more abundant than previously recognized. Contrary to naive expectations, the symmetry triangle appears to encompass important elements of symmetry and "not all is lost" inside it. Although, the examples considered were presented in the framework of a bosonic model, it is important to emphasize that these symmetry concepts are applicable to any many-body problem (bosons and fermions) endowed with an algebraic structure. Examples of many-body Hamiltonians with fermionic PDS and QDS are known [48–50]. The PDS algorithms discussed, for constructing Hamiltonians with PDS, are applicable to any semi-simple algebra and can be extended to coupled algebraic structure, $G_1 \times G_2$ [51, 52]. Attempts are under way to extend the PDS notion to Bose-Fermi symmetries and supersymmetries [53].

An important virtue of the PDS algorithms is their ability to incorporate and provide a selection criterion for higher-order terms [54]. Such terms are needed for an accurate description of the data and for extensions of ab-initio and beyond-mean-field methods to larger systems, which necessitate a strategy to deal with A-body effective interactions and proliferation of parameters. On one hand, the PDS approach allows more flexibility by relaxing the constrains of an exact dynamical symmetry (DS). On the other hand, the PDS approach picks particular symmetry-breaking terms which do not destroy results previously obtained with a DS for a segment the spectrum. The PDS construction is implemented order by order, yet the scheme is non-perturbative in the sense that the non-solvable states can experience strong symmetry-breaking. These virtues generate an efficient tool which can greatly enhance the scope of algebraic modeling of dynamical systems.

Correlated quantum many-body systems often display an astonishing regular excitation patterns which raises a fundamental question, namely, how simplicity emerges out of complexity in such circumstances. The simple patterns show up amidst a complicated environment of other states. It is natural to associate the "simple" states with a symmetry that protects their purity and special character. This symmetry, however, is shared by only a subset of states, and is broken in the remaining eigenstates of the same Hamiltonian. It thus appears that realistic quantum many-body Hamiltonians can accommodate simultaneously eigenstates with different symmetry character. The symmetry in question cannot be exact but only partial or "apparent". These are precisely the defining ingredients of PDS and QDS. These novel concepts of symmetries can thus offer a possible clue in addressing the "simplicity out of complexity" challenge.

Acknowledgments

I thank P. Van Isacker, J.E. García Ramos, C. Kremer, N. Pietralla and M. Macek for a fruitful collaboration on the topics discussed and F. Iachello, P. Cejnar and R.F. Casten for insightful discussions. This work is supported by the Israel Science Foundation.

Journal of Physics: Conference Series **597** (2015) 012003 doi:10.1088/1742-6596/597/1/012003

References

[1] Bohm A, Néeman Y and Barut A O eds 1988 *Dynamical Groups and Spectrum Generating Algebras* (Singapore: World Scientific)
[2] Iachello F and Arima A 1987 *The Interacting Boson Model* (Cambridge: Cambridge Univ. Press)
[3] Iachello F and Levine R D 1994 *Algebraic Theory of Molecules* (Oxford: Oxford Univ. Press)
[4] Alhassid Y and Leviatan A 1992 *J. Phys. A* **25** L1265
[5] García-Ramos J E, Leviatan A and Van Isacker P 2009 *Phys. Rev. Lett.* **102** 112502
[6] Leviatan A 2011 *Prog. Part. Nucl. Phys.* **66** 93
[7] Bahri C and Rowe D J 2000 *Nucl. Phys. A* **662** 125
[8] Rowe D J 2004 *Nucl. Phys. A* **745** 47
[9] Turner P S and Rowe D J 2005 *Nucl. Phys. A* **756** 333
[10] Rosensteel G and Rowe D J 2005 *Nucl. Phys. A* **759** 92
[11] Bohr A and Mottelson B R 1975 *Nuclear Structure II* (Benjamin, Reading, MA)
[12] Ginocchio J N and Kirson M W 1980 *Phys. Rev. Lett.* **44** 1744
[13] Dieperink A E L, Scholten O and Iachello F 1980 *Phys. Rev. Lett.* **44** 1747
[14] Warner D D and Casten R F 1983 *Phys. Rev. C* **28**, 1798
[15] Lipas P O, Toivonen P and Warner D D 1985 *Phys. Lett. B* **155** 295
[16] Casten R F and Warner D D 1983 *Prog. Part. Nucl. Phys.* **9** 311
[17] Leviatan A 1996 *Phys. Rev. Lett.* **77** 818
[18] Leviatan A and Sinai I 1999 *Phys. Rev. C* **60** 061301
[19] Casten R F, Cakirli R B, Blaum K and Couture A 2014 *Phys. Rev. Lett.* **113** 112501
[20] Leviatan A, Novoselsky A and Talmi I 1986 *Phys. Lett. B* **172** 144
[21] Cejnar P, Jolie J and Casten R F 2010 *Rev. Mod. Phys.* **82** 2155
[22] Van Isacker P 1999 *Phys. Rev. Lett.* **83** 4269
[23] Leviatan A and Van Isacker P 2002 *Phys. Rev. Lett.* **89** 222501
[24] Cejnar P and Jolie J 1998 *Phys. Lett. B* **420** 241; *ibid* 1998 *Phys. Rev. E* **58** 387
[25] Kremer C, Beller J, Leviatan A, Pietralla N, Rainovski G, Trippel R and Van Isacker P 2014 *Phys. Rev. C* **89** 041302(R)
[26] McCutchan E A, Zamfir N V and Casten R F 2004 *Phys. Rev. C* **69** 064306
[27] Rowe D J , Rochford P and Repka J 1988 *J. Math. Phys. (NY)* **29** 572
[28] Rowe D J 2004 in *Computational and Group-Theoretical Methods in Nuclear Physics*, Escher J, Castaños O, Hirsch J G, Pittel S and Stoitcheva G eds 2004 (Singapore: World Scientific)
[29] Rowe D J 2009 private communication
[30] Van Isacker P 2014 *Nuclear Physics News* **24** 23
[31] Macek M, Dobeš J and Cejnar P 2009 *Phys. Rev. C* **80** 014319; *ibid* 2010 **82** 014308
[32] Van Isacker P 2014 private communication
[33] Whelan N , Alhassid Y and Leviatan A 1993 *Phys. Rev. Lett.* **71** 2208
[34] Leviatan A and Whelan N D 1996 *Phys. Rev. Lett.* **77** 5202
[35] Macek M, Dobeš J, Stránský P and Cejnar P 2010 *Phys. Rev. Lett.* **105** 072503
[36] Bonatsos D, McCutchan E A and Casten R 2010 *Phys. Rev. Lett.* **104** 022502
[37] Alhassid Y and Whelan N 1991 Phys. Rev. Lett. **67** 816
[38] Carr L ed. 2010 *Understanding Quantum Phase Transitions* (CRC press)
[39] Emary C and Brandes T 2003 *Phys. Rev. Lett.* **90** 044101; *ibid* 2003 *Phys. Rev. E* **67** 066203
[40] Macek M and Leviatan A 2011 *Phys. Rev. C* **84** 041302(R)
[41] Leviatan A and Macek M 2012 *Phys. Lett. B* **714** 110
[42] Macek M and Leviatan A 2014 *Ann. Phys. (NY)* **351** 302
[43] Leviatan A 2006 *Phys. Rev. C* **74** 051301(R)
[44] Leviatan A 2007 *Phys. Rev. Lett.* **98** 242502
[45] Hatch R L and Levit S 1982 *Phys. Rev. C* **25** 614
[46] Peres A 1984 *Phys. Rev. Lett.* **53** 1711
[47] Stránský P, Hruška P and Cejnar P 2009 *Phys. Rev. E* **79** 066201
[48] Escher J and Leviatan A 2000 *Phys. Rev. Lett.* **84** 1866; *ibid* 2002 *Phys. Rev. C* **65** 054309
[49] Rosensteel G and Rowe D J 2003 *Phys. Rev. C* **67** 014303
[50] Van Isacker P and Heinze S 2008 *Phys. Rev. Lett.* **100** 052501; *ibid* 2014 *Ann. Phys. (NY)* **349** 73
[51] Talmi I 1997 *Phys. Lett. B* **405** 1
[52] Leviatan A and Ginocchio J N 2000 *Phys. Rev. C* **61** 024305
[53] Van Isacker P, Jolie J, Thomas T and Leviatan A 2014 work in progress
[54] Leviatan A, García-Ramos J E and Van Isacker P 2013 *Phys. Rev. C* **87** 021302(R)

Journal of Physics: Conference Series **597** (2015) 012004 doi:10.1088/1742-6596/597/1/012004

Reflection positivity for the circle group

K-H Neeb[1] and G Ólafsson[2]

[1] Department Mathematik, FAU Erlangen-Nürnberg, Cauerstrasse 11, 91058-Erlangen, Germany

[2] Department of mathematics, Louisiana State University, Baton Rouge, LA 70803, USA

E-mail: neeb@math.fau.de, olafsson@math.lsu.edu

Abstract. In this note we characterize those unitary one-parameter groups $(U_t^c)_{t\in\mathbb{R}}$ which admit euclidean realizations in the sense that they are obtained by the analytic continuation process corresponding to reflection positivity from a unitary representation U of the circle group. These are precisely the ones for which there exists an anti-unitary involution J commuting with U^c. This provides an interesting link with the modular data arising in Tomita–Takesaki theory. Introducing the concept of a positive definite function with values in the space of sesquilinear forms, we further establish a link between KMS states and reflection positivity on the circle.

1. Introduction

In this note we continue our investigations of the mathematical foundations of *reflection positivity*, a basic concept in constructive quantum field theory ([10, 16, 17, 13, 14, 15]). Originally, reflection positivity, also called Osterwalder–Schrader positivity, arises as a requirement on the euclidean side to establish a duality between euclidean and relativistic quantum field theories ([27]). It is closely related to "Wick rotations" or "analytic continuation" in the time variable from the real to the imaginary axis.

The underlying fundamental concept is that of a *reflection positive Hilbert space*, introduced in [24]. This is a triple $(\mathcal{E}, \mathcal{E}_+, \theta)$, where $\mathcal{E}$ is a Hilbert space, $\theta : \mathcal{E} \to \mathcal{E}$ is a unitary involution and $\mathcal{E}_+$ is a closed subspace of $\mathcal{E}$ which is θ-positive in the sense that the hermitian form $\langle \theta u, v\rangle$ is positive semidefinite on $\mathcal{E}_+$. [1] Let $\mathcal{N} := \{v \in \mathcal{E}_+ : \langle \theta v, v\rangle = 0\}$, write $\widehat{\mathcal{E}}$ for the Hilbert space completion of the quotient $\mathcal{E}_+/\mathcal{N}$ and $q : \mathcal{E}_+ \to \widehat{\mathcal{E}}, v \mapsto \widehat{v}$ for the canonical map. If $T : \mathcal{D}(T) \subseteq \mathcal{E}_+ \to \mathcal{E}_+$ is a linear or antilinear operator with $T(\mathcal{N} \cap \mathcal{D}(T)) \subseteq \mathcal{N}$, then there exists a well-defined operator $\widehat{T} : \mathcal{D}(\widehat{T}) \subseteq \widehat{\mathcal{E}} \to \widehat{\mathcal{E}}$ defined by $\widehat{T}(\widehat{v}) = \widehat{Tv}$ for $v \in \mathcal{D}(T)$. The $\widehat{\ }$ is the *OS-quantization functor*.

To see how this relates to group representations, let us call a triple (G, H, τ) a *symmetric Lie group* if G is a Lie group, τ is an involutive automorphism of G and H is an open subgroup of the group G^τ of τ-fixed points. Then the Lie algebra $\mathfrak{g}$ of G decomposes into τ-eigenspaces $\mathfrak{g} = \mathfrak{h} \oplus \mathfrak{q}$ and we obtain the *Cartan dual Lie algebra* $\mathfrak{g}^c = \mathfrak{h} \oplus i\mathfrak{q}$. If (G, H, τ) is a symmetric Lie group and $(\mathcal{E}, \mathcal{E}_+, \theta)$ a reflection positive Hilbert space, then we say that a unitary representation $\pi : G \to \mathcal{E}$ is *reflection positive with respect to* (G, H, τ) if the following three conditions hold:

(RP1) $\pi(\tau(g)) = \theta\pi(g)\theta$ for every $g \in G$.

[1] We write $\langle v, w\rangle$ for the scalar product in a complex Hilbert space and we assume that it is linear in the *first* argument.

(RP2) $\pi(h)\mathcal{E}_+ = \mathcal{E}_+$ for every $h \in H$.

(RP3) There exists a subspace $\mathcal{D} \subseteq \mathcal{E}_+ \cap \mathcal{E}^\infty$, dense in $\mathcal{E}_+$, such that $\mathrm{d}\pi(X)\mathcal{D} \subset \mathcal{D}$ for all $X \in \mathfrak{q}$.

A typical source of reflection positive representations are the representations $(\pi_\varphi, \mathcal{H}_\varphi)$ obtained via GNS construction (cf. Theorem A.4) from τ-invariant positive definite functions $\varphi\colon G \to B(\mathcal{V})$, respectively the kernel $K(x,y) = \varphi(xy^{-1})$, where $\mathcal{V}$ is a Hilbert space (cf. Appendix A). If $G_+ \subseteq G$ is an open subset with $G_+H = G_+$, then φ is called *reflection positive for* (G, G_+, τ) if the kernel $Q(x,y) = \varphi(x\tau(y)^{-1})$ is positive definite on G_+. For $\mathcal{E} = \mathcal{H}_\varphi$, the subspace $\mathcal{E}_+$ is generated by the functions $K(\cdot, y)v$, $y \in G_+, v \in \mathcal{V}$, and $\widehat{\mathcal{E}}$ identifies naturally with the Hilbert space $\mathcal{H}_Q \subseteq \mathcal{V}^{G_+}$ (cf. [24, Prop.1.11], [21, Ex. 5.17a]). If the kernels $\langle Q(x,y)v, v\rangle$ are smooth for v in a dense subspace of $\mathcal{V}$, then (RP1-3) are readily verified.

If π is a reflection positive representation, then $\pi^c_H(h) := \widehat{\pi(h)}$ defines a unitary representation of H on $\widehat{\mathcal{E}}$. However, we would like to have a unitary representation π^c of the simply connected Lie group G^c with Lie algebra $\mathfrak{g}^c$ on $\widehat{\mathcal{E}}$ extending π^c_H in such a way that the derived representation is compatible with the operators $i\widehat{\mathrm{d}\pi(X)}$, $X \in \mathfrak{q}$, that we obtain from (RP3) on a dense subspace of $\widehat{\mathcal{E}}$. If such a representation exists, then we call $(\pi, \mathcal{E})$ a *euclidean realization* of the representation $(\pi^c, \widehat{\mathcal{E}})$ of G^c. Sufficient conditions for the existence of π^c have been developed in [21]. The prototypical pair (G, G^c) consists of the euclidean motion group $\mathbb{R}^d \rtimes \mathrm{O}_d(\mathbb{R})$ and the simply connected covering of the Poincaré group $\mathcal{P}^\uparrow_+ = \mathbb{R}^d \rtimes \mathrm{SO}_{1,d}(\mathbb{R})_0$.

In [25] we studied reflection positive one-parameter groups and hermitian contractive semigroups as one key to reflection positivity for more general symmetric Lie groups and their representations. Here a crucial point is that, for every unitary one-parameter group $U^c_t = e^{itH}$ with $H \geq 0$ on the Hilbert space $\mathcal{V}$, we obtain by $\varphi(t) := e^{-|t|H}$ a $B(\mathcal{V})$-valued function on $\mathbb{R}$ which is reflection positive for $(\mathbb{R}, \mathbb{R}_+, -\mathrm{id}_\mathbb{R})$ and which leads to a euclidean realization of U^c. From this we derive that all representations of the $ax+b$-group, resp., the Heisenberg group which satisfy the positive spectrum condition for the translation group, resp., the center, possess natural euclidean realizations.

The present note grew out of the attempt to extract the representation theoretic aspects from the discussion of reflection positivity for the circle group $\mathbb{T}$ in [18]. Here we continue our project by exploring reflection positive functions $\varphi\colon \mathbb{T} \to B(\mathcal{V})$ for $(\mathbb{T}, \mathbb{T}^+, \tau)$, where $\tau(z) = z^{-1}$ and $\mathbb{T}^+$ is a half circle. This leads us naturally to anti-unitary involutions, an aspect that did not show up for triples (G, G_+, τ), where G_+ is a semigroup. We start in Section 2 with a discussion of the integral representation of reflection positive operator-valued functions for $(\mathbb{T}, \mathbb{T}^+, \tau)$ due to Klein and Landau ([18]) and in Section 3 we characterize those unitary one-parameter groups $(U^c, \mathcal{H})$ which admit euclidean realizations in this context as those for which there exists an anti-unitary involution J commuting with U^c (Theorem 3.4). Any such pair (J, U^c) with $U^c_t = e^{itH}$ can be encoded in the pair (J, Δ), where $\Delta = e^{-\beta H}$ is positive selfadjoint with $J\Delta J = \Delta^{-1}$, a relation well-known from Tomita-Takesaki theory. Such pairs are closely linked to real standard subspaces, a connection discussed in Section 4. In Section 5 we finally explain how all this connects to KMS states for C^*-dynamical systems $(\mathcal{A}, \mathbb{R}, \alpha)$ and conclude with a short discussion of perspectives. To establish the connection with KMS states, we need the concept of a positive definite function on a group G with values in the space $\mathrm{Sesq}(\mathcal{V})$ of sesquilinear forms on a real or complex vector space $\mathcal{V}$. This concept is briefly developed in Appendix A, where we explain in particular how the GNS construction works in this context. Appendix B provides some tools to obtain integral representations of positive definite functions on convex sets which sharpen some known results in this context.

We hope that this note will prove useful in the further development of the representation theoretic side of reflection positivity under the presence of anti-unitary involutions which occur in many recent constructions in Quantum Field Theory (see in particular [7, p. 627], [2], [6]).

2. Reflection positivity on the circle group $\mathbb{T}$

Let $G := \mathbb{T}_\beta := \mathbb{R}/\beta\mathbb{Z}$, where $\beta > 0$, so that $\mathbb{T}_\beta$ is a circle of length β. We write $[t] := t + \beta\mathbb{Z}$ for the image of t in $\mathbb{T}_\beta$ and write $\mathbb{T}_\beta^+ := \{[t] \in \mathbb{T}_\beta : 0 < t < \beta/2\}$ for the corresponding semicircle. We further fix the involutive automorphism $\tau_\beta(z) = z^{-1}$ given by inversion in $\mathbb{T}_\beta$. In the following we identify functions on $\mathbb{T}_\beta$ with β-periodic functions on $\mathbb{R}$.

Definition 2.1. Let $\mathcal{V}$ be a Hilbert space. A weak operator continuous function $\varphi: \mathbb{T}_\beta \to B(\mathcal{V})$ is called *reflection positive* w.r.t. $(\mathbb{T}_\beta, \mathbb{T}_\beta^+, \tau_\beta)$ if and only if it is positive definite and the kernel $(\varphi(t+s))_{0<t,s<\beta/2}$ is positive definite, which is equivalent to the positive definiteness of the kernel

$$Q_\varphi(t,s) := \varphi\Big(\frac{t+s}{2}\Big), \quad 0 < t,s < \beta \quad \text{on }]0,\beta[. \tag{1}$$

Remark 2.2. The reflection positivity of φ ensures that the corresponding β-periodic GNS-representation of $\mathbb{R}$ on the reproducing kernel Hilbert space $\mathcal{E} := \mathcal{H}_\varphi \subseteq \mathcal{V}^{\mathbb{T}_\beta} \subseteq \mathcal{V}^{\mathbb{R}}$ by $\pi_\varphi(t)f := f(\cdot + t)$ is reflection positive with respect to $(\theta f)(x) = f(-x)$ and the closed subspace $\mathcal{E}_+$ generated by $\varphi(-t)v$, $t \in \mathbb{T}_\beta^+$, $v \in \mathcal{V}$ (cf. Proposition A.4). Here $\tau_\beta(x) = -x$ and $H = \{0\}$. Conditions (RP1/2) are obvious, and (RP3) is satisfied because the restriction of φ to $]0,\beta[$ is automatically smooth (Theorem 2.4). For the basis element $X = 1 \in \mathbf{L}(\mathbb{T}_\beta) \cong \mathbb{R}$, the operator $\widehat{d\pi(X)}$ acts on $\widehat{\mathcal{E}} \subseteq \mathcal{V}^{\mathbb{T}_\beta^+}$ by $\frac{d}{dt}$.

Example 2.3. Basic examples of reflection positive functions are given by

$$f_\lambda(t) = e^{-t\lambda} + e^{-(\beta-t)\lambda} = 2e^{-\beta\lambda/2}\cosh((\tfrac{\beta}{2} - t)\lambda)$$

for $0 \le t \le \beta$ and $\lambda \ge 0$. The corresponding kernel Q_{f_λ} is positive definite on all of $\mathbb{R}$ because f_λ is a Laplace transform of the positive measure $\delta_\lambda + e^{-\beta\lambda}\delta_{-\lambda}$. A direct calculation shows that the Fourier series of the β-periodic extension of f_λ to $\mathbb{R}$ (also denoted f_λ) is given by

$$f_\lambda(t) = \sum_{n\in\mathbb{Z}} c_n e^{2\pi i n t/\beta} \quad \text{with} \quad c_n = \frac{2\beta\lambda(1-e^{-\beta\lambda})}{(\lambda\beta)^2 + (2\pi n)^2}. \tag{2}$$

As $c_n \ge 0$ for every $n \in \mathbb{Z}$, the function f_λ on $\mathbb{T}_\beta$ is positive definite. This shows that f_λ is reflection positive.

Theorem 2.4. ([18, Thm. 3.3]) *A β-periodic weak operator continuous function $\varphi: \mathbb{R} \to B(\mathcal{V})$ is reflection positive w.r.t. $(\mathbb{T}_\beta, \mathbb{T}_\beta^+, \tau_\beta)$ if and only if there exists a* Herm$(\mathcal{V})_+$*-valued measure μ_+ on $[0,\infty[$ such that*

$$\varphi(t) = \int_0^\infty e^{-t\lambda} + e^{-(\beta-t)\lambda}\, d\mu_+(\lambda) \quad \text{for} \quad 0 \le t \le \beta. \tag{3}$$

Then the measure μ_+ is uniquely determined by φ.

Proof. If φ is given by (3), then the kernel $\varphi(\frac{x+y}{2})$ is clearly positive definite. To see that φ is a positive definite function of $\mathbb{T}_\beta$, we use the Fourier expansion $f_\lambda(t) = \sum_n c_n(\lambda)e^{2\pi i n t/\beta}$ from (2) to obtain

$$\varphi(t) = \sum_{n\in\mathbb{Z}} e^{2\pi i n t/\beta} \int_0^\infty c_n(\lambda)\, d\mu_+(\lambda).$$

Now the positivity of the operators $\int_0^\infty c_n(\lambda)\, d\mu_+(\lambda)$ shows that φ is positive definite, hence reflection positive.

If, conversely, $\varphi: [0,\beta] \to B(\mathcal{V})$ is reflection positive, then it can be written as a Laplace transform $\varphi = \mathcal{L}(\mu)$ of a Herm$(\mathcal{V})_+$-valued measure μ on $\mathbb{R}$ ([11, Thm. 18.8]; see also

Theorem B.3 below). Since the reflection in $\beta/2$ is the composition of the reflection $r(t) = -t$ in 0 and the translation by β, the function φ is also symmetric with respect to $\frac{\beta}{2}$. This symmetry requirement is equivalent to $r_*\mu = e_{-\beta}\mu$ for $e_\beta(\lambda) = e^{\beta\lambda}$, so that φ can be written as in (3). □

Remark 2.5. (a) It is often more convenient to work with the measure μ on $\mathbb{R}$ defined by

$$d\mu(\lambda) = d\mu_+(\lambda) + e^{\beta\lambda}d\mu_+(-\lambda), \quad \text{which satisfies} \quad r_*\mu = e_{-\beta}\mu \quad \text{for} \quad r(\lambda) = -\lambda. \tag{4}$$

We thus obtain the description $\varphi(t) = \int_{\mathbb{R}} e^{-t\lambda}\, d\mu(\lambda) = \mathcal{L}(\mu)(t)$ of φ as the Laplace transform $\mathcal{L}(\mu)$. The existence of these integrals for $0 \leq t \leq \beta$ only requires that the measure μ is finite. We then have $\mu(\mathbb{R}) = \varphi(0) = \varphi(\beta)$ by (4).

(b) In view of (4), the measure $\nu := e_{-\beta/2}\mu$ is symmetric. For the Fourier transform $\widehat{\mu}(z) = \int_{\mathbb{R}} e^{iz\lambda}\, d\mu(\lambda)$ which is defined on the closed strip $\overline{\mathcal{D}_\beta} = \{z \in \mathbb{C} \colon 0 \leq \operatorname{Im} z \leq \beta\}$, we thus obtain the relations

$$\widehat{\mu}(i\beta - z) = \widehat{\mu}(z), \quad z \in \overline{\mathcal{D}_\beta}, \qquad \text{and} \qquad \widehat{\nu}(t) = \widehat{\nu}(-t) = \widehat{\mu}\Big(t + i\frac{\beta}{2}\Big), \quad t \in \mathbb{R}. \tag{5}$$

Remark 2.6. (The representation U^c on $\widehat{\mathcal{E}}$) With the integral representation from Theorem 2.4, we can make the unitary one-parameter group U^c of the dual group $(\mathbb{T}_\beta)^c \cong \mathbb{R}$ on $\widehat{\mathcal{E}}$ more explicit. From Remark 2.2, we recall that $\widehat{\mathcal{E}}$ can be identified with the reproducing kernel Hilbert space corresponding to the kernel $\varphi(\frac{x+y}{2})$ on the real interval $[0,\beta]$ (cf. (1)).

(a) In view of Theorem B.3, $\widehat{\mathcal{E}}$ can be identified with the vector-valued L^2-space $L^2(\mathbb{R},\mu;\mathcal{V})$ with the scalar product $\langle \xi, \eta\rangle = \int_{\mathbb{R}}\langle d\mu(\lambda)\xi(\lambda), \eta(\lambda)\rangle$. On this space we have a natural unitary one-parameter group defined by

$$(U^c_t f)(\lambda) := e^{it\lambda}f(\lambda) \quad \text{for} \quad t, \lambda \in \mathbb{R}.$$

Since $L^2(\mathbb{R},\mu;\mathcal{V})$ contains the constant functions, we obtain a bounded operator $j \colon \mathcal{V} \to L^2(\mathbb{R},\mu;\mathcal{V}), j(v)(\lambda) := v$. Then $j(\mathcal{V})$ is U^c-cyclic in $L^2(\mathbb{R},\mu;\mathcal{V})$ ([22, Lemma III.8]), and we have

$$\langle U^c_t j(v), j(w)\rangle = \int_{\mathbb{R}} e^{it\lambda}\langle d\mu(\lambda)v, w\rangle = \langle \varphi(-it)v, w\rangle \quad \text{for} \quad v, w \in \mathcal{V}, t \in \mathbb{R}, \tag{6}$$

where we have used the analytic continuation of φ to the strip $\{z \in \mathbb{C} \colon 0 \leq \operatorname{Re} z \leq \beta\}$ that follows from the integral representation (Theorem 2.4) and Theorem B.3. Therefore $(U^c, L^2(\mathbb{R},\mu;\mathcal{V}))$ can be identified with the vector-valued GNS representation corresponding to the positive definite function $\widehat{\mu}(t) = \varphi(-it)$ on $\mathbb{R}$ obtained by analytic continuation from φ (cf. Proposition A.4). If $(Hf)(\lambda) = \lambda f(\lambda)$ is the infinitesimal generator of $U^c_t = e^{itH}$, then, for every $v \in \mathcal{V}$, we have $v \in \mathcal{D}(e^{-\beta H/2})$ because $\int_{\mathbb{R}} e^{-\beta\lambda}\langle d\mu(\lambda)v, v\rangle = \langle\varphi(\beta)v, v\rangle$ is finite (Lemma B.4). For $0 \leq t \leq \beta$ and $v, w \in \mathcal{V}$, we now obtain $\langle\varphi(t)v, w\rangle = \int_{\mathbb{R}} e^{-t\lambda}\langle d\mu(\lambda)v, w\rangle = \langle e^{-tH}v, w\rangle$, resulting in the dilation formula

$$\varphi(t) = j^* e^{-tH} j \quad \text{for} \quad 0 \leq t \leq \beta. \tag{7}$$

Further, $(Jf)(\lambda) := e^{-\beta\lambda/2}f(-\lambda)$, is a unitary involution on $L^2(\mathbb{R},\mu;\mathcal{V})$ with $JHJ = -H$, and

$$R := e^{\beta H/2}J = Je^{-\beta H/2}, \quad (Rf)(\lambda) = f(-\lambda)$$

is an involution with domain $\mathcal{D}(R) = \mathcal{D}(e^{-\beta H/2})$ and $\mathcal{V} \subseteq \operatorname{Fix}(R) := \{v \in \mathcal{D}(R) \colon Rv = v\}$.

(b) Alternatively, we can use $\widetilde{j} \colon \mathcal{V} \to L^2(\mathbb{R},\mu;\mathcal{V}), \widetilde{j}(v)(\lambda) = e^{-\beta\lambda/4}v$ to obtain

$$\langle U^c_t\widetilde{j}(v), \widetilde{j}(w)\rangle = \int_{\mathbb{R}} e^{it\lambda}e^{-\beta\lambda/2}\langle d\mu(\lambda)v, w\rangle = \Big\langle \varphi\Big(\frac{\beta}{2} - it\Big)v, w\Big\rangle,$$

which exhibits U^c as the GNS representation of the symmetric positive definite function $\psi(t) := \varphi\ \frac{\beta}{2} - it$.

3. Existence of euclidean realizations

The following proposition provides various characterizations of unitary one-parameter groups with reflection symmetry. As we shall see below, these are precisely the ones with a euclidean realization from $(\mathbb{T}_\beta, \mathbb{T}_\beta^+, \tau_\beta)$.

Proposition 3.1. *For a unitary one-parameter group $(U_t)_{t\in\mathbb{R}}$ on $\mathcal{H}$ with spectral measure $E\colon \mathfrak{B}(\mathbb{R}) \to B(\mathcal{H})$, the following are equivalent:*

(i) *There exists an anti-unitary involution J on $\mathcal{H}$ with $JU_tJ = U_t$ for $t \in \mathbb{R}$.*

(ii) *For $\mathcal{H}_\pm := E(\mathbb{R}_\pm^\times)\mathcal{H}$, the unitary one-parameter groups $U_t^+ := U_t|_{\mathcal{H}_+}$ and $U_t^- := U_{-t}|_{\mathcal{H}_-}$ are unitarily equivalent.*

(iii) *The unitary one-parameter group $(U, \mathcal{H})$ is equivalent to a GNS representation $(\pi_\psi, \mathcal{H}_\psi)$, where $\psi\colon \mathbb{R} \to B(\mathcal{V})$ is a symmetric positive definite function.*

(iv) *There exists a unitary involution R on $\mathcal{H}$ with $RU_tR = U_{-t}$ for $t \in \mathbb{R}$.*

Proof. Every cyclic unitary one-parameter group is isomorphic to $L^2(\mathbb{R}, \mu)$ with $(U_tf)(\lambda) = e^{it\lambda}f(\lambda)$. Then $Kf = \overline{f}$ is an anti-unitary involution. Decomposing into cyclic subspaces, we thus obtain an anti-unitary involution K on $\mathcal{H}$ satisfying $KU_tK = U_{-t}$ for $t \in \mathbb{R}$. As (i)-(iv) hold for the trivial representation on the subspace $\mathcal{H}_0 = E(\{0\})\mathcal{H}$ of fixed points, we may w.l.o.g. assume that $\mathcal{H}_0 = \{0\}$, so that $\mathcal{H} = \mathcal{H}_+ \oplus \mathcal{H}_-$.

(i) $\Rightarrow$ (ii): The operator $W := KJ$ is unitary and satisfies $WU_tW^{-1} = U_{-t}$ for $t \in \mathbb{R}$. Therefore $W\mathcal{H}_\pm = \mathcal{H}_\mp$ and we obtain a unitary intertwining operator $(U^+, \mathcal{H}_+) \to (U^-, \mathcal{H}_-)$.

(ii) $\Rightarrow$ (iii): We write $\mathcal{H} = \mathcal{K} \oplus \mathcal{K}$ with $U_t = W_t \oplus W_{-t}$ and $(W, \mathcal{K}) \cong (U^+, \mathcal{H}_+)$. Then the representations on both summands are disjoint. Put $\mathcal{V} := \{(v, v)\colon v \in \mathcal{K}\}$ and let $\mathcal{W} \subseteq \mathcal{H}$ be the closed U-invariant subspace generated by $\mathcal{V}$. Since $\mathcal{W}$ is invariant under all spectral projections, it contains the projection of $\mathcal{V}$ onto both factors, so that $\mathcal{W} = \mathcal{H}$. The corresponding positive definite function $\psi(t) := P_\mathcal{V} U_t P_\mathcal{V}^*$ obtained from the orthogonal projection $P_\mathcal{V}\colon \mathcal{H} \to \mathcal{V}$ satisfies

$$\langle \psi(t)(v, v), (w, w)\rangle = \langle (W_tv, W_{-t}v), (w, w)\rangle = \langle (W_t + W_{-t})v, w\rangle \quad \text{for} \quad v, w \in \mathcal{K},$$

so that ψ is symmetric. Hence (iii) follows from Proposition A.4.

(iii) $\Rightarrow$ (iv): If ψ is symmetric, then $R\colon \mathcal{H}_\psi \to \mathcal{H}_\psi, (Rf)(t) := f(-t)$ is a unitary involution with the required properties because the kernel $Q(t, s) = \psi(t - s)$ satisfies $Q(-t, -s) = Q(t, s)$ (cf. [23, Rem. II.4.5(c)]).

(iv) $\Rightarrow$ (ii) follows from $R(\mathcal{H}_\pm) = \mathcal{H}_\mp$.

(ii) $\Rightarrow$ (i): As above, we write $\mathcal{H} = \mathcal{K} \oplus \mathcal{K}$ with $U_t = W_t \oplus W_{-t}$ and $(W, \mathcal{K}) \cong (U^+, \mathcal{H}_+)$ and let $K\colon \mathcal{K} \to \mathcal{K}$ be an anti-unitary involution with $KU_tK = U_{-t}$ for $t \in \mathbb{R}$. Then $J(v, w) := (Kw, Kv)$ has the required properties. □

Remark 3.2. Let $U_t = e^{itH}$ be a continuous unitary one-parameter group on the complex Hilbert space $\mathcal{H}$ with infinitesimal generator H and J an anti-unitary involution with $JU_tJ = U_{\varepsilon t}$ for $\varepsilon \in \{\pm 1\}$. Then $JHJ = -\varepsilon H$. If the operator H is non-negative, then so is JHJ, which can only happen for $\varepsilon = -1$.

Definition 3.3. We have seen in Remark 2.2 that every weakly operator continuous reflection positive function $\varphi\colon \mathbb{T}_\beta \to B(\mathcal{V})$ leads to a reflection positive unitary representation $(U, \mathcal{E})$ of $\mathbb{T}_\beta$ (Remark 2.2). If $(U^c, \widehat{\mathcal{E}})$ is the corresponding unitary one-parameter group from Remark 2.6, then we call $(U, \mathcal{E})$ a *euclidean realization of* U^c. We have already seen in Remark 2.6 that it can be obtained by the GNS construction from the positive definite function $\mathbb{R} \to B(\mathcal{V}), t \mapsto \varphi(it)$ or the symmetric function $\varphi(it + \frac{\beta}{2})$. In this sense it is obtained from U by an analytic continuation process.

At this point it is a natural question which unitary one-parameter groups $(U^c, \mathcal{H})$ have a euclidean realization in the sense of Definition 3.3. This can now be stated in terms of the conditions discussed in Proposition 3.1:

Theorem 3.4. (Realization Theorem) *A unitary one-parameter group $(U_t^c)_{t\in\mathbb{R}}$ on a Hilbert space $\mathcal{H}$ has a euclidean realization in terms of a reflection positive representation of $(\mathbb{T}_\beta, \mathbb{T}_{\frac{\beta}{2}}^+, \theta)$ if and only if there exists an anti-unitary involution J on $\mathcal{H}$ commuting with U^c.*

Proof. Since the assertion is trivial for the trivial representation, we may assume that there are no non-zero fixed vectors, i.e., $\mathcal{H}^{U^c} = \{0\}$.

From Proposition 3.1 we know that the existence of an anti-unitary involution commuting with U^c is equivalent to the realizability of U^c by the GNS construction from a symmetric positive definite function. If U^c has a euclidean realization as in Definition 3.3, then $\psi(t) := \varphi(\frac{\beta}{2} + it)$ is such a function (Remark 2.6(b)).

Suppose, conversely, that there exists an anti-unitary involution J on $\mathcal{H}$ commuting with U^c. As in the proof of Proposition 3.1, we write $\mathcal{H} = \mathcal{V} \oplus \mathcal{V}$ with $U_t^c = e^{itA} \oplus e^{-itA}$ for a positive selfadjoint operator A on $\mathcal{V}$. For $\beta > 0$, we consider the bounded operator $j\colon \mathcal{V} \to \mathcal{H}, j(v) = (v, e^{-\frac{\beta}{2}A}v)$. Since the projection $P(v_1, v_2) = (v_1, 0)$ onto the first component is $E(\mathbb{R}_+^\times)$ for the spectral measure E of U^c (here we use that $E(\{0\}) = 0$), the closed U^c-invariant subspace generated by $j(\mathcal{V})$ is adapted to the decomposition $\mathcal{H} = \mathcal{V} \oplus \mathcal{V}$. Hence its cyclicity follows from the fact that the range of $e^{-\frac{\beta}{2}A}$ is dense in $\mathcal{V}$. As $j(\mathcal{V})$ is cyclic,

$$\psi\colon \mathbb{R} \to B(\mathcal{V}), \quad \psi(t) := j^* U_t^c j = j^* e^{itH} j \quad \text{for} \quad U_t^c = e^{itH},$$

is a positive definite function for which the GNS representation is equivalent to $(U^c, \mathcal{H})$ (Proposition A.4). Now $j^*(w_1, w_2) = w_1 + e^{-\frac{\beta}{2}A} w_2$ yields

$$\psi(t) = e^{itA} + e^{-\frac{\beta}{2}A} e^{-itA} e^{-\frac{\beta}{2}A} = e^{itA} + e^{(-it-\beta)A}.$$

Since $A \geq 0$, $\varphi(z) := \psi(iz) = e^{-zA} + e^{-(\beta-z)A}$ defines a bounded operator for $0 \leq \operatorname{Re} z \leq \beta$. For $0 \leq t \leq \beta$, we thus obtain the positive definite function

$$\varphi(t) = e^{-tA} + e^{-(\beta-t)A} = j^* e^{-tH} j, \quad 0 \leq t \leq \beta, \quad \text{satisfying} \quad \varphi(\beta - t) = \varphi(t). \tag{8}$$

Hence φ defines a reflection positive function on $\mathbb{T}_\beta$ for which $\varphi(-it) = \psi(t)$, so that by (6) the corresponding representation of the dual group $(\mathbb{T}_\beta)^c \cong \mathbb{R}$ is equivalent to U^c. □

Remark 3.5. (a) In [25, Prop. 6.1], we have shown that a unitary one-parameter group $(U_t)_{t\in\mathbb{R}}$ with $U_t = e^{itH}$ has a euclidean realization in terms of the triple $(\mathbb{R}, \mathbb{R}_+, -\mathrm{id}_\mathbb{R})$ if and only if $H \geq 0$, and then $\varphi(t) := e^{-|t|H}$ is a corresponding reflection positive function.

(b) For the function φ from (8), a $\mathrm{Herm}(\mathcal{V})_+$-valued measure with $\mathcal{L}(\mu) = \varphi$ is obtained as in (4) from the spectral measure E_+ of A by $d\mu(\lambda) = dE_+(\lambda) + e^{\beta\lambda} dE_+(-\lambda)$. Here $e^{-tH} = e^{-tA} \oplus e^{tA}$, so that $j(v) = (v, e^{-\beta A/2}v) \in \mathcal{D}(e^{\beta H/2})$.

(c) Let $U_t^c = e^{itH}$ be a unitary one-parameter group on $\mathcal{H}$ and J be a unitary involution on $\mathcal{H}$ with $JHJ = -H$. Then $R := Je^{-\beta H/2}$ is an unbounded involution with $\mathcal{D}(R) = \mathcal{D}(e^{-\beta H/2})$. We further assume that $j\colon \mathcal{V} \to \mathcal{H}$ is a bounded operator with cyclic range contained in $\mathrm{Fix}(R) = \{v \in \mathcal{D}(R)\colon Rv = v\}$. Then

$$\varphi(t) := j^* e^{-tH} j \in B(\mathcal{V}), \quad 0 \leq t \leq \beta \tag{9}$$

defines a function on $[0, \beta]$ for which the kernel $\varphi(\frac{t+s}{2})$ is positive definite, and we further have

$$\begin{aligned}\varphi(t) &= j^* e^{-tH} j = j^* R^* e^{-tH} R j = j^* e^{-\beta H/2} J e^{-tH} J e^{-\beta H/2} j \\ &= j^* e^{-\beta H/2} e^{tH} e^{-\beta H/2} j = j^* e^{-(\beta-t)H} j = \varphi(\beta - t).\end{aligned}$$

Therefore φ defines a reflection positive function on $\mathbb{T}_\beta$. In view of (9), it is the Laplace transform of the measure j^*Ej, where E is the spectral measure of H.

4. Standard subspaces

To connect reflection positivity on $\mathbb{T}_\beta$ with the modular theory of von Neumann algebras, we now take a closer look at *standard real subspaces* of a complex Hilbert space $\mathcal{H}$.

Definition 4.1. A closed real subspace $V \subseteq \mathcal{H}$ is said to be *standard* if $V \cap iV = \{0\}$ and $\overline{V + iV} = \mathcal{H}$. Then we define the corresponding *Tomita operator* on the dense subspace $V_{\mathbb{C}} := V + iV$ of $\mathcal{H}$ by $S_V(x + iy) := x - iy$ for $x, y \in V$ (cf. [6], [20]).

Lemma 4.2. *If $V \subseteq \mathcal{H}$ is a standard real subspace and $S := S_V$, then*

(i) *S is a closed densely defined antilinear involution.*

(ii) *If $\Delta := S^*S$ and $S = J\Delta^{1/2}$ is the polar decomposition of S, then Δ is a positive selfadjoint, J is an anti-unitary involution and $J\Delta J = \Delta^{-1}$.*

(iii) *$S^* = JSJ$ is the complex conjugation with respect to the standard subspace $J(V)$.*

(iv) *The real orthogonal complement of V w.r.t. $\mathrm{Re}\langle\cdot,\cdot\rangle$ is $iJ(V)$. In particular, $\mathcal{H} = V \oplus iJ(V)$ as a real Hilbert space.*

(v) *$J(V) = V^{\perp,\omega}$ for the symplectic form $\omega := \mathrm{Im}\langle\cdot,\cdot\rangle$.*

Proof. (i), (ii): For $x, y \in V$ and $z := x + iy$, we have

$$\|(z, S(z))\|^2 = \|x + iy\|^2 + \|x - iy\|^2 = 2(\|x\|^2 + \|y\|^2), \tag{10}$$

so that the graph of S is closed because $V \oplus V$ is complete, which in turn follows from the closedness of V. We conclude that S is a closed densely defined operator on $\mathcal{H}$, so that S has a polar decomposition as in (ii). Since $\mathrm{im}(S) = V_{\mathbb{C}}$ is dense, J is an isometry. With the same arguments as in [4, Prop. 2.5.11], it now follows that J is an anti-unitary involution satisfying $J\Delta J = \Delta^{-1}$.

(iii) This further leads to $S^* = \Delta^{1/2}J = JSJ$, which shows that $S^* = S_{JV}$ is the complex conjugation corresponding to the standard subspace $J(V)$.

(iv) For $w \in \mathcal{H}$, we have the following chain of equivalences:

$$\begin{aligned}\mathrm{Re}\langle iV, w\rangle = \{0\} &\Leftrightarrow (\forall v \in V_{\mathbb{C}})\, \mathrm{Re}\langle v, w\rangle = \mathrm{Re}\langle Sv, w\rangle = \mathrm{Re}\langle w, Sv\rangle \\ &\Leftrightarrow (\forall v \in V_{\mathbb{C}})\langle v, w\rangle = \langle w, Sv\rangle \;\Leftrightarrow\; w \in \mathcal{D}(S^*)\ \&\ S^*w = w \;\Leftrightarrow\; w \in J(V).\end{aligned}$$

(v) follows immediately from (iv) because $iV^{\perp,\omega}$ is the orthogonal complement w.r.t. $\mathrm{Re}\langle\cdot,\cdot\rangle$. □

Remark 4.3. By Lemma 4.2, the scalar product $\langle\cdot,\cdot\rangle$ is real-valued on $V \times V$ if and only if $V \subseteq J(V)$, which is equivalent to $J(V) = V$. This happens only if $S^* = S$, and this in turn is equivalent to $\Delta = \mathbf{1}$, i.e., $J = S$.

Lemma 4.4. *Let Δ be an injective positive selfadjoint operator on $\mathcal{H}$ and J an anti-unitary involution with $J\Delta J = \Delta^{-1}$. Then*

(i) *$S := J\Delta^{1/2} \colon \mathcal{D}(\Delta^{1/2}) \to \mathcal{H}$ is a closed antilinear involution and $V := \{v \in \mathcal{D}(S) \colon Sv = v\}$ is a standard real subspace of $\mathcal{H}$ with $S_V = S$.*

(ii) *The selfadjoint operator $H := \log \Delta$ satisfies $JHJ = -H$ and $U_{(t,\varepsilon)} := e^{itH}J^\varepsilon = \Delta^{it}J^\varepsilon$ defines an anti-unitary representation of the direct product group $\mathbb{R} \times \mathbb{Z}/2\mathbb{Z}$.*

Proof. That S is an involution follows from $S^2 = J\Delta^{1/2}J\Delta^{1/2} = \Delta^{-1/2}\Delta^{1/2} = \mathrm{id}_{\mathcal{D}(S)}$. Further, the closedness of the selfadjoint operator $\Delta^{1/2}$ implies that S is closed. Now (i) follows from (10), and (ii) is clear. □

Example 4.5. (Modular data of von Neumann algebras) Tomita–Takesaki theory ([4, §2.5]) starts with a cyclic separating vector Ω for a von Neumann algebra $\mathcal{M} \subseteq B(\mathcal{H})$. Then $V := \{M\Omega : M^* = M \in \mathcal{M}\}$ is a standard real subspace of $\mathcal{H}$, and for $S = S_V$, $\Delta := S^*S$ is the *modular operator* and $\alpha_t(M) := \Delta^{-it}M\Delta^{it}$ defines a one-parameter group of automorphisms of $\mathcal{M}$ (the *modular group*). Further, the *modular involution* J satisfies $J\mathcal{M}J = \mathcal{M}'$.

The preceding two lemmas imply that we have a one-to-one correspondence between the following data.

(a) standard closed subspaces $V \subseteq \mathcal{H}$.
(b) pairs (Δ, J), where $0 < \Delta = \Delta^*$ and J is an anti-unitary involution satisfying $J\Delta J = \Delta^{-1}$.
(c) pairs (H, J), where $H = H^*$ and J is an anti-unitary involution satisfying $JHJ = -H$.
(d) strongly continuous anti-unitary representations of $\mathbb{R} \times \mathbb{Z}/2\mathbb{Z} \cong \mathbb{R}^\times$.

Remark 4.6. Let $U_t^c = e^{itH}$ be a unitary one-parameter group on $\mathcal{H}$ and J be an anti-unitary involution on $\mathcal{H}$ with $JHJ = -H$. Then $S := Je^{-\beta H/2}$ is an unbounded antilinear involution with $\mathcal{D}(S) = \mathcal{D}(e^{-\beta H/2})$. We further assume that $\mathcal{V}$ is a real vector space and that $j: \mathcal{V} \to \mathcal{H}$ is a linear map with cyclic range contained in $\mathrm{Fix}(S) = \{v \in \mathcal{D}(S) : Sv = v\}$. As the operators U_t^c commute with S, $\mathrm{Fix}(S)$ is a closed U^c-invariant subspace. Now $\varphi(z)(v,w) := \langle e^{-zH}j(v), j(w)\rangle$ defines for $0 \leq \mathrm{Re}\, z \leq \beta$ an element $\varphi(z) \in \mathrm{Sesq}(\mathcal{V})$. It is easy to see that the kernel $\varphi(\frac{z+\overline{w}}{2})$ is positive definite. For $v, w \in \mathcal{V}$, we further have:

$$\begin{aligned}
\varphi(z)(v,w) &= \langle e^{-zH}j(v), j(w)\rangle = \langle e^{-zH}Sj(v), Sj(w)\rangle = \langle e^{-zH}Je^{-\beta H/2}j(v), Je^{-\beta H/2}j(w)\rangle \\
&= \langle Je^{\overline{z}H}e^{-\beta H/2}j(v), Je^{-\beta H/2}j(w)\rangle = \langle e^{-\beta H/2}j(w), e^{\overline{z}H}e^{-\beta H/2}j(v)\rangle \\
&= \langle j(w), e^{-(\beta-\overline{z})H}j(v)\rangle = \overline{\varphi(\beta - \overline{z})(v,w)}. \qquad (11)
\end{aligned}$$

This means that, for $0 \leq \mathrm{Re}\, z \leq \beta$, the relation $\varphi(\beta - \overline{z}) = \overline{\varphi(z)}$ holds in $\mathrm{Sesq}(\mathcal{V})$. From that we derive in particular that $\mathrm{Re}\,\varphi$ defines a reflection positive function $\mathbb{T}_\beta \to \mathrm{Herm}(\mathcal{V})$. Note that all functions $\varphi^{v,v}(t) := \varphi(t)(v,v)$ are real-valued, hence reflection positive. The function φ itself admits a β-periodic extension to $\mathbb{R}$ if and only if $\varphi(0)$ is real-valued on $V \times V$. Since the scalar product of $\mathcal{H}$ is in general not real-valued on $V \times V$, the form $\varphi(0)$ need not be symmetric

5. KMS states

Let $(\mathcal{A}, \mathbb{R}, \alpha)$ be a C^*-dynamical system, i.e., $\alpha: \mathbb{R} \to \mathrm{Aut}(\mathcal{A})$ is a homomorphism defining a continuous $\mathbb{R}$-action on the C^*-algebra $\mathcal{A}$. We recall from [3, Props. 5.3.3, 5.3.7] that an α-invariant state ω of $\mathcal{A}$ is a *KMS state at value* $\beta > 0$ if, for all $A, B \in \mathcal{A}$, there exists a bounded holomorphic function $F_{A,B} : \mathcal{D}_\beta \to \mathbb{C}$ extending continuously to $\overline{\mathcal{D}_\beta}$, such that

$$F_{A,B}(t) = \omega(A\alpha_t(B)) \quad \text{and} \quad F_{A,B}(t + i\beta) = \omega(\alpha_t(B)A) \quad \text{for} \quad t \in \mathbb{R}. \qquad (12)$$

The following lemma links our concept of a positive definite function with values in sesquilinear forms to invariant states of $\mathcal{A}$ (Definition A.3).

Lemma 5.1. *Let $(\mathcal{A}, G, \alpha)$ be a C^*-dynamical system and ω be a G-invariant state of $\mathcal{A}$. Then*

$$\psi: G \to \mathrm{Sesq}(\mathcal{A}), \quad \psi(g)(A,B) := \omega(B^*\alpha_g(A))$$

is a positive definite function.

Proof. Let $(\pi_\omega, U, \mathcal{H}_\omega, \Omega)$ be the cyclic covariant representation with $U_g\Omega = \Omega$ for $g \in G$ ([4, Cor. 2.3.17]). For the $\mathbb{R}$-equivariant linear map $j: \mathcal{A} \to \mathcal{H}_\omega, j(A) = \pi(A)\Omega$ with dense range we then have

$$\psi(g)(A,B) = \langle \pi(B^*\alpha_g(A))\Omega, \Omega\rangle = \langle U_g\pi(A)\Omega, \pi(B)\Omega\rangle = \langle U_g j(A), j(B)\rangle,$$

Therefore the assertion follows from Proposition A.4. □

Proposition 5.2. *For $G = \mathbb{R}$, an invariant state ω is a KMS state if and only if ψ extends to a function $\overline{\mathcal{D}_\beta} \to \mathrm{Sesq}(\mathcal{A})$ which is pointwise continuous and pointwise holomorphic on $\mathcal{D}_\beta$ on $\mathcal{A} \times \mathcal{A}$ and satisfies*

$$\psi(t + i\beta)(A, B) = \psi(-t)(B^*, A^*) = \overline{\psi(t)(A^*, B^*)} \quad \textit{for} \quad t \in \mathbb{R}. \tag{13}$$

Proof. The fact that ψ is positive definite implies in particular that $\psi(-t) = \psi(t)^*$:

$$\psi(-t)(A, B) = \omega(B^*\alpha_{-t}(A)) = \omega(\alpha_t(B^*)A) = \overline{\omega(A^*\alpha_t(B))} = \psi(t)^*(A, B).$$

Comparing with (12), we see that $\psi(t)(A, B) = F_{B^*,A}(t)$, so that the assertion follows from $\omega(\alpha_t(A)B^*) = \omega(A\alpha_{-t}(B^*)) = \psi(-t)(B^*, A^*)$ and [3, Props. 5.3.3, 5.3.7]. □

Remark 5.3. (a) By analytic continuation, the relation (13) implies that

$$\psi(i\beta - z)(A, B) = \psi(z)(B^*, A^*) = \overline{\psi(-\overline{z})(A^*, B^*)} \quad \text{for} \quad z \in \overline{\mathcal{D}_\beta}.$$

For $\varphi(t) := \psi(it)$, we obtain in particular

$$\varphi(\beta - t)(A, B) = \varphi(t)(B^*, A^*) = \overline{\varphi(t)(A^*, B^*)} \quad \text{for} \quad 0 \leq t \leq \beta.$$

Using the notation from Appendix A, we define $\varphi^{A,B}(t) := \varphi(t)(A, B)$. Then $\varphi^{A,A}(t) := \varphi(t)(A, A)$ is real for $A \in \mathcal{A}$ and that $\varphi^{A,A^*}(\beta - t) = \varphi^{A,A^*}(t)$. For $A = A^*$, it follows in particular that the function $\varphi^{A,A}$ on $[0, \beta]$ defines a reflection positive function on $\mathbb{T}_\beta$.

(b) In the situation of the proof of Proposition 5.2, for $G = \mathbb{R}$, Lemma B.4 implies that $\pi(\mathcal{A})\Omega \subseteq \mathcal{D}(e^{-\beta H/2})$ and that

$$\varphi(t)(A, B) = \langle e^{-tH}\pi(A)\Omega, \pi(B)\Omega\rangle \quad \text{for} \quad 0 \leq t \leq \beta.$$

(c) In the context of Tomita–Takesaki theory (Example 4.5), we put $U_t := \Delta^{-it}$ and $H = -\log \Delta$, so that $\Delta^{1/2} = e^{-H/2}$, which corresponds to $\beta = 1$. For the state $\omega(M) := \langle M\Omega, \Omega\rangle$ and $\psi(t)(A\Omega, B\Omega) = \langle U_t A\Omega, B\Omega\rangle = \varphi(-it)(A\Omega, B\Omega)$, we obtain from (11) for $A = A^*, B = B^*$

$$\psi(t + i)(A\Omega, B\Omega) = \varphi(1 - it)(A\Omega, B\Omega) = \overline{\varphi(-it)}(A\Omega, B\Omega) = \overline{\psi(t)}(A\Omega, B\Omega),$$

which is the KMS condition (13) for $\beta = 1$ ([3, Prop. 5.3.10]).

6. Perspectives

The main outcome of the present note is that it clarifies the connection between reflection positivity for the triple $(\mathbb{T}_\beta, \mathbb{T}_\beta^+, \tau_\beta)$ and the modular data (J, Δ) arising naturally in Tomita–Takesaki theory, resp., the theory of KMS states. We hope that this will serve as a basis for a deeper understanding of reflection positivity for higher dimensional groups.

Relativistic KMS states: One interesting direction is to connect the relativistic KMS states for a d-dimensional space time introduced by J. Bros and D. Buchholz in [5] to reflection positivity for the group $\mathbb{R}^d$ (see also [9] for $d = 2$). Here the strip $\mathcal{D}_\beta \subseteq \mathbb{C}$ is replaced by tube domain

$$\mathcal{T}_\beta := \{z \in \mathbb{C}^d : \operatorname{Im} z \in V_+ \cap (\beta e - V_+)\},$$

where V_+ is the open forward light cone and $e \in V_+$ is a timelike vector of unit length. One expects a duality between $G^c = \mathbb{R}^d$ and $G = \mathbb{T}_\beta \times \mathbb{R}^{d-1}$ and a suitable domain $G^+ \subseteq G$ for reflection positivity.

ax+b-group and generalizations: There are still several aspects of reflection positivity for the $ax+b$-group $G \cong \mathbb{R} \rtimes \mathbb{R}^\times$ (the affine group of the real line) that are not covered by the discussion in [25]. Here the only non-trivial involution is given by $\tau(b,a) = (-b,a)$. In [25] we have seen how reflection positivity for (G, G^+, τ), where $G^+ = \{(b,a) : b > 0, a > 0\}$, leads to euclidean realizations of those representations π^c of $G^c \cong \mathbb{R} \rtimes \mathbb{R}^\times_+$ satisfying a positive spectrum condition for translations. For any anti-unitary involution J satisfying $J\pi(b,a)J = \pi(-b,a)$, these representations extend naturally by $\pi(0,-1) := J$ to anti-unitary representations of the non-connected group $\mathbb{R} \rtimes \mathbb{R}^\times$. Such representations arise naturally in the context of Borchers' triples ([8], [2]), where they actually extend to anti-unitary representations of $G^c := \mathbb{R}^d \rtimes_\gamma \mathbb{R}^\times \cong (\mathbb{R}^2 \times \mathbb{R}^{d-2}) \rtimes \mathrm{SO}_{1,1}(\mathbb{R})$ and $J = \pi(0,0,-1)$ acts on Minkowski space $\mathbb{R}^d \cong \mathbb{R}^2 \oplus \mathbb{R}^{d-2}$ by $J\pi(b,c,\mathbf{1})J = \pi(-b,c,\mathbf{1})$, where $\mathbb{R}^2$ is 2-dimensional Minkowski space and $\mathbb{R}^{d-2}$ is space-like. The dual group is the subgroup $G \cong (\mathbb{R}^2 \times \mathbb{R}^{d-2}) \rtimes \mathrm{SO}_2(\mathbb{R})$ of the d-dimensional motion group. This should provide a natural framework for extending the reflection positivity for the circle discussed here in which one can also treat relativistic KMS states. Clearly, the crucial case to be understood is $d = 2$. On $G \cong \mathbb{R}^d \rtimes \mathrm{SO}_2(\mathbb{R})$, the involution induced by the time reflection θ is given by $\tau(b,a) = (\theta(b), a^{-1})$ and likewise on G^c. Note that this involution commutes with the action of J.

Reflection positivity on the side of G^c: Classically, reflection positivity is a condition imposed in the euclidean model where we have a unitary representation $(\pi, \mathcal{E})$ of the group G, whereas on the dual side we have a representation $(\pi^c, \widehat{\mathcal{E}})$ which has no specific reflection symmetry. However, in the theory of modular inclusions, one encounters anti-unitary representations of the $ax+b$-group corresponding to a modular pair (J, Δ) and a unitary one-parameter group U which satisfy the commutation relations

$$\Delta^{it} U(s) \Delta^{-it} = U(e^{-2\pi t} s) \quad \text{and} \quad JU(s)J = U(-s) \quad \text{for} \quad t, s \in \mathbb{R}.$$

In [6, Thm. 3.2] it is shown that, in this context, the infinitesimal generator H of $U(s) = e^{isH}$ satisfies $H \geq 0$ if and only if $U(s)V \subseteq V$ holds for $s \geq 0$, where V is the standard real subspace specified by (J, Δ) (cf. Lemma 4.4). This situation has a remarkable similarity with the reflection positivity for $(\mathbb{R}, \mathbb{R}_+, -\mathrm{id}_\mathbb{R})$, where we have a unitary one-parameter group $(U_t)_{t \in \mathbb{R}}$ of $\mathcal{E}$ whose positive part acts by isometries on the subspace $\mathcal{E}_+$ and the spectrum of the generator of the dual one-parameter group U^c is positive ([25]). However, here the subspace V is real, $\langle Jv, v \rangle \geq 0$ for $v \in V$ and $(U(s))_{s \geq 0}$ acts by real isometries on V.

More general reflection positive functions: For a symmetric Lie group (G, τ), the notion of a reflection positive function $\varphi \colon G \to \mathbb{C}$ has been introduced with the idea that these should be positive definite functions corresponding to unitary representations U on a reflection positive Hilbert space $(\mathcal{E}, \mathcal{E}_+, \theta)$ such that $\varphi(g) = \langle \pi(g)v, v \rangle$ holds for a θ-fixed vector $v \in \mathcal{E}_+$. Then reflection positivity with respect to a domain $G^+ \subseteq G$ corresponds to $\pi(G^+)^{-1} v \subseteq \mathcal{E}_+$ (cf. [24]) and is encoded in the positive definiteness of the kernel $\varphi(g\tau(h)^{-1})$ on G^+. For $(\mathbb{R}, \mathbb{R}_+, -\mathrm{id}_\mathbb{R})$ and $(\mathbb{T}_\beta, \mathbb{T}_\beta^+, \tau_\beta)$, this implies that φ is real-valued (cf. Theorem 2.4 for $\dim \mathcal{V} = 1$). But this condition is too restrictive to cover the β-periodic functions φ^{A,A^*} from Remarks 4.6 (see also Remark 5.3). This suggests to work with a more general concept where we give up the θ-invariance of v, so that φ need no longer be τ-invariant. Extending $\pi \colon G \to \mathrm{U}(\mathcal{E})$ to a representation $\widehat{\pi}$ of the group $G_\tau := G \rtimes \{\mathbf{1}, \tau\}$ by $\pi(\mathbf{1}, \tau) := \theta$, the condition $\pi(G^+)^{-1} v \subseteq \mathcal{E}_+$ corresponds to the positive definiteness of the kernel $\langle \theta\pi(h^{-1})v, \pi(g^{-1})v \rangle = \langle \widehat{\pi}(g\tau(h)^{-1}, \tau)v, v \rangle$, which can be expressed in terms of the positive definite function $\widehat{\varphi}(g) := \langle \widehat{\pi}(g)v, v \rangle$ on the non-connected group G_τ. It is therefore desirable to obtain an explicit description of the so specified functions for $(\mathbb{R}, \mathbb{R}_+, -\mathrm{id}_\mathbb{R})$ and $(\mathbb{T}_\beta, \mathbb{T}_\beta^+, \tau_\beta)$ and thus a generalization of the corresponding results in [24] for $\mathbb{R}$ and Theorem 2.4 for $\mathbb{T}_\beta$. Of course, it would be of particular interest to see if the functions φ^{A,A^*} from Remarks 4.6 fall into this larger class.

Acknowledgments
K.-H. Neeb was supported by DFG-grant NE 413/7-2, Schwerpunktprogramm "Darstellungstheorie" and Gestur Ólafsson was supported by NSF grant DMS-1101337 and the Emerging Fields Project "Quantum Geometry" of the University of Erlangen.

Appendix A. Form-valued kernels

Classically, reproducing kernels arise from Hilbert spaces $\mathcal{H}$ of functions $f: X \to \mathbb{C}$ (X a set) for which the evaluations $f \mapsto f(x)$ are continuous, hence representable by elements $K_x \in \mathcal{H}$ by $f(x) = \langle f, K_x \rangle$, and then $K(x,y) := K_y(x) = \langle K_y, K_x \rangle$ is called the *reproducing kernel* of the space $\mathcal{H}$. Then K determines $\mathcal{H}$ uniquely. Accordingly, we write $\mathcal{H}_K \subseteq \mathbb{C}^X$ for the Hilbert space determined by K. It is a classical result that a kernel function $K: X \times X \to \mathbb{C}$ is the reproducing kernel of some Hilbert space if and only if it is *positive definite* in the sense that, for any finite collection $x_1, \ldots, x_n \in X$, the matrix $(K(x_j, x_k))_{1 \leq j,k \leq n}$ is positive semidefinite (cf. [1], [23, Ch. 1]). There is a natural generalization to Hilbert spaces $\mathcal{H}$ of functions with values in a Hilbert space $\mathcal{V}$, i.e., $\mathcal{H} \subseteq \mathcal{V}^X$. Then $K_x(f) = f(x)$ is a linear operator $K_x: \mathcal{H} \to \mathcal{V}$ and we obtain a kernel $K(x,y) := K_x K_y^* \in B(\mathcal{V})$ with values in the bounded operators on $\mathcal{V}$. However, there are also situations where one would like to deal with kernels whose values are unbounded operators, so that one has to generalize this context further. As we shall explain below, the concept of a positive definite kernel with values in the space $\mathrm{Sesq}(\mathcal{V})$ of sesquilinear complex-valued forms on $\mathcal{V}$ provides a natural context to deal with all relevant cases. [2]

For a real or complex vector space $\mathcal{V}$, we write $\mathcal{V}^\sharp$ for the complex vector space of antilinear maps $\mathcal{V} \to \mathbb{C}$.

Definition A.1. Let X be a set. We call a map $K: X \times X \to \mathrm{Sesq}(\mathcal{V})$ *a positive definite kernel* if the associated scalar-valued kernel

$$K^\flat: (X \times \mathcal{V}) \times (X \times \mathcal{V}) \to \mathbb{C}, \quad K^\flat((x,v),(y,w)) := K(x,y)(w,v)$$

is positive definite. For elements f of the corresponding reproducing kernel Hilbert space $\mathcal{H}_{K^\flat} \subseteq \mathbb{C}^{X \times \mathcal{V}}$, we then have

$$f(y,w) = \langle f, K^\flat_{y,w} \rangle \quad \text{with} \quad K^\flat_{y,w}(x,v) := K^\flat((x,v),(y,w)) = K(x,y)(w,v).$$

Therefore $v \mapsto K^\flat_{x,v}$ is linear, and this implies that $f(x,\cdot)$ is antilinear. We identify $\mathcal{H}_{K^\flat}$ with a subspace of $(\mathcal{V}^\sharp)^X$ by identifying $f \in \mathcal{H}_{K^\flat}$ with the function $f^\sharp: X \to \mathcal{V}^\sharp, f^\sharp(x) := f(x,\cdot)$. We call

$$\mathcal{H}_K := \{f^\sharp : f \in \mathcal{H}_{K^\flat}\} \subseteq (\mathcal{V}^\sharp)^X$$

the *(vector-valued) reproducing kernel space associated to* K. The elements

$$K_{x,v} := (K^\flat_{x,v})^\sharp = K(\cdot, x)(v, \cdot), \quad x \in X, v \in \mathcal{V},$$

then form a dense subspace of $\mathcal{H}_K$ with

$$\langle K_{y,w}, K_{x,v} \rangle = K^\flat((x,v),(y,w)) = K(x,y)(w,v). \tag{A.1}$$

[2] If $\mathcal{V}$ is a real space, then $\mathrm{Sesq}(\mathcal{V})$ is the space of bilinear maps $\mathcal{V} \times \mathcal{V} \to \mathbb{C}$, and if $\mathcal{V}$ is complex, it stands for those maps $\mathcal{V} \times \mathcal{V} \to \mathbb{C}$ which are linear in the first and antilinear in the second argument. The uniqueness of sesquilinear extension leads to $\mathrm{Sesq}(\mathcal{V}) = \mathrm{Sesq}(\mathcal{V}_\mathbb{C})$ for a real space $\mathcal{V}$. If $\mathcal{V}$ is complex, then polarization shows that every $\varphi \in \mathrm{Sesq}(\mathcal{V})$ is uniquely determined by its values on the diagonal.

Example A.2. (a) If $K\colon X\times X\to B(\mathcal{V})$ is an operator-valued kernel, where $\mathcal{V}$ is a complex Hilbert space, then we obtain a $\mathrm{Sesq}(\mathcal{V})$-valued kernel by $Q(x,y)(v,w) := \langle K(x,y)v,w\rangle$. We have a natural inclusion $B(\mathcal{V})\hookrightarrow \mathrm{Sesq}(\mathcal{V}), A\mapsto \langle A\cdot,\cdot\rangle$, whose range is the space of continuous sesquilinear forms. All the functions $f^\sharp\colon X\to\mathcal{V}^\sharp$ in $\mathcal{H}_Q$ take values in continuous functionals, hence can be identified with $\mathcal{V}$-valued functions. This leads to the realization $\mathcal{H}_K\hookrightarrow\mathcal{V}^X$ (cf. [23, Ch. 1]).

(b) For a one-point set $X=\{*\}$, a positive definite $\mathrm{Sesq}(\mathcal{V})$-valued kernel is simply a positive semidefinite $K\in\mathrm{Sesq}(\mathcal{V})$. The corresponding Hilbert space $\mathcal{H}_K\subseteq\mathcal{V}^\sharp$ is generated by the elements $K_v=K(v,\cdot)$ with $\langle K_v,K_w\rangle = K(v,w)$.

In particular, if $\mathcal{V}$ is a Hilbert space, then the natural inclusion $\mathcal{V}\hookrightarrow\mathcal{V}^\sharp, v\mapsto\langle v,\cdot\rangle$, corresponds to the kernel $K(v,w)=\langle v,w\rangle$.

(c) If $\mathcal{V}=\mathbb{C}$, then $\mathrm{Sesq}(\mathcal{V})\cong\mathbb{C}$ and $\mathrm{Sesq}(\mathcal{V})$-valued kernels are complex-valued kernels.

(d) If $\mathcal{A}$ is a C^*-algebra and $\omega\in\mathcal{A}^*$ a positive functional, then $K_\omega(A,B) := \omega(B^*A)$ is a positive semidefinite sesquilinear kernel for which the corresponding Hilbert space $\mathcal{H}_{K_\omega}\subseteq\mathcal{A}^\sharp$ can be obtained from the GNS representation $(\pi_\omega,\mathcal{H}_\omega,\Omega)$ ([4, Cor. 2.3.17]) by the embedding

$$\Gamma\colon\mathcal{H}_\omega\to\mathcal{A}^\sharp,\quad \Gamma(\xi)(A) := \langle\xi,\pi(A)\Omega\rangle\quad\text{because}\quad\langle\pi(A)\Omega,\pi(B)\Omega\rangle=\omega(B^*A)=K_\omega(A,B).$$

Note that $\mathcal{A}$ has a natural representation on $\mathcal{A}^\sharp$ by $(A.\beta)(B) := \beta(A^*B)$ and that Γ is equivariant with respect to this representation.[3]

Definition A.3. Let G be a group. A function $\varphi\colon G\to\mathrm{Sesq}(\mathcal{V})$ is said to be *positive definite* if the $\mathrm{Sesq}(\mathcal{V})$-valued kernel $K(g,h) := \varphi(gh^{-1})$ is positive definite.

Proposition A.4. (GNS-construction) (a) *Let $\varphi\colon G\to\mathrm{Sesq}(\mathcal{V})$ be a positive definite function. Then $(\pi_\varphi(g)f)(h) := f(hg)$ defines a unitary representation of G on the reproducing kernel Hilbert space $\mathcal{H}_\varphi\subseteq(\mathcal{V}^\sharp)^G$ with kernel $\varphi(gh^{-1})$ and the range of the map $j\colon\mathcal{V}\to\mathcal{H}_\varphi, j(v)(g) := \varphi(g)(v,\cdot)$ is a cyclic subspace, i.e., $\pi(G)j(\mathcal{V})$ spans a dense subspace of $\mathcal{H}$. We then have*

$$\varphi(g)(v,w)=\langle\pi_\varphi(g)j(v),j(w)\rangle\quad\text{for}\quad g\in G, v,w,\in\mathcal{V}.$$

(b) *If, conversely, $(\pi,\mathcal{H})$ is a unitary representation of G and $j\colon\mathcal{V}\to\mathcal{H}$ a linear map whose range is cyclic, then*

$$\varphi\colon G\to\mathrm{Sesq}(\mathcal{V}),\quad\varphi(g)(v,w) := \langle\pi(g)j(v),j(w)\rangle$$

is a $\mathrm{Sesq}(\mathcal{V})$-valued positive definite function and $(\pi,\mathcal{H})$ is unitarily equivalent to $(\pi_\varphi,\mathcal{H}_\varphi)$.

Proof. (cf. [23, Sect. 3.1]) (a) For the kernel $K(g,h) := \varphi(gh^{-1})$ and $v\in\mathcal{V}$, the right invariance of the kernel K implies that $\pi(g)K_{h,v}=K_{hg^{-1},v}$, and from that one easily derive the invariance of $\mathcal{H}_\varphi$ under right translations and the unitarity of their restrictions to $\mathcal{H}_\varphi$. Then $j(v)=\varphi(\cdot)(v,\cdot)=K_{\mathbf{1},v}$ and $\pi(g)j(v)=K_{g^{-1},v}$ imply that $j(\mathcal{V})\subseteq\mathcal{H}_\varphi$ is cyclic. Finally we note that

$$\langle\pi(g)j(v),j(w)\rangle=\langle K_{g^{-1},v},K_{\mathbf{1},w}\rangle=K(g)(v,w)=\varphi(g)(v,w).$$

(b) The positive definiteness of φ follows easily from the relation $\varphi(gh^{-1})(v,w) = \langle\pi(h)^{-1}j(v),\pi(g)^{-1}j(w)\rangle$. Since $j(\mathcal{V})$ is cyclic, the map

$$\Gamma(\xi)(g)(v) := \langle\xi,\pi(g)^{-1}j(v)\rangle$$

defines an injection $\mathcal{H}\hookrightarrow(\mathcal{V}^\sharp)^G$ whose range is the subspace $\mathcal{H}_\varphi$ and which is equivariant with respect to the right translation action π_φ. □

[3] This realization of the Hilbert space $\mathcal{H}_\omega$ has the advantage that we can see its elements as elements of the space $\mathcal{A}^\sharp$ (see [23] for many applications of this perspective). Usually, $\mathcal{H}_\omega$ is obtained as the Hilbert completion of a quotient of $\mathcal{A}$ by a left ideal which leads to a much less concrete space.

Journal of Physics: Conference Series **597** (2015) 012004 doi:10.1088/1742-6596/597/1/012004

Remark A.5. If $\mathcal{V}$ is a Hilbert space and j is continuous, then we have the adjoint operator $j^*\colon \mathcal{H} \to \mathcal{V}$ is well-defined and we obtain the $B(\mathcal{V})$-valued positive definite function $\varphi(g) := j^*\pi(g)j$ which can be used to realize $\mathcal{H}$ in $\mathcal{V}^G$ (cf. Example A.2(a)).

Appendix B. Integral representations

For a more concrete realization of unitary representations associated to positive definite functions in L^2-spaces, integral representations are of crucial importance. The following result is a straightforward generalization of Bochner's Theorem for locally compact abelian groups. Here we write $\mathrm{Sesq}(\mathcal{V})_+ \subseteq \mathrm{Sesq}(\mathcal{V})$ for the convex cone of positive semidefinite forms.

Proposition B.1. *Let G be a locally compact abelian group. If $\varphi\colon G \to \mathrm{Sesq}(\mathcal{V})$ is a positive definite function for which all functions $\varphi^{v,w} := \varphi(\cdot)(v,w), v,w \in \mathcal{V}$, are continuous, then there exists a uniquely determined $\mathrm{Sesq}(\mathcal{V})_+$-valued Borel measure μ on the locally compact group $\widehat{G}$ such that $\widehat{\mu}(g) := \int_{\widehat{G}} \chi(g)\, d\mu(\chi) = \varphi(g)$ holds for every $g \in G$ pointwise on $\mathcal{V} \times \mathcal{V}$.*

Proof. First, Bochner's Theorem for scalar-valued positive definite functions yields for every $v \in \mathcal{V}$ a finite positive measure μ^v on $\widehat{G}$ such that $\varphi^{v,v} = \widehat{\mu^v}$. By polarization, we obtain complex measures $\mu^{v,w}$, $v, w \in \mathcal{V}$, on $\widehat{G}$ with $\varphi^{v,w} = \widehat{\mu^{v,w}}$. Then the collection $(\mu^{v,w})_{v,w\in\mathcal{V}}$ of complex measures on $\widehat{G}$ defines a $\mathrm{Sesq}(\mathcal{V})_+$-valued measure by $\mu(\cdot)(v,w) := \mu^{v,w}$ for $v, w \in \mathcal{V}$, and this measure satisfies $\widehat{\mu} = \varphi$. □

Remark B.2. Suppose that E is the spectral measure on the character group $\widehat{G}$ for which the continuous unitary representation $(\pi, \mathcal{H})$ is represented by $\pi(g) = \int_{\widehat{G}} \chi(g)\, dE(\chi)$. Then, for $v \in \mathcal{H}$, the positive definite function $\pi^v(g) := \langle \pi(g)v, v\rangle$ is the Fourier transform of the measure $E^{v,v} = \langle E(\cdot)v, v\rangle$. This establishes a close link between spectral measures and the representing measures in the preceding proposition.

We say that a subset $\mathcal{D}$ in the real vector space E (contained in $E_{\mathbb{C}} = E + iE$) is *finitely open* if, for each finite-dimensional subspace $F \subseteq E$, the intersection $F \cap \mathcal{D}$ is open in F. If $\mathcal{V}$ is a Hilbert space, then a function $f\colon T_{\mathcal{D}} := \mathcal{D} + iE \to E$ on the corresponding tube domain is called *Gateaux holomorphic* if the restriction to each tube domain $T_{F\cap\mathcal{D}}$ is holomorphic. We write $\mathcal{O}_G(T_{\mathcal{D}}, \mathcal{V})$ for the space of all Gateaux holomorphic $\mathcal{V}$-valued functions on $T_{\mathcal{D}}$. For the theory of holomorphic functions on domains in infinite dimensional spaces we refer to Hervé's monograph [12] or [23] for the connections to representation theory.

Theorem B.3. (Laplace transforms and positive definite kernels) *Let E be a real vector space and $\mathcal{D} \subseteq E$ a non-empty convex finitely open subset. Let $\mathcal{V}$ be a Hilbert space and $\varphi\colon \mathcal{D} \to B(\mathcal{V})$ such that*

(L1) *the kernel $Q_\varphi(x,y) = \varphi\big(\frac{x+y}{2}\big)$ is positive definite.*

(L2) *φ is weak operator continuous on very line segment in $\mathcal{D}$, i.e., all functions $t \mapsto \langle \varphi(x+th)v, v\rangle$, $v \in \mathcal{V}$, are continuous on $\{t \in \mathbb{R}\colon x + th \in \mathcal{D}\}$.*

Then the following assertions hold:

(i) *There exists a unique $\mathrm{Herm}^+(\mathcal{V})$-valued measure μ on the smallest σ-algebra on E^* for which all point evaluations are continuous such that*

$$\varphi(x) = \mathcal{L}(\mu)(x) := \int_{E^*} e^{-\lambda(x)}\, d\mu(\lambda) \quad \text{for} \quad x \in \mathcal{D}.$$

(ii) *The map*

$$\mathcal{F}\colon L^2(E^*, \mu; \mathcal{V}) \to \mathcal{O}_G(T_{\mathcal{D}}, \mathcal{V}), \quad \langle \mathcal{F}(f)(z), v\rangle := \langle f, e_{-\frac{\overline{z}}{2}} v\rangle$$

is unitary onto the reproducing kernel space $\mathcal{H}_\varphi := \mathcal{H}_{Q_\varphi}$ corresponding to the kernel associated to φ. It intertwines the unitary representation

$$(\pi(u)f)(\alpha) := e^{i\alpha(u)}f(\alpha) \quad \text{on} \quad L^2(E^*,\mu) \quad \text{and} \quad (\widetilde{\pi}(u)f)(z) := f(z-2iu) \quad \text{on} \quad \mathcal{H}_\varphi.$$

(iii) *φ extends to a unique Gateaux holomorphic function $\widehat{\varphi}$ on the tube domain $T_\mathcal{D}$ which is positive definite in the sense that the kernel $\widehat{\varphi}(\frac{z+\overline{w}}{2})$ is positive definite.*

Proof. (i) follows from [11, Thm. 18.8] and (ii) from [26, Thm. I.7], provided we show that (L2) implies that, for every trace class operator $S \in B_1(\mathcal{V})$, the function $x \mapsto \operatorname{tr}(\varphi(x)S)$ is continuous on line segments in $\mathcal{D}$. To this end, we may assume that $E = \mathbb{R}$ and $\mathcal{D} =]a,b[$. Since $B_1(\mathcal{H})$ is spanned by positive operators with trace 1, we may w.l.o.g. assume that $S = S^* \geq 0$ with $\operatorname{tr} S = 1$. We may further assume that $\dim \mathcal{V} = \infty$; otherwise the assertion is trivial. Then there exists an orthogonal sequence $(v_n)_{n\in\mathbb{N}}$ in $\mathcal{V}$ with $Sv = \sum_n \langle v, v_n\rangle v_n$. This leads to

$$\operatorname{tr}(\varphi(x)S) = \sum_n \varphi_n(x) \quad \text{with} \quad \varphi_n(x) := \langle \varphi(x)v_n, v_n\rangle.$$

By assumption, all functions φ_n are continuous. Applying [26, Thm. I.7] to the case $\mathcal{V} = \mathbb{C}$, we obtain positive measures ν_n on $\mathbb{R}$ with with $\varphi_n = \mathcal{L}(\nu_n)$ on $]a,b[$. Let $\nu := \sum_n \nu_n$. Then

$$\mathcal{L}(\nu)(x) = \sum_n \mathcal{L}(\nu_n)(x) = \sum_n \varphi_n(x) = \operatorname{tr}(\varphi(x)S),$$

and the continuity (actually the analyticity) of $\mathcal{L}(\nu)$ on $]a,b[$ follows from [23, Cor. V.4.4]. □

The preceding theorem generalizes in an obvious way to $\mathrm{Sesq}(\mathcal{V})$-valued functions, where the corresponding measure μ has avalues in the cone $\mathrm{Sesq}(\mathcal{V})_+$. One can use the same arguments as in the proof of Bochner's Theorem.

The following lemma sharpens the "technical lemma" in [19, App. A].

Lemma B.4. *Let $U_t = e^{itH}$ be a unitary one-parameter group on $\mathcal{H}$, E the corresponding spectral measure, $v \in \mathcal{H}$, $E^v := \langle E(\cdot)v, v\rangle$, $\beta > 0$ and $\varphi(t) := \langle U_t v, v\rangle = \int_\mathbb{R} e^{it\lambda}\, dE^v(\lambda)$. Then the following are equivalent:*

(i) *There exists a continuous function ψ on $\overline{\mathcal{D}_\beta}$, holomorphic on $\mathcal{D}_\beta$, such that $\psi|_\mathbb{R} = \varphi$.*

(ii) *$\mathcal{L}(E^v)(\beta) = \int_\mathbb{R} e^{-\beta\lambda}\, dE^v(\lambda) < \infty$.*

(iii) *$v \in \mathcal{D}(e^{-\frac{\beta}{2}H})$.*

Proof. That (i) implies (ii) follows from [28, p. 311]. If, conversely, (ii) is satisfied, then $\psi(z) := \mathcal{L}(E^v)(-iz)$ is defined on $\overline{\mathcal{D}_\beta}$, holomorphic on $\mathcal{D}_\beta$ and $\psi|_\mathbb{R} = \varphi$. Finally, the equivalence of (ii) and (iii) follows immediately from the definition of $e^{-\frac{\beta}{2}H}$ in terms of the spectral measure E. □

Lemma B.5. (Criterion for the existence of $\mathcal{L}(\mu)(x)$) *Let $\mathcal{V}$ be a Hilbert space and μ be a finite* $\mathrm{Herm}(\mathcal{V})_+$*-valued Borel measure on $\mathbb{R}$, so that we can consider its Laplace transform $\mathcal{L}(\mu)$, taking values in* $\mathrm{Herm}(\mathcal{V})$, *whenever the integral*

$$\operatorname{tr}(\mathcal{L}(\mu)(x)S) = \int_\mathbb{R} e^{-\lambda x}\, d\mu^S(\lambda) \quad \textit{for} \quad d\mu^S(\lambda) = \operatorname{tr}(d\mu(\lambda)S),$$

exists for every positive trace class operator S on $\mathcal{V}$. This is equivalent to the finiteness of the integrals $\mathcal{L}(\mu^v)(x)$ for every $v \in \mathcal{V}$, where $d\mu^v(\lambda) = \langle d\mu(\lambda)v, v\rangle$.

Proof. For $x \in E$, the existence of $\mathcal{L}(\mu)(x)$ implies the finiteness of the integrals $\mathcal{L}(\mu^v)(x)$ for $v \in \mathcal{V}$. Suppose, conversely, that all these integrals are finite. Then we obtain by polarization a hermitian form

$$\beta(v, w) := \int_{\mathbb{R}} e^{-\lambda x} \langle d\mu(\lambda)v, w\rangle$$

on $\mathcal{V}$. We claim that β is continuous. As $\mathcal{V}$ is in particular a Fréchet space, it suffices to show that, for every $w \in \mathcal{V}$, the linear functional $\lambda(v) := \beta(v, w)$ is continuous ([29, Thm. 2.17]).

The linear functionals $\lambda_n(v) := \int_{-n}^{n} e^{-\lambda x} \langle d\mu(\lambda)v, w\rangle$ are continuous because μ is a bounded measure and the functions e_x are bounded on bounded intervals. By the Monotone Convergence Theorem, combined with the Polarization Identity, $\lambda_n \to \lambda$ holds pointwise on $\mathcal{V}$, and this implies the continuity of λ ([29, Thm. 2.8]).

For a positive trace class operators $S = \sum_n \langle \cdot, v_n\rangle v_n$ with $\operatorname{tr} S = \sum_n \|v_n\|^2 < \infty$, we now obtain

$$\mathcal{L}(\mu^S)(x) = \sum_n \mathcal{L}(\mu^{v_n})(x) = \sum_n \beta(v_n, v_n) \leq \|\beta\| \sum_n \|v_n\|^2 < \infty. \qquad \square$$

References

[1] Aronszajn N 1950 Theory of reproducing kernels *Trans. Amer. Math. Soc.* **68** 337–404
[2] Borchers H-J 1992 The CPT-Theorem in two-dimensional theories of local observables *Comm. Math. Phys.* **143** 315–332
[3] Bratteli O and D W Robinson 1996 Operator Algebras and Quantum Statistical Mechanics II 2nd ed. *Texts and Monographs in Physics* (Berlin: Springer-Verlag)
[4] Bratteli O and D W Robinson 2002 Operator Algebras and Quantum Statistical Mechanics I 2nd ed. *Texts and Monographs in Physics* (Berlin: Springer-Verlag)
[5] Bros J and Buchholz D 1994 Towards a relativistic KMS-condition *Nuclear Phys.* B **429:2** 291–318
[6] Brunetti R, Guido D and Longo R 2002 Modular localization and Wigner particles *Rev. Math. Phys.* **14** 759–785
[7] Buchholz D and Summers S J 2004 An algebraic characterization of vacuum states in Minkowski space. III. Reflection maps *Comm. Math. Phys.* **246:3** 625–641
[8] Buchholz D. Lechner G, and Summers S J 2011 Warped convolutions, Rieffel deformations and the construction of quantum field theories *Comm. Math. Phys.* **304:1** 95–123
[9] Gérard C and Jäkel C 2006 On the relativistic KMS condition for the $P(\varphi)_2$ model (arXiv:math-ph/0609088)
[10] Glimm J and Jaffe 1981 *A Quantum Physics – A Functional Integral Point of View* (New York: Springer-Verlag)
[11] Glöckner H 2003 Positive Definite Functions on Infinite Dimensional Convex Cones *Mem. Amer. Math. Soc.* **789** (Rhode Island: Amer. Math. Soc.)
[12] Hervé M 1989 *Analyticity in infinite dimensional spaces* (Berlin: de Gruyter)
[13] Jaffe A and Ritter G 2008 Reflection positivity and monotonicity *J. Math. Phys.* **49:5** 052301 10
[14] Jaffe A and Ritter G 2007 Quantum field theory on curved backgrounds. I. The euclidean functional integral *Comm. Math. Phys.* **270** 545–572
[15] Jaffe A and Ritter G 2007 Quantum field theory on curved backgrounds. II. Spacetime symmetries (arXiv:hep-th/0704.0052v1)
[16] Jorgensen P E T and Ólafsson G 1998 Unitary representations of Lie groups with reflection symmetry *J. Funct. Anal.* **158** 26–88
[17] Jorgensen P E T and Ólafsson G 2000 Unitary representations and Osterwalder-Schrader duality, in: *The Mathematical Legacy of Harish-Chandra, Doran R S and Varadarajan V S, eds., Proc. Symp. in Pure Math.* **68** (Rhode Island: Amer. Math. Soc.)
[18] Klein A and Landau L 1981 Periodic Gaussian Osterwalder–Schrader positive processes and the two-sided Markov property on the circle *Pac. J. Math.* **94:2** 341–367
[19] Klein A and Landau L 1983 From the Euclidean group to the Poincaré group via Osterwalder-Schrader positivity *Comm. Math. Phys.* **87** 469–484
[20] Longo R 2008 Real Hilbert subspaces, modular theory, SL(2, R) and CFT, in: *Von Neumann Algebras in Sibiu , 33-91, Theta Ser. Adv. Math.* **10** (Bucharest: Theta)
[21] Merigon S, Neeb K H and Olafsson G 2014 Integrability of unitary representations on reproducing kernel spaces (*Preprint* arXiv:math.RT.1406.2681)
[22] Neeb K H 1998 Operator valued positive definite kernels on tubes *Monatshefte für Math.* **126** 125–160

[23] Neeb K H 2000 Holomorphy and Convexity in Lie Theory *Expositions in Mathematics* **28** (Berlin: de Gruyter Verlag)

[24] Neeb K H and Ólafsson G 2014 Reflection positivity and conformal symmetry *J. Funct. Anal.* **266** 2174–2224

[25] Neeb K H and Ólafsson G 2013 Reflection positive one-parameter groups and dilations to appear in: *Complex Analysis and Operator Theory* (arXiv:math.RT.1312.6161)

[26] Neeb K H and Ørsted B 2002 Representations in L^2-spaces on infinite-dimensional symmetric cones *J. Funct. Anal.* **190** 133–178

[27] Osterwalder K and Schrader R 1973 Axioms for Euclidean Green's functions. 1 *Comm. Math. Phys.* **31** 83–112

[28] Richter H 1966 Wahrscheinlichkeitstheorie, 2nd edition *Grundlehren der math. Wiss.* **86** (Berlin: Springer Verlag)

[29] Rudin W 1973 *Functional Analysis* (New York: McGraw Hill)

Comments on supersymmetric quantum field theories

Yuji Tachikawa
Department of Physics, Faculty of Science,
University of Tokyo, Bunkyo-ku, Tokyo 133-0022, Japan, and
Institute for the Physics and Mathematics of the Universe,
University of Tokyo, Kashiwa, Chiba 277-8583, Japan

E-mail: yuji.tachikawa@ipmu.jp

Abstract. New mathematical conjectures have often been extracted from the analysis of supersymmetric quantum field theories. Here we would like to illustrate how this extraction process is performed.

1. Mathematical conjectures from QFTs: the classic example

The prime example where a big mathematical conjecture was obtained from the analysis of supersymmetric quantum field theories (QFTs) was the discovery of the mirror symmetry in the early 1990s. Let us quickly recall what this symmetry is.

Given a Calabi-Yau manifold X, there is an associated 2d supersymmetric QFT $Q(X)$:

$$X \mapsto Q(X) \tag{1}$$

In the space of 2d supersymmetric QFTs, there is an involutive automorphism σ

$$Q \mapsto \sigma(Q) \mapsto \sigma^2(Q) = Q. \tag{2}$$

Note that there is not yet a precise mathematical definition of what a 2d supersymmetric QFT is, or what this automorphism σ is. However, the quantum field theorists knew these concepts well enough, to find the following. Namely, given a Calabi-Yau X, it often happens that there is another Calabi-Yau Y such that

$$\sigma(Q(X)) = Q(Y). \tag{3}$$

Properties of X and Y related in a way very mysterious to those who do not know the background QFT material. For example, the complex structure of X is related to the symplectic structure of Y.

By now there is a big mathematical industry studying the relation between X and Y. However, mathematicians still do not talk directly about $Q(X)$, as if it is the one-who-must-not-be-named.

2. Mathematical conjectures from QFTs: a recent one

There are many other minor such events where QFT led to new mathematical relations. For example, consider the following expression

$$I(u,v;x,y) = \oint \frac{\prod_{\pm\pm\pm} \Gamma_{p,q}(tu^{\pm1}v^{\pm1}z^{\pm1}) \prod_{\pm\pm\pm} \Gamma_{p,q}(tx^{\pm1}y^{\pm1}z^{\pm1})}{\prod_{\pm} \Gamma_{p,q}(t^2z^{\pm2}) \prod_{\pm} \Gamma_{p,q}(z^{\pm2})} \frac{dz}{2\pi iz} \tag{4}$$

where

$$\Gamma_{p,q}(z) = \prod_{j,k\geq 0} \frac{1 - z^{-1}p^{j+1}q^{k+1}}{1 - zp^jq^k} \tag{5}$$

is the elliptic gamma function.

This integral $I(u,v;x,y)$ is obviously symmetric under the exchanges $u \leftrightarrow v$, $x \leftrightarrow y$. However, it is also symmetric under the exchange $u \leftrightarrow x$. On the physics side, this symmetry was conjectured in [1] in 2009. On the math side, this symmetry was proved completely independently in [2] in 2009.

3. Previous formalisms of QFTs

What is this thing called QFT, that touches many parts of mathematics? Those who practice QFT, including the author, feel they know what that is. However, those feelings are not enough to convey it to the diverse audience of this conference series. We clearly need some formalism, and we want some solid starting point. At this point, a reader might wonder that in fact there have been many such formalisms. Examples include Wightman axioms, algebraic quantum field theories, topological quantum field theories, and vertex operator algebras. They indeed capture some of the aspects of QFTs, but none of them is comprehensive enough to even state what the mirror symmetry is, or what the Seiberg-Witten theory is.

Such a formalism has not been written down anywhere yet, but the author thinks it should not be impossible to do so. After all, the axioms of a group, that look so straightforward today, took many years to be straightened out.

There is one other thing to be pointed out. In the study of QFT, in whatever formalization mentioned so far, or in particle physics community in general, people tend to study each individual QFT, Q_1, Q_2, ... one by one. For example, the whole experimental high energy particle physics can be said to be the quest to find exactly which QFT Q_{SM} describes elementary particles, and thus, the universe.

But this is like studying groups one by one. There is nothing wrong with that, particularly when there is a few particularly interesting groups / QFTs. We should, however, also study group homomorphism $G \to H$, representations of groups, action of groups on spaces, the quotient space G/H, etc. Similarly, we should study the interrelation among QFTs, the relation of QFTs with other mathematical objects, etc.

4. A formalism of QFTs

From now on, a QFT is always assumed to be four-dimensional and $\mathcal{N} = 2$ supersymmetric, unless otherwise specified. A few basic formal axioms of QFTs are the following:

- Given a QFT Q and a 4d manifold X, we can form the partition function $Z_Q(X) \in \mathbb{C}$.
- Given two QFTs Q_1 and Q_2, we can form a new QFT $Q_1 \times Q_2$.
- This product is commutative, associative, has a unit $\bullet$, i.e. $Q \times \bullet = Q$.
- The product is compatible with taking the partition function $Z_{Q_1\times Q_2}(X) = Z_{Q_1}(X)Z_{Q_2}(Y)$.

We also need a concept of G-symmetric QFTs:

- Given a G-symmetric QFT Q and a 4d manifold X with G connection, we can form the partition function $Z_Q(X) \in \mathbb{C}$.
- Given a G-symmetric QFT Q and a homomorphism $\varphi : H \to G$, we can regard Q as an H-symmetric QFT.
- Given a G_1-symmetric QFT Q_1 and a G_2-symmetric QFT G_2, their product $Q_1 \times Q_2$ is $G_1 \times G_2$-symmetric, and $Z_{Q_1 \times Q_2}(X) = Z_{Q_1}(X) Z_{Q_2}(X)$.
- Finally, the unit $\bullet$ is G-symmetric for any G.

From this extended viewpoint we can think of an unqualified QFT as an $\{id\}$-symmetric QFT. Note that these formal properties are very much like spaces with G action:

- Given a space X with G action and a homomorphism $\varphi : H \to G$, we can regard that X has H action.
- Given a space X_1 with G_1 action and a space X_2 with G_2 action, their product $X \times Y$ has $G_1 \times G_2$ action.
- A point $\bullet$ has trivial G action for any G.

Now, recall that given a space X with $G \times H$ action, X/G is a space with H action. Similarly, given a QFT Q that is $G \times H$-symmetric, we can form a new QFT $Q /\!\!/\!\!/ G$, which is H-symmetric. Note that the G symmetry is gone. Usually this operation is called coupling to the gauge group G. With the formal operations so far, we already have interesting QFTs:

$$\bullet /\!\!/\!\!/ G. \tag{6}$$

They are called pure $\mathcal{N} = 2$ gauge theories with gauge group G in the standard physics literature. For a four-manifold M, a slightly modified version of

$$Z_{\bullet /\!\!/\!\!/ \mathrm{SU}(2)}(M) \tag{7}$$

is the Donaldson invariant, as shown in [3].

Another basic construction is the following. Given a symplectic representation R of G, there is a G-symmetric QFT we denote by Hyp(R):

$$R \mapsto \mathrm{Hyp}(R). \tag{8}$$

This is usually called the free hypermultiplet in R, and has the following formal properties:

$$\mathrm{Hyp}(R \oplus R') = \mathrm{Hyp}(R) \times \mathrm{Hyp}(R'), \qquad \mathrm{Hyp}(0) = \bullet \tag{9}$$

where 0 stands for a trivial, zero-dimensional representation.

Now we know there are QFTs of the form $\mathrm{Hyp}(R) /\!\!/\!\!/ G$, which are usually called $\mathcal{N} = 2$ supersymmetric gauge theories. For example, take $G = \mathrm{U}(1)$, and $R = V \oplus V^*$, where $V \simeq \mathbb{C}$ is a standard 1-dimensional representation of U(1). Let

$$Q' = \mathrm{Hyp}(R) /\!\!/\!\!/ \mathrm{U}(1). \tag{10}$$

A version of its partition function, $Z_{Q'}(M)$, is the Seiberg-Witten invariant. These invariants are concrete objects but rather deep, and will not be discussed further in this note.

5. Mathematical objects associated to QFTs

Let us discuss two easier objects to discuss, associated to a given G-symmetric QFT Q. These are the Higgs branch $\mathcal{M}_{\text{Higgs}}(Q)$ and the superconformal index $\text{SCI}(Q)$.

The Higgs branch $\mathcal{M}_{\text{Higgs}}(Q)$ is a Hyperkähler space with G action, and the superconformal index $\text{SCI}(Q)$ is a class function on G that is a formal power series in p, q, t. They preserve formal properties under the multiplication of the QFTs:

$$\mathcal{M}_{\text{Higgs}}(Q_1 \times Q_2) = \mathcal{M}_{\text{Higgs}}(Q_1) \times \mathcal{M}_{\text{Higgs}}(Q_2), \qquad \text{SCI}(Q_1 \times Q_2) = \text{SCI}(Q_1) \times \text{SCI}(Q_2). \tag{11}$$

For the gauging $Q /\!\!/\!\!/ G$, we have

$$\mathcal{M}_{\text{Higgs}}(Q /\!\!/\!\!/ G) = \mathcal{M}_{\text{Higgs}}(Q) /\!\!/\!\!/ G \tag{12}$$

where $/\!\!/\!\!/ G$ is the hyperkähler quotient construction and

$$\text{SCI}(Q /\!\!/\!\!/ G) = \left(\frac{1}{\Gamma_{p,q}(t)\Gamma'_{p,q}(1)}\right)^r \frac{1}{|W_G|} \oint \frac{\text{SCI}(Q)(z)}{\prod_\alpha \Gamma_{p,q}(z^\alpha)\Gamma_{p,q}(t^2 z^\alpha)} \prod_{i=1}^r \frac{dz_i}{2\pi\sqrt{-1}z_i} \tag{13}$$

where α runs over the roots of G. Here we restricted $z \in G$ to lie in the Cartan torus $z \in \text{U}(1)^r \subset G$.

For the free hypermultiplets $Q = \text{Hyp}(R)$, we have

$$\mathcal{M}_{\text{Higgs}}(\text{Hyp}(R)) = R, \qquad \text{SCI}(\text{Hyp}(R))(z) = \prod_w \Gamma_{p,q}(tz^w) \tag{14}$$

where z is in the Cartan of G and w runs over the weights of R.

Now, given a symplectic representation R of $G \times H$, we can form $Q = \text{Hyp}(R) /\!\!/\!\!/ G$ that is H-symmetric. For this QFT, we have

$$\mathcal{M}_{\text{Higgs}}(\text{Hyp}(R) /\!\!/\!\!/ G) = R /\!\!/\!\!/ G, \tag{15}$$

which is a hyperkähler space with H action. Similarly, we have

$$\text{SCI}(\text{Hyp}(R) /\!\!/\!\!/ G)(y) = \oint \frac{\prod_{w \oplus v} \Gamma_{p,q,t}(z^w y^v)}{\prod_\alpha \Gamma_{p,q}(z^\alpha)\Gamma_{p,q}(t^2 z^\alpha)} \prod_{i=1}^r \frac{dz_i}{2\pi z_i}, \tag{16}$$

which is the so-called elliptic beta integral. They are both well-studied in mathematics.

Note that we start from a group G and its representation R, that are both well established mathematical objects. We then pass to $\text{Hyp}(R) /\!\!/\!\!/ G$, which is not well formulated yet. Then we pass back to $\mathcal{M}_{\text{Higgs}}(\text{Hyp}(R) /\!\!/\!\!/ G)$ or $\text{SCI}(\text{Hyp}(R) /\!\!/\!\!/ G)$, both of which are again well-established mathematical-physical objects. How do we get something new?

6. Gaiotto's construction

The key is that there are a more transcendental construction coming from six dimensions. Let G be a simply-laced group. Then it is believed that there is a 6-dimensional supersymmetric QFT $\mathcal{S}_G$ with very good properties, called the 6d $\mathcal{N}$=(2,0) theory. Given a k-punctured Riemann surface C and a 4d manifold X, define

$$Z_G[C](X) = \mathcal{S}_G(X \times C). \tag{17}$$

This gives a 4d QFT $\mathcal{S}_G[C]$ depending on C.

With k-points, $\mathcal{S}_G[C]$ is G^k symmetric. For example,

$\mathcal{S}_G[$  $]$, $\mathcal{S}_G[$ 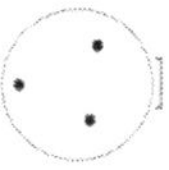 $]$ (18)

are G^2, G^3 symmetric, respectively. Not only that, $\mathcal{S}_G[C]$ depends only on the topology of C: you can exchange points without changing the theory. More mathematically, there is an S_k action on k marked points on C. Therefore, when C has k marked points, $\mathcal{S}_G[C]$ is $S_k \ltimes G^k$ symmetric. For example, $\mathcal{S}_G[$  $]$ is $S_3 \ltimes G^3$ symmetric.

Now, we can connect two punctures of two Riemann surfaces:

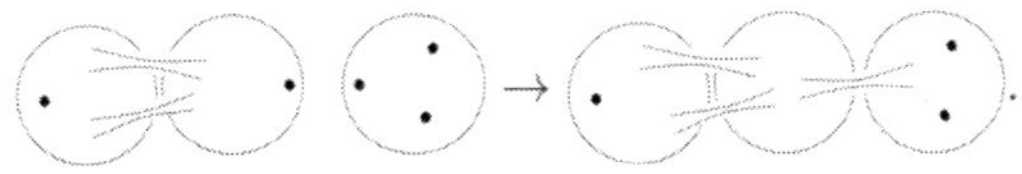 (19)

Correspondingly, we have the crucial property

$\mathcal{S}_G[$  $] = (\mathcal{S}_G[$  $] \times \mathcal{S}_G[$ 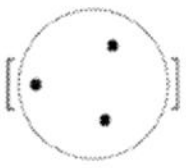 $])/\!\!/\!\!/G$ (20)

where the gauging G is performed with respect to $G_{\text{diag}} \to G \times G$ associated to the two punctures connected. This was first formulated by Gaiotto in [4] and opened up a new direction in the study of supersymmetric quantum field theories.

We can say that the assignment

$$C \mapsto \mathcal{S}_G[C] \tag{21}$$

maps the operations among Riemann surfaces to the operations among QFTs. For those who know the axioms of 2d topological QFT, this can be phrased as follows: a usual 2d topological QFT takes values in the monoidal category of vector spaces, whereas this topological QFT $\mathcal{S}_G$ takes values in (a higher version of) the monoidal category of 4d supersymmetric QFTs.

7. Gaiotto's construction and mathematical conjectures

Now, consider the equation

$\mathcal{S}_G[$  $] = (\mathcal{S}_G[$ 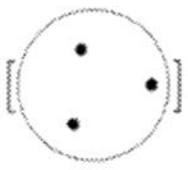 $] \times \mathcal{S}_G[$ 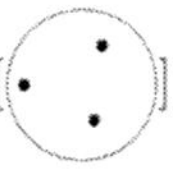 $])/\!\!/\!\!/G.$ (22)

Note that on the left hand side, G^4 are manifestly interchangeable, by the formal property. There is an action of $S_4 \ltimes G^4$. On the right hand side, however, we started from two objects with $S_3 \ltimes G^3$ symmetry. But by connecting two G, we only have the symmetry

$$S_2 \ltimes ((S_2 \ltimes G^2) \times (S_2 \ltimes G^2)). \tag{23}$$

This means that something nontrivial is going on. This non-triviality, however, happens within the category of QFTs.

We can get something nontrivial happening in something well defined, by applying $\mathcal{M}_{\text{Higgs}}$ or SCI. For this, let us take $G = \mathrm{SU}(2)$. Then it is known that

$\mathrm{Hyp}(V_1 \otimes V_2 \otimes V_3) = \mathcal{S}_{\mathrm{SU}(2)}[$ 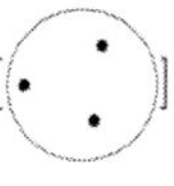 $]$ (24)

where $V_i \simeq \mathbb{C}^2$ is the defining representation of SU(2). The left hand side correctly has $\mathrm{SU}(2)^3$ action, with S_3 permuting them. We now have

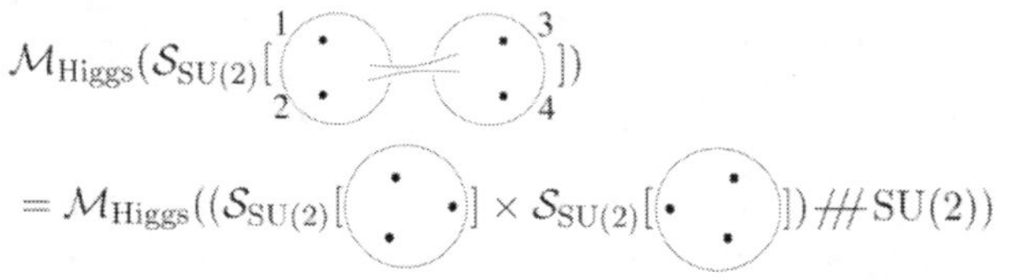

$$= (V_1 \otimes V_2 \otimes V \oplus V \otimes V_3 \otimes V_4)///\mathrm{SU}(2) \quad (25)$$

From the left hand side, it should have S_4 permuting four SU(2)s. The right hand side does not obviously have this symmetry. In fact, the right hand side is the ADHM construction of the minimal nilpotent orbit of $\mathrm{SO}(8)_{\mathbb{C}}$. There is an outer automorphism S_3 acting on SO(8), that provides S_4 permuting SU(2)s.

We also have

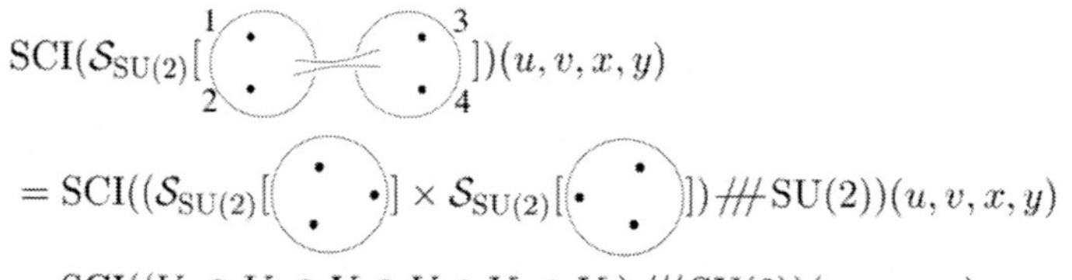

$$\begin{aligned} &= \mathrm{SCI}((V_1 \otimes V_2 \otimes V \oplus V \otimes V_3 \otimes V_4)/\!\!/\!\!/\mathrm{SU}(2))(u,v,x,y) \\ &= \frac{1}{2\Gamma_{p,q}(t)\Gamma'_{p,q}(1)} I(u,v;x,y) \end{aligned} \quad (26)$$

where $I(u,v;x,y)$ is the hyperelliptic beta integral defined in (4) From the left hand side, we see that the right hand side should have the S_4 symmetry permuting u,v,x,y. This was how it was conjectured on the physics side in [1]. On the math side, this symmetry was proved completely independently in [2], as already mentioned.

There are more operations, extracting concrete, well-defined mathematical physical objects from QFTs. For example, given a G-symmetric QFT Q, there is a vertex operator algebra $W(Q)$ that has $\hat{g}$ affine Lie algebra as a vertex subalgebra. Further, $W(Q/\!\!/\!\!/G)$ is given by something like the Drinfeld-Sokolov reduction of $W(Q)$ with respect to $\hat{g}$, and $W(\mathrm{Hyp}(R))$ is the standard symplectic boson VOA. This operation was introduced in [5]. Then we have

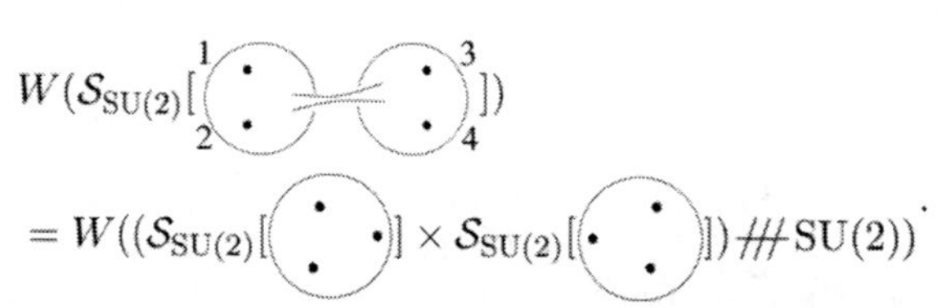

$$= W((V_1 \otimes V_2 \otimes V \oplus V \otimes V_3 \otimes V_4)/\!\!/\!\!/\mathrm{SU}(2)) \quad (27)$$

This is a VOA that can explicitly be written down, and has four $\widehat{\mathfrak{su}(2)}$ affine Lie subalgebra. But the existence of S_4 action permuting four $\widehat{\mathfrak{su}(2)}$ has not been proved. In fact the final VOA is conjectured to be just $\widehat{\mathfrak{so}(8)}_{-2}$, and any mathematician reading this note should consult [5] and prove it.

Yet another object one can associate to a G-symmetric Q is the Nekrasov partition function $Z_{\mathrm{Nekrasov}}(Q)$, which is an element in the equivariant cohomology of the moduli space of

Journal of Physics: Conference Series **597** (2015) 012005 doi:10.1088/1742-6596/597/1/012005

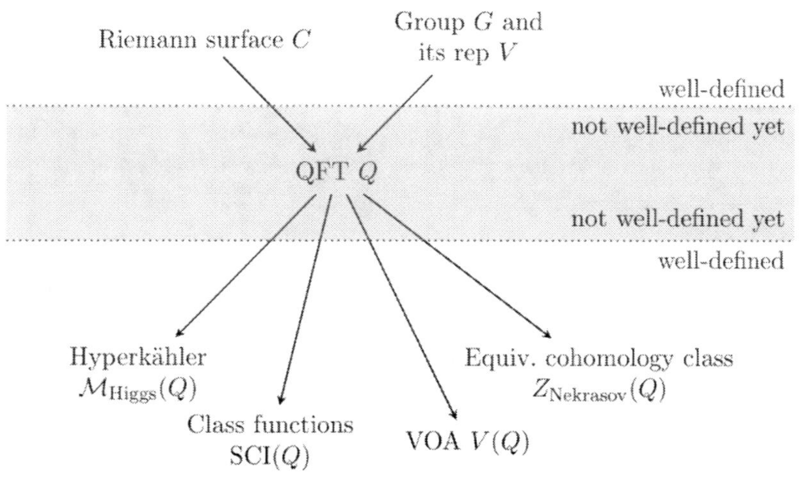

Figure 1. Relations of various objects discussed in the note.

G instantons. $Z_{\mathrm{Nekrasov}}(\mathrm{Hyp}(R))$ is determined by the index bundle of the Dirac operator associated to the representation R of G in the instanton background, and $Z_{\mathrm{Nekrasov}}(Q/\!/\!/G)$ is also computable, given $Z_{\mathrm{Nekrasov}}(Q)$.

There are more formal properties satisfied by $\mathcal{S}_G[C]$ that depends on the complex structure of C, which was not explained in this note. The ability to permute four points on

$Z_{\mathrm{Nekrasov}}(\mathcal{S}_G[$ 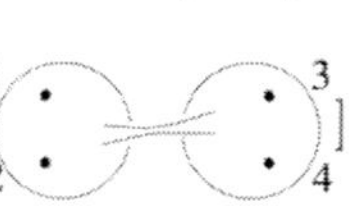 $]),$ (28)

when carefully analyzed, turns out to imply that there should be a $W(G)$-algebra action on the equivariant cohomology of the moduli space of G instantons.

This was how L. Fernando Alday, Davide Gaiotto and the author conjectured the relation between 4d gauge theory and 2d conformal field theory in [6], although the argument presented here was streamlined with lots of hindsight today. That conjecture, which has recently proved by mathematicians, was the main reason the author was awarded a prize in this conference.

8. Conclusions

As a summary, in Fig. 1, the relations of various objects discussed in this note are summarized in a picture. The steps are simple to understand: we pass from a well-defined realm to a not-well-defined realm, where some properties are 'known'. We then pass back to a well-defined realm again, producing precisely-formulated mathematical conjectures.

The author believes that many properties of QFT that are used to derive new mathematical conjectures are formalizable. Of course, there will remain some 'deus ex machina' in the process. For example, the existence of the operation

$$Q \mapsto Q/\!/\!/G \tag{29}$$

almost contains the solution to the mass-gap problem. This is because of the following. Seiberg and Witten have already shown in [7] that, given some basic properties of $\bullet /\!/\!/ G$, we can easily deduce that the pure non-supersymmetric G gauge theory has a mass gap, which is one of the Clay Millenium Problems.

Similarly, the existence of the 6d $\mathcal{N} = (2,0)$ theory $\mathcal{S}_G$ is another 'deus ex machina'. But, assuming that, the map $C \mapsto \mathcal{S}_G[C]$ can be constructed rather formally, and many of the properties follow straightforwardly, in a way understandable to mathematicians and mathematical physicists. So, we will be able to rigorously show that the symmetry $u \leftrightarrow x$ of the hyperelliptic beta integral $I(u, v; x, y)$ in (4) follows from the existence of $\mathcal{S}_{\mathrm{SU}(2)}$.

The author hopes that in this way, mathematicians will be able to see which part of the analysis done by quantum field theorists is truly deep and which part is more mundane.

Acknowledgments

The author would like to thank Luis Fernando Alday and Davide Gaiotto, for the fruitful collaboration that led to [6]. After all, the author was the last one to join the collaboration: they needed a Mathematica code to compute Nekrasov's partition function, which the author happened to have written as a project for my master's thesis. Both Fernando and Davide moved on to other projects, and the author is the only one out of three who is still working on it. The work of YT is supported in part by JSPS Grant-in-Aid for Scientific Research No. 25870159, and in part by WPI Initiative, MEXT, Japan at IPMU, the University of Tokyo.

References

[1] Gadde A, Pomoni E, Rastelli L and Razamat S S 2010 *JHEP* **03** 032 (*Preprint* 0910.2225)
[2] van de Bult F J 2011 *Ramanujan J.* **25** 1–20 (*Preprint* 0909.4793)
[3] Witten E 1988 *Commun. Math. Phys.* **117** 353
[4] Gaiotto D 2012 *JHEP* **1208** 034 (*Preprint* 0904.2715)
[5] Beem C, Lemos M, Liendo P, Peelaers W, Rastelli L and van Rees B C 2013 (*Preprint* 1312.5344)
[6] Alday L F, Gaiotto D and Tachikawa Y 2010 *Lett. Math. Phys.* **91** 167–197 (*Preprint* 0906.3219)
[7] Seiberg N and Witten E 1994 *Nucl. Phys.* **B426** 19–52 (*Preprint* hep-th/9407087)

Symmetric coordinates in solids: magnetic Bloch oscillations

Joshua Zak

Department of Physics, Technion, Israel Institute of Technology, Haifa, 32000, Israel

E-mail: zak@physics.technion.ac.il

Abstract. There has recently been a revival of the Bloch theory of energy bands in solids. This revival was caused, on one hand, by the discovery of topological insulators and the discovery of graphene, and, on the other end, by a very efficient new technique that was developed for creating artificial solids. These are the cold atoms in optical lattices. Last year geometric phases were measured in energy bands of cold atoms in a one-dimensional optical lattice by using Bloch oscillations. These phases are related to the Wyckoff positions, or the symmetry centers in the Bravais lattice. In this lecture a theoretical frame is developed for magnetic Bloch oscillations, meaning oscillations in the presence of a magnetic field. The theory is based on the kq-representation and the symmetric coordinates in solids. It is shown that for a Bloch electron in a magnetic field the orbit quasi-center is a conserved quantity. This is similar to the conservation of the quasi-momentum for an electron in a periodic potential. When an electric field is turned on, the orbit quasi-center oscillates in a similar way to the Bloch oscillations in the absence of a magnetic field. But there is a difference because the magnetic Brillouin zone is different. It depends on the strength of the magnetic field. An analogy is drawn between Bloch oscillations and magnetic Bloch oscillations. By using the magnetic translations it is indicated that a magnetic Wannier-Stark ladder appears in the spectrum of a Bloch electron in crossed magnetic and electric fields. The geometric phases for magnetic Bloch oscillations should be magnetic field dependent.

1. Introduction

The derivation of magnetic Bloch oscillation will be carried out in this talk by using the basic operators in quantum mechanics (x and p being the coordinate and momentum operators)

$$\exp\left(\frac{i}{\hbar}pa\right) \text{ and } \exp\left(ix\frac{2\pi}{a}\right) \tag{1}$$

where a is an arbitrary constant. It can be checked that the operators commute

$$\left[e^{\frac{i}{\hbar}pa}, e^{ix\frac{2\pi}{a}}\right] = 0 \tag{2}$$

These exponentiated operators have a long history. They appear already in the book by Hermann Weyl "The Theory of Groups and Quantum Mechanics" (1928) in connection with the Weyl-Heisenberg group [1]. In the same year, 1928, they appear in Felix Bloch Ph.D. thesis (under Heisenberg) for building the Bloch Theory of Solids [2]. In 1932 von Neumann uses them in his book "Mathematical Foundations of Quantum Mechanics" [3]. In 1960 Julian Schwinger uses them in the construction of finite Hilbert spaces [4]. In 1964 they are used in constructing the magnetic translation group [5, 6]. Yakov Zel'dovich uses them in the definition of quasi-energy in 1967 [7]. Finally, in 1967 they are used in the kq-representation [8, 9, 10]. It should be pointed out that there is no classical analog for the commutation relation (2) of these operators: there are no two classical functions, one of x, $f(x)$, and one of p, $g(p)$, that give a vanishing Poisson bracket.

Paul Dirac in "Recollections of an Exciting Era" [11] draws attention to the correspondence between a quantum commutator and a classical Poisson bracket. They differ by a constant. This correspondence does not hold for the exponentiated operators in Eq. (1). According to Noether's theorem [12] in classical physics, continuous symmetries lead to conservation laws. In quantum mechanics Wigner has shown that any transformation, continuous or discrete, that leaves the Hamiltonian unchanged can be expressed by a unitary or anti-unitary operator, which is a conserved quantity [13]. The difference of conserved quantities in classical and quantum mechanics can best be seen by comparing the motion of a free particle with the motion in a periodic potential. For a free particle the Hamiltonian H does not depend on the coordinate x and any continuous translation will not change H. The linear momentum p is conserved both in classical and quantum mechanics. For a particle in a periodic potential the Hamiltonian remains invariant under discrete translations by the period a of the periodic potential. As a consequence of this discrete symmetry there is no conserved quantity in classical mechanics. But in quantum mechanics this invariance of the Hamiltonian under discrete translations leads to the important conservation law of the quasi-momentum (or the Bloch momentum) k.

As was pointed out above the exponentiated operators in Eq. (1) define the kq-representation [8, 9, 10, 14]. The eigenvalues of these operators, k and q, are the symmetric coordinates in solids (in reference [10] these coordinates are called "natural coordinates"). One should compare the k and q in solids with the spherical coordinates r, θ, and φ in atomic physics. In the latter case the potential depends only on the absolute value r of the radius vector $\vec{r}$, while θ and φ appear in the conserved angular momentum $\vec{L}$. By choosing spherical coordinates one simplifies the solution of the problem significantly. In a crystalline solid the potential is a function of the quasi-coordinate q only, $V(q)$, while the quasi-momentum k is a conserved quantity. The operators x and p are connected to k and q in the following way [14]

$$x = i\frac{\partial}{\partial k} + q \quad , \quad p = -i\hbar\frac{\partial}{\partial q} \tag{3}$$

The wave function $C(k,q)$ in the kq-representation is given by the wave function $\psi(x)$ in the x-representation as follows [14]

$$C(k,q) = \left(\frac{a}{2\pi}\right)^{1/2} \sum_n e^{ikan} \psi(q - na) \tag{4}$$

Here a is like in Eq. (1). The effectiveness of the use of the kq-representation is best demonstrated in the derivation of the acceleration theorem for a Bloch electron in an electric field E. The Hamiltonian for this problem in the kq-representation is [14]

$$H = -\frac{\hbar^2}{2m}\frac{\partial^2}{\partial q^2} + V(q) - eE\left(i\frac{\partial}{\partial k} + q\right) = H_0 - eE\left(i\frac{\partial}{\partial k} + q\right) \tag{5}$$

where H_0 is the Bloch Hamiltonian for $E = 0$. From this Hamiltonian the acceleration theorem for the quasi-momentum follows immediately (one needs only to use the simple quantum mechanics rule that the time rate of change of k is given by the commutator of the Hamiltonian with k)

$$\hbar\frac{\mathrm{d}k}{\mathrm{d}t} = i[H,k] = eE \tag{6}$$

Here H is the Hamiltonian in Eq. (5). Eq. (6) is exact [15]! It should be pointed that the existing derivations of the acceleration theorem (6) are not entirely elementary and its validity is restricted to weak electric fields and for k in a small part of the Brillouin zone [16]. For cold atoms in optical lattices the driving force is different from the one in Eq. (6), e.g. gravity [17] or a magnetic field gradient acting on the spin of the atom [18]. A description of the properties of ultracold bosonic and fermionic quantum gases can be found in the review article [19]. The celebrated acceleration theorem in Eq. (6) is of central importance in the Bloch theory of energy bands in solids, in general, and in Bloch oscillations, in particular [20]. In this lecture a similar equation to Eq. (6) is developed for the vector $\vec{k}^{(B)}$ which labels the eigenvalues of the magnetic translations for a Bloch electron in a magnetic field $\vec{B}$. It is shown how $\vec{k}^{(B)}$ is related to the orbit quasi-center $\vec{q}_0$. Using the connection between $\vec{k}^{(B)}$ and $\vec{q}_0$, an elementary equation is developed for the orbit quasi-center $\vec{q}_0$ of a Bloch electron in crossed magnetic and electric fields. It is shown that Bloch-like oscillations may exist also in the presence of a magnetic field $\vec{B}$.

2. Motion in Magnetic and Electric Fields

For a particle of charge e in a periodic potential $V(x, y)$, a magnetic field in the z-direction, and an electric field in the xy-plane, the Hamiltonian is

$$H = \frac{1}{2m}\left(\vec{p} - \frac{e}{c}\vec{A}\right)^2 + V(x, y) - e\vec{E}\cdot\vec{r} \equiv H_0^{(B)} - e\vec{E}\cdot\vec{r} \tag{7}$$

In what follows, we shall use the symmetric gauge for the vector potential $\vec{A} = \frac{1}{2}\vec{B}\times\vec{r}$. For a free particle in a magnetic field $\vec{B}$ the conserved quantity is known to be the orbit center $\vec{r}_0$ [21]:

$$x_0 = x + \tfrac{c}{eB}\left(p_y - \tfrac{eB}{2c}x\right) \quad , \quad y_0 = y - \tfrac{c}{eB}\left(p_x + \tfrac{eB}{2c}y\right) \tag{8}$$

There is a simple picture of this conservation law for a charged particle moving in the plane perpendicular to the magnetic field $\vec{B}$: this particle moves in a circle with a fixed orbit center $\vec{r}_0$. This holds both in classical and quantum mechanics. The difference is that in quantum mechanics $\vec{r}_0$ is an operator with components x_0 and y_0 that do not commute [21]. When in addition to the magnetic field there is also a periodic potential $V(x, y)$ (the Hamiltonian of the problem is $H_0^{(B)}$ in Eq. (7)), the symmetry of the problem is then the magnetic translation group [5, 6]. This group has a subgroup of commuting magnetic translations [22]

$$T\left(\vec{b}_1\right) = \exp\left[\tfrac{i}{\hbar}\left(\vec{p} + \tfrac{e}{2c}\vec{B}\times\vec{r}\right)\cdot\vec{b}_1\right] \quad , \quad T\left(\vec{b}_2\right) = \exp\left[\tfrac{i}{\hbar}\left(\vec{p} + \tfrac{e}{2c}\vec{B}\times\vec{r}\right)\cdot\vec{b}_2\right] \tag{9}$$

where $\vec{b}_1$ and $\vec{b}_2$ are multiples of the periods $\vec{a}_1$ and $\vec{a}_2$ of the periodic potential $V(x, y)$

$$\vec{b}_1 = N_1\vec{a}_1 \quad , \quad \vec{b}_2 = N_2\vec{a}_2 \quad , \quad \vec{b} = m_1\vec{b}_1 + m_2\vec{b}_2 \tag{10}$$

Here m_1, m_2 are integers running from $-\infty$ to ∞, and where N_1, N_2 are positive integers that appear in the rationality condition on the magnetic field B [22]

$$eBb_1b_2 = hc \tag{11}$$

Having in mind the commutation relation of the operators x_0 and y_0 in Eq. (8)

$$[x_0, y_0] = -i\frac{\hbar c}{eB} \tag{12}$$

it is easy to check that the magnetic translations $T\left(\vec{b}_1\right)$ and $T\left(\vec{b}_2\right)$ in Eq. (9) commute; where we have also used the rationality condition in Eq. (11).

Journal of Physics: Conference Series **597** (2015) 012006 doi:10.1088/1742-6596/597/1/012006

3. Orbit Quasi-center

We are now in a position to define the orbit quasi-center $\vec{q}_0$. Since the magnetic translations in Eq. (9) commute, they have simultaneous eigenfunctions and eigenvalues. We shall denote their eigenvalues in the following way (we use Eq. (8)):

$$T\left(\vec{b}_1\right) = \exp\left(-\frac{ieB}{\hbar c} y_0 b_1\right) \sim \exp\left(ik_x^{(B)} b_1\right) \tag{13}$$

$$T\left(\vec{b}_2\right) = \exp\left(\frac{ieB}{\hbar c} x_0 b_2\right) \sim \exp\left(ik_y^{(B)} b_2\right) \tag{14}$$

In Eqs. (13) and (14) x_0 and y_0 are the operators of the orbit center $\vec{r}_0$, and the magnetic $\vec{k}^{(B)}$-vector labels the eigenvalues of the magnetic translations $T\left(\vec{b}_1\right)$ and $T\left(\vec{b}_2\right)$. The $\vec{k}^{(B)}$-vector has no direct physical meaning. But we can label the eigenvalues of the exponentiated operators in the Eqs. (13), (14) also by the orbit quasi-center $\vec{q}_0$

$$\exp\left(-\frac{ieB}{\hbar c} y_0 b_1\right) \sim \exp\left(-\frac{ieB}{\hbar c} q_{0y} b_1\right) \tag{15}$$

$$\exp\left(\frac{ieB}{\hbar c} x_0 b_2\right) \sim \exp\left(\frac{ieB}{\hbar c} q_{0x} b_2\right) \tag{16}$$

Here q_{0x} and q_{0y} are the x and y-components of the orbit quasi-center $\vec{q}_0$. Eqs. (15), (16) give the definition of the orbit quasi-center $\vec{q}_0$. The reason we call it quasi-center is because it is defined mod $\vec{b}$ in Eq. (10). This is in line with the definition of the quasi-momentum k in the Bloch theory, k is defined modulo the vector of the reciprocal lattice. Since the magnetic translations $T\left(\vec{b}_1\right)$ and $T\left(\vec{b}_2\right)$ commute with the Hamiltonian $H_0^{(B)}$ in Eq. (7), $\vec{q}_0$ is a conserved quantity for a Bloch electron in a magnetic field $\vec{B}$. This is in full analogy with the Bloch theory: for a free particle, the conserved quantity is the momentum p, and when the crystal potential is turned on, the conserved quantity is the quasi-momentum k.

The concept of the orbit quasi-center $\vec{q}_0$ is consistent with the classical Pippard network [23]: a charged particle in a magnetic field moves on a fixed circle with a definite center $\vec{r}_0$; in the presence of a periodic potential, the conserved quantity is $\vec{q}_0$, which is defined mod $\vec{b}$ in Eq. (10) and the particle moves then on a set of circular orbits with centers on the lattice $\vec{b}$, which is the Pippard network [23].

The next step is to consider how $\vec{q}_0$ varies in time when the electric field is turned on. We have

$$\frac{\mathrm{d}}{\mathrm{d}t} T\left(\vec{b}\right) = \frac{i}{\hbar}\left[H, T\left(\vec{b}\right)\right] = -\frac{i}{\hbar}\left[e\vec{E}\cdot\vec{r}_0, T\left(\vec{b}\right)\right] \tag{17}$$

because $T\left(\vec{b}\right)$ commutes with $H_0^{(B)}$ in Eq. (7) and with the components of the operator $\vec{p} + \frac{e}{2c}\vec{B}\times\vec{r}$ that appear in the brackets in Eq. (8) (also, in the Hamiltonian H in Eq. (7)). Having in mind the commutation relation for the operators x_0 and y_0 in Eq. (12), we find by using Eq. (17) that

$$\dot{\vec{q}}_0 = \frac{c\vec{E}\times\vec{B}}{B^2} \tag{18}$$

where $\vec{q}_0$ is the orbit quasi-center in the magnetic field $\vec{B}$, namely, $\vec{q}_0 = \vec{r}_0 \mod \vec{b}$ in Eq. (10). One can check that the same equation as in Eq. (18) holds also for the orbit center $\vec{r}_0$, when $V(x,y) = 0$ in Eq. (7)

$$\dot{\vec{r}}_0 = \frac{c\vec{E}\times\vec{B}}{B^2} \tag{19}$$

The latter equation is well known in classical electrodynamics, where it is called the equation for the drift velocity [24]. Since on the right hand side of Eq. (19) there is a constant vector, the orbit center $\vec{r}_0$ will grow with time indefinitely. The situation is different for the orbit quasi-center $\vec{q}_0$. The latter being defined mod $\vec{b}$, it will oscillate with time. We call this the magnetic Bloch oscillations of the orbit quasi-center $\vec{q}_0$.

4. Magnetic Bloch Oscillations

There is an analogy between the Eq. (6) for the quasi-momentum k and Eq. (18) for the orbit quasi-center $\vec{q}_0$. In both cases, we have oscillations. This analogy goes much further. For a charged particle in a magnetic field (for the Hamiltonian $H_0^{(B)}$ in Eq. (7) with $V(x, y) = 0$) the energy spectrum consists of sharp Landau levels [21]. In what follows, we shall consider the simplest case of the rationality condition as given by Eq. (11). Under this condition, when a periodic potential is introduced, each Landau level broadens into a magnetic energy band [22] (the more general case of the rationality condition is considered in the calculation of the quantum Hall effect [25]). The energy levels inside the magnetic Bloch band are labeled by the orbit quasi-center $\vec{q}_0$. When the magnetic field $\vec{B} = 0$, the energy levels inside a Bloch band are labeled by the quasi momentum k. As was mentioned in the introduction the acceleration theorem in Eq. (6) forms the basis for the Bloch oscillations for a Bloch electron in an electric field. A similar theorem can be derived for the magnetic $\vec{k}^{(B)}$-vector which labels the eigenvalues of the magnetic translations in Eqs. (13) and (14) when a magnetic field $\vec{B}$ is present. The magnetic translations commute with $H_0^{(B)}$ in Eq. (7). For calculating the time derivative of the magnetic $\vec{k}^{(B)}$-vector we have to find the commutator of the magnetic translation with the electric field energy $-e\vec{E} \cdot \vec{r}$ in Eq. (7). The result is exactly like in Eq. (6) for the quasi-momentum k

$$\hbar \frac{\mathrm{d}\vec{k}^{(B)}}{\mathrm{d}t} = e\vec{E} \tag{20}$$

By comparing Eqs. (13), (14) with Eqs. (15), (16) respectively we find the following relation between the orbit quasi-center $\vec{q}_0$ and the magnetic $\vec{k}^{(B)}$

$$\hbar\vec{k}^{(B)} = \frac{e}{c}\vec{B} \times \vec{q}_0 \tag{21}$$

From Eqs. (20) and (21) we find again Eq. (18) for the orbit quasi-center, which is the main result of the talk [26].

Let us also show that the Hamiltonian in Eq. (7) has a spectrum containing a ladder structure like in the case of the Wannier-Stark ladder [14]. This follows from the following observation. Let ψ_ϵ be an eigenstate of the Hamiltonian in Eq. (7) corresponding to the energy ϵ. We consider the case when the electric field points in the x-direction. The electric field energy in the Hamiltonian (7) is then

$$-eE_x x \tag{22}$$

Consider the magnetic translation (see Eq. (9))

$$T(m\vec{a}_1) = \exp\left(-\frac{ieB}{\hbar c} y_0 a_1 m\right) \tag{23}$$

Its commutation with H in Eq. (7) for the electric field energy in Eq. (22) is

$$T^+(ma_1)\, H T(m\vec{a}_1) = H + ma_1 eE_x \tag{24}$$

Table 1. Conserved quantities and the influence if an electric force on them. $V(x,y)$: periodic potential; $\vec{B}$: magnetic field; $\vec{p}$: momentum; $\vec{r}_0$: orbit center; $\vec{k}$: quasi-momentum; $\vec{q}_0$: orbit quasi-center; $\vec{E}$: electric field. Remark: as explained in the text for neutral atoms $\vec{E}$ and $\vec{B}$ can be created artificially.

	Free space	$V(x,y)$	$\vec{B}$	$\vec{B}$ and $V(x,y)$
Conserved quantities	$\vec{p}$	$\vec{k}$	$\vec{r}_0$	$\vec{q}_0$
Influence of an electric force on them	$\dot{\vec{p}} = e\vec{E}$	$\hbar\dot{\vec{k}} = e\vec{E}$	$\dot{\vec{r}}_0 = \frac{c\vec{E}\times\vec{B}}{B^2}$	$\dot{\vec{q}}_0 = \frac{c\vec{E}\times\vec{B}}{B^2}$

The result in Eq. (24) was obtained in a way similar to the results in Eqs. (17) and (18). This shows that if ψ_ϵ is an eigenfunction of H in Eq. (7) with the eigenvalue ϵ, then the function $T(ma_1)\psi_\epsilon$ is an eigenfunction of H with the eigenvalue $\epsilon + ma_1eE_x$

$$HT(m\vec{a}_1)\psi_\epsilon = (\epsilon + ma_1eE_x)T(ma_1)\psi_\epsilon \tag{25}$$

for $m = 0, \pm1, \pm2, \ldots$. This shows that the magnetic Wannier-Stark ladder is contained in the spectrum of H. The argument of Eq. (25) was originally used for indicating that a Stark ladder might exist when there is no magnetic field [27].

At the beginning, we pointed out how one mimics an electric force on neutral atoms in optical lattices [17, 18]. There are also ways to mimic artificial vector gauge potentials for neutral particles in an optical lattice [28, 29]. The motion of a charged particle in crossed magnetic and electric fields is more complicated than given by Eq. (19). The particle itself moves on a curve called trochoid [24]. It is the orbit center $\vec{r}_0$ that satisfies the simple equation (19). Eq. (18) is a quantum-mechanical result which is obtained when a periodic potential is turned on. There is no classical analog for Eq. (18).

In Table 1, we summarize the conserved quantities we deal with, and how they are influenced by the "electric" force. The quotation marks are used to indicate that for neutral atoms one uses substitutes for the electric force.

We have started the talk with the exponentiated operators in Eq. (1). The eigenvalues of these operators are the symmetric coordinates, the quasi-momentum k and the quasi-coordinate q, for a Bloch electron in a solid. They are the kq-representaion. The orbit quasi-center $\vec{q}_0$, that was used for the magnetic Bloch oscillations, is defined by the Eqs. (15) and (16). By changing notations

$$x_0 = \tfrac{cP}{eB} \quad , \quad y_0 = X \quad , \quad b_2 = a \quad , \quad b_1 = \tfrac{hc}{eBa} \tag{26}$$

where b_1 is obtained from the rationality condition in Eq. (11), the operators in Eqs. (15) and (16) turn into the exponentiated operators of the kq-representation in Eq. (1)

$$\exp\left(\tfrac{ieB}{\hbar c}x_0b_2\right) \to \exp\left(\tfrac{i}{\hbar}Pa\right) \quad , \quad \exp\left(-\tfrac{ieB}{\hbar c}y_0b_1\right) \to \exp\left(-iX\tfrac{2\pi}{a}\right) \tag{27}$$

The operators X and P form the phase plane of the orbit center $\vec{r}_0$. As was mentioned in the introduction, von Neumann [3] used the exponentiated operators in Eq. (1) or (27) to construct a complete set of functions

$$e^{ix\frac{2\pi}{a}m}e^{\frac{i}{\hbar}pan}\psi_0(x) \tag{28}$$

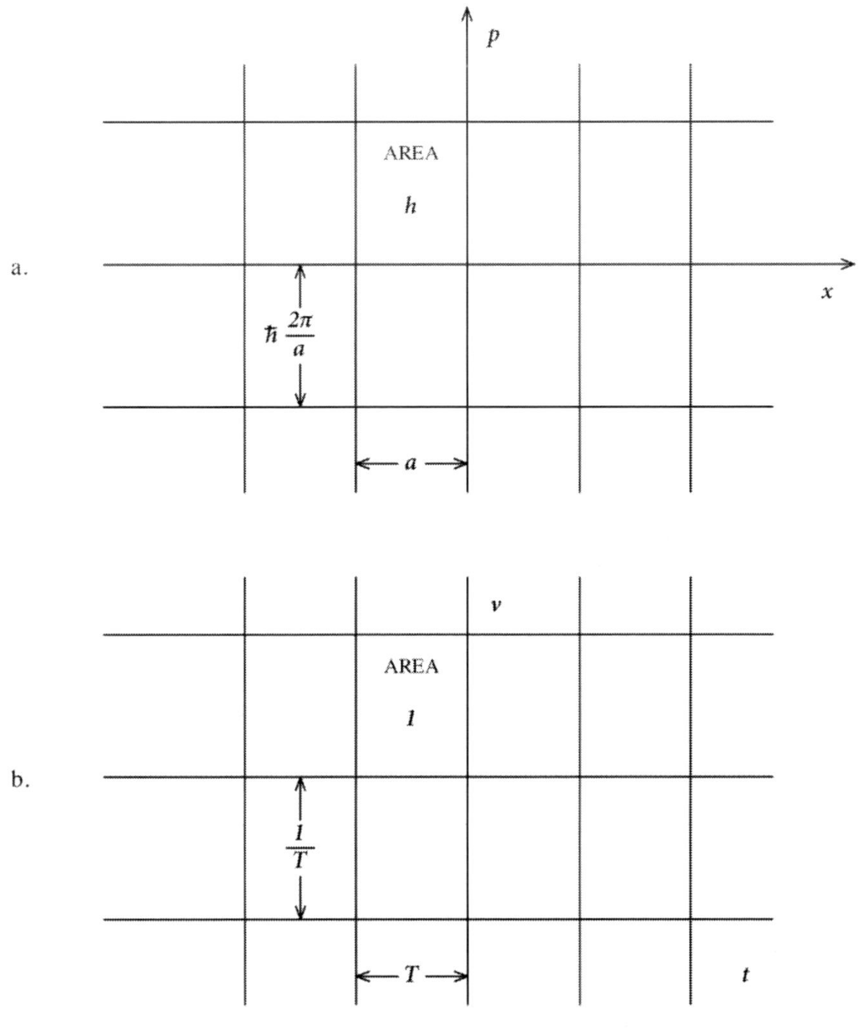

Figure 1. a. von Neumann lattice b. Gabor lattice

where m and n run from $-\infty$ to $+\infty$ and $\psi_0(x)$ is the ground state of the harmonic oscillator. This set is located on a lattice in phase plane as shown in Figure 1 (this figure contains also the lattice in frequency ν and time t plane [30]). In physics this lattice is called the von Neumann lattice, in signal processing it is called the Gabor lattice. What we show here is that this lattice in phase plane xp is closely related to the Pippard network [23] for a Bloch electron in a magnetic field in the coordinate plane x_0y_0 of the orbit center $\vec{r}_0$ in Eq. (8). The points inside the unit cell of area h (the Planck constant) of the von Neumann lattice are related to the orbit quasi-center q_{0x}, q_{0y}, while the lattice points in Figure 1.a are related to b_1 and b_2 in Eq. (26). The complete set of functions in Eq. (28) is very widely used in signal processing [31], where x and p is replaced by time t and frequency ν. It should be pointed out that von Neumann was right when he claimed that the set in Eq. (28) is complete. It turns out however that it is overcomplete by one member [32].

In summary, one deals here with the problem of a Bloch particle in crossed electric and magnetic fields. It is an old problem that has recently attracted wide renewed interest in view of the new techniques of cold atoms in optical lattices subject to synthetic electric and magnetic fields. Our focus is on magnetic Bloch oscillations. For deriving the main result given by Eq. (18), we use the magnetic translations [5, 6]. Their eigenvalues were originally labeled by a magnetic $\vec{k}^{(B)}$-vector, which has no direct physical meaning. As is well known by now, this $\vec{k}^{(B)}$-vector is related to the orbit quasi-center [22]. The main result in Eq. (18) replaces the quasi-momentum acceleration theorem for the case of zero magnetic field. A crucial condition for the use of commuting magnetic translations is the rationality condition given by Eq. (11). They quantize the magnetic flux in the simplest case. Eq. (18) cannot be obtained for an irrational flux because then there are no commuting magnetic translations, Eq. (9). It is pointed out that the Hamiltonian (7) in combination with its symmetry of magnetic translations contains the spectrum of a ladder structure, like in the case of the Wannier-Stark ladder for zero magnetic field. In our case we choose the electric field in the x-direction, Eq. (22), which simplifies to a great extent the derivation of the equations leading to the magnetic Wannier-Stark ladder. In the latter case the experimental observation of the Wannier-Stark levels [20] came much later than their prediction [27]. The theoretical calculation of their lifetime seems to remain a challenging problem [33].

References

[1] Weyl H 1950 *The Theory of Groups and Quantum Mechanics* (New York: Dover)
[2] Bloch F 1928 *Zeitschrift fuer Physik* **52** 555
[3] von Neumann J 1955 *Mathematical Foundations of Quantum Mechanics* (Princeton: Princeton University Press)
[4] Schwinger J 1960 *Proceedings of the National Academy of Sciences USA* **46** 570
[5] Brown E 1964 *Physical Review* **133** A1038
[6] Zak J 1964 *Physical Review* **134** A1602, A1607
[7] Zel'dovich Y B 1967 *Soviet Physics JETP* **51** 1492
[8] Zak J 1967 *Physcial Review Letters* **19** 1385
[9] Zak J 1969 *Physics Today* **22** 64
[10] Zak J 1970 *Physics Today* **23** 51
[11] Dirac P 1972 *Recollections of an Exciting Era* (Enrico Fermi School of Physics, Course LVII, Varenna)
[12] Noether E 1918 *Nachrichten von der Gesellschaft der Wissenschaften zu Goettingen* **2** 235
[13] Wigner E P 1959 *Group Theory* (New York: Academic Press)
[14] Zak J 1972 *Solid State Physics* vol 27 (New York: Academic Press)
[15] Levanda M and Fleurov V 2001 *Annals of Physics* **292** 199
[16] Kittel C 1987 *Quantum Theory of Solids* (New York: Wiley)
[17] Poli N, Wang F Y, Tarallo M G, Alberti A, Prevedelli M and Tino G 2011 *Physical Review Letters* **106** 038501
[18] Atala M, Aidelsburger M, Barreiro J T, Abanin D, Kitagaro T, Demler E and Bloch I 2013 *Nature Physics* **9** 795
[19] Bloch I 2005 *Nature Physics* **1** 23
[20] Voisin P, Bleuse J, Bouche C, Gillard S, Alibert C and Regreny A 1988 *Physical Review Letters* **61** 1639
[21] Johnson M H and Lippmann B A 1949 *Physical Review* **76** 828
[22] Dana I and Zak J 1983 *Physical Review* **28** 811
[23] Pippard A B 1964 *Philos. Trans. R. Soc. London Ser. A* **256** 317
[24] Jackson J D 1999 *Classical Electrodynamics* (Hoboken, New Jersey: John Wiley & Sons, Inc.) p. 586
[25] Thouless D J, Kohmoto M, Nightingale M P and den Nijs M 1982 *Physical Review Letters* **42** 405
[26] Zak J 2014 *EPL* **105** 67001
[27] Wannier G H 1962 *Reviews of Modern Physics* **34** 645
[28] Jimenez-Garcia K, Blanc L J L, Williams R A, Beeler M C, Perry A R and Spielman I B 2012 *Physical Review Letters* **108** 225303

[29] Struck J, Olschlager C, Weinberg M, Hanke P, Simonet J, Eckardt A, Lewenstein M, Sengstock K and Windpassinger P 2012 *Physical Review Letters* **108** 225304
[30] Gabor D 1946 *J. Inst. Elect. Eng.* **93** 429
[31] Janssen A J E M 1988 *Philips Journal of Research* **43** 23
[32] Perelomov A P 1971 *Theoretical and Mathematical Physics* **6** 156
[33] Avron J E 1982 *Annals of Physics (N.Y.)* **143** 33

Bi-fractional Wigner functions

S Agyo, C Lei and A Vourdas

Department of Computing, University of Bradford, Bradford BD7 1DP, United Kingdom

E-mail: C.Lei1@bradford.ac.uk

Abstract. Two fractional Fourier transforms are used to define bi-fractional displacement operators, which interpolate between displacement operators and parity operators. They are used to define bi-fractional coherent states. They are also used to define the bi-fractional Wigner function, which is a two-parameter family of functions that interpolates between the Wigner function and the Weyl function. Links to the extended phase space formalism are also discussed.

1. Introduction

Wigner and Weyl functions and also P and Q functions play a central role in the general area of phase space methods [1, 2]. They are quantities which describe pseudo-probabilities and correlations associated to quantum states. There have also been other more general quantities, which interpolate among them, and provide a more general formalism. In this general context in a recent publication [3], we presented a two-parameter interpolation between displacement and parity operators (or similarly between Weyl and Wigner functions). The displacement and parity operators are related through a two-dimensional Fourier transform, which we replace with fractional Fourier transforms [4, 5, 6, 7]. This leads to a two-parameter family of operators which we call bi-fractional displacement operators, and which include as special cases the displacement and parity operators.

In the present paper we review briefly and extend this work, by discussing links with the extended phase space method in refs [8, 9, 10, 11, 12]. The latter uses the fact that the Wigner function $W(x,p)$ is related to the Weyl function $\widetilde{W}(X,P)$ through a two-dimensional Fourier transform. x,p are position and momentum entering in the Wigner function, and X,P are position and momentum increments entering in the Weyl function. One Fourier transform involves the variables x,P and the other the p,X. These Fourier transforms lead to uncertainty relations between the x,P variables (and also the X,p variables), which have been studied in [13]. An 'extended Wigner function' (and other 'extended quantities') that depend on x,p,X,P have been introduced in the extended phase space $x-p-X-P$. In this language, the two variables of the bi-fractional displacement operator studied here, are in the direction defined by the angle ϕ_1 in the $x-P$ plane, and also in the direction defined by the angle ϕ_2 in the $p-X$ plane (see fig1). Using the bi-fractional displacement operator we get a two-parameter family of functions which include as special cases the Wigner and Weyl functions.

In section 2, we discuss briefly fractional Fourier transforms, in order to define the notation. In section 3, we introduce the bifractional displacement operators. In section 4, we act with them on the vacuum to get bi-fractional coherent states. In section 5, we introduce the 'bi-fractional Wigner functions', which is a two parameter family of functions, that includes Wigner and Weyl functions as special cases. In section 6, we link the present work with the extended phase space formalism. We conclude in section 7, with a discussion of our results.

2. Fractional Fourier Transform

The fractional Fourier Transform has been studied extensively for a long time. It is a generalisation of the Fourier transform given by

$$f_\phi(\alpha) = \exp(i\phi\mathcal{A}^\dagger\mathcal{A})f(\alpha); \quad \mathcal{A} = \frac{\alpha + \partial_\alpha}{\sqrt{2}}; \quad \mathcal{A}^\dagger = \frac{\alpha - \partial_\alpha}{\sqrt{2}} \tag{1}$$

This is the differential form and it is equivalent to the following integral form

$$f_\phi(\alpha) = \int \mathcal{K}_\phi(\alpha,\beta)f(\beta)d\beta, \tag{2}$$

where $\mathcal{K}_\phi(\alpha,\beta)$ is the kernel

$$\mathcal{K}_\phi(\alpha,\beta) = \begin{cases} \left[\frac{1+i\cot\phi}{2\pi}\right]\exp\left[\frac{-i(\alpha^2+\beta^2)\cot\phi}{2} + \frac{i\alpha\beta}{\sin\phi}\right], & \text{if } \phi \neq n\pi \\ \delta(\alpha-\beta), & \text{if } \phi = 2n\pi \\ \delta(\alpha+\beta), & \text{if } \phi = (2n+1)\pi \end{cases} \tag{3}$$

It is easily seen that for $\phi = \pi/2$ we get the Fourier transform, and for $\phi = \pi$ the parity operator with respect to the origin which we denote as $\mathcal{P}(0,0)$. We note that,

$$\int d\beta\, \mathcal{K}_{\phi_1}(\alpha,\beta)\mathcal{K}_{\phi_2}(\beta,\gamma) = \mathcal{K}_{\phi_1+\phi_2}(\alpha,\gamma) \tag{4}$$

The two-dimensional fractional Fourier transform is given by

$$f_{\phi_1,\phi_2}(\alpha,\beta) = \int \mathcal{K}_{\phi_1}(\beta,\alpha')\mathcal{K}_{\phi_2}(\alpha,-\beta')f(\alpha',\beta')d\alpha'd\beta' \tag{5}$$

3. Bi-fractional Displacement operator

We consider the harmonic oscillator in the Hilbert space, $\mathcal{H}$ with $\hat{x}$ and $\hat{p}$ as the position and momentum operators. We define the displacement operator as

$$\mathcal{D}(X,P) = \exp(i\sqrt{2}P\hat{x} - i\sqrt{2}X\hat{p}). \tag{6}$$

The bi-fractional displacement operator is defined as

$$\mathcal{B}_{\phi_1,\phi_2}(\alpha,\beta) = [\cos(\phi_1-\phi_2)]^{1/2}\int \mathcal{K}_{\phi_2}(\beta,X)\mathcal{K}_{\phi_1}(\alpha,-P)\mathcal{D}(X,P)dX\,dP. \tag{7}$$

$\mathcal{B}_{\phi_1,\phi_2}(\alpha,\beta)$ is a continuous function of ϕ_1,ϕ_2. There is an important difference between Eqs(5),(7). In Eq.(5) α',β' are independent variables. In Eq.(7) X,P are dual quantum variables, multiplied by $\hat{x}$, $\hat{p}$ which do not commute. Integrations should be performed very carefully taking into account the ordering of the operators $\hat{x}$, $\hat{p}$. A result of this is the prefactor $[\cos(\phi_1-\phi_2)]^{1/2}$ which is important for unitarity.

In [3] we have proved that the $\mathcal{B}_{\phi_1,\phi_2}(\alpha,\beta)$ are elements of the group $\boldsymbol{\mathcal{HW}} \rtimes \boldsymbol{\mathcal{SU}(1,1)}$ which is the semidirect product of the Heisenberg Weyl group $\boldsymbol{\mathcal{HW}}$ of displacements, by the $\boldsymbol{\mathcal{SU}(1,1)}$ group of squeezing transformations. They are 'special elements' in the sense that not every element of this group can be written as $\mathcal{B}_{\phi_1,\phi_2}(\alpha,\beta)$.

A special case of $\mathcal{B}_{\phi_1,\phi_2}(\alpha,\beta)$ is the parity operator with respect to the point (α,β), which is defined as

$$\mathcal{P}(\alpha,\beta) = \mathcal{D}\left(\frac{\alpha}{2},\frac{\beta}{2}\right)\mathcal{P}(0,0)\mathcal{D}^{\dagger}\left(\frac{\alpha}{2},\frac{\beta}{2}\right) = \mathcal{D}(\alpha,\beta)\mathcal{P}(0,0) \tag{8}$$

The parity operator is related to the displacement operator through a two-dimensional Fourier transform

$$\begin{aligned}\mathcal{P}(x,p) &= \frac{1}{2\pi}\int \mathcal{D}(X,P)\exp\left[i(pX - Px)\right] dX\, dP \\ &= \int \mathcal{D}(X,P)\, \mathcal{K}_{\frac{\pi}{2}}(p,X)\, \mathcal{K}_{\frac{\pi}{2}}(x,-P) dX\, dP\end{aligned} \tag{9}$$

For $\phi_1 = \phi_2 = \frac{\pi}{2}$, we get

$$\mathcal{B}_{\frac{\pi}{2},\frac{\pi}{2}}(\alpha,\beta) = \mathcal{P}(\alpha,\beta) \tag{10}$$

The bi-fractional displacement operator $\mathcal{B}_{\phi_1,\phi_2}(\alpha,\beta)$ is unitary:

$$[\mathcal{B}_{\phi_1,\phi_2}(\alpha,\beta)]^{\dagger} = \mathcal{B}_{-\phi_1,-\phi_2}(-\alpha,-\beta) \tag{11}$$

Special cases are:

$$\begin{aligned}&\mathcal{B}_{0,0}(\alpha\,\beta) = \mathcal{D}(\beta,-\alpha) \\ &\mathcal{B}_{\frac{\pi}{2},\frac{\pi}{2}}(\alpha,\beta) = \mathcal{P}(\alpha,\beta) \\ &\mathcal{B}_{\pi,\pi}(\alpha,\beta) = \mathcal{D}(-\beta,\alpha)\end{aligned} \tag{12}$$

The relationship between two bi-fractional displacement operators with different variables is

$$\begin{aligned}\mathcal{B}_{\psi_1+\phi_1,\psi_2+\phi_2}(\alpha,\beta) &= \frac{|\cos(\psi_1+\phi_1-\psi_2-\phi_2)|^{1/2}}{|\cos(\psi_1-\psi_2)|^{1/2}} \\ &\times \int d\sigma d\omega \mathcal{K}_{\phi_2}(\beta,\beta')\mathcal{K}_{\phi_1}(\alpha,\alpha')\mathcal{B}_{\psi_1,\psi_2}(\alpha',\beta')\end{aligned} \tag{13}$$

Acting with the displacement operator on both sides of the bifractional displacement operator we get

$$\begin{aligned}\mathcal{D}(\nu,\xi)\mathcal{B}_{\phi_1,\phi_2}(\alpha,\beta)\mathcal{D}^{\dagger}(\nu,\xi) &= \mathcal{B}_{\phi_1,\phi_2}(-\alpha-2\nu\sin\phi_1,\beta+2\xi\sin\phi_2) \\ &\times \exp\left[i(2\alpha\nu\cos\phi_1+\nu^2\sin 2\phi_1)\right] \\ &\times \exp\left[i(2\beta\xi\cos\phi_2+\xi^2\sin 2\phi_2)\right]\end{aligned} \tag{14}$$

4. Bifractional coherent states

Acting with $\mathcal{B}_{\phi_1,\phi_2}(\alpha,\beta)$ on the vacuum $|0\rangle$ we get the generalized coherent states

$$|\alpha,\beta;\phi_1,\phi_2\rangle = \mathcal{B}_{\phi_1,\phi_2}(\alpha,\beta)\,|0\rangle \tag{15}$$

In the special case when $\phi_1 = \phi_2 = 0$, and $\phi_1 = \phi_2 = \frac{\pi}{2}$ we get Glauber coherent states:

$$|-\beta,\alpha;0,0\rangle = |\alpha,\beta;\frac{\pi}{2},\frac{\pi}{2}\rangle = |\alpha,\beta\rangle \tag{16}$$

They are eigenstates of the creation and annihilation operators

$$\begin{aligned} b(\phi_1, \phi_2) &= \mathcal{B}_{\phi_1,\phi_2}(0,0)a\,\mathcal{B}_{\phi_1,\phi_2}(0,0) \\ b^\dagger(\phi_1, \phi_2) &= \mathcal{B}_{\phi_1,\phi_2}(0,0)a^\dagger\mathcal{B}_{\phi_1,\phi_2}(0,0), \end{aligned} \tag{17}$$

but they have novel non-trivial properties with respect to $a, a^\dagger$. They also satisfy the resolution of the identity

$$\frac{1}{2\pi}\int |\alpha,\beta;\phi_1,\phi_2\rangle\,\langle\alpha,\beta;\phi_1,\phi_2|\,d\alpha d\beta = \mathbf{1} \tag{18}$$

In [3] we have studied the Bargmann functions of these coherent states. Also the whole formalism of coherent state quantization, can be studied using the bifractional coherent states.

5. Bi-fractional Wigner functions

We consider the trace of a trace-class operator Θ with $\mathcal{B}_{\phi_1,\phi_2}(\alpha,\beta)$:

$$\begin{aligned} \mathcal{V}_{\phi_1,\phi_2}(\alpha,\beta|\Theta) &= \mathrm{Tr}[\Theta\mathcal{B}_{\phi_1,\phi_2}(\alpha,\beta)] \\ &= |\cos(\phi_1-\phi_2)|^{1/2}\int \mathcal{K}_{\phi_2}(\beta,X) \\ &\times\ \mathcal{K}_{\phi_1}(\alpha,-P)\widetilde{\mathcal{W}}(X,P|\Theta)dXdP \end{aligned} \tag{19}$$

This is a generalization of the Wigner and Weyl functions. Indeed

$$\begin{aligned} &\mathcal{V}_{0,0}(\alpha,\beta|\Theta) = \widetilde{\mathcal{W}}(\beta,-\alpha|\Theta) \\ &\mathcal{V}_{\frac{\pi}{2},\frac{\pi}{2}}(\alpha,\beta|\Theta) = \mathcal{W}(\alpha,\beta|\Theta) \\ &\mathcal{V}_{\pi,\pi}(\alpha,\beta|\Theta) = \widetilde{\mathcal{W}}(-\beta,\alpha|\Theta) \end{aligned} \tag{20}$$

In the special case that $\phi_1 = \phi_2 = 0$, the variable α becomes the variable P, the variable β becomes the variable X (see fig.1), and $\mathcal{V}_{\phi_1,\phi_2}(\alpha,\beta|\Theta)$ is the Weyl function. In the special case that $\phi_1 = \phi_2 = \frac{\pi}{2}$, the variable α becomes the variable x, the variable β becomes the variable p, and $\mathcal{V}_{\phi_1,\phi_2}(\alpha,\beta|\Theta)$ is the Wigner function. In general, the function $\mathcal{V}_{\phi_1,\phi_2}(\alpha,\beta|\Theta)$ is a two-parameter interpolation between the Wigner and Weyl functions.

When $\Theta = \rho$ is a density matrix, $\mathcal{V}_{\phi_1,\phi_2}(\alpha,\beta|\rho)$ is the bi-fractional Wigner function for the state described by ρ. As example, we consider the superposition of coherent states

$$|s\rangle = \mathcal{N}[|\alpha_0,\beta_0\rangle + |-\alpha_0,-\beta_0\rangle] \tag{21}$$

where $\mathcal{N}$ is a normalization factor. In figs 2, 3 we plot $\Re[\mathcal{V}_{\frac{\pi}{4},\frac{\pi}{4}}(\alpha,\beta|\Theta)]$ and $|\mathcal{V}_{\frac{\pi}{4},\frac{\pi}{4}}(\alpha,\beta|\Theta)|$ for this state with $\alpha_0 = 1.8$ and $\beta_0 = 0$. In the Weyl function the auto-terms are in the middle and the cross-terms are in the 'wings', while the opposite is true in the Wigner function. Here we have both auto-terms and cross-terms in the wings.

6. Links with the extended phase space formalism

The extended phase space formalism uses the fact that the Wigner function $W(x,p)$ and the Weyl function $\widetilde{W}(X,P)$ are related through a two-dimensional Fourier transform. The Wigner function describes quantum noise, and the Weyl function quantum correlations. Starting from a function $g(u)$ with $u \in \mathbb{R}^N$, and its Fourier tranform $\widetilde{g}(v)$ with $v \in \mathbb{R}^N$, we can introduce (in any context) 'a Wigner function' $W(u,v)$ (and other quantities) in the $u-v$ phase space, which is $\mathbb{R}^{2N}$. The extended phase space formalism introduces extended Wigner functions that

Figure 1. The α direction in the $x-P$ plane within the extended phase space $x-p-X-P$. Also the β direction in the $p-X$ plane. For clarity, the two planes are drawn separately, but they belong to the same four-dimensional space.

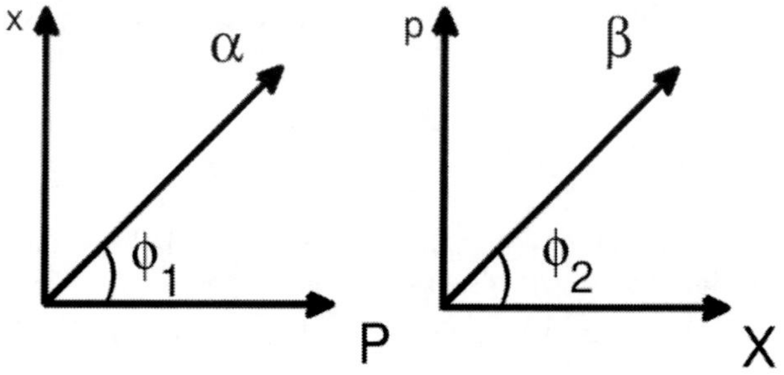

Figure 2. $\Re[\mathcal{V}_{\frac{\pi}{4},\frac{\pi}{4}}(\alpha,\beta|\Theta)]$ for the state of Eq.(21), with $\alpha_0 = 1.8$ and $\beta_0 = 0$.

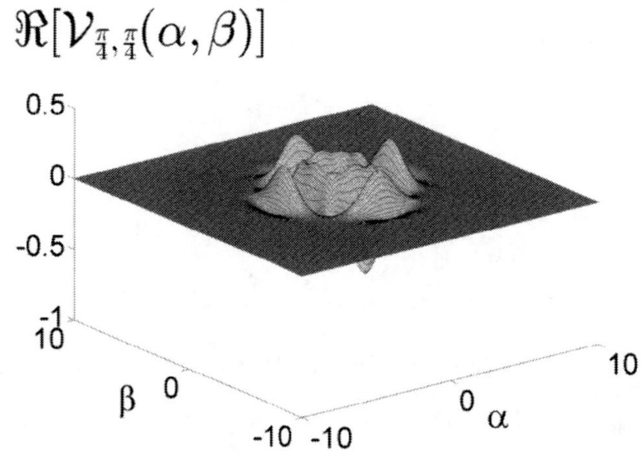

depend on four variables (x, p, X, P) and studies transformations in the extended phase space $x-p-X-P$. The $x-P$ plane links quantum noise in the x-direction, with quantum correlations in the P-direction. Uncertainty relations in the $x-P$ plane have been studied in [13]. They are related to the usual uncertainties in the case of pure states, but they are different for mixed states. Analogous comments can be made for the $p-X$ plane.

In the extended phase space $x-p-X-P$ we can introduce the following dispacement operator of Weyl functions:

$$\begin{aligned}\mathfrak{D}(A_1, A_2, B_1, B_2) &= \exp[(B_2 - A_2)\partial_P] \exp\left(-i\frac{A_1 + B_1}{2}P\right) \exp[(B_1 - A_1)\partial_X] \\ &\times \exp\left(i\frac{A_2 + B_2}{2}X\right)\end{aligned} \quad (22)$$

Figure 3. $|\mathcal{V}_{\frac{\pi}{4},\frac{\pi}{4}}(\alpha,\beta|\Theta)|$ for the state of Eq.(21), with $\alpha_0 = 1.8$ and $\beta_0 = 0$.

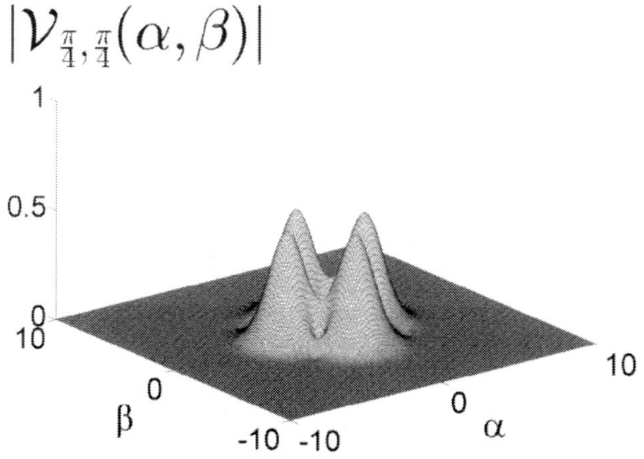

Acting with this on the Weyl function $\widetilde{W}(X, P|\Theta)$ we get [9]

$$\mathfrak{D}(A_1, A_2, B_1, B_2)\widetilde{W}(X, P|\Theta) = \exp\left(-i\frac{A_1 + B_1}{2}P + i\frac{A_2 + B_2}{2}X - iA_1B_2 + iA_2B_1\right) \times \widetilde{W}(X + B_1 - A_1, P - A_2 + B_2) \quad (23)$$

We call Θ' the operator with this Weyl function:

$$\widetilde{W}(X, P|\Theta') = \mathfrak{D}(A_1, A_2, B_1, B_2)\widetilde{W}(X, P|\Theta) \quad (24)$$

In this language, the present paper considers quantities in the general $\alpha - \beta$ plane (fig.1), which becomes the $X - P$ plane in the special case $\phi_1 = \phi_2 = 0$, and the $x - p$ plane in the special case $\phi_1 = \phi_2 = \frac{\pi}{2}$. It then studies relations between the various quantities. For example, we show that $\mathcal{V}_{\phi_1,\phi_2}(\alpha, \beta|\Theta')$ is related to $\widetilde{W}(X, P|\Theta)$ as follows:

$$\begin{aligned}\mathcal{V}_{\phi_1,\phi_2}(\alpha, \beta|\Theta') &= |\cos(\phi_1 - \phi_2)|^{1/2} \int \mathcal{K}_{\phi_2}(\beta, X)\mathcal{K}_{\phi_1}(\alpha, -P) \\ &\times [\mathfrak{D}(A_1, A_2, B_1, B_2)\widetilde{W}(X, P|\Theta)]dXdP \\ &= |\cos(\phi_1 - \phi_2)|^{1/2} \int \mathcal{K}_{\phi_2}(\beta, X + \mu_2)\exp(i\nu_2 X) \\ &\times \mathcal{K}_{\phi_1}(\alpha, -P - \mu_1)\exp(-i\nu_1 P)\,\widetilde{W}(X, P|\Theta)dXdP. \end{aligned} \quad (25)$$

Here

$$\mu_2 = A_1 - B_1; \quad \nu_2 = \frac{A_2 + B_2}{2}; \quad \mu_1 = A_2 - B_2; \quad \nu_1 = \frac{A_1 + B_1}{2}. \quad (26)$$

We next introduce the displaced kernel of the fractional Fourier transform

$$\mathfrak{K}_\phi(x, y|\kappa, \lambda) = \exp(i\kappa y)\exp(\lambda\partial_y)\mathcal{K}_\phi(x, y) = \exp(i\kappa y)\mathcal{K}_\phi(x, y + \lambda), \quad (27)$$

and rewrite Eq.(25) in the compact form

$$\begin{aligned} \mathcal{V}_{\phi_1,\phi_2}(\alpha,\beta|\Theta') &= |\cos(\phi_1-\phi_2)|^{1/2}\int \mathfrak{K}_{\phi_2}(\beta,X|\nu_2,\mu_2)\mathfrak{K}_{\phi_1}(\alpha,-P|-\nu_1,-\mu_1) \\ &\times \widetilde{\mathcal{W}}(X,P|\Theta)dXdP \end{aligned} \tag{28}$$

Other similar relations can also be proved. In particular symplectic $Sp(4,R)$ transformations in the $x-p-X-P$ extended phase space require further study.

7. Discussion

We have studied the bi-fractional displacement operators $\mathcal{B}_{\phi_1,\phi_2}(\alpha,\beta)$. They include as special cases, both the displacement operators and the parity operators (Eq.(12)). We stress the importance of the prefactor $[\cos(\phi_1-\phi_2)]^{1/2}$ in Eq.(7), for unitarity. This is related to the fact that in Eq.(7) X, P are dual quantum variables, multiplied by $\hat{x}, \hat{p}$ which do not commute.

Using them we have defined the bi-fractional coherent states in Eq.(15). We have also defined the bi-fractional Wigner functions in Eq.(19), which include as special cases, both the Wigner and Weyl functions (Eq.(20)).

We have interpreted the formalism within the extended phase space, which relates quantum noise with correlations. The bi-fractional displacement operators $\mathcal{B}_{\phi_1,\phi_2}(\alpha,\beta)$, are displacement operators in the $\alpha-\beta$ plane of fig.1. The formalism provides an interpolation between various quantities in phase space.

For further work we suggest the application of these ideas to finite quantum systems with variables in $\mathbb{Z}_d$ (the integers modulo d). We expect this generalization to be easier for prime d, in which case $\mathbb{Z}_d$ is a field. Overall, the merit of the formalism is that different quantities appear to be special cases of a single quantity.

References

[1] Zachos C K, Fairlie D B and Curtright T L 2005 *Quantum Mechanics in Phase Space* (Singapore: World Scientific)
[2] Schleich W P 2001 *Quantum Optics in Phase Space* (New York: Wiley)
[3] Agyo S, Lei C and Vourdas A 2015 *Phys. Lett. A* **379** 255
[4] Namias V 1980 *J. Inst. Math. Applic.* **25** 241
[5] McBride A C and Kerr F H 1987 *IMA J. Appl. Math* **39** 159
[6] Bailey D H and Swarztrauber P N 1991 *SIAM Review* **33** 389
[7] Ozaktas H M, Zalevsky Z and Kutay M A 2001 *The Fractional Fourier Transform* (New York: Wiley)
[8] Chountasis S and Vourdas A 1998 *Phys. Rev. A* **58** 1794
[9] Chong C C and Vourdas A 2001 *J. Phys. A* **34** 9849
[10] Chountasis S and Vourdas A 1999 *J. Phys. A* **32** 6949
[11] Ponomarenko S A and Wolf E 2001 *Phys. Rev. A* **63** 062106
[12] Agarwal G S and Ponomarenko S A 2003 *Phys. Rev. A* **67** 032103
[13] Vourdas A 2004 *Phys. Rev. A* **69** 022108

Representations of conformal Galilei algebra with integer spin and an application

N Aizawa[1], **R Chandrashekar**[2] **and J Segar**[3]

[1] Department of Mathematics and Information Sciences, Osaka Prefecture University, Nakamozu Campus, Sakai, Osaka 599-8531, Japan

[2] Department of Physics, National Chung Hsing University, Taichung 40227, Taiwan

[3] Department of Physics, Ramakrishna Mission Vivekananda College, Mylapore, Chennai 600 004, India

E-mail: aizawa@mi.s.osakafu-u.ac.jp

Abstract. Conformal Galilei algebra is a class of non-semisimple Lie algebras. Each member of the class is labelled by two parameters d (positive integer) and ℓ (positive integer or positive half-integer). We investigate the lowest weight representations of the conformal Galilei algebra for $d = 1$ and any integer ℓ. We start with the Verma modules and then extract all the irreducible lowest weight modules. This is done by searching and explicitly constructing singular vectors. As an application of our construction of singular vectors, we obtain the partial differential equations which are symmetric by kinematical transformation generated by the conformal Galilei algebra.

1. Introduction

Our aim in the present work is to study representation theory and its application to a particular class of non-semisimple Lie algebras. The class is called conformal Galilei algebra (CGA) and can be regarded as a nonrelativistic version of conformal algebras [1]. Each member of the class is labelled by two parameters d (positive integer) and ℓ (positive integer or positive half-integer). CGA has a vast variety of physical applications. However such studies of CGA was limited to lower values of ℓ (see e.g. [10] for references on $\ell = 1/2$ and $\ell = 1$ CGA). It is in the last couple of years that more focus has been directed towards CGA with higher values of ℓ [2, 3, 4, 5, 6, 7, 8, 9]. In this work we focus on $d = 1$ CGA for any integer ℓ, and study the lowest weight Verma module in some detail. This shall be done by computing the Kac determinant and by the construction of singular vectors. As an application of our results, we are able to obtain partial differential equations which are symmetric under the kinematical transformations generated by the CGA. The present work is a continuation of our previous works on the lowest (or highest) weight representations of CGA done in Refs. [10, 11] in which $d = 1$ CGA with half-integer ℓ are studied. We remark a very recent work [12] discussing a similar problem. In Ref. [12] a classification of all finite weight modules over the $d = 1$ CGA for any ℓ has been done. We would like to be more explicit in the present work. Namely, we start with the Verma modules and show explicitly how to arrive at the irreducible modules. We give the structure of the irreducible modules through a concrete formula.

The plan of this article is as follow: We give the definition of $d = 1$ CGA in the next section. The Verma modules over the $d = 1$ CGA are introduced in §3 and their reducibility is studied by singular vectors. Those singular vectors are also used to obtain symmetric partial differential equations in §4. Finally, a classification of all irreducible lowest weight modules is given in §5.

2. Preliminary: definition of $d = 1$ CGA

Throughout this paper we denote the $d = 1$ CGA by $\mathfrak{g}_\ell$. The complex Lie algebra $\mathfrak{g}_\ell$ for a fixed integer ℓ has the elements [1]:

$$D,\ H,\ C,\ P_n\ (n = 0, 1, \ldots, 2\ell). \tag{1}$$

Their nonvanishing commutators are given by

$$\begin{array}{lll} [D,H] = H, & [D,C] = -C, & [C,H] = 2D, \\ [H,P_n] = -nP_{n-1}, & [D,P_n] = (\ell - n)P_n, & [C,P_n] = (2\ell - n)P_{n+1}. \end{array} \tag{2}$$

One may see from this that $\langle\, P_0, P_1, \ldots, P_{2\ell}\,\rangle$ is an Abelian ideal of $\mathfrak{g}_\ell$ so that $\mathfrak{g}_\ell$ is *not* semisimple. It is known that this Lie algebra has no central extensions [13]. The subalgebra spanned by $\langle\, H, D, C\,\rangle$ is isomorphic to $so(2,1) \simeq sl(2,\mathbb{R}) \simeq su(1,1)$. The Abelian subalgebra spanned by $\langle\, P_n\,\rangle_{n=0,1,\ldots,2\ell}$ carries the spin ℓ representation of the $sl(2,\mathbb{R})$ subalgebra.

One may introduce the algebraic anti-involution $\omega : \mathfrak{g}_\ell \to \mathfrak{g}_\ell$ by

$$\omega(D) = D, \quad \omega(H) = C, \quad \omega(P_n) = P_{2\ell-n}. \tag{3}$$

It is not difficult to verify that ω satisfies the required relations:

$$\omega([X,Y]) = [\omega(Y), \omega(X)], \qquad \omega^2(X) = X \quad {}^\forall X \in \mathfrak{g}_\ell. \tag{4}$$

Let us define the degree of the generators based on their commutators with respect to D:

$$\deg(D) = 0, \quad \deg(H) = 1, \quad \deg(C) = -1, \quad \deg(P_n) = \ell - n. \tag{5}$$

With respect to the sign of the degree one may define the triangular decomposition of $\mathfrak{g}_\ell$:

$$\mathfrak{g}_\ell = \mathfrak{g}_\ell^+ \oplus \mathfrak{g}_\ell^0 \oplus \mathfrak{g}_\ell^-, \tag{6}$$

where

$$\begin{aligned} \mathfrak{g}_\ell^+ &= \langle\, H,\ P_0,\ P_1,\ \ldots,\ P_{\ell-1}\,\rangle \\ \mathfrak{g}_\ell^0 &= \langle\, D,\ P_\ell\,\rangle \\ \mathfrak{g}_\ell^- &= \langle\, C,\ P_{\ell+1},\ P_{\ell+2},\ \ldots,\ P_{2\ell}\,\rangle \end{aligned}$$

Eqn. (6) is a decomposition of $\mathfrak{g}_\ell$ as a direct sum of the vector spaces.

3. Lowest weight Verma modules and their irreducibility

In this section we study the lowest weight Verma modules over $\mathfrak{g}_\ell$. First let us define the Verma modules. We assume the existence of the lowest weight vector $|\delta, p\rangle$ defined by

$$D\,|\delta,p\rangle = \delta\,|\delta,p\rangle\,, \qquad P_\ell\,|\delta,p\rangle = p\,|\delta,p\rangle\,, \qquad X\,|\delta,p\rangle = 0, \quad {}^\forall X \in \mathfrak{g}_\ell^-. \tag{7}$$

The Verma module over $\mathfrak{g}_\ell$ is defined, as usual (see e.g. [14]), as a module induced from $|\delta,p\rangle$: $V_\ell^{\delta,p} = U(\mathfrak{g}_\ell^+)\,|\delta,p\rangle$ where $U(\mathfrak{g}_\ell^+)$ is the enveloping algebra of $\mathfrak{g}_\ell^+$. In the sequel we assume that $p \neq 0$. The case of $p = 0$ is discussed in [16].

We choose the following basis of $V_\ell^{\delta,p}$:

$$H^k P_{\ell-1}^{m_1} P_{\ell-2}^{m_2} \cdots P_0^{m_\ell} |\delta, p\rangle . \tag{8}$$

This is an eigenvector of D with the eigenvalue $\delta + N$ where

$$N = k + \sum_{i=1}^{\ell} i\ m_i, \tag{9}$$

and we refer to N as the *level* as usual. Then $V_\ell^{\delta,p}$ has a grading structure according to the value of N :

$$V_\ell^{\delta,p} = \bigoplus_{N=0}^{\infty} (V_\ell^{\delta,p})_N, \qquad (V_\ell^{\delta,p})_N = \{\ |v\rangle \in V_\ell^{\delta,p} \mid D\,|v\rangle = (\delta+N)\,|v\rangle\ \}. \tag{10}$$

One may define an inner product for the vectors in $V_\ell^{\delta,p}$. Let $|x\rangle$ and $|y\rangle$ be any two vectors in $V_\ell^{\delta,p}$:

$$|x\rangle = X\,|\delta,p\rangle\,, \qquad |y\rangle = Y\,|\delta,p\rangle\,, \quad X, Y \in U(\mathfrak{g}_\ell^+).$$

We define the inner product of $|x\rangle$ and $|y\rangle$ by

$$\langle x \mid y\rangle = \langle \delta, p|\,\omega(X) Y\,|\delta,p\rangle\,, \quad \langle \delta,p \mid \delta,p\rangle = 1. \tag{11}$$

Next we define the Kac determinant at level N (see e.g. [15]). Let $|v_1\rangle\,, |v_2\rangle\,, \ldots, |v_r\rangle$ be a set of basis of the subspace $(V_\ell^{\delta,p})_N$. We consider the matrix $(\langle v_i \mid v_j\rangle)$ whose entries are the inner products of the basis of $(V_\ell^{\delta,p})_N$. The determinant of this matrix is called the Kac determinant at level N : $\Delta_N^{(\ell)} = \det(\langle v_i \mid v_j\rangle)$. Our first result is an explicit formula of the Kac determinant which is helpful to know the reducibility of $V_\ell^{\delta,p}$.

Proposition 1 *The Kac determinants $\Delta_N^{(\ell)}$ at level N of $\mathfrak{g}_\ell$ are given as follows (up to overall sign):*

$$(i) \quad \Delta_N^{(1)} = \left(\prod_{m=0}^{N} m! \right)^2 (2p)^{N(N+1)}, \tag{12}$$

$$(ii) \quad \Delta_N^{(\ell)} = (\ell+1)^2\, p^2\, \delta_{N1}, \quad (\ell \geq 2). \tag{13}$$

The proof is found in [16]. (i) shows that if $\ell = 1$ then the Verma module $V_1^{\delta,p}$ has no singular vectors so that it is irreducible. (ii) shows that if $\ell \geq 2$ then there exist no singular vectors in the level $N = 1$ subspace of $V_\ell^{\delta,p}$. As shown in Proposition 3 and 4 there exist singular vectors in the subspace of other levels. Hence we have shown the following:

Proposition 2

(i) $V_\ell^{\delta,p}$ is irreducible if $\ell = 1$.

(ii) $V_\ell^{\delta,p}$ is reducible if $\ell \geq 2$.

Proposition 3 *Following are the singular vectors in $V_\ell^{\delta,p}$ for $\ell \geq 2$.*

$$(\mathcal{S}^{(2n)})^k\,|\delta,p\rangle\,, \quad (\mathcal{S}^{(2n+1)})^k\,|\delta,p\rangle\,, \quad k = 1, 2, \ldots \tag{14}$$

where n takes the value of a positive integer. The maximal value of n is determined by ℓ and N in such a way that $\mathcal{S}^{(2n)}$ and $\mathcal{S}^{(2n+1)}$ given below are well-defined.

$$\mathcal{S}^{(2n)} = p\, P_{\ell-2n} + \sum_{j=1}^{n-1} a_j\, P_{\ell-2n+j}\, P_{\ell-j} + \frac{(-1)^n}{2}\,\frac{\ell!\,(\ell+2n)!}{((\ell+n)!)^2}\, P_{\ell-n}^2, \tag{15}$$

$$\begin{aligned}\mathcal{S}^{(2n+1)} &= p^2\, P_{\ell-2n-1} - \frac{\ell+2n+1}{(n+\frac{1}{2})(\ell+1)}\, P_{\ell-1}\mathcal{S}^{(2n)} \\ &\quad - \frac{n-\frac{1}{2}}{n+\frac{1}{2}}\,\frac{\ell+2n+1}{\ell+1}\, p\, P_{\ell-1}\, P_{\ell-2n} - \sum_{j=1}^{n-1} b_j\, P_{\ell-2n+j}\, P_{\ell-j-1},\end{aligned} \tag{16}$$

where

$$a_j = (-1)^j\, \frac{\ell!\,(\ell+2n)!}{(\ell+j)!\,(\ell+2n-j)!}, \qquad b_j = (-1)^j\, \frac{n-\frac{1}{2}-j}{n+\frac{1}{2}}\, \frac{p\,\ell!\,(\ell+2n+1)!}{(\ell+j+1)!\,(\ell+2n-j)!}.$$

Proposition 4 *Following are the singular vectors in $V_\ell^{\delta,p}$ for $\ell \geq 2$.*

$$(\mathcal{T}^{(2n+1)})^k\, |\delta, p\rangle, \quad k = 1, 2, \ldots \tag{17}$$

where n takes the value of a positive integer. The maximal value of n is determined by ℓ and N in such a way that $\mathcal{T}^{(2n+1)}$ given below is well-defined.

$$\begin{aligned}\mathcal{T}^{(2n+1)} &= P_{\ell-1}\left((\ell+2)\, P_{\ell-1}^2 - 2p\,(\ell+1)\, P_{\ell-2}\right)^n \\ &\quad + \sum_{j=0}^{n-1} c_j\, P_{\ell-1}^{2(n-j-1)}\, P_{\ell-2}^j\, P_{\ell-3} + \sum_{j=1}^{n} d_j\, P_{\ell-1}^{2n-2j+1}\, P_{\ell-2}^j,\end{aligned} \tag{18}$$

where

$$c_j = \binom{n-1}{j}\left(-\frac{2p(\ell+1)}{\ell+2}\right)^j c_0, \quad d_1 = -p(\ell+1)\,(\ell+2)^{n-1}.$$

Other coefficients c_0, d_j ($j \leq 1 \leq n$) are determined by the relations

$$\begin{aligned}&(j+1)\,(\ell+2)\, d_{j+1} + (2n-2j+1)\, p\,(\ell+1) d_j + (\ell+3)\, c_{j-1} \\ &\quad + \binom{n}{j}(-2)^j\,(p(\ell+1))^{j+1}\,(\ell+2)^{n-j} = 0, \quad (j = 1, 2, \ldots, n)\end{aligned} \tag{19}$$

with $d_{n+1} = 0$.

Proposition 3, 4 are proved by verifying that the vectors (14) and (17) satisfies the definition of singular vectors (see [16]).

4. Differential equations symmetric under the group generated by $\mathfrak{g}_\ell$

The singular vectors obtained in the previous section can be used to derive partial differential equations having particular symmetries. The symmetries are generated by $\mathfrak{g}_\ell$, i.e., the symmetry group is the exponentiation of $\mathfrak{g}_\ell$, and the partial differential equations are invariant under the change of independent variables, i.e., the kinematical symmetries, caused by the group. This can be done by applying the method developed for real semisimple Lie groups in [17]. In this section, before presenting our results, we give a brief review of the method with suitable modification for the present case.

The basic idea of [17] is to realize the Verma modules in a space of C^∞-class functions. Then the generators of $\mathfrak{g}_\ell$ will be differential operators and the singular vector will be converted to differential equations.

Let G be a complex semisimple Lie group and $\mathfrak{g}$ its Lie algebra. The Lie algebra $\mathfrak{g}$ has the triangular decomposition $\mathfrak{g} = \mathfrak{g}^+ \oplus \mathfrak{g}^0 \oplus \mathfrak{g}^-$. The corresponding decomposition of G is denoted by $G = G_+ G_0 G_-$. Consider the space of C^∞-class functions on G having the property called *right covariance*:

$$C_\Lambda = \{ \ f \in C^\infty(G) \mid f(gxg_-) = e^{\Lambda(H)} f(g) \ \}, \tag{20}$$

where $\Lambda \in \mathfrak{g}^*$ (algebra dual to $\mathfrak{g}_\ell$), $g \in G$, $H \in \mathfrak{g}^0$, $x = e^H \in G_0$, $g_- \in G_-$. Because of the right covariance, the functions of C_Λ are actually function on G/B with $B = G_0 G_-$, or on G_+. We keep using the same notations for the restricted representation space of functions on G_+. Then the Lie algebra $\mathfrak{g}_\ell$ is realized by the standard left action of $\mathfrak{g}_\ell$ on C_Λ :

$$\pi_L(X) f(g) = \frac{d}{d\tau} f(e^{-\tau X} g)\bigg|_{\tau=0}, \qquad X \in \mathfrak{g},\ g \in G. \tag{21}$$

We introduce the right action of $\mathfrak{g}$ on C_Λ by the standard formula:

$$\pi_R(X) f(g) = \frac{d}{d\tau} f(g e^{\tau X})\bigg|_{\tau=0}, \qquad X \in \mathfrak{g},\ g \in G. \tag{22}$$

Using the right covariance we can show that the function $f \in C_\Lambda$ has the properties of lowest weight vector:

$$\pi_R(H) f(g) = \Lambda(H) f(g), \quad \pi_R(X) f(g) = 0, \qquad H \in \mathfrak{g}^0,\ X \in \mathfrak{g}^-. \tag{23}$$

This allows us to realize the Verma module $V^\Lambda \simeq U(\mathfrak{g}^+) v_0$ with the lowest weight vector v_0 in terms of the function in C_Λ and differential operators $\pi_R(X)$, $X \in \mathfrak{g}^+$.

Now suppose that the Verma module V^Λ has a singular vector. The general structure of a singular vector is

$$v_s = \mathcal{P}(X_1, X_2, \ldots, X_s) v_0, \quad X_k \in \mathfrak{g}^+. \tag{24}$$

where $\mathcal{P}$ denotes a homogeneous polynomial in its variables. It is shown in [17] that if the operator $\pi_R(\mathcal{P})$ has a nontrivial kernel

$$\pi_R(\mathcal{P}) \psi = 0, \tag{25}$$

for some function ψ on G_+, then the differential equation (25) has the kinematical symmetry generated by $\mathfrak{g}_\ell$.

Now let us apply this scheme to our case. We parametrize an element of G_+ as $g = \exp(tH) \exp\left(\sum_{n=0}^{\ell-1} x_n P_n \right)$. Then the right action of $\mathfrak{g}_\ell^+$ yields

$$\pi_R(P_n) = \frac{\partial}{\partial x_n}, \qquad \pi_R(H) = \frac{\partial}{\partial t} + \sum_{j=1}^{\ell-1} j x_j \frac{\partial}{\partial x_{j-1}}. \tag{26}$$

From (25), Proposition 3 and Proposition 4 we obtain the following hierarchies of partial differential equations.

Proposition 5 *The following equations have the kinematical symmetry generated by* $\mathfrak{g}_\ell$.

$$\left(p\frac{\partial}{\partial x_{\ell-2n}} + \sum_{j=1}^{n-1} a_j \frac{\partial^2}{\partial x_{\ell-2n+j}\partial x_{\ell-j}} + \frac{(-1)^n}{2}\frac{\ell!(\ell+2n)!}{((\ell+n)!)^2}\frac{\partial^2}{\partial x_{\ell-n}^2} \right)^k \psi(x) = 0, \tag{27}$$

$$\left(p^2\frac{\partial}{\partial x_{\ell-2n-1}} - \frac{\ell+2n+1}{(n+\frac{1}{2})(\ell+1)}\frac{\partial}{\partial x_{\ell-1}}\left(\sum_{j=1}^{n-1} a_j\frac{\partial^2}{\partial x_{\ell-2n+j}\partial x_{\ell-j}} + \frac{(-1)^n}{2}\frac{\ell!(\ell+2n)!}{((\ell+n)!)^2}\frac{\partial^2}{\partial x_{\ell-n}^2}\right) \right.$$
$$\left. - \frac{\ell+2n+1}{\ell+1}p\frac{\partial^2}{\partial x_{\ell-1}x_{\ell-2n}} - \sum_{j=1}^{n-1} b_j \frac{\partial^2}{\partial x_{\ell-2n+j}x_{\ell-j-1}} \right)^k \psi(x) = 0, \tag{28}$$

$$\left(\frac{\partial}{\partial x_{\ell-1}}\left((\ell+2)\frac{\partial^2}{\partial x_{\ell-1}^2} - 2p(\ell+1)\frac{\partial}{\partial x_{\ell-2}}\right)^n \right.$$
$$\left. + \sum_{j=0}^{n-1} c_j \left(\frac{\partial}{\partial x_{\ell-1}}\right)^{2(n-j-1)}\left(\frac{\partial}{\partial x_{\ell-2}}\right)^j \frac{\partial}{\partial x_{\ell-3}} + \sum_{j=1}^{n} d_j\left(\frac{\partial}{\partial x_{\ell-1}}\right)^{2n-2j+1}\left(\frac{\partial}{\partial x_{\ell-2}}\right)^j \right)^k \psi(x) = 0. \tag{29}$$

We have obtained highly nontrivial differential equations. To have a close look at the equations, we give examples of the hierarchies of equations for $n = 1, 2$. For $n = 1$ the equations (27)-(29) read as follows:

$$\left(p\frac{\partial}{\partial x_{\ell-2}} - \frac{\ell+2}{2(\ell+1)}\frac{\partial^2}{\partial x_{\ell-1}^2} \right)^k \psi(x) = 0, \tag{30}$$

$$\left(p^2\frac{\partial}{\partial x_{\ell-3}} - \frac{\ell+3}{\ell+1}p\frac{\partial^2}{\partial x_{\ell-2}\partial x_{\ell-1}} + \frac{(\ell+2)(\ell+3)}{3(\ell+1)^2}\frac{\partial^3}{\partial x_{\ell-1}^3} \right)^k \psi(x) = 0, \tag{31}$$

$$\left((\ell+2)\frac{\partial^3}{\partial x_{\ell-1}^3} - 3(\ell+1)p\frac{\partial^2}{\partial x_{\ell-1}\partial x_{\ell-2}} + \frac{3(\ell+1)^2}{\ell+3}p^2\frac{\partial}{\partial x_{\ell-3}} \right)^k \psi(x) = 0. \tag{32}$$

The corresponding equations for $n = 2$ are given by

$$\left(p\frac{\partial}{\partial x_{\ell-4}} - \frac{\ell+4}{\ell+1}\frac{\partial^2}{\partial x_{\ell-3}\partial x_{\ell-1}} + \frac{(\ell+3)(\ell+4)}{2(\ell+1)(\ell+2)}\frac{\partial^2}{\partial x_{\ell-2}^2} \right)^k \psi(x) = 0, \tag{33}$$

$$\left(p^2\frac{\partial}{\partial x_{\ell-5}} - \frac{\ell+5}{\ell+1}p\frac{\partial^2}{\partial x_{\ell-4}\partial x_{\ell-1}} + \frac{(\ell+4)(\ell+5)}{5(\ell+1)(\ell+2)}p\frac{\partial^2}{\partial x_{\ell-3}\partial x_{\ell-2}} \right.$$
$$\left. + \frac{2(\ell+4)(\ell+5)}{5(\ell+1)^2}\frac{\partial^3}{\partial x_{\ell-3}\partial x_{\ell-1}^2} - \frac{(\ell+3)(\ell+4)(\ell+5)}{5(\ell+1)^2(\ell+2)}\frac{\partial^3}{\partial x_{\ell-2}\partial x_{\ell-1}} \right)^k \psi(x) = 0, \tag{34}$$

$$\left((\ell+2)^2\frac{\partial^5}{\partial x_{\ell-1}^5} - 5(\ell+1)(\ell+2)p\frac{\partial^4}{\partial x_{\ell-2}\partial x_{\ell-1}^3} + \frac{3(\ell+1)^2(\ell+2)}{\ell+3}p^2\frac{\partial^3}{\partial x_{\ell-1}^2\partial x_{\ell-3}} \right.$$
$$\left. + 6(\ell+1)^2p^2\frac{\partial^3}{\partial x_{\ell-1}\partial x_{\ell-2}^2} - \frac{6(\ell+1)^3}{\ell+3}p^3\frac{\partial^2}{\partial x_{\ell-2}\partial x_{\ell-3}} \right)^k \psi(x) = 0. \tag{35}$$

5. Irreducible lowest weight modules over $\mathfrak{g}_\ell$

We have shown that the Verma modules over $\mathfrak{g}_\ell$ are reducible in many cases (Proposition 2). It is known that the Verma module is, in a sense, the largest lowest weight module. That is, one can derive all irreducible lowest weight modules starting from the Verma module $V_\ell^{\delta,p}$. Here we give a classification of all irreducible lowest weight modules over $\mathfrak{g}_\ell$:

Theorem 1 *All irreducible lowest weight modules over $\mathfrak{g}_\ell$ are listed as follows:*

(i) the Verma module $V_1^{\delta,p}$ for $\ell = 1$.

(ii) the quotient module $V_\ell^{(\ell)}$ for $\ell \geq 2$. This module is infinite dimensional with the basis vectors $H^k P_{\ell-1}^m \left|u_0^{(\ell)}\right\rangle$ where k, m are nonnegative integers. See Lemma 2 for the definition of the lowest weight vector $\left|u_0^{(\ell)}\right\rangle$.

Theorem 1 coincides with the results in [12] and the statement is stronger than our previous result [16] where the complete classification is not given. Theorem 1 (i) is already mentioned in Proposition 2. So we give an outline of the proof of (ii) below. The proof of Lemma 1, 2 is found in [16].

We consider the quotient module $V_\ell^{\delta,p}/\mathcal{I}^{(2)}$ where $\mathcal{I}^{(2)}$ is the largest $\mathfrak{g}_\ell$-submodule of $V_\ell^{\delta,p}$. Since there is no singular vectors in the $N = 1$ subspace $(V_\ell^{\delta,p})_1$, $\mathcal{I}^{(2)}$ will be induced by the singular vector in the $N = 2$ subspace.

Lemma 1 *There exists precisely one (up to overall constant) singular vector in the $N = 2$ subspace $(V_\ell^{\delta,p})_2$. This singular vector is given by*

$$\left|v_s^{(2)}\right\rangle = (2p(\ell+1)P_{\ell-2} - (\ell+2)P_{\ell-1}^2)\,|\delta, p\rangle . \tag{36}$$

Define $\mathcal{I}^{(2)} = U(\mathfrak{g}_\ell^+)\left|v_s^{(2)}\right\rangle$, then $\mathcal{I}^{(2)}$ is the largest $\mathfrak{g}_\ell$-submodule in $V_\ell^{\delta,p}$. Let $\left|u_0^{(2)}\right\rangle$ be the lowest weight vector in $V_\ell^{\delta,p}/\mathcal{I}^{(2)}$. Then

$$\begin{aligned} &D\left|u_0^{(2)}\right\rangle = \delta\left|u_0^{(2)}\right\rangle, \quad P_\ell\left|u_0^{(2)}\right\rangle = p\left|u_0^{(2)}\right\rangle, \\ &X\left|u_0^{(2)}\right\rangle = 0,\ X \in \mathfrak{g}_\ell^-, \quad P_{\ell-2}\left|u_0^{(2)}\right\rangle = \frac{\ell+2}{2p\,(\ell+1)}P_{\ell-1}^2\left|u_0^{(2)}\right\rangle . \end{aligned} \tag{37}$$

It follows that the basis of $V_\ell^{\delta,p}/\mathcal{I}^{(2)}$ is given by

$$H^k\, P_{\ell-1}^{m_1}\, P_{\ell-3}^{m_3} \cdots P_0^{m_\ell}\left|u_0^{(2)}\right\rangle , \tag{38}$$

This is an eigenvector of D with the eigenvalue $\delta + k + m_1 + \sum_{i=3}^{\ell} i\,m_i$. We define the level $N^{(2)}$ in the quotient space $V_\ell^{\delta,p}/\mathcal{I}^{(2)}$ by $N^{(2)} = k + m_1 + \sum_{i=3}^{\ell} i\,m_i$.

One can show that there exists no singular vector in $N^{(2)} = 1, 2$ subspaces of $V_\ell^{\delta,p}/\mathcal{I}^{(2)}$ and precisely one singular vector exists in $N^{(2)} = 3$ subspace. Using the singular vector in $N^{(2)} = 3$ subspace one can obtain the largest $\mathfrak{g}_\ell$-submodule $\mathcal{I}^{(3)}$ in $V_\ell^{\delta,p}/\mathcal{I}^{(2)}$. Then we consider the quotient module $V_\ell^{\delta,p}/\mathcal{I}^{(2)}/\mathcal{I}^{(3)}$ and search singular vectors in the quotient module. In fact, one can repeat this process until we arrive at $V_\ell^{\delta,p}/\mathcal{I}^{(2)}/\cdots/\mathcal{I}^{(\ell)}$. This is assured by the next lemma:

Lemma 2 *Suppose that we have arrived at the quotient space* $V_\ell^{(\lambda)} := V_\ell^{\delta,p}/\mathcal{I}^{(2)}/\cdots/\mathcal{I}^{(\lambda)}$, $(2 \leq \lambda \leq \ell)$. *Namely, we have the quotient space with the basis*

$$H^k P_{\ell-1}^{m_1} P_{\ell-\lambda-1}^{m_{\lambda+1}} \cdots P_0^{m_\ell} \left|u_0^{(\lambda)}\right\rangle, \tag{39}$$

where $\left|u_0^{(\lambda)}\right\rangle$ *is the lowest weight vector in* $V_\ell^{(\lambda)}$, *defined by*

$$\begin{aligned} & D\left|u_0^{(\lambda)}\right\rangle = \delta\left|u_0^{(\lambda)}\right\rangle, \quad P_\ell\left|u_0^{(\lambda)}\right\rangle = p\left|u_0^{(\lambda)}\right\rangle, \quad X\left|u_0^{(\lambda)}\right\rangle = 0, \ X \in \mathfrak{g}_\ell^-, \\ & P_{\ell-a}\left|u_0^{(\lambda)}\right\rangle = \frac{p\,(\ell+a)!}{a!\,\ell!\,(p(\ell+1))^a} P_{\ell-1}^a \left|u_0^{(\lambda)}\right\rangle, \quad (a = 2, 3, \ldots, \lambda). \end{aligned} \tag{40}$$

Then

(i) $V_\ell^{(\lambda)}$ *is the graded vector space:*

$$\begin{aligned} & V_\ell^{(\lambda)} = \bigoplus_{N^{(\lambda)}=0}^{\infty} (V_\ell^{(\lambda)})_{N^{(\lambda)}}, \quad (V_\ell^{(\lambda)})_{N^{(\lambda)}} = \{\ |v\rangle \in V_\ell^{(\lambda)} \mid D\,|v\rangle = (\delta + N^{(\lambda)})\,|v\rangle\ \}, \\ & N^{(\lambda)} = k + m_1 + \sum_{i=\lambda+1}^{\ell} i m_i. \end{aligned} \tag{41}$$

(ii) The subspace $(V_\ell^{(\lambda)})_{N^{(\lambda)}}$ *has a nonvanishing Kac determinant if* $1 \leq N^{(\lambda)} \leq \lambda$. *The Kac determinant is given by (up to a sign factor)*

$$\Delta_{N^{(\lambda)}} = (p\,(\ell+1))^{N^{(\lambda)}(N^{(\lambda)}+1)} \prod_{k=0}^{N^{(\lambda)}} k!\,(N^{(\lambda)} - k)!. \tag{42}$$

This implies that there exists no singular vectors in the level $N^{(\lambda)}$ *subspaces if* $1 \leq N^{(\lambda)} \leq \lambda$.

(iii) If $2 \leq \lambda \leq \ell - 1$, *then there exists precisely one (up to overall constant) singular vector in the* $N^{(\lambda)} = \lambda + 1$ *subspace of* $V_\ell^{(\lambda)}$. *The singular vector is given by*

$$\left|v_s^{(\lambda+1)}\right\rangle = (\,(\lambda+1)!\,p^\lambda(\ell+1)^\lambda P_{\ell-\lambda-1} - (\ell+2)(\ell+3)\cdots(\ell+\lambda+1)P_{\ell-1}^{\lambda+1}\,)\left|u_0^{(\lambda)}\right\rangle. \tag{43}$$

Now we consider the module $V_\ell^{(\ell)} = V_\ell^{\delta,p}/\mathcal{I}^{(2)}/\cdots/\mathcal{I}^{(\ell)}$. The basis of this space is $H^k P_{\ell-1}^m \left|u_0^{(\ell)}\right\rangle$ where $\left|u_0^{(\ell)}\right\rangle$ is the lowest weight vector defined by (40) with $\lambda = \ell$. It is not difficult to verify that there exits no singular vector in $V_\ell^{(\ell)}$ so that $V_\ell^{(\ell)}$ is irreducible.

One may wonder what happens if we take a quotient by $\mathfrak{g}_\ell$-submodule which is not largest. It can be shown that this does not produce any new irreducible modules. We explain this by taking the Verma module $V_\ell^{\delta,p}$ as an example. First let us note the following fact:

Lemma 3 *All singular vectors in* $V_\ell^{\delta,p}$ *are independent of* H.

It is an easy task to prove this lemma. We assume that the singular vector is kth order in H, then demand that the vector is annihilated by $P_{\ell+1}$. It results that all the terms which is kth order in H must disappear from the singular vector. This reduces the order in H by one. Repeating the same argument we prove the lemma.

Next let $|w^N\rangle$ be a singular vector in level N $(2 < N \leq \ell)$ subspace of $V_\ell^{\delta,p}$.

Lemma 4 *In the quotient space* $V_\ell^{(N-1)}$

(i) $|w^N\rangle$ *is reduced to the zero vector if* $|w^N\rangle$ *is independent of* $P_{\ell-N}$

(ii) $|w^N\rangle$ *is reduced to singular vector given in (43) if* $|w^N\rangle$ *depends on* $P_{\ell-N}$

This lemma is proved as follows. From Lemma 3, $|w^N\rangle$ is a linear combination of $P_{\ell-1}^{m_1} P_{\ell-2}^{m_2} \cdots P_{\ell-N}^{m_N} |\delta, p\rangle$ so that $|w^N\rangle$ is annihilated by any $P_n \in \mathfrak{g}_\ell^-$. The condition $C|w^N\rangle = 0$ requires some relations among the coefficients of the linear combination. In the quotient space $V_\ell^{(N-1)}$ we have additional relations (40). Substituting (40) into $|w^N\rangle$ and using the relation from $C|w^N\rangle = 0$ one may verify that the lemma is true.

Now consider the quotient module $V_\ell^{\delta,p}/\mathcal{W}$ where $\mathcal{W} = U(\mathfrak{g}_\ell^+)|w^N\rangle$. The singular vectors in the level M $(2 \le M \le N-1)$ subspaces of $V_\ell^{\delta,p}$ are also singular vectors in $V_\ell^{\delta,p}/\mathcal{W}$. Thus one can consider the quotient module $V_\ell^{\delta,p}/\mathcal{W}/\mathcal{I}^{(2)}/\cdots/\mathcal{I}^{(N-1)}$. In the case of Lemma 4 (i) the contribution from $\mathcal{W}$ disappears, while in the case of Lemma 4 (ii) taking the quotient by $\mathcal{W}$ has the same effect as taking the quotient by $\mathcal{I}^{(N)}$. Therefore we arrive at the result.

Lemma 5

(i) $V_\ell^{\delta,p}/\mathcal{W}/\mathcal{I}^{(2)}/\cdots/\mathcal{I}^{(N-1)} \simeq V_\ell^{(N-1)}$ *if* $|w^N\rangle$ *is independent of* $P_{\ell-N}$

(ii) $V_\ell^{\delta,p}/\mathcal{W}/\mathcal{I}^{(2)}/\cdots/\mathcal{I}^{(N-1)} \simeq V_\ell^{(N)}$ *if* $|w^N\rangle$ *depends on* $P_{\ell-N}$

Lemma 5 means that taking the quotient by $\mathcal{W}$ does not produce any new irreducible module.

Finally, let $|z^N\rangle$ be be a singular vector in level N $(> \ell)$ subspace of $V_\ell^{\delta,p}$. Then by the arguments similar to Lemma 4, 5 one may come to the next lemma:

Lemma 6 *In the quotient space* $V_\ell^{(\ell)}$*, any singular vector* $|z^N\rangle$ *is reduced to the zero vector. Hence* $V_\ell^{\delta,p}/\mathcal{Z}/\mathcal{I}^{(2)}/\cdots/\mathcal{I}^{(\ell)} \simeq V_\ell^{(\ell)}$ *where* $\mathcal{Z} = U(\mathfrak{g}_\ell^+)|z^N\rangle$.

Thus we have proved that taking a quotient of $V_\ell^{\delta,p}$ by the $\mathfrak{g}_\ell$-submodule which is not the largest does not produce any new irreducible modules. Though we have discussed the case of the Verma module $V_\ell^{\delta,p}$, the same arguments hold for the quotient module $V_\ell^{(\lambda)}$ and, the quotient by the $\mathfrak{g}_\ell$-submodule does not produce any new irreducible module if the submodule is not the largest.

Acknowledgments
The work of N.A. is supported by the grants-in-aid from JSPS (Contract No.26400209). R. C. was financially supported through the MOST grants 102-2811-M-005-025 and 102-2628-M-005-001-MY4 in Taiwan.

References

[1] Negro J, del Olmo M A and Rodríguez-Marco A 1997 *J. Math. Phys.* **38** 3786
[2] Galajinsky A and Masterov I 2011 *Phys. Lett.* **B702** 265
[3] Galajinsky A and Masterov I 2013 *Nucl. Phys.* **B866** 212
[4] Galajinsky A and Masterov I 2013 *Phys. Lett.* **B723** 190
[5] Andrzejewski K, Gonera J and Maslanka P 2012 *Phys. Rev.* **D86** 065009
[6] Andrzejewski K, Gonera J, Kosinski P and Maslanka P 2013 *Nucl. Phys.* **B876** 309
[7] Andrzejewski K and Gonera J 2013 *Phys. Lett.* **B721** 319–322
[8] Andrzejewski K and Gonera J 2013 *Phys. Rev.* **D88** 065011
[9] Andrzejewski K, Gonera J and Kijanka-Dec A 2013 *Phys. Rev.* **D87** 065012
[10] Aizawa, N, Isaac P S and Kimura Y 2012 *Int. J. Math.* **23** 1250118
[11] Aizawa N, Kimura Y and Segar J 2013 *J. Phys. A: Math. Theor.* **46** 405204
[12] Lü R, Mazorchuk V and Zhao K 2014 *J. Pure Appl. Algebra* **218** 1885
[13] Martelli D and Tachikawa Y 2010 *J. High Energy Phys.* JHEP05(2010)091
[14] Dixmier J 1977 *Enveloping algebras* (Amsterdam: North Holland)

[15] Kac V and Raina A K 1987 *Bombay Lectures on Highest Weight Representations of Infinite Dimensional Lie Algebras* (Singapore: World Scientific)

[16] Aizawa N, Chandrashekar R and Segar J 2014 Lowest weight representations, singular vectors and invariant equations for a class of conformal Galilei algebras *Preprint* arXiv:1408.4842 [math-ph]

[17] Dobrev V K 1988 *Rep. Math. Phys.* **25** 159

Quaternionic Hardy spaces in the open unit ball and half space and Blaschke products

Daniel Alpay[1], **Fabrizio Colombo**[2] **and Irene Sabadini**[2]

[1] Department of Mathematics, Ben-Gurion University of the Negev, Beer-Sheva 84105 Israel
[2] Politecnico di Milano, Dipartimento di Matematica, Via E. Bonardi, 9 20133 Milano, Italy

E-mail: dany@math.bgu.ac.il, fabrizio.colombo@polimi.it, irene.sabadini@polimi.it

Abstract. The Hardy spaces $H_2(\mathbb{B})$ and $H_2(\mathbb{H}_+)$, where $\mathbb{B}$ and $\mathbb{H}_+$ denote, respectively, the open unit ball of the quaternions and the half space of quaternions with positive real part, as well as Blaschke products, have been intensively studied in a series of papers where they are used as a tool to prove other results in Schur analysis. This paper gives an overview on the topic, collecting the various results available.

1. Introduction

In this paper we give an overview of the results available on the Hardy spaces $H_2(\Omega)$ where Ω is either the open unit ball of the quaternions or the half space of quaternions with positive real part. These spaces have been studied in a series of papers, see [1, 2, 3, 4, 5, 6] as a tool to prove other results in Schur analysis and the purpose of this survey is to collect them in one paper. For the proofs of the various results, we refer the reader to the original sources.

We will work in the framework of slice hyperholomorphic functions. Several functions spaces have been considered in this setting, e.g. the Bergman spaces, see [11, 12, 13], Besov, Bloch and Dirichlet spaces, see [9], the Fock space see [7] while other Hardy spaces $H_p(\mathbb{B})$ are in [14]. To define the class of slice hyperholomorphic functions, we need some terminology that is introduced below.

By $\mathbb{H}$ we denote the algebra of real quaternions, namely the set of elements of the form $p = x_0 + x_1 i + x_2 j + x_3 k$ where i, j, k is the standard basis of quaternions. Let $\mathbb{S}$ be the set of purely imaginary quaternions with norm 1. Any $I \in \mathbb{S}$ is such that $I^2 = -1$, so we can consider the complex plane of elements of the form $x + Iy$ where $x, y \in \mathbb{R}$. Let $\Omega \subseteq \mathbb{H}$ be an open set and let $f : \Omega \to \mathbb{H}$ be a real differentiable function. Let $I \in \mathbb{S}$ and let f_I be the restriction of f to the complex plane $\mathbb{C}_I$. We say that f is a (left) slice hyperholomorphic function in Ω if, for every $I \in \mathbb{S}$, f_I satisfies

$$\frac{1}{2}\left(\frac{\partial}{\partial x} + I\frac{\partial}{\partial y}\right) f_I(x + Iy) = 0.$$

An analogous definition can be given for right slice hyperholomorphic functions. The class of (left) slice hyperholomorphic functions includes, in particular, converging power series with quaternionic coefficients written on the right. For the basic notions and main properties of these functions we refer the reader to the books [10, 16]. The class of slice hyperholomorphic functions is not closed, in general, under the pointwise multiplication. However, slice hyperholomorphic

functions can be multiplied, at least on axially symmetric s-domains, in order to obtain a function of the same kind using the so-called $\star$-product. We recall that axially symmetric s-domains Ω are domains intersecting the real line, which remain connected when intersected with any complex plane $\mathbb{C}_I$ and such that whenever an element $x + Iy \in \Omega$ all the elements of the form $x + Jy \in \Omega$ when J varies in $\mathbb{S}$. In the special case in which Ω is a ball with center at the origin and if f, g are two slice hyperholomorphic functions they can be written as $f(p) = \sum_{k=0}^{\infty} p^k f_k$, $g(p) = \sum_{k=0}^{\infty} p^k g_k$. Their $*$-product is $(f \star g)(p) = \sum_{k=0}^{\infty} p^k (\sum_{r=0}^{k} f_r g_{k-r})$, thus it coincides with the product defined in [15].
Pointwise multiplication and slice multiplication can be related as in the following result, see e.g. [10, Proposition 4.3.22]:

Proposition 1.1 *Let $U \subseteq \mathbb{H}$ be an axially symmetric s-domain, $f, g : U \to \mathbb{H}$ be slice hyperholomorphic functions and let us assume that $f(p) \neq 0$. Then*

$$(f \star g)(p) = f(p)g(f(p)^{-1}pf(p)), \tag{1}$$

for all $p \in U$. If $f(p) = 0$ then $(f \star g)(p) = 0$.

Note that the transformation $p \to f(p)^{-1}pf(p)$ is clearly a rotation in $\mathbb{H}$, since $|p| = |f(p)^{-1}pf(p)|$.
Note also that if $(f \star g)(p) = 0$ then either $f(p) = 0$ or $g(f(p)^{-1}pf(p)) = 0$.
This latter fact is very well known for polynomials with quaternionic coefficients, see [17], for which it is also well known that they might have spheres of zeros. Given a nonreal quaternion α, it can be written as $\alpha = \alpha_0 + \alpha_1 i + \alpha_2 j + \alpha_3 k = a + Ib$ where $a = \alpha_0$, $I = (\alpha_1 i + \alpha_2 j + \alpha_3 k)/|\alpha_1 i + \alpha_2 j + \alpha_3 k| \in \mathbb{S}$, $b = |\alpha_1 i + \alpha_2 j + \alpha_3 k|$. The sphere defined by α is the set $[\alpha] = \{a + Jb \,|\, J \in \mathbb{S}\}$; it is the set of elements satisfying

$$p^2 - 2ap + (a^2 + b^2) = p^2 - 2\mathrm{Re}(\alpha)p + |\alpha|^2 = (p - \alpha) \star (p - \bar{\alpha}) = 0.$$

This discussion suggests that the definition of multiplicity of a zero has to be given in appropriate way:

Definition 1.2 *We say that the* multiplicity of the spherical zero $[\alpha]$ *of a function $Q(p)$ is m if m is the maximum of the integers r such that $(p^2 + 2\mathrm{Re}(\alpha)p + |\alpha|^2)^r$ divides $Q(p)$.*
Let $\alpha_j \in \mathbb{H} \setminus \mathbb{R}$ and let

$$Q(p) = (p - \alpha_1) \star \ldots \star (p - \alpha_n) \star g(p) \quad \alpha_{j+1} \neq \bar{\alpha}_j, \quad j = 1, \ldots, n-1, \quad g(p) \neq 0. \tag{2}$$

We say that α_1 is a zero of Q of multiplicity 1 *if $\alpha_j \notin [\alpha_1]$ for $j = 2, \ldots, n$.*
We say that α_1 is a zero of Q of multiplicity $n \geq 2$ *if $\alpha_j \in [\alpha_1]$ for all $j = 2, \ldots, n$.*
If $\alpha_j \in \mathbb{R}$ we can repeat the notion of multiplicity of α_1 where (2) holds by removing the assumption $\alpha_{j+1} \neq \bar{\alpha}_j$.

The definition of multiplicity, in the case of a real zero, coincides with the standard notion of multiplicity since, in this case, the $\star$-product reduces to the pointwise product. Note that if a function has a sphere of zeros at $[\alpha]$ with multiplicity n, at most one point on $[\alpha]$ can have higher multiplicity.
As a consequence of Proposition 1.1 one has (see Corollary 3.3 in [4]):

Corollary 1.3 *If $\lim_{r\to 1} |f(re^{I\theta})| = 1$, for all I fixed in $\mathbb{S}$, then*

$$\lim_{r\to 1} |(f \star g)(re^{I\theta})| = |g(e^{I'\theta})|,$$

a.e. for $\theta \in [0, 2\pi)$ (and $I' \in \mathbb{S}$ depends on θ, I and f).

2. The Hardy space of the unit ball

The quaternionic Hardy space $H_2(\mathbb{B})$ of the unit ball $\mathbb{B}$ is defined as the space of square summable (left) slice regular power series, see [3]:

$$H_2(\mathbb{B}) = \left\{ f(p) = \sum_{k=0}^{\infty} p^k f_k : \|f\|^2_{H_2(\mathbb{B})} := \sum_{k=0}^{\infty} |f_k|^2 < \infty \right\}.$$

The space $H_2(\mathbb{B})$ is a right quaternionic Hilbert space if equipped with the inner product

$$\langle f, g \rangle = \sum_{k=0}^{\infty} \bar{g}_k f_k \quad \text{if} \quad f(p) = \sum_{k=0}^{\infty} p^k f_k, \ g(p) = \sum_{k=0}^{\infty} p^k g_k. \tag{3}$$

If f is as in (3) then for a fixed $I \in \mathbb{S}$ we have

$$\begin{aligned} \int_0^{2\pi} |f(re^{I\theta})|^2 d\theta &= \int_0^{2\pi} \left(\sum_{j,k=0}^{\infty} r^{k+j} \overline{f}_k e^{I(j-k)\theta} f_j \right) d\theta \\ &= \sum_{j,k=0}^{\infty} r^{k+j} \overline{f}_k \left(\int_0^{2\pi} e^{I(j-k)\theta} d\theta \right) f_j = 2\pi \cdot \sum_{n=0}^{\infty} r^{2n} |f_n|^2. \end{aligned}$$

This last formula implies that the norm in $H_2(\mathbb{B})$ can be equivalently defined as, see [4]:

$$\|f\|^2_{H_2(\mathbb{B})} = \sup_{0 \le r < 1} \frac{1}{2\pi} \int_0^{2\pi} |f(re^{I\theta})|^2 d\theta \tag{4}$$

where the value of the integral on the right is the same for each $I \in \mathbb{S}$. Observe that the supremum in (4) can be replaced by the limit as r tends to one. The integrand in (4) depends on $I \in \mathbb{S}$, so one could have put in the definition of the norm also the supremum with respect to $I \in \mathbb{S}$, however since the integral does not depend on the choice of I the supremum with respect to $I \in \mathbb{S}$ is not needed.

Remark 2.1 The space $H_2(\mathbb{B})$ can be alternatively characterized as the reproducing kernel Hilbert space with reproducing kernel

$$k_{H_2(\mathbb{B})}(p, q) = \sum_{n=0}^{\infty} p^n \overline{q}^n. \tag{5}$$

The function $k_{H_2(\mathbb{B})}(\cdot, q)$ belongs to $H_2(\mathbb{B})$ for every $q \in \mathbb{B}$ and the fact that $k_{H_2(\mathbb{B})}(p, q)$ is a reproducing kernel means for any function $f \in H_2(\mathbb{B})$ as in (3),

$$\langle f, k_{H_2(\mathbb{B})}(\cdot, q) \rangle_{H_2(\mathbb{B})} = \sum_{k=0}^{\infty} q^k f_k = f(q). \tag{6}$$

An important feature of slice hyperholomorphic functions which is important in the sequel is the following formula:

Theorem 2.2 *(Representation Formula) Let $\Omega \subseteq \mathbb{H}$ be an axially symmetric s-domain and let $f : \Omega \to \mathbb{H}$ be a slice regular function. The following equality holds for all $q = x + Iy \in \Omega$, $J, K \in \mathbb{S}$:*

$$f(x + Iy) = (J - K)^{-1}[Jf(x + Jy) - Kf(x + Ky)] + I(J - K)^{-1}[f(x + Jy) - f(x + Ky)]. \tag{7}$$

This formula implies that if we have two points on a same sphere, the value of the function at any other point on the sphere is uniquely determined by the values in the first two points. Moreover we have:

Proposition 2.3 *The finite collection of functions* $\{k_{H_2(\mathbb{B})}(\cdot, q_i)\}$ *based on distinct points* $q_1, \dots, q_k \in \mathbb{B}$ *is (right) linearly independent in* $H_2(\mathbb{B})$ *if and only if none three of these points belong to the same 2-sphere.*

The linear dependence of three functions $k_{H_2(\mathbb{B})}(\cdot, p_i)$ based on equivalent points explains in an alternative way that the restriction of any function $f \in H_2(\mathbb{B})$ to any 2-sphere is completely determined by the values of f at any two points of this sphere.
The following result, see [1, Theorem 3.3], [3, Theorem 6.2] shows that Schur functions are contractions on $H_2(\mathbb{B})$:

Theorem 2.4 *Let* $S : \mathbb{B} \to \mathbb{H}$. *The following are equivalent:*

(i) S *is slice regular on* $\mathbb{B}$ *and* $|S(p)| \leq 1$ *for all* $p \in \mathbb{B}$.

(ii) The operator M_S *of left* $\star$*–multiplication by* S

$$M_S : f \mapsto S \star f \tag{8}$$

is a contraction on $H_2(\mathbb{B})$, *that is,* $\|S \star f\|_{H_2(\mathbb{B})} \leq \|f\|_{H_2(\mathbb{B})}$ *for all* $f \in H_2(\mathbb{B})$.

The matrix-valued version of this result is in [4, Theorem 4.6]. In analogy to the complex case we introduce the Hardy space $H_\infty(\mathbb{B})$ of bounded slice regular functions on $\mathbb{B}$ with norm $\|S\|_\infty = \sup_{p\in\mathbb{B}} |S(p)| < \infty$ and the space $\mathcal{M}(H_2(\mathbb{B}))$ of bounded multipliers, that is, the functions $S : \mathbb{B} \to \mathbb{H}$ such that the operator M_S of left $\star$–multiplication (8) is bounded on $H_2(\mathbb{B})$. By its definition, the set of slice hyperholomorphic functions on $\mathbb{B}$ which are bounded by 1 and denoted by $\mathcal{R}(\mathbb{B}, \overline{\mathbb{B}})$ is the closed unit ball of $H_\infty(\mathbb{B})$, see [1, Corollary 3.5]. Moreover, we have the following consequence of Theorem 2.4.

Corollary 2.5 $H_\infty(\mathbb{B}) = \mathcal{M}(H_2(\mathbb{B}))$ *and* $\|S\|_\infty = \|M_S\|$ *for every* $S \in H_\infty(\mathbb{B})$.

2.1. Blaschke products

We now recall the notion of Blaschke factors and products, see [5, 4, 6]. According to the fact that the zeros of a slice hyperholomorphic functions can also be spheres, we will introduce two different types of Blaschke factors.

Definition 2.6 *Let* $a \in \mathbb{H}$, $|a| < 1$. *The function*

$$B_a(p) = (1 - p\bar{a})^{-\star} \star (a - p)\frac{\bar{a}}{|a|} \tag{9}$$

is called Blaschke factor at a.

Remark 2.7 Let $\lambda(p) = 1 - p\bar{a}$ and let us apply formula (1) to the products $\lambda^c(p) \star \lambda(p)$ and $\lambda^c(p) \star (a - p)$. We have

$$\begin{aligned} B_a(p) &= (\lambda^c(p) \star \lambda(p))^{-1}\lambda^c(p) \star (a - p)\frac{\bar{a}}{|a|} = (\lambda^c(p)\lambda(\tilde{p}))^{-1}\lambda^c(p)(a - \tilde{p})\frac{\bar{a}}{|a|} \\ &= \lambda(\tilde{p})^{-1}(a - \tilde{p})\frac{\bar{a}}{|a|} = (1 - \tilde{p}\bar{a})^{-1}(a - \tilde{p})\frac{\bar{a}}{|a|}, \end{aligned} \tag{10}$$

where $\tilde{p} = \lambda^c(p)^{-1}p\lambda^c(p)$. Formula (10) gives the Blaschke factor $B_a(p)$ in terms of the pointwise multiplication.

The following property extends the analogous property in the complex case:

Theorem 2.8 *Let $a \in \mathbb{H}$, $|a| < 1$. The Blaschke factor $B_a(q)$ has the following properties:*

(i) it takes the unit ball $\mathbb{B}$ to itself;

(ii) it takes the boundary of the unit ball to itself;

(iii) it has a unique zero for $p = a$.

In [4, Theorem 5.6] we proved the following result:

Theorem 2.9 *Let $\{a_j\} \subset \mathbb{B}$, $j = 1, 2, \dots$ be a sequence of nonzero quaternions such that $[a_i] \neq [a_j]$ if $i \neq j$ and assume that $\sum_{j\geq 1}(1 - |a_j|) < \infty$. Then the function*

$$B(p) := \Pi^{\star}_{j\geq 1}(1 - p\bar{a}_j)^{-\star} \star (a_j - p)\frac{\bar{a}_j}{|a_j|}, \tag{11}$$

where $\Pi^{\star}$ denotes the $\star$-product, converges uniformly on the compact subsets of $\mathbb{B}$.

The Blaschke factor vanishing at the sphere $[a]$ is given in the following definition:

Definition 2.10 *Let $a \in \mathbb{H}$, $|a| < 1$. The function*

$$B_{[a]}(p) = (1 - 2\text{Re}(a)p + p^2|a|^2)^{-1}(|a|^2 - 2\text{Re}(a)p + p^2) \tag{12}$$

is called Blaschke factor at the sphere $[a]$.

The definition of $B_{[a]}(p)$ does not depend on the choice of the point a that identifies the 2-sphere. Indeed all the elements in the sphere $[a]$ have the same real part and module. We also have, see [6]:

Proposition 2.11 *The $\star$-inverse of B_a and $B_{[a]}$ are $B_{\bar{a}^{-1}}$, $B_{[a^{-1}]}$ respectively.*

The following result is the analogue of Theorem 2.9:

Proposition 2.12 *A Blaschke product having zeros at the set of spheres*

$$Z = \{([c_1], \nu_1), ([c_2], \nu_2), \dots\}$$

where $c_j \in \mathbb{B}$, the sphere $[c_j]$ is a zero of multiplicity ν_j, $j = 1, 2, \dots$ and $\sum_{j\geq 1} \nu_j(1 - |c_j|) < \infty$ is given by

$$\prod_{j\geq 1}(B_{[c_j]}(p))^{\nu_j}.$$

By Theorem 2.9 and 2.12 we can prove the following general result, see [4, 6]:

Theorem 2.13 *A Blaschke product having zeroes at the set*

$$Z = \{(a_1, n_1), \dots, ([c_1], m_1), \dots\}$$

where $a_j \in \mathbb{B}$, a_j have respective multiplicities $n_j \geq 1$, $a_j \neq 0$ for $j = 1, 2, \dots$, $[a_i] \neq [a_j]$ if $i \neq j$, $c_i \in \mathbb{B}$, the spheres $[c_j]$ have respective multiplicities $m_j \geq 1$, $j = 1, 2, \dots$, $[c_i] \neq [c_j]$ if $i \neq j$ and

$$\sum_{i,j\geq 1}\Big(n_i(1 - |a_i|) + 2m_j(1 - |c_j|)\Big) < \infty \tag{13}$$

is of the form

$$\prod_{i\geq 1}(B_{[c_i]}(p))^{m_i} \prod_{i\geq 1}^{\star} \prod_{j=1}^{\star n_i} (B_{\alpha_{ij}}(p)),$$

where $n_j \geq 1$, $\alpha_{11} = a_1$ and α_{ij} are suitable elements in $[a_i]$, and if $\alpha_{ij} \in \mathbb{H} \setminus \mathbb{R}$ $\alpha_{i\,j+1} \neq \overline{\alpha_{ij}}$, for $j = 2, 3, \dots$.

In the case in which one has to construct a Blaschke product having a zero at a_i with multiplicity n_i by prescribing the factors $(p - a_{i1}) \star \cdots \star (p - a_{in_i})$, $a_{ij} \in [a_i]$ for all $j = 1, \ldots, n_i$, the factors in the Blaschke product must be chosen accordingly, see [4, 6].

Definition 2.14 *A Blaschke product of the form*

$$B(p) = \prod_{i=1}^{r} (B_{[c_i]}(p))^{m_i} \prod_{i=1}^{\star s} \prod_{j=1}^{\star n_i} (B_{a_{ij}}(p)), \tag{14}$$

is said to have degree $d = \sum_{i=1}^{r} 2m_i + \sum_{j=1}^{s} n_j$.

Let us denote by $\mathcal{H}(B)$ the quaternionic Hilbert space with reproducing kernel K_B. We have, see [6]:

Theorem 2.15 *Let $B(p)$ be a Blaschke product as in (14). Then* $\dim(\mathcal{H}(B)) = \deg B$.

Theorem 2.16 *Let B_a be a Blaschke factor. The operator*

$$M_a \,:\, f \mapsto B_a \star f$$

is an isometry from $\mathbf{H}_2(\mathbb{B})$ into itself.

The following problem is the simplest Beurling-Lax type problem in the present setting and we show below how to solve it:

Problem 2.17 *Given N points $a_1, \ldots, a_N \in \mathbb{B}$, and M spheres $[c_1], \ldots, [c_M]$ in $\mathbb{B}$ such that the spheres $[a_1], \ldots [a_N], [c_1], \ldots, [c_M]$ are pairwise non-intersecting, find all $f \in \mathbf{H}_2(\mathbb{B})$ such that*

$$f(a_i) = 0, \quad i = 1, \ldots, N, \tag{15}$$

and

$$f([c_j]) = 0, \quad j = 1, \ldots, M. \tag{16}$$

Theorem 2.18 *There is a Blaschke product B such that the solutions of Problem 2.17 are the functions $f = B \star g$, when g runs through $\mathbf{H}_2(\mathbb{B})$.*

3. The Hardy space of the half-space

Let us consider the half-space $\mathbb{H}_+$ of the quaternions q such that $\mathrm{Re}(q) > 0$ and set $\Pi_{+,I} = \mathbb{H}_+ \cap \mathbb{C}_I$. We will denote by f_I the restriction of a function f defined on $\mathbb{H}_+$ to $\Pi_{+,I}$. We define

$$\mathbf{H}_2(\Pi_{+,I}) = \{f \text{ slice hyperholomorphic in } \mathbb{H}_+ \,:\, \int_{-\infty}^{+\infty} |f_I(Iy)|^2 dy < \infty\},$$

where $f(Iy)$ denotes the nontangential value of f at Iy. Note that these value exist almost everywhere, in fact any $f \in \mathbf{H}_2(\Pi_{+,I})$ when restricted to a complex plane $\mathbb{C}_I$ can be written as $f_I(x + Iy) = F(x + Iy) + G(x + Iy)J$ where J is any element in $\mathbb{S}$ orthogonal to I, and F, G are $\mathbb{C}_I$-valued holomorphic functions. Since the nontangential values of F and G exist almost everywhere at Iy, also the nontangential value of f exists at Iy a.e. on $\Pi_{+,I}$, see [2].

Remark 3.1 In alternative, we could have defined $\mathbf{H}_2(\Pi_{+,I})$ as the set of slice hyperholomorphic functions f such that $\sup_{x>0}\int_{-\infty}^{+\infty}|f_I(x+Iy)|^2dy<\infty$. However note that $f_I(x+Iy)=F(x+Iy)+G(x+Iy)J$, see the above discussion, and so $|f_I(x+Iy)|^2=|F(x+Iy)|^2+|G(x+Iy)|^2$. Thus, from the result in the complex case, we have

$$\begin{aligned}\sup_{x>0}\int_{-\infty}^{+\infty}|f_I(x+Iy)|^2dy &= \sup_{x>0}\int_{-\infty}^{+\infty}|F(x+Iy)|^2dy+\sup_{x>0}\int_{-\infty}^{+\infty}|G(x+Iy)|^2dy\\ &= \int_{-\infty}^{+\infty}|F(Iy)|^2dy+\int_{-\infty}^{+\infty}|G(Iy)|^2dy \qquad (1)\\ &= \int_{-\infty}^{+\infty}|f_I(Iy)|^2dy.\end{aligned}$$

In $\mathbf{H}_2(\Pi_{+,I})$ we define the scalar product

$$\langle f,g\rangle_{\mathbf{H}_2(\Pi_{+,I})}=\int_{-\infty}^{+\infty}\overline{g_I(Iy)}f_I(Iy)dy,$$

where $f_I(Iy)$, $g_I(Iy)$ denote the nontangential values of f,g at Iy on $\Pi_{+,I}$. This scalar product gives the norm

$$\|f\|_{\mathbf{H}_2(\Pi_{+,I})}=\left(\int_{-\infty}^{+\infty}|f_I(Iy)|^2dy\right)^{\frac{1}{2}},$$

(which is finite by our assumptions).

Proposition 3.2 *Let f be slice hyperholomorphic in $\mathbb{H}_+$ and assume that $f\in\mathbf{H}_2(\Pi_{+,I})$ for some $I\in\mathbb{S}$. Then for all $J\in\mathbb{S}$ the following inequalities hold*

$$\frac{1}{2}\|f\|_{\mathbf{H}_2(\Pi_{+,I})}\le\|f\|_{\mathbf{H}_2(\Pi_{+,J})}\le 2\|f\|_{\mathbf{H}_2(\Pi_{+,I})}.$$

An immediate consequence of this result is:

Corollary 3.3 *A function $f\in\mathbf{H}_2(\Pi_{+,I})$ for some $I\in\mathbb{S}$ if and only if $f\in\mathbf{H}_2(\Pi_{+,J})$ for all $J\in\mathbb{S}$.*

We now introduce the Hardy space of the half space $\mathbb{H}_+$, see [2, Definition 4.4]:

Definition 3.4 *We define $\mathbf{H}_2(\mathbb{H}_+)$ as the space of slice hyperholomorphic functions on $\mathbb{H}_+$ such that*

$$\sup_{I\in\mathbb{S}}\int_{-\infty}^{+\infty}|f(Iy)|^2dy<\infty. \qquad (2)$$

We have:

Proposition 3.5 *The function*

$$k(p,q)=(\bar p+\bar q)(|p|^2+2\mathrm{Re}(p)\bar q+\bar q^2)^{-1} \qquad (3)$$

is slice hyperholomorphic in p and $\bar q$ on the left and on the right, respectively in its domain of definition, i.e. for $p\notin[\bar q]$. The restriction of $\frac{1}{2\pi}k(p,q)$ to $\mathbb{C}_I\times\mathbb{C}_I$ coincides with $k_{\Pi_+}(z,w)$. Moreover we have the equality:

$$k(p,q)=(|q|^2+2\mathrm{Re}(q)p+p^2)^{-1}(p+q). \qquad (4)$$

The function $k(p,q)$ can also be constructed by taking the left $\star$-inverse with respect to the variable p or the right $\star$-inverse with respect to the variable q, that is

$$k(p,q) = (|q|^2 + 2\mathrm{Re}(q)p + p^2)^{-1}(p+q) = (p+\bar{q})^{-\star}$$

or

$$k(p,q) = (\bar{p}+\bar{q})(|p|^2 + 2\mathrm{Re}(p)\bar{q} + \bar{q}^2)^{-1} = (p-\bar{q})^{-\star_r}.$$

Proposition 3.6 *The kernel* $\frac{1}{2\pi}k(p,q)$ *is reproducing, i.e. for any* $f \in \mathbf{H}_2(\mathbb{H}_+)$

$$f(p) = \int_{-\infty}^{\infty} \frac{1}{2\pi} k(p, Iy) f(Iy) dy.$$

For some computations, it is useful to know that the kernel $k(p,q)$ satisfies (see [2, Proposition 4.7])

$$pk(p,q) + k(p,q)\bar{q} = 1.$$

We know that if $\{\phi_n(z)\}$ is an orthonormal basis for $\mathbf{H}_2(\Pi_{+,I})$, for some $I \in \mathbb{S}$, then

$$k(z,w) = \sum_{n=1}^{\infty} \phi_n(z)\overline{\phi_n(w)}, \tag{5}$$

and so the kernel $k(z,w)$ is positive definite. In [2, Proposition 4.8] we have proved the following:

Proposition 3.7 *Let* $\{\phi_n(z)\}$ *be an orthonormal basis for* $\mathbf{H}_2(\Pi_{+,I})$, *for some* $I \in \mathbb{S}$, *and let* $\{\Phi_n(q)\} = \{\mathrm{ext}(\phi_n(z))\}$ *be the sequence of the slice hyperholomorphic extensions of its elements. Then* $\{\Phi_n(q)\}$ *is an orthonormal basis for* $\mathbf{H}_2(\mathbb{H}_+)$, *and*

$$k(p,q) = \sum_{n=0}^{\infty} \Phi_n(p)\overline{\Phi_n(q)}.$$

Example 3.8 As an example of decomposition using an orthonormal basis, we consider

$$\Phi_n(p) = \sqrt{2}(p+1)^{-n-1}(p-1)^n.$$

We have

$$k(p,q) = \sum_{n=0}^{\infty} \Phi_n(p)\overline{\Phi_n(q)} = 2\sum_{n=0}^{\infty}(p+1)^{-n-1}(p-1)^n(\bar{q}-1)^n(\bar{q}+1)^{-n-1}.$$

We now introduce the Blaschke factors in the half space $\mathbb{H}_+$, see [2].

Definition 3.9 *For* $a \in \mathbb{H}_+$ *set*

$$b_a(p) = (p+\bar{a})^{-\star} \star (p-a).$$

The function $b_a(p)$ *is called Blaschke factor at* a *in the half space* $\mathbb{H}_+$.

Remark 3.10 The function $b_a(p)$ is defined outside the sphere $[-a]$ as it can be easily seen by rewriting it as

$$b_a(p) = (p^2 + 2\mathrm{Re}(a)p + |a|^2)^{-1}(p+a) \star (p-a) = (p^2 + 2\mathrm{Re}(a)p + |a|^2)^{-1}(p^2 - a^2)$$

and it has a zero for $p = a$.

We have the following result which characterizes the convergence of a Blaschke product:

Theorem 3.11 *Let* $\{a_j\} \subset \mathbb{H}_+$, $j = 1, 2, \ldots$ *be a sequence of quaternions such that* $\sum_{j\geq 1} \mathrm{Re}(a_j) < \infty$. *Then the function*

$$B(p) := \Pi^{\star}_{j\geq 1}(p + \bar{a}_j)^{-\star} \star (p - a_j), \tag{6}$$

converges uniformly on the compact subsets of $\mathbb{H}_+$.

As in the unit ball case, we have two kinds of Blaschke factors. In fact, products of the form

$$b_a(p) \star b_{\bar{a}}(p) = ((p + \bar{a})^{-\star} \star (p - a)) \star ((p + a)^{-\star} \star (p - \bar{a}))$$

can be written as

$$b_a(p) \star b_{\bar{a}}(p) = (p^2 + 2\mathrm{Re}(a)p + |a|^2)^{-1}(p^2 - 2\mathrm{Re}(a)p + |a|^2),$$

and they admit the sphere $[a]$ as set of zeros. Thus if we want to construct a Blaschke product vanishing at some assigned spheres, it is convenient to introduce the following:

Definition 3.12 *For* $a \in \mathbb{H}_+$ *set*

$$b_{[a]}(p) = (p^2 + 2\mathrm{Re}(a)p + |a|^2)^{-1}(p^2 - 2\mathrm{Re}(a)p + |a|^2).$$

The function $b_a(p)$ *is called Blaschke factor at the sphere* $[a]$ *in the half space* $\mathbb{H}_+$.

Note that the definition is well posed since it does not depend on the choice of the point a. As a consequence of Theorem 3.11 we have:

Corollary 3.13 *Let* $\{c_j\} \subset \mathbb{H}_+$, $j = 1, 2, \ldots$ *be a sequence of quaternions such that* $\sum_{j\geq 1} \mathrm{Re}(c_j) < \infty$. *Then the function*

$$B(p) := \Pi_{j\geq 1}(p^2 + 2\mathrm{Re}(c_j)p + |c_j|^2)^{-1}(p^2 - 2\mathrm{Re}(c_j)p + |c_j|^2), \tag{7}$$

converges uniformly on the compact subsets of $\mathbb{H}_+$.

Thus we can prove the following [2, Theorem 4.14]:

Theorem 3.14 *A Blaschke product having zeros at the set*

$$Z = \{(a_1, \mu_1), (a_2, \mu_2), \ldots, ([c_1], \nu_1), ([c_2], \nu_2), \ldots\}$$

where $a_j \in \mathbb{H}_+$, a_j *have respective multiplicities* $\mu_j \geq 1$, $[a_i] \neq [a_j]$ *if* $i \neq j$, $c_i \in \mathbb{H}_+$, *the spheres* $[c_j]$ *have respective multiplicities* $\nu_j \geq 1$, $j = 1, 2, \ldots$, $[c_i] \neq [c_j]$ *if* $i \neq j$ *and*

$$\sum_{i,j\geq 1} \Big(\mu_j(1 - |a_j|) + 2\nu_i(1 - |c_i|)\Big) < \infty$$

is given by

$$\prod_{i\geq 1}(b_{[c_i]}(p))^{\nu_i} \prod_{j\geq 1}^{\star} \prod_{k=1}^{\star\mu_j}(b_{a_{jk}}(p))^{\star\mu_j},$$

where $a_{11} = a_1$ *and* $a_{jk} \in [a_j]$ *are such that* $\alpha_{j+1} \neq \bar{\alpha}_j$, $j = 1, \ldots, n-1$, *if* $\alpha_j \in \mathbb{H} \setminus \mathbb{R}$, $k = 1, 2, 3, \ldots, \mu_j$.

We conclude this section by proving that the operator of multiplication by a Blaschke factor is an isometry. In the proof we are in need of the notion of conjugate of a function f. Given a slice hyperholomorphic function f consider its restriction to a complex plane $\mathbb{C}_I$ and write it, as customary, in the form $f_I(z) = F(z) + G(z)J$ where J is an element in $\mathbb{S}$ orthogonal to I and F, G are $\mathbb{C}_I$-valued holomorphic functions. Define $f^c(p) = \text{ext}(\overline{F(\bar{z})} - G(z)J)$ as the unique extension of the function $\overline{F(\bar{z})} - G(z)J$, see e.g. [10]. Note that if $f(p) = \sum_{n\geq 0} p^n a_n$ then $f^c(p) \sum_{n\geq 0} p^n \bar{a}_n$. We have the following, see [2]:

Lemma 3.15 *Let* $f \in \mathbf{H}_2(\mathbb{H}_+)$. *Then* $\|f\|_{\mathbf{H}_2(\mathbb{H}_+)} = \|f^c\|_{\mathbf{H}_2(\mathbb{H}_+)}$.

Theorem 3.16 *Let* b_a *be a Blaschke factor. The operator*

$$M_{b_a} \, : \, f \mapsto b_a \star f$$

is an isometry from $\mathbf{H}_2(\mathbb{H}_+)$ *into itself.*

More generally, a function S slice hyperholomorphic in the right-half-plane will be such that M_S is a contraction from the Hardy space of the right half-plane into itself if and only it is bounded by one in modulus there. The operator range $\sqrt{I - M_S M_S^*}$ with the lifted norm is then the associated de Branges Rovnyak space. For more information see [2], and [8].

References

[1] Alpay D, Bolotnikov V, Colombo F and Sabadini I 2014 Self-mappings of the quaternionic unit ball: multiplier properties, Schwarz-Pick inequality, and Nevanlinna–Pick interpolation problem (to appear in *Indiana Univ. Math. J.*)

[2] Alpay D, Colombo F, Lewkowicz I and Sabadini I 2014 Realizations of slice hyperholomorphic generalized contractive and positive functions (to appear in *Milan J. Math.* DOI 10.1007/s00032-014-0231-9)

[3] Alpay D, Colombo F and Sabadini I 2012 *Integral Equations and Operator Theory* **72** 253

[4] Alpay D, Colombo F and Sabadini I 2013 *J. Anal. Math.* **121** 87

[5] Alpay D, Colombo F and Sabadini I 2014 *J. Geom. Anal.* **24** 843

[6] Alpay D, Colombo F and Sabadini I 2014 in *Hypercomplex Analysis: New Perspectives and Applications* Trends in Mathematics (Basel: Birkhauser) 19

[7] Alpay D, Colombo F. Sabadini I and Salomon G 2014 in *Hypercomplex Analysis: New Perspectives and Applications* Trends in Mathematics (Basel: Birkhauser) 43

[8] de Branges L and Rovnyak J 1966 in *Perturbation theory and its applications in quantum mechanics* ed. C Wilcox (New York: Wiley) 295

[9] Castillo Villalba C M P, Colombo F, Gantner J and Gonzalez-Cervantes J O 2014 Bloch, Besov and Dirichlet spaces of slice hyperholomorphic functions (to appear in Complex Analysis and Operator Theory DOI 10.1007/s11785-014-0380-4)

[10] Colombo F, Sabadini I and Struppa D C 2011 *Noncommutative functional calculus. Theory and applications of slice hyperholomorphic functions* Progress in Mathematics Vol 289 (Basel: Birkhauser)

[11] Colombo F, Gonzalez-Cervantes J O, Luna-Elizarraras M E, Sabadini I and Shapiro M V 2013 in *Advances in Hypercomplex Analysis, Springer INdAM Series* **1** 39

[12] Colombo F, Oscar González-Cervantes J O and Sabadini I 2012 *Compl. Var. Ell. Eq.* **57** 825

[13] Colombo F, González-Cervantes J O and Sabadini I 2013 *Compl. Var. Ell. Eq.* **58** 1355

[14] de Fabritiis C, Gentili G and Sarfatti G *Quaternionic Hardy Spaces* (preprint 2014)

[15] Fliess M 1974 *J. Math. Pures Appl.* **53** 197

[16] Gentili G, Stoppato C and Struppa D C 2013 *Regular Functions of a Quaternionic Variable*, Springer Monographs in Mathematics (Berlin: Springer)

[17] Lam T Y 1991 *A First Course in Noncommutative Rings* Graduate Texts in Mathematics Vol. **123** (New York: Springer-Verlag)

Journal of Physics: Conference Series **597** (2015) 012010 doi:10.1088/1742-6596/597/1/012010

Complexity analysis of human physiological signals based on case studies

Maia Angelova, Philip Holloway and Jason Ellis
Northumbria University, Newcastle upon Tyne NE2 1XE, UK
E-mail: maia.angelova@northumbria.ac.uk

Abstract. This work focuses on methods for investigation of physiological time series based on complexity analysis. It is a part of a wider programme to determine non-invasive markers for healthy ageing. We consider two case studies investigated with actigraphy: (a) sleep and alternations with insomnia, and (b) ageing effects on mobility patterns. We illustrate, using these case studies, the application of fractal analysis to the investigation of regulation patterns and control, and change of physiological function. In the first case study, fractal analysis techniques were implemented to study the correlations present in sleep actigraphy for individuals suffering from acute insomnia in comparison with healthy controls. The aim was to investigate if complexity analysis can detect the onset of adverse health-related events. The subjects with acute insomnia displayed significantly higher levels of complexity, possibly a result of too much activity in the underlying regulatory systems. The second case study considered mobility patterns during night time and their variations with age. It showed that complexity metrics can identify change in physiological function with ageing. Both studies demonstrated that complexity analysis can be used to investigate markers of health, disease and healthy ageing.

1. Introduction
In the past few decades there has been growing interest for modelling physiological systems using complex systems approaches. Here, a complex system is defined as a system featuring: (a) a large number of interactive components; (b) aggregate activity that is non-linear; and (c) exhibits self-organisation. There is a growing interest in the development of new techniques to analyse and describe the dynamics of physiological systems, especially for distinguishing between the dynamics of healthy and impaired systems, or more importantly predicting the onset of adverse health related events. Many of these techniques revolve around complexity analysis and are based upon fractals. The object of investigation is considered to be complex if it has a high fractal dimension. The term fractal, first used by Mandelbrot [1], describes an object that displays self-similarity across multiple scales, known as multi-scale invariance.

This invariance is also known as power-law scaling and is easily observed in natural objects, living organisms and groups of organisms. Lipsitz *et al* [2] showed a loss in the fractal structures of the dendritic arbor between a young and old male indicating a possible loss in fractality associated with the ageing process. The theory of fractals and self-similarity is not restricted to geometric objects or images, it can also be applied to time-series where the dynamics of the time series exhibits self-similarity in the time-domain.

The dynamical systems approach to physiological time series allows to study the dynamics of the regulating mechanisms underlying the physiological process. A large number of metrics have been introduced (see for examples [3] and the references therein), based on signal and spectral analysis, from which conclusions can be made about the dynamics of the underlying mechanisms. In this paper, we will illustrate, using two case studies, the application of fractal analysis to the investigation of regulation and control and change of physiological function.

The aim of this paper is to illustrate the concepts of complexity analysis via applications to two sets of health data. The first case study is based on sleep and alternation with insomnia [4], and the second one is focused on the mobility patterns and how the complexity metrics reflect the change of physiological function with age [5]. Both studies are based on actigraphy data. Lichstein *et al* [6] demonstrated actigraphy to be a satisfactory objective measure of sleep for 4 out of 5 sleep parameters, which motivated the data collecting method in the second case study.

This paper is organised as follows: Section 2 introduces the data for two case studies: sleep data and mobility data obtained via actigraphy, Section 3 introduces the complexity methods and techniques arising from fractal analysis, Section 4 shows the findings and Section 5 presents the results and final conclusions.

2. Data

For both case studies the data was collected with actiwatch. The device measures the amount of motor movement made during the day and night. It was worn at all times throughout the day and night. The actiwatch unit was fully waterproof and as such, did not need to be taken off. For these case studies, each dataset consists of activity counts, summated at one minute epochs for a period of two weeks.

For the first case study [4], measurements for 21 healthy subjects aged 23 to 65 (mean-40, standard deviation-16) and 26 patients with acute insomnia aged 18 to 64 (mean-32, standard deviation-12) were recorded. For the purpose of this study, only night-time periods, between the hours of 11pm and 6am (420 minutes), were analysed and daytime measurements were excluded.

For the second case study [5], the night-time periods of the actigraphy were considered. All recordings were grouped and analysed according to the age groups 20-30, 30-40, 60-70 and 80-90.

3. Methods

The dynamics of many physiological systems have been shown to contain fractal structures and these structures have also been known to deteriorate with both age and health problems [2, 7, 8]. In a frequency domain, the power-law relation of a time-series can be written as

$$S(f) \propto f^{-\beta}, \tag{1}$$

where f is the frequency, $S(f)$ is the spectral power and β is the scaling parameter (generally $0 \leq \beta \leq 2$) [9]. Long-range correlations exist in a time series if β is around 1, such correlations are a sign of self-similarity across time.

Complexity has been shown to exist in many physiological systems, most notably in heart-rate variability (HRV) [10, 11, 12] and has been shown to decline in patients with cardiac diseases [13]. Further, Lipsitz and Goldberger [2] proposed that there is a reduction in the complexity of a physiological or behavioural system with the onset of age or disease. High complexity is often found in systems in which fluctuations are paramount to healthy behaviour, for example heart rate variability. Goldberger suggested that a high level of complexity could imply variance with a continuous adaptability to the subject's environment [14].

3.1. Detrended Fluctuation Analysis

DFA was first introduced by Peng *et al* [15] to analyse long-range correlations in DNA sequences. In the literature, such correlations are also known as $1/f$-scaling. This type of analysis has been used to show that the complexity of heart rate dynamics increased after 8 weeks of aerobic training [16] and has been applied to HRV to quantify sleep [17].

DFA is a method of determining, statistically, the self-affinity of a signal. The exponent gained quantifies the correlation properties of the signal in order to identify complex behaviour. The algorithm works by first integrating a time series, then splitting the signal into equally sized boxes. A least squares line is then fit to the data in each box in order to detrend the series. The root-mean-squared deviation is calculated to show the typical fluctuations of the series, $F(n)$,

$$F(n) = \sqrt{\frac{1}{N}\sum_{k=1}^{N}[y(k) - y_n(k)]^2}, \tag{2}$$

where N denotes the length of the data, n is the box size, $y(k)$ is the integrated time series and $y_n(k)$ is the local trend. This algorithm is then repeated for every box to provide a relationship between the average local fluctuations, $F(n)$, and the box size, n. For our studies, $F(n)$ was calculated for box sizes of 5-420 minutes. A log-log plot is then used to deduce whether there is a linear relationship, indicating the presence of self-similar scaling, denoted by the α parameter. The correlations of the time series are then presented as $n^{-\alpha}$, where the parameter α estimates the type of correlation present.

An exponent between $0.75 \leq \alpha \leq 1.25$ is indicative of $1/f$-type scaling. This signifies the presence of long-range correlations and hidden-hierarchical structures such as self-organised criticality.

DFA has previously been applied to investigate the heart rate dynamics of patients during sleep to distinguish sleep stages [18], and has been shown to have applications for the detection of obstructive sleep apnea [19, 20]. Ivanov *et al* [21] applied DFA to investigate the correlations present in the heart beat dynamics during sleep and wake periods for three groups: healthy individuals, individuals having suffered heart-failure, and cosmonauts. The results in [21] illustrated a clear decline in complexity during sleep for the healthy individuals and cosmonauts, and an increase in complexity during sleep for the individuals who have suffered heart-failure, compared with the healthy controls.

3.2. Magnitude and Sign Analysis

Any long-range correlated time-series can be divided into two sub-series formed by the magnitude and sign of each increment [22, 23]. The DFA quantifies the linear fractal characteristics related to two-point correlations, while the magnitude and sign analyses (MSA) method is applied to investigate the long-range nonlinear properties that may exist in the data [22]. The MSA consists of the following steps: (i) calculation of the incremental value between successive values; (ii) decomposition of this incremental series into a magnitude series and sign series; (iii) subtraction of the magnitude and sign series from their respective means in order to avoid artificial trends; (iv) integration of the magnitude and sign series in order to allow for a more accurate calculation of the DFA exponent; and (v) DFA analysis to obtain α for both series.

The DFA exponents obtained from the magnitude and sign series are denoted α^{mag} and α^{sign} respectively. Fig 1 illustrates the MSA method for one night of actigraphy.

Positive correlations in the magnitude series ($\alpha^{mag} > 0.5$) indicate that an increment with a large magnitude is more likely to be followed by an increment with a large magnitude, and similarly for increments with small magnitudes. They could also be a reliable marker of long-term nonlinear properties [24]. Anti-correlation in the sign series indicates that a positive increment is more likely to be followed by a negative increment and vice versa. MSA is a complementary method to DFA as it distinguishes long-range correlations (similar to DFA) but also quantifies the nonlinear properties as well as the temporal organisation of the series.

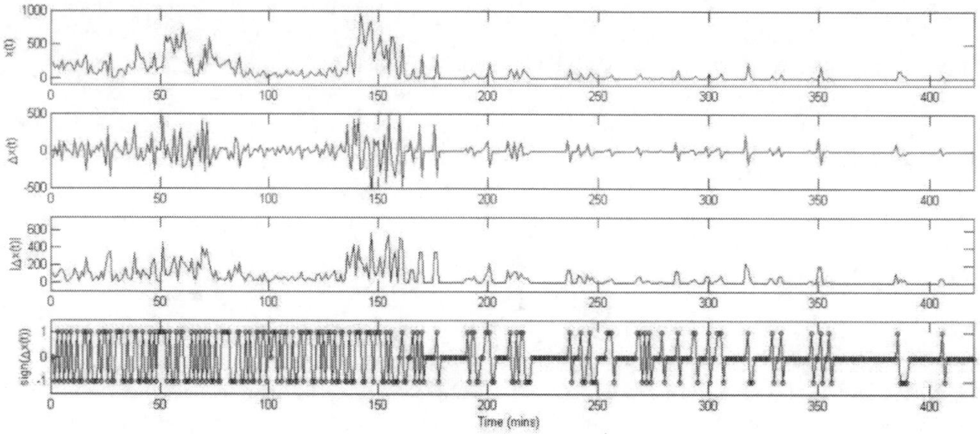

Figure 1. A typical time-series for one night of actigraphy (top), the incremental series of the original signal (second), the magnitude of the increments (third), and the sign series of the increments (bottom).

3.3. Power Spectral Analysis

Power Spectral Analysis (PSA) is another method used to study time series. It is based on power spectrum calculation from the data (3.3), then logarithms are taken of both the frequency and the power to produce a linear relationship, the slope of which can indicate the presence of scaling or self-similarity, β. Both parameters β and α are used to calculate the correlations in a time series, and are linearly related [26],

$$\beta = 2\alpha - 1. \tag{3}$$

Random uncorrelated white noise does not have long-range correlations resulting in $\beta \simeq 0$. A random-walk process which only has short-range correlations between successive points has a β parameter of around 2. Similar to DFA, $\beta \simeq 1$ identifies the presence of long-range correlations. Table 1 illustrates the relationship between DFA and PSA scaling exponents, α, and, β, as well as the type of noise each value represents.

Correlation	DFA (α)	PSA (β)
White noise	1/2	0
Pink noise	1	1
Brownian noise	3/2	2

Table 1. Scaling exponents α and β for different type of noise.

Both DFA and PSA have been proven useful in the study of physiologic complexity with a loss in complexity shown in cases with heart conditions [13, 11]. Both methods have also very recently been implemented to study the transition of scaling behaviour across sleep and wake periods, also using actigraphy data, concluding that both methods could be exploited to detect sleep/wake transitions [27]. Here, DFA is primarily used to analyse the complexity of actigraphy during night-time hours via the α-scaling exponent. The β exponent is used to verify the results.

3.4. Recurrence Quantitative Analysis
Recurrence Quantitative Analysis (RQA) contains small and large scale structures which can be used as a qualitative measure on the dynamics of the system. Single, isolated recurrence points can occur if states are rare. A diagonal line (running parallel to the main diagonal) occurs when a segment of the trajectory runs parallel to another segment. The greater the percentage of these diagonals suggests a higher determinism in the system (Fig. 2).

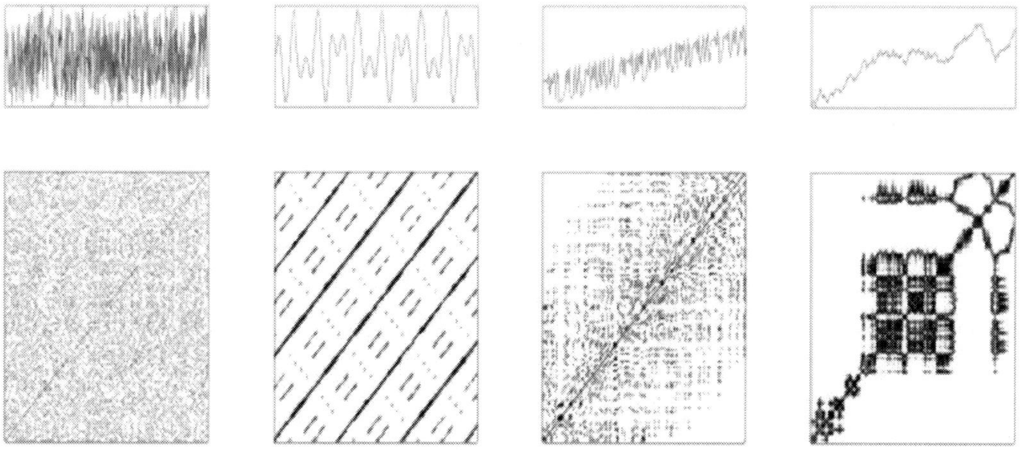

Figure 2. Illustration of the steps in RQA analysis.

3.5. Variability Analysis
Variability measures the extent to which data points in a series diverge from their mean value. Common measures of variability include range, variance and standard deviation. The most common use of variability in physiology is in the quantification of HRV, where a high degree of variability reflects healthy system function. Here a standard deviation (std) is used as a crude method of finding the activity of an individual during the night. High std value is considered as a sign of more activity during the night. Therefore, similar to calculating complexity, a lower variability is expected in those who do not suffer adverse health related events, such as insomnia.

For the purpose of this study, it is important to show that there is a correlation between the complexity and variability of night-time movements. Lipsitz [7] highlighted that there is a difference between the two. A relationship does often exist between complexity and variability, one key example being the complexity and HRV of heart beat dynamics, both of which decline with age and disease.

4. Results
4.1. Sleep Study
Activity distributions were constructed using the raw actigraph data for two case studies: (a) healthy sleepers and acute insomniacs and (b) healthy individuals from age groups 20-90 to investigate their mobility function. These indicate the distribution of the number of movements made in a given minute. Healthy sleepers tend much closer to zero movements per minute with quite a small tail-off whereas acute insomniacs show far greater dispersion towards a higher number of movements per minute.

The DFA parameter α was calculated for each night for every individual from both groups, these were then averaged to give each person a score. Variability was calculated alongside

DFA in order to examine whether a linear relationship exists between the two. The results are presented in Fig 3. A linear regression line is fit to the data and a positive correlation can be seen between the DFA exponent and the variability of the time-series thus justifying our hypothesis that a more complex signal would be indicative of a disturbed night's sleep. The R^2 value, known as the coefficient of determination, represents how well the regression line fits the data. For the healthy individuals (black circles), the regression line fits 70% of the data showing good correlation between the α-exponent from DFA and the variability. Thus, a positive linear relationship exists for the group. However, the group with acute insomnia showed more dispersion around their regression line, but a positive correlation between the complexity and variability of the night-time movements was still observed.

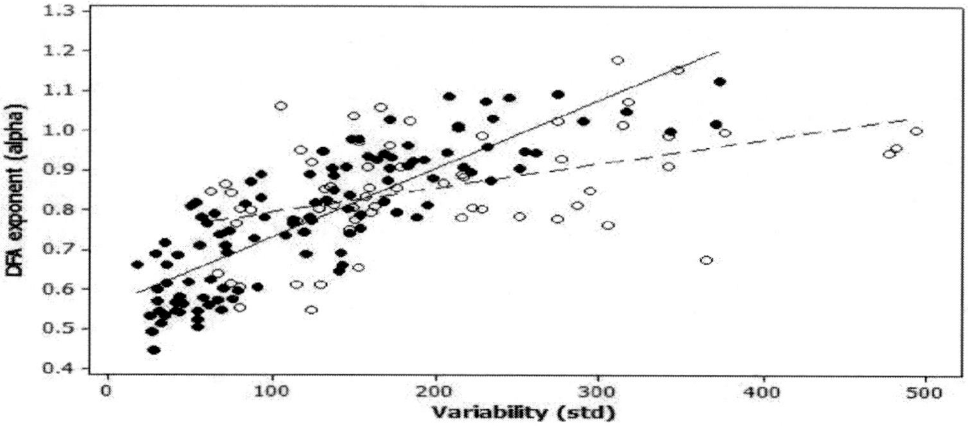

Figure 3. Scatter plot to show the linear relationship between the DFA parameter α and variability, calculated for each night, for both healthy sleepers (solid circles) and acute insomniacs (hollow circles). The solid and broken lines illustrate a correlation between the DFA parameter and the variability measures respectively.

Fig 4 shows the results gained from DFA analysis for a typical night of actigraphy from a healthy individual and one with acute insomnia, both individuals aged 25. A clear difference can be seen in the scaling parameter α, with a higher correlation seen in the individual with acute insomnia.

Figure 5 refers to a 25 year old female suffering from insomnia. It demonstrates a high complexity with a β value of 1.08. The results show similar findings to those gained from power spectral analysis.

These results are consistent across the whole dataset with those with acute insomnia exhibiting higher correlations during sleep than healthy subjects. From our data, 73% of all individuals with acute insomnia fell within the complex ($1/f$-scaling) region, $0.75 \leq \alpha \leq 1.25$, compared with just 33% of normal, healthy sleepers. Also, out of every individual who fell within this region, over 70% suffered from acute insomnia. In addition, significance testing was carried out on the two groups to look for significant differences in the complexity scores. A Mann-Whitney U test gave a p-value of 0.0015 indicating a highly significant difference between the complexities of healthy sleepers (mean 0.73) and acute insomniacs (mean 0.84).

The night-to-night variability of sleep complexity was calculated for both healthy sleepers and those with insomnia. These results demonstrate slight night-to-night fluctuations in the complexity for both groups, however none of these fluctuations are significant.

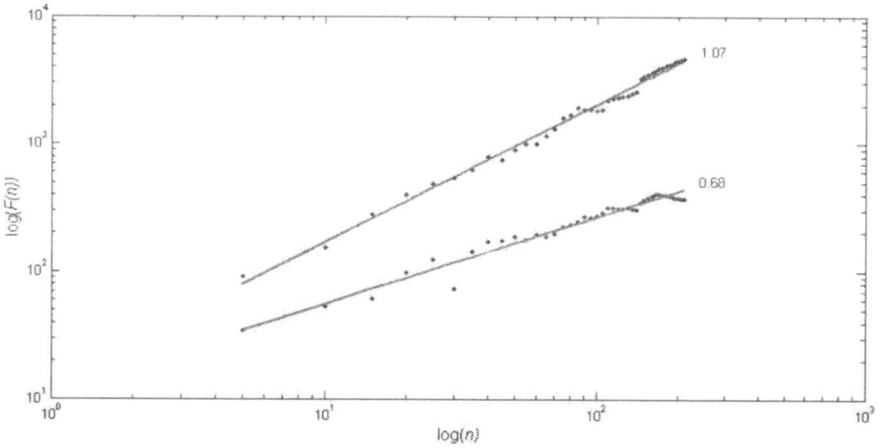

Figure 4. Log-log plot of the root-mean squared deviation, $F(n)$, versus the box size, n (5-420 mins) for a healthy sleeper, $\alpha = 0.68$, and an individual with acute insomnia, $\alpha = 1.07$.

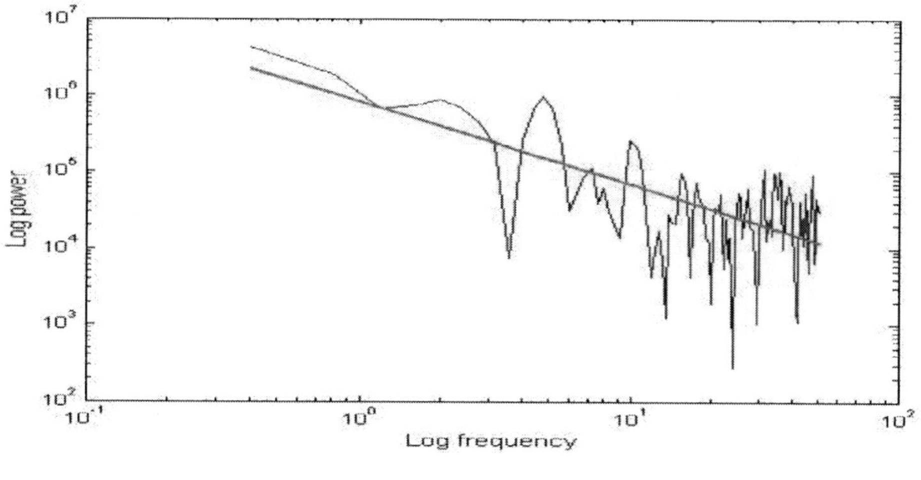

Figure 5. Log-transformed PS from one night of actigraphy for female, aged 25, with acute insomnia, $\beta = 1.08, \alpha = 1.04$, indicating high complexity.

4.2. Mobility Study

The second case study investigated PS, DFA and RQA for healthy individuals aged from 20 to 90. It showed that a threshold is apparent in many of the analyses present in the 60-70 group. A dynamical RQA measure, the determinism of the activity, is shown on Fig 6. The results for complexity metrics, obtained by RQA for the mobility case study, are shown in Table 2. The peak levels of the activity and trapping time show sharp change in value in the 60-70 age group region. Thus, the second case study also shows that complexity could prove to be a very good tool for the analysis of change of physiological function with age.

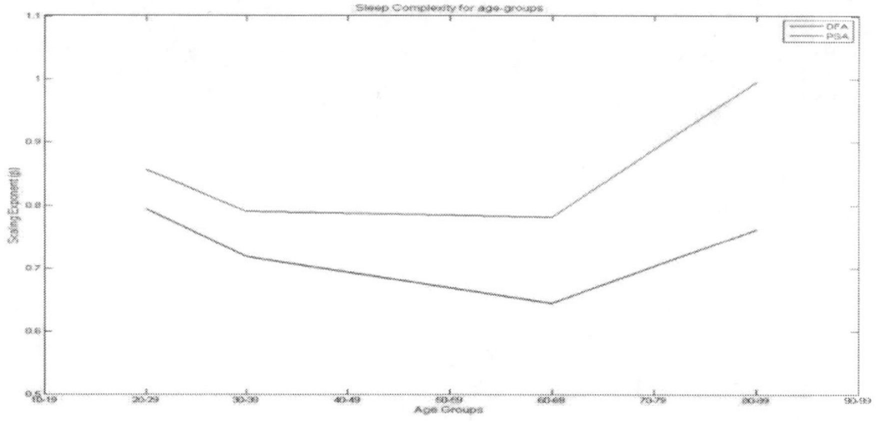

Figure 6. RQA plots showing treshold values for complexity metrics.

	20-30	30-40	60-70	80-90
Determinism	69.76	68.11	66.15	70.69
LMAX	6.34	6.15	5	4.26
TT	10.29	9.15	7.70	6.51

Table 2. Complexity measures for mobility study obtained by RQA

5. Discussion and Conclusions

This study showed that complexity analysis, applied to actigraphy data, can differentiate between individuals with healthy functions and the onset of health-related events. In the first case study, our results suggest that this is the case with normal sleepers demonstrating significantly less complexity (with less correlation) in night-time movements than those with acute insomnia. In the second case study our results showed a clear treshold in change of physiological functions for the 60-70 age group.

Due to the nature of the data within these studies, each time-series may contain a considerable amount of zero-recordings - resulting from no movement made within a one minute epoch. This could potentially affect the accuracy of the DFA parameter α. Chen *et al* [33] investigated the effect of nonstationarities on DFA and in particular the effects of a segmented time-series with different local properties. It was reported that for nonstationary time-series, segments with high positive correlations will dominate. Therefore, there is a possibility that the results discussed here are a slight exaggeration of the true correlations that exist, especially in the case of the healthy group who had a higher degree of zero-movements. We also cannot ignore the probability that some of the activity seen during the night could be the result of conscious movements.

Fractal techniques such as DFA, PSA and RQA have never been applied to actigraphy of sleep before with the hypothesis of high correlations being a marker of sleep-related health issues (in this case, acute insomnia), despite some studies showing an apparent increase in night-time movement complexity with other health problems, e.g. previous heart-failure [21]. The results gained are very promising and could provide a useful non-invasive markers for the identification of of health-related conditions.

Acknowledgments
This work was partly supported by European FP7 project 247541 MATSIQEL, Marie Curie Actions-IRSES.

References
[1] Mandelbrot BB 1983 *The Fractal Geometry of Nature* (New York: Freeman) 1–58
[2] Lewis A, Lipsitz LA and Goldberger AL 1992 *JAMA* **267**(13) 1806-1809
[3] Dauwels J, Vialatte F and Cichocki A 2010 *Current Alzheimer Research* **7**(6) 487–505
[4] Holloway P, Angelova M, Lombardo S, St Clair Gibson A and Lee D Ellis J 2014 *J R Soc Interface* 20131112
[5] Holloway P, Angelova M, St Clair Gibson A, Lombardo S, Lee D, Ellis J, Lambert and Rauch L 2012 Modelling techniques for analysis of human activity patterns. Intelligent Systems IS12, Proceedings of the 6th IEEE International Conference, 6-8 September 2012 Sofia, 275–280
[6] Lichstein KL, Stone KC, Donaldson J, Nau SD, Soeffing JP, Murray D, Lester KW and Aguillard RN 2006 *SLEEP* **29** (2) 232–239
[7] Lipsitz LA 2002 *J. Gerontol. Biol. Sci.* **57A** (3) B115–B125
[8] Melo RC, Santos MDB, Silva E, Quitrio RJ, Moreno MA, Reis MS, Verzola IA, Oliveira L, Martins LEB, Gallo-Junior L and Catai AM 2005 *Braz. J. Med. Biol. Res.* **38** 1331-1338
[9] Blackledge JM 2005 *Digital Image Processing: Mathematical and computational methods* (West Sussex: Horwood Publishing)
[10] Akselrod S, Gordon D, Ubel FA, Shannon DC, Barger AC and Cohen RJ 1981 *Science, New Science* **213** (4504) 220–222
[11] Ivanov PCh, Amaral LAN, Goldberger AL, Havlin S, Rosenblum MG, Struzik ZR and Stanley HE 1999 *Nature* **399** 461–465
[12] Valencia JF 2009 *IEEE Eng. Med. Biol. Mag.* **28** 72–78
[13] Goldberger AL Amaral LA, Hausdorff JM, Ivanov P, Peng CK and Stanley HE 2002 *PNAS USA* **99** 2466–2472
[14] Goldberger AL, Moody GB and Costa MD 2012 available online at http://physionet.org/tutorials/cv/
[15] Peng CK, Buldyrev SV, Havlin S, Simons M, Stanley HE and Goldberger AL 1994 *Phys. Rev. E.* **49** 1685–1689
[16] Tulppo MP, Hautala AJ, Makikallio TH, Laukkanen RT, Nissila S, Hughson RL and Huikuri HV 2003 *J. Appl. Physiol.* **95** 364-372
[17] Yazawa T, Asai I, Shimoda Y and Katsuyama T 2010 *WCECS* **2** 921–925
[18] Kantelhardt JW, Ashkenazy Y, Ivanov PCh, Bunde A, Havlin S, Penzel T, Peter J-H and Stanley HE 2002 *Phys. Rev. E* **65** (5) 051908
[19] Mietus JE, Peng CK, Ivanov PCh and Goldberger AL 2000 *Comput. Cardiol.* **27** 753–756
[20] Penzel T, Kantelhardt JW, Becker HF, Peter JH and Bunde A 2003 *Comput. Cardiol.* **30** 307–310
[21] Ivanov PCh, Bunde A, Amaral LAN, Havlin S, Fritsch-Yelle J, Baevsky RM, Stanley HE and Goldberger AL 1999 *Europhys. Lett.* **48** (5) 594-600
[22] Ashkenazy Y, Ivanov PCh, Havlin S, Peng C-K, Goldberger AL and Stanley HE 2001 *Phys. Rev. Lett.* **86** (9) 1900–1903
[23] Ashkenazy Y, Ivanov PCh, Havlin S, Peng CK, Yamamoto Y, Goldberger AL and Stanley HE 2000 *Comput. Cardiol.* **27** 139–142
[24] Ashkenazy Y, Havlin S, Ivanov PCh, Peng CK, Schulte-Frohlinde V and Stanley HE 2003 *Physica A* **323** 19–41
[25] Buysse DJ., Cheng Y, Germain A., Moul DE, Franzen P, Fletcher M and Monk TH 2010 *Sleep Med.* **11** 56–71
[26] Peng CK and Havlin S, Stanley HE and Goldberger AL 1995 *Chaos* **5** 82–87
[27] Wohlfahrt P, Kantelhardt JW, Zinkhan M, Schumann AY, Penzel T, Fietze I, Pillmann F and Stang A 2013 *EPL* **103** 68002
[28] Hu K, Ivanov PCh, Chen Z, Hilton MF, Stanley HE and Shea SA 2004 *Physica A* **337** 307–318
[29] Schneider-Helmert D 1987 *Sleep* **10** (5) 452–462
[30] Jurysta F, Lanquart JP, Sputaels V, Dumont M, Migeotte PF, Leistedt S, Linkowski P and van de Borne P 2009 *Clin. Neurophysiol.* **120** (6) 1054–1060
[31] Sadeh A, Hauri PJ, Kripke DF and Lavie P 1995 *Sleep* **18** 288-302
[32] Ivanov PCh, Hu K, Hilton MF, Shea SA and Stanley HE 2007 *PNAS* **104** (52) 20702–20707
[33] Chen Z, Ivanov PCh, Hu K and Stanley HE 2002 *Phys. Rev. E* **65** (4)

Journal of Physics: Conference Series **597** (2015) 012011 doi:10.1088/1742-6596/597/1/012011

The orthogonal projection on slice functions on the quaternionic sphere

Nicola Arcozzi[1] **and Giulia Sarfatti**[2]

Dipartimento di Matematica, Università di Bologna, Piazza di Porta San Donato 5, 40126 Bologna, Italy.

E-mail: nicola.arcozzi@unibo.it, giulia.sarfatti@unibo.it

Abstract. We study the L^p norm of the orthogonal projection from the space of quaternion valued L^2 functions to the closed subspace of *slice* L^2 functions.

The aim of this short note is to study the orthogonal projection Π from the space of quaternion valued L^2 functions to its closed subspace of *slice* L^2 functions. In particular, to compute the norm of the projection operator we will first show that we can write Π in terms of a quaternionic *slice* Poisson kernel.

Let $\mathbb{H} = \mathbb{R} + \mathbb{R}i + \mathbb{R}j + \mathbb{R}k$ denote the non commutative 4-dimensional real algebra of quaternions and let $\mathbb{B} = \{q \in \mathbb{H} : |q| < 1\}$ be the unit ball in $\mathbb{H}$. Its boundary $\partial\mathbb{B}$ contains elements of the form $q = e^{It}$, $I \in \mathbb{S}$, $t \in \mathbb{R}$, where $\mathbb{S} = \{q \in \mathbb{H} : q^2 = -1\}$ is the two dimensional sphere of imaginary units in $\mathbb{H}$. We endow $\partial\mathbb{B}$ with the measure $d\Sigma\left(e^{It}\right) = d\sigma(I)dt$, which is naturally associated with the Hardy space $H^2(\mathbb{B})$ of the slice regular functions on $\mathbb{B}$, see [1]. We here normalize the measures so to have $\Sigma(\partial\mathbb{B}) = \sigma(\mathbb{S}) = 1$. The functions we are concerned with here satisfy algebraic conditions and size conditions. A function $f : \partial\mathbb{B} \to \mathbb{H}$ is a *slice function* if for any $I, J \in \mathbb{S}$

$$f(e^{Jt}) = \frac{1}{2}\left[f(e^{It}) + f(e^{-It})\right] + \frac{JI}{2}\left[f(e^{-It}) - f(e^{It})\right]. \tag{1}$$

That is, slice functions are affine in the $\mathbb{S}$ variable:

$$f(e^{Jt}) = a(t) + Jb(t), \tag{2}$$

where $a, b : \partial\mathbb{B} \to \mathbb{H}$ are functions depending on t alone. Condition (2) has an algebraic nature. Equation (1) also provides a formula to extend a function f_I defined on a slice $\partial\mathbb{B}_I = \partial\mathbb{B}\cap(\mathbb{R}+\mathbb{R}I)$ to a slice function $f := \mathrm{ext}(f_I)$ defined on the entire sphere $\partial\mathbb{B}$.

Slice functions were introduced in the more general setting of real alternative algebras by Ghiloni and Perotti in [5]. A notable class of slice functions is provided by the restriction to the unit sphere in $\mathbb{H}$ of slice regular functions. Such functions have been object of intensive research

[1] Partially supported by INDAM-GNAMPA and by the PRIN project Real and Complex Manifolds of the Italian MIUR.

[2] Partially supported by INDAM-GNSAGA, by the PRIN project Real and Complex Manifolds and by the FIRB project Differential Geometry and Geometric Function Theory of the Italian MIUR.

since the seminal work by Gentili and Struppa [3]. A concrete example is given by converging power series of the form $e^{It} \mapsto \sum_{n\in\mathbb{Z}} e^{Int}a_n$, with quaternionic coefficients a_n.

We refer to [4] for results concerning slice regular functions and to [2] for functional analysis results in the quaternionic setting.

We denote by $L_s = L_s(\partial\mathbb{B})$ the class of the measurable slice functions on $\partial\mathbb{B}$, which has a natural structure of right linear space on $\mathbb{H}$.

We will denote by $L^p = L^p(\partial\mathbb{B})$, $1 \le p \le \infty$, the space of the functions $\varphi : \partial\mathbb{B} \to \mathbb{H}$ such that $\int_{\partial\mathbb{B}} |\varphi|^p d\Sigma =: \|\varphi\|_p^p < \infty$, and let $L_s^p := L^p \cap L_s$. It is not difficult to prove that L_s^2 is a closed subspace of L^2 (see Proposition 5 below), hence we have an orthogonal projection operator $\Pi : L^2 \to L_s^2$: Π is self-adjoint, surjective, and $\Pi^2 = \Pi$.

Let

$$\|\Pi\|_{p,p} = \sup_{L^2 \ni \varphi \neq 0} \frac{\|\Pi\varphi\|_p}{\|\varphi\|_p}.$$

Obviously, $\|\Pi\|_{2,2} = 1$.

Our main goal here is computing the norm of $\Pi : L^p \to L_s^p$.

Theorem 1. *(i)* $\|\Pi\|_{2,2} = 1$, $\|\Pi\|_{\infty,\infty} = \frac{4}{3}$.

(ii) Let $2 \le p \le \infty$. *Then,* $\|\Pi\|_{p,p} \le 2\left(\frac{2p-2}{3p-2}\right)^{\frac{p-1}{p}}$.

(iii) If $1 \le p, q \le \infty$ *and* $\frac{1}{p} + \frac{1}{q} = 1$, *then* $\|\Pi\|_{p,p} = \|\Pi\|_{q,q}$.

The general theory implies that Π is a contraction on L^2 and the equality in (iii). The inequality in (ii) gives the right constant for $p = \infty$, but not for $p = 2$. Finding the exact value of $\|\Pi\|_{p,p}$ for all p's might shed some light on the geometry of slice regular functions, and we think it is an interesting open problem. The study presented in this paper began when discussing some problems related to Preprint [6].

On route to the proof of Theorem 1, we shall write Π as an explicit integral operator.

Let $\varphi \in L^2$. The measure $d\Sigma$ we are considering is the product of the (normalization of the) standard Lebesgue measure on the unit circle times the standard area element on the two sphere. Hence, φ belongs to the L^2 space of the unit circle $\partial\mathbb{B}_I$ for almost every $I \in \mathbb{S}$, and we can expand φ as a power series on almost each slice of the form

$$\varphi(e^{It}) = \sum_{n=-\infty}^{\infty} e^{Int} a_n(I).$$

The coefficients depend on the imaginary unit $I \in \mathbb{S}$.

Proposition 2. *Let* $\varphi \in L^2$. *Then the orthogonal projection of* φ *to the space of slice functions is*

$$\Pi\varphi(e^{Jt}) = \sum_{n=-\infty}^{\infty} e^{Jnt}\tilde{a}_n$$

for any $e^{Jt} \in \partial\mathbb{B}$, *where the coefficients are given by the integral means*

$$\tilde{a}_n = \int_{\mathbb{S}} d\sigma(I) a_n(I)$$

for any $n \in \mathbb{Z}$.

Proof. Let $f(e^{It}) = \sum_{n\in\mathbb{Z}} e^{Int}\alpha_n \in L^2_s$. Taking into account that, for any measurable function $g : \partial\mathbb{B} \to \mathbb{H}$, $\int_\pi^{2\pi} g(e^{It})dt = \int_0^\pi g(e^{-It})dt$, which implies that

$$\int_{\partial\mathbb{B}} g(e^{It})d\Sigma(e^{It}) = \frac{1}{2\pi}\int_0^{2\pi}\int_{\mathbb{S}} g(e^{It})d\sigma(I)dt,$$

we can write

$$\begin{aligned}\langle f, \varphi\rangle_{L^2} &= \frac{1}{2\pi}\int_0^{2\pi} dt \int_{\mathbb{S}} d\sigma(I)\overline{\varphi(e^{It})}f(e^{It}) = \int_{\mathbb{S}} d\sigma(I) \sum_{n=-\infty}^{\infty} \overline{a_n(I)}\alpha_n \\ &= \sum_{n=-\infty}^{\infty} \overline{\int_{\mathbb{S}} d\sigma(I)a_n(I)}\alpha_n = \sum_{n=-\infty}^{\infty} \overline{\tilde{a}_n}\alpha_n = \langle f, \tilde{\varphi}\rangle_{L^2}\end{aligned}$$

where $\tilde{\varphi}(e^{Jt}) = \sum_{n\in\mathbb{Z}} e^{Jnt}\tilde{a}_n$ is a slice function. Hence $\Pi\varphi = \tilde{\varphi}$. □

We observe first that functions f in L^2_s extend as power series to the interior of the unit ball:

$$f(re^{It}) = \sum_{n=-\infty}^{+\infty} r^{|n|}e^{Int}\widehat{f}(n), \text{ where } \widehat{f}(n) = \frac{1}{2\pi}\int_{-\pi}^{\pi} e^{-nIs}f(e^{Is})ds \text{ is independent of } I. \tag{3}$$

For $0 \le r < 1$, let P^I_r be the Poisson kernel on the unit circle of the complex plane $\mathbb{R}+\mathbb{R}I \subset \mathbb{H}$:

$$P^I_r(t) = \frac{1-r^2}{|1-re^{It}|^2} = \sum_{n=-\infty}^{+\infty} r^{|n|}e^{Int}.$$

We point out that $P^I_r = P_r$ is independent of the imaginary unit I. In the following Lemma, we use the notation P^I_r to stress the role of the imaginary unit in formula (4).

Lemma 3. *For any $\varphi \in L^2$, the extension to the interior of $\mathbb{B}$ of its projection is given by*

$$\Pi\varphi(re^{It}) = \int_{\partial\mathbb{B}} K(re^{It}, e^{Js})\varphi(e^{Js})d\Sigma(e^{Js}),$$

where

$$K(re^{It}, e^{Js}) = \frac{1}{2}\left[P^J_r(t+s) + P^J_r(t-s)\right] + \frac{IJ}{2}\left[P^J_r(t+s) - P^J_r(t-s)\right]. \tag{4}$$

Proof. Fix $J \in \mathbb{S}$ such that the restriction φ_J of φ to $\partial\mathbb{B}_J$ belongs to the L^2 space of the circle $\partial\mathbb{B}_J$, and consider the harmonic extension of φ_J to the interior of the disc $\mathbb{B}_J$,

$$\varphi_J(re^{Js}) = \sum_{n=-\infty}^{\infty} r^{|n|}e^{Jns}a_n(J).$$

The coefficients of the projection $\Pi\varphi$, recalling Proposition 2, are given by

$$\tilde{a}_n = \int_{\mathbb{S}} d\sigma(J)a_n(J) = \int_{\mathbb{S}} d\sigma(J)\frac{1}{2\pi}\int_0^{2\pi} e^{-Jns}\varphi(e^{Js})ds.$$

Hence, for any $I \in \mathbb{S}, t \in \mathbb{R}$,

$$\begin{aligned}\Pi\varphi(re^{It}) &= \sum_{n=-\infty}^{+\infty} r^{|n|}e^{Int}\tilde{a}_n = \int_{\mathbb{S}} d\sigma(J)\frac{1}{2\pi}\int_0^{2\pi}\sum_{n=-\infty}^{+\infty} r^{|n|}e^{Int}e^{-Jns}\varphi(e^{Js})ds \\ &= \int_{\partial\mathbb{B}} K(re^{It}, e^{Js})\varphi(e^{Js})d\Sigma(e^{Js})\end{aligned}$$

where

$$K(re^{It}, e^{Js}) = \sum_{n=-\infty}^{+\infty} r^{|n|}e^{Int}e^{-Jns}.$$

To conclude, notice that $K(re^{It}, e^{Js})$ is a slice function with respect to the variable e^{It}, hence it can be written as

$$\begin{aligned}K(re^{It}, e^{Js}) &= \frac{1}{2}\left(K(re^{Jt}, e^{Js}) + K(re^{-Jt}, e^{Js})\right) + \frac{IJ}{2}\left(K(re^{-Jt}, e^{Js}) - K(re^{Jt}, e^{Js})\right) \\ &= \frac{1}{2}\left(\sum_{n=-\infty}^{+\infty} r^{|n|}e^{Jn(t-s)} + \sum_{n=-\infty}^{+\infty} r^{|n|}e^{Jn(-t-s)}\right) \\ &\quad + \frac{IJ}{2}\left(\sum_{n=-\infty}^{+\infty} r^{|n|}e^{Jn(-t-s)} - \sum_{n=-\infty}^{+\infty} r^{|n|}e^{Jn(t-s)}\right) \\ &= \frac{1}{2}\left(P_r^J(t-s) + P_r^J(-t-s))\right) + \frac{IJ}{2}\left(P_r^J(-t-s) - P_r^J(t-s)\right).\end{aligned}$$

□

Corollary 4. *We have that* $\Pi\varphi(re^{It}) = A(r,t) + IB(r,t)$, *where*

$$A(r,t) = \frac{1}{2\pi}\int_{-\pi}^{\pi}\frac{1}{2}\left[P_r(t+s) + P_r(t-s)\right]\left\{\int_{\mathbb{S}}\varphi(e^{Js})d\sigma(J)\right\}ds$$

and

$$B(r,t) = \frac{1}{2\pi}\int_{-\pi}^{\pi}\frac{1}{2}\left[P_r(t+s) - P_r(t-s)\right]\left\{\int_{\mathbb{S}}J\varphi(e^{Js})d\sigma(J)\right\}ds.$$

In A and B the function φ enters, respectively, through its "imaginary mean" and "imaginary first moment" on each copy of $\mathbb{S}$ inside $\partial\mathbb{B}$.

Passing in the limit as $r \to 1^-$, using the fact that the classical Poisson kernel as linear operator tends to the Dirac delta function, we obtain -at least when $\varphi \in C = C(\partial\mathbb{B}, \mathbb{H})$ is continuous on $\partial\mathbb{B}$-:

$$\Pi\varphi(e^{tI}) = \int_{\mathbb{S}}(1 - IJ)\varphi(e^{Jt})d\sigma(J). \tag{5}$$

By density of C in L^p ($p < \infty$), the integral formula for Π extends to L^p if Π is bounded on L^p. Since Π is bounded on L^2, by inclusion it is defined on L^p's at least when $2 \leq p \leq \infty$. Formula (5) is in this case true for *a.e.* e^{tI} in $\partial\mathbb{B}$.

Consider the case $p = \infty$ of the theorem first. Taking the essential supremum, we have that

$$\|\Pi\varphi\|_\infty \leq \int_{\mathbb{S}}|1 - IJ|d\sigma(J)\|\varphi\|_\infty, \tag{6}$$

Journal of Physics: Conference Series **597** (2015) 012011 doi:10.1088/1742-6596/597/1/012011

with equality if $\varphi(e^{Js}) = |1 - I_0 J|(1 - I_0 J)^{-1}$ for some $I_0 \in \mathbb{S}$. Using spherical coordinates on $\mathbb{S}$ such that $I = (0, 0, 1)$ and $J = (\sin t \cos s, \sin t \sin s, \cos t)$ with $s \in (0, 2\pi)$ and $t \in (0, \pi)$, we get that $|1 - IJ| = 2\sin(t/2)$, $d\sigma(J) = \sin(t)dtds$, and hence that

$$\int_{\mathbb{S}} |1 - IJ| d\sigma(J) = 4/3, \tag{7}$$

thus concluding the proof of Part (i) of Theorem 1. Similarly, when $2 \le p < \infty$ we have, with $\frac{1}{p} + \frac{1}{q} = 1$,

$$\|\Pi\varphi\|_p \le \left(\int_{\mathbb{S}} |1 - IJ|^q d\sigma(J)\right)^{1/q} \|\varphi\|_p. \tag{8}$$

We have used Hölder's inequality in (5) and the fact that the first factor on the right of (8) is independent of I. A simple calculation with the same spherical coordinates on $\mathbb{S}$ used to obtain (7) gives

$$\left(\int_{\mathbb{S}} |1 - IJ|^q d\sigma(J)\right)^{1/q} = 2\left(\frac{2}{q+2}\right)^{1/q} = 2\left(\frac{2p-2}{3p-2}\right)^{\frac{p-1}{p}},$$

and the proof of Theorem 1 is finished.

Proposition 5. *L^2_s is a closed subspace of L^2.*

Proof. Let $\{f_n\}_n$ be a sequence in L^2_s converging in L^2 norm to $f \in L^2$. We want to show that the difference $f(e^{Jt}) - (a(t) + Jb(t))$, where $a(t) = \frac{1}{2}[f(e^{It}) + f(e^{-It})]$ and $b(t) = \frac{I}{2}[f(e^{-It}) - f(e^{It})]$ for any $I \in \mathbb{S}$, equals zero in L^2 norm. For any $n \in \mathbb{N}$ we have

$$\begin{aligned} &\|f(e^{Jt}) - (a(t) + Jb(t))\|_2 \\ &\le \|f(e^{Jt}) - f_n(e^{Jt})\|_2 + \|f_n(e^{Jt}) - a_n(t) - Jb_n(t)\|_2 + \|a_n(t) + Jb_n(t) - a(t) - Jb(t)\|_2, \end{aligned}$$

with obvious notation. The first summand tends to 0 as n goes to ∞ and the second summand vanishes for any n since $f_n \in L^2_s$. The last term can be bounded as

$$\|a_n(t) + Jb_n(t) - a(t) - Jb(t)\|_2 \le \|a_n(t) - a(t)\|_2 + \|b_n(t) - b(t)\|_2,$$

where we can choose any slice $\partial\mathbb{B}_I$ to compute both quantities on the right side. The fact that f_n converges to f in L^2 norm yields that, for σ-almost every I, $\int_0^{2\pi} |f_n(e^{It}) - f(e^{It})|^2 dt$ tends to 0 as n goes to ∞, and hence, that both $\|a_n(t) - a(t)\|_2$ and $\|b_n(t) - b(t)\|_2$ tend to 0 as n goes to ∞. Therefore we conclude that $f \in L^2_s$. □

We end the note with some remarks. It is a peculiar fact that $1 < \|\Pi\|_{\infty,\infty} < \infty$. Many natural projection operators defined on L^2 spaces extend to contractions to L^p for $1 < p < \infty$. This is the case with restriction operators $\varphi \mapsto \chi_E\varphi$, where χ_E is a measurable characteristic function; or, closer to the operator we are considering here, the projection onto the space of functions endowed with some symmetry. For instance, let $L_{\mathbb{S}}$ be the space of the functions φ on $\partial\mathbb{B}$ which are measurable and such that $\varphi(e^{It}) = a(t)$ depends on t alone, set $L^p_{\mathbb{S}} = L^p \cap L_{\mathbb{S}}$, and let $\Pi_{\mathbb{S}} : L^2 \to L^2_{\mathbb{S}}$ be the orthogonal projection. We might write

$$\Pi_{\mathbb{S}}\varphi(e^{It}) = \int_{SO(3)} \varphi(e^{RIt}) dh(R),$$

where $SO(3)$ is the orthogonal group, acting on $\mathbb{S}$ as $I \mapsto RI$, for any $R \in SO(3)$, and dh is its (bi-invariant) Haar measure, normalized to have $h(SO(3)) = 1$. Alternatively, we might ask φ to be covariant under rotations,

$$\varphi(e^{RIt}) = R\varphi(e^{It}). \tag{9}$$

(If $\varphi = e^{Ks}$, then $R\varphi = e^{RKs}$.) The projection from L^2 to the subspace of the functions satisfying (9) is

$$\Pi'_{\mathbb{S}}\varphi(e^{It}) = \int_{SO(3)} R^{-1}\varphi(e^{R(I)t})dh(R). \tag{10}$$

Observe that there is a formal analogy between (9) and (10), and the integrals appearing in Corollary 4.

At the other extreme, some projection operators on L^2 are unbounded in L^∞. The prototype is the conjugate function operator in one complex variable, and an avatar of it in our context might be

$$\varphi(e^{It}) = \sum_{n=-\infty}^{\infty} e^{Int}a_n(I) \mapsto \tilde{\Pi}\varphi(e^{It}) = \sum_{n=0}^{\infty} e^{Int}a_n(I).$$

Tipically, projections of this kind can be expressed in terms of singular integral operators.

The projection Π onto slice functions seems to be intermediate between these two families of operators. It shares with the first the fact that it is linked with a geometric-algebraic invariance property. This might "explain" the boundedness in L^∞. On the other hand, the invariance can not be simply stated in measure-theoretic terms, and this might "explain" why it is not a contraction of L^∞: cancellations in the integrals play a role, and this causes the L^p norm of the operator to grow.

To have a hint as to which kind of cancellations are involved, it is instructive to consider the relation:

$$\int_{\mathbb{S}} |\Pi\varphi(e^{It})|^2 d\sigma(I) = \left|\int_{\mathbb{S}} \varphi(e^{It})d\sigma(I)\right|^2 + \left|\int_{\mathbb{S}} I\varphi(e^{It})d\sigma(I)\right|^2,$$

which follows from Corollary 4. If the function $\varphi(e^{Jt}) = a(t) + Jb(t)$ is already slice, then the first integral vanishes for $a = 0$, while the second does for $b = 0$. This is somehow a measure of the amount of cancellation which is going on in the integrals.

References

[1] Arcozzi N and Sarfatti G 2014 Invariant metrics for the quaternionic Hardy space *J. Geom. Anal.* online first
[2] Brackx F, Delanghe R and Sommen F 1982 *Clifford analysis* (Boston, MA: Pitman)
[3] Gentili G and Struppa D C 2007 A new theory of regular functions of a quaternionic variable *Adv. Math.* **216** 279-301.
[4] Gentili G, Stoppato C and Struppa D C 2013 *Regular functions of a quaternionic variable* (Berlin: Springer)
[5] Ghiloni R and Perotti A 2011 Slice regular functions on real alternative algebras *Adv. Math.* **226** 1662-91
[6] Sarfatti G 2014 Quaternionic Hankel operators and approximation by slice regular functions *Preprint*

Journal of Physics: Conference Series **597** (2015) 012012 doi:10.1088/1742-6596/597/1/012012

On the raising and lowering difference operators for eigenvectors of the finite Fourier transform

M K Atakishiyeva[1] **and N M Atakishiyev**[2]

[1] Facultad de Ciencias Universidad Autónoma del Estado de Morelos 62250 Cuernavaca, Morelos, México

[2] Instituto de Matemáticas, Unidad Cuernavaca Universidad Nacional Autónoma de México 62210 Cuernavaca, Morelos, México

E-mail: `natig@matcuer.unam.mx`

Abstract. We construct explicit forms of raising and lowering difference operators that govern eigenvectors of the finite (discrete) Fourier transform. Some of the algebraic properties of these operators are also examined.

1. Introduction

We choose to set out by recalling first some mathematical aspects of the classical Fourier transform (FT). The orthonormalized wave functions $\psi_n(x)$ of the linear harmonic oscillator in non-relativistic quantum mechanics,

$$\int_{\mathbb{R}} \psi_m(x)\, \psi_n(x)\, dx = \delta_{mn}\,, \qquad m, n = 0, 1, 2, \dots, \tag{1.1}$$

are explicitly given as

$$\psi_n(x) := c_n^{-1}\, H_n(\xi)\, \exp\left(-\xi^2/2\right), \qquad c_n = \sqrt{\sqrt{\pi}\, 2^n\, n!}\,, \tag{1.2}$$

where $H_n(\xi)$ are the classical Hermite polynomials and $\xi := \sqrt{m\,\omega/\hbar}\, x$ is a dimensionless coordinate (see, for example, [1]). In quantum mechanics they emerge as eigenfunctions of the Hamiltonian $\mathbf{H}$ for the linear harmonic oscillator,

$$\mathbf{H}\, \psi_n(x) \equiv \frac{\hbar\,\omega}{2}\left(\xi^2 - \frac{d^2}{d\xi^2}\right)\psi_n(x) = \hbar\,\omega\,(n + 1/2)\,\psi_n(x)\,, \tag{1.3}$$

which is a self-adjoint differential operator of the second order. The wave functions $\psi_n(x)$ (we recall that the functions $H_n(x)e^{-x^2/2}$ are usually referred to as *Hermite functions* in the mathematical literature) represent an important explicit example of an orthonormal and complete system in the Hilbert space $L^2(\mathbb{R}, dx)$ of square-integrable functions on the full real line $x \in \mathbb{R}$. It is further well known that the wave functions of the linear harmonic oscillator $\psi_n(x)$ possess the simple transformation property with respect to the Fourier transform: *they are also eigenfunctions of the Fourier transform, associated with the eigenvalues* i^n, that is,

$$(\mathcal{F}\,\psi_n)\,(x) \equiv \frac{1}{\sqrt{2\pi}} \int_{\mathbb{R}} e^{\,\mathrm{i}\,x\,y}\, \psi_n(y)\, dy = \mathrm{i}^{\,n}\, \psi_n(x)\,. \tag{1.4}$$

Assume now for the moment that we are not aware of this important mathematical fact and are interested in finding eigenfunctions of the FT operator $\mathcal{F}$ explicitly. Since mutually commuting operators have the same set of eigenfunctions, one may solve this problem by defining such a self-adjoint differential operator with simple spectrum of distinct eigenvalues, that commutes with the FT operator (1.4). Then the eigenfunctions of that differential operator can be found by solving a corresponding to this case differential equation and they will be at same time the eigenfunctions of the FT operator (1.4). So in this way one reduces a problem of finding eigenfunctions of the FT operator $\mathcal{F}$ to one of solving some differential equation.

Indeed, to illustrate this point let us start with the first-order differential operator $\frac{d}{dx}$ and evaluate its action on (1.4):

$$\frac{d}{dx}\int_{\mathbb{R}} e^{\,\mathrm{i}\,x\,y}\, f(y)\, dy = \mathrm{i}\int_{\mathbb{R}} e^{\,\mathrm{i}\,x\,y}\, y\, f(y)\, dy\,, \tag{1.5}$$

where $f(x) \in L^2(\mathbb{R}, dx)$ and for the sake of simplicity from now on we write x instead of the dimensionless variable ξ. Consequently, from the right side of (1.5) one deduces that the next step should be to evaluate

$$x\int_{\mathbb{R}} e^{\,\mathrm{i}\,x\,y}\, f(y)\, dy = -\,\mathrm{i}\int_{\mathbb{R}}\left(\frac{d\, e^{\,\mathrm{i}\,x\,y}}{dy}\right) f(y)\, dy = \mathrm{i}\int_{\mathbb{R}} e^{\,\mathrm{i}\,x\,y}\,\frac{d\, f(y)}{dy}\, dy\,, \tag{1.6}$$

upon integrating by parts the middle term in (1.6). From (1.5) and (1.6) it thus follows that

$$\left(x \pm \frac{d}{dx}\right)\int_{\mathbb{R}} e^{\,\mathrm{i}\,x\,y}\, f(y)\, dy = \pm\,\mathrm{i}\int_{\mathbb{R}} e^{\,\mathrm{i}\,x\,y}\left(y \pm \frac{d}{dy}\right) f(y)\, dy\,. \tag{1.7}$$

In the operator form these identities can be represented as

$$\mathbf{a}\,\mathcal{F} = \mathrm{i}\,\mathcal{F}\,\mathbf{a}\,, \qquad \mathbf{a}^{\dagger}\,\mathcal{F} = -\,\mathrm{i}\,\mathcal{F}\mathbf{a}^{\dagger}\,, \tag{1.8}$$

where

$$\mathbf{a} := \frac{1}{\sqrt{2}}\left(x + \frac{d}{dx}\right)\,, \qquad \mathbf{a}^{\dagger} := \frac{1}{\sqrt{2}}\left(x - \frac{d}{dx}\right)\,, \tag{1.9}$$

are the lowering and raising first-order differential operators, which obey the standard Heisenberg commutation relation

$$\left[\mathbf{a}\,,\,\mathbf{a}^{\dagger}\right] := \mathbf{a}\,\mathbf{a}^{\dagger} \,-\, \mathbf{a}^{\dagger}\,\mathbf{a} \equiv \left[\frac{d}{dx}\,,\,x\right] = \mathbf{I}\,. \tag{1.10}$$

From the identities (1.8) it follows now at once that the self-adjoint second-order differential operators $\mathbf{a}\,\mathbf{a}^{\dagger}$ and $\mathbf{a}^{\dagger}\mathbf{a}$ commute with the FT operator (1.4), i.e., $[\mathbf{a}\,\mathbf{a}^{\dagger}, \mathcal{F}] = 0$ and $[\mathbf{a}^{\dagger}\mathbf{a}\,, \mathcal{F}] = 0$. So we are naturally led to the desired differential operator of the second order,

$$\mathbf{H} = \frac{1}{2}\Big(\mathbf{a}\,\mathbf{a}^{\dagger} \,+\, \mathbf{a}^{\dagger}\,\mathbf{a}\Big) = \frac{1}{2}\Big(x^2 - \frac{d^2}{dx^2}\Big)\,, \tag{1.11}$$

which commutes with the Fourier transform operator $\mathcal{F}$ and represents, in effect, the well-known Hamiltonian (1.3) for the linear harmonic oscillator in quantum mechanics.

On the other hand, there exist a finite (or discrete) analogue of the integral Fourier transform (1.4). We recall that the *finite (discrete) Fourier transform* (FFT) based on points N is represented by an $N \times N$ unitary symmetric matrix $\mathbf{\Phi}^{(N)} = \left(\Phi^{(N)}_{m,n}\right)$ with elements (see, for example, [2, 3])

$$\Phi^{(N)}_{m,n} := \frac{1}{\sqrt{N}}\,\exp\left(\frac{2\pi\mathrm{i}}{N}\,m\,n\right) \equiv \frac{1}{\sqrt{N}}\,q^{m\,n}\,, \quad q := e^{\frac{2\pi\mathrm{i}}{N}}\,, \tag{1.12}$$

where $m,\, n \in \{0, 1, \ldots, N-1\}$. Given a complex valued vector $\vec{y}$ with components $\{y_k\}_{k=0}^{N-1}$, one can compute another vector $\vec{z}$ with components

$$z_m = \sum_{n=0}^{N-1} \Phi_{m,n}^{(N)} y_n , \tag{1.13}$$

referred to as *the finite Fourier transform* of the vector $\vec{y}$. Those vectors $\vec{f}^{\,(k)}$, which are solutions of the standard equations

$$\sum_{n=0}^{N-1} \Phi_{m,n}^{(N)} f_n^{(k)} = \lambda_k f_m^{(k)}, \qquad k \in \{0, 1, \ldots, N-1\} , \tag{1.14}$$

then represent eigenvectors of the FFT operator $\mathbf{\Phi}^{(N)}$, associated with the eigenvalues λ_k. Since the fourth power of $\mathbf{\Phi}^{(N)}$ is the unit matrix, the only four distinct eigenvalues among λ_k's are ± 1 and $\pm \mathrm{i}$.

Although there exists a good deal of discussion in the literature on eigenvectors of the FFT (see, for example [4]-[11] and references therein), the problem of deriving eigenvectors of FFT analytically still remains to be solved. So our objective is to find, in complete analogy with the continuous case (1.11), such a *self-adjoint difference operator with distinct eigenvalues*, which commutes with the FFT. The ability to solve a difference equation for eigenvectors thus obtained will enable one to define an analytical form of the desired set of eigenvectors for the finite Fourier transform.[1]

This presentation is a brief account of the first development along those lines, discussed above: namely, we have succeeded in constructing the explicit forms of raising and lowering difference operators that govern eigenvectors of the finite Fourie transform.

The layout of the paper is as follows. Section 2 collects those known facts about eigenvectors of the FFT , which are needed in section 3 for deriving some explicit forms of difference operators, commuting with the FFT $\mathbf{\Phi}^{(N)}$. In section 4 we construct an explicit form of the difference equation for eigenvectors of the FFT. Section 5 closes the paper with a brief remark, which outline some further research directions of interest.

Throughout this exposition we employ standard notations of the theory of special functions, see, for example, [12] or [13].

2. Eigenvectors of the FFT and their properties

It is a simple matter to verify that the matrix elements of $\mathbf{\Phi}^{(N)}$ are, by their definition, formally N-periodic, that is,

$$\Phi_{m,m'}^{(N)} = \Phi_{m+N,m'}^{(N)} = \Phi_{m,m'+N}^{(N)} . \tag{2.1}$$

This means that required explicit forms of the all eigenvectors $\vec{f}^{\,(n)}$ of the FFT operator $\mathbf{\Phi}^{(N)}$ should exhibit the same periodicity property

$$f_k^{(n)} = f_{k+N}^{(n)} , \qquad n, k = 0, 1, ..., N-1 . \tag{2.2}$$

We now turn to the question of defining how such N-periodic vectors $\vec{y}$ (which are not necessarily eigenvectors of the FFT operator $\mathbf{\Phi}^{(N)}$) transform when their components y_k are

[1] We emphasize from the outset that the idea of employing such approach for deriving eigenvectors of the FFT is not new (see, for example, [4, 5]). However, it seems that there never was consistent attempt to carry out this program explicitly for arbitrary N.

multiplied by powers of $q^{\pm k}$ and acted upon by the shift operators $\mathbf{T}^{(\pm)}$. These relations are building blocks of the rest of this work.

Let us define first two operators $\mathbf{Q}^{(\pm)}$ with matrix elements $Q_{kl}^{(\pm)} := q^{\pm k}\delta_{kl}$, where $q = e^{2\pi \mathrm{i}/N} \equiv e^{\mathrm{i}\theta_N}$, $\theta_N := 2\pi/N$. Then, by this definition, $\mathbf{Q}^{(\pm)}\vec{y} = \vec{z}^{(\pm)}$, where $z_k^{(\pm)} = \sum_{l=0}^{N-1} Q_{kl}^{(\pm)} y_l = \sum_{l=0}^{N-1} q^{\pm k}\delta_{kl}\, y_l = q^{\pm k}\, y_k$. This means that under the action of the operators $\mathbf{Q}^{(\pm)}$ the components y_k of an arbitrary vector $\vec{y}$ get multiplied by the factors $q^{\pm k}$, respectively. Consequently, for their linear combinations $\mathbf{C} := \frac{1}{2}\left(\mathbf{Q}^{(+)} + \mathbf{Q}^{(-)}\right)$ and $\mathbf{S} := \frac{1}{2\mathrm{i}}\left(\mathbf{Q}^{(+)} - \mathbf{Q}^{(-)}\right)$ one obtains that

$$\left(\mathbf{C}\,\vec{y}\right)_k = \frac{1}{2}\left(q^{k} + q^{-k}\right) y_k = \cos(k\,\theta_N)\, y_k\,,$$
$$\left(\mathbf{S}\,\vec{y}\right)_k = \frac{1}{2\mathrm{i}}\left(q^{k} - q^{-k}\right) y_k = \sin(k\,\theta_N)\, y_k\,. \tag{2.3}$$

The next step is to define a pair of the shift operators $\mathbf{T}^{(\pm)}$ with matrix elements $T_{kl}^{(\pm)} := \delta_{k\pm1,\,l}$, where $\delta_{-1,\,l} \equiv \delta_{N-1,\,l}$ and $\delta_{N,\,l} \equiv \delta_{0,\,l}$. Then

$$\left(\mathbf{T}^{(\pm)}\,\vec{y}\right)_k = \sum_{l=0}^{N-1} T_{kl}^{(\pm)}\, y_l = \sum_{l=0}^{N-1} \delta_{k\pm1,\,l}\, y_l = y_{k\pm1}\,, \tag{2.4}$$

where $y_{-1} \equiv y_{N-1}$ and $y_N \equiv y_0$.

Lemma. Components of any N-periodic vector $\vec{y}$ satisfy the following identities:

$$\sum_{k=0}^{N-1} q^{j(k\pm1)}\, y_k = \sum_{k=0}^{N-1} q^{j\,k}\, y_{k\mp1}\,, \qquad j = 0, 1, 2, ..., N-1\,. \tag{2.5}$$

Proof. Begin with the left side of the first identity in (2.5),

$$\sum_{k=0}^{N-1} q^{j\,(k+1)}\, y_k = \sum_{m=1}^{N} q^{j\,m}\, y_{m-1} = \sum_{m=1}^{N-1} q^{j\,m}\, y_{m-1} + q^{j\,N}\, y_{N-1} = \sum_{m=0}^{N-1} q^{j\,m}\, y_{m-1}\,, \tag{2.6}$$

in view of the evident relation $q^{Nj} = 1$. The second identity in (2.5) is argued similarly. So the lemma is proved. □

The operators $\mathbf{Q}^{(\pm)}$ and $\mathbf{T}^{(\pm)}$ are actually interconnected through the FFT operator $\mathbf{\Phi}^{(N)}$ in the following way.

Proposition. The intertwining relations

$$\mathbf{Q}^{(\pm)}\,\mathbf{\Phi}^{(\mathbf{N})} = \mathbf{\Phi}^{(\mathbf{N})}\,\mathbf{T}^{(\mp)}\,, \qquad \mathbf{T}^{(\pm)}\,\mathbf{\Phi}^{(\mathbf{N})} = \mathbf{\Phi}^{(\mathbf{N})}\,\mathbf{Q}^{(\pm)}\,, \tag{2.7}$$

for the operators $\mathbf{Q}^{(\pm)}$ and $\mathbf{T}^{(\pm)}$ with the FFT operator $\mathbf{\Phi}^{(N)}$ are respectively valid.

Proof. Two relations on the left in (2.7) follow directly from the lemma, if one takes into account the definition of the shift operators $\mathbf{T}^{(\pm)}$. So let us consider now matrix elements of two identities on the right in (2.7):

$$\left(\mathbf{T}^{(\pm)}\,\mathbf{\Phi}^{(\mathbf{N})}\right)_{m\,n} = \sum_{k=0}^{N-1} T_{m\,k}^{(\pm)}\,\Phi_{k\,n}^{(N)} = \sum_{k=0}^{N-1} \delta_{m\pm1,\,k}\,\Phi_{k\,n}^{(N)} = \Phi_{m\pm1,\,n}^{(N)}$$
$$= \frac{1}{\sqrt{N}}\, q^{(m\pm1)\,n} = q^{\pm n}\,\Phi_{m,\,n}^{(N)} = \sum_{l=0}^{N-1} \Phi_{m,\,l}^{(N)}\, q^{\pm l}\,\delta_{l\,n} = \left(\mathbf{\Phi}^{(\mathbf{N})}\,\mathbf{Q}^{(\pm)}\right)_{m\,n}\,.$$

This completes the proof of the proposition. □

Notice that if a vector $\mathbf{y}^{(n)}$ with the components $\{y_j^{(n)}\}_{j=0}^{N-1}$ is N-periodic, then all m vectors $\mathbf{z}^{(n;\,m)}$, which have, by definition, components $\{z_j^{(n;\,m)}\}_{j=0}^{N-1} := \{q^{mj}\, y_j^{(n)}\}_{j=0}^{N-1}$, m is an arbitrary integer number, are also N-periodic. Therefore, applying each line in (2.7) m times in succession, one arrives at the following corollary to the proposition:

$$q^{\pm mj} \sum_{k=0}^{N-1} \Phi_{j,k}^{(N)}\, y_k^{(n)} = \sum_{k=0}^{N-1} \Phi_{j,k}^{(N)}\, \mathbf{T}_{\mp}^{m}\, y_k^{(n)} ,$$

$$\mathbf{T}_{\pm}^{m} \sum_{k=0}^{N-1} \Phi_{j,k}^{(N)}\, y_k^{(n)} = \sum_{k=0}^{N-1} \Phi_{j,k}^{(N)}\, q^{\pm mk}\, y_k^{(n)} , \tag{2.8}$$

where m is an integer.

3. Difference operators, commuting with $\mathbf{\Phi}^{(N)}$

We are now in a position to discuss explicit forms of shift (difference) operators, which commute with the finite Fourier transform operator $\mathbf{\Phi}^{(N)}$. Making use of (2.7), we evaluate first, in analogy with (1.5), that

$$\frac{1}{2}\left(\mathbf{T}^{(+)} + \mathbf{T}^{(-)}\right) \mathbf{\Phi}^{(\mathbf{N})} = \frac{1}{2}\, \mathbf{\Phi}^{(\mathbf{N})} \left(\mathbf{Q}^{(+)} + \mathbf{Q}^{(-)}\right) = \mathbf{\Phi}^{(\mathbf{N})}\, \mathbf{C} . \tag{3.1}$$

Since on the other hand, in view of the same proposition (2.7),

$$\mathbf{C}\, \mathbf{\Phi}^{(\mathbf{N})} = \frac{1}{2}\left(\mathbf{Q}^{(+)} + \mathbf{Q}^{(-)}\right) \mathbf{\Phi}^{(\mathbf{N})} = \frac{1}{2}\, \mathbf{\Phi}^{(\mathbf{N})} \left(\mathbf{T}^{(+)} + \mathbf{T}^{(-)}\right), \tag{3.2}$$

one readily concludes from (3.1) and (3.2) that the operator $\mathbf{U}^{(1)}$,

$$\mathbf{U}^{(1)} := \mathbf{C} + \frac{1}{2}\left(\mathbf{T}^{(+)} + \mathbf{T}^{(-)}\right), \tag{3.3}$$

with matrix elements $U_{kl}^{(1)} = \cos k\theta_N\, \delta_{kl} + \frac{1}{2}\left(\delta_{k+1,\,l} + \delta_{k-1,\,l}\right)$, does commute with the FFT operator $\mathbf{\Phi}^{(N)}$:

$$\mathbf{U}^{(1)}\, \mathbf{\Phi}^{(N)} = \mathbf{\Phi}^{(N)}\, \mathbf{U}^{(1)} .$$

Moreover, one verifies that in the continuous limit as $N \to \infty$

$$\lim_{N\to\infty} \frac{N}{\pi}\left(\mathbf{I} - \frac{1}{2}\mathbf{U}^{(1)}\right) = \frac{1}{2}\left(x^2 - \frac{d^2}{dx^2}\right), \tag{3.4}$$

where x is dimensionless coordinate and $\mathbf{I}$ is the $N \times N$ unit matrix (consult [4, 8] for a detailed discussion of how to treat such convergence of a discrete operator to a continuum one in a rigorous mathematical manner).

To facilitate ease of comparing the operator $\mathbf{U}^{(1)}$ with other previously considered in the literature operators, which commute with $\mathbf{\Phi}^{(N)}$, we display its matrix form:

$$\left(U_{m,m'}^{(1)}\right) = \frac{1}{2}\begin{pmatrix} 2 & 1 & 0 & 0 & \cdots & 0 & 0 & 1 \\ 1 & 2\cos\frac{2\pi}{N} & 1 & 0 & \cdots & 0 & 0 & 0 \\ 0 & 1 & 2\cos\frac{4\pi}{N} & 1 & \cdots & 0 & 0 & 0 \\ \cdots & \cdots & \cdots & \cdots & \cdots & \cdots & \cdots & \cdots \\ 0 & 0 & 0 & 0 & \cdots & 1 & 2\cos\frac{4\pi}{N} & 1 \\ 1 & 0 & 0 & 0 & \cdots & 0 & 1 & 2\cos\frac{2\pi}{N} \end{pmatrix}.$$

This matrix has repeatedly surfaced in the literature on the finite Fourier transform. For instance, it coincides with the matrix S in [5], see formula (12) therein. An eigenvalue problem for the matrix (or difference operator) $\mathbf{U^{(1)}}$ probably first occurs in [6] and it is therefore known as Harper's equation, whereas the combination $\frac{2\pi}{N}\left(\mathbf{I}-\frac{1}{2}\mathbf{U^{(1)}}\right)$ represents the Hamiltonian of the so-called *Harper function harmonic oscillator* [7, 8].

However, the operator $\mathbf{U^{(1)}}$ fails to serve to our purposes for generic N (*in spite of the fact that it does commute with the* $\mathbf{\Phi}^{(N)}$ *and reduces to the linear harmonic oscillator Hamiltonian (1.3) in the continuous limit as* $N\to\infty$) because it has multiple eigenvalues (in particular, when N is divisible by four, it has two zero eigenvalues).

Before proceeding to the problem of finding another difference operator in question (*with distinct eigenvalues for generic* N), we call attention to the following two observations. Firstly, one may generalize $\mathbf{U^{(1)}}$ and derive a family of operators $\mathbf{U^{(k)}}$,

$$\mathbf{U^{(k)}} := \frac{1}{2}\left[\left(\mathbf{T}^{(+)}\right)^k + \left(\mathbf{T}^{(-)}\right)^k + \left(\mathbf{Q}^{(+)}\right)^k + \left(\mathbf{Q}^{(-)}\right)^k\right], \tag{3.5}$$

where k is a positive integer. Then it is not hard to verify, by using relations (2.7), that all operators $\mathbf{U^{(k)}}$, $k=1,2,3,...$, commute with the finite Fourier transform $\mathbf{\Phi}^{(N)}$. Secondly, since $\mathbf{U^{(1)}}$ commutes with $\mathbf{\Phi}^{(N)}$, so does its square, $\left(\mathbf{U^{(1)}}\right)^2$. Combining these two facts, one therefore concludes that the operator

$$\mathbf{V_1} := \left(\mathbf{U^{(1)}}\right)^2 - \frac{1}{2}\mathbf{U^{(2)}} \tag{3.6}$$

also does commute with the finite Fourier transform $\mathbf{\Phi}^{(N)}$. The reason for picking up such a combination of the operators $\mathbf{U^{(1)}}$ and $\mathbf{U^{(2)}}$ in (3.6) is that it still contains only linear terms in the shift operators $\mathbf{T}^{(\pm)}$. Indeed, by definition (3.6), one has

$$\mathbf{V_1} = \left(\mathbf{U^{(1)}}\right)^2 - \frac{1}{2}\mathbf{U^{(2)}} = \frac{1}{4}\Big[\left(\mathbf{T}^{(+)}+\mathbf{T}^{(-)}+\mathbf{Q}^{(+)}+\mathbf{Q}^{(-)}\right)^2$$

$$-\left(\mathbf{T}^{(+)}\right)^2-\left(\mathbf{T}^{(-)}\right)^2-\left(\mathbf{Q}^{(+)}\right)^2-\left(\mathbf{Q}^{(-)}\right)^2\Big] \equiv \mathbf{I} + \cos\frac{\pi}{N}\,\mathbf{V_2}, \tag{3.7}$$

upon taking into account the readily verified commutation relations

$$\mathbf{T}^{(+)}\mathbf{Q}^{(\pm)} = q^{\pm 1}\mathbf{Q}^{(\pm)}\mathbf{T}^{(+)}, \qquad \mathbf{T}^{(-)}\mathbf{Q}^{(\pm)} = q^{\mp 1}\mathbf{Q}^{(\pm)}\mathbf{T}^{(-)} \tag{3.8}$$

between the operators $\mathbf{T}^{(\pm)}$ and $\mathbf{Q}^{(\pm)}$. The operator $\mathbf{V_2}$ in (3.7) is equal to

$$\mathbf{V_2} = \left(\cos\frac{\pi}{N}\,\mathbf{C} - \sin\frac{\pi}{N}\,\mathbf{S}\right)\mathbf{T}^{(+)} + \left(\cos\frac{\pi}{N}\,\mathbf{C} + \sin\frac{\pi}{N}\,\mathbf{S}\right)\mathbf{T}^{(-)}, \tag{3.9}$$

or, in the matrix form,

$$\left((\mathbf{V_2})_{m,m'}\right) = \frac{1}{2}\begin{pmatrix} 0 & \cos\frac{\pi}{N} & 0 & 0 & \cdots & 0 & 0 & \cos\frac{\pi}{N} \\ \cos\frac{\pi}{N} & 0 & \cos\frac{3\pi}{N} & 0 & \cdots & 0 & 0 & 0 \\ 0 & \cos\frac{3\pi}{N} & 0 & \cos\frac{5\pi}{N} & \cdots & 0 & 0 & 0 \\ \cdots & \cdots & \cdots & \cdots & \cdots & \cdots & \cdots & \cdots \\ 0 & 0 & 0 & 0 & \cdots & \cos\frac{5\pi}{N} & 0 & \cos\frac{3\pi}{N} \\ \cos\frac{\pi}{N} & 0 & 0 & 0 & \cdots & 0 & \cos\frac{3\pi}{N} & 0 \end{pmatrix}.$$

In the continuous limit as $N \to \infty$ one obtains that

$$\lim_{N\to\infty} \frac{N}{\pi}\left(\mathbf{I} - \frac{1}{2}\mathbf{V_2}\right) = x^2 - \frac{d^2}{dx^2}\,. \tag{3.10}$$

Examples (3.5) and (3.6) just reflect the fact that there are infinitely many possibilities for constructing difference operators, which commute with the finite Fourier transform $\mathbf{\Phi}^{(N)}$. So we need to find among them *a difference operators with the simplest spectrum of distinct eigenvalues for generic* N. Clearly, in the continuous limit as $N \to \infty$ this operator should reduce to the second-order differential equation (1.3) for the Hermite functions (1.2).

As we shall establish in the next section, a required difference operator is actually comprised of the two above-introduced operators $\mathbf{U}^{(2)}$ and $\mathbf{V_1}$ (or, equivalently, $\mathbf{V_2}$).

4. Difference equation for eigenvectors of the FFT

It is a remarkable fact that an appropriate difference equation for the eigenvectors $\mathbf{f}^{(n)}$ of the FFT, associated with the eigenvalues $\lambda_n = \mathrm{i}^{\,n}$, can be constructed in complete analogy with the continuous case (that is, the Hermite functions (1.2)). To bring out this fact we now address ourselves to a study of difference analogues of the lowering $\mathbf{a}$ and raising $\mathbf{a}^\dagger$ differential operators (1.9), which factorize the operator $\mathbf{H}$ in (1.11),

$$\mathbf{H} = \mathbf{a}^\dagger\mathbf{a} + \frac{1}{2}\mathbf{I}. \tag{4.1}$$

Recall also that very often it is more convenient to work directly with the number operator $\mathbf{N}$, defined as

$$\mathbf{N} := \mathbf{H} - \frac{1}{2}\mathbf{I} = \mathbf{a}^\dagger\mathbf{a}. \tag{4.2}$$

Then from the commutation relation (1.10) it follows at once that

$$\mathbf{N}\,\mathbf{a} = \mathbf{a}(\mathbf{N}-\mathbf{I}), \qquad \mathbf{N}\,\mathbf{a}^\dagger = \mathbf{a}^\dagger(\mathbf{N}+\mathbf{I}), \tag{4.3}$$

and the above identities confirm algebraically that the operators $\mathbf{a}$ and $\mathbf{a}^\dagger$ do lower and raise, respectively, the eigenvalues of the number operator $\mathbf{N}$ (or, equivalently, of the Hamiltonian $\mathbf{H}$). So in that way one naturally arrives at the conclusion that the Fourier transform (1.4) can be equivalently represented in another operator form as

$$\mathcal{F} = \exp\left(\frac{\pi \mathrm{i}}{2}\,\mathbf{N}\right). \tag{4.4}$$

Of course, from this operator form the identities (1.8) now follow at once, upon taking into account the commutation relations (4.3).

To construct difference analogues of the operators $\mathbf{a}$ and $\mathbf{a}^\dagger$, which are evidently building blocks of the continuous case (4.1), we may proceed as in the previous section. So let us consider first two operators $\mathcal{X}$ and $\mathcal{D}$ $(N \geq 3)$:

$$\mathcal{X} := \frac{1}{2\mathrm{i}}\sqrt{\frac{N}{2\pi}}\left(\mathbf{Q}^{(+)} - \mathbf{Q}^{(-)}\right) \equiv \sqrt{\frac{N}{2\pi}}\,\mathbf{S}, \qquad \mathcal{D} := \frac{1}{2}\sqrt{\frac{N}{2\pi}}\left(\mathbf{T}^{(+)} - \mathbf{T}^{(-)}\right). \tag{4.5}$$

From the intertwining relations (2.7) it follows then that

$$\mathcal{D}\,\mathbf{\Phi}^{(N)} = \mathrm{i}\,\mathbf{\Phi}^{(N)}\mathcal{X}, \qquad \mathcal{X}\,\mathbf{\Phi}^{(N)} = \mathrm{i}\,\mathbf{\Phi}^{(N)}\mathcal{D}. \tag{4.6}$$

Also, one readily verifies that the operators $\mathcal{X}$ and $\mathcal{D}$ are normalized in such a way that the limit relations

$$\lim_{N\to\infty} \mathcal{X} = x, \qquad \lim_{N\to\infty} \mathcal{D} = \frac{d}{dx}, \tag{4.7}$$

for these two operators do hold.

Remark. It is important to observe that the action of the difference operator $\mathcal{D}$ on the product of two functions is more complicated than the Leibniz product rule for the derivative $\frac{d}{dx}$. Indeed, it is not hard to verify that for two arbitrary functions $u(j)$ and $v(j)$ one has

$$u(j+1)\,v(j+1) - u(j-1)\,v(j-1) = u(j+1)[v(j+1) - v(j-1)] + u(j+1)\,v(j-1)$$

$$-u(j-1)[v(j-1) - v(j+1)] - u(j-1)\,v(j+1) = [u(j+1) + u(j-1)]\,[v(j+1) - v(j-1)]$$
$$+[u(j+1) - u(j-1)]\,[v(j+1) + v(j-1)] - [u(j+1)\,v(j+1) - u(j-1)\,v(j-1)]\,.$$

Consequently,

$$u(j+1)\,v(j+1) - u(j-1)\,v(j-1) = \frac{1}{2}\,[u(j+1) + u(j-1)]\,[v(j+1) - v(j-1)]$$

$$+\frac{1}{2}\,[u(j+1) - u(j-1)]\,[v(j+1) + v(j-1)].$$

Taking into account the definition of the operator $\mathcal{D}$ in (4.5), one concludes that

$$\mathcal{D}[u(j)\,v(j)] = [\mathcal{A}u(j)]\;[\mathcal{D}v(j)] + [\mathcal{D}u(j)]\;[\mathcal{A}v(j)]\,, \tag{4.8}$$

where $\mathcal{A}$ is *the averaging difference operator*, defined as (*cf.* formula (21.6.3) in [14])

$$\mathcal{A} := \frac{1}{2}\left(\mathbf{T}^{(+)} + \mathbf{T}^{(-)}\right). \tag{4.9}$$

The next natural step is to evaluate that the commutation relation between the operators $\mathcal{X}$ and $\mathcal{D}$ is of the form

$$[\,\mathcal{D}, \mathcal{X}\,] = \frac{N}{\pi}\,\sin\frac{\pi}{N}\,\mathbf{V_2}\,, \tag{4.10}$$

where $\mathbf{V_2}$ is defined in (3.9). This commutation relation is a discrete analogue of the Heisenberg commutation relation (1.10) for the operators x and $\frac{d}{dx}$ and on account of (3.10) the (4.10) reduces to the latter in the continuous limit as $N \to \infty$.

To imitate the steps (1.5)–(1.7) of the continuous case, one evaluates now, by employing each of the identities in (4.6) twice, that

$$\mathcal{D}^{\mathbf{2}}\,\mathbf{\Phi}^{(N)} = -\,\mathbf{\Phi}^{(N)}\mathcal{X}^{\mathbf{2}}\,, \qquad \mathcal{X}^{\mathbf{2}}\,\mathbf{\Phi}^{(N)} = -\,\mathbf{\Phi}^{(N)}\mathcal{D}^{\mathbf{2}}\,. \tag{4.11}$$

This means that a difference operator $\mathcal{H}$, defined as

$$\mathcal{H} := \frac{1}{2}\left(\mathcal{X}^{\mathbf{2}} - \mathcal{D}^{\mathbf{2}}\right), \tag{4.12}$$

does commute with the FFT operator $\mathbf{\Phi}^{(N)}$, that is, $\mathcal{H}\,\mathbf{\Phi}^{(N)} = \mathbf{\Phi}^{(N)}\,\mathcal{H}$. Since in the continuous limit as $N \to \infty$ the (4.12) reduces to the linear harmonic oscillator Hamiltonian $\mathbf{H}/\hbar\,\omega$ (because of the limit relations (4.7)), the difference operator $\mathcal{H}$ can serve as a required discrete analogue of the continuous case (1.3).

Difference analogues of the lowering and raising operators $\mathbf{a}$ and $\mathbf{a}^\dagger$ in (4.1) may be now defined as

$$\mathbf{b}_N := \frac{1}{\sqrt{2}}\,(\mathcal{X}+\mathcal{D}) = \frac{1}{2}\sqrt{\frac{N}{\pi}}\,\left[\mathbf{S} + \frac{1}{2}\Big(\mathbf{T}^{(+)} - \mathbf{T}^{(-)}\Big)\right], \tag{4.13a}$$

$$\mathbf{b}_N^\dagger := \frac{1}{\sqrt{2}}\,(\mathcal{X}-\mathcal{D}) = \frac{1}{2}\sqrt{\frac{N}{\pi}}\,\left[\mathbf{S} - \frac{1}{2}\Big(\mathbf{T}^{(+)} - \mathbf{T}^{(-)}\Big)\right]. \tag{4.13b}$$

We also display the matrix form of the operators $\mathbf{b}_N$ and $\mathbf{b}_N^\dagger$, respectively:

$$\Big((\mathbf{b}_N)_{m,m'}\Big) = \frac{1}{4}\sqrt{\frac{N}{\pi}}\begin{pmatrix} 0 & 1 & 0 & 0 & \cdots & 0 & 0 & -1 \\ -1 & 2\sin\frac{2\pi}{N} & 1 & 0 & \cdots & 0 & 0 & 0 \\ 0 & -1 & 2\sin\frac{4\pi}{N} & 1 & \cdots & 0 & 0 & 0 \\ \cdots & \cdots & \cdots & \cdots & \cdots & \cdots & \cdots & \cdots \\ 0 & 0 & 0 & 0 & \cdots & -1 & -2\sin\frac{4\pi}{N} & 1 \\ 1 & 0 & 0 & 0 & \cdots & 0 & -1 & -2\sin\frac{2\pi}{N} \end{pmatrix},$$

$$\Big((\mathbf{b}_N^\dagger)_{m,m'}\Big) = \frac{1}{4}\sqrt{\frac{N}{\pi}}\begin{pmatrix} 0 & -1 & 0 & 0 & \cdots & 0 & 0 & 1 \\ 1 & 2\sin\frac{2\pi}{N} & -1 & 0 & \cdots & 0 & 0 & 0 \\ 0 & 1 & 2\sin\frac{4\pi}{N} & -1 & \cdots & 0 & 0 & 0 \\ \cdots & \cdots & \cdots & \cdots & \cdots & \cdots & \cdots & \cdots \\ 0 & 0 & 0 & 0 & \cdots & 1 & -2\sin\frac{4\pi}{N} & -1 \\ -1 & 0 & 0 & 0 & \cdots & 0 & 1 & -2\sin\frac{2\pi}{N} \end{pmatrix}.$$

As a direct consequence of (4.10), these difference analogues of $\mathbf{a}$ and $\mathbf{a}^\dagger$ satisfy the following commutation relation

$$\left[\mathbf{b}_N,\,\mathbf{b}_N^\dagger\right] \equiv [\mathcal{D},\,\mathcal{X}\,] = \frac{N}{\pi}\,\sin\frac{\pi}{N}\,\mathbf{V_2} \tag{4.14}$$

and by construction they reduce to $\mathbf{a}$ and $\mathbf{a}^\dagger$ in the continuous limit as $N \to \infty$.

This arresting analogy with the continuous case still develops to an even greater extent since by using definitions (4.13a) and (4.13b) it is straightforward to verify that

$$\mathbf{b}_N^\dagger\,\mathbf{b}_N = \frac{1}{2}\,(\mathcal{X}-\mathcal{D})\,(\mathcal{X}+\mathcal{D}) = \frac{1}{2}\Big(\mathcal{X}^2 - \mathcal{D}^2 + \mathcal{X}\mathcal{D} - \mathcal{D}\mathcal{X}\Big)$$

$$= \mathcal{H} - \frac{1}{2}\Big[\mathcal{D},\,\mathcal{X}\Big] = \mathcal{H} - \frac{1}{2}\left[\mathbf{b}_N,\,\mathbf{b}_N^\dagger\right]. \tag{4.15}$$

So the difference operators $\mathbf{b}_N$ and $\mathbf{b}_N^\dagger$, introduced above in (4.13), do factorize the difference operator $\mathcal{H}$ in the form

$$\mathcal{H} = \mathbf{b}_N^\dagger\,\mathbf{b}_N + \frac{1}{2}\left[\mathbf{b}_N,\,\mathbf{b}_N^\dagger\right] = \frac{1}{2}\Big(\mathbf{b}_N\,\mathbf{b}_N^\dagger + \mathbf{b}_N^\dagger\,\mathbf{b}_N\Big), \tag{4.16}$$

which closely resembles the symmetrized expression (1.11) for $\mathbf{H}$ in the continuous case. Therefore the operator $\mathcal{N} := \mathbf{b}_N^\dagger\,\mathbf{b}_N$ can be regarded as a difference analogue of the number operator $\mathbf{N}$. Observe that since the difference between the operators $\mathcal{N}$ and $\mathcal{H}$ is just proportional to the operator $\mathbf{V_2}$,

$$\mathcal{N} \equiv \mathcal{H} - \frac{1}{2}\left[\mathbf{b}_N,\,\mathbf{b}_N^\dagger\right] = \mathcal{H} - \frac{N}{2\pi}\,\sin\frac{\pi}{N}\,\mathbf{V_2}, \tag{4.16'}$$

the former operator $\mathcal{N}$ commutes with the FFT operator $\mathbf{\Phi}^{(N)}$ as well (see (3.6) and (3.7)).

Finally, it turns out that thus introduced difference operators $\mathbf{b}_N$ and $\mathbf{b}_N^\dagger$ have "proper" commutation relations with the FFT operator $\mathbf{\Phi}^{(N)}$: from intertwining relations (4.6) it follows at once that

$$\mathbf{b}_N\,\mathbf{\Phi}^{(N)} = \mathrm{i}\,\mathbf{\Phi}^{(N)}\,\mathbf{b}_N\,, \qquad \mathbf{\Phi}^{(N)}\,\mathbf{b}_N^\dagger = \mathrm{i}\,\mathbf{b}_N^\dagger\,\mathbf{\Phi}^{(N)}. \tag{4.17}$$

Consequently, the difference operators $\mathbf{b}_N$ and $\mathbf{b}_N^\dagger$ in (4.13) actually represent n-lowering and n-raising operators, respectively, for the eigenvectors $\mathbf{f}^{(n)}$ of the FFT, associated with the eigenvalues $\lambda_n = \mathrm{i}^n$. Thus,

$$\mathbf{b}_N\,\mathbf{f}^{(n)} = \mu_n^{(N)}\,\mathbf{f}^{(n-1)}\,, \qquad \mathbf{b}_N^\dagger\,\mathbf{f}^{(n)} = \nu_{n+1}^{(N)}\,\mathbf{f}^{(n+1)}. \tag{4.18}$$

where $\mu_n^{(N)}$ rand $\nu_n^{(N)}$ are some constant factors, which may depend on n.

Remark. For arbitrary integer N it is not difficult to construct explicitly the "ground eigenvector" $\{f_m^{(0)}\}_{m=0}^{N-1}$ of the FFT (1.12) with the eigenvalue $\lambda_0 = \mathrm{i}^0 = 1$. Indeed, this eigenvector has the form $\{f_m^{(0)}\} := c_0^{(N)}\{\sqrt{N}+1, 1, 1, ..., 1\}$, which is a consequence of the fact the sum of elements in the k-th row of FFT-matrix $\left(\Phi_{m,m'}^{(N)}\right)$, $k = 1, 2, 3, ..., N-1$, actually represents a geometric progression

$$S_N(x) := 1 + x + x^2 + ... + x^{N-1} = \frac{1-x^N}{1-x}$$

in the variable $x = q^k$ and therefore all of $S_N(q^k)$ are equal to zero for any integer value of $k = 1, 2, 3, ..., N-1$ (recall that $q^N = 1$ by definition (1.12) of q). Thus,

$$\sum_{m'=0}^{N-1}\Phi_{m,m'}^{(N)}\,f_{m'}^{(0)} = \frac{c_0^{(N)}}{\sqrt{N}}\begin{pmatrix} 1 & 1 & 1 & \cdots & 1 & 1 \\ 1 & q & q^2 & \cdots & q^{N-2} & q^{N-1} \\ 1 & q^2 & q^4 & \cdots & q^{2(N-2)} & q^{2(N-1)} \\ \cdots & \cdots & \cdots & \cdots & \cdots & \cdots \\ 1 & q^{N-1} & q^{2(N-1)} & \cdots & q^{(N-2)(N-1)} & q^{(N-1)^2} \end{pmatrix}\begin{pmatrix} \sqrt{N}+1 \\ 1 \\ 1 \\ \cdots \\ 1 \end{pmatrix}$$

$$= \frac{c_0^{(N)}}{\sqrt{N}}\begin{pmatrix} \sqrt{N}+1+N-1 \\ \sqrt{N}+S_n(q) \\ \sqrt{N}+S_n(q^2) \\ \cdots \\ \sqrt{N}+S_n(q^{N-1}) \end{pmatrix} = c_0^{(N)}\begin{pmatrix} \sqrt{N}+1 \\ 1 \\ 1 \\ \cdots \\ 1 \end{pmatrix} = f_m^{(0)}.$$

To normalize this "ground eigenvector" $\{f_m^{(0)}\}$ to one, it is sufficient to choose the constant $c_0^{(N)} = [2\sqrt{N}(\sqrt{N}+1)]^{-1/2}$.

It follows at once from (4.17) that the difference operator $\mathcal{B}^{(N)} := \mathbf{b}_N^\dagger\,\mathbf{b}_N$ commutes with the FFT operator $\mathbf{\Phi}^{(N)}$, that is,

$$\left[\mathcal{B}^{(N)},\,\mathbf{\Phi}^{(N)}\right] = 0\,.$$

We thus conclude that a required *difference analogue* of the second-order differential equation $\mathbf{N}\,\psi_n(x) = n\,\psi_n(x)$, which is associated with the continuous case, has the form

$$\mathcal{B}^{(N)}\,\mathbf{f}^{(n)} = \lambda_n^{(N)}\,\mathbf{f}^{(n)}, \qquad \lambda_n^{(N)} = \mu_n^{(N)}\,\nu_n^{(N)}, \tag{4.19}$$

where

$$\mathcal{B}^{(N)} = \frac{1}{2}\,(\mathcal{X}-\mathcal{D})\,(\mathcal{X}+\mathcal{D}) = \frac{N}{4\pi}\left(I - \mathbf{S_2} - 2\sin\frac{\pi}{N}\,\mathbf{R_2}\right) \tag{4.20}$$

and $\mathbf{S}_2$ and $\mathbf{R}_2$ are defined in (3.5) and (3.8), respectively. Observe that a formal solution of the difference equation (4.19) may be constructed in terms of the *ground* eigenvector $\mathbf{f}^{(0)}$, governed by, on account of (4.13), a difference equation

$$\mathbf{b}_N\,\mathbf{f}^{(0)} = \frac{1}{4}\sqrt{\frac{N}{\pi}}\left[\mathbf{T}_+ - \mathbf{T}_- + 2\,\mathbf{S}\right]\mathbf{f}^{(0)} = 0\,. \tag{4.21}$$

Then any other eigenvector $\mathbf{f}^{(n)}$ with an arbitrary n is found as

$$\mathbf{f}^{(n)} = \frac{1}{\prod_{k=1}^{n}\nu_k^{(N)}}\left(\mathbf{b}_N^{\dagger}\right)^n\mathbf{f}^{(0)}\,, \tag{4.22}$$

in conformity with the well-known formula

$$\psi_n(x) = \frac{1}{\sqrt{n!}}\left(\mathbf{a}^{\dagger}\right)^n\psi_0(x) \tag{4.23}$$

for the classical Hermite functions (1.2).

5. Concluding remarks

To summarize, we have constructed the explicit forms of raising and lowering difference operators that govern eigenvectors of the finite Fourie transform. The main algebraic properties of these raising and lowering difference operators have been examined in detail. We hope that this particular knowledge will help us to achieve the same degree of analytic tractability with the finite Fourier transform, as we observe in the case of its continuous counterpart, briefly outlined in Introduction.

Acknowledgments

We are grateful to V.B. Matveev for illuminating discussions. The participation of MKA in this work has been partially supported by the SEP-CONACyT project 168104 "Operadores integrales y pseudodiferenciales en problemas de fisíca matemática" and by the *Proyecto de Redes* PROMEP (México). NMA has been partially supported by the CONACyT project CB220692 "Óptica Matemática".

References

[1] Landau L D and Lifshitz E M 1991 *Quantum Mechanics (Non-relativistic Theory)* (Oxford: Pergamon Press)
[2] Auslander L and Tolimieri R 1979 Is computing with the finite Fourier transform pure or applied mathematics? *Bull. Amer. Math. Soc.* **1** 847–897
[3] Terras A 1999 *Fourier Analysis on Finite Groups and Applications* (Cambridge: Cambridge University Press)
[4] Grünbaum F A 1982 The eigenvectors of the discrete Fourier transform: a version of the Hermite functions *J. Math. Anal. Appl.* **88** 355–363
[5] Dickinson B W and Steiglitz K 1982 Eigenvectors and functions of the discrete Fourier transform *IEEE Transactions on Acoustics, Speech, and Signal Processing* **30** 25–31
[6] Harper P G 1955 Single band motion of conduction electrons in a uniform magnetic field *Proc. Phys. Soc. A* **68** 874–878
[7] Barker L 2000 The discrete fractional Fourier transform and Harper's equation *Mathematika* **47** 281–297
[8] Barker L 2001 Continuum quantum systems as limits of discrete systems III Operators *J. Math. Phys.* **42** 4653–4668
[9] Mehta M L 1987 Eigenvalues and eigenvectors of the finite Fourier transform *J. Math. Phys.* **28** 781-785
[10] Matveev V B 2001 Intertwining relations between the Fourier transform and discrete Fourier transform, the related functional identities and beyond *Inverse Problems* **17** 633–657
[11] Atakishiyev N M 2006 On q-extensions of Mehta's eigenvectors of the finite Fourier transform *Int. J. Mod. Phys. A* **21** 4993–5006

[12] Gasper G and Rahman M 2004 *Basic Hypergeometric Functions* (Cambridge: Cambridge University Press)
[13] Andrews G E Askey R and Roy R 1999 *Special Functions* (Cambridge: Cambridge University Press)
[14] Ismail M E H 2005 *Classical and Quantum Orthogonal Polynomials in One Variable* (Cambridge: Cambridge University Press)

A curved Hénon–Heiles system and its integrable perturbations

Ángel Ballesteros, Alfonso Blasco and Francisco J Herranz
Departamento de Física, Universidad de Burgos, E-09001 Burgos, Spain
E-mail: angelb@ubu.es, ablasco@ubu.es, fjherranz@ubu.es

Abstract. The constant curvature analogue on the two-dimensional sphere and the hyperbolic space of the integrable Hénon–Heiles Hamiltonian $\mathcal{H}$ given by

$$\mathcal{H} = \frac{1}{2}(p_1^2 + p_2^2) + \Omega\left(q_1^2 + 4q_2^2\right) + \alpha\left(q_1^2 q_2 + 2q_2^3\right),$$

where Ω and α are real constants, is revisited. The resulting integrable curved Hamiltonian, $\mathcal{H}_\kappa$, depends on a parameter κ which is just the curvature of the underlying space and allows one to recover $\mathcal{H}$ under the smooth flat/Euclidean limit $\kappa \to 0$. This system can be regarded as an integrable cubic perturbation of a specific curved 1 : 2 anisotropic oscillator, which was already known in the literature. The Ramani series of potentials associated to $\mathcal{H}_\kappa$ is fully constructed, and corresponds to the curved integrable analogues of homogeneous polynomial perturbations of $\mathcal{H}$ that are separable in parabolic coordinates. Integrable perturbations of $\mathcal{H}_\kappa$ are also fully presented, and they can be regarded as the curved counterpart of integrable rational perturbations of the Euclidean Hamiltonian $\mathcal{H}$. It will be explicitly shown that the latter perturbations can be understood as the 'negative index' counterpart of the curved Ramani series of potentials. Furthermore, it is shown that the integrability of the curved Hénon–Heiles Hamiltonian $\mathcal{H}_\kappa$ is preserved under the simultaneous addition of curved analogues of 'positive' and 'negative' families of Ramani potentials.

1. Introduction

The Hénon–Heiles Hamiltonian

$$H = \frac{1}{2}(p_1^2 + p_2^2) + \frac{1}{2}(q_1^2 + q_2^2) + \lambda\left(q_1^2 q_2 - \frac{1}{3}q_2^3\right)$$

was introduced in [1] in order to model a Newtonian axially-symmetric galactic system. Nevertheless, it was soon considered as the paradigm of a two-dimensional (2D) system that exhibited chaotic behaviour (see, for instance, [2, 3, 4]). Later on, when the following generalization containing adjustable parameters was introduced

$$\mathcal{H} = \frac{1}{2}(p_1^2 + p_2^2) + \Omega_1 q_1^2 + \Omega_2 q_2^2 + \alpha\left(q_1^2 q_2 + \beta q_2^3\right),$$

it was proven that the only Liouville-integrable [5] members of this family of generalized Hénon–Heiles Hamiltonians were given by *three* specific choices of the real parameters Ω_1, Ω_2, α and β (see [6, 7, 8, 9, 10, 11, 12, 13, 14, 15, 16, 17, 18] and references therein):

Journal of Physics: Conference Series **597** (2015) 012013 doi:10.1088/1742-6596/597/1/012013

- The Sawada–Kotera system, given by $\beta = 1/3$ and $\Omega_1 = \Omega_2 = \Omega$:

$$\mathcal{H} = \frac{1}{2}(p_1^2 + p_2^2) + \Omega\left(q_1^2 + q_2^2\right) + \alpha\left(q_1^2 q_2 + \frac{1}{3}q_2^3\right). \tag{1}$$

- The Korteweg–de Vries (KdV) system, with $\beta = 2$ and (Ω_1, Ω_2) arbitrary parameters:

$$\mathcal{H} = \frac{1}{2}(p_1^2 + p_2^2) + \Omega_1 q_1^2 + \Omega_2 q_2^2 + \alpha\left(q_1^2 q_2 + 2q_2^3\right). \tag{2}$$

- The Kaup–Kuperschdmit system, with $\beta = 16/3$ and $\Omega_2 = 16\Omega_1 = 16\Omega$:

$$\mathcal{H} = \frac{1}{2}(p_1^2 + p_2^2) + \Omega\left(q_1^2 + 16q_2^2\right) + \alpha\left(q_1^2 q_2 + \frac{16}{3}q_2^3\right). \tag{3}$$

Beyond the former integrable cases, it is worthy to emphasize that there exists a very interesting family of integrable homogeneous potentials, deeply connected to the particular KdV system (2) that arises when $\Omega_2 = 4\Omega_1$. These are the so-called Ramani series of integrable potentials [19, 20], which can be freely superposed by preserving integrability, since they are just the polynomial potentials on the Euclidean plane that can be separated in parabolic coordinates [11, 21]. Furthermore, this classical separability property underlies the fact that a large collection of *integrable rational perturbations* can be added to the Ramani potentials by preserving the integrability of the complete Hamiltonian (see [21, 22, 23, 24, 25] and references therein).

In this paper we firstly review the integrable *curved* analogue on the 2D sphere $\mathbf{S}^2$ and the hyperbolic (or Lobachevski) space $\mathbf{H}^2$ of the *flat* KdV Hénon–Heiles Hamiltonian (2) with $\Omega_2 = 4\Omega_1$, which has been recently presented in [26], together with the full curved counterpart of the integrable Ramani series of potentials. In this approach, all the results depend on the Gaussian curvature κ of the underlying space in an explicit form, so that all the flat/Euclidean expressions can be recovered performing the zero-curvature limit (contraction) $\kappa \to 0$ from the curved expressions. Alternatively, the curvature κ can also be understood as a *deformation parameter* providing the curved systems as deformed versions from the flat/Euclidean ones that preserve the integrability of the former. Secondly, we present new integrable perturbations of the curved KdV system, thus generalizing the results of [26].

The structure of the paper is as follows. In the next section, we review all the *flat* integrable Hamiltonian background, that is, the properties and structure of the KdV Hénon–Heiles Hamiltonian $\mathcal{H}$ (hereafter with $\Omega_2 = 4\Omega_1$) and the Ramani series of potentials on the Euclidean plane $\mathbf{E}^2$. In section 3, we construct the known integrable rational perturbations of $\mathcal{H}$ as Ramani potentials with *negative* indices. In section 4, we briefly review the ambient (or Weierstrass) and Beltrami (projective) canonical variables for $\mathbf{S}^2$ and $\mathbf{H}^2$, which are needed in the curved framework. The resulting curved KdV Hénon–Heiles Hamiltonian $\mathcal{H}_\kappa$ and curved Ramani potentials are addressed in section 5. Finally, section 6 is devoted to present the new integrable perturbations of $\mathcal{H}_\kappa$, that can be understood as the 'negative index' counterpart of the curved Ramani potentials, as well as the result that the superposition of all the curved Ramani terms does preserve the integrability of the system.

2. An integrable Hénon–Heiles system on the plane

Le us consider the following tuning $\Omega_2 = 4\Omega_1 = 4\Omega$ in the KdV Hénon–Heiles Hamiltonian (2) defined on $\mathbf{E}^2$:

$$\mathcal{H} = \frac{1}{2}(p_1^2 + p_2^2) + \Omega(q_1^2 + 4q_2^2) + \alpha\left(q_1^2 q_2 + 2q_2^3\right), \tag{4}$$

where (q_1, q_2) are Cartesian coordinates and (p_1, p_2) their conjugate momenta satisfying the usual canonical Poisson bracket $\{q_i, p_j\} = \delta_{ij}$. This system is known to be endowed with a constant of motion *quadratic* in the momenta and given by

$$\mathcal{I} = p_1(q_1p_2 - q_2p_1) + q_1^2\left(2\Omega q_2 + \frac{\alpha}{4}(q_1^2 + 4q_2^2)\right), \tag{5}$$

that is, $\{\mathcal{H}, \mathcal{I}\} = 0$. Hence $\mathcal{H}$ is integrable in the Liouville sense. Notice that the Hamiltonian (4) can be regarded as an integrable cubic perturbation added to the $1:2$ oscillator with frequencies $(\omega, 2\omega)$ once the identification $\omega^2 = 2\Omega$ is performed.

It is worth making a carefully analysis of the potentials composing both the Hamiltonian (4) and its invariant (5). For this, let us recall that the so-called *Ramani series of integrable potentials* consists of homogeneous polynomial potentials of degree n given by [19, 20]

$$\mathcal{V}_n(q_1, q_2) = \sum_{i=0}^{[\frac{n}{2}]} 2^{n-2i}\binom{n-i}{i} q_1^{2i} q_2^{n-2i}, \qquad n = 1, 2, \ldots \tag{6}$$

In this respect, we remark that the quadratic potential (the $1:2$ oscillator) and the cubic potential in $\mathcal{H}$, say $\mathcal{V}_2$ and $\mathcal{V}_3$, are just the second- and the third-order Ramani potentials, respectively. Moreover, the integral $\mathcal{I}$ contains the linear $\mathcal{V}_1$ and the quadratic $\mathcal{V}_2$ Ramani potentials; namely

$$\mathcal{V}_1 = 2q_2, \qquad \mathcal{V}_2 = q_1^2 + 4q_2^2, \qquad \mathcal{V}_3 = 4q_1^2q_2 + 8q_2^3.$$

In general, it can be straightforwardly proven that a Hamiltonian $\mathcal{H}_n$ containing the Ramani potential $\mathcal{V}_n$ is Liouville integrable, and its integral of the motion $\mathcal{L}_n$ involves the $\mathcal{V}_{n-1}$ potential, namely

$$\mathcal{H}_n = \frac{1}{2}(p_1^2 + p_2^2) + \alpha_n\mathcal{V}_n, \qquad \mathcal{L}_n = p_1(q_1p_2 - q_2p_1) + \alpha_n q_1^2\mathcal{V}_{n-1}, \qquad \{\mathcal{H}_n, \mathcal{L}_n\} = 0. \tag{7}$$

The formula (6) requires the definition of the 0-th order Ramani potential as a trivial constant $\mathcal{V}_0 := 1$, that is, the first Hamiltonian system within the Ramani series reads

$$\mathcal{H}_1 = \frac{1}{2}(p_1^2 + p_2^2) + \alpha_1(2q_2), \qquad \mathcal{L}_1 = p_1(q_1p_2 - q_2p_1) + \alpha_1 q_1^2.$$

A crucial mathematical property of the Ramani potentials is the fact that they can be freely superposed without losing integrability [20, 24, 25]. More explicitly:

Proposition 1. *The Hamiltonian written in Cartesian canonical variales* (p_1, p_2, q_1, q_2) *as*

$$\mathcal{H}_{(M)} = \frac{1}{2}\left(p_1^2 + p_2^2\right) + \sum_{n=1}^{M}\alpha_n\mathcal{V}_n = \frac{1}{2}\left(p_1^2 + p_2^2\right) + \sum_{n=1}^{M}\sum_{i=0}^{[\frac{n}{2}]}\alpha_n 2^{n-2i}\binom{n-i}{i} q_1^{2i}q_2^{n-2i}, \tag{8}$$

where $M \in \mathbb{N}^+$ *and* α_n *are arbitrary real constants, is endowed with the following integral of the motion*

$$\begin{aligned}\mathcal{L}_{(M)} &= p_1(q_1p_2 - q_2p_1) + q_1^2\sum_{n=1}^{M}\alpha_n\mathcal{V}_{n-1}\\ &= p_1(q_1p_2 - q_2p_1) + q_1^2\left(\sum_{n=1}^{M}\sum_{i=0}^{[\frac{n-1}{2}]}\alpha_n 2^{n-1-2i}\binom{n-1-i}{i}q_1^{2i}q_2^{n-1-2i}\right).\end{aligned} \tag{9}$$

Therefore, the relationship between the KdV Hénon–Heiles Hamiltonian $\mathcal{H}$ (4) and its constant of motion $\mathcal{I}$ (5) with the Hamiltonian $\mathcal{H}_{(M)}$ (8) and the integral $\mathcal{L}_{(M)}$ (9) comes out as a byproduct of proposition 1, since by setting

$$M = 3, \qquad \alpha_1 = 0, \qquad \alpha_2 = \Omega, \qquad \alpha_3 = \alpha/4, \tag{10}$$

we obtain that

$$\mathcal{H} \equiv \mathcal{H}_{(3)} = \frac{1}{2}(p_1^2 + p_2^2) + \alpha_2 \mathcal{V}_2 + \alpha_3 \mathcal{V}_3,$$
$$\mathcal{I} \equiv \mathcal{L}_{(3)} = p_1(q_1 p_2 - q_2 p_1) + q_1^2 (\alpha_2 \mathcal{V}_1 + \alpha_3 \mathcal{V}_2).$$

3. Integrable rational perturbations of a KdV Hénon–Heiles system on the plane

It is worth stressing that the Ramani potentials can be extended in order to provide integrable rational perturbations of the Hamiltonian (4) by starting from the 0-th order Ramani potential and going 'backwards', that is, by considering *negative* indices n.

In order to make this statement explicit, let us start from the trivial Hamiltonian defined by the 0-th order potential $\mathcal{V}_0 := 1$. This is clearly integrable as Poisson-commutes with the integral $\mathcal{L}_0$ given by

$$\mathcal{H}_0 = \frac{1}{2}(p_1^2 + p_2^2) + \alpha_0 \mathcal{V}_0, \qquad \mathcal{L}_0 = p_1(q_1 p_2 - q_2 p_1) + \alpha_0 1. \tag{11}$$

Therefore, according to (7), it seems natural to define

$$\mathcal{L}_0 := p_1(q_1 p_2 - q_2 p_1) + \alpha_0 q_1^2 \mathcal{V}_{-1}, \qquad \text{where} \qquad \mathcal{V}_{-1} := \frac{1}{q_1^2} \equiv \frac{\mathcal{V}_0}{q_1^2}.$$

It is quite remarkable that $\mathcal{V}_{-1}$ is just a Rosochatius or Winternitz potential [27, 28, 29]. From it, we can construct the corresponding $n = -1$ potential which is again integrable; namely

$$\mathcal{H}_{-1} = \frac{1}{2}(p_1^2 + p_2^2) + \alpha_{-1}\mathcal{V}_{-1}, \quad \mathcal{L}_{-1} = p_1(q_1 p_2 - q_2 p_1) + \alpha_{-1} q_1^2 \mathcal{V}_{-2}, \quad \mathcal{V}_{-2} := -\frac{2q_2}{q_1^4} \equiv -\frac{\mathcal{V}_1}{q_1^4}.$$

In this way, the complete series of rational perturbations of the Hamitonian (4), which can be understood as the 'negative' counterparts of the Ramani potentials (6), are found to be [21, 23]:

$$\mathcal{H}_{-n} = \frac{1}{2}(p_1^2 + p_2^2) + \alpha_{-n}\mathcal{V}_{-n}, \qquad \mathcal{L}_{-n} = p_1(q_1 p_2 - q_2 p_1) + \alpha_{-n} q_1^2 \mathcal{V}_{-(n+1)},$$
$$\mathcal{V}_{-n} = (-1)^{n+1} \frac{\mathcal{V}_{n-1}}{q_1^{2n}}, \qquad \{\mathcal{H}_{-n}, \mathcal{L}_{-n}\} = 0, \qquad n = 1, 2 \ldots \tag{12}$$

Nevertheless, for the sake of clarity, let us consider real parameters with positive indices, λ_n, defined by

$$\lambda_n := (-1)^{n+1} \alpha_{-n}, \qquad n = 1, 2 \ldots \tag{13}$$

which allow us to rewrite (12) as

$$\mathcal{H}_{-n} = \frac{1}{2}(p_1^2 + p_2^2) + \lambda_n \frac{\mathcal{V}_{n-1}}{q_1^{2n}}, \qquad \mathcal{L}_{-n} = p_1(q_1 p_2 - q_2 p_1) - \lambda_n \frac{\mathcal{V}_n}{q_1^{2n}}, \qquad n = 1, 2 \ldots \tag{14}$$

Moreover, the Ramani potentials (6), the 0-th potential $\mathcal{V}_0 = 1$ and the rational perturbations (14) can be freely superposed leading to an integrable Hamiltonian that generalizes the results given in proposition 1 as follows.

Proposition 2. [25] *The Hamiltonian given by*

$$\begin{aligned}\mathcal{H}_{(M,R)} &= \frac{1}{2}\left(p_1^2+p_2^2\right)+\sum_{n=1}^{M}\alpha_n\mathcal{V}_n+\alpha_0\mathcal{V}_0+\sum_{n=1}^{R}\lambda_n\frac{\mathcal{V}_{n-1}}{q_1^{2n}}\\ &= \frac{1}{2}\left(p_1^2+p_2^2\right)+\sum_{n=1}^{M}\sum_{i=0}^{[\frac{n}{2}]}\alpha_n 2^{n-2i}\binom{n-i}{i}q_1^{2i}q_2^{n-2i}+\alpha_0\\ &\quad+\sum_{n=1}^{R}\sum_{i=0}^{[\frac{n-1}{2}]}\lambda_n 2^{n-1-2i}\binom{n-1-i}{i}\frac{q_2^{n-1-2i}}{q_1^{2(n-i)}},\end{aligned}\tag{15}$$

where α_n, α_0 and λ_n are arbitrary real constants, is integrable for any indices $M, R \in \mathbb{N}^+$. The corresponding integral of the motion reads

$$\mathcal{L}_{(M,R)} = p_1(q_1p_2-q_2p_1)+q_1^2\sum_{n=1}^{M}\alpha_n\mathcal{V}_{n-1}+\alpha_0\mathcal{V}_0-\sum_{n=1}^{R}\lambda_n\frac{\mathcal{V}_n}{q_1^{2n}},$$

where $\mathcal{V}_n$ are given in (6) and $\mathcal{V}_0 = 1$.

For instance, if we set $M = 3$ and $R = 4$ we obtain the following integrable generalization of the KdV Hénon–Heiles Hamiltonian (4):

$$\begin{aligned}\mathcal{H}_{(3,4)} &= \frac{1}{2}\left(p_1^2+p_2^2\right)+\alpha_1\mathcal{V}_1+\alpha_2\mathcal{V}_2+\alpha_3\mathcal{V}_3+\alpha_0\mathcal{V}_0+\lambda_1\frac{\mathcal{V}_0}{q_1^2}+\lambda_2\frac{\mathcal{V}_1}{q_1^4}+\lambda_3\frac{\mathcal{V}_2}{q_1^6}+\lambda_4\frac{\mathcal{V}_3}{q_1^8}\\ &= \frac{1}{2}\left(p_1^2+p_2^2\right)+\alpha_1\left(2q_2\right)+\alpha_2\left(q_1^2+4q_2^2\right)+\alpha_3\left(4q_1^2q_2+8q_2^3\right)+\alpha_0\\ &\quad+\lambda_1\frac{1}{q_1^2}+\lambda_2\frac{2q_2}{q_1^4}+\lambda_3\frac{q_1^2+4q_2^2}{q_1^6}+\lambda_4\frac{4q_1^2q_2+8q_2^3}{q_1^8},\end{aligned}$$

which Poisson-commutes with

$$\begin{aligned}\mathcal{L}_{(3,4)} &= p_1(q_1p_2-q_2p_1)+q_1^2\left(\alpha_1\mathcal{V}_0+\alpha_2\mathcal{V}_1+\alpha_3\mathcal{V}_2\right)+\alpha_0\mathcal{V}_0\\ &\quad-\left(\lambda_1\frac{\mathcal{V}_1}{q_1^2}+\lambda_2\frac{\mathcal{V}_2}{q_1^4}+\lambda_3\frac{\mathcal{V}_3}{q_1^6}+\lambda_4\frac{\mathcal{V}_4}{q_1^8}\right).\end{aligned}$$

Recall that the λ_1-potential behaves as a centrifugal barrier on $\mathbf{E}^2$ when $\lambda_1 > 0$ [29].

4. Ambient and Beltrami canonical variables

In order to achieve the generalization of the above results to the 2D sphere $\mathbf{S}^2$ and hyperbolic space $\mathbf{H}^2$, let us consider the one-parameter family of 3D real Lie algebras $\mathfrak{so}_\kappa(3)$ with commutation relations and Casimir invariant given by [29, 30]:

$$[J_{12}, J_{01}] = J_{02}, \qquad [J_{12}, J_{02}] = -J_{01}, \qquad [J_{01}, J_{02}] = \kappa J_{12},\tag{16}$$

$$\mathcal{C} = J_{01}^2 + J_{02}^2 + \kappa J_{12}^2,\tag{17}$$

where κ is a real parameter. The 2D homogeneous space $\mathrm{SO}_\kappa(3)/\mathrm{SO}(2)$, where $\mathrm{SO}_\kappa(3)$ is the Lie group of $\mathfrak{so}_\kappa(3)$ and $\mathrm{SO}(2)=\langle J_{12}\rangle$, has constant Gaussian curvature equal to κ. This generic family of homogeneous space comprises the three relevant cases with constant curvature:

$\kappa>0$: Sphere	$\kappa=0$: Euclidean plane	$\kappa<0$: Hyperbolic space
$\mathbf{S}^2=\mathrm{SO}(3)/\mathrm{SO}(2)$	$\mathbf{E}^2=\mathrm{ISO}(2)/\mathrm{SO}(2)$	$\mathbf{H}^2=\mathrm{SO}(2,1)/\mathrm{SO}(2)$

These 2D spaces can be embedded in $\mathbb{R}^3=(x_0,x_1,x_2)$ where the *ambient* or Weierstrass coordinates must satisfy

$$x_0^2+\kappa(x_1^2+x_2^2)=1.$$

Next, if we apply a central projection with pole $(0,0,0)\in\mathbb{R}^3$ from $(x_0,x_1,x_2)\in\mathbb{R}^3$ to the 2D projective space, we obtain the *Beltrami* coordinates $\mathbf{q}=(q_1,q_2)\in\mathbb{R}^2$ given by

$$x_0=\frac{1}{\sqrt{1+\kappa\mathbf{q}^2}},\qquad \mathbf{x}=\frac{\mathbf{q}}{\sqrt{1+\kappa\mathbf{q}^2}},\qquad \mathbf{q}=\frac{\mathbf{x}}{x_0},\tag{18}$$

such that $\mathbf{x}=(x_1,x_2)$, the conjugate Beltrami momenta are $\mathbf{p}=(p_1,p_2)$ and hereafter we denote

$$\mathbf{q}^2=q_1^2+q_2^2,\qquad \mathbf{p}^2=p_1^2+p_2^2,\qquad \mathbf{q}\cdot\mathbf{p}=q_1p_1+q_2p_2.$$

A symplectic realization of $\mathfrak{so}_\kappa(3)$ (16) in terms of the Beltrami canonical variables $(\mathbf{q},\mathbf{p})$ turns out to be [28, 29, 30]

$$J_{0i}=p_i+\kappa(\mathbf{q}\cdot\mathbf{p})q_i,\quad i=1,2;\qquad J_{12}=q_1p_2-q_2p_1.\tag{19}$$

In this framework, the curved kinetic energy $\mathcal{T}_\kappa$ for a particle moving on these spaces comes from the Casimir (17) under the above realization:

$$\mathcal{T}_\kappa\equiv\frac{1}{2}\mathcal{C}=\frac{1}{2}(J_{01}^2+J_{02}^2+\kappa J_{12}^2)=\frac{1}{2}\left(1+\kappa\mathbf{q}^2\right)\left(\mathbf{p}^2+\kappa(\mathbf{q}\cdot\mathbf{p})^2\right).\tag{20}$$

Notice that the flat/Euclidean limit $\kappa\to 0$ of the above expressions leads to

$$x_0=1,\qquad \mathbf{x}=\mathbf{q},\qquad J_{0i}=p_i,\qquad J_{12}=q_1p_2-q_2p_1,\qquad \mathcal{T}=\frac{1}{2}\mathbf{p}^2.$$

5. A KdV Hénon–Heiles system on the sphere and the hyperbolic space

In this section we summarize the construction of the curved counterpart of the KdV Hénon–Heiles system (4), that has been recently presented in [26] by making use of of ambient and Beltrami dynamical variables. Such construction requires to obtain, firstly, the curved Ramani potentials $\mathcal{V}_{\kappa,n}$ and, secondly, their superposition, so generalizing proposition 1 to the curved case. In this way, the definition of the curved integrable KdV Hénon–Heiles Hamiltonian $\mathcal{H}_\kappa$ comes out as a byproduct, and the main object in the construction here presented turns out to be the curved Ramani potentials.

Proposition 3. [26] *The Ramani potentials on the sphere* $\mathbf{S}^2$ *and the hyperbolic space* $\mathbf{H}^2$ *are defined in Beltrami coordinates* (q_1,q_2) *(18) as*

$$\mathcal{V}_{\kappa,n}=\left(\frac{1+\kappa\mathbf{q}^2}{1-\kappa q_2^2}\right)^2\sum_{i=0}^{[\frac{n}{2}]}2^{n-2i}\binom{n-i}{i}\left(\frac{q_1}{\sqrt{1+\kappa\mathbf{q}^2}}\right)^{2i}\left(1-\frac{i}{n-i}\left[\frac{\kappa q_1^2}{1+\kappa\mathbf{q}^2}\right]\right)\left(\frac{q_2}{1+\kappa\mathbf{q}^2}\right)^{n-2i}\tag{21}$$

with $n = 1, 2\ldots$ Each Ramani Hamiltonian

$$\mathcal{H}_{\kappa,n} = \mathcal{T}_\kappa + \alpha_n \mathcal{V}_{\kappa,n},$$

is integrable, as is endowed with a constant of motion $\mathcal{L}_{\kappa,n}$ which is quadratic in the momenta

$$\mathcal{L}_{\kappa,n} = J_{01}J_{12} + \alpha_n \frac{q_1^2}{1+\kappa\mathbf{q}^2}\mathcal{V}_{\kappa,n-1}, \qquad \{\mathcal{H}_{\kappa,n}, \mathcal{L}_{\kappa,n}\} = 0, \tag{22}$$

where $\mathcal{V}_{\kappa,0}$ is defined by

$$\mathcal{V}_{\kappa,0} := \frac{(1+\kappa q_2^2)(1+\kappa\mathbf{q}^2)}{(1-\kappa q_2^2)^2}, \tag{23}$$

and J_{01}, J_{12} and $\mathcal{T}_\kappa$ are the functions given by (19) and (20).

We stress that the flat limit $\kappa \to 0$ of the above expressions leads to the Euclidean Ramani potential $\mathcal{V}_n$ (6), the Hamiltonian $\mathcal{H}_n$ and its integral of motion $\mathcal{L}_n$ (7) along with $\mathcal{V}_0 = 1$. Recall also that, under the flat limit, Beltrami coordinates reduce to Cartesian ones. However, note that in the curved case $\mathcal{V}_{\kappa,0}$ is no longer a trivial potential. We also remark that the quadratic Ramani Hamiltonian, $\mathcal{H}_{\kappa,2} = \mathcal{T}_\kappa + \alpha_2\mathcal{V}_{\kappa,2}$, is just the *superintegrable* curved $1:2$ oscillator, formerly introduced in [31] and further studied in [29, 30].

In terms of the ambient coordinates (x_0, x_1, x_2) (18), the curved Ramani potentials (21) and (23) turn out to be

$$\mathcal{V}_{\kappa,0} = \frac{1-\kappa x_1^2}{(x_0^2-\kappa x_2^2)^2}, \quad \mathcal{V}_{\kappa,n} = \frac{1}{(x_0^2-\kappa x_2^2)^2}\sum_{i=0}^{[\frac{n}{2}]} 2^{n-2i}\binom{n-i}{i}x_1^{2i}\left(1-\frac{i}{n-i}\kappa x_1^2\right)(x_0x_2)^{n-2i} \tag{24}$$

which affords their parametrization in any coordinate system under the appropriate change of variables.

Furthermore, as in the Euclidean case, the curved Ramani potentials can be freely superposed. Therefore, the expressions (8) and (9) given in proposition 1 can be generalized to the curved case as follows.

Proposition 4. [26] *The Hamiltonian formed by the linear superposition of the curved Ramani potentials (21) and given by*

$$\mathcal{H}_{\kappa,(M)} = \mathcal{T}_\kappa + \sum_{n=1}^{M}\alpha_n\mathcal{V}_{\kappa,n}, \qquad M \in \mathbb{N}^+, \tag{25}$$

Poisson-commutes with the function

$$\mathcal{L}_{\kappa,(M)} = J_{01}J_{12} + \frac{q_1^2}{1+\kappa\mathbf{q}^2}\sum_{n=1}^{M}\alpha_n\mathcal{V}_{\kappa,n-1},$$

where J_{01}, J_{12} and $\mathcal{T}_\kappa$ are the functions given by (19) and (20).

As a straightforward consequence, the curved counterpart of the Hénon–Heiles KdV Hamiltonian (4) on $\mathbf{S}^2$ and $\mathbf{H}^2$ along with its integral (5), written in Beltrami variables, can be obtained from proposition 4 for the particular case $\mathcal{H}_{\kappa,(3)}$ by setting (10); namely

$$\mathcal{H}_\kappa = \mathcal{T}_\kappa + \mathcal{V}_\kappa = \mathcal{T}_\kappa + \Omega\,\mathcal{V}_{\kappa,2} + \frac{\alpha}{4}\mathcal{V}_{\kappa,3},$$
$$\mathcal{V}_\kappa = \Omega\frac{q_1^2(1+\kappa q_2^2)+4q_2^2}{(1-\kappa q_2^2)^2} + \alpha\frac{q_1^2q_2(1+\kappa\mathbf{q}^2-\frac{1}{2}\kappa q_1^2)+2q_2^3}{(1-\kappa q_2^2)^2(1+\kappa\mathbf{q}^2)}. \tag{26}$$

Journal of Physics: Conference Series **597** (2015) 012013 doi:10.1088/1742-6596/597/1/012013

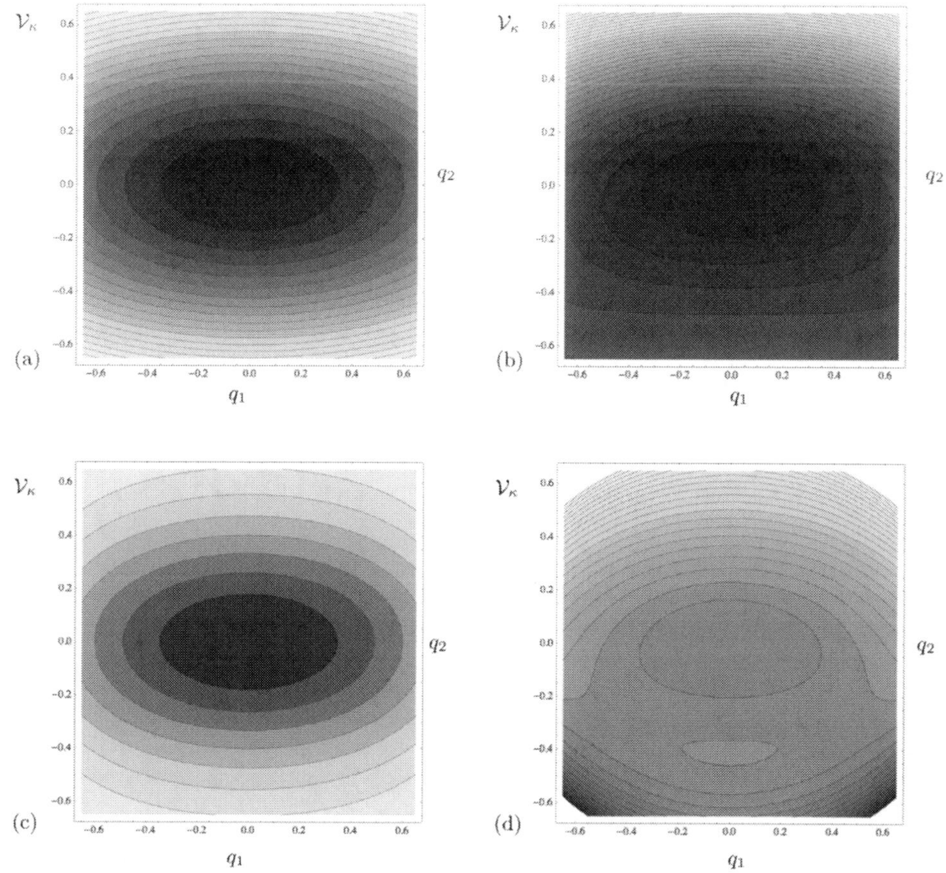

Figure 1. Level plots for the potential $\mathcal{V}_\kappa$ (26) in Beltrami projective coordinates (q_1, q_2) for four different values of the constants (κ, Ω, α) involved. Values of the potential function $\mathcal{V}_\kappa$ are coded as follows: dark blue represents low values, and higher values are red-shifted. Figure (a) corresponds to $(\kappa, \Omega, \alpha) = (0, 1, 0)$, (b) to $(0, 1, 2)$, (c) to $(-1, 1, 0)$ and (d) to $(-1, 1, 2)$.

And, therefore, the Hamiltonian $\mathcal{H}_\kappa$ Poisson-commutes with the corresponding integral of the motion that comes from $\mathcal{L}_{\kappa,(3)}$:

$$\mathcal{I}_\kappa = J_{01}J_{12} + \frac{q_1^2}{1+\kappa\mathbf{q}^2}\left(\Omega\,\mathcal{V}_{\kappa,1} + \frac{\alpha}{4}\,\mathcal{V}_{\kappa,2}\right)$$
$$= (p_1 + \kappa(\mathbf{q}\cdot\mathbf{p})q_1)(q_1p_2 - q_2p_1) + \frac{q_1^2}{1+\kappa\mathbf{q}^2}\left(\Omega\,\frac{2q_2(1+\kappa\mathbf{q}^2)}{(1-\kappa q_2^2)^2} + \alpha\,\frac{q_1^2(1+\kappa q_2^2)+4q_2^2}{4(1-\kappa q_2^2)^2}\right).$$

Level plots for the potential $\mathcal{V}_\kappa$ (26) are shown in figure 1, where the Euclidean ($\kappa = 0$) and hyperbolic ($\kappa < 0$) KdV Hénon–Heiles potentials are represented and compared. We stress that both the curvature κ and the constant α can be considered as deformation parameters that preserve the integrability of the initial Hamiltonian. Case (a) represents the flat 1 : 2 (superintegrable) anharmonic oscillator potential on $\mathbf{E}^2$ with no Hénon–Heiles term, in particular with $\kappa = 0$, $\Omega = 1$ and $\alpha = 0$. In case (b) the Hénon–Heiles cubic term is added on $\mathbf{E}^2$ through $\alpha = 2$ (we keep $\Omega = 1$), while the underlying space is still flat ($\kappa = 0$). Here we see that the Hénon–Heiles term breaks the $q_2 \to -q_2$ symmetry of the anharmonic oscillator. The (superintegrable) curved 1 : 2 anharmonic oscillator potential on $\mathbf{H}^2$ with no Hénon–Heiles term ($\kappa = -1$, $\Omega = 1$ and $\alpha = 0$) is represented in case (c). Now we realize that the non-vanishing curvature modifies the values of the potential, although its general shape around the origin is quite similar to the flat case. Finally, when the curved Hénon–Heiles term is added ($\kappa = -1$, $\Omega = 1$ and $\alpha = 2$) the plot (d) is obtained on $\mathbf{H}^2$. Note that although superintegrability is broken when the Hénon–Heiles term is considered, both (b) and (d) potentials always generate integrable motions on the hyperbolic plane.

6. Integrable perturbations of a curved KdV Hénon–Heiles system

In this last section we present, as new results, the curved analogues on $\mathbf{S}^2$ and $\mathbf{H}^2$ of the 'negative' counterparts of the Ramani potentials (12) and of their integrable superposition (15).

As in section 3, let us start from the curved Hamiltonian $\mathcal{H}_{\kappa,0}$ corresponding to the potential $\mathcal{V}_{\kappa,0}$ (23), which, in contrast with the flat case $\mathcal{V}_0 = 1$, is now a non-trivial function:

$$\mathcal{H}'_{\kappa,0} = \mathcal{T}_\kappa + \alpha_0 \mathcal{V}_{\kappa,0} = \mathcal{T}_\kappa + \alpha_0 \frac{(1+\kappa q_2^2)(1+\kappa \mathbf{q}^2)}{(1-\kappa q_2^2)^2} .$$

This Hamiltonian is, in fact, integrable as it Poisson-commutes with the function

$$\mathcal{L}'_{\kappa,0} = J_{01}J_{12} + \alpha_0 \frac{2\kappa\, q_1^2 q_2}{(1-\kappa q_2^2)^2} .$$

According to (22), let us define

$$\mathcal{L}'_{\kappa,0} := J_{01}J_{12} + \alpha_0 \frac{q_1^2}{1+\kappa\mathbf{q}^2}\, \mathcal{V}'_{\kappa,-1}, \qquad \mathcal{V}'_{\kappa,-1} := \kappa \frac{2q_2(1+\kappa\mathbf{q}^2)}{(1-\kappa q_2^2)^2} \equiv \kappa \mathcal{V}_{\kappa,1}.$$

By induction, it can be easily shown that

$$\mathcal{V}'_{\kappa,-n} = \kappa\, \mathcal{V}_{\kappa,n}, \qquad n = 1,2,\ldots$$

and they vanish when $\kappa = 0$. Consequently, under this procedure we have obtained curved Beltrami potentials with negative index, but they are just proportional to (21) and, therefore, we have not obtained any new result concerning integrable perturbations.

However, it turns out that the curved analogue of the rational perturbations (12) can be obtained by starting from the trivial (free geodesic) curved Hamiltonian with a constant potential α_0 (as in (11)) which is clearly integrable; explicitly

$$\mathcal{H}_{\kappa,0} = \mathcal{T}_\kappa + \alpha_0, \qquad \mathcal{L}_{\kappa,0} = J_{01}J_{12} + \alpha_0, \qquad \{\mathcal{H}_{\kappa,0}, \mathcal{L}_{\kappa,0}\} = 0.$$

If we again assume that recurrence (22) should hold, we are led to define

$$\mathcal{L}_{\kappa,0} := J_{01}J_{12} + \alpha_0 \frac{q_1^2}{1+\kappa\mathbf{q}^2}\, \mathcal{V}_{\kappa,-1}, \qquad \mathcal{V}_{\kappa,-1} := \frac{1+\kappa\mathbf{q}^2}{q_1^2}.$$

Surprisingly enough, we find that $\mathcal{V}_{\kappa,-1} \equiv 1/x_1^2$ is just the curved Rosochatius or Winternitz potential, which corresponds to a noncentral oscillator potential on the sphere with center (in ambient coordinates) located at $O_1 = (0,1,0)$ (see [28, 29, 31, 32] for a detailed discussion).

Now, if we construct the Hamiltonian $\mathcal{H}_{\kappa,-1} = \mathcal{T}_\kappa + \alpha_{-1}\mathcal{V}_{\kappa,-1}$, the potential $\mathcal{V}_{\kappa,-2}$ is then obtained through the corresponding integral of motion:

$$\mathcal{L}_{\kappa,-1} = J_{01}J_{12} + \alpha_{-1} \frac{q_1^2}{1+\kappa\mathbf{q}^2}\, \mathcal{V}_{\kappa,-2}, \quad \text{where} \quad \mathcal{V}_{\kappa,-2} := -\frac{2q_2(1+\kappa\mathbf{q}^2)}{q_1^4} = -\frac{(1-\kappa q_2^2)^2}{q_1^4}\, \mathcal{V}_{\kappa,1} .$$

Table 1. The Ramani potentials for $n = \{0, \pm1, \pm2, \pm3\}$ on $\mathbf{E}^2$ in Cartesian coordinates $\mathbf{q}$ (6) and (12) along with their curved counterpart on $\mathbf{S}^2$ and $\mathbf{H}^2$ in ambient coordinates (24) and (28) such that $x_0^2 + \kappa\mathbf{x}^2 = 1$. Recall that $x_0 = 1$ and $\mathbf{x} = \mathbf{q}$ when $\kappa = 0$.

$\mathbf{E}^2$: Cartesian coordinates $\mathbf{q}$	$\mathbf{S}^2$ and $\mathbf{H}^2$: Ambient coordinates $(x_0, \mathbf{x})$
$\mathcal{V}_{-3} = \dfrac{q_1^2 + 4q_2^2}{q_1^6}$	$\mathcal{V}_{\kappa,-3} = \dfrac{x_1^2(1-\kappa x_1^2) + 4x_0^2 x_2^2}{x_1^6}$
$\mathcal{V}_{-2} = -\dfrac{2q_2}{q_1^4}$	$\mathcal{V}_{\kappa,-2} = -\dfrac{2x_0 x_2}{x_1^4}$
$\mathcal{V}_{-1} = \dfrac{1}{q_1^2}$	$\mathcal{V}_{\kappa,-1} = \dfrac{1}{x_1^2}$
$\mathcal{V}_0 = 1$	$\mathcal{V}_{\kappa,0} = \dfrac{1-\kappa x_1^2}{(x_0^2 - \kappa x_2^2)^2}$
$\mathcal{V}_1 = 2q_2$	$\mathcal{V}_{\kappa,1} = \dfrac{2x_0 x_2}{(x_0^2 - \kappa x_2^2)^2}$
$\mathcal{V}_2 = q_1^2 + 4q_2^2$	$\mathcal{V}_{\kappa,2} = \dfrac{x_1^2(1-\kappa x_1^2) + 4x_0^2 x_2^2}{(x_0^2 - \kappa x_2^2)^2}$
$\mathcal{V}_3 = 4q_1^2 q_2 + 8q_2^3$	$\mathcal{V}_{\kappa,3} = \dfrac{4x_0 x_1^2 x_2(1 - \frac{1}{2}\kappa x_1^2) + 8x_0^3 x_2^3}{(x_0^2 - \kappa x_2^2)^2}$

From this, the integrable curved Ramani potentials $\mathcal{V}_{\kappa,-n}$ can be defined as follows.

Proposition 5. *Let us define*

$$\mathcal{V}_{\kappa,-1} = \frac{1 + \kappa\mathbf{q}^2}{q_1^2}, \qquad \mathcal{V}_{\kappa,-m} = (-1)^{m+1}\frac{(1-\kappa q_2^2)^2}{q_1^{2m}}\left(1 + \kappa\mathbf{q}^2\right)^{m-2}\mathcal{V}_{\kappa,m-1}, \tag{27}$$

with $m = 2, 3, \ldots$ *and* $\mathcal{V}_{\kappa,m-1}$ *given by (21). The Hamiltonian defined by* $\mathcal{H}_{\kappa,-n} = \mathcal{T}_\kappa + \alpha_{-n}\mathcal{V}_{\kappa,-n}$ *(*$n = 1, 2, \ldots$*) Poisson-commutes with*

$$\mathcal{L}_{\kappa,-n} = J_{01}J_{12} + \alpha_{-n}\frac{q_1^2}{1+\kappa\mathbf{q}^2}\mathcal{V}_{\kappa,-(n+1)}\,.$$

Consequently, the generalization of the Euclidean rational perturbations (12) to a constant curvature framework is achieved. In ambient coordinates (18), the potentials (27) read

$$\mathcal{V}_{\kappa,-1} = \frac{1}{x_1^2}, \qquad \mathcal{V}_{\kappa,-m} = (-1)^{m+1}\frac{(x_0^2 - \kappa x_2^2)^2}{x_1^{2m}}\mathcal{V}_{\kappa,m-1}, \qquad m = 2, 3, \ldots \tag{28}$$

with $\mathcal{V}_{\kappa,m-1}$ given in (24). These results are illustrated in table 1 by writing the first Ramani potentials with positive and negative indices on $\mathbf{E}^2$, $\mathbf{S}^2$ and $\mathbf{H}^2$. We also remark that proposition 5 can be written in terms of coefficients with positive indices, λ_n, through the definition (13).

Finally, the potentials $\mathcal{V}_{\kappa,-n}$ can also be added to the Hamiltonian (25) leading to the following full integrable curved superposition, that constitutes the main result of this paper. We write it in ambient coordinates as follows.

Theorem 6. *The Hamiltonian formed by the linear superposition of the curved Ramani potentials (24) and (28)*

$$\mathcal{H}_{\kappa,(M,R)} = \mathcal{T}_\kappa + \sum_{n=1}^{M}\alpha_n\mathcal{V}_{\kappa,n} + \alpha_0\mathcal{V}_{\kappa,0} + \lambda_1\frac{1}{x_1^2} + (x_0^2 - \kappa x_2^2)^2\sum_{m=2}^{R}\lambda_m\frac{\mathcal{V}_{\kappa,m-1}}{x_1^{2m}}\,,$$

where α_n, α_0, λ_1, λ_m are arbitrary real constants and $M, R \in \mathbb{N}^+$, is integrable. Its constant of the motion is given by the function

$$\mathcal{L}_{\kappa,(M,R)} = J_{01}J_{12} + x_1^2 \sum_{n=1}^{M} \alpha_n \mathcal{V}_{\kappa,n-1} + \alpha_0 x_1^2 \kappa \mathcal{V}_{\kappa,1} - \lambda_1 \frac{2x_2x_0}{x_1^2} - (x_0^2 - \kappa x_2^2)^2 \sum_{m=2}^{R} \lambda_m \frac{\mathcal{V}_{\kappa,m}}{x_1^{2m}} .$$

In this way, we have obtained the generalization of proposition 2 to $\mathbf{S}^2$ and $\mathbf{H}^2$. The above result can be straightforwardly written in terms of Beltrami variables through (18).

Finally, we would like to mention that the construction of the curved counterpart of the KdV Hénon–Heiles Hamiltonian (2) with arbitrary parameters (Ω_1, Ω_2), as well as of the Sawada–Kotera (1) and the Kaup–Kuperschdmit (3) Hénon–Heiles systems is currently in progress.

Acknowledgments
This work was partially supported by the Spanish MINECO under grant MTM2013-43820-P and by Junta de Castilla y León under grant BU278U14.

References

[1] Hénon M and Heiles C 1964 *Astron. J.* **69** 73
[2] Tabor M 1989 *Chaos and integrability in nonlinear dynamics* (New York: Wiley)
[3] Gutzwiller M C 1990 *Chaos in Classical and Quantum Mechanics* (*Interdisciplinary Applied Mathematics* vol 1) (New York: Springer)
[4] Boccaletti D and Pucacco G 2004 *Theory of Orbits* (Berlin: Springer)
[5] Perelomov A M 1990 *Integrable systems of classical mechanics and Lie algebras* (Basel: Birhäuser)
[6] Bountis T, Segur H and Vivaldi F 1982 *Phys. Rev.* A **25** 1257
[7] Chang Y F, Tabor M and Weiss J 1982 *J. Math. Phys.* **23** 531
[8] Grammaticos B, Dorizzi B and Padjen R 1982 *Phys. Lett.* A **89** 111
[9] Hietarinta J 1983 *Phys. Rev.* A **28** 3670
[10] Fordy A P 1983 *Phys. Lett.* A **97** 21
[11] Wojciechowski S 1984 *Phys. Lett. A* **100** 277
[12] Sahadevan R and Lakshmanan M 1986 *J. Phys. A: Math. Gen.* **19** L949
[13] Fordy A P 1990 *Physica* D **52** 204
[14] Sarlet W 1991 *J. Phys. A: Math. Gen.* **24** 5245
[15] Ravoson V, Gavrilov L and Caboz R 1993 *J. Math. Phys.* **34** 2385
[16] Lakshmanan M and Sahadevan R 1993 *Phys. Rep.* **224** 1
[17] Conte R, Fordy A P and Pickering A 1993 *Physica* D **69** 33
[18] Conte R, Musette M and Verhoeven C 2005 *Theor. Math. Phys.* **144** 888
[19] Ramani A, Dorizzi B and Grammaticos B 1982 *Phys. Rev. Lett.* **49** 1539
[20] Hietarinta J 1987 *Phys. Rep.* **147** 87
[21] Ferapontov E V and Fordy A P 1997 *J. Geom. Phys.* **21** 169
[22] Hone A N W, Novikov V and Verhoeven C 2006 *Inv. Probl.* **22** 2001
[23] Hone A N W, Novikov V and Verhoeven 2008 *Phys. Lett.* A **372** 1440
[24] Blasco A 2009 *Integrability of non-linear Hamiltonian systems with N degrees of freedom* PhD Thesis (Burgos: Burgos University)
[25] Ballesteros A and Blasco A 2010 *Ann. Phys.* **325** 2787
[26] Ballesteros A, Blasco A, Herranz F J and Musso 2014 arXiv:1411.2033
[27] Fris J, Mandrosov V, Smorodinsky Y A, Uhlir M and Winternitz P 1965 *Phys. Lett.* **16** 354
[28] Ballesteros A and Herranz F J 2007 *J. Phys. A: Math. Theor.* **40** F51
[29] Ballesteros A, Herranz F J and Musso F 2013 *Nonlinearity* **26** 971
[30] Ballesteros A, Blasco A, Herranz F J and Musso F 2014 *J. Phys. A: Math. Theor.* **47** 345204
[31] Rañada M F and Santander M 1999 *J. Math. Phys.* **40** 5026
[32] Ballesteros A, Herranz F J, Santander M and Sanz-Gil T 2003 *J. Phys. A: Math. Theor.* **36** L93

Exactly solvable deformations of the oscillator and Coulomb systems and their generalization

A Ballesteros[1], A Enciso[2], F J Herranz[1], O Ragnisco[3] and D Riglioni[4]

[1] Departamento de Física, Universidad de Burgos, E-09001 Burgos, Spain

[2] Instituto de Ciencias Matemáticas, CSIC, Nicolás Cabrera 13-15, E-28049 Madrid, Spain

[3] Dipartimento di Matematica e Fisica, Università di Roma Tre and Istituto Nazionale di Fisica Nucleare sezione di Roma Tre, Via Vasca Navale 84, I-00146 Roma, Italy

[4] Centre de Recherches Mathématiques, Université de Montréal, H3T 1J4 2920 Chemin de la tour, Montreal, Canada

E-mail: angelb@ubu.es, aenciso@icmat.es, fjherranz@ubu.es, ragnisco@fis.uniroma3.it, riglioni@crm.umontreal.ca

Abstract. We present two maximally superintegrable Hamiltonian systems $\mathcal{H}_\lambda$ and $\mathcal{H}_\eta$ that are defined, respectively, on an N-dimensional spherically symmetric generalization of the Darboux surface of type III and on an N-dimensional Taub–NUT space. Afterwards, we show that the quantization of $\mathcal{H}_\lambda$ and $\mathcal{H}_\eta$ leads, respectively, to exactly solvable deformations (with parameters λ and η) of the two basic quantum mechanical systems: the harmonic oscillator and the Coulomb problem. In both cases the quantization is performed in such a way that the maximal superintegrability of the classical Hamiltonian is fully preserved. In particular, we prove that this strong condition is fulfilled by applying the so-called conformal Laplace–Beltrami quantization prescription, where the conformal Laplacian operator contains the usual Laplace–Beltrami operator on the underlying manifold plus a term proportional to its scalar curvature (which in both cases has non-constant value). In this way, the eigenvalue problems for the quantum counterparts of $\mathcal{H}_\lambda$ and $\mathcal{H}_\eta$ can be rigorously solved, and it is found that their discrete spectrum is just a smooth deformation (in terms of the parameters λ and η) of the oscillator and Coulomb spectrum, respectively. Moreover, it turns out that the maximal degeneracy of both systems is preserved under deformation. Finally, new further multiparametric generalizations of both systems that preserve their superintegrability are envisaged.

1. Introduction

It is well known that if we consider a natural classical Hamiltonian system on the N-dimensional (ND) Euclidean space

$$\mathcal{H} = \mathcal{T}(\mathbf{p}) + \mathcal{U}(\mathbf{q}), \tag{1}$$

the harmonic oscillator potential $\mathcal{U}(\mathbf{q}) = \omega^2\mathbf{q}^2$ and the Coulomb potential $\mathcal{U}(\mathbf{q}) = -k/|\mathbf{q}|$ define two *maximally superintegrable* (MS) systems (in the Liouville sense), since both systems are endowed with $(2N-1)$ functionally independent and globally defined integrals of the motion. In the first case such integrals are provided by the components of the Demkov–Fradkin tensor [1, 2], and in the second one by the angular momenta together with the N components of the Runge–Lenz vector (see e.g. [3] and references therein). At the classical dynamical level, the footprint of superintegrability consists in the fact that all bounded trajectories of these two systems are closed ones, a fact which is diretly related with Bertrand's theorem [4]. Moreover, when the quantization of these systems is performed it is found that such superintegrability implies that their spectrum exhibits maximal degeneracy due to a superabundance of quantum integrals of the motion.

In this paper we review two spherically symmetric deformations of the oscillator and Coulomb systems that define two new MS systems [5, 6]. As a consequence, their quantization [7, 8, 9] is shown to present maximal degeneracy in the spectra. At a first sight, the existence of such deformations could seem impossible since the only spherically symmetric potentials on the Euclidean space that are MS are just the oscillator and the Coulomb ones. Therefore, the addition of any radial perturbation on these systems leads to superintegrability breaking and thus to a lack of maximal degeneracy in the spectra, a fact that is very well known in quantum perturbation theory. However, as we shall see, such superintegrable perturbations can be obtained if *both the potential and the kinetic energy* are simultaneously deformed in a very precise way. Explicitly, the Hamiltonian (1) will be smoothly deformed into

$$\mathcal{H}_\mu(\mathbf{q},\mathbf{p}) = \mathcal{T}_\mu(\mathbf{q},\mathbf{p}) + \mathcal{U}_\mu(\mathbf{q}), \tag{2}$$

where μ can be regarded as a (generic) *deformation parameter* in such a manner that we will be no longer working on the *flat* Euclidean space, but on a suitable curved space with metric and kinetic energy given by

$$\mathrm{d}s_\mu^2 = \sum_{i,j=1}^N g_{ij}(\mathbf{q})\mathrm{d}q_i\mathrm{d}q_j, \qquad \mathcal{T}_\mu(\mathbf{q},\mathbf{p}) = \frac{1}{2}\sum_{i,j=1}^N g^{ij}(\mathbf{q})p_ip_j.$$

This fact will provide additional interesting geometric features to the systems we will deal with. In particular, we will see that the curved/deformed generalization of the Demkov–Fradkin tensor and of the Runge–Lenz vector do exist, and will be the essential tool to prove the MS property of the deformed systems.

We recall that the quantization problem on curved spaces is clearly a non-trivial one, since the kinetic energy term $\mathcal{T}_\mu(\mathbf{q},\mathbf{p})$ is a function of both positions and momenta that creates severe ordering ambiguities. Nevertheless, we shall explicitly show that a quantization of $\mathcal{H}_\mu$ (2) that preserves the MS property is achieved through the *conformal Laplacian quantization* (see [8, 9, 10, 11, 12, 13] and references therein):

$$\hat{\mathcal{H}}_{\mathrm{c},\mu} = \hat{\mathcal{T}}_{\mathrm{c},\mu} + \mathcal{U}_\mu = -\frac{\hbar^2}{2}\Delta_{\mathrm{c},\mu} + \mathcal{U}_\mu = -\frac{\hbar^2}{2}\left(\Delta_{\mathrm{LB},\mu} - \frac{(N-2)}{4(N-1)}R_\mu\right) + \mathcal{U}_\mu\,, \tag{3}$$

where R_μ is the scalar curvature on the underlying ND curved manifold $\mathcal{M}_\mu$, the operator $\Delta_{\mathrm{c},\mu}$ is the conformal Laplacian [14] and $\Delta_{\mathrm{LB},\mu}$ is the usual Laplace–Beltrami operator on $\mathcal{M}_\mu$, *i.e.*,

$$\Delta_{\mathrm{LB},\mu} = \sum_{i,j=1}^N \frac{1}{\sqrt{g}}\partial_i\sqrt{g}g^{ij}\partial_j\,,$$

where g^{ij} is the inverse of the metric tensor g_{ij} and g is the corresponding determinant. The limit $\mu\to 0$ gives rise to the quantization of the flat Hamiltonian (1) with $\mathrm{d}s^2 = \mathrm{d}\mathbf{q}^2$ since

$$\lim_{\mu\to 0} R_\mu = 0, \qquad \lim_{\mu\to 0}\Delta_{\mathrm{c},\mu} = \lim_{\mu\to 0}\Delta_{\mathrm{LB},\mu} = \Delta = \nabla^2, \qquad \lim_{\mu\to 0}\hat{\mathcal{H}}_{\mathrm{c},\mu} = -\frac{\hbar^2}{2}\nabla^2 + \mathcal{U}\,.$$

We also recall that the quantization (3) can be related through a similarity transformation to the Hamiltonian obtained by means of the so-called *direct Schrödinger quantization* prescription on conformally flat spaces [7, 8]

$$\hat{\mathcal{H}}_\mu = \hat{\mathcal{T}}_\mu + \mathcal{U}_\mu = -\frac{\hbar^2}{2f_\mu(r)^2}\Delta + \mathcal{U}_\mu\,,$$

where $f_\mu(r) = f_\mu(|\mathbf{q}|)$ is the conformal factor of the metric on $\mathcal{M}_\mu$ written as $\mathrm{d}s_\mu^2 = f_\mu(r)^2\mathrm{d}\mathbf{q}^2$. In this case, the scalar curvature reads

$$R_\mu = -(N-1)\left(\frac{(N-4)f_\mu'(r)^2 + f_\mu(r)\left(2f_\mu''(r) + 2(N-1)r^{-1}f_\mu'(r)\right)}{f_\mu(r)^4}\right). \qquad (4)$$

In the next two sections, we review the exactly solvable deformations of the ND isotropic oscillator $\hat{\mathcal{H}}_{c,\lambda}$ [8] and the Coulomb system $\hat{\mathcal{H}}_{c,\eta}$ [9], correspondingly. New results are sketched in the last section by presenting the only possible multiparametric spherically symmetric generalizations of the above systems which are MS with *quadratic* integrals of motion, that is, the most generic deformations that can be endowed, respectively, with a generalized Demkov–Fradkin tensor and with a Runge–Lenz N-vector.

2. An exactly solvable deformation of the oscillator system

The ND classical Hamiltonian system given by

$$\mathcal{H}_\lambda(\mathbf{q},\mathbf{p}) = \mathcal{T}_\lambda(\mathbf{q},\mathbf{p}) + \mathcal{U}_\lambda(\mathbf{q}) = \frac{\mathbf{p}^2}{2(1+\lambda\mathbf{q}^2)} + \frac{\omega^2\mathbf{q}^2}{2(1+\lambda\mathbf{q}^2)}, \qquad (5)$$

where λ and ω are real parameters and $\mathbf{q},\mathbf{p} \in \mathbb{R}^N$ are canonical coordinates and momenta, was proven in [5] to be MS. The kinetic energy $\mathcal{T}_\lambda(\mathbf{q},\mathbf{p})$ can be interpreted as the one generating the geodesic motion of a particle with unit mass on a conformally flat space with metric and (non-constant) scalar curvature (4) given by

$$\mathrm{d}s_\lambda^2 = (1+\lambda\mathbf{q}^2)\mathrm{d}\mathbf{q}^2, \qquad R_\lambda(\mathbf{q}) = -\lambda\,\frac{(N-1)\left(2N + 3\lambda(N-2)\mathbf{q}^2\right)}{(1+\lambda\mathbf{q}^2)^3}.$$

Such a curved space is, in fact, an ND spherically symmetric generalization $\mathcal{M}_\lambda$ [15, 16] of the Darboux surface of type III [17, 18, 19]. The limit $\lambda \to 0$ of the above expressions leads to the well known results concerning the (flat) ND isotropic harmonic oscillator with frequency ω:

$$\mathcal{H} = \frac{1}{2}\mathbf{p}^2 + \frac{1}{2}\omega^2\mathbf{q}^2, \qquad \mathrm{d}s^2 = \mathrm{d}\mathbf{q}^2, \qquad R = 0.$$

The remarkable point is that $\mathcal{H}_\lambda$ is a MS Hamiltonian, a fact that can be stated as follows.

Proposition 1. [5, 6] *(i) The Hamiltonian $\mathcal{H}_\lambda$ (5) is endowed with the following constants of motion ($m = 2,\dots,N$):*

• $(2N-3)$ *angular momentum integrals:*

$$C^{(m)} = \sum_{1\le i<j\le m}(q_ip_j - q_jp_i)^2, \qquad C_{(m)} = \sum_{N-m<i<j\le N}(q_ip_j - q_jp_i)^2, \qquad C^{(N)} = C_{(N)}. \qquad (6)$$

• N^2 *integrals which form the ND curved/deformed Demkov–Fradkin tensor ($i,j = 1,\dots,N$):*

$$I_{\lambda,ij} = p_ip_j - \left(2\lambda\mathcal{H}_\lambda(\mathbf{q},\mathbf{p}) - \omega^2\right)q_iq_j, \qquad \mathcal{H}_\lambda = \frac{1}{2}\sum_{i=1}^N I_{\lambda,ii}\,.$$

(ii) Each of the three sets $\{\mathcal{H}_\lambda, C^{(m)}\}$, $\{\mathcal{H}_\lambda, C_{(m)}\}$ ($m = 2,\dots,N$) and $\{I_{\lambda,ii}\}$ ($i = 1,\dots,N$) is formed by N functionally independent functions in involution.

(iii) The set $\{\mathcal{H}_\lambda, C^{(m)}, C_{(m)}, I_{\lambda,ii}\}$ *for* $m = 2,\ldots,N$ *with a fixed index* i *is constituted by* $(2N-1)$ *functionally independent functions.*

Let us now consider the standard definitions for the quantum positions $\hat{\mathbf{q}} = (\hat{q}_1,\ldots,\hat{q}_N)$, momenta $\hat{\mathbf{p}} = (\hat{p}_1,\ldots,\hat{p}_N)$ and $\nabla = (\frac{\partial}{\partial q_1},\ldots,\frac{\partial}{\partial q_N})$ operators $(i,j = 1,\ldots,N)$:

$$\hat{q}_i\,\psi(\mathbf{q}) = q_i\,\psi(\mathbf{q}), \qquad \hat{p}_i\,\psi(\mathbf{q}) = -\mathrm{i}\hbar\,\frac{\partial\,\psi(\mathbf{q})}{\partial q_i}, \qquad [\hat{q}_i,\hat{p}_j] = \mathrm{i}\hbar\,\delta_{ij}, \qquad \mathbf{q}\cdot\nabla = \sum_{i=1}^{N} q_i\frac{\partial}{\partial q_i}\,.$$

If we now apply the conformal Laplacian quantization prescription (3) to $\mathcal{H}_\lambda$ (5) we find:

Proposition 2. [8] *Let* $\hat{\mathcal{H}}_{c,\lambda}$ *be the quantum Hamiltonian given by*

$$\hat{\mathcal{H}}_{c,\lambda} = -\frac{\hbar^2}{2}\Delta_{\mathrm{LB},\lambda} + \frac{\omega^2\mathbf{q}^2}{2(1+\lambda\mathbf{q}^2)} - \hbar^2\lambda(N-2)\left(\frac{2N+3\lambda(N-2)\mathbf{q}^2}{8(1+\lambda\mathbf{q}^2)^3}\right),$$
$$\text{with}\qquad \Delta_{\mathrm{LB},\lambda} = \frac{1}{(1+\lambda\mathbf{q}^2)}\Delta + \frac{\lambda(N-2)}{(1+\lambda\mathbf{q}^2)^2}\,(\mathbf{q}\cdot\nabla)\,. \tag{7}$$

(i) $\hat{\mathcal{H}}_{c,\lambda}$ *commutes with the* $(2N-3)$ *quantum angular momentum operators* $(m = 2,\ldots,N)$

$$\hat{C}^{(m)} = \sum_{1\le i<j\le m}(\hat{q}_i\hat{p}_j - \hat{q}_j\hat{p}_i)^2, \qquad \hat{C}_{(m)} = \sum_{N-m<i<j\le N}(\hat{q}_i\hat{p}_j - \hat{q}_j\hat{p}_i)^2, \qquad \hat{C}^{(N)} = \hat{C}_{(N)}\,, \tag{8}$$

as well as with the N^2 *curved/deformed Demkov–Fradkin operators given by*

$$\hat{I}_{c,\lambda,ij} = \hat{p}_i\hat{p}_j - (N-2)\frac{\mathrm{i}\hbar\lambda}{2(1+\lambda\hat{\mathbf{q}}^2)}(\hat{q}_i\hat{p}_j + \hat{q}_j\hat{p}_i) + \frac{(N-2)\hbar^2\lambda^2\hat{q}_i\hat{q}_j}{(1+\lambda\hat{\mathbf{q}}^2)^2}\left(1 - \frac{N-2}{4}\right)$$
$$-\frac{(N-2)\hbar^2\lambda}{2(1+\lambda\hat{\mathbf{q}}^2)}\delta_{ij} - 2\lambda\hat{q}_i\hat{q}_j\hat{\mathcal{H}}_{c,\lambda}(\hat{\mathbf{q}},\hat{\mathbf{p}}) + \omega^2\hat{q}_i\hat{q}_j,$$

with $i,j = 1,\ldots,N$ *and such that* $\hat{\mathcal{H}}_{c,\lambda} = \frac{1}{2}\sum_{i=1}^{N}\hat{I}_{c,\lambda,ii}$.
(ii) Each of the three sets $\{\hat{\mathcal{H}}_{c,\lambda}, \hat{C}^{(m)}\}$, $\{\hat{\mathcal{H}}_{c,\lambda}, \hat{C}_{(m)}\}$ $(m = 2,\ldots,N)$ *and* $\{\hat{I}_{c,\lambda,ii}\}$ $(i = 1,\ldots,N)$ *is formed by* N *algebraically independent commuting observables.*
(iii) The set $\{\hat{\mathcal{H}}_{c,\lambda}, \hat{C}^{(m)}, \hat{C}_{(m)}, \hat{I}_{c,\lambda,ii}\}$ *for* $m = 2,\ldots,N$ *with a fixed index* i *is formed by* $(2N-1)$ *algebraically independent observables.*
(iv) $\hat{\mathcal{H}}_{c,\lambda}$ *is formally self-adjoint on the space* $L^2(\mathcal{M}_\lambda)$, *associated with the underlying Darboux III space, defined by*

$$\langle\Psi|\Phi\rangle_{c,\lambda} = \int_{\mathcal{M}_\lambda}\overline{\Psi(\mathbf{q})}\,\Phi(\mathbf{q})\,(1+\lambda\mathbf{q}^2)^{N/2}\,\mathrm{d}\mathbf{q}.$$

The complete solution to the eigenvalue problem along with the corresponding eigenfunctions for the case of *positive* deformation parameter λ is summarized in the following statement.

Theorem 3. [8] *Let* $\hat{\mathcal{H}}_{c,\lambda}$ *be the quantum Hamiltonian (7) with* $\lambda > 0$. *Then:*
(i) The continuous spectrum of $\hat{\mathcal{H}}_{c,\lambda}$ *is given by* $[\frac{\omega^2}{2\lambda},\infty)$. *There are no embedded eigenvalues and its singular spectrum is empty.*
(ii) $\hat{\mathcal{H}}_{c,\lambda}$ *has an infinite number of eigenvalues, all of which are contained in* $(0,\frac{\omega^2}{2\lambda})$. *Their only accumulation point is* $\frac{\omega^2}{2\lambda}$ *which is the bottom of the continuous spectrum.*

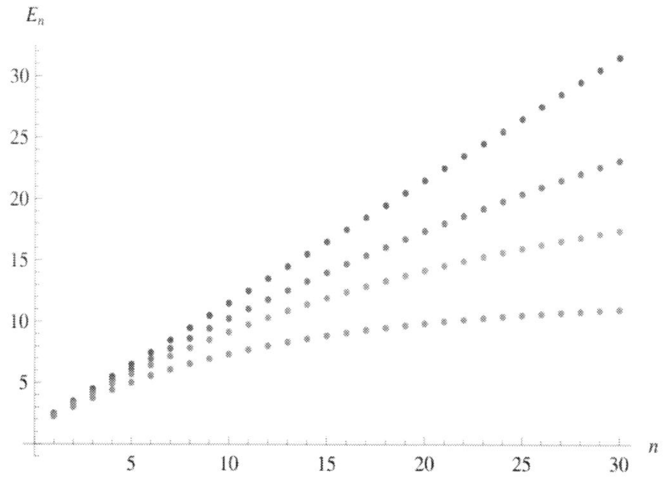

Figure 1. The discrete spectrum $E_{\lambda,n}$ (9) for $0 \leq n \leq 25$, $N = 3$, $\hbar = \omega = 1$ and $\lambda = \{0, 0.01, 0.02, 0.04\}$ starting from the upper dot (straight) line corresponding to the isotropic harmonic oscillator with $\lambda = 0$, that is, $E_{0,n}$. In the same order, $E_{\lambda,0} = \{1.5, 1.48, 1.46, 1.41\}$ and $E_{\lambda,\infty} = \{\infty, 50, 25, 12.5\}$.

(iii) All the discrete eigenvalues of $\hat{\mathcal{H}}_{c,\lambda}$ are of the form

$$E_{\lambda,n} = -\lambda\hbar^2\left(n + \frac{N}{2}\right)^2 + \hbar\left(n + \frac{N}{2}\right)\sqrt{\hbar^2\lambda^2\left(n + \frac{N}{2}\right)^2 + \omega^2}, \qquad n \in \mathbb{N}. \tag{9}$$

(iv) The eigenfunction $\Psi_{c,\lambda}$ of $\hat{\mathcal{H}}_{c,\lambda}$ with eigenvalue $E_{\lambda,n}$ is given by

$$\Psi_{c,\lambda}(\mathbf{q}) = (1 + \lambda\mathbf{q}^2)^{(2-N)/4}\prod_{i=1}^{N}\exp\{-\beta^2 q_i^2/2\}H_{n_i}(\beta q_i), \qquad \beta = \sqrt{\frac{\Omega}{\hbar}},$$

where H_{n_i} are Hermite polynomials with $n_i \in \mathbb{N}$ such that $n_1 + \cdots + n_N = n$ and the deformed frequency Ω is defined by

$$\Omega = \sqrt{\omega^2 - 2\lambda E_{\lambda,n}}\,.$$

Moreover, the bound states of this system satisfy

$$E_{\lambda,\infty} = \lim_{n\to\infty} E_{\lambda,n} = \frac{\omega^2}{2\lambda}, \qquad \lim_{n\to\infty}(E_{\lambda,n+1} - E_{\lambda,n}) = 0.$$

The discrete spectrum (9) is depicted in figure 1 as a function of n for several values of λ. As it was remarked in the introduction, the spectrum turns out to be maximally degenerate since it can be described as a function of just one quantum number $n \in \mathbb{N}$. By taking into account the definition of n, the number of degenerate states $D(E_{\lambda,n})$ for a given energy level $E_{\lambda,n}$ corresponds to all the possible combinations of $\{n_i \in \mathbb{N}\}$ obeying to the constraint $\sum_{i=1}^{N} n_i = n$, namely

$$D(E_{\lambda,n}) = \frac{(n + N - 1)!}{n!(N - 1)!},$$

which for $N = 3$ reduces to the well known expression $D(E_{\lambda,n}) = (n+2)(n+1)/2$ for the degeneracies of the isotropic oscillator. Therefore, the curved system has a similar set of integrals of the motion as the undeformed one and, as a consequence, it exhibits the same degeneracy.

3. An exactly solvable deformation of the Coulomb system

Now we consider the ND Hamiltonian system given by

$$\mathcal{H}_\eta = \mathcal{T}_\eta(\mathbf{q},\mathbf{p}) + \mathcal{U}_\eta(\mathbf{q}) = \frac{|\mathbf{q}|}{2(\eta+|\mathbf{q}|)}\,\mathbf{p}^2 - \frac{k}{\eta+|\mathbf{q}|}, \tag{10}$$

where η and k are real parameters. The metric and scalar curvature of the underlying manifold $\mathcal{M}_\eta$ turns out to be

$$\mathrm{d}s_\eta^2 = \left(1+\frac{\eta}{|\mathbf{q}|}\right)\mathrm{d}\mathbf{q}^2, \qquad R_\eta = \eta(N-1)\,\frac{4(N-3)r+3\eta(N-2)}{4r(\eta+r)^3}, \qquad r=|\mathbf{q}|=\sqrt{\mathbf{q}^2}\,.$$

We remark that the system (10) is directly related to a reduction [20] of the geodesic motion on the Taub–NUT space [21, 22, 23, 24, 25, 26, 27, 28, 29, 30, 31]. In fact, this system can be regarded as an η-deformation of the ND Euclidean Coulomb problem with coupling constant k, since the limit $\eta \to 0$ yields

$$\mathcal{H} = \frac{1}{2}\,\mathbf{p}^2 - \frac{k}{|\mathbf{q}|}, \qquad \mathrm{d}s^2 = \mathrm{d}\mathbf{q}^2, \qquad R = 0.$$

Remarkably enough, the Hamiltonian $\mathcal{H}_\eta$ turns out to be a MS classical system, and this result can be summarized as follows.

Proposition 4. [6] *(i) The Hamiltonian $\mathcal{H}_\eta$ (10) is endowed with the $(2N-3)$ angular momentum integrals (6) and Poisson-commutes with the $\mathcal{R}_{\eta,i}$ components $(i=1,\dots,N)$ of the Runge–Lenz N-vector given by*

$$\mathcal{R}_{\eta,i} = \sum_{j=1}^N p_j(q_jp_i - q_ip_j) + \frac{q_i}{|\mathbf{q}|}\,(\eta\mathcal{H}_\eta(\mathbf{q},\mathbf{p})+k)\,.$$

(ii) The set $\{\mathcal{H}_\eta, C^{(m)}, C_{(m)}, \mathcal{R}_{\eta,i}\}$ with $m=2,\dots,N$ and a fixed index i is formed by $(2N-1)$ functionally independent functions.

We also recall that the classical system $\mathcal{H}_\eta$ has been fully solved in [32]. Next the quantum counterpart of (10) can be obtained by applying (3), and reads:

Proposition 5. [9] *(i) The quantum Hamiltonian $\hat{\mathcal{H}}_{\mathrm{c},\eta}$ given by*

$$\hat{\mathcal{H}}_{\mathrm{c},\eta} = -\frac{\hbar^2}{2}\Delta_{\mathrm{LB},\eta} - \frac{k}{\eta+|\mathbf{q}|} + \hbar^2\eta(N-2)\,\frac{4(N-3)|\mathbf{q}|+3\eta(N-2)}{32|\mathbf{q}|(\eta+|\mathbf{q}|)^3},$$

$$\text{with}\qquad \Delta_{\mathrm{LB},\eta} = \frac{|\mathbf{q}|}{\eta+|\mathbf{q}|}\,\Delta - \frac{\eta(N-2)}{2|\mathbf{q}|(|\mathbf{q}|+\eta)^2}\,(\mathbf{q}\cdot\nabla), \tag{11}$$

commutes with the $(2N-3)$ quantum angular momentum operators (8) as well as with the following N Runge–Lenz operators $(i=1,\dots,N)$:

$$\hat{\mathcal{R}}_{\mathrm{c},\eta,i} = \frac{1}{2}\sum_{j=1}^N\left(\hat{p}_j + \mathrm{i}\hbar\,\eta\,\frac{(N-2)\hat{q}_j}{4(\eta+|\hat{\mathbf{q}}|)\,\hat{\mathbf{q}}^2}\right)(\hat{q}_j\hat{p}_i - \hat{q}_i\hat{p}_j)$$
$$+\frac{1}{2}\sum_{j=1}^N(\hat{q}_j\hat{p}_i - \hat{q}_i\hat{p}_j)\left(\hat{p}_j + \mathrm{i}\hbar\,\eta\,\frac{(N-2)\hat{q}_j}{4(\eta+|\hat{\mathbf{q}}|)\,\hat{\mathbf{q}}^2}\right) + \frac{\hat{q}_i}{|\hat{\mathbf{q}}|}\left(\eta\hat{\mathcal{H}}_{\mathrm{c},\eta}(\mathbf{q},\mathbf{p})+k\right).$$

(ii) Each of the three sets $\{\hat{\mathcal{H}}_{c,\eta}, \hat{C}^{(m)}\}$, $\{\hat{\mathcal{H}}_{c,\eta}, \hat{C}_{(m)}\}$ $(m = 2, \dots, N)$ *and* $\{\hat{\mathcal{R}}_{c,\eta,i}\}$ $(i = 1, \dots, N)$ *is formed by* N *algebraically independent commuting operators.*
(iii) The set $\{\hat{\mathcal{H}}_{c,\eta}, \hat{C}^{(m)}, \hat{C}_{(m)}, \hat{\mathcal{R}}_{c,\eta,i}\}$ *for* $m = 2, \dots, N$ *with a fixed index* i *is formed by* $(2N-1)$ *algebraically independent operators.*
(iv) $\hat{\mathcal{H}}_{c,\eta}$ *is formally self-adjoint on the Hilbert space* $L^2(\mathcal{M}_\eta)$ *with the scalar product*

$$\langle \Psi | \Phi \rangle_{c,\eta} = \int_{\mathcal{M}_\eta} \overline{\Psi(\mathbf{q})}\, \Phi(\mathbf{q}) \left(1 + \frac{\eta}{|\mathbf{q}|}\right)^{N/2} d\mathbf{q}.$$

For a *positive* value of the deformation parameter η, the complete solution of the eigenvalue problem for this quantum mechanical deformed Coulomb problem is the following.

Theorem 6. [9] *Let* $\hat{\mathcal{H}}_{c,\eta}$ *be the quantum Hamiltonian (11) with* $k > 0$ *and* $\eta > 0$*. Then:*
(i) The continuous spectrum of $\hat{\mathcal{H}}_{c,\eta}$ *is given by* $[0, \infty)$*. There are no embedded eigenvalues and the singular spectrum is empty.*
(ii) $\hat{\mathcal{H}}_{c,\eta}$ *has an infinite number of eigenvalues* $E_{\eta,n,l}$*, depending only on the sum* $(n+l)$ *and accumulating at* 0*.*
(iii) The eigenvalues $E_{\eta,n,l}$ *of* $\hat{\mathcal{H}}_{c,\eta}$ *are of the form*

$$E_{\eta,n,l} = \frac{-k^2}{\hbar^2\left(n+l+\frac{N-1}{2}\right)^2 + k\eta + \sqrt{\hbar^4\left(n+l+\frac{N-1}{2}\right)^4 + 2\,\hbar^2 k\eta\left(n+l+\frac{N-1}{2}\right)^2}}, \tag{12}$$

such that the radial eigenfunction $\Phi_{c,\eta}(r)$ *of* $\hat{\mathcal{H}}_{c,\eta}$ *with eigenvalue* $E_{\eta,n,l}$ *reads*

$$\Phi_{c,\eta}(r) = \left(1 + \frac{\eta}{r}\right)^{\frac{2-N}{4}} r^l \exp\left(-\frac{Kr}{\hbar^2\left(n+l+\frac{N-1}{2}\right)}\right) L_n^{2l+N-2}\left(\frac{2Kr}{\hbar^2\left(n+l+\frac{N-1}{2}\right)}\right),$$

where L_n^α *are generalized Laguerre polynomials and the deformed coupling constant* K *reads*

$$K = k + \eta\, E_{\eta,n,l}.$$

Since $\hat{\mathcal{H}}_{c,\eta}$ is a Hamiltonian with radial symmetry, its complete eigenfunction is so given by $\Psi_{c,\eta} = \Phi_{c,\eta}(r)Y_{\boldsymbol{l}}(\boldsymbol{\theta})$ where $Y_{\boldsymbol{l}}(\boldsymbol{\theta})$ denotes the usual hyperspherical harmonics, $\boldsymbol{\theta} = (\theta_1, \dots, \theta_{N-1})$ and $\boldsymbol{l}$ is a vector of $N-1$ quantum numbers $\boldsymbol{l} = (l_2, \dots, l_{N-1}, l_N = l)$ such that (see (8))

$$\hat{C}^{(m)}Y_{\boldsymbol{l}}(\boldsymbol{\theta}) = \hbar^2 l_m(l_m + m - 2)Y_{\boldsymbol{l}}(\boldsymbol{\theta}), \qquad l_m - l_{m-1} \geq 0, \qquad m = 2, \dots N.$$

Notice also that the bound states of this system satisfy

$$\lim_{n,l \to \infty} E_{\eta,n,l} = 0, \qquad \lim_{\mathfrak{n} \to \infty} \left(E_{\eta,\mathfrak{n}+1} - E_{\eta,\mathfrak{n}}\right) = 0, \qquad \mathfrak{n} = n + l.$$

As expected, the limit $\eta \to 0$ of $E_{\eta,n,l}$ provides the well known formula for the standard Coulomb eigenvalues $E_{0,n,l}$

$$E_{0,n,l} = -\frac{k^2}{2\hbar^2\left(n+l+\frac{N-1}{2}\right)^2}.$$

And we find that the perturbative series for the eigenvalues of the deformed system $\hat{\mathcal{H}}_{c,\eta}$ (11) reads

$$E_{\eta,n,l} = E_{0,n,l} + \eta\, \frac{k^3}{2\hbar^4\left(n+l+\frac{N-1}{2}\right)^4} - \eta^2\, \frac{5k^4}{8\hbar^6\left(n+l+\frac{N-1}{2}\right)^6} + O(\eta^3).$$

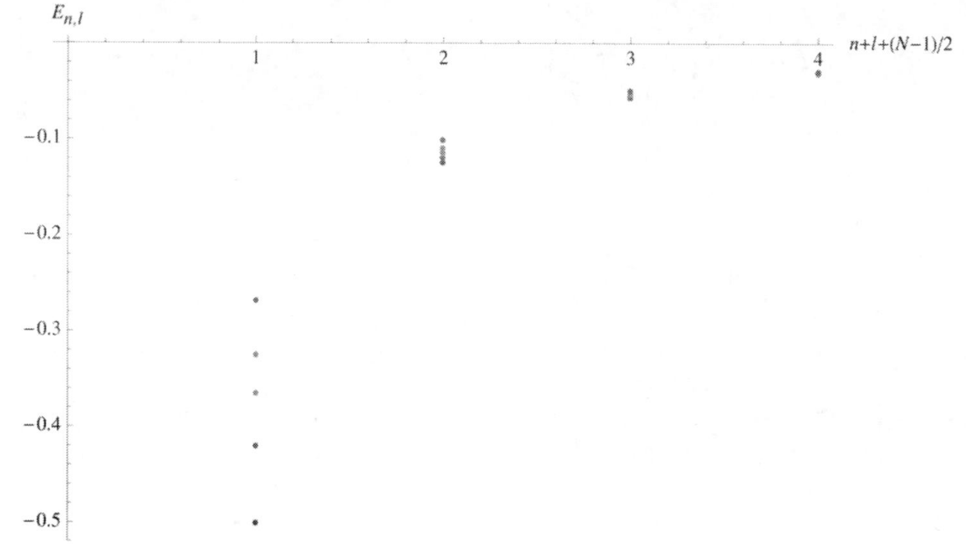

Figure 2. Discrete spectrum (12) for the fundamental and the three first excited states of the Hamiltonian $\hat{\mathcal{H}}_{c,\eta}$ (11) when $\eta = \{0, 0.2, 0.4, 0.6, 1\}$ with $\hbar = k = 1$ and $N \geq 3$. Note that the effect of the η deformation is quite strong for the fundamental state, since it comes from the shift $r \to r + \eta$ in the usual Coulomb potential.

In figure 2 the eigenvalues of the fundamental and of the first three excited states are plotted for different values of the deformation parameter η.

As we can see from (12) the spectrum is maximally degenerate as, again, it depends on a unique principal quantum number $\mathfrak{n} = n + l$. The degeneracy $D(E_{\eta,\mathfrak{n}})$ of a given energy level $E_{\eta,\mathfrak{n}}$ can be computed straightforwardly by taking into account that the cardinality $D(L_l)$ given by the set of the hyperspherical harmonics $\{Y_l(\boldsymbol{\theta})\}$ having the same quantum number l and such that $\hat{C}^{(N)}\{Y_l(\boldsymbol{\theta})\} = \hbar^2 l(l + N - 2)\{Y_l(\boldsymbol{\theta})\}$ reads [33]

$$D(L_l) = \frac{(2l + N - 2)(l + N - 3)!}{l!(N-2)!}.$$

From it we obtain that

$$D(E_{\eta,\mathfrak{n}}) = \sum_{l=0}^{\mathfrak{n}} D(L_l) = \frac{(2\mathfrak{n} + N - 1)(\mathfrak{n} + N - 2)!}{\mathfrak{n}!(N-1)!}.$$

In particular, for $N = 3$ we obtain $D(E_{\eta,\mathfrak{n}}) = (\mathfrak{n} + 1)^2$, which coincides with the degeneracy of the energy levels of the undeformed Coulomb problem.

4. Generalization

So far we have reviewed some *specific* exactly solvable deformations of the oscillator and Coulomb potentials, which can be regarded as the most natural MS deformations beyond constant curvature. Nevertheless, there are more possible generalizations within this framework that preserves the classical MS property and that would lead to other exactly solvable deformed oscillator and Coulomb systems. These arise within the classification of Bertrand Hamiltonians formerly introduced in [34] and further developed in [35, 36, 37]. Such systems are MS and their underlying Bertrand spaces are spherically symmetric ones. If we require to keep *quadratic* integrals of motion, so generalizing the Demkov–Fradkin tensor and the Runge–Lenz N-vector, it can be shown that there only exists *one* possible generalization of the deformations of the oscillator and Coulomb systems here studied that depends on *two* deformation parameters.

In particular, the two-parameter MS deformation of the oscillator system turns out to be

$$\mathcal{H}_{\lambda,\xi}(\mathbf{q},\mathbf{p})=\mathcal{T}_{\lambda,\xi}(\mathbf{q},\mathbf{p})+\mathcal{U}_{\lambda,\xi}(\mathbf{q})=\frac{(1-\xi\mathbf{q}^4)^2\mathbf{p}^2}{2(1+\lambda\mathbf{q}^2+\xi\mathbf{q}^4)}+\frac{\omega^2\mathbf{q}^2}{2(1+\lambda\mathbf{q}^2+\xi\mathbf{q}^4)},$$

where ξ is a real parameter. Obviously, the limit $\xi\to 0$ gives rise to the Hamiltonian $\mathcal{H}_\lambda$ (5) The underlying manifold $\mathcal{M}_{\lambda,\xi}$ is endowed with a conformally flat metric given by

$$\mathrm{d}s^2_{\lambda,\xi}=\frac{(1+\lambda\mathbf{q}^2+\xi\mathbf{q}^4)}{(1-\xi\mathbf{q}^4)^2}\,\mathrm{d}\mathbf{q}^2.$$

And the corresponding scalar curvature (4) reads

$$R_{\lambda,\xi}(r)=-\frac{(N-1)}{(1+\lambda r^2+\xi r^4)^3}\Big\{N\big(2+3\lambda r^2+6\xi r^4+\lambda\xi r^6\big)\big(\lambda+3\lambda\xi r^4+2\xi r^2(3+\xi r^4)\big)$$
$$-6r^2(\lambda^2-4\xi)(1-\xi r^4)^2\Big\},$$

where recall that $r=|\mathbf{q}|=\sqrt{\mathbf{q}^2}$.

As far as the Coulomb system is concerned, the resulting two-parameter MS deformation is given by

$$\mathcal{H}_{\eta,\zeta}=\mathcal{T}_{\eta,\zeta}(\mathbf{q},\mathbf{p})+\mathcal{U}_{\eta,\zeta}(\mathbf{q})=\frac{(1-\zeta\mathbf{q}^2)^2|\mathbf{q}|}{2(\eta+|\mathbf{q}|+\eta\zeta\mathbf{q}^2)}\,\mathbf{p}^2-\frac{k(1+\zeta\mathbf{q}^2)}{(\eta+|\mathbf{q}|+\eta\zeta\mathbf{q}^2)},$$

which generalizes the one-parameter Hamiltonian $\mathcal{H}_\eta$ (10). Hence the metric of the underlying spherically symmetric space $\mathcal{M}_{\eta,\zeta}$ and its scalar curvature are found to be

$$\mathrm{d}s^2_{\eta,\zeta}=\frac{(\eta+|\mathbf{q}|+\eta\zeta\mathbf{q}^2)}{(1-\zeta\mathbf{q}^2)^2|\mathbf{q}|}\,\mathrm{d}\mathbf{q}^2,$$

$$R_{\eta,\zeta}(r)=-\frac{(N-1)}{4r(\eta+r+\eta\zeta r^2)^3}\Big\{6\eta(1-\zeta r^2)^2\Big(\eta+r\big(2+\zeta r(6\eta+r[2+\eta\zeta r])\big)\Big)$$
$$-N\big(3\eta+r(4+\eta\zeta r[6-\zeta r^2])\big)\big(\eta-\zeta r^2(6\eta+r[4+3\eta\zeta r])\big)\Big\}.$$

It is worth stressing that $\mathcal{M}_{\eta,\zeta}$ turns out to be the ND spherically symmetric generalization of the Darboux surface of type IV [17, 18, 19] constructed in [15, 16].

Consequently, by applying the conformal Laplacian quantization (3) to the above two-parameter Hamiltonians, new exactly solvable systems, $\hat{\mathcal{H}}_{\mathrm{c},\lambda,\xi}$ and $\hat{\mathcal{H}}_{\mathrm{c},\eta,\zeta}$, would be obtained as deformations of the oscillator and Coulomb systems. Their solution would generalize the results presented in theorems 3 and 6. Work on this line is currently in progress.

Acknowledgments
This work was partially supported by the Spanish MINECO through the Ramón y Cajal program (A.E.) and under grants MTM2013-43820-P (A.B and F.J.H.) and FIS2011-22566 (A.E.), by the Spanish Junta de Castilla y León under grant BU278U14 (A.B., A.E. and F.J.H.), by the ICMAT Severo Ochoa under grant SEV-2011-0087 (A.E.), and by a postdoctoral fellowship from the Laboratory of Mathematical Physics of the CRM, Université de Montréal (D.R.).

References
[1] Demkov Y N 1959 *Soviet Phys. JETP* **9** 63
[2] Fradkin D M 1965 *Amer. J. Phys.* **33** 207
[3] Ballesteros A, Enciso A, Herranz F J and Ragnisco O 2009 *Commun. Math. Phys.* **290** 1033
[4] Bertrand J 1873 *C.R. Acad. Sci. Paris* **77** 849
[5] Ballesteros A, Enciso A, Herranz F J and Ragnisco O 2008 *Physica D* **237** 505
[6] Ballesteros A, Enciso A, Herranz F J, Ragnisco O and Riglioni D 2011 *SIGMA* **7** 048
[7] Ballesteros A, Enciso A, Herranz F J, Ragnisco O and Riglioni D 2011 *Phys. Lett. A* **375** 1431
[8] Ballesteros A, Enciso A, Herranz F J, Ragnisco O and Riglioni D 2011 *Ann. Phys.* **326** 2053
[9] Ballesteros A, Enciso A, Herranz F J, Ragnisco O and Riglioni D 2014 *Ann. Phys.* **351** 540
[10] Wald R M 1984 *General Relativity* (Chicago: The University of Chicago Press)
[11] Liu Z J and Qian M 1992 *Trans. Amer. Math. Soc.* **331** 321
[12] Landsman N P 1998 *Mathematical Topics Between Classical and Quantum Mechanics* (New York: Springer)
[13] Michel J P, Radoux F and Silhan J 2014 *SIGMA* **10** 016
[14] Baer C and Dahl M 2003 *Geom. Funct. Anal.* **13** 483
[15] Ballesteros A, Herranz F J and Ragnisco O 2007 *Phys. Lett. B* **652** 376
[16] Ballesteros A, Enciso A, Herranz F J and Ragnisco O 2009 *Ann. Phys.* **324** 1219
[17] Koenigs G 1972 *Leçons sur la thèorie gènèrale des surfaces* vol 4 ed G Darboux (New York: Chelsea) p 368
[18] Kalnins E G, Kress J M, Miller W Jr and Winternitz P 2003 *J. Math. Phys.* **44** 5811
[19] Grosche C, Pogosyan G S and Sissakian A N 2007 *Phys. Part. Nuclei* **38** 525
[20] Iwai T and Katayama N 1994 *J. Phys. A: Math. Gen.* **27** 3179
[21] Manton N S 1982 *Phys. Lett. B* **110** 54
[22] Atiyah M F and Hitchin N J 1985 *Phys. Lett. A* **107** 21
[23] Gibbons G W and Manton N S 1986 *Nucl. Phys. B* **274** 183
[24] Fehér L G and Horváthy P A 1987 *Phys. Lett. B* **183** 182
[25] Gibbons G W and Ruback P J 1988 *Comm. Math. Phys.* **115** 267
[26] Iwai T and Katayama N 1995 *J. Math. Phys.* **36** 1790
[27] Iwai T, Uwano Y and Katayama N 1996 *J. Math. Phys.* **37** 608
[28] Bini D, Cherubini C and Jantzen R T 2002 *Class. Quantum Grav.* **19** 5481
[29] Bini D, Cherubini C, Jantzen R T and Mashhoon B 2003 *Class. Quantum Grav.* **20** 457
[30] Gibbons G W and Warnick C M 2007 *J. Geom. Phys.* **57** 2286
[31] Jezierski J and Lukasik M 2007 *Class. Quantum Grav.* **24** 1331
[32] Latini D and Ragnisco O 2014 *preprint* arXiv:1411.3571
[33] Atkinson K and Han W 2012 *Spherical Harmonics and Approximations on the Unit Sphere: An Introduction Lecture Notes in Mathematics* **2044** (New York: Springer) p 16
[34] Perlick V 1992 *Class. Quantum Grav.* **9** 1009
[35] Ballesteros A, Enciso A, Herranz F J and Ragnisco O 2008 *Class. Quantum Grav.* **25** 165005
[36] Riglioni D 2013 *J. Phys. A: Math. Theor.* **46** 265207
[37] Post S and Riglioni D 2014 *preprint* arXiv:1410.4495

Constructing the quantum Hall system on the Grassmannians $\mathbf{Gr}_2(\mathbb{C}^N)$

F Ballı, A Behtash, S Kürkçüoğlu and G Ünal

Department of Physics, Middle East Technical University, Dumlupinar Boulevard, 06800, Cankaya, Ankara, Turkey

E-mail: fatih.balli@metu.edu.tr, alireza.behtash@metu.edu.tr, kseckin@metu.edu.tr, ugonul@metu.edu.tr

Abstract. In this talk, we give the formulation of Quantum Hall Effects (QHEs) on the complex Grassmann manifolds $\mathbf{Gr}_2(\mathbb{C}^N)$. We set up the Landau problem in $\mathbf{Gr}_2(\mathbb{C}^N)$, solve it using group theoretical techniques and provide the energy spectrum and the eigenstates in terms of the $SU(N)$ Wigner $\mathcal{D}$-functions for charged particles on $\mathbf{Gr}_2(\mathbb{C}^N)$ under the influence of abelian and non-abelian background magnetic monopoles or a combination of these thereof. For the simplest case of $\mathbf{Gr}_2(\mathbb{C}^4)$ we provide explicit constructions of the single and many-particle wavefunctions by introducing the Plücker coordinates and show by calculating the two-point correlation function that the lowest Landau level (LLL) at filling factor $\nu = 1$ forms an incompressible fluid. Finally, we heuristically identify a relation between the $U(1)$ Hall effect on $\mathbf{Gr}_2(\mathbb{C}^4)$ and the Hall effect on the odd sphere S^5, which is yet to be investigated in detail, by appealing to the already known analogous relations between the Hall effects on $\mathbb{C}P^3$ and $\mathbb{C}P^7$ and those on the spheres S^4 and S^8, respectively. The talk is given by S. Kürkçüoğlu at the Group 30 meeting at Ghent University, Ghent, Belgium in July 2014 and based on the article by F.Ballı, A.Behtash, S.Kürkçüoğlu, G.Ünal [1].

1. Introduction

A 4-dimensional generalization of the quantum Hall effect (QHE) was introduced by Hu and Zhang in [2]. They treat the Landau problem on S^4 for charged particles carrying an additional $SU(2)$ degree of freedom which are under the influence of an $SU(2)$ background gauge field. In the thermodynamic limit, the multi-particle problem in the lowest Landau level (LLL) with filling factor $\nu = 1$ may be seen as an incompressible 4-dimensional quantum Hall liquid as demonstrated by these authors. Appearance of massless chiral bosons at the edge of a 2-dimensional quantum Hall droplet [3, 4, 5, 6] generalizes to this setting. It is found that among the edge excitations of this 4-dimensional quantum Hall droplet not only photons and gravitons but also other massless higher spin states occur.

Further developments took place after the work of Hu and Zhang. Other higher-dimensional generalizations of QHE to a variety of manifolds including complex projective spaces $\mathbb{C}P^N$, S^8, S^3, the Flag manifold $\frac{SU(3)}{U(1)\times U(1)}$, as well as quantum Hall systems based on higher dimensional fuzzy spheres have been investigated by several authors [7, 8, 9, 10, 11]. Nair and Karabali examined the QHE on $\mathbb{C}P^N$ [7] and solved the Landau problem on $\mathbb{C}P^N$ using the coset realization of $\mathbb{C}P^N$ over $SU(N+1)$ and performing a suitable restriction of the Wigner $\mathcal{D}$-functions on the latter to obtain the wave functions and the energy spectrum for charged particles under the influence of both $U(1)$ abelian and/or non abelian $SU(N)$ gauge field backgrounds. The degeneracy in each LL is identified with the dimension of the irreducible representation (IRR) to which the wave functions belong. A close connection between the Hall effects on $\mathbb{C}P^3$ and $\mathbb{C}P^7$ with abelian backgrounds and those on the spheres S^4 and S^8 with $SU(2)$ and $SO(8)$ backgrounds, respectively [7, 8, 11] has been revealed in these investigations. These models possess several features which make them interesting in their own right and worthy for further study.

Here we focus on the results of our formulation of QHE on the complex Grassmannians $\mathbf{Gr}_k(\mathbb{C}^N)$ as reported in our article [1]. $\mathbf{Gr}_k(\mathbb{C}^N)$ are generalizations of complex projective spaces $\mathbb{C}P^N$ and share many of their features, such as being a Kähler manifold. Plücker embedding of $\mathbf{Gr}_k(\mathbb{C}^N)$ into $\mathbb{C}P^{\binom{N}{k}-1}$ is quite useful in capturing several of these features. For the case $k = 2$, the Plücker embedding describes $\mathbf{Gr}_2(\mathbb{C}^N)$ as a projective algebraic hypersurface in $\mathbb{C}P^N$. For $\mathbf{Gr}_k(\mathbb{C}^4)$ this becomes the well-known Klein Quadric in $\mathbb{C}P^5$ [12]. Employing group theoretical techniques we solved the Landau problem on $\mathbf{Gr}_2(\mathbb{C}^N)$ and developed the physics of at the LLL at the filling factor $\nu = 1$. In this proceedings article, we provide a description of the essential features of our methods and give our main results and refer the reader to our article [1] for a full discussion.

It is also worthwhile to remark that Landau problem on two and higher dimensional spaces have close and striking connections to string physics, D-branes and stringy matrix models and to the structure fuzzy spaces such as the fuzzy sphere S_F^2 and fuzzy complex projective spaces $\mathbb{C}P_F^N$. These connections are studied at various levels of sophistication in the literature [13, 14, 15]. Fuzzy spaces arise as quantized versions of their parent manifolds and they are described by finite-dimensional matrix algebras which tend to the algebra of functions over the parent manifolds under a suitable mapping such as the diagonal coherent state map. Quantum field theories are formulated over fuzzy spaces as matrix models with finite degrees of freedom, while preserving the symmetries of the parent space, which makes them appealing for QFT applications (see, [16] and references therein). Construction of fuzzy spaces using geometric quantization methods yields Hilbert spaces $\mathcal{H}_N$ of wave-functions which are holomorphic sections of $U(1)$ bundles over the commutative parent manifold and the matrix algebras Mat_N of linear transformations on $\mathcal{H}_N$'s form the fuzzy spaces [15]. Observables on the fuzzy spaces belong to this matrix algebra. It has been observed that the LLL in Landau problems over S^2, $\mathbb{C}P^N$ in $U(1)$ backgrounds define Hilbert spaces which are identical to $\mathcal{H}_N$ as they are also holomorphic sections of $U(1)$ bundles over these spaces. Similar structural relations between S_F^4 and QHE on S^4 also exists [15]. Building upon this connection, observables of QHE problem are also contemplated as linear transformations in Mat_N acting on $\mathcal{H}_N$. From this angle, we see that there appears almost an immediate connection of our findings for QHE problem on $\mathbf{Gr}_2(\mathbb{C}^N)$ to fuzzy Grassmann spaces which are discussed in some detail in the literature [17, 18, 19].

Landau problem on S^2 , S^4 and in higher dimensions, which may be of interest in the context of string theory, have descriptions in terms of strings interacting with D-branes [13, 14]. In the 2-dimensional case, one considers a $D2$-brane wrapped around an S^2 and with N $D0$-branes dissolved on it. Stack of K $D6$- branes extending in directions perpendicular to $D2$-brane are then moved to the center of the $D2$-brane. Due to Hanany-Witten effect [20], K fundamental strings stretch between a $D2$-brane and $D6$-branes. Each $D0$-brane provides a magnetic flux quantum over the world volume of $D2$-brane while the end points of the string on $D2$-brane play the role of charged particles under the world-volume gauge field. Low energy excitations of this system are described by the QHE system on S^2 with K playing the number of charged particles, N being the magnetic flux and the ratio $\frac{K}{N} = \nu$ being the filling factor. In this picture, background magnetic field may be described as the density of $D0$-branes on the $D2$-brane and $D0$-brane may be viewed to form an incompressible fluid. An alternative point of view is obtained by describing the background magnetic field in terms of a combination of $D0$-

branes and flux due to a background 2-form field $B_{\mu\nu}$. In a similar manner, Fabinger [14] was able to argue that QHE on S^4 describes the low energy dynamics of a configuration of strings interacting with D-branes, where one now wraps a stack of $D4$-branes on S^4 and spreads $D0$-branes on it. Moving flat infinite $D4$-branes to the center of S^4 gives once again fundamental strings connecting the branes at the center and those forming the S^4. Low energy dynamics of this configuration turns out to be the QHE of Hu and Zhang on S^4. Alternatively, one may develop another interpretation of the latter in terms of a certain number of $D0$-branes expanded into a fuzzy four sphere S_F^4 [21]. We consider the possibility that these connections between string physics, fuzzy geometries and QHE systems over two and higher dimensional compact manifolds may be further exploited to give a description of QHE on $\mathbf{Gr}_2(\mathbb{C}^4)$ in terms of strings-D-branes configuration, although it may prove very hard to address the stability of the latter. Nevertheless, we hope that our results may be preliminarily conceived as a low energy limit of such a strings-D-branes configuration.

2. Review of QHE on $\mathbb{C}P^1$

In this section we give a short summary of the formulation of quantum Hall problem on $\mathbb{C}P^1$ as a warm up for the developments in the subsequent sections. Treatment of QHE on $\mathbb{C}P^1 \equiv S^2$ is originally due to Haldane [22]. Karabali and Nair [7] have provided a reformulation which is adaptable to higher dimensional spaces which we follow here.

Landau problem on $\mathbb{C}P^1$ can be viewed as electrons on a two-sphere under the influence of a Dirac monopole sitting at the center. We take up the task to construct the Hamiltonian for a single electron under the influence of this Dirac monopole. We use the fact that the functions on the group manifold of $SU(2) \equiv S^3$ may be expanded in terms of the Wigner-$\mathcal{D}$ functions $\mathcal{D}^{(j)}_{L_3 R_3}(g)$ where g is an $SU(2)$ group element and j is an integral or a half-odd integral number labeling the IRR of $SU(2)$. The subscripts L_3 and R_3 are the eigenvalues of the third component of the left- and right-invariant vector fields on $SU(2)$.Throughout this article we sometimes denote the left and right invariant vector fields of $SU(N)$ and their eigenvalues by L_i and R_i, respectively, which one is meant will be clear from the context. The left- and right-invariant vector fields on $SU(2)$ satisfy

$$[L_i\,, L_j] = -\varepsilon_{ijk} L_k\,, \quad [R_i\,, R_j] = \varepsilon_{ijk} R_k\,, \quad [L_i\,, R_j] = 0\,. \tag{1}$$

Functions and sections of bundles over $\mathbb{C}P^1$ may be obtained from the Wigner-$\mathcal{D}$ functions on $SU(2)$ by a suitable restriction of the latter. The coset realization of $\mathbb{C}P^1$ is

$$\mathbb{C}P^1 \equiv S^2 = \frac{SU(2)}{U(1)}\,. \tag{2}$$

This implies that the sections of $U(1)$ bundle over $\mathbb{C}P^1$ should fulfill

$$\mathcal{D}(g e^{iR_3\theta}) = e^{i\frac{n}{2}\theta}\mathcal{D}(g)\,, \tag{3}$$

where n is an integer. This condition is satisfied by the functions of the form $\mathcal{D}^{(j)}_{L_3 \frac{n}{2}}(g)$. Eigenvalue $\frac{n}{2}$ of R_3 corresponds to the strength of the Dirac monopole sitting at the center of the sphere and $\mathcal{D}^{(j)}_{L_3 \frac{n}{2}}(g)$ are the desired wave functions as will argue so shortly. Let us observe that $\mathcal{D}^{(j)}_{L_3 0}(g)$ simply correspond to the spherical harmonics on S^2, which are nothing but the wave functions for electrons on a sphere with no background magnetic field.

In the presence of a magnetic field B, the Hamiltonian must involve covariant derivatives whose commutator is proportional to B. We take this commutator as $[D_+\,, D_-] = B$. We observe that the covariant derivatives $D_\pm$ may be identified by the right invariant vector fields $R_\pm = R_1 \pm iR_2$, as

$$D_\pm = \frac{1}{\sqrt{2}\ell} R_\pm\,, \tag{4}$$

where ℓ denotes the radius of the sphere. Note that $[R_+ , R_-] = 2R_3$, for the eigenvalue $\frac{n}{2}$ of R_3 we have

$$B = \frac{n}{2\ell^2} , \tag{5}$$

for the magnetic monopole field with the magnetic charge $\frac{n}{2}$ in accordance with the Dirac quantization condition. Magnetic flux through the sphere is $2\pi n$.

In terms of the covariant derivatives the Hamiltonian is given as

$$\begin{aligned} H &= \frac{1}{2M}(D_+D_- + D_-D_+) \\ &= \frac{1}{2M\ell^2}(\sum_{i=1}^{3} R_i^2 - R_3^2) , \end{aligned} \tag{6}$$

where M is the mass of the particle. We have $\sum_{i=1}^3 R_i^2 = \sum_{i=1}^3 L_i^2 = j(j+1)$, with j labeling the irreducible representations of $SU(2)$. To have $\frac{n}{2}$ occuring as one of the possible eigenvalues of R_3, we need $j = \frac{1}{2}n + q$ where q is an integer. Spectrum of the Hamiltonian is therefore

$$\begin{aligned} E_{q,n} &= \frac{1}{2M\ell^2}\left((\frac{n}{2} + q)(\frac{n}{2} + q + 1) - \frac{n^2}{4}\right) \\ &= \frac{B}{2M}(2q+1) + \frac{q(q+1)}{2M\ell^2} \end{aligned} \tag{7}$$

Eigenfunctions of this Hamiltonian are now clearly seen to be $\mathcal{D}^{(j)}_{L_3 \frac{n}{2}}(g)$ as we have noted earlier. Integer q is readily interpreted as the Landau level (LL) index. The ground state, that is the Lowest Landau Level (LLL), is at $q = 0$ and has the energy $\frac{B}{2M}$. The LLL is separated from the higher LL by finite energy gaps. Degeneracy at an LL controlled by the left invariant vector fields L_i since they commute with the covariant derivatives $[L_i , D_j] = 0$. Each LL is $(2j + 1 = n + 1 + 2q)$-fold degenerate. These are the wavefunctions $\mathcal{D}^{(j)}_{L_3 \frac{n}{2}}(g)$ at a given LL with L_3 eigenvalues ranging from $-j$ to j.

Wave functions $\mathcal{D}^{(j)}_{L_3 \frac{n}{2}}(g)$ may be given in explicit form by choosing a suitable coordinate system. We refer the reader to the original literature [7] where this is done in detail. In [7] it is also shown that the LLL form an incompressible liquid by computing the two-point correlation function for the wave-function density. We investigate this crucial property of the LLL for our case in section ...

3. Landau Problem on the Grassmannian $\mathbf{Gr}_2(\mathbb{C}^4)$

We start with recalling a few facts about the Grassmannians and their geometry. Complex Grassmannians $\mathbf{Gr}_k(\mathbb{C}^N)$ are the set of all k-dimensional linear subspaces of the vector space $\mathbb{C}^N$ with the complex dimension $k(N-k)$. They are smooth and compact complex manifolds and admit Kähler structures. Grassmannians are homogeneous spaces and have the coset realization

$$\mathbf{Gr}_k(\mathbb{C}^N) = \frac{SU(N)}{S[U(N-k) \times U(k)]} \sim \frac{SU(N)}{SU(N-k) \times SU(k) \times U(1)} . \tag{8}$$

We see that $\mathbf{Gr}_1(\mathbb{C}^N) \equiv \mathbb{C}P^N$ and $\mathbf{Gr}_2(\mathbb{C}^4)$ is therefore the simplest Grassmannian that is not a projective space.

To set up and solve the Landau problem on $\mathbf{Gr}_2(\mathbb{C}^4)$, we contemplate that $SU(4)$ Wigner $\mathcal{D}$-functions may be suitably restricted to obtain the harmonics and local sections of bundles over $\mathbf{Gr}_2(\mathbb{C}^4)$. Let $g \in SU(4)$ and let us denote the left- and the right-invariant vector fields on $SU(4)$ by L_α and R_α $(\alpha : 1, \cdots, 15)$; they fulfill the Lie algebra commutation relations for $SU(4)$. We introduce the Wigner-$\mathcal{D}$ functions on $SU(4)$ as

$$g \to \mathcal{D}^{(p,q,r)}_{L^{(1)} L_3^{(1)} L^{(2)} L_3^{(2)} L_{15}; R^{(1)} R_3^{(1)} R^{(2)} R_3^{(2)} R_{15}}(g) \tag{9}$$

where (p, q, r) are three integers labeling the irreducible representations of $SU(4)$, and the subscripts denote the relevant quantum numbers for the left- and right- rotations. In particular, the left and right generators of $SU(2) \times SU(2)$ subgroup are labeled by $L_\alpha \equiv (L_i^{(1)}, L_i^{(2)})$ and $R_\alpha \equiv (R_i^{(1)}, R_i^{(2)})$ $(i : 1, 2, 3, \alpha : 1, \cdots, 6)$ with corresponding $SU(2) \times SU(2)$ quadratic Casimirs $\mathcal{C}_2^L = L^{(1)}(L^{(1)} + 1) + L^{(2)}(L^{(2)} + 1)$, $\mathcal{C}_2^R = R^{(1)}(R^{(1)} + 1) + R^{(2)}(R^{(2)} + 1)$.

Hamiltonian on $\mathbf{Gr}_2(\mathbb{C}^4)$ may be written down as

$$\begin{aligned} H &= \frac{1}{2M\ell^2} \sum_{\alpha=7}^{14} R_\alpha^2 \\ &= \frac{1}{2M\ell^2} \left(\mathcal{C}_2(p, q, r) - \mathcal{C}_2^R - R_{15}^2 \right) , \end{aligned} \tag{10}$$

where $\mathcal{C}_2(p, q, r)$ is the quadratic Casimir of $SU(4)$ in the IRR (p, q, r) with the eigenvalue

$$\mathcal{C}_2(p, q, r) = \frac{3}{8}(r^2 + p^2) + \frac{1}{2}q^2 + \frac{1}{8}(2pr + 4pq + 4qr + 12p + 16q + 12r). \tag{11}$$

The dimension of the IRR (p, q, r) is

$$\dim(p, q, r) = \frac{1}{12}(p + q + 2)(p + q + r + 3)(q + r + 2)(p + 1)(q + 1)(r + 1). \tag{12}$$

Coset realization of $\mathbf{Gr}_2(\mathbb{C}^4)$ signals that, there can be both abelian and non-abelian background gauge fields corresponding to the gauging of the $U(1)$ and one or both of the $SU(2)$ subgroups.

3.1. $U(1)$ *gauge field background*

In this case we are concerned with the branching

$$SU(4) \supset SU(2) \times SU(2) \times U(1) . \tag{13}$$

and obtaining the wave functions with the $U(1)$ background gauge field, requires us to restrict $\mathcal{D}^{(p,q,r)}$ in such a way that they transform trivially (i.e. singlets) under the right action of $SU(2) \times SU(2)$, and carry a right $U(1)$ charge (R_{15} eigenvalue).

$SU(4)$ IRR (p, q, r) has the following branching in Young tableaux notation, which keeps the $SU(2) \times SU(2)$ in the singlet representation,

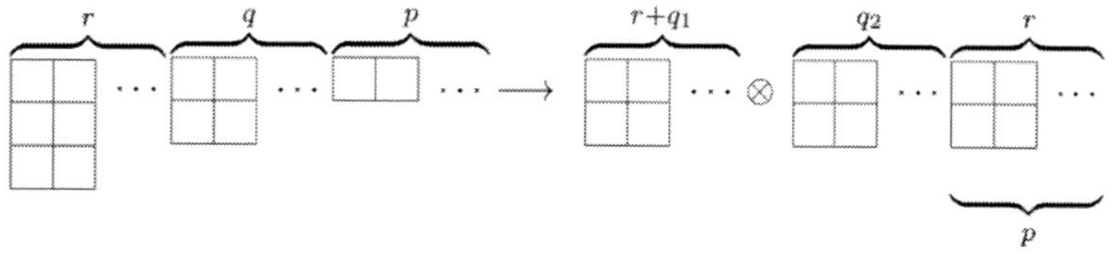

(14)

Here we have introduced the splitting $q = q_1 + q_2$ in the representation in order to handle the partition of columns labeled by q in the branching. We see that a trivial representation of $SU(2) \times SU(2)$ may be obtained if and only if p is equal to r.

It is possible to show that [1]

$$R_{15} = \frac{n}{\sqrt{2}} = \frac{q_1 - q_2}{\sqrt{2}}\,, \tag{15}$$

with $n \in \mathbb{Z}$ being the $U(1)$ charge of the branching.

Putting these facts together energy spectrum corresponding to the Hamiltonian (10) turns out to be

$$E = \frac{1}{2M\ell^2}\left(p^2 + 3p + np + 2q_2^2 + 4q_2 + 2pq_2 + 2n(1+q_2)\right)\,. \tag{16}$$

The LLL energy at a fixed monopole background n is obtained for $q_2 = p = 0$ and it is

$$E_{LLL} = \frac{n}{M\ell^2} = \frac{2B}{M}\,, \tag{17}$$

with the degeneracy $\dim(0\,,n\,,0) = \frac{1}{12}(n+1)(n+2)^2(n+3)$. In (17), $B = \frac{n}{2\ell^2}$ is the field strength of the $U(1)$ magnetic monopole.

The wave functions corresponding to this energy spectrum are

$$\mathcal{D}^{(p\,,q_1+q_2\,,p)}_{L^{(1)}L_3^{(1)}L^{(2)}L_3^{(2)}L_{15}\,;0\,,0\,,0\,,0\,,\frac{n}{\sqrt{2}}}(g) \equiv \mathcal{D}^{(p\,,[\frac{q+n}{2}]+[\frac{q-n}{2}]\,,p)}_{L^{(1)}L_3^{(1)}L^{(2)}L_3^{(2)}L_{15}\,;0\,,0\,,0\,,0\,,\frac{n}{\sqrt{2}}}(g)\,. \tag{18}$$

The degeneracy of each Landau level is given by the dimension of the IRR $(p\,,q\,,p)$ in equation (12). This means that the set of left quantum numbers $\{L^{(1)}\,,L_3^{(1)}\,,L^{(2)}\,,L_3^{(2)}\,,L_{15}\}$ can take on $\dim(p\,,q_1+q_2\,,p)$ different values as a set.

For the many-body problem in which all the states of LLL are filled with the filling factor $\nu = 1$, in the thermodynamic limit $\ell \to \infty, \mathcal{N} \to \infty$ we obtain a finite spatial density of particles

$$\rho = \frac{\mathcal{N}}{\frac{\pi^4\ell^8}{12}} \underset{\ell\to\infty\,,\mathcal{N}\to\infty}{\longrightarrow} \frac{n^4}{\pi^4\ell^8} = \left(\frac{2B}{\pi}\right)^4\,, \tag{19}$$

where we have used $\mathcal{N} = \dim(0\,,n\,,0) = \frac{1}{12}(n+1)(n+2)^2(n+3)$ for the number of fermions in the LLL with $\nu = 1$, and $\mathrm{vol}(\mathbf{Gr}_2(\mathbb{C}^4)) = \frac{\pi^4\ell^8}{12}$.

We note that the case $n = 0$ simply reduces the Wigner-$\mathcal{D}$ functions to the harmonics on $\mathbf{Gr}_2(\mathbb{C}^4)$ corresponding to the wave functions of a particle on $\mathbf{Gr}_2(\mathbb{C}^4)$ with vanishing monopole background.

3.2. Single $SU(2)$ gauge field and $U(1)$ gauge field background

In this case we need to restrict to $\mathcal{D}^{(p,q,r)}$, which transform as a singlet under one or the other $SU(2)$ in the right action of $SU(2) \times SU(2)$, and carry a $U(1)$ charge. There are a range of possibilities within the branching (13) as given in the following Young tableaux decomposition:

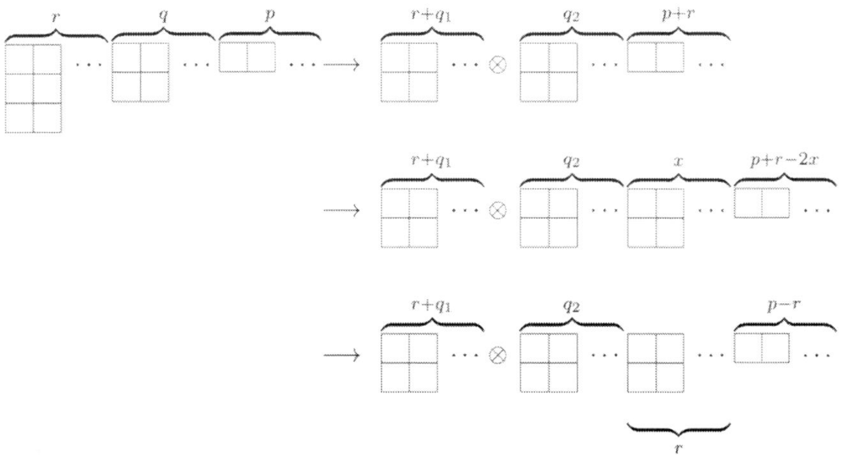

(20)

(21)

(22)

Here we have assumed that $p > r$, split $q_1 + q_2 = q$ and introduced the integer x $(0 \leq x \leq r)$ to conveniently represent the generic case. Eigenvalues of R_{15} take the form

$$R_{15} = \frac{n}{\sqrt{2}} = \frac{1}{2\sqrt{2}} \left(2(q_1 - q_2) - (p - r)\right) . \tag{23}$$

We observe from the Young tableaux that the first $SU(2)$ in the branching remains a singlet while the second may take on values over the range

$$R^{(1)} = 0 , \quad R^{(2)} = \frac{p - r}{2} , \cdots , \frac{r + p}{2} . \tag{24}$$

$R^{(2)}$ values are integers since n being an integer restricts $p - r$ to even integers.

Some algebra shows that the energy spectrum is

$$E = \frac{1}{2M\ell^2} \left(2q_2^2 + 2q_2(n + R^{(2)} + m + 2) + n(R^{(2)} + m + 2) + (R^{(2)} + m)(2 + m) \right) , \tag{25}$$

while the LLL energy at fixed background charges $R^{(2)}$ and n is obtained for $q_2 = m = 0$ and reads

$$E_{LLL} = \frac{1}{2M\ell^2} \left(n(R^{(2)} + 2) + 2R^{(2)} \right) . \tag{26}$$

Wave functions with this energy spectrum may be given in the form

$$\mathcal{D}^{(p\,,q_1+q_2\,,r)}_{L^{(1)} L_3^{(1)} L^{(2)} L_3^{(2)} L_{15}\,; 0\,, 0\,, R^{(2)}\,, R_3^{(2)}\,, \frac{n}{\sqrt{2}}}(g) . \tag{27}$$

In the thermodynamic limit, for pure $SU(2)$ background $(n = 0\,, R^{(1)} = 0\,, R^{(2)} \neq 0)$, $R^{(2)}$ should scale in the thermodynamic limit as $R^{(2)} \sim \ell^2$. The number of fermions in the LLL with $\nu = 1$ is $\mathcal{N} = \dim(R^{(2)}\,, 0\,, R^{(2)}) \approx \frac{\left(R^{(2)}\right)^5}{6}$ and the corresponding spatial density is $\rho \approx \frac{\left(R^{(2)}\right)^4}{\pi^4 \ell^8}$, which is finite. For case when both $U(1)$ and $SU(2)$ backgrounds are present, we may choose either one of n or $R^{(2)}$ to scale like ℓ^2. Taking $n \sim \ell^2$ and $R^{(2)}$ to be finite in thermodynamic limit, we get $\rho \sim \frac{n^4}{2\pi^4 \ell^8 R^{(2)}}$, which is also finite.

Finally we remark that interchanging the Young tableaux of two $SU(2)$'s amounts to interchanging $R^{(1)}$ and $R^{(2)}$ in (24), and also a flip in the sign of the $U(1)$ charge. In the relevant formulas above, one can compensate for these changes by replacing $R^{(2)}$ with $R^{(1)}$ and substituting $|n|$ for n.

The treatment of the most general case, with $SU(2) \times SU(2)$ gauge field background and generalization of the results of this section to the Landau problem on $\mathbf{Gr}_2(\mathbb{C}^N)$ are somewhat lengthy and omitted here. Interested readers can find a full discussion of these in our article [1].

4. Local Form of the Wave Functions and the Gauge Fields

Plücker coordinates for $\mathbf{Gr}_k(\mathbb{C}^N)$ are constructed out of a projective embedding, the so-called Plücker embedding $\mathbf{Gr}_k(\mathbb{C}^n) \hookrightarrow \mathbf{P}\left(\bigwedge^k \mathbb{C}^n\right)$ [12, 24]. For $\mathbf{Gr}_2(\mathbb{C}^4)$ this construction involves the projective space $\mathbf{P}\left(\mathbb{C}^4 \wedge \mathbb{C}^4\right) \equiv \mathbb{C}P^5$. We introduce two sets of complex coordinates $v_\alpha\,, w_\alpha$ $(\alpha = 1\,, \cdots, 4)$ and take the fully antisymmetric basis for the exterior product space $\mathbb{C}^4 \wedge \mathbb{C}^4$ in the form

$$P_{\alpha\beta} = \frac{1}{\sqrt{2}}(v_\alpha w_\beta - v_\beta w_\alpha)\,. \tag{28}$$

$P_{\alpha\beta}$ may be seen as the homogenous coordinates on $\mathbb{C}P^5$ with the identification $P_{\alpha\beta} \sim \lambda P_{\alpha\beta}$ where $\lambda \in U(1)$ and $\sum_{\alpha,\beta}^4 |P_{\alpha\beta}|^2 = 1$. Plücker embedding of $\mathbf{Gr}_2(\mathbb{C}^4)$ in $\mathbb{C}P^5$ is given by the homogeneous condition

$$\varepsilon_{\alpha\beta\gamma\delta} P_{\alpha\beta} P_{\gamma\delta} = P_{12}P_{34} - P_{13}P_{24} + P_{14}P_{23} = 0\,. \tag{29}$$

In fact this condition defines the Klein quadric Q_4 in $\mathbb{C}P^5$, which is complex analytically equivalent to $\mathbf{Gr}_2(\mathbb{C}^4)$.

It is possible to use $P_{\alpha\beta}$ to parametrize the columns of $g \in SU(4)$ in the IRR $(0,1,0)$. We choose a parametrization of the form $g_{N6} = P_N := P_{\alpha\beta}$, $g_{N5} = \varepsilon_{NM} P^*_M = \varepsilon_{\alpha\beta\gamma\delta} P^*_{\gamma\delta}$ with $N \equiv [\alpha\beta]$, $N = 1\,, \cdots, 6$ and $\alpha\beta = (12, 13, 14, 23, 24, 34)$.

Wave functions in the $U(1)$ background gauge field are the sections of $U(1)$ bundle over $\mathbf{Gr}_2(\mathbb{C}^4)$. They satisfy the gauge transformation property

$$\mathcal{D}^{(0\,,q_1+q_2\,,0)}(gh) = \mathcal{D}^{(0\,,q_1+q_2\,,0)}(g e^{i\lambda_{15}\theta}) = \mathcal{D}^{(0\,,q_1+q_2\,,0)}(g) e^{i\frac{n}{\sqrt{2}}\theta}\,. \tag{30}$$

In the IRR $(0,1,0)$ we have $\lambda_{15} = \frac{1}{\sqrt{2}} diag(0,0,0,0,-1,1)$. Using this fact and (30), we infer that

$$\mathcal{D}^{(0\,,1\,,0)}(g) \sim P_{\alpha\beta}\,. \tag{31}$$

Since $(0,q,0)$ IRR is the q-fold symmetric tensor product of the IRR $(0,1,0)$ we find that

$$\mathcal{D}^{(0\,,q_1+q_2\,,0)}(g) \sim P_{\alpha_1\beta_1} P_{\alpha_2\beta_2} \cdots P_{\alpha_{q_1}\beta_{q_1}} P^*_{\gamma_1\delta_1} P^*_{\gamma_2\delta_2} \cdots P^*_{\gamma_{q_2}\delta_{q_2}}\,. \tag{32}$$

LLL wave functions are those with $q_2 = 0$ and take the form

$$\mathcal{D}^{(0\,,q_1\,,0)}_{L^{(1)} L_3^{(1)} L^{(2)} L_3^{(2)} L_{15}\,;\,0,0,0,0,\frac{n}{\sqrt{2}}}(g) \sim P_{\alpha_1\beta_1} P_{\alpha_2\beta_2} \cdots P_{\alpha_{q_1}\beta_{q_1}}\,, \tag{33}$$

which are holomorphic in the Plücker coordinates.

At filling factor $\nu = 1$ LLL has $\mathcal{N} = \dim(0,1,0) = \frac{1}{12}(n+1)(n+2)^2(n+3)$ number of particles. Its wave-function is given in terms of the Slater determinant as

$$\Psi_{MP} = \frac{1}{\sqrt{\mathcal{N}!}} \varepsilon^{\Lambda_1\Lambda_2\cdots\Lambda_n} \Psi_{\Lambda_1}(P^{(1)}) \Psi_{\Lambda_2}(P^{(2)}) \cdots \Psi_{\Lambda_{\mathcal{N}}}(P^{(\mathcal{N})})\,. \tag{34}$$

Journal of Physics: Conference Series **597** (2015) 012015 doi:10.1088/1742-6596/597/1/012015

Here $P^{\mathbf{i}}$ denotes the ith position fixed in the Hall fluid and correspondingly $\Psi_{\Lambda_j}(P^{\mathbf{i}})$ refers to the wave function of the jth particle located at the position $P^{\mathbf{i}}$. For a one-particle wave function in our notation is

$$\Psi_{\Lambda_i}(P^{\mathbf{i}}) \equiv \Psi^{\mathbf{i}}_{\alpha\beta} \sim P^{\mathbf{i}}_{\alpha\beta} \,. \tag{35}$$

LLL wave function given in (33) is then

$$\Psi_{\Lambda_i}(P^{\mathbf{i}}) \equiv \Psi^{\mathbf{i}}_{\Lambda_i} \sim (P^{\mathbf{i}}_{\alpha\beta})^n \,. \tag{36}$$

Two point correlation function is of the form

$$\Omega(1,2) = \int |\Psi_{MP}|^2 d\mu(3) d\mu(4) \cdots d\mu(\mathcal{N}) = |\Psi^{\mathbf{1}}|^2 |\Psi^{\mathbf{2}}|^2 - |\Psi^{*\mathbf{1}}_{\Lambda} \Psi^{\mathbf{2}}_{\Lambda}|^2 . \tag{37}$$

wher $d\mu(i)$ denote integration measure of integration on $\mathbf{Gr}_2(\mathbb{C}^4)$ in the coordinates of the ith particle.

Taking the normalized coordinate chart $\gamma_i := \frac{P_{\alpha\beta}}{P_{12}}$ where $P_{12} \neq 0$ and using the notation $\vec{X} = \vec{\gamma}\ell$, yields after some algebra the result in the thermodynamic limit as

$$\Omega(1,2) \approx 1 - e^{-2B\left(\vec{x}^{\mathbf{1}} - \vec{x}^{\mathbf{2}}\right)^2} e^{-2B\ell^2\left(\det \Gamma^{\mathbf{1}} - \det \Gamma^{\mathbf{2}}\right)^2} , \tag{38}$$

where we have used $n = 2B\ell^2$ and introduced $\Gamma^{\mathbf{i}} := \begin{pmatrix} \gamma^{\mathbf{i}}_2 & \gamma^{\mathbf{i}}_1 \\ \gamma^{\mathbf{i}}_4 & \gamma^{\mathbf{i}}_3 \end{pmatrix}$. This result shows the two-point function of the particles located at the positions $\vec{x}^{\mathbf{1}}, \vec{x}^{\mathbf{2}}$ on $\mathbf{Gr}_2(\mathbb{C}^4)$, is extracted from that of the particles on $\mathbb{C}P^5$ at the positions $\vec{X}^{\mathbf{1}}, \vec{X}^{\mathbf{2}}$ by a restriction of these particles to the algebraic variety determined by $X^{\mathbf{i}}_5 \equiv \ell \det \Gamma^{\mathbf{i}}$, as expected. It is apparent from the form of $\Omega(1,2)$ that the probability of finding two particles at the same point goes to zero, signalling the incompressibility of the Hall fluid.

A short calculation show that the $U(1)$ gauge field $A = -\frac{in}{\sqrt{2}} \mathrm{Tr}\left(\lambda^{15}_{(6)} g^{-1} dg\right)$ may be written in terms of Plücker coordinates as

$$A = -in P^*_N dP_N \,, \tag{39}$$

and the associated field strength $F = dA$ is

$$F = -in dP^*_N \wedge dP_N \,. \tag{40}$$

We note that F is an antisymmetic, gauge invariant, and closed two-form on $\mathbf{Gr}_2(\mathbb{C}^4)$ and as such it is proportional to the Kähler two-form over $\mathbf{Gr}_2(\mathbb{C}^4)$. It is known from very general considerations [25] that the integral of F over a non-contractable two surface Σ in $\mathbf{Gr}_2(\mathbb{C}^4)$ is an integral multiple of 2π:

$$\frac{1}{2\pi} \int_\Sigma F = n \,. \tag{41}$$

In the present context, this result signals an analogue of the Dirac quantization condition with $\frac{n}{2}$ identified as the magnetic monopole charge and $B = \frac{n}{2\ell^2}$.

5. Final Remarks

We have presented some of our essential results on the formulation of the quantum Hall problem on $\mathbf{Gr}_2(\mathbb{C}^N)$. A group theoretical approach has been used to obtain the energy spectrum and the wave function for the Landau problem on $\mathbf{Gr}_2(\mathbb{C}^N)$ and this is supplemented by the local description of the LLL physics at $\nu = 1$ using Plücker coordinates.

It is worthwhile to briefly discuss the following observation about the QHE on $\mathbf{Gr}_2(\mathbb{C}^4)$ with $U(1)$ background. The isomorphisms $Spin(6) \cong SU(4)$ and $Spin(4) \cong SU(2) \times SU(2)$, indicate that the Stiefel manifold $\mathbf{St}_2(\mathbb{R}^6) \equiv \frac{Spin(6)}{Spin(4)}$ forms the principal $U(1)$ fibration [23]

$$U(1) \longrightarrow \mathbf{St}_2(\mathbb{R}^6) \longrightarrow \mathbf{Gr}_2(\mathbb{C}^4)\,. \tag{42}$$

There are also a family of fibrations $\mathbf{St}_{k-1}(\mathbb{R}^{n-1}) \longrightarrow \mathbf{St}_k(\mathbb{R}^n) \longrightarrow S^{n-1}$, which for $k=2$ and $n=6$ becomes

$$S^4 \longrightarrow \mathbf{St}_2(\mathbb{R}^6) \longrightarrow S^5\,. \tag{43}$$

Putting these facts together, we see that $\mathbf{Gr}_2(\mathbb{C}^4)$ has the local structure $\frac{S^5 \times S^4}{U(1)}$. Thus, we think that the QHE on S^5 with the S^4 fibers, associated to a $SO(5)$ gauge field background, may be seen as a QHE on $\mathbf{Gr}_2(\mathbb{C}^4)$ with a $U(1)$ background gauge field. The natural question to answer here is then what we mean by QHE on S^5. This problem may be treated by generalizing the formulation of the QHE on the 3-sphere [9]

$$S^3 = \frac{SU(2) \times SU(2)}{SU(2)_{diag}} \cong \frac{Spin(4)}{Spin(3)}\,, \tag{44}$$

which selects the constant background gauge field as the spin connection. In a construction generalizing this to the QHE on S^5,

$$S^5 = \frac{SO(6)}{SO(5)} = \frac{Spin(6)}{Spin(5)}\,, \tag{45}$$

one will be naturally selecting a constant $SO(5)$ background gauge field taking it again as the spin connection. Such a choice of the gauge field appears to be consistent with our heuristic argument. Our observation is inspired by and bears a resemblance to the relation between the QHE on $\mathbb{C}P^7$ and S^8. The former can be realized locally as $\frac{S^8 \times S^7}{U(1)}$, while the latter forms the base of the 3rd Hopf map $S^7 \longrightarrow S^{15} \longrightarrow S^8$, and S^{15} is a $U(1)$ bundle over $\mathbb{C}P^7$ [8].

Acknowledgments

F. B., S. K., and G.Ü. acknowledge the support of TÜBİTAK under the project No.110T738. S.K. also acknowledges the support of TÜBA-GEBIP program of the Turkish Academy of Sciences.

References

[1] Balli F, Behtash A, Kurkcuoglu S and Unal G 2014 Quantum Hall Effect on the Grassmannians $\mathbf{Gr}_2(\mathbb{C}^N)$ *Phys. Rev.* D **89** 105031
[2] Hu J and Zhang S.C 2001 A Four Dimensional Generalization of the Quantum Hall Effect *Science* **294** 823
[3] Ezawa Z F 2003 *Quantum Hall Effects: Field Theoretical Approach and Related Topics* (Singapore: World Scientific)
[4] Halperin B I 1982 Quantized Hall conductance, current-carrying edge states, and the existence of extended states in a two-dimensional disordered potential *Phys. Rev.* B **25** 2185
[5] Wen X G 1990 Electrodynamical Properties of Gapless Edge Excitations in the Fractional Quantum Hall States *Phys. Rev. Lett.* **64** 2206
[6] Stone M 1990 Schur functions, chiral bosons, and the quantum-Hall-effect edge states *Phys. Rev. B* **42** 8399
[7] Karabali D and Nair V P 2002 Quantum Hall Effect in Higher Dimensions *Nucl. Phys. B* **641** 533
[8] Bernevig B A, Chern C H, Hu J P, Toumbas N and Zhang S C 2003 The Eight Dimensional Quantum Hall Effect and the Octonions *Phys. Rev. Lett.* **91** 236803
[9] Nair V P and Randjbar-Daemi S 2004 Quantum Hall effect on S^3, edge states and fuzzy S^3/Z_2 *Nucl. Phys. B* **679** 447
[10] Daoud M and Jellal A 2008 Quantum Hall Effect on the Flag Manifold $\mathbf{F}_2$ *Int. J. Mod. Phys.* **A23** 3129

[11] Hasebe K and Kimura Y 2004 Dimensional Hierarchy in Quantum Hall Effects on Fuzzy Spheres *Phys. Lett. B* **602** 255
[12] Ward R S and Wells R O 1991 *Twistor Geometry and Field Theory* (Cambridge: Cambridge University Press)
[13] Bernevig B A, Brodie J H, Susskind L and Toumbas N 2001 How Bob Laughlin tamed the giant graviton from Taub - NUT space *J. High Energy Phys.* JHEP0102200103
[14] Fabinger M 2002, Higher dimensional quantum Hall effect in string theory *J. High Energy Phys.* JHEP02052002037
[15] Karabali D, Nair V P, Randjbar-Daemi S 2005 *Fields to Strings: Circumnavigating Theoretical Physics*, ed Shifman M, Vainshtein A and Wheater J (Singapore: World Scientific)
[16] Balachandran A P, Kurkcuoglu S and Vaidya S 2007 *Lectures on fuzzy and fuzzy SUSY physics* (Singapore: World Scientific)
[17] Dolan B P and Jahn O 2003 Fuzzy complex Grassmannian spaces and their star products *Int. J. Mod. Phys.* A **18** 1935
[18] Murray S and Sämann C 2008 Quantization of Flag Manifolds and their Supersymmetric Extensions *Adv. Theor. Math. Phys.* **12** 641
[19] Huet I and Murray S 2012 Polarization Tensors for the fuzzy Grassmannians $Gr^F_{2;n}$ and Grassmannian harmonics on $Gr_{2;n}$ (*Preprint* hep-th/1204.3347)
[20] Hanany A and Witten E 1997 Type IIB superstrings, BPS monopoles and three-dimensional gauge dynamics *Nucl. Phys. B* **492** 152
[21] Ramgoolam S 2001 On spherical harmonics for fuzzy spheres in diverse dimensions *Nucl. Phys.* B **610** 461
[22] Haldane F D M 1983 Fractional Quantization of the Hall Effect: A Hierarchy of Incompressible Quantum Fluid States *Phys. Rev. Lett.* **51** 605
[23] Uskov D B and Rau A R P 2006 Geometric phase for $SU(N)$ through fiber bundles and unitary integration *Phys. Rev.* A **74** 030304
[24] Griffiths P and Harris J 1994 *Principles of Algebraic Geometry* (New York: John Wiley Sons)
[25] Nair V P 2005 *Quantum Field Theory: A Modern Perspective* (New York: Springer)

On equations of motion on Siegel-Jacobi spaces generated by linear Hamiltonians in the generators of the Jacobi group

Stefan Berceanu

National Institute for Physics and Nuclear Engineering, Department of Theoretical Physics, PO BOX MG-6, Bucharest-Magurele, Romania

E-mail: `Berceanu@theory.nipne.ro`

Abstract. It is proved that the equations of classical motion and the quantum evolution on the Siegel-Jacobi disk generated by a Hamiltonian linear in the generators of the Jacobi group G_1^J obtained by the Wei-Norman method and a method used in the context of Berezin's quantization are identical. In a certain set of variables the motion on the Siegel disk and $\mathbb{C}$ are decoupled. The geometric significance and the meaning in the context of coherent states of this coordinates are emphasized.

1. Introduction

The Jacobi group G_n^J - the semidirect product of the Heisenberg group H_n with $\mathrm{Sp}(n,\mathbb{R})$ - is intensively studied in mathematics [1, 2, 3]. The Jacobi group G_n^J is relevant in physics, in particular in quantum mechanics [4], being responsible [5] for the squeezed states in quantum optics [6, 7, 8]. To the Jacobi group [9] it is associated the so called Siegel-Jacobi ball [1, 2, 3, 10, 11], the partially bounded homogeneous Kähler manifold $\mathcal{D}_n^J \approx \mathbb{C}^n \times \mathcal{D}_n$, where the the Siegel ball $\mathcal{D}_n = \mathrm{Sp}(n,\mathbb{R})/\mathrm{U}(n)$ admits a realization as a bounded domain [12]. To the Jacobi group we have attached coherent states [13] based on the Siegal-Jacobi spaces [14, 9, 3, 15]. Different aspects of quantum mechanics via coherent states and geometry on the Siegel-Jacobi disk $\mathcal{D}_1^J$ are emphasized in [16]. The equations of motion on the Siegel-Jacobi ball and Siegel-Jacobi upper half plane generated by linear Hamiltonians in the generators of the Jacobi group are studied in [14, 15, 17] using a method developed in [18, 19, 17] in the context of Berezin's approach to quantization [20, 21] for coherent states [13]. On the other side, equations of motion on the Siegel-Jacobi disk are studied in [22] by the Wei-Norman method [23] applied to the quasienergy operator [24] in the context of Lie systems of differential equations [25, 26]. The Wei-Norman equations for G_1^J were studied in [27, 28].

In this note I underline the equivalence between the equations of motion on the Siegel-Jacobi disk $\mathcal{D}_1^J$ (also called extended Poincaré disk, $\mathfrak{M}$) generated by a hermitian Hamiltonian linear in the generators of the Jacobi group G_1^J obtained with apparently different methods - the Wei-Norman method [22] and Berezin's method [14, 17, 15, 29]. The equivalence of the equations of motion on the Siegel-Jacobi disk is proved in a set of variables - called FC-variables - in which the motion on $\mathbb{C}$ and $\mathcal{D}_1$ are decoupled. The FC-transform has a special geometric interpretation and also has significance in the context of coherent states. If $(z,w) \in \mathbb{C} \times \mathcal{D}_1$ are local coordinates on the Siegel-Jacobi disk [14], it was proved in [17, 15] that under the FC-transform, $FC : (\eta, w) \rightarrow (z = \eta - w\bar{\eta}, w)$ the Kähler to form $\omega_{\mathcal{D}_1^J}(z, w)$ becomes just the sum of $\omega_{\mathbb{C}}(\eta)$ and $\omega_{\mathcal{D}_1}(w)$, recalling the fundamental conjecture for homogeneous Kähler manifolds [30, 31]. The FC-transform expresses also the change of variables from the normalized to the unnormalized coherent state vectors [16]. The equations of motion on $\mathfrak{M}$ associated to a hermitian Hamiltonian linear in the generators of the Jacobi group G_1^J obtained in [22] are a particular case of equations of motion on the Siegel-Jacobi spaces attached to a hermitian Hamiltonian linear in the generators of G_n^J [17].

The paper is laid out as follows. §2 recalls the method to obtain what we call Berezin's equations of motion on a homogeneous Kähler manifold $M = G/H$ generated by a hermitian Hamiltonian linear in the generators of the Lie group G if the Lie algebra $\mathfrak{g}$ admits a certain holomorphic differential representation [18, 19, 32, 17]. §3 briefly summarizes the Wei-Norman method [23] to solve linear differential equations attached to linear operators in the generators of a Lie algebra $\mathfrak{g}$. §4 recalls the commutation relations of the generators of the Jacobi algebra $\mathfrak{g}_1^J := \mathfrak{h}_1 \rtimes \mathfrak{su}(1,1)$ and the operators which appear in the unitary representation of the Jacobi group G_1^J. §5 presents the quasienergy operator attached to the time-dependent Schrödinger equation on the Siegel-Jacobi disk $\mathcal{D}_1^J$. A hermitian Hamiltonian linear in the generators of the Jacobi group G_1^J is considered. Using the results from [33, 34] and passing from the complex representation adopted in [14, 17] to the real representation of the Jacobi group G_1^J used in [22], with the technique of §3, the equations of motion on $\mathfrak{M}$ from [22] are reobtained and shown to be identical with those obtained by Berezin's method [17]. For completeness, the equations of motion on the Siegel-Jacobi ball [15] are briefly recalled in §5. The main contribution of the paper is contained in Proposition 2. More details are presented in [29].

Conventions. The Hilbert space $\mathfrak{H}$ considered in this paper is endowed with a scalar product $< \cdot, \cdot >$ antilinear in the first argument. $\mathbb{R}$, $\mathbb{C}$ and $\mathbb{N}$ denotes the field of real, complex numbers, respectively the set of non-negative integers. We denote the imaginary unit $\sqrt{-1}$ by i, and the Real and Imaginary part of a complex number by $\Re$ and respectively $\Im$, i.e. for $z \in \mathbb{C}$ we have $z = \Re z + \mathrm{i}\Im z$, and $\bar{z} = \Re z - \mathrm{i}\Im z$. Also we use the notation $cc(z) := \bar{z}$ for $z \in \mathbb{C}$ and $cc(A) = A^\dagger$ for an operator A. We denote by $M_n(\mathbb{F})$ the set of $n \times n$ matrices with entries in the field $\mathbb{F}$. If $A \in M_n(\mathbb{F})$, then A^t ($A^\dagger$) denotes the transpose (respectively, the hermitian conjugate) of A. I denotes the unit operator, while $\mathbb{I}_n$ denotes the unit matrix of $M_n(\mathbb{F})$. If $A \in M_n(\mathbb{F})$, we denote by $A^s := \frac{1}{2}(A + A^t)$. We use Einstein convention that repeated indices are implicitly summed over. We denote the differential by d. If π is an unitary irreducible representation of a Lie group G with Lie algebra $\mathfrak{g}$ on a complex separable Hilbert space $\mathfrak{H}$, then we denote for the derived representation $\boldsymbol{X} := \mathrm{d}\,\pi(X)$, $X \in \mathfrak{g}$. If X, Y are elements in the lie algebra $\mathfrak{g}$, then $\mathrm{ad}X(Y) = [X, Y]$.

2. Berezin's equations of motion via coherent states

In the group theoretic approach to coherent states [13] it is considered the triplet $(G, \pi, \mathfrak{H})$, where π is a continuous, unitary, irreducible representation of the Lie group G on the separable complex Hilbert space $\mathfrak{H}$.

The normalized (unnormalized) vectors $\underline{e}_x$ (respectively, e_z), based on the homogeneous manifold M=G/H, supposed to be Kählerian, are defined as

$$\underline{e}_x = \exp(\sum_{\phi \in \Delta_+} x_\phi \boldsymbol{X}_\phi^+ - \bar{x}_\phi \boldsymbol{X}_\phi^-) e_0, \; e_z = \exp(\sum_{\phi \in \Delta_+} z_\phi \boldsymbol{X}_\phi^+) e_0. \tag{1}$$

Above e_0 is the extremal weight vector of the representation π, Δ_+ are the positive roots of the Lie algebra $\mathfrak{g}$ of G, and $X_\phi, \phi \in \Delta$, are the generators. $\boldsymbol{X}_\phi^+$ ($\boldsymbol{X}_\phi^-$) corresponds to the positive (respectively, negative) generators [13, 32].

We denote by FC [16] (from *fundamental conjecture* [30, 31]) the change of variables $x \rightarrow z$ in formula (1) such that

$$\underline{e}_x = \tilde{e}_z,\ \tilde{e}_z := < e_z, e_z >^{-\frac{1}{2}} e_z,\ z = FC(x). \tag{2}$$

By a *dequantization procedure*, the motion on the classical phase space - a homogeneous Kähler manifold $M = G/H$ - can be described by the local equations of motion [20, 21] $\dot{z}_\alpha = \mathrm{i}\{\mathcal{H}, z_\alpha\},\ \alpha \in \Delta_+$, where $\mathcal{H}$ is the classical Hamiltonian (*the covariant symbol*)

$$\mathcal{H} = < e_z, e_z >^{-1} < e_z, \boldsymbol{H} e_z > \tag{3}$$

attached to the quantum Hamiltonian $\boldsymbol{H}$, and the Poisson bracket $\{,\}$ is introduced using the inverse of the balanced metric [35] matrix $g_{\alpha\bar{\beta}} = \frac{\partial^2}{\partial z_\alpha \partial \bar{z}_\beta} \ln < e_z, e_z >$. The time-dependent Schrödinger equation is expressed as

$$\boldsymbol{H}(t)\psi(t) = \mathrm{i}\hbar \frac{\mathrm{d}\,\psi(t)}{\mathrm{d}t}. \tag{4}$$

A linear Hamiltonian in the generators $\boldsymbol{X}_\lambda$ of the group of symmetry G is considered

$$\boldsymbol{H} = \sum_{\lambda \in \Delta} \epsilon_\lambda \boldsymbol{X}_\lambda. \tag{5}$$

We look for the solution of the Schrödinger equation of motion (4) generated by the Hamiltonian (5) as

$$\psi(t) = \mathrm{e}^{\mathrm{i}\phi} \tilde{e}_z. \tag{6}$$

Let us suppose that the differential action corresponding to the operator $\boldsymbol{X}_\lambda$ in (5) can be expressed in a local system of coordinates as a holomorphic first order differential operator with polynomial coefficients,

$$\mathbb{X}_\lambda = P_\lambda + \sum_{\beta \in \Delta_+} Q_{\lambda,\beta} \frac{\partial}{\partial z_\beta},\ \lambda \in \Delta. \tag{7}$$

It can be proven [18, 19, 17] that:

Proposition 1. *If the generators of the group of symmetry admit the representation* (7), *then the classical motion and quantum evolution on* $M = G/H$ *generated by the hermitian linear Hamiltonian* (5) *are solutions of the first order differential equations*

$$\mathrm{i}\,\dot{z}_\alpha = \sum_{\lambda \in \Delta} \epsilon_\lambda Q_{\lambda,\alpha},\ \alpha \in \Delta_+. \tag{8}$$

3. Wei-Norman method

Let us consider a Lie algebra $\mathfrak{g}$ with the generators $\{X_i\}_{i=1,\ldots,n}$ and the linear operator

$$A(t) = \sum_{i=1}^{n} \epsilon_i(t) X_i, \tag{9}$$

where $\epsilon_i(t)$ are scalar functions of t. Let us attach to the linear operators U and A the differential equation

$$\frac{\mathrm{d}U(t)}{\mathrm{d}t} = A(t)U(t),\quad U(0) = I. \tag{10}$$

In the Wei-Norman method [23] the solution of the equation (10) is searched in the form of product of exponentials

$$U(t) = \prod_{i=1}^{n} \exp(\xi_i(t)X_i). \tag{11}$$

Then the functions ξ satisfy a first order differential equation which depends only on the Lie algebra $\mathfrak{g}$ and the $\epsilon(t)$-s:

$$\eta\dot{\xi} = \epsilon, \tag{12}$$

where η is obtained by the formulae

$$\left(\prod_{j=1}^{r} \exp(\xi_j \boldsymbol{X}_j)\right) \boldsymbol{X}_i \left(\prod_{j=r}^{1} \exp(-\xi_j \boldsymbol{X}_j)\right) = \sum_{k=1}^{n} \eta_{ki}\boldsymbol{X}_k,\ i, r = 1, 2, \ldots, n. \tag{13}$$

In our calculation, what we need is the following formula

$$U^{-1}(t)\dot{U}(t) = \sum_{i=1}^{n} \dot{\xi}_i \boldsymbol{Y}_i, \quad \boldsymbol{Y}_i = \left(\prod_{k=n}^{i+1} [\exp(-\xi_k \mathrm{ad}\boldsymbol{X}_k)]\right)\boldsymbol{X}_i, \tag{14}$$

obtained [29] with technique presented in [23] and the Baker-Hausdorff formula [36]:

$$\begin{aligned} e^X Y e^{-X} = e^{\mathrm{ad}X}Y &= \sum_{n=0}^{\infty} \frac{(\mathrm{ad}X)^n}{n!} Y \\ &= Y + [X, Y] + \frac{1}{2!}[X, [X, Y]] + \cdots + \frac{1}{n!}[\underbrace{[X, [X, \ldots, [}_{n \text{ brackets}} X, Y \underbrace{] \ldots]}_{n}] + \ldots. \end{aligned}$$

4. The Jacobi Lie algebra $\mathfrak{g}_1^J$ and the Lie Jacobi group G_1^J

The (6-dimensional) Jacobi algebra $\mathfrak{g}_1^J$ is the semi-direct sum $\mathfrak{g}_1^J := \mathfrak{h}_1 \rtimes \mathfrak{su}(1,1)$, where $\mathfrak{h}_1$ is the 3-dimensional Heisenberg algebra generated by the boson creation (respectively, annihilation) operators $a^\dagger$ (a), $\mathfrak{su}(1,1)$ has the generators $K_{0,+,-}$, verifying the non-trivial commutation relations [14]

$$[a, a^\dagger] = I, \tag{15a}$$

$$[K_0, K_\pm] = \pm K_\pm\ ,\quad [K_-, K_+] = 2K_0, \tag{15b}$$

$$\left[K_0,\ a^\dagger\right] = \frac{1}{2}a^\dagger, [K_0, a] = -\frac{1}{2}a. \tag{15c}$$

To the Heisenberg group H_1 it is attached the unitary displacement operator

$$D(\alpha) := \exp(\alpha\boldsymbol{a}^\dagger - \bar{\alpha}\boldsymbol{a}) \tag{16a}$$

$$= \exp(-\frac{1}{2}|\alpha|^2)\exp(\alpha\boldsymbol{a}^\dagger)\exp(-\bar{\alpha}\boldsymbol{a}) \tag{16b}$$

$$= \exp(\frac{1}{2}|\alpha|^2)\exp(-\bar{\alpha}\boldsymbol{a})\exp(\alpha\boldsymbol{a}^\dagger). \tag{16c}$$

We denote by S the unitary irreducible positive discrete series representation D_+^k of the group SU(1,1) with Bargmann index k [37] and we introduce the notation $\underline{S}(z) = S(w)$, where

$$\underline{S}(z) := \exp(z\boldsymbol{K}_+ - \bar{z}\boldsymbol{K}_-),\ z \in \mathbb{C}; \tag{17a}$$

$$S(w) = \exp(w\boldsymbol{K}_+)\exp(\rho\boldsymbol{K}_0)\exp(-\bar{w}\boldsymbol{K}_-) \tag{17b}$$

$$= \exp(-\bar{w}\boldsymbol{K}_-)\exp(-\rho\boldsymbol{K}_0)\exp(w\boldsymbol{K}_+),\ |w| < 1; \tag{17c}$$

$$w = \frac{z}{|z|}\tanh(|z|),\quad \rho = \ln(1 - w\bar{w}), z \neq 0, \tag{17d}$$

and $w = 0$ for $z = 0$ in (17d). We attach to the Jacobi group G_1^J the unitary operator [38] defined on the Siegel-Jacobi disk $\mathcal{D}_1^J$ [14]:

$$T(\xi) = D(\alpha)S(w), \quad \mathcal{D}_1^J \ni \xi = (\alpha, w) \in \mathbb{C} \times \mathcal{D}_1. \tag{18}$$

In [14] we have introduced unnormalized Perelomov vectors e_ξ defined in the points $\xi = (\alpha, w)$ of the Jacobi disk $\mathcal{D}_1^J$.

The real base [2, 4] used in [22] and the base (15) of the Lie algebra $\mathfrak{g}_1^J$ are related by the relations:

$$\boldsymbol{N}_1 = \boldsymbol{a} + \boldsymbol{a}^\dagger; \quad \boldsymbol{N}_2 = \mathrm{i}(\boldsymbol{a} - \boldsymbol{a}^\dagger); \tag{19a}$$

$$\boldsymbol{K}_1 = \frac{1}{2}(\boldsymbol{K}_+ + \boldsymbol{K}_-); \quad \boldsymbol{K}_2 = \frac{1}{2\mathrm{i}}(\boldsymbol{K}_+ - \boldsymbol{K}_-). \tag{19b}$$

In [22] it was considered instead of the representation (18) the equivalent representation

$$T(\xi) = D(x, y)S(u, v), \quad \xi = (u, v, x, y) \in \mathfrak{M}, \quad \text{where} \tag{20a}$$

$$D(x, y) = \exp(\mathrm{i}\, y\boldsymbol{N}_1 + \mathrm{i}\, x\boldsymbol{N}_2), \quad S(u, v) = \exp(\mathrm{i}\, k_1\boldsymbol{K}_1 + \mathrm{i}\, k_2\boldsymbol{K}_2), \tag{20b}$$

$$k_1 = \frac{v}{2s}\ln\frac{1+s}{1-s}, \quad k_2 = \frac{u}{2s}\ln\frac{1+s}{1-s}, \quad s = (u^2 + v^2)^{\frac{1}{2}}, \ |s| < 1. \tag{20c}$$

The parameters in the representations (20b) and (18) are related by the relations

$$\alpha = x + \mathrm{i}\, y; \quad w = u + \mathrm{i}\, v. \tag{21}$$

5. Equations of motion on the Siegel-Jacobi disk $\mathcal{D}_1^J$

To any linear operator $\boldsymbol{A}$ defined on $\mathfrak{H}$, we attach [4, 33, 34] the operator $\hat{\boldsymbol{A}}(\xi)$ based on the Siegel-Jacobi disk $\mathcal{D}_1^J$

$$\hat{\boldsymbol{A}}(\xi) := T^{-1}(\xi)\boldsymbol{A}T(\xi), \quad \mathcal{D}_1^J \ni \xi = (\alpha, w) \in \mathbb{C} \times \mathcal{D}_1. \tag{22}$$

As in [22], we consider the following family of unitary operators

$$U(\xi, \varphi) := \exp(-\mathrm{i}\,\varphi)T(\xi), \tag{23}$$

where $\xi \in \mathcal{D}_1^J$ and φ is a real phase. Let $\tau := \frac{t}{\hbar}$. In [22] it was introduced the quasienergy operator $\boldsymbol{E} := \mathrm{i}\,\frac{\mathrm{d}}{\mathrm{d}\tau} - \boldsymbol{H}$ [24]. With (22), we get for $\hat{\boldsymbol{E}}(\xi, \varphi) := U(-\xi, -\varphi)\boldsymbol{E}U(\xi, \varphi)$ the expression

$$\hat{\boldsymbol{E}}(\xi, \varphi) = \frac{\mathrm{d}\varphi}{\mathrm{d}\tau}\boldsymbol{I} + \mathrm{i}\, T(\xi)^{-1}\dot{T}(\xi) - \hat{\boldsymbol{H}}(\xi). \tag{24}$$

We have the relations [33, 34]:

$$\hat{\boldsymbol{a}}(\alpha, w) = r(\boldsymbol{a} + w\boldsymbol{a}^\dagger) + \alpha, \quad r = (1 - w\bar{w})^{-\frac{1}{2}}, \tag{25}$$

$$\hat{\boldsymbol{K}}_0(\alpha, w) = r^2[\bar{w}\boldsymbol{K}_- + \left(1 + |w|^2\right)\boldsymbol{K}_0 + w\boldsymbol{K}_+] + r\Re[\alpha(\boldsymbol{a}^\dagger + \bar{w}\boldsymbol{a})] + \frac{1}{2}|\alpha|^2, \tag{26}$$

$$\hat{\boldsymbol{K}}_-(\alpha, w) = r^2[\boldsymbol{K}_- + 2w\boldsymbol{K}_0 + w^2\boldsymbol{K}_+] + \alpha r(\boldsymbol{a} + w\boldsymbol{a}^\dagger) + \frac{1}{2}\alpha^2. \tag{27}$$

We express (18) as product of exponentials of the generators as in (11), where

$$\boldsymbol{X}_1 = \boldsymbol{I}; \ \boldsymbol{X}_2 = \boldsymbol{a}^\dagger; \ \boldsymbol{X}_3 = \boldsymbol{a}; \ \boldsymbol{X}_4 = \boldsymbol{K}_+; \ \boldsymbol{X}_5 = \boldsymbol{K}_0; \ \boldsymbol{X}_6 = \boldsymbol{K}_-, \tag{28}$$

$$\xi_1 = -\frac{1}{2}|z|^2;\ \xi_2 = z;\ \xi_3 = -\bar{z};\ \xi_4 = w;\ \xi_5 = \ln(1 - w\bar{w});\ \xi_6 = -\bar{w}. \tag{29}$$

We apply (14) to the operator $T(\xi)$ (18) in the variables $(z, w) \in \mathcal{D}_1^J$. With the notation:

$$\begin{aligned}
\boldsymbol{Y}_2 &= \mathrm{e}^{-\xi_6 \mathrm{ad}\boldsymbol{X}_6} e^{-\xi_5 \mathrm{ad}\boldsymbol{X}_5} e^{-\xi_4 \mathrm{ad}\boldsymbol{X}_4} e^{-\xi_3 \mathrm{ad}\boldsymbol{X}_3} \boldsymbol{X}_2,\\
\boldsymbol{Y}_3 &= \mathrm{e}^{-\xi_6 \mathrm{ad}\boldsymbol{X}_6} e^{-\xi_5 \mathrm{ad}\boldsymbol{X}_5} e^{-\xi_4 \mathrm{ad}\boldsymbol{X}_4} \boldsymbol{X}_3,\\
\boldsymbol{Y}_4 &= \mathrm{e}^{-\xi_6 \mathrm{ad}\boldsymbol{X}_6} e^{-\xi_5 \mathrm{ad}\boldsymbol{X}_5} \boldsymbol{X}_4,\\
\boldsymbol{Y}_5 &= \mathrm{e}^{-\xi_6 \mathrm{ad}\boldsymbol{X}_6} \boldsymbol{X}_5,
\end{aligned}$$

we obtained the expressions:

$$\begin{aligned}
&\boldsymbol{Y}_1 = \boldsymbol{I};\ \boldsymbol{Y}_2 = -\xi_3 + \mathrm{e}^{-\frac{\xi_5}{2}}(\boldsymbol{a}^\dagger - \xi_6 \boldsymbol{a});\ \boldsymbol{Y}_3 = (\mathrm{e}^{\frac{\xi_5}{2}} - \xi_4\xi_6 \mathrm{e}^{-\frac{\xi_5}{2}})\boldsymbol{a} + \xi_4 \mathrm{e}^{-\frac{\xi_5}{2}}\boldsymbol{a}^\dagger;\\
&\boldsymbol{Y}_4 = \mathrm{e}^{-\xi_5}(\boldsymbol{K}_+ + 2\xi_4\boldsymbol{K}_0 + \xi_4^2\boldsymbol{K}_-);\ \boldsymbol{Y}_5 = \boldsymbol{K}_0 + \xi_4\boldsymbol{K}_-;\ \boldsymbol{Y}_6 = \boldsymbol{K}_-.
\end{aligned} \tag{30}$$

With (30) and (14) we obtain for $T^{-1}\dot{T}(z, w)$ the value:

$$T^{-1}\dot{T}(z, w) = \mathrm{i}\,\Im(z\dot{\bar{z}}) + r[(\dot{z} - \dot{\bar{z}}w)\boldsymbol{a}^\dagger - cc] + [\dot{w}r^2\boldsymbol{K}_+ - cc] + 2\,\mathrm{i}\,\Im(\dot{w}\bar{w})r^2\boldsymbol{K}_0. \tag{31}$$

In the notation of [14, 15], we consider a hermitian Hamiltonian linear in the generators of the Jacobi group G_1^J:

$$\boldsymbol{H}_0 = \epsilon_a \boldsymbol{a} + \bar{\epsilon}_a \boldsymbol{a}^\dagger + \epsilon_0 \boldsymbol{K}_0 + \epsilon_+ \boldsymbol{K}_+ + \epsilon_- \boldsymbol{K}_-,\quad \bar{\epsilon}_+ = \epsilon_-,\ \epsilon_0 = \bar{\epsilon}_0. \tag{32}$$

With equations (25)-(27), we calculate $\hat{\boldsymbol{H}}_0(\xi)$, $\xi = (\alpha, w) \in \mathbb{C} \times \mathcal{D}_1$:

$$\begin{aligned}
\hat{\boldsymbol{H}}_0(\alpha, w) &= I_0 + C_1\boldsymbol{a}^\dagger + \bar{C}_1\boldsymbol{a} + C_0\boldsymbol{K}_0 + C_+\boldsymbol{K}_+ + \bar{C}_+\boldsymbol{K}_-, \quad \text{where:}\\
I_0 &= \epsilon_a\alpha + \bar{\epsilon_a}\bar{\alpha} + \frac{1}{2}(\epsilon_0|\alpha|^2 + \epsilon_-\alpha^2 + \epsilon_+\bar{\alpha}^2),\\
\frac{C_1}{r} &= \bar{\epsilon}_a + \epsilon_a w + \frac{\epsilon_0}{2}(\alpha + \bar{\alpha}w) + \epsilon_-\alpha w + \epsilon_+\bar{\alpha},\\
\frac{C_0}{r^2} &= \epsilon_0(1 + |w|^2) + 2(\epsilon_- w + \epsilon_+\bar{w}),\\
\frac{C_+}{r^2} &= \epsilon_0 w + \epsilon_- w^2 + \epsilon_+.
\end{aligned} \tag{33}$$

In [22] the Hamiltonian (32) was written down in real coordinates as:

$$\boldsymbol{H}_0 = 2\varepsilon_0\boldsymbol{K}_0 + 2\varepsilon_1\boldsymbol{K}_1 + 2\varepsilon_2\boldsymbol{K}_2 + 2\nu_1\boldsymbol{N}_1 + 2\nu_2\boldsymbol{N}_2. \tag{34}$$

The correspondence of the coefficients of the Hamiltonians (34) and (32) is

$$\epsilon_a = \nu_1 + \mathrm{i}\,\nu_2;\ \epsilon_0 = 2\varepsilon_0;\ \epsilon_+ = \varepsilon_1 - \mathrm{i}\,\varepsilon_2. \tag{35}$$

Now we introduce (31) and (33) into (24), we make the change of variables (21) and operators

(19) and we get:

$$\begin{aligned}
\hat{\boldsymbol{E}}(u,v,x,y) &= G_0\boldsymbol{I} + G_1\boldsymbol{N}_1 + G_2\boldsymbol{N}_2 + H_0\boldsymbol{K}_0 + H_1\boldsymbol{K}_1 + H_2\boldsymbol{K}_2, \quad \text{where:}\\
G_0 &= \dot{\varphi} + y\dot{x} - x\dot{y} - 2(\nu_1 x - \nu_2 y) - \varepsilon_0(x^2+y^2) - \varepsilon_1(x^2-y^2) + 2\varepsilon_2 xy,\\
-\frac{G_1}{r} &= (1+u)\dot{y} - \dot{x}v + \nu_1(1+u) - \nu_2 v + \varepsilon_0[x(1+u)+yv]\\
&\quad + \varepsilon_1[x(1+u)-yv] - \epsilon_2[y(1+u)+xv],\\
\frac{G_2}{r} &= -(1-u)\dot{x} + \dot{y}v + \nu_1 v + \nu_2(u-1) + \varepsilon_0[y(1-u)+xv]\\
&\quad + \varepsilon_1[xv + y(u-1)] + \varepsilon_2[x(u-1)-yv],\\
-\frac{H_0}{2r^2} &= \dot{v}u - \dot{u}v + \varepsilon_0(1+u^2+v^2) + 2(\varepsilon_1 u - \varepsilon_2 v),\\
-\frac{H_1}{2r^2} &= \dot{v} + \varepsilon_0 u + \varepsilon_1(u^2-v^2+1) - 2\varepsilon_2 uv,\\
\frac{H_2}{2r^2} &= -\dot{u} + \varepsilon_0 v + 2\varepsilon_1 uv + \varepsilon_2(u^2-v^2-1).
\end{aligned}$$

Identifying the coefficients of $\boldsymbol{N}_1$, $\boldsymbol{N}_2$, and respectively $\boldsymbol{K}_1$, $\boldsymbol{K}_2$, we get the equations of motion in the real coordinates (u,v,x,y) and (38)

$$\begin{aligned}
\dot{x} &= -\varepsilon_2 x + (\varepsilon_0 - \varepsilon_1)y - \nu_2,\\
\dot{y} &= -(\varepsilon_0 + \varepsilon_1)x + \varepsilon_2 y - \nu_1;
\end{aligned} \tag{36}$$

$$\begin{aligned}
\dot{u} &= 2v(\varepsilon_1 u + \varepsilon_0) - \varepsilon_2(1-u^2+v^2),\\
\dot{v} &= 2u(\varepsilon_2 v - \varepsilon_0) - \varepsilon_1(1+u^2-v^2);
\end{aligned} \tag{37}$$

$$-\frac{H_0}{2} = \varepsilon_0 + \varepsilon_1 u - \varepsilon_2 v, \qquad G_0 = \dot{\varphi} - (\nu_1 x - \nu_2 y). \tag{38}$$

We have proved:

Proposition 2. *Let us consider $\Phi \in \mathfrak{H}$ such that $\boldsymbol{K}_0\Phi = k\Phi$, $k \in \mathbb{R}$. Then*

$$\Psi(\xi,\varphi) = U(\xi,\varphi)\Phi = \mathrm{e}^{-\mathrm{i}\varphi}T(\xi)\Phi, \quad \xi = (u,v,x,y) \in \mathfrak{M}, \tag{39}$$

is a solution of (4) *for the hermitian Hamiltonian* (34) *on the Siegel-Jacobi disk $\mathcal{D}^J_1$. $x,y \in \mathbb{R}$ verify* (36), *$(u,v) \in \mathcal{D}_1$ verify* (37), *while the phase φ in* (23) *verifies the differential equation:*

$$\dot{\varphi} = \nu_1 x - \nu_2 y + 2k(\varepsilon_0 + \varepsilon_1 u - \varepsilon_2 v). \tag{40}$$

In the complex variable $w = u + \mathrm{i}\,v$ the motion (37) *on the Siegel disc $\mathcal{D}_1$ is described by the Riccati equation*

$$\mathrm{i}\,\dot{w} = \epsilon_+ + \epsilon_0 w + \epsilon_- w^2. \tag{41}$$

The equations of motion (36) *in the complex variable $\eta = x + \mathrm{i}\,y$ become*

$$\mathrm{i}\,\dot{\eta} = \bar{\epsilon}_a + \epsilon_+\bar{\eta} + \frac{\epsilon_0}{2}\eta, \tag{42}$$

In conclusion, we have proved that the Riccati equation (41) on $\mathcal{D}_1$ obtained with the Wei-Norman method coincides with Berezin's equation of motion (4.8b) or (4.10b) in [15], with the difference of notation $\epsilon_+ \leftrightarrow \epsilon_- = \bar{\epsilon}_+$. The equation (42) for $\eta \in \mathbb{C}$ obtained with the Wei-Norman method is just Berezin's equation of motion (4.10a) in [15], with the correspondence

$\epsilon_a \leftrightarrow \bar{\epsilon}_a$, $\epsilon_+ \leftrightarrow \epsilon_- = \bar{\epsilon}_+$. *The quantum and classical Berezin's equations of motion on the Siegel-Jacobi disk determined by a hermitian Hamiltonian, linear in the generators of Jacobi group G^J_1, expressed in the FC-coordinates, are the same as the equations obtained applying the Wei-Norman method. The equations of motion* (41) *on* $\mathcal{D}_1$ *and* (42) *on* $\mathbb{C}$ *and* (36) *on* $\mathbb{R}^2$, *determined by the hermitian Hamiltonian* (32) *or* (34) *linear in the generators of the Jacobi group G^J_1 obtained with the Wei-Norman method are a particular case of the Berezin's equations of motion* (45a) *on* $\mathcal{D}_n$, (46) *on* $\mathbb{C}^n$ *and respectively* (48) *on* $\mathbb{R}^{2n}$, *determined by the hermitian Hamiltonian* (43), *linear in the generators of the Jacobi group G^J_n*. See [29] for a comparison of phases in the two approaches. In the present paper we have proved Proposition 2 by brute-force calculation in the case of the Jacobi group $G^J_1 = H_1 \rtimes \mathrm{SU}(1,1)$, but we believe that the identity of the results in the two methods is true in more general situations, for some semidirect products of Lie groups.

Appendix: Berezin's Equations of motion on the Siegel-Jacobi ball

In [9, 17] we have defined unnormalized coherent state vectors $e_{z,W}$, where $z \in \mathbb{C}^n$, and $W \in M(n,\mathbb{C}) : W = W^t, \mathbb{I}_n - W\bar{W} > 0$ describes a point in the noncompact hermitian symmetric space $\mathcal{D}_n = \mathrm{Sp}(n,\mathbb{R})/\mathrm{U}(n)$, realized as a homogenous bounded domain.

We consider a hermitian Hamiltonian linear in the generators of the group G^J_n [17]

$$\boldsymbol{H} = \epsilon_i \boldsymbol{a}_i + \bar{\epsilon}_i \boldsymbol{a}^\dagger_i + \epsilon^0_{ij}\boldsymbol{K}^0_{ij} + \epsilon^-_{ij}\boldsymbol{K}^-_{ij} + \epsilon^+_{ij}\boldsymbol{K}^+_{ij}, \tag{43}$$

$$\epsilon^\dagger_0 = \epsilon_0;\ \epsilon_- = \epsilon^t_-;\ \epsilon_+ = \epsilon^t_+;\ \epsilon^\dagger_+ = \epsilon_-. \tag{44}$$

With the technique recalled in §2, we have proved in [17] that:

Proposition 3. *a) The differential equations for* $(W,z) \in \mathcal{D}^J_n$ *are*

$$\mathrm{i}\dot{W} = \epsilon_- + (W\epsilon_0)^s + W\epsilon_+ W,\ W \in \mathcal{D}_n, \tag{45a}$$

$$\mathrm{i}\dot{z} = \epsilon + W\bar{\epsilon} + \frac{1}{2}\epsilon^t_0 z + W\epsilon_+ z.\ z \in \mathbb{C}^n, \tag{45b}$$

b) Under the FC transform, $z = \eta - W\bar{\eta}$, *the differential equations in the variables* $\eta \in \mathbb{C}^n$ *and* $W \in \mathcal{D}_n$ *become independent:* W *verifies* (45a) , *while* η *verifies*

$$\mathrm{i}\dot{\eta} = \epsilon + \epsilon_-\bar{\eta} + \frac{1}{2}\epsilon^t_0\eta,\ \eta \in \mathbb{C}^n. \tag{46}$$

c) The linear system of differential equations associated with matrix Riccati equation (45a) *reads*

$$\begin{pmatrix} \dot{X} \\ \dot{Y} \end{pmatrix} = h_c \begin{pmatrix} X \\ Y \end{pmatrix},\ h_c = \begin{pmatrix} -\mathrm{i}(\frac{\epsilon_0}{2})^t & -\mathrm{i}\epsilon_- \\ \mathrm{i}\epsilon_+ & \mathrm{i}\frac{\epsilon_0}{2} \end{pmatrix} \in \mathfrak{sp}(n,\mathbb{R})_{\mathbb{C}},\ W = X/Y \in \mathcal{D}_n. \tag{47}$$

d) In (46) *we introduce* $\eta = \xi - \mathrm{i}\zeta$, $\xi,\zeta \in \mathbb{R}^n$ *and we put* $\epsilon = b + \mathrm{i}a$, $\epsilon_- = m + \mathrm{i}n$, $\epsilon^t_0/2 = p + \mathrm{i}q$ *where* $a, b \in \mathbb{R}^n$, $m,n,p,q \in M(n,\mathbb{R})$. *The first order complex differential equation equation* (46) *is equivalent with a system of first order real differential equations with real coefficients, which we write as*

$$\dot{Z} = h_r Z + F,\ Z = \begin{pmatrix} \xi \\ \zeta \end{pmatrix},\ F = \begin{pmatrix} a \\ b \end{pmatrix},\ h_r = \begin{pmatrix} n+q & m-p \\ m+p & -n+q \end{pmatrix} \in \mathfrak{sp}(n,\mathbb{R}). \tag{48}$$

Acknowledgments

The author expresses his gratitude to the organizers of the Colloquium on Group Theoretical Methods in Physics for the opportunity to deliver a talk at XXX meeting in Ghent, Belgium. This research was conducted in the framework of the ANCS project program PN 09 37 01 02/2009 and the UEFISCDI - Romania program PN-II Contract No. 55/05.10.2011. The author is indebted to the unknown referee for the suggested corrections.

References

[1] Eichler M and Zagier D 1985 *The Theory of Jacobi Forms* Progress in Mathematics **55** (Boston: Birkhäuser)
[2] Berndt R and Schmidt R 1998 *Elements of the Representation Theory of the Jacobi Group* Progress in Mathematics **163** (Basel: Birkhäuser)
[3] Berceanu S and Gheorghe A 2011 On the geometry of Siegel-Jacobi domains *Int. J. Geom. Methods Mod. Phys.* **8** 1783–98 (*Preprint* mathDG/1011.3317)
[4] Berceanu S and Gheorghe A 2008 Applications of the Jacobi group to Quantum Mechanics *Romanian J. Phys.* **53** 1013–21 (*Preprint* math.DG/0812.0448)
[5] Berceanu S 2008 *AIP Conf. Proc., Geometric Methods in Physics* 1191 ed P Kielanowski, S T Ali et al pp 21-29 (*Preprint* mathDG/0910.5563)
[6] Mandel L and Wolf E 1995 *Optical Coherence and Quantum Optics* (Cambridge: Cambridge University Press)
[7] Ali S T, Antoine J P and Gazeau J P 2000 *Coherent States, Wavelets, and their Generalizations* (New York: Springer-Verlag)
[8] Drummond P D and Ficek D 2004 *Quantum Squeezing* (Berlin: Springer)
[9] Berceanu S 2008 *Perspectives in Operator Algebra and Mathematical Physics* ed F P.Boca, R Purice et al (Bucharest: The Theta Foundation) pp 1-25 (*Preprint* mathDG/060404381)
[10] Yang J H 2002 The method of orbits for real Lie groups *Kyungpook Math. J.* **42** 199–272
[11] Yang J H 2010 Invariant metrics and Laplacians on the Siegel-Jacobi disk *Chin. Ann. Math.* **31B** 85–100
[12] S. Helgason 1978 *Differential Geometry, Lie Groups and Symmetric Spaces* (New York: Academic Press)
[13] Perelomov A M 1986 *Generalized Coherent States and their Applications* (Springer: Berlin)
[14] Berceanu S 2006 A holomorphic representation of the Jacobi algebra *Rev. Math. Phys.* **18** 163-99; — 2012 Errata *Rev. Math. Phys.* **24** 1292001 p 2 (*Preprint* math.DG/0408219v3)
[15] Berceanu S 2013 Consequences of the fundamental conjecture for the motion on the Siegel-Jacobi disk *Int. J. Geom. Methods Mod. Phys.* **10** 1250076 p 18 (*Preprint* mathDG/1110.5469v2)
[16] Berceanu S 2014 Coherent states and geometry on the Siegel-Jacobi disk *Int. J. Geom. Methods Mod. Phys.* 1450035 p 25 (*Preprint* mathDG/1307.4219v2)
[17] Berceanu S 2012 A convenient coordinatization of Siegel-Jacobi domains *Rev. Math. Phys.* **24** 1250024 p 38 (*Preprint* mathDG/1204.5610)
[18] Berceanu S and Gheorghe A 1992 On equations of motion on hermitian symmetric spaces *J. Math. Phys.* **33** 998–1007
[19] Berceanu S and Boutet de Monvel L 1993 Linear dynamical systems, coherent state manifolds, flows and matrix Riccati equation *J. Math. Phys.* **34** 2353–71
[20] Berezin F A 1975 Quantization in complex symmetric spaces *Izv. Akad. Nauk SSSR Ser. Mat.* **39** 363–402, 472
[21] Berezin F A 1978 Models of Gross-Neveu type are quantization of a Classical Mechanics with a nonlinear phase space *Commun. Math. Phys.* **63** 131–53
[22] Gheorghe A 2013 Quantum and classical Lie systems for extended symplectic groups *Romanian J. Phys.* **58** 1436–45
[23] Wei J and Norman E 1964 On global representations of the solutions of linear differential equations as a product of exponentials *Proc. AMS* **15** 327–334
[24] J. Howland S 1974 Stationary scattering theory for time-dependent Hamiltonians *Math. Ann.* **207** 315–35
[25] S. Lie 1880 Allgemeine Untersuchungen über Differentialgleichungen, die eine continuirliche endliche Gruppe gestatten *Math. Ann.* **6** 441–528
[26] Lie S and Scheffers G 1893 *Vorlesungen Über Continuierliche Gruppen mit Geometrischen und Anderen Anwendungen* (Leipzig: Teubner)
[27] Cariñena J, Grabowski J and Marmo G 2000 *Lie-Scheffers Systems: a Geometric Approach* (Naples: Bibliopolis)
[28] Cariñena J F, Grabowski J and Ramos A 2001 Reduction of time-dependent systems admitting a superposition principle *Acta Appl. Math.* **66** 67–87
[29] Berceanu S 2014 Wei-Norman and Berezin's equations of motion on the Siegel-Jacobi disk *Preprint* mathDG/1403.6594
[30] Vinberg E B and Gindikin S G 1976 Kählerian manifolds admitting a transitive solvable automorphism group *Math. Sb.* **74 (116)** 333–51
[31] Dorfmeiser J and Nakajima K 1988 The fundamental conjecture for homogeneous Kähler manifolds *Acta Mathematica* **161** 23–70
[32] Berceanu S 2005 *Advances in Operator Algebras and Mathematical Physics* ed F. Boca, O. Bratteli et al (Bucharest: The Theta Foundation) pp 1–24 (*Preprint* math.DG/060404381v2)
[33] Berceanu S 2011 Generalized coherent states based on Siegel-Jacobi disk *Romanian J. Phys.* **56** 856–67

[34] Berceanu S and Gheorghe A 2011 Matrix elements of the Jacobi group *Romanian J. Phys.* **56** 1056–68
[35] Berceanu S 2014 Bergman representative coordinates on the Siegel-Jacobi disk *Preprint* mathDG/1409.0368
[36] Magnus W 1954 On the exponential solution of differential equations for a linear operator *Comm. Pure Appl. Math.* **7** 649–73
[37] Bargmann V 1947 Irreducible unitary representations of the Lorentz group *Ann. of Math.* **48** 568–640
[38] Stoler P 1970 Equivalence classes of minimum uncertainty packets *Phys. Rev.* D **1** 3217–3219

Painlevé IV Hamiltonian systems and coherent states

D Bermudez[1,3], A Contreras-Astorga[2,3] and DJ Fernández C[3]

[1]Department of Physics of Complex Systems, Weizmann Institute of Science, Rehovot 76100, Israel

[2]Department of Mathematics and Actuarial Science, Indiana University Northwest, 3400 Broadway, Gary IN 46408, USA

[3]Departamento de Física, Cinvestav, A.P. 14-740, 07000 México D.F., Mexico

E-mail: david.bermudez@weizmann.ac.il, aloncont@iun.edu, david@fis.cinvestav.mx

Abstract. Schrödinger Hamiltonians with third-order differential ladder operators are linked to the Painlevé IV equation. Some of these appear from applying SUSY QM to the harmonic oscillator. Departing from them, we will build coherent states as eigenstates of the annihilation operator, then as displaced versions of the extremal states, both involving the third-order ladder operators, and finally as displaced extremal states using linearized ladder operators. To each Hamiltonian corresponds two families of coherent states for fixed ladder operators: one in the infinite dimension subspace associated with the oscillator spectrum and another in the finite dimension one generated by the eigenstates created by SUSY QM.

1. Introduction

For a long time, coherent states (CS) for physical systems of different nature have been successfully derived through the several available definitions [1–10]. For instance, if the Hilbert space $\mathcal{H}$ is of finite dimension they have been built up through the group theoretical construction of Perelomov [6]. On the other hand, for systems as the harmonic oscillator, such that $\mathcal{H}$ is generated by a discrete infinite set of orthogonal eigenstates of the Hamiltonian, they can be derived using Barut-Girardello's definition, as eigenstates of the annihilation operator [5]. It would be important to address the subject for systems of mixed type, for which $\mathcal{H}$ turns out to be the direct sum of a finite dimension subspace and an infinite dimension one. Let us note that a subset of Painlevé IV Hamiltonian systems, called in this way because they are determined by solutions of the Painlevé IV equation [11], has been recently identified [12–16]. In this work we will explore the coherent states construction for systems in such a subset.

In order to do that, this paper has been organized as follows. In Section 2, the second-order polynomial Heisenberg algebras (PHA) and the general one-dimensional Schrödinger Hamiltonians ruled by them will be studied. In Section 3 the family of kth order supersymmetric (SUSY) partners of the harmonic oscillator Hamiltonian will be constructed, while in Section 4 the subfamily of Painlevé IV Hamiltonian systems, ruled as well by second-order PHA, will be identified. In Section 5 the coherent states construction for the third-order differential ladder operators generating the second-order algebra will be addressed. In Section 6 the same will be done for a pair of linearized annihilation and creation operators built from the third-order ones. Finally, Section 7 contains our conclusions.

2. Second order PHA

Systems ruled by a second-order polynomial Heisenberg algebra possess three generators $\{H, \mathcal{L}^-, \mathcal{L}^+\}$ which satisfy [13–16]:

$$[H, \mathcal{L}^\pm] = \pm\mathcal{L}^\pm, \tag{1}$$

$$[\mathcal{L}^-, \mathcal{L}^+] = N_3(H+1) - N_3(H) = P_2(H), \tag{2}$$

$$N_3(H) \equiv \mathcal{L}^+\mathcal{L}^- = (H-\varepsilon_1)(H-\varepsilon_2)(H-\varepsilon_3). \tag{3}$$

The simplest realization arises for H taking Schrödinger's form

$$H = -\frac{1}{2}\frac{\mathrm{d}^2}{\mathrm{d}x^2} + V(x), \tag{4}$$

where the potential $V(x)$ is to be determined, while $\mathcal{L}^+$ is a third-order differential ladder operator factorized as

$$\mathcal{L}^+ = L_a^+ L_b^+, \tag{5}$$

$$L_a^+ = \frac{1}{\sqrt{2}}\left[-\frac{\mathrm{d}}{\mathrm{d}x} + f(x)\right], \tag{6}$$

$$L_b^+ = \frac{1}{2}\left[\frac{\mathrm{d}^2}{\mathrm{d}x^2} + g(x)\frac{\mathrm{d}}{\mathrm{d}x} + h(x)\right]. \tag{7}$$

By solving the resulting system of equations we arrive at

$$f = g + x, \tag{8}$$

$$h = \frac{g'}{2} - \frac{g^2}{2} - 2xg - x^2 + a, \tag{9}$$

$$V = \frac{x^2}{2} - \frac{g'}{2} + \frac{g^2}{2} + xg + \varepsilon_1 - \frac{1}{2}, \tag{10}$$

where $g(x)$ satisfies

$$g'' = \frac{g'^2}{2g} + \frac{3}{2}g^3 + 4xg^2 + 2(x^2-a)g + \frac{b}{g}, \tag{11}$$

$$a = \varepsilon_2 + \varepsilon_3 - 2\varepsilon_1 - 1, \qquad b = -2(\varepsilon_2 - \varepsilon_3)^2, \tag{12}$$

which is the Painlevé IV equation [11]. Thus, if we find a solution $g(x)$ to Eq. (11), for given extremal energies $\varepsilon_i, i = 1, 2, 3$, then the potential $V(x)$ and the associated third-order ladder operators $\mathcal{L}^\pm$ become determined. This is the reason to call Painlevé IV Hamiltonian system to those ruled by the second-order PHA of Eqs. (1-3), with H and $\mathcal{L}^\pm$ given by Eqs. (4-12).

3. Harmonic oscillator SUSY partners

As is well known, a generic k-th order SUSY transformation applied to the harmonic oscillator produces a new Hamiltonian [17–20] (see also [21])

$$H_k = -\frac{1}{2}\frac{\mathrm{d}^2}{\mathrm{d}x^2} + V_k(x), \tag{13}$$

whose potential (indeed a family) is expressed in terms of the initial one $V_0(x) = x^2/2$ as

$$V_k(x) = \frac{x^2}{2} - \frac{\mathrm{d}^2}{\mathrm{d}x^2}\{\ln W[u(x,\epsilon_0), \ldots, u(x,\epsilon_{k-1})]\}, \tag{14}$$

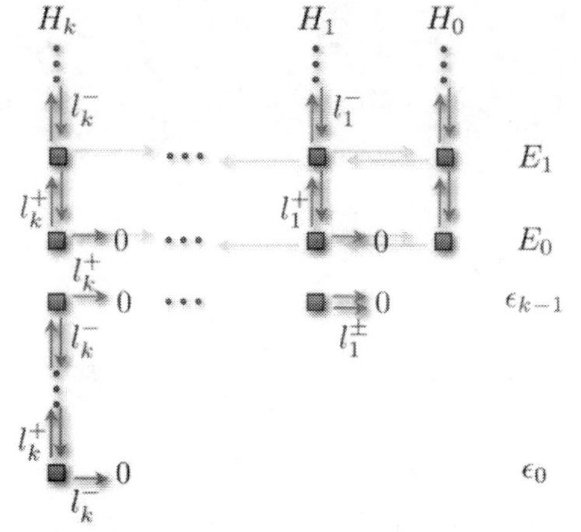

Figure 1. Action of the operators $l_j^\pm$ onto the eigenvectors of $H_j, j = 1, \ldots, k$.

where $u(x, \epsilon_j), j = 0, \ldots, k-1$ are k seed solutions of the stationary Schrödinger equation associated to $\epsilon_0 < \ldots < \epsilon_{k-1} < E_0 = \frac{1}{2}$, which are given by

$$u(x,\epsilon) = e^{-\frac{x^2}{2}} \left[{}_1F_1\left(\tfrac{1-2\epsilon}{4}, \tfrac{1}{2}, x^2\right) + 2\nu x \frac{\Gamma(\frac{3-2\epsilon}{4})}{\Gamma(\frac{1-2\epsilon}{4})} \, {}_1F_1\left(\tfrac{3-2\epsilon}{4}, \tfrac{3}{2}; x^2\right) \right]. \tag{15}$$

Note that $V_k(x)$ is non-singular if $|\nu_{k-j}| < 1$ for j odd and $|\nu_{k-j}| > 1$ for j even, $j = 1, 2, \ldots, k$. The spectrum of H_k becomes:

$$\mathrm{Sp}\,[H_k] = \{\epsilon_0, \epsilon_1, \ldots, \epsilon_{k-1}, E_n = n + 1/2, n = 0, 1, \ldots\}. \tag{16}$$

The family of potentials $V_k(x)$ of Eq. (14), in general, is generated by $2k$ free parameters: the k factorization energies $\epsilon_0, \ldots, \epsilon_{k-1}$ and the corresponding constants $\nu_0, \ldots, \nu_{k-1}$.

4. Non-trivial realizations of PIV Hamiltonian systems

Among the family of potentials $V_k(x)$ of Eq. (14), ruled in general by a $2k$th order PHA, there is a subfamily having associated also a second-order PHA which consists of Hamiltonians generated by connected seed solutions [12–16] (see also [20])

$$u(x, \epsilon_{k-j-1}) = (a^-)^j u(x, \epsilon_{k-1}), \qquad \epsilon_{k-j-1} = \epsilon_{k-1} - j, \quad j = 0, \ldots, k-1, \tag{17}$$

where the only free seed $u(x, \epsilon_{k-1})$ is such that $\epsilon_{k-1} < E_0 = 1/2$ and $|\nu_{k-1}| < 1$. As a consequence, the Hamiltonian H_k has also third-order differential ladder operators $l_k^\pm$ such that

$$[H_k, l_k^\pm] = \pm l_k^\pm, \tag{18}$$

$$[l_k^-, l_k^+] = N(H_k + 1) - N(H_k), \tag{19}$$

$$N(H_k) = l_k^+ l_k^- = \left(H_k - \tfrac{1}{2}\right)(H_k - \epsilon_0)(H_k - \epsilon_{k-1} - 1). \tag{20}$$

It turns out that now $\mathcal{H} = \mathcal{H}_{\text{iso}} \oplus \mathcal{H}_{\text{new}}$, where the action of $l_k^\pm$ on the basis of eigenvectors $|n^k\rangle$ of H_k in $\mathcal{H}_{\text{iso}}$ (such that $H_k|n^k\rangle = E_n|n^k\rangle$, $n = 0, 1, \ldots$) becomes

$$l_k^-|n^k\rangle = \sqrt{(E_n - E_0)(E_n - \epsilon_0)(E_n - \epsilon_0 - k)}\,|n - 1^k\rangle, \tag{21}$$

$$l_k^+|n^k\rangle = \sqrt{(E_{n+1} - E_0)(E_{n+1} - \epsilon_0)(E_{n+1} - \epsilon_0 - k)}\,|n + 1^k\rangle. \tag{22}$$

On the other hand, on the eigenstates $|\epsilon_j^k\rangle$ of H_k in $\mathcal{H}_{\rm new}$ (such that $H_k|\epsilon_j^k\rangle = \epsilon_j|\epsilon_j^k\rangle$, $j = 0,\dots k-1$) this action turns out to be:

$$l_k^-|\epsilon_j^k\rangle = \sqrt{(\epsilon_j - E_0)(\epsilon_j - \epsilon_0)(\epsilon_j - \epsilon_0 - k)}|\epsilon_{j-1}^k\rangle, \tag{23}$$

$$l_k^+|\epsilon_j^k\rangle = \sqrt{(\epsilon_{j+1} - E_0)(\epsilon_{j+1} - \epsilon_0)(\epsilon_{j+1} - \epsilon_0 - k)}|\epsilon_{j+1}^k\rangle. \tag{24}$$

An scheme representing this action is given in Figure 1.

5. PIV coherent states

Once the systems under study, and their annihilation and creation operators $l_k^\pm$, have been identified, let us build the corresponding coherent states [22, 23].

5.1. Barut-Girardello coherent states

First, let us look for them as eigenstates of l_k^- on each of the two subspaces, namely:

$$l_k^-|z^k\rangle = z|z^k\rangle. \tag{25}$$

On the subspace $\mathcal{H}_{\rm iso}$ we have

$$|z_{\rm iso}^k\rangle = \sum_{n=0}^{\infty} c_n|n^k\rangle. \tag{26}$$

Then we get,

$$|z_{\rm iso}^k\rangle = c_0(z,z)\sum_{n=0}^{\infty}\frac{z^n}{\sqrt{n!}}\sqrt{\frac{\Gamma(E_0-\epsilon_0+1)\Gamma(E_0-\epsilon_0-k+1)}{\Gamma(E_0-\epsilon_0+1+n)\Gamma(E_0-\epsilon_0-k+1+n)}}|n^k\rangle, \tag{27}$$

where the normalization factor $c_0(z,z)$ is obtained from the function

$$c_0(a,b) \equiv [{}_0F_2(E_0-\epsilon_0+1, E_0-\epsilon_0-k+1; a^*b)]^{-1/2}. \tag{28}$$

The properties of the CS $|z_{\rm iso}^k\rangle$ of Eq. (27) are the following:

– The so-called reproducing kernel turns out to be:

$$\langle z_{\rm iso}^{\prime k}|z_{\rm iso}^k\rangle = \frac{c_0(z',z')c_0(z,z)}{c_0^2(z',z)}. \tag{29}$$

A plot illustrating its modulus for $z' = -5+i$, $\epsilon_0 = -4$, $k = 4$ is shown in Figure 2.
– There is continuity of the CS $|z_{\rm iso}^k\rangle$ in z, i.e., $\||z_{\rm iso}^{\prime k}\rangle - |z_{\rm iso}^k\rangle\| \to 0$ when $z \to z'$.
– The restriction of the identity operator $\mathbb{I}$ on $\mathcal{H}_{\rm iso}$ admits the following decomposition:

$$\int |z_{\rm iso}^k\rangle\langle z_{\rm iso}^k|\,\mu_1(z)\mathrm{d}z = \mathbb{I}|_{\mathcal{H}_{\rm iso}}, \tag{30}$$

with the positive definite measure μ_1 given by [24] ($r = |z|$)

$$\mu_1(z) \equiv \frac{G_{0\ 3}^{3\ 0}\left(0, E_0-\epsilon_0, E_0-\epsilon_0-k\,|\,r^2\right)}{\pi c_0^2(z,z)\Gamma(E_0-\epsilon_0+1)\Gamma(E_0-\epsilon_0-k+1)}, \tag{31}$$

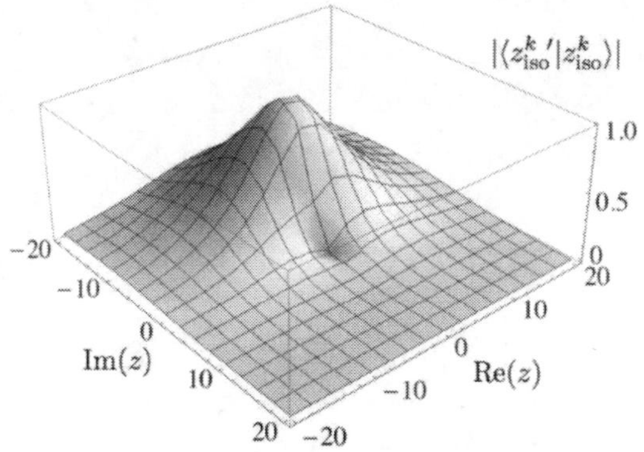

Figure 2. Modulus of the reproducing kernel of Eq. (29) for $z' = -5 + i$, $\epsilon_0 = -4$, $k = 4$.

where G is a Meijer G function.

– A CS $|z_{\rm iso}^k\rangle$ evolves into another CS of the same kind,

$$U(t)|z_{\rm iso}^k\rangle = e^{-iH_k t}|z_{\rm iso}^k\rangle = e^{-iE_0 t}|z(t)_{\rm iso}^k\rangle, \quad z(t) = ze^{-it}. \tag{32}$$

– The mean energy value $\langle H_k\rangle_{\rm iso} = \langle z_{\rm iso}^k|H_k|z_{\rm iso}^k\rangle$ acquires the form:

$$\langle H_k\rangle_{\rm iso} = \frac{1}{2} + \frac{r^2\,{}_0F_2(E_0-\epsilon_0+2, E_0-\epsilon_0-k+2; r^2)}{(E_0-\epsilon_0+1)(E_0-\epsilon_0-k+1)\,{}_0F_2(E_0-\epsilon_0+1, E_0-\epsilon_0-k+1; r^2)}. \tag{33}$$

– State probability. For a system in a CS $|z_{\rm iso}^k\rangle$, the probability $p_n^{\rm iso}(z) \equiv |\langle n^k|z_{\rm iso}^k\rangle|^2$ that an energy measurement gives the value E_n can be written as:

$$p_n^{\rm iso}(z) = c_0^2(z,z)\frac{|z|^{2n}}{n!}\frac{\Gamma(E_0-\epsilon_0+1)\Gamma(E_0-\epsilon_0-k+1)}{\Gamma(E_0-\epsilon_0+1+n)\Gamma(E_0-\epsilon_0-k+1+n)}. \tag{34}$$

On the other hand, on $\mathcal{H}_{\rm new}$ the only eigenstate of l_k^- is the extremal state $|\epsilon_0^k\rangle$ with eigenvalue $z = 0$, i.e., there is no family of CS in $\mathcal{H}_{\rm new}$ generated through this definition.

5.2. Perelomov coherent states

Now, let us look for the CS as the action of the "displacement" operator (which is not unitary)

$$D(z) = \exp\left(-\tfrac{1}{2}|z|^2\right)\exp\left(zl_k^+\right)\exp\left(-z^*l_k^-\right), \tag{35}$$

onto the extremal states of the system.

On $\mathcal{H}_{\rm new}$ we calculate $|z_{\rm new}^k\rangle = N_z' D(z)|\epsilon_0^k\rangle$, with N_z' being a normalization factor, leading to:

$$|z_{\rm new}^k\rangle \;=\; N(z,z)\left[\sum_{j=0}^{k-1}\sqrt{(E_0-\epsilon_0-j)_j(k-j)_j}\,\frac{z^j}{\sqrt{j!}}|\epsilon_j^k\rangle\right], \tag{36}$$

where $N(z,z)$ is obtained from the function

$$N(a,b) \;\equiv\; \left[\sum_{j=0}^{k-1}\frac{(a^*b)^j}{j!}(E_0-\epsilon_0-j)_j(k-j)_j\right]^{-1/2}. \tag{37}$$

The properties of the CS $|z^k_{\rm new}\rangle$ of Eq. (36) are the following:
– The reproducing kernel is now:

$$\langle z'^k_{\rm new}|z^k_{\rm new}\rangle = \frac{N(z',z')N(z,z)}{N^2(z',z)}. \tag{38}$$

– Once again there is continuity in the label z, since $\big\||z'^k_{\rm new}\rangle - |z^k_{\rm new}\rangle\big\| \to 0$ when $z \to z'$.
– The restriction of the identity operator on $\mathcal{H}_{\rm new}$ is expressed as

$$\int |z^k_{\rm new}\rangle\langle z^k_{\rm new}|\,\mu_2(z){\rm d}z = \mathbb{I}|_{\mathcal{H}_{\rm new}}, \tag{39}$$

where the positive definite measure μ_2 reads now:

$$\mu_2(z) \equiv \frac{G^{1\ 2}_{2\ 1}\left({}^{-k,\epsilon_0-E_0}_{\quad 0}\ \big|\ r^2\right)}{\pi N^2(z,z)\Gamma(E_0-\epsilon_0)\Gamma(k)}. \tag{40}$$

– A CS $|z^k_{\rm new}\rangle$ evolves into another CS of the same kind,

$$U(t)|z^k_{\rm new}\rangle = e^{-iH_kt}|z^k_{\rm new}\rangle = e^{-i\epsilon_0t}|z(t)^k_{\rm new}\rangle, \quad z(t) = ze^{-it}. \tag{41}$$

– The mean energy $\langle H_k\rangle_{\rm new} = \langle z^k_{\rm new}|H_k|z^k_{\rm new}\rangle$ now is given by

$$\langle H_k\rangle_{\rm new} = \epsilon_0 + N^2(z,z)\left[\sum_{j=0}^{k-1} j(E_0-\epsilon_0-j)_j(k-j)_j\frac{|z|^{2j}}{j!}\right]. \tag{42}$$

– State probability. For a system in a CS $|z^k_{\rm new}\rangle$, the probability $p^{\rm new}_j(z) \equiv |\langle\epsilon^k_j|z^k_{\rm new}\rangle|^2$ that an energy measurement gives the value ϵ_j is now:

$$p^{\rm new}_j(z) = N^2(z,z)(E_0-\epsilon_0-j)_j(k-j)_j\frac{|z|^{2j}}{j!}. \tag{43}$$

On the other hand, on $\mathcal{H}_{\rm iso}$ the only normalizable CS of kind $D(z)|0^k\rangle$ appears for $z=0$, i.e., we do not get a family of Perelomov coherent states on $\mathcal{H}_{\rm iso}$ but just the single CS $|0^k\rangle$.

6. Linearized PIV coherent states

We have derived families of CS on $\mathcal{H}_{\rm iso}$ and $\mathcal{H}_{\rm new}$ by using Barut-Girardello's definition for the former and Perelomov's one for the latter. It would be important to produce CS in both subspaces using just one of these two definitions. This can be done by introducing new annihilation and creation operators whose actions onto the eigenstates of H_k are simpler than the $l^\pm_k$ ones [23].

Thus, let us introduce the so-called linearized annihilation and creation operators as follows:

$$\ell^+_k \equiv \sigma(H_k)l^+_k, \qquad \ell^-_k \equiv \sigma(H_k+1)l^-_k, \tag{44}$$

$$\sigma(H_k) = [(H_k-\epsilon_0)(H_k-\epsilon_0-k)]^{-1/2}, \tag{45}$$

which act on the eigenvectors of H_k in a simplified way:

$$\ell^-_k|n^k\rangle = \sqrt{n}\,|n-1^k\rangle, \tag{46}$$

$$\ell^+_k|n^k\rangle = \sqrt{n+1}\,|n+1^k\rangle, \tag{47}$$

$$\ell^+_k|\epsilon^k_j\rangle = (1-\delta_{j,k-1})\sqrt{\epsilon_{j+1}-E_0}\,|\epsilon^k_{j+1}\rangle, \tag{48}$$

$$\ell^-_k|\epsilon^k_j\rangle = (1-\delta_{j,0})\sqrt{\epsilon_j-E_0}\,|\epsilon^k_{j-1}\rangle. \tag{49}$$

We construct the CS using Perelomov's definition which involves these ladder operators $\ell^\pm_k$.

6.1. Linearized PIV coherent states on $\mathcal{H}_{\rm iso}$

On $\mathcal{H}_{\rm iso}$ the operators $\{H_k, \ell_k^-, \ell_k^+\}$ satisfy the Heisenberg-Weyl algebra. Let us use the displacement operator

$$\mathcal{D}(z) = \exp\left(-\frac{1}{2}|z|^2\right)\exp\left(z\ell_k^+\right)\exp\left(-z^*\ell_k^-\right). \tag{50}$$

Hence, the linearized CS $|z_{\rm isol}^k\rangle = \mathcal{D}(z)|0^k\rangle$ acquire the form

$$|z_{\rm isol}^k\rangle = \exp\left(-\frac{|z|^2}{2}\right)\sum_{n=0}^{\infty}\frac{z^n}{\sqrt{n!}}|n^k\rangle. \tag{51}$$

Several properties of the CS $|z_{\rm isol}^k\rangle$ are the same as for the standard CS, namely:

– The reproducing kernel becomes:

$$\langle z_{\rm isol}'^k|z_{\rm isol}^k\rangle = \exp\left(-\frac{|z|^2}{2} - \frac{|z'|^2}{2} + z'^* z\right). \tag{52}$$

– There is continuity in the label z, i.e., $\left\||z_{\rm isol}'^k\rangle - |z_{\rm isol}^k\rangle\right\| \to 0$ when $z \to z'$.

– The restriction of the identity operator on $\mathcal{H}_{\rm iso}$ decomposes into

$$\frac{1}{\pi}\int |z_{\rm isol}^k\rangle\langle z_{\rm isol}^k|\, {\rm d}\,{\rm Re}(z)\, {\rm d}\,{\rm Im}(z) = \mathbb{I}|_{\mathcal{H}_{\rm iso}}. \tag{53}$$

– There is temporal stability in the CS $|z_{\rm isol}^k\rangle$, namely,

$$U(t)|z_{\rm isol}^k\rangle = e^{-iE_0 t}|z(t)_{\rm isol}^k\rangle, \qquad z(t) = ze^{-it}. \tag{54}$$

– The mean energy $\langle H_k\rangle_{\rm isol} = \langle z_{\rm isol}^k|H_k|z_{\rm isol}^k\rangle$ becomes the standard one,

$$\langle H_k\rangle_{\rm isol} = |z|^2 + E_0. \tag{55}$$

– State probability. For a system in a CS $|z_{\rm isol}^k\rangle$, the probability $p_n^{\rm isol}(z) = |\langle n^k|z_{\rm isol}^k\rangle|^2$ that an energy measurement gives the value E_n is now a Poisson distribution:

$$p_n^{\rm isol}(z) = e^{-|z|^2}\frac{|z|^{2n}}{n!}. \tag{56}$$

Although the previous properties of the CS $|z_{\rm isol}^k\rangle$ are the same as for the standard CS, this does not mean that the associated wavefunctions are Gaussian, which is illustrated in Figure 3, where the modulus of the CS wavefunction (blue curve) is plotted for $\nu_3 = 0$, $\epsilon_3 = -5/2$, $z = 1.5$ with $\langle H_k\rangle_{\rm isol} = 11/4$. The corresponding potential $V_4(x)$ (purple curve) becomes a rational extension of the harmonic oscillator, where the eigenfunctions $\langle x|n^k\rangle$ are expressed in terms of exceptional Hermite polynomials [25, 26].

The Barut-Girardello CS of Eq. (27) and the linearized ones of Eq. (51) coexist in the same subspace $\mathcal{H}_{\rm iso}$, thus they can be compared to each other. In Figure 4 the mean energy values $\langle H_k\rangle_{\rm iso}$ and $\langle H_k\rangle_{\rm isol}$ as functions of $r = |z|$ for both sets of CS are plotted. In purple we can observe the slow increasing behavior of $\langle H_k\rangle_{\rm iso}$ for the Barut-Girardello CS (see Eq. (33)); on the other hand, for the linearized PIV CS on $\mathcal{H}_{\rm iso}$ it is seen in blue the standard half parabola with a minimum $\langle H_k\rangle_{\rm isol} = 1/2$ at $r = 0$ (see Eq. (55)). The system under consideration corresponds to a SUSY partner of the harmonic oscillator generated for $k = 4$ with four connected energy levels at $\epsilon_0 = -11/2$, $\epsilon_1 = -9/2$, $\epsilon_2 = -7/2$ and $\epsilon_3 = -5/2$.

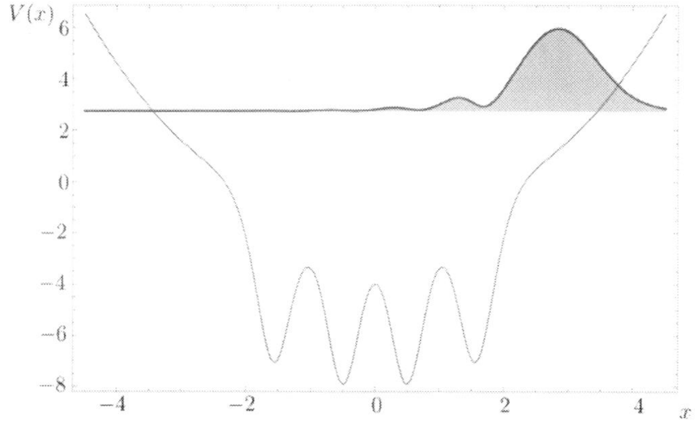

Figure 3. Modulus of the wavefunction $\langle x|z^k_{\rm isol}\rangle$ for $\nu_3 = 0$, $\epsilon_3 = -5/2$, $z = 1.5$, with $\langle H_k\rangle_{\rm isol} = 11/4$.

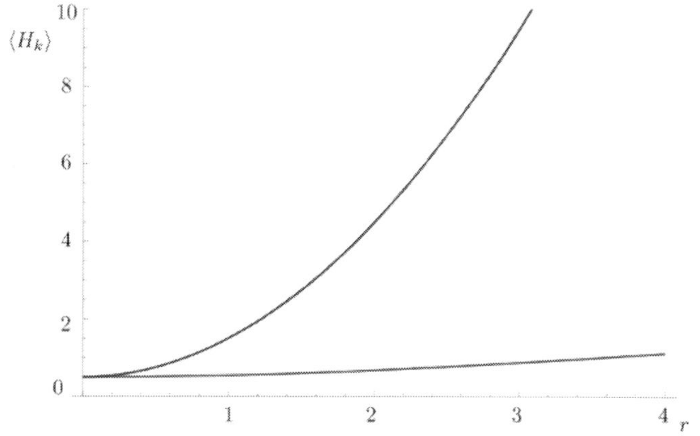

Figure 4. Mean energies $\langle H_k\rangle_{\rm iso}$ and $\langle H_k\rangle_{\rm isol}$ as functions of $r = |z|$ for the Barut-Girardello CS (purple curve) and the linearized PIV CS (blue curve) on $\mathcal{H}_{\rm iso}$, Eqs. (33) and (55) respectively. The system under consideration corresponds to a SUSY partner of the harmonic oscillator for $k = 4$ and $\epsilon_0 = -11/2$.

6.2. Linearized PIV coherent states on $\mathcal{H}_{\rm new}$

On $\mathcal{H}_{\rm new}$ the operators $\{H_k, \ell_k^-, \ell_k^+\}$ do not satisfy the Heisenberg-Weyl algebra. However, we will use the same displacement operator of Eq. (50) to act on the extremal state $|\epsilon_0^k\rangle \in \mathcal{H}_{\rm new}$. Thus, the linearized CS $|z^k_{\rm newl}\rangle = C_z' D(z)|\epsilon_0^k\rangle$ become now

$$|z^k_{\rm newl}\rangle = C(z,z)\sum_{j=0}^{k-1}\frac{(iz)^j}{j!}\sqrt{\frac{1}{\Gamma(E_0-\epsilon_0-j)}}|\epsilon_j^k\rangle, \quad (57)$$

where

$$C(a,b) = \left[\sum_{j=0}^{k-1}\frac{(a^*b)^j}{(j!)^2}\frac{1}{\Gamma(E_0-\epsilon_0-j)}\right]^{-1/2}. \quad (58)$$

The properties of $|z^k_{\text{newl}}\rangle$ are now non-standard:
– The reproducing kernel is given by:

$$\langle z'^k_{\text{newl}}|z^k_{\text{newl}}\rangle = \frac{C(z',z')C(z,z)}{C^2(z',z)}. \tag{59}$$

– There is once again continuity in the label z, namely, $\| |z'^k_{\text{newl}}\rangle - |z^k_{\text{newl}}\rangle \| \to 0$ when $z \to z'$.
– The resolution of the identity on $\mathcal{H}_{\text{new}}$ turns out to be:

$$\int_{\mathbb{C}} |z^k_{\text{newl}}\rangle\langle z^k_{\text{newl}}| \, \mu_3(z)\text{d}z = \mathbb{I}|_{\mathcal{H}_{\text{new}}}, \tag{60}$$

where the positive definite measure μ_3 becomes now:

$$\mu_3(z) = \frac{f_3(r)}{\pi C^2(z,z)}, \tag{61}$$

$$f_3(r) = G^{2\,1}_{1\,2}\left({}^{\epsilon_0 - E_0}_{0,0} \,\middle|\, r^2 \right) = \Gamma^2(E_0+1-\epsilon_0)U(E_0+1-\epsilon_0, 1; r^2). \tag{62}$$

– Temporal stability. A CS $|z^k_{\text{newl}}\rangle$ evolves once again in a CS,

$$U(t)|z^k_{\text{newl}}\rangle = e^{-i\epsilon_0 t}|z(t)^k_{\text{newl}}\rangle, \quad z(t) = ze^{-it}. \tag{63}$$

– The mean energy $\langle H_k\rangle_{\text{newl}} = \langle z^k_{\text{newl}}|H_k|z^k_{\text{newl}}\rangle$ becomes now:

$$\langle H_k\rangle_{\text{newl}} = \epsilon_0 + C^2(z,z)\sum_{j=0}^{k-1} \frac{j|z|^{2j}}{(j!)^2}\frac{1}{\Gamma(E_0-\epsilon_0-j)}. \tag{64}$$

– State probability. For a system in a CS $|z^k_{\text{newl}}\rangle$, the probability $p_j^{\text{newl}}(z) = |\langle \epsilon^k_j|z^k_{\text{newl}}\rangle|^2$ that an energy measurement gives the value ϵ_j is not longer a Poisson distribution:

$$p_j^{\text{newl}}(z) = \frac{|z|^{2j}}{(j!)^2}\frac{C^2(z,z)}{\Gamma(E_0-\epsilon_0-j)}. \tag{65}$$

Once again, the wavefunction $\langle x|z^k_{\text{newl}}\rangle$ is not Gaussian, as can be seen in the example of Figure 5 (blue curve) built for $\nu_3 = 0$, $\epsilon_3 = -5/2$, $z = 2e^{-1.59i}$, with $\langle H_k\rangle_{\text{newl}} = -3.0923$.

In Figure 6 the mean energy values $\langle H_k\rangle_{\text{new}}$ and $\langle H_k\rangle_{\text{newl}}$ as functions of $r = |z|$ are shown for the Perelomov CS (purple curve) and the linearized PIV CS (blue curve) on $\mathcal{H}_{\text{new}}$ (see Eqs. (42) and (64) respectively) with parameters $\epsilon_0 = -11/2$ and $k = 4$. The minimum for both curves corresponds to ϵ_0 when $r = 0$, while the maximum becomes ϵ_3 when $r \to \infty$. It can be seen that $\langle H_k\rangle_{\text{newl}}$ for the linearized PIV CS on $\mathcal{H}_{\text{new}}$ grows slower than for the Perelomov ones.

7. Conclusions

In this paper we have constructed several CS for a subfamily of Painlevé IV Hamiltonian systems which are connected with the PIV equation and have third-order differential ladder operators. The corresponding Hilbert space is of type $\mathcal{H} = \mathcal{H}_{\text{iso}} \oplus \mathcal{H}_{\text{new}}$, thus on each subspace we have derived two independent sets of CS.

As for a criterion for choosing a certain type of CS, this depends on the physical situation we are dealing with. For instance, if we are studying the semiclassical limit of the system the CS in $\mathcal{H}_{\text{iso}}$ would be appropriate, since their mean energy value can be made arbitrarily large. However, if we want to characterize a "more quantum" system's behavior it would be better to use either the CS in $\mathcal{H}_{\text{new}}$ or those in $\mathcal{H}_{\text{iso}}$ for $|z|$ small.

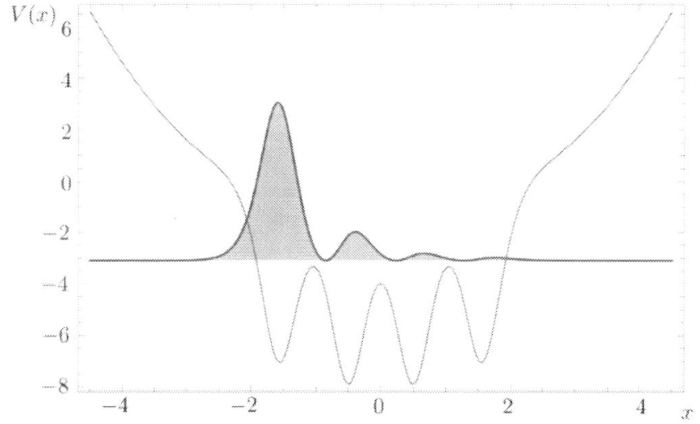

Figure 5. Modulus of the wavefunction $\langle x|z^k_{\text{newl}}\rangle$ for $\nu_3 = 0$, $\epsilon_3 = -5/2$, $z = 2e^{-1.59i}$ with $\langle H_k\rangle_{\text{newl}} = -3.0923$.

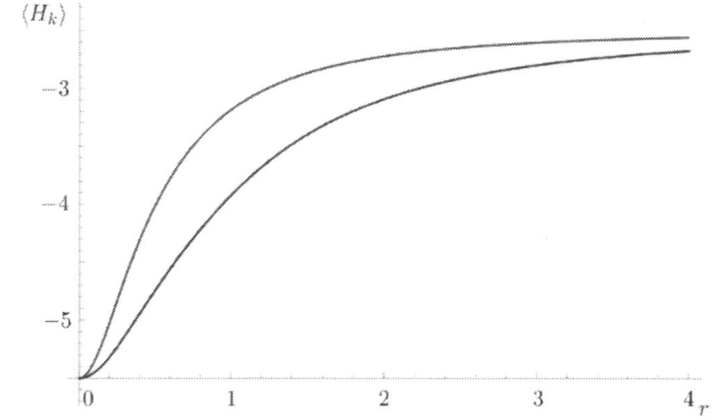

Figure 6. Mean energies $\langle H_k\rangle_{\text{new}}$ and $\langle H_k\rangle_{\text{newl}}$ as functions of r for the Perelomov CS (purple curve) and the linearized PIV CS (blue curve) on $\mathcal{H}_{\text{new}}$, Eqs. (42) and (64) respectively. The system under consideration corresponds to a SUSY partner of the harmonic oscillator for $k = 4$ and $\epsilon_0 = -11/2$.

Acknowledgments

The authors acknowledge the support of Conacyt (Mexico), Project 152574. DB and ACA also acknowledge Conacyt fellowships 207672 and 207577.

References

[1] Glauber R J 1963 The quantum theory of optical coherence *Phys. Rev.* **130** 2529–39
[2] Glauber R J 1963 Coherent and incoherent states of the radiation field *Phys. Rev.* **131** 2766–88
[3] Klauder J R 1963 Continuous representation theory. I. Postulates of continuous representation theory *J. Math. Phys.* **4** 1955–58
[4] Klauder J R 1963 Continuous representation theory. II. Generalized relation between quantum and classical dynamics *J. Math. Phys.* **4** 1958–73
[5] Barut A O and Girardello L 1971 New "coherent" states associated with non compact groups *Commun. Math. Phys.* **21** 41–55
[6] Perelomov A 1986 *Generalized coherent states and their applications* (Berlin: Springer-Verlag)

[7] Gazeau J P and Klauder J R 1999 Coherent states for systems with discrete and continuous spectrum *J. Phys. A: Math. Gen.* **32** 123–32

[8] Ali S T, Antoine J P and Gazeau J P 2000 *Coherent states, wavelets and their generalizations* (New York: Springer-Verlag)

[9] Quesne C 2001 Generalized coherent states associated with the C_λ-extended oscillator *Ann. Phys.* **293** 147–88

[10] Ben Geloun J, Hnybida J and Klauder J R 2012 Coherent states for continuous spectrum operators with non-normalizable fiducial states *J. Phys. A: Math. Theor.* **45** 085301

[11] Conte R and Musette M 2008 *The Painlevé handbook* (Dordrecht: Springer)

[12] Andrianov A, Cannata F, Ioffe M and Nishnianidze D 2000 Systems with higher-order shape invariance: spectral and algebraic properties *Phys. Lett. A* **266** 341–49

[13] Carballo J M, Fernández D J, Negro J and Nieto L M 2004 Polynomial Heisenberg algebras *J. Phys. A: Math. Gen.* **37** 10349–62

[14] Bermudez D and Fernández D J 2011 Supersymmetric quantum mechanics and Painlevé IV equation *SIGMA* **7** 025

[15] Bermudez D and Fernández D J 2011 Non-hermitian Hamiltonians and the Painlevé IV equation with real parameters *Phys. Lett. A* **375** 2974–78

[16] Bermudez D and Fernández D J 2014 Supersymmetric quantum mechanics and Painlevé equations *AIP Conf. Proc.* **1575** 50–88

[17] Fernández D J, Hussin V and Nieto L M 1994 Coherent states for isospectral oscillator Hamiltonians *J. Phys. A: Math. Gen.* **27** 3547–64

[18] Fernández D J, Nieto L M and Rosas-Ortiz O 1995 Distorted Heisenberg algebra and coherent states for isospectral oscillator Hamiltonians *J. Phys. A: Math. Gen.* **28** 2693–708

[19] Fernández D J and Hussin V 1999 Higher-order SUSY, linearized nonlinear Heisenberg algebras and coherent states *J. Phys. A: Math. Gen.* **32** 3603-19

[20] Andrianov A A and Sokolov A V 2003 Nonlinear sypersymmetry in quantum mechanics: algebraic properties and differential representation *Nucl. Phys. B* **660** 25–50

[21] Fernández D J, Hussin V and Rosas-Ortiz O 2007 Coherent states for Hamiltonians generated by supersymmetry *J. Math. Phys. A: Math. Theor.* **40** 6491–511

[22] Bermudez D, Contreras-Astorga A and Fernández D J 2014 Painlevé IV coherent states *Ann. Phys.* **350** 615-634

[23] Bermudez D, Contreras-Astorga A and Fernández D J 2014 Linearized coherent states for Hamiltonian systems with two equidistant ladder spectra *J. Phys.: Conf. Ser.* **512** 012018

[24] Sixdeniers J M and Penson K A 2000 On the completeness of coherent states generated by binomial distribution *J. Phys. A: Math. Gen.* **33** 2907–16

[25] Marquette I and Quesne C 2013 New families of superintegrable systems from Hermite and Laguerre exceptional orthogonal polynomials *J. Math. Phys.* **54** 042102

[26] Gómez-Ullate D, Grandati Y and Milson R 2014 Rational extensions of the quantum harmonic oscillator and exceptional Hermite polynomials. *J. Phys. A: Math. Theor.* **47** 015203

Boundary conditions, semigroups, quantum jumps, and the quantum arrow of time

Arno Bohm
Department of Physics, University of Texas at Austin, Austin TX 78712, USA
E-mail: bohm@physics.utexas.edu

Abstract. Experiments on quantum systems are usually divided into preparation of states and the registration of observables. Using the traditional mathematical methods (the Hilbert space and Schwartz space of distribution theory), it is not possible to distinguish mathematically between observables and states. The Hilbert space as well as Schwartz space boundary conditions for the dynamical equations lead by mathematic theorems (Stone-von Neumann) to unitary group with $-\infty < t < \infty$. But in the experimental set-up, one clearly distinguishes between the preparation of a state and the registration of an observable in that state. Furthermore, a state must be prepared first before an observable can be measured in this state (causality). This suggests time asymmetric boundary conditions for the dynamical equations of quantum theory. Such boundary conditions have been provided by Hardy space in the Lax-Phillips theory for electromagnetic and acoustic scattering phenomena. The Paley-Wiener theorem for Hardy space then leads to semi-group and time asymmetry in quantum physics. It introduces a finite "beginning of time " t_0 for a time asymmetric quantum theory, which have been observed as an ensemble of finite times $t_0^{(i)}$, the onset times of dark periods in the quantum jump experiments on a single ion.

1. Introduction

The choice of the right boundary conditions for the dynamical equation of quantum physics (e.g. Schrödinger or Heisenberg equation) is as important as the equation, or even more so. The boundary condition specify the space of allowed solutions. The first boundary condition introduced by von Neumann was the Hilbert space boundary condition; it was followed by the Schwartz space boundary condition which justifies the Dirac formalism. Both (Hilbert space $\mathcal{H}$ and Schwartz space Φ) lead by mathematical theorems [1][2] to (unitary for $\mathcal{H}$) group evolutions with the time extending over $\infty < t < \infty$. But heuristic arguments (causality) requires a beginning of time, namely a time t_0 at which the quantum state of a (scattering) experiment has been prepared and after which the registration (or detection) of an observable can begin. Thus time t in the quantum theory should extend over $t_0 < t < \infty$. This beginning of time t_0 for an ensemble of quanta, is observed as an ensemble of finite beginnings of time $\{t_0^{(i)}\} \longleftrightarrow t_0$, and t_0 is thus the time at which the state ϕ^+ describing this ensemble has been prepared, and after which the observable can be registered by a detector (i.e. can be detected). The existence of such time t_0 described by an ensemble of times $\{t_0^{(i)}\}$ has been established in experiments on single quantum systems (like the onset times $\{t_0^{(i)}\}$ of the i-th dark period in Dehmelt's and others [3] quantum jump experiments on a single ion). The existence of this t_0 suggests that one replaces the Hilbert space boundary condition, as well as the Schwartz space boundary condition for the dynamical equations, by some other boundary conditions for the dynamical (Schrödinger or Heisenberg) equations, which do not lead to the standard unitary group evolution of [1] or to the group evolution in the Schwartz space Φ [2], with t extending over $-\infty < t < \infty$.

Such a pair of new boundary condition for the dynamical equation is provided by the Hardy spaces boundary condition (cf. also in the Lax-Phillips theory [4]). Using this pair of Hardy space boundary conditions for the solutions of the dynamical equation (Schrödinger and Heisenberg equation) of quantum theory leads, by another mathematical theorem, the Paley-Wiener theorem, to a quantum mechanical semi-group evolution with $t_0 < t < \infty$. The Hardy space boundary condition thus leads to time asymmetry of a new quantum theory with finite "beginnings of time" $t_0^{(i)}$. This ensemble of individual beginnings of time for an individual micro-system is exactly what is observed by the onset times $t_0^{(i)}$ of the dark-period in numerous quantum jump experiments [3]. The Hardy space axiom, using a pair of spaces, Φ_- for the states and Φ_+ for the observables, also provides the means to distinguish mathematically between (accelerator prepared) states and (detector registered) observables, in scattering experiments like e.g. Fig.1 below.

The main new aspects of the Hardy space boundary conditions are:

1.) the finite beginnings of time $\{t_0^{(i)}\}$ and
2.) the equality of inverse width $\hbar/\Gamma$ of the Breit-Wigner resonance (e.g, corresponding to a pole of the S-matrix) and the lifetime τ for its exponentially decaying state, which fulfill $\tau = \hbar/\Gamma$.

The Hardy space axiom provides the most promising mathematical theory for quantum physics. It combines resonance scattering and decay phenomena into one unified theory of scattering and decay phenomena. This time asymmetric quantum theory can be extended into the relativistic domain by replacing the non-relativistic energy E by the relativistic invariant mass squared: $E \to s^2 = E^2 - \vec{p}^2$ [5].

2. Boundary Conditions and Dynamical Equations

Quantum physics separates roughly into two areas which are characterized by the two groups that govern the space-time motion, the Galileo group for non-relativistic motions, and the Poincare group for motion in relativistic space-time. For the non-relativistic theory which we shall discuss here only, one usually works with the solutions of the dynamical differential equation:

The Schrödinger equation

$$i\hbar \frac{\partial \phi^+(t)}{\partial t} = H\,\phi^+(t) \qquad (1)$$

for the state vector $\phi^+(t)$.
or
One could also consider the von Neumann equation for a more complicated state described by the density operator $\rho(t)$:

$$i\hbar \frac{\partial \rho(t)}{\partial t} = [H, \rho(t)] \qquad (2)$$

Or the Heisenberg equation

$$i\hbar \frac{\partial A(t)}{\partial t} = -[H, A(t)] \qquad (3)$$

for the observables (operator) $A(t)$

And one could consider the Heisenberg equation for the special case of the simple observable $A_{\psi^-} = |\psi^-(t)\rangle\langle\psi^-(t)|$; then one can use in place of A_ψ the "observable vector" ψ^- and in place of (3) the "vector-Heisenberg" equation:

$$i\hbar \frac{\partial}{\partial t}\psi^-(t) = -H\psi^-(t) \qquad (4)$$

We shall use for our discussion here the simplest situation: the Schrödinger equation (1) for the state vector $\phi(t)$; and for the observable $A(t)$, we restrict ourself to the Heisenberg equation (4) using the "observable vector" $\psi^-(t)$ of the special case $A_{\psi^-} = |\psi^-(t)\rangle\langle\psi^-(t)|$. The state vector ϕ^+ of (1), as well as the "observable" vector ψ^- of (4), represent physical apparatuses in laboratory experiments. This is depicted in Fig.1 for the example of the preparation of the "simplest" state $\phi^+(t)$ and the registration of the "simplest" observable $|\psi^-\rangle\langle\psi^-|$, in terms of detector counts of the $K^0 \to \pi^+ \pi^-$. (Remark: For a relativistic theory, one replaces the Lippmann-Schwinger eigenkets by the eigenkets labeled by the eigenvalues of relativistic 4-velocity operator $P_\mu M^{-1}$ ($\mu = 0, 1, 2, 3$), where $M^2 = P^\mu P_\mu$ is the mass-square operator and P_μ is the 4-momentum operator [5].)

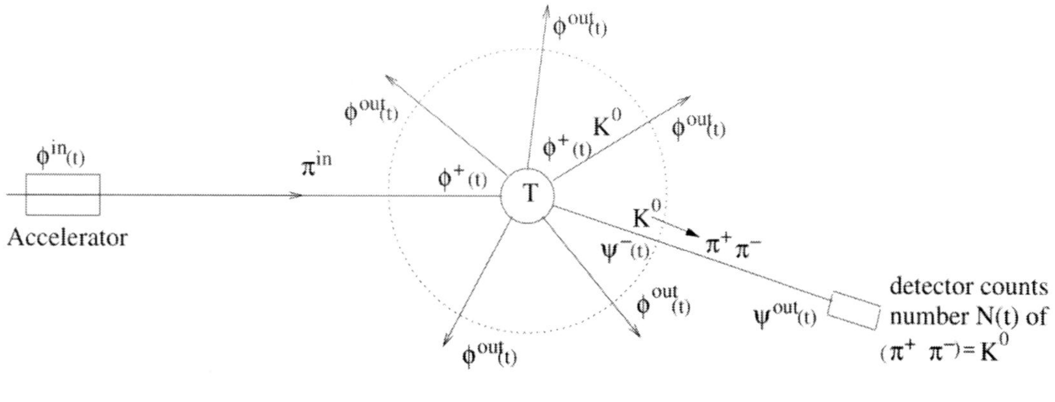

Figure 1. Preparation and decay of K^0 in the reaction

$$\pi^{in}\, p\ (\mathrm{T}) \longrightarrow \Lambda\ (\mathrm{T}')\ K^0_S\ (10^{-23}\ \mathrm{s}) \hookrightarrow \pi^+ \pi^-\ (10^{-10}\mathrm{s})$$

The theoretical predictions which need to be compared with the experimental data (given by $N(t)$, the detector counts in Fig.1), are the Born probabilities:

$$\begin{array}{cccc} & \text{Heisenberg} & & \text{Schrödinger picture} \\ \mathcal{P}_\rho(A(t)) \equiv & \mathrm{Tr}(A(t)\,\rho) & = & \mathrm{Tr}(A\,\rho(t)) \end{array} \tag{5}$$

In the special case (discussed here for simplicity) the probability to detect the observable $A_{\psi^-} = |\psi^-(t)\rangle\langle\psi^-(t)|$ in the prepared state $|\phi^+(t)\rangle\langle\phi^+(t)|$, is given by

$$\mathcal{P}_{\phi^+}(\psi^-(t)) = \mathrm{Tr}(|\psi^-(t)\rangle\langle\psi^-(t)\,||\phi^+\rangle\langle\phi^+| = |\langle\psi^-|\phi^+(t)\rangle|^2. \tag{6}$$

To solve a differential equation (like e.g, the Schrödinger eq. (1) or the Heisenberg eq. (4)) requires the choice of a **Boundary Condition** (that means one needs to choose the mathematical spaces, to which our solutions of the equations (1) or of eq. (4) should belong). The right choice of a Boundary Condition is very important for quantum physics. The boundary condition specifies the mathematical spaces which the solutions of the differential equation (1)-(4) need to belong to. In particular the boundary conditions specify the topological properties of the spaces of solution (e.g, the definition of convergence of infinite sequences in the spaces of solutions of (1) or of (4)). This choice of a specific boundary conditions has important consequences for the physical properties described by the theory, like, e.g, the existence of Dirac kets (not possible under Hilbert space boundary condition but possible under Schwartz space axiom). Or the mathematical meaning of Lippmann-Schwinger kets [6] (they do not exist as functional in the dual of the Schwartz space), or the mathematical requirements on the spaces containing Gamow vectors with exponential time evolution [8], and the possible relation of the Gamow vectors of the lifetime τ to the Breit-Wigner resonances (of Lorentzian line-shape) [9].

The boundary conditions are usually formulated as the choice of the mathematical spaces to which the solution of the dynamical differential equation (1)-(4) must belong. For instance, one has chosen in the past as admitted solutions of the dynamical equation (1) the Hilbert space $\mathcal{H}$, and one chooses as admitted solution for ψ of equation (4) also the Hilbert space boundary condition: $\phi(t) \in \mathcal{H}$, $\psi(t) \in \mathcal{H}$. One could equally well have chosen the Schwartz space (7d) as boundary condition for the dynamical equations; then one would obtain, as a bonus, that the Dirac kets are continuous antilinear functionals on the Schwartz space Φ, i.e. the scattering states would be mathematically defined as elements of the Schwartz space $|E\rangle \in \Phi^\times$. But in Schwartz space one would still not obtain Gamow kets [8] or Lippmann-Schwinger kets [6], and one could not distinguish in Schwartz space Φ between states ϕ^+ and observables ψ^-.

Since it is pretty obvious that the state $\phi^+(t)$ in Fig.1 has totally different physical meaning than the observable ψ^-, one should expect that the $\phi^+(t)$ of (1) and the $\psi^-(t)$ of (4), should not both be from the Hilbert space $\mathcal{H}$ and also not from the Schwartz space Φ.

Our goal will therefore be to arrive at new boundary conditions for the dynamical equation (1), (4) which:

1. mathematically distinguish between state ϕ and observable ψ
2. lead to a mathematical theory which unifies resonance and decay phenomena.

In the historical development of quantum theory, the following boundary conditions have been used for the dynamical equation (1) and (4):

(i) Hilbert space boundary condition of von Neumann ("Hilbert space axiom"):

Set of state vectors $\{\phi\}$ = Set of observables $\{\psi\} \div \mathcal{H}$= mathematical Hilbert space (meaning the use of Lebesgue integrals which most physisists ignore.) (7a)

(ii) Schwartz space boundary condition will provide us with the wonderful Dirac formalism, with the Dirac kets being $|E\rangle \in \Phi^\times$ (continuous, antilinear functionals on the Schwartz space Φ). Using the Schwartz space axiom, every vector $\phi \in \Phi$ has a basis vector expansion

$$\phi = \int_0^\infty |E\rangle\langle E|\phi\rangle\,, \tag{7b}$$

but convergence is defined not by one norm as in $\mathcal{H}$: $\phi_n \xrightarrow{\text{in } \mathcal{H}} \phi$, iff $||\phi_n - \phi|| \to 0$, but by a countable number of norms

$$\phi_n \xrightarrow{\Phi} \phi \quad \text{iff} \quad (\phi_n, (A+1)^p \phi_n) \to 0 \text{ for every power } p = 0, 1, 2 \cdots \tag{7c}$$

The operator A in the definition of the countable number of scalar products is a distinguished observable in Φ, e.g. the Hamiltonian or the Nelson operator in more complicated systems. And in Schwartz space, states fulfilling (1) or (2) and observables fulfilling (3) or (4) will be mathematically identified

$$\underset{\text{fulfilling (1)}}{\text{Set of state vectors } \{\phi\}} = \underset{\text{fulfilling (4)}}{\text{Set of observables } \{\psi\}} \div \Phi = \text{Schwartz space } \Phi\,. \tag{7d}$$

This would identify the physically different entities, accelerator prepared state $\{\phi^+\}$ and registered observable $\{\psi^-\}$ detected by the counter in Fig.1, and describe them by the same mathematical Schwartz space Φ.

Furthermore, from the boundary condition (7a), as well as from (7d) follows (by the Stone-von Neumann theorem [1] for the Hilbert space $\mathcal{H}$, and by a similar theorem for the Schwartz space Φ [2]), that the time evolution in both cases is given by *a group evolution*, similar to the unitary group:

The solutions of the Schrödinger equation (1) under the Hilbert space condition $\phi \in \mathcal{H}$ are: (Stone-von Neumann)

$$\phi(t) = U^{\dagger}(t)\phi = e^{-iHt/\hbar}\phi\,, \quad \text{with} \quad -\infty < t < +\infty \quad \text{for} \quad \phi \in \mathcal{H}\,. \tag{8a}$$

And solution of (1) under the Schwartz space boundary condition $\phi \in \Phi$ are also given by:

$$\phi(t) = U^{\dagger}_{\Phi}(t)\,\phi = e^{-iHt/\hbar}\phi\,; \quad -\infty < t < \infty \quad \text{for} \quad \phi \in \Phi\,. \tag{8b}$$

($U^{\dagger}_{\Phi}(t) = U^{\dagger}(t)\,|_{\Phi}$ is the restriction of $U^{\dagger}$ to the Schwartz subspace $\Phi \subset \mathcal{H}$.)

Similar results hold for the solutions of the Heisenberg equation under the Hilbert boundary conditions:

$$\psi(t) = U(t)\,\psi = e^{iHt/\hbar}\psi, \text{ with } \quad -\infty < t < +\infty \quad \text{for}\, \psi \in \mathcal{H}\,, \tag{9a}$$

and for the Schwartz space boundary condition $\psi \in \Phi$:

$$\psi(t) = U_{\Phi}(t)\,\psi = e^{iHt/\hbar}\psi\,, \quad -\infty < t < \infty \quad \text{for}\, \psi \in \Phi\,. \tag{9b}$$

These result: the (unitary) group evolution (8a) and (8b) for the state $\phi(t)$, as well as (9a) and (9b) for the observable $\psi(t)$, have been well-known for the Hilbert space axiom (Stone-von Neumann theorem [1]). The group evolution (8b) and (9b) have also been proven for the Schwartz space boundary condition $\psi\,,\ \phi \in \Phi$. [2]

As a consequence of these theorems (Stone-von Neumann and Schwartz space) it follows, that if one uses the *Hilbert space boundary condition*, or if one uses the *Schwartz space boundary condition* (Dirac's formalism), one predicts the Born probability $\mathcal{P}_\rho(A(t)) = |\langle\psi^-(t)\,|\phi^+\rangle|^2$ to detect an observable $A(t) = |\psi^-(t)\rangle\langle\psi^-(t)|$ in the state ϕ^+ for <u>all</u> time t: $-\infty < t < +\infty$.

This contradicts our feeling of *causality*, since $-\infty < t < +\infty$ in (9a) (9b) means that the probability (6) to detect the observable $A(t) = |\psi^-(t)\rangle\langle\psi^-(t)\,|$ in the prepared state $\phi^+(t_0)$ is also *predicted*, for times of the past i.e. for $t < t_0$, where t_0 is the time in the future of t at which the state ϕ^+ will be prepared. Such prediction does not make physical sense.

Thus, accepting the Hilbert space or the Schwartz space boundary condition (7a) or (7d) would lead to a conflict with our feeling for causality, because:
A detector which is represented by the observable ψ^- (i.e. by $|\psi^-(t)\rangle\langle\psi^-(t)\,|$ or by an operator $A(t)$, as in Fig.1) cannot detect anything connected to this scattering process until after the (finite) time t_0, at which the accelerator has been turned on and a state ϕ^+ has been prepared by the preparation part of the experiment, as seen from Fig.1. Thus, our standard quantum theory which makes predictions for all times $-\infty < t < +\infty$ (like the theory based on the Hilbert space axiom (9a) and also the theory based on the Schwartz space axiom (9b)) can not be quite right.

At best such boundary condition (7a)(7d) would provide an inaccurate, and deficient theoretical description of quantum processes, as shown and discussed using the experiment of Fig.1, because predictions for $-\infty < t < +\infty$ would contradict the causality principle, which asserts that:

> A state ϕ^+ needs to be prepared first, by a time t_0, *before* an observable (detector) $|\psi^-(t)\rangle\langle\psi^-(t)\,|\rightarrow|\psi^{out}\rangle\langle\psi^{out}\,|$ can be measured in the state ϕ^+ at times $t \geq t_0$ by detector counts, $N(t)/N \sim |\langle\psi^-(t)\,|\phi(t_0)\rangle|^2$: There cannot be any $K^0_S \rightarrow \pi^+\,\pi^-$ decays detected, before the π^{in} (Fig.1) hits the target T. Event by event (i), the K^0 has to be produced by the finite time $t^{(i)}_0$ at the target T before $\pi^+\,\pi^-$ can be registered by the detector $|\psi^-(t)\rangle\langle\psi^-(t)\,|$ at times $t > t^{(i)}_0$. (10a)

The beginning of time t_0 for an ensemble of i micro-systems, is of course not a specific time on the clocks in the laboratory where the experiment takes place, because this experiment involves an **ensemble** of quantum projectiles that come into being at an **ensemble** of beginnings of time t_0^i; it usually also involves an ensemble of target particles T. Thus the time t_0, after which the decay products can be counted, would also be an ensemble of times; unless one can do experiments on single particles, like Dehmelt's quantum jump experiment [3]. The lifetime is a weighted average of individual time intervals with an ensemble of "beginnings of time" $t_0^{(i)}$ and an ensemble of (usually) different "individual lifetime $\tau^{(i)}$. This ensemble of beginnings of time $t_0^{(i)}$ is the finite "beginning of time" t_0 for the quantum system (particle) on the metastable level, that thereafter decays.

Thus, the well-known unitary-group time evolution (8a) and (8b) of standard quantum mechanics (7a) or (7d) needs to be replaced by a semigroup time evolution $t_0 < t < \infty$, where the **finite** semi-group time t_0 represents the ensemble of the individual beginning of time $t_0^{(i)}$ for each individual micro-object, $i = 1, 2 \cdots$: $\{t_0^{(i)}\} = t_0$ of the prepared state ϕ^+ [10],

$$\phi(t) = e^{-iH(t-t_0)}\phi(t_0)\,, \quad t > t_0\,. \tag{10b}$$

In the quantum jump experiment [3], the lifetime is a weighted average of individual lifetimes $\tau^{(i)}$ of single ions (i) on a meta-stable level. This is the way how the lifetime for the long-lifetime state in the quantum jumps experiments were measured [3]. The exponential decay law for the time evolution of this ensemble of **single** meta-stable (excited) states has been observed with sufficient accuracy in [3] to justify the interpretation that this ensemble of "single meta-stable ion states" obeys the exponential law: The i-th meta-stable particle has a finite beginning of time t_0 which is observed as the onset times of the dark periods. The ensemble of these onset times of the dark periods $\{t_0^{(i)}\}$ is the finite time $t_0(=0)$ representing the ensemble $\{t_0^{(i)}\}$ of onset times of dark periods. In order to accommodate this ensemble of finite beginnings of time $\{t_0^{(i)}\}$, one needs a state (operator or state vector) with a finite beginning of time t_0, representing the ensembles of the beginnings of time for the dark periods $\{t_0^{(i)}\} \equiv t_0 \equiv 0 \leq t < \infty$. Here $t_0 = 0$ represents the ensemble observed as onset times of the dark period, starting with the i-th quantum jump onto the metastable level at $\{t_0^{(i)}\}$.

Since the standard Hilbert space (or the Schwartz space) theory for the dynamical equations leads (by the mathematical theorem [1]) to the unitary group evolution $U(t)$ and $U^\dagger(t)$ with $-\infty < t < \infty$, these spaces (the Hilbert space or the Schwartz space) can **not** be used for boundary conditions that lead to a causal evolution with a finite beginning of time t_0, as it is observed by the onset times of the dark periods in [3].

In order to obtain a quantum theory with a finite beginning of time t_0, and which describes a state of an ensemble with finite beginning of times $\{t_0^{(i)}\} \equiv t_0(=0)$ as it has been observed in [3], one needs new boundary conditions for the dynamical equations, which replace the standard (Hilbert space (7a), or also the Schwartz space (7d)) boundary conditions for the dynamical equations. The new mathematical boundary condition must lead to a finite beginning of time t_0, that represents the time at which the state has been prepared and after which the observable can be registered by the detectors. Since quantum theory deals with an ensemble of micro-particles, we can expect that this beginning of time t_0 is observed as an ensemble of times $t_0^{(i)}$, like the onset times (on the clock in the lab) of the dark periods, with durations that obey the exponential law, as shown in the experiment of [3].

Summarizing: Causality suggests that the Born probabilities of the observable $A(t) = |\,\psi^-(t)\rangle\langle\psi^-(t)\,|$ in the prepared state ϕ^+ make sense only after a finite time, the preparation time t_0 of the state. This finite t_0 is observed as an ensemble of finite times $\{t_0^{(i)} = \text{finite}\} \equiv t_0$.

The Born probabilities

$$\mathcal{P}_{\phi^+}(|\psi^-(t)\rangle\langle\psi^-(t)|) = |\langle\psi^-(t)|\phi^+(t_0)\rangle|^2 = |\langle\psi^-(t)|\phi^+\rangle|^2 , \tag{11}$$

are physically defined only for time $t > t_0$.

Here t_0 is the *time at which the state* ϕ^+ *is prepared and after which the observable* ψ^- *can be registered (i.e. detected) in the state* ϕ^+. Since a quantum state represents an ensemble of micro systems in the lab, the beginning of time t_0 represents (usually) an ensemble of finite times, $t_0 \leftrightarrow \{t_0^{(i)}\}$; these $t_0^{(i)}$ are different times on a clock in the lab, but they represent the same time t_0 for the metastable state. These $t_0^{(i)}$ have been observed as the onset times of the dark periods in Dehmelt's and others quantum jump experiments [3] with single ions in a Paul trap; where each excited single-ion (i) of the ensemble, has its own "beginning of time $t_0^{(i)}$ " by the clocks in the lab [10].

3. Lippmann-Schwinger Kets and Hardy Space Boundary Conditions

3.1. From Time Symmetric to Time Asymmetric Boundary Conditions

The above described observations [3] suggest, that there is a finite quantum mechanical beginning of time; therefore one must solve the dynamical differential equation not under the standard Hilbert -or the Schwartz space boundary conditions, because the Hilbert space axiom and the Schwartz space axioms (due to the Stone-von Neumann theorem) does not allow for a finite beginning of time t_0 [1]. The insight provided by the experiments on single ion [3] indicates that one needs *new boundary conditions* which distinguish mathematically between the space of states and the space of observables [10] and this leads to finite "beginnings of time" $t_0^{(i)}$, and therewith to a *semi-group evolution* with $\{t_0^{(i)}\} \cong t_0 \leq t < \infty$.

It is incredible that a mathematical theorem (Stone-von Neumann for the Hilbert space which requires Lebesgue integrals that most physicists do not even use [3]), could have distracted us from the time asymmetry in the quantum world.

In quantum physical experiments, the time asymmetry is an expression of causality: a state needs to be prepared first before an observable can be registered in the state, as it is displayed in Fig.1.

To obtain such causal (or time-asymmetric) solutions of the dynamical equations:

$$\psi^-(t) = \mathcal{U}(t - t_0)\,\psi(t_0) \quad \text{with the beginning of time } t_0 : \quad t_0 \leq t < +\infty , \tag{12}$$

one does not require new dynamical equations; but one needs to use *new boundary conditions for the solutions* of the standard dynamical equations (1)-(4). That means one has to look for new mathematical spaces for the state vectors $\{\phi^+\}$ (representing the preparation apparatus (accelerator)) and for new mathematical spaces describing the observables $\{\psi^-\}$ (representing the registration apparatus i.e, the detector in the scattering experiment of Fig.1).

Since the standard boundary conditions for the dynamical equation (1) or (4), which use

$$\text{the Hilbert space axiom:} \qquad \{\phi\} = \{\psi\} \div \mathcal{H} \quad \text{of von Neumann,} \tag{13a}$$

or which use:

$$\text{the Schwartz space axiom:} \quad \{\phi\} = \{\psi\} \div \Phi \quad \text{of the Dirac formulation,} \tag{13b}$$

lead to the time symmetric (unitary) group evolution (8a)(8b), (9a)(9b) with $-\infty < t < +\infty$, we must look for new boundary conditions which respect causality, in place of the Hilbert space and the Schwartz space axioms (7a) and (7d) for the two dynamical differential equations (1) and (4). These new boundary conditions, for the Schrödinger equation of the states $\{\phi^+\}$ and for the Heisenberg equation of the observables $\{\psi^-\}$, need to differ from each other.

Furthermore, since the $\{\phi^+\}$ represents the prepared states, and the $\{\psi^-\}$ represents the detected observable, as it is suggested by the arrangement of the scattering experiments in Fig.1, it is more natural to use two different mathematical spaces: One space, say Φ_-, will represent the space of states $\{\phi^+\}$ and the other space, say Φ_+, will represent the space of observables $\{\psi^-\}$. The question then is: Which mathematical spaces or which pair of mathematical spaces should be chosen for these two sets of physical entities, the set of states $\{\phi^+\}$ obeying (1) and the set of observables $\{\psi^-\}$ obeying (2)?

The best answer at this time in history, based on experimental results in quantum resonance scattering and decay phenomena is, that these spaces are the pair of Hardy spaces (Hardy spaces have been used before by Lax-Phillips [4] in classical scattering theory, and also by Y. Strauss *et al* [11] in quantum scattering).

The new Hardy space boundary conditions for the dynamical equations (1)-(4) has the advantage that it will employ different mathematical spaces for the two different physical entities, one space Φ_- for the states $\{\phi^+\}$ and the other space which we call Φ_+ for the observables $\{\psi^-\}$. There is no reason to use one and the same mathematical space Φ (e.g. the Schwartz space) to represent for such different physical entities as the set of states $\{\phi^+\}$ representing the preparation apparatus, and the set of observables $\{\psi^-\}$ representing the detector.

Thus, the new rule that relates the physical quantities to the mathematical representation space is given by a new axiom of quantum theory, the Hardy space axiom.

3.2. The Hardy Space Axiom

The set of state vectors $\{\phi^+\}$ representing the accelerator-prepared states (see Fig.1 above) and fulfilling the Schrödinger equation (1), is the Hardy space Φ_- of the lower complex energy semi-plane (2nd sheet of the analytic S-matrix).

$$\text{physically prepared in-states} \quad \{\phi^+\} \doteqdot \Phi_- \quad \begin{array}{l}\text{mathematical Hardy space for the}\\ \text{solutions of the Schrödinger eq.(1)}\end{array} \tag{14a}$$

The set of observable vectors $\{\psi^-\}$ representing the detector-registered observables (Fig.1) and fulfilling the Heisenberg equation (4), is the Hardy space Φ_+ of the upper complex energy semi-plane (2nd sheet of the analytic S-matrix):

$$\text{experimentally detected out-observables} \quad \{\psi^-\} \doteqdot \Phi_+ \quad \begin{array}{l}\text{mathematical Hardy space for the}\\ \text{solutions of the Heisenberg eq.(4)}\end{array} \tag{14b}$$

Here we have used a notation that labels the prepared in-state vector (in-state of a scattering experiment in Fig.1) by ϕ^+, as it is the custom in scattering theory; this label $^+$ has its origin in the standard notation for the in-state kets $|E^+\rangle$ of the Lippmann-Schwinger equation (as will be defined below in (16) [6] [7]). For the out-vectors representing the detector, we use the notation ψ^- (which here represents the <u>observable</u> $|\psi^-\rangle\langle\psi^-|$). The observable-vector ψ^- needs to obey the Heisenberg equation (4). The label $^-$ for the observable ψ^- and the label $^+$ for the in-state ϕ^+ have their origin in the standard notation for the Lippmann-Schwinger kets (see below in equation (16) (17)).

This miss-match (14a) and (14b)) of the notation for vectors and for spaces:

$$\phi^+ \in \Phi_- \quad \text{(lower complex plane),} \quad \text{for the prepared in- state} \tag{15a}$$

$$\psi^- \in \Phi_+ \quad \text{(upper complex plane)} \quad \text{for the detected observable} \tag{15b}$$

Journal of Physics: Conference Series **597** (2015) 012018 doi:10.1088/1742-6596/597/1/012018

has its origin in the different conventions used in physics for the vectors ϕ^+ and ψ^-, and the convention used in mathematics for the Hardy space Φ_- and the Hardy space Φ_+.

The notation used for the famous Lippmann-Schwinger kets [6] in the standard physics literature (c.f equation (17) below), results in the mismatch of $\pm$ in (15a) and (15b) between the state vectors ϕ^+ and their mathematical space $\{\phi^+\} = \Phi_-$ (in the mathematicians' convention for Hardy spaces), and for the Lippmann-Schwinger kets and the labels $\mp$ for the dual spaces $\Phi_-^\times$, $\Phi_+^\times$ (which are the dual space of the Hardy spaces in (14a) and (14b)).

Therefore, we want to give the heuristic Lippmann-Schwinger ket [6] a mathematical meaning, by defining them as the antilinear continuous functionals on the Hardy space (14a) and (14b):

$$|E^+\rangle \in \Phi_-^\times = \text{dual of the space } \Phi_- \quad \text{and} \quad |E^-\rangle \in \Phi_+^\times = \text{dual of the space } \Phi_+ . \qquad (15c)$$

The Lippmann-Schwinger kets $|\,E^+\rangle$ and $|\,E^-\rangle$ were initially introduced by the Lippmann-Schwinger equation of (16) below. They have with (15c) obtained the mathematical meaning as the continuous antilinear functionals on the Hardy space Φ_- and Φ_+, respectively.

Our Hardy space Φ_- is realized by the "smooth Hardy" function (the intersection of Hardy class $\mathcal{H}^2_-$ with the Schwartz space S). In analogy to the components $\langle E\,|\phi\rangle$ of the Dirac basis vector expansion (7b), the components of the Hardy space vectors (14a) (14b) are very well-behaved functions

$$\phi^+(E) = \langle {}^+E\,|\phi^+\rangle \in (\mathcal{H}^2_- \cap S)_{\mathbb{R}_+} \text{ on } \mathbb{C}_- . \qquad (15d)$$

This means $\phi^+(E)$ are smooth functions that can also be continued into the lower complex energy semiplane $\mathbb{C}_-$ second sheet of the S-matrix. This definition is compatible with the physicists convention for the Lippmann-Schwinger kets [6]. The energy wave function $\phi^+(E) = \langle {}^+E\,|\phi^+\rangle$ of the accelerator prepared in-states ϕ^+ are analytic function in the lower complex energy plane $\mathbb{C}_-$; for this $\mathbb{C}_-$ one choose the second sheet of the S-matrix (where the resonance poles of the S-matrix are located).

Similarly, the Hardy space Φ_+ is realized by the smooth Hardy function

$$\psi^-(E) = \langle {}^-E\,|\psi^-\rangle \in (\mathcal{H}^2_+ \cap S)_{\mathbb{R}_+} \text{ on } \mathbb{C}_+ , \qquad (15e)$$

i.e. smooth functions on the upper complex energy semi-plane $\mathbb{C}_+$ (second sheet of the S-matrix). As a consequence, their complex conjugates $\overline{\psi^-(E)} = \langle \psi^-\,|E^-\rangle$ are also analytical functions on the second sheet of the S-matrix lower complex semi-plane, which is important for the analytic continuation of the S-matrix (e.g. to the places where the resonance poles of the S-matrix are located).

From (15d),(15e) follows that the product of these smooth Hardy functions $\langle\psi^-\,|E^-\rangle\langle {}^+E\,|\phi^+\rangle$ can be analytically continued into the lower complex semi-plane second sheet of the S-matrix element $S(E)$, where the resonance poles of the S-matrix are located. (in general there are also other interesting features on the second sheet of the S-matrix, e.g, higher order poles which will not be discussed in this paper.) First order poles of the S-matrix correspond to Breit-Wigner resonances.

The Hardy space boundary conditions (14a) and (14b) (or (15d) and (15e)) for the dynamical equation (1) and (4) were already foreshadowed by the Lippmann-Schwinger equations (16), though Hardy space mathematicians [4] and physicists using Lippmann-Schwinger equations did not seem to have been aware of each other.

The Lippmann-Schwinger kets [6] were postulated earlier (in analogy to the Dirac kets) as the in-plane wave "states" $|E^+\rangle$, and the out-plane wave "states" $|E^-\rangle$ which:

1. are energy "eigenkets" of the "exact Hamiltonian $H = K + V$; and of the "interaction-free Hamiltonian K

$$H\,|E^\pm\rangle = E\,|E^\pm\rangle , \qquad K\,|E\rangle = E\,|E\rangle$$

that fulfill the Lippmann-Schwinger equations [6]:

$$|E^{\pm}\rangle = |E\rangle + \frac{1}{E - K \pm i\epsilon} V |E^{\pm}\rangle \qquad H = K + V\,, \tag{16}$$

2. whose formal "solutions" are given by:

$$|E \pm i\epsilon\rangle = |E^{\pm}\rangle = |E\rangle + \frac{1}{E - H \pm i\epsilon} V |E\rangle = \Omega^{\pm}|E\rangle, \quad \epsilon \to +0\,. \tag{17}$$

Now, with (15d) and (15e), the Lippmann-Schwinger ket $|E^{\pm}\rangle$ have been given a mathematical meaning. They are defined as the continuous antilinear functionals on the pair of the Hardy spaces Φ_- and Φ_+, $|E^{\pm}\rangle \in \Phi_{\mp}^{\times}$. These functionals $F_E = |E^{\pm}\rangle$ are also the generalized eigenvectors F_E with eigenvalue E of the Hamiltonian H; i.e

$$\langle H\phi^+|E^+\rangle \equiv \langle \phi^+ |\, H^{\times}\, |E^+\rangle = E\langle \phi^+|E^+\rangle\,, \tag{18a}$$

and

$$\langle H\psi^-|E^-\rangle \equiv \langle \psi^- |\, H^{\times}\, |E^-\rangle = E\langle \psi^-|E^-\rangle\,. \tag{18b}$$

($H^{\times}$ in (18a)(18b) is the definition of the operator $H^{\times}$ conjugate to H in Φ_- and Φ_+.)

In the mathematics literature, one uses the notation $\mathcal{H}^2_-$ for the Hardy *class* function (L^2-integrable) of the lower complex semi-plane, and the notation $\mathcal{H}^2_+$ for the Hardy class function (L^2-integrable) of the upper complex semi-plane. The energy wave functions that we are interested in here for the quantum mechanical scattering and resonance theory, are not the L^2-integrable function classes analytic in the lower complex plane $f_-(E) \subset \mathcal{H}^2_-$ (Hardy class functions in $\mathbb{C}_-$) or the L^2-integrable function classes analytic in the upper complex plane $f_+(E) \subset \mathcal{H}^2_+$ (Hardy class in $\mathbb{C}_+$) [4].

For the theory of resonance scattering and decay phenomena, we use the space of smooth Hardy functions (15d) and (15e). They are defined as the space of (L^2-integrable) Hardy functions $\mathcal{H}^2_-$ which have also the properties of Schwartz functions S on the lower complex energy semi-plane $\mathbb{C}_-$ (second sheet of the S-matrix) [10]:

$$\phi^+(E) = \langle {}^+E|\phi^+\rangle \in \left(\mathcal{H}^2_- \cap S\right)_{\mathbb{R}_+}\,. \tag{15d}$$

Here $\mathcal{H}^2_- \cap S$ is the intersection of $\mathcal{H}^2_-$ with the Schwartz space S. Similarly we choose the set of functions:

$$\psi^-(E) = \langle {}^-E|\psi^-\rangle \in \left(\mathcal{H}^2_+ \cap S\right)_{\mathbb{R}_+} \tag{15e}$$

on the upper complex energy semi-plane $\mathbb{C}_+$ (second sheet of the S-matrix).

The $\mathcal{H}^2_+ \cap S$ are the spaces of smooth Hardy function $\langle {}^-E\,|\psi^-\rangle \in \mathcal{H}^2_+ \cap S$ on the upper complex energy plane, and the $\langle {}^+E\,|\phi^+\rangle \in \mathcal{H}^2_- \cap S$ are the smooth Hardy functions on the lower complex plane (second sheet of the S-matrix), i.e. the energy wave function $\langle {}^+E|\phi^+\rangle$ and $\langle {}^-E|\psi^-\rangle$ are the smooth Hardy space functions that can be analytically continued into the lower (for ϕ^+) and upper (for ψ^-) complex energy plane, second sheet of the S-matrix. Therefore the product $\langle \psi^-|E^-\rangle\langle {}^+E\,|\phi^+\rangle = \overline{\langle {}^-E|\psi^-\rangle}\langle {}^+E\,|\phi^+\rangle$ are smooth Hardy functions that can be analytically continued into the lower complex energy plane (second sheet of the S-matrix), where (as depicted in Fig.2 at z_R) the resonance poles of the S-matrix $S(E)$ (and other interesting physical entities) are located.

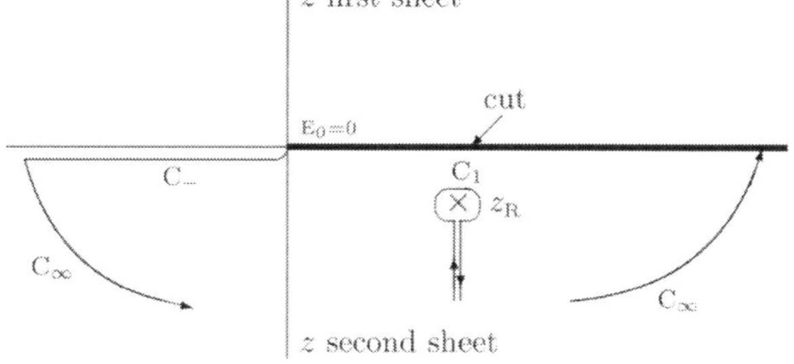

Figure 2: First upper complex plane and second sheet lower complex plane with cut and resonance pole of the S-matrix at $z_R = E_R - i\Gamma/2$, $\tau = \hbar/\Gamma$= lifetime.

To simplify the notation, we have so far omitted the additional quantum number j, j_3, η in the energy kets for which one usually choose the angular momentum j and its component j_3 and uses η for the other degeneracy quantum numbers. Using the complete set of observables H, J^2, J_3, $\widehat{\eta}$, the energy eigenkets for the spaces $\Phi^{\times}_{\mp}$ are $|E^{\pm}\rangle = |E, j, j_3, \eta^{\pm}\rangle$ and the energy wave function $\langle^{\pm}E\,|\phi^{\pm}\rangle = \langle^{\pm}E, j, j_3, \eta\,|\phi^{\pm}\rangle$ are given by the Hardy functions (15d) and (15e). (H, J^2, J_3, $\widehat{\eta}$ are a complete system of commuting observables and η denote additional species quantum numbers, not important here.)

The Hardy space kets (i.,e continuous antilinear functionals of $\Phi^{\times}_{-}$) $|E^{+}\rangle = |E, j, j_3, \eta^{+}\rangle$ are taken as basis systems for a Dirac basis ket expansion (using the "nuclear spectral theorem" for the "nuclear" linear topological space Φ_{-}) of the in-state vectors $\phi^{+} \in \Phi_{-}$ (*representing the preparation apparatus, e.g. the accelerator*, in the scattering experiment of Fig.1):

$$\phi^{+} = \sum_{j,j_3,\eta} \int_0^{\infty} dE|E, j, j_3, \eta^{+}\rangle\langle^{+}E, j, j_3, \eta|\phi^{+}\rangle = \int_0^{\infty} dE|E^{+}\rangle\langle^{+}E|\phi^{+}\rangle = \int dE\,|E^{+}\rangle\phi^{+}(E) \tag{19}$$

Nuclear space here means that Φ_{-} fulfills the conditions of the nuclear spectral theorem which means the conditions for a Dirac basis vector expansion like (19), which the Schwartz -and the Hardy space do.

Similarly, the $|E^{-}\rangle = |E, j, j_3, \eta^{-}\rangle$ are taken as basis systems for out-vectors $\{\psi^{-}\}$ (the ψ^{-} *represent the observables* $|\psi^{-}\rangle\langle\psi^{-}|$ registered (or "detected") by the *detector* of Fig.1):

$$\psi^{-} = \sum_{j,j_3,\eta} \int_0^{\infty} dE'|E', j, j_3, \eta^{-}\rangle\langle^{-}E', j, j_3, \eta|\psi^{-}\rangle = \int_0^{\infty} dE'|E'^{-}\rangle\langle^{-}E'|\psi^{-}\rangle = \int dE'\,|E'^{-}\rangle\psi^{-}(E') \tag{20}$$

4. The Time Evolution Semigroup

The Lippmann-Schwinger ket expansion (19) and (20) were originally postulated in analogy to the Dirac basis vector expansion, just replacing the Dirac kets by the Lippmann-Schwinger ket (16) (17), which were also just postulated [6] . The difference between (19),(20) and the Dirac ket expansions is, that the original Dirac ket expansion is justified as the "nuclear spectral theorem" for the Schwartz-Rigged Hilbert space Φ [12].

To justify the Lippmann-Schwinger ket expansions (19) and (20), and to provide a mathematic definition of the Lippmann-Schwinger kets, required the construction of a *new pair of Rigged Hilbert spaces*, which are based on the pair of "smooth"-Hardy functions (15d) and (15e) of the upper complex energy plane 1st sheet, and lower complex energy plane [2] second sheet of the analytic S-matrix and along the positive real energy axis, cf. Fig.2.

Thus

the in-states $\phi^{+} \in \Phi_{-}$	and the out-observables $\psi^{-} \in \Phi_{+}$	(21a)
have the energy wave functions	have the energy wave functions	
$\phi^{+}(E) = \langle^{+}E\|\phi^{+}\rangle \in \left(\mathcal{H}^2_{-} \cap S\right)_{\mathbb{R}_+}$	$\psi^{-}(E) = \langle^{-}E\,\|\psi^{-}\rangle \in \left(\mathcal{H}^2_{+} \cap S\right)_{\mathbb{R}_+}$	(21b)

In (21b), $\mathcal{H}^2_\pm$ is the standard notation of the Hardy class functions (L^2-integrable) on the lower complex semiplane (for $\phi^+(E)$), and on the upper complex semi-plane (for $\psi^-(E)$), second sheet of the S-matrix. The function spaces $(\mathcal{H}^2_\pm \cap S)_{\mathbb{R}_+}$ are the intersections of $\mathcal{H}^2_\pm$ with the Schwartz function space S restricted to the positive energy. $\mathcal{H}^2_\pm \cap S$ are the spaces of Hardy functions of the lower ($-$) (in (15d)) and the upper ($+$) (in (15e)) complex energy plane, second sheet of the S-matrix; thus $\langle\psi^-|^-E\rangle\langle^+E|\phi^+\rangle \in (\mathcal{H}^2_- \cap S)$ on the second sheet lower complex E-plane, where the resonance poles of the S-matrix are located.

The Dirac kets $|E\rangle$ are mathematically defined as continuous antilinear functionals on the **Schwartz** space Φ: $|E\rangle \in \Phi^\times$ (the space of "tempered distribution"). This is as far as the distribution theory using the Schwartz space could go. The shortcoming of the Schwartz space is that the states $\{\phi\}$ and the observables $\{\psi\}$ can not be mathematically distinguished from each other since both are mathematically described by the same Schwartz space cf (8b) and (9b). But physically, it is clear that states and observables are different physical entities, the accelerator defines the states and the detector defines the observables.

The new idea, drawn from the *pair* of phenomenological Lippmann-Schwinger kets (16), is to relate the set of prepared in-states ϕ^+ (experimentally defined by the preparation apparatus (e.g, accelerator etc of Fig.1) to the mathematical Hardy class space $\mathcal{H}^2_-$; (this then leads to the smooth Hardy functions (15d)).And also to relate the observable ψ^- to the smooth Hardy functions (15e).

This is expressed by *the Hardy space axiom* of time-asymmetric quantum theory (15d) and (15e) and in terms of the abstract vector spaces:

$$\text{Set of accelerator prepared states } \{\phi^+\} \doteqdot \Phi_- = \text{ Hardy space on } \mathbb{C}_-\text{(lower complex energy plane, 2nd sheet of the } S\text{-matrix).} \tag{14a}$$

$$\text{Set of detector registered observables } \{\psi^-\} \doteqdot \Phi_+ = \text{ Hardy space on } \mathbb{C}_+\text{(upper complex energy plane, 2nd sheet of the } S\text{-matrix).} \tag{14b}$$

The accelerator prepared energy wave functions $\phi^+(E)$ are the smooth Hardy functions (15d), and the detector registered energy wave function are the smooth Hardy functions $\psi^-(E)$ (15e).

Taking as the new axiom of quantum physics, the Hardy space axiom (14a),(14b) or (15d), (15e), the Schrödinger and the Heisenberg equation have to be solved under these Hardy space boundary conditions.

The great improvement of the Hardy space axiom (14a), (14b) over the standard Hilbert space axiom (7a) or the Schwartz space axiom (7d) is that the Hilbert space and the Schwartz space boundary condition for the dynamical equations lead to the unitary group evolution (8a) and (9a), and (8b) and (9b), for which t extends over $-\infty < t < \infty$ (by the Stone-von Neumann theorem *etc*).

In contrast, the Hardy space boundary conditions (14a), (14b) leads (by *the Paley-Wiener theorem* [13] for Hardy functions) to two different *semi-groups evolutions*, one for the accelerator prepared states $\phi^+ \in \Phi_-$ obeying the Schrödinger equation, and the other for the detector-registered observable $\psi^- \in \Phi_+$ obeying the Heisenberg equation. Their time evolutions are given by the two semigroups:

$$\begin{array}{cc} \text{for solution of} & \text{for solution of} \\ \text{the Schrödinger equation} & \text{the Heisenberg equation} \\ \text{for states in the space } \Phi_- & \text{for observables in the space } \Phi_+ \\ \phi^+(t) = e^{-iHt/\hbar}\phi^+ & \psi^-(t) = e^{iHt/\hbar}\psi^- \\ \text{with } 0 \le t < \infty & \text{with } 0 \le t < \infty \end{array} \tag{22}$$

Remember that the change in time evolution from the unitary group evolution (8a), (8b) and (9a), (9b) to the semigroup in (22), is due to our change from the Hilbert space and from the Schwartz space boundary condition for the dynamical equations to the new Hardy space boundary conditions (14a) and (14b). As a consequence of the change for the boundary condition from (the Hilbert space with the Stone-von Neumann $\to -\infty < t < +\infty$) to (the Hardy space with the Paley-Wiener $\to 0 < t < +\infty$), we obtain a **new quantum theory** with strictly causal time evolution:

The probability to detect the time evolved observable $\psi^-(t)$ in the state ϕ^+ is predicted as the Born probability:

$$\begin{aligned} \mathcal{P}_{\phi^+}(\psi^-(t)) &= |\langle\psi^-(t)|\phi^+\rangle|^2 = |\langle e^{iHt/\hbar}\psi^-|\phi^+\rangle|^2 \\ &= |\langle\psi^-|e^{-iHt/\hbar}|\phi^+\rangle|^2 = |\langle\psi^-|\phi^+(t)\rangle|^2 \quad \underline{\text{for } t \geq 0 \text{ only}}\,. \end{aligned} \tag{23}$$

This mathematical prediction is a consequence of the new Hardy space axioms (14a) and (14b) from which (22) and thus (23) follows as a consequence of the Paley-Wiener theorem [13].

This is a new theoretical prediction which follows from (14a),(14b). It is in agreement with the experimental observation that the detector counts $N(t)/N$:

$$N(t)/N \sim \text{probability for detected observable } \mathcal{P}_{\phi^+}(\psi^-(t)), \text{ are detected only for } t \geq t_0 = 0\,. \tag{24}$$

This result (24) also establishes agreements of the calculated probabilities (23) and the experimental data $N(t)/N$. In (24), t_0 is the time at which the state ϕ^+ had been prepared by the accelerator in Fig.1, and after which the registration of the observable makes sense. The theoretical prediction (23) follows from the Hardy space boundary condition in perfect agreement with our general feeling about causality. But the result cannot be obtained from the standard Hilbert space or Schwartz space axioms (8a)-(9b), which "predict the probabilities for all time $-\infty < t < \infty$. The Hardy space prediction (23) is in agreement with causality (24) whereas the Hilbert space and Schwartz space predictions (8a)-(9b) are not causal (they also predict for the past $-\infty < t < 0$). The question then is: how can one see such beginnings of time t_0 ?

The agreement between the physicists' set of state vectors $\{\phi^+\}$ and the mathematical space Φ_-, as well as the agreement between the physicists' set of observables $\{\psi^-\}$ and the mathematical space Φ_+ is perfect, except for the mix-up in the notation of $\phi^+ \in \Phi_-$ and $\psi^- \in \Phi_+$ of the + and − signs.

The reason for this "mis-labeling" between the ± labels of the pair $\phi^+ \in \Phi_-$ and $\psi^- \in \Phi_+$ are just different notational conventions; the Lippmann-Schwinger kets $|E^\pm\rangle$ were labeled by physicists [6][7] such that the energy wave function $\langle^+E|\phi^+\rangle$ and the $\langle\psi^-|E^-\rangle = \overline{\langle^-E|\psi^-\rangle}$ are analytic function on the lower complex plane (2nd sheet of the S-matrix), where resonance poles of the S-matrix and other interesting physical properties (singularities) are located.

The new statement of ***physical contents*** is the Hardy space boundary condition (14a) and (14b) which are expressed in terms of the energy wave functions (on the second sheet of the S-matrix below the cut from $0 \leq E < +\infty$). These wave functions have the property of the smooth Hardy functions

$$\langle^+E|\phi^+\rangle \in \left(\mathcal{H}^2_- \cap S\right)_{\mathbb{R}_+} \quad \text{and} \quad \langle^-E|\psi^-\rangle = \overline{\langle\psi^-|E^-\rangle} \in \left(\mathcal{H}^2_+ \cap S\right)_{\mathbb{R}_+}\,. \tag{25}=(15d), (15e)$$

This is an axiom which one can only conjecture if one looks at many aspects of resonance and decay phenomena. The result of such search process was that

> the prepared state $\phi^+ \in \Phi_-$ and the registered observable $\psi^- \in \Phi_+$are represented by Hardy spaces of the lower and upper complex energy plane (2nd sheet of the S-matrix $S_j(E)$); thus $|E^+\rangle \in \Phi_-^\times$ and $|E^-\rangle \in \Phi_+^\times$ are continuous antilinear functionals on the dual of the Hardy spaces Φ_- describing states, and Φ_+ describing observables. (26)

The mathematical entities which in quantum mechanics had been associated with the prepared "state" ϕ^+ fulfilling eq.(1) and the detected observable $|\psi^-\rangle\langle\psi^-|$ fulfilling eq.(4) are represented in the new mathematical theory by a pair of (mathematically well-defined) Hardy spaces, Φ_- for the state ϕ^+, and Φ_+ for the observable ψ^- whose energy wave functions are given by the smooth Hardy functions (14a)(14b), (15d)(15e).

The Hilbert space, and Schwartz space theory with Dirac kets, have done a fair job in the description of scattering processes, but they did not succeed to provide an adequate description of resonance and decay phenomena. For that one needed the Lippmann-Schwinger kets and also Gamow vectors with exponential time evolution for the decaying states, and one needed energy wave function $\phi^+(E) = \langle^+E|\phi^+\rangle$, $\overline{\psi^-(E)} = \overline{\langle^-E|\psi^-\rangle}$ which can be analytically continued into the complex energy plane (e.g. second sheet of the S-matrix where the resonance poles are located). And all this could be done so far with the "smooth Hardy function (25).

If one wants a theory that unifies the Breit-Wigner resonance of width Γ with an exponentially decaying state of lifetime $\tau = \hbar/\Gamma$ (as an exact relation), one needs the Hardy space axiom (14a) (14b) (15d) (15e). Also the Hardy space axiom leads to a strictly causal theory for the experimental predictions (24) with (23).

In order to achieve all this, new boundary condition of the dynamical equations of quantum mechanics were required: instead of solving the dynamical equation under the standard Hilbert space -(or Schwartz space)- boundary conditions, one has to use the (pair of) Hardy spaces. Then one obtains, again from a mathematical theorem of Paley-Wiener the time asymmetric dynamical evolution (22), (23) which agrees with our intuition of causality and it explains experimental observations like the quantum jump experiments [3].

The two dynamical equations of quantum mechanics (e.g. (1) and (4)) are usually considered as describing two pictures (Schrödinger or Heisenberg) of the same physical situation. This is indeed the case if one chooses the same Hilbert space boundary conditions as boundary condition of these differential equations. In most experimental situations (e.g. the scattering experiment of Fig.1), however, (for instance the scattering experiment of Fig.1), this is not justified: detected observable $A = |\psi^-\rangle\langle\psi^-|$ and prepared state $\rho = |\phi^+\rangle\langle\phi^+|$ are utterly different from each other. Therefore, the standard axiom (e.g. Hilbert space but also Schwartz space axiom (for the mathematical Dirac kets)which predicts the Stone-von Neumann unitary group) do not apply.

A new boundary condition, which was foreshadowed by the Lippmann-Schwinger equation, is the Hardy space boundary conditions (14a)(14b). It was conjectured from the requirement that, the width of the resonance $\Gamma = \hbar/\tau$ the inverse lifetime of decaying state. This equality of resonance width $\Gamma = \hbar/\tau$ inverse lifetime has recently been experimentally confirmed to high accuracy [14].

References

[1] Stone M H 1932 *Ann. Math.* **33** No. 3 643; Neumann V J 1932 *Ann. Math.* **33** No. 3 567
[2] Bohm A and Gadella M 1989 *Dirac Kets, Gamow Vectors and Gelfand Triplet* Lecture Note in Physics Vol **348** (New York and Berlin: Springer-Verlag), specifically Ch4 P.82
[3] Dehmelt H, Nagourney W and Sandberg J 1986 *Phys. Rev. Lett.* **56** 2797; Sauter Th, Neuhauser W, Blatt R and Toschek P E 1986 *Phys. Rev. Lett.* **57** 1696; Bergquist J C, Hulet R G, Itano W M and Wineland D J 1986 *Phys. Rev. Lett.* **57** 1699; Peik E, Hollemann G and Walther H 1994 *Phys. Rev.* A **49** 402; Zhao X, Yu N, Dehmelt H and Nagourney W 1995 *Phys. Rev.* A **51** 4483
[4] Lax P D and Phillips R S 1967 *Scattering Theory* 1st Edition (New York: Academic Press)
[5] Bohm A, Kaldass H and Wickramasekara S 2003 *Fortsch. Phys.* **51** No 6 569-603, 604-634
[6] Lippmann B A and Schwinger J 1950 *Phys. Rev.* **79** 469480
[7] The same convention is also used in Bohm A 1986, 1993, 2001 *Quantum Mechanics* (Springer-Verlag) Chapter XV and XXI
[8] Gamow G 1928 *Z. Phys.* **51** 204-12
[9] Breit G and Wigner E P 1936 *Phys. Rev.* **49** 519
[10] Bohm A, Kielanoski P and Wickramasekara S 2006 *Ann. Phys.* **32** 2299

[11] Strauss Y, Horwitz L P and Eisenberg E 2000 *J. Math. Phys.* **41** No. 12 8050; Strauss Y 2003 *Int. J. Theor. Phys.* **42** 2285

[12] Gelfand I M and Vilenkin G F 1967 *Generalised Functions* Vol IV (New York: Accademic Press) Chapter 1

[13] Paley R and Wiener N 1934 *Fourier Transform in the Complex Domain* (New York: American Mathematical Society)

[14] Oates C W, Vogel K R and Hall J L 1996 *Phys. Rev. Lett.* **76** 2866; Volz U, Majerus M, Liebel H, Schmitt A and Schmoranzer A 1996 *Phys. Rev. Lett.* **76** 2862

Spin–orbit coupling in low-symmetry systems

L L Boyle
Vine Cottage, Patrixbourne, Kent CT4 5BX, UK

E-mail: Laurence.Boyle@gmail.com

Abstract. Some of the complications which arise in labelling angular-momentum states in low-symmetry systems are discussed. In particular the assignment of irreducible representation labels for spin–orbit coupled states can be quite tricky because of the existence of automorphisms of the symmetry group of the system which affect the labelling of the representations if the definitions of the symmetry elements are changed. A detailed study is made for molecules of C_4 symmetry as the automorphism group structure is particularly rich especially when the number of electrons is odd.

1. Introduction

In high-symmetry cases quantum numbers can be correlated with labels of irreducible representations. In the ideal case this correlation derives from the eigenvalues of one or more operators. When two states correspond to the same irreducible representation, however, an additional quantum number is used which indicates the position of the state in the energetical sequence. Such subsidiary quantum numbers are well-known and sometimes useful. A notable example occurs in atomic systems in the presence of a linear magnetic field since component operators can be defined which have eigenvalues which can be used to define subsidiary quantum numbers. In the presence of an electric, molecular or crystal field, however, the situation becomes more interesting.

Low symmetries have been thought by some to be of little interest because of the small number of irreducible representations of the corresponding symmetry groups. However, in some cases this leads to interesting consequences for the definition of the states (as defined by irreducible representations) in terms of quantum numbers. To illustrate the problem the well-known case of spin–orbit coupling in atomic [1,2] and molecular systems [3] will be examined.

In the high-symmetry case of atoms, orbital states and even-electron spin states are described by irreducible representations of $O(3)^+$ while odd-electron spin states are described by irreducible representations of SU(2). Spin–orbit coupled states derive from weighted sums of products of spin states and orbital states and are constructed in the context of $O(3)^+$ for even-electron atoms and SU(2) for odd-electron atoms.

For atoms an orbital quantum number, L, is defined by the dimensionality of the orbital representation, 2L+1, while the spin quantum number, S, is defined by the dimensionality, 2S+1,of the spin representation. Multiplication of the spin and orbital irreducible representations leads in general to a reducible representation. When this is reduced to a sum of irreducible representations one obtains the spin–orbit quantum number, J, from the dimensionality, 2J+1, of each component irreducible representation. J turns out to be quantised (since consecutive values differ by 1 unit) and bounded by the limits $L+S \geq J \geq |L-S|$. In low-symmetry systems, however, the (2J+1)-fold degeneracy is lifted

Journal of Physics: Conference Series **597** (2015) 012019 doi:10.1088/1742-6596/597/1/012019

and it is the purpose of this contribution to assess the extent to which this lifting of the degeneracy has any effect on the physical properties of the states.

2. Orbital states

One well-known example which illustrates the effect of the lifting of the degeneracy is that of the use of magnetic quantum numbers. These are really only relevant to the properties of orbital and spin angular momenta in a linear magnetic field which can be internal or external. In such a scenario there is a unique direction, z, related to the field and the orbital magnetic quantum numbers, L_z, are related to the characters for the smallest rotation about that direction. In the case of an atom, $L \geq L_z \geq -L$, and, by analogy, $S \geq S_z \geq -S$, while for a linear molecule $L_z = \pm L$ and $S_z = \pm S$.

The point group C_4 will be chosen as a suitable example of a low-symmetry system. The character table for this group may be written

C_4	**E**	$\mathbf{A}^2$	**A**	$\mathbf{A}^3$
Γ_1	1	1	1	1
Γ_2	1	1	−1	−1
Γ_3	1	−1	i	−i
Γ_3*	1	−1	−i	i

where the symmetry elements are written at the tops of the columns and the representation labels at the left-hand side of the rows of characters. The generator **A** is a rotation operator such that $\mathbf{A}\phi = \phi + \frac{1}{2}\pi$ where ϕ is an angle expressed in radians. The effect of **A** on the wave function $\psi = \exp(iL_z\phi)$ is to produce $\mathbf{A}\psi = \exp\{iL_z(\phi + \frac{1}{2}\pi)\}$. This can be factorised as $\mathbf{A}\psi = \exp(i\pi L_z/2)\,\psi$ so that the eigenvalue, $\exp(i\pi L_z/2)$, which is the imaginary i raised to the power L_z, defines the character for the single generating element **A** and hence the irreducible representation spanned by ψ is identified. The result is that when $|L_z|\text{mod}4 = 0$ or 2 the representation is Γ_1 or Γ_2 respectively. When $|L_z|\text{mod}4 = 1$ or 3 the representation is Γ_3 or Γ_3^* respectively if $L_z > 0$ and Γ_3^* or Γ_3 respectively if $L_z < 0$.

3. Spin states

The discussion of the effect of a molecular field on spin states differs from the above only when the number of electrons is odd. Then the double group, which is a subgroup of SU(2) must be used. In the case of C_4 the double group is isomorphic to the cyclic group C_8 for which the character table is:-

	C_8	**E**	$\mathbf{A}^4$	$\mathbf{A}^2$	$\mathbf{A}^6$	**A**	$\mathbf{A}^5$	$\mathbf{A}^3$	$\mathbf{A}^7$
C_4		**E**		$\mathbf{A}^2$		**A**		$\mathbf{A}^3$	
Γ_1	Γ_1	1	1	1	1	1	1	1	1
Γ_2	Γ_2	1	1	1	1	−1	−1	−1	−1
Γ_3	Γ_3	1	1	−1	−1	i	i	−i	−i
Γ_3*	Γ_3*	1	1	−1	−1	−i	−i	i	i
	Γ_4	1	−1	i	−i	$\sqrt{i}$	$-\sqrt{i}$	$i\sqrt{i}$	$-i\sqrt{i}$
	Γ_4*	1	−1	−i	i	$-i\sqrt{i}$	$i\sqrt{i}$	$-\sqrt{i}$	$\sqrt{i}$
	Γ_5	1	−1	i	−i	$-\sqrt{i}$	$\sqrt{i}$	$-i\sqrt{i}$	$i\sqrt{i}$
	Γ_5*	1	−1	−i	i	$i\sqrt{i}$	$-i\sqrt{i}$	$\sqrt{i}$	$-\sqrt{i}$

In the top line of this table the generator **A** is now of order 8 and the correlation of elements with those of the single group C_4 discussed in the preceding section is effected by arranging the columns so that the single group is effectively embedded in the double group.

For odd-electron spin states the only relevant irreducible representations are the lower four in the table, *i. e.* Γ_4, Γ_4*, Γ_5 and Γ_5*, which in this context are often referred to as double-valued representations. They can be characterised by their characters under the generating element **A** of C_8

which are respectively the first, seventh, fifth and third powers of $\sqrt{i}$ which is $(i + 1)/\sqrt{2} = \exp(i\pi/2)$. When $2|S_z|\text{mod}8 = 1$ or 7 the representation is Γ_4 or Γ_4^* respectively if $S_z > 0$ and respectively Γ_3^* or Γ_3 if $S_z < 0$. When $2|S_z|\text{mod}8 = 3$ or 5 the representation is Γ_5^* or Γ_5 respectively if $S_z > 0$ and Γ_5 or Γ_5^* respectively if $S_z < 0$. For even-electron spin states the results are identical to those for L_z states of the same numerical value as S_z.

4. Spin-orbit coupled states

The spin–orbit coupled states are characterised by a quantum number J_z which is the sum of L_z and S_z. The rules for assigning irreducible representation labels to J_z should parallel those for the corresponding determination for S_z.

In the model case of atomic symmetry the spacing pattern of the spin–orbit coupled states is obtained by assuming that the spin–orbit coupling term in the Hamiltonian operator is $\zeta\mathbf{L}\cdot\mathbf{S}$ where ζ is a spin–orbit coupling constant and $\mathbf{L}\cdot\mathbf{S}$ is the scalar product of the axial-vector angular-momentum operators **L** and **S**. To evaluate this contribution to the energy one expands $\mathbf{L}\cdot\mathbf{S}$ as $\frac{1}{2}(\mathbf{J}^2 - \mathbf{L}^2 - \mathbf{S}^2)$ and then by assuming that the eigenvalues of $\mathbf{J}^2$, $\mathbf{L}^2$ and $\mathbf{S}^2$ are respectively J(J+1), L(L+1) and S(S+1) the contribution of the term $\zeta\mathbf{L}\cdot\mathbf{S}$ to the energy of the atomic state is $\frac{1}{2}\zeta\{J(J+1) - L(L+1) - S(S+1)\}$. Since the energy of a set of states is usually largely dependent on L and S, the spacing pattern is determined by the J-values of the states. The number of levels in the pattern will be 2L+1 if $L \leq S$ or 2S+1 if $S \leq L$. The spacings between energetically-adjacent states will form an arithmetic progression from which information can be deduced leading to the identification of L, S and J for each state. Identification of a J_z-value requires the states to be split using a small external linear magnetic field.

In linear molecules the external field is essentially present as an internal field and two types of field need to be distinguished. A purely electric field retains the degeneracy of complex-conjugate pairs of states. These would be states characterised by ± pairs of J_z-values. Such degeneracies would be lifted by magnetic fields.

In the low-symmetry case of molecules of point-group symmetry C_4 the physical effect of spin–orbit coupling is such that states with values of J differing by an integral multiple of 4 have the same symmetry and so can interact. The evidence for this interaction is derived from spectroscopy. This interaction of states with the same symmetry is known as quenching since the mixing of states of lower J_z-value with the states of higher J_z-value reduces the interaction with the magnetic field.

5. The effect of automorphisms

In all cases where representation labels are being assigned to physical states one must examine the automorphism group of the symmetry group to see what additional information needs to be specified to fix a label [4]. In the case of the cyclic group C_4 discussed in the preceding section, the only automorphism is that which interchanges the rotation **A** with its inverse $\mathbf{A}^{-1} = \mathbf{A}^3$. The choice of the sense of direction of the angle ϕ is purely one of convention and is difficult to relate to the macroscopic choice of the north or south pole of a magnet. Inevitably this means that different authors have used different conventions and so the labelling of a state as Γ_3 or Γ_3^* needs to be interpreted with caution and consistency.

The double group of C_4 is particularly interesting from the point of view of its automorphisms.

A	$\mathbf{A}^7$	$\mathbf{A}^5$	$\mathbf{A}^3$
Γ_3	Γ_3*	Γ_3	Γ_3*
Γ_3*	Γ_3	Γ_3*	Γ_3
Γ_4	Γ_4*	Γ_5	Γ_5*
Γ_4*	Γ_4	Γ_5*	Γ_5
Γ_5	Γ_5*	Γ_4	Γ_4*
Γ_5*	Γ_5	Γ_4*	Γ_4

These only affect the single-valued representations Γ_3 and Γ_3* and the four double-valued representations. The columns of the table above show the four possible cases. These are characterised by the power of the generating element **A** to which **A** is transformed by the automorphism. The first column gives the irreducible representation which will be changed and the subsequent columns give the transformed sets. The automorphism $\mathbf{A}\rightarrow\mathbf{A}^7$ produces complex conjugation while the automorphism $\mathbf{A}\rightarrow\mathbf{A}^5$ only affects the double-valued representations and can be regarded as the second generator of the automorphism group which is isomorphic to Klein's *Vierergruppe*.

The effect on the J_z quantum numbers is non-trivial. Replacing **A** by $\mathbf{A}^3$, for instance, would relabel a Γ_4 ($J_z = {}^1\!/_2$ or $-{}^7\!/_2$) state as a Γ_5* ($J_z = {}^3\!/_2$ or $-{}^5\!/_2$). In fact the four possible values, (${}^3\!/_2$, ${}^1\!/_2$, $-{}^1\!/_2$, $-{}^3\!/_2$) of J_z for a $J = {}^3\!/_2$ state, which would be degenerate in the parent atomic symmetry, would all have separate symmetry labels and the relationship between the label to be chosen by convention and the explicit form of an electronic wave function for each would depend on the model chosen to describe the Hamiltonian operator and its eigenfunctions for the system.

6. Conclusion

The aim of this contribution is to indicate how the study of the automorphisms of a group in certain systems of low symmetry is important for establishing conventions for the labelling of electronic states. Due to the existence of automorphisms which have no effect on the labelling of states in symmetries which can be regarded as supergroups of the low-symmetry groups discussed, these problems have not been explicitly addressed hitherto.

Spin–orbit coupling was chosen as one well-known example which would illustrate the relevant points. Couplings involving rotational or vibrational angular momenta would also serve as interesting examples for future work.

References

[1] Herzberg G 1936 *Atomspektren und Atomstruktur: eine Einführung für Chemiker, Physiker und Physikochemiker* (Dresden: Theodor Steinkopff)

[2] Wybourne B G 1965 *Spectroscopic Properties of the Rare Earths* (New York: Interscience)

[3] Herzberg G 1966 *Electronic Spectra and Electronic Structure of Polyatomic Molecules* (New York: Van Nostrand Reinhold)

[4] Boyle L L 2014 *J. Phys.: Conf. Ser.* **512** 012019

Relativistic resonance and decay phenomena

Hai V Bui

Department of Physics, University of Texas at Austin, Austin TX 78712, USA

E-mail: `viethai@physics.utexas.edu`

Abstract. The exact relation $\tau = \hbar/\Gamma$ between the width Γ of a resonance and the lifetime τ for the decay of this resonance could not be obtained in standard quantum theory based on the Hilbert space or Schwartz space axiom in non-relativistic physics as well as in the relativistic regime. In order to obtain the exact relation, one has to modify the Hilbert space axiom or the Schwartz space axiom and choose new boundary conditions based on the Hardy space axioms in which the space of the states and the space of the observables are described by two different Hardy spaces. As consequences of the new Hardy space axioms, one obtains, instead of the symmetric time evolution for the states and the observables, asymmetrical time evolutions for the states and observables which are described by two semi-groups. A relativistic resonance obeying the exponential time evolution can be described by a relativistic Gamow vector, which is defined as superposition of the exact out-plane wave states with a Breit-Wigner energy distribution of the width Γ.

1. Introduction

Resonance and decaying processes have been studied for a long time in scattering experiments. Although resonance and decaying state are studied in two difference processes, resonance in the scattering process and decaying state in the decay process, they are believed to be two aspects of the same phenomenon. It means that the lifetime of the decaying state is mathematically inverse width of the resonance. This belief came first in non-relativistic scattering theory [1] using the Hilbert space axiom for which an approximate relation between lifetime of the decaying state and the width of resonance was obtained. Khalfin [2] showed that in the standard quantum theory (based on Hilbert space axiom), there exists no vector with the exponential time evolution, i.e. it is impossible to obtain the exact relation between the width of the resonance and the lifetime of the decaying state in the quantum theory using the Hilbert space axiom. Such a vector, called Gamow vector, with an exponential time evolution property was introduced by Gamow [3]. However, the Gamow vector has not been favored by physicists since there are no complex eigenvalues for a self-adjoint Hamiltonian H of physical system in the conventional quantum physics.

The time evolution of the solutions of the dynamical equations (Schrödinger equation for the state in the Schrödinger picture or Heisenberg equation for the observable in the Heisenberg picture) extends over $-\infty < t < \infty$. This time-symmetric evolution is a result of the famous mathematical theorem (Stone-von Neumann theorem [4]) for the Hilbert space. The time-symmetric evolution of the observable (in the Heisenberg picture) or the state (in the Schrödinger picture) leads to the result, that the Born probability to detect the observable in the state exists for all time, $-\infty < t < \infty$. This means that the detectors can detect the observables (or products of decay) for all the time, even before the apparatus are turned on (e.g. an accelerator

Journal of Physics: Conference Series **597** (2015) 012020 doi:10.1088/1742-6596/597/1/012020

in the scattering experiment). This is in contrast to the causality condition for the scattering experiment since the detector can not detect anything until the experiment is prepared and the apparatus (e.g. accelerator) are turned on.

In order to obtain the unified theory of resonances and decaying states, one has to modify the Hilbert space and also the Schwartz space boundary condition for the space of the states and the space of the observables, and use a new boundary condition, called Hardy space axioms, which distinguishes between the space of the states and the space of the observables. The space of states and the space of observables are mathematically described by two different Hardy spaces, Hardy space on the lower and on the upper complex energy plane second sheet of the S-matrix, respectively. The result of the Hardy space axioms is that, one can get an exact relation between the width of the resonance and the lifetime of corresponding decaying state. Furthermore, one predicts that the probability to detect an observable in a state exists only after a finite time $t_0(=0)$ at which the state has been prepared. In relativistic regime, the states and the observables can be represented by the two different Poincaré semi-groups, one is the forward light cone semi-group for the observables and another is the backward light cone semi-group for the states.

2. Overview about Resonance and Decaying State

A resonance is defined as an intermediate state in scattering processes, e.g,:

$$a + b \rightarrow R \rightarrow c + d + \cdots \tag{1}$$

In the scattering process (1), two beams of particles, i.e, a and b, are accelerated through an accelerator and collide (for example, head-on collisions in electron-positron or proton-anti proton experiments), or just one beam of particles a is accelerated by the accelerator and hits the target b which is at rest. The resonance R is formed after the collisions and then decays into the products c, $d \cdots$ which are registered by detectors.

A resonance R is characterized by a mass M_R and a width Γ_R. The mass M_R and width Γ_R of the resonance R are experimentally measured by fitting the partial cross section $\sigma_j(W)$ to the Breit-Wigner line shape with a slowly varying background $B(W)$. The partial cross section $\sigma_j(W)$ is a function of the center-of-mass energy:

$$\sigma_j(W) \sim \left| \frac{r}{W - (M_R - i\Gamma_R/2)} + B(W) \right|^2 . \tag{2}$$

Here, W is the square root of the center-of-mass energy s, $W = \sqrt{s}$.

The width Γ_R of a resonance R can theoretically be related by the decay rate Γ of resonance R to final states f using quantum field theory [5]:

$$\Gamma = \frac{1}{2m} \sum_f \int d\Pi_f \, |\mathcal{M}(R \rightarrow f)|^2 . \tag{3}$$

Here, $d\Pi_f$ is the relativistically invariant phase space of the final states f and $\mathcal{M}(R \rightarrow f)$ is a scattering amplitude of resonance R which decays into the final states f. The accuracy of the value of decay rate depends on the approximation used in calculation of scattering amplitude $\mathcal{M}(R \rightarrow f)$, i.e, the order in the Feynman diagrams used to calculate the matrix $\mathcal{M}(R \rightarrow f)$.

A decaying state D appears in decay process:

$$D \rightarrow \eta + \eta' \cdots \tag{4}$$

In the decay experiment, the decay products $\eta, \eta', \cdots$, are counted by detectors which may be constructed to observe some specific particles.

A decaying state D is characterized by a mass m_D and a lifetime τ_D. The lifetime τ_D of the decaying state D is experimentally obtained by fitting the counting rate $N_\eta/N(t)$ of the product η in the time interval Δt_i to the exponential law for a partial decay rate of product η, $R_\eta(t)$:

$$R_\eta(t) \approx \frac{1}{N}\frac{\Delta N_\eta(t_i)}{\Delta t_i} \sim e^{-\frac{t}{\tau}} . \tag{5}$$

Here, N is the total number of decay products of the decaying state D, i.e, it is sum of all partial decay product η, $\eta' \cdots$ which are counted by the detectors:

$$N = \sum_{\text{all } \eta \text{ decay products}} N_\eta . \tag{6}$$

The observed partial decay rate $\frac{1}{N}\frac{\Delta N_\eta(t_i)}{\Delta t_i}$ is fitted by (5) to the exponential time evolution $e^{-\frac{t}{\tau}}$. From this fit, the lifetime τ_D of the decaying state D is determined. Physicists believe that the resonance and decaying state are the same entities, i.e, they are two different aspects of the same phenomenon and the lifetime τ can be calculated as inverse width Γ, and vice versa:

$$\tau = \frac{\hbar}{\Gamma} . \tag{7}$$

The relation (7) between the lifetime τ and the width Γ was first obtained in non-relativistic quantum physics in the zero order of probability to detect the observable ψ in the state $\phi(t)$ (by Goldberger and Watson) using Weisskopf-Wigner methods [1]:

$$\begin{aligned} |\langle\psi|\phi(t)\rangle|^2 = \mathcal{P}_{\phi(t)}(\psi) &\sim e^{-\frac{\Gamma t}{\hbar}}(1 + \tfrac{\Gamma t}{2} + \tfrac{3\Gamma^2 t^2}{28} + \tfrac{\Gamma^3 t^3}{84} + ...) \\ &\sim e^{-\frac{\Gamma t}{\hbar}} = e^{-\frac{t}{\tau}} \quad \text{where} \quad \tau \approx \hbar/\Gamma . \end{aligned} \tag{8}$$

There is a problem in the standard (Hilbert space) quantum theory that the exact relation between the lifetime τ and the width Γ in (7) can not be obtained because the exponential time evolution is not possible in the Hilbert space (Khalfin's theorem) [2]. In contrast to this theorem based on the Hilbert space, experimental data show that the decay probability of quantum systems is in good agreement with the exponential law, if the decaying state can be isolated from the background, e.g, in K^0_S-decay experiment in the relativistic regime [6].

Another problem in standard quantum theory is that Born's probability to detect an observable $\Lambda(t) = |\psi(t)\rangle\langle\psi(t)|$ in the state ϕ:

$$\mathcal{P}_\phi(\Lambda(t)) = \text{Tr}(|\psi(t)\rangle\langle\psi(t)||\phi\rangle\langle\phi|) = |\langle\psi(t)|\phi\rangle|^2 = |\langle e^{-iHt}\psi(0)|\phi\rangle|^2, \quad -\infty < t < +\infty \tag{9}$$

is theoretically predicted for all time $-\infty < t < +\infty$. This prediction (9) is a mathematical consequence (by Stone-von Neumann theorem [4]) of the Hilbert space boundary condition for the solutions of the Schrödinger equation for the state ϕ or of the Heisenberg equation for the observable. However, the prediction (9) violates the causality, which states that a state ϕ must be prepared first (e.g, by an accelerator) before the observable $\psi(t)$ (for decay products) can be measured (e.g, by detectors) in the scattering experiment.

3. Hardy space boundary condition for the quantum system

The causality principle suggests that in the scattering experiment, the detectors are only able to detect the observables (e.g. decay products) after the state has been prepared (e.g. after the accelerator is turned on and the beam has hit the fixed target, or two accelerated beams are guided to make head-on collision). This means that the Born probability to detect an observable

$\Lambda(t) = |\psi(t)\rangle\langle\psi(t)|$ in the state ϕ can *experimentally* not be for all time $-\infty < t < +\infty$, but only for the time t after the finite time t_0 at which the state has been prepared, i.e, for $t \geq t_0$. The experimental Born probability to detect an observable $\psi(t)$ in the state ϕ:

$$\mathcal{P}_\phi(\psi(t))) = |\langle\psi(t)|\phi\rangle|^2 = |\langle e^{-iH(t-t_0)}\psi(t_0)|\phi\rangle|^2, \tag{10}$$

exits only for the time t after the time t_0, $t_0 \leq t < +\infty$. The finite time t_0 can mathematically be set as zero, $t_0 = 0$, then the probability (10) can be rewritten as

$$\mathcal{P}_\phi(\psi(t))) = |\langle\psi(t)|\phi\rangle|^2 = |\langle e^{-iHt}\psi|\phi\rangle|^2, \quad \text{for the time } 0 \leq t < +\infty. \tag{11}$$

From (11), it is clear that the time evolution of the observable $\psi(t)$ can not be described by the time-symmetric (unitary) group which is a consequence of the Stone-von Neumann theorem for the solutions of the Heisenberg equation using the Hilbert space or the Schwartz space axiom. Hence, the time evolution needs to be given by the semi-group with the finite starting time t_0 (which can be chosen as $t_0 = 0$):

$$\psi(t) = e^{-iHt}\psi, \quad \text{for the time } 0 \leq t < +\infty. \tag{12}$$

Theoretically, this semi-group time evolution (12) can not be obtained from the conventional quantum mechanics using the Hilbert space axiom or the Schwartz space axiom. But the semi-group time evolution like (12) can be obtained using the Hardy space axioms [7].
The Hardy space axiom asserts that:

The space of in-states $\{\phi^+\}$ is the Hardy space Φ_- of the lower complex s-plane $\mathbb{C}_-$ on second sheet of S-matrix:

$$\{\phi^+\} = \Phi_- \subset \mathcal{H} \subset \Phi_-^\times. \tag{13a}$$

Space of observables $\{\psi^-\}$ is the Hardy space Φ_+ of the upper complex s-plane $\mathbb{C}_+$ on second sheet of S-matrix:

$$\{\psi^-\} = \Phi_+ \subset \mathcal{H} \subset \Phi_+^\times. \tag{13b}$$

Here, we use the superscriptions $^+$ in ϕ^+ to describe the prepared in-states and $^-$ in ψ^- to describe the registered observables. But the subscriptions $_-$ and $_+$ in in $\Phi_\mp$ to describe the Hardy space of the lower and upper complex s-plane $\mathbb{C}_\mp$ on the second sheet of S-matrix, respectively.

These Hardy spaces $\Phi_\pm$ in (13) are mathematically represented by the smooth Hardy functions rather than the L^2-integrable functions of the invariance mass $s = p^\mu p_\mu$ used in the Hilbert space. The Hardy space axiom requires that the functions of $s = p^\mu p_\mu$ in the upper and lower half-plane of the second sheet of the s-surface are smooth Hardy functions [8]:

$$\text{for the observable } \psi^- \in \Phi^+ \iff \psi^-(s) = \langle^-\hat{\boldsymbol{p}}\, j_3[sj], \eta|\psi^-\rangle \in \tilde{\mathcal{S}} \cap \mathcal{H}_+^2|_{\mathbb{R}_{s_0}} \otimes \tilde{\mathcal{S}}(\mathbb{R}^3), \tag{14a}$$

$$\text{for the in-state } \phi^+ \in \Phi^- \iff \phi^+(s) = \langle^+\hat{\boldsymbol{p}}\, j_3[sj], \eta|\phi^+\rangle \in \tilde{\mathcal{S}} \cap \mathcal{H}_-^2|_{\mathbb{R}_{s_0}} \otimes \tilde{\mathcal{S}}(\mathbb{R}^3). \tag{14b}$$

Here, the space $\tilde{\mathcal{S}} \cap \mathcal{H}_\pm^2|_{\mathbb{R}_{s_0}}$ are not the spaces of L^2-integrable energy wave functions but are the intersection between the Schwartz function space $\tilde{\mathcal{S}}$ and Hardy class function $\mathcal{H}_\pm^2$ of the lower and upper complex energy plane (for the analytic S-matrix), respectively. $\mathbb{R}_{s_0}$ means that the value of $s = p_\mu p^\mu$ is restricted to the physical values, $s_0 = (m_1 + m_2 + \cdots + m_n)^2 < s < \infty$. The space $\tilde{\mathcal{S}}(\mathbb{R}^3)$ is the Schwartz function space of the space-components of the 4-velocity vector $\hat{\boldsymbol{p}} = \boldsymbol{p}/s$.

4. Representations of Relativistic Resonance and Decaying State

A relativistic particle of the invariant mass square $s = m_p^2$ and spin j_p is described by the unitary irreducible representation (UIR) of the Poincaré group $\mathcal{P}$ labeled by invariant mass squared $s = m_p^2$ and spin j_p, (s, j_p) and other quantum numbers η_p (species quantum number, isospin, lepton number etc) [9]:

$$\mathcal{P} = \{(\Lambda, x) \mid \Lambda \in \overline{SO(3,1)},\ \det\Lambda = +1,\ \Lambda_0^0 \geq 0\}\,. \tag{15}$$

Instead of using the momentum eigenvectors $|\, \boldsymbol{p}\, j_3[sj], \eta\rangle$ as basis vectors of the irreducible representation space of [s,j], one can as well use the three space-components of the 4-velocity $\widehat{\boldsymbol{p}} = \boldsymbol{p}/s$ and label the eigenvectors as $|\widehat{\boldsymbol{p}}\, j_3[s\, j], \eta\rangle$. These labels can also be used for the basis vectors of semi-group which will be discussed later. The 4-velocity kets $|\,\widehat{\boldsymbol{p}}\, j_3[sj], \eta\rangle$ are the eigenkets of: A conjugate operator $M^\times$ of the mass operator $M = (P_\mu P^\mu)^{1/2}$ with eigenvalue $\sqrt{s}$, a conjugate operator $\widehat{P}_\mu^\times$ of the 4-velocity operator $\widehat{P}_\mu = P_\mu M^{-1}$ with eigenvalue $\widehat{p}_\mu$, a total angular momentum operator (spin) j where the eigenvalue of $-\widehat{w}^2 = -\widehat{w}^\mu \widehat{w}_\mu$ is $j(j+1)$, and a third-component of the total angular momentum operator $U(L(\widehat{p}))\widehat{w}_3 U^{-1}(L(\widehat{p}))$ with eigenvalue j_3 [10]

$$\begin{aligned}
&M^\times\, |\widehat{\boldsymbol{p}}\, j_3[sj], \eta\rangle = \sqrt{s}\, |\widehat{\boldsymbol{p}}\, j_3[sj], \eta\rangle\,, \\
&\widehat{P}_\mu^\times\, |\widehat{\boldsymbol{p}}\, j_3[sj], \eta\rangle = \widehat{p}_\mu\, |\widehat{\boldsymbol{p}}\, j_3[sj], \eta\rangle\,, \\
&-\widehat{w}^2\, |\widehat{\boldsymbol{p}}\, j_3[sj], \eta\rangle = j(j+1)\, |\widehat{\boldsymbol{p}}\, j_3[sj], \eta\rangle\,, \\
&U(L(\widehat{p}))\widehat{w}_3 U^{-1}(L(\widehat{p}))\, |\widehat{\boldsymbol{p}}\, j_3[sj], \eta\rangle = j_3\, |\widehat{\boldsymbol{p}}\, j_3[sj], \eta\rangle\,.
\end{aligned} \tag{16}$$

Here $s = p_\mu p^\mu$ is invariant mass squared, $\widehat{p}_\mu = p_\mu/\sqrt{s}$ is 4-velocity with $\widehat{p}^0 = \widehat{E}(\widehat{\boldsymbol{p}}) = \sqrt{1 + \widehat{\boldsymbol{p}}^2}$, and space-components $\widehat{\boldsymbol{p}} \in \mathcal{R}^3$, the Pauli-Lubanski operator is $\widehat{w} = \epsilon^{\mu\nu\rho\sigma}\, \widehat{P}_\nu\, J_{\rho\sigma}$, and η is particle species number. Both operator $\widehat{P}_\mu^\times$ and $M^\times$ are conjugate operators of $\widehat{P}_\mu$ and M, respectively, in the dual Schwartz space $\Phi^\times$ which are unique extensions of $\widehat{P}_\mu^\dagger$ and $M^\dagger$ in the Hilbert space $\mathcal{H}$ to the dual Schwartz space $\Phi^\times$ (continuous antilinear functionals) of the Schwartz space Φ..

The 4-velocity kets $|\widehat{\boldsymbol{p}}\, j_3[sj], \eta\rangle$ are elements of the dual space $\Phi^\times$. The Schwartz space Φ is a dense subspace of Hilbert space $\mathcal{H}$ and then one obtains the Schwartz space triplet (or Rigged Hilbert Space (RHS)): $\Phi \subset \mathcal{H} \subset \Phi^\times$.

The normalization of free 4-velocity eigenkets is chosen as

$$\langle \widehat{\boldsymbol{p}}\, j_3[sj], \eta\, |\widehat{\boldsymbol{p}}'\, j_3'[s'j'], \eta'\rangle = 2\,\widehat{E}(\widehat{\boldsymbol{p}})\, \delta(\widehat{\boldsymbol{p}} - \widehat{\boldsymbol{p}}')\, \delta(s - s')\, \delta_{j_3 j_3'}\, \delta_{jj'} \delta_{\eta,\eta'}\,, \tag{17}$$

where $\widehat{E}(\widehat{\boldsymbol{p}}) \equiv E(\boldsymbol{p})/\sqrt{s} = \sqrt{1 + \widehat{\boldsymbol{p}}^2} = \gamma = \frac{1}{\sqrt{1-v^2}}$.

If the relativistic particle system consists of 2-particles, which each of which is represented by an unitary irreducible representation (UIR) of Poincaré group labeled by the mass squared s_i, spin j_i, $i = 1, 2$ (and by the other quantum numbers collectively denoted by quantum number η- (including particle species numbers)), the direct product space of the 2-particle system $\mathcal{H}_{12} = \mathcal{H}_1 \otimes \mathcal{H}_2$, can be reduced into a direct sum of two UIR spaces of different mass and spin, $\mathcal{H}_{12} = \mathcal{H}_1^\eta(s_1, j_1) \oplus \mathcal{H}_2^\eta(s_2, j_2)$ [10] [11].

In the center of mass system of two decay products of masses m_1 and m_2, any vector $\phi \in \Phi \subset \mathcal{H} \subset \Phi^\times$ can be expanded as a continuous linear superposition of the 4-velocity kets which is analogy to the Dirac's basis vector expansion for momentum kets (mathematically given by Nuclear Spectral Theorem for Rigged Hilbert Space (RHS) [7]):

$$\phi = \sum_{jj_3\eta} \int_{(m_1+m_2)^2}^{\infty} ds \int \frac{d^3\widehat{\boldsymbol{p}}}{2\widehat{p}^0}\, |\widehat{\boldsymbol{p}}\, j_3[sj], \eta\rangle \langle \widehat{\boldsymbol{p}}\, j_3[sj], \eta\, |\phi\rangle\,. \tag{18}$$

The interacting in- and out-states basis kets $|\,\widehat{\boldsymbol{p}}\,j_3\,[sj]\eta, n^{\pm}\rangle$ are obtained from the free-interaction basis kets by the Moeller wave operators $\Omega^{\pm}$ [12]:

$$|\widehat{\boldsymbol{p}}\,j_3[sj],\eta^{\pm}\rangle = \Omega^{\pm}\,|\widehat{\boldsymbol{p}}\,j_3[sj],\eta\rangle\,. \tag{19}$$

The in-state $|\widehat{\boldsymbol{p}}\,j_3[sj],\eta^{+}\rangle$ and out-state $|\widehat{\boldsymbol{p}}\,j_3[sj],\eta^{-}\rangle$ basis kets are elements of dual spaces: $|\,\widehat{\boldsymbol{p}}\,j_3[sj],\eta^{\pm}\rangle \in \Phi^{\times}_{\mp}$ of the Hardy spaces $\Phi_{\mp}$, respectively. By the Hardy space axiom of quantum theory, these Hardy spaces $\Phi_{\mp}$ contain the spaces of the prepared in-state ϕ^{+} and detected or observed "out-state" ψ^{-}: $\phi^{+} \in \Phi_{-}$ and $\psi^{-} \in \Phi_{+}$, respectively. We restrict the representation space of $[s,\ j]$ to the case in which 4-velocity $\widehat{p}_{\mu} = p_{\mu}/\sqrt{s}$ is real while the mass $\sqrt{s}$ and therewith 4-momentum p^{μ} are complex. This is a restriction of generality but may be sufficient to accommodate to resonance and decay phenomena. This is called "*minimally complex*" representation for the 4-velocity ket [10].

In the *minimally complex* representation, the kets $|\widehat{\boldsymbol{p}}\,j_3[sj],\eta^{\pm}\rangle$ are eigenkets of the conjugate operator $\widehat{P}^{\times}_{\mu}$ of the 4 velocity operator $\widehat{P}_{\mu} = P_{\mu}\,M^{-1}$ with real eigenvalue $\widehat{p}_{\mu} = p_{\mu}/m$ and the conjugate operator $M^{\times}$ of the mass operator $M = (P_{\mu}\,P^{\mu})^{1/2}$ with the complex eigenvalue $\sqrt{s}$, respectively:

$$\begin{aligned} M^{\times}\,|\widehat{\boldsymbol{p}}\,j_3[sj],\eta^{\pm}\rangle &= \sqrt{s}\,|\widehat{\boldsymbol{p}}\,j_3[sj],\eta^{\pm}\rangle\,, \\ \widehat{P}^{\times}_{\mu}\,|\widehat{\boldsymbol{p}}\,j_3[sj],\eta^{\pm}\rangle &= \widehat{p}_{\mu}\,|\widehat{\boldsymbol{p}}\,j_3[sj],\eta^{\pm}\rangle\,. \end{aligned} \tag{20}$$

The operator $M^{\times}$ and $\widehat{P}^{\times}_{\mu}$ are conjugate operators of M and $\widehat{P}_{\mu}$ in the dual spaces $\Phi^{\times}_{\pm}$ of the Hardy spaces $\Phi_{\pm}$, respectively, which are unique extensions of $M^{\dagger}$ and $\widehat{P}^{\dagger}_{\mu}$ in the Hilbert space $\mathcal{H}$ to the dual spaces $\Phi^{\times}_{\pm}$.

As a consequence of (20), the kets $|\,\widehat{\boldsymbol{p}}\,j_3[sj],\eta^{\pm}\rangle$ are also the eigenkets of the conjugate operators $(P^{\mu})^{\times}$ of the exact 4-momentum operator P^{μ} with the eigenvalue $\widehat{p}^{\mu}\sqrt{s}$:

$$(P^{\mu})^{\times}\,|\widehat{\boldsymbol{p}}\,j_3[sj],\eta^{\pm}\rangle = \widehat{p}^{\mu}\sqrt{s}\,|\widehat{\boldsymbol{p}}\,j_3[sj],\eta^{\pm}\rangle\,, \quad \text{where} \quad \widehat{p}^{\mu} = (\widehat{p}^{0},\widehat{\boldsymbol{p}}) = (\gamma,\gamma\boldsymbol{v})\,. \tag{21}$$

Working in the rest frame of the system ($\widehat{\boldsymbol{p}} = 0$), the interacting in- and out-kets $|\widehat{\boldsymbol{p}} = 0\,j_3\,[sj]\eta, n^{\pm}\rangle$ can be described as formal solutions of the Lippmann-Schwinger equation. Therefore, in analogy to the Lippmann-Schwinger equation in non-relativistic scattering, the "solution" of the Lippmann-Schwinger equation [13] in the relativistic case is

$$|\mathbf{0}\,j_3\,[sj]\eta, n^{\pm}\rangle = \left(1 + \frac{1}{s - H \pm i\epsilon}V\right)|\mathbf{0}\,j_3\,[sj]\eta, n\rangle\,. \tag{22}$$

The in- and out-kets in the lab frame can then be obtained from the rest frame kets by the rotation-free Lorentz "boost transformation, $\mathcal{U}(L(\widehat{p}))$:

$$|\widehat{\boldsymbol{p}}\,j_3[sj],\eta^{\pm}\rangle = (\mathcal{U}^{\dagger}(L(\widehat{p}))^{\times}\,|\mathbf{0}\,j_3\,[sj]\eta, n^{\pm}\rangle\,. \tag{23}$$

Here $L(\widehat{p})$ is the rotation-free Lorentz boost which is a function of the real 4-velocity $\widehat{p}^{\mu}$. The conjugate operator $(\mathcal{U}^{\dagger}(L(\widehat{p}))^{\times}$ acts on the dual space $\Phi^{\times}_{\pm}$ and $\mathcal{U}^{\dagger}$ is the restriction of the unitary operator $U^{\dagger}$ to the Hardy spaces $\Phi_{\pm}$.

Corresponding to the two distinct Hardy spaces $\Phi_{\pm}$ for the out-observables ψ^{-} and in-states ϕ^{+} (according to the Hardy space axiom (13)), there are two semi-groups of the Poincaré transformation. One is into the forward light cone for the Hardy space Φ_{+} of the upper complex s-plane $\mathbb{C}_{+}$ on second sheet of S-matrix [14]:

$$\mathcal{P}_{+} = \{(\Lambda, x)\mid \Lambda \in \overline{SO(3,1)},\ \det\Lambda = +1,\ \Lambda^{0}_{0} \geq 0,\ x^2 = t^2 - \mathbf{x} \geq 0, t \geq 0\}\,. \tag{24a}$$

Another semi-group of the Poincaré transformation is into the backward light cone for the Hardy space Φ_- of the lower complex s-plane $\mathbb{C}_+$ on second sheet of S-matrix:

$$\mathcal{P}_- = \{(\Lambda, x) \mid \Lambda \in \overline{SO(3,1)},\ \det\Lambda = +1,\ \Lambda_0^0 \geq 0,\ x^2 = t^2 - \mathbf{x} \geq 0, t \leq 0\}\,. \tag{24b}$$

The transformations of the 4-velocity kets $|\hat{\boldsymbol{p}}\, j_3[sj], \eta^\mp\rangle \in \Phi_\pm^\times$ under inhomogeneous Lorentz transformation (Λ, x) are given by the operators:

$$\mathcal{U}_+^\times(\Lambda, x)\ |\hat{\boldsymbol{p}}\, j_3[sj], \eta^-\rangle = e^{-ip.x} \sum_{j_3'} D^j_{j_3 j_3'}(W(\Lambda^{-1}, p))\ |\Lambda^- 1\hat{\boldsymbol{p}}\, j_3[sj], \eta^-\rangle \text{ for } t \geq 0, x^2 \geq 0\,, \tag{25a}$$

$$\mathcal{U}_-^\times(\Lambda, x)\ |\hat{\boldsymbol{p}}\, j_3[sj], \eta^+\rangle = e^{-ip.x} \sum_{j_3'} D^j_{j_3 j_3'}(W(\Lambda^{-1}, p))\ |\Lambda^- 1\hat{\boldsymbol{p}}\, j_3[sj], \eta^+\rangle \text{ for } t \leq 0, x^2 \geq 0\,. \tag{25b}$$

Here, the subscriptions $\pm$ of the operators $\mathcal{U}_\pm^\times$ represent the transformations of the 4-velocity kets $|\hat{\boldsymbol{p}}\, j_3[sj], \eta^\mp\rangle$ under the corresponding semi-groups of Poincaré transformation into the forward light cone (25a) and into the the backward light cone (25b), respectively.

5. Relativistic Gamow Vector and Time Asymmetric Evolution

The probability to detect the out-observable ψ^- in the in-state ϕ^+ can be given in term of reduced S-matrix element or j-th partial S-matrix $S_j^{\eta\eta'}(s)$:

$$\begin{aligned}(\psi^-, \phi^+) &= \textstyle\sum_j \int_{s_0}^\infty ds \sum_{j_3} \int \frac{d^3\hat{p}}{2\hat{p}^0}\ \langle\psi^-\ |\hat{p}\, j_3[sj]\eta^-\rangle S_j^{\eta\eta'}(s)\langle^+\hat{p}\, j_3[sj]\eta'\,|\phi^+\rangle \\ &= \textstyle\sum_j (\psi^-, \phi^+)_j\,.\end{aligned} \tag{26}$$

Here, $(\psi^-, \phi^+)_j$ is the j-th partial probability which responds to the j-th partial S-matrix $S_j^{\eta\eta'}(s)$:

$$(\psi^-, \phi^+)_= \int_{s_0}^\infty ds \sum_{j_3} \int \frac{d^3\hat{p}}{2\hat{p}^0}\ \langle\psi^-\ |\hat{p}\, j_3[sj]\eta^-\rangle S_j^{\eta\eta'}(s)\langle^+\hat{p}\, j_3[sj]\eta'\,|\phi^+\rangle\,. \tag{27}$$

The j-th partial S-matrix $S_j^{\eta\eta'}(s)$ in (26) or (27) is then connected to the scattering amplitude of the j-th partial wave $a_j(s)$ of the scattering process:

$$S_j^{\eta\eta'}(s) = \begin{cases} 2\,i\,a_j(s) + 1 & \text{for elastic scattering } \eta = \eta' \\ 2\,i\,a_j^\eta(s) & \text{for reaction from } \eta' \text{ into channel } \eta \end{cases} \tag{28}$$

In S-matrix theory [15], a resonance is defined as a first order pole on 2nd sheet of the j-th partial S-matrix $S_j(s)$ at the complex value s_R. For simplicity, we assume that there is only one first-order pole $s_R = M_R - i\Gamma/2$ of j-th partial S-matrix $S_j(s)$ which corresponds to one resonance in the scattering process (1). In other words, there is just one resonance created in the collisions in the scattering experiment. Then, we can expand the j-th partial S-matrix $S_j(s)$ as a series of the first-order pole $s_R = M_R - i\Gamma/2$:

$$S_j(s) = \frac{R}{s - s_R} + R_0 + R_1(s - s_R) + \cdots \tag{29}$$

Therefore, the j-th partial S-matrix of the probability can be expressed in terms of a contour integral C around the pole s_R of $S_j(s)$ and a background which does not depend on the pole s_R:

$$\begin{aligned}(\psi^-, \phi^+)_j &= \textstyle\int_{s_0}^\infty ds \sum_{j_3} \int \frac{d^3\hat{p}}{2\hat{p}^0}\ \langle\psi^-\ |\hat{p}\, j_3[sj]\eta^-\rangle S_j^{\eta\eta'}(s)\langle^+\hat{p}\, j_3[sj]\eta'\,|\phi^+\rangle \\ &= \textstyle\sum_{j_3} \int \frac{d^3\hat{p}}{2\hat{p}^0} \int_{s_0}^{-\infty_{II}} ds\, \langle\psi^-\ |\hat{p}\, j_3[sj]\eta^-\rangle S_{II}(s)\langle^+\hat{p}\, j_3[sj]\eta'\,|\phi^+\rangle + \\ &\qquad \textstyle + \sum_{j_3} \int \frac{d^3\hat{p}}{2\hat{p}^0} \oint_C ds\, \langle\psi^-\ |\hat{p}\, j_3[sj]\eta^-\rangle \frac{R}{s-s_R}\langle^+\hat{p}\, j_3[sj]\eta'\,|\phi^+\rangle \\ &= \textstyle\langle\psi^-\,|\phi^{bg}\rangle + \sum_{j_3} \int \frac{d^3\hat{p}}{2\hat{p}^0} \int_{-\infty_{II}}^\infty ds\, \langle\psi^-\ |\hat{p}\, j_3[sj]\eta^-\rangle \frac{R}{s-s_R}\langle^+\hat{p}\, j_3[sj]\eta'\,|\phi^+\rangle\,.\end{aligned} \tag{30}$$

Here, the term

$$|\phi^{bg}\rangle = \sum_{j_3} \int \frac{d^3\hat{p}}{2\hat{p}^0} \int_{s_0}^{-\infty_{II}} ds \ |\hat{p}\, j_3[sj]\eta^-\rangle S_{II}(s) \langle^+ \hat{p}\, j_3[sj]\eta'|\phi^+\rangle , \tag{31}$$

does not relate to the resonance pole s_R and therefore represents the non-resonant background term in the scattering experiment.

One can then define a relativistic Gamow ket (or 4-velocity eigenket at a pole $s = s_R$) as superposition of the exact "out-plane waves" $|\hat{p}\, j_3[s\, j]\eta^-\rangle$ with the Breit-Wigner energy distribution $\frac{1}{s-s_R}$ [16]:

$$\psi^G_{s_R} = |\hat{p}\, j_3[s_R\, j]\eta^-\rangle = \frac{i}{2\pi} \int_{-\infty_{II}}^{\infty} ds \ |\hat{p}\, j_3[sj]\eta^-\rangle \frac{1}{s - s_R} . \tag{32}$$

The relativistic Gamow kets $\psi^G_{s_R}$ are transformed by the Poincaré semi-group into the forward light cone $\boldsymbol{\mathcal{P}}_+$ (24a). Moreover, the relativistic Gamow kets are eigenvectors of the 4-momentum operators P^μ with complex eigenvalues $\sqrt{s_R}$:

$$P^\times_\mu \ |\hat{p}\, j_3[s_R\, j]\eta^-\rangle = \sqrt{s_R}\ \hat{p}_\mu \ |\hat{p}\, j_3[s_R\, j]\eta^-\rangle \quad \text{with} \quad \hat{p}_\mu = (\hat{p}^0, \hat{\mathbf{p}}) = (\gamma, \gamma\boldsymbol{v}) . \tag{33}$$

Since the relativistic Gamow kets $\psi^G_{s_R}$ are transformed under the Poincaré semi-group of forward light cone $\boldsymbol{\mathcal{P}}_+$, the space-time translation (I, x) of the Gamow ket $\psi^G_{s_R}(t)$ is then given by:

$$\begin{aligned} \mathcal{U}^\times_+(I,x)\ |\hat{\mathbf{p}}\, j_3[s_R j], \eta^-\rangle &= e^{-ip.x}\ |\hat{\mathbf{p}}\, j_3[s_R j], \eta^-\rangle \\ &= e^{-i\gamma\sqrt{s_R}(t - \boldsymbol{x}.\boldsymbol{v})}\ |\hat{\mathbf{p}}\, j_3[s_R j], \eta^-\rangle \quad \text{for} \quad t \geq 0, x^2 \geq 0 . \end{aligned} \tag{34}$$

Working in the rest frame $\hat{\boldsymbol{p}} = 0$, a time evolution of the relativistic Gamow kets $\psi^G_{s_R}(t')$ at time $t' = t\gamma$ (proper time) is given by

$$\begin{aligned} \psi^G_{s_R}(t') &= e^{-iH^\times t'/\hbar}\ |\hat{\mathbf{p}} = 0\, , j_3\, [s_R\, j]\eta^-\rangle = e^{-i\gamma\sqrt{s_R}\, t}\ |\hat{\mathbf{p}} = 0\, , j_3\, [s_R\, j]\eta^-\rangle \\ &= e^{-i\sqrt{s_R}\, t'}\ |\hat{\mathbf{p}} = 0\, , j_3\, [s_R\, j]\eta^-\rangle = e^{-iM_R t'/\hbar} e^{-(\Gamma_R/2)t'/\hbar}\ |\hat{\mathbf{p}} = 0\, , j_3\, [s_R\, j]\eta^-\rangle \end{aligned} \tag{35}$$

and is valid only for $t' \geq 0$. The probability to detect an observable $\psi^-(t)$ in the Gamow ket $\psi^G_{s_R}(t)$ in the rest frame is then given by

$$\begin{aligned} |\langle\psi^-(t)\, |\hat{\mathbf{p}} = 0\, , j_3\, [s_R\, j]\eta^-\rangle|^2 &= |\langle e^{iHt/\hbar}\psi^-\ |\hat{\mathbf{p}} = 0 j_3\, [s_R\, j]\eta^-\rangle|^2 \\ &= e^{-\Gamma_R\, t/\hbar}\ |\langle\psi^-\ |\hat{\mathbf{p}} = 0\, , j_3\, [s_R\, j]\eta^-\rangle|^2 , \end{aligned} \tag{36}$$

and only for $t \geq 0$ because of (34). Hence, we get for the Gamow state an exact exponential decay with a lifetime τ which is exactly given by the width $\Gamma = \Gamma_R$ of the corresponding scattering resonance:

$$\tau = \frac{\hbar}{\Gamma} . \tag{37}$$

Summarizing: The resonances in the relativistic scattering process or unstable, decaying states in relativistic regime are described by the relativistic Gamow kets $\psi^G_{s_R}$ defined in (32). These Gamow kets $\psi^G_{s_R}$ are associated with the first-order pole of the S-matrix at $s_R = M_R - i\Gamma_R/2$. As a result of (35), the time evolution of the relativistic Gamow kets $\psi^G_{s_R}$ is not given by an unitary group for which the time t extends for all time $-\infty < t < \infty$, but by the semi-group (34) for which the time t extends over $0 \leq t < \infty$.

Journal of Physics: Conference Series **597** (2015) 012020 doi:10.1088/1742-6596/597/1/012020

Acknowledgment

We gratefully appreciate the hospitality of the Ghent University, Belgium.

References

[1] Goldberger M and Watson K 1964 *Collision theory* (New York: Wiley)
[2] Khalfin L A 1958 *Sov. Phys. JETP* **6** 1053-63
[3] Gamow G 1928 *Z. Phys.* **51** 204-12
[4] Stone M H 1932 *Ann. Math.* **33** No.3 643; Neumann V J 1932 *Ann. Math.* **33** No.3 567
[5] Peskin M E and Schroeder D V 1995 *An Introduction to Quantum Field Theory* (Boston: Addison Wesley)
[6] Gibbon L K 1997 *Phys. Rev. D* **55** 6625
[7] Bohm A and Gadella M 1989 *Dirac Kets, Gamow Vectors and Gelfand triplet* Lecture Note in Physics **348** (Berlin: Springer)
[8] Bohm A, Kaldass H and Wickramasekara S 2003 *Fortschr. Phys.* **51** 569
[9] Wigner E P 1939 *Ann. Math* **40** 149; Bargmann V and Wigner E P 1984 *Proc. Natl. Acad. Sci. USA* **34** 211
[10] Bohm A and Kaldass H 1999 *Phys. Rev.* A **60** 4606
[11] Joos H 1962 *Fortschr. Physik* **10** 65; Macfarlane A J 1962 *Rev. Mod. Phys.* **34** 41
[12] Weinberg S 1995 *The Quantum Theory of Fields* **1** (Cambridge: Cambridge University Press)
[13] Lippmann B A and Schwinger J 1950 *Phys. Rev.* **79** 469480
[14] Bohm A, Kaldass H and Wickramasekara S 2003 *Fortschr. Phys.* **51** 604
[15] Eden R J, Landshoff P V, Olive D I and Polkinghorne J C 1966 *The Analytic S-matrix* (Cambridge: Cambridge University Press)
[16] Bohm A and Yoshihiro S 2005 *Phys. Rev. D* **71** 085018

Generalized harmonic functions and unitary representations of low dimensional Lie groups

R Campoamor-Stursberg [1] **and M Rausch de Traubenberg** [2]
[1]I.M.I-U.C.M, Plaza de Ciencias 3, E-28040 Madrid, Spain
[2]IPHC-DRS, UdS, CNRS, IN2P3; 23 rue du Loess, Strasbourg, 67037 Cedex, France
E-mail: [1]`rutwig@ucm.es`, [2]`Michel.Rausch@iphc.cnrs.fr`

Abstract. The unitary representations of the three dimensional simple Lie groups are reconsidered from the perspective of harmonic functions acting on certain manifolds related to differential realisations of the groups themselves. By means of contractions of Lie groups, the procedure is also applied to the group E_2 of rotations-translations in two dimensions.

1. Introduction

Besides their fundamental importance in Analysis and Geometry, harmonic functions constitute an indispensable tool for many branches of Classical Physics, specially in connection with potential theory, providing an appropriate formalism to describe electrostatic, magnetic or gravitational potentials, as well as in Quantum physics, where harmonic functions play a relevant role in the operators techniques and the symmetry analysis of quantum systems, hence for the representation theory of these groups [1, 2, 3, 4, 5, 6, 7, 8, 9, 10]. Their generalisation to higher dimensions is a natural consequence of these applications, particularly in the context of relativistic systems [2, 11, 12, 13, 14]. As representative examples of this connection we have the relationship between unitary representations of pseudo-orthogonal groups $SO(p,q)$ and harmonic functions, *i.e.* eigenfunctions of the Laplacian, on the $(n-1)-$sphere $\mathbb{S}^{n-1}$ or the hyperboloids (see [15] for references). Harmonic functions have also been considered for other types of simple Lie groups, as $SU(3)$ in the context of elementary particles [16, 17] or the group $Sp(6,\mathbb{R})$ in the quantum mechanics of three bodies [18].

In this work we propose an alternative construction of the representations of the three dimensional semisimple Lie groups (complex and real) from the point of view of harmonic functions. To proceed, we identify $SL(2,\mathbb{C})$ (resp. $SL(2,\mathbb{R}), SU(2)$) with the complex unit three-sphere $\mathbb{S}^3_{\mathbb{C}}$ (resp. with the hyperboloid of signature $(2,2)$ $\mathbb{H}_{2,2}$ or with three-sphere $\mathbb{S}^3$) that we describe by means of a specific system of coordinates. Removing appropriate regions from these manifolds (either cylinders or circles), these systems of coordinates give rise to a bijective correspondence. Using these systems of coordinates enables us to obtain differential realisations of the Lie algebras $\mathfrak{sl}(2,\mathbb{C}), \mathfrak{su}(2)$ and $\mathfrak{sl}(2,\mathbb{R})$ acting respectively on the spaces $\mathbb{S}^3_{\mathbb{C}} \setminus \Big(\mathcal{C}_2 \times \mathcal{C}_2\Big), \mathbb{H}_{2,2} \setminus \mathbb{S}^1$ or $\mathbb{S}^3 \setminus \Big(\mathbb{S}^1 \times \mathbb{S}^1\Big)$, thus allowing to construct all unitary representations in terms of harmonic functions.

In analogous way, we consider an alternative construction of unitary representations of the group of rotations-translations in two dimensions E_2 on the cone with one point removed

$\mathcal{C}_{2,2} \setminus \{0\}$, which can be seen as a singular limit of $\mathbb{H}_{2,2} \setminus \mathbb{S}^1$. It has to be emphasised that all these constructions are related by natural operations such as taking real forms or considering contractions of Lie algebras. We also observe that in the case of $SL(2,\mathbb{Z})$ and E_2, their first homotopy group being isomorphic to $\mathbb{Z}$, the representations are constructed in some special cases (more precisely, when the eigenvalues of the semisimple element or the spin do not belong to $\frac{1}{2}\mathbb{N}$) on a suitable covering space of the considered manifold. The proposed approach can be sumarised in the following diagram:

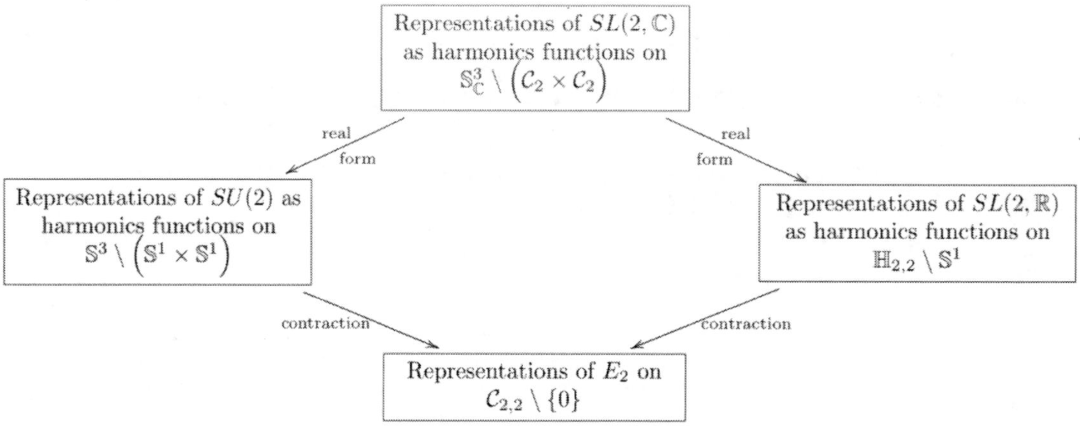

Alternative constructions of the representations considered in this paper can be found for example in [19, 20, 21]. We also observe that in [22], a unified construction of the Lie groups $SO(3), SO(1,2)$ and E_2, was developed, however basing on a different Ansatz and techniques as those used in this work.

The contents of the paper is the following. In Section 2 we explicitly parameterise the Lie groups $SL(2,\mathbb{C}), SL(2,\mathbb{R})$ and $SU(2)$. Considering appropriate systems of coordinates a bijective correspondence deduced from the paremeterisation is obtained after removal of some regions on the previous manifolds. Section 3 is devoted to an explicit construction of a differential realisation of the Lie algebras $\mathfrak{sl}(2,\mathbb{C})$, $\mathfrak{sl}(2,\mathbb{R})$ and $\mathfrak{su}(2)$. In Section 4 we construct explicitly unitary representations of $\mathfrak{sl}(2,\mathbb{C})$, $\mathfrak{sl}(2,\mathbb{R})$ and $\mathfrak{su}(2)$ as harmonic functions on the corresponding manifold. A special emphasis on $\mathfrak{sl}(2,\mathbb{R})$ is given. Finally, Section 5 focuses on the extension of this approach for the description of unitary representations of E_2.

2. Parameterisation of the Lie groups

In this section we construct explicit systems of coordinates for the Lie groups $SL(2,\mathbb{C}), SU(2), SL(2,\mathbb{R})$. It is shown that removing appropriate regions of the corresponding manifolds allows us to obtain bijective parameterisations. These will be central for the construction of unitary representations of the corresponding Lie groups.

The Lie group $SL(2,\mathbb{C})$ is the set of two-by-two complex unimodular matrices

$$\begin{aligned} SL(2,\mathbb{C}) &= \left\{ \begin{pmatrix} \alpha & \beta \\ \gamma & \delta \end{pmatrix} , \ \alpha,\beta,\gamma,\delta \in \mathbb{C} \ , \ \alpha\delta - \beta\gamma = 1 \right\} , \\ &\cong \ \mathbb{S}^3_{\mathbb{C}} = \left\{ z_1, z_2, z_3, z_4 \in \mathbb{C}, \ z_1^2 + z_2^2 + z_3^2 + z_4^2 = 1 \right\} \subset \mathbb{C}^4, \end{aligned}$$

with $\alpha = z_1 - iz_2, \delta = z_1 + iz_2, \beta = z_4 - iz_3, \gamma = -z_4 - iz_3$. An adapted system of coordinates on $\mathbb{S}^3_{\mathbb{C}}$ can be given as follows, writing $z_1^2 + z_2^2 + z_3^2 + z_4^2 = z_+ \bar{z}'_+ + z_- \bar{z}'_-$ We define $\Theta = \vartheta_0 + i\vartheta_1$, $\Phi_\pm = \varphi_{\pm 0} + i\varphi_{\pm 1} \in \mathbb{C}$

and set

$$\begin{aligned} z_+ &= \cos(\Theta)e^{i\Phi_+} , \\ z_- &= \sin(\Theta)e^{i\Phi_-} , \\ z'_+ &= \cos(\bar\Theta)e^{i\bar\Phi_+} , \\ z'_- &= \sin(\bar\Theta)e^{i\bar\Phi_-} . \end{aligned} \tag{1}$$

with

$$(\vartheta_0, \varphi_{+0}, \varphi_{-0}, \vartheta_1, \varphi_{+1}, \varphi_{-1}) \in [0, \frac{\pi}{2}] \times [0, 2\pi[\times[0, 2\pi[\times\mathbb{R}^3 .$$

It can be shown that although (1) covers $\mathbb{S}^3_\mathbb{C}$, the correspondence is not one-to-one. However, removing the two cylinders $\mathcal{C}_2$

$$\begin{aligned} \Theta &= 0 , \quad (\varphi_{+0}, \varphi_{+1}) \in [0, 2\pi[\times\mathbb{R} , \\ \Theta &= \frac{\pi}{2} , \quad (\varphi_{-0}, \varphi_{-1}) \in [0, 2\pi[\times\mathbb{R} . \end{aligned}$$

and denoting $\mathcal{I}^3_\mathbb{C} = \Big\{(\vartheta_0, \varphi_{+0}, \varphi_{-0}, \vartheta_1, \varphi_{+1}, \varphi_{-1}) \in [0, \frac{\pi}{2}] \times [0, 2\pi[\times[0, 2\pi[\times\mathbb{R}^3 \text{ ,s.t. } (\vartheta_0, \vartheta_1) \neq (0, 0), (\pi/2, 0)\Big\}$, we have a bijection between $\mathcal{I}^3_\mathbb{C}$ and $\mathbb{S}^3_\mathbb{C} \setminus \Big(\mathcal{C}_2 \times \mathcal{C}_2\Big)$ [23].

The group $SL(2, \mathbb{R})$ is the set of two-by-two real unimodular matrices

$$\begin{aligned} SL(2, \mathbb{R}) &= \left\{ \begin{pmatrix} \alpha & \beta \\ \gamma & \delta \end{pmatrix} , \quad \alpha, \beta, \gamma, \delta \in \mathbb{R} , \quad \alpha\delta - \beta\gamma = 1 \right\} , \\ &\cong \mathbb{H}_{2,2} = \left\{ x_1, x_2, x_3, x_4 \in \mathbb{R}, \ x_1^2 + x_2^2 - x_3^2 - x_4^2 = 1 \right\} \subset \mathbb{R}^4. \end{aligned}$$

Since $SL(2, \mathbb{R})$ is a real form on $SL(2, \mathbb{C})$, we consider the real form of $\mathbb{S}^3_\mathbb{C}$ parameterised using

$$\Theta = i\vartheta_1 = i\rho , \ \Phi_+ = \varphi_{+0} = \varphi_+ , \Phi_- = \phi_{-0} = \varphi_- , \tag{2}$$

such that (1) reduces to

$$\zeta_+ = \cosh\rho e^{i\varphi_+} , \ \zeta_- = \sinh\rho e^{i\varphi_-} , \tag{3}$$

with $(\rho, \varphi_+, \varphi_-) \in \mathbb{R}_+ \times [0, 2\pi[\times[0, 2\pi[$. We obviously have

$$|\zeta_+|^2 - |\zeta_-|^2 = 1 .$$

The variables $\zeta_\pm$ clearly parameterise the hyperboloid $\mathbb{H}_{2,2}$. If we remove the circle

$$\rho = 0 , \quad \varphi_+ \in [0, 2\pi[,$$

that is $|\zeta_+|^2 = 1$, from $\mathbb{H}_{2,2}$, we have a bijection between $\mathcal{I}_{2,2} = \mathbb{R}^*_+ \times [0, 2\pi[\times[0, 2\pi[$ and $\mathbb{H}_{2,2} \setminus \mathbb{S}^1$. Further it can be seen that the first homotopy group of $\mathbb{H}_{2,2} \setminus \mathbb{S}^1$ is isomorphic to $\mathbb{Z} \times \mathbb{Z}$ [23]. Several coverings of $\mathbb{H}_{2,2} \setminus \mathbb{S}^1$ that will be relevant in the sequel can be defined (for other coverings see [23]):

- the $(p, 1)$−sheeted covering $\widetilde{\mathbb{H}_{2,2} \setminus \mathbb{S}^1}^{(p,1)}$ with $\rho \in \mathbb{R}^*_+, \ 0 \leq \varphi_+ < 2p\pi, \ 0 \leq \varphi_- < 2\pi$,
- the $(\infty, 1)$−sheeted covering $\widetilde{\mathbb{H}_{2,2} \setminus \mathbb{S}^1}^{(\infty,1)}$ with $\rho \in \mathbb{R}^*_+, \varphi_+ \in \mathbb{R}, \ 0 \leq \varphi_- < 2\pi$.

The Lie group $SU(2)$ is the set of two-by-to unitary complex matrices of determinant one

$$\begin{aligned} SU(2) &= \left\{U = \begin{pmatrix} \alpha & \beta \\ -\bar{\beta} & \bar{\alpha} \end{pmatrix} , \alpha, \beta \in \mathbb{C} \, , |\alpha|^2 + |\beta|^2 = 1\right\} \\ &\cong \mathbb{S}^3 = \left\{x_1, x_2, x_3, x_4 \in \mathbb{R}, x_1^2 + x_2^2 + x_3^2 + x_4^2 = 1\right\} \subset \mathbb{R}^4 \, . \end{aligned}$$

Since $SU(2)$ is a real form of $SL(2,\mathbb{C})$ we consider the real form of $\mathbb{S}^3_{\mathbb{C}}$ parametrised by

$$\Theta = \vartheta_0 = \theta \, , \Phi_+ = \varphi_{+0} = \varphi_+ \, , \Phi_- = \varphi_{-0} = \varphi_- \, , \tag{4}$$

which gives in (1)

$$w_+ = \cos\theta e^{i\varphi_+} \, , \ w_- = \sin\theta e^{i\varphi_-} \, ,$$

leading to

$$|w_+|^2 + |w_-|^2 = 1 \, ,$$

with $(\theta, \varphi_+, \varphi_-) \in [0, \pi/2] \times [0, 2\pi[\times[0, 2\pi[$. The variables $w_\pm$ clearly parameterise the sphere $\mathbb{S}^3$. Now, removing two circles from $\mathbb{S}^3$

$$\begin{aligned} \theta &= 0 \, , \ \varphi_+ \in [0, 2\pi[\, , \\ \theta &= \frac{\pi}{2} \, , \ \varphi_- \in [0, 2\pi[\, , \end{aligned}$$

that is, the circles $|w_+|^2 = 1$ and $|w_-|^2 = 1$, we have a bijection between $\mathcal{I}_3 =]0, \pi/2[\times[0, 2\pi[\times[0, 2\pi[$ and $\mathbb{S}^3 \setminus \left(\mathbb{S}^1 \times \mathbb{S}^1\right)$ [23].

Note finally that the manifolds $\mathbb{S}^3 \setminus \left(\mathbb{S}^1 \times \mathbb{S}^1\right)$ and $\mathbb{H}_{2,2} \setminus \mathbb{S}^1$ are real forms of the manifold $\mathbb{S}^3_{\mathbb{C}} \setminus \left(\mathcal{C}_2 \times \mathcal{C}_2\right)$ [23]. The real manifolds can be endowed with a scalar product and a Laplacian related to a scalar product and Laplacian on $\mathbb{S}^3 \setminus \left(\mathbb{S}^1 \times \mathbb{S}^1\right)$. In this note we only consider the case of $SL(2,\mathbb{R})$. The Laplacian is given by [23]

$$\Delta = -\frac{1}{\cosh\rho \sinh\rho} \partial_\rho (\cosh\rho \sinh\rho \partial_\rho) + \frac{1}{\cosh^2\rho} \partial^2_{\varphi_+} - \frac{1}{\sinh^2\rho} \partial^2_{\varphi_-} \, . \tag{5}$$

The scalar product firstly defined on $\mathbb{H}_{2,2} \setminus \mathbb{S}^1$ extends naturally to the covering considered above. On $\widetilde{\mathbb{H}_{2,2} \setminus \mathbb{S}^1}^{(p,1)}$ we have

$$(f, g)_{(p,1)} = \frac{1}{p} \frac{1}{(2\pi)^2} \int\limits_0^{+\infty} \cosh\rho \sinh\rho \, \mathrm{d}\rho \int\limits_0^{2p\pi} \mathrm{d}\varphi_+ \int\limits_0^{2\pi} \mathrm{d}\varphi_- \bar{f}(\rho, \varphi_+, \varphi_-) g(\rho, \varphi_+, \varphi_-) \, , \tag{6}$$

and on $\widetilde{\mathbb{H}_{2,2} \setminus \mathbb{S}^1}^{(\infty,1)}$ it reduces to

$$(f, g)_{(\infty,1)} = \frac{1}{(2\pi)^2} \int\limits_0^{+\infty} \cosh\rho \sinh\rho \, \mathrm{d}\rho \int\limits_{-\infty}^{+\infty} \mathrm{d}\varphi_+ \int\limits_0^{2\pi} \mathrm{d}\varphi_- \bar{f}(\rho, \varphi_+, \varphi_-) g(\rho, \varphi_+, \varphi_-) \, . \tag{7}$$

3. Differential realisations of the Lie groups

The parameterisations of Section 2 enable us to obtain explicit differential realisations of the Lie algebras $\mathfrak{sl}(2,\mathbb{C}), \mathfrak{sl}(2,\mathbb{R})$ and $\mathfrak{su}(2)$ which define a left action respectively on $\mathbb{S}^3_{\mathbb{C}} \setminus \Big(\mathcal{C}_2 \times \mathcal{C}_2\Big), \mathbb{H}_{2,2} \setminus \mathbb{S}^1$ and $\mathbb{S}^3 \setminus \Big(\mathbb{S}^1 \times \mathbb{S}^1\Big)$.

For the Lie algebra $\mathfrak{sl}(2,\mathbb{C})$ the realisation is explicitly given by

$$
\begin{aligned}
L_+ &= \frac{1}{4}e^{i(-\varphi_{-0}-i\varphi_{-1}+\varphi_{+0}+i\varphi_{+1})}\Big(\tan(\vartheta_0+i\vartheta_1)\big(-i\partial_{\varphi_{+0}}-\partial_{\varphi_{+1}}\big)+ \\
&\quad \partial_{\vartheta_0}-i\partial_{\vartheta_1}+\cot(\vartheta_0+i\vartheta_1)\big(-i\partial_{\varphi_{-0}}-\partial_{\varphi_{-1}}\big)\Big) \\
&= \frac{1}{2}e^{i(\Phi_+-\Phi_-)}\Big(-i\tan\Theta\partial_{\Phi_+}+\partial_\Theta-i\cot\Theta\partial_{\Phi_-}\Big)\,, \\
L_- &= \frac{1}{4}e^{i(\varphi_{-0}+i\varphi_{-1}-\varphi_{+0}-i\varphi_{+1})}\Big(\tan(\vartheta_0+i\vartheta_1)\big(-i\partial_{\varphi_{+0}}-\partial_{\varphi_{+1}}\big)+ \\
&\quad -\partial_{\vartheta_0}+i\partial_{\vartheta_1}+\cot(\vartheta_0+i\vartheta_1)\big(-i\partial_{\varphi_{-0}}-\partial_{\varphi_{-1}}\big)\Big) \\
&= \frac{1}{2}e^{i(\Phi_--\Phi_+)}\Big(-i\tan\Theta\partial_{\Phi_+}-\partial_\Theta-i\cot\Theta\partial_{\Phi_-}\Big)\,, \\
L_0 &= -\frac{i}{4}\Big(\partial_{\varphi_{+0}}-i\partial_{\varphi_{+1}}-\partial_{\varphi_{-0}}+i\partial_{\varphi_{-1}}\Big)=-\frac{i}{2}\Big(\partial_{\Phi_+}-\partial_{\Phi_-}\Big)\,,
\end{aligned}
\tag{8}
$$

and

$$
\begin{aligned}
\bar{L}_+ &= \frac{1}{4}e^{i(-\varphi_{-0}+i\varphi_{-1}+\varphi_{+0}-i\varphi_{+1})}\Big(\tan(\vartheta_0-i\vartheta_1)\big(-i\partial_{\varphi_{+0}}+\partial_{\varphi_{+1}}\big)+ \\
&\quad \partial_{\vartheta_0}+i\partial_{\vartheta_1}+\cot(\vartheta_0-i\vartheta_1)\big(-i\partial_{\varphi_{-0}}+\partial_{\varphi_{-1}}\big)\Big) \\
&= \frac{1}{2}e^{i(\bar\Phi_+-\bar\Phi_-)}\Big(-i\tan\bar\Theta\partial_{\bar\Phi_+}+\partial_{\bar\Theta}-i\cot\bar\Theta\partial_{\bar\Phi_-}\Big)\,, \\
\bar{L}_- &= \frac{1}{4}e^{i(\varphi_{-0}-i\varphi_{-1}-\varphi_{+0}+i\varphi_{+1})}\Big(\tan(\vartheta_0-i\vartheta_1)\big(-i\partial_{\varphi_{+0}}+\partial_{\varphi_{+1}}\big)+ \\
&\quad -\partial_{\vartheta_0}-i\partial_{\vartheta_1}+\cot(\vartheta_0-i\vartheta_1)\big(-i\partial_{\varphi_{-0}}+\partial_{\varphi_{-1}}\big)\Big) \\
&= \frac{1}{2}e^{i(\bar\Phi_--\bar\Phi_+)}\Big(-i\tan\bar\Theta\partial_{\bar\Phi_+}-\partial_{\bar\Theta}-i\cot\bar\Theta\partial_{\bar\Phi_-}\Big)\,, \\
\bar{L}_0 &= -\frac{i}{4}\Big(\partial_{\varphi_{+0}}+i\partial_{\varphi_{+1}}-\partial_{\varphi_{-0}}-i\partial_{\varphi_{-1}}\Big)=-\frac{i}{2}\Big(\partial_{\bar\Phi_+}-\partial_{\bar\Phi_-}\Big)\,,
\end{aligned}
\tag{9}
$$

such that we have the $\mathfrak{sl}(2,\mathbb{C})$ commutation relations

$$
\begin{gathered}
\Big[L_0,L_\pm\Big]=\pm L_\pm\,,\quad \Big[L_+,L_-\Big]=2L_0\,, \\
\Big[\bar L_0,\bar L_\pm\Big]=\pm \bar L_\pm\,,\quad \Big[\bar L_+,\bar L_-\Big]=2\bar L_0\,, \\
\Big[L_a,\bar L_b\Big]=0\,.
\end{gathered}
$$

The differential realisation of $\mathfrak{sl}(2,\mathbb{R})$ is obtained using (8) and (9) through the real form (2)

$$
\begin{aligned}
J_+ &= \frac{1}{2}e^{i(\varphi_+-\varphi_-)}\Big(-i\tanh(\rho)\partial_{\varphi_+}-\partial_\rho+i\coth\rho\partial_{\varphi_-}\Big)\,, \\
J_- &= \frac{1}{2}e^{i(\varphi_--\varphi_+)}\Big(-i\tanh(\rho)\partial_{\varphi_+}+\partial_\rho+i\coth\rho\partial_{\varphi_-}\Big)\,, \\
J_0 &= -\frac{i}{2}\Big(\partial_{\varphi_+}-\partial_{\varphi_-}\Big)\,,
\end{aligned}
\tag{10}
$$

and satisfies the $\mathfrak{sl}(2,\mathbb{R})$ commutation relations

$$
\Big[J_0,J_\pm\Big]=\pm J_\pm\,,\quad \Big[J_+,J_-\Big]=-2J_0\,.
$$

Finally, the differential realisation of $\mathfrak{su}(2)$ is obtained using (8) and (9) through the real form (4) and is given by

$$\begin{aligned}
R_+ &= \frac{1}{2}e^{i(\varphi_+-\varphi_-)}\Big(-i\tan\theta\partial_{\varphi_+}+\partial_\theta-i\cot\theta\partial_{\varphi_-}\Big) , \\
R_- &= \frac{1}{2}e^{i(\varphi_--\varphi_+)}\Big(-i\tan\theta\partial_{\varphi_+}-\partial_\theta-i\cot\theta\partial_{\varphi_-}\Big) , \\
R_0 &= -\frac{i}{2}\Big(\partial_{\varphi_+}-\partial_{\varphi_-}\Big) ,
\end{aligned}$$

and satisfies the $\mathfrak{su}(2)$ commutations relations

$$\big[R_0,R_\pm\big]=\pm R_\pm \ , \big[R_+,R_-\big]=2R_0 \ .$$

4. Representations as harmonic functions for the three dimensional simple Lie groups

In this section we show that all unitary representations of the Lie groups $SL(2,\mathbb{C}), SL(2,\mathbb{R})$ and $SU(2)$ can be obtained from their spinor(s) representation(s). The case of $SU(2)$ is trivial and it is not difficult to see that all unitary representations can be defined as harmonic functions on $\mathbb{S}^3\setminus\Big(\mathbb{S}^1\times\mathbb{S}^1\Big)$. This case will not be considered here (see [23] for the details).

Irreducible representations of $SL(2,\mathbb{C})$ have been studied by Gel'fand and can be given in terms of homogenous functions in $\mathbb{C}^2$ [2, 24, 25, 26, 27] (see [28] for an English translation of [24]). Unitary representations are characterised by two numbers ℓ_0 and ℓ_1 and correspond either to $\ell_0\in\frac{1}{2}\mathbb{N}$ and $\ell_1=i\sigma,\sigma\in\mathbb{R}$ for the principal series or to $\ell_0=0$ and $0<\ell_1\le 1$ for the complementary series. We have shown that we can extend the Gel'fand formulæ for all unitary representations such that they are living on $\mathbb{S}^3_{\mathbb{C}}\setminus\Big(\mathcal{C}_2\times\mathcal{C}_2\Big)$ and are harmonic. Since the expressions are complicated and not very enlightening we refer to [23] for explicit expressions.

Unitary representations of $SL(2,\mathbb{R})$ have been studied by various authors [6, 8, 9, 29, 30] and divide into two types: the discrete series (which is bounded either from below or above) and the continuous (principal and supplementary) series, which are unbounded. We have shown that all unitary representations of $SL(2,\mathbb{R})$ can be defined as harmonic functions of $\mathbb{H}_{2,2}\setminus\mathbb{S}^1$ (or one of his coverings), but in this note we only consider the case of discrete series. The discrete series is characterised by $s>0$. For the discrete series bounded from below we have

$$\begin{aligned}
\mathcal{D}_s^+=\Bigg\{\Psi^+_{s,n} &= \sqrt{\frac{2\Gamma(n+2s)}{\Gamma(-1+2s)\Gamma(n+1)}}\bar{\zeta}_+{}^{-2s-n}\bar{\zeta}_-{}^n \\
&= \sqrt{\frac{2\Gamma(n+2s)}{\Gamma(-1+2s)\Gamma(n+1)}}e^{-in\varphi_-+i(2s+n)\varphi_+}\cosh^{-2s-n}\rho\sinh^n\rho\ ,\quad n\in\mathbb{N}\Bigg\} ,
\end{aligned}$$

while for discrete series bounded from above

$$\begin{aligned}
\mathcal{D}_s^-=\Bigg\{\Psi^-_{s,n} &= \sqrt{\frac{2\Gamma(n+2s)}{\Gamma(-1+2s)\Gamma(n+1)}}\zeta_+^{-2s-n}\zeta_-^n \\
&= \sqrt{\frac{2\Gamma(n+2s)}{\Gamma(-1+2s)\Gamma(n+1)}}e^{-i(2s+n)\varphi_++in\varphi_-}\cosh^{-2s-n}\rho\sinh^n\rho\ ,\quad n\in\mathbb{N}\Bigg\} ,
\end{aligned}$$

which satisfy

$$\left\{\begin{aligned}
J_+\Psi^+_{s,n} &= \sqrt{(n+1)(n+2s)}\Psi^+_{s,n+1} , \\
J_-\Psi^+_{s,n} &= \sqrt{n(n+2s-1)}\Psi^+_{s,n-1} , \\
J_0\Psi^+_{s,n} &= (n+s)\Psi^+_{s,n} ,
\end{aligned}\right.
\qquad
\left\{\begin{aligned}
J_+\Psi^-_{s,n} &= -\sqrt{n(n+2s-1)}\Psi^-_{s,n-1} \\
J_-\Psi^-_{s,n} &= -\sqrt{(n+1)(n+2s)}\Psi^-_{s,n+1} , \\
J_0\Psi^-_{s,n} &= -(n+s)\Psi^-_{s,n} .
\end{aligned}\right.$$

Note that $\mathcal{D}_s^+$ and $\mathcal{D}_s^-$ are isomorphic representations. In order that the expressions above are well defined they have to be defined on some covering of $\mathbb{H}_{2,2}\backslash\mathbb{S}^1$. In particular, when $2s = p/q \in \mathbb{Q}$, the representation $\mathcal{D}_s^\pm$ is defined on $\widetilde{\mathbb{H}_{2,2}\setminus\mathbb{S}^1}^{(q,1)}$ and corresponds to a representation of the q−sheeted covering of $SL(2,\mathbb{R})$ and when $s \in \mathbb{R}\backslash\mathbb{Q}$ the representation $\mathcal{D}_s^\pm$ is defined on $\widetilde{\mathbb{H}_{2,2}\setminus\mathbb{S}^1}^{(\infty,1)}$ and corresponds to a representation of the universal covering of $SL(2,\mathbb{R})$. Using (5), a direct computation gives

$$\Delta\Psi^\pm_{s,m} = -4s(s-1)\Psi^\pm_{s,m} = -4Q\Psi^\pm_{s,m} ,$$

where $Q = J_0^2 - 1/2(J_+J_- + J_-J_+)$ is the Casimir operator of $\mathfrak{sl}(2,\mathbb{R})$ and the functions $\Psi^\pm_{s,m}$ are harmonic functions defined on $\mathbb{H}_{2,2}\setminus\mathbb{S}^1$ (or one of his coverings).

Now we check unitarity of the representations using the scalar product (6) or (7). In order to proceed we have to check the convergence of the various integrals, since we are integrating on non-compact manifolds. Using hypergeometric functions, a routine computation shows that in all cases, after integration upon the angles and performing a change of variables, the integrals involving ρ reduce to

$$I_{s,n} = \int\limits_1^{+\infty} r^{-4s-2n+1}(r^2-1)^n \mathrm{d}r ,$$

with $n \geq 0$. These integrals converge if $s > 1/2$ and give

$$I_{s,n} = \frac{1}{2}\frac{\Gamma(n+1)\Gamma(2s-1)}{\Gamma(2s+n)} .$$

In particular, this means that when $2s = p/q, 2s' = p'/q'$, introducing q'' as the least common multiple of q and q' (6), provides the following identity when $s, s' > 1/2$

$$(\Psi^{\varepsilon'}_{s',n'}, \Psi^{\varepsilon}_{s,n})_{q''} = \delta^{\varepsilon'\varepsilon}\delta_{s's}\delta_{n'n} ,$$

and when either s or s' is an irrational number, when $s, s' > 1/2$ holds we have from (7) that

$$(\Psi^{\varepsilon'}_{s',n'}, \Psi^{\varepsilon}_{s,n})_{q''} = \delta^{\varepsilon'\varepsilon}\delta_{n'n}\delta(s-s') .$$

Observe that when $s \in \mathbb{R}$, the eigenvalues of the Casimir operator Q are continuous hence the eigenfunctions $\Psi^\pm_{s,n}$ cannot be normalised.

5. Extension of the formalism for the group of rotations-translations in two dimensions

Interestingly all the formalism we have applied for simple Lie groups extends for the group E_2 of rotation translations in two-dimensions. This group can be, for instance, obtained by an Inönü-Wigner contraction of the group $SU(2)$ or $SL(2,\mathbb{R})$. Studying a singular limit of the hyperboloid $\mathbb{H}_{2,2} \cong SL(2,\mathbb{R})$ enables us to describe the cone $\mathcal{C}_{2,2}$ of equation

$$x_1^2 + x_2^2 - x_3^2 - x_4^2 = 0 ,$$

which is strongly related to E_2 [23]. To have an explicit realisation of unitary representations of E_2 we have to proceed in several steps. Firstly, considering a singular limit of (3) leads to a bijective parameterisation of the cone with one point removed $\mathcal{C}\setminus\{0\}$. Secondly, considering the same singular limit for (10) leads to a differential realisation of $\mathfrak{e}_2$ (the Lie algebra of E_2) and an explicit realisation of all unitary representations of $\mathfrak{e}_2$ on $\mathcal{C}_{2,2}\setminus\{0\}$. Next, to define an invariant scalar product we observe that $\mathcal{C}_{2,2}\setminus\{0\}$ can be continuously deformed (is homeomorphic) to $\mathbb{R}\times\mathbb{S}^1\times\mathbb{S}^1$ and since $\mathbb{S}^1$ is the one-point compactification of $\mathbb{R}$, we have explicitly constructed unitary representations of $\mathfrak{e}_2$ on $\mathbb{S}^1\times\mathbb{S}^1\times\mathbb{S}^1$, parameterised by $\Psi, \varphi_+, \varphi_- \in [0, 2\pi[$. With this realisation the generators of $\mathfrak{e}_2$ are given by

$$\begin{aligned} P_+ &= -\frac{1}{2}e^{i(\varphi_+-\varphi_-)}\partial_\Psi , \\ P_- &= \frac{1}{2}e^{i(\varphi_- - \varphi_+)}\partial_\Psi , \\ J &= -\frac{i}{2}(\partial_{\varphi_+} - \partial_{\varphi_-}) , \end{aligned}$$

with commutations relations

$$[J, P_\pm] = \pm P_\pm \ , [P_+, P_-] = 0 \ .$$

Unitary representations of $\mathfrak{e}_2$ are characterised by two numbers $p \in \mathbb{R}$ and $-1/2 < s \le 1/2$ and are given by

$$\mathcal{D}_{p,s} = \left\{ \Lambda_{p,s,n} = e^{i2p\Psi} e^{i(n+s)\varphi_+ - in\varphi_-} \ , n \in \mathbb{Z} \right\},$$

such that

$$\begin{aligned} P_+ \Lambda_{p,s,n} &= -ip\Lambda_{p,s,n+1} \ , \\ P_- \Lambda_{p,s,n} &= ip\Lambda_{p,s,n-1} \ , \\ J\Lambda_{p,s,n} &= (s+n)\Lambda_{p,s,n} \ . \end{aligned}$$

As for the $SL(2,\mathbb{R})$ case, the functions Λ are defined on appropriate coverings. These coverings will be given by their parameterisation. Two cases are considered. The first one corresponds to $2s = r/q$ with $\Psi \in \mathbb{R}, \varphi_+ \in [0, 2q\pi[, \varphi_- \in [0, 2\pi[$, and second case to $s \in \mathbb{R} \setminus \mathbb{Q}$ with $\Psi, \varphi_+ \in \mathbb{R}, \varphi_- \in [0, 2\pi[$.

Now, if f and g are two functions defined on an appropriate covering of $\mathbb{S}^1 \times \mathbb{S}^1 \times \mathbb{S}^1$, we define the scalar product

$$(f, g)_q = \frac{1}{q} \frac{1}{(2\pi)^2} \int\limits_{-\infty}^{+\infty} \mathrm{d}\Psi \int\limits_0^{2q\pi} \mathrm{d}\varphi_+ \int\limits_0^{2\pi} \mathrm{d}\varphi_- \bar{f}(\Psi, \varphi_+, \varphi_-) g(\Psi, \varphi_+, \varphi_-) \ ,$$

in the first case and

$$(f, g)_\infty = \frac{2}{(2\pi)^2} \int\limits_{-\infty}^{+\infty} \mathrm{d}\Psi \int\limits_{-\infty}^{+\infty} \mathrm{d}\varphi_+ \int\limits_0^{2\pi} \mathrm{d}\varphi_- \bar{f}(\Psi, \varphi_+, \varphi_-) g(\Psi, \varphi_+, \varphi_-) \ ,$$

in the second case. It is then straightforward to check that for $s, s' \in \mathbb{Q}$, *i.e.*, $2s = r/q, 2s' = r'/q'$ (q'' being the least common multiple of q, q'), we have

$$(\Lambda_{p', \frac{r'}{2q'}, n'}, \Lambda_{p, \frac{r}{2q}, n})_{q''} = \delta_{nn'} \delta_{\frac{r}{q} \frac{r'}{q'}} \delta(p - p') \ ,$$

while for s or s' a real number (but not a rational), we get

$$(\Lambda_{p', s', n'}, \Lambda_{p, s, n})_\infty = \delta_{nn'} \delta(s' - s) \delta(p - p') \ .$$

The functions $\Lambda_{p,s,n}$ are hence orthonormal, and the representations $\mathcal{D}_{p,s}$ are unitary.

6. Final remarks

Unitary representations of the three dimensional simple Lie groups have been constructed in terms of generalised harmonic functions acting on appropriate smooth manifolds related to differential realisations. Using the well known relation of these groups with the group of rotations-translations in two dimensions via Inönü-Wigner contractions, the argument is extended to the construction of unitary multiplets of the latter. This procedure provides an alternative perspective concerning the central role of spinor representations. In principle, up to the restrictions expected in the integration on non-compact manifolds, the method can be formally proposed for higher rank simple Lie groups. The corresponding analysis for SU(3) in the context of the Elliott model or the pseudo-orthogonal groups SO($5 - p, p$) in the frame of kinematical groups constitute physically motivating situations to further enlarge the Ansatz to the case of subduced representations or its application in reductions chains of symmetry groups.

Acknowledgments
The authors express their gratitude to Marcus Slupinski for fruitful discussions and valuable suggestions that improved the manuscript. This work was partially supported by the research project MTM2010-18556 of the MICINN (Spain).

References
[1] Talman J D 1968 *Special Functions, A Group Theoretic Approach* (New York: Benjamin)
[2] Gel'fand I M, Minlos R A and Shapiro Z Y 1963 *Representations of the Rotation and Lorentz Groups and their Applications* (Oxford: Pergamon Press)
[3] Berezin F and Gel'fand I M 1962 *Amer. Math. Soc. Translations 21* 193
[4] Helgason S 1981 *Topics in Harmonic Analysis on Homogeneous Spaces* (Boston: Birkhauser)
[5] Floyd L W 1985 *Tokyo J. Math.* **8** 99
[6] Bargmann V 1947 *Annals Math.* **48** 568–640
[7] Bargmann V 1962 *Rev. Mod. Phys.* **34** 829–845
[8] Gel'fand I, Graev M and Vilenkin 1966 *Generalized Functions, Vol.5* (New York: Academic)
[9] Lang S 1975 $SL_2(R)$ (Addison-Wesley Publishing Company)
[10] Warner G 1972 *Harmonic Analysis on Semisimple Lie Groups II* (New York: Springer)
[11] Fischer J, Niederle J and Raczka R 1966 *J. Math. Phys.* **7** 816
[12] M Huszár M 1985 *Acta Phys. Hungarica* **58** 175
[13] Mukunda N W 1967 *J. Math. Phys.* **8** 50
[14] Mukunda N W 1968 *J. Math. Phys.* **9** 417
[15] Strichartz R 1973 *J. Funct. Anal* **12** 341
[16] Bég M A B and Ruegg H 1965 *J. Math. Phys.* **6** 677
[17] Nelson T 1967 *J. Math. Phys.* **8** 857–863
[18] Marsh S and Buck B 1982 *J. Phys.* **A15** 2337–2348
[19] Vilenkin N and Klimyk A 1994 *Representation of Lie groups and special functions. Recent advances. Transl. from the Russian by V. A. Groza a. A. A. Groza.* (Dordrecht: Kluwer Academic Publishers)
[20] Vilenkin N and Klimyk A 1993 *Representation of Lie groups and special functions. Volume 2: Class I representations, special functions, and integral transforms. Transl. from the Russian by V. A. Groza a. A. A. Groza.* (Dordrecht: Kluwer Academic Publishers)
[21] Vilenkin N and Klimyk A 1992 *Representation of Lie groups and special functions. Volume 3: Classical and quantum groups and special functions. Transl. from the Russian by V. A. Groza a. A. A. Groza.* (Dordrecht: Kluwer Academic Publishers Group)
[22] Negro J, del Olmo M and Rodriguez-Marco A 2002 *J. Phys.* **A35** 2283 (*Preprint* `quant-ph/0110152`)
[23] Campoamor-Stursberg R and Rausch de Traubenberg M 2014 (*Preprint* `1404.4705`)
[24] Naĭmark M A 1958 *The linear representations of the Lorentz group* (Линейные представления группы Лоренца) (Moskow: Fizmatgiz)
[25] Ginsburg V A and Tamm I Y 1947 *JETP* **17** 227
[26] Harish-Chandra 1947 *Proc. R. Soc. Lond. A* **189** 372
[27] Stoyanov D T and Todorov I 1968 *J. Math. Phys.* **9** 2146–2167
[28] Naĭmark M A 1964 *Linear representations of the Lorentz group* International series of monographs in pure and applied mathematics (Pergamon Press)
[29] Wybourne B G 1974 *Classical Groups for Physicists* (New York: John Wiley & Sons)
[30] Ui H 1970 *Prog. Theor. Phys.* **44** 689–702

Journal of Physics: Conference Series **597** (2015) 012022 doi:10.1088/1742-6596/597/1/012022

Quantum physics and signal processing in rigged Hilbert spaces by means of special functions, Lie algebras and Fourier and Fourier-like transforms

E Celeghini[1,2], **M A del Olmo**[2]

[1]Dipartmento di Fisica, Universitá di Firenze, I50019 Sesto Fiorentino, Firenze, Italy
[2]Departamento de Física Teórica, Universidad de Valladolid, E-47005, Valladolid, Spain

E-mail: celeghini@fi.infn.it, olmo@fta.uva.es

Abstract. Quantum physics and signal processing in the line $\mathbb{R}$ are strictly related to Fourier transform and Weyl-Heisenberg algebra. We discuss here the addition of a new discrete variable that measures the degree of the Hermite functions and allows to obtain the projective algebra $io(2)$. A rigged Hilbert space is found and a new discrete basis in $\mathbb{R}$ obtained. All operators defined on $\mathbb{R}$ are shown to belong to the universal enveloping algebra of $io(2)$ allowing, in this way, their algebraic treatment. Introducing in the half-line a Fourier-like transform, the procedure is extended to $\mathbb{R}^+$ and can be easily generalized to $\mathbb{R}^n$ and to spherical cohordinate systems.

1. Introduction

The starting points of quantum mechanics in the line $\mathbb{R}$ and its wave functions counterpart $L^2(\mathbb{R})$ are the Weyl-Heisenberg algebra $h(1) \equiv \langle X, P, \mathbb{I}\rangle$ [1] that defines position and momentum and the Fourier transform FT [2] that connects them. This is true also in signal processing, also if there the operatorial structure is scarcely taken in consideration.

We consider here a more predictive approach, adding to $h(1)$ a new operator N related to the degree of the Hermite functions [3] that became the transition matrices between discrete and continuous bases in quantum space [4]. A projective algebra $io(2)$ [6, 7] is constructed and a rigged Hilbert space (RHS)[5] found, where N defines a new discrete basis in $\mathbb{R}$ and $L^2(\mathbb{R})$. Moreover the spaces of the operators defined on $\mathbb{R}$, $\{\mathcal{O}[\mathbb{R}]\}$, and on $L^2(\mathbb{R})$, $\{\mathcal{O}[L^2(\mathbb{R})]\}$, are isomorphic to the universal enveloping algebra UEA$[io(2)]$.

Three bases are indeed obtained, two continuous -$\{|x\rangle\}$ and $\{|p\rangle\}$ related to position X and momentum P- and a new one discrete, $\{|n\rangle\}$, connected to the Hermite functions. As in a Hilbert space all bases have the same cardinality, the mathematical structure is a RHS [5], the true space of quantum mechanics. A state in quantum mechanics was indeed originally defined as a ray in a RHS, an intricate concept involving a Gelfand triple $\phi \subseteq \mathcal{H} \subseteq \phi'$, where $\mathcal{H}$ is a Hilbert space, ϕ (dense subset of $\mathcal{H}$) is the "*ket*" space and ϕ' (dual of ϕ) is the "*bra*" space. However quickly RHS has been considered an unnecessary complication as all results can be found (also if by means of elaborated limits on functions with compact support) in the Hilbert space obtained representing, by means of the axiom of choice, each entire ray with a vector of norm one and phase zero.

We attempt to show here that RHS are not so complicated to justify its elimination from all recent textbooks and can be more powerful than standard Hilbert spaces because they allow the operatorial description (and the inclusion inside the same algebra) of variables with different cardinality. Hoping that it could convince more than a general discussion, that can be found in [5], we discuss here the 1-dimensional case.

The projective algebra $io(2) \equiv \langle X, P, N, \mathbb{I} \rangle$ above introduced is isomorphic to the so called algebra of the harmonic oscillator $\langle H, a, a^\dagger, \mathbb{I} \rangle$ [6, 7], usually described inside the UEA$[h(1)]$. As the ∞-dimensional UEA$[io(2)]$ is a complicated object, a set of relevant subspaces emerges in $\mathbb{R}$ and $\{\mathcal{O}[\mathbb{R}]\}$ as well as in $L^2(\mathbb{R})$ and $\{\mathcal{O}[L^2(\mathbb{R})]\}$. In particular, as ∞-many copies of $io(2)$ are contained into UEA$[io(2)]$, ∞-many operators that close $io(2)$ are found.

Note that in $L^2(\mathbb{R})$ (but not in $\mathbb{R}$, $\{\mathcal{O}[\mathbb{R}]\}$ and $\{\mathcal{O}[L^2(\mathbb{R})]\}$) the subspaces derived from the algebra can be obtained also from the fractional Fourier transform (FrFT) [8]. Indeed the eigenvectors of the FrFT are, as the ones of FT, the Hermite functions and for each $k \in \mathbb{N}$ a FrFT can be constructed such that its eigenvalues are the kth roots of unity, each one corresponding to one subspace.

The results stated in $\mathbb{R}$ can be easily extended to orthogonal and pseudo-orthogonal spaces of any dimension by means of a tensorial construction.

Finally to exhibit that the approach in not peculiar of $\mathbb{R}$, the discussion is repeated on the half line $\mathbb{R}^+$. For this purpose, starting from the Hermite functions and the relations between the Hermite functions and the generalized Laguerre functions, two Fourier-like transforms are constructed each one related to a different definition of conjugation of bases.

2. The line $\mathbb{R}$

To describe the line we start from the unitary irreducible representations of the translation group T^1

$$P\,|p\rangle \;=\; p\,|p\rangle\,, \qquad\qquad U^p(x)\,|p\rangle \;=\; e^{-\mathrm{i}px}\,|p\rangle\,.$$

The regular representation $\{|p\rangle\}$ $(-\infty < p < \infty)$ is such that

$$\langle\, p\,|\,p'\,\rangle \;=\; \sqrt{2\pi}\;\delta(p-p')\,, \qquad\qquad \frac{1}{\sqrt{2\pi}}\int_{-\infty}^{+\infty} |p\rangle\, dp\, \langle p| \;=\; \mathbb{I}$$

and the conjugate basis $\{|x\rangle\}$, defined by the operator X, is obtained from the FT

$$|x\rangle := \left[\frac{1}{\sqrt{2\pi}}\int_{-\infty}^{+\infty} dp\; e^{-\mathrm{i}px}\right]|p\rangle\;, \qquad \langle\, x\,|\,x'\,\rangle \;=\; \sqrt{2\pi}\;\delta(x-x')\;, \qquad \frac{1}{\sqrt{2\pi}}\int_{-\infty}^{+\infty} |x\rangle\, dx\, \langle x| \;=\; \mathbb{I}\;.$$

The operators X and P close, together with $\mathbb{I}$, the Weyl-Heisenberg algebra [1]. At this point, we move to a non standard approach related to the ray representations [6, 7, 9] of the inhomogeneous orthogonal algebra $io(2)$. In addition to X and P (which spectra have the cardinality of the continuum $\aleph_1$) an operator N (with spectrum of cardinality $\aleph_0$), related to the index n of the Hermite polynomials $H_n(x)$, is indeed introduced. As the Hermite functions

$$\psi_n(x) := \frac{e^{-x^2/2}}{\sqrt{2^n n! \sqrt{\pi}}}\; H_n(x)\,,$$

are a basis of the space of real/complex square integrable functions on the line $L^2((-\infty, \infty)) \equiv L^2(\mathbb{R})$ [2]:

$$\int_{-\infty}^{\infty} \psi_n(x)\; \psi_{n'}(x)\; dx = \delta_{n,n'}\;, \qquad\qquad \sum_{n=0}^{\infty} \psi_n(x)\; \psi_n(x') \;=\; \delta(x-x')\,.$$

The basic idea is now to introduce the set of vectors $\{|n\rangle\}$

$$|n\rangle := (2\pi)^{-1/4} \int_{-\infty}^{\infty} dx \; \psi_n(x)\, |x\rangle \qquad n \in \mathbb{N}\,, \tag{1}$$

that, by inspection, is an orthonormal and complete set in $\mathbb{R}$

$$\langle\, n \,|\, n' \,\rangle = \delta_{n\,n'}\,, \qquad \sum_{n=0}^{\infty} |n\rangle\,\langle n| = \mathbb{I}\,.$$

Also $\{|n\rangle\}$ is thus a discrete basis in the real line $\mathbb{R}$, i.e. $\mathbb{R} \equiv \{|p\rangle\} \equiv \{|x\rangle\} \equiv \{|n\rangle\}$.

Relations among the three bases are easily established, as $\{\psi_n(x)\}$ are eigenvectors of FT,

$$[\text{FT}]\; \psi_n(x) = \left[\frac{1}{\sqrt{2\pi}} \int_{-\infty}^{\infty} dx\; e^{\,\mathrm{i}px}\right] \psi_n(x) = \mathbf{i}^n\; \psi_n(p)\,, \tag{2}$$

$$|\,x\,\rangle = (2\pi)^{1/4} \sum_{n=0}^{\infty} \psi_n(x)\,|\,n\rangle\,, \qquad |p\,\rangle = \left[\frac{1}{\sqrt{2\pi}} \int_{-\infty}^{\infty} dx\, e^{\,\mathrm{i}px}\right] |x\,\rangle\,,$$

$$|n\rangle = \mathbf{i}^n\; (2\pi)^{-1/4} \int_{-\infty}^{\infty} dp\; \psi_n(p)\; |p\rangle\,, \qquad |\,p\,\rangle = (2\pi)^{1/4} \sum_{n=0}^{\infty} \mathbf{i}^n\; \psi_n(p)\;|\,n\rangle\,.$$

For an arbitrary vector $|f\rangle \in \mathbb{R}$ we thus have

$$|f\rangle = \frac{1}{\sqrt{2\pi}} \int_{-\infty}^{+\infty} dx\; f(x)\; |x\rangle = \frac{1}{\sqrt{2\pi}} \int_{-\infty}^{+\infty} dp\; f(p)\; |p\rangle = \sum_{n=0}^{\infty} f_n\; |n\rangle\,,$$

$$f(x) := \langle x|f\rangle = (2\pi)^{1/4} \sum_{n=0}^{\infty} \psi_n(x)\; f_n\,, \qquad f(p) := \langle p|f\rangle = (2\pi)^{1/4} \sum_{n=0}^{\infty} (-\mathbf{i})^n\; \psi_n(p)\; f_n\,,$$

$$f_n := \langle n|f\rangle = (2\pi)^{-1/4} \int_{-\infty}^{+\infty} dx\; \psi_n(x)\; f(x) = \mathbf{i}^n (2\pi)^{-1/4} \int_{-\infty}^{+\infty} dp\; \psi_n(p)\; f(p)$$

and the wave functions $f(x)\,, f(p)$ and the sequence $\{f_n\}$ describe $|f\rangle$ in the three bases.

All seems trivial, but $\{|n\rangle\}$ has the cardinality of the natural numbers $\aleph_0$ and, as all bases in a Hilbert space have the same cardinality, the structure we have constructed (the quantum space on the line $\mathbb{R}$) is not a Hilbert space but a rigged Hilbert space.

In this way, by means of FT and Hermite functions, we went back to the foundations of quantum mechanics, where a physical state was defined in a Hilbert space as a ray and not as a vector, as we are used considering RHS as an unnecessary complication. However the problem is apparent also in the simple case of the harmonic oscillator: the algebraic description consider $H\,(\equiv N + 1/2)\,,\ a\,,\ a^\dagger$ and $\mathbb{I}$. But $X = \frac{1}{\sqrt{2}}(a + a^\dagger)$ and $P \equiv -\mathrm{i}\, D_x = \frac{-\mathrm{i}}{\sqrt{2}}(a - a^\dagger)$ do not belong to the algebra because they are not operators in a Hilbert space. In a RHS, instead, X and P are standard operators also if their spectra are continuous and we have no problems to consider them as generators of a Lie algebra together with the number operator N [4, 5]. This allows us to include all differential operators inside the Lie algebra and to extend the set of operators defined in the universal enveloping algebra.

3. Algebra of the line $\mathbb{R}$

To describe the structure of the operators on the line we introduce now in $L^2(\mathbb{R})$ the operators X, $D_x(=\mathbf{i}P)$, N and $\mathbb{I}$

$$X\psi_n(x) := x\,\psi_n(x)\,, \quad D_x\psi_n(x) := \psi_n'(x)\,, \quad N\psi_n(x) := n\,\psi_n(x)\,, \quad \mathbb{I}\,\psi_n(x) := \psi_n(x).$$

Recurrence relations of Hermite polynomials

$$H_n'(x) \;=\; 2\,n\,H_{n-1}(x), \qquad H_n'(x) - 2\,x\,H_n(x) \;=\; H_{n+1}(x)$$

can be rewritten as

$$a\;\psi_n(x) \;=\; \sqrt{n}\;\psi_{n-1}(x), \qquad a^\dagger\;\psi_n(x) \;=\; \sqrt{n+1}\;\psi_{n+1}(x)\,,$$

where the operators a and $a^\dagger$ are defined in terms of the operators X and P

$$a := \frac{1}{\sqrt{2}}\,(X + \mathbf{i}P)\;\;, \qquad a^\dagger := \frac{1}{\sqrt{2}}\,(X - \mathbf{i}P)\;\;.$$

The algebra contains the rising and lowering operators on the Hemite functions [10, 11]

$$[N, a^\dagger] = a^\dagger, \qquad [N, a] = -a\;, \qquad [a, a^\dagger] = \mathbb{I}\;, \qquad [\mathbb{I}, \bullet] = 0 \tag{3}$$

and is isomorphic to the projective algebra $io(2)$ [6, 7]:

$$[N, X] = -\mathbf{i}P\;, \qquad [N, P] = \mathbf{i}X\;, \qquad [X, P] = \mathbf{i}\,\mathbb{I}\;, \qquad [\mathbb{I}, \bullet] = 0\,. \tag{4}$$

Because of Eq.(1), $\mathbb{R}$ and $L^2(\mathbb{R})$) are isomorphic and, as discussed in [9], they are representations with zero value of the Casimir operator of the projective algebra (4) (or equivalently (3)),

$$C \equiv (X^2 - D_x^2)/2 - (N + 1/2)\,\mathbb{I} \;=\; \{a, a^\dagger\}/2 - (N + 1/2)\,\mathbb{I} = 0. \tag{5}$$

On the representations we can assume $\mathbb{I} = 1$ and write

$$C\,|n\rangle \;=\; \Big[\{a, a^\dagger\}/2 - N - 1/2\Big]\,|n\rangle \;=\; 0$$

that in $L^2(\mathbb{R})$ can be alternatively written as

$$C\;\psi_n(x) \;=\; \left[(X^2 - D_x^2)/2 - N - 1/2\right]\;\psi_n(x) \;=\; 0\;.$$

Eq.(5) gives indeed the operatorial identity that defines $\mathbb{R}$ and $L^2(\mathbb{R})$

$$N \;\equiv\; \left(X^2 - D_x^2 - 1\right)/2 \;\equiv\; \{a, a^\dagger\}/2 - 1/2\,, \tag{6}$$

by inspection equivalent to the Hermite differential equation:

$$H_n''(x) - 2\,x\,H_n'(x) + 2\,n\,H_n(x) = 0\,.$$

4. Universal enveloping algebra of $io(2)$

The representations are irreducible, so that, on both spaces $L^2(\mathbb{R})$ and $\mathbb{R}$, all operators of the UEA$[io(2)]$ are defined and an isomorphism exists between the UEA$[io(2)]$ and the space of the operators $\{\mathcal{O}[L^2(\mathbb{R})]\}$ and $\{\mathcal{O}[\mathbb{R}]\}$:

$$\{\mathcal{O}[L^2(\mathbb{R})]\} \equiv UEA[io(2)] \equiv \{\mathcal{O}[\mathbb{R}]\} , \tag{7}$$

i.e. each operator $\mathcal{O}$ can be written

$$\mathcal{O} = \sum c_{\alpha\,\beta\,\gamma}\, X^\alpha\, D_x{}^\beta\, N^\gamma = \sum d_{\alpha\,\beta\,\gamma}\, a^{\dagger\alpha}\, N^\beta\, a^\gamma.$$

From the analytical point of view, an ordered monomial $X^\alpha\, D_x{}^\beta\, N^\gamma \in$ UEA$[io(2)]$ is an order β differential operator but, because of the operatorial identity (6), we have

$$D_x^2 \equiv X^2 - 2N - 1,$$

and any operator in $\{\mathcal{O}[L^2(\mathbb{R})]\}$ and in $\{\mathcal{O}(\mathbb{R})\}$ can be written

$$\mathcal{O} = f_0(X)\, g_0(N) + f_1(X)\, D_x\, g_1(N)$$

that on the basis vector $\psi_n(x)$ becomes

$$\mathcal{O}\, \psi_n(x) = f_0(x)\, g_0(n)\, \psi_n(x) + f_1(x)\, g_1(n)\, \psi'_n(x).$$

5. Substructures in the line

The UEA$[io(2)]$ is a rich structure and it contains ∞-many $io(2)$ algebras that allow to construct an intricate structure of subspaces. Moreover for any $k \in \mathbb{N}$ we have

$$UEA[io(2)] \supset \oplus_{q=0}^{k-1}\, io_{k,q}(2) ,$$

where k and q are parameters that define each copy $io_{k,q}(2)$ of the algebra $io(2)$. Indeed, starting from n and k , we can define two other integers $r :=$ Quotient$[n,k]$ and $q :=$ Mod$[n,k]$ so that $n = k\,r + q$ and the corresponding operators R and Q are, of course, diagonal on $\{|k\,r+q\rangle\}$

$$R\,|kr+q\rangle = r\,|kr+q\rangle , \qquad Q\,|kr+q\rangle = q\,|kr+q\rangle .$$

By inspection, the operators $A^\dagger_{k,q}$, $A_{k,q} \in$ UEA$[io(2)]$

$$A^\dagger_{k,q} := (a^\dagger)^k \frac{\sqrt{N+k-q}}{\sqrt{k\prod_{j=1}^k (N+j)}} , \qquad A_{k,q} := \frac{\sqrt{N+k-q}}{\sqrt{k\prod_{j=1}^k (N+j)}} (a)^k \tag{8}$$

are defined on the whole set $\{|k\,r+q\rangle\}$ and give

$$A^\dagger_{k,q}\,|k\,r+q\rangle = \sqrt{r+1}\,|k(r+1)+q\rangle , \qquad A_{k,q}\,|k\,r+q\rangle = \sqrt{r}\,|k(r-1)+q\rangle .$$

Each couple (k,q) gives thus a representation with $C=0$ of the algebra $io(2)$, we denote $io_{k,q}(2)$:

$$[R, A^\dagger_{k,q}] = +A^\dagger_{k,q} \qquad [R, A_{k,q}] = -A_{k,q} \qquad [A_{k,q}, A^\dagger_{k,q}] = \mathbb{I} \qquad [\mathbb{I}, \bullet] = 0.$$

In particular, for $k=4$ and $0 \le q < 4$, we can define

$$\mathbb{R}_{4,q} := \{|4r+q\rangle\} , \qquad L^2_{4,q}(\mathbb{R}) := \{\psi_{4\,r+q}(x)\} \qquad (r = 0,1,2,\ldots) ,$$

so that the operators $\{\mathcal{O}[L^2_{4,q}(\mathbb{R})]\}$ and $\{\mathcal{O}[\mathbb{R}_{4,q}]\}$ acting inside each of them belong to UEA$[io_{4,q}(2)]$

$$\{\mathcal{O}[L^2(\mathbb{R}_{4,q})]\} \equiv UEA[io_{4,q}(2)] \equiv \{\mathcal{O}[\mathbb{R}_{4,q}]\} .$$

The spaces $L^2(\mathbb{R})$ and $\mathbb{R}$ can be thus divided in four subspaces each one supporting the corresponding representation of $io_{4,q}(2)$:

$$L^2(\mathbb{R}) = \oplus_{q=0}^{3} L^2_{4,q}(\mathbb{R}) = \oplus_{q=0}^{3} \{\psi_{4r+q}(x)\} , \tag{9}$$

$$\mathbb{R} = \oplus_{q=0}^{3} \mathbb{R}_{4,q} = \oplus_{q=0}^{3} \{|4r+q\rangle\}.$$

These results are general as Eqs. (8) allow to write for all $k \in \mathbb{N}$, $0 \leq q < k$ and $r = 0, 1, 2, \ldots$

$$L^2_{k,q}(\mathbb{R}) := \{\psi_{kr+q}(x)\} , \qquad \mathbb{R}_{k,q} := \{|kr+q\rangle\} ,$$

$$\{\mathcal{O}[L^2(\mathbb{R}_{k,q})]\} \equiv UEA[io_{k,q}(2)] \equiv \{\mathcal{O}[\mathbb{R}_{k,q}]\},$$

$$L^2(\mathbb{R}) = \oplus_{q=0}^{k-1} L^2_{k,q}(\mathbb{R}) = \oplus_{q=0}^{k-1} \{\psi_{kr+q}(x)\} , \qquad \mathbb{R} = \oplus_{q=0}^{k-1} \mathbb{R}_{k,q} = \oplus_{q=0}^{k-1} \{|kr+q\rangle\}.$$

Eq.(9) can also be obtained from FT (2) that divides $L^2(\mathbb{R})$ in four subspaces each one corresponding to one eigenvalue of FT (but does not give any information about the operators). Besides its generalization to all $k \in \mathbb{N}$ allows to divide $L^2(\mathbb{R})$ in k subspaces (again disregarding the operators) and can be obtained from the fractional Fourier transform (FrFT) [8]:

$$[\text{FrFT}]_\alpha \, \psi_n(x) = e^{\mathbf{i}\alpha n} \, \psi_n(x') \qquad \alpha \in \mathbb{C} .$$

Indeed specializing α to $\alpha = 2\pi/k$ we get

$$[\text{FrFT}]_{2\pi/k} \, \psi_n(x) = e^{\mathbf{i}\frac{2\pi n}{k}} \, \psi_n(x') , \tag{10}$$

$$[\text{FrFT}]_{2\pi/k} \, f(x) := \frac{1}{\sqrt{2\pi}} \int_{-\infty}^{\infty} dx \, e^{ixx'} \, e^{2\pi\mathbf{i}(1/k-1/4)N} \, f(x) = [\text{FT}] \left[e^{2\pi\mathbf{i}(1/k-1/4)N} \right] f(x)$$

where N –diagonal on each $\psi_n(x)$– on the whole space is rewritten as $N \equiv (X^2 - D_x^2 - 1)/2$. As an example, for $k = 3$ Eq. (10) gives

$$[\text{FrFT}]_{2\pi/3} \, \psi_n(x) = e^{i\frac{2\pi n}{3}} \, \psi_n(x') ,$$

that, similarly to Eq.(9), exhibits three subspaces of $L^2(\mathbb{R})$

$$L^2(\mathbb{R}) = L^2_{3,0}(\mathbb{R}) \oplus L^2_{3,1}(\mathbb{R}) \oplus L^2_{3,2}(\mathbb{R}) = \{\psi_{3r}\} \oplus \{\psi_{3r+1}\} \oplus \{\psi_{3r+2}\} ,$$

but, like FT, does not give any information about the operators acting inside each of them .

6. The half-line $\mathbb{R}^+$

The results obtained on the line $\mathbb{R}$ can be rewritten in $\mathbb{R}^+$, the vector space defined by the operator Y , with eigenvectors $\{|y\rangle\}$ where y is into the open set $(0 < y < \infty)$:

$$Y\,|y\rangle = y\,|y\rangle , \qquad \langle y\,|\,y'\rangle = \delta(y - y') , \qquad \int_0^\infty |y\rangle\, dy\, \langle y| = \mathbb{I}.$$

The generalized Laguerre polynomials $L_n^\alpha(y)$ $(n \in \mathbb{N})$ play now the role of the Hermite polynomials. For each value of α an alternative discrete basis $\{|n\rangle\}$ is obtained as follows. From $L_n^\alpha(\mathrm{y})$ the normalized generalized Laguerre functions $M_n^\alpha(y)$ are

$$M_n^\alpha(y) := \sqrt{\frac{\Gamma(n+1)}{\Gamma(n+\alpha+1)}}\ y^{\alpha/2}\ e^{-y/2}\ L_n^\alpha(y)\ ,$$

that satisfy

$$\int_0^\infty M_n^\alpha(y)\ M_m^\alpha(y)\ dy\ =\ \delta_{nm}\ , \qquad \sum_{n=0}^\infty M_n^\alpha(y)\ M_n^\alpha(y')\ =\ \delta(y-y')$$

so that, like the set $\{\psi_n(x)\}$ in $L^2(\mathbb{R})$, the set $\{M_n^\alpha(x)\}$ is -for fix and arbitrary α- a basis in $L^2(\mathbb{R}^+)$ [2]. Again, like in (1), we define -for $n \in \mathbb{N}$- the vector $|n\rangle$

$$|n\rangle\ :=\ \int_0^\infty dy\ M_n^\alpha(y)\ |y\rangle$$

and we have

$$\langle n|n'\rangle = \delta_{nm}\ , \qquad \sum_{n=0}^\infty |n\rangle\langle n|\ =\ \mathbb{I}\ ,$$

i.e. $\{|n\rangle\}_{n\in\mathbb{N}}$ is an orthonormal basis of the half-line $\mathbb{R}^+$ and $M_n^\alpha(y)$ are the transformations matrices that relate the two bases $\{|y\rangle\}$ and $\{|n\rangle\}$:

$$M_n^\alpha(y)\ =\ \langle\, y\,|\,n\,\rangle\ =\ \langle\, n\,|\,y\,\rangle .$$

Like in the line, an arbitrary vector $|f\rangle \in \mathbb{R}^+$ can be written

$$|f\rangle = \int_0^\infty dy\, f(y)\, |y\rangle\ =\ \sum_{n=0}^\infty |n\rangle\, f_n\ ,$$

$$f(y) := \langle y|f\rangle = \sum_{n=0}^\infty M_n^\alpha(y)\ f_n\ , \qquad f_n := \langle n|f\rangle = \int_0^\infty dy\ M_n^\alpha(y)\ f(y)\ ,$$

so that the wave function $f(y)$ and the sequence $\{f_n\}$ describe the vector $|f\rangle$ in the two bases.

As one discrete basis $\{|n\rangle\}$ and one continuous one $\{|y\rangle\}$ have been found also the half-line is a RHS. The third basis will be constructed in Sect. 9.

7. Algebra of the half-line

We introduce now the associated algebra. For $L_n^\alpha(y)$ the recurrence relations are

$$y\, L_n^\alpha(y)' + (n+\alpha+1-y)\, L_n^\alpha(y) = (n+1)\, L_{n+1}^\alpha(y)\ ,$$

$$-y\, L_n^\alpha(y)' + n\, L_n^\alpha(y) = (n+\alpha)\, L_{n-1}^\alpha(y)$$

that, defining the operators Y, D_y, $\mathbb{N}$ and $\mathbb{I}$,

$$Y\, M_n^\alpha(y) := y\, M_n^\alpha(y)\, , \quad D_y\, M_n^\alpha(y) := M_n^\alpha(y)'\ , \quad N\, M_n^\alpha(y) := n\, M_n^\alpha(y)\ , \quad \mathbb{I}\, M_n^\alpha(y) := M_n^\alpha(y)\ ,$$

can be rewritten

$$\begin{aligned} J_+ \, M_n^\alpha(y) &:= \left(Y \, D_y + N + 1 + \frac{\alpha - Y}{2} \right) M_n^\alpha(y) = \sqrt{(n+1)(n+\alpha+1)} \, M_{n+1}^\alpha(y) \, , \\ J_- \, M_n^\alpha(y) &:= \left(-Y \, D_y + N + \frac{\alpha - Y}{2} \right) M_n^\alpha(y) = \sqrt{n(n+\alpha)} \, M_{n-1}^\alpha(y) \, . \end{aligned} \tag{11}$$

By inspection we see that $J_\pm = J_\mp{}^\dagger$, i.e. $J_\mp$ is the hermitian conjugate operator of $J_\pm$. Note that, unlike the full line, to define the rising and lowering operators the operator N is needed.

To complete eqs. (11), we introduce now $J_3 := N + (\alpha+1)/2$

$$J_3 \, M_n^\alpha(y) = [N + (\alpha+1)/2] \, M_n^\alpha(y) = (n + (\alpha+1)/2) \, M_n^\alpha(y) \, , \tag{12}$$

that, together with $J_\pm$, close the $su(1,1)$ algebra

$$[J_3, J_\pm] = \pm J_\pm \, , \qquad [J_+, J_-] = -2 \, J_3$$

of which, as shown by Eqs.(11, 12), $\{M_n^\alpha(y)\}$ are an unitary irreducible representation. Furtermore, as $\{|n\rangle\}$ is isomorphic to $\{M_n^\alpha(y)\}$, we have

$$\begin{aligned} J_+ \, |n\rangle &= \sqrt{(n+1)(n+\alpha+1)} \, |n+1\rangle \, , \\ J_3 \, |n\rangle &= (n + (\alpha+1)/2) \, |n\rangle \, , \\ J_- \, |n\rangle &= \sqrt{n(n+\alpha)} \, |n-1\rangle . \end{aligned}$$

The Casimir operator $\mathcal{C}$ is :

$$\mathcal{C} = \left(J_3^2 - \frac{1}{2}\{J_+, J_-\} \right) = \frac{\alpha^2 - 1}{4} \, ,$$

that, remembering Eqs.(11, 12), can be written as

$$\left[Y \, D_y{}^2 + D_y + N + \frac{\alpha+1}{2} - \frac{\alpha^2}{2Y} - \frac{Y}{4} \right] M_n^\alpha(y) = 0 \, , \tag{13}$$

reformulation of the associated Laguerre differential equation:

$$y \, L_n^\alpha(y)'' + (\alpha + 1 - y) \, L_n^\alpha(y)' + n \, L_n^\alpha(y) = 0 \, .$$

Eq.(13) defines the operatorial identity of the half-line, similar to Eq.(6) of the line,

$$N \equiv -Y \, D_y{}^2 - D_y - \frac{\alpha+1}{2} + \frac{\alpha^2}{4Y} + \frac{Y}{4} . \tag{14}$$

8. UEA and substructures on the half-line

Also in the half line, the representation is irreducible so that the operators acting on the $L^2(\mathbb{R}^+)$ and $\mathbb{R}^+$ belong to the UEA$[su(1,1)]$, i.e. can be written as

$$\mathcal{O} = \sum c_{\alpha\,\beta\,\gamma} \, J_+{}^\alpha \, J_3{}^\beta \, J_-{}^\gamma = \sum d_{\alpha\,\beta\,\gamma} \, Y^\alpha \, D_y{}^\beta \, N^\gamma$$

and the space of the operators $\{\mathcal{O}[L^2(\mathbb{R}^+)]\}$ and $\{\mathcal{O}[\mathbb{R}^+]\}$) are isomorphic to the UEA$[su(1,1)]$;

$$\{\mathcal{O}[L^2(\mathbb{R}^+)]\} \equiv UEA[su(1,1)] \equiv \{\mathcal{O}[\mathbb{R}^+]\}.$$

The monomials $J_+{}^\alpha \, J_3{}^\beta \, J_-{}^\gamma \in$ UEA$[su(1,1)]$ look to be differential operators of order $\alpha+\gamma$ but the identity (14) can also be read

$$D_y{}^2 \equiv -\frac{1}{Y}\left(D_y + N + \frac{\alpha+1}{2} - \frac{\alpha^2}{2Y} - \frac{Y}{4}\right)$$

so that all the off-diagonal operators can be written as zero or one order differential operators and all the diagonal ones as zero order or equivalent to Eq.(13).

Also the UEA$[su(1,1)]$ contains, in analogy with the UEA$[io(2)]$, ∞-many $su(1,1)$, $su_{k,q}(1,1)$. We can define $L^2_{k,q}(\mathbb{R}^+) := \{M_{kr+q}(\mathbb{R}^+)\}$ and $\mathbb{R}^+_{k,q} := \{|kr+q\rangle\}$ so that for each $k \in \mathbb{N}$ the spaces $L^2(\mathbb{R}^+)$ and $\mathbb{R}^+$ are sums of k subspaces

$$L^2(\mathbb{R}^+) = \oplus_{q=0}^{k-1} L_{k,q}(\mathbb{R}^+) \,, \qquad \mathbb{R}^+ = \oplus_{q=0}^{k-1} R^+_{k,q} \,.$$

The same subspaces structure is found in the spaces of operators $\{\mathcal{O}[L^2(\mathbb{R}^+)]\}$ and $\{\mathcal{O}(\mathbb{R}^+)\}$, i.e. we have

$$\{\mathcal{O}[L^2(\mathbb{R}^+_{k,q})]\} \equiv UEA[su_{k,q}(1,1)] \equiv \{\mathcal{O}[\mathbb{R}^+_{k,q}]\} \,.$$

9. Fourier-like transforms on $\mathbb{R}^+$

Finally to find the conjugate basis of $\{|y\rangle\}$, we need something that plays in $\mathbb{R}^+$ the role of the FT in $\mathbb{R}$, i.e. an integral transform that has as eigenvectors the $\{M_n^\alpha(y)\}$. Remembering the fact that Hermite functions are eigenstates of FT and the well known relations [3]

$$\psi_{2n}(x) = (-1)^n (x^2)^{1/4} M_n^{-1/2}(x^2) \,, \qquad \psi_{2n+1}(x) = (-1)^n x (x^2)^{-1/4}\, M_n^{+1/2}(x^2)$$

two transforms $\mathrm{T}^\pm$ are found

$$[\mathrm{T}^+] \; f(y) := \left[\frac{(-1)^n}{\sqrt{2\pi}} \int_0^\infty dy \frac{\sin[\sqrt{yy'}]}{(yy')^{1/4}}\right] \; f(y) \,,$$

$$[\mathrm{T}^-] \; f(y) := \left[\frac{(-1)^n}{\sqrt{2\pi}} \int_0^\infty dy \frac{\cos[\sqrt{yy'}]}{(yy')^{1/4}}\right] \; f(y) \,,$$

that have $M_n^{\pm 1/2}(y)$ as eigenvectors and $(-1)^n$ as eigenvalues

$$[\mathrm{T}^\pm] \; M_n^{\pm 1/2}(y) = (-1)^n \; M_n^{\pm 1/2}(y').$$

In this way, we find two alternative conjugate basis of $\{|y\rangle\}$

$$|q\rangle^+ = \frac{(-1)^n}{\sqrt{2\pi}} \int_0^\infty dy \; \frac{\sin[\sqrt{qy}]}{(qy)^{1/4}} \, |y\rangle \quad , \qquad |q\rangle^- = \frac{(-1)^n}{\sqrt{2\pi}} \int_0^\infty dy \; \frac{\cos[\sqrt{qy}]}{(qy)^{1/4}} \, |y\rangle$$

and two different related substructures

$$L^2(\mathbb{R}^+) = L_0^2(\mathbb{R}^+)^\pm \oplus L_1^2(\mathbb{R}^+)^\pm = \{M_{2n}^{\pm 1/2}\} \oplus \{M_{2n+1}^{\pm 1/2}\} \qquad \forall n \in \mathbb{N}. \tag{15}$$

Like in $\mathbb{R}$, the subspaces of Eq.(15) (but again not the operators acting on them) can be generalized introducing the fractional transforms $[\mathrm{FrT}^\pm]$

$$[\mathrm{FrT}^\pm]_{2\pi/k} \; f(y) := [T^\pm] \left[e^{2\pi \mathrm{i}(1/k - 1/2)N} \right] \; f(y) \,,$$

where N is given by the identity (14) and is diagonal on $M_n^{\pm 1/2}(y)$

$$[\mathrm{FrT}^\pm]_{2\pi/k} \; M_n^{\pm 1/2}(y) = e^{\,\mathrm{i}2\pi n/k} \; M_n^{\pm 1/2}(y').$$

10. Conclusions

Rigged Hilbert spaces are shown to be more predictive than Hilbert spaces, both in quantum physics and signal processing in optics and informatics, as operators of different cardinality can be considered together.

In rigged Hilbert spaces continuous and discrete bases exist with special functions as transformation matrices between them.

In a RHS, operators of different cardinality can be together generators of a Lie algebra and/or elements of an universal enveloping algebra.

An elaborate algebraic structure is found, inside both the quantum space and the space of operators defined on it. In particular, an infinite set of substructures emerges both in the space of the states and in the space of operators acting on them.

An alternative definition of integral transform, as an operator that has special functions as eigenvectors, has been introduced allowing -unlike for the case of $\mathbb{R}$- to give for $\mathbb{R}^+$ two alternative definitions of conjugate basis.

References

[1] Coutinho S C 1995 *A Primer of Algebraic D-modules* (Cambridge Mass.: Cambridge University Press)
[2] Folland G 1992 *Fourier Analysis and its Applications* (Belmont, California: Wadsworth)
[3] Olver F et al. 2010 *NIST Handbook of Mathematical Functions* (Cambridge Mass.: Cambridge University Press)
[4] Celeghini E and del Olmo M A 2013 *Annals of Physics* **335** 78-85
[5] Böhm A 1978 *The Rigged Hilbert Space and Quantum Mechanics* (Berlin: Springer-Verlag)
[6] Hamermesh M 1962 *Group Theory* (Reading Mass.: Addison-Wesley)
[7] Bargmann V 1954 *Ann. of Math.* **59** 1-46
[8] Ozaktas H M, Zalevsky Z and Alper Kutay M 2001 *The Fractional Fourier Transform* (Chichester: Wiley)
[9] Celeghini E and Tarlini M 1981 *Nuovo Cimento* **65B** 172-180
[10] Bangerezako G 1999 *J. Comput. Appl. Math.* **107** 219-33
[11] Lorente M 2001 *J. Phys. A: Math Gen.* **34** 569-88

Lie groups, algebraic special functions and Jacobi polynomials

Enrico Celeghini[1,2], **Mariano A del Olmo**[2] **and Miguel A Velasco**[3]

[1]Dipartimento di Fisica, Università di Firenze and INFN–Sezione di Firenze
I50019 Sesto Fiorentino, Firenze, Italy

[2]Departamento de Física Teórica and IMUVA, Universidad de Valladolid,
E-47011, Valladolid, Spain

[3]CIEMAT, Madrid, Spain

E-mail: celeghini@fi.infn.it, olmo@fta.uva.es, miguelangel.velasco@ciemat.es

Abstract. We present an attempt of classification of special functions in terms of a set of formal properties mainly based on their "ladder symmetry" algebras. Thus, the "algebraic special functions" are defined. We discuss, here, the case related with the Jacobi polynomials. We construct the relevant objets associated with the corresponding symmetry group $SU(2,2)$: generators, subgroups, invariants and representations. Harmonic analyses based on significant subgroups for the Jacobi functions are developed.

1. Introduction

In a recent series of papers [1, 2, 3] we have revisited the connection between special functions, in particular the classical orthogonal polynomials, with Lie groups, differential equations and Hilbert spaces. The revision started from the Hermite polynomials in the quantum harmonic oscillator to arrive to the study of the symmetries of quantum systems where, in many cases, orthogonal polynomials are involved in the construction of bases of L^2 spaces [4, 5, 6, 7].

Our approach is based on ideas of Wigner (symmetry group) [8, 9] and of Truesdell ("functions with additional properties") [10]. A lot of work has been done in this direction since the seminal book by Wigner, in particular we can mention Vilenkin [11, 12, 13] and Miller [14, 15]. Both have made an impressive work in the study of the symmetries of special functions.

Starting from Miller, Vilenkin and Truesdell we have defined the "algebraic special functions" (ASF) as those functions with the following set of formal properties: they admit a set of ladder operators that span a Lie algebra $\mathcal{G}$ =Lie(G), they support a unitary irreducible representation (UIR) of G and they are bases of Hilbert spaces.

The main facts associated to these ASF are: the defining second order differential equation of the corresponding family of ASF can be obtained by the factorization method [16, 17] from the Casimir of each possible subalgebra of $\mathcal{G}$ or from the Casimir of the full algebra $\mathcal{G}$ associated to a specific representation; the set of operators acting on the Hilbert space is homomorphic to the Universal Enveloping Algebra (UEA) built on $\mathcal{G}$; and the exponential map defines all possible changes of bases in the corresponding Hilbert space. So, they are well adapted to the quantum mechanics framework.

Journal of Physics: Conference Series **597** (2015) 012023 doi:10.1088/1742-6596/597/1/012023

We have got some interesting results that we can summarize as follows:

(i) for Hermite, Laguerre and Legendre polynomials we have obtained symmetry Lie groups of rank one. Thus, the Heisenberg-Weyl group $H(1)$ for algebraic Hermite functions

$$K_n(x) = \frac{e^{-x^2/2}}{\sqrt{2^n n! \sqrt{\pi}}} H_n(x), \qquad \forall n \in \mathbb{N},$$

where $H_n(x)$ are the Hermite polynomials. However the Lie group $SU(1,1)$ describes the Legendre polynomials $P_n(x)$ and also the algebraic Laguerre functions

$$M_n(x) = e^{-x/2} L_n(x), \qquad \forall n \in \mathbb{N},$$

where $L_n(x)$ are the Laguerre polynomials [2].

(ii) Groups of rank two are obtained for the associated Legendre polynomials, spherical harmonics, Gegenbauer polynomials and Jacobi polynomials $J_n^{(\alpha,\alpha)}(x)$. For instance, for the associated Legendre polynomials $P_l^m(x)$ we get the algebraic associated Legendre functions

$$T_l^m(x) = \sqrt{\frac{(l-m)!}{(l+m)!}}\, P_l^m(x), \qquad l \in \mathbb{N},\ m \in \mathbb{Z},\ 0 \le |m| \le l,$$

that as well as the spherical harmonics support a particular UIR of $SO(3,2)$ [1].

We present in this paper a case whose symmetry group has rank three: the Jacobi polynomials $J_n^{(\alpha,\beta)}(x)$. As we have noted in the previous cases the rank of the group is related with the number of independent parameters labeling the orthogonal polynomials. The idea is to connect an operator to each label, such that the polynomials are eigenvectors of it. These operators will span the Cartan subalgebra of the corresponding symmetry Lie algebra.

Some of the results described here have been obtained by Miller many years ago [18, 19] in connection with the wave equation in four dimensions and the Gaussian hypergeometric functions ${}_2F_1(\alpha,\beta,\gamma;x)$, generalization of the Jacobi polynomials $J_n^{(\alpha,\beta)}(x)$ The conformal group $SU(2,2)$ found by Miller is recovered here. Later, in Ref. [20] Floreanini and Vinet construct, following the approach of Miller and the results of Kalnins and Miller [21], a representation of the euclidean algebra in four dimensions $e(4)$ supported by functions of three variables that include the Jacobi polynomials. In these cases the problem of the coexistence of variations of the parameters α,β,γ,n and derivatives in the variable x was solved in a creative manner by means of the introduction of new variables and new functions that include the hypergeometric functions and the Jacobi polynomials, respectively. Thus, the ladder operators act on the space of these auxiliary functions instead on the space of the hypergeometric functions or of Jacobi polynomials where only the corresponding recurrence relations are defined.

In our approach we realize all recurrence relations as true operators in the vector space of Jacobi polynomials $J_n^{(\alpha,\beta)}(x)$ associating the parameters n,α,β to operators, that have the parameters as eigenvalues. We can thus define the operators in the space of the corresponding orthogonal polynomials and describe the algebraic structure in the vector space of these polynomials.

The realization of the ladder operators acting on the Jacobi polynomials allows us to construct objects associated to a Lie group: generators, invariants, subgroups, universal enveloping algebra, representations, etc.

Different subspaces of the Jacobi functions are found to support unitary irreducible representations of subgroups of $SU(2,2)$. In particular, it looks interesting for physical

Journal of Physics: Conference Series **597** (2015) 012023 doi:10.1088/1742-6596/597/1/012023

applications that the SU(2,2)-UIR can be splitted in two subrepresentations dividing the Jacobi functions in bosonic ($SO(3,2)$) and fermionic ($Spin(3,2)$) ones, stressing the relation of the algebraic Jacobi functions (AJF) with the Wigner d_j-matrices [22, 23].

We also present harmonic analysis approaches based on the groups $SU(1,1)$, $SO(3,2)$ and $SU(2,2)$, respectively. In this way subsets of AJF are bases of Hilbert spaces. The operators acting on the associated spaces of square integrable functions belong to the corresponding universal enveloping algebra.

2. Algebraic Jacobi functions

The Jacobi polynomials, $J_n^{(\alpha,\beta)}(x)$, are polynomials of degree $n \in \mathbb{N}$ with $\alpha, \beta \in \mathbb{R}$ and $\alpha > -1$, $\beta > -1$ [24, 25]. Firstly we take three new (discrete) variables (j, m, q) instead of (n, α, β) to label them. The relation between these two set of parameters is

$$j := n + \frac{\alpha+\beta}{2}, \qquad m := \frac{\alpha+\beta}{2}, \qquad q := \frac{\alpha-\beta}{2}.$$

Next we include a x-depending factor related with the integration measure of the Jacobi polynomials. Hence we obtain the called "algebraic Jacobi functions" (AJF)

$$\mathcal{J}_j^{m,q}(x) := \sqrt{\frac{\Gamma(j+m+1)\,\Gamma(j-m+1)}{\Gamma(j+q+1)\,\Gamma(j-q+1)}} \left(\frac{1-x}{2}\right)^{\frac{m+q}{2}} \left(\frac{1+x}{2}\right)^{\frac{m-q}{2}} J_{j-m}^{(m+q,m-q)}(x). \tag{1}$$

For our group-theoretical purposes we impose to (j, m, q) that

$$j \geq |m|, \qquad j \geq |q|, \qquad 2j \in \mathbb{N}, \quad j - m \in \mathbb{N}, \quad j - q \in \mathbb{N}, \tag{2}$$

resulting that (j, m, q) are all together integers or half-integers. The conditions (2) rewritten in terms of the parameters (n, α, β) are different from the original ones. However the change is motivated by the introduction of the normalization inside the functions and by the algebra structure requirements (2). Moreover the AJF verify that $\mathcal{J}_j^{m,q}(x) = 0$ for $j < |q|$ (see eq. (1)) and they can be extended to $j < |m|$ by considering a limit procedure

$$\hat{\mathcal{J}}_j^{m,q}(x) := \lim_{\varepsilon\to 0} \mathcal{J}_j^{m+\varepsilon,q}(x) = \begin{cases} \mathcal{J}_j^{m,q}(x) & \forall\,(j,m,q) \text{ verifying conditions (2)} \\ 0 & j < |m| \end{cases}.$$

Henceforth we remove the hat to the AJF $\hat{\mathcal{J}}_j^{m,q}(x)$. They present additional symmetries hidden inside the Jacobi polynomials:

$$\begin{aligned} \mathcal{J}_j^{m,q}(x) &= \mathcal{J}_j^{q,m}(x), & \mathcal{J}_j^{m,q}(x) &= (-1)^{l-m}\,\mathcal{J}_j^{m,-q}(-x), \\ \mathcal{J}_j^{m,q}(x) &= (-1)^{l-q}\,\mathcal{J}_j^{-m,q}(-x), & \mathcal{J}_j^{m,q}(x) &= (-1)^{m+q}\,\mathcal{J}_j^{-m,-q}(x). \end{aligned}$$

The AJF for m and q fixed verify the orthogonality relation

$$\int_{-1}^{1} \mathcal{J}_j^{m,q}(x)\;(j+1/2)\;\mathcal{J}_{j'}^{m,q}(x)\,dx = \delta_{j\,j'} \tag{3}$$

and the completeness relation

$$\sum_{j=sup(|m|,|q|)}^{\infty} \mathcal{J}_j^{m,q}(x)\;(j+1/2)\;\mathcal{J}_j^{m,q}(y) = \delta(x-y). \tag{4}$$

The differential Jacobi equation satisfied by the Jacobi polynomials $J_n^{(\alpha,\beta)}(x)$

$$\left[(1-x^2)\frac{d^2}{dx^2} - ((\alpha+\beta+2)x + (\alpha-\beta))\frac{d}{dx} + n(n+\alpha+\beta+1)\right] J_n^{(\alpha,\beta)}(x) = 0$$

can be rewritten in terms of the AJF $\mathcal{J}_j^{m,q}(x)$

$$\left[-(1-x^2)\,\frac{d^2}{dx^2} + 2\,x\frac{d}{dx} + \frac{2\,m\,q\,x + m^2 + q^2}{1-x^2} - j(j+1)\right] \mathcal{J}_j^{m,q}(x) = 0\,, \tag{5}$$

where the symmetry under the interchange $m \Leftrightarrow q$ is evident.

An important fact is that the AJFs (1) are essentially the elements of the Wigner d_j-rotation matrices [22, 23] after the change of variable $x = \cos\beta$. Indeed, the relation

$$d^j_{qm}(\beta) = \sqrt{\frac{(j+m)!\,(j-m)!}{(j+q)!\,(j-q)!}}\,(\sin\beta/2)^{m-q}\,(\cos\beta/2)^{m+q}\,J_{j-m}^{(m-q,m+q)}(\cos\beta) \tag{6}$$

allows to write

$$d^j_{qm}(\beta) = \mathcal{J}_j^{m,-q}(\cos\beta).$$

3. Symmetry algebras of the algebraic Jacobi functions

Let us start by introducing, not only the operators X and D_x of the configuration space

$$X\,f(x) = x\,f(x), \qquad D_x\,f(x) = f'(x), \qquad [X, D_x]\,f(x) = -f(x),$$

but three other diagonal operators J, M and Q on the AJF

$$J\,\mathcal{J}_j^{m,q}(x) = j\,\mathcal{J}_j^{m,q}(x), \quad M\,\mathcal{J}_j^{m,q}(x) = m\,\mathcal{J}_j^{m,q}(x), \quad Q\,\mathcal{J}_j^{m,q}(x) = q\,\mathcal{J}_j^{m,q}(x). \tag{7}$$

The procedure consists in transforming the differential-difference equations and difference equations verified by the Jacobi polynomials (that one can find in Refs. [24, 25, 26]) in terms of ladder operators. Since there are many recurrence relations, we start considering the equations (18.9.15) and (18.9.16) of Ref. [24]

$$\begin{aligned}
\frac{d}{dx}J_n^{(\alpha,\beta)}(x) &= \frac{1}{2}(n+\alpha+\beta+1)\,J_{n-1}^{(\alpha+1,\beta+1)}(x)\,,\\
\frac{d}{dx}\left[(1-x)^\alpha(1+x)^\beta J_n^{(\alpha,\beta)}(x)\right] &= -2(n+1)(1-x)^{\alpha-1}(1+x)^{\beta-1}\,J_{n+1}^{(\alpha-1,\beta-1)}(x),
\end{aligned}$$

that in terms of the AJF become

$$A_\pm\,\mathcal{J}_j^{m,q}(x) = \sqrt{(j\mp m)\,(j\pm m+1)}\;\mathcal{J}_j^{m\pm1,q}(x), \tag{8}$$

where

$$A_\pm := \pm\sqrt{1-X^2}\,D_x + \frac{1}{\sqrt{1-X^2}}\,(XM+Q). \tag{9}$$

By inspection only the parameter m of AJF changes and varies in ±1. Both operators $A_\pm$ together with $A_3 := M$ close a $su(2)$ Lie algebra (denoted by $su_A(2)$) that commutes with J and Q

$$[A_3, A_\pm] = \pm A_\pm\,, \qquad [A_+, A_-] = 2A_3.$$

In this way the set of AJF such that $2j \in \mathbb{N}$, $j - m \in \mathbb{N}$ and $-j \leq m \leq j$ support the $(2j+1)$-dimensional UIR of the Lie group $SU_A(2)$ independently from the value of q.

Taking into account the differential realization (9) of $A_\pm$ and the Casimir, $\mathcal{C}_A$, of $su_A(2)$

$$[\mathcal{C}_A - J(J+1)]\ \mathcal{J}_j^{m,q}(x) \equiv \left[A_3^2 + \frac{1}{2}\{A_+, A_-\} - J(J+1)\right]\ \mathcal{J}_j^{m,q}(x) = 0\ , \tag{10}$$

we recover the Jacobi differential equation (5) in operator terms

$$\left[-(1-X^2)D_x^2 + 2XD_x + \frac{1}{1-X^2}(2XMQ + M^2 + Q^2) - J(J+1)\right] \mathcal{J}_j^{m,q}(x) = 0, \tag{11}$$

which can also be obtained by two factorized equations that reproduce the Jacobi equation

$$\begin{aligned} &[A_+\,A_- - (J+M)\,(J-M+1)]\ \mathcal{J}_j^{m,q}(x) = 0\ ,\\ &[A_-\,A_+ - (J-M)\,(J+M+1)]\ \mathcal{J}_j^{m,q}(x) = 0\ . \end{aligned} \tag{12}$$

Note that eqs. (10) and (12) are particular cases of a general rule: the defining differential equation can be recovered applying to the ASF the second order Casimir operator of any involved algebra and sub-algebra as well as any diagonal product of ladder operators.

Now taking into account the symmetry $m \Leftrightarrow q$ of $\mathcal{J}_j^{m,q}(x)$, we construct the algebra $su_B(2)$ of operators $B_\pm, B_3$, that change q leaving j and m unchanged, from $A_\pm$ and A_3 by the interchange

$$(A_\pm, A_3) \quad \overset{M \Leftrightarrow Q}{\Longleftrightarrow} \quad (B_\pm, B_3).$$

Obviously the counterparts of the expressions (8)–(12) for $B_\pm, B_3$, are obtained by the changes $m \Leftrightarrow q$ and $M \Leftrightarrow Q$ in these expressions.

Since the operators $A_\pm$ and A_3 commute with $B_\pm$ and B_3 we get, on the space of $\mathcal{J}_j^{m,q}(x)$ with j fixed, the symmetry algebra

$$su_A(2) \oplus su_B(2).$$

So, the set of $\mathcal{J}_j^{m,q}(x)$ (for fixed j and $-j \leq m \leq j$, $-j \leq q \leq j$) determines a UIR of $SU_A(2) \otimes SU_B(2)$.

4. $su(1,1)$-ladder operators

Proceeding in a similar way, from the remaining difference and differential-difference relations we can obtain eight new infinitesimal generators. In a first step we construct two ones ($C_\pm$ such that $C_\pm^\dagger = C_\mp$) that determine a $su(1,1)$ algebra ($su_C(1,1)$):

$$\begin{aligned} C_+ &:= +\frac{(1+X)\sqrt{1-X}}{\sqrt{2}}\,D_x - \frac{1}{\sqrt{2(1-X)}}\,(X\,(J+1) - (J+1+M+Q))\,,\\ C_- &:= -\frac{(1+X)\sqrt{1-X}}{\sqrt{2}}\,D_x - \frac{1}{\sqrt{2(1-X)}}\,(X\,J - (J+M+Q))\,. \end{aligned} \tag{13}$$

They act on the space $\{\mathcal{J}_j^{m,q}\}$ (for j, m, q integer and half-integer such that $j \geq |m|, |q|$)

$$\begin{aligned} C_+\,\mathcal{J}_j^{m,q}(x) &= \sqrt{(j+m+1)(j+q+1)}\ \mathcal{J}_{j+1/2}^{m+1/2,\,q+1/2}(x),\\ C_-\,\mathcal{J}_j^{m,q}(x) &= \sqrt{(j+m)(j+q)}\ \mathcal{J}_{j-1/2}^{m-1/2,\,q-1/2}(x). \end{aligned} \tag{14}$$

They change all the parameters (j, m, q) by $\pm 1/2$. Moreover

$$[C_+, C_-] = -2C_3, \qquad [C_3, C_\pm] = \pm C_\pm, \qquad C_3 := J + \frac{1}{2}(M+Q) + \frac{1}{2}.$$

The Casimir $\mathcal{C}_C$ of $su_C(1,1)$ is

$$\mathcal{C}_C \mathcal{J}_j^{m,q}(x) \equiv \left[C_3^2 - \frac{1}{2}\{C_+, C_-\} \right] \mathcal{J}_j^{m,q}(x) = \frac{1}{4}\left[(m-q)^2 - 1 \right] \mathcal{J}_j^{m,q}(x).$$

The Jacobi equation (11) can be derived from $\mathcal{C}_C$

$$\begin{aligned} &\left[\mathcal{C}_C - \frac{1}{4}(M-Q)^2 + \frac{1}{4} \right] \mathcal{J}_j^{m,q}(x) \\ &\qquad \equiv \left[C_3^2 - \tfrac{1}{2}\{C_+, C_-\} - \tfrac{1}{4}(M-Q)^2 + 1/4 \right] \mathcal{J}_j^{m,q}(x) = 0, \end{aligned} \tag{15}$$

and also from

$$\begin{aligned} &[C_+ C_- - (J+M)(J+Q)] \, \mathcal{J}_j^{m,q}(x) = 0, \\ &[C_- C_+ - (J+1+M)(J+1+Q)] \, \mathcal{J}_j^{m,q}(x) = 0. \end{aligned}$$

Analyzing (15) we see that since $(m-q) = 0, \pm 1, \pm 2, \pm 3, \cdots$ the IR of su(1, 1) with Casimir $\mathcal{C}_C = (m-q)^2/4 - 1/4 = -1/4, 0, 3/4, 2, 15/4, \cdots$ are obtained. Hence, the set of AJF supports many infinite-dimensional UIR of the discrete series of $SU_C(1,1)$ [27].

We can find six more ladder operators, $D\pm, E\pm, F\pm$, whose explicit differential expressions are easily obtained from those of $C_\pm$ (13) according to the changes suggested by

$$\begin{aligned} D_\pm(X, D_x, J, M, Q) &= C_\pm(-X, -D_x, J, M, -Q), \\ E_\pm(X, D_x, J, M, Q) &= C_\pm(-X, -D_x, J, -M, Q), \\ F_\pm(X, D_x, J, M, Q) &= -C_\pm(X, D_x, J, -M, -Q). \end{aligned} \tag{16}$$

All the relations where the operators $C_\pm$ are involved can be rewritten for these new operators taking into account (16).

The generators $A_\pm, B_\pm, C_\pm, D_\pm, E_\pm, F_\pm, J, M, Q$ close the Lie algebra $su(2,2)$ [3]. Its quadratic Casimir has the form

$$\begin{aligned} \mathcal{C}_{su(2,2)} &= \tfrac{1}{2}\left(\{A_+, A_-\} + \{B_+, B_-\} - \{C_+, C_-\} - \{D_+, D_-\} - \{E_+, E_-\} - \{F_+, F_-\} \right) \\ &\qquad + 2J(J+1) + M^2 + Q^2 + \frac{1}{2} \equiv -\frac{3}{2}. \end{aligned}$$

Once more taking into account the differential realization of the operators involved in the Casimir we recover the Jacobi equation.

The AJF support a UIR of the group $SU(2,2)$ with the value -3/2 of $\mathcal{C}_{su(2,2)}$ such that the integer and half-integer values of (j, m, q) are putted all together.

A more detailed discussion can be found in Ref. [3]

5. Symmetries for fermion and boson states

The operators $A_\pm$ ($B_\pm$) change only a label m (q) of ± 1 but the operators $C_\pm, D_\pm, E_\pm, F_\pm$ change all the three labels (j, m, q) by $\pm 1/2$. Hence, we can obtain operators that only change j by ± 1 and leave the other two parameters invariant by composing the action of pairs of operators, for instance $F_\pm C_\pm$ (or $C_\pm F_\pm$, $D_\pm E_\pm$, $E_\pm D_\pm$). They are second order differential operators but they can be reduced to first order ones when act on $\mathcal{J}_j^{m,q}$ by means of the Jacobi equation. We have seen that the $\mathcal{J}_j^{m,q}$ have $j \geq |m|$ and $j \geq |q|$, but now we have to consider separately specific values of m and q.

Thus we define two conjugate hermitian operators

$$K_+ := F_+ C_+ \frac{1}{\sqrt{(J+1)^2 - Q^2}}, \qquad K_- := F_- C_- \frac{1}{\sqrt{J^2 - Q^2}},$$

that can be written as

$$\begin{aligned} K_+ &:= \left(-(1-X^2)\,D_x + X\,(J+1) + \frac{MQ}{J+1}\right) \frac{J+1}{\sqrt{(J+1)^2 - Q^2}}, \\ K_- &:= \left((1-X^2)\,D_x + X\,J + \frac{MQ}{J}\right) \frac{J}{\sqrt{J^2 - Q^2}}, \end{aligned} \tag{17}$$

with the condition that they act on $\mathcal{J}_j^{m,q}$ with $j \geq |m| > |q|$. Their explicit action on $\mathcal{J}_j^{m,q}$ is

$$\begin{aligned} K_+\, \mathcal{J}_j^{m,q}(x) &= \sqrt{(j+1)^2 - m^2}\, \mathcal{J}_{j+1}^{m,q}(x), \\ K_-\, \mathcal{J}_j^{m,q}(x) &= \sqrt{j^2 - m^2}\, \mathcal{J}_{j-1}^{m,q}(x). \end{aligned}$$

The operators $K_\pm$ together with $K_3 := J + 1/2$ close the Lie algebra $su_K(1,1)$. Hence $\{\mathcal{J}_j^{m,q};\ m, q \text{ fixed}, j \geq |m| > |q|\}$ is a basis of the $SU(1,1)$–UIRs with Casimir $\mathcal{C}_K = m^2 - 1/4$.

In the case $|m| < |q|$ it is enough to interchange $M \Leftrightarrow Q$ (or $m \Leftrightarrow q$) in $K_\pm$. However, when $|m| = |q|$ the action of K_- is not well defined in eq. (17) for $j = |m| = |q|$, but we can extend its definition to this case by considering the limit

$$K_- := \lim_{\epsilon \to 0} \left[\left((1-X^2)\,D_x + X\,J + \frac{(M+\epsilon)(Q+\epsilon)}{J}\right) \frac{J}{\sqrt{J^2 - (Q+\epsilon)^2}} \right].$$

In any case the $\mathcal{J}_j^{m,q}$ determine, for fixed m and q, a UIR of $SU(1,1)$. Thus, if $|m| \geq |q|$ we have eqs. (5) with $j = |m|, |m|+1, |m|+2 \ldots$ and Casimir $\mathcal{C}_K = m^2 - 1/4$, while for $|m| < |q|$ we have to exchange q and m everywhere. These $SU(1,1)$–UIRs contain only states with integer or half-integer values of its labels. From a physical point of view the Wigner d_j-matrices (6) do not mix integer and half-integer spins, i.e. bosons and fermions.

These $SU(1,1)$ groups cannot, in general, be extended to larger groups except when $q = 0$ or $m = 0$. Then, the $\mathcal{J}_j^{m,0}$ (with $j, m \in \mathbb{Z}$ and $j \geq |m|$) are related to the associated Legendre polynomials $P_{n+\alpha}^{\alpha}$ and support a UIR of $SO(3,2)$. In this case we get not only $K_\pm$ but also $A_\pm$ that allow us to obtain the whole algebra $so(3,2)$ described in [1]. For $\mathcal{J}_j^{0,q}$ the results are similar.

The "fermions" states $\{\mathcal{J}_j^{m,\pm 1/2}\}$ and $\{\mathcal{J}_j^{\pm 1/2,q}\}$ are related to the same algebra $so(3,2)$ but to a representation of its covering group $Spin(3,2)$ [28] with all the parameters half-integer.

6. L^2–functions spaces and $\mathcal{J}_j^{m,q}(x)$

It is well known that the orthogonal polynomials are bases on L^2–functions spaces [29]. In this case according to the relations (3) and (4) the set $\{\mathcal{J}_j^{m,q}(x) \mid m,q \text{ fixed}\}_{j=\sup(|m|,|q|)}^{\infty}$ is a basis of the L^2–functions defined in the interval $\mathbb{E} = (-1,1)$. Moreover the AJF are transition matrices between continuous and discrete bases in representation spaces.

Effectively, let us consider the simplest case. From expressions (3) and (4) we are able to consider j and x as conjugate variables on the same space after fixed m and q. Let $\{|j,(m,q)\rangle;\ m,q \text{ fixed},\ j = \sup(|m|,|q|)\}$ be the set of eigenvectors of K_3 and $\mathcal{C}_K$

$$\begin{aligned} K_3\ |j,(m,q)\rangle &= (j+\tfrac{1}{2})\ |j,(m,q)\rangle, \\ \mathcal{C}_K\ |j,(m,q)\rangle &= (m^2-\tfrac{1}{4})\ |j,(m,q)\rangle. \end{aligned} \tag{18}$$

The action of $K_\pm$ on these vectors gives

$$\begin{aligned} K_+\ |j,(m,q)\rangle &= \sqrt{(j+1)^2-m^2}\ |j+1,(m,q)\rangle, \\ K_-\ |j,(m,q)\rangle &= \sqrt{j^2-m^2}\ |j-1,(m,q)\rangle. \end{aligned} \tag{19}$$

So the set $\{|j,(m,q)\rangle\}$ is a basis of the space support of the IUR of $SU(1,1)$ determined by expressions (18) and (19), i.e.

$$\langle j,(m,q)|j',(m,q)\rangle = \delta_{j\,j'}, \qquad \sum_{j=sup(|m|,|q|)}^{\infty} |j,(m,q)\rangle\ \langle j,(m,q)| = \mathcal{I}. \tag{20}$$

The orthogonality and completeness of $\{|j,(m,q)\rangle\}$ (20) and the AJF allows us to define the vectors

$$|x,(m,q)\rangle := \sum_{j=sup(|m|,|q|)}^{\infty} |j,(m,q)\rangle\ \sqrt{j+1/2}\ \mathcal{J}_j^{m,q}(x)$$

which are a basis of the space $\mathbb{E} = (-1,1)$

$$\langle x,(m,q)|x',(m,q)\rangle = \delta(x-x'), \qquad \int_{-1}^{+1} |x,(m,q)\rangle\, dx\, \langle x,(m,q)| = \mathcal{I}\,.$$

This implies that the $\{\mathcal{J}_j^{m,q}(x)\}$ are the transition matrices between the two bases, i.e.

$$\mathcal{J}_j^{m,q}(x) = \frac{1}{\sqrt{j+1/2}}\ \langle x,(m,q)|j,(m,q)\rangle = \frac{1}{\sqrt{j+1/2}}\ \langle j,(m,q)|x,(m,q)\rangle\ ,$$

and

$$|j,(m,q)\rangle = \int_{-1}^{+1} |x,(m,q)\rangle\ \sqrt{j+1/2}\ \mathcal{J}_j^{m,q}(x)\, dx\,.$$

Note that there is a basis associated to each triple $(m,q,\mathfrak{g} \in SU(1,1))$.

The case of two discrete variables, (j,m) or (j,q), is related to the group $SO(3,2)$ and has been discussed in [1], so it will be not reconsidered here.

The case of three variables (j,m,q) corresponds to $SU(2,2)$ and now both m and q are modified by the group action. Like in the previous case let us consider the set of eigenvectors of J, M, Q and also of the Casimir operators $\mathcal{C}_A$, $\mathcal{C}_B$, $\mathcal{C}_C$, $\mathcal{C}_D$, $\mathcal{C}_E$ and $\mathcal{C}_F$

$$\{|j,m,q\rangle\ ;\ m,q \in \mathbb{Z}/2,\ j \geq |m|,\ j \geq |q|,\ j-m \in \mathbb{N},\ j-q \in \mathbb{N}\},$$

where $\mathbb{Z}/2 = \{0, \pm 1/2, \pm 1, \pm 3/2, \ldots\}$. Hence the action, for instance, of the operators $C_\pm$ on these vectors is easy to compute obtaining

$$C_+|j,m,q\rangle = \sqrt{(j+m+1)(j+q+1)}\,|j+1/2, m+1/2, q+1/2\rangle,$$
$$C_-|j,m,q\rangle = \sqrt{(j+m)(j+q)}\,|j-1/2, m-1/2, q-1/2\rangle,$$

and similarly for the other generators of $su(2,2)$. These vectors support the IUR of $SU(2,2)$ above described. So

$$\langle j,m,q|j',m',q'\rangle = \delta_{j\,j'}\,\delta_{m\,m'}\,\delta_{q,q'}, \qquad \sum_{j,m,q} |j,m,q\rangle\,\langle j,m,q| = \mathcal{I}.$$

We can define a new set of vectors

$$\{|x,m,q\rangle\ ;\ x\in\mathbb{E}=(-1,1)\subset\mathbb{R},\ q\in\mathbb{Z}/2,\ m-q\in\mathbb{Z}\}$$

in terms of the vectors $|j,m,q\rangle$ and the AJF $\mathcal{J}_j^{m,q}(x)$ by

$$|x,m,q\rangle := \sum_{j=sup(|m|,|q|)}^{\infty} |j,m,q\rangle\,\sqrt{j+1/2}\,\mathcal{J}_j^{m,q}(x).$$

Orthonormality and completeness are easily obtained

$$\langle x,m,q|x',m',q'\rangle = \delta(x-x')\,\delta_{m\,m'}\,\delta_{q\,q'}, \qquad \sum_{m,q}\int_{-1}^{+1} |x,m,q\rangle\,dx\,\langle x,m,q| = \mathcal{I}.$$

The space in this case can be identified as $\mathbb{E}\times\mathbb{Z}\times\mathbb{Z}/2$ and is the direct sum of the spaces $\mathbb{E}_{m,q}$ (associated to the configuration space $E=(-1,1)\subset\mathbb{R}$) with m and q fixed,

$$\mathbb{E}\times\mathbb{Z}\times\mathbb{Z}/2 = \bigcup_{m-q\in\mathbb{Z}}\ \bigcup_{q\in\mathbb{Z}/2} \mathbb{E}_{m,q},$$

where $\mathbb{Z}\times\mathbb{Z}/2$ is related to the set of pairs $\{(m-q,q)\}$ since m and q are together integer or half-integer.

The $\{\mathcal{J}_j^{m,q}(x)\}$ are the transition matrices between the two bases $\{|j,m,q\rangle\}$ and $\{|x,m,q\rangle\}$

$$\mathcal{J}_j^{m,q}(x) = \frac{1}{\sqrt{j+1/2}}\,\langle x,m,q|j,m,q\rangle = \frac{1}{\sqrt{j+1/2}}\,\langle j,m,q|x,m,q\rangle\,.$$

Like in Ref. [1, 2] the role of the AJF as transition matrices reflects the fact that the algebra generators can be seen as differential operators in $\mathbb{E}\times\mathbb{Z}\times\mathbb{Z}/2$ or algebraic operators in the space of labels $\mathbb{N}/2\times\mathbb{N}\times\mathbb{N}$ related to the set of triplets $\{(j, j-m, j-q)\}$. This allows to make explicit the Lie algebra structure in contrast with Ref. [11, 12, 14].

An arbitrary vector $|f\rangle \in L^2(\mathbb{E},\mathbb{Z},\mathbb{Z}/2)$ can be written as

$$|f\rangle = \sum_{m,q=-\infty}^{\infty}\int_{-1}^{+1} dx\,|x,m,q\rangle\,f^{m,q}(x) = \sum_{m,q=-\infty}^{+\infty}\ \sum_{j=sup(|m|,|q|)}^{\infty} |j,m,q\rangle\,f_j^{m,q},$$

where

$$f^{m,q}(x) = \langle x,m,q|f\rangle = \sum_{j=sup(|m|,|q|)}^{\infty} \sqrt{j+1/2}\,\mathcal{J}_j^{m,q}(x)\,f_j^{m,q},$$
$$f_j^{m,q} = \langle j,m,q|f\rangle = \int_{-1}^{+1} dx\,\sqrt{j+1/2}\,\mathcal{J}_j^{m,q}(x)\,f^{m,q}(x)\,.$$

So, the L^2–functions, $f^{m,q}(x)$, defined on $(\mathbb{E},\mathbb{Z},\mathbb{Z}/2)$ can be developped as

$$f^{m,q}(x) = \sum_{m,q=-\infty}^{\infty} \sum_{j=sup(|m|,|q|)}^{\infty} \sqrt{j+1/2}\,\mathcal{J}_j^{m,q}(x)\,f_j^{m,q}.$$

Note that since $\{\mathcal{J}_j^{m,q}\}$ is a basis of a UIR of $SU(2,2)$ and, at the same time, a basis of the L^2–functions defined on $(\mathbb{E},\mathbb{Z},\mathbb{Z}/2)$ then $L^2(\mathbb{E},\mathbb{Z},\mathbb{Z}/2)$ supports the same UIR of $SU(2,2)$. This implies that every change of basis in $L^2(\mathbb{E},\mathbb{Z},\mathbb{Z}/2)$ is related to an element $\mathfrak{g}$ of $SU(2,2)$ and that the operators acting on $L^2(\mathbb{E},\mathbb{Z},\mathbb{Z}/2)$ can be written inside the UEA$[su(2,2)]$.

7. Conclusions

The algebraic special functions are relevant since they constitute a bridge between second order differential equations in one dimension and Lie algebras. Moreover they allow to establish a homorphism between the UEA of the corresponding symmetry Lie algebra and the vector space of the operators defined on the L^2–functions. In particular the algebraic Jacobi functions coincide with the Wigner d_j-matrices, which play an important role in Quantum Mechanics. This fact enhances the interest of the algebraic special functions in physical applications.

References

[1] Celeghini E and del Olmo M A 2013 *Ann. of Phys.* **333** 90
[2] Celeghini E and del Olmo M A 2013 *Ann. of Phys.* **335** 78
[3] Celeghini E, del Olmo M A and Velasco M A 2014 Lie groups of Jacobi polynomials and Wigner d-matrices (*Preprint* arXiv: 1402.5217)
[4] Calzada J A, Negro J and del Olmo M A 2006 *J. Math. Phys.* **47** 043511
[5] Calzada J A, Kuru S, Negro J and del Olmo M A 2008 *J. Phys. A: Math. Theor.* **41** 255201
[6] Calzada J A, Kuru S, Negro J and del Olmo M A 2012 *Ann. Phys.* **327** 808
[7] Celeghini E, Kuru S, Negro J and del Olmo M A 2013 *Ann. Phys.* **332** 27
[8] Wigner E P 1955 *The application of group theory to the special functions of mathematical physics* (Princeton: Princeton Univ. Press)
[9] Talman J D 1968 *Special functions: a group theoretic approach* (New York: Benjamin)
[10] Truesdell C 1948 *Annals in Math. Studies* vol **18** (Princeton: Princeton Univ. Press)
[11] Vilenkin N Ja 1968 *Special Functions and the Theory of Group Representations* (Providence: Amer. Math. Soc.)
[12] Vilenkin N Ja and Klimyk A U 1991, 1992, 1993 *Representation of Lie Groups and Special Functions* vol 1, 2, 3 (Dordrecht: Kluwer)
[13] Vilenkin N Ja and Klimyk A U 1995 *Representation of Lie Groups and Special Functions: Recent Advances* (Dordrecht: Kluwer)
[14] Miller Jr W 1968 *Lie Theory and Special Functions* (New York: Academic Press)
[15] Miller Jr W 1977 *Symmetry and Separation of Variables* (Reading, MA: Addison-Wesley)
[16] Schrödinger E 1941 *Proc. Roy. Irish Acad.* **A46**, 183; **A47** 53
[17] Infeld L and Hull T E 1951 *Rev. Mod. Phys.* **23** 21
[18] Miller Jr W 1972 *SIAM J. Math. Anal.* **3** 31; *J. Math. Phys.* **13** 1393
[19] Miller Jr W 1973 *SIAM J. Math. Anal.* **4** 314; *SIAM J. Appl. Math.* **25** 226
[20] Floreanini R and Vinet L 1996 *J. Math. Phys.* **37** 4135
[21] Kalnins E G and Miller W 1989 *Rocky Mountain J.* **19** 223
[22] Wigner E P 1927 *Zeitschrift für Physik* **43** 624
[23] Biedenharn L C and Louck J D 1981 *Angular Momentum in Quantum Mechanics* (Reading, MA: Addison-Wesley)

Journal of Physics: Conference Series **597** (2015) 012023 doi:10.1088/1742-6596/597/1/012023

[24] Olver F W J, Lozier D W, Boisvert R F and Clark C W (edit.) 2010 *NIST Handbook of Mathematical Functions* (New York: Cambridge Univ. Press)
[25] Luke Y L 1969 *The Special Functions and Their Approximations* Vol.1 (San Diego: Academic Press)
[26] Abramowitz M and Stegun I 1972 *Handbook of Mathematical Functions with Formulas, Graphs and Mathematical Tables* (San Diego: Dover)
[27] Bargmann V 1947 *Ann. of Math.* **48** 368
[28] Lawson H B and M.L. Michelsohn M L 1989 *Spin Geometry* (Princeton: Princeton Univ. Press).
[29] Celeghini E and del Olmo M A 2014 Quantum physics and signal processing in rigged Hilbert spaces by means of special functions, Lie algebras and Fourier and Fourier-like transforms (to be publish in J. Phys.: Conf. Ser. 2015; *Preprint* arXiv: 1411.3263)

Warped product of Hamiltonians and extensions of Hamiltonian systems

Claudia Maria Chanu, Luca Degiovanni and Giovanni Rastelli

Dipartimento di Matematica, Università di Torino, via Carlo Alberto 10, 10123, Torino, Italia

E-mail: `claudiamaria.chanu@unito.it`, `luca.degiovanni@unito.it`, `giovanni.rastelli@unito.it`

Abstract. The aim of this article is to set the theory of extensions of Hamiltonian systems, developed in a series of previous papers, in the framework of warped products of Hamiltonian systems. Some illustrative examples and plots are given.

1. Introduction

In Sec. 2 the definition of warped product of Riemannian manifolds is recalled. This geometric concept suggests the dynamic idea of warped product of Hamiltonian systems. Some properties are introduced as well as two examples of application: the geodesic Hamiltonian on the torus $\mathbb{T}^2$, the Jacobi-Calogero and Wolfes systems. In Sec. 3 we motivate and introduce the extension procedure (developed in a series of papers [4]-[9]) which transforms natural Hamiltonians with N degrees of freedom into $(N+1)$-dimensional ones possessing a non-trivial first integral of higher degree in the momenta. The interesting fact is that the new integral can be computed in a simple recursive way. The extended Hamiltonians perfectly fit in the framework of warped products. Moreover, the geometric necessary conditions for the extension procedure are deeply related with the geometry of warped products of Riemannian manifolds. The fundamental examples of Hamiltonian systems which illustrate the extension procedure are the well-known Tremblay-Turbiner-Winternitz (TTW) system [11] , [19], (that is a generalization of the Jacobi -Calogero and Wolfes systems) and the anisotropic caged oscillator [11], [20]. The dynamics of the TTW system and of the geodesic flow on $\mathbb{T}^2$ as warped product of Hamiltonians is sketched in three figures. The superintegrability of the TTW Hamiltonian is then proved in a new way thanks to its identification as an extended Hamiltonian. Moreover, it is straightforward to see that the extension of an N-dimensional superintegrable systems is an $N+1$-dimensional superintegrable system. Then, the extension procedure allows the construction of new superintegrable systems from known ones [7]. In Sec. 4 the extension procedure is summarized in a simple algorithm.

2. Warped products

Given n Riemannian manifolds (Q_k, ds_k^2), $(k=1,\dots,n)$, their warped product is the Riemannian manifold $Q = \times_{k=1}^n Q_k$, with metric $ds^2 = \alpha^k ds_k^2$, where $\alpha^k : Q \to \mathbb{R}$ or $\mathbb{C}$. A simple example of a non-trivial warped product of Riemannian manifolds is the Euclidean plane with metric $ds^2 = dr^2 + r^2 d\theta^2$ in polar coordinates (r,θ) which can be interpreted as warped product of $(\mathbb{R}^1)^+ \times \mathbb{S}^1$ with $\alpha^1 = 1$, $\alpha^2 = r^2$.

Given a Riemannian manifold, the possibility of writing it as warped product of lower dimensional manifolds is characterized by several theorems involving its geodesic completeness, curvature, etc. (see for example [18]).

The concept of warped product can be straightforwardly generalized to Hamiltonian systems. Let H_k be n non-degenerate Hamiltonians on T^*Q_k with canonical coordinates (q^{i_k}, p_{j_k}). Let α^k be n functions defined on $Q = \times_{k=1}^n Q_k \to \mathbb{R}$ (we do not consider here the more general case of α_k defined on the whole T^*Q). Let

$$H = \alpha^k H_k \tag{1}$$

be a Hamiltonian in $\times_{k=1}^n T^*Q_k = T^*Q$ with the obvious canonical coordinates.

The Hamiltonian vector field (on T^*Q_k) generated by H_k is

$$X_k = \frac{\partial H_k}{\partial p_{i_h}}\frac{\partial}{\partial q^{i_h}} - \frac{\partial H_k}{\partial q^{i_h}}\frac{\partial}{\partial p_{i_h}},$$

hence, the Hamiltonian vector field (on Q) generated by H is given by

$$\begin{aligned} X_H &= \alpha^k\frac{\partial H_k}{\partial p_{i_h}}\frac{\partial}{\partial q^{i_h}} - \alpha^k\frac{\partial H_k}{\partial q^{i_h}}\frac{\partial}{\partial p_{i_h}} - H_k\frac{\partial \alpha^k}{\partial q^{i_h}}\frac{\partial}{\partial p_{i_h}} \\ &= \alpha^k X_k - H_k\frac{\partial \alpha^k}{\partial q^{i_h}}\frac{\partial}{\partial p_{i_h}} \end{aligned}$$

Therefore, the functions

$$\alpha^k = \frac{dt_k}{d\tau}, \tag{2}$$

determine a position-dependent time-scaling and we can write

$$\frac{dq^{i_k}}{d\tau} = \frac{d\hat{q}^{i_k}}{d\tau}, \tag{3}$$

meaning that the Hamilton equations for the configuration coordinates on each Q_k coincide with the Hamilton equations for the configuration coordinates on Q rescaled for the Hamiltonian parameter τ.

An example of such a rescaling (for $k = 1$) is provided by the classical Jacobi transform, mapping a natural Hamiltonian $H_1 = G + V - E = \frac{1}{2}g^{ij}p_ip_j + V(q^h) - E$ $(E \in \mathbb{R})$ in the geodesic Hamiltonian $H = \frac{G}{E-V}$ whose metric is conformal to the original one. In this case $\alpha^1 = (E-V)^{-1}$ and the two Hamiltonian vector fields are parallel on the submanifold $H_1 = 0$.

Moreover, the Hamilton equations for the momenta on Q

$$\frac{dp_{i_k}}{d\tau} = \frac{d\hat{p}_{i_k}}{d\tau} - \frac{\partial \alpha^h}{\partial q^{i_k}}H_h. \tag{4}$$

mean that the momentum components of the Hamiltonian vector field of H on T^*Q coincide with the Hamiltonian fields of each H_k rescaled according to the parameter τ and modified by a term in each point equal to the linear combination of the differentials of the α^k with coefficients given by the energies in those points of the non-interacting systems H_k. We may say that in H the interaction among the H_k is determined in each point by the energies H_k, assumed free, oriented according to the differentials of the α^k. The α^k themselves determine the time-scaling among the Hamiltonian parameters of each H_k and that of H.

For any function f we have

$$\frac{df}{d\tau} = \{H, f\} = \{\alpha^k, f\}H_k + \alpha^k\{H_k, f\},$$

where $\{,\}$ are the canonical Poisson brackets on T^*Q. Hence, if $f = H_m$ then

$$\frac{dH_m}{d\tau} = -H_k\{H_m, \alpha^k\}, \tag{5}$$

where $\{H_m, \alpha^k\}$ represents the "time-variation" of the function α^k with respect to the Hamiltonian parameter of the system H_m assumed non-interacting with the other systems. We may therefore say that each system of Hamiltonian H_m is interacting with the whole system H with an exchange of energy determined by (5).

If the Hamiltonians $H_k = \frac{1}{2}g_k^{i_k j_k} p_{i_k} p_{j_k} + V(q^{i_k})$ are natural, then also their warped product H, as defined here, is a natural Hamiltonian, with the warped product of the (Q_k, g_k) as configuration manifold.

2.1. *Example: the geodesic Hamiltonian on $\mathbb{T}^2$*

The Torus $\mathbb{T}^2$ with the metric inherited from $\mathbb{E}^3$ may be represented by

$$\begin{aligned} x &= (R + r\cos u)\cos v, \\ y &= (R + r\cos u)\sin v, \\ z &= r\sin u, \end{aligned}$$

$(r, R \in \mathbb{R})$ with metric tensor

$$g_{uu} = r^2, \quad g_{uv} = 0, \quad g_{vv} = (R + r\cos u)^2.$$

The torus is then $\mathbb{S}^1 \times \mathbb{S}^1$ with metric tensor given by the warped product $ds^2 = r^2 du^2 + (R + r\cos u)^2 dv^2$. The geodesic flow has Hamiltonian

$$H = \frac{1}{2r^2}p_u^2 + \frac{1}{2(R + r\cos u)^2}p_v^2,$$

that is the warped product of $H_1 = \frac{1}{2}p_u^2$, $H_2 = \frac{1}{2}p_v^2$ with $\alpha^1 = \frac{1}{r^2}$, $\alpha^2 = \frac{1}{(R+r\cos u)^2}$. To both H_1 and H_2 can be added scalar potentials depending on u and v respectively. In Figure 1 a segment of geodesic is shown, together with its projection on the spaces $u = p_u = 0$, $v = p_v = 0$, respectively. The graphs of the percentage time variations of H_1 and H_2 are also drawn. The function H_2 is a constant of the motion for H.

2.2. *Example: the Jacobi-Calogero and Wolfes Hamiltonians*

Let three points (x^1, x^2, x^3) with unitary masses on a line be interacting with scalar potential

$$V_{JC} = \frac{k}{(x^1 - x^2)^2} + \frac{k}{(x^2 - x^3)^2} + \frac{k}{(x^3 - x^1)^2},$$

or

$$V_W = \frac{k}{(x^1 + x^2 - 2x^3)^2} + \frac{k}{(x^2 + x^3 - 2x^1)^2} + \frac{k}{(x^3 + x^1 - 2x^2)^2}.$$

These functions are respectively known as Jacobi-Calogero's (V_{JC}) and Wolfes' (V_W) potentials, and they are more generally considered with additional harmonic terms or added together (see

Journal of Physics: Conference Series **597** (2015) 012024 doi:10.1088/1742-6596/597/1/012024

[3] and references therein). The Jacobi-Calogero potential is a two-body interaction, while the Wolfes is a three-body interaction. Both potentials are integrable and superintegrable in both their classical and quantum formulations. The corresponding Hamiltonians share the same inertial term

$$\frac{1}{2}\left(p_1^2+p_2^2+p_3^2\right),$$

and the coordinates (x^i) can be considered as Cartesian coordinates in $\mathbb{R}^3$ so that the Hamiltonians of the systems are functions on $T^*\mathbb{R}^3$. When written in cylindrical coordinates (r,ϕ,z) with axis z passing trough the points $(0,0,0)$ and $(1,1,1)$, the Hamiltonian of the two systems become

$$H=\frac{1}{2}(p_r^2+p_z^2)+\frac{1}{r^2}\left(\frac{1}{2}p_\phi^2+V(\phi)\right),$$

where $V(\phi)=k_I\sin^{-2}(3\phi)$ for the Jacobi-Calogero and $V(\phi)=k_{II}\cos^{-2}(3\phi)$ for the Wolfes potential, and k_I, k_{II} are suitable constants. After the remarkable observation that the two potentials coincide under a $\frac{\pi}{6}$ rotation around the axis, we see that H, for every $V(\phi)$, is the warped product of the two Hamiltonians

$$H_1=\frac{1}{2}(p_r^2+p_z^2),\quad H_2=\frac{1}{2}p_\phi^2+V(\phi),$$

with $\alpha^1=1$, $\alpha^2=r^{-2}$. It is easy to check that H_2 is always a first integral of H.

3. Extensions of Hamiltonian systems

Other interesting examples fitting into this framework are two families of superintegrable Hamiltonian systems with two degrees of freedom, intensively studied in the last few years, whose potential depends on a rational parameter k :

- the TTW-system [12, 13, 19]

$$H=\frac{1}{2}p_r^2+\omega^2r^2+\frac{1}{r^2}\left(\frac{1}{2}p_\psi^2+\frac{a}{\sin^2 k\psi}+\frac{b}{\cos^2 k\psi}\right)$$

- the anisotropic caged oscillator, [20, 11]

$$H=\frac{1}{2}p_x^2+\frac{1}{2}p_y^2+\omega^2x^2+k\omega^2y^2+\frac{a}{x^2}+\frac{b}{y^2}.$$

The interest for these systems arises from the fact that they are maximally superintegrable for any rational value of the parameter k, in their classical as well as quantum realizations. Their quantum version has been intensively studied. Moreover, the reduction of the three-dimensional Jacobi-Calogero and Wolfes systems on the (r,ϕ) plane coincide with the TTW-system for $a=k_I$, $b=\omega=0$ and $a=\omega=0$, $b=k_{II}$, respectively.

The previous systems have Hamiltonians of type

$$H=\frac{1}{2}p_u^2+f(u)+\alpha^2(u)\left(\frac{1}{2}g^{ij}p_ip_j+V(q^i)\right) \tag{6}$$

i.e., they are constructed by merging an N-dimensional natural Hamiltonian $H_2=L=\frac{1}{2}g^{ij}p_ip_j+V(q^i)$ with a 1-dimensional natural Hamiltonian $H_1=\frac{1}{2}p_u^2+f(u)$ through the functions $\alpha^1=1$, $\alpha^2=\alpha^2(u)$.

Since these two examples are both superintegrable systems (i.e., they admit a third independent first integral K other than H and H_2, associated with the separation of variables), it is worth to investigate if it is their structure of warped products the responsible of the existence of K. Namely, we search for conditions on $\alpha^2(u)$, g^{ij}, $V(q^i)$, $f(u)$ such that H admits a polynomial in the momenta first-integral K, of degree determined by the rational parameter k. Therefore, in the following we will not assume separation of variables, that is essential in the previous proofs of the superintegrability of these systems, and focus on the existence of a single constant of motion. The results, proved in several papers, are summarized below.

Roughly speaking, given a natural Hamiltonian $L = \frac{1}{2}g^{ij}p_ip_j + V(q^i)$ with N degrees of freedom, we call *extensions of* L the Hamiltonians with $N+1$ degrees of freedom of the form

$$H = \frac{1}{2}p_u^2 + f(u) + \alpha^2(u)L,$$

such that, starting from a first order differential operator of the form

$$p_u + w\gamma(u)X_L,$$

where X_L is the Hamiltonian vector field of L and w a suitable constant, a first integral K can be constructed by applying iteratively the operator to a non-trivial function $G(q^i, p_i)$. A series of results of increasing generality exposed in [3]-[9], ended in the following Theorem, proved in [9]:

	$c = 0$	$c \neq 0$
$\gamma =$	$-Au$	$\dfrac{1}{T_\kappa(cu)}$
$\alpha^2 = -\dfrac{m^2}{n^2}\gamma' =$	$\dfrac{m^2}{n^2}A$	$\dfrac{m^2 c}{n^2 S_\kappa^2(cu)}$
$f =$	$\dfrac{m^2}{n^2}L_0A^2u^2 + \dfrac{\omega}{A^2u^2}$	$\omega T_\kappa^2(cu)$

Table 1. Functions involved in the modified (m, n)-extension of L (A, ω constants)

Theorem 1. *Let L be a natural Hamiltonian such that there exist two constants c, L_0, not both zero, such that equation*

$$X_L^2(G) = -2(cL + L_0)G \tag{7}$$

admits a non trivial solution G (i.e., $X_L(G) \neq 0$) and let γ, α^2 and f be defined by Table 1 where the functions

$$S_\kappa(x) = \begin{cases} \dfrac{\sin\sqrt{\kappa}x}{\sqrt{\kappa}} & \kappa > 0 \\ x & \kappa = 0 \\ \dfrac{\sinh\sqrt{|\kappa|}x}{\sqrt{|\kappa|}} & \kappa < 0 \end{cases} \qquad T_\kappa(x) = \begin{cases} \dfrac{\tan\sqrt{\kappa}x}{\sqrt{\kappa}} & \kappa > 0 \\ x & \kappa = 0 \\ \dfrac{\tanh\sqrt{|\kappa|}x}{\sqrt{|\kappa|}} & \kappa < 0 \end{cases}$$

$$C_\kappa(x) = \begin{cases} \cos\sqrt{\kappa}x & \kappa > 0 \\ 1 & \kappa = 0 \\ \cosh\sqrt{|\kappa|}x & \kappa < 0 \end{cases}$$

are known as Trigonometric Tagged Functions *and resume in a single function the standard trigonometric and hyperbolic functions [15].*

Then, the Hamiltonian H depending on the rational parameter m/n, called modified (m,n)-extension of L,

$$H = \frac{1}{2}p_u^2 + f(u) + \alpha^2(u)L = \frac{1}{2}p_u^2 + \frac{m^2}{n^2}L_0\gamma^2 + \frac{\omega}{\gamma^2} - \frac{m^2}{n^2}\gamma' L$$

admits a first integral of the form

$$K_{2s,n} = \left(\left(p_u + \frac{2s}{n^2}\gamma X_L\right)^2 + 2\omega\gamma\right)^s G_n$$

for $m = 2s$ and

$$K_{4s+2,2n} = \left(\left(p_u + \frac{4s+2}{4n^2}\gamma X_L\right)^2 + 2\omega\gamma\right)^{2s+1} G_{2n}$$

for $m = 2s+1$, where G_n and G_{2n} are the n-th and $2n$-th terms of the recursion

$$G_1 = G, \qquad G_{n+1} = X_L(G)\,G_n + \frac{1}{n}G\,X_L(G_n).$$

The extended Hamiltonian H is therefore based on different Riemannian manifolds according to the values of κ (which can be arbitrarily chosen), and of c, (which is determined by the curvature of the metric tensor g [6]). When L is a one-dimensional Hamiltonian, the metric tensor g is not defined and the possible extensions H and their configuration manifolds can be summarized by Table 2 in the case $c \neq 0$, while for $c = 0$ the extension is

$$H = \frac{1}{2}p_u^2 + \frac{m^2}{n^2}L_0A^2u^2 + \frac{\omega}{A^2u^2} + \frac{m^2}{n^2}AL,$$

and is based on either the Euclidean or the Minkowski plane, according to the sign of A.

When L is not one-dimensional, the curvature of its base manifold appears into the integrability conditions of (8) [6]. For example, when L is based on $\mathbb{S}^2$, and G is assumed to be independent from the momenta, then c must be equal to the curvature of the sphere (see Theorem 2 below). For $\kappa = 1$, $\alpha^2 = m^2n^{-2}c\sin^{-2}u$ and H is based on a Riemannian manifold, which is S^3 only for the trivial case $m = n$. For $\kappa = 0$, $\alpha^2 = m^2n^{-2}cu^{-2}$ and the base of H is $\mathbb{E}^3$ for $m = n$. For κ negative we have $\alpha^2 = m^2n^{-2}c\sin h^{-2}u$ and the base of H is again a Riemannian manifold, conformally flat only for $m = n$. When L is based on the pseudosphere $\mathbb{H}^2$, whose curvature is negative, c must be negative and the extensions are based on a Lorentzian manifold, whose geometry is again determined by the sign of κ.

An intrinsic characterization of extended Hamiltonians systems is given in [6] and requires the existence on the base manifold of H of a particular conformal Killing vector.

The general TTW system and the caged anisotropic oscillator can be written in form of extensions on any two-dimensional manifold of constant curvature (for superintegrable systems with higher-order symmetries on flat and curved manifolds see also [1, 2, 10, 14, 16, 17]). This has been done for the Euclidean plane in [9] and reported (for the TTW system) as an example below. Hence, by Theorem 1, this allows to compute directly the additional first integral and prove superintegrability in a constructive way.

	$c=1$	Manifold
$\kappa=1$	$H=\frac{1}{2}p_u^2+\omega\tan^2 u+\frac{m^2}{n^2\sin^2 u}L$	Sphere
$\kappa=0$	$H=\frac{1}{2}p_u^2+\omega u^2+\frac{m^2}{n^2u^2}L$	Euclidean plane
$\kappa=-1$	$H=\frac{1}{2}p_u^2+\omega\tanh^2 u+\frac{m^2}{n^2\sinh^2 u}L$	Pseudo-sphere
	$c=-1$	
$\kappa=1$	$H=\frac{1}{2}p_u^2+\omega\tan^2 u-\frac{m^2}{n^2\sin^2 u}L$	de Sitter
$\kappa=0$	$H=\frac{1}{2}p_u^2+\omega u^2-\frac{m^2}{n^2u^2}L$	Minkowski plane
$\kappa=-1$	$H=\frac{1}{2}p_u^2+\omega\tanh^2 u-\frac{m^2}{n^2\sinh^2 u}L$	anti-de Sitter

Table 2. The possible (m,n)-extensions of $L=\frac{1}{2}p_v^2+V(v)$ and their configuration manifolds

3.1. Example: The Tremblay-Turbiner-Winternitz system
By setting $k=\frac{m}{n}$, $u=r$, $v=k\psi$, the complete TTW system in the Euclidean plane becomes

$$H=\frac{1}{2}p_u^2+\frac{1}{u^2}\left(\frac{m^2}{2n^2}p_v^2+\frac{a}{\sin^2 v}+\frac{b}{\cos^2 v}\right)+\omega u^2.$$

From [9] we know that H can be written as

$$H=\frac{1}{2}p_u^2+\frac{(2m)^2}{n^2u^2}\left(\frac{1}{2}p_q^2+\frac{c_1+c_2\cos q}{\sin^2 q}\right)+\omega u^2,$$

where $q=2v$ and $c_1=\frac{a+b}{2k^2}$ $c_2=\frac{a-b}{2k^2}$. The factorized Hamiltonian $L(q,p_q)$ fulfills the hypotheses of Theorem 1 for $c=1$, $L_0=0$ and $G=p_q\sin q$. The coefficient of L is obtained for $\kappa=0$, that is $\gamma=1/u$. The corresponding first integral is of the form

$$\bar{K}_{2m,n}=\left(\left(p_u+\frac{2m}{n^2u}X_L\right)^2+2\frac{\omega}{u}\right)^m G_n.$$

We show in Figures 2 and 3 the trajectories of the particular case $k_2=0$ of the TTW system with harmonic term. In each figure the first two graphs show the relative change (w.r.t. the initial condition) of H_1 and H_2. The two following plots represent the projection of the integral curves in the subspaces $p_u=u=0$ and $p_v=v=0$, respectively. The last one displays the trajectory of the system in the plane (u,v) with axes $x=u\cos v$, $y=y\sin v$. These curves are drawn in the non-Euclidean space with metric $g^{uu}=1$, $g^{vv}=\frac{m^2}{u^2}$, $g^{uv}=0$, where m is now the rational parameter. If the v coordinate is reparametrized into mv, the system can be represented in the Euclidean space, but the graph become self-overlapping for $m<\frac{1}{2}$. Since m is rational, the trajectories always close.

The two-dimensional reductions of the Jacobi-Calogero and Wolfes systems are clearly members of the TTW family, then, they are extended Hamiltonian systems. On the contrary, the Hamiltonian of the geodesic flow of $\mathbb{T}^2$ of Example 2.1 is not an extended Hamiltonian.

The key point of the method is the existence of solutions of (7), this implies global geometric properties of the Riemannian configuration manifold of L.

As instance, in the simplest case when we assume $G = G(q^i)$, independent of the momenta, (7) splits into

$$\nabla^i\nabla^j G + cg^{ij}G = 0,$$
$$\nabla_i V\nabla^i G - 2(cV + L_0)G = 0.$$

The solution $G(q^i)$ may depend on up to $N+1$ parameters a_i.

Theorem 2. *The function $G(q^i)$ solution of (7) satisfies*

$$\nabla^i\nabla^j G + cg^{ij}G = 0, \tag{8}$$

and depends on $N+1$ parameters, and we say that G is complete, iff g is of constant curvature c.

These are the only non trivial solutions on a two-dimensional manifold.

Families of solutions depending on less than $N+1$ parameters can be found on manifolds of non-constant curvature of dimension ≥ 3.

Remark 1. Equation (8) is fundamental in theory of warped manifolds [18]. For example, the existence of non-trivial solutions of (8) on a complete Riemannian manifold of metric g with $c \neq 0$, implies that g has constant non zero curvature. Moreover, in the neighbourhood of non-stationary points of the geodesic flow of any Riemannian manifold, the existence of solutions implies that g itself has the structure of warped product, and that there exist local coordinates $(u^1, \ldots, u^n)$ such that

$$ds^2 = (G'(u^n))^2 f_{ij}(u^h)du^i du^j + (du^n)^2,$$

where the f_{ij} depend on $(u^1, \ldots, u^{n-1})$ and G' is the ordinary derivative of G with respect to u^n ([18], Lemma 1.2). This form of the metric tensor is a necessary condition for writing a Hamiltonian based on the manifold as extended Hamiltonian.

4. Conclusions

The extension procedure, when it occurs, maps separable systems into separable systems and (maximally) superintegrable systems into (maximally) superintegrable systems. Indeed, if L admits separation of variables, then also its extension H does, but separation of L it is not a requirement of the method. Moreover, if the Hamiltonian L is superintegrable with $2n-1$ first integrals, then its extension H is superintegrable with $(2n-1)+2 = 2(n+1)-1$ first integrals (the new integrals being the Hamiltonian H itself and the first integral K). Therefore, the extension of superintegrable systems is a powerful method for building new superintegrable systems from old ones [7].

We can resume the extension procedure, here for the case of $G = G(q^i)$, in a simple algorithm

1 - Consider L on a (pseudo-)Riemannian manifold of metric g and constant curvature c

2 - Solve equation $\nabla^i\nabla^j G + cg^{ij}G = 0$ for the functions $G(q^i; a_1, \ldots, a_{n+1})$

3 - Solve equation $\nabla_i V\nabla^i G - 2(cV + L_0)G = 0$ for the potential V

4 - Determine the extension H in the two cases $c = 0$ or $c \neq 0$.

5 - Compute the additional first integral as described in Theorem 1.

The dynamics of extended Hamiltonian systems can be understood within the framework of warped products of Hamiltonian systems, since any extension is a warped product of Hamiltonians.

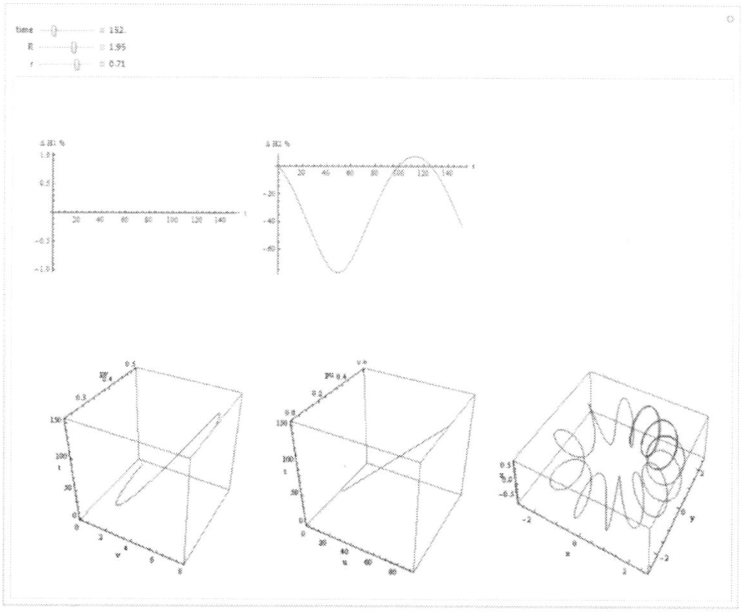

Figure 1. Geodesic on the Torus

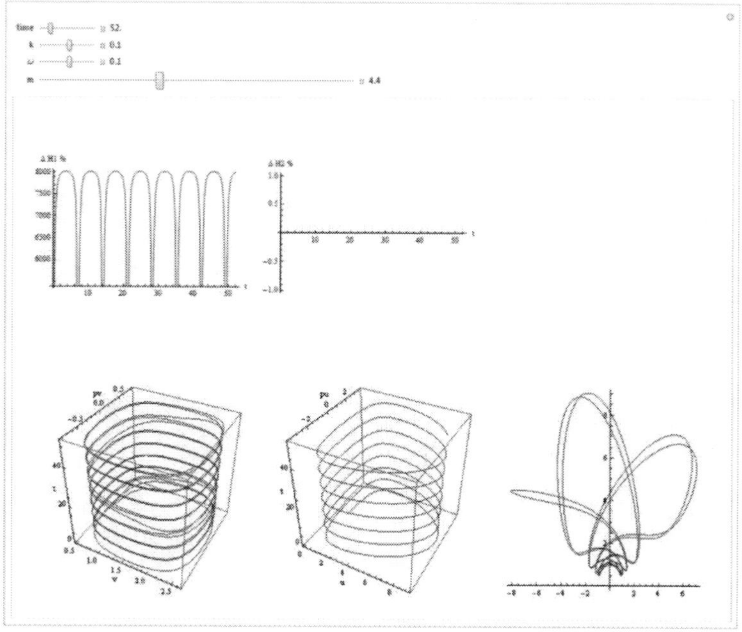

Figure 2. Tremblay-Turbiner-Winternitz system A

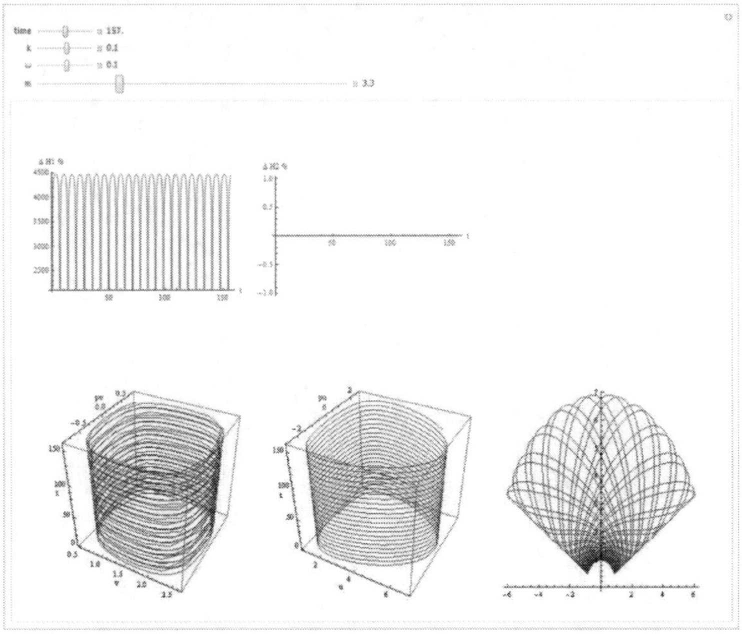

Figure 3. Tremblay-Turbiner-Winternitz system B

Acknowledgments
We thank Peter Petersen for indicating the fundamental article [18].

References
[1] Ballesteros A, Herranz F J, Santander M and Sanz-Gil T 2003 *J. Phys. A: Math. Gen.* **36** L93
[2] Ballesteros A and Herranz F J 2009 *J. Phys. A: Math. Theor.* **42** 245203
[3] Chanu C, Degiovanni L and Rastelli G 2008 *J. Math. Phys.* **49** 112901
[4] Chanu C, Degiovanni L and Rastelli G 2011 *SIGMA* **7** 038
[5] Chanu C, Degiovanni L and Rastelli G 2011 *J. Math. Phys.* **52** 032903
[6] Chanu C, Degiovanni L and Rastelli G 2012 *J. Phys.: Conf. Ser.* **343** 012101
[7] Chanu C, Degiovanni L and Rastelli G 2012 *SIGMA* **8** 070
[8] Chanu C, Degiovanni L and Rastelli G 2013 *J. Math. Phys.* **55** 122703
[9] Chanu C, Degiovanni L and Rastelli G 2014 *J. Math. Phys.* **55** 122701
[10] Kalnins E G, Kress J M, Pogosyan G S and Miller W 2001 *J. Phys. A: Math. Gen.* **34** 4705
[11] Kalnins E G, Kress J M and Miller W Jr 2010 *SIGMA* **6** 066
[12] Kalnins E G, Kress J M and Miller W Jr 2013 *J. Phys.* A: *Math. Theor.* **46** 022002
[13] Maciejewski A J, Przybylska M and Yoshida H 2010 *J. Phys.* A: *Math. Theor.* **43** 382001
[14] Marquette I 2009 *J. Math. Phys.* **50** 122102
[15] Rañada M F and Santander M 1999 *J. Math. Phys.* **40** 50265057
[16] Rañada M F 2014 *J. Phys. A* **47** 165203
[17] Rodriguez M A, Tempesta P and Winternitz P 2008 *Phys. Rev. E* **78** 046608
[18] Tashiro Y 1965 *Trans. Amer. Math. Soc.* **117** 251275
[19] Tremblay F, Turbiner A V and Winternitz P 2009 *J. Phys.* A **42** 242001
[20] Evans N W and Verrier P E 2008 *J. Math. Phys.* **49** 092902

The Dicke model as the contraction limit of a pseudo-deformed Richardson-Gaudin model

Pieter W Claeys[1,2], Stijn De Baerdemacker[1,2], Mario Van Raemdonck[1,2,3] and Dimitri Van Neck[1,2]

[1] Ghent University, Center for Molecular Modeling, Technologiepark 903, 9052 Ghent, Belgium
[2] Ghent University, Department of Physics and Astronomy, Proeftuinstraat 86, 9000 Ghent, Belgium
[3] Ghent University, Department of Inorganic and Physical Chemistry, Krijgslaan 281 (S3), 9000 Ghent, Belgium

E-mail: stijn.debaerdemacker@ugent.be

Abstract. The Dicke model is derived in the contraction limit of a pseudo-deformation of the quasispin algebra in the $su(2)$-based Richardson-Gaudin models. Likewise, the integrability of the Dicke model is established by constructing the full set of conserved charges, the form of the Bethe Ansatz state, and the associated Richardson-Gaudin equations. Thanks to the formulation in terms of the pseudo-deformation, the connection from the $su(2)$-based Richardson-Gaudin model towards the Dicke model can be performed adiabatically.

1. Introduction

The interaction of a single quantized mode of electromagnetic radiation (photons) with a two-state system, such as a nuclear spin or two-level atom, can be modeled by means of the Rabi Hamiltonian [1]. Although simple and physically transparent in its formulation, the Rabi Hamiltonian does not offer a known exact eigenstate in general, due to the 'counter-rotating' interaction terms in the Hamiltonian (We kindly refer the reader to a recent discussion on the solvability and integrability of the Rabi Hamiltonian [2]). In the resonance regime however, these fast-frequency counter-rotating terms average out with respect to the collective two-level oscillations, and can be neglected in first-order perturbation (the so-called Rotating Wave Approximation (RWA)). This gives rise to the Jaynes-Cummings (JC) Hamiltonian of quantum electrodynamics [3]. Due to the RWA, the JC Hamiltonian exhibits an additional symmetry, which allows for the decomposition of the Hilbert space into irreducible representations (irreps) conserving the total number of (bosonic and atomic) excitations. For the two-level ($s = 1/2$ quasispin) formulation of the JC model, these irreps remain two-dimensional and therefore reduce to two-level mixing models. One of the successes of theoretical and experimental quantum mechanics is that such simple theoretical models have been validated in sophisticated experiments situated in cavity quantum electrodynamics [4], circuit quantum electrodynamics [5], etc.

A direct consequence of the additional symmetry is that the two-level JC model exhibits two linearly independent conserved operators, *i.e.* the Hamiltonian and the operator counting the total number of excitations. As the model has two degrees of freedom (the electromagnetic photon mode and the two-level system), the model is *integrable* by definition [6], and supports a Bethe Ansatz product wavefunction [7]. This result can be generalized to general $(2s+1)$-level systems, also referred to as Tavis-Cummings models [8], or a collections of inequivalent $(2s+1)$-level systems interacting via a single mediating bosonic mode, called the Dicke model [9]. The integrability of the Dicke model has been established first by Gaudin [7], by means of an infinite dimensional Schwinger representation of the central spin in the Gaudin magnet. Later, the full set of conserved charges of the Dicke model has been derived by Dukelsky et. al. [10], by mapping one of the $su(2)$ quasispin copies of the Richardson-Gaudin conserved charges onto the bosonic mode [11]. As a Bethe Ansatz integrable model, the conserved charges and Bethe Ansatz state of the Dicke model can also be obtained from the Algebraic Bethe Ansatz method [12], however a direct and simplified derivation of the Bethe-Ansatz state with the corresponding Bethe equations solution is feasible using a commutator scheme [13].

For the derivation of the conserved charges of the Dicke model from those of the Richardson-Gaudin model [7, 10], a contraction of one of the quasispin algebras in the model was required. This contraction maps the $su(2)$ quasispin directly onto a bosonic Heisenberg-Weyl algebra $hw(1)$. Recently, it has been shown how the Bethe Ansatz states in the Richardson-Gaudin (RG) solution [14, 15] of the reduced Bardeen-Cooper-Schriefer (BCS) Hamiltonian for conventional superconductivity [16] can be connected adiabatically to a product state of generalized bosons [17] employing a pseudo deformation of the quasispin algebra. The product structure of the Bethe Ansatz state is reminiscent of the projected BCS approximation, in which the superconducting state is approximated by a condensate of collective Cooper pairs [16, 18]. However, in general the RG variables (or rapidities) parametrizing the generalized quasispin creation operators in the Bethe Ansatz state differ from each other, which is not reconcilable with the concept of a true condensate. The pseudo deformation maps the hard-core bosonic $su(2)$ quasispin algebra of the BCS pairing Hamiltonian into a genuinely bosonic $hw(1)$ by means of a continuous pseudo-deformation parameter. The reduced BCS Hamiltonian remains integrable along the path of deformation, allowing for an adiabatic and injective mapping of the exact Bethe Ansatz states into a condensate of (orthogonal) bosonic modes [19]. Accordingly, the coupled set of RG equations of integrability reduce to a single decoupled equation, equivalent to the secular equation of the particle-particle Tamm-Dancoff Approximation (*pp*-TDA) for the elementary pairing modes for the reduced BCS Hamiltonian [20]. Reversely, the pseudo deformation enables one to numerically reconstruct the solution of the coupled set of non-linear RG equations from the simpler decoupled secular *pp*-TDA equation, by adiabatically reintroducing the Pauli principle in the integrable model [17, 21].

In the present manuscript, we will reassess the derivation of the conserved charges of the Dicke from those of the Richardson-Gaudin (RG) models in the framework of the pseudo-deformed quasispin algebra. This will enable an adiabatic connection of the Dicke model with the $su(2)$ based Richardson-Gaudin models. The connection can be made consistently on the level of the Hamiltonian, the conserved charges, the Bethe Ansatz state, as well as the Bethe (or Richardson-Gaudin) equations, and sheds light on how the Dicke model can be embedded within the larger class of Richardson-Gaudin integrable models.

2. Richardson-Gaudin models

The conserved charges of the Richardson-Gaudin [11] models are parametrized by

$$R_i = S_i^0 + g\sum_{k\neq i}^{m}[X_{ik}\tfrac{1}{2}(S_k^\dagger S_i + S_i^\dagger S_k) + Z_{ik}S_i^0 S_k^0] \tag{1}$$

with the set of operators $\{S_i^\dagger, S_i, S_i^0\}$ $(i = 1 \dots m)$ spanning a set of m mutually independent $su(2)_i$ quasispin algebras

$$[S_i^0, S_k^\dagger] = \delta_{ik} S_k^\dagger, \qquad [S_i^0, S_k] = -\delta_{ik} S_k, \qquad [S_i^\dagger, S_k] = 2\delta_{ik} S_k^0, \tag{2}$$

with irreps $|s_i, \mu_i\rangle$, and the X and Z antisymmetric matrices obeying the Gaudin algebra

$$X_{ij}X_{jk} - X_{ik}(Z_{ij} + Z_{jk}) = 0, \qquad \forall i \neq j \neq k \neq i. \tag{3}$$

As soon as X and Z span the Gaudin algebra, the conserved charges commute mutually

$$[R_i, R_j] = 0, \qquad \forall i, j, \tag{4}$$

and therefore define a RG integrable model. Because the conserved charges (1) are in involution, they have a common set of eigenstates. These eigenstates are given by means of the Bethe Ansatz

$$|\psi\rangle = \prod_{\alpha=1}^{N} \left(\sum_{i=1}^{m} X_{i\alpha} S_i^\dagger \right) |\theta\rangle \tag{5}$$

with N the number of excitations, and $|\theta\rangle = \otimes_{i=1}^{m} |s_i, -s_i\rangle$ the tensor product of the lowest-weight irreps of each $su(2)_i$ copy. The notation $X_{i\alpha}$ signifies that the m-dimensional matrix X has been extended to an $(m+N)$-dimensional matrix which still obeys the Gaudin algebra (3) and the antisymmetry condition. For state (5) to be an eigenstate of the conserved charges (1), the extended Z (and X) matrices need to satisfy the following Bethe Ansatz equations

$$1 + g\sum_{i=1}^{m} Z_{i\alpha} s_i - g \sum_{\beta \neq \alpha} Z_{\beta\alpha} = 0, \qquad \forall \alpha = 1 \dots N. \tag{6}$$

Up to this point, the Gaudin algebra has not been specified, so these results are fully independent of the particular realisation of X and Z. There are several parametrizations existing in the literature [22, 11], which can be classified according to the following copy-independent relation

$$X_{ij}^2 - Z_{ij}^2 = c, \qquad \forall i \neq j, \tag{7}$$

with c a constant for all $i \neq j$. The widely used $c = 0$ parametrization gives rise to the rational model, whereas the $c > 0$ and $c < 0$ are known as the trigonometric and hyperbolic models respectively[1]. In the present manuscript, we will mainly focus on the trigonometric case, which can be parametrized in terms of a set of m real variables $\{\eta_i\}$ [11]

$$X_{ij} = \frac{\sqrt{(1+\eta_i^2)(1+\eta_j^2)}}{\eta_i - \eta_j}, \qquad Z_{ij} = \frac{1 + \eta_i\eta_j}{\eta_i - \eta_j} \tag{8}$$

and $c = 1$, but equivalent parametrizations exist [10].

[1] although the trigonometric and hyperbolic case can be transformed into one another by introducing an imaginary coupling constant $g \to ig$

3. Pseudo deformed quasi-spin algebra

The pseudo deformed quasispin algebra $su(2)$ is given by [17]

$$[S^0(\xi), S^\dagger(\xi)] = S^\dagger(\xi), \quad [S^0(\xi), S(\xi)] = -S(\xi), \quad [S^\dagger(\xi), S(\xi)] = 2\left(\xi S^0(\xi) + (\xi - 1)s\right) \tag{9}$$

with ξ the pseudo-deformation parameter and s the original ($\xi = 1$) $su(2)$ irrep label. It can be easily verified that the limits $\xi = 1$ and $\xi = 0$ give rise to the original $su(2)$ algebra and a (unnormalised) bosonic $hw(1)$ algebra respectively. The nomenclature *pseudo* deformation is chosen because this algebra can be reduced to a canonical $su(2)$ algebra

$$[A^0(\xi), A^\dagger(\xi)] = A^\dagger(\xi), \qquad [A^0(\xi), A(\xi)] = -A(\xi), \qquad [A^\dagger(\xi), A(\xi)] = 2A^0(\xi), \tag{10}$$

with

$$A^\dagger(\xi) = \frac{1}{\sqrt{\xi}} S^\dagger(\xi), \quad A(\xi) = \frac{1}{\sqrt{\xi}} S(\xi), \quad A^0(\xi) = S^0(\xi) + \left(1 - \frac{1}{\xi}\right) s \tag{11}$$

except for the $\xi = 0$ limit. For this limit, the following operators

$$b^\dagger = \sqrt{\tfrac{1}{2s}}\, S^\dagger(0), \quad b = \sqrt{\tfrac{1}{2s}}\, S(0), \quad b^\dagger b = S^0(0) \tag{12}$$

close the $hw(1)$ algebra

$$[b^\dagger b, b^\dagger] = b^\dagger, \quad [b^\dagger b, b] = -b \quad [b, b^\dagger] = 1. \tag{13}$$

The representations of the $\{A^\dagger(\xi), A(\xi), A^0(\xi)\}$ algebra are labeled by $s(\xi) = s(1)/\xi$. The physical interpretation of this is that the effective degeneracy of the quasispin algebra is gradually increased with decreasing ξ, reaching infinity for the full $\xi = 0$ contraction limit. It should be noted that only discrete values of $\xi_n = \frac{2s}{n}$ (with $n = 2s, 2s+1, \ldots$) give rise to unitary irreps. Nevertheless, this is not an obstacle because the theory of RG integrability does not depend on matrix representations (with integer dimensions), and the parameter ξ can be regarded as a continuous variable. This is illustrated by the following construction. Because the set of generators $\{A^\dagger(\xi), A(\xi), A^0(\xi)\}$ span an $su(2)$ algebra, the following set of conserved charges are equally in involution

$$R_i(\xi) = A_i^0(\xi) + g\xi \sum_{k \neq i}^{m} [X_{ik} \tfrac{1}{2}(A_k^\dagger(\xi) A_i(\xi) + A_i^\dagger(\xi) A_k(\xi)) + Z_{ik} A_i^0(\xi) A_k^0(\xi)], \tag{14}$$

provided the matrices X and Z fulfill the Gaudin algebra (3). Note that the coupling constant g in eq. (1) has been renormalized as $g\xi$. The Bethe Ansatz state

$$|\psi\rangle = \prod_{\alpha=1}^{N} \left(\sum_{i=1}^{m} X_{i\alpha} A_i^\dagger(\xi) \right) |\theta\rangle \tag{15}$$

is again an eigenstate of the conserved charges (14) if the pseudo-deformed RG equations

$$1 + g \sum_{i=1}^{m} Z_{i\alpha} \xi s_i(\xi) - g\xi \sum_{\beta \neq \alpha} Z_{\beta\alpha} = 0, \qquad \forall \alpha = 1 \ldots N, \tag{16}$$

are solved. Although the irrep labels $s(\xi)$ approach infinity for $\xi \to 0$, the value of $\xi s(\xi)$ remains constant over the range $\xi \in [0, 1]$, so $\lim_{\xi \to 0} [\xi s(\xi)] = s$. Consequently, the problem of solving the set of pseudo-deformed RG equations (16) is perfectly well defined for any value of the parameter $\xi \in [0,1]$, and is not restricted to discrete values $\xi_n = \frac{2s}{n}$ for unitary irreps. The $\xi \to 0$ limit offers a particular case, because the conserved charges (up to a diverging constant) become purely bosonic

$$R_i(\xi=0) = b_i^\dagger b_i + g\sum_{k\neq i}^{m}[X_{ik}\sqrt{s_i s_k}(b_i^\dagger b_k + b_k^\dagger b_i) - Z_{ik}(s_i b_k^\dagger b_k + s_k b_i^\dagger b_i)] \tag{17}$$

and the pseudo-deformed RG equations become completely decoupled

$$1 + g\sum_{i=1}^{m} Z_{i\alpha}s_i = 0, \qquad \forall\alpha = 1\ldots N, \tag{18}$$

which is equivalent to a set of independent secular equations of the *pp*-TDA. This construction forms the backbone of the numerical RG solver method introduced in [17, 21]. In this approach, the RG equations are solved in the tractable full contraction limit (18), and are then adiabatically brought into the original form (6) by tuning $\xi \to 1$.

4. Dicke model derived from Richardson-Gaudin models

The pseudo-deformation in the previous section was used to transform all $su(2)_i$ copies ($i \in [1\ldots m]$) in the system into a bosonic $hw(1)_i$. Because the JC and Dicke models contain only one bosonic mode, it would be interesting to see whether an equivalent scheme can be used to obtain the Dicke model from the $\otimes_{i=0}^{m} su(2)$ RG systems by deforming a single quasispin copy. Denote this special copy by $i=0$, and consider the conserved charge

$$R_0(\xi) = A_0^0(\xi) + g\sum_{k\neq 0}^{m}[\tfrac{1}{2}X_{0k}(A_0^\dagger(\xi)S_k + S_k^\dagger A_0(\xi)) + Z_{0k}A_0^0(\xi)S_k^0]. \tag{19}$$

In contrast to the pseudo-deformed conserved charges (14), the coupling constant will be renormalized as $g = \sqrt{\frac{2\xi}{s_0 G^2}\frac{G^2}{\hbar\omega}}$, with G a finite, ξ-independent, constant and $\hbar\omega$ the energy of the electromagnetic field. The other conserved charges are given by

$$\begin{aligned} R_i(\xi) = S_i^0 &+ g\textstyle\sum_{k\neq 0,i}^{m}[\tfrac{1}{2}X_{ik}(S_i^\dagger S_k + S_k^\dagger S_i) + Z_{ik}S_i^0 S_k^0] \\ &+ g[\tfrac{1}{2}X_{i0}(S_i^\dagger A_0(\xi) + A_0^\dagger(\xi)S_i) + Z_{i0}A_0^0(\xi)S_i^0]. \end{aligned} \tag{20}$$

From here on, we will employ the trigonometric realization (8) of the Gaudin algebra (3). Without loss of generality, the X_{0k} and Z_{0k} matrix elements can be evaluated in the $\eta_0 \to \infty$ limit, giving rise to the parametrization

$$X_{0k} = \lim_{\eta_0\to\infty}\frac{\sqrt{1+\eta_0^2}\sqrt{1+\eta_k^2}}{\eta_0-\eta_k} = \sqrt{1+\eta_k^2}, \qquad Z_{0k} = \lim_{\eta_0\to\infty}\frac{1+\eta_0\eta_k}{\eta_0-\eta_k} = \eta_k. \tag{21}$$

Using this realization and the definition of the Gaudin algebra (3), it is straightforward to show that

$$X_{ik} = \frac{X_{i0}X_{0k}}{Z_{i0}+Z_{0k}} = \frac{\sqrt{1+\eta_i^2}\sqrt{1+\eta_k^2}}{\eta_i-\eta_k}, \qquad Z_{ik} = \sqrt{X_{ik}^2-1} = \frac{1+\eta_i\eta_k}{\eta_i-\eta_k}, \tag{22}$$

which is indeed a standard trigonometric parametrization (8) of the Gaudin algebra (3). The set of parameters $\{\eta_k\}$ can be chosen freely, so we renormalize them as $\eta_k = -\sqrt{\frac{\xi}{2s_0G^2}}\varepsilon_k$, with $\{\varepsilon_k\}$

a set of free, ξ-independent, variables. It is noteworthy that the X_{0k} and Z_{0k} matrix elements take the form

$$X_{0k} = 1 + \tfrac{\xi}{4s_0G^2}\varepsilon_k^2 + \mathcal{O}(\xi^2), \qquad Z_{0k} = -\sqrt{\tfrac{\xi}{2s_0G^2}}\varepsilon_k, \tag{23}$$

in the $\xi \to 0$ contraction limit, which is related to the parametrization proposed by Dukelsky et. al. [10] in their derivation of the Dicke model. Substituting the expressions of X_{0k} and Z_{0k} (eqs. (21)) in the conserved charge, and taking the proper $\xi \to 0$ limit, one obtains the Hamiltonian of the Dicke model (up to a divergent constant)

$$H_{\text{Dicke}} = \hbar\omega R_0(\xi \to 0) = \hbar\omega b^\dagger b + \sum_{k=1}^{m} \varepsilon_k S_k^0 + G\sum_{k=1}^{m}(b^\dagger S_k + S_k^\dagger b), \tag{24}$$

where the $i = 0$ notation has been omitted for the bosonic operators. A similar procedure leads to the other conserved charges

$$\hbar\omega R_i(\xi \to 0) = (\hbar\omega - \varepsilon_i)S_i^0 + \sum_{k\neq i}^{m} \frac{2G^2}{\varepsilon_k - \varepsilon_i}[\tfrac{1}{2}(S_i^\dagger S_k + S_k^\dagger S_i) + S_i^0 S_k^0] - G(S_i^\dagger b + b^\dagger S_i). \tag{25}$$

To obtain an expression for the Bethe Ansatz state in the contraction limit, one needs a parametrization of the matrix elements $X_{0\alpha}$ and $X_{k\alpha}$. The former can be chosen according to eqs. (21) with $\eta_\alpha = -\sqrt{\frac{\xi}{2s_0G^2}}x_\alpha$, from which the latter can be derived

$$X_{k\alpha} = \frac{X_{k0}X_{0\alpha}}{Z_{k0} + Z_{0\alpha}} = \frac{\sqrt{1+\eta_k^2}\sqrt{1+\eta_\alpha^2}}{\eta_k - \eta_\alpha}. \tag{26}$$

Substituting these parametrisations for $X_{0\alpha}$ and $X_{k\alpha}$ into the Bethe Ansatz state eq. (5), and taking the $\xi \to 0$ limit, one obtains

$$|\psi\rangle = \left(\frac{2s_0}{\xi}\right)^{\frac{N}{2}} \prod_{\alpha=1}^{N}\left(b^\dagger - G\sum_{k=1}^{m}\frac{S_k^\dagger}{\varepsilon_k - x_\alpha}\right)|\theta\rangle, \tag{27}$$

where the factor $(\frac{2s_0}{\xi})^{\frac{N}{2}}$ can be absorbed by the normalization of the state.

Finally, the RG equations for the Dicke model can be obtained likewise. Starting from the RG equations (6) with the $i = 0$ quasispin copy replaced by a pseudo deformed $su(2)$,

$$1 + gZ_{0\alpha}s_0(\xi) + g\sum_{k=1}^{m} Z_{k\alpha}s_k - g\sum_{\beta\neq\alpha} Z_{\beta\alpha} = 0, \qquad \forall\alpha = 1\ldots N. \tag{28}$$

and substituting the proper expressions for $Z_{k\alpha}$ and $Z_{\beta\alpha}$

$$Z_{k\alpha} = \frac{1 + \eta_k\eta_\alpha}{\eta_k - \eta_\alpha}, \qquad Z_{\beta\alpha} = \frac{1 + \eta_\beta\eta_\alpha}{\eta_\beta - \eta_\alpha}, \tag{29}$$

we obtain the RG equations for the Dicke Hamiltonian

$$(\hbar\omega - x_\alpha) - 2G^2\sum_{k=1}^{m}\frac{s_k}{\varepsilon_k - x_\alpha} + 2G^2\sum_{\beta\neq\alpha}^{N}\frac{1}{x_\beta - x_\alpha} = 0, \quad \forall\alpha = 1\ldots N. \tag{30}$$

5. Conclusions

In conclusion, we have derived the RG integrability of the Dicke model in the contraction limit of a single pseudo deformed quasispin algebra copy in the $\otimes_{i=0} su(2)_i$ RG system. In addition to the Dicke Hamiltonian, the additional conserved charges for the integrability have been obtained, as well as the correct form for the Bethe Ansatz state and the RG equations which the rapidities in the Bethe Ansatz need to satisfy. It would be interesting to investigate how the Bethe Ansatz eigenstates of the trigonometric RG Hamiltonian are adiabatically connected to the Bethe Ansatz eigenstates of the Dicke model. This will constitute the subject of future investigations.

Acknowledgments

PWC and SDB acknowledge financial support from FWO-Vlaanderen as pre-doctoral and post-doctoral fellows respectively.

References

[1] Grynberg G, Aspect A and Fabre C 2010 *Introduction to Quantum Optics* (Cambridge: Cambridge University Press)
[2] Batchelor M T and Zhou H Q 2014 (*Preprint* `quant-ph/1408.3816`)
[3] Jaynes E and Cummings F W 1963 *Proc. IEEE* **51** 89
[4] Brune M, Schmidt-Kaler F, Maali A, Dreyer J, Hagley E, Raimond J M and Haroche S 1996 *Phys. Rev. Lett.* **76** 1800
[5] Wallraff A, Schuster D I, Blais A, Frunzio L, Huang R S, Majer J, Kumar S, Girvin S M and Schoelkopf R J 2004 *Nature* **398** 162
[6] Caux J S and Mossel J 2011 *J. Stat. Mech.: Theor. Exp.* P02023
[7] Gaudin M 1976 *J. Phys. (Paris)* **37** 1087
[8] Tavis M and Cummings F W 1968 *Phys. Rev.* **170** 379
[9] Dicke R H 1954 *Phys. Rev.* **93** 99
[10] Dukelsky J, Dussel G G, Esebbag C and Pittel S 2004 *Phys. Rev. Lett.* **93** 050403
[11] Ortiz G, Somma R, Dukelsky J and Rombouts S 2005 *Nucl. Phys. B* **707** 421
[12] Babelon O and Talalaev D 2007 *J. Stat. Mech.: Theor. Exp.* P06013
[13] Tsyplyatyev O, von Delft J and Loss D 2010 *Phys. Rev. B* **82** 092203
[14] Richardson R W 1963 *Phys. Lett.* **3** 277
[15] Richardson R W and Sherman N 1964 *Nucl. Phys.* **52** 221
[16] Bardeen J, Cooper L N and Schrieffer J 1957 *Phys. Rev.* **108** 1175
[17] De Baerdemacker S 2012 *Phys. Rev. C* **86** 044332
[18] Cooper L N 1956 *Phys. Rev.* **104** 1189
[19] De Baerdemacker S 2011 *Proceedings of the 28th International Colloquium on Group Theoretical Methods in Physics (GROUP28)* (*J. Phys. Conf. Ser.* vol 284) ed Angelova M, Zakrzewski W, Hussin V and Piette B p 012020
[20] Ring P and Schuck P 2004 *The Nuclear Many-Body Problem* 3rd ed (Berlin: Springer)
[21] Van Raemdonck M, De Baerdemacker S and Van Neck D 2014 *Phys. Rev. B* **89** 155136
[22] Dukelsky J, Esebbag E and Schuck P 2001 *Phys. Rev. Lett.* **87** 066403

Unique characterization of the Fourier transform in the framework of representation theory

H De Bie[1], R Oste[2] and J Van der Jeugt[2]

[1]Department of Mathematical Analysis, Faculty of Engineering and Architecture – Ghent University, Galglaan 2, 9000 Gent, Belgium

[2]Department of Applied Mathematics, Computer Science and Statistics, Faculty of Sciences – Ghent University, Krijgslaan 281, 9000 Gent, Belgium

E-mail: Hendrik.DeBie@UGent.be, Roy.Oste@UGent.be, Joris.VanderJeugt@UGent.be

Abstract. In this paper we elaborate upon the investigation initiated in [3] of typical and distinctive properties of the Fourier transform (FT), in particular the crucial role played by the Howe dual pair $(O(m), \mathfrak{sl}_2)$. We prove in detail a result on the unique characterization of the FT making extensive use of a representation of the Lie algebra $\mathfrak{sl}_2$. As an example, we consider the case $m = 1$. We refer to [3] for a detailed study involving the derivation of a class of operators portraying FT symmetry properties.

1. Introduction

Due to its useful properties the classical Fourier transform (FT) has many applications in a whole range of areas such as harmonic analysis and signal processing. Over $\mathbb{R}^m$, the FT is given by the integral transform

$$\mathcal{F}[f](y) = \frac{1}{(2\pi)^{m/2}} \int_{\mathbb{R}^m} \mathrm{e}^{i\langle x,y\rangle} f(x)\,\mathrm{d}x,$$

where $\langle x, y\rangle = \sum_{j=1}^{m} x_j y_j$ denotes the standard inner product of two m-dimensional vectors $x = (x_1, \ldots, x_m)$ and $y = (y_1, \ldots, y_m)$.

In [3], we initiated the investigation of typical and distinctive properties of the FT. This led to the question as to what suffices to uniquely characterize the Fourier transform. Hereto, we consider in particular invariance and symmetry properties governed by the dual pair $O(m)$ and $\mathfrak{sl}_2$. This is motivated by the fact that the FT can be expressed as an operator exponential containing orthogonally invariant operators that generate a realization of $\mathfrak{sl}_2$. Other means of characterizing the FT include for instance the interaction of the FT with the convolution product, which is used in [1] to obtain a uniqueness result. Our approach allows us to make explicit use of the representation theory of the Lie algebra $\mathfrak{sl}_2$.

Now, a natural course of actions is the following: We start from a select set of FT properties determined by the underlying symmetry and we set out to determine the class of all operators also satisfying these. This problem necessarily has at least one solution, namely the FT. Two possible scenarios then arise depending on the imposed properties. If the FT turns out to be the sole solution, this gives us a unique characterization the FT. To obtain other transforms we can then relax the prescribed properties to less strict ones. At the other end of the spectrum is the case where we find an overabundance of solutions. We can then investigate how we can reduce this class of solutions by imposing additional properties. In this way we arrive at a (preferably finite) set of interesting transforms. On top of that, by continuing this process it ultimately gives us a list of properties, the combination of which is exclusively satisfied by the FT.

The explicit derivation of operators having Fourier-like symmetry properties is the subject of [3]. Now we focus on a specific result which uniquely characterizes the FT. Moreover, we elaborate upon the case $m = 1$ where the approach of [3] does not yield any new operators.

The paper is organized as follows. In Section 2, we discuss the properties and symmetries of the FT relevant to our cause. In Section 3, we prove a uniqueness result for the FT. In Section 4, we consider in detail the one-dimensional case.

2. Properties

We will start by elaborating upon what makes the FT so useful. Our goal is then to determine a class of operators T having these specific properties.

A major property of the Fourier transform consists of its interaction with differential operators. In particular, we have for $j = 1, \ldots, m$:

$$\begin{aligned} \mathcal{F} \circ \partial_{x_j} &= -i\, y_j \circ \mathcal{F} \\ \mathcal{F} \circ x_j &= -i\, \partial_{y_j} \circ \mathcal{F} \end{aligned} \tag{1}$$

We want to relax these interactions as prescribing them uniquely characterizes the FT up to a multiplicative factor. They give rise to the following symmetries for $j = 1, \ldots, m$:

$$\mathcal{F} \circ \left(x_j^2 - \partial_{x_j}^2\right) = \left(y_j^2 - \partial_{y_j}^2\right) \circ \mathcal{F} \tag{2}$$

where one recognises the Hamiltonian of the quantum harmonic oscillator.

Now, the FT also exhibits a symmetry with respect to the orthogonal group $O(m)$. The relations (2) combined with the invariance under $O(m)$ yield

$$\mathcal{F} \circ (-\Delta_x + |x|^2) = (-\Delta_y + |y|^2) \circ \mathcal{F} \tag{3}$$

which is compatible with $O(m)$. Here the Laplace operator Δ_x and the norm squared $|x|^2$ are defined as

$$\Delta_x := \sum_{j=1}^{m} \partial_{x_j}^2, \qquad |x|^2 := \sum_{j=1}^{m} x_j^2.$$

Also

$$\mathcal{F} \circ (\Delta_x + |x|^2) = -(\Delta_y + |y|^2) \circ \mathcal{F}. \tag{4}$$

This is related to an alternative formulation of the FT as an operator exponential:

$$\mathcal{F} = e^{i\frac{\pi}{4}(-\Delta_x + |x|^2 - m)}. \tag{5}$$

The operators that occur here highlight a connection with the Lie algebra $\mathfrak{sl}_2$ as they satisfy

$$\left[\Delta_x/2, |x|^2/2\right] = \mathbb{E}_x + \tfrac{m}{2} \qquad \left[\mathbb{E}_x + \tfrac{m}{2}, |x|^2/2\right] = |x|^2 \qquad \left[\mathbb{E}_x + \tfrac{m}{2}, \Delta_x/2\right] = -\Delta_x. \tag{6}$$

Here we denoted by $\mathbb{E}_x$ the Euler operator, defined as

$$\mathbb{E}_x := \sum_{j=1}^{m} x_j \partial_{x_j}.$$

An operator T will have the same symmetries (3) and (4) if it satisfies

$$\begin{aligned} T \circ \Delta_x &= -|y|^2 \circ T \\ T \circ |x|^2 &= -\Delta_y \circ T, \end{aligned} \tag{7}$$

which are reminiscent of (1). These are called the Helmholtz relations because if (7) holds for an integral transform

$$Tf(y) = \frac{1}{(2\pi)^{m/2}} \int_{\mathbb{R}^m} K(x,y) f(x) \, \mathrm{d}x,$$

then its kernel will satisfy

$$\frac{1}{(2\pi)^{m/2}} \int_{\mathbb{R}^m} (\Delta_y K(x,y) + |x|^2 K(x,y)) f(x) \, \mathrm{d}x = 0 \qquad (\forall f \in \mathcal{S}(\mathbb{R}^m)). \tag{8}$$

Here $\mathcal{S}(\mathbb{R}^m)$ denotes the space of smooth and rapidly decreasing functions on $\mathbb{R}^m$, also called the Schwartz space, which is dense in $L^2(\mathbb{R}^m)$. Note that T satisfying the Helmholtz relations, together with the commutation relations (6), implies that one also has

$$T \circ (\mathbb{E}_x + \tfrac{m}{2}) = -(\mathbb{E}_x + \tfrac{m}{2}) \circ T.$$

Another property of the FT is that it is an automorphism on $\mathcal{S}(\mathbb{R}^m)$. A decomposition of $\mathcal{S}(\mathbb{R}^m)$ with respect to both the orthogonal $O(m)$ symmetry and the algebraic $\mathfrak{sl}_2$ symmetry is given by the generalized Hermite functions [2]

$$\phi_{j,k,\ell} := 2^j j! \, L_j^{\frac{m}{2}+k-1}(|x|^2) \, H_k^{(\ell)} \, \mathrm{e}^{-|x|^2/2}. \tag{9}$$

Here $j, k \in \mathbb{Z}_{\geq 0}$, $L_j^{\frac{m}{2}+k-1}$ is the Laguerre polynomial and $\{ H_k^{(\ell)} \mid \ell = 1, ..., \dim(\mathcal{H}_k) \}$ is a basis for the space of spherical harmonics of degree k, denoted by $\mathcal{H}_k := \ker \Delta_x \cap \mathcal{P}_k$, with $\mathcal{P}_k$ the space of homogeneous polynomials of degree k. The basis (9) forms eigenfunctions of the FT with action given by

$$\mathcal{F}(\phi_{j,k,\ell}) = \mathrm{e}^{i\frac{\pi}{2}(2j+k)} \phi_{j,k,\ell}.$$

Note that this action is independent of the index ℓ of the Hermite functions, which is precisely due to the orthogonal symmetry of the FT.

Another way to write the Hermite functions (see [2]) is

$$\phi_{j,k,\ell} = \left(-\frac{\Delta_x}{2} - \frac{|x|^2}{2} + \mathbb{E}_x + \frac{m}{2} \right)^j H_k^{(\ell)} \, \mathrm{e}^{-|x|^2/2}. \tag{10}$$

The operator that acts on the Hermite functions gives rise to another realization of the Lie algebra $\mathfrak{sl}_2$ by means of

$$h = -\frac{\Delta_x}{2} + \frac{|x|^2}{2}, \quad e = -\frac{\Delta_x}{4} - \frac{|x|^2}{4} + \frac{1}{2}(\mathbb{E}_x + \tfrac{m}{2}), \quad f = \frac{\Delta_x}{4} + \frac{|x|^2}{4} + \frac{1}{2}(\mathbb{E}_x + \tfrac{m}{2}), \tag{11}$$

which satisfy

$$[h, e] = 2e, \qquad [h, f] = -2f, \qquad [e, f] = h.$$

For every $k \in \mathbb{Z}_{\geq 0}$ and $\ell \in \{1, \ldots, \dim(\mathcal{H}_k)\}$, the set of Hermite functions $\{ \phi_{j,k,\ell} \mid j \in \mathbb{Z}_{\geq 0} \}$ then forms a basis for the positive discrete series representation of $\mathfrak{sl}_2$ with lowest weight $k + m/2$. This can be seen from their action on the basis (9) which is given by

$$h \, \phi_{j,k,\ell} = \left(2j + k + \tfrac{m}{2}\right) \phi_{j,k,\ell}, \tag{12}$$

and

$$e\,\phi_{j,k,\ell} = \frac{1}{2}\,\phi_{j+1,k,\ell}, \qquad f\,\phi_{j,k,\ell} = -j\,(2j-2+m+2k)\,\phi_{j-1,k,\ell}.$$

Returning to the operator formulation (5) of the Fourier transform, this can now be rewritten in terms of h as

$$\mathcal{F} = e^{i\frac{\pi}{4}(-\Delta_x+|x|^2-m)} = e^{i\frac{\pi}{2}(h-\frac{m}{2})}.$$

We see that, up to a complex phase factor, the FT is an element of $\exp(\mathfrak{sl}_2)$. Now, we want to find out if any other elements of $\exp(\mathfrak{sl}_2)$ give rise to an operator that satisfies the properties we considered. The operators (11) allow us to prove a unique characterization of the Fourier transform. This is the subject of the following section.

3. Uniqueness result

Theorem 1. *Let T be an operator that satisfies the properties:*

(i) the Helmholtz relations

$$\begin{aligned} T\circ\Delta_x &= -|y|^2\circ T \\ T\circ|x|^2 &= -\Delta_y\circ T \end{aligned}$$

(ii) $T\,\phi_{j,k,\ell} = \mu_{j,k}\,\phi_{j,k,\ell}$ *with* $\mu_{j,k}\in\mathbb{C}$

(iii) $T^4 = \mathrm{id}$

(iv) $T = A\exp(z)$ *with* $A\in\mathbb{C}$ *and* $z\in\mathfrak{sl}_2 = \mathrm{span}\{\Delta_x, |x|^2, [\Delta_x, |x|^2]\}$.

Then T equals, up to multiplication by an integer power of i, the classical Fourier transform or its inverse.

Proof. To prove this result we will use the realization of $\mathfrak{sl}_2$ by means of the operators (11) and their relation with the basis (9) to our advantage. A general element of $\mathfrak{sl}_2$ can then be written as

$$z = B\,h + C\,e + D\,f$$

with complex numbers $B, C, D\in\mathbb{C}$.

In terms of the operators h, e, f, the Helmholtz property translates to the (anti-)commutation relations

$$T\circ h = h\circ T, \qquad T\circ e = -e\circ T, \qquad T\circ f = -f\circ T.$$

Hence, in order for $T = A\exp(z)$ to satisfy the Helmholtz relations, we must have $z\in\mathfrak{sl}_2$ such that

$$\exp(z)\,h\,\exp(-z) = h, \qquad \exp(z)\,e\,\exp(-z) = -e, \qquad \exp(z)\,f\,\exp(-z) = -f.$$

Using $\exp(X)\,Y\exp(-X) = \exp(\mathrm{ad}_X)\,Y$ with $\mathrm{ad}_X Y = [X,Y]$ (Lemma 5.3 of [8]) and

$$[z,h] = -C\,h + 2D\,e, \qquad [z,e] = B\,h - 2D\,f, \qquad [z,f] = -2B\,e + 2C\,f$$

to work out the left-hand sides, these conditions give rise to

$$(1+4CDX)\,h + (-4BCX-2CY)\,e + (-4BDX+2DY)\,f = h \tag{13}$$

$$(-2BDX-DY)\,h + \big(1+(4B^2+2CD)X+2BY\big)\,e + (-2D^2X)\,f = -e \tag{14}$$

$$(-2BCX+CY)\,h + (-2C^2X)\,e + \big(1+(4B^2+2CD)X-2BY\big)\,f = -f \tag{15}$$

Journal of Physics: Conference Series **597** (2015) 012026 doi:10.1088/1742-6596/597/1/012026

with

$$X = \sum_{n=1}^{\infty} \frac{(4B^2+4CD)^{n-1}}{(2n)!}, \qquad Y = \sum_{n=0}^{\infty} \frac{(4B^2+4CD)^{n}}{(2n+1)!}.$$

We find requirements for B, C, D by equating the coefficients of the operators h, e, f in the left-hand sides with those in the right-hand sides.

From the coefficients of f in (14) and e in (15) we conclude that when $X \neq 0$, we must have $C = 0$ and $D = 0$. Similarly, if $X = 0$, we deduce from the coefficient of e in (14) that $Y \neq 0$. This in turn implies again that $C = 0$ and $D = 0$ (from the coefficients of e and f in (13)). Hence, we necessarily have $C = 0$ and $D = 0$. Note that this immediately covers $T = \exp(z)$ upholding property (ii), as for $z = B\,h$ we have using (12)

$$T\,\phi_{j,k,\ell} = A\exp\big(B(2j+k+\tfrac{m}{2})\big)\,\phi_{j,k,\ell}. \tag{16}$$

Plugging $C = 0$ and $D = 0$ into (13)–(15), the only non-trivial coefficient equations yield

$$-1 = 1 + 4B^2\sum_{n=1}^{\infty}\frac{(4B^2)^{n-1}}{(2n)!} + 2B\sum_{n=0}^{\infty}\frac{(4B^2)^n}{(2n+1)!} = \exp(2B),$$

and

$$-1 = \exp(-2B).$$

Both of which lead to the condition

$$2B = i(\pi + 2\pi n) \iff B = i\frac{\pi}{2}(2n+1),$$

for some integer n. This completely fixes all viable elements $z \in \exp(\mathfrak{sl}_2)$.

Now, in order for property (iii) to hold, the eigenvalues of T^4 must all equal 1. Together with the expression (16) for the eigenvalues of $T = A\exp(z)$ this leads to $A^4\exp(i\pi m) = 1$, which gives $A = \exp\big(i\frac{\pi}{2}(-\frac{m}{2}+N)\big)$ for some integer N.

In this way we arrive at the form

$$T = \exp\left(i\frac{\pi}{2}\left((2n+1)\left(h-\frac{m}{2}\right)+nm+N\right)\right) \qquad (n, N \in \mathbb{Z})$$

Now, the operator $h - m/2$ has integer eigenvalues and for integer k we have $i^k = i^{k \bmod 4}$. Hence, for n an even integer, T is precisely the operator exponential form of the Fourier transform, up to multiplication by an integer power of i, namely i^{nm+N}. Likewise, for n odd T equals the inverse Fourier transform

$$\exp\left(-i\frac{\pi}{2}\left(h-\frac{m}{2}\right)\right),$$

up to an integer power of i. □

Property (iv) in the previous theorem is rather restrictive. In order to see whether the other properties suffice to uniquely characterize the FT, we investigate what happens if we omit it. The objective is thus to determine all operators

$$T\colon \mathcal{S}(\mathbb{R}^m) \to \mathcal{S}(\mathbb{R}^m)$$

that satisfy

(i) the Helmholtz relations

$$\begin{aligned} T \circ \Delta_x &= -|y|^2 \circ T \\ T \circ |x|^2 &= -\Delta_y \circ T \end{aligned}$$

(ii) $T\phi_{j,k,\ell} = \mu_{j,k}\,\phi_{j,k,\ell}$ with $\mu_{j,k} \in \mathbb{C}$

(iii) $T^4 = \mathrm{id}$

As opposed to the previous result, the solution to this problem will in general not be limited to the FT and its inverse. To determine these other solutions, we make explicit use of the fact that we already know of one specific solution, namely the FT, and use this to rewrite the requirements. As the FT is an automorphism on the space $\mathcal{S}(\mathbb{R}^m)$, we propose to look for $\widetilde{T}$ such that $\widetilde{T}[\mathcal{F}[f]]$ has the desired properties and then define the operator T by $T[f] = \widetilde{T}[\mathcal{F}[f]]$, or thus

$$T = \widetilde{T} \circ \mathcal{F}. \tag{17}$$

This leads to the following result:

Proposition 2. *Any operator* $\widetilde{T}$ *of the form*

$$\widetilde{T} = \exp\left(i\tfrac{\pi}{2}F\right), \tag{18}$$

with F *an operator that*

- *commutes with (the generators of)* $\mathfrak{sl}_2 = \mathrm{span}\{\Delta_x, |x|^2, [\Delta_x, |x|^2]\}$
- *has integer eigenvalues on the functions* $\{\phi_{j,k,\ell}\}$ *(independent of* ℓ*)*

will yield an operator T *by* (17) *that satisfies the properties (i)–(iii).*

The problem is thus reduced to finding suitable operators F, the requirements for which are more clear-cut than those for $\widetilde{T}$. Natural candidates for F are operators in the center of $\mathcal{U}(\mathfrak{sl}_2)$, where $\mathcal{U}(\mathfrak{sl}_2)$ denotes the universal enveloping algebra of $\mathfrak{sl}_2$. The center is the subspace consisting of those elements that commute with all of $\mathcal{U}(\mathfrak{sl}_2)$, or adequately with the elements of $\mathfrak{sl}_2$. This subspace is finitely generated by the Casimir element [4, 5, 6]. For a realization of $\mathfrak{sl}_2$ by means of a triplet h, e, f satisfying

$$[h, e] = 2e, \qquad [h, f] = -2f, \qquad [e, f] = h,$$

the Casimir element has as explicit expression:

$$\Omega = h^2 + 2ef + 2fe = h^2 + 2h + 4fe. \tag{19}$$

Moreover, as the Casimir is a diagonal operator on the representation space, the condition of $\{\phi_{j,k,\ell}\}$ being eigenfunctions is immediately fulfilled for F an operator in the center of $\mathcal{U}(\mathfrak{sl}_2)$. Hence, the only remaining requirement on F is that its eigenvalues must be integers. To resolve this we resort to the notion of integer-valued polynomials, see [3].

Except for one particular case, this is a fruitful approach yielding a complete class of operator solutions. We elaborate on this special case in the next section. For the general study hereof we again refer to [3].

Journal of Physics: Conference Series **597** (2015) 012026 doi:10.1088/1742-6596/597/1/012026

4. The one-dimensional case

We now consider the specific case where $m = 1$. Instead of dealing with vectors $x = (x_1, \ldots, x_m)$ and $y = (y_1, \ldots, y_m)$, we now denote by x and y the single coordinate $x \equiv x_1$ and $y \equiv y_1$. In this case, the orthogonal group $O(1)$ becomes a finite group of order 2. The orthogonal symmetry of the FT then comes down to commuting with the sign inversion operator

$$\mathcal{F}\big[f(-x)\big](y) = \mathcal{F}\big[f(x)\big](-y).$$

Moreover for $m = 1$, the operators in our realization of $\mathfrak{sl}_2$ reduce to

$$\Delta_x := \frac{\mathrm{d}^2}{\mathrm{d}x^2}, \qquad |x|^2 := x^2, \qquad \mathbb{E}_x := x\frac{\mathrm{d}}{\mathrm{d}x}.$$

As the Laplace operator only has one term, the space of polynomials contained in its kernel is built up out of only two parts, namely the space $\mathcal{H}_0$ which is generated by 1, and $\mathcal{H}_1$ generated by x. The basis (9) then consists of

$$\phi_{j,0} = 2^j j!\, L_j^{-1/2}(x^2)\,\mathrm{e}^{-x^2/2}, \qquad \phi_{j,1} = 2^j j!\, L_j^{1/2}(x^2)\, x\,\mathrm{e}^{-x^2/2}, \tag{20}$$

for $j \in \mathbb{Z}_{\geq 0}$. Note that there is no index ℓ as the spaces $\mathcal{H}_0$ and $\mathcal{H}_1$ are one-dimensional. These are in fact the Hermite functions (up to a scalar factor), as the generalized Laguerre polynomials are related to the Hermite polynomials as follows

$$H_{2j}(x) = (-1)^j 2^{2j} j! L_j^{-1/2}(x^2), \quad H_{2j+1}(x) = (-1)^j 2^{2j+1} j! x L_j^{-1/2}(x^2).$$

Indeed, the basis (20) can be unified as

$$\psi_k(x) = \left(i\frac{\sqrt{2}}{2}\right)^k H_k(x)\,\mathrm{e}^{-x^2/2}, \tag{21}$$

where we have slightly adjusted the factor in front. The action of the FT on (20) is given by

$$\mathcal{F}(\phi_{j,0}) = (-1)^j \phi_{j,0}, \qquad \mathcal{F}(\phi_{j,1}) = (-1)^j i\, \phi_{j,1}.$$

On (21) this becomes

$$\mathcal{F}(\psi_k) = i^k \psi_k.$$

This in fact corresponds to unifying two representations of $\mathfrak{sl}_2$ as one representation of the Lie superalgebra $\mathfrak{osp}(1|2)$. Indeed, introducing the operators

$$b^+ = i\frac{\sqrt{2}}{2}\left(x - \frac{\mathrm{d}}{\mathrm{d}x}\right) \quad \text{and} \quad b^- = -i\frac{\sqrt{2}}{2}\left(x + \frac{\mathrm{d}}{\mathrm{d}x}\right), \tag{22}$$

they satisfy the relations

$$\big[\{b^-, b^+\}, b^\pm\big] = \pm 2b^\pm,$$

where $\{a, b\} = ab + ba$ denotes the anti-commutator. Hence b^+ and b^- generate a realization of the Lie superalgebra $\mathfrak{osp}(1|2)$, [7]. The even (or "bosonic") elements of this algebra

$$h = \frac{1}{2}\{b^-, b^+\}, \qquad e = \frac{1}{4}\{b^+, b^+\}, \qquad f = -\frac{1}{4}\{b^-, b^-\}.$$

generate an even subalgebra isomorphic with $\mathfrak{sl}_2$. Working out the anti-commutators, these are precisely the operators (11) we considered earlier. For instance, using (22),

$$2h = \{b^-, b^+\} = \frac{1}{2}\left(x + \frac{\mathrm{d}}{\mathrm{d}x}\right)\left(x - \frac{\mathrm{d}}{\mathrm{d}x}\right) + \frac{1}{2}\left(x - \frac{\mathrm{d}}{\mathrm{d}x}\right)\left(x + \frac{\mathrm{d}}{\mathrm{d}x}\right) = -\frac{\mathrm{d}^2}{\mathrm{d}x^2} + x^2$$

The action on the eigenfunctions is as follows

$$b^+ \psi_k = \psi_{k+1}, \qquad b^- \psi_k = k\, \psi_{k-1},$$

while

$$h\,\psi_k = (k + \tfrac{1}{2})\,\psi_k, \quad e\,\psi_k = \frac{1}{2}\,\psi_{k+2}, \quad f\,\psi_k = -\frac{1}{2}k\,(k-1)\,\psi_{k-2}.$$

In line with (10), we can write

$$\psi_k = (b^+)^k \psi_0$$

Now, for the basis functions (20) we have

$$h\,\phi_{j,0} = \left(2j + \tfrac{1}{2}\right)\phi_{j,0}, \qquad h\,\phi_{j,1} = \left(2j + \tfrac{3}{2}\right)\phi_{j,1}.$$

The sets $\{\, \phi_{j,0} \mid j \in \mathbb{Z}_{\geq 0} \,\}$ and $\{\, \phi_{j,1} \mid j \in \mathbb{Z}_{\geq 0} \,\}$ thus form a basis for the positive discrete series representation of $\mathfrak{sl}_2$ with lowest weights $1/2$ and $3/2$ respectively. Together they span the same space as $\{\, \psi_k \mid k \in \mathbb{Z}_{\geq 0} \,\}$. This last set forms a basis for the irreducible representation of the Lie superalgebra $\mathfrak{osp}(1|2)$ with lowest weight $1/2$.

Returning to the objective we had in mind we look at the Casimir element of $\mathfrak{sl}_2$ for $m = 1$. However, this element turns out to be a scalar constant as it is given by

$$\Omega_1 = h^2 + 2h + 4fe = -\frac{3}{4}.$$

Hence, looking back at Proposition 2, an operator F in the center of $\mathcal{U}(\mathfrak{sl}_2)$ will, by relation (18), yield the classical FT up to a multiplicative scalar factor.

In this case there is another set we can consider, as besides $\mathfrak{sl}_2$ we also have a realization of $\mathfrak{osp}(1|2)$. The center of $\mathcal{U}(\mathfrak{osp}(1|2))$ is finitely generated by another Casimir element:

$$C = \frac{1}{4} + \frac{1}{2}\left[b^-, b^+\right] + h^2 + 2h + 4fe.$$

We see that C differs from the Casimir element of $\mathfrak{sl}_2$, given by (19), by an additional term. This extra term is related to the Scasimir element (see [4])

$$S = \frac{1}{2}(b^- b^+ - b^+ b^- - 1),$$

which is a square root of the Casimir operator. Using (22), for our realization of $\mathfrak{osp}(1|2)$ this element becomes

$$S = \frac{1}{4}\left(x + \frac{\mathrm{d}}{\mathrm{d}x}\right)\left(x - \frac{\mathrm{d}}{\mathrm{d}x}\right) - \frac{1}{4}\left(x - \frac{\mathrm{d}}{\mathrm{d}x}\right)\left(x + \frac{\mathrm{d}}{\mathrm{d}x}\right) - \frac{1}{2} = \frac{1}{4} + \frac{1}{4} - \frac{1}{2} = 0,$$

and hence $C = S^2 = 0$. In this way we again find no solutions other than the classical FT.

These results can also be explained in another fashion. In the one-dimensional case, the requirement for the kernel of an integral transform in order to satisfy the Helmholtz relations, (8), reduces to

$$\frac{\mathrm{d}^2}{\mathrm{d}y^2} K + x^2\, K = 0.$$

This differential equation has as general solution

$$K = C_1 e^{ixy} + C_2 e^{-ixy}, \qquad C_1, C_2 \in \mathbb{C}$$

where we recognize the kernel of the Fourier transform and that of its inverse. For $m = 1$, the Helmholtz relations thus suffice to uniquely characterize the FT or its inverse.

In dimension $m \geq 2$, the Casimir element Ω_m becomes

$$\Omega_m = h^2 + 2h + 4fe = \left(\mathbb{E}_x + \tfrac{m-2}{2}\right)^2 - |x|^2 \Delta_x - 1.$$

As in this case Ω_m does not reduce to a trivial operator, by Proposition 2 we do find solutions that differ from the FT or its inverse. We again refer to [3] for a detailed study.

References

[1] Alesker S, Artstein-Avidan S, Faifman D and Milman V 2010 A characterization of product preserving maps with applications to a characterization of the Fourier transform *Illinois J. Math.* **54** 1115–1132

[2] Coulembier K, De Bie H and Sommen F 2011 Orthogonality of the Hermite polynomials in superspace and Mehler type formulae *Proc. London Math. Soc.* **103** 786–825

[3] De Bie H, Oste R and Van der Jeugt J 2014 Generalized Fourier transforms and unique characterization of the Fourier transform (arXiv:1411.6856)

[4] Frappat L, Sciarrino A and Sorba P 2000 *Dictionary on Lie algebras and superalgebras* (San Diego, CA: Academic Press Inc.)

[5] Humphreys J 1972 *Introduction to Lie algebras and representation theory* (Springer, New York)

[6] Jacobson N 2013 *Lie algebras* (Courier Dover Publications)

[7] Lievens S, Stoilova N I and Van der Jeugt J 2008 The Paraboson Fock Space and Unitary Irreducible Representations of the Lie Superalgebra $\mathfrak{osp}(1|2n)$ *Commun. Math. Phys.* **281** 805–826

[8] Miller W 1973 *Symmetry groups and their applications* (New York: Academic Press) p 160

Rank-3 root systems induce root systems of rank 4 via a new Clifford spinor construction

Pierre-Philippe Dechant

Department of Mathematics, University of Durham,
Institute for Particle Physics Phenomenology, Ogden Centre for Fundamental Physics,
Department of Physics, University of Durham, South Rd, Durham DH1 3LE, UK and
Department of Mathematics, York Centre for Complex Systems Analysis, University of York,
Heslington, York YO10 5GG, UK

E-mail: ppd22@cantab.net

Abstract. In this paper, we show that via a novel construction every rank-3 root system induces a root system of rank 4. Via the Cartan-Dieudonné theorem, an even number of successive Coxeter reflections yields rotations that in a Clifford algebra framework are described by spinors. In three dimensions these spinors themselves have a natural four-dimensional Euclidean structure, and discrete spinor groups can therefore be interpreted as 4D polytopes. In fact, we show that these polytopes have to be root systems, thereby inducing Coxeter groups of rank 4, and that their automorphism groups include two factors of the respective discrete spinor groups trivially acting on the left and on the right by spinor multiplication. Special cases of this general theorem include the exceptional 4D groups D_4, F_4 and H_4, which therefore opens up a new understanding of applications of these structures in terms of spinorial geometry. In particular, 4D groups are ubiquitous in high energy physics. For the corresponding case in two dimensions, the groups $I_2(n)$ are shown to be self-dual, whilst via a similar construction in terms of octonions each rank-3 root system induces a root system in dimension 8; this root system is in fact the direct sum of two copies of the corresponding induced 4D root system.

1. Introduction

Root systems are useful mathematical abstractions, which are polytopes that generate reflection (Coxeter) groups. Certain families of root systems exist in any dimension, whereas others – exceptional ones – only exist as accidental structures in specific dimensions. Root systems in different dimensions are largely thought to be independent of each other (with the exception of sub-root systems). The Freudenthal-Tits magic square [1] makes some non-trivial connections, but geometric insight as to why these should exist is scarce. In this paper, we present a novel connection between root systems in different dimensions that has a geometric origin.

Clifford's Geometric Algebra provides a mathematical framework that generalises the more familiar vector space and matrix methods. In this setup, orthogonal transformations are encoded (in fact, doubly covered) by versors, which are the Clifford geometric product of several unit vectors, via a sandwiching prescription. In particular, the rotations (i.e. the special orthogonal group) are doubly covered by geometric products of an even number of unit vectors, resulting in spinors, or 'rotors'. These elements in the even-subalgebra can themselves have a Euclidean structure and can thus be reinterpreted as vectors in a different space. Here, we show that such a construction can induce a root system (in the same or another dimension) from a given root system. This systematises the observations made in [2, 3, 4], and opens up a new – spinorial – view of the geometry of root systems (where by spinorial we mean something related to the double cover of the rotations i.e. special orthogonal transformations), with interesting applications particularly concerning the interplay of three and four dimensions, and notably the exceptional root systems D_4, F_4 and H_4: D_4, the root system related to $SO(8)$ with exceptional triality symmetry, F_4, the largest crystallographic group in 4D, and the largest non-crystallographic group, H_4.

The rest of this paper is organised as follows. Section 2 introduces Coxeter groups and root systems, and Section 3 gives the necessary Clifford algebra background. The Induction Theorem is stated and proven in Section 4, and the self-duality of the two-dimensional case is discussed in Section 5. We extend our approach to a related octonionic construction in Section 6. Conclusions are given in Section 7.

2. Coxeter Groups

Definition 2.1 (Coxeter group). *A Coxeter group is a group generated by some involutive generators $s_i, s_j \in S$ subject to relations of the form $(s_i s_j)^{m_{ij}} = 1$ with $m_{ij} = m_{ji} \geq 2$ for $i \neq j$.*

The finite Coxeter groups have a geometric representation where the involutions are realised as reflections at hyperplanes through the origin in a Euclidean vector space $\mathcal{E}$ (essentially just the classical reflection groups). In particular, let $(\cdot|\cdot)$ denote the inner product in $\mathcal{E}$, and λ, $\alpha \in \mathcal{E}$.

Definition 2.2 (Reflections and roots). *The generator s_α corresponds to the reflection*

$$s_\alpha : \lambda \rightarrow s_\alpha(\lambda) = \lambda - 2\frac{(\lambda|\alpha)}{(\alpha|\alpha)}\alpha \tag{1}$$

at a hyperplane perpendicular to the root vector α.

The action of the Coxeter group is to permute these root vectors, and its structure is thus encoded in the collection $\Phi \in \mathcal{E}$ of all such roots, which form a root system:

Definition 2.3 (Root system). *Root systems are defined by the two axioms*

(i) Φ only contains a root α and its negative, but no other scalar multiples: $\Phi \cap \mathbb{R}\alpha = \{-\alpha, \alpha\}$ $\forall\, \alpha \in \Phi$.

(ii) Φ is invariant under all reflections corresponding to vectors in Φ: $s_\alpha \Phi = \Phi$ $\forall\, \alpha \in \Phi$.

Root systems and their associated Coxeter groups describe polyhedral symmetries such as the symmetries of the Platonic Solids, and are also central in Lie Theory [5, 6]. They have manifold practical applications, for instance to crystals and quasicrystals [7, 8, 9], the structure of viruses [10] and other polyhedral objects [11, 12, 13, 14], as well as high energy theory [15, 16, 17] and singularities [18, 19]. A recent characterisation and visualisation of the relevant groups in two and three dimensions in a conformal framework was given in [20, 21, 22]. Some of the 4-dimensional root systems have unusual accidental properties, such as triality in D_4 (or better known in the physics literature as the triality between vector and spinor representations of $SO(8)$, important in showing the equivalence of the Green-Schwarz and Ramond-Neveu-Schwarz strings), the largest non-crystallographic Coxeter group H_4 (which contains the GUT group $A_4 = SU(5)$ as a subgroup), or the largest symmetry group F_4 of a lattice in four dimensions (which also contains the little groups of String and M-Theory, D_4 and B_4, respectively). These peculiarities are discussed in more detail elsewhere [2, 4] – here, we stress the general nature of our argument without the need to discuss individual root systems and their applications.

3. Geometric Algebra

The study of Clifford algebras and Geometric Algebra originated with Grassmann's, Hamilton's and Clifford's geometric work [23, 24, 25]. However, the geometric content of the algebras was soon lost when interesting algebraic properties were discovered in mathematics, and Gibbs advocated the use of the hybrid system of vector calculus in physics. When Clifford algebras resurfaced in physics in the context of quantum mechanics, it was purely for their algebraic properties, and this continues in particle physics to this day. Thus, it is widely thought that Clifford algebras are somehow intrinsically quantum mechanical in nature. The original geometric meaning of Clifford algebras has been revived in the work of David Hestenes [26, 27, 28]. Here, we follow an exposition along the lines of [29].

In a manner reminiscent of complex numbers carrying both real and imaginary parts in the same algebraic entity, one can consider the geometric product of two vectors defined as the sum of their scalar (inner/symmetric) product and wedge (outer/ exterior/antisymmetric) product

$$ab \equiv a \cdot b + a \wedge b. \tag{2}$$

The wedge product is the outer product introduced by Grassmann, as an antisymmetric product of two vectors, which naturally defines a plane. Unlike the constituent inner and outer products, the geometric product is invertible, as a^{-1} is simply given by $a^{-1} = a/(a^2)$. This leads to many algebraic simplifications over standard vector space techniques, and also feeds through to the differential structure of the theory, with Green's function methods that are not achievable with vector calculus methods. This geometric product can be extended to the product of more vectors via associativity and distributivity, resulting in higher grade objects called multivectors. There are a total of 2^n elements in the algebra, since it truncates at grade n multivectors due to the scalar nature of the product of parallel vectors and the antisymmetry of orthogonal vectors. Essentially, a Clifford algebra is a deformation of the exterior algebra by a quadratic form, and for a Geometric Algebra this is the metric of space(time).

The geometric product provides a very compact and efficient way of handling reflections in any number of dimensions, and thus by the Cartan-Dieudonné theorem also rotations [30]. For a unit vector n, we consider the reflection of a vector a in the hyperplane orthogonal to n. Thanks to the geometric product in Clifford algebra the two terms in Eq. (1) combine into a single term and thus a 'sandwiching prescription':

Theorem 3.1 (Reflections). *In Geometric Algebra, a vector 'a' transforms under a reflection in the (hyper-)plane defined by a unit normal vector 'n' as*

$$a' = -nan. \tag{3}$$

This is a remarkably compact and simple prescription for reflecting vectors in hyperplanes. More generally, higher grade multivectors of the form $M = ab \dots c$ (so-called versors) transform similarly ('covariantly'), as $M = ab \dots c \to \pm nannbn \dots ncn = \pm nab \dots cn = \pm nMn$. Even more importantly, from the Cartan-Dieudonné theorem, rotations are the product of successive reflections. For instance, compounding the reflections in the hyperplanes defined by the unit vectors n and m results in a rotation in the plane defined by $n \wedge m$.

Proposition 3.2 (Rotations). *In Geometric Algebra, a vector 'a' transforms under a rotation in the plane defined by $n \wedge m$ via successive reflection in hyperplanes determined by the unit vectors 'n' and 'm' as*

$$a'' = mnanm =: Ra\tilde{R}, \tag{4}$$

where we have defined $R = mn$ and the tilde denotes the reversal of the order of the constituent vectors $\tilde{R} = nm$.

Journal of Physics: Conference Series **597** (2015) 012027 doi:10.1088/1742-6596/597/1/012027

Theorem 3.3 (Rotors and spinors). *The object $R = mn$ generating the rotation in Eq. (4) is called a rotor. It satisfies $\tilde{R}R = R\tilde{R} = 1$. Rotors themselves transform single-sidedly under further rotations, and thus form a multiplicative group under the geometric product, called the rotor group. Since R and $-R$ encode the same rotation, the rotor group is a double-cover of the special orthogonal group, and is thus essentially the Spin group. Objects in Geometric Algebra that transform single-sidedly are called spinors, so that rotors are normalised spinors.*

Corollary 3.4 (Discrete spinor groups). *Discrete spinor groups are of even order.*

Higher multivectors transform in the above covariant, double-sided way as $MN \rightarrow (RM\tilde{R})(RN\tilde{R}) = RM\tilde{R}RN\tilde{R} = R(MN)\tilde{R}$.

The Geometric Algebra of three dimensions Cl(3) spanned by three orthogonal unit vectors e_1, e_2 and e_3 contains three bivectors e_1e_2, e_2e_3 and e_3e_1 that square to -1, as well as the highest grade object $e_1e_2e_3$ (trivector and pseudoscalar), which also squares to -1.

$$\underbrace{\{1\}}_{1 \text{ scalar}} \quad \underbrace{\{e_1, e_2, e_3\}}_{3 \text{ vectors}} \quad \underbrace{\{e_1e_2 = Ie_3, e_2e_3 = Ie_1, e_3e_1 = Ie_2\}}_{3 \text{ bivectors}} \quad \underbrace{\{I \equiv e_1e_2e_3\}}_{1 \text{ trivector}}. \tag{5}$$

Theorem 3.5 (Quaternions and spinors of Cl(3)). *The unit spinors $\{1, -Ie_1, -Ie_2, -Ie_3\}$ of* Cl(3) *are isomorphic to the quaternion algebra $\mathbb{H}$.*

This completes the background that we shall need for our proof of the Induction Theorem.

4. Induction Theorem

In this section, we show that every root system of rank 3 induces a root system in dimension 4.

Proposition 4.1 ($O(4)$-structure of spinors and quaternions). *The space of* Cl(3)*-spinors and quaternions have a 4D Euclidean signature.*

Proof. For quaternions, this is given via conjugation defined by $\bar{q} = q_0 - q_ie_i$, as $(p,q) = \frac{1}{2}(\bar{p}q + p\bar{q})$, $|q|^2 = \bar{q}q = q_0^2 + q_1^2 + q_2^2 + q_3^2$. For a spinor $R = a_0 + a_1Ie_1 + a_2Ie_2 + a_3Ie_3$, the norm is given by $R\tilde{R} = a_0^2 + a_1^2 + a_2^2 + a_3^2$ (or via Theorem 3.5), and the inner product is $(R_1, R_2) = \frac{1}{2}(R_1\tilde{R}_2 + R_2\tilde{R}_1)$. □

Lemma 4.2 (Discrete quaternion groups give root systems). *Any finite subgroup G of even order in $\mathbb{H}$ is a root system.*

Proof. This is stated and proven in [31]. □

Lemma 4.3 (Rank-3 Coxeter groups and finite even quaternion groups). *The spinors defined from any rank-3 Coxeter group are isomorphic to an even subgroup of the quaternions.*

Proof. From Corollary 3.4, the spinor group generated by a Coxeter group is discrete and even. Because of Theorem 3.5, for a Coxeter group of rank 3 this spinor group is isomorphic to a finite even order quaternion group. □

Lemma 4.4 (Discrete spinor groups in 3D give 4D root systems). *A discrete group of spinors in three dimensions is a four-dimensional root system.*

Proof. Due to Lemma 4.3, the discrete spinor group is even and isomorphic to an even quaternion group. From Lemma 4.2 it follows that this is a root system. □

Theorem 4.5 (Induction Theorem). *Any rank-3 root system induces a root system of rank 4.*

Proof. A root system in three dimensions gives rise to corresponding Coxeter reflections (Section 2), acting in Geometric Algebra as given by Eq. (3). An even number of successive reflections yields spinors via Theorem 3.3, and from Corollary 3.4, this group is even. Via Lemma 4.4, this even spinor group yields a root system in four dimensions.

Alternatively, one can check the root system axioms from Definition 2.3 directly. The first is satisfied, since trivially for an element in a discrete spinor group its negative is also in the discrete spinor group by Corollary 3.4. One can easily check the axiom of closure of the root system under reflections by using Eq. (1) in 4D using the Euclidean inner product in Proposition 4.1. For spinors R_1 and R_2 this amounts to $R_2 \to R_2' = R_2 - 2(R_1, R_2)/(R_1, R_1)R_1 = R_2 - ((R_1\tilde{R}_2 + R_2\tilde{R}_1)R_1/(R_1\tilde{R}_1) = -R_1\tilde{R}_2R_1/(R_1\tilde{R}_1)$. Closure of the root system is thus ensured by closure of the spinor group. This also has very interesting consequences for the automorphism group of these root systems, which contains two factors of the spinor group acting from the left and the right [4] (in this sense, the above closure under reflections amounts to a certain twisted conjugation). □

Theorem 4.6 (Automorphism groups of the induced root systems). *The automorphism groups of these induced root systems contain two factors of the respective spinor group (i.e. the root system itself), acting on the left and on the right (trivially, in the spinorial picture, as group multiplication in the spinor group, by group closure).*

Theorem 4.7 (Non-existence of a reduction theorem). *Not every rank-4 root system can be induced by a rank-3 root system.*

Proof. A counterexample is provided by $I_2(4) \times A_1 \times A_1$. □

In the above proof no reference was made to any particular root system, and it is thus valid in generality. However, the number of such root systems is limited, so that one can list the induced root systems on a case-by-case basis. Table 1 contains the list of 4D root systems that are induced by 3D root systems, as well as the results of later sections. It is particularly noteworthy that the exceptional four-dimensional root systems arise via our Clifford spinor construction; in particular, the fact that irreducible root systems arise in this way. One might speculate philosophically whether their existence in some sense hinges on the accidentalness of this 3D Clifford spinor construction.

Example 1. *The simple roots of $A_1 \times A_1 \times A_1$ can be taken as $\alpha_1 = e_1$, $\alpha_2 = e_2$ and $\alpha_3 = e_3$. Closure of these under reflections via Eq. (3) gives $(\pm 1, 0, 0)$ and permutations thereof, which are the 6 vertices of the root system, the octahedron. Combining two reflections yields a spinor, so forming rotors according to $R_{ij} = \alpha_i\alpha_j$ gives, e.g. $R_{11} = \alpha_1^2 = 1 \equiv (1, 0, 0, 0)$, or $R_{23} = \alpha_2\alpha_3 = e_2e_3 = Ie_1 \equiv (0, 1, 0, 0)$, where we denote components as given in Theorem 3.5. Explicit calculation of all cases generates the 8 permutations of $(\pm 1, 0, 0, 0)$. When interpreted as a 4D polytope, these are the vertices of the 16-cell, which is the root system of $A_1 \times A_1 \times A_1 \times A_1$.*

Example 2. *Other specific examples are the inductions $A_3 \to D_4$ (cuboctahedron to 24-cell), $B_3 \to F_4$ and $H_3 \to H_4$ (icosidodecahedron to 600-cell) [2, 3, 4]. This demonstrates how exceptional phenomena in dimension four could arise from regular root systems (i.e. A_n and B_n for $n = 3$), and hints that a spinorial view of their geometry can shed light on many of their applications.*

5. Spinors in dimension two

The space of spinors $\psi = a + be_1e_2 \equiv a + bI$ in two-dimensional Euclidean space is also two-dimensional, and has a natural Euclidean structure given by $\psi\tilde{\psi} = a^2 + b^2$. This induces a rank-2 root system from any rank-2 root system in a similar way to the construction above. However, this construction does not yield any new root systems by the following theorem.

Journal of Physics: Conference Series **597** (2015) 012027 doi:10.1088/1742-6596/597/1/012027

Theorem 5.1 (Self-duality of $I_2(n)$). *Two-dimensional root systems are self-dual under the Clifford spinor construction.*

Proof. A 2D root vector $\alpha_i = a_1e_1 + a_2e_2$ is in bijection with a spinor by $\alpha_i \to \alpha_1\alpha_i = e_1\alpha_i = a_1 + a_2e_1e_2 = a_1 + a_2I$ (taking $\alpha_1 = e_1$ without loss of generality). This is the same as forming a spinor between those two root vectors. The infinite family of two-dimensional root systems $I_2(n)$ is therefore self-dual. The order of the Coxeter group $|W|$ matches the number of roots $|\Phi|$, even for the well-known crystallographic cases in 2D: for instance, for the A_n family one has the general formulae $|W| = (n+1)!$ and $|\Phi| = n(n+1)$, with equality for $n = 2$, as $A_2 = I_2(3)$. For $B_2 = I_2(4)$, $|W| = 2^n n!$ and $|\Phi| = 2n^2$, such that equality holds for $n = 2$. For $G_2 = I_2(6)$, one also has $|W| = |\Phi| = 12$. □

6. An octonion construction: eight-dimensional root systems

The crucial fact that our 3D and 2D (nD) constructions depended on was that Clifford algebra allowed us to construct a group of spinors with 2^{n-1} components, that – when thought of as a set of vectors in 2^{n-1}-dimensional space – fulfilled the first axiom of a root system, i.e. that α and $-\alpha$ are contained in the set, by construction. This 2^{n-1}-dimensional space also had a Euclidean metric. The second part of the proof that these sets are in fact root systems was then to show closure under reflections in these roots as defined by the Euclidean metric. For instance, this is no longer the case in 4D, where the space of spinors is 8-dimensional. However, here we follow an analogous construction in terms of octonions, that does yield root systems in 8D.

The octonions $\mathbb{O}$ are a generalisation of the quaternions, complex and real numbers. It is a (non-associative division) algebra with one scalar and seven imaginary units that satisfy certain commutativity properties. In analogy with complex and quaternionic conjugation, there is an octonionic conjugation, which defines an 8-dimensional Euclidean metric.

In order to mimic our Clifford construction above, we take the simple roots of the 3D root systems to be along the first three imaginary units i, j and k, i.e. the simple roots of A_1^3 would be $\alpha_1 = i$, $\alpha_2 = j$ and $\alpha_3 = k$. We then consider the closure of this set of octonions under octonion multiplication (here 16 octonions, namely the 8 unit octonions and their negatives). Again this gives a set (group) of octonions which contains $\pm\alpha_i$ by construction. It is also straightforward to show that these sets of octonions are closed under reflections with respect to the 8-dimensional Euclidean metric, as this reflection essentially amounts to octonion multiplication and the above procedure of taking the closure under octonion multiplication therefore guarantees closure under reflections. The resulting set of vectors is therefore a root system in eight dimensions (here A_1^8). However, the resulting root system in each case is just the direct sum of two copies of the corresponding induced 4D root system, e.g. H_3 induces H_4 in 4D and $H_4 \oplus H_4$ in 8D. The argument in the paper is general and does not make reference to any particular root system. Table 1 therefore summarises the results for each of the limited number of root systems for illustrative purposes.

7. Conclusions

We have shown how via a Clifford spinor construction, any root system of rank 3 induces a root system in four dimensions. This was done in full generality without reference to any particular root system. However, since the number of root systems in 3D is finite, one gets a concrete list of cases. One finds that the 4D root systems induced in this way contain mostly the exceptional root systems that generate the exceptional Coxeter groups D_4 (triality), F_4 (largest crystallographic group in 4D) and H_4 (largest non-crystallographic group). This construction therefore offers a novel perspective on exceptional phenomena in four dimensions and the peculiar structure of their automorphism groups, in terms of spinorial geometry.

Table 1. Summary of the induced root systems in two, four and eight dimensions.

start root system	induced root system
$I_2(n)$	$I_2(n)$
$A_1 \oplus I_2(n)$	$I_2(n) \oplus I_2(n)$
A_3	D_4
B_3	F_4
H_3	H_4
$A_1 \oplus I_2(n)$	$I_2(n) \oplus I_2(n) \oplus I_2(n) \oplus I_2(n)$
A_3	$D_4 \oplus D_4$
B_3	$F_4 \oplus F_4$
H_3	$H_4 \oplus H_4$

Via an analogous construction in terms of octonions, each such 3D root system also induces a root system in eight dimensions; however, this root system is reducible and in fact consists of two copies of the induced 4D root system in each case. In the two-dimensional case, root systems (i.e. $I_2(n)$) were shown to be self-dual. This spinorial view sheds light on the peculiarities of root systems, in particular certain rank-4 root systems and their automorphism groups (see [2, 4]), and more generally, opens up a new field of study in the spinorial geometry of root systems. This could be particularly interesting for applications in high energy physics, where the rank-4 groups are ubiquitous.

Acknowledgments

I would like to thank my family and friends for their support and David Hestenes, Eckhard Hitzer, Anthony Lasenby, Joan Lasenby, Reidun Twarock, Mike Hobson and Céline Bœhm for helpful discussions.

References

[1] Baez J C 2002 *Bull. Amer. Math. Soc.* **39** 145–205
[2] Dechant P P 2013 *Advances in Applied Clifford Algebras* **23** 301–321 (*Preprint* 1205.1451)
[3] Dechant P P 2014 *Advances in Applied Clifford Algebras* **24** 89–108 (*Preprint* 1207.5005)
[4] Dechant P P 2013 *Acta Crystallographica Section A: Foundations of Crystallography* **69** 592–602
[5] Kac V G 1994 *Infinite-Dimensional Lie Algebras* (*Progress in mathematics* vol 44) (Cambridge: Cambridge University Press)
[6] Fuchs J and Schweigert C 2003 *Symmetries, Lie Algebras and Representations* (Cambridge: Cambridge University Press)
[7] Patera J and Twarock R 2002 *Journal of Physics A: Mathematical and General* **35** 1551–1574
[8] Koca M, Koca N O and Koç R 1998 *Turkish Journal of Physics* **22** 421–436
[9] Dechant P P, Boehm C and Twarock R 2013 *Journal of Mathematical Physics* **54** 093508
[10] Dechant P P, Boehm C and Twarock R 2012 *Journal of Physics A: Mathematical and Theoretical* **45** 285202 (*Preprint* 1110.5219)
[11] Twarock R 2002 *Phys. Lett. A* **300** 437–444
[12] Dechant P P, Wardman J, Keef T and Twarock R 2014 *Acta Crystallographica Section A* **70** 162–167
[13] Koca M, Al-Ajmi M and Koç R 2007 *Journal of Mathematical Physics* **48** 113514
[14] Koca M 2011 *Journal of Physics Conference Series* **284** 012040 (*Preprint* 1006.3149)
[15] Gross D J, Harvey J A, Martinec E J and Rohm R 1985 *Nucl.Phys.* **B256** 253
[16] Damour T, Henneaux M and Nicolai H 2002 *Physical Review Letters* **89** 221601 (*Preprint* hep-th/0207267)
[17] Koca M, Koc R and Al-Barwani M 2001 *Journal of Physics A: Mathematical and General* **34** 11201
[18] Shcherbak O P 1988 *Russian Mathematical Surveys* **43** 149
[19] Henneaux M, Persson D and Spindel P 2008 *Living Reviews in Relativity* **11** 1 (*Preprint* 0710.1818)

[20] Hestenes D 2002 *Point Groups and Space Groups in Geometric Algebra* (Boston: Birkhäuser) pp 3–34
[21] Hestenes D and Holt J W 2007 *Journal of Mathematical Physics* **48** 023514
[22] Hitzer E and Perwass C 2010 *Advances in Applied Clifford Algebras* **20**(3) 631–658
[23] Grassmann H 1844 *Die lineale Ausdehnungslehre* (Leipzig: Otto Wigand)
[24] Hamilton W R 1844 *Philos. Mag.* **25** 489
[25] Clifford W 1878 *American Journal of Mathematics* **1** pp. 350–358
[26] Hestenes D 1966 *Space-Time Algebra* (New York: Gordon and Breach)
[27] Hestenes D and Sobczyk G 1984 *Clifford algebra to geometric calculus: a unified language for mathematics and physics* Fundamental theories of physics (Dordrecht: Reidel)
[28] Hestenes D 1999 *New foundations for classical mechanics; 2nd ed.* Fundamental theories of physics (Dordrecht: Kluwer)
[29] Doran C and Lasenby A N 2003 *Geometric Algebra for Physicists* (Cambridge: Cambridge University Press)
[30] Garling D J H 2011 *Clifford Algebras: An Introduction* (Cambridge: Cambridge University Press)
[31] Humphreys J E 1990 *Reflection groups and Coxeter groups* (Cambridge: Cambridge University Press)

Noncommutative supergeometry and quantum supergroups

Axel de Goursac

Chargé de Recherches FNRS, IRMP, Université Catholique de Louvain, Chemin du Cyclotron 2, 1348 Louvain-la-Neuve, Belgium

E-mail: axelmg@melix.net

Abstract. This is a review of concepts of noncommutative supergeometry - namely Hilbert superspace, C*-superalgebra, quantum supergroup - and corresponding results. In particular, we present applications of noncommutative supergeometry in harmonic analysis of Lie supergroups, non-formal deformation quantization of supermanifolds, quantum field theory on noncommutative spaces; and we give explicit examples as deformation of flat superspaces, noncommutative supertori, solvable topological quantum supergroups.

Introduction

Differential supergeometry has its origin [1, 2] in the second half of the 20th century. Its objects are supermanifolds involving, besides the usual commuting coordinates, also anticommuting coordinates. The corresponding algebras of functions are then $\mathbb{Z}_2$-graded commutative. This graded point of view is very useful in various domains of mathematics such as differential geometry, spinor geometry, Clifford algebras, and also in physics for fermionic or supersymmetric theories, or for the BRST formalism in gauge theories.

Besides, noncommutative geometry is a growing field of mathematics [3] whose essential principle lies in the duality between spaces and commutative algebras, so that the properties of spaces can be characterized algebraically. Then, a noncommutative algebra can be seen as corresponding to some "noncommutative space". This very rich way of thinking allows generalizing classical notions and theorems of usual geometry, and it is even possible to prove some new results for differential geometry in this more general noncommutative framework. Moreover, noncommutative geometry is the appropriate framework for quantization and also a challenging tool to build new models in quantum physics. It has in particular deep links with quantum gravity, quantum Hall effect and string theories [4, 5].

Putting together these two ingredients – supergeometry and noncommutativity – gives rise to noncommutative supergeometry (NCSG). Noncommutative supergeometry deals with (noncommutative) $\mathbb{Z}_2$-graded algebras that correspond to "noncommutative superspaces" in the point of view of noncommutative geometry described above. It should be the right framework for fermionic or supersymmetric quantum field theories on noncommutative spacetime, or BRST formalism for noncommutative spaces... Early results [6, 7] concerning an application of NCSG to quantum field theory on the Moyal space also motivated the author to develop this program of NCSG at a mathematical level as well as to look at physical applications.

Of course, there is already an extensive literature on graded algebras (see e.g. [8]), useful for NCSG at the algebraic level. Some work in the direction of noncommutative Q-manifolds was also performed in [9, 10]. However, at the analytic level, NCSG was really initiated in [11] with the definition of its basic objects: Hilbert superspaces and C*-superalgebras that endow noncommutative superspaces with topological properties. In this paper, we review these analytic notions, as well as the corresponding notion of symmetry – quantum supergroup – defined and studied in the Fréchet setting [12]. We stress the relevance of this definition of Hilbert superspace, rather than the standard one, by presenting an extended Stone–von Neumann theorem valid in any signature. At the operator algebra level, a universal deformation formula that extends Rieffel's deformation quantization to the graded setting is valid in the category of C*-superalgebras, underlying its relevance. We also look at an application of NCSG to quantum field theory.

1. Classical supergeometry

In this section, we recall basics of supergeometry (see [1, 2, 13, 14, 15, 16] for the sheaf-theoretic approach and [17, 18, 19]) for the concrete approach).

A *supermanifold* M of dimension $m|n$ is the data of a smooth manifold of dimension m called its body and denoted $\mathbb{B}M$ together with a sheaf on $\mathbb{B}M$ of $\mathbb{Z}_2$-graded commutative algebras; and this locally ringed space is locally isomorphic to $\mathcal{C}^\infty_{\mathbb{R}^m} \otimes \bigwedge \mathbb{R}^n$. We denote by $\mathcal{C}^\infty(M)$ the algebra of global sections of the sheaf corresponding to $\mathbb{B}M$. It is a $\mathbb{Z}_2$-graded commutative Fréchet algebra. The simplest example is the locally ringed space $\mathcal{C}^\infty_{\mathbb{R}^m} \otimes \bigwedge \mathbb{R}^n$ itself, called *superspace* of dimension $m|n$ and denoted by $\mathbb{R}^{m|n}$, for which we have

$$\mathcal{C}^\infty(\mathbb{R}^{m|n}) \simeq \mathcal{C}^\infty(\mathbb{R}^m) \otimes \bigwedge \mathbb{R}^n. \tag{1}$$

To obtain explicit expressions, let us fix the canonical basis $(\xi^i)_{1\leq i\leq n}$ of $\mathbb{R}^n$. A smooth superfunction f on the superspace $\mathbb{R}^{m|n}$, i.e. a section $f \in \mathcal{C}^\infty(\mathbb{R}^{m|n})$, can be decomposed as

$$\sum_I f_I(x)\xi^I,$$

for any $x \in \mathbb{R}^m = \mathbb{B}\mathbb{R}^{m|n}$, and where the sum is taken on all ordered subsets I of $\{1,\dots,n\}$ with $\xi^I := \prod_{i\in I}\xi^i = \xi^{i_1}\xi^{i_2}\cdots\xi^{i_k}$ if $I = \{i_1,\dots,i_k\}$ (ordered), and $\xi^\emptyset = 1$ by convention; the coefficients f_I being smooth functions in $\mathcal{C}^\infty(\mathbb{R}^m)$.

Owing to the equivalence of this definition with the one of the concrete approach, we will note in the following that $f(x,\xi) = \sum_I f_I(x)\xi^I$ for any $(x,\xi) \in \mathbb{R}^{m|n}$, even if a rigorous definition of this notation involves functors of points. For any two (ordered) subsets $I = \{i_1,\dots,i_l\}$ and $J = \{j_1,\dots,j_\ell\}$ of $\{1,\dots,n\}$ we define $\varepsilon(I,J)$ to be zero if I and J overlap; if $I \cap J = \emptyset$, we set $\varepsilon(I,J)$ to the parity of the list $(i_1,\dots,i_k,j_1,\dots,j_\ell)$, defined as -1 raised to the number of transpositions needed to put it in increasing order. It satisfies

$$\varepsilon(I,J) = (-1)^{|I||J|}\varepsilon(J,I), \qquad \varepsilon(I,J\cup K) = \varepsilon(I,J)\varepsilon(I,K) \text{ if } J\cap K = \emptyset. \tag{2}$$

As a consequence, we can perform explicit products of superfunctions $\xi^I.\xi^J = \varepsilon(I,J)\xi^{I\cup J}$ for $I \cap J = \emptyset$. We recall the Lebesgue–Berezin *integration* for superfunctions:

$$\int_{\mathbb{R}^{m|n}} \mathrm{d}x\mathrm{d}\xi\, f(x,\xi) := \int_{\mathbb{R}^m} \mathrm{d}x\, f_{\{1,\dots,n\}}(x). \tag{3}$$

Journal of Physics: Conference Series **597** (2015) 012028 doi:10.1088/1742-6596/597/1/012028

2. Philosophy of NCSG

The duality between spaces and commutative algebras is transverse to numerous domains of Mathematics. At the topological level, the Gelfand–Naïmark theorem states that there is a contravariant correspondence between the category of locally compact Hausdorff spaces and the one of commutative C*-algebras: $X \mapsto \mathcal{C}(X) = \{X \to \mathbb{C} \text{ continuous}\}$ defining an equivalence of categories. Then, the category of noncommutative topological spaces can be defined as the dual of the category of C*-algebras [3].

Omitting topological conditions and in view of the $\mathbb{Z}_2$-graded commutative structure of the functions $\mathcal{C}^\infty(M)$ on a supermanifold M, we can define *noncommutative superspaces* as associated to $\mathbb{Z}_2$-graded (noncommutative) algebras. So the difference with noncommutative geometry is simply a $\mathbb{Z}_2$-grading, but we will see that this grading induces important features. This algebraic definition of noncommutative superspaces is very simple and gives rise to a well-known object (see for example [8] on graded algebras). Note that geometric objects were considered and constructed for noncommutative superspaces [7] such as the noncommutative analogues of vector fields (graded derivations), de Rham forms (bigraded noncommutative differential calculus based on graded derivations), connections and covariant derivatives (noncommutative ε-connections), curvatures, gauge transformations... These tools were actually defined for graded algebras with a more general grading and a bicharacter (like color algebras), opening the subject of *noncommutative graded* (or *color*) *geometry* [7].

However, topological structures like C*-algebras play a crucial role in noncommutative geometry. For NCSG, we introduced in [11] two topological notions, that of Hilbert superspaces and of C*-superalgebra that we recall here.

The standard notion of Hilbert superspace [15, 20], also used in [21] consists of the data of a $\mathbb{Z}_2$-graded Hilbert space $\mathcal{H} = \mathcal{H}_0 \oplus \mathcal{H}_1$ with hermitian positive definite scalar product $(-,-)$ and together with an inner product defined by

$$\langle \varphi, \psi \rangle := i^{|\varphi||\psi|}(\varphi, \psi). \tag{4}$$

This inner product is of degree 0 and superhermitian, i.e. $\forall x, y \in \mathcal{H}$, $\overline{\langle x, y \rangle} = (-1)^{|x||y|}\langle y, x \rangle$. The notion of Hilbert superspace presented below is a generalization of the one associated to (4). Indeed, the inner product can be of degree 1 below and the link between $\langle -,- \rangle$ is by a more general operator J, not only $J(\psi) = i^{|\psi|}\psi$.

Definition 2.1 ([11]) A *Hilbert superspace* of parity $n \in \mathbb{Z}_2$ is a complex $\mathbb{Z}_2$-graded vector space $\mathcal{H} = \mathcal{H}_+ \oplus \mathcal{H}_-$ endowed with a superhermitian (sesquilinear) homogeneous inner product $\langle -,- \rangle$ of degree n, such that there exists a *fundamental symmetry* J of degree n i.e. an endomorphism J of $\mathcal{H}$ satisfying $\forall x, y \in \mathcal{H}$,

- $J^2(x) = (-1)^{(n+1)|x|}x$, and $\langle J(x), J(y) \rangle = \langle x, y \rangle$,
- $(x,y)_J := \langle x, J(y) \rangle$ defines a hermitian positive definite scalar product on $\mathcal{H}$ for which $\mathcal{H}$ is complete.

The Hilbert topology of a Hilbert superspace $\mathcal{H}$ is actually independent of the choice of the fundamental symmetry J [22], so the space of continuous endomorphisms $\mathcal{B}(\mathcal{H})$ is canonically defined. In parity 1, a Hilbert superspace $(\mathcal{H}, \langle -,- \rangle)$ is a Krein space, while a Hilbert superspace of parity 0 corresponds to the orthogonal sum of the two Krein spaces $(\mathcal{H}_+, \langle -,- \rangle)$ and $(\mathcal{H}_-, i\langle -,- \rangle)$. This category of Hilbert superspaces is stable for the direct sum and the tensor product. The next Example produces Hilbert superspaces in the sense of Definition 2.1 but not in the sense of (4) for $n > 0$.

Example 2.2 ([11, 22]) In view of the decomposition (1), we set $L^2(\mathbb{R}^{m|n}) = L^2(\mathbb{R}^m)\otimes\bigwedge\mathbb{R}^n$. It is a Hilbert superspace of parity n (mod 2) with the usual $\mathbb{Z}_2$-grading and the superhermitian scalar product

$$\langle f_1, f_2\rangle = \int_{\mathbb{R}^{m|n}} \mathrm{d}x\mathrm{d}\xi\ \overline{f_1(x,\xi)}f_2(x,\xi) \tag{5}$$

by using the Berezin integration (3). A fundamental symmetry is given by the analog of the Hodge operation $J(f)(x,\xi) = \sum_I \varepsilon(I,\complement I)f_I(x)\xi^{\complement I}$, where ε was defined in (2) and $\complement I$ denotes the complement of the subset I in $\{1,\dots,n\}$.

For any bounded operator $T\in\mathcal{B}(\mathcal{H})$, there exists a *superadjoint* $T^\dagger\in\mathcal{B}(\mathcal{H})$ defined by

$$\forall x,y\in\mathcal{H}\quad:\quad \langle T^\dagger(x),y\rangle = (-1)^{|T||x|}\langle x,T(y)\rangle. \tag{6}$$

The next Definition corresponds to the standard notion of real form in [15].

Definition 2.3 ([11]) A *superinvolution* on a complex $\mathbb{Z}_2$-graded algebra $\mathbf{A}$ is an antilinear map $\dagger:\mathbf{A}\to\mathbf{A}$ homogeneous of degree 0 such that

$$\forall a,b\in\mathbf{A}\quad:\quad (a^\dagger)^\dagger = a,\qquad (ab)^\dagger = (-1)^{|a||b|}b^\dagger a^\dagger.$$

For example, $\mathcal{B}(\mathcal{H})$ is a $\mathbb{Z}_2$-graded algebra with a superinvolution given by the superadjoint.

A *C*-superalgebra* is a complex $\mathbb{Z}_2$-graded algebra with superinvolution, faithfully represented on a Hilbert superspace $\mathcal{H}$ by $\rho:\mathbf{A}\to\mathcal{B}(\mathcal{H})$ (with ρ of degree 0 and compatible with superinvolutions), and such that $\mathbf{A}$ is closed for the operator norm topology of $\mathcal{B}(\mathcal{H})$ via ρ.

This concrete definition allows directly to define the tensor product of two C*-superalgebras. Spectral properties of C*-superalgebras are currently under study. Morphisms of C*-superalgebras are homogeneous homomorphisms of degree 0 compatible with superinvolutions. Owing to the beginning of this section, we can define the category of *noncommutative topological superspaces* as the dual category of C*-superalgebras.

Example 2.4 ([11]) The space $L^\infty(\mathbb{R}^{m|n}) = L^\infty(\mathbb{R}^m)\otimes\bigwedge\mathbb{R}^n$ is a supercommutative C*-superalgebra, with usual complex conjugation $\overline{f(x,\xi)} := \sum_I \overline{f_I(x)}\xi^I$ as superinvolution, represented by multiplication operators on the Hilbert superspace $L^2(\mathbb{R}^{m|n})$. The choice of J as in Example 2.2 leads to the C*-norm $\|f\| := \sum_I \|f_I\|_\infty$. But the involution * (attached to the scalar product $\langle -, J(-)\rangle$) does not stabilize $L^\infty(\mathbb{R}^{m|n})$ so that it is not a C*-algebra. Noncommutative examples of C*-superalgebras will be given in Section 4.

Symmetries of noncommutative spaces are quantum groups [23, 24]. In the present graded context, we call the symmetries of noncommutative superspaces quantum supergroups. This notion already exists in the literature at the algebraic level (see e.g. [24]) but in the present program, we want to extend this notion at the topological level. Natural objects to consider would be C*-quantum supergroups (with a C*-superalgebra structure) and it is currently under study. A structure going in this direction is that of Fréchet quantum supergroups.

Definition 2.5 ([12]) A *Fréchet quantum supergroup* is a $\mathbb{Z}_2$-graded Fréchet space with a graded topological tensor product and a continuous homogeneous of degree 0 Hopf algebra structure.

A *representation* of a Fréchet quantum supergroup H is a Fréchet $\mathbb{Z}_2$-graded H-comodule algebra with continuous homogeneous of degree 0 coaction.

Example 2.6 The simplest and well-known example corresponds to the Abelian supergroup $\mathbb{R}^{m|n}$. It can be seen in particular as a Fréchet quantum supergroup but commutative by considering $H = \mathcal{C}^\infty(\mathbb{R}^{m|n})$, which is a Fréchet-Hopf algebra for the usual pointwise product, unit and for the usual coproduct, counit and antipode given by, for any $z_i \in \mathbb{R}^{m|n}$,

$$\Delta(f)(z_1, z_2) = f(z_1 + z_2), \qquad \varepsilon(f) = f(0), \qquad S(f)(z) = f(-z).$$

We will see noncommutative examples of Fréchet quantum supergroups in Section 4.

3. Application to harmonic analysis of Lie supergroups

Let us present the Heisenberg supergroup. We endow the superspace $\mathbb{R}^{2m|2n}$ with an even symplectic superform, associated to the matrix

$$\omega = \begin{pmatrix} 0 & \mathbb{1}_m & 0 & 0 \\ -\mathbb{1}_m & 0 & 0 & 0 \\ 0 & 0 & \mathbb{1}_{r+2s} & 0 \\ 0 & 0 & 0 & -\mathbb{1}_r \end{pmatrix}$$

of size $(2m+2n)$ in the canonical basis, with $2n = 2r+2s$. the odd part of this symplectic form corresponds to a symmetric form and one has to choose its signature $(r+2s, r)$. The Heisenberg supergroup is the supergroup $G = \mathbb{R}^{2m|2n} \times \mathbb{R}^{1|0}Z$ (with Z an even generator) defined as usual by its Lie superalgebra relations

$$\forall a, a' \in \mathbb{R}^{2m|2n},\ \forall t, t' \in \mathbb{R}^{1|0} \quad : \quad [a + tZ, a' + t'Z] = \omega(a, a')Z,$$

where $[-,-]$ denotes the graded Lie bracket.

Proposition 3.1 ([11, 22]) *Kirillov's Orbit Method applied to the Heisenberg supergroup G yields a Hilbert superspace $\mathcal{H}_\theta := L^2(\mathbb{R}^{m|r}) \otimes Hol(\mathbb{C}^{0|s})$, with elements $\varphi(q, \xi, \zeta) = \sum_{I,J} \varphi_{IJ}(q)\xi^I\zeta^J$ ($q \in \mathbb{R}^m$, $\xi \in \mathbb{R}^{0|r}$, $\zeta \in \mathbb{C}^{0|s}$) and superhermitian inner product*

$$\langle \varphi, \psi \rangle = (2i)^s \int \mathrm{d}q\mathrm{d}\xi\mathrm{d}\zeta\mathrm{d}\overline{\zeta}\ \overline{\varphi(q, \xi, \zeta)}\psi(q, \xi, \zeta)e^{\frac{i}{\theta}\zeta\overline{\zeta}}. \tag{7}$$

Moreover, this method also produces a representation U_θ of G on $\mathcal{H}_\theta$

$$U_\theta(g)\varphi(q_0, \xi_0, \zeta_0) = e^{\frac{2i}{\theta}\left(t + (\frac{1}{2}q - q_0)p + (\frac{1}{2}\xi - \xi_0)\eta + \frac{1}{2}(\frac{1}{2}\zeta - \zeta_0)\overline{\zeta}\right)}\varphi(q_0 - q, \xi_0 - \xi, \zeta_0 - \zeta)$$

for $g = (q, p, \xi, \eta, \zeta, \overline{\zeta}, t) \in G$ with $t \in \mathbb{R}$, $p \in \mathbb{R}^m$, $\eta \in \mathbb{R}^{0|r}$ in the real polarization; U_θ generalizing the usual Schrödinger representation (corresponding to the case $r = s = 0$). This representation U_θ is superunitary, namely

$$\forall g \in G,\ \forall \varphi, \psi \in \mathcal{H}_\theta \quad : \quad \langle U_\theta(g)\varphi, U_\theta(g)\psi \rangle = \langle \varphi, \psi \rangle.$$

In case of positive signature $r = 0$, $n = s$, then the standard inner product (4) corresponds to (7) for which a fundamental symmetry is given by $J(\varphi) = \sum_I (-i)^{|I|}\varphi_I\zeta^I$. However, in case of mixed signature $r \neq 0$, the standard inner product (4) does not correspond to (7) for which any fundamental symmetry has to be more general than a multiplication by a power of i. By using superunitarity induced by (7) and the notion of Hilbert superspace of Definition 2.1, one obtains an extended Stone–von Neumann theorem in any signature [22], identifying the superunitary dual of G with irreducible representations of the Clifford algebra. On the contrary, the standard notion of Hilbert superspace associated to (4) leads to an empty dual superunitary in mixed signature [25, Theorem 5.8], which is not satisfactory.

Theorem 3.2 ([22]) *Any superunitary representation of G with fixed character $e^{\frac{2i}{\theta}t}$ decomposes as an orthogonal direct sum of the irreducible superunitary representations U_θ.*

So we see that this structure of Hilbert superspace is more adapted than the standard one [15, 20] of (4) to the analytic point of view of unitary representations of Lie supergroups.

4. Application to non-formal deformation quantization

Contrary to the formal case extensively studied and classified in Refs. [26, 27, 28, 29, 30], there are only few available examples of non-formal deformation quantization of Lie groups in the smooth non-graded setting. Rieffel [31] built the deformation of Abelian Lie groups and the associated universal deformation formula (UDF). This was also recently extended to (non-Abelian) Kählerian Lie groups [32, 33] and to the case of $SL(2,\mathbb{R})$ [34] and of $SU(1,n)$ [35]. Star-exponentials [36, 37] associated to these deformations were computed in the non-formal setting in [38, 39] and such deformations were linked to Hilbert algebras and multipliers in [40]. Non-formal deformation quantization was also extended to the complex case [41, 42] and to Abelian p-adic groups [43].

Let us examine the deformation quantization of the Heisenberg supergroup $G = \mathbb{R}^{2m|n} \times \mathbb{R}^{1|0}Z$. By using the superunitary representation of G from Proposition 3.1 as well as the symmetric structure of its coadjoint orbit $\mathbb{R}^{2m|n}$, we have constructed a non-formal deformation quantization of $\mathbb{R}^{2m|n}$ invariant under the symmetry G.

Theorem 4.1 ([11]) *The star-product defined by*

$$(f_1 \star_\theta f_2)(z) = \frac{1}{\pi^{2m}\theta^{2m-n}} \int \mathrm{d}z_1 \mathrm{d}z_2\ K_\theta(z_1,z_2) f_1(z+z_1) f_2(z+z_2), \tag{8}$$

for $z=(x,\xi)\in\mathbb{R}^{2m|n}$, *with supergroup cocycle* $K_\theta(z_1,z_2) = e^{\frac{-2i}{\theta}\omega(z_1,z_2)}$, *on the space*

$$\mathcal{B}^1(\mathbb{R}^{2m|n}) := \{f\in\mathcal{C}^\infty(\mathbb{R}^{2m|n}),\ \forall\alpha\in\mathbb{N}^{2m},\ \sup_{x\in\mathbb{R}^{2m}} \sum_I |\partial_x^\alpha f_I(x)| < +\infty\} \quad \simeq \quad \mathcal{B}(\mathbb{R}^{2m})\otimes \bigwedge \mathbb{R}^n,$$

is associative, G-invariant and satisfies the tracial property

$$\int \mathrm{d}z\ (f_1\star_\theta f_2)(z) = \int \mathrm{d}z\ f_1(z) f_2(z).$$

Moreover, there is a Weyl-type quantization map $\Omega_\theta : \mathcal{B}^1(\mathbb{R}^{2m|n}) \to \mathcal{B}(\mathcal{H}_\theta)$, *compatible with the star-product and G-equivariant, i.e.*

$$\Omega_\theta(f_1\star_\theta f_2) = \Omega_\theta(f_1)\Omega_\theta(f_2), \qquad \Omega_\theta(L^*_{g^{-1}} f) = U_\theta(g)\Omega_\theta(f)U_\theta(g)^{-1}.$$

The space $\mathcal{B}^1(\mathbb{R}^{2m|n})$, endowed with the star-product, is a Fréchet algebra isomorphic to the tensor product of the Moyal–Weyl algebra $\mathcal{B}(\mathbb{R}^{2m})$ and of the Clifford algebra $Cl(n,\mathbb{C})$. Note that the space $\mathcal{B}^1(\mathbb{R}^{2m|n})$ is part of a pseudodifferential calculus adapted to the superspace $\mathbb{R}^{2m|n}$ for which an oscillatory integral is also defined.

An analogous construction was also performed [44] with a different cocycle for the *Poincaré supergroup* in $2|2$ dimensions, i.e. the supergroup defined by even generators H, E, F, odd generators Γ, Ξ and Lie superalgebra relations

$$[H,E]=2E,\quad [H,F]=-2F,\quad [H,\Gamma]=\Gamma,\quad [H,\Xi]=-\Xi,\quad [\Gamma,\Gamma]=-2E,\quad [\Xi,\Xi]=-2F.$$

For the Heisenberg supergroup, a universal deformation formula was also built, namely the use of $\star_\theta$, can also deform any algebra on which the supergroup $\mathbb{R}^{2m|n}$ is acting. Let us describe it. We consider an action ρ of the supergroup $\mathbb{R}^{2m|n}$ on a Fréchet algebra $(\mathbf{A}, |\cdot|_j)$ satisfying the conditions:

- $\forall z_1,z_2\in\mathbb{R}^{2m|n}$, $\rho_{z_1+z_2} = \rho_{z_1}\rho_{z_2}$ and $\rho_0 = \mathrm{id}$.
- $\forall z\in\mathbb{R}^{2m|n}$, $\forall a,b\in\mathbf{A}$, $\rho_z(ab) = \rho_z(a)\rho_z(b)$, and ρ_z is linear.

- By writing $z = (x, \xi) \in \mathbb{R}^{2m|n}$, we can expand the action as: $\rho_{(x,\xi)}(a) = \sum_I \rho_x(a)_I \xi^I$; $\forall a \in \mathbf{A}$, $\forall I$, $x \mapsto \rho_x(a)_I$ is $\mathbf{A}$-valued and continuous.
- The action is subisometric, i.e. there exists a constant $C > 0$ such that $\forall a \in \mathbf{A}$, $\forall I$, $\forall j$, $\exists k$, $\forall x \in \mathbb{R}^{2m}$, $|\rho_x(a)_I|_j \leq C|a|_k$.

The star-product (8) can be directly extended to $\mathbf{A}$-valued superfunctions $\mathcal{B}^1_{\mathbf{A}}(\mathbb{R}^{2m|n})$. Note that this space is Fréchet for the seminorms $|f|_{j,\alpha} = \sup_{x\in\mathbb{R}^{2m}}\{\sum_I |\partial_x^\alpha f_I(x)|_j\}$. We recall that the set of *smooth vectors* of $\mathbf{A}$ for the action ρ is defined as

$$\mathbf{A}^\infty = \{a \in \mathbf{A}, \quad \rho^a := z \mapsto \rho_z(a) \text{ is smooth on } \mathbb{R}^{2m|n}\}.$$

This set $\mathbf{A}^\infty$ is dense in $\mathbf{A}$, and for any $a \in \mathbf{A}^\infty$, the map ρ^a lies in $\mathcal{B}^1_{\mathbf{A}^\infty}(\mathbb{R}^{2m|n})$, so that we can form the star-product of ρ^a and ρ^b, for a and b smooth vectors.

Theorem 4.2 ([11]) *The expression*

$$a \star_\theta b := (\rho^a \star_\theta \rho^b)(0),$$

for $a, b \in \mathbf{A}^\infty$, defines an associative product on $\mathbf{A}^\infty$. If $\mathbf{A}$ is a C-superalgebra and ρ is compatible with the degree and the superinvolution, then $(\mathbf{A}^\infty, \star_\theta)$ can be completed to another C*-superalgebra denoted by $\mathbf{A}_\theta$.*

Therefore, the category of C*-superalgebras is stable by this deformation quantization, which stresses the interest of this structure for noncommutative supergeometry. Note that if $\mathbf{A}$ is a C*-algebra on which $\mathbb{R}^{2m|n}$ acts, the deformation $(\mathbf{A}^\infty, \star_\theta)$ cannot be completed into a C*-algebra in general. The *universal deformation formula* (UDF) of Theorem 4.2 is a generalization of Drinfeld twists [45, 46] to the non-formal setting.

This UDF thus produces a large class of (deformed) noncommutative superspaces. For example, as soon as $\mathbb{R}^{2m|n}$ acts on a trivial compact supermanifold, one can deform its algebra of continuous superfunctions to a noncommutative C*-superalgebra [11]. Noncommutative supertori appear as particular cases of this deformation.

Definition 4.3 ([11]) The *noncommutative supertorus* of dimension $2m|n$ and signature (p, q) (with $n = p + q$) is the C*-superalgebra generated by the even elements U_j, V_j $(1 \leq j \leq m)$ and by the odd elements Γ_k, Ξ_ℓ $(1 \leq k \leq p,\ 1 \leq \ell \leq q)$ satisfying the relations

$$\begin{aligned} &U_j^\dagger U_j = U_j U_j^\dagger = V_j^\dagger V_j = V_j V_j^\dagger = \mathbb{1}, \qquad \Gamma_k^\dagger = \Gamma_k, \qquad \Xi_\ell^\dagger = \Xi_\ell,\\ &U_j V_j = e^{2i\pi\theta} V_j U_j, \qquad (\Gamma_k)^2 = i\theta, \qquad (\Xi_\ell)^2 = -i\theta, \end{aligned}$$

the other relations being trivial. It coincides with a deformation of the usual supertorus of dimension $2m|n$, with deformation parameter θ.

By using the UDF of Theorem 4.2, we can also construct interesting examples of Fréchet quantum supergroups. Inspired by [47] in the non-graded case, we consider an action $\pi : \mathbb{R}^{1|0} \to Sp(\mathbb{R}^{2m|n}, \omega)$ and $G' := \mathbb{R}^{1|0} \ltimes_\pi \mathbb{R}^{2m|n}$. G' is a solvable Lie supergroup.

Theorem 4.4 ([12]) *The space $H := \mathcal{C}^\infty(\mathbb{R}^{1|0})\hat{\otimes}\mathcal{B}^1(\mathbb{R}^{2m|n})$, endowed with the usual Hopf algebra structure of G' except for the product, which is given by the deformation*

$$f_1 \star f_2(t, z) := \frac{1}{\pi^{2m} t^{2m-n}} \int \mathrm{d}z_1 \mathrm{d}z_2\ f_1(t, z + z_1) f_2(t, z + z_2) K_t(z_1, z_2),$$

for any $t \in \mathbb{R}^{1|0} \simeq \mathbb{R}$ and $z \in \mathbb{R}^{2m|n}$, is a Fréchet quantum supergroup. An analogue of the multiplicative unitary operator $W \in \mathcal{L}(H\hat{\otimes}H)$ can be defined for this quantum supergroup by

$$W(f_1 \otimes f_2) = (\Delta f_1) \star (1 \otimes f_2).$$

It satisfies the pentagonal equation $W_{12}W_{13}W_{23} = W_{23}W_{12}$ as well as the superunitarity condition with respect to a Hilbert superspace structure of $L^2(G' \times G')$.

A similar construction of Fréchet quantum supergroup was performed [12] for connected Lie supergroups with supertoral subgroups.

5. Application to QFT

We give here an application where a NCSG is useful to understand a renormalizable noncommutative quantum field theory. Let us consider the Moyal space [48, 49] given by the deformation quantization of $\mathbb{R}^{2m}$. It turns out that the $\phi^{\star_\theta 4}$ theory on the Moyal space is not renormalizable because of a new divergence called UV-IR mixing [50], generic to noncommutative spaces [51]. The first solution of this problem was found by H. Grosse and R. Wulkenhaar by adding a *harmonic term* to the standard $\phi^{\star_\theta 4}$ *action functional*

$$S_{GW}[\phi] = \int \Big(\frac{1}{2}(\partial_\mu\phi)^2 + \frac{2\Omega^2}{\theta^2}x^2\phi^2 + \frac{M^2}{2}\phi^2 + \lambda\,\phi^{\star_\theta 4}\Big)\mathrm{d}x, \tag{9}$$

which becomes a renormalizable model ($m = 1, 2$) to all orders [52, 53, 54] for $\Omega \neq 0$. This model possesses new interesting properties concerning its vacuum configurations [55], its Connes–Kreimer algebra [56], its symmetries [57, 58], its commutative limit [59], its beta function [60] and finally its solvability for $\theta \to \infty$ [61]. Note that there exists now another renormalizable real scalar theory on the Moyal space [62, 63]. A gauge theory associated to (9) was constructed in [64, 65]. It exhibits similar features [66, 67] and is a candidate to renormalizability. See [68, 69, 70] for the study of the BRST symmetry.

We want to give an interpretation of the harmonic term in (9). It turns out that the action (9) can be reformulated in terms of the commutator and anticommutator

$$S_{GW}[\phi] = \int \Big(\frac{1}{2}\big|[\frac{i}{2}x_\mu, \phi]_{\star_\theta}\big|^2 + \frac{\Omega^2}{2}\big|\{\frac{i}{2}x_\mu, \phi\}_{\star_\theta}\big|^2 + \frac{M^2}{2}\phi^2 + \lambda\,\phi^{\star_\theta 4}\Big)\mathrm{d}x.$$

This is a sign of a graded symmetry. Actually, this model can be interpreted in the setting of deformation quantization of the superspace $\mathbb{R}^{2m|1}$. We consider the natural differential calculus [7] for this graded algebra associated to the graded derivations $\mathfrak{d} \in \big\{[\frac{i}{2}x_\mu, -]_{\star_\theta}, [\frac{i}{2}x_\mu\xi, -]_{\star_\theta}\big\}_{\mu\in\{1,..,2m\}}$.

Proposition 5.1 ([6, 11]) *The standard $\Phi^{\star_\theta 4}$ action of the deformation quantization of $\mathbb{R}^{2m|1}$, after integration with respect to the odd variable ξ and identification of the even and odd fields $\phi_1(x) = b\phi_0(x)$ ($b \in \mathbb{R}$) in $\Phi(x,\xi) = \phi_0(x) + \phi_1(x)\xi$, gives exactly rise to the renormalizable Grosse–Wulkenhaar action (9):*

$$tr\Big(\frac{1}{2}\sum_{\mathfrak{d}}|\mathfrak{d}(\Phi)|^2 + \frac{M^2}{2}\Phi^2 + \Lambda\Phi^{\star_\theta 4}\Big) = S_{GW}[\phi_0]$$

with $\Omega^2 = \frac{b^4\theta^2}{16}$ and $\lambda = \Lambda(1 + \frac{b^4\theta^2}{16})$.

In this setting, the Langmann–Szabo duality [71] corresponds to the switch of grading and the gauge action of [64] can be directly derived as the Yang–Mills action of the deformation of $\mathbb{R}^{2m|1}$ with differential calculus associated to the graded derivations $\mathfrak{d}$.

This shows that the NCSG of the deformation of $\mathbb{R}^{2m|1}$ is likely to be related to the renormalizability of the noncommutative quantum field theory (9). This interpretation could be useful to obtain renormalizable quantum field theories on other noncommutative spaces.

Acknowledgments
This work was supported by the Belgian Interuniversity Attraction Pole (IAP) within the framework "Dynamics, Geometry and Statistical Physics" (DYGEST), and by the Max Planck Institut für Mathematik Bonn.

References

[1] Berezin F and Leites D 1976 *Soviet Maths Doklady* **16** 1218–1222
[2] Kostant B 1977 *Graded manifolds, graded Lie theory and prequantization* Lecture Notes in Mathematics 570 (Berlin: Springer)
[3] Connes A 1994 *Noncommutative Geometry* (San Diego, New York, London: Academic Press)
[4] Seiberg N and Witten E 1999 *JHEP* **09** 032 (*Preprint* hep-th/9908142)
[5] Douglas M R and Nekrasov N A 2001 *Rev. Mod. Phys.* **73** 977–1029 (*Preprint* hep-th/0106048)
[6] de Goursac A 2010 *SIGMA* **6** 048 (*Preprint* 1003.5788)
[7] de Goursac A, Masson T and Wallet J C 2012 *J. Noncommut. Geom.* **6** 343–387 (*Preprint* 0811.3567)
[8] Nastasescu C and van Oystaeyen F 1982 *Graded ring theory* (Amsterdam-New York: North-Holland Publishing Co.)
[9] Schwarz A 1999 *Lett. Math. Phys.* **50** 309
[10] Schwarz A 2003 *Nucl. Phys.* **B650** 475–496
[11] Bieliavsky P, de Goursac A and Tuynman G 2012 *J. Funct. Anal.* **263** 549–603 (*Preprint* 1011.2370)
[12] de Goursac A 2015 *Pacif. J. Math.* **273** 169–195 (*Preprint* 1105.2420)
[13] Leites D A 1980 *Russian Math. Surveys* **35** 1–64
[14] Manin Y I 1997 *Gauge Field Theory and Complex Geometry* Grundlehren der Mathematischen Wissenschaften, **289** (Berlin: Springer-Verlag)
[15] Deligne P and Morgan J W 1999 *Notes on supersymmetry (after Joseph Bernstein)* (Quantum Fields and Strings: A Course for Mathematicians, pp. 41-98)
[16] Carmeli C, Caston L and Fioresi R 2011 *Mathematical foundations of supersymmetry* (EMS Series of Lectures in Mathematics, Europ. Math. Soc.)
[17] DeWitt B 1984 *Supermanifolds* (Cambridge: Cambridge University Press)
[18] Rogers A 2007 *Supermanifolds, Theory and Applications* (Singapore: World Scientific Publishing)
[19] Tuynman G M 2005 *Supermanifolds and Supergroups* (Amsterdam: Kluwer Academic Publishers)
[20] Carmeli C, Cassinelli G, Toigo A and Varadarajan V S 2006 *Commun. Math. Phys.* **263** 217–258
[21] Alldridge A, Hilgert J and Laubinger M 2013 *J. London Math. Soc.* **87** 561–585
[22] de Goursac A and Michel J P *in progress*
[23] Woronowicz S L 1987 *Commun. Math. Phys.* **111** 613
[24] Majid S 1995 *Foundations of quantum group theory* (Cambridge: Cambridge University Press)
[25] Salmasian H 2010 *Commun. Math. Phys.* **297** 189–227
[26] Bayen F, Flato M, Fronsdal C, Lichnerowicz A and Sternheimer D 1978 *Ann. Phys.* **11** 61–151
[27] Omori H, Maeda Y and Yoshioka A 1991 *Adv. Math.* **85** 224–255
[28] Lecomte P, Michor P and Schicketanz H 1992 *J. Pure Appl. Algebra* **77** 87–102
[29] Fedosov B 1994 *J. Differential Geom.* **40** 213–238
[30] Kontsevich M 2003 *Lett. Math. Phys.* **66** 157–216 (*Preprint* q-alg/9709040)
[31] Rieffel M A 1993 *Mem. Amer. Math. Soc.* **106** R6
[32] Bieliavsky P 2002 *J. Sympl. Geom.* **1** 269–320 (*Preprint* math/0010004)
[33] Bieliavsky P and Gayral V *to appear in Mem. Amer. Math. Soc.* (*Preprint* 1109.3419)
[34] Bieliavsky P, Detournay S and Spindel P 2009 *Commun. Math. Phys.* **289** 529–559 (*Preprint* 0806.4741)
[35] Korvers S 2014 *PhD thesis, UCLouvain*
[36] Cahen M, Flato M, Gutt S and Sternheimer D 1985 *J. Geom. Phys.* **2** 35–49
[37] Arnal D and Cortet J C 1990 *J. Funct. Anal.* **92** 103–175
[38] Bieliavsky P, Gayral V, de Goursac A and Spinnler F 2014 *in Developments and Retrospectives in Lie Theory: Geometric and Analytic Methods (Mason, Penkov, Wolf Eds), Springer, Dev. Math.* **37** 41–76 (*Preprint* 1311.1871)
[39] Bieliavsky P, de Goursac A, Maeda Y and Spinnler F *in progress*
[40] de Goursac A (*Preprint* 1410.3434)
[41] Omori H, Maeda Y, Miyazaki N and Yoshioka A 2000 *RIMS Kokyuroku* **1150** 128–132
[42] Garay M, de Goursac A and van Straten D 2014 *Ann. Phys.* **342** 83–102 (*Preprint* 1309.0437)
[43] Gayral V and Jondreville D (*Preprint* 1409.3349)
[44] Bieliavsky P, de Goursac A and Michel J P *in progress*
[45] Drinfeld V G 1989 *Leningrad Math. J.* **1** 1419–1457

[46] Giaquinto A and Zhang J 1998 *J. Pure Appl. Algebra* **128** 133
[47] Rieffel M A 1993 *Contemp. Math.* **145** 465–491
[48] Gracia-Bondia J M and Varilly J C 1988 *J. Math. Phys.* **29** 869–879
[49] Cagnache E, D'Andrea F, Martinetti P and Wallet J C 2011 *J. Geom. Phys.* **61** 1881
[50] Minwalla S, Van Raamsdonk M and Seiberg N 2000 *JHEP* **02** 020 (*Preprint* hep-th/9912072)
[51] Gayral V 2005 (*Preprint* hep-th/0507208)
[52] Grosse H and Wulkenhaar R 2003 *JHEP* **12** 019 (*Preprint* hep-th/0307017)
[53] Grosse H and Wulkenhaar R 2005 *Commun. Math. Phys.* **256** 305–374 (*Preprint* hep-th/0401128)
[54] Rivasseau V, Vignes-Tourneret F and Wulkenhaar R 2006 *Commun. Math. Phys.* **262** 565–594 (*Preprint* hep-th/0501036)
[55] de Goursac A, Tanasa A and Wallet J C 2008 *Eur. Phys. J.* **C53** 459–466 (*Preprint* 0709.3950)
[56] Tanasa A and Vignes-Tourneret F 2008 *J. Noncommut. Geom.* **2** 125–139 (*Preprint* 0707.4143)
[57] de Goursac A and Wallet J C 2011 *J. Phys. A: Math. Theor.* **44** 055401 (*Preprint* 0911.2645)
[58] Hounkonnou M N and Samary D O 2010 *J. Phys. A* **A43** 155202 (*Preprint* 0909.4562)
[59] de Goursac A 2013 *Ann. H. Poincare* **14** 2025–2043 (*Preprint* 1207.6208)
[60] Disertori M, Gurau R, Magnen J and Rivasseau V 2007 *Phys. Lett.* **B649** 95–102 (*Preprint* hep-th/0612251)
[61] Grosse H and Wulkenhaar R 2014 (*Preprint* 1406.7755)
[62] Gurau R, Magnen J, Rivasseau V and Tanasa A 2009 *Commun. Math. Phys.* **287** 275–290 (*Preprint* 0802.0791)
[63] Tanasa A 2010 *SIGMA* **6** 047 (*Preprint* 1003.4877)
[64] de Goursac A, Wallet J C and Wulkenhaar R 2007 *Eur. Phys. J.* **C51** 977–987 (*Preprint* hep-th/0703075)
[65] Grosse H and Wohlgenannt M 2007 *Eur. Phys. J.* **C52** 435–450 (*Preprint* hep-th/0703169)
[66] de Goursac A, Wallet J C and Wulkenhaar R 2008 *Eur. Phys. J.* **C56** 293–304 (*Preprint* 0803.3035)
[67] Cagnache E, Masson T and Wallet J C 2011 *J. Noncommut. Geom.* **5** 39 (*Preprint* 0804.3061)
[68] Blaschke D N, Grosse H and Schweda M 2007 *Europhys. Lett.* **79** 61002 (*Preprint* 0705.4205)
[69] Blaschke D N, Grosse H, Kronberger E, Schweda M and Wohlgenannt M 2010 *Eur. Phys. J.* **C67** 575–582 (*Preprint* 0912.3642)
[70] Blaschke D N, Grosse H and Wallet J C 2013 *JHEP* **1306** 038 (*Preprint* 1302.2903)
[71] Langmann E and Szabo R J 2002 *Phys. Lett.* **B533** 168–177 (*Preprint* hep-th/0202039)

Geometry of surfaces associated to Grassmannian sigma models

L Delisle[1], **V Hussin**[2,3] **and W J Zakrzewski**[4]

[1] Institut de mathématiques de Jussieu–Paris Rive Gauche, UP7D–Campus des Grands Moulins, Bâtiment Sophie Germain, Cases 7012, 75205 Paris Cedex 13.
[2] Département de mathématiques et de statistique, Université de Montréal, C. P. 6128, Succ. Centre-ville, Montréal (Québec) H3C 3J7, Canada.
[3] Centre de recherches mathématiquxes, Université de Montréal, C. P. 6128, Succ. Centre-ville, Montréal (Québec) H3C 3J7, Canada.
[4] Department of mathematical sciences, University of Durham, Durham DH1 3LE, United Kingdom.

E-mail: laurent.delisle@imj-prg.fr, hussin@dms.umontreal.ca, w.j.zakrzewski@durham.ac.uk

Abstract. We investigate the geometric characteristics of constant Gaussian curvature surfaces obtained from solutions of the $G(m, n)$ sigma model. Most of these solutions are related to the Veronese sequence. We show that we can distinguish surfaces with the same Gaussian curvature using additional quantities like the topological charge and the mean curvature. The cases of $G(1, n) = \mathbb{C}P^{n-1}$ and $G(2, n)$ are used to illustrate these characteristics.

1. Introduction

In recent papers [1,2], we have classified some relevant solutions of the Grassmannian $G(m, n)$ sigma model that are associated to constant Gaussian curvature surfaces in $su(n)$. In our construction, we have found, among others, some non-equivalent solutions with the same constant Gaussian curvature. In the non-holomorphic case [2], we have considered a particular set of solutions starting from the knowledge of the corresponding solutions of $G(1, n) = \mathbb{C}P^{n-1}$, the so-called Veronese solutions. In the holomorphic case [1], we presented some conjectures and constructed some solutions which are not related to the Veronese ones. These results clarified and extended results obtained elsewhere [3-9].

In this contribution, we aim to show that some of the surfaces that have the same constant Gaussian curvature and correspond to non-equivalent solutions of $G(m, n)$, up to a gauge transformation, may be distinguished by other geometric characteristics such as the topological charge and the mean curvature. The case of $G(2, n)$ will be discussed in detail to show how this works out.

In Section 2, we discuss the $G(m, n)$ sigma model and we define the geometric quantities for the surfaces associated to $G(m, n)$. In particular, we recall a class of solutions of the model that lead to surfaces with constant Gaussian curvature and explain the relation with the Veronese sequence. We compute explicitly the additional geometric characteristics of these surfaces. In Section 3, we show how these quantities could explain the differences between surfaces of same constant Gaussian curvature. We re-visit the case of $G(1, n) = \mathbb{C}P^{n-1}$ and give some general results for $G(m,n)$. We show, in the case of $G(2,n)$, how different solutions with the same Gaussian curvature can be distinguished by calculating their topological charge and/or their mean curvature. We conclude this section with the case of non-Veronese holomorphic solutions. Section 4 presents our conclusions and future outlook.

2. Surfaces associated to solutions of $G(m,n)$

2.1. *The model*

The two-dimensional $G(m,n)$ sigma model is a field theory [10] defined on the complex plane $\mathbb{C}$ which has the Grassmannian manifold $G(m,n)$ as its target space:

$$G(m,n) \cong \frac{U(n)}{U(m) \times U(n-m)}, \quad n > m, \tag{1}$$

where $U(k)$ is the set of $k \times k$ unitary matrices. The field $Z(x_+, x_-)$ defined on an open and simply connected subset Ω of $\mathbb{C}$ thus takes values in $G(m,n)$. The elements Z of $G(m,n)$ are parametrized by $n \times m$ matrices and satisfy $Z^\dagger Z = \mathbb{I}_m$. Moreover, they correspond to critical points of the energy functional defined via the Lagrangian density

$$\mathcal{L}(Z) = \frac{1}{2}\mathrm{Tr}\left[(D_+Z)^\dagger D_+Z + (D_-Z)^\dagger D_-Z\right], \tag{2}$$

where $D_\pm\Lambda = \partial_\pm\Lambda - \Lambda(Z^\dagger\partial_\pm Z)$ are the covariant derivatives, $\partial_\pm = \partial_{x_\pm}$ and (x_+, x_-) are complex local coordinates on Ω. We consider the case of $\Omega = \mathbb{C}$ and require the energy of these fields to be finite. To achieve this we have to impose the boundary conditions $D_\pm Z \longrightarrow 0$ as $|x_\pm| \longrightarrow \infty$. With such boundary conditions, the complex plane $\mathbb{C}$ is compactified into the two-sphere S^2 via the stereographic projection and, as a consequence, the fields Z are harmonic maps [11] from S^2 into the Grassmann manifold $G(m,n)$.

Using the variation of the energy, we deduce the Euler-Lagrange equations of the model given as $(\pm \longleftrightarrow \mp)$

$$D_+D_-Z + Z(D_-Z)^\dagger D_-Z = 0, \quad Z^\dagger Z = \mathbb{I}_m. \tag{3}$$

The finite energy solutions of these equations are fully known in the $G(1,n) \cong \mathbb{C}P^{n-1}$ case [11]. They are given by

$$Z_i = \frac{P_+^i f}{|P_+^i f|}, \quad i = 0, 1, \cdots, n-1, \tag{4}$$

where $f = f(x_+) \in \mathbb{C}^n$ is holomorphic and P_+ is an orthogonalizing operator defined recursively as

$$P_+^0 f = f, \quad P_+ f = \partial_+ f - \frac{f^\dagger\partial_+ f}{|f|^2} f, \quad P_+^i f = P_+(P_+^{i-1} f), \quad P_+^n f = 0. \tag{5}$$

For the $G(m,n)$ model with $m \geq 2$, the complete set of solutions is not known, but we can use the solutions (4) of the $G(1,n)$ model to construct particular classes of them:

$$Z_{(i_1,i_2,\cdots,i_m)} = \left(\frac{P_+^{i_1} f}{|P_+^{i_1} f|}, \frac{P_+^{i_2} f}{|P_+^{i_2} f|}, \cdots, \frac{P_+^{i_m} f}{|P_+^{i_m} f|}\right), \quad 0 \leq i_1 < i_2 < \cdots < i_m \leq n-1. \tag{6}$$

A convenient way to reformulate the model in a gauge-invariant way involves using orthogonal projectors [10]. Indeed, for $G(m,n)$, we define a rank m hermitian orthogonal projector $\mathbb{P}$ as

$$\mathbb{P} = ZZ^\dagger. \tag{7}$$

This projector satisfies

$$\mathbb{P}^2 = \mathbb{P}^\dagger = \mathbb{P}, \quad \mathrm{Tr}(\mathbb{P}) = m. \tag{8}$$

The Lagrangian density (2) and the Euler-Lagrange equations (3) can be rewritten in an equivalent way as

$$\mathcal{L}(\mathbb{P}) = \frac{1}{2}\mathrm{Tr}(\partial_+\mathbb{P}\partial_-\mathbb{P}), \quad [\partial_+\partial_-\mathbb{P}, \mathbb{P}] = 0, \quad \mathbb{P}^2 = \mathbb{P}. \tag{9}$$

A solution of the type (6) leads to a projector of the form:

$$\mathbb{P}_\beta = \sum_{j=0}^{n-1} \beta_j \mathbb{P}_j, \quad \mathbb{P}_j = Z_j Z_j^\dagger = \frac{P_+^j f (P_+^j f)^\dagger}{|P_+^j f|^2}, \tag{10}$$

where β is a $n-$column vector such that $\beta_j = 0$ or 1 for all j and $\sum_{j=0}^{n-1} \beta_j = m$.

The key, in constructing surfaces from the solutions of the $G(m,n)$ model, is to observe that the Euler-Lagrange equations (9) may be rewritten as a conservation law:

$$\partial_+\mathbf{L} - \partial_-\mathbf{L}^\dagger = 0, \quad \mathbf{L} = [\partial_-\mathbb{P}, \mathbb{P}]. \tag{11}$$

Then, using the Poincaré lemma and the fact that Ω is simply connected [12], we may define a surface $\mathbf{X} \in su(n)$ (the set of $n \times n$ hermitian and traceless matrices) via its tangent space as

$$d\mathbf{X} = \mathbf{L}^\dagger dx_+ + \mathbf{L} dx_- \tag{12}$$

or explicitly

$$\partial_+\mathbf{X} = [\partial_+\mathbb{P}, \mathbb{P}], \quad \partial_-\mathbf{X} = -[\partial_-\mathbb{P}, \mathbb{P}]. \tag{13}$$

2.2. Lagrangian and topological densities

Let us recall that the topological density is defined by [10]

$$\mathcal{Q}(\mathbb{P}) = \frac{1}{2}\mathrm{Tr}\left[(D_+Z)^\dagger D_+ Z - (D_-Z)^\dagger D_- Z\right] = \frac{1}{2}\mathrm{Tr}\left[\mathbb{P}[\partial_-\mathbb{P}, \partial_+\mathbb{P}]\right], \tag{14}$$

showing, in particular, that for a holomorphic (or anti-holomorphic) solution, which satisfies $D_-Z = 0$ (or $D_+Z = 0$), it coincides with the Lagrangian density (up to a sign).

Hence, for the solutions of type (10), we get the following explicit expressions for the Lagrangian density (9) and the topological density (14):

$$\mathcal{L}(\mathbb{P}_\beta) = \frac{1}{2}\sum_{j=1}^{n-1}(\beta_{j-1} - \beta_j)^2 \frac{|P_+^j f|^2}{|P_+^{j-1} f|^2}, \quad \mathcal{Q}(\mathbb{P}_\beta) = \frac{1}{2}\sum_{j-1}^{n-1}(\beta_{j-1} - \beta_j)\frac{|P_+^j f|^2}{|P_+^{j-1} f|^2}. \tag{15}$$

Let us exhibit some properties of the topological charge. First we rewrite the topological density as

$$\mathcal{Q}(\mathbb{P}_\beta) = \mathcal{Q}\left(\sum_{j=0}^{n-1}\beta_j\mathbb{P}_j\right) = \frac{1}{2}\sum_{j=0}^{n-1}\beta_j\left(\frac{|P_+^{j+1} f|^2}{|P_+^j f|^2} - \frac{|P_+^j f|^2}{|P_+^{j-1} f|^2}\right) = \sum_{j=0}^{n-1}\beta_j \mathcal{Q}(\mathbb{P}_j), \tag{16}$$

showing that $\mathcal{Q}$ is a purely additive quantity. Furthermore, using the topological property

$$\mathcal{Q}(\mathbb{P}_j) = \frac{|P_+^{j+1} f|^2}{|P_+^j f|^2} - \frac{|P_+^j f|^2}{|P_+^{j-1} f|^2} = \partial_+\partial_- \ln(|P_+^j f|^2), \tag{17}$$

the topological density takes the compact form

$$\mathcal{Q}(\mathbb{P}_\beta) = \partial_+\partial_- \ln \prod_{j=0}^{n-1} |P_+^j f|^{\beta_j}. \tag{18}$$

2.3. Mean and Gaussian curvatures

In order to extract some geometric properties of the surface $\mathbf{X}$ defined as in (12), we introduce a scalar product on the Lie algebra $su(n)$:

$$\langle A, B\rangle = \frac{1}{2}\mathrm{Tr}(AB), \quad A, B \in su(n). \tag{19}$$

The first fundamental form [12] of the surface $\mathbf{X}$ is given by:

$$\mathbf{I} = \langle d\mathbf{X}, d\mathbf{X}\rangle = g_{++}dx_+^2 + 2g_{+-}dx_+dx_- + g_{--}dx_-^2, \tag{20}$$

where $g_{\mu\nu}$ are the components of the metric tensor and are given by

$$g_{\pm\pm} = \langle \partial_\pm \mathbf{X}, \partial_\pm \mathbf{X}\rangle = -\langle \partial_\pm \mathbb{P}, \partial_\pm \mathbb{P}\rangle, \quad g_{\pm\mp} = \langle \partial_\pm \mathbf{X}, \partial_\mp \mathbf{X}\rangle = \langle \partial_\pm \mathbb{P}, \partial_\mp \mathbb{P}\rangle. \tag{21}$$

Since we are interested in solutions $\mathbb{P}$ given in (10), we can show that our surfaces are conformal maps and that the metric components take the form:

$$g_{\pm\pm} = 0, \quad g_{+-} = g_{-+} = \frac{1}{2}\mathrm{Tr}(\partial_+\mathbb{P}\partial_-\mathbb{P}). \tag{22}$$

Note that the expression for g_{+-} is identical to the expression for the Lagrangian density (9). Using the Brioschi formula [12], we see that the Gaussian curvature $\mathcal{K}$ of the surface $\mathbf{X}$ associated to the solution $\mathbb{P}$:

$$\mathcal{K} = -\frac{1}{g_{+-}}\partial_+\partial_- \ln g_{+-}. \tag{23}$$

Let us now calculate the expression for the mean curvature $\mathcal{H}$ [12] associated to solutions of the model. As we have shown the considered surfaces are conformal maps and we know that the expression for the mean curvature is given as

$$\mathcal{H} = \frac{1}{2}\mathrm{Tr}(\mathbf{II}(\mathbf{I}^{-1})), \tag{24}$$

where $\mathbf{II}$ is the second fundamental form defined as

$$\mathbf{II} = \langle \partial_+^2\mathbf{X}, N\rangle dx_+^2 + 2\langle \partial_+\partial_-\mathbf{X}, N\rangle dx_+dx_- + \langle \partial_-^2\mathbf{X}, N\rangle dx_-^2 = -\langle d\mathbf{X}, dN\rangle. \tag{25}$$

In the above expression, N is a normal unit vector to the surface $\mathbf{X}$ and thus satisfies $\langle d\mathbf{X}, N\rangle = 0$. Using the conformal property of the surfaces the mean curvature is given by

$$\mathcal{H} = \frac{\langle \partial_+\partial_-\mathbf{X}, N\rangle}{g_{+-}}. \tag{26}$$

Due to the expression (13) and from the Euler-Lagrange equations (9), we easily get :

$$\partial_+\partial_-\mathbf{X} = [\partial_+\mathbb{P}, \partial_-\mathbb{P}] = [\partial_+\mathbf{X}, \partial_-\mathbf{X}]. \tag{27}$$

We may thus define a unit normal vector N to the surface $\mathbf{X}$ as

$$N = \frac{[\partial_+\mathbf{X}, \partial_-\mathbf{X}]}{\|[\partial_+\mathbf{X}, \partial_-\mathbf{X}]\|} = \frac{\partial_+\partial_-\mathbf{X}}{\|\partial_+\partial_-\mathbf{X}\|} \in su(n) \tag{28}$$

remembering that $\langle d\mathbf{X}, N\rangle = 0$. Hence, we see that the mean curvature $\mathcal{H}$ is given by

$$\mathcal{H} = \frac{\|\partial_+\partial_-\mathbf{X}\|}{g_{+-}} = 2\frac{\|[\partial_+\mathbb{P}, \partial_-\mathbb{P}]\|}{\mathrm{Tr}(\partial_+\mathbb{P}\partial_-\mathbb{P})}. \tag{29}$$

3. Solutions of $G(m,n)$, their associated surfaces and geometric characteristics

3.1. Special case of $\mathbb{C}P^{n-1}$ and the Veronese sequence

In the late 80's, Bolton and *al.*[11] fully classified constant Gaussian curvature surfaces $\mathbf{X}$ associated to the solutions (4) of the $\mathbb{C}P^{n-1}$ model. Indeed, the set of all solutions is obtained from the Veronese holomorphic curve f defined as

$$f(x_+) = \left(1, \sqrt{\binom{n-1}{1}}x_+, \cdots, \sqrt{\binom{n-1}{r}}x_+^r, \cdots, x_+^{n-1}\right)^T. \tag{30}$$

This holomorphic curve satisfies the following identity, which will prove to be useful in the rest of the paper:

$$\frac{|P_+^i f|^2}{|P_+^{i-1} f|^2} = \frac{\alpha_{i,n}}{(1+|x|^2)^2} \tag{31}$$

where

$$\alpha_{i,n} = i(n-i), \quad i = 1, 2, \cdots, n-1. \tag{32}$$

In this case, the Gaussian curvature $\mathcal{K}$ takes the form

$$\mathcal{K}(Z_i) = \frac{4}{r_i(1,n)}, \quad r_i(1,n) = r_i = n-1+2i(n-1-i). \tag{33}$$

We see that the quantity r_i admits an obvious symmetry given by

$$r_i = r_{n-1-i}, \tag{34}$$

which shows that some surfaces associated to non-equivalent solutions Z_i and Z_{n-1-i} have the same value of constant Gaussian curvature. Note that when n is odd, we omit $i = \frac{n-1}{2}$ since $Z_i = Z_{n-1-i}$.

Thus we need other geometric quantities to differentiate them. In this case, the topological density is sufficient since we have

$$\mathcal{Q}(Z_i) = \frac{q_i(1,n)}{2(1+|x|^2)^2}, \quad q_i(1,n) = q_i = n-1-2i, \tag{35}$$

where q_i satisfy the relation:

$$q_{n-1-i} = -q_i. \tag{36}$$

To go further, we express the quantity $\alpha_{i,n}$ in terms of the r's and q's given, respectively, in (33) and (35). We get

$$\alpha_{m+j,n} = \frac{1}{2}\left[r_m + (2j-1)q_m\right] - j(j-1), \tag{37}$$

which will help us to express the geometric expressions in terms of the r's and the q's. For example, we have

$$\alpha_{m,n} = \frac{1}{2}\left[r_m - q_m\right], \quad \alpha_{m+1,n} = \frac{1}{2}\left[r_m + q_m\right]. \tag{38}$$

The mean curvature $\mathcal{H}_i$, associated to the solution Z_i, is constant and is given by

$$\mathcal{H}_i = 2\frac{\sqrt{\alpha_{i,n}^2 - \alpha_{i,n}\alpha_{i+1,n} + \alpha_{i+1,n}^2}}{\alpha_{i,n} + \alpha_{i+1,n}} = \frac{\sqrt{r_i^2 + 3q_i^2}}{r_i}. \tag{39}$$

This geometric quantity is not necessary to differentiate surfaces of $\mathbb{C}P^{n-1}$ and of equal Gaussian curvature, but it will become relevant for higher dimensional Grassmannians.

3.2. Geometric characteristics of the Veronese curves of $G(m,n)$

For the general Grassmannian $G(m,n)$, we take the solution $\mathbb{P}_\beta$ given in (10) with f chosen as the Veronese holomorphic curve (30). The Lagrangian and topological densities are then given by

$$\mathcal{L}(\mathbb{P}_\beta) = \frac{r_\beta(m,n)}{2(1+|x|^2)^2}, \quad \mathcal{Q}(\mathbb{P}_\beta) = \frac{q_\beta(m,n)}{2(1+|x|^2)^2}, \tag{40}$$

where

$$r_\beta(m,n) = \sum_{j=1}^{n-1}(\beta_{j-1}-\beta_j)^2\alpha_{j,n}, \quad q_\beta(m,n) = \sum_{j=1}^{n-1}(\beta_{j-1}-\beta_j)\alpha_{j,n}. \tag{41}$$

Due to the fact that β_j are 0 or 1, we easily deduce the following expressions:

$$r_\beta(m,n) - q_\beta(m,n) = 2\sum_{j=1}^{n-1}\beta_j\alpha_{j,n} - 2\sum_{j=1}^{n-1}\beta_j\beta_{j-1}\alpha_{j,n}, \tag{42}$$

$$r_\beta(m,n) + q_\beta(m,n) = 2\sum_{j=1}^{n-1}\beta_{j-1}\alpha_{j,n} - 2\sum_{j=1}^{n-1}\beta_j\beta_{j-1}\alpha_{j,n}. \tag{43}$$

These expressions are slightly different from the ones obtained in the $\mathbb{C}P^{n-1}$ case. Indeed, they exhibit interactions between consecutive projectors $\mathbb{P}_{j-1}$ and $\mathbb{P}_j$ in the general expression of $\mathbb{P}_\beta$, which is absent in the $\mathbb{C}P^{n-1}$ case.

Let us recall that the Gaussian curvature of the surface associated to the solution $\mathbb{P}_\beta$ is given by

$$\mathcal{K} = \frac{4}{r_\beta(m,n)}. \tag{44}$$

Moreover, we can also calculate the numerator of the mean curvature expression given in (29) and we get

$$\|[\partial_+\mathbb{P}_\beta, \partial_-\mathbb{P}_\beta]\|^2 \propto \sum_{j=1}^{n-1}(\beta_{j-1}-\beta_j)^2\alpha_{j,n}((\beta_{j-1}-\beta_j)^2\alpha_{j,n} - \frac{1}{2}(\beta_j-\beta_{j+1})^2\alpha_{j+1,n} - \frac{1}{2}(\beta_{j-2}-\beta_{j-1})^2\alpha_{j-1,n}). \tag{45}$$

For example, the holomorphic solutions, in the $G(m,n)$ case, which are described by $\beta_i = 1$ for $i = 0, 1, \cdots, m-1$ and $\beta_i = 0$ for $i = m, m+1, \cdots, n-1$ lead to

$$r_\beta^{(hol)}(m,n) = q_\beta^{(hol)}(m,n) = \alpha_{m,n} = m(n-m), \quad \mathcal{H}_\beta^{(hol)}(m,n) = 2. \tag{46}$$

Let us illustrate, in the following subsection, the case of $G(2,n)$ and the need to use further geometric characteristics to distinguish surfaces with the same Gaussian curvature.

3.2.1. The case of $G(2,n)$: some examples In this case, we have already computed the Gaussian curvature for some surfaces associated to non-holomorphic solutions of $G(2,n)$. For example, we have found [2] that

$$r_{2,3}(2,7) = r_{0,5}(2,7) = 22. \tag{47}$$

If we now compute the topological charge we find that

$$q_{2,3}(2,7) = q_{0,5}(2,7) = 2. \tag{48}$$

This means that we need other quantities to differentiate the geometry of these two surfaces.

The explicit forms of the quantities $r_\beta(2,n)$ and $q_\beta(2,n)$ are obtained directly from (42) and (43). Indeed, we have to distinguish two cases: when $\beta_j = \beta_{j+1} = 1$ (**interaction**), we get

$$q_{j,j+1}(2,n) = 2(n-2-2j), \tag{49}$$
$$r_{j,j+1}(2,n) = 2(n-2+j(n-2-j)) = q_{j,j+1}(2,n) + 2\alpha_j; \tag{50}$$

and when $\beta_j = \beta_k = 1$ for $k > j+1$ (**absence of interaction**), we get

$$q_{j,k}(2,n) = 2(n-1-j-k), \tag{51}$$
$$r_{j,k}(2,n) = 2(n-1+j(n-1-j)+k(n-1-k)) = q_{j,k}(2,n) + 2\alpha_j + 2\alpha_k. \tag{52}$$

Finally, we have (note that, for simplicity, we have set $\alpha_i = \alpha_{i,n}$)

$$\mathcal{H}_{i,i+1} = 2\frac{\sqrt{\alpha_i^2 + \alpha_{i+2}^2}}{\alpha_i + \alpha_{i+2}}, \tag{53}$$
$$\mathcal{H}_{i,i+2} = 2\frac{\sqrt{\alpha_i^2 - \alpha_i\alpha_{i+1} + \alpha_{i+1}^2 - \alpha_{i+1}\alpha_{i+2} + \alpha_{i+2}^2 - \alpha_{i+2}\alpha_{i+3} + \alpha_{i+3}^2}}{\alpha_i + \alpha_{i+1} + \alpha_{i+2} + \alpha_{i+3}}, \tag{54}$$
$$\mathcal{H}_{i,j>i+2} = 2\frac{\sqrt{\alpha_i^2 - \alpha_i\alpha_{i+1} + \alpha_{i+1}^2 + \alpha_j^2 - \alpha_j\alpha_{j+1} + \alpha_{j+1}^2}}{\alpha_i + \alpha_{i+1} + \alpha_j + \alpha_{j+1}}. \tag{55}$$

The mean curvature is constant in each cases and it can be used to differentiate the geometry of the surfaces associated to $\mathbb{P}_{2,3}$ and $\mathbb{P}_{0,5}$ in $G(2,7)$. Indeed, we have

$$\mathcal{H}_{2,3} = \frac{2\sqrt{61}}{11}, \quad \mathcal{H}_{0,5} = \frac{4\sqrt{7}}{11}. \tag{56}$$

In Table 1, we give the examples of $G(2,n)$ with $n = 4, 5, 6$.

Table 1. The $G(2,4)$, $G(2,5)$ and $G(2,6)$ models

(i,j)	$r_{i,j}$	$q_{i,j}$	$\mathcal{H}_{i,j}$
$(0,1)$	4	4	2
$(1,2)$	6	0	$\sqrt{2}$
$(0,2)$	10	2	$\sqrt{\frac{2}{5}}$

(i,j)	$r_{i,j}$	$q_{i,j}$	$\mathcal{H}_{i,j}$
$(0,1)$	6	6	2
$(1,2)$	10	2	$\frac{2\sqrt{13}}{5}$
$(0,2)$	16	4	$\frac{\sqrt{7}}{4}$
$(0,3)$	14	2	$\frac{2\sqrt{11}}{7}$
$(0,4)$	8	0	$\sqrt{2}$
$(1,3)$	20	0	$\frac{1}{\sqrt{5}}$

(i,j)	$r_{i,j}$	$q_{i,j}$	$\mathcal{H}_{i,j}$
$(0,1)$	8	8	2
$(1,2)$	14	4	$\frac{\sqrt{106}}{7}$
$(2,3)$	16	0	$\sqrt{2}$
$(0,2)$	22	6	$\frac{\sqrt{58}}{11}$
$(0,3)$	22	4	$\frac{7\sqrt{2}}{11}$
$(0,4)$	18	2	$\frac{\sqrt{74}}{9}$
$(0,5)$	10	0	$\sqrt{2}$
$(1,3)$	30	2	$\frac{\sqrt{2}}{3}$
$(1,4)$	26	0	$\frac{7\sqrt{2}}{13}$

3.2.2. The case of $G(2,n)$: more general results In this section, we give a partial answer to the following question: If $q_{i,j} = q_{k,l}$ and $r_{i,j} = r_{k,l}$, is the mean curvature sufficient to differentiate the surfaces associated to $\mathbb{P}_{i,j}$ and $\mathbb{P}_{k,l}$?

To answer this question, we have to consider three cases.

The first case arises when $j = i+1$ and $l = k+1$. In this case, $q_{i,i+1} = q_{k,k+1}$ and so $k = i$ and we are, thus, dealing with the same solution. We may exclude this case.

The second one is corresponds to $j > i+1$ and $l > k+1$. In this case, the condition $q_{i,j} = q_{k,l}$ leads to $i+j = k+l$ which is equivalent to $l = i+j-k$. Then one wants $r_{ij} = r_{kl}$ which implies that

$$r_{i,j} - r_{k,l} = r_{i,j} - r_{k,i+j-k} = 4(i-k)(j-k) = 0 \iff i = k \quad \text{or} \quad j = k. \tag{57}$$

But if $i = k$, then $j = l$ and so we are dealing with the same solution. Moreover, when $j = k$, we have $i = l$, but this contradicts the original assumption that $j > i+1$. So we may ignore this case too.

The third case is the one we have encountered for different grids: $j = i+1$ and $l > k+1$. The constraint $q_{i,i+1} = q_{k,l}$ leads to $l_{k,i} = 2i - k + 1$ and together with $l > k+1$ implies $i > k$. Solving the constraint $r_{i,i+1} = r_{k,l}$, we get an expression for the dimension n which is given by

$$n_{k,i} = 3i + 1 - 4k + \frac{2k(1+k)}{1+i}, \quad i > k. \tag{58}$$

So let us now look for a couple (k,i) such that n_{ki} is an integer. Once we have found such a couple, we must check that it satisfies the constraints $l_{k,i} = 2i - k + 1 < n$ and $i < n-1$.

We have

- $n_{0,i} = 3i + 1 \in \mathbb{N}, \quad i > 0$: This shows that the projectors $\mathbb{P}_{i,i+1}$ and $\mathbb{P}_{0,2i+1}$ have the same Gaussian curvature and topological charge. But can their corresponding mean curvatures be different? If they are the same, we have

$$\left(\frac{\mathcal{H}_{0,2i+1}}{\mathcal{H}_{i,i+1}}\right)^2_{n=3i+1} = \frac{2 + i + 3i^2 + i^3 + 2i^4}{2 - 6i + i^2 + 8i^3 + 4i^4} = 1 \iff i = 1. \tag{59}$$

  However, the case $i = 1$ can be easily understood since, in this case, $n_{0,1} = 4$ and we are thus comparing the projectors $\mathbb{P}_{1,2}$ and $\mathbb{P}_{0,3}$ which correspond to the same solution of the $G(2,4)$ model. So we see that, in this case, the mean curvatures are different.
- $n_{1,i} = -3 + 3i + \frac{4}{1+i} \in \mathbb{N}, \quad i > 1$: This shows that the only admissible i is $i = 3$. In this case, we have $n_{1,3} = 7$ and we are comparing the projectors $\mathbb{P}_{3,4}$ and $\mathbb{P}_{1,6}$. Using the completeness relation, this is the same as comparing the projectors $\mathbb{P}_{2,3}$ and $\mathbb{P}_{0,5}$ which is consistent with the discussion of subsection 3.2.1. As we already know, the mean curvatures are different.
- $n_{2,i} = -7 + 3i + \frac{12}{1+i} \in \mathbb{N}, \quad i > 2$: Putting all the constraints together, this shows that the only admissible i are $i = 11$ and $i = 5$. In the first case, we get $n_{2,11} = 27$ and we are comparing the projectors $\mathbb{P}_{11,12}$ and $\mathbb{P}_{2,21}$. The second possibility leads to $n_{2,5} = 10$ and we are, then, comparing the projectors $\mathbb{P}_{5,6}$ and $\mathbb{P}_{2,9}$. As before, we can show that the mean curvatures are different.

Let us make some further comments. Indeed, if we set $l = n-1$, then we have that $n = 2i + 2 - k$ which can then be compared with the formula for $n_{k,i}$. Doing so, we obtain $i = k-1$ or $i = 1+2k$. The case $i = k-1$ must be rejected due to the constraint that $i > k$. This means that we are left with the unique choice $i = 1+2k$. Thus we have $n_{k,1+2k} = 4+3k$

and we are comparing surfaces associated to the projectors $\mathbb{P}_{1+2k,2+2k}$ and $\mathbb{P}_{k,3+3k}$. This general result is consistent with the above examples. Furthermore, we can show that

$$\left(\frac{\mathcal{H}^{(4+3k)}_{k,3+3k}}{\mathcal{H}^{(4+3k)}_{1+2k,2+2k}}\right)^2 = \frac{9+18k+18k^2+9k^3+2k^4}{9+36k+49k^2+24k^3+4k^4} = 1 \iff k = -\frac{9}{2}, -2, -1, 0. \tag{60}$$

Since $k > 0$, the mean curvatures are different. Note that the case $k = 0$ is to be rejected as previously discussed.

In Table 2, we give some possible values of i. This table contains all the examples mentioned

Table 2. Some possible values of i in the expression for $n_{k,i}$

i	$n_{k,i}$	$l_{k,i}$	Comments
$2k(1+k)-1$	$6k^2+2k-1$	$4k^2+3k-1$	$k>0$
$k(1+k)-1$	$k(3k-1)$	$2k^2+k-1$	$k>1$
$2k+1$	$3k+4$	$3(k+1)$	$k\geq 0$
$2k-1$	$3k-1$	$3k-1$	N/A

above. The above discussion works for $k = 3, 4$, but for $k = 5$ new cases arise. Indeed, there are two of them: $n_{5,19} = 41$ with $l_{5,19} = 34$ and $n_{5,14} = 27$ with $l_{5,14} = 24$. The later one can be explained using the fact that $2k(1+k)$ in the expression of $n_{k,i}$ is always divisible by 4. We have summarized all this information in Table 3. We conjecture that, that as k increases, the

Table 3. Values of i for which $2k(1+k)$ is divisible by 4

i	$n_{k,i}$	$l_{k,i}$	Comments
$m(2m+3)$	$6m^2+m+1$	$4m(m+1)$	$k=2m+1 \quad m\geq 2$
$m(2m+1)-1$	$6m^2-5m+2$	$4m^2-1$	$k=2m \quad m\geq 2$

number of cases increases too and it is totally dependent on the prime decomposition of the factor $2k(1+k)$ in the expression for $n_{k,i}$.

3.3. Non-Veronese holomorphic solutions

We have conjectured [1] that we can construct a holomorphic solution in $G(m,n)$ of constant Gaussian curvature $\mathcal{K} = \frac{4}{r}$ for all integer values of $r = 1, 2, \cdots, \alpha_{m,n}$. The maximal value $r = \alpha_{m,n} = m(n-m)$ is obtained from the Veronese holomorphic curve (30) and its $m-1$ consecutive derivatives. The values of $r = 1, 2, \cdots, m(n-1-m)$ are obtained from the natural immersion of $G(m, n-1)$ into $G(m,n)$. The other values are not obtained from such immersions nor from the Veronese curve. Furthermore, for the same value of the missing r, we may find non-equivalent solutions $Z_i = \hat{Z}_i\hat{L}_i$, $i = 1, 2$. As an example, in the $G(2,5)$ case, we have obtained [1] two non-equivalent holomorphic solutions corresponding to $r = 5$ parametrized by

$$\hat{Z}_1^T = \begin{pmatrix} 1 & 0 & \sqrt{5}x_+ & \sqrt{5}x_+^2 & 0 \\ 0 & 1 & \sqrt{5}x_+^2 & \frac{7}{\sqrt{5}}x_+^3 & \frac{1}{\sqrt{5}}x_+^3 \end{pmatrix}, \quad \hat{Z}_2^T = \begin{pmatrix} 1 & 0 & x_+ & \frac{1}{\sqrt{5}}x_+^2 & 0 \\ 0 & 1 & 2x_+ & \frac{7}{\sqrt{5}}x_+^2 & \sqrt{5}x_+^3 \end{pmatrix}. \tag{61}$$

Of course, we know that, in the holomorphic case, the Lagrangian density corresponds to the topological one. Since the Lagrangian density of these two solutions is the same this is also the case for the topological density. Hence, the mean curvature will distinguish them. Indeed, using $Z = \hat{Z}\hat{L}$ that satisfies $\hat{Z}^\dagger\hat{Z} = (\hat{L}\hat{L}^\dagger)^{-1}$, we get $\mathbb{P} = \hat{Z}(\hat{Z}^\dagger\hat{Z})^{-1}\hat{Z}^\dagger$. In this case, the mean curvature is not a constant, *i.e.* $\mathcal{H}_i = \mathcal{H}_i(|x|^2)$. Moreover, we can show that $\mathcal{H}_1 \neq \mathcal{H}_2$. Indeed from (29), we get

$$\left(\frac{\mathcal{H}_1}{\mathcal{H}_2}\right)^2 = \frac{\mathcal{P}_1(|x|^2)}{\mathcal{P}_2(|x|^2)}, \quad \mathcal{P}_2(y) = y^6\mathcal{P}_1\left(\frac{1}{y}\right) = 25 + 110y + 285y^2 + 428y^3 + 355y^4 + 150y^5 + 25y^6.$$

4. Conclusions and Outlook

In this paper, we have studied various geometric properties of surfaces constructed from the solutions of the two-dimensional $G(m,n)$ Sigma model. The aim of our work was to see whether we can differentiate surfaces of equal constant Gaussian curvature by involving also the topological density and the mean curvature. This problem originated from the complete classification of constant Gaussian curvature surfaces [11] associated to Veronese holomorphic curves for $\mathbb{C}P^{n-1}$. In this classification, some non-equivalent solutions had, due to a symmetry property, identical Gaussian curvatures. In these cases, the topological charges had different values.

We have not fully solved the problem which has turned out to be more complicated than originally envisaged. So here we report where we are at present. We have generalized our previous results to more general Grassmannians. We have obtained explicit expressions for the Gaussian curvature, the topological density and the mean curvature for these solutions. We have shown, in the case of $G(2,n)$, that some non-equivalent solutions may have identical Gaussian curvature and also identical topological densities. This has led us to consider the second fundamental form of these surfaces and we have computed their mean curvature. Some partial results for the $G(2,n)$ case show that the mean curvature is sufficient to distinguish surfaces with identical Gaussian curvatures and topological charges. Furthermore, we have shown that the projectors having this property are all of the form $\mathbb{P}_{i,i+1}$ and $\mathbb{P}_{k,l}$ with $l > k+1$.

The case of holomorphic solutions which are not of the Veronese type and that cannot be obtained from immersions of lower dimensional Grassmannians is really challenging since we have no general formula for these cases. Our results are complete for the $G(2,5)$ model where we have shown that all solutions of equal Gaussian curvature and topological density have distinct mean curvature.

Acknowledgements

This work has been supported in part by research grants from Natural sciences and engineering research council of Canada (NSERC). Laurent Delisle acknowledges a NSERC Postdoctoral fellowship.

References

[1] Delisle L, Hussin V and Zakrzewski W J 2013 *Journal of Geometry and Physics* **66**, 24
[2] Delisle L, Hussin V and Zakrzewski W J 2013 *Journal of Geometry and Physics* **71** 1
[3] Li Z-Q and Yu Z-H 1999 *Manuscripta Math.* **100** 305
[4] Jiao X and Peng J 2003 *Transaction of the American Mathematical Society* **355** 3715
[5] Jiao X and Peng J 2004 *Differential geometry and its applications* **20** 267
[6] Jiao X and Peng J 2010 *Front. Math. China* **5** 297
[7] Fei J, Jiao X and Xu X 2011 *Proc. Indian Acad. Sci.* **121** 181
[8] Peng C K and Xu X W 2014 *Israel Journal of Mathematics* **202** 1
[9] Peng C K and Xu X W 2014 *Journal de Mathématiques Pures et Appliquées*, to be published
[10] Zakrzewski W J 1989 *Low Dimensional Sigma Models* (Bristol: Adam Hilger)
[11] Bolton J, Jensen G R, Rigoli M and Woodward L M 1988 *Math. Ann.* **279** 599
[12] Hélein F 2001 *Constant mean curvature surfaces, harmonic maps and integrable systems* (ETH Zürich: Birkhäuser)

Journal of Physics: Conference Series **597** (2015) 012030 doi:10.1088/1742-6596/597/1/012030

On two subgroups of U(n), useful for quantum computing

Alexis De Vos[1] **and Stijn De Baerdemacker**[2]

[1]Cmst, Vakgroep elektronika en informatiesystemen, Imec v.z.w., Universiteit Gent, Sint Pietersnieuwstraat 41, B - 9000 Gent, Belgium

[2]Center for Molecular Modeling, Vakgroep fysica en sterrenkunde, Universiteit Gent, Technologiepark 903, B - 9052 Gent, Belgium

[2]Ghent Quantum Chemistry Group, Vakgroep anorganische en fysische chemie, Universiteit Gent, Krijgslaan 281 - S3, B - 9000 Gent, Belgium

E-mail: `alex@elis.UGent.be`, `Stijn.DeBaerdemacker@UGent.be`

Abstract. As two basic building blocks for any quantum circuit, we consider the 1-qubit `PHASOR` circuit $\Phi(\theta)$ and the 1-qubit `NEGATOR` circuit $N(\theta)$. Both are roots of the `IDENTITY` circuit. Indeed: both $\Phi(0)$ and $N(0)$ equal the 2×2 unit matrix. Additionally, the `NEGATOR` is a root of the classical `NOT` gate. Quantum circuits (acting on w qubits) consisting of controlled `PHASOR`s are represented by matrices from ZU(2^w); quantum circuits consisting of controlled `NEGATOR`s are represented by matrices from XU(2^w). Here, ZU(n) and XU(n) are subgroups of the unitary group U(n): the group XU(n) consists of all $n \times n$ unitary matrices with all $2n$ line sums (i.e. all n row sums and all n column sums) equal to 1 and the group ZU(n) consists of all $n \times n$ unitary diagonal matrices with first entry equal to 1. Any U(n) matrix can be decomposed into four parts: $U = \exp(i\alpha)\, Z_1 X Z_2$, where both Z_1 and Z_2 are ZU(n) matrices and X is an XU(n) matrix. We give an algorithm to find the decomposition. For $n = 2^w$ it leads to a four-block synthesis of an arbitrary quantum computer.

1. Introduction

The unitary group U(n) is important for quantum computing, because all quantum circuits acting on w qubits can be represented by a member of the unitary group U(2^w). For example, all quantum circuits, acting on a single qubit, are represented by a matrix from U(2). The simplest U(2) matrix is the 2×2 unit matrix. It represents the `IDENTITY` gate or `I` gate:

$$\begin{pmatrix} 1 & 0 \\ 0 & 1 \end{pmatrix} = \mathtt{I}\ .$$

This gate is trivial, as it performs no action on the qubit: the output qubit equals the input qubit. Within U(2), the `I` matrix has a lot of square roots: four diagonal matrices

$$\begin{pmatrix} 1 & 0 \\ 0 & 1 \end{pmatrix},\ \begin{pmatrix} -1 & 0 \\ 0 & 1 \end{pmatrix},\ \begin{pmatrix} 1 & 0 \\ 0 & -1 \end{pmatrix},\ \text{and}\ \begin{pmatrix} -1 & 0 \\ 0 & -1 \end{pmatrix}\ ,$$

as well as an infinity of 'anti-diagonal' matrices:

$$\sqrt{-1}\begin{pmatrix} 0 & e^{i\chi} \\ -e^{-i\chi} & 0 \end{pmatrix}\ ,$$

where χ is an arbitrary real number. From this set, we choose two elements:

$$\begin{pmatrix} 1 & 0 \\ 0 & -1 \end{pmatrix} \text{ and } \begin{pmatrix} 0 & 1 \\ 1 & 0 \end{pmatrix} .$$

The latter matrix results from the choices $\sqrt{-1} = -i$ and $\chi = \pi/2$ and represents the `NOT` gate or `X` gate [1]. The former matrix represents the `Z` gate. Whereas the `X` gate is a classical computer gate, inverting the incoming bit, the `Z` gate is a true quantum gate.

We now introduce a generalization of both the `X` and the `Z` matrix:

$$\begin{aligned} N(t) &= (1-t)\,\mathtt{I} + t\,\mathtt{X} \\ \Phi(t) &= (1-t)\,\mathtt{I} + t\,\mathtt{Z}\,, \end{aligned}$$

where t is an interpolation parameter [2]. We can easily prove that these matrices are unitary, iff t is of the form $(1-e^{i\theta})/2$, resulting in

$$\begin{aligned} N(\theta) &= \tfrac{1}{2}\,(1+e^{i\theta})\,\mathtt{I} + \tfrac{1}{2}\,(1-e^{i\theta})\,\mathtt{X} \\ \Phi(\theta) &= \tfrac{1}{2}\,(1+e^{i\theta})\,\mathtt{I} + \tfrac{1}{2}\,(1-e^{i\theta})\,\mathtt{Z}\,. \end{aligned}$$

We thus have constructed two 1-dimensional subgroups of the 4-dimensional group U(2):

$$\begin{aligned} N(\theta) &= \frac{1}{2}\begin{pmatrix} 1+e^{i\theta} & 1-e^{i\theta} \\ 1-e^{i\theta} & 1+e^{i\theta} \end{pmatrix} \\ \Phi(\theta) &= \begin{pmatrix} 1 & 0 \\ 0 & e^{i\theta} \end{pmatrix} . \end{aligned}$$

The gate represented by the matrix $N(\theta)$, we call the `NEGATOR` gate. It thus constitutes a generalization of the `NOT` gate. The gate represented by the matrix $\Phi(\theta)$, we call the `PHASOR` gate. We use the following symbols for these quantum gates:

$$-\boxed{N(\theta)}- \quad \text{and} \quad -\boxed{\Phi(\theta)}- \;,$$

respectively. In the literature [3] [4] [5] [6], some of these gates have a particular notation:

$$\begin{aligned} N(0) &= \mathtt{I} \\ N(\pi/4) &= \mathtt{W} \\ N(\pi/2) &= \mathtt{V} \\ N(\pi) &= \mathtt{X} \\ N(2\pi) &= \mathtt{I} \\[1ex] \Phi(0) &= \mathtt{I} \\ \Phi(\pi/4) &= \mathtt{T} \\ \Phi(\pi/2) &= \mathtt{S} \\ \Phi(\pi) &= \mathtt{Z} \\ \Phi(2\pi) &= \mathtt{I}\,. \end{aligned}$$

In particular, the `V` gate is known as 'the square root of `NOT`' [7] [8] [9] [10].

The subgroup of all $N(\theta)$ matrices, we denote by XU(2); the subgroup of all $\Phi(\theta)$ matrices, we denote by ZU(2). These two subgroups of U(2) have three interesting properties.

Journal of Physics: Conference Series **597** (2015) 012030 doi:10.1088/1742-6596/597/1/012030

(i) Their intersection is minimal, as it equals U(2)'s trivial subgroup consisting of merely one 2×2 matrix, i.e. the identity matrix $\mathtt{I}$.

(ii) Their closure is maximal, as it equals U(2) itself. Indeed, an arbitrary member of U(2), i.e.

$$U = e^{i\alpha} \begin{pmatrix} \cos(\varphi)\, e^{i\psi} & \sin(\varphi)\, e^{i\chi} \\ -\sin(\varphi)\, e^{-i\chi} & \cos(\varphi)\, e^{-i\psi} \end{pmatrix} \tag{1}$$

can be synthesized by cascading merely **NEGATOR**s and **PHASOR**s. The reader may verify the following two matrix decompositions:

$$\begin{aligned} U &= e^{i\alpha+i\varphi+i\psi} \begin{pmatrix} 1 & 0 \\ 0 & i\,e^{-i\psi-i\chi} \end{pmatrix} \frac{1}{2} \begin{pmatrix} 1+e^{-i2\varphi} & 1-e^{-i2\varphi} \\ 1-e^{-i2\varphi} & 1+e^{-i2\varphi} \end{pmatrix} \begin{pmatrix} 1 & 0 \\ 0 & -i\,e^{-i\psi+i\chi} \end{pmatrix} \\ U &= e^{i\alpha-i\varphi+i\psi} \begin{pmatrix} 1 & 0 \\ 0 & -i\,e^{-i\psi-i\chi} \end{pmatrix} \frac{1}{2} \begin{pmatrix} 1+e^{i2\varphi} & 1-e^{i2\varphi} \\ 1-e^{i2\varphi} & 1+e^{i2\varphi} \end{pmatrix} \begin{pmatrix} 1 & 0 \\ 0 & i\,e^{-i\psi+i\chi} \end{pmatrix} . \end{aligned} \tag{2}$$

Because, moreover, any phase factor can be decomposed as

$$\begin{pmatrix} e^{i\beta} & 0 \\ 0 & e^{i\beta} \end{pmatrix} = \begin{pmatrix} 0 & 1 \\ 1 & 0 \end{pmatrix} \begin{pmatrix} 1 & 0 \\ 0 & e^{i\beta} \end{pmatrix} \begin{pmatrix} 0 & 1 \\ 1 & 0 \end{pmatrix} \begin{pmatrix} 1 & 0 \\ 0 & e^{i\beta} \end{pmatrix} ,$$

we conclude that U equals the cascade of three **NEGATOR**s and three **PHASOR**s:

$$\begin{aligned} U &= N(\pi)\,\Phi(\alpha+\varphi+\psi)\,N(\pi)\,\Phi(\alpha+\varphi-\chi+\pi/2)\,N(-2\varphi)\,\Phi(-\psi+\chi-\pi/2) \\ U &= N(\pi)\,\Phi(\alpha-\varphi+\psi)\,N(\pi)\,\Phi(\alpha-\varphi-\chi-\pi/2)\,N(2\varphi)\,\Phi(-\psi+\chi+\pi/2) . \end{aligned}$$

(iii) We note that the two subgroups are each other's Hadamard conjugate:

$$\mathrm{ZU}(2) = H\ \mathrm{XU}(2)\ H \quad \text{and} \quad \mathrm{XU}(2) = H\ \mathrm{ZU}(2)\ H ,$$

where H denotes the Hadamard matrix:

$$H = \frac{1}{\sqrt{2}} \begin{pmatrix} 1 & 1 \\ 1 & -1 \end{pmatrix} .$$

2. Decomposition of an arbitrary unitary matrix

Two-qubit circuits are represented by matrices from U(4). We may apply either the **NEGATOR** gate or the **PHASOR** gate from the previous section to either the first qubit or the second qubit. Here are two examples:

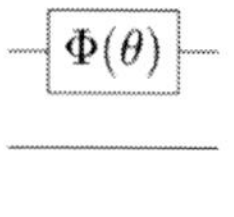

and
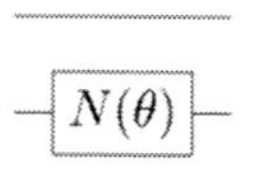

,

i.e. a **PHASOR** acting on the first qubit and a **NEGATOR** acting on the second qubit, respectively. These circuits are represented by the 4×4 unitary matrices

$$\begin{pmatrix} 1 & 0 & 0 & 0 \\ 0 & 1 & 0 & 0 \\ 0 & 0 & e^{i\theta} & 0 \\ 0 & 0 & 0 & e^{i\theta} \end{pmatrix} \quad \text{and} \quad \frac{1}{2} \begin{pmatrix} 1+e^{i\theta} & 1-e^{i\theta} & 0 & 0 \\ 1-e^{i\theta} & 1+e^{i\theta} & 0 & 0 \\ 0 & 0 & 1+e^{i\theta} & 1-e^{i\theta} \\ 0 & 0 & 1-e^{i\theta} & 1+e^{i\theta} \end{pmatrix} ,$$

respectively.

Journal of Physics: Conference Series **597** (2015) 012030 doi:10.1088/1742-6596/597/1/012030

However, we also introduce more sophisticated gates: the so-called 'controlled PHASORs' and 'controlled NEGATORs'. Two examples are

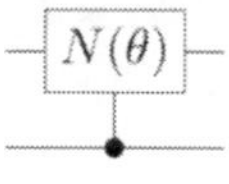

and

,

i.e. a positive-polarity controlled NEGATOR acting on the first qubit, controlled by the second qubit, and a negative-polarity controlled PHASOR acting on the second qubit, controlled by the first qubit, respectively. The former symbol is read as follows: 'if the second qubit equals 1, then the NEGATOR acts on the first qubit; if, however, the second qubit equals 0, then the NEGATOR is inactive, i.e. the first qubit undergoes no change'. The latter symbol is read as follows: 'if the first qubit equals 0, then the PHASOR acts on the second qubit; if, however, the first qubit equals 1, then the PHASOR is inactive, i.e. the second qubit undergoes no change'. The matrices representing these circuit examples are:

$$\begin{pmatrix} 1 & 0 & 0 & 0 \\ 0 & \frac{1}{2}(1+e^{i\theta}) & 0 & \frac{1}{2}(1-e^{i\theta}) \\ 0 & 0 & 1 & 0 \\ 0 & \frac{1}{2}(1-e^{i\theta}) & 0 & \frac{1}{2}(1+e^{i\theta}) \end{pmatrix} \text{ and } \begin{pmatrix} 1 & 0 & 0 & 0 \\ 0 & e^{i\theta} & 0 & 0 \\ 0 & 0 & 1 & 0 \\ 0 & 0 & 0 & 1 \end{pmatrix},$$

respectively.

We now give two examples of a 3-qubit circuit:

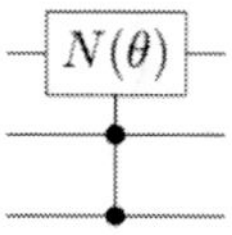

and

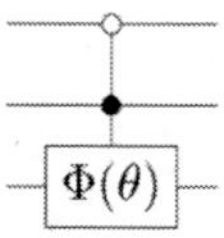

,

i.e. a positive-polarity controlled NEGATOR acting on the first qubit and a mixed-polarity controlled PHASOR acting on the third qubit. The 8×8 matrices representing these circuit examples are:

$$\begin{pmatrix} 1 & 0 & 0 & 0 & 0 & 0 & 0 & 0 \\ 0 & 1 & 0 & 0 & 0 & 0 & 0 & 0 \\ 0 & 0 & 1 & 0 & 0 & 0 & 0 & 0 \\ 0 & 0 & 0 & \frac{1}{2}(1+e^{i\theta}) & 0 & 0 & 0 & \frac{1}{2}(1-e^{i\theta}) \\ 0 & 0 & 0 & 0 & 1 & 0 & 0 & 0 \\ 0 & 0 & 0 & 0 & 0 & 1 & 0 & 0 \\ 0 & 0 & 0 & 0 & 0 & 0 & 1 & 0 \\ 0 & 0 & 0 & \frac{1}{2}(1-e^{i\theta}) & 0 & 0 & 0 & \frac{1}{2}(1+e^{i\theta}) \end{pmatrix} \text{ and } \begin{pmatrix} 1 & 0 & 0 & 0 & 0 & 0 & 0 & 0 \\ 0 & 1 & 0 & 0 & 0 & 0 & 0 & 0 \\ 0 & 0 & 1 & 0 & 0 & 0 & 0 & 0 \\ 0 & 0 & 0 & e^{i\theta} & 0 & 0 & 0 & 0 \\ 0 & 0 & 0 & 0 & 1 & 0 & 0 & 0 \\ 0 & 0 & 0 & 0 & 0 & 1 & 0 & 0 \\ 0 & 0 & 0 & 0 & 0 & 0 & 1 & 0 \\ 0 & 0 & 0 & 0 & 0 & 0 & 0 & 1 \end{pmatrix},$$

respectively. We note the following properties:

- the former matrix has all eight row sums and all eight column sums equal to 1;
- the latter matrix is diagonal and has upper-left entry equal to 1.

Because the multiplication of two square matrices with all line sums equal to 1 automatically yields a third square matrix with all line sums equal to 1, we can easily demonstrate that an arbitrary quantum circuit like

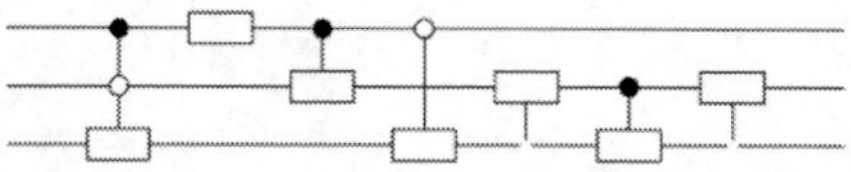

,

consisting merely of uncontrolled NEGATORs and controlled NEGATORs is represented by a $2^w \times 2^w$ unitary matrix with all line sums equal to 1. The $n \times n$ unitary matrices with all line sums equal to 1 form a group XU(n), subgroup of U(n). We thus can say that an arbitrary NEGATOR circuit is represented by an XU(2^w) matrix. A laborious proof [11] demonstrates that the converse theorem is also valid: any member of XU(2^w) can be synthesized by an appropriate string of (un)controlled NEGATORs.

Because the multiplication of two diagonal square matrices yields a third diagonal square matrix and because the multiplication of two unitary matrices with first entry equal to 1 yields a third unitary matrix with first entry equal to 1, we can easily demonstrate that an arbitrary quantum circuit like

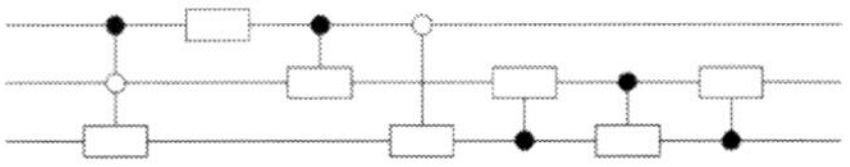

,

consisting merely of uncontrolled PHASORs and controlled PHASORs is represented by a $2^w \times 2^w$ unitary diagonal matrix with first entry equal to 1. The $n \times n$ unitary diagonal matrices with upper-left entry equal to 1 form a group ZU(n), subgroup of U(n). We thus can say that an arbitrary PHASOR circuit is represented by a ZU(2^w) matrix. The converse theorem is also valid: any member of ZU(2^w) can be synthesized by an appropriate string of (un)controlled PHASORs.

We summarize: the study of NEGATOR and PHASOR circuits leads to the introduction of two subgroups of the unitary group U(n):

- the subgroup XU(n), consisting of all $n \times n$ unitary matrices with all of their $2n$ line sums are equal to 1;
- the subgroup ZU(n), consisting of all $n \times n$ diagonal unitary matrices with upper-left entry equal to 1.

These subgroups have properties similar to the three properties of XU(2) and ZU(2), discussed in the previous section.

(i) The intersection of XU(n) and ZU(n) is minimal, as it is the trivial subgroup consisting of a single matrix, i.e. the $n \times n$ unit matrix.

(ii) The closure of XU(n) and ZU(n) is U(n). Indeed, any U(n) matrix U can be decomposed as

$$U = e^{i\alpha}\, Z_1 X_1 Z_2 X_2 Z_3 ... Z_{p-1} X_{p-1} Z_p \ , \tag{3}$$

with $p \leq n$ and where all Z_j are ZU(n) matrices and all X_j are XU(n) matrices [12]. Because we have the identity

$$\mathrm{diag}(a, a, a, a, a, ..., a, a) = \ P_0\ \mathrm{diag}(1, a, 1, a, 1, ..., 1, a)\ P_0^{-1}\ \mathrm{diag}(1, a, 1, a, 1, ..., 1, a) \ ,$$

where a is a short-hand notation for $e^{i\alpha}$ and P_0 is the circulant permutation matrix

$$\begin{pmatrix} 0 & 1 & 0 & 0 & ... & 0 & 0 \\ 0 & 0 & 1 & 0 & ... & 0 & 0 \\ 0 & 0 & 0 & 1 & ... & 0 & 0 \\ \vdots & & & & & & \\ 0 & 0 & 0 & 0 & ... & 0 & 1 \\ 1 & 0 & 0 & 0 & ... & 0 & 0 \end{pmatrix} ,$$

we can conclude that U equals the product of $n+1$ or less XU(n) members and $n+2$ or less ZU(n) members.

(iii) We have that

$$\mathrm{ZU}(n) \subset H_n\ \mathrm{XU}(n)\ H_n \ \text{ and } \ \mathrm{XU}(n) \supset H_n\ \mathrm{ZU}(n)\ H_n\ ,$$

where H_n is an $n \times n$ complex Hadamard matrix [13]. In particular, the group H_n ZU(n) H_n is the subgroup of XU(n) consisting of all circulant XU(n) matrices. For arbitrary n, one may choose for H_n the $n \times n$ discrete Fourier transform F_n:

$$F_n = \begin{pmatrix} 1 & 1 & 1 & 1 & \dots & 1 \\ 1 & \omega & \omega^2 & \omega^3 & \dots & \omega^{n-1} \\ 1 & \omega^2 & \omega^4 & \omega^6 & \dots & \omega^{2(n-1)} \\ \vdots & & & & & \\ 1 & \omega^{n-1} & \omega^{2(n-1)} & \omega^{3(n-1)} & \dots & \omega^{(n-1)(n-1)} \end{pmatrix} ,$$

where ω is the primitive n th root of unity. For $n = 2^w$, also the real Hadamard matrix $H^{\otimes w}$ (Kronecker product of 2×2 Hadamard matrices H) is a possible choice:

$$\begin{pmatrix} 1 & 1 \\ 1 & -1 \end{pmatrix} \otimes \begin{pmatrix} 1 & 1 \\ 1 & -1 \end{pmatrix} \otimes \dots \otimes \begin{pmatrix} 1 & 1 \\ 1 & -1 \end{pmatrix} ,$$

with w factors in the tensor product.

Because of the identity

$$\begin{aligned} \mathrm{diag}(1, a_2, a_3, ..., a_{n-1}, a_n) &= \mathrm{diag}(1, a_2, 1, ..., 1, 1)\, \mathrm{diag}(1, 1, a_3, ..., 1, 1)\ ... \\ &\quad \mathrm{diag}(1, 1, 1, ..., a_{n-1}, 1)\ \mathrm{diag}(1, 1, 1, ..., 1, a_n)\ , \end{aligned}$$

it is clear that the group ZU(n) is isomorphic to the direct product $\mathrm{U}(1)^{n-1}$ and thus is an $(n-1)$-dimensional Lie group. The group XU(n) is isomorphic to its conjugate

$$F_n\, \mathrm{XU}(n)\, F_n^{-1}\ .$$

One can verify that, if X is an XU(n) matrix, then the matrix $F_n X F_n^{-1}$ is an $n \times n$ unitary matrix with first row $1, 0, 0, ..., 0$ and thus also first column $1, 0, 0, ..., 0$. Conversely, it is possible to prove that a matrix of the form

$$X = F_n \begin{pmatrix} 1 & \mathbf{0}_{1\times(n-1)} \\ \mathbf{0}_{(n-1)\times 1} & Y \end{pmatrix} F_n^{-1}\ , \tag{4}$$

where $\mathbf{0}_{a\times b}$ denotes the $a \times b$ zero matrix and Y is an arbitrary member of U($n-1$), is an XU(n) matrix [11]. Thus XU(n) is isomorphic to U($n-1$) and therefore is an $(n-1)^2$-dimensional subgroup of the n^2-dimensional group U(n). We summarize by noting the beautiful symmetry

$$\mathrm{ZU}(n) \cong \mathrm{U}(1)^{n-1} \ \text{ and } \ \mathrm{XU}(n) \cong \mathrm{U}(n-1)^1\ .$$

In the past, properties of the subgroup XU(n) of U(n) have not been studied. Here, we note in particular the group chain

$$\mathrm{P}(n) \subset \mathrm{XU}(n) \subset \mathrm{U}(n)\ ,$$

where P(n) denotes the finite group of $n \times n$ permutation matrices. As all classical reversible circuits (acting on w bits) are represented by a matrix from P(2^w), we may say that the group XU(2^w) represents computers, situated 'between' classical computers and full-fledged quantum computers.

Decomposition (3) should be compared with other factorizations of U(n), described in the literature. Diţă [14] has proposed a product $D_nO_nD_{n-1}O_{n-1}...D_2O_2D_1$, where all D_j are diagonal unitary matrices and all O_j are orthogonal matrices. Reck et al. [15] have derived a decomposition of the form $T_1T_2...T_pD$, where all T_j are 2-parameter unitary matrices (describing a beam splitter plus phase shifter), D is a diagonal matrix, and $p \leq n(n-1)/2$. Rowe et al. [16] have applied two mutually commuting subgroups of U(n). Finally, Patera and Zassenhaus [17] have introduced the generalized Pauli group $\mathcal{P}_n$. We note that, for odd n, this finite group is generated by the permutation matrix P_0 together with a particular member of ZU(n), i.e. diag$(1, \omega, \omega^2, ..., \omega^{n-1})$, where ω is the primitive n th root of unity.

3. Short decomposition of an arbitrary unitary matrix

In Reference [18], it is conjectured that a shorter decomposition (3) exists: with $p \leq 2$. Thus an arbitrary member U of U(n) may be decomposed as

$$U = \exp(i\alpha)\, Z_1 X Z_2 \ , \tag{5}$$

with X a member of XU(n) and both Z_1 and Z_2 member of ZU(n). For $n = 2$, the conjecture is proved by eqns (2). For an arbitrary unitary matrix U with $n > 2$, a recent (non-constructive) proof is provided by Idel and Wolf [19], based on symplective topology. For a given unitary matrix U, De Vos and De Baerdemacker [18] provide a numerical procedure (reminiscent of Sinkhorn's construction [20] of a doubly stochastic matrix) in order to find the scalar $\exp(i\alpha)$, as well as the three matrices Z_1, X, and Z_2. For thousands of examples (from $3 \leq n \leq 32$), the algorithm converges to a solution. We conjecture that the algorithm always converges.

Thus any quantum circuit looks like

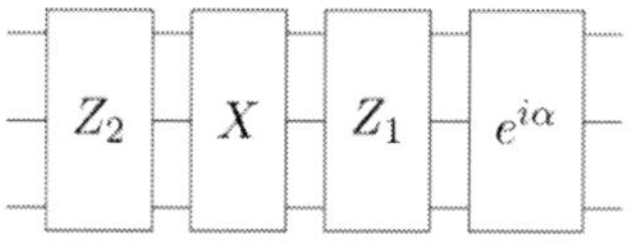

,

i.e. the cascade of an overall phase factor, an input section consisting merely of (un)controlled `PHASOR`s, a core section consisting merely of (un)controlled `NEGATOR`s, and an output section consisting merely of (un)controlled `PHASOR`s. In turn, the phase factor $e^{i\alpha}$ may be decomposed into two `NEGATOR` circuits (i.e. two classical cyclic-shift circuits [21] [22]) and two uncontrolled `PHASOR`s. We note that this circuit decomposition is not unique, because the matrix decomposition (5) is not unique. See Appendix.

Instead of synthesizing a quantum circuit with X and Z building-blocks, one can also opt for X and Hadamard blocks. Indeed, introducing the circulant XU(2^w) matrices

$$\begin{aligned} X_1 &= H^{\otimes w}\, Z_1\, H^{\otimes w} \\ X_2 &= H^{\otimes w}\, Z_2\, H^{\otimes w} \ , \end{aligned}$$

we have

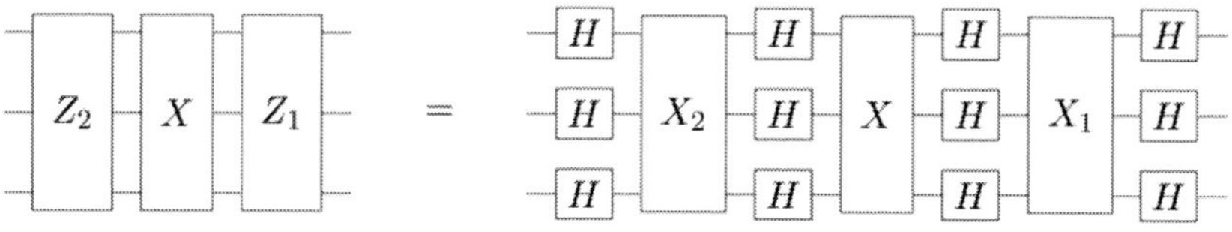

.

Thus any quantum computer (up to an overall phase) consists of three (or less) `NEGATOR` circuits and $4w$ (or less) `HADAMARD` gates.

By applying alternatingly (4) and (5), again and again, i.e. for $n = 2^w$, $n = 2^w - 1$, $n = 2^w - 2$, ..., and finally for $n = 2$, we find yet another decomposition of an arbitrary quantum computer [19] [23]:

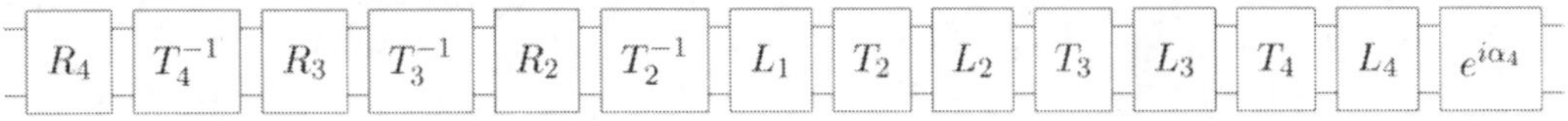

Here the circuits T_2, T_3, ..., and T_{2^w} implement the constant block-diagonal matrices

$$T_j = \begin{pmatrix} \mathbf{1}_{2^w - j} & \\ & F_j \end{pmatrix} ,$$

where $\mathbf{1}_a$ is the $a \times a$ unit matrix and F_j is the $j \times j$ discrete Fourier transform. The $2^w - 1$ circuits R_j and the 2^w circuits L_j are

- either $\mathrm{ZU}(2^w)$ circuits
- or circulant $\mathrm{XU}(2^w)$ circuits.

4. Conclusion

Like the qubit being a quantum counterpart of the classical bit, the `NEGATOR` gate is the quantum counterpart of the classical `NOT` gate and the controlled `NEGATOR` is the quantum counterpart of the classical controlled `NOT` (a.k.a. the `TOFFOLI` gate). Any quantum circuit can be built from (un)controlled `NEGATOR`s, supplemented with either (un)controlled `PHASOR` gates or `HADAMARD` gates.

Appendix

The matrix decomposition (5) is not unique. For $n = 2$, eqns (2) show two different solutions. E.g. the 2×2 discrete Fourier transform (a.k.a. the Hadamard transform)

$$F_2 = \frac{1}{\sqrt{2}} \begin{pmatrix} 1 & 1 \\ 1 & -1 \end{pmatrix}$$

has two and only two decompositions:

$$\frac{1-i}{\sqrt{2}} \begin{pmatrix} 1 & 0 \\ 0 & i \end{pmatrix} \frac{1}{2} \begin{pmatrix} 1+i & 1-i \\ 1-i & 1+i \end{pmatrix} \begin{pmatrix} 1 & 0 \\ 0 & i \end{pmatrix}$$

and

$$\frac{1+i}{\sqrt{2}} \begin{pmatrix} 1 & 0 \\ 0 & -i \end{pmatrix} \frac{1}{2} \begin{pmatrix} 1-i & 1+i \\ 1+i & 1-i \end{pmatrix} \begin{pmatrix} 1 & 0 \\ 0 & -i \end{pmatrix} .$$

However, if, in (1), either $\cos(\varphi)$ or $\sin(\varphi)$ is zero, decompositions (2) constitute an infinity of decompositions. This is confirmed by the identities

$$\begin{pmatrix} e^{ix} & 0 \\ 0 & e^{iy} \end{pmatrix} = e^{ix} \begin{pmatrix} 1 & 0 \\ 0 & e^{-ix+iz} \end{pmatrix} \begin{pmatrix} 1 & 0 \\ 0 & 1 \end{pmatrix} \begin{pmatrix} 1 & 0 \\ 0 & e^{iy-iz} \end{pmatrix} \quad \text{and}$$

$$\begin{pmatrix} 0 & e^{ix} \\ e^{iy} & 0 \end{pmatrix} = e^{iz} \begin{pmatrix} 1 & 0 \\ 0 & e^{iy-iz} \end{pmatrix} \begin{pmatrix} 0 & 1 \\ 1 & 0 \end{pmatrix} \begin{pmatrix} 1 & 0 \\ 0 & e^{ix-iz} \end{pmatrix} ,$$

where, in both cases, z may take any value.

If $n > 2$, then an infinity of decompositions of the matrix U arises whenever the matrix contains a row with n 1 zeroes and thus automatically also a column with n 1 zeroes (provided that row is not the first row and that column is not the first column). If no such row (and column) is present, there may be either a finite or an infinite number of decompositions. The former is illustrated by the example of the 3×3 discrete Fourier transform:

$$F_3 = \frac{1}{\sqrt{3}} \begin{pmatrix} 1 & 1 & 1 \\ 1 & \omega & \omega^2 \\ 1 & \omega^2 & \omega \end{pmatrix} ,$$

where ω is the primitive cubic root of unity, i.e. $-\frac{1}{2} + i\frac{\sqrt{3}}{2}$. Indeed, for this matrix six and only six decompositions are possible:

$$-i\omega \begin{pmatrix} 1 & 0 & 0 \\ 0 & 1 & 0 \\ 0 & 0 & \omega^2 \end{pmatrix} \frac{1}{2\sqrt{3}} \begin{pmatrix} \sqrt{3}-i & \sqrt{3}-i & 2i \\ \sqrt{3}-i & 2i & \sqrt{3}-i \\ 2i & \sqrt{3}-i & \sqrt{3}-i \end{pmatrix} \begin{pmatrix} 1 & 0 & 0 \\ 0 & 1 & 0 \\ 0 & 0 & \omega^2 \end{pmatrix} ,$$

$$-i\omega \begin{pmatrix} 1 & 0 & 0 \\ 0 & \omega^2 & 0 \\ 0 & 0 & 1 \end{pmatrix} \frac{1}{2\sqrt{3}} \begin{pmatrix} \sqrt{3}-i & 2i & \sqrt{3}-i \\ 2i & \sqrt{3}-i & \sqrt{3}-i \\ \sqrt{3}-i & \sqrt{3}-i & 2i \end{pmatrix} \begin{pmatrix} 1 & 0 & 0 \\ 0 & \omega^2 & 0 \\ 0 & 0 & 1 \end{pmatrix} ,$$

$$-i \begin{pmatrix} 1 & 0 & 0 \\ 0 & \omega & 0 \\ 0 & 0 & \omega \end{pmatrix} \frac{1}{2\sqrt{3}} \begin{pmatrix} 2i & \sqrt{3}-i & \sqrt{3}-i \\ \sqrt{3}-i & \sqrt{3}-i & 2i \\ \sqrt{3}-i & 2i & \sqrt{3}-i \end{pmatrix} \begin{pmatrix} 1 & 0 & 0 \\ 0 & \omega & 0 \\ 0 & 0 & \omega \end{pmatrix} ,$$

$$i\omega^2 \begin{pmatrix} 1 & 0 & 0 \\ 0 & \omega & 0 \\ 0 & 0 & 1 \end{pmatrix} \frac{1}{2\sqrt{3}} \begin{pmatrix} \sqrt{3}+i & \sqrt{3}+i & -2i \\ -2i & \sqrt{3}+i & \sqrt{3}+i \\ \sqrt{3}+i & -2i & \sqrt{3}+i \end{pmatrix} \begin{pmatrix} 1 & 0 & 0 \\ 0 & 1 & 0 \\ 0 & 0 & \omega \end{pmatrix} ,$$

$$i\omega^2 \begin{pmatrix} 1 & 0 & 0 \\ 0 & 1 & 0 \\ 0 & 0 & \omega \end{pmatrix} \frac{1}{2\sqrt{3}} \begin{pmatrix} \sqrt{3}+i & -2i & \sqrt{3}+i \\ \sqrt{3}+i & \sqrt{3}+i & -2i \\ -2i & \sqrt{3}+i & \sqrt{3}+i \end{pmatrix} \begin{pmatrix} 1 & 0 & 0 \\ 0 & \omega & 0 \\ 0 & 0 & 1 \end{pmatrix} ,$$

and

$$i \begin{pmatrix} 1 & 0 & 0 \\ 0 & \omega^2 & 0 \\ 0 & 0 & \omega^2 \end{pmatrix} \frac{1}{2\sqrt{3}} \begin{pmatrix} -2i & \sqrt{3}+i & \sqrt{3}+i \\ \sqrt{3}+i & -2i & \sqrt{3}+i \\ \sqrt{3}+i & \sqrt{3}+i & -2i \end{pmatrix} \begin{pmatrix} 1 & 0 & 0 \\ 0 & \omega^2 & 0 \\ 0 & 0 & \omega^2 \end{pmatrix} .$$

The latter is illustrated by the decomposition of the 4×4 Fourier matrix:

$$\begin{aligned} F_4 &= \frac{1}{2} \begin{pmatrix} 1 & 1 & 1 & 1 \\ 1 & i & -1 & -i \\ 1 & -1 & 1 & -1 \\ 1 & -i & -1 & i \end{pmatrix} \\ &= 1 \begin{pmatrix} 1 & 0 & 0 & 0 \\ 0 & a & 0 & 0 \\ 0 & 0 & 1 & 0 \\ 0 & 0 & 0 & -a \end{pmatrix} \frac{1}{2} \begin{pmatrix} 1 & -ia & 1 & ia \\ 1/a & 1 & -1/a & 1 \\ 1 & ia & 1 & -ia \\ -1/a & 1 & 1/a & 1 \end{pmatrix} \begin{pmatrix} 1 & 0 & 0 & 0 \\ 0 & i/a & 0 & 0 \\ 0 & 0 & 1 & 0 \\ 0 & 0 & 0 & -i/a \end{pmatrix} \\ &= \frac{1}{b} \begin{pmatrix} 1 & 0 & 0 & 0 \\ 0 & b & 0 & 0 \\ 0 & 0 & -1 & 0 \\ 0 & 0 & 0 & b \end{pmatrix} \frac{1}{2} \begin{pmatrix} b & 1 & -b & 1 \\ 1 & i/b & 1 & -i/b \\ -b & 1 & b & 1 \\ 1 & -i/b & 1 & i/b \end{pmatrix} \begin{pmatrix} 1 & 0 & 0 & 0 \\ 0 & b & 0 & 0 \\ 0 & 0 & -1 & 0 \\ 0 & 0 & 0 & b \end{pmatrix} , \end{aligned}$$

where both a and b are allowed to have any value on the unit circle of the complex plane.

The Fourier matrices F_n not only are important in quantum computing [1], but also in quantum optics. In the latter application, they constitute the canonical form of the $n \times n$ unitary matrices with all entries having the same modulus (i.e. having modulus $1/\sqrt{n}$). Constant-modulus unitary matrices are important for multiports (particle beam splitters) [24].

References

[1] Nielsen M and Chuang I 2000 *Quantum Computation and Quantum Information* (Cambridge: Cambridge University Press)
[2] De Vos A and De Baerdemacker S 2014 Matrix calculus for classical and quantum circuits *A.C.M. Journal on Emerging Technologies in Computing Systems* **11** 9
[3] Wille R and Drechsler R 2010 *Towards a Design Flow for Reversible Logic* (Dordrecht: Springer)
[4] Sasanian Z and Miller D 2011 Transforming MCT circuits to NCVW circuits *Proc. 3 rd Int. Workshop on Reversible Computation (Gent)* pp 163–174
[5] Selinger P 2015 Efficient Clifford+T approximations of single-qubit operators *Quantum Information & Computation* **15** 159
[6] Amy M, Maslov D and Mosca M 2013 Polynomial-time T-depth optimization of Clifford+T circuits via matroid partitioning *IEEE Trans. on Computer-Aided Design of Integrated Circuits and Systems* **33** 1486
[7] Deutsch D 1992 Quantum computation *Physics World* **5** 57
[8] Galindo A and Martín-Delgado M 2002 Information and computation: classical and quantum aspects *Review of Modern Physics* **74** 347
[9] De Vos A, De Beule J and Storme L 2009 Computing with the square root of `NOT` *Serdica Journal of Computing* **3** 359
[10] Vandenbrande S, Van Laer R and De Vos A 2012 The computational power of the square root of `NOT` *Proc. 10 th Int. Workshop on Boolean Problems (Freiberg)* pp 257–262
[11] De Vos A and De Baerdemacker S 2013 The `NEGATOR` as a basic building block for quantum circuits *Open Systems & Information Dynamics* **20** 1350004
[12] De Vos A and De Baerdemacker S 2014 The decomposition of U(n) into XU(n) and ZU(n) *Proc. 44 th Int. Symposium on Multiple-Valued Logic (Bremen)* pp 173–177
[13] Tadej W and Życzkowski K 2006 A concise guide to complex Hadamard matrices *Open Systems & Information Dynamics* **13** 133
[14] Diţă P 2003 Factorization of unitary matrices *Journal of Physics A: Mathematical and General* **36** 2781
[15] Reck M, Zeilinger A, Bernstein H and Bertani P 1994 Experimental realization of any discrete unitary operator *Physical Review Letters* **73** 58
[16] Rowe D, Sanders B and de Guise H 1999 Representations of the Weyl group and Wigner functions for SU(3) *Journal of Mathematical Physics* **40** 3604
[17] Patera J and Zassenhaus H 1988 The Pauli matrices in n dimensions and finest gradings of simple Lie algebras of type A_{n-1} *Journal of Mathematical Physics* **29** 667
[18] De Vos A and De Baerdemacker S 2014 Scaling a unitary matrix *Open Systems & Information Dynamics* **21** 1450013
[19] Idel M and Wolf M 2014 Sinkhorn normal form for unitary matrices (`arXiv:math-ph` 1408.5728)
[20] Sinkhorn R 1964 A relationship between arbitrary positive matrices and doubly stochastic matrices *The Annals of Mathematical Statistics* **35** 876
[21] Beth T and Rötteler M 2001 Quantum algorithms: applicable algebra and quantum physics In: Alber G, Beth T, Horodecki M, Horodecki P, Horodecki R, Rötteler M, Weinfurter H, Werner R and Zeilinger A 2001 *Quantum Information* (Berlin: Springer Verlag) pp 96–150
[22] De Vos A 2010 *Reversible Computing* (Weinheim: Wiley–VCH Verlag) p 49
[23] De Vos A and De Baerdemacker S 2014 The synthesis of a quantum circuit *Proc. 11 th Int. Workshop on Boolean Problems (Freiberg)* pp 129–136
[24] Mattle K, Michler M, Weinfurter H, Zeilinger A and Zukowski M 1995 Non-classical statistics at multiport beam splitters *Applied Physics B* **60** S111

Supersymmetric harmonic oscillator and nonlinear supercoherent states

Erik Díaz-Bautista and David J Fernández C
Physics Department, Cinvestav, A.P. 14-740, 07000 México D.F., México
E-mail: ediaz@fis.cinvestav.mx, david@fis.cinvestav.mx

Abstract. We find and analyze the (nonlinear) supercoherent states associated with the generalized nonlinear supersymmetric annihilation operator (SAO) of the supersymmetric harmonic oscillator. We discuss as well the uncertainty relation for a special case in order to compare our results with those obtained for the linear supercoherent states.

1. Introduction

For the standard harmonic oscillator with angular frequency ω, the commutation relations between the Hamiltonian and the annihilation and creation operators $\hat{H}, \hat{a}, \hat{a}^\dagger$ generate the well known Heisenberg-Weyl algebra:

$$[\hat{H}, \hat{a}] = -\hbar\omega\hat{a}, \quad [\hat{H}, \hat{a}^\dagger] = \hbar\omega\hat{a}^\dagger, \quad [\hat{a}, \hat{a}^\dagger] = \hat{1}. \tag{1}$$

However, if a generalized form for the annihilation and creation operators is employed, which respects the commutation relations with the Hamiltonian, it is possible to establish a connection with the so-called polynomial deformations of the Heisenberg algebra.

Following the above ideas, the analysis of the algebraic commutation relations between the Hamiltonian $\hat{H}_s$ of the SUSY harmonic oscillator [1, 2],

$$\hat{H}_s = \omega \begin{pmatrix} \hat{a}^\dagger\hat{a} & 0 \\ 0 & \hat{a}\hat{a}^\dagger \end{pmatrix}, \tag{2}$$

and the corresponding creation $\hat{A}^\dagger$ and annihilation operators $\hat{A}$ of the system requires the knowledge of the explicit form of these operators. However, as it was shown in [3, 4], this form is not unique.

Recently [5], it was introduced a general expression for $\hat{A}$ and its eigenvectors $|Z\rangle$ with complex eigenvalues z were found (called supercoherent states) which become expressed in terms of the standard harmonic oscillator coherent states [6, 7].

Although the expression found in [5] for $\hat{A}$ is very general, it is still not unique. In this work, we will consider some of its deformations $\hat{Q}$, which maintain however the structure given in [5]. We will analyze also a particular deformation of the SAO and we will find the explicit form of its eigenvectors $|\mathcal{X}\rangle$ with complex eigenvalue X. Due to the deformation assumed, the eigenstates of $\hat{Q}$ turn out to be expressed in terms of nonlinear coherent states, associated to deformed Lie algebras (called nonlinear algebras) [8, 9, 10, 11]. Therefore, it is natural to call the states $|\mathcal{X}\rangle$ nonlinear supercoherent states.

2. Standard coherent states (CS)

A standard coherent state $|\alpha\rangle$ can be defined as an eigenstate of the annihilation operator $\hat{a}$ with complex eigenvalue α, i.e.,

$$\hat{a}|\alpha\rangle = \alpha|\alpha\rangle, \quad \alpha \in \mathbb{C}, \quad \langle\alpha|\alpha\rangle = 1. \tag{3}$$

In the Fock basis, the normalized coherent states read

$$|\alpha\rangle = e^{-\frac{1}{2}|\alpha|^2} \sum_{n=0}^{\infty} \frac{\alpha^n}{\sqrt{n!}} |n\rangle. \tag{4}$$

There are other definitions that can be used to build this type of quantum states, which are equivalent to each other for the harmonic oscillator; nevertheless, in this paper we will use the previous definition.

3. Nonlinear coherent states (NLCS)

The nonlinear coherent states $|z\rangle_f$ can be defined as eigenstates of the deformed annihilation operator $\tilde{a} = f(\hat{N})\hat{a}$ [12], i.e.,

$$\tilde{a}|z\rangle_f = f(\hat{N})\hat{a}|z\rangle_f = z|z\rangle_f, \quad z \in \mathbb{C}, \tag{5}$$

$f(\hat{N})$ being a well behaved real function of the number operator $\hat{N} = \hat{a}^\dagger\hat{a}$. These states depend strongly on how many Fock states $|n\rangle$ are annihilated by $\tilde{a}$, which in turn depend on the explicit form of the function $f(\hat{N})$ since

$$\tilde{a}|n\rangle = \sqrt{n} f(n-1)|n-1\rangle, \quad n = 0, 1, 2, \ldots \tag{6}$$

For instance, for $f(\hat{N}) = \hat{N} + \hat{1}$ it turns out that $f(n-1) = n \neq 0$, $\forall\ n = 1, 2, \ldots$ which implies that $\tilde{a}|0\rangle = 0$. Thus:

$$|\alpha\rangle_{nl} = [{}_0F_2(1,1;r^2)]^{-1/2} \sum_{n=0}^{\infty} \frac{\alpha^n}{\Gamma(n+1)\sqrt{\Gamma(n+1)}} |n\rangle, \tag{7}$$

where $r = |\alpha|$ and ${}_pF_q$ is the generalized hypergeometric function

$${}_pF_q(a_1, \ldots, a_p, b_1, \ldots, b_q; x) = \frac{\Gamma(b_1)\ldots\Gamma(b_q)}{\Gamma(a_1)\ldots\Gamma(a_p)} \sum_{n=0}^{\infty} \frac{\Gamma(a_1+n)\ldots\Gamma(a_p+n)}{\Gamma(b_1+n)\ldots\Gamma(b_q+n)} \frac{x^n}{n!}. \tag{8}$$

On the other hand, for $f(\hat{N}) = \hat{N}$ we have that $f(0) = 0$. This assumption implies that $\tilde{a}|0\rangle = \tilde{a}|1\rangle = 0$. Hence, the corresponding NLCS become now

$$|\alpha\rangle_{NL} = \left[{}_0F_2(1,2;r^2)\right]^{-1/2} \sum_{n=0}^{\infty} \frac{\alpha^n}{\Gamma(n+1)\sqrt{\Gamma(n+2)}} |n+1\rangle. \tag{9}$$

This result indicates that the contribution of the ground state is removed from $|\alpha\rangle_{NL}$ since this eigenstate $|0\rangle$ is isolated from the remaining ones due to $\tilde{a}|0\rangle = \tilde{a}^\dagger|0\rangle = 0$.

Note that, if the form of the function $f(\hat{N})$ is chosen appropriately, it is possible to isolate more energy eigenstates.

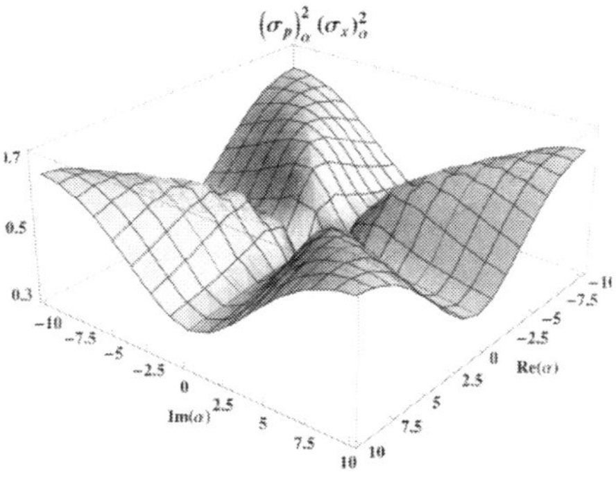

Figure 1. Heisenberg uncertainty relation $(\sigma_x)^2_\alpha(\sigma_p)^2_\alpha$ as function of α for the nonlinear coherent states of Eq. (7).

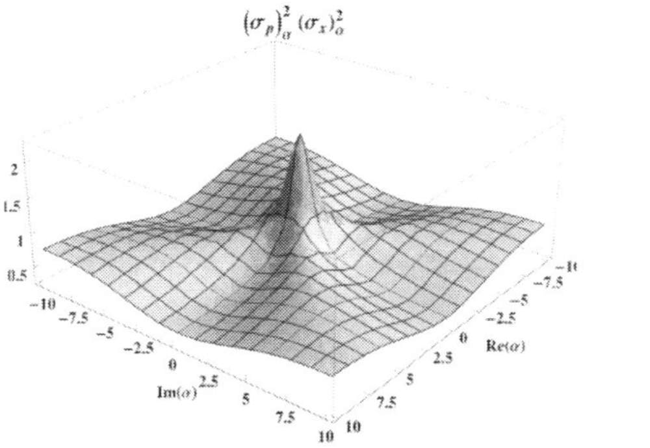

Figure 2. Heisenberg uncertainty relation $(\sigma_x)^2_\alpha(\sigma_p)^2_\alpha$ as function of α for the nonlinear coherent states of Eq. (9).

3.1. Heisenberg uncertainty relation

Knowing the explicit form of the standard coherent states $|\alpha\rangle$ and the nonlinear ones $|\alpha\rangle_{nl}$ and $|\alpha\rangle_{NL}$, the corresponding Heisenberg uncertainty relations can be straightforwardly calculated.

For the standard coherent states, the mean values of the operators $\hat{x}$, $\hat{p}$, $\hat{x}^2$, $\hat{p}^2$ become (with $\hbar = m = \omega = 1$):

$$\begin{aligned} \langle\hat{x}\rangle &= \sqrt{2}\mathrm{Re}(\alpha), & \langle\hat{x}^2\rangle &= \frac{1}{2} + 2[\mathrm{Re}(\alpha)]^2, \\ \langle\hat{p}\rangle &= \sqrt{2}\mathrm{Im}(\alpha), & \langle\hat{p}^2\rangle &= \frac{1}{2} + 2[\mathrm{Im}(\alpha)]^2. \end{aligned} \tag{10}$$

Hence, the Heisenberg uncertainty relation turns out to be given by $(\sigma_x)^2_\alpha(\sigma_p)^2_\alpha = 1/4$.

For the nonlinear coherent states $|\alpha\rangle_{nl}$, the mean values of the operators $\hat{x}$, $\hat{p}$, $\hat{x}^2$, $\hat{p}^2$ are:

$$\begin{aligned} \langle\hat{x}\rangle_{nl} &= \sqrt{2}\mathrm{Re}(\alpha)\left[\frac{{}_0F_2(1,2;r^2)}{{}_0F_2(1,1;r^2)}\right], & \langle\hat{x}^2\rangle_{nl} &= \frac{1}{2} + [\mathrm{Re}(\alpha)]^2\beta(r) + [\mathrm{Im}(\alpha)]^2\sigma(r), \\ \langle\hat{p}\rangle_{nl} &= \sqrt{2}\mathrm{Im}(\alpha)\left[\frac{{}_0F_2(1,2;r^2)}{{}_0F_2(1,1;r^2)}\right], & \langle\hat{p}^2\rangle_{nl} &= \frac{1}{2} + [\mathrm{Re}(\alpha)]^2\sigma(r) + [\mathrm{Im}(\alpha)]^2\beta(r), \end{aligned} \tag{11}$$

where

$$\beta(r) = \left[\frac{{}_0F_2(2,2;r^2)}{{}_0F_2(1,1;r^2)}\right] + \frac{1}{2}\left[\frac{{}_0F_2(1,3;r^2)}{{}_0F_2(1,1;r^2)}\right], \quad \sigma(r) = \left[\frac{{}_0F_2(2,2;r^2)}{{}_0F_2(1,1;r^2)}\right] - \frac{1}{2}\left[\frac{{}_0F_2(1,3;r^2)}{{}_0F_2(1,1;r^2)}\right]. \tag{12}$$

The explicit expression for the Heisenberg uncertainty relation is (see Figure 1):

$$(\sigma_x)^2_\alpha(\sigma_p)^2_\alpha = \left[\frac{1}{2} - [\mathrm{Re}(\alpha)]^2\tau(r) + [\mathrm{Im}(\alpha)]^2\sigma(r)\right]\left[\frac{1}{2} - [\mathrm{Im}(\alpha)]^2\tau(r) + [\mathrm{Re}(\alpha)]^2\sigma(r)\right], \tag{13}$$

where

$$\tau(r) = 2\left[\frac{{}_0F_2(1,2;r^2)}{{}_0F_2(1,1;r^2)}\right]^2 - \beta(r). \tag{14}$$

For the nonlinear coherent states $|\alpha\rangle_{NL}$ the mean values of the operators $\hat{x}$, $\hat{p}$, $\hat{x}^2$, $\hat{p}^2$ are:

$$\begin{aligned} \langle\hat{x}\rangle_{NL} &= \sqrt{2}\mathrm{Re}(\alpha)\left[\frac{{}_0F_2(2,2;r^2)}{{}_0F_2(1,2;r^2)}\right], & \langle\hat{x}^2\rangle_{NL} &= \frac{3}{2}+[\mathrm{Re}(\alpha)]^2\left[\frac{{}_0F_2(2,3;r^2)}{{}_0F_2(1,2;r^2)}\right], \\ \langle\hat{p}\rangle_{NL} &= \sqrt{2}\mathrm{Im}(\alpha)\left[\frac{{}_0F_2(2,2;r^2)}{{}_0F_2(1,2;r^2)}\right], & \langle\hat{p}^2\rangle_{NL} &= \frac{3}{2}+[\mathrm{Im}(\alpha)]^2\left[\frac{{}_0F_2(2,3;r^2)}{{}_0F_2(1,2;r^2)}\right]. \end{aligned} \tag{15}$$

The expression for the Heisenberg uncertainty relation is now (see Figure 2) [13]:

$$(\sigma_x)^2_\alpha(\sigma_p)^2_\alpha = \left[\frac{3}{2}-[\mathrm{Re}(\alpha)]^2\rho(r)\right]\left[\frac{3}{2}-[\mathrm{Im}(\alpha)]^2\rho(r)\right], \tag{16}$$

where

$$\rho(r) = 2\left[\frac{{}_0F_2(2,2;r^2)}{{}_0F_2(1,2;r^2)}\right]^2 - \left[\frac{{}_0F_2(2,3;r^2)}{{}_0F_2(1,2;r^2)}\right]. \tag{17}$$

4. Nonlinear supercoherent states

The generalized form for the SAO $\hat{A}$ [5] and its deformation are respectively:

$$\hat{A} = \begin{pmatrix} k_1\hat{a} & k_2 \\ k_3\hat{a}^2 & k_4\hat{a} \end{pmatrix}, \quad \hat{Q} = \begin{pmatrix} k_1 f(\hat{N})\hat{a} & k_2 \\ k_3[f(\hat{N})\hat{a}]^2 & k_4 f(\hat{N})\hat{a} \end{pmatrix}, \tag{18}$$

where $k_i \in \mathbb{C}$ and $f(\hat{N}) = \hat{N}+\hat{1}$ or $f(\hat{N}) = \hat{N}$. We will label $\hat{Q} = \hat{A}'$ when $f(\hat{N}) = \hat{N}+\hat{1}$, and $\hat{Q} = \hat{A}''$ when $f(\hat{N}) = \hat{N}$. Note even that $\hat{Q} = \hat{A}$ when $f(\hat{N}) = \hat{1}$.

Both linear and nonlinear supercoherent states are defined in the generic way as

$$\hat{Q}|\mathcal{X}\rangle = X|\mathcal{X}\rangle, \quad |\mathcal{X}\rangle = \begin{pmatrix} \sum_{n=0}^{\infty} a_n|n\rangle \\ \sum_{n=1}^{\infty} c_n|n-1\rangle \end{pmatrix}, \quad X \in \mathbb{C}. \tag{19}$$

From this equation, after several calculations and simplifications, an overall matrix relationship is found as

$$K\begin{pmatrix} \tilde{a}_{n+1} \\ \tilde{c}_{n+1} \end{pmatrix} = \begin{pmatrix} \tilde{a}_n \\ \tilde{c}_n \end{pmatrix}, \quad K = \begin{pmatrix} k_1 & k_2 \\ k_3 & k_4 \end{pmatrix}, \tag{20}$$

where the quantities $\tilde{a}_n$ and $\tilde{c}_n$ are defined as

$$\tilde{a}_n = \beta_n\sqrt{n!}X^{1-n}a_n, \quad \tilde{c}_n = \beta_{n-1}\sqrt{(n-1)!}X^{1-n}c_n, \tag{21}$$

with

$$X = \begin{cases} z, & \text{for } \hat{Q}=\hat{A} \\ Y, & \text{for } \hat{Q}=\hat{A}' \\ Z, & \text{for } \hat{Q}=\hat{A}'' \end{cases}, \quad \beta_n = \begin{cases} 1, & \text{for } \hat{Q}=\hat{A} \text{ and } n\geq 1 \\ n!, & \text{for } \hat{Q}=\hat{A}' \text{ and } n\geq 1 \\ (n-1)!, & \text{for } \hat{Q}=\hat{A}'' \text{ and } n\geq 2 \end{cases}. \tag{22}$$

As can be seen, both the linear and nonlinear supercoherent states depend on the eigenvalues $\psi_\pm$ of the matrix K. This induces a classification of the supercoherent states $|\mathcal{X}\rangle$ into three different families as follows: degenerate ($\psi_+ = \psi_- \equiv \psi \neq 0$); singular ($\psi_+\psi_- = 0$); generic (everything else).

Furthermore, by rewriting a_n and c_n in terms of $\psi_\pm$, the eigenstates of $\hat{A}$ turn out to be expressed in terms of the CS in (4) while those of $\hat{A}'$ and $\hat{A}''$ are determined by the ones of

(7) or (9), when choosing $f(\hat{N}) = \hat{N} + \hat{1}$ or $f(\hat{N}) = \hat{N}$ respectively. Then we will identify the respective supercoherent states as follows:

$$|\mathcal{X}\rangle = \begin{cases} |Z\rangle & \text{for } \hat{Q} = \hat{A} \\ |\mathcal{Y}\rangle & \text{for } \hat{Q} = \hat{A}' \\ |\mathcal{Z}\rangle & \text{for } \hat{Q} = \hat{A}'' \end{cases} . \tag{23}$$

4.1. Supercoherent states classification

4.1.1. Generic. The general solution for the matrix Eq. (20) is

$$|\mathcal{X}\rangle = \mathcal{B}_1|\mathcal{X}_A\rangle + \mathcal{B}_2|\mathcal{X}_C\rangle, \tag{24}$$

where

$$\mathcal{B}_1 = \begin{cases} a_0/k_1 & \text{for } \hat{Q} = \hat{A}, \hat{A}' \\ a_1/k_1 & \text{for } \hat{Q} = \hat{A}'' \end{cases}, \quad \mathcal{B}_2 = \begin{cases} c_1/(Xk_1) & \text{for } \hat{Q} = \hat{A}, \hat{A}' \\ c_2/(Xk_1) & \text{for } \hat{Q} = \hat{A}'' \end{cases}, \tag{25a}$$

$$|\mathcal{X}_A\rangle = \frac{1}{\psi_+ - \psi_-} G_A \begin{pmatrix} |\chi_+\rangle \\ |\chi_-\rangle \end{pmatrix}, \quad |\mathcal{X}_C\rangle = \frac{1}{\psi_+ - \psi_-} G_C \begin{pmatrix} |\chi_+\rangle \\ |\chi_-\rangle \end{pmatrix}, \tag{25b}$$

with

$$\chi_\pm \equiv X\psi_\pm^{-1}, \quad |\chi_\pm\rangle = \begin{cases} |\beta_\pm\rangle & \text{for } \hat{Q} = \hat{A} \\ |\varphi_\pm\rangle_{nl} & \text{for } \hat{Q} = \hat{A}' \\ |\phi_\pm\rangle_{NL} & \text{for } \hat{Q} = \hat{A}'' \end{cases}, \tag{26a}$$

and

$$G_A = \begin{pmatrix} \psi_+(\psi_+ - k_4) & -\psi_-(\psi_- - k_4) \\ k_3X & -k_3X \end{pmatrix}, \tag{26b}$$

$$G_C = \begin{pmatrix} k_2\psi_+\psi_- & -k_2\psi_+\psi_- \\ X[k_1\psi_+ - (k_1^2 + k_2k_3)] & -X[k_1\psi_- - (k_1^2 + k_2k_3)] \end{pmatrix}. \tag{26c}$$

Besides the set $\{|\mathcal{X}_A\rangle, |\mathcal{X}_C\rangle\}$, we can choose a new basis for the state space formed by the elements

$$|\mathcal{X}_\pm\rangle = \begin{pmatrix} k_2\psi_\pm|\chi_\pm\rangle \\ (\psi_\pm - k_1)X|\chi_\pm\rangle \end{pmatrix}, \tag{27}$$

whereby it is possible to pass from a parameter space $\{k_1, k_2, k_3, k_4\}$ to a new one, formed by $\{\psi_+, \psi_-, k_1, k_2\}$.

4.1.2. Degenerate. Explicitly the supercoherent states are

$$|\mathcal{X}\rangle = \mathcal{B}_1|\mathcal{X}_A^d\rangle + \mathcal{B}_2|\mathcal{X}_C^d\rangle, \tag{28}$$

where

$$|\mathcal{X}_A^d\rangle = G_A^{(d)} \begin{pmatrix} |\chi\rangle \\ |\chi'\rangle \end{pmatrix}, \quad |\mathcal{X}_C^d\rangle = G_C^{(d)} \begin{pmatrix} |\chi\rangle \\ |\chi'\rangle \end{pmatrix}, \tag{29}$$

with

$$|\chi'\rangle = \frac{d}{d\chi}|\chi\rangle, \tag{30a}$$

$$G_A^{(d)} = \begin{pmatrix} k_1 & -(\psi - k_4)\chi \\ 0 & -k_3\chi^2 \end{pmatrix}, \qquad G_C^{(d)} = \psi \begin{pmatrix} 0 & -k_2\chi \\ k_1\chi & -(\frac{k_4-k_1}{2})\chi^2 \end{pmatrix}. \tag{30b}$$

4.1.3. Singular. The corresponding supercoherent states are now

$$|\mathcal{X}_s\rangle = \begin{pmatrix} k_1|\chi\rangle \\ k_3\chi|\chi\rangle \end{pmatrix}, \quad \chi \equiv \frac{X}{(k_1 + k_4)}. \tag{31}$$

4.2. Superposition and uncertainties

Consider a superposition of states $|\mathcal{X}_\pm\rangle$ with parameters η and λ as follows:

$$|\mathcal{X}_m\rangle = \cos\eta|\mathcal{X}_+\rangle + e^{i\lambda}\sin\eta|\mathcal{X}_-\rangle = \begin{pmatrix} \gamma_{1+}|\chi_+\rangle + \gamma_{1-}|\chi_-\rangle \\ \gamma_{2+}|\chi_+\rangle + \gamma_{2-}|\chi_-\rangle \end{pmatrix}, \tag{32}$$

where $\gamma_{1\pm}$ and $\gamma_{2\pm}$ are given by

$$\begin{aligned} \gamma_{1+} &= k_2\psi_+\cos\eta, & \gamma_{1-} &= k_2\psi_- e^{i\lambda}\sin\eta, \\ \gamma_{2+} &= (\psi_+ - k_1)\cos\eta, & \gamma_{2-} &= (\psi_- - k_1)e^{i\lambda}\sin\eta. \end{aligned} \tag{33}$$

The mean value of an arbitrary observable $\hat{c}$ is then

$$\langle\hat{c}\rangle = \frac{\langle\mathcal{X}_m|\hat{c}\mathcal{X}_m\rangle}{\langle\mathcal{X}_m|\mathcal{X}_m\rangle}, \tag{34a}$$

$$\langle\mathcal{X}_m|\hat{c}\mathcal{X}_m\rangle = \Gamma^+\langle\chi_+|\hat{c}\chi_+\rangle + \Gamma^-\langle\chi_-|\hat{c}\chi_-\rangle + 2\mathrm{Re}(\Gamma^{+-}\langle\chi_+|\hat{c}\chi_-\rangle), \tag{34b}$$

with

$$\Gamma^+ = |\gamma_{1+}|^2 + |\gamma_{2+}X|^2, \quad \Gamma^- = |\gamma_{1-}|^2 + |\gamma_{2-}X|^2, \quad \Gamma^{+-} = \gamma_{1+}^*\gamma_{1-} + \gamma_{2+}^*\gamma_{2-}|X|^2. \tag{35}$$

This allows us to find the expressions for the uncertainties of the operators $\hat{x}$ and $\hat{p}$ and their squares.

4.2.1. Uncertainties for $\hat{Q} = \hat{A}$. The expressions for the uncertainties are [5]:

$$\sigma_x^2 = \frac{\Gamma^+\frac{1}{2}[(\beta_+ + \beta_+^*)^2 + 1]e^{|\beta_+|^2} + \Gamma^-\frac{1}{2}[(\beta_- + \beta_-^*)^2 + 1]e^{|\beta_-|^2} + 2\mathrm{Re}[\Gamma^{+-}\frac{1}{2}[(\beta_- + \beta_+^*)^2 + 1]e^{\beta_+^*\beta_-}]}{\Gamma^+e^{|\beta_+|^2} + \Gamma^-e^{|\beta_-|^2} + 2\mathrm{Re}[\Gamma^{+-}e^{\beta_+^*\beta_-}]}$$
$$- \left(\frac{\Gamma^+\frac{1}{\sqrt{2}}(\beta_+ + \beta_+^*)e^{|\beta_+|^2} + \Gamma^-\frac{1}{\sqrt{2}}(\beta_- + \beta_-^*)e^{|\beta_-|^2} + 2\mathrm{Re}[\Gamma^{+-}\frac{1}{\sqrt{2}}(\beta_- + \beta_+^*)e^{\beta_+^*\beta_-}]}{\Gamma^+e^{|\beta_+|^2} + \Gamma^-e^{|\beta_-|^2} + 2\mathrm{Re}[\Gamma^{+-}e^{\beta_+^*\beta_-}]}\right)^2 \tag{36}$$

$$\sigma_p^2 = \frac{\Gamma^+\frac{1}{2}[-(\beta_+ - \beta_+^*)^2 + 1]e^{|\beta_+|^2} + \Gamma^-\frac{1}{2}[-(\beta_- - \beta_-^*)^2 + 1]e^{|\beta_-|^2} + 2\mathrm{Re}[\Gamma^{+-}\frac{1}{2}[-(\beta_- - \beta_+^*)^2 + 1]e^{\beta_+^*\beta_-}]}{\Gamma^+e^{|\beta_+|^2} + \Gamma^-e^{|\beta_-|^2} + 2\mathrm{Re}[\Gamma^{+-}e^{\beta_+^*\beta_-}]}$$
$$- \left(\frac{\Gamma^+\frac{-i}{\sqrt{2}}(\beta_+ - \beta_+^*)e^{|\beta_+|^2} + \Gamma^-\frac{-i}{\sqrt{2}}(\beta_- + \beta_-^*)e^{|\beta_-|^2} + 2\mathrm{Re}[\Gamma^{+-}\frac{-i}{\sqrt{2}}(\beta_- + \beta_+^*)e^{\beta_+^*\beta_-}]}{\Gamma^+e^{|\beta_+|^2} + \Gamma^-e^{|\beta_-|^2} + 2\mathrm{Re}[\Gamma^{+-}e^{\beta_+^*\beta_-}]}\right)^2 \tag{37}$$

where the quantities in Eq. (35) are evaluated taking $X = z$.

4.2.2. Uncertainties for $\hat{Q} = \hat{A}'$. For this case, the expressions for the uncertainties are:

$$\begin{aligned}\sigma_x^{2(nl)} &= \Delta_1^{-1}\Bigg\{\Gamma^+\left[[\mathrm{Re}(\varphi_+)]^2\left[\frac{{}_0F_2(2,2;|\varphi_+|^2)}{{}_0F_2(1,1;|\varphi_+|^2)}+\frac{1}{2}\frac{{}_0F_2(1,3;|\varphi_+|^2)}{{}_0F_2(1,1;|\varphi_+|^2)}\right]+[\mathrm{Im}(\varphi_+)]^2\left[\frac{{}_0F_2(2,2;|\varphi_+|^2)}{{}_0F_2(1,1;|\varphi_+|^2)}-\frac{1}{2}\frac{{}_0F_2(1,3;|\varphi_+|^2)}{{}_0F_2(1,1;|\varphi_+|^2)}\right]+\frac{1}{2}\right]\\
&+\Gamma^-\left[[\mathrm{Re}(\varphi_-)]^2\left[\frac{{}_0F_2(2,2;|\varphi_-|^2)}{{}_0F_2(1,1;|\varphi_-|^2)}+\frac{1}{2}\frac{{}_0F_2(1,3;|\varphi_-|^2)}{{}_0F_2(1,1;|\varphi_-|^2)}\right]+[\mathrm{Im}(\varphi_-)]^2\left[\frac{{}_0F_2(2,2;|\varphi_-|^2)}{{}_0F_2(1,1;|\varphi_-|^2)}-\frac{1}{2}\frac{{}_0F_2(1,3;|\varphi_-|^2)}{{}_0F_2(1,1;|\varphi_-|^2)}\right]+\frac{1}{2}\right]\\
&+2\mathrm{Re}\Bigg(\Gamma^{+-}\frac{1}{2\sqrt{{}_0F_2(1,1;|\varphi_+|^2)}\sqrt{{}_0F_2(1,1;|\varphi_-|^2)}}\\
&\left[\frac{(\varphi_-^2+\varphi_+^{*2})}{2}[{}_0F_2(1,3;\varphi_+^*\varphi_-)]+2\varphi_+^*\varphi_-[{}_0F_2(2,2;\varphi_+^*\varphi_-)]+{}_0F_2(1,1;\varphi_+^*\varphi_-)\right]\Bigg)\Bigg\}\\
&-\Delta_1^{-2}\Bigg\{\sqrt{2}\Gamma^+\mathrm{Re}(\varphi_+)\left[\frac{{}_0F_2(1,2;|\varphi_+|^2)}{{}_0F_2(1,1;|\varphi_+|^2)}\right]+\sqrt{2}\Gamma^-\mathrm{Re}(\varphi_-)\left[\frac{{}_0F_2(1,2;|\varphi_-|^2)}{{}_0F_2(1,1;|\varphi_-|^2)}\right]\\
&+\sqrt{2}\mathrm{Re}\left(\Gamma^{+-}(\varphi_-+\varphi_+^*)\frac{[{}_0F_2(1,2;\varphi_+^*\varphi_-)]}{\sqrt{{}_0F_2(1,1;|\varphi_+|^2)}\sqrt{{}_0F_2(1,1;|\varphi_-|^2)}}\right)\Bigg\}^2 \qquad (38)\end{aligned}$$

$$\begin{aligned}\sigma_p^{2(nl)} &= \Delta_1^{-1}\Bigg\{\Gamma^+\left[[\mathrm{Re}(\varphi_+)]^2\left[\frac{{}_0F_2(2,2;|\varphi_+|^2)}{{}_0F_2(1,1;|\varphi_+|^2)}-\frac{1}{2}\frac{{}_0F_2(1,3;|\varphi_+|^2)}{{}_0F_2(1,1;|\varphi_+|^2)}\right]+[\mathrm{Im}(\varphi_+)]^2\left[\frac{{}_0F_2(2,2;|\varphi_+|^2)}{{}_0F_2(1,1;|\varphi_+|^2)}+\frac{1}{2}\frac{{}_0F_2(1,3;|\varphi_+|^2)}{{}_0F_2(1,1;|\varphi_+|^2)}\right]+\frac{1}{2}\right]\\
&+\Gamma^-\left[[\mathrm{Re}(\varphi_-)]^2\left[\frac{{}_0F_2(2,2;|\varphi_-|^2)}{{}_0F_2(1,1;|\varphi_-|^2)}-\frac{1}{2}\frac{{}_0F_2(1,3;|\varphi_-|^2)}{{}_0F_2(1,1;|\varphi_-|^2)}\right]+[\mathrm{Im}(\varphi_-)]^2\left[\frac{{}_0F_2(2,2;|\varphi_-|^2)}{{}_0F_2(1,1;|\varphi_-|^2)}+\frac{1}{2}\frac{{}_0F_2(1,3;|\varphi_-|^2)}{{}_0F_2(1,1;|\varphi_-|^2)}\right]+\frac{1}{2}\right]\\
&+2\mathrm{Re}\Bigg(\Gamma^{+-}\frac{1}{2\sqrt{{}_0F_2(1,1;|\varphi_+|^2)}\sqrt{{}_0F_2(1,1;|\varphi_-|^2)}}\\
&\left[-\frac{(\varphi_-^2+\varphi_+^{*2})}{2}[{}_0F_2(1,3;\varphi_+^*\varphi_-)]+2\varphi_+^*\varphi_-[{}_0F_2(2,2;\varphi_+^*\varphi_-)]+{}_0F_2(1,1;\varphi_+^*\varphi_-)\right]\Bigg)\Bigg\}\\
&-\Delta_1^{-2}\Bigg\{\sqrt{2}\Gamma^+\mathrm{Im}(\varphi_+)\left[\frac{{}_0F_2(1,2;|\varphi_+|^2)}{{}_0F_2(1,1;|\varphi_+|^2)}\right]+\sqrt{2}\Gamma^-\mathrm{Im}(\varphi_-)\left[\frac{{}_0F_2(1,2;|\varphi_-|^2)}{{}_0F_2(1,1;|\varphi_-|^2)}\right]\\
&+\sqrt{2}\mathrm{Re}\left(-i\Gamma^{+-}(\varphi_--\varphi_+^*)\frac{[{}_0F_2(1,2;\varphi_+^*\varphi_-)]}{\sqrt{{}_0F_2(1,1;|\varphi_+|^2)}\sqrt{{}_0F_2(1,1;|\varphi_-|^2)}}\right)\Bigg\}^2 \qquad (39)\end{aligned}$$

where $X = Y$ in Eq. (35) and

$$\Delta_1 = \Gamma^+ + \Gamma^- + 2\mathrm{Re}\left(\Gamma^{+-}\left[\frac{{}_0F_2(1,1;\varphi_+^*\varphi_-)}{\sqrt{{}_0F_2(1,1;|\varphi_+|^2)}\sqrt{{}_0F_2(1,1;|\varphi_-|^2)}}\right]\right). \qquad (40)$$

4.2.3. Uncertainties for $\hat{Q} = \hat{A}''$. Similarly, the expressions for the uncertainties are now:

$$\begin{aligned}\sigma_x^{2(NL)} &= \Delta_2^{-1}\Bigg\{\Gamma^+\left[[\mathrm{Re}(\phi_+)]^2\left[\frac{{}_0F_2(2,3;|\phi_+|^2)}{{}_0F_2(1,2;|\phi_+|^2)}\right]+\frac{3}{2}\right]+\Gamma^-\left[[\mathrm{Re}(\phi_-)]^2\left[\frac{{}_0F_2(2,3;|\phi_-|^2)}{{}_0F_2(1,2;|\phi_-|^2)}\right]+\frac{3}{2}\right]\\
&+2\mathrm{Re}\Bigg(\Gamma^{+-}\frac{1}{2\sqrt{{}_0F_2(1,2;|\phi_+|^2)}\sqrt{{}_0F_2(1,2;|\phi_-|^2)}}\\
&\left[\frac{(\phi_-+\phi_+^*)^2}{2}[{}_0F_2(2,3;\phi_+^*\phi_-)]+3[{}_0F_2(1,2;\phi_+^*\phi_-)]\right]\Bigg)\Bigg\}\\
&-\Delta_2^{-2}\Bigg\{\sqrt{2}\Gamma^+\mathrm{Re}(\phi_+)\left[\frac{{}_0F_2(2,2;|\phi_+|^2)}{{}_0F_2(1,2;|\phi_+|^2)}\right]+\sqrt{2}\Gamma^-\mathrm{Re}(\phi_-)\left[\frac{{}_0F_2(2,2;|\phi_-|^2)}{{}_0F_2(1,2;|\phi_-|^2)}\right]\\
&+\sqrt{2}\mathrm{Re}\left(\Gamma^{+-}(\phi_-+\phi_+^*)\frac{[{}_0F_2(2,2;\phi_+^*\phi_-)]}{\sqrt{{}_0F_2(1,2;|\phi_+|^2)}\sqrt{{}_0F_2(1,2;|\phi_-|^2)}}\right)\Bigg\}^2 \qquad (41)\end{aligned}$$

$$\sigma_p^{2(NL)} = \Delta_2^{-1}\Bigg\{\Gamma^+\left[[\mathrm{Im}(\phi_+)]^2\left[\frac{{}_0F_2(2,3;|\phi_+|^2)}{{}_0F_2(1,2;|\phi_+|^2)}\right]+\frac{3}{2}\right]+\Gamma^-\left[[\mathrm{Im}(\phi_-)]^2\left[\frac{{}_0F_2(2,3;|\phi_-|^2)}{{}_0F_2(1,2;|\phi_-|^2)}\right]+\frac{3}{2}\right]$$
$$+2\mathrm{Re}\Bigg(\Gamma^{+-}\frac{1}{2\sqrt{{}_0F_2(1,2;|\phi_+|^2)}\sqrt{{}_0F_2(1,2;|\phi_-|^2)}}$$
$$\left[-\frac{(\phi_- - \phi_+^*)^2}{2}[{}_0F_2(2,3;\phi_+^*\phi_-)]+3[{}_0F_2(1,2;\phi_+^*\phi_-)]\right]\Bigg)\Bigg\}$$
$$-\Delta_2^{-2}\Bigg\{\sqrt{2}\Gamma^+\mathrm{Im}(\phi_+)\left[\frac{{}_0F_2(2,2;|\phi_+|^2)}{{}_0F_2(1,2;|\phi_+|^2)}\right]+\sqrt{2}\Gamma^-\mathrm{Im}(\phi_-)\left[\frac{{}_0F_2(2,2;|\phi_-|^2)}{{}_0F_2(1,2;|\phi_-|^2)}\right]$$
$$+\sqrt{2}\mathrm{Re}\left(-i\Gamma^{+-}(\phi_- - \phi_+^*)\frac{[{}_0F_2(2,2;\phi_+^*\phi_-)]}{\sqrt{{}_0F_2(1,2;|\phi_+|^2)}\sqrt{{}_0F_2(1,2;|\phi_-|^2)}}\right)\Bigg\}^2 \tag{42}$$

with $X = Z$ in Eq. (35) and

$$\Delta_2 = \Gamma^+ + \Gamma^- + 2\mathrm{Re}\left(\Gamma^{+-}\left[\frac{{}_0F_2(1,2;\phi_+^*\phi_-)}{\sqrt{{}_0F_2(1,2;|\phi_+|^2)}\sqrt{{}_0F_2(1,2;|\phi_-|^2)}}\right]\right). \tag{43}$$

5. A particular case

To analyze the behavior of the supercoherent states (linear and nonlinear) the following particular values for the parameters k_i are taken: $k_1 = k_4 = 1$, $k_2 = \cos\theta$ and $k_3 = \sin\theta$. Thus, the deformed SAO in (18) takes the generic form:

$$\hat{Q} = \begin{pmatrix} f(\hat{N})\hat{a} & \cos\theta \\ \sin\theta[f(\hat{N})\hat{a}]^2 & f(\hat{N})\hat{a} \end{pmatrix}, \tag{44}$$

meanwhile the superposition (32) (with $\eta = \lambda = \pi/4$) becomes

$$|\mathcal{X}_\theta\rangle = \frac{1}{\sqrt{2}}(|\mathcal{X}_{+,\theta}\rangle + |\mathcal{X}_{-,\theta}\rangle e^{i\pi/4}). \tag{45}$$

On the other hand, the eigenvalues $\psi_\pm$ of the matrix K of Eq. (20) become now:

$$\psi_{\pm,\theta} = 1 \pm \sqrt{\frac{1}{2}\sin(2\theta)}, \tag{46}$$

so that for $0 < \theta < \pi/2$ both $\psi_\pm$ are real while for $\pi/2 < \theta < \pi$ they turn out to be complex. By substituting equations (45) and (46) in the expressions (36)-(42), the Heisenberg uncertainty relation for each form of the operator $\hat{Q}$ is found.

5.1. Heisenberg uncertainty relation for $\hat{Q} = \hat{A}$.

The Heisenberg uncertainty relation associated with the linear supercoherentes states of [5] is shown in Figures 3 and 4. For z real (left plot) the uncertainty reaches a maximum value equal to 0.83 at $|z| \sim 0.5$ for real eigenvalues $\psi_\pm$ ($0 < \theta < \pi/2$), while it shows a growing behavior as $|z|$ increases for complex eigenvalues $\psi_\pm$ ($\pi/2 < \theta < \pi$). For complex z (right plot), the uncertainty relation behaves similarly as for the real case.

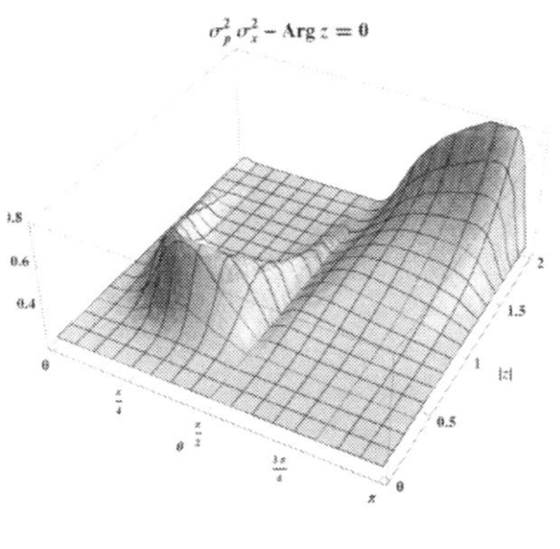

Figure 3. Heisenberg uncertainty relation $\sigma_x^2\sigma_p^2$ (Eqs. (36) and (37)) for the supercoherent states $|Z\rangle$ with z real.

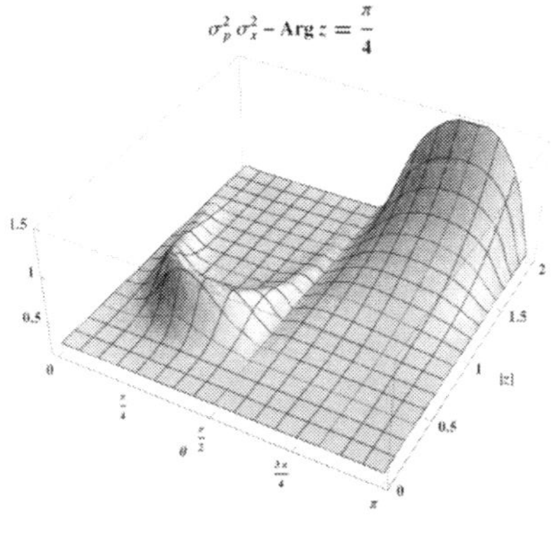

Figure 4. Heisenberg uncertainty relation $\sigma_x^2\sigma_p^2$ (Eqs. (36) and (37)) for the supercoherent states $|Z\rangle$ with z complex.

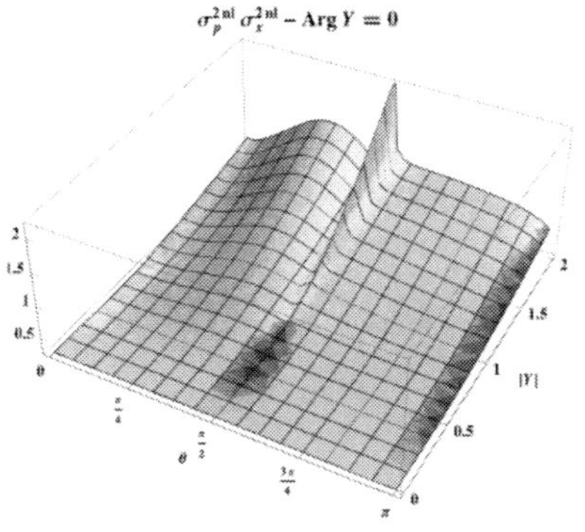

Figure 5. Heisenberg uncertainty relation $\sigma_x^{2(nl)}\sigma_p^{2(nl)}$ (Eqs. (38) and (39)) for the nonlinear supercoherent states $|\mathcal{Y}\rangle$ with Y real.

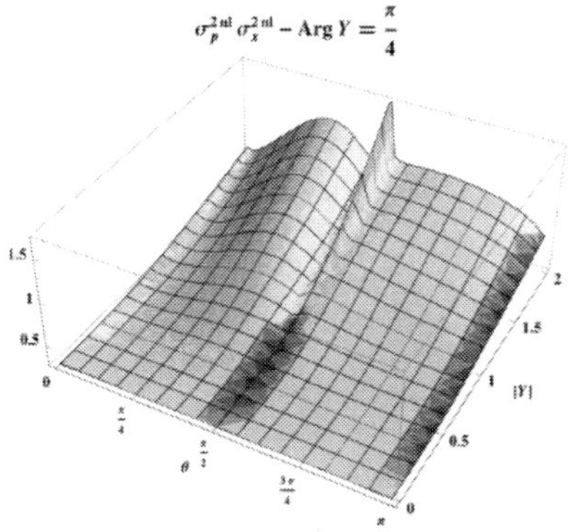

Figure 6. Heisenberg uncertainty relation $\sigma_x^{2(nl)}\sigma_p^{2(nl)}$ (Eqs. (38) and (39)) for the nonlinear supercoherent states $|\mathcal{Y}\rangle$ with Y complex.

5.2. Heisenberg uncertainty relation for $\hat{Q} = \hat{A}'$.

Figures 5 and 6 show the Heisenberg uncertainty relation $\sigma_x^{2(nl)}\sigma_p^{2(nl)}$ for the nonlinear supercoherent states $|\mathcal{Y}\rangle$ of the operator $\hat{A}'$. For Y real (left plot), for both real $(0 < \theta < \pi/2)$ as well as complex $(\pi/2 < \theta < \pi)$ eigenvalues $\psi_\pm$, the uncertainty starts from a minimum value and grows slowly. In the region of degenerate eigenvalues, for $\theta = \pi/2$, the uncertainty reaches a maximum value. For complex Y (right plot), the uncertainty relation behaves similarly as for the real case.

5.3. Heisenberg uncertainty relation for $\hat{Q} = \hat{A}''$.

Figures 7 and 8 show the uncertainty relations $\sigma_x^{2(NL)}\sigma_p^{2(NL)}$ for the nonlinear supercoherent states $|\mathcal{Z}\rangle$ of the operator $\hat{A}''$. For Z real (left plot), for both real $(0 < \theta < \pi/2)$ as well as complex $(\pi/2 < \theta < \pi)$ eigenvalues $\psi_\pm$, the uncertainty starts from a value equal to 2.25, then it decreases smoothly and grows again. In the region of degenerate eigenvalues, for $\theta = \pi/2$ eigenvalues $\psi_\pm$, the uncertainty reaches a minimum value close to 0.25. For complex Z (right

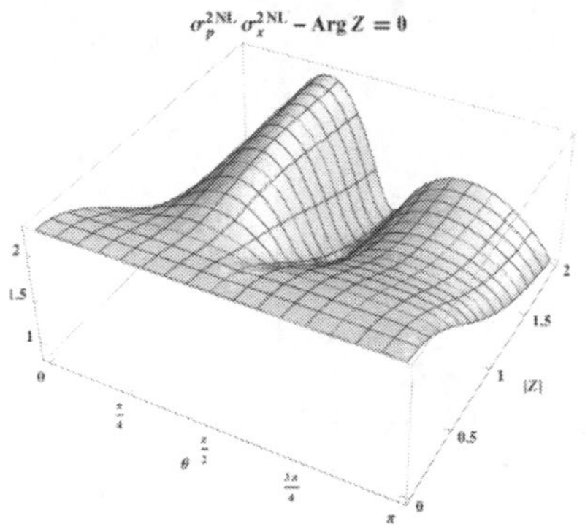

Figure 7. Heisenberg uncertainty relation $\sigma_x^{2(NL)}\sigma_p^{2(NL)}$ (Eqs. (41) and (42)) for nonlinear supercoherent states $|\mathcal{Z}\rangle$ with Z real.

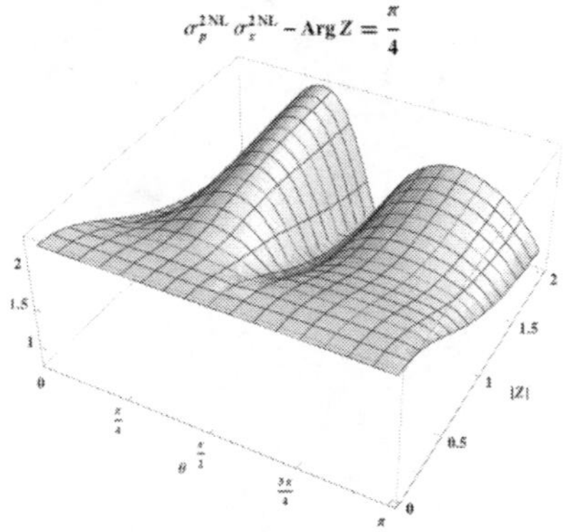

Figure 8. Heisenberg uncertainty relation $\sigma_x^{2(NL)}\sigma_p^{2(NL)}$ (Eqs. (41) and (42)) for nonlinear supercoherent states $|\mathcal{Z}\rangle$ with Z complex.

plot), the uncertainty relation behaves similarly as for the real case.

As shown in the Figures above, the behavior of the Heisenberg uncertainty relation for the supercoherent states $|Z\rangle$ becomes more involved than for the nonlinear ones, $|\mathcal{Y}\rangle$ and $|\mathcal{Z}\rangle$. Moreover, due to the nonlinear supercoherent states $|\mathcal{Z}\rangle$ are expressed in terms of the nonlinear coherent states $|\alpha\rangle_{NL}$, which isolate the eigenstate $|0\rangle$, the corresponding uncertainty relation seems to start from a different value than its counterparts associated to $\hat{A}$ and $\hat{A}'$.

6. Conclusion

The introduction of a deformation in the form of the supersymmetric annihilation operator $\hat{A}$ given in [5] allowed us to find interesting results related with the nonlinear supercoherent states. As it was noted, the form of the operator $\tilde{a} = f(\hat{N})\hat{a}$ enabled us to isolate the ground state contribution, which has interesting implications for the properties of the nonlinear supercoherent states and their associated Heisenberg uncertainty relation.

Let us note that the results shown in this paper obey the Heisenberg uncertainty principle ($\sigma_x^2\sigma_p^2 \geq 1/4$), which ensures that the states constructed here are consistent with quantum theory.

Finally, the freedom we have to choose the form of the function $f(\hat{N})$ will allow to consider a more general algebraic study, focused on the commutation relations of the operators $\{\hat{H}_s, \hat{A}, \hat{A}^\dagger\}$ of the supersymmetric harmonic oscillator and the possible relation with the polynomial deformations of the Heisenberg algebra, as it happens with the operators $\{\hat{H}, \hat{a}, \hat{a}^\dagger\}$ for the standard harmonic oscillator.

Acknowledgments

The authors acknowledge the support of Conacyt, Project **152574**. ED also acknowledges the Conacyt fellowship 290649.

References

[1] Wess J and Zumino B 1974 *Nucl. Phys. B* **70** 39
[2] Witten E 1981 *Nucl. Phys. B* **188** 513
[3] Aragone C and Zypman F 1986 *J. Phys. A: Math. Gen.* **19** 2267

Journal of Physics: Conference Series **597** (2015) 012031 doi:10.1088/1742-6596/597/1/012031

[4] Bérubé-Lauzière Y and Hussin V 1993 *J. Phys. A: Math. Gen.* **26** 6271
[5] Kornbluth M and Zypman F 2013 *J. Math. Phys.* **54** 012101
[6] Glauber R J 1963 *Phys. Rev. Lett.* **10** 84; 1963 *Phys. Rev.* **130** 2529; ibid 2766.
[7] Glauber R J 1965 in *Quantum Optics and Electronics*, Les Houches, eds. C. DeWitt, A. Blandin and C. Cohen-Tannoudji (New York: Gordon and Breach)
[8] Granovoskii Ya I, Lutzenko I M and Zhedanov A S 1992 *Ann. Phys.* **217** 1
[9] Vinet L and Letourneau P 1995 *Ann. Phys.* **243** 144
[10] Junker G and Roy P 1997 *Phys. Lett. A* **232** 155; ibid 1999 *A* **257** 113
[11] Karassiov V P 1994 *J. Phys. A: Math. Gen.* **27** 153
[12] Hong-Yi Fan and Hai-Ling Cheng 2002 *Commun. Theor. Phys.* **37** 655; 2000 *Jo. Opt. B: Quantum Semiclass. Opt.* **2**, issue **1**, 65
[13] Fernández D J, Hussin V and Nieto L M 1994 *J. Phys. A: Math. Gen.* **27** 3547

Invariant differential operators for non-compact Lie groups: the SO*(12) case

V K Dobrev

Institute for Nuclear Research and Nuclear Energy, Bulgarian Academy of Sciences, 72 Tsarigradsko Chaussee, 1784 Sofia, Bulgaria

E-mail: dobrev@inrne.bas.bg

Abstract. In the present paper we continue the project of systematic construction of invariant differential operators on the example of the non-compact algebra $so^*(12)$. We give the main multiplets of indecomposable elementary representations. Due to the recently established parabolic relations the multiplet classification results are valid also for the algebra $so(6,6)$ with suitably chosen maximal parabolic subalgebra.

1. Introduction

Invariant differential operators play very important role in the description of physical symmetries. In a recent paper [1] we started the systematic explicit construction of invariant differential operators. We gave an explicit description of the building blocks, namely, the parabolic subgroups and subalgebras from which the necessary representations are induced. Thus we have set the stage for study of different non-compact groups.

In the present paper we focus on the algebra $so^*(12)$. The algebras $so^*(4n)$ are form a subclass of the class of algebras, which we call 'conformal Lie algebras' in [2], which have very similar properties to the canonical conformal algebras of Minkowski space-time. In our subclass we have the algebras: $so^*(4)$, $so^*(8)$, $so^*(12)$, ... However the first two cases are reduced to well known conformal algebras due to the coincidences: $so^*(4) \cong so(3) \oplus so(2,1)$, $so^*(8) \cong so(6,2)$. Thus, the algebra $so^*(12)$ is the lowest nontrivial member of our subclass.

This paper is a sequel of [1] and [3] and due to the lack of space we refer to these papers for motivations and extensive list of literature on the subject.

2. Preliminaries

Let G be a semisimple non-compact Lie group, and K a maximal compact subgroup of G. Then we have an Iwasawa decomposition $G = KA_0N_0$, where A_0 is abelian simply connected vector subgroup of G, N_0 is a nilpotent simply connected subgroup of G preserved by the action of A_0. Further, let M_0 be the centralizer of A_0 in K. Then the subgroup $P_0 = M_0A_0N_0$ is a minimal parabolic subgroup of G. A parabolic subgroup $P = MAN$ is any subgroup of G which contains a minimal parabolic subgroup.

The importance of the parabolic subgroups comes from the fact that the representations induced from them generate all (admissible) irreducible representations of G [4–6].

Let ν be a (non-unitary) character of A, $\nu \in \mathcal{A}^*$, let μ fix an irreducible representation D^μ of M on a vector space V_μ.

We call the induced representation $\chi = \mathrm{Ind}_P^G(\mu \otimes \nu \otimes 1)$ an *elementary representation* of G [7]. Their spaces of functions are:

$$\mathcal{C}_\chi = \{\mathcal{F} \in C^\infty(G, V_\mu) \mid \mathcal{F}(gman) = e^{-\nu(H)} \cdot D^\mu(m^{-1})\mathcal{F}(g)\} \tag{1}$$

where $a = \exp(H) \in A$, $H \in \mathcal{A}$, $m \in M$, $n \in N$. The representation action is the *left* regular action:

$$(\mathcal{T}^\chi(g)\mathcal{F})(g') = \mathcal{F}(g^{-1}g'), \quad g, g' \in G. \tag{2}$$

For our purposes we need to restrict to *maximal* parabolic subgroups P, so that rank $A = 1$. Thus, for our representations the character ν is parameterized by a real number d, called the conformal weight or energy.

An important ingredient in our considerations are the *highest/lowest weight representations* of $\mathcal{G}$. These can be realized as (factor-modules of) Verma modules V^Λ over $\mathcal{G}^{\mathbb{C}}$, where $\Lambda \in (\mathcal{H}^{\mathbb{C}})^*$, $\mathcal{H}^{\mathbb{C}}$ is a Cartan subalgebra of $\mathcal{G}^{\mathbb{C}}$, weight $\Lambda = \Lambda(\chi)$ is determined uniquely from χ [8, 9].

Actually, since our ERs will be induced from finite-dimensional representations of $\mathcal{M}$ (or their limits) the Verma modules are always reducible. Thus, it is more convenient to use *generalized Verma modules* $\tilde{V}^\Lambda$ such that the role of the highest/lowest weight vector v_0 is taken by the space $V_\mu v_0$. For the generalized Verma modules (GVMs) the reducibility is controlled only by the value of the conformal weight d. Relatedly, for the intertwining differential operators only the reducibility w.r.t. non-compact roots is essential.

One main ingredient of our approach is as follows. We group the (reducible) ERs with the same Casimirs in sets called *multiplets* [9, 10]. The multiplet corresponding to fixed values of the Casimirs may be depicted as a connected graph, the vertices of which correspond to the reducible ERs and the lines between the vertices correspond to intertwining operators. The explicit parametrization of the multiplets and of their ERs is important for understanding of the situation.

In fact, the multiplets contain explicitly all the data necessary to construct the intertwining differential operators. Actually, the data for each intertwining differential operator consists of the pair (β, m), where β is a (non-compact) positive root of $\mathcal{G}^{\mathbb{C}}$, $m \in I\!N$, such that the BGG [11] Verma module reducibility condition (for highest weight modules) is fulfilled:

$$(\Lambda + \rho, \beta^\vee) = m, \quad \beta^\vee \equiv 2\beta/(\beta, \beta). \tag{3}$$

When (3) holds then the Verma module with shifted weight $V^{\Lambda - m\beta}$ (or $\tilde{V}^{\Lambda - m\beta}$ for GVM and β non-compact) is embedded in the Verma module V^Λ (or $\tilde{V}^\Lambda$). This embedding is realized by a singular vector v_s determined by a polynomial $\mathcal{P}_{m,\beta}(\mathcal{G}^-)$ in the universal enveloping algebra $(U(\mathcal{G}_-))\, v_0$, $\mathcal{G}^-$ is the subalgebra of $\mathcal{G}^{\mathbb{C}}$ generated by the negative root generators [12]. More explicitly, [9], $v^s_{m,\beta} = \mathcal{P}^m_\beta\, v_0$ (or $v^s_{m,\beta} = \mathcal{P}^m_\beta\, V_\mu v_0$ for GVMs). Then there exists [9] an intertwining differential operator

$$\mathcal{D}^m_\beta : \mathcal{C}_{\chi(\Lambda)} \longrightarrow \mathcal{C}_{\chi(\Lambda - m\beta)} \tag{4}$$

given explicitly by:

$$\mathcal{D}^m_\beta = \mathcal{P}^m_\beta(\widehat{\mathcal{G}^-}) \tag{5}$$

where $\widehat{\mathcal{G}^-}$ denotes the *right* action on the functions $\mathcal{F}$, cf. (1).

3. The non-compact Lie algebras $so^*(12)$

The group $G = SO^*(2n)$ consists of all matrices in $SO(2n,\mathbb{C})$ which commute with a real skew-symmetric matrix times the complex conjugation operator C :

$$SO^*(2n) \doteq \{\ g \in SO(2n,\mathbb{C}) \mid J_n C g = g J_n C\} \tag{6}$$

The Lie algebra $\mathcal{G} = so^*(2n)$ is given by:

$$\begin{aligned} so^*(2n) &\doteq \{\ X \in so(2n,\mathbb{C}) \mid J_n C X = X J_n C\} = \\ &= \{\ X = \begin{pmatrix} a & b \\ -\bar{b} & \bar{a} \end{pmatrix} \mid a,b \in gl(n,\mathbb{C}),\ \ {}^t a = -a,\ \ b^\dagger = b\ \} . \end{aligned} \tag{7}$$

$\dim_R \mathcal{G} = n(2n-1)$, $\mathrm{rank}\,\mathcal{G} = n$.

The Cartan involution is given by: $\Theta X = -X^\dagger$. Thus, $\mathcal{K} \cong u(n)$:

$$\mathcal{K} = \{\ X = \begin{pmatrix} a & b \\ -b & a \end{pmatrix} \mid a,b \in gl(n,\mathbb{C}),\ \ {}^t a = -a = -\bar{a},\ \ b^\dagger = b = \bar{b}\ \} . \tag{8}$$

Thus, $\mathcal{G} = so^*(2n)$ has discrete series representations and highest/lowest weight representations. The complimentary space $\mathcal{P}$ is given by:

$$\mathcal{P} = \{\ X = \begin{pmatrix} a & b \\ b & -a \end{pmatrix} \mid a,b \in gl(n,\mathbb{C})\ ,\ \ {}^t a = -a = \bar{a},\ \ b^\dagger = b = -\bar{b}\ \ \} . \tag{9}$$

$\dim_R \mathcal{P} = n(n-1)$. The split rank is $r \equiv [n/2]$. The subalgebras $\mathcal{N}_0^\pm$ which form the root spaces of the root system $(\mathcal{G},\mathcal{A}_0)$ are of real dimension $n(n-1) - [n/2]$.

The maximal parabolic subalgebras have $\mathcal{M}$-factors as follows [1]:

$$\mathcal{M}_j^{\max} = so^*(2n-4j) \oplus su^*(2j)\ , \quad j = 1,\ldots,r\ . \tag{10}$$

The $\mathcal{N}^\pm$ factors in the maximal parabolic subalgebras have dimensions: $\dim(\mathcal{N}_j^\pm)^{\max} = j(4n-6j-1)$.

For even $n = 2r$ we choose a *maximal* parabolic $\mathcal{P} = \mathcal{MAN}$ such that $\mathcal{A} \cong so(1,1)$, $\mathcal{M} = \mathcal{M}_r^{\max} = su^*(n)$. We note also that

$$\mathcal{K}^{\mathbb{C}} \cong u(1)^{\mathbb{C}} \oplus sl(n,\mathbb{C}) \cong \mathcal{A}^{\mathbb{C}} \oplus \mathcal{M}^{\mathbb{C}} \tag{11}$$

Thus, the factor $\mathcal{M}$ has the same finite-dimensional (nonunitary) representations as the finite-dimensional (unitary) representations of the semi-simple subalgebra of $\mathcal{K}$. The property (11) distinguishes the class we called 'conformal Lie algebras' [2], to which class the algebras $so^*(4r)$ belong.

Further we restrict to our case of study $\mathcal{G} = so^*(12)$.

We label the signature of the ERs of $\mathcal{G}$ as follows:

$$\chi = \{n_1, n_2, n_3, n_4, n_5\,;\, c\}\ , \qquad n_j \in \mathbb{Z}_+\ , \quad c = d - \tfrac{15}{2} \tag{12}$$

where the last entry of χ labels the characters of $\mathcal{A}$, and the first five entries are labels of the finite-dimensional (nonunitary) irreps of $\mathcal{M} = su^*(6)$ when all $n_j > 0$ or limits of the latter when some $n_j = 0$.

Below we shall use the following conjugation on the finite-dimensional entries of the signature:

$$(n_1, n_2, n_3, n_4, n_5)^* \doteq (n_5, n_4, n_3, n_2, n_1) . \tag{13}$$

The ERs in the multiplet are related also by intertwining integral operators introduced in [13]. These operators are defined for any ER, the general action being:

$$\begin{aligned} G_{KS} &: \mathcal{C}_\chi \longrightarrow \mathcal{C}_{\chi'} , \\ \chi &= \{ n_1, \ldots, n_5 ; c \} , \qquad \chi' = \{ (n_1, \ldots, n_5)^* ; -c \}. \end{aligned} \tag{14}$$

Further, we need the root system of the complexification $\mathcal{G}^{\mathbb{C}} = so(12, \mathbb{C})$. The positive roots are given standardly as:

$$\begin{aligned} \alpha_{ij} &= \epsilon_i - \epsilon_j , \quad 1 \leq i < j \leq 6 , \\ \beta_{ij} &= \epsilon_i + \epsilon_j , \quad 1 \leq i < j \leq 6 \end{aligned} \tag{15}$$

where ϵ_i are standard orthonormal basis: $(\epsilon_i, \epsilon_j) = \delta_{ij}$. The compact roots are α_{ij} - they form (by restriction) the root system of the semisimple part of $\mathcal{K}^{\mathbb{C}} \cong su^*(6)^{\mathbb{C}}$, while the roots β_{ij} are noncompact.

Further, we give the correspondence between the signatures χ and the highest weight Λ. The connection is through the Dynkin labels:

$$m_i \equiv (\Lambda + \rho, \alpha_i^\vee) = (\Lambda + \rho, \alpha_i) , \quad i = 1, \ldots, 5, \tag{16}$$

where $\Lambda = \Lambda(\chi)$, ρ is half the sum of the positive roots of $\mathcal{G}^{\mathbb{C}}$. The explicit connection is:

$$n_i = m_i , \quad c = - \tfrac{1}{2}(m_1 + 2m_2 + 3m_3 + 4m_4 + 2m_5 + 3m_6) \tag{17}$$

Finally, we remind that according to [3] the above considerations are applicable also for the algebra $so(6,6)$ with parabolic $\mathcal{M}$-factor $sl(6, I\!R)$.

4. Main multiplets of SO*(12)

The main multiplets are in 1-to-1 correspondence with the finite-dimensional irreps of $so^*(12)$, i.e., they are labelled by the six positive Dynkin labels $m_i \in I\!N$.

The number of ERs/GVMs in the corresponding multiplets is (cf. [3]):

$$|W(\mathcal{G}^{\mathbb{C}}, \mathcal{H}^{\mathbb{C}})| \,/\, |W(\mathcal{K}^{\mathbb{C}}, \mathcal{H}^{\mathbb{C}})| = |W(so(12, \mathbb{C}))| \,/\, |W(sl(6, \mathbb{C}))| = 32 \tag{18}$$

where $\mathcal{H}$ is a Cartan subalgebra of both $\mathcal{G}$ and $\mathcal{K}$.

The signatures of the 32 ERs/GVMs in a main multiplet are given in the following pair-wise manner:

$$\begin{aligned}
\chi_0^\pm &= \{(m_1, m_2, m_3 m_4, m_5)^\pm;\ \pm\tfrac{1}{2}(m_1 + 2m_2 + 3m_3 + 4m_4 + 2m_5 + 3m_6)\} \\
\chi_a^\pm &= \{(m_1, m_2, m_3, m_{4,6}, m_5)^\pm;\ \pm\tfrac{1}{2}(m_1 + 2m_2 + 3m_3 + 4m_4 + 2m_5 + m_6)\} \\
\chi_b^\pm &= \{(m_1, m_2, m_{34}, m_6, m_{45})^\pm;\ \pm\tfrac{1}{2}(m_1 + 2m_2 + 3m_3 + 2m_4 + 2m_5 + m_6)\} \\
\chi_c^\pm &= \{(m_1, m_2, m_{35}, m_6, m_4)^\pm;\ \pm\tfrac{1}{2}(m_1 + 2m_2 + 3m_3 + 2m_4 + m_6)\} \\
\chi_{c'}^\pm &= \{(m_1, m_{23}, m_4, m_6, m_{35})^\pm;\ \pm\tfrac{1}{2}(m_1 + 2m_2 + m_3 + 2m_4 + 2m_5 + m_6)\} \\
\chi_d^\pm &= \{(m_1, m_{23}, m_{45}, m_6, m_{34})^\pm;\ \pm\tfrac{1}{2}(m_1 + 2m_2 + m_3 + 2m_4 + m_6)\} \\
\chi_{d'}^\pm &= \{(m_{12}, m_3, m_4, m_6, m_{25})^\pm;\ \pm\tfrac{1}{2}(m_1 + m_3 + 2m_4 + 2m_5 + m_6)\} \\
\chi_e^\pm &= \{(m_1, m_{24}, m_5, m_{4,6}, m_3)^\pm;\ \pm\tfrac{1}{2}(m_1 + 2m_2 + m_3 + m_6)\} \\
\chi_{e'}^\pm &= \{(m_{12}, m_3, m_{45}, m_6, m_{24})^\pm;\ \pm\tfrac{1}{2}(m_1 + m_3 + 2m_4 + m_6)\} \\
\chi_{e''}^\pm &= \{(m_2, m_3, m_4, m_6, m_{15})^\pm;\ \pm\tfrac{1}{2}(-m_1 + m_3 + 2m_4 + 2m_5 + m_6)\} \\
\chi_f^\pm &= \{(m_{12}, m_{34}, m_5, m_{4,6}, m_{23})^\pm;\ \pm\tfrac{1}{2}(m_1 + m_3 + m_6)\} \\
\chi_{f'}^\pm &= \{(m_2, m_3, m_{45}, m_6, m_{14})^\pm;\ \pm\tfrac{1}{2}(-m_1 + m_3 + 2m_4 + m_6)\} \\
\chi_{f''}^\pm &= \{(m_1, m_{24,6}, m_5 m_4, m_3)^\pm;\ \pm\tfrac{1}{2}(m_1 + 2m_2 + m_3 - m_6)\} \\
\chi_g^\pm &= \{(m_{13}, m_4, m_5, m_{34,6}, m_2)^\pm;\ \pm\tfrac{1}{2}(m_1 - m_3 + m_6)\} \\
\chi_{g'}^\pm &= \{(m_{12}, m_{34,6}, m_5, m_4, m_{23})^\pm;\ \pm\tfrac{1}{2}(m_1 + m_3 - m_6)\} \\
\chi_{g''}^\pm &= \{(m_2, m_{34}, m_5, m_{4,6}, m_{13})^\pm;\ \pm\tfrac{1}{2}(-m_1 + m_3 + m_6)\}
\end{aligned} \tag{19}$$

where $(k_1, k_2, k_3, k_4, k_5)^- = (k_1, k_2, k_3, k_4, k_5)$, $(k_1, k_2, k_3, k_4, k_5)^+ = (k_1, k_2, k_3, k_4, k_5)^*$. They are given explicitly in Fig. 1. The pairs $\Lambda^\pm$ are symmetric w.r.t. to the bullet in the middle of the figure - this represents the Weyl symmetry realized by the Knapp-Stein operators (14): $G_{KS} : \mathcal{C}_{\chi^\mp} \longleftrightarrow \mathcal{C}_{\chi^\pm}$.

Matters are arranged so that in every multiplet only the ER with signature χ_0^- contains a finite-dimensional nonunitary subrepresentation in a finite-dimensional subspace $\mathcal{E}$. The latter corresponds to the finite-dimensional irrep of $so*(12)$ with signature $\{m_1, \ldots, m_6\}$. The subspace $\mathcal{E}$ is annihilated by the operator G^+, and is the image of the operator G^-. The subspace $\mathcal{E}$ is annihilated also by the intertwining differential operator acting from χ_0^- to χ_a^-. When all $m_i = 1$ then $\dim \mathcal{E} = 1$, and in that case $\mathcal{E}$ is also the trivial one-dimensional UIR of the whole algebra $\mathcal{G}$. Furthermore in that case the conformal weight is zero: $d = \frac{15}{2} + c = \frac{15}{2} - \frac{1}{2}(m_1 + 2m_2 + 3m_3 + 4m_4 + 2m_5 + 3m_6)_{|m_i=1} = 0$.

In the conjugate ER χ_0^+ there is a unitary discrete series subrepresentation in an infinite-dimensional subspace $\mathcal{D}$. It is annihilated by the operator G^-, and is the image of the operator G^+.

All the above is valid also for the algebra $so(6,6)$, cf. [3]. However, the latter algebra does not have highest/lowest weight representations while the algebra $so^*(12)$ has highest/lowest weight series representations.

Thus, for $so^*(12)$ the ER with signature χ_0^+ contains both a holomorphic discrete series representation and a conjugate anti-holomorphic discrete series representation. The direct sum of the holomorphic and the antiholomorphic representations spaces form the invariant subspace $\mathcal{D}$ mentioned above. Note that the corresponding lowest weight GVM is infinitesimally equivalent only to the holomorphic discrete series, while the conjugate highest weight GVM is infinitesimally equivalent to the anti-holomorphic discrete series.

In Fig. 1 we use the notation: $\Lambda^\pm = \Lambda(\chi^\pm)$. Each intertwining differential operator is represented by an arrow accompanied by a symbol i_{jk} encoding the root α_{jk} and the number

$m_{\alpha_{jk}}$ which is involved in the BGG criterion. This notation is used to save space, but it can be used due to the fact that only intertwining differential operators which are non-composite are displayed, and that the data β, m_β, which is involved in the embedding $V^\Lambda \longleftrightarrow V^{\Lambda - m_\beta, \beta}$ turns out to involve only the m_i corresponding to simple roots, i.e., for each β, m_β there exists $i = i(\beta, m_\beta, \Lambda) \in \{1, \ldots, 5\}$, such that $m_\beta = m_i$. Hence in Fig. 1. the data α_{jk}, $m_{\alpha_{jk}}$ is represented by i_{jk} on the arrows.

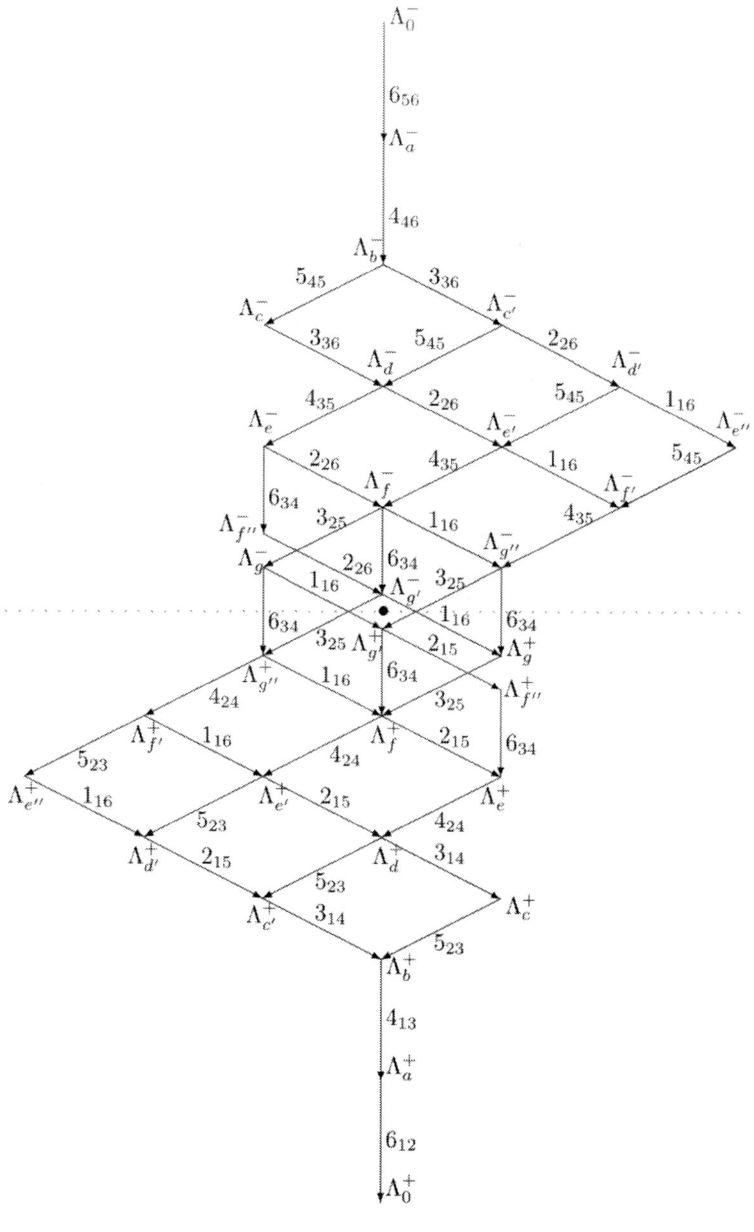

Fig. 1. $SO^*(12)$ main multiplets

Acknowledgments
The author would like to thank the Organizers for the kind hospitality and invitation to present a talk at the XXX International Colloquium on Group Theoretical Methods in Physics (Ghent, July 2014).

References
[1] Dobrev V K 2008 *Rev. Math. Phys.* **20** (hep-th/0702152) 407–449
[2] Dobrev V K 2009 *J. Phys. A: Math. Theor.* **42** 285203 (arXiv:0812.2690 [hep-th])
[3] Dobrev V K 2013 *J. High Energy Phys.* 02 015 (arXiv:1208.0409)
[4] Langlands R P 1988 On the classification of irreducible representations of real algebraic groups *Math. Surveys and Monographs* Vol. 31 (New York: American Mathematical Society)
[5] Zhelobenko D P 1974 *Harmonic Analysis on Semisimple Complex Lie Groups* (Moscow: Nauka)
[6] Knapp A W and Zuckerman G J 1977 Classification theorems for representations of semisimple groups, in: *Lecture Notes in Math.* Vol. 587 (Berlin: Springer) 138–159; *Ann. Math.* **116** (1982) 389–501
[7] Dobrev V K, Mack G, Petkova V B, Petrova S G and Todorov I T 1976 *Rep. Math. Phys.* **9** 219–246; Harmonic Analysis on the n-Dimensional Lorentz Group and Its Applications to Conformal Quantum Field Theory *Lecture Notes in Physics* Vol. 63 (Berlin: Springer)
[8] Harish-Chandra 1966 *Ann. Math.* **116** 1–111
[9] Dobrev V K 1988 *Rep. Math. Phys.* **25** 159–181
[10] Dobrev V K 1985 *Lett. Math. Phys.* **9** 205–211
[11] Bernstein I N, Gel'fand I M and Gel'fand S I 1971 *Funkts. Anal. Prilozh.* **5** (1) 1–9; English translation: *Funct. Anal. Appl.* **5** 1–8
[12] Dixmier J 1977 *Enveloping Algebras* (New York: American Mathematical Society)
[13] Knapp A W and Stein E M 1971 *Ann. Math.* **93** 489–578; 1980 *Inv. Math.* **60** 9–84

Quantum confinement in hydrogen bond of DNA and RNA

C S dos Santos[1], **E Drigo Filho**[1] **and R M Ricotta**[2]
[1]Instituto de Biociências, Letras e Ciências Exatas, IBILCE-UNESP
Rua Cristovão Colombo 2265, 15054-000 São José do Rio Preto ,SP, Brazil
[2]Faculdade de Tecnologia de São Paulo, FATEC/SP-CEETPS-UNESP
Praça Fernando Prestes 30, 01124-060 São Paulo, SP, Brazil
E-mail: carlos.sisantos@catolica.edu.br, drigo@ibilce.unesp.br, regina@fatecsp.br

Abstract. The hydrogen bond is a fundamental ingredient to stabilize the DNA and RNA macromolecules. The main contribution of this work is to describe quantitatively this interaction as a consequence of the quantum confinement of the hydrogen. The results for the free and confined system are compared with experimental data. The formalism to compute the energy gap of the vibration motion used to identify the spectrum lines is the Variational Method allied to Supersymmetric Quantum Mechanics.

1. Introduction

One of the most important interactions to stabilize the DNA and RNA macromolecules is the hydrogen bond between pairs of nitrogenous bases. It promotes a link among the complementary pairs of bases, maintaining the double helix structure. In the DNA the pairs of bases are guanine-cytosine and adenine-thymine; in the RNA the thymine is substituted by uracil, [1]. In a recent work, the hydrogen bonds present in proteins and in smaller molecules were studied by considering the hydrogen quantum confinement in the NH and OH molecular groups, [2]. In the present work, we extended the results for the nucleic acids, which present the NH molecular group.

The hydrogen bond is present in various systems, [3], and it is classified as a donor-acceptor interaction. From the experimental point of view, it is possible to identify if a particular molecular group is forming an hydrogen bond due to the change in the vibrational spectrum. This change can be observed by different experimental techniques, as for instance the infrared spectrum analysis.

The main point of this work is to show that the hydrogen bond can be treated as the quantum confinement of the hydrogen, which would change the transitions related to the NH group present in nucleic acids. As usual, the Morse potential, [4], is used to simulate the vibration of the NH molecular group, [5]-[11]. The parameters used are the ones found in literature for these groups without the hydrogen interaction, i.e., without confinement. The geometrical restrictions due to the hydrogen bonds are obtained from standard experimental results for this kind of systems, [3]. The calculation of the energy levels is made through the variational method with test functions obtained from Supersymmetric Quantum Mechanics formalism, SQM. This procedure has alrealdy been used in different quantum problems, with and without confinement, [12]-[15],

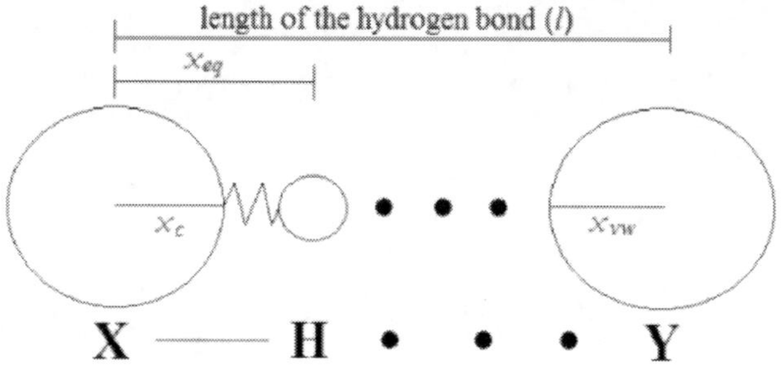

Figure 1. Schematic one-dimensional model, where x_c is the covalent radius of the atom donor, x_{vw} is the van der Waals radius of the atom and acceptor and x_{eq} is the equilibrium distance of covalent bond. X is the donor atom and Y represents the atom acceptor.

including specific studies involving the Morse potential, [16], which corroborates its applicability in the present case.

In section 2 the adopted model for the hydrogen bonds is presented. We then present a review of the SQM formalism in section 3. In Section 4 the specific mathematical formalism concerning the Morse potential is developed, including the identification of the wave functions to be used in the Variational Method. The numerical results, with the appropriate parameters of the NH group are presented as well as the comparison with the experimental data are in section 5. In section 6 the conclusions are discussed.

2. Model

Figure 1 shows a schematic model of a hydrogen bonding between the donor atom X and the acceptor atom Y, here denoted by X-H$\cdots$Y. The hydrogen bond is represented by H$\cdots$Y and covalent bond is represented by X-H. The X and Y atoms are considered as hard spheres and the hydrogen is treated as a material point, since the electron cloud is shared with the donor and the acceptor atoms. The origin of the coordinate system is located at the center of the atom X. Any distance used is computed from this point and it is considered center-to-center. The covalent radius is adopted by the condition of impenetrability of the donor atom X; the employment of the van der Waals radius is justified because it is a measure of the volume excluded by the acceptor atom Y, i.e., the hydrogen cannot penetrate this region.

Thus, the hydrogen nuclei may oscillate within a region determined between a minimum distance (x_{min}), given by the covalent radius (x_c) of atom X, and a maximum distance (x_{max}), given by the length of the hydrogen bond ($l_{x\cdots y}$) subtracted by the van der Waals radius (x_{vw}) of the atom Y, i.e.:

$$x_{max} = l_{X\cdots Y} - x_{vw} \quad (1)$$

The vibrational energy of this quantum system is suitably described by the confined one-dimensional Morse potential, widely known to well describe oscillations of diatomic molecules, [4], [12]-[14], [16]. The energy absorbed/emitted is obtained from the difference between the energies of the ground state and the first excited state. Since the confined Morse potential is non-exactly solvable, the energies of those states evaluated by using the variational method associated to the formalism of SQM.

3. Supersymmetric Quantum Mechanics, SQM

SQM is a well known algebraic method commonly used to exploit different aspects of non-relativistic quantum mechanical systems. It is particularly efficient to solve exactly all the quantum potential problems by means of the Hamiltonian factorization, thus providing the entire spectrum of wave functions and respective energies and a hierarchy of Hamiltonians, all related by the supersymmetric algebra, [17].

Using the superalgebra, a given Hamiltonian H_1, which has the energy $E_0^{(1)}$ and wave function $\Psi_0^{(1)}(x)$ for the lowest state, can be factorized in terms of bosonic operators $A_1^{\pm}$, such that the Schrödinger equation can be rewritten (in $\hbar = c = 1$ units) as

$$H_1 = -\frac{d^2}{dx^2} + V_1(x) = A_1^+ A_1^- + E_0^{(1)} \tag{2}$$

where

$$A_1^- \Psi_0^{(1)}(x) = 0. \tag{3}$$

The following realization of the bosonic operators in terms of the superpotential $w_1(x)$

$$A_1^{\pm} = \left(\mp\frac{d}{dx} + w_1(x)\right) \tag{4}$$

results in the Riccati equation

$$w_1^2 - \frac{d}{dx}w_1 = V_1(x) - E_0^{(1)} \tag{5}$$

and defines the ground state wave function in terms of the superpotential

$$\Psi_0^{(1)}(x) = Nexp(-\int_0^x w_1(\bar{x})d\bar{x}). \tag{6}$$

It can be shown that a n-members hierarchy of Hamiltonians can be generated, all related by the superalgebra, [17],

$$H_n = A_n^+ A_n^- + E_0^{(n)} \quad , \tag{7}$$

where

$$A_n^{\pm} = \left(\mp\frac{d}{dx} + w_n(x)\right) \tag{8}$$

with simple relations connecting the eigenvalues and eigenfunctions of the n-members,

$$\Psi_n^{(1)} = A_1^+ A_2^+ ... \psi_0^{(n+1)} \quad , \quad E_n^{(1)} = E_0^{(n+1)}. \tag{9}$$

In equation (9), the upper index between parentheses refers to the hierarchy Hamiltonian member and the lower index to the level within the hierarchy.

At this point, it should be remarked that if the potential is non-exactly solvable, such as in a confined system, an approximation technique is needed and the variational method has already appeared as fully appropriate, [12].

4. Mathematical Formalism

The one-dimensional Morse potential, [4], defined in terms of the parameters D_e, the dissociation energy of molecule; β, the parameter related to the width of potential well and x_{eq}, the inter-nuclear equilibrium distance, is given by

$$V_M = D_e(e^{-2\beta(x-x_{eq})} - 2e^{-\beta(x-x_{eq})}). \tag{10}$$

By treating the molecule as an oscillator of frequency ν, we can Taylor expand the Morse potential around the equilibrium position to find, in second order, the β parameter given by

$$\beta = (\frac{2\pi^2\mu}{D_e})^{\frac{1}{2}}\nu \tag{11}$$

and fix it through experimental data. With the change of variables, $y = \beta x$, the dimensionless Schrödinger equation for a molecule subject to this potential is given by

$$-\frac{d^2\Psi(y)}{dy^2} + \Lambda^2(e^{-2(y-y_{eq})} - 2e^{-(y-y_{eq})})\Psi(y) = \epsilon\Psi(y) \tag{12}$$

with

$$\Lambda^2 = \frac{2\mu D_e}{\beta^2\hbar^2} \tag{13}$$

and

$$\epsilon = \frac{2\mu E}{\beta^2\hbar^2} \tag{14}$$

where μ is the reduced mass of the system. The Hamiltonian operator in equation (12) is given by

$$\hat{H}_M = -\frac{d^2}{dy^2} + \Lambda^2(e^{-2(y-y_{eq})} - 2e^{-(y-y_{eq})}) \tag{15}$$

Since the Schrödinger equation is exactly solvable, we identify $\hat{H}_M$ with H_1 of equation (2) and proceed the process of factorization, calculating the whole hierarchy. The result is, [12]

$$V_{n+1}(y) = \Lambda^2(e^{-2(y-y_{eq})} - 2e^{-(y-y_{eq})}) + 2n\Lambda e^{-(y-y_{eq})}$$

$$w_{n+1}(y) = -\Lambda e^{-(y-y_{eq})} + (\lambda - \frac{2n+1}{2})$$

$$\epsilon_0^{(n+1)} = -(\Lambda - \frac{2n+1}{2})^2. \tag{16}$$

In addition we can calculate exactly the wave functions. The ground state wave function is evaluated by using equation (6) and is given by

$$\Psi_0^{(1)}(y) \propto exp\left(-\Lambda e^{-(y-y_{eq})} - y(\Lambda - \frac{1}{2})\right) \tag{17}$$

This wave function is then used to evaluate, through the SQM algebra, the wave function for the excited states by using equation (9); the first excited state is given by

$$\Psi_1^{(1)}(y) \propto -2\left(\Lambda e^{-(y-y_{eq})} - \Lambda + 1\right) exp\left(-\Lambda e^{-(y-y_{eq})} - y(\Lambda - \frac{3}{2})\right) \tag{18}$$

The wave functions given in (17) and (18) are the ground state and the first excited state wave functions of the exact system, non-confined or with no hydrogen bond. We refer to them as $\Psi_{n(non-confined)}$, $n = 0, 1$. As our aim is the lowest state and the first excited state of the confined system, with the hydrogen bond, a potential barrier is introduced, i.e., $V_1(y_{max}) = V_1(y_{min}) = \infty$, meaning that these wave functions must be zero outside the barrier.

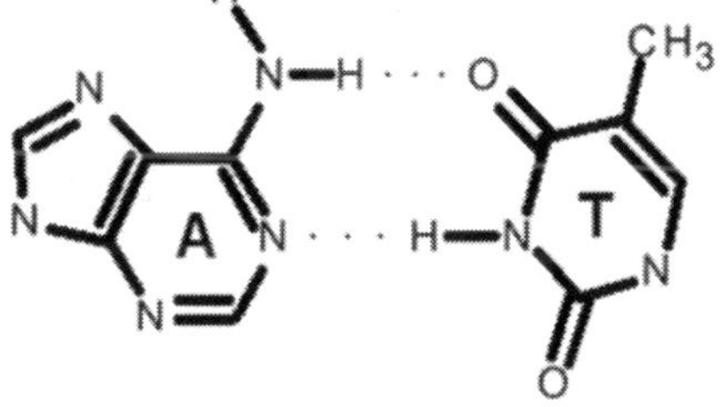

Figure 2. Schematic model hydrogen bonds present in the AT pair in DNA.

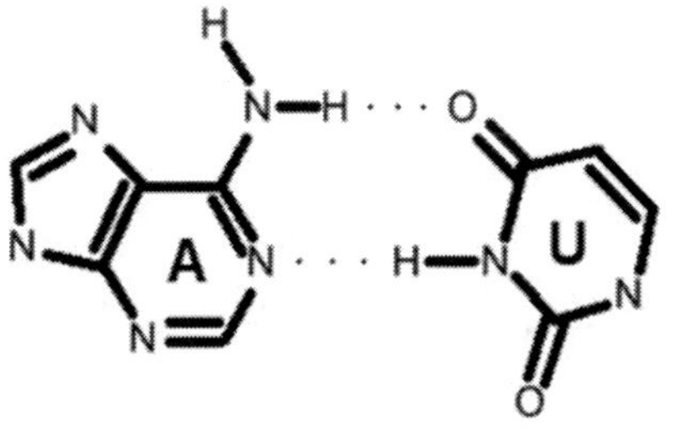

Figure 3. Schematic model hydrogen bonds present in the AU pair in RNA.

Consequently, the probability of finding the hydrogen outside the confinement region must be zero.

As the system is no longer exact, it is necessary to use an approximative method, which we choose to be the variational method associated to SQM. From the SQM point of view, this is equivalent to introduce two additional terms in the superpotential w_1, associated to the superalgebra, [13], [14]. Adopting this procedure, the wave functions to be used in the variational method are given by

$$\Psi_n^{(1)}{}_{(confined)} \propto \Psi_n^{(1)}{}_{(non-confined)} \cdot (y_{max} - y)(y - y_{min}) \quad n = 0, 1 \ , \tag{19}$$

with the corresponding energies obtained by solving

$$E_n = \frac{\int_{y_{min}}^{y_{max}} \Psi_n^* \hat{H}_M \Psi_n dy}{\int_{y_{min}}^{y_{max}} \Psi_n^* \Psi_n dy} \tag{20}$$

where $\hat{H}_M$ is given by equation (15) and Ψ_n are the trial wave functions given by equation (19), which depend on (17) and (18). The integration limits are the covalent radius of the atom for which the hydrogen is chemically bound (x_{min}) and the van der Waals radius complementary of hydrogen bond (x_{max}). The results of energy are presented in cm^{-1} units (inverse wavelength), the same as the ones from spectroscopic results.

5. Results

The hydrogen bonds studied involve the NH group. The necessary parameters, presented in Tables 1 and 2 below, were obtained experimentally, as indicated.

Table 1. Parameters used for the calculations of energy to the NH group, [18].

	NH
Equilibrium distance (x_{eq})	1.0362 Å
Harmonic vibrational wavenumber (ω)	$3282cm^{-1}$
Energy dissociation (D_e)	3.47 eV
Reduced mass (μ)	$1.5614.10^{-27}kg$

Table 2. Covalent radius and van der Waals radius used for the nitrogen and oxygen atoms [19], [20].

	N	O
Covalent radius (x_c)	0.71 Å	0.66 Å
van der Waals radius (x_{vw})	1.55Å	1.52 Å

5.1. Results for NH group

5.1.1. Non-confined case

Using the values given in Table 1 the value of the Morse potential β parameter, determined by the relation (11) for NH group is $\beta = 2.3168.10^{10}m^{-1}$. The value of the constant Λ for NH group, given by equation (13), is $\Lambda = 17.0534$.

In the situation free of any confinement the calculation of the transition from the first excited state to the ground state is made according to equation (16). In this case, in spectroscopic units, this value corresponds to $E_1 - E_0 = 3089cm^{-1}$ and should be found in spectroscopy. In studies by infrared spectroscopy with polypeptides it is observed an absorption band in $3090cm^{-1}$ and it has been described as a vibration frequency of the NH molecular group, [21]. This value is found very close to the calculated and can be identified, according to the model proposed, as vibration of NH without the formation of hydrogen bonding.

5.1.2. Confined case

When the NH group form molecular hydrogen bonding the vibration frequency is shifted. An interesting system where you can observe this effect are in the nucleic acids, DNA and RNA, because between the base pairs there are hydrogen bonds that allow the maintenance of the structure of the macromolecule. Here we analyze the vibrations of the NH group in hydrogen bonding between base pairs adenine-thymine (AT) and adenine-uracil (AU). In these base pairs two hydrogen bonds are formed, one with oxygen and the other with nitrogen. The schematic models are shown in Figures 2 and 3.

The lengths of the hydrogen bonds of AT and AU present in the DNA and RNA molecules, respectively, are well known in the literature and are obtained by X-ray diffraction, [22]. The distances are given between the donor atom and acceptor atom, because by X-ray diffraction it is not possible to determine the hydrogen positions. The lengths of the hydrogen bonds are shown in Table 3 below.

Table 3. Length of hydrogen bonds in nucleic acids, [22].

	AT	AU
(N...N	2.927 Å	2.820Å
(N...O)	2.861 Å	2.950Å

These lengths are used in equation (1) to determine the hydrogen maximum limit of oscillation.

Solving equation (20) through the variational method, we found the transition energies for each bond; they are summarized in Table 4. These values are compared with results given in literature obtained experimentally.

Table 4. Energy levels of the NH group in the hydrogen bonds between the base pairs AT and AU. E_0 is the ground state energy, E_1 is the energy of the first excited state and $\Delta E_{1\to 0} = E_1 - E_0$, in (cm^{-1}) units.

		E_0	E_1	$\Delta E_{1\to 0}$
AT	(N$\cdots$N)	-26330	-23122	3218
	(N$\cdots$O)	-26314	-23039	3275
AU	(N$\cdots$N)	-26225	-22794	3431
	(N$\cdots$O)	-26340	-23175	3165

Femtosecond vibrational spectroscopy with DNA oligomers realized with an alternating sequence of 23 adenine and thymine base pairs, [23], shows that there are three absorption peaks in the vibrational spectrum interesting to remark.

One is between 3000 and $3100cm^{-1}$ and two around $3250cm^{-1}$, one being located between 3200 and $3250cm^{-1}$, related to the N...N hydrogen bond in the AT pair, and the other between 3250 e $3300cm^{-1}$, related to the N...O hydrogen bond in the AT pair. These last two peaks can be identified, according to our model, as the vibration of the hydrogen bond in NH, with a deviation between 0.56%-0.98% and 0.76%-0.77% respectively to the two ranges. The first peak, located around 3000 and $3100cm^{-1}$, was found to be in fact at $3075cm^{-1}$, and can be associated, according to our model ($3089cm^{-1}$), with NH group without hydrogen bond, with a deviation of 0.46%.

Thus, these experimental results of the AT of DNA and AU of RNA oligomers show perfect agreement to the results obtained through our method.

Other infrared spectroscopy studies with RNA oligomers, [24], show the vibration spectrum when the NH participates in hydrogen bonds with some explicit numerical values. One of the peaks of the spectrum is located at $3185cm^{-1}$, very close to $3165cm^{-1}$ relative to the binding (N$\cdots$O) calculated by our method, with a deviation of only 0.63%. A peak between 3050 e $3100cm^{-1}$ can also be observed, which can de associated to the non-confined situation, with no hydrogen bond of NH group of our model, ($3089cm^{-1}$).

6. Conclusion

In this work the quantum confinement of hydrogen present when hydrogen bonds are stablished was studied through the variational method associated to SQM. The vibration spectrum of NH group was evaluated in the situations of non-confinement, i.e., with no hydrogen bonds and when there is confinement, i.e., when the hydrogen bonds are formed. We focused our attention to the case of DNA and RNA oligomers, which present the NH group. As a result we conclude that the presence of hydrogen bonds promote significant changes in the vibration spectrum of NH group. The comparison with experimental data reported in the literature showed perfect agreement, with percentage errors smaller than 1%, which reenforces the applicability of the proposal.

We conclude that the variational method associated to SQM was successfully applied to the calculation of the vibrational spectrum of NH molecular group with the principle of quantum confinement of hydrogen bonds. The treatment proposed here can be extended to the study of other molecular groups, as FH and CH, involved in hydrogen bonds.

References

[1] Yang M, Szyc L and Elsaesser T 2011 *J. Phys. Chem.* B **115** 1262
[2] Santos C S, Drigo Filho E and Ricotta R M Quantum Confinement in Hydrogen Bond, submitted for publication.
[3] Jeffrey G A 1997 *An Introduction to Hydrogen Bonding* (Oxford: Oxford University Press)
[4] Morse P M 1929 *Phys. Rev.* **34** 57
[5] Leviel L and Marechal Y 1971 *J. Chem. Phys.* **54** 1104
[6] Rekik N, Oujia B and Wójcik M J 2008 *J. Chem. Phys.* **352** 65
[7] Noureddine I, Houcine G and Brahim O 2012 *Intern. J. Quantum Chem.* **112** 1006
[8] Houcine G, Noureddine I and Brahim O 2012 *Intern. J. Quantum Chem.* **112** 1373
[9] Goryainov S V 2012 *J. Phys. B: Atom. Mol. Opt. Phys.* **407** 4233
[10] Marechal Y and Witkowski A 1968 *J. Chem. Phys.* **48** 3697
[11] Blaise P, Wojcik M J and Rousseau O H 2005 *J. Chem. Phys.* **122** 64306
[12] Drigo Filho E and Ricotta R M 2000 *Phys. Lett.* A **296** 296
[13] Drigo Filho E and Ricotta R M 2002 *Phys. Lett.* A **299** 137
[14] Silva F R and Drigo Filho E 2010 *Chem. Phys. Lett.* **498** 198
[15] Varshni Y P 2004 *Mod. Phys. Lett.* A **19** 2757
[16] Ley-Koo E, Mateos-Cortés S and Villa-Torres G 1998 *Intern. J. Quantum Chem. Phys.* **58** 23
[17] Sukumar C V 1985 *J. Phys. A: Math. Gen.* **18** L57
[18] Janoschek R 2001 *Pure and Appl. Chem.* **73** 1521
[19] Cordero B 2008 *Dalton Transactions* **21** 2832
[20] Bondi A 1964 *J. Phys. Chem.* **68** 441
[21] Badger R M and Rubalcava H 1954 *Proc. Nat. Acad. Sci.* **40** 12
[22] Saenger W 1983 *Principles of Nucleic Acid Structure* (Berlin: Springer-Verlag)
[23] Szyc L, Yang M, Dreyer J, Nibbering E T J and Elsaesser T 2008 *J. Chem. Phys.* **357** 36
[24] Woutersen S and Cristalli C 2004 *J. Chem. Phys.* **11** 5381

Fundamental solution of k-hyperbolic harmonic functions in odd spaces

Sirkka-Liisa Eriksson and Heikki Orelma

Department of Mathematics, Tampere University of Technology, P.O.Box 553, FI-33101 Tampere, Finland

E-mail: Sirkka-Liisa.Eriksson@tut.fi, Heikki.Orelma@tut.fi

Abstract. We study k-hyperbolic harmonic functions

$$\Delta u - \frac{k}{x_n}\frac{\partial u}{\partial x_n} = 0,$$

in the upper half space $\mathbb{R}_+^{n+1} = \{(x_0, x_1, ..., x_n) \mid x_i \in \mathbb{R},\ x_n > 0 \text{ for } i = 0, ..., n\}$. The operator $x_n^{\frac{2k}{n-1}}\left(\Delta u - \frac{k}{x_n}\frac{\partial u}{\partial x_n}\right)$ is the Laplace-Beltrami operator with respect to the Riemannian metric $ds_k^2 = x_n^{-\frac{2k}{n-1}}(dx_0^2 + dx_1^2 + ... + dx_n^2)$. In case $k = n - 1$ the Riemannian metric is the hyperbolic distance of Poincaré upper half space. The proposed functions are connected to the axially symmetric potentials studied notably by Weinstein, Huber and Leutwiler. We present the fundamental solution in case n is even using the hyperbolic metric. The main tool is the transformation of k-hyperbolic harmonic functions to eigenfunctions of the hyperbolic Laplace operator.

1. Introduction

We are studying function theory connected to the hyperbolic Riemannian metric

$$ds^2 = \frac{dx_0^2 + dx_1^2 + ... + dx_n^2}{x_n^2}$$

in the Poincaré upper half space

$$\mathbb{R}_+^{n+1} = \{(x_0, x_1, ..., x_n) \mid x_i \in \mathbb{R},\ x_n > 0 \text{ for } i = 0, ..., n\}.$$

This metric is interesting, since it is invariant under Möbius transformation mapping upper half space onto itself (see [7]). Moreover, Leutwiler noticed in [9] and [10] that the power function x^m ($m \in \mathbb{N}_0$), calculated using Clifford algebras, is a conjugate gradient of a hyperbolic harmonic function f, that is

$$\Delta_h f = x_n^2 \Delta f - (n-1)\, x_n \frac{\partial f}{\partial x_n} = 0$$

where as usual

$$\Delta h = \frac{\partial^2 h}{\partial x_0^2} + ... + \frac{\partial^2 h}{\partial x_n^2}.$$

Journal of Physics: Conference Series **597** (2015) 012034 doi:10.1088/1742-6596/597/1/012034

and Δ_h is the hyperbolic Laplace-Beltrami operator with respect to the preceding metric. He started to study these type of functions.

In this paper, we study k-hyperbolic harmonic functions. Their function theory is investigated by the first author in [3] and for $k = n-1$ for example in [5]. The Riemannian metric connected to k-hyperbolic harmonic functions is

$$ds_k^2 = \frac{dx_0^2 + dx_1^2 + ... + dx_n^2}{x_n^{\frac{2k}{n-1}}}$$

and they satisfy the equation

$$x_n \Delta h - k \frac{\partial h}{\partial x_n} = 0$$

(see [13]). Weinstein introduced axially symmetric potential theory in [14] and his motivation was that if $p < 0$ is an integer then an axially symmetric harmonic function in $2-p$-dimensional space satisfies the preceding equation in the meridian plane (see for example [8]). It is known that k-hyperbolic harmonic functions are forming also a harmonic space in the sense of Contantinescu and Cornea [2] but this result is not giving any kind of concrete presentations for k-harmonic functions.

We also need to consider the generalized Weinstein equation

$$x_n^2 \Delta h - k x_n \frac{\partial h}{\partial x_n} + lh = 0 \tag{1}$$

in an open domain whose closure is contained in the upper half space. Earlier this equation has been researched for example by Leutwiler and Akim in [1] and in [11]. Our general technical assumption is that the constants $l, k \in \mathbb{R}$ satisfy $4l \leq (k+1)^2$. Our idea is that we transfer solutions of this equation to the solutions of Laplace-Beltrami equation of the hyperbolic metric in the Poincaré upper half space. In the main result, we present the fundamental solution of the equation (1) in terms of the hyperbolic distance function in case n is even. Earlier these results have been presented in $\mathbb{R}^3$ (see [6]).

2. Preliminaries

We recall the properties of the hyperbolic distance that we need later on (see the proof for example in [12]).

Theorem 2.1 *The hyperbolic distance $d_h(x, a)$ between the points $x = (x_0, ..., x_n)$ and $a = (a_0, ..., a_n)$ in $\mathbb{R}_+^{n+1}$ may be computed as*

$$d_h(x, a) = \operatorname{arcosh} \lambda(x, a),$$

where

$$\begin{aligned}\lambda(x, a) &= \frac{(x_0 - a_0)^2 + ... + (x_{n-1} - a_{n-1})^2 + x_n^2 + a_n^2}{2x_n a_n} \\ &= \frac{|x - a|^2 + |x - \hat{a}|^2}{4x_n a_n}.\end{aligned}$$

In above $\hat{a} = (a_0, ..., a_{n-1}, -a_n)$ and $|x - a|$ is the usual Euclidean distance.

We also apply the simple calculation rules of the hyperbolic distance stated next.

Lemma 2.2 *If* $x = (x_0, x_1, ..., x_n)$ *and* $a = (a_0, a_1, ..., a_n)$ *are points in* $\mathbb{R}^{n+1}_+$ *then*

$$|x-a|^2 = 2x_n a_n (\lambda(x,a) - 1), \tag{2}$$

$$|x-\hat{a}|^2 = 2x_n a_n (\lambda(x,a) + 1), \tag{3}$$

$$\frac{|x-a|^2}{|x-\hat{a}|^2} = \frac{\lambda(x,a)-1}{\lambda(x,a)+1} = \tanh^2(\frac{d_h(x,a)}{2}), \tag{4}$$

where $\widehat{a} = (a_0, .., a_{n-1}, -a_n)$.

We also note the relation between the Euclidean and hyperbolic balls.

Proposition 2.3 *The hyperbolic ball* $B_h(a, r_h)$ *with the hyperbolic center* $a = (a_0, ..., a_n)$ *and the radius* r_h *is the same as the Euclidean ball with the Euclidean center*

$$c_a(r_h) = a = (a_0, ..., a_n \cosh r_h)$$

and the Euclidean radius $r_e = a_n \sinh r_h$.

Definition 2.4 *Let* $\Omega \subset \mathbb{R}^{n+1}_+$ *be open. A twice continuously differentiable function* $f : \Omega \to \mathbb{R}$ *is* $k-$*hyperbolic harmonic if*

$$\Delta f - \frac{k}{x_n}\frac{\partial f}{\partial x_n} = 0$$

for any $x \in \Omega$.

An important tool for handling k-hyperbolic harmonic functions is their transformation to eigenfunctions of the hyperbolic Laplace operator, stated next.

Lemma 2.5 *Let* Ω *be an open set contained in* $\mathbb{R}^{n+1}_+$. *A function* $f : \Omega \to \mathbb{R}$ *satisfies the Weinstein equation (1) if and only if the function* $g(x) = x_n^{\frac{n-k-1}{2}} f(x)$ *satisfies the following equation*

$$\Delta_h g + \frac{1}{4}\left(n^2 - (k+1)^2 + 4l\right) g = 0. \tag{5}$$

Especially, a function $f : \Omega \to \mathbb{R}$ *is* k*-hyperbolic harmonic if and only if the function* $g(x) = x_n^{\frac{n-k-1}{2}} f(x)$ *satisfies the following equation*

$$\Delta_h g + \frac{1}{4}\left(n^2 - (k+1)^2\right) g = 0. \tag{6}$$

3. The fundamental solution in $\mathbb{R}^{n+1}_+$ when n is even

We are looking for a fundamental solutions of the equation (6) connected to the hyperbolic Laplace operator. The hyperbolic Laplace of functions depending on λ is computed in [4] as follows.

Lemma 3.1 *If* f *is twice continuously differentiable depending only on* $\lambda = \lambda(x, e_n)$ *then*

$$\Delta_h f(x) = \left(\lambda^2 - 1\right)\frac{\partial^2 f}{\partial \lambda^2} + (n+1)\lambda\frac{\partial f}{\partial \lambda}.$$

It is easier to compute the solutions in terms of the hyperbolic distance.

Theorem 3.2 *If f is twice continuously differentiable depending only on $r_h = d_h(x, e_n) =$ arcosh $\lambda(x, e_n)$ then the hyperbolic Laplace in $\mathbb{R}^{n+1}_+$ is given by*

$$\Delta_h f(r_h) = \frac{\partial^2 f}{\partial r_h^2} + n \coth r_h \frac{\partial f}{\partial r_h}.$$

Proof. Using $\lambda(x, e_n) = \cosh r_h$, we compute

$$\frac{\partial r_h}{\partial \lambda} = \frac{1}{\sinh r_h}$$

and

$$\frac{\partial^2 r_h}{\partial \lambda^2} = -\frac{\cosh r_h}{\sinh^3 r_h}.$$

Hence applying the chain rule we obtain

$$\begin{aligned}
\frac{\partial f(r_h)}{\partial \lambda} &= \frac{\partial f}{\partial r_h}\frac{\partial r_h}{\partial \lambda} = \frac{\partial f}{\partial r_h}\frac{1}{\sinh r_h},\\
\frac{\partial^2 f(\lambda)}{\partial \lambda^2} &= \frac{\partial^2 f}{\partial r_h^2}\left(\frac{\partial r_h}{\partial \lambda}\right)^2 + \frac{\partial f}{\partial r_h}\frac{\partial^2 r_h}{\partial \lambda^2}\\
&= \frac{\partial^2 f}{\partial r_h^2}\frac{1}{\sinh^2 r_h} - \frac{\partial f}{\partial r_h}\frac{\cosh r_h}{\sinh^3 r_h},
\end{aligned}$$

completing the proof by the preceding Lemma. ■

In case n even, the preceding equation may be solved inductively. The crucial step is the following result.

Lemma 3.3 *Let $s \in \mathbb{N}_0$. If f is a solution of the equation*

$$\frac{\partial^2 f}{\partial r_h^2} + 2s \coth r_h \frac{\partial f}{\partial r_h} = \gamma f \tag{7}$$

depending on $r_h = d_h(x, e_n)$ then the function $g(r_h) = \frac{1}{\sinh r_h}\frac{\partial f}{\partial r_h}(r_h)$ satisfies the equation

$$\frac{\partial^2 g}{\partial r_h^2} + 2(s+1)\coth r_h \frac{\partial g}{\partial r_h} = (\gamma - 1 - 2s)\, g.$$

Proof. We just compute

$$\begin{aligned}
\frac{\partial^2 g}{\partial r_h^2} &= \left(-\frac{1}{\sinh r_h} + 2\frac{\cosh^2 r_h}{\sinh^3 r_h}\right)\frac{\partial f}{\partial r_h} - 2\frac{\cosh r_h}{\sinh^2 r_h}\frac{\partial^2 f}{\partial r_h^2} + \frac{1}{\sinh r_h}\frac{\partial^3 f}{\partial r_h^3},\\
2(s+1)\frac{\partial g}{\partial r_h}\frac{\cosh r_h}{\sinh r_h} &= -\frac{2(s+1)\cosh^2 r_h}{\sinh^3 r_h}\frac{\partial f}{\partial r_h} + \frac{2(s+1)\cosh r_h}{\sinh^2 r_h}\frac{\partial^2 f}{\partial^2 r_h}.
\end{aligned}$$

Since $\frac{\partial^2 f}{\partial r_h^2} = \gamma f - 2s \coth r_h \frac{\partial f}{\partial r_h}$ we conclude

$$\begin{aligned}
\frac{1}{\sinh r_h}\frac{\partial^3 f}{\partial r_h^3} &= \gamma g - 2s\frac{\cosh r_h}{\sinh^2 r_h}\frac{\partial^2 f}{\partial r_h^2} + 2s\frac{1}{\sinh^3 r_h}\frac{\partial f}{\partial r_h}\\
&= \gamma g - 2s\frac{\cosh r_h}{\sinh^2 r_h}\frac{\partial^2 f}{\partial r_h^2} + 2s\frac{\cosh^2 r_h}{\sinh^3 r_h}\frac{\partial f}{\partial r_h} - 2s\frac{1}{\sinh r_h}\frac{\partial f}{\partial r_h}
\end{aligned}$$

and therefore
$$\frac{\partial^2 g}{\partial r_h^2}+2\left(s+1\right)\frac{\partial g}{\partial r_h}\frac{\cosh r_h}{\sinh r_h}=\left(\gamma-1-2s\right)g.$$

■

Applying the previous result inductively, we are able to solve our main equation.

Proposition 3.4 *Let A be the operator defined by $A=\frac{1}{\sinh(r_h)}\frac{\partial}{\partial r_h}$. If f is a solution of the equation*
$$\frac{\partial^2 f}{\partial r_h^2}=\gamma f$$
then $g=A^m f$ satisfies the equation
$$\frac{\partial^2 g}{\partial r_h^2}+2m\frac{\partial g}{\partial r_h}\frac{\cosh r_h}{\sinh r_h}=\left(\gamma-m^2\right)g.$$

Proof. Applying inductively the previous lemma, we may deduce the result as follows
$$\begin{aligned}\frac{\partial^2 g}{\partial r_h^2}+2m\frac{\partial g}{\partial r_h}\frac{\cosh r_h}{\sinh r_h}&=\left(\gamma-\sum_{j=0}^{m-1}\left(1+2j\right)\right)g\\&=\left(\gamma-\left(m+\left(m-1\right)m\right)\right)g=\left(\gamma-m^2\right)g.\end{aligned}$$

■

Corollary 3.5 *If f is a solution of the equation*
$$\frac{\partial^2 f}{\partial r_h^2}=(\gamma+m^2)f$$
then $g=A^m f$ satisfies the equation
$$\frac{\partial^2 g}{\partial r_h^2}+2m\frac{\partial g}{\partial r_h}\frac{\cosh r_h}{\sinh r_h}=\gamma g.$$

We are ready to state our main result.

Theorem 3.6 *Let $\gamma=\frac{(k+1)^2-4l}{4}-m^2$, $\alpha=\sqrt{(k+1)^2-4l}$ and $n=2m$. The solution of the equation*
$$\frac{\partial^2 g}{\partial r_h^2}+2m\frac{\partial g}{\partial r_h}\frac{\cosh r_h}{\sinh r_h}=\gamma g$$
depending on the hyperbolic distance $r_h=d_h\left(x,a\right)$ with a pole at $x=a$ is
$$F_{n,k,l}\left(r_h\right)=\frac{s_{n,k,l}\left(r_h\right)}{\sinh^{n-1}r_h}$$
where $s_{n,k,l}\left(r_h\right)=\sum_{j=-m+1}^{j=m-1}a_{j,n,l}\cosh\left(\frac{r_n(\alpha+2j)}{2}\right)$ for some constant $a_{j,n,l}$ defined inductively by
$$\begin{aligned}a_{j,n+2,l}&=\frac{a_{(j-1),n}\left(\alpha+2j-2n\right)-a_{(j+1),n,l}\left(\alpha+2\left(j+n\right)\right)}{4}\\a_{j,n,0}&=0\qquad \text{if } j<1-m \text{ or } j>m-1,\\a_{0,2,0}&=1\end{aligned}$$

Journal of Physics: Conference Series **597** (2015) 012034 doi:10.1088/1742-6596/597/1/012034

and

$$s_{n,k,l}(0) = \frac{(-1)^{m-1}(n-1)!!}{2}.$$

Moreover, the function

$$H_{n,k}(x,a) = (-1)^{m-1} x_n^{\frac{k+1-n}{2}} a_n^{\frac{k+1-n}{2}} F_{n,k,0}(d_h(a,x))$$

is k-hyperbolic with respect to the both variables a and x.

Proof. Using the preceding corollary we obtain that the function $G_{n,k,l}(r_h) = CA^m f$ is the eigenfunction corresponding to the value γ when we choose

$$f(r_h) = \begin{cases} \sinh\left(\frac{\alpha r_h}{2}\right), & k \neq -1, \\ r_h, & k = -1. \end{cases}$$

Setting $C = \frac{2}{\alpha}$, we deduce that

$$F_{2,k,l}(r_h) = \frac{\cosh\left(\frac{\alpha r_h}{2}\right)}{\sinh r_h}$$

is the eigenfunction corresponding to the value γ and the result holds when $m = 1$ and $a_{0,2,0} = 1$. Assume that the result holds for some $n = 2m$. Then

$$F_{n+2,k,l}(r_h) = \frac{\sum_{j=-m+1}^{j=m-1} a_{j,n,l}(\alpha+2j)\sinh\left(\frac{r_n(\alpha+2j)}{2}\right)\sinh r_h}{2\sinh^{n+1} r_h} - \frac{(n-1)\sum_{j=-m+1}^{j=m-1} a_{j,n,l}\cosh\left(\frac{r_n(\alpha+2j)}{2}\right)\cosh r_n}{\sinh^{n+1} r_h}. \tag{8}$$

Applying the hyperbolic identities, we obtain

$$F_{n+2,k,l}(r_h) = \frac{\sum_{j=1-m}^{j=m-1} a_{j,n,l}(\alpha+2(j-n+1))\cosh\left(\frac{r_n(|k+1|+2(j+1))}{2}\right)}{4\sinh^{n+1} r_h} - \frac{\sum_{j=1-m}^{j=m-1} a_{j,n,l}(\alpha+2(j+n-1))\cosh\left(\frac{r_n(\alpha+2(j-1))}{2}\right)}{4\sinh^{n+1} r_h}.$$

Setting $a_{j,n,l} = 0$ if $j < 1-m$ or $j > m-1$, we obtain

$$a_{j,n+2,l} = \frac{a_{(j-1),n}(\alpha+2j-2n) - a_{(j+1),n}(\alpha+2(j+n))}{4}$$

completing the proof of the first assertion. The value of $s_{n,k,l}(0)$ follows inductively from (8). ■

The preceding function leads to the fundamental solution. For the proof we need the Green's identity (see for example [1]).

Lemma 3.7 *Let $R \subset \mathbb{R}_+^{n+1}$ be an open set with the smooth boundary contained $\mathbb{R}_+^{n+1}$ and denote the volume element corresponding to the Riemannian metric*

$$ds^2 = \frac{dx_0^2 + dx_1^2 + ... + dx_n^2}{x_n^{\frac{2k}{n-1}}}$$

by $dx_{(k)} = x_n^{-\frac{k(n+1)}{n-1}} dx$, *the surface elements by* $d\sigma_{(k)} = x_n^{-\frac{kn}{n-1}} d\sigma$ *and the outer normal* $\frac{\partial u}{\partial m_k} = x_n^{\frac{k}{n-1}} \frac{\partial u}{\partial m}$ *where* m *is the outer normal to the the surface* ∂R. *Then the Laplace-Beltrami operator is*

$$\Delta_k = x_n^{\frac{2k}{n-1}} \left(\Delta - \frac{k}{x_n} \frac{\partial}{\partial x_n} \right)$$

and

$$\int_R \left(u \Delta_k v dx_{(k)} - v \Delta_k u dx_{(k)} \right) = \int_{\partial R} \left(u \frac{\partial v}{\partial m_k} - v \frac{\partial u}{\partial m_k} \right) d\sigma_{(k)}$$

for any functions u *and* v *that are twice continuously differentiable functions in an open neighborhood containing the closure* $\overline{R}$ *of* R.

As usual, we obtain setting $v = 1$ the result

Proposition 3.8 *If* $R \subset \mathbb{R}_+^{n+1}$ *be an open set and* $f : R \to \mathbb{R}$ *is* k*-hyperbolic harmonic then*

$$\int_{\partial B_h(y,\rho)} \frac{\partial u}{\partial m_k} d\sigma_{(k)} = 0$$

for any $B_h(y, \rho)$ *with* $\overline{B_h(y, \rho)} \subset R$.

We need to prove that the preceding function $H_{n,k}(y, a)$ is Lebesgue integrable and that its normal derivative has some nice limiting property.

Lemma 3.9 *The function* $H_{n,k}(y, a)$ *is Lebesgue integrable in the hyperbolic ball* $B_h(a, R_h)$ *and*

$$\int_{B_h(a,R_h)} |F_{n,k,0}(d_h(a, y))| \frac{dy}{y_n^{\frac{2kn}{n-1}}} \leq M \sinh^2 r_h$$

for some positive $M > 0$.

Proof. It is enough to prove the statement for $a = (0, ..., 1)$. We recall that

$$\frac{s_{n,k}(d_h(y, a))}{\sinh^{n-1} d_h(y, a)} = \frac{s_{n,k}(d_h(y, a))}{|y - \cosh R_h a|^{n-1}}$$

in $B_h(a, R_h)$. Since $s_{n,k}(d_h(y, a)) y_n^{-\frac{2kn}{n-1}}$ is a continuous function, we only need to consider the integral

$$\int_{B_h(a,R_h)} \sinh^{1-n} d_h(y, a) dy = \int_{B_e(\cosh R_h a, \sinh R_h)} \frac{dy}{|y - \cosh R_h a|^{n-1}},$$

since $\sinh d_h(y, a) = |y - \cosh R_h a|$ by Proposition 2.3. Hence we have

$$\int_{B_e(\cosh R_h a, R_h)} \sinh^{1-n} d_h(y, a) dy = \int_0^{\sinh r_h} r \int_{\partial B_e(\cosh R_h a, 1)} dy dr = \frac{\omega_n \sinh^2 r_h}{2},$$

where ω_n is the surface measure of the unit ball in $\mathbb{R}^{n+1}$. ∎

Lemma 3.10 *Let $\Omega \subset \mathbb{R}_{+}^{n+1}$ be open. Let u be continuous in Ω. Denote $H_{n,k}(y,x) = (-1)^{\frac{n}{2}-1} y_n^{\frac{k+1-n}{2}} x_n^{\frac{k+1-n}{2}} F_{n,k,0}(y,x)$. Then*

$$\lim_{R_h\to 0} \frac{\int_{\partial B_h(x,R_h)} u \frac{\partial H_{n,k}}{\partial m_k} d\sigma_{(k)}}{2^{\frac{n+2}{2}} \pi^{\frac{n}{2}}} = -u(x).$$

for any hyperbolic balls $B_h(x,r_h)$ satisfying $\overline{B_h(x,r_h)} \subset \Omega$.

Proof. We abbreviate $F_{n,k,0} = F_{n,k}$. Using Proposition 2.3 we infer that in $\partial B_h(x,R_h)$ the outside pointing normal at y is

$$m = (\nu_0, \nu_1, ..., \nu_n) = \frac{(y_0 - x_0, y_1 - x_1, ..., y_n - x_n \cosh R_h)}{x_n \sinh R_h}.$$

Denote $r_h = d_h(y,x)$. We first compute

$$\begin{aligned}
(-1)^{\frac{n}{2}-1} \frac{\partial H_{n,k}}{\partial m_k} &= (-1)^{\frac{n}{2}-1} y_n^{\frac{k}{n-1}} \frac{\partial H_{n,k}}{\partial m} = (-1)^{\frac{n}{2}-1} y_n^{\frac{k}{n-1}} (m, \operatorname{grad} H_{n,k}) \\
&= y_n^{\frac{(n-1)(k+1-n)+2k}{2(n-1)}} x_n^{\frac{k+1-n}{2}} \frac{\partial F_{n,k}}{\partial r_h} \sum_{i=0}^{n} \nu_i \frac{\partial r_h}{\partial y_i} \\
&+ \frac{k+1-n}{2} y_n^{\frac{(n-1)(k-1-n)+2k}{2(n-1)}} x_n^{\frac{k+1-n}{2}} \nu_n F_{n,k}.
\end{aligned}$$

Applying Lemma 2.1 we infer

$$\frac{\partial r_h}{\partial y_i} = \frac{\partial r_h}{\partial y_i} = \frac{\partial \arccos \lambda(y,x)}{\partial y_i} = \frac{y_i - x_i - x_n(\cosh r_h - 1)\delta_{in}}{y_n x_n \sinh r_h}$$

and therefore we conclude

$$\sum_{i=0}^{n} \nu_i \frac{\partial r_h}{\partial y_i} = \frac{1}{y_n}.$$

Hence we have

$$\begin{aligned}
(-1)^{\frac{n}{2}-1} \frac{\partial H_{n,k}}{\partial m_k}(y) &= y_n^{\frac{(n-1)(k-1-n)+2k}{2(n-1)}} x_n^{\frac{k+1-n}{2}} \frac{\partial F_{n,k}}{\partial r_h} + \\
&\frac{k+1-n}{2} y_n^{\frac{(n-1)(k-1-n)+2k}{2(n-1)}} x_n^{\frac{k+1-n}{2}} \nu_n F_{n,k} \\
&= y_n^{\frac{(n-1)(k-1-n)+2k}{2(n-1)}} x_n^{\frac{k+1-n}{2}} \frac{s'_{k,n}(r_h)}{\sinh^{n-1} r_h} - \\
&(n-1) y_n^{\frac{(n-1)(k-1-n)+2k}{2(n-1)}} x_n^{\frac{k+1-n}{2}} \frac{s_{n,k}(r_h) \cosh r_h}{\sinh^n r_h} + \\
&\frac{k+1-n}{2} y_n^{\frac{(n-1)(k-1-n)+2k}{2(n-1)}} x_n^{\frac{k+1-n}{2}} \nu_n F_{n,k}.
\end{aligned}$$

Since $B_h(x,R_h) = B(x_e, x_n \sin R_h)$ for $x_e = (x_0, x_1, ..., x_n \cosh R_h)$ we obtain

$$\lim_{R_h \to 0} \frac{1}{\omega_n x_n^n \sinh^n R_h} \int_{\partial B_h(x,R_h)} y_n^{\frac{(n-1)(k-1-n)+2k}{2(n-1)}} x_n^{\frac{k+1+n}{2}} \sinh R_h s'_{k,n}(R_h) d\sigma_{(k)} = 0.$$

Similarly, we compute that

$$\lim_{R_h\to 0}\frac{1}{\omega_n x_n^n\sinh^n R_h}\int_{\partial B_h(x,R_h)}\frac{u(y)\,(y_n - x_n\cosh R_h)\,s_{n,k}(R_h)}{x_n}d\sigma_{(k)} = 0.$$

Since $\omega_n = (-1)^{\frac{n}{2}-1}\frac{2^{\frac{n+2}{2}}\pi^{\frac{n}{2}}}{s_{n,k}(0)}$ when n is even, manipulating the last integral we obtain

$$\begin{aligned}
&\lim_{R_h\to 0}\frac{-(n-1)(-1)^{\frac{n}{2}-1}}{\omega_n x_n^n\sinh^n R_h}\int_{\partial B_h(x,R_h)}u(y)\,y_n^{\frac{(n-1)(k-1-n)+2k}{2(n-1)}}x_n^{\frac{k+1+n}{2}}s_{n,k}(r_h)\cosh R_h d\sigma_{(k)}\\
&=\lim_{r_h\to 0}\frac{-(n-1)\,x_n^{\frac{k+1+n}{2}}s_{n,k}(0)\cosh R_h}{2^{\frac{n+2}{2}}\pi^{\frac{n}{2}}x_n^n\sinh^n R_h}\int_{\partial B_h(x,R_h)}u(y)\frac{(-1)^{\frac{n}{2}-1}s_{n,k}(r_h)}{y_n^{\frac{k+1+n}{2}}s_{n,k}(0)}d\sigma\\
&=\lim_{r_h\to 0}\frac{-(n-1)\,x_n^{\frac{k+1+n}{2}}\cosh R_h}{\omega_n x_n^n\sinh^n R_h}\int_{\partial B_h(x,R_h)}u(y)\frac{s_{n,k}(r_h)}{y_n^{\frac{k+1+n}{2}}s_{n,k}(0)}d\sigma\\
&=-(n-1)\,u(x),
\end{aligned}$$

completing the proof. ■

Our most important result is the following formula.

Theorem 3.11 *Let $\Omega\subset\mathbb{R}_+^{n+1}$ be open and $B_h(y,\rho)$ a hyperbolic ball with the hyperbolic center y and the hyperbolic radius ρ satisfying $\overline{B_h(y,\rho)}\subset\Omega$. If u is a twice continuously differentiable function in Ω and $x\in B_h(y,\rho)$ then*

$$\begin{aligned}
u(x) &= \frac{1}{2^{\frac{n+2}{2}}\pi^{\frac{n}{2}}(n-1)}\int_{\partial B_h(y,\rho)}\left(H_{n,k}(y,x)\frac{\partial u(y)}{\partial m_k} - u(y)\frac{\partial H_{n,k}(y,x)}{\partial m_k}\right)d\sigma_{(k)}(y)\\
&-\frac{1}{2^{\frac{n+2}{2}}\pi^{\frac{n}{2}}(n-1)}\int_{B_h(y,\rho)}\Delta_k u(y)\,H_{n,k}(y,x)\,dx_{(k)}
\end{aligned}$$

where $d\sigma_{(k)}$, $dx_{(k)}$ and $\frac{\partial}{\partial m_k}$ are the same as in Lemma 3.7.

Proof. Denote $B_h(y,\rho)=B$ and pick a hyperbolic ball such that $\overline{B_h(x,R_h)}\subset B$. Since $H_{h,k}$ is k-hyperbolic harmonic we obtain

$$\begin{aligned}
\int_{B\setminus B_h(x,R_h)}H_{n,k}\Delta_k u dx_{(k)} &= \int_{\partial R}\left(H_{n,k}\frac{\partial u}{\partial m_k} - u\frac{\partial H_{n,k}}{\partial m_k}\right)d\sigma_{(k)}\\
&-\int_{\partial B_h(x,R_h)}(u\frac{\partial H_{n,k}}{\partial m_k} + H_{n,k}\frac{\partial u}{\partial m_k})d\sigma_{(k)}.
\end{aligned}$$

Since $\frac{\partial u}{\partial m_k}$ and $y_n^{\frac{k+1-n}{2}-\frac{2kn}{n-1}}x_n^{\frac{k+1-n}{2}}s_{n,k}(d_h(x,y))$ are bounded we obtain

$$\int_{\partial B_h(x,R_h)}\left|H_{n,k}(x,y)\frac{\partial u}{\partial m_k}\right|d\sigma_{(k)}(y)\le\frac{M}{\sinh^{n-1}R}\int_{\partial B_h(x,R_h)}d\sigma = M\omega_n\sinh R_h$$

and therefore

$$\lim_{R_h\to 0}\int_{\partial B_h(x,R_h)}\left|H_{n,k}(x,y)\frac{\partial u}{\partial m_k}\right|d\sigma_{(k)}(y)=0.$$

Applying Lemma 3.10 we conclude the result. ■

Corollary 3.12 *If $\phi\in C_0^\infty(\mathbb{R}_+^{n+1})$ and $\rho>0$ the radius such that supp(ϕ) $\subset B_h(y,\rho)$ then*

$$\phi(x) = -\frac{1}{2^{\frac{n+2}{2}}\pi^{\frac{n}{2}}(n-1)}\int_{B_h(y,\rho)}\Delta_k\phi(y)\,H_{n,k}(y,x)\,dx_{(k)}.$$

Conclusion

We have found the fundamental k-hyperbolic solution in odd dimensional spaces. Earlier these results have only been proved in $\mathbb{R}^3_+$ (see [6]). Using this result in consecutive papers, we will obtain kernels in hyperbolic function theory based on k-hyperbolic functions and Cauchy type integral formulas.

References

[1] Akin Ö and Leutwiler H 1994 On the invariance of the solutions of the Weinstein equation under Möbius transformations. In Classical and modern potential theory and applications (Chateau de Bonas, 1993) *NATO Adv. Sci. Inst. Ser. C Math. Phys. Sci.* **430** (Dordrecht: Kluwer) 19–29
[2] Constantinescu C and Cornea A 1972 *Potential theory on harmonic spaces* (Berlin: Springer)
[3] Eriksson-Bique S-L 2003 k-hypermonogenic functions *Progress in analysis* (Singapore: World Scientific) 337–48
[4] Eriksson S-L 2010 Hyperbolic Extensions of Integral formulas *Adv. Appl. Clifford Alg.* **20** pp 575–86
[5] Eriksson S-L and Leutwiler H 2009 Hyperbolic harmonic functions and their function theory *Potential theory and Stochastics in Albac* (Theta series in advanced mathematics) pp 85–100
[6] Eriksson S-L and Orelma H 2014 Hyperbolic Laplace Operator and the Weinstein Equation in $\mathbb{R}^3_+$ *Adv. in Appl. Clifford Algebras* **24** 109–24
[7] Hua L-K 1981 *Starting with the unit circle* (Berlin: Springer)
[8] Huber A 1954 A Uniqueness of Generalized Axially Symmetric Potentials *Ann. of Math.* **60** 351–368
[9] Leutwiler H 1992 Modified Clifford analysis *Complex Variables* **1** 153–71
[10] Leutwiler H 1992 Modified quaternionic analysis *Complex Variables* **20** 19–51
[11] Leutwiler H 1987 Best constants in the Harnack inequality for the Weinstein equation *Aequationes Math.* **34** 304–15
[12] Leutwiler H 2004 Appendix: Lecture notes of the course "*Hyperbolic harmonic functions and their function theory*" *Clifford algebras and potential theory, Univ. Joensuu Dept. Math. Rep. Ser.* **7** (Joensuu: Univ. Joensuu) pp 85–109
[13] Orelma H 2010 New Perspectives in Hyperbolic Function Theory *Doctoral Dissertation* (Tampere University of Technology 892)
[14] Weinstein A 1953 Generalized Axially Symmetric Potential Theory *Bull. Amer. Math. Soc.* **59** 20–38

Factorization method for the truncated harmonic oscillator

D J Fernández C[1] and V S Morales-Salgado[2]

Departamento de Física, Cinvestav, A.P. 14-740, *07000 México D.F. Mexico*

E-mail: [1]`david@fis.cinvestav.mx`, [2]`vmorales@fis.cinvestav.mx`

Abstract. Factorization procedures of first and second order are used to generate Hamiltonians with known spectra departing from the harmonic oscillator with an infinite potential barrier. Certain systems obtained in a straightforward way through said method possess differential ladder operators of both types, third and fourth order. Since systems with this kind of operators are linked with the Painlevé IV and V equations respectively, several solutions of these non-linear second-order differential equations will be simply found.

1. Introduction

The idea of factorizing a differential operator in Quantum Mechanics yields a mighty technique for solving Schrödinger equation [1, 2, 3, 4]. This method can be presented in several ways, supplying us with interesting new results concerning the solvability of quantum problems [5, 6, 7, 8, 9, 10, 11, 12, 13, 14, 15, 16] and the spectral design of new potentials [11, 13, 15].

When applied to the harmonic oscillator, this procedure deforms the potential as well as the underlying algebra of the system in a specific way [18, 17, 10, 19]. These deformations of the harmonic oscillator algebra, which are called Polynomial Heisenberg Algebras (PHA), turn out to be characterized by differential ladder operators of order $m+1$, with the commutator between them being polynomials of degree m in the Hamiltonian. While systems realizing the second order PHA are connected to the Painlevé IV (PIV) equation, those realizing the third order algebras are linked to the Painlevé V (PV) equation [21, 20, 24, 23, 22, 17, 25].

These results have been used recently to produce plenty of non-singular solutions to the PIV and PV equations [14, 26, 27]. With the aim of studying some singular solutions supplied by said connections, we will continue the analysis now for the harmonic oscillator with an infinite potential barrier at $x = 0$ [7, 28, 29], which we will call truncated harmonic oscillator. Caution must be employed to avoid that singularities appear inside the domain of the potential.

The organization of this work is the following: in Section 2 we will review the factorization method as needed in the rest of the paper as well as the way in which the Hamiltonians built from the harmonic oscillator yields realizations of the second and third order PHA's. We will also present the procedure which connects these systems with the PIV and PV equations to produce solutions of these non-linear second-order differential equations. Section 3 is devoted to analyze the harmonic oscillator with an infinite potential barrier at the origin, its first and second order transformations, and the systems thus obtained. In Section 4 we will obtain several solutions to the PIV equation by using the extremal states of the supersymmetric partners of the truncated harmonic oscillator. Thus far, Sections 2-4 comprise a review of a previous work which can be found in [28, 29]. However, Section 5 contains original results on how to obtain several solutions to the PV equation respectively by using said extremal states. Finally, in Section 6 we will summarize the results attained.

2. Factorization method

Let us suppose the existence of a one-dimensional Schrödinger Hamiltonian $H = -\frac{1}{2}\frac{d^2}{dx^2} + V(x)$, together with a k-th order differential operator A^+. Then, for another one-dimensional Schrödinger Hamiltonian $\tilde{H} = -\frac{1}{2}\frac{d^2}{dx^2} + \tilde{V}(x)$ which is "intertwined" with H through the relation

$$\tilde{H}A^+ = A^+ H,$$

it is true that solutions of the stationary Schrödinger equation $\tilde{H}\phi_n(x) = E_n\phi_n(x)$ are connected to those of $H\psi_n(x) = E_n\psi_n(x)$ by

$$\phi_n(x) \equiv C_n A^+ \psi_n(x),$$

where C_n is a normalization constant. We will call $\tilde{H}$ and H supersymmetric partners, and we will assume that both are Hermitian so that the potentials V and $\tilde{V}$ are real.

If the new k-th order differential operator $A \equiv (A^+)^\dagger$ is introduced, then there is also an intertwining relation involving A, which can be written as $HA = A\tilde{H}$. Even more, the products of A and A^+ become factorized as follows

$$A^+A = \prod_{i=1}^{k}(\tilde{H} - \epsilon_i), \quad AA^+ = \prod_{i=1}^{k}(H - \epsilon_i).$$

For $k = 1$ the intertwining operator A and the new potential $\tilde{V}$ are determined by a solution to the stationary Schrödinger equation $Hu = \epsilon u$ as follows:

$$\begin{aligned} A^+ &= \frac{1}{\sqrt{2}}\left[-\frac{d}{dx} + [\ln(u)]'\right], \\ \tilde{V} &= V - [\ln(u)]'', \end{aligned}$$

where ϵ is called *factorization energy* and u *transformation function* or *seed solution*.

For a second order intertwining operator ($k = 2$) the procedure is fixed by a pair of transformation functions u_i, which are also solutions of $Hu_i = \epsilon_i u_i$, $i = 1, 2$. In a similar fashion as for $k = 1$ now we obtain

$$\begin{aligned} A^+ &= \frac{1}{2}\left\{\frac{d^2}{dx^2} - [\ln W(u_1, u_2)]'\frac{d}{dx} + \frac{1}{2}\left([\ln W(u_1, u_2)]'' + [\ln W(u_1, u_2)]'^2\right) - 2V + \epsilon_1 + \epsilon_2\right\}, \\ \tilde{V} &= V - [\ln W(u_1, u_2)]'', \end{aligned}$$

where $W(u_1, u_2)$ is the Wronskian of the two seed solutions u_1, u_2.

The general case follows straightforwardly by induction from the previous cases [31, 32, 33, 34, 35, 36, 37]. As an example of this general procedure next we apply the method to the harmonic oscillator.

2.1. Harmonic oscillator

For the specific harmonic oscillator Hamiltonian

$$H = -\frac{1}{2}\frac{d^2}{dx^2} + \frac{x^2}{2},$$

a transformation of order k leads to a Hamiltonian $\tilde{H}$ having at most k new levels created at the positions defined by the factorization energies ϵ_j, $j = 1, .., k$.

In such a case, the eigenfunctions $\phi_n(x)$ of $\tilde{H}$ associated to the eigenvalues $E_n = n + \frac{1}{2}$ of H are

$$\phi_n(x) = \frac{A^+\psi_n(x)}{\sqrt{(E_n - \epsilon_1)...(E_n - \epsilon_k)}},$$

while the ones $\phi_{\epsilon_j}(x)$ associated to the new eigenvalues ϵ_j, are

$$\phi_{\epsilon_j} \propto \frac{W(u_1, .., u_{j-1}, u_{j+1}, ..., u_k)}{W(u_1, .., u_k)}.$$

The k transformation functions u_j, $j = 1, ..., k$ are chosen from the general solutions of the stationary Schrödinger equation $-u_j'' + Vu_j = \epsilon_j u_j$ with $V = \frac{x^2}{2}$:

$$u_j(x) = e^{-x^2/2}\left[{}_1F_1\left(\frac{1-2\epsilon_j}{4}, \frac{1}{2}, x^2\right) + 2\nu_j \frac{\Gamma(\frac{3-2\epsilon_j}{4})}{\Gamma(\frac{1-2\epsilon_j}{4})} x\, {}_1F_1\left(\frac{3-2\epsilon_j}{4}, \frac{3}{2}, x^2\right)\right]. \quad (1)$$

In order to have a non-singular transformation, the Wronskian $W(u_1, .., u_k)$ must have no zeros on the real axis. This is achieved by demanding $\epsilon_k < \epsilon_{k-1} < ... < \epsilon_1 < E_0 = 1/2$ and $|\nu_1| < 1$, $|\nu_2| > 1$, $|\nu_3| < 1$,... so that the new potential

$$\tilde{V}(x) = \frac{x^2}{2} - [\ln W(u_1, ..., u_k)]''$$

won't have singularities in its domain.

Hamiltonians $\tilde{H}$ obtained by this procedure possess natural ladder operators, $L^\pm = A^+a^\pm A$. Since $a^\pm$ are the usual first order ladder operators of the harmonic oscillator, then $L^\pm$ are differential operators of order $2k+1$. The set of operators $\{\tilde{H}, L^+, L^-\}$ realizes the so-called Polynomial Heisenberg Algebras (PHA), which in general are described by the commutators

$$[\tilde{H}, L^\pm] = \pm L^\pm,$$
$$[L^-, L^+] = P_{2k}(\tilde{H}),$$

where $P_{2k}(\tilde{H})$ is a polynomial of degree $2k$ on the Hamiltonian $\tilde{H}$, which defines the system as a deformation of the harmonic oscillator.

In particular, systems with third order ladder operators $L^\pm$ realize a second order PHA. By factorizing them as $L^+ = L_1^+L_2^+$ and $L^- = L_2^-L_1^-$ where $L_{1,2}^- = (L_{1,2}^+)^\dagger$ and $L_1^+ = \frac{1}{\sqrt{2}}\left[-\frac{d}{dx} + f(x)\right]$, $L_2^+ = \frac{1}{2}\left[\frac{d^2}{dx^2} + g(x)\frac{d}{dx} + h(x)\right]$, we find at the end the following set of equations:

$$f(x) = x + g(x),$$
$$h = \tfrac{g'}{2} - \tfrac{g^2}{2} - 2xg - x^2 + \varepsilon_2 + \varepsilon_3 - 2\varepsilon_1 - 1,$$
$$V = \tfrac{x^2}{2} - \tfrac{g'}{2} + \tfrac{g^2}{2} + xg + \varepsilon_1 - \tfrac{1}{2},$$
$$\tfrac{d^2g}{dx^2} = \tfrac{1}{2g}\left(\tfrac{dg}{dx}\right)^2 + \tfrac{3}{2}g^3 + 4xg^2 + 2(x^2 - a)g + \tfrac{b}{g}.$$

The last equation can be recognized as the Painlevé IV (PIV) differential equation with constant parameters $a = \varepsilon_2 + \varepsilon_3 - 2\varepsilon_1 - 1$ and $b = -2(\varepsilon_2 - \varepsilon_3)^2$. Thus, we get a connection between second order PHA's and the PIV equation. We can either solve such a non-linear differential equation in order to realize a second order PHA, or take a known realization of a second order PHA and obtain then specific solutions to the PIV equation.

The last is done by using the factorization

$$N \equiv L^+L^- = \left(\tilde{H} - \varepsilon_1\right)\left(\tilde{H} - \varepsilon_2\right)\left(\tilde{H} - \varepsilon_3\right), \tag{2}$$

which indicates that there are three extremal states ϕ_λ, with $\lambda = \varepsilon_1, \varepsilon_2, \varepsilon_3$, corresponding to the eigenvalues ε_1, ε_2 and ε_3 respectively such that $L^-\phi_\lambda = L_2^-L_1^-\phi_\lambda = 0$ and thus $N\phi_\lambda = 0$. In particular,

$$L_1^-\phi_\lambda = \frac{1}{\sqrt{2}}\left[\frac{d}{dx} + f(x)\right]\phi_\lambda = 0$$

satisfies the extremal state condition. Without lost of generality, we denote by ϕ_{ε_1} such a state, but let us keep in mind that there are three possible identifications. Finally, let us recall that $f(x) = x + g(x)$ to obtain

$$g(x) = -x - [\ln\phi_{\varepsilon_1}]', \tag{3}$$

which is a solution to the PIV equation with fixed parameters a, b.

On the other hand, systems with fourth order ladder operators $L^+ = L_1^+L_2^+$ and $L^- = L_2^-L_1^-$ realize a third order PHA [10, 25]. Choosing $L_1^+ = \frac{1}{2}\left[\frac{d^2}{dx^2} + g_1(x)\frac{d}{dx} + h_1(x)\right]$, $L_2^+ = \frac{1}{2}\left[\frac{d^2}{dx^2} + g(x)\frac{d}{dx} + h(x)\right]$ and $L_{1,2}^- = (L_{1,2}^+)^\dagger$ we find this other set of equations:

$$g_1(x) = -g(x) - x,$$
$$g(x) = \frac{x}{w-1},$$
$$\frac{d^2w}{dz^2} = \left(\frac{1}{2w} + \frac{1}{w-1}\right)\left(\frac{dw}{dz}\right)^2 - \frac{1}{z}\frac{dw}{dz} + \frac{(w-1)^2}{z^2}\left(aw + \frac{b}{w}\right) + c\frac{w}{z} + d\frac{w(w+1)}{w-1}. \tag{4}$$

Equation (4) is recognized as the Painlevé V (PV) differential equation, with $z = x^2$ and $a = \frac{(\varepsilon_1-\varepsilon_2)^2}{2}$, $b = -\frac{(\varepsilon_3-\varepsilon_4)^2}{2}$, $c = \frac{\varepsilon_1+\varepsilon_2-\varepsilon_3-\varepsilon_4-1}{2}$, $d = -\frac{1}{8}$ being fixed parameters.

This time we are concerned with the factorization

$$N \equiv L^+L^- = \left(\tilde{H} - \varepsilon_1\right)\left(\tilde{H} - \varepsilon_2\right)\left(\tilde{H} - \varepsilon_3\right)\left(\tilde{H} - \varepsilon_4\right) \tag{5}$$

which shows that there are four extremal states such that $L^-\phi_\lambda = L_2^-L_1^-\phi_\lambda = 0$ and thus $N\phi_\lambda = 0$, $\lambda = \varepsilon_1, \varepsilon_2, \varepsilon_3, \varepsilon_4$ corresponding to the eigenvalues ε_1, ε_2, ε_3 and ε_4 respectively.

In a similar fashion as the previous case, but with lengthier calculations, without lost of generality we can write $g_1 = (\ln[W(\phi_{\varepsilon_3}, \phi_{\varepsilon_4})])'$ thus obtaining $g(x) = -x - (\ln[W(\phi_{\varepsilon_3}, \phi_{\varepsilon_4})])'$. The use of two extremal states in this construction enables to fix now six different identifications for ϕ_{ε_3} and ϕ_{ε_4} leading to g.

Finally, through the change of variable $x = \sqrt{z}$, the function

$$w(z) = 1 + \frac{\sqrt{z}}{g(\sqrt{z})}.$$

produces the solution to the PV equation with fixed parameters a, b, c and d [21, 37].

Journal of Physics: Conference Series **597** (2015) 012035 doi:10.1088/1742-6596/597/1/012035

3. Truncated harmonic oscillator

Adding an infinite potential barrier to the previous potential leads to modify the boundary condition at $x = 0$. The Hamiltonian of this system is

$$H_0 = -\frac{1}{2}\frac{d^2}{dx^2} + V_0, \qquad V_0 = \begin{cases} \frac{x^2}{2} & \text{if } x > 0, \\ \infty & \text{if } x \leq 0. \end{cases}$$

While both functions (and energies)

$$\psi_n(x) \propto x\, e^{-x^2/2}\, {}_1F_1\Big(-n, \frac{3}{2}, x^2\Big), \qquad E_n = 2n + \frac{3}{2},$$

$$\chi_n(x) \propto e^{-x^2/2}\, {}_1F_1\Big(-n, \frac{1}{2}, x^2\Big), \qquad \mathcal{E}_n = 2n + \frac{1}{2},$$

satisfy $H_0\psi_n = E_n\psi_n$ and $H_0\chi_n = \mathcal{E}_n\chi_n$ respectively, only the ψ_n satisfy the boundary condition at the origin making each E_n an eigenvalue of H_0, which is not the case for $\mathcal{E}_n$.

In order to implement the factorization method, with a major control on the singularities of the new potentials, we will use transformation functions as in equation (1) but with a definite parity, i.e., we will proceed by cases taking $u_j(x) = xe^{-x^2/2}\,{}_1F_1\left(\frac{3-2\epsilon_j}{4}, \frac{3}{2}, x^2\right)$ as the odd transformation function and $u_j(x) = e^{-x^2/2}\,{}_1F_1\left(\frac{1-2\epsilon_j}{4}, \frac{1}{2}, x^2\right)$ as the even one.

3.1. Intertwining operator with $k = 1$

Implementing a first order intertwining $H_1A^+ = A^+H_0$ using an odd transformation function u induces new Hamiltonians H_1 with potentials of the form

$$V_1 = V_0 + 1 + \frac{1}{x^2} - \left\{ \ln\left[{}_1F_1\left(\frac{3-2\epsilon}{4}, \frac{3}{2}, x^2\right)\right]\right\}'', \tag{6}$$

with eigenfunctions and eigenvalues given by $\phi_n \propto A^+\psi_n$ and $E_n = 2n + \frac{3}{2}$. For $\epsilon < \frac{3}{2}$ there is no new level of H_1 at ϵ, while in the limit $\epsilon = \frac{3}{2}$ one can erase the level $E_0 = \frac{3}{2}$.

On the other hand, the use of an even transformation function u produces new Hamiltonians H_1 with potentials of the form

$$V_1 = V_0 + 1 - \left\{ \ln\left[{}_1F_1\left(\frac{1-2\epsilon}{4}, \frac{1}{2}, x^2\right)\right]\right\}''. \tag{7}$$

In this case the eigenfunctions and eigenvalues of H_1 become $\phi_n \propto A^+\chi_n$ and $\mathcal{E}_n = 2n + \frac{1}{2}$. Again, for $\epsilon < \frac{1}{2}$ there is no new level of H_1 at ϵ, and with the choice $\epsilon = \frac{1}{2}$ one can erase the level $E_0 = \frac{1}{2}$.

In any case, for both of these first order intertwinings the new Hamiltonians are isospectral to H_0, up to a shift in the ground state energy. This will not hold anymore for second order intertwinings.

3.2. Intertwining operator with $k = 2$

A second order intertwining $H_2A^+ = A^+H_0$, where $H_2 = -\frac{1}{2}\frac{d^2}{dx^2} + V_2$, requires to use two transformation functions, u_1 and u_2 associated to ϵ_1 and ϵ_2 respectively, as described in Section 2. Without lost of generality, let us suppose that $\epsilon_2 < \epsilon_1$. We have found four non-equivalent combinations for these seed solutions [28, 29]. The results obtained are to be described in what follows.

If both u_1 and u_2 are odd then the new potential takes the form

$$V_2 = V_0 + \frac{3}{x^2} + 2 - [\ln \omega_1(x)]'',$$

where the function $\omega_1(x)$ is free of zeros in the interval $(0,\infty)$ as long as $\epsilon_1, \epsilon_2 < \frac{3}{2}$ or $\frac{3+4n}{2} < \epsilon_1, \epsilon_2 < \frac{3+4(n+1)}{2}$. The set of eigenvalues of H_2 is given by $\{E_n = 2n + \frac{3}{2}, \quad n = 0, 1, ...\}$ with corresponding eigenfunctions $\phi_n(x) \propto A^+\psi_n(x)$. As a limiting case, when one of the factorization energies takes the value E_n, then this level is erased from the set of eigenvalues.

In the case that u_1 and u_2 are both even it is found that

$$V_2 = V_0 + \frac{1}{x^2} + 2 - [\ln \omega_2(x)]'',$$

where now the function $\omega_2(x)$ will not have zeros in $(0,\infty)$ whenever $\epsilon_1, \epsilon_2 < \frac{1}{2}$ or $\frac{1+4n}{2} < \epsilon_1, \epsilon_2 < \frac{1+4(n+1)}{2}$. In general, H_2 has eigenvalues $\mathcal{E}_n = 2n + \frac{1}{2}$, $n = 0, 1, ...,$ with $\phi_n(x) \propto A^+\chi_n(x)$ being their corresponding eigenfunctions. However when one of the factorization energies becomes $\mathcal{E}_n$, this level is missing from the set of eigenvalues.

For u_1 being odd and u_2 even with $\frac{1+4n}{2} < \epsilon_2 < \epsilon_1 < \frac{3+4n}{2}$, the new potential becomes

$$V_2 = V_0 + 2 - [\ln \omega_3(x)]''.$$

It does not have any singularity in the domain $(0,\infty)$ other than the original one at $x = 0$. The eigenfunctions transform as $\phi_n(x) \propto A^+\psi_n(x)$ with eigenvalues $E_n = 2n + \frac{3}{2}$. The limiting case of this choice occurs when $\epsilon_1 = E_n$, which erases the level E_n. In addition, for $\epsilon_2 \neq \mathcal{E}_n$ the level ϵ_2 is added with $\phi_{\epsilon_2} \propto \frac{u_1}{W(u_1,u_2)}$.

Finally, when u_1 is chosen even and u_2 odd, the new potential turns out to be

$$V_2 = V_0 + 2 - [\ln \omega_4(x)]'',$$

where the function $\omega_4(x)$ does not have zeros in $(0,\infty)$ whenever $\epsilon_2 < \epsilon_1 < \frac{1}{2}$ or $\frac{3+4n}{2} < \epsilon_2 < \epsilon_1 < \frac{5+4n}{2}$. The eigenfunctions of H_2 are $\phi_n(x) \propto A^+\psi_n$ with eigenvalues $E_n = 2n + \frac{3}{2}$. With the choice $\epsilon_2 = E_n$ we can erase precisely this level. For $\epsilon_1 \neq \mathcal{E}_n$ there is an extra level at ϵ_1 with eigenfunction $\phi_{\epsilon_1} \propto \frac{u_2}{W(u_1,u_2)}$.

4. Solutions to the Painlevé IV equation

When performing a first order transformation on the truncated harmonic oscillator the supersymmetric partners H_1 possess natural third-order ladder operators $L^\pm = A^+a^\pm A$, thus we require just to identify the three extremal states to connect with specific solutions to the PIV equation. From equation (2) we know that the extremal states of H_1 are related to the eigenvalues ϵ, $\epsilon + 1$, $\frac{1}{2}$, and they are given by

$$\phi_{\varepsilon_1} \propto \frac{1}{u}, \qquad \phi_{\varepsilon_2} \propto A^+a^+u, \qquad \phi_{\varepsilon_3} \propto A^+e^{\frac{-x^2}{2}}. \qquad (8)$$

Once again, one chooses a definite parity for $u(x)$ in order to control better the position of the singularities introduced by the transformation. For an odd transformation function, equation (3) gives a solution to the PIV equation associated to the first extremal state of equation (8) of the form

$$g_1 = \frac{1}{x} - 2x + \left(1 - \frac{2}{3}\epsilon\right) x \frac{{}_1F_1(\frac{7-2\epsilon}{4}; \frac{5}{2}; x^2)}{{}_1F_1(\frac{3-2\epsilon}{4}; \frac{3}{2}; x^2)}.$$

Meanwhile, for an even transformation function this procedure gives a solution to the PIV equation associated also to the first extremal state of equation (8) of the form

$$g_1 = -2x + (1-2\epsilon)\, x\, \frac{{}_1F_1(\frac{5-2\epsilon}{4}; \frac{3}{2}; x^2)}{{}_1F_1(\frac{1-2\epsilon}{4}; \frac{1}{2}; x^2)}.$$

Once g_1 is obtained, independently of the parity of $u(x)$ the other two solutions to the PIV equation are obtained through the expressions

$$g_2 = -g_1 - 2x - 2\left[\frac{x+(2\epsilon-x^2)(g_1+x)+(g_1+x)^3}{x^2-2\epsilon-1-(g_1+x)^2}\right],$$

$$g_3 = -\frac{g_1'+2}{g_1+2x}.$$

Now, a second order transformation generates Hamiltonians H_2 having natural fifth-order ladder operators $L^{\pm} = A^{+}a^{\pm}A$. Thus we need to employ the reduction theorem described in [14] which ensures that if $u_2(x) = a^{-}u_1(x)$ and $\epsilon_2 = \epsilon_1 - 1$, then the system has as well third order ladder operators $l^{\pm}$ such that

$$l^{+}l^{-} = (H_2 - \epsilon_1 + 1)(H_2 - \epsilon_1 - 1)(H_2 - 1/2).$$

Therefore, the sought extremal states corresponding to the eigenvalues $\epsilon_1 - 1$, $\epsilon_1 + 1$, $\frac{1}{2}$ are

$$\phi_{\varepsilon_1} \propto \frac{u_1}{W[u_1, u_2]}, \qquad \phi_{\varepsilon_2} \propto A^{+}a^{+}u_1, \qquad \phi_{\varepsilon_3} \propto A^{+}e^{\frac{-x^2}{2}}.$$

From the procedure described previously, based on equation (3), we get now the following solutions to the PIV equation

$$\begin{aligned}
g_1 &= -x - \alpha + 2\left[\frac{x+\alpha}{x^2+1-2\epsilon_1-\alpha^2}\right], \\
g_2 &= g_1 + \frac{2\alpha^2 - 2x^2 + 2(2\epsilon_1+1)}{\alpha - g_1 - x}, \\
g_3 &= \frac{(x+\alpha)g_1^2 + \left[2\epsilon_1 - 1 + (x+\alpha)^2\right]g_1 + (2\epsilon_1-3)(x+\alpha)}{(x+\alpha)^2 + (x+\alpha)g_1 + 2\epsilon_1 - 1},
\end{aligned}$$

where $\alpha = \frac{u_1'}{u_1}$.

5. Solutions to the Painlevé V equation

As for the PV equation, let us recall first that the truncated harmonic oscillator Hamiltonian has a pair of second-order ladder operators $(a^{+})^2$ and $(a^{-})^2$ which connect in a natural way the eigenfunctions of H_0.

In this way, for a first-order transformation the Hamiltonian H_1 possesses also natural fourth order ladder operators $L^{\pm} = A^{+}(a^{\pm})^2A$, which satisfy the conditions described in the paragraphs after equation (5). Using now the six different identifications of the extremal states ϕ_{ε_3} and ϕ_{ε_4} supplies us with several explicit solutions to the PV equation:

$$w_1 = 1 + \frac{2\sqrt{2z}(1+2\epsilon-z+\sqrt{2z}\alpha)}{\sqrt{2z}(2+z) - 4(1+z)\alpha + 2\sqrt{2z}\alpha^2},$$

$$w_2 = \frac{\left(2\alpha+\sqrt{2z}\right)\left(4\alpha - 4\epsilon\sqrt{2z} + \sqrt{2}z^{3/2} - 2\sqrt{2}\alpha^2\sqrt{z} - 4\sqrt{2z}\right)}{\left(2\alpha-\sqrt{2z}-\right)\left(4\alpha - 4\epsilon\sqrt{2z} + \sqrt{2}z^{3/2} - 2\sqrt{2}\alpha^2\sqrt{z} + 4\sqrt{2z}\right)},$$

$$w_3 = 1 + \frac{\sqrt{2z}\left(8\epsilon z - 2z^2 + 4\alpha^2 z + 4\sqrt{2z}\alpha - 16\right)}{8\alpha - 2\sqrt{2}z^{3/2}\left(\alpha^2 + 2\epsilon - 3\right) + 4\alpha z\left(\alpha^2 + 2\epsilon - 1\right) - 8\sqrt{2z}\left(\alpha^2 + \epsilon - 1\right) + \sqrt{2}z^{5/2} - 2\alpha z^2},$$

$$w_4 = \frac{2\sqrt{2} - 2\alpha\sqrt{z} - \sqrt{2}z}{2\sqrt{2} - 2\alpha\sqrt{z} + \sqrt{2}z},$$

$$w_5 = \frac{2\alpha + \sqrt{2}z}{2\alpha - \sqrt{2}z},$$

$$w_6 = -\frac{-4\alpha + \sqrt{2}z^{3/2} + 2\sqrt{2}\left(\alpha^2 - 1\right)\sqrt{z} + 4\alpha z}{4\alpha - 2\sqrt{2}\sqrt{z}\left(\alpha^2 + 2\epsilon - 2\right) + \sqrt{2}z^{3/2}},$$

where again we have $\alpha = \frac{u'}{u}$.

Finally, for a second-order transformation the natural ladder operators $L^{\pm} = A^{+}(a^{\pm})^2 A$ of H_2 are now of sixth order. So, we find ourselves in the need of a slight generalization of the reduction theorem employed above. Such generalization requires now that $u_2 = (a^-)^2 u_1$, with $\epsilon_2 = \epsilon_1 - 2$, thus providing us with the fourth order ladder operators $l^{\pm}$ for H_2 such that

$$l^+ l^- = \left(H_2 - \frac{1}{2}\right)\left(H_2 - \frac{3}{2}\right)(H_2 - \epsilon_1 + 2)(H_2 - \epsilon_1 - 2).$$

Analytic expressions for solutions to the PV equation obtained through this method, with a second-order transformation and an arbitrary factorization energy ϵ_1, are too long to be depicted here. However, some simple examples for specific values of ϵ_1 can be shown.

If we set $\varepsilon_1 = \epsilon_1 - 2$, $\varepsilon_2 = \frac{3}{2}$, $\varepsilon_3 = \frac{1}{2}$, $\varepsilon_4 = \epsilon_1 + 2$, we get:

$$w(z) = \frac{3 - z}{2} \quad \text{with} \quad \epsilon_1 = \frac{3}{2} \quad \text{and } u_1(x) \text{ odd}$$

For the identification $\varepsilon_1 = \frac{1}{2}$, $\varepsilon_2 = \epsilon_1 - 2$, $\varepsilon_3 = \frac{3}{2}$, $\varepsilon_4 = \epsilon_1 + 2$, it is obtained:

$$w(z) = \frac{1 - z}{2} \quad \text{with} \quad \epsilon_1 = \frac{1}{2} \quad \text{and } u_1(x) \text{ even}$$

Finally, for the permutation $\varepsilon_1 = \epsilon_1 - 2$, $\varepsilon_2 = \epsilon_1 + 2$, $\varepsilon_3 = \frac{1}{2}$, $\varepsilon_4 = \frac{3}{2}$, we arrive at:

$$w(z) = \frac{1 + 2z - z^2}{4 + 4z} \quad \text{with} \quad \epsilon_1 = \frac{3}{2} \quad \text{and } u_1(x) \text{ even}$$

6. Conclusions

In this work we have obtained sundry supersymmetric partners of the harmonic oscillator with an infinite potential barrier. In particular those generated from the first-order technique turned out to be isospectral to the truncated harmonic oscillator, while those obtained from second-order SUSY offered greater possibilities for spectral design, e.g., it is possible to erase one or two consecutive levels in the energy spectrum. It is also possible to add a new level to the original spectrum in almost any position on the energy axis.

For each parity choice of the transformation functions we have found the factorization energy domain allowing the non-singular first and second order supersymmetric transformations in $(0, \infty)$. It must be noted that, for the initial singular potential the first-order transformations with $u(x)$ even and the second-order ones with both $u_1(x)$ and $u_2(x)$ even behave in a peculiar way, since they transform the eigenfunctions of the harmonic oscillator which are not physical solutions of the original singular potential into those which are eigenfunctions of the new Hamiltonian and vice versa.

Even more, a simple and direct procedure to obtain several explicit solutions to the Painlevé IV and V equations was implemented using the extremal states associated to the supersymmetric partners of the harmonic oscillator with an infinite potential barrier at the origin.

Acknowledgments

The authors acknowledge the support of Conacyt (México), Project 152574. VSMS also acknowledges the Conacyt fellowship 243374.

References

[1] Infeld L 1941 *Phys. Rev.* **59** 737
[2] Infeld L and Hull T E 1951 *Rev. Mod. Phys.* **23** 21
[3] Dirac P A M 1957 *The Principles of Quantum Mechanics* (Oxford: Clarendon)
[4] Mielnik B 1984 *J. Math. Phys.* **25** 3387
[5] Cooper F, Khare A and Sukhatme U 1995 *Phys. Rep.* **251** 267
[6] Junker G and Roy P 1998 *Ann. Phys.* **270** 155
[7] Márquez I F, Negro J and Nieto L M 1998 *J. Phys. A: Math. Gen.* **31** 4115
[8] Quesne C and Vansteenkiste N 1999 *Helv. Phys. Acta* **72** 71
[9] Mielnik B and Rosas-Ortiz O 2004 *J. Phys. A: Math. Gen.* **37** 10007
[10] Carballo J M, Fernández D J, Negro J and Nieto L M 2004 *J. Phys. A: Math. Gen.* **37** 10349
[11] Contreras-Astorga A and Fernández D J 2008 *J. Phys. A: Math. Theor.* **41** 475303
[12] Marquette I 2009 *J. Math. Phys.* **50** 095202
[13] Fernández D J 2010 *AIP Conf. Proc.* **1287** 3
[14] Bermúdez D and Fernández D J 2011 *SIGMA* **7** 025
[15] Andrianov A A and Ioffe M V 2012 *J. Phys. A: Math. Theor.* **45** 503001
[16] Gómez-Ullate D, Grandati Y and Milson R 2013 (*Preprint* arXiv:1306.5143 [math-ph])
[17] Mateo J and Negro J 2008 *J. Phys. A: Math. Theor.* **41** 045204
[18] Fernández D J and Hussin V 1999 *J. Phys. A, Math. Gen.* **32** 3603
[19] Fernández A J, Negro J and Nieto L M 2004 *Phys. Lett. A* **324** 139
[20] Veselov A P and Shabat A B 1993 *Funct. Anal. Appl.* **27** 81
[21] Adler V E 1994 *Physica D* **73** 335
[22] Dubov S Y, Eleonskii V M and Kulagin N E 1994 *Chaos* **4** 47
[23] Bassom A P, Clarkson P A and Hicks A C 1995 *Stud. Appl. Math.* **95** 1
[24] Andrianov A A, Cannata F, Ioffe M and Nishnianidze D 2000 *Phys. Lett. A* **266** 341
[25] Bermúdez D, Fernández D J and Negro J 2014 (*Preprint* Cinvestav)
[26] Bermúdez D, Fernández D J 2011 *Phys. Lett. A* **375** 2974
[27] Bermúdez D 2012 *SIGMA* **8** 069
[28] Fernández D J and Morales-Salgado V S 2014 *J. Phys. A: Math. Theor.* **47** 035304
[29] Fernández D J and Morales-Salgado V S 2014 *J. Phys: Conf. Ser.* **512** 012024
[30] Aoyama H, Sato M and Tanaka T 2001 *Nucl. Phys. B* **619** 105
[31] Fernández D J and Fernández-García N 2005 *AIP Conference Proceedings* **744** 236
[32] Andrianov A A, Ioffe M V and Spiridonov V P 1993 *Phys. Lett. A* **174** 273
[33] Andrianov A A, Ioffe M V, Cannata F and Dedonder J P 1995 *Int. J. Mod. Phys. A* **10** 2683
[34] Bagrov V G and Samsonov B F 1997 *Phys. Part. Nucl.* **28** 374
[35] Fernández D J, Hussin V and Mielnik B 1998 *Phys. Lett. A* **244** 309
[36] Cariñena J F, Ramos A and Fernández D J 2001 *Ann. Phys.* **292** 42
[37] Iwasaki K, Kimura H, Shimomura S and Yoshida M 1991 *From Gauss to Painlevé* (Braunschweig: Friederich Vieweg and Sohn)

Auxiliary functions for rational ODEs with 2 and 3 singular values

G Filipuk

Institute of Mathematics, University of Warsaw, ul. Banacha 2, 02-097 Warsaw, Poland

E-mail: `filipuk@mimuw.edu.pl`

Abstract. We examine auxiliary functions which are related to certain second order nonlinear differential equations with 2 and 3 singular values. The auxiliary functions satisfy first order linear differential equations. It is shown that after imposing suitable conditions on the coefficients of the linear equations and on the form of the auxiliary functions one can reduce the nonlinear differential equations to the (modified) third and fifth Painlevé equations and the fourth Painlevé equation.

1. Introduction

The singularity structure of a given nonlinear differential equation is in general very complicated. Fixed singularities occur at points where the equation itself is singular in some sense. The location of the singularities may also depend on the initial conditions and such singularities are called movable. The coefficients of the equation are not necessarily singular at such points. It is known from the theory of ordinary differential equations [9] that linear equations do not have movable singularities. Nonlinear equations in general may have movable poles, algebraic branch points, essential singularities and others (see [9] for numerous examples of possible singularities).

Classes of second-order rational ordinary differential equations, admitting certain families of formal algebraic series solutions, is considered in [1, 2, 3, 10]. For all solutions of these equations, it is shown that any movable singularity that can be reached by analytic continuation along a finite-length curve is an algebraic branch point (or, in particular cases, a pole). As it is shown in [1, 3], this yields in particular the known fact that the only movable singularities of solutions of the Painlevé equations (P_I)—(P_{VI}) are poles for generic values of parameters. This is known as the Painlevé property. In the technical part of the proof certain auxiliary functions with certain properties appear and play an important role.

Solutions of the Painlevé equations are often referred to as the Painlevé transcendents or nonlinear special functions and have numerous and significant applications in mathematics and mathematical physics. The Painlevé equations are irreducible and satisfy second order nonlinear differential equations. The third, fourth and fifth Painlevé equations are given respectively by equations (10), (13), (15). There exist also modified versions of the third and fifth Painlevé equations (e.g., (6)) which are related to the original ones by a change of variables. For more information on the Painlevé equations the reader is referred to the book [5].

There are several proofs of the meromophic nature of the Painleveé equations (see, for instance, [8, 7, 6, 5, 10] and [1, 3] with further references therein). In [1, 2, 3, 10] the proof of the fact that any movable singularity that can be reached by analytic continuation along a rectifiable curve is an algebraic branch point relies on the Cauchy theorem for certain systems of differential equations in the complex domain and the existence of certain auxiliary functions (which are not unique) which allow constructing such systems and which remain bounded as the singularities are approached. These functions satisfy first-order linear ODEs with some potentially singular terms that can be removed. Each singular value of the equation should be treated separately and, hence, one needs to consider several bounded functions. Note that the study of similar auxiliary functions for the Painlevé equations was also done in [4] (and in the references therein).

The construction of auxiliary functions in [4, 3, 10] may look a bit artificial and therefore, in this paper we are interested to understand the construction of auxiliary functions for the Painlevé type equations in general. We consider several second-order rational ordinary differential equations of the form (3), (7), (11) with 2 singular values $w = 0, \infty$ and equations (14), (16) with 3 singular values $w = 0, 1, \infty$. In each case we analyse the auxiliary functions

$$V(z) = E(z,w)w'^2 + F(z,w)w' + G(z,w) \tag{1}$$

which satisfy the first order linear equations of the form

$$V' = P(z,w)V + Q(z,w)w' + R(z,w) \tag{2}$$

and which are related to the nonlinear differential equations for $w(z)$. We show that under suitable conditions on functions E, F, G, P, Q, R the nonlinear equations are reduced to the Painlevé equations. The most important conditions on the functions are that they are rational and P, Q and R are holomorphic at singular values of the nonlinear equation. We get the (modified) third and fifth Painlevé equations and the fourth Painlevé equation. By a similar argument as in Lemma 4.3 [10] one immediately gets that under those conditions the function $V(z)$ remains bounded as $w(z)$ stays away from singular values and thus the results of the present paper can be used to improve the proof of the Painlevé property for the Painlevé equation [10] if combined with the methods in [1, 2, 3]. Moreover, we show that there is always a freedom in the choice of P for the Painlevé equations and discuss several examples.

2. Main results

2.1. The modified third Painlevé equation

Let $w(z)$ be defined by the equation

$$w'' = \frac{w'^2}{w} + \sum_{n=-1}^{3} a_n(z)w^n, \tag{3}$$

where $' = d/dz$ and $a_n(z)$ are analytic functions of z. Singular values of this equation are $w = 0$ and $w = \infty$. Take

$$E(z,w) = \frac{1}{w^2} \tag{4}$$

in (1). The choice of E is unique if we assume that the function V satisfies equation (2).

Assume that the function G satisfying

$$\frac{\partial G}{\partial w} = -\frac{2}{w^2}\sum_{n=-1}^{3} a_n(z)w^n$$

is rational in w. This yields

$$G(z,w) = g(z) + \frac{a_{-1}(z)}{w^2} + \frac{2a_0(z)}{w} - 2a_2(z)w - a_3(z)w^2 \tag{5}$$

and $a_1(z) = 0$ (to avoid the logarithmic term). Here $g(z)$ is an arbitrary function.

By a simple calculation the function P satisfies

$$P = w\left(F + w\frac{\partial F}{\partial w}\right).$$

Since we want this function to be analytic at singular values $w(z) = 0$ and $w(z) = \infty$ and also be non-constant, there is a freedom of the choice. The function F should be rational in w, so one can assume that

$$P = \frac{p_1(z)}{w+1} + \frac{p_2(z)}{(w+1)^2}.$$

The functions $p_1(z) \neq 0$ and $p_2(z) \neq 0$ are to be determined from other conditions. By solving a differential equation

$$w\left(F + w\frac{\partial F}{\partial w}\right) = \frac{p_1(z)}{w+1} + \frac{p_2(z)}{(w+1)^2}$$

for $F(z, w)$ we get

$$F = \frac{f(z) - p_1(z)/(w+1) + \log(w/(w+1))(p_1 + p_2)}{w}.$$

The function F is rational if the condition $p_2(z) = -p_1(z)$ is imposed and it is given by

$$F(z, w) = \frac{p_2(z)}{w(w+1)} + \frac{f(z)}{w},$$

where $f(z)$ is an arbitrary function. The coefficient Q in the differential equation (2) is given by

$$Q = wF^2 - w^2 F F_w + F_z,$$

where $F_w = \partial F/\partial w$ and $F_z = \partial F/\partial z$. From the proof in [3] we observe that the functions P, Q and R in (2) should be holomorphic at $w = 0$ and $w = \infty$ and the function Q has a second order zero at $w = \infty$. Observe that with our choice of P as above, the function P is automatically holomorphic at $w = 0$ and $w = \infty$. Other conditions give the following system of differential equations:

$$\begin{aligned}
&p_2'(z) + f'(z) = 0, \quad f'(z) = 0,\\
&p_2(z)a_{-1}(z) + a_{-1}(z)f(z) + a_{-1}'(z) = 0,\\
&p_2(z)a_0(z) + a_0(z)f(z) + 2a_0'(z) = 0,\\
&a_3(z)f(z) - a_3'(z) = 0, \quad a_2(z)f(z) - 2a_2'(z) = 0,
\end{aligned}$$

which is solved by

$$\begin{aligned}
f(z) = C_1, \quad p_2(z) = C_2, \quad a_{-1}(z) = \delta e^{(-C_2 - C_1)z},\\
a_0(z) = \beta e^{(-C_2 - C_1)z/2}, \quad a_3(z) = \gamma e^{C_1 z}, \quad a_2 = \alpha e^{C_1 z/2},
\end{aligned}$$

where C_1 and C_2 are arbitrary constants. Finally, taking $C_1 = 2a$ and $C_2 - C_1 = 2b$, we get that the equation (3) has the following form:

$$w'' = \frac{w'^2}{w} + \gamma e^{2az} w^3 + \alpha e^{az} w^2 + \beta e^{bz} + \frac{\delta e^{2bz}}{w},$$

where a, b, α, β, γ, δ are arbitrary constants. In particular, if $C_1 = 0$, $C_2 = 2$ $(a = 0,\ b = 1)$ we get the modified third Painlevé equation given by

$$w'' = \frac{w'^2}{w} + \alpha w^2 + \beta + e^{2z}\left(\gamma w^3 + \frac{\delta}{w}\right). \tag{6}$$

For the modified third Painlevé equation we get the auxiliary function given by

$$V(z) = \frac{w'^2}{w} - \frac{2w'}{w(w+1)} - \gamma w^2 - 2\alpha w + \frac{2\beta e^z}{w} + \frac{\delta e^{2z}}{w^2} + g(z).$$

We can summarize the results in this section as follows.

Theorem 2.1 *Assume the following conditions for the coefficients of (1) and (2):*

(i) the function E satisfies (4);

(ii) the function G satisfying

$$\frac{\partial G}{\partial w} = -\frac{2}{w^2}\sum_{n=-1}^{3} a_n(z)w^n$$

is rational in w and is given by (5);

(iii) the function F satisfying

$$w\left(F + w\frac{\partial F}{\partial w}\right) = P := \frac{g_1(z)}{w+1} + \frac{g_2(z)}{(w+1)^2}$$

is rational in w, where $g_1(z)$ and $g_2(z)$ are functions to be determined from other conditions;

(iv) the functions P, Q and R in (2) are holomorphic at $w = 0$ and $w = \infty$;

(v) the function Q has a second order zero at $w = \infty$.

Then equation (3) has the following form:

$$w'' = \frac{w'^2}{w} + \gamma e^{2az}w^3 + \alpha e^{az}w^2 + \beta e^{bz} + \frac{\delta e^{2bz}}{w},$$

where a, b, α, β, γ, δ are arbitrary constants.

We remark that the function P should have at least a second order pole at $w = -1$ and by taking more terms in the expression of P we can get a similar statement. If the function P is either constant or has the first order pole, then the function F is not rational. We also note that we can choose another function P with a pole of second order at $w = 1$ and get the same modified Painlevé equation. Moreover, this function is equal (up to the arbitrary function $g(z)$) to the function U in [7].

2.2. The third Painlevé equation

Let the function $w(z)$ satisfy the equation

$$w'' = \frac{w'^2}{w} - \frac{w'}{z} + \sum_{n=-1}^{3} a_n(z)w^n, \tag{7}$$

where $' = d/dz$ and $a_n(z)$ are analytic functions of z as before. Singular values of this equation are $w = 0$ and $w = \infty$. We take

$$E(z,w) = \frac{z^2}{w^2} \tag{8}$$

in (1).

Theorem 2.2 *Assume condition (8) and the following conditions for the coefficients of (1) and (2):*

(i) the function G satisfying

$$\frac{\partial G}{\partial w} = -\frac{2z^2}{w^2}\sum_{n=-1}^{3} a_n(z)w^n$$

is rational in w;

(ii) the function F satisfying

$$\frac{w}{z^2}\left(F + w\frac{\partial F}{\partial w}\right) = P := \frac{p_1(z)}{w-1} + \frac{p_2(z)}{(w-1)^2}$$

with $p_1(z) \neq 0$, $p_2(z) \neq 0$ is rational in w;

(iii) the functions P, Q and R in (2) are holomorphic at $w = 0$ and $w = \infty$;

(iv) the function Q has a second order zero at $w = \infty$.

Then equation (7) has the following form:

$$w'' = \frac{w'^2}{w} - \frac{w'}{z} + \gamma z^{2a}w^3 + \alpha z^{a-1}w^2 + \beta z^b + \frac{\delta z^{2b+2}}{w}, \tag{9}$$

where a, b, α, β, γ, δ are arbitrary constants.

Proof 1 *Condition 1 implies $a_1(z) = 0$ since*

$$G(z,w) = g(z) + \frac{a_{-1}(z)z^2}{w^2} + \frac{2z^2a_0}{w} - 2z^2a_1(z)\log w - 2z^2a_2(z)w - z^2a_3(z)w^2.$$

In this case the function G does not have logarithmic terms. As in the previous section, we could choose $P \neq 0$ equal to any function which is holomorphic at $w = 0$ and $w = \infty$ such that the differential equation for F has a rational solution. If $P(z,w) = p_1(z)/(w-1)$, then F has a logarithmic term if $p_1(z) \neq 0$. Hence, if we choose P as in the theorem, then

$$F(z,w) = \frac{f(z) + z^2((p_1(z) - p_2(z))\log((w-1)/w) - p_2(z)/(w-1))}{w}.$$

Condition 2 implies $p_1(z) = p_2(z)$ and the function

$$F(z,w) = \frac{f(z)}{w} - \frac{z^2p_1(z)}{w(w-1)},$$

where $f(z)$, $p_1(z)$ are the functions to be determined from other conditions. We have already chosen the function P such that it is automatically holomorphic at $w = 0$ and $w = \infty$. Substituting into

$$Q = \frac{z^2w^2F_z - w^4FF_w - w^3F^2 - zw^2F}{z^2w^2}$$

and to the expression for the function R, which is cumbersome so we omit it, we analyse other conditions in the theorem on Q and R. Their holomorphicity at $w = 0$ and $w = \infty$ gives the following system of differential equations:

$$\begin{aligned}
&z^2p_1(z) - f(z) + z^3p_1'(z) + zf'(z) = 0,\\
&zf'(z) - f(z) = 0,\ \ 2z^2a_2'(z) - a_2(z)f(z) + 4za_2(z) = 0,\\
&2za_{-1}(z) + z^2p_1(z)a_{-1}(z) + a_{-1}(z)f(z) + z^2a_{-1}'(z),\\
&4za_0(z) + z^2p_1(z)a_0(z) + a_0(z)f(z) + 2z^2a_0'(z) = 0,\\
&z^2a_3'(z) - a_3(z)f(z) + 2za_3(z) = 0,
\end{aligned}$$

which is solved by

$$f(z) = az, \ p_1(z) = b/z, \ a_{-1}(z) = \delta z^{-2-a-b},$$
$$a_0(z) = \beta z^{-2-a/2-b/2}, \ a_3(z) = \gamma z^{a-2}, \ a_2 = \alpha z^{a/2-2}.$$

This yields the statement of the theorem after renaming the constants $a \to 2(1+a), b \to -2(3+a+b)$. We remark that one can take more terms in the expression of P and get a similar statement.

In particular, if $b=-1$, $a=0$ in (9), we get the third Painlevé equation given by

$$w'' = \frac{w'^2}{w} - \frac{w'}{z} + \frac{1}{z}\left(\alpha w^2 + \beta\right) + \gamma w^3 + \frac{\delta}{w}, \tag{10}$$

where α, β, γ, δ are arbitrary complex parameters. In this case the auxiliary function $V(z)$ is given by

$$V(z) = \frac{z^2 w'^2}{w^2} + \frac{2z(w+1)w'}{w(w-1)} - \gamma z^2 w^2 - 2\alpha z w + \frac{2\beta z}{w} + \frac{\delta z^2}{w^2} + g(z).$$

One can immediately see that the function coincides with the function U in [10] up to the arbitrary function $g(z)$.

2.3. The fourth Painlevé equation

Assume that the function $w(z)$ satisfies an equation of the form

$$w'' = \frac{w'^2}{2w} + \sum_{n=-1}^{3} a_n(z) w^n \tag{11}$$

with singular values $w=0$ and $w=\infty$.

Theorem 2.3 *Let*

$$E(z,w) = \frac{1}{w}$$

in (1). We impose the following conditions on the coefficients of (1) and (2):

(i) the function G satisfying

$$\frac{\partial G}{\partial w} = -\frac{2}{w}\sum_{n=-1}^{3} a_n(z) w^n$$

is rational in w;

(ii) the function F satisfying

$$\frac{1}{2}F + w\frac{\partial F}{\partial w} = P := \frac{p_1(z)}{w+1} + \frac{p_2(z)}{(w+1)^2}$$

is rational in w with $p_1(z) \neq 0$ and $p_2(z) \neq 0$;

(iii) the functions P, Q and R are holomorphic at $w=0$ and $w=\infty$;

(iv) the function Q has a second order zero at $w=\infty$.

Then equation (11) has the following form:

$$w'' = \frac{w'^2}{2w} + a_3 w^3 + \left(a_2 - \frac{4a_3 cz}{3}\right) w^2 + \frac{1}{6}\left(cz(2a_3 cz - 3a_2) - 12\alpha\right) w + \frac{\beta}{w}, \tag{12}$$

where a_2, a_3 and c are arbitrary constants.

Proof 2 *Condition 1 implies* $a_0(z) = 0$ *since* $G(z,w) = g(z) + 2a_{-1}(z)/w - 2a_0(z)\log w - 2a_1(z)w - a_2(z)w^2 - 2a_3(z)w^3/3$. *We further get* $p_2(z) = -2p_1(z)$ *since*

$$F(z,w) = \frac{f(z) + \mathrm{arctg}\sqrt{w}(2p_1(z) + p_2(z)) + p_2(z)\sqrt{w}/(w+1)}{\sqrt{w}}.$$

Thus, the function

$$F(z,w) = -\frac{2p_1(z)}{w+1},$$

where $p_1(z)$ *is to be determined from other conditions given by*

$$p_1'(z) = 0, \;\; a_{-1}'(z) = 0, \;\; a_3'(z) = 0$$
$$4p_1(z)a_3(z) + 3a_2'(z) = 0, \;\; 2a_1'(z) + p_1(z)a_2(z) = 0.$$

Solving for coefficients $a_i(z)$ *and* $p_1(z)$ *we get the statement of the theorem.*

In particular, if $a_3 = 3/2$, $a_2 = 0$, $c = -2$ in (12) we get the fourth Painlevé equation given by

$$w'' = \frac{w'^2}{2w} + \frac{3}{2}w^3 + 4zw^2 + 2(z^2 - \alpha)w + \frac{\beta}{w}, \tag{13}$$

where α and β are arbitrary complex parameters. The auxiliary function is given by

$$V(z) = \frac{w'^2}{w} + \frac{4w'}{w+1} + \frac{2\beta}{w} - 4(z^2 - \alpha)w - 4zw^2 - w^3 + g(z).$$

One can analise the auxiliary functions in [10] and [5] to see that up to a choice of P and a constant functions our construction gives essentially the same auxiliary functions.

2.4. The fifth Painlevé equation

Assume that the function $w(z)$ satisfies

$$w'' = \left(\frac{1}{2w} + \frac{1}{w-1}\right)w'^2 - \frac{w'}{z} + \sum_{n=-1}^{3} a_n(z)w^n + \frac{b(z)}{w-1} \tag{14}$$

with singular values $w = 0$, 1 and $w = \infty$. We choose the function

$$V(z) = \frac{z^2 w'^2}{w(w-1)^2} + F(z,w)w' + G(z,w)$$

which satisfies (2). Similarly to the previous subsections we state the following theorem.

Theorem 2.4 *Let the function* G *satisfying*

$$\frac{\partial G}{\partial w} = -\frac{2z^2}{w(w-1)^2}\left(\sum_{n=-1}^{3} a_n(z)w^n + \frac{b(z)}{w-1}\right)$$

be rational in w. *Take the function* F *in the form*

$$F(z,w) = -\frac{2z^2 p(z)}{3(w-1)(w+1)^2}$$

which satisfies

$$\frac{(w-1)(F(3w-1)+2w(w-1)F_w)}{2z^2} = P := \frac{p(z)}{(w+1)} - \frac{10p(z)}{3(w+1)^2} + \frac{8p(z)}{3(w+1)^3}.$$

If the functions P, Q and R are holomorphic at $w=0$, 1 and $w=\infty$ and the function Q has a second order zero at $w=\infty$, then the coefficients in (14) are given by

$$b(z) = 2a_{-1}(z) + a_0(z) - a_2(z) - 2a_3(z), \quad p(z) = \frac{c}{z}, \quad a_3(z) = -\frac{a_2(z)}{2},$$

$$a_2(z) = -\frac{2\alpha}{z^2},$$

$$a_1(z) = \frac{\delta z^{c/6} + \gamma z^{c/12} + \beta + \alpha}{z^2},$$

$$a_0(z) = \frac{2\delta z^{c/6} - 2\beta}{z^2}, \quad a_{-1}(z) = \frac{\beta}{z^2},$$

where c, α, β, γ, δ are arbitrary. In particular, if $c = 12$, we get the fifth Painlevé equation given by

$$w'' = \left(\frac{1}{2w} + \frac{1}{w-1}\right)w'^2 - \frac{w'}{z} + \frac{(w-1)^2}{z^2}\left(\alpha w + \frac{\beta}{w}\right) + \gamma\frac{w}{z} + \delta\frac{w(w+1)}{w-1}. \tag{15}$$

The proof is similar to the previous sections.

Next we remark about the dependence of the resulting equation on the choice of P. Indeed, we make certain general assumptions (like the holomorphicity at certain points) on the coefficients in the proof of the theorem. Therefore, we expect that certain changes in the choice of P might lead to a different result. For example, the choice in [10] was that $F = -4z/(w^2-1)$. In this case one should take P in the form $P = 2(w-1)^2/(w+1)^2$. However, this function is holomorphic at $w = 0, 1, \infty$ and one can show that equation (14) is still (15). We could have chosen $P(z,w) = p(z) + p_1(z)/(w+1) + p_2(z)(w+1)^2$ and get (15) as well.

2.5. The modified fifth Painlevé equation

If the function $w(z)$ satisfies

$$w'' = \left(\frac{1}{2w} + \frac{1}{w-1}\right)w'^2 + \sum_{n=-1}^{3} a_n(z)w^n + \frac{b(z)}{w-1} \tag{16}$$

with singular values $w = 0$, 1 and $w = \infty$, and the function $V(z)$ is given by

$$V(z) = \frac{w'^2}{w(w-1)^2} + F(z,w)w' + G(z,w),$$

we have

Theorem 2.5 *Let the function G satisfying*

$$\frac{\partial G}{\partial w} = -\frac{2}{w(w-1)^2}\left(\sum_{n=-1}^{3} a_n(z)w^n + \frac{b(z)}{w-1}\right)$$

be rational in w. Take the function F in the form

$$F(z,w) = -\frac{2p(z)}{3(w-1)(w+1)^2}$$

which satisfies

$$\frac{1}{2}(w-1)\left(F(3w-1)+2w(w-1)\frac{\partial F}{\partial w}\right) = P := \frac{p(z)}{(w+1)} - \frac{10p(z)}{3(w+1)^2} + \frac{8p(z)}{3(w+1)^3}.$$

If the functions P, Q and R are holomorphic at $w = 0, 1$ and $w = \infty$ and the function Q has a second order zero at $w = \infty$, then equation (16) takes the following form:

$$w'' = \left(\frac{1}{2w} + \frac{1}{w-1}\right)w'^2 + (w-1)^2\left(\alpha w + \frac{\beta}{w}\right) + \gamma e^{cz/12} w + \delta e^{cz/6}\frac{w(w+1)}{w-1},$$

where c, α, β, γ, δ are arbitrary. In particular, if $c = 12$, we get the modified fifth Painlevé equation.

Proof 3 *The proof is computational. The function F is rational provided $b(z) = 2a_{-1}(z) + a_0(z) - a_2(z) - 2a_3(z)$ and $a_2(z) = -2a_3$. Moreover, one gets*

$$p(z) = c,\ \ a_{-1}(z) = \beta,\ \ a_3(z) = \alpha,\ \ a_0(z) = -2\beta + 2\delta e^{cz/6},$$
$$a_1(z) = \alpha + \beta + \delta e^{cz/6} + \gamma e^{cz/12}.$$

3. Discussion

In this paper we analised the principles of the construction of the auxiliary bounded functions related to the Painlevé equations. Auxiliary functions satisfy the property that they are analytic around every zero and pole for the (modified) third and fourth Painlevé equations and additionally at every 1-point for the (modified) fifth Painlevé equation. We showed that in particular, our construction yields the functions considered in [5, 7, 10] in the proofs of the Painlevé property of the Painlevé equations. In principle, similar computations can be done for the sixth Painlevé equation with 4 singular values $w = 0, 1, z, \infty$, but the calculations become cumbersome and more involved.

Acknowledgments

The author is grateful to Professor R. Halburd (University College London, UK) for illuminating discussions and helpful comments and suggestions. Calculations were partially obtained in the Interdisciplinary Centre for Mathematical and Computational Modelling (ICM), Warsaw University, within grant nr G34-18. Research is partially supported by Polish MNiSzW Grant No N N201 397937.

References

[1] Filipuk G and Halburd R 2009 *J. Math. Phys.* **50** 023509
[2] Filipuk G and Halburd R 2009 *Comput. Methods Funct. Theory* **9** 551
[3] Filipuk G and Halburd R 2009 *Stud. App. Math.* **123** 17
[4] Golubev V 1912 *Mat. Sbornik* **28** 323 (in Russian)
[5] Gromak V, Laine I and Shimomura S 2002 *Painlevé Differential Equations in the Complex Plane* (Berlin: de Gruyter)
[6] Hinkkanen A and Laine I 2001 *Rep. Univ. Jyväskylä Dep. Math. Stat.* **83** 133
[7] Hinkkanen A and Laine I 2001 *J. Anal. Math.* **85** 323
[8] Hinkkanen A and Laine I 1999 *J. Anal. Math.* **79** 345
[9] Ince E 1956 *Ordinary Differential Equations* (New York: Dover)
[10] Shimomura S 2003 *Japan. J. Math.* **29** 159

Journal of Physics: Conference Series **597** (2015) 012037 doi:10.1088/1742-6596/597/1/012037

Integral quantizations with POVM and some applications

R Fresneda[1] **and J-P Gazeau**[2]

[1] Federal University of ABC (UFABC) Santo André - SP - Brasil . CEP 09210-580

[2] APC, Université Paris Diderot, Sorbonne Paris-Cité 75205 Paris Cedex 13, France

[2] Centro Brasileiro de Pesquisas Fisicas Rua Xavier Sigaud 150, 22290-180 - Rio de Janeiro, RJ, Brazil

E-mail: `rodrigo.fresneda@ufabc.edu.br, gazeau@apc.univ-paris7.fr`

Abstract. We present quantizations of functions (or distributions) on a measure space viewed as a classical set. They are based on positive operator-valued measures. To illustrate the large range of potentialities of the method, we develop examples where the classical sets are the group SU(2), the (projective) Weyl-Heisenberg group, and the affine group. Applications to quantum cosmology and to the quantization of constraints are outlined.

1. General framework for quantization

Quantization of a "classical" object X, such as a phase space in mechanics, can be formulated as a linear map $f \mapsto A_f$ from the space of $\mathcal{C}(X)$ of complex-valued functions $f(x)$ on X to a space $\mathcal{A}(\mathcal{H})$ of linear operators A_f in some complex Hilbert space $\mathcal{H}$ (we omit here the necessary domain considerations). One further requires that

(i) the constant function $f = 1$ be mapped to the identity operator,

(ii) real functions be mapped to (essentially) self-adjoint or at least to symmetric operators A_f in $\mathcal{A}(\mathcal{H})$.

This broad definition can be applied to specific physical problems by adding appropriate structure to X and to $\mathcal{C}(X)$, such as measure, topology, manifold structure, closure under algebraic operations, etc. Moreover, one is free to define which functions or operators are observables, whose spectra have a physical meaning in terms of measurement.

Finally, given an appropriate structure, like a measure or a Poisson structure, one can define an unambiguous classical limit $\check{f} \mapsto f$ of quantum observables through X-labeled expectation values of A_f defining a new function, $\check{f}(x)$ (namely, "lower symbols")

We describe in Section 2 a general quantization procedure based on (normalized) positive operator-valued measures (POVM). We illustrate the method in Section 3 with the cases where the classical set X has a group structure, like SU(2), the projective Weyl-Heisenberg group, and the affine group. Section 4 is an application of the Weyl-Heisenberg integral quantization to simple constraints in the complex plane, showing the difference between two approaches.

2. Integral quantization with POVM

Integral quantization [1, 2, 3, 4, 5] is a generalization of the Berezin-Toeplitz-Klauder (or "anti-Wick" or standard coherent state) quantization. It can be formulated as follows.

One starts from a measure space $(X, \mathfrak{B}, \nu)$ (or (X, ν) for short), where $\mathfrak{B}$ is the σ-algebra of ν-measurable subsets, a separable Hilbert space $\mathcal{H}$, and an X-labelled family of positive semi-definite and unit trace operators, i.e. density operators, on $\mathcal{H}$,

$$X \ni x \mapsto \rho(x) \in \mathcal{L}(\mathcal{H}), \quad \rho(x) > 0, \quad \mathrm{tr}(\rho(x)) = 1, \tag{1}$$

resolving the identity I on $\mathcal{H}$,

$$\int_X \rho(x)\, \mathrm{d}\nu(x) = I, \quad \text{in a weak sense.} \tag{2}$$

One then equips X with a suitable topology, so the normalized positive operator-valued measure (POVM) $\mathfrak{m}_\rho$ on the corresponding σ-algebra $\mathfrak{B}(X)$ of Borel sets is defined through the following map of

$$\mathfrak{B}(X) \ni \Delta \mapsto \mathfrak{m}_\rho(\Delta) = \int_\Delta \rho(x)\, \mathrm{d}\nu(x) \tag{3}$$

into the convex cone $\mathcal{L}^+(\mathcal{H})$ of positive bounded operators on $\mathcal{H}$. As a result, the resolution of the identity produces the X-labeled family of probability distributions

$$X \ni x_0, x \mapsto p_{x_0}(x) = \mathrm{tr}\,(\rho(x_0)\, \rho(x)) \tag{4}$$

on (X, ν). The expression $p_{x_0}(x)$ measures the degree of localization of x w.r.t. x_0, in a sense determined by the family $\rho(x)$, and vice versa due to the symmetry $p_{x_0}(x) = p_x(x_0)$ on the measure space (X, ν). In the particular case where $\rho(x)$ is rank-one projector operator (i.e. "pure coherent state"),

$$\rho(x) = |x\rangle\, \langle x|\,, \quad \langle x\, |x\rangle = 1\,, \tag{5}$$

one has

$$p_{x_0}(x) = |\, \langle x_0\, |x\rangle\, |^2\,, \tag{6}$$

From the fact that any density operator ρ is Hilbert-Schmidt, with norm $\|\rho\| = \sqrt{\mathrm{tr}\rho\rho^\dagger} = \sqrt{\mathrm{tr}\rho^2}$, one can introduce the associated distance

$$d_{\mathrm{HS}}(x, x') = \|\rho(x) - \rho(x')\| = \sqrt{\mathrm{tr}(\rho(x) - \rho(x'))^2}\,. \tag{7}$$

This object forces any pair of points in X to be finitely separated since we have

$$d_{\mathrm{HS}}(x, x') = \sqrt{\mathrm{tr}\,((\rho(x))^2 + (\rho(x'))^2 - 2\rho(x)\rho(x'))} \leq \sqrt{2}\sqrt{1 - \mathrm{tr}(\rho(x)\rho(x'))} \leq \sqrt{2}\,. \tag{8}$$

The integral POVM quantization is the linear map

$$f \mapsto A_f = \int_X \rho(x)\, f(x)\, \mathrm{d}\nu(x)\,. \tag{9}$$

This should be understood in terms of the sesquilinear form

$$B_f(\psi_1, \psi_2) = \int_X \langle\psi_1|\, \rho(x)\, |\psi_2\rangle\; f(x)\, \mathrm{d}\nu(x)\,, \tag{10}$$

defined on a dense subspace of $\mathcal{H}$. There is a caveat: if f is real and at least semi-bounded, Friedrich extension of B_f univocally defines a self-adjoint operator. However, if f is not semi-bounded, there is no natural choice of a self-adjoint operator associated with B_f. We must give more details about the Hilbert space and operator domains, see for instance [6].

Self-adjointness is essential in the Copenhagen-Dirac-von Neumann interpretation of quantum physics, since the spectral theorem for bounded or unbounded self-adjoint operators is the key for (sharp) quantum measurement. Hence, if A_f is self-adjoint, its spectrum, which is determined by its projector-valued (PV) spectral measure E_f corresponding to the integral representation

$$A_f = \int_{\mathbb{R}} \lambda \, \mathrm{d}E_f(\lambda)\,, \tag{11}$$

might have a remote connection with its classical spectrum $\{f(x)\,,\, x \in X\}$ appearing in the integral representation (9). While the POVM used for quantization - and built from a family $\rho(x)$ resolving the identity with respect to the fixed measure ν - should be considered as a frame to analyse such functions on X, the PV measure in the integral representation (11) is proper to the quantum observable and to functions of it. However, there are simple examples (see below for position and momentum in the Weyl-Heisenberg case) where classical and quantum spectra can be considered as identical regardless of the difference between their respective PV and POVM. Moreover, the frame $x \mapsto \rho(x)$ itself may be associated to a specific system to be quantized. A nice pedagogical example (the seastar) is presented in Chap. 11 of [2].

2.1. *Semi-classical aspects*

Some of the properties of the operator A_f can be grasped by examining the lower (Lieb) or covariant (Berezin) symbol, which in the present generalization reads as

$$A_f \mapsto \check{f}(x) := \mathrm{tr}(\rho(x)\, A_f)\,. \tag{12}$$

This function could be viewed as a sort of Wigner function [1]. However, it has a real probabilistic content, since it is the local averaging of the classical function f with respect to the probability distribution $\mathrm{tr}(\rho(x)\rho(x'))$

$$f(x) \mapsto \check{f}(x) = \int_X f(x')\, \mathrm{tr}(\rho(x)\rho(x'))\, \mathrm{d}\nu(x')\,. \tag{13}$$

The Bargmann-Segal-like map $f \mapsto \check{f}$ above is in general a regularization of the function f. Indeed, depending on the topology on X, the original f can be singular. It can be even a distribution (see below).

2.2. *Quantum measurement: sharp or unsharp?*

In the quantum context for which A_f is a self-adjoint operator viewed as an observable for a given system, with spectral decomposition (11), and given a density operator $\rho_m = \sum_i q_i \left|\phi_i\right\rangle \left\langle\phi_i\right|$ describing the mixed state of an ensemble, one interprets the quantity

$$\mathrm{tr}\left(\rho_m \mathrm{d}E_f(\lambda)\right) \tag{14}$$

à la Copenhagen-Dirac-von Neumann, that is, as the probability of measuring the spectral value λ. Now, the expectation value of the measurement of the observable A_f when the system is in the state ρ_m is given by the alternative expressions

$$\mathrm{tr}\left(\rho_m A_f\right) = \int_{\mathbb{R}} \lambda\, \mathrm{tr}\left(\mathrm{d}E_f(\lambda)\right) \qquad \text{(PV, sharp)} \tag{15}$$

$$= \int_X f(x)\, \mathrm{tr}(\rho_m \rho(x))\, \mathrm{d}\nu(x) \qquad \text{(POVM, unsharp)}\,. \tag{16}$$

The latter is the average of the original f with respect to the probability density $p_m(x) := \mathrm{tr}(\rho_m \rho(x))$. Hence, it might be viewed as an unsharp measurement, possibly through some marginal integration [7, 8]. Note that when the ρ_m is chosen as the element $\rho_m = \rho(x_0)$ of the family of density operators (1), the measurement (15) has the averaging interpretation given by (13).

3. Covariant integral quantization with UIR of a group

3.1. The material

Let G be a Lie group with left Haar measure $\mathrm{d}\mu(g)$, and let $g \mapsto U(g)$ be a unitary irreducible representation (UIR) of G in a Hilbert space $\mathcal{H}$. Consider a bounded operator M on $\mathcal{H}$ such that the operator

$$R := \int_G \mathsf{M}(g)\,\mathrm{d}\mu(g)\,, \quad \mathsf{M}(g) := U(g)\,\mathsf{M}\,U^{\dagger}(g)\,, \tag{17}$$

is defined in a weak sense. From the left invariance of $\mathrm{d}\mu(g)$, R commutes with all operators $U(g)$, $g \in G$, and so from Schur's Lemma, $R = c_{\mathsf{M}} I$ with

$$c_{\mathsf{M}} = \int_G \mathrm{tr}\left(\rho_0\,\mathsf{M}(g)\right)\,\mathrm{d}\mu(g)\,, \tag{18}$$

where the unit trace positive operator ρ_0 is chosen in order to make the integral convergent. The resolution of the identity follows:

$$\int_G \mathsf{M}(g)\,\mathrm{d}\nu(g) = I\,, \quad \mathrm{d}\nu(g) := \mathrm{d}\mu(g)/c_{\mathsf{M}}\,. \tag{19}$$

For a square-integrable UIR U for which ρ is an "admissible" density operator, $c(\eta) = \int_G \mathrm{d}\mu(g)\,|\mathrm{tr}(\rho U(g))|^2 < \infty$, the resolution of the identity is then obeyed by the family

$$\rho(g) = U(g)\rho U^{\dagger}(g) \tag{20}$$

This allows for a *covariant* integral quantization of complex-valued functions on the group

$$f \mapsto A_f = \int_G \rho(g)\,f(g)\,\mathrm{d}\nu(g)\,, \tag{21}$$

which is covariant in the sense that

$$U(g) A_f U^{\dagger}(g) = A_{\mathsf{U}(g)f}\,, \tag{22}$$

where $g \mapsto \mathsf{U}(g)$ is the regular representation of G for $f \in L^2(G, \mathrm{d}\mu(g))$

$$(\mathsf{U}(g)f)(g') := f(g^{-1}g')\,. \tag{23}$$

The semiclassical portrait and the corresponding classical limit are given via the lower symbol

$$\check{f}(g) := \int_G tr(\rho(g)\,\rho(g'))\,f(g')d\nu\,(g')\;, \tag{24}$$

a generalization of the Berezin transform on G.

3.2. Illustration with SU(2)

A proper rotation in space $\mathcal{R}(\omega,\hat{n}) : \vec{r} \mapsto \vec{r'}$, where $\hat{n}$ a unit vector $\hat{n}$ and ω a rotation angle $0 \leq \omega < 2\pi$, can be defined as

$$\vec{r'} = \mathcal{R}(\omega,\hat{n}) \cdot \vec{r} = \vec{r} \cdot \hat{n}\, \hat{n} + \cos\omega\, \hat{n} \times (\vec{r} \times \hat{n}) + \sin\omega\, (\hat{n} \times \vec{r}) \,. \tag{25}$$

In scalar-vector quaternionic form,

$$(0,\vec{r'}) = \xi(0,\vec{r})\bar{\xi}\,,\ \xi := \left(\cos\frac{\omega}{2}, \sin\frac{\omega}{2}\,\hat{n}\right) ,$$

with $q = (q_0, \vec{q}) \in \mathbb{H}$ and

$$qq' := (q_0 q'_0 - \vec{q}\cdot\vec{q'}, q'_0\vec{q} + q_0\vec{q'} + \vec{q}\times\vec{q'}) \,.$$

The (quaternionic) conjugate of $q = (q_0, \vec{q})$ is $\bar{q} = (q_0, -\vec{q})$, the squared norm is $\|q\|^2 = q\bar{q}$, and the inverse of a nonzero quaternion is $q^{-1} = \bar{q}/\|q\|^2$. Unit quaternions, i.e., quaternions with norm 1, form the multiplicative subgroup isomorphic to SU(2).

The correspondence between the canonical basis of $\mathbb{H} \simeq \mathbb{R}^4$, $(e_0 \equiv 1, e_1, e_2, e_3)$, and the Pauli matrices is

$$e_0 \leftrightarrow \sigma_0\,,\ e_a \leftrightarrow (-1)^{a+1} i\sigma_a\,,\ a = 1,2,3\,. \tag{26}$$

Unit quaternions constitute the three-sphere S^3. In hypersherical coordinates (α,θ,ϕ) of S^3 one has

$$\mathrm{S}^3 \ni \xi = \|q\|\,(\cos\alpha, \sin\alpha\,\hat{n})\,,\quad 0 \leq \alpha, \theta \leq \pi\,,\ 0 \leq \phi < 2\pi\,,$$
$$\dot{\xi} = \sin^2\alpha\,\sin\theta\,\mathrm{d}\alpha\,\mathrm{d}\theta\,\mathrm{d}\phi\,,\quad \int_{\mathrm{SU}(2)} \dot{\xi} = 2\pi^2\,.$$

The unit ball $\mathbb{B}$ in $\mathbb{R}^3$ parametrizes the set of 2×2 complex density matrices ρ. For general 2×2 complex density matrices, $\|\vec{a}\| \leq 1$, this becomes

$$\rho = \frac{1}{2}(1 + \vec{a}\cdot\vec{\sigma}) = \frac{1}{2}(1 - i(0,\vec{r})) \equiv \frac{1}{2}(1 - i\underset{\sim}{\mathbf{r}}) \equiv \rho_{\underset{\sim}{\mathbf{r}}}\,,\ r_i = (-1)^{i+1} a_i\,. \tag{27}$$

If $\|\vec{a}\| = 1$, i.e. $\vec{a} \in S^2$ ("Bloch sphere" in this context), with spherical coordinates (θ,ϕ), then ρ is the pure state $\rho = |\theta,\phi\rangle\langle\theta,\phi|$. For $\xi \in \mathrm{SU}(2)$, one defines the family of density matrices labeled by ξ:

$$\rho(\xi) := \xi\rho\bar{\xi} = \frac{1}{2}(1 - i\xi\underset{\sim}{\mathbf{r}}\bar{\xi})\,. \tag{28}$$

It is then straightforward to prove the resolution of the identity

$$\int_{\mathrm{SU}(2)} \rho(\xi)\,\frac{\dot{\xi}}{\pi^2} = I\,, \tag{29}$$

which allows us to quantize functions on $SU(2)$ or S^3 along the linear map

$$f(\xi) \mapsto A_f = \int_{\mathrm{SU}(2)} f(\xi)\,\rho(\xi)\,\frac{\dot{\xi}}{\pi^2} \in \mathrm{M}(2,\mathbb{C})\,. \tag{30}$$

The symbols corresponding to the Pauli matrices can be obtained via the Hopf map $S^3 \to S^2$, where the Hopf fibration is understood in terms of the transitive action of rotations on S^2.

Let us fix the North pole unit vector $\left(0,\hat{k}\right) \in S^2$. Then $\chi = \xi\left(0,\hat{k}\right)\bar{\xi}$ is the image of the rotation by the unit quaternion ξ. The components of χ correspond to the operators

$$A_{\chi_1} = \frac{a_3}{3}\sigma_1\,, \; A_{\chi_2} = -\frac{a_3}{3}\sigma_2\,, \; A_{\chi_3} = \frac{a_3}{3}\sigma_3\,. \tag{31}$$

With the redefinitions $2\omega = \beta$, $\psi_2 - \psi_1 = \gamma + \frac{\pi}{2}$, $\psi_1 + \psi_2 - 2\pi = -\alpha + \frac{\pi}{2}$, the algebra of Euler angle operators becomes

$$[A_\beta, A_\alpha] = iA_\gamma\,, \; [A_\gamma, A_\beta] = iA_\alpha\,, \; [A_\alpha, A_\gamma] = iA_\beta\,, \tag{32}$$

i.e., the Lie algebra of SU(2), as we also had with the quantization of the Hopf map. The Euler parametrization of SU(2) seems to be privileged from an algebraic point of view.

3.3. Weyl-Heisenberg covariant integral quantization

In the absence of square-integrability over G, there exists a definition of square-integrable covariant coherent states with respect to a left coset manifold $X = G/H$, with H a closed subgroup of G, equipped with a quasi-invariant measure ν [2]. For a global Borel section $\sigma : X \to G$ of the group, let ν_σ be the unique quasi-invariant measure defined by

$$\mathrm{d}\nu_\sigma(x) = \lambda(\sigma(x), x)\mathrm{d}\nu(x)\,, \tag{33}$$

where $\lambda(g, x)\mathrm{d}\nu(x) = \mathrm{d}\nu(g^{-1}x)$, $(\forall g \in G)$. Let U be a UIR which is square integrable mod(H) and ρ mod(H) an admissible density operator, i.e. $c_\eta := \int_X \mathrm{tr}\left(\rho\,\rho_\sigma(x)\right)\mathrm{d}\nu_\sigma(x) < \infty$. With $\rho_\sigma(x) = U(\sigma(x))\rho U(\sigma(x))^\dagger$, we have the resolution of the identity and the resulting quantization

$$f \mapsto A_f = \frac{1}{c_\eta}\int_X f(x)\,\rho_\sigma(x)\,\mathrm{d}\nu_\sigma(x) = I\,. \tag{34}$$

In the case of the Weyl-Heisenberg group, X is the coset space $G_{HW}/$(phase subgroup) $\sim \mathbb{C}$, while the measure is the Lebesgue measure in the complex plane $\mathrm{d}^2z = \mathrm{d}z \wedge \mathrm{d}\bar{z}/2i$. The triplet $\{a, a^\dagger, I\}$ generates the Weyl-Heisenberg algebra characterized by the canonical commutation rule $\left[a, a^\dagger\right] = I$. To each $z \in \mathbb{C}$ corresponds the (unitary) displacement operator $D(z)$, encoding the CCR

$$\mathbb{C} \ni z \mapsto D(z) = e^{za^\dagger - \bar{z}a}\,, \; D(z)D(z') = e^{\frac{1}{2}(z\bar{z}' - \bar{z}z')}D(z+z')\,.$$

Let $\varpi(z)$ be a function on the complex plane defining a bounded operator $\mathsf{M} := \int_{\mathbb{C}} D(z)\,\varpi(z)\frac{\mathrm{d}^2z}{\pi}$ and such that $\varpi(0) = 1$. The family of displaced operators $\mathsf{M}(z) := D(z)\mathsf{M}D(z)^\dagger$ resolves the identity

$$\int_{\mathbb{C}} \mathsf{M}(z)\,\frac{\mathrm{d}^2z}{\pi} = I\,, \tag{35}$$

and the resulting quantization map is given by

$$f \mapsto A_f = \int_{\mathbb{C}} f(z)\,\mathsf{M}(z)\,\frac{\mathrm{d}^2z}{\pi}\,, \quad A_{f(z-z_0)} = D(z_0)A_{f(z)}D(z_0)^\dagger\,. \tag{36}$$

For $q = \frac{\sqrt{2}}{2}\left(z + \bar{z}\right)$ and $p = \frac{\sqrt{2}}{2i}\left(z - \bar{z}\right)$, one has

$$A_qA_p - A_pA_q = i\left[a, a^+\right] = I\,, \tag{37}$$

i.e., the canonical commutation rule is preserved by the set of quantizations parametrized by functions $\varpi(z)$. In particular, the Cahill-Glauber function [9] $\varpi_s(z) = e^{s|z|^2/2}$ for $s = 0$ corresponds to the Wigner-Weyl integral quantization, while the cases $s = -1, 1$ correspond, respectively, to the CS (anti-normal) and normal quantization ($s = 1$ should be understood in an asymptotic sense). Only for $s \leq -1$ is the operator $\mathsf{M} = \int_{\mathbb{C}} D(z)\, \varpi(z) \frac{\mathrm{d}^2 z}{\pi}$ a density operator, which gives rise to a POVM.

The choice of weight function $\varpi(z)$ determines different orderings. For the simplest one involving the classical q and p, we have

$$A_{qp} = A_q A_p - \frac{i}{2} + \partial_{\bar{z}}^2 \varpi\big|_{z=0} - \partial_z^2 \varpi\big|_{z=0} - \left(\partial_{\bar{z}} \varpi\big|_{z=0}\right)^2 + \left(\partial_z \varpi\big|_{z=0}\right)^2 . \tag{38}$$

For instance, with $\varpi(z) = (cz + 1)e^{s|z|^2}$, $c, s \in \mathbb{C}$, $\mathrm{Re}\, s < 1$, we have $A_{qp} = A_q A_p - \frac{i}{2} + c^2$, and putting $c = e^{i\pi/4}/\sqrt{2}$ leads to the so-called xp-quantization $A_{qp} = A_q A_p$. For ϖ chosen *real* and *even*, one has $A_z = a\,, \quad A_{\overline{f(z)}} = A^\dagger_{f(z)}$. Only for real even $\varpi(z)$ one arrives at the correct energy spectrum of the harmonic oscillator. In general,

$$A_{|z|^2} = a^\dagger a + \frac{1}{2} - \partial_z \partial_{\bar{z}} \varpi\big|_{z=0} + a\, \partial_z \varpi\big|_{z=0} - a^\dagger\, \partial_{\bar{z}} \varpi\big|_{z=0} . \tag{39}$$

3.4. Affine integral quantization

3.4.1. The material As the Wigner-Weyl integral quantization, based on the Weyl-Heisenberg group, is natural for a classical motion on the line, affine or wavelet quantization, based on the affine group, is natural for a classical motion on the half-line.

Set X is the upper half-plane $\Pi_+ := \{(q,p)\,|\, p \in \mathbb{R}\,,\, q > 0\}$ equipped with measure $\mathrm{d}q\mathrm{d}p$. It is the phase space for the motion on the half-line $q > 0$. Equipped with the multiplication $(q,p)(q_0,p_0) = (qq_0, p_0/q + p)$, $q \in \mathbb{R}^*_+$, $p \in \mathbb{R}$, X is viewed as the affine group $\mathrm{Aff}_+(\mathbb{R})$ of the real line. The affine group $\mathrm{Aff}_+(\mathbb{R})$ has two non-equivalent UIR, $U_\pm$. Both are square integrable. The UIR $U_+ \equiv U$ is carried by the Hilbert space $\mathcal{H} = L^2(\mathbb{R}^*_+, dx)$ in the form

$$U(q,p)\psi(x) = (e^{ipx}/\sqrt{q})\psi(x/q) . \tag{40}$$

The unit-norm state $\psi \in L^2(\mathbb{R}^\dagger_+, \mathrm{d}x) \cap L^2(\mathbb{R}^\dagger_+, \mathrm{d}x/x)$ ("fiducial vector") produces all *wavelets*, or equivalently, CS defined as $|q,p\rangle = U(q,p)|\psi\rangle$:

$$\rho = |\psi\rangle\langle\psi| \to U(q,p) M U^\dagger(q,p) \equiv \rho(q,p) . \tag{41}$$

This yields the crucial resolution of the identity

$$\int_{\Pi_+} \frac{\mathrm{d}q\mathrm{d}p}{2\pi c_{-1}} |q,p\rangle\langle q,p| = I\,, \qquad c_\gamma := \int_0^\infty \mathrm{d}x |\psi(x)|^2/x^{2+\gamma} . \tag{42}$$

The covariant integral that arises from the resolution of the identity is

$$A_f = \int_{\Pi_+} \frac{\mathrm{d}q\mathrm{d}p}{2\pi c_{-1}} f(q,p) |q,p\rangle\langle q,p| , \tag{43}$$

and $U(g) A_f U^\dagger(g) = A_{\mathsf{U}(g)f}$, $\mathsf{U}(g)f)(g') := f(g^{-1}g')$. Quantization is canonical (up to a multiplicative constant) for q and p:

$$A_p = P = -i\partial/\partial x\,, \; A_{q^\beta} = (c_{\beta-1}/c_{-1})\, Q^\beta , \tag{44}$$

where $Qf(x) = xf(x)$.

The quantization of the kinetic energy yields

$$A_{p^2} = P^2 + KQ^{-2}, \quad K = K(\psi) = \int_0^\infty u\mathrm{d}u(\psi'(u))^2/c_{-1}. \tag{45}$$

Thus affine integral quantization forbids a quantum free particle moving on the positive line to reach the origin. Furthermore, the operator $P^2 = -d^2/dx^2$ in $L^2(\mathbb{R}_+^*, \mathrm{d}x)$ is not essentially self-adjoint, whereas the above regularized operator, defined on the domain of smooth compactly supported functions, is essentially self-adjoint for $K \geq 3/4$. Therefore, quantum dynamics of the free motion is consistent.

3.4.2. Affine Integral Quantization for FLRW Quantum Cosmology FLRW models filled with a barotropic fluid with equation of state $\mathrm{p} = w\rho$ and resolving the Hamiltonian constraint lead to a model of a singular universe, or equivalently, of a particle moving on the half-line $(0, \infty)$ with reduced variable Hamiltonian.

$$h(q,p) = \alpha(w)p^2 + 6\tilde{k}q^{\beta(w)}, \quad q > 0, \quad \{q,p\} = 1, \tag{46}$$

with $\tilde{k} = (\int \mathrm{d}\omega)^{2/3}k$, $\alpha(w) = 3(1-w)^2/32$ and $\beta(w) = 2(3w+1)/(3(1-w))$. The constant k is 0, -1 or 1 (in suitable units of inverse area) depending on whether the universe is flat, open or closed.

Assume a closed universe with radiation content : $w = 1/3$ and $k = +1$. Affine quantization on $\mathbb{R}_+^*$ with a fiducial vector like $\psi(x) \propto \exp(-(\alpha(\nu)x + \beta(\nu)/x)$, with parameter $\nu > 0$, yields the quantum Hamiltonian [10]

$$A_h = H = \frac{1}{24}P^2 + \frac{a_P^2 K(\nu)}{24}\frac{1}{Q^2} + 6\frac{a_P^2}{\sigma^2}\frac{c_1}{c_{-1}}Q^2, \tag{47}$$

where a_P is a Planck area. For $K(\nu) \geq 3/4$, wavelet quantization removes the quantum singularity and then the quantum evolution is well-defined, contrary to what happens in canonical quantization

The quantum states and their dynamics have a phase space representation through lower symbols. For a state $|\phi\rangle$:

$$\Phi(q,p) = \langle q,p|\phi\rangle/\sqrt{2\pi}. \tag{48}$$

The associated probability distribution on phase space is

$$\varrho_\phi(q,p) = \frac{1}{2\pi c_{-1}}|\langle q,p|\phi\rangle|^2. \tag{49}$$

With (energy) eigenstates of some quantum Hamiltonian H at our disposal, we can compute the time evolution

$$\varrho_\phi(q,p,t) := \frac{1}{2\pi c_{-1}}|\langle q,p|e^{-iHt}|\phi\rangle|^2 \tag{50}$$

for any state ϕ.

In general the lower symbol $\check{f}(q,p)$ differs from its classical counterpart $f(q,p)$: it is a quantum-corrected effective observable. Thus, computing the lower symbol of the Hamiltonian leads to the semiclassical Friedmann equation for the scale factor $a(t)$:

$$\left(\frac{\dot{a}}{a}\right)^2 + \frac{kc^2}{a^2} + c^2a_P^2(1-w)^2\frac{\nu}{128}\frac{1}{V^2} = \frac{8\pi G}{3c^2}\rho. \tag{51}$$

Note that the repulsive potential depends explicitly on the volume. This excludes non-compact universes from quantum modelling.

Finally, the singularity resolution is confirmed: as the singularity is approached ($a \to 0$), the repulsive potential grows faster ($\sim a^{-6}$) than the density of the fluid ($\sim a^{-3(1+w)}$) and so at some point the two terms become equal and the contraction is brought to a halt.

The form of the repulsive potential does not depend on the state of fluid filling the universe: the singularity is avoided thanks to the quantum geometry.

4. Integral quantization of constraints

4.1. Two approaches

First approach. On a general level, suppose that (X, ν) is a smooth n-dimensional manifold on which is defined the space $\mathcal{D}'(X)$ of distributions as the topological dual of compactly supported n-forms on X [12]. Some of these distributions, e.g. $\delta(u(x)) = \delta \circ u(x)$, express geometrical constraints, e.g. $u(x) = 0$. Given a POVM integral quantization determined by a family of density operators $x \mapsto \rho(x)$ resolving the identity in some Hilbert space $\mathcal{H}$, one extends the map $f \mapsto A_f$ to these objects. We thus obtain a quantum version, $A_{\delta\circ u}$ of these constraints.

Second approach. Another viewpoint in the quantization of constraints, such as the above $u(x) = 0$, is that of Dirac's [13]. It consists of determining the kernel of the operator A_u, which in our case is obtained from the integral quantization map $u \mapsto A_u$.

Lower symbols. One completes both approaches with the semi-classical analysis provided by their respective lower symbols $\check{\delta \circ u}$ and $\check{u} = 0$. In the sequel, illustrative examples are provided by the complex plane with Lebesgue measure $(X, \nu) = (\mathbb{C}, \mathrm{d}z/\pi)$.

4.2. Example of constraints in the plane

In this example, $(X, \nu) = (\mathbb{C}, \mathrm{d}^2z/\pi)$, and we adopt the integral quantization based on the Weyl-Heisenberg group as it was described in Subsection 3.3. For instance, we can choose as weight functions the Cahill-Glauber function $\varpi_s(z) = e^{s|z|^2/2}$ for $s \leq -1$. This choice yields the family of density operators which are Weyl-Heisenberg displaced thermal states resolving the identity,

$$\rho_t(z) := D(z)\rho_t D(z)^\dagger \,, \; D(z) = e^{za^\dagger - \bar{z}a} \,, \quad \rho_t := (1-t)\sum_{n=0}^{\infty} t^n \,|n\rangle\langle n| \,, \; 0 \leq t := \frac{s+1}{s-1} < 1 \,. \tag{52}$$

$$\int_{\mathbb{C}} \rho_t(z)\, \frac{\mathrm{d}^2 z}{\pi} = I \,. \tag{53}$$

The density operator ρ_t is viewed as a thermal state when $t = e^{-\hbar\omega/kT}$. Zero temperature, or equivalently $t = 0$, corresponds to the projector $\rho_0 = |0\rangle\langle 0|$ and the corresponding $\rho_0(z)$ reduces to the rank-one CS projector $|z\rangle\langle z|$. Note that the Weyl-Wigner integral quantization , and the related Wigner transform, correspond to the value $t = -1$ (i.e., $s = 0$ or $T = i\hbar\omega/\pi$ is purely imaginary). For this particular case in which ρ_{-1} loses its probabilistic meaning, $\rho_{-1} = 2\mathsf{P}$, where $\mathsf{P} = \sum_{n=0}^{\infty}(-1)^n |n\rangle\langle n|$ is the parity operator.

Let $f(z) = 0$ be an equation defining some subset of the complex plane, like points, curves, etc. The first approach towards quantization yields the operator

$$\delta \circ f \mapsto A_{\delta\circ f} = \int_{\mathbb{C}} \delta\left(f\left(z\right)\right)\, \rho_t(z)\, \frac{\mathrm{d}^2 z}{\pi} \,, \tag{54}$$

whose lower symbol is $\check{\delta \circ f}(z) = \int_{\mathbb{C}} \delta\left(f\left(z'\right)\right) \langle z|\, \rho_t(z')\, |z\rangle \, \frac{\mathrm{d}^2 z'}{\pi}$.

The constraint can then be understood in terms of the kernel of the operator. In the **Dirac quantization** procedure we directly quantize the function $f(z)$

$$f \mapsto A_f = \int_{\mathbb{C}} f(z)\, \rho_t(z)\, \frac{\mathrm{d}^2 z}{\pi}, \tag{55}$$

We then analyze $\ker A_f$ through the zeros of the lower symbol $\check{f}(z) = \int_{\mathbb{C}} f(z')\, \mathrm{tr}(\rho_t(z)\rho_t(z'))\, \frac{\mathrm{d}^2 z'}{\pi}$.

Both methods are obviously not mathematically equivalent, except for a few cases. They are possibly *physically* equivalent, i.e. indistinguishable in terms of measurement.

Here are two simple examples. they are handled with CS, $\rho_0(z) = |z\rangle\langle z|$, for the sake of simplicity. A more thorough investigation can be found in [11].

4.2.1. The point: $f(z) = z - z_0 = 0$ Here we study the constraint $f(z) = z - z_0 = 0$. In the first approach, we have

$$A_{\pi\delta\circ f} = \int \pi\delta(z - z_0)\, |z\rangle\langle z|\, \frac{\mathrm{d}^2 z}{\pi} = |z_0\rangle\langle z_0|\,. \tag{56}$$

The lower symbol is purely Gaussian and concentrates about the point z_0.

$$\delta \check{\circ} f(z) = |\langle z|\, z_0\rangle|^2 = e^{-|z-z_0|^2}\,. \tag{57}$$

In the second approach one has

$$A_f = \int_{\mathbb{C}} (z - z_0)\, |z\rangle\langle z|\, \frac{\mathrm{d}^2 z}{\pi} = a - z_0 I\,. \tag{58}$$

where a is the lowering operator

$$a\,|z\rangle = z\,|z\rangle\,. \tag{59}$$

Therefore, $|v\rangle \in \ker A_f$ is such that $|v\rangle = \lambda\,|z_0\rangle$, $\lambda \in \mathbb{C}$. The lower symbol of A_f is identical to the original f

$$\check{f}(z) = z - z_0 = f(z)\,. \tag{60}$$

which means that classical and semi-classical constraints are the same. Therefore the two methods already differ on this elementary example since the first one gives a smooth Gaussian portrait of the point z_0.

Note that the lower symbol of the projector $P_{z_0} = |z_0\rangle\langle z_0|$ corresponding to the element of the kernel is the same as (57)

$$\check{P}_{z_0} = e^{-|z-z_0|^2}\,. \tag{61}$$

It is only at this level that the two methods yield similar results.

4.2.2. The half-line $f(z) = P(\gamma, J) = \gamma - \gamma_0 = 0$ The constraint is given by $f(z) = P(J,\gamma) = \gamma - \gamma_0 = 0$. The first approach gives

$$\begin{aligned} A_{\pi\delta\circ f} &= \int \pi\delta(\gamma - \gamma_0)\, |z\rangle\langle z|\, \frac{\mathrm{d}^2 z}{\pi} \\ &= \sum_{n,n'\geq 0} |e_n\rangle\langle e_{n'}|\, \frac{\Gamma\left(\frac{n+n'}{2} + 1\right)}{\sqrt{n!n'!}}\, \frac{1}{2} e^{i(n-n')\gamma_0}\,. \end{aligned}$$

Its lower symbol is given by

$$\begin{aligned} \pi\delta \,\check{\circ}\, f(z) &= \int_0^\infty \exp\left[-\left|r\,e^{i\gamma} - r'\,e^{i\gamma_0}\right|^2\right] r'\,\mathrm{d}r' \\ &= \frac{\sqrt{\pi}}{2}\, e^{-r^2\sin^2(\gamma-\gamma_0)}\,\mathrm{erfc}(r\,\cos(\gamma-\gamma_0))\,. \end{aligned} \tag{62}$$

The second approach yields

$$\begin{aligned} A_f &= \int (\gamma-\gamma_0)\,|z\rangle\langle z|\,\frac{\mathrm{d}z}{\pi} = \pi I + i \sum_{\substack{n,n'\geq 0 \\ n\neq n'}} \frac{\Gamma\left(\frac{n+n'}{2}+1\right)}{\sqrt{n!n'!}}\,\frac{1}{n-n'}\,|e_n\rangle\langle e_{n'}| - \gamma_0 I \\ &= A_{\daleth} - \gamma_0 I \end{aligned}$$

where $A_{\daleth}$ is the angle operator. In this case the spectrum of operator is continuous with support on $[0,2\pi]$, so the kernel is not properly defined in the Hilbert space, except in a distributional sense. The lower symbol of A_f is the Fourier series

$$\begin{aligned} \check{A}_f &= \langle J,\gamma|\,A_{\daleth}\,|J,\gamma\rangle - \gamma_0 = \pi + i\,e^{-J}\sum_{n\neq n'} \frac{\Gamma\left(\frac{n+n'}{2}+1\right)}{n!n'!}\,\frac{z^{n'}\,\bar{z}^n}{n'-n} - \gamma_0 \\ &= \pi - 2\sum_{q=1}^{\infty} d_q(\sqrt{J})\,\frac{\sin q\gamma}{q} - \gamma_0\,, \end{aligned} \tag{63}$$

where the function $d_q(r)$ is given by:

$$d_q(r) = e^{-r^2} r^q\,\frac{\Gamma(\frac{q}{2}+1)}{\Gamma(q+1)}\,{}_1F_1\left(\frac{q}{2}+1; q+1; r^2\right)\,. \tag{64}$$

It can be shown that this positive function is bounded by 1.

For large J this lower symbol tends to the Fourier series of the 2π-periodic angle function $\daleth(\gamma) = \gamma$ for $\gamma \in [0,2\pi)$:

$$\langle J,\gamma|\,A_{\daleth}\,|J,\gamma\rangle \approx \pi - 2\sum_{q=1}^{\infty}\frac{1}{q}\,\sin q\gamma\,. \tag{65}$$

Such a behavior is understood in terms of the classical limit of these quantum objects. Indeed, by restoring physical dimensions to our formulas, we know that the quantity $|z|^2 = J$ should appear divided by the Planck constant $\hbar$. Hence, the limit $J \to \infty$ in our previous expressions can also be considered as the classical limit $\hbar \to 0$.

5. Conclusion

The main idea put forward is a map between classical observables, functions on a set, and quantum observables, operators on a Hilbert space. In general this is not well-defined, because of the appearance of unbounded operators, domain issues, etc. It is possible to bypass such subtleties by working with the lower symbols of operators, which live in the classical setting, but carry some information from the noncommutative and probabilistic quantum setting.

Besides the freedom allowed by integral quantization, the advantages of the method with regard to other quantization procedures are manifold:

(i) Only a minimal amount of restrictions have to be imposed on the classical objects being quantized, which allows us to consider the quantization of more general objects, such as distributions.

(ii) Once a choice of a (positive) operator-valued measure has been made, which must be consistent with experiment, there is no more ambiguity (no ordering problem). To each classical object corresponds one and only one quantum object.

(iii) The method produces in essence a regularizing effect.

(iv) The method, through POVM choices, offers the possibility of keeping a full probabilistic content. In particular, the Weyl-Wigner integral quantization is not defined by a POVM.

Acknowledgments
J.P. Gazeau thanks the CBPF, the CNPq, and the UFABC for financial support. R. Fresneda thanks CNPq, FAPESP and CAPES for financial support.

References
[1] Bergeron H and Gazeau J P 2014 Integral quantizations with two basic examples *Annals of Physics (NY)* **344** 43-68 (arXiv:1308.2348 [quant-ph, math-ph])
[2] Ali S T, Antoine J P and Gazeau J P 2013 *Coherent states, wavelets and their generalizations* 2d edition, Theoretical and Mathematical Physics (New York: Springer)
[3] Bergeron H, Curado E M F, Gazeau J P and Rodrigues Ligia M C S 2014 Quantizations from (P)OVM's, Proceedings of the 8th Symposium on Quantum Theory and Symmetries, El Colegio Nacional, Mexico City, 5-9 August, 2013, Ed. K.B. Wolf *J. Phys.: Conf. Ser.* (arXiv: 1310.3304 [quant-ph, math-ph])
[4] Baldiotti M, Fresneda R and Gazeau J P 2014 Three examples of covariant integral quantization *Proceedings of Science* ICMP 2013
[5] Gazeau J P and Heller B *POVM Quantization, submitted*
[6] Bergeron H, Gazeau J P, Siegl P and Youssef A 2010 *Eur. Phys. Lett.* **92** 60003; Bergeron H, Siegl P and Youssef A 2012 *J. Phys. A: Math. Theor.* **45** 244028
[7] Busch P, Grabowski M and Lahti P J 1995 *Operational Quantum Physics* (Berlin: Springer-Verlag)
[8] Holevo A S 2011 Probabilistic and statistical aspects of quantum theory Edizioni della Normale
[9] Cahill K E and Glauber R 1969 Ordered expansion in boson amplitude operators *Phys. Rev.* **177** 1857-1881
[10] Bergeron H, Dapor A, Gazeau J P and Małkiewicz P 2014 Smooth big bounce from affine quantization *Phys. Rev. D* **89** 083522 (arXiv:1305.0653 [gr-qc])
[11] Baldiotti M, Fresneda R and Gazeau J P, About Dirac & Dirac constraint quantizations *submitted*
[12] Grosser M 2008 A note on distribution spaces on manifolds *Novi Sad Math.* **38** 121–128
[13] Dirac P A M 2001 *Lectures on Quantum Mechanics* (New York: Dover)

Phase transitions in small-world systems: application to functional brain networks

B R Gadjiev and T B Progulova

Dubna International University for Nature, Society and Man,
19 Universitetskaya street, Dubna, 141980, Russia

E-mail: gadjiev@uni-dubna.ru

Abstract. In the present paper the problem of symmetry breaking in the systems with a small-world property is considered. The obtained results are applied to the description of the functional brain networks. Origin of the entropy of fractal and multifractal small–world systems is discussed. Applying the maximum entropy principle the topology of these networks has been determined. The symmetry of the regular subgroup of a small-world system is described by a discrete subgroup of the Galilean group. The algorithm of determination of this group and transformation properties of the order parameter have been proposed. The integer basis of the irreducible representation is constructed and a free energy functional is introduced. It has been shown that accounting the presence of random connections leads to an integro-differential equation for the order parameter. For q-exponential distributions an equation of motion for the order parameter takes the form of a fractional differential equation. We consider the system that is described by a two-component order parameter and discuss the features of the spatial distribution of solutions.

1. Introduction

Recently Eguiluz et al. [1] have presented a method of construction of functional brain networks proceeding from the results of functional magnetic resonance imaging measurements in a human. In these experiments, a magnetic resonance activity of certain parts of brain (so-called voxels) is measured at each discrete time step. By $x_i(t)$ we denote a voxel's activity at the instant of time t. It was proposed to consider that two voxels are functionally linked if the value of their temporal correlation exceeds a certain positive value r_c independent of the value of their anatomical connection. The correlation coefficient between any pairs of voxels x_i and x_j is calculated as

$$r(i,j) = \frac{\langle x_i(t)x_j(t)\rangle - \langle x_i(t)\rangle\langle x_j(t)\rangle}{\sigma(x_i(t))\sigma(x_j(t))} \quad (1)$$

where $\sigma^2(x_l(t)) = \langle x_l^2(t)\rangle - \langle x_l(t)\rangle^2$, brackets $\langle ... \rangle$ represents temporal averages and $x_l(t)$ is the blood oxygenation level dependent signal of the voxel i in case of brain scanning data.

The elements of the correlation matrix determine the value of correlations among various parts of the cerebral cortex. Using highly correlated nodes Eguiluz et al. [1] have constructed a network and determined that the degree distribution of the obtained network has the form $p(k){\sim}k^{-\gamma}$, where $\gamma \approx 2$ (figure 1) [1]. It has been also shown that these networks possess a small-world structure, a community structure and are fractals [2].

However it should be stressed that the degree distributions for $r_c = 0.6$ and $r_c = 0.7$ are actually described by the q-exponentional distribution

$$p(k) = c\left[1 - (1-q)\frac{|k-k_0|^\alpha}{\kappa_c}\right]^{-\frac{q}{q-1}} \tag{2}$$

and for $r_c = 0.7$, $\alpha \approx 1$ whereas for $r_c = 0.6$, $\alpha > 1$ [3]. For $r_c = 0.5$ the form of the degree distribution of the correlation network is characteristic of stretched exponential probability distributions [3]. Fitting with the help of formula (2) is actually performed by using the maximum likelihood method [5].

In paper [6] the correlation network obtained from functional magnetic resonance imaging measurements is compared with that of the derived from the numerical simulation of 2D Ising model at various temperatures. Near the critical temperature a striking similarity in statistical properties of these two networks are observed and it makes them indistinguishable. The similarity of these networks allows to suppose that a collective dynamics is inherent in a human brain.

On the other hand it also follows that the dynamics of the brain functioning in such systems takes place near the critical point and generates spatio-temporal structures.

To study the dynamics generating spatio-temporal structures in brain we develop the theory of the Landau – Ginzburg type for the systems with a small-world structure.

We first introduce an algorithm to generate small-world networks. Then applying the maximum entropy principle and accounting multifractality of the system we determine a degree distribution for such systems.

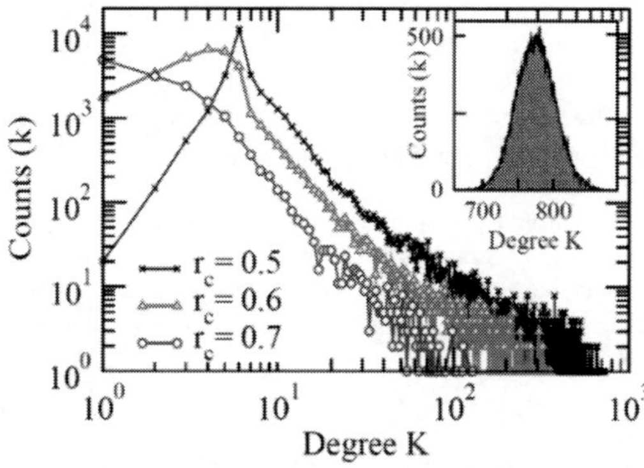

Figure 1. Degree distribution for three values of the correlation threshold. The inset depicts the degree distribution for an equivalent randomly connected network. Figure taken from [1].

Using the principle of least action we derive the equation of motion for the order parameter representing spatio-temporal structures near the critical point. We also consider a specific example.

2. Topology of functional brain network with a small-world property

The algorithm constructing a small-world network was for the first time proposed by Watts and Strogatz [7]. At the initial instant of time there is one-dimensional lattice of N nodes with periodic boundary conditions where each link connecting a vertex to one of its k nearest neighbours in the clockwise sense is left in place with probability $1-p$, and with probability p is reconnected to a randomly chosen other vertex. Long range connections are therefore introduced. As a result a network structure with a small-world property emerges. However this network is not scale-free that is it has degree distributions with non-power law forms [8].

A human brain consists of $\sim 10^{10}$ neurons, each of which is connected with other neurons by $\sim 2 \cdot 10^4$ links. As is shown in [1] the degree distributions of functional brain networks in the log-log scale have a rectilinear region. Then another construction of a small-world network is a more convenient model for a human brain. We proceed from a closed system of N nodes with periodic boundary conditions where each node is connected with the neighbors by n links. A new edge is added to this system at each instant of time, one of the ends of this edge being connected with one of the nodes of the regular lattice with probability $1/N$ while the other end of the edge being connected with that in accordance with the preferential attachment principle $k_i(t)/\sum_j k_j(t)$. Thus the rate of the connectivity $k_i(t)$ change of the node (without taking into account the contribution of the initial regular graph edges) is determined by two contributions and represented by the equation

$$\frac{\partial k_i(t)}{\partial t} = \frac{A}{N} + A\frac{k_i(t)}{\sum_j k_j(t)}. \tag{3}$$

Taking into account that at the instant t a full network connectivity is $\sum_j k_j = 2t$ and the change in the total degree of the network at one time step is $\Delta k = \sum_i\big(k_i(t) - k_i(t-1)\big) = 2$ we obtain $A = 1$. Then equation (3) takes the form

$$\frac{\partial k_i(t)}{\partial t} = \frac{1}{N} + \frac{k_i(t)}{\sum_j k_j(t)}. \tag{4}$$

The solution of this equation has the form

$$k_i(t) = 2t/N + ct^{1/2}, \tag{5}$$

where c is a constant of integration which can be determined from the condition $\sum_i k_i(t) = 2t$. The degree distribution of this network is described by the q-exponential distribution in the form [9]

$$p_{st}(k) = Z^{-1}\left(1 - \frac{1}{\beta}(1-q)(k-k_0)^2\right)^{\frac{q}{1-q}}. \tag{6}$$

Here q is a measure of complexity of the system. The value $q = 1$ corresponds to the degree distribution in the form of the Gaussian distribution emerging at large temporal shaping of such system. At $q \neq 1$ and for large enough k the network is characterized by the degree distribution emerging at the earlier steps of forming a network. Distribution (6) describes the results of processing of functional magnetic resonance imaging measurement data $r_c = 0.6$ [1] pretty well.

The presented algorithm shows that order and disorder are inherent in small-world systems.

However it should be taken into consideration that the structure of these networks may be homogeneous, fractal and multifractal.

To derive a distribution function for such systems we use the maximum entropy principle.

The famous Boltzmann-Gibbs entropy which is determined as

$$S = \sum_{i=1}^{W} s_i = \sum_{i=1}^{W} p_i \ln p_i \tag{7}$$

is characteristic for homogeneous structures, where the distribution function $p(x)$ is normalized to unity.

We consider the quantity

$$G_I = -\int_0^1 p \ln p \, dp. \tag{8}$$

Fractional generalization of this integral gives

$$\Gamma(\alpha) G_I = -\int_0^1 p^{\alpha} \ln p \, dp, \tag{9}$$

where $\Gamma(\alpha)$ is a gamma function.

For admissible systems when $S = \sum_i s_i$ in case of a fractal structure we have $s_i = p_i^{\alpha} \ln p_i$ presenting the Shafee entropy [8]. It is clear that at $\alpha = 1$ we obtain the Boltzmann – Gibbs entropy.

A multifractal is a mixture of fractals. Suppose that dimensions of fractals in a multifractal are distributed uniformly that is $f(\alpha) = 1/(\alpha - 1)$. Then multiplying (9) by $f(\alpha)$ and integrating over α we obtain

$$G_M = -\int_0^1 \frac{p^\alpha - p}{\alpha - 1} dp. \tag{10}$$

Consequently for admissible systems we have

$$G_{Tsallis} = \sum_{i=1}^{W} \frac{p_i - p_i^\alpha}{\alpha - 1}, \tag{11}$$

representing the Tsallis entropy [7]. If $\alpha \to 1$ the Tsallis entropy coincides with that of Boltzmann – Gibbs.

These results can be generalized if we consider a multifractal structure as a mixture of fractal substructures. We introduce the distribution function $f(\alpha)$ allowing to introduce an equation for determination of the systems entropy

$$g(p) \sim \int_{-\ln p}^{\mu} e^{-y} f(y) dy. \tag{12}$$

In particular if $f(\alpha) = \delta(\alpha - 1)$, $f(\alpha) = \delta(\alpha - \theta)$ and $f(\alpha) = 1/(\alpha - 1)$ we obtain the Boltzmann – Gibbs, Shafee and Tsallis entropies correspondingly. It should be stressed that if $f(\alpha)$ has the form of a power-law function $f(y) = y^{\alpha - 1}$ we obtain the Anteneodo – Plastino entropy [4]

$$S_\gamma = \sum_{i=1}^{W} s_\gamma(p_i), \tag{13}$$

where

$$s_\gamma(p_i) = \Gamma\left(\frac{\gamma+1}{\gamma}, -\ln p_i\right) - p_i \Gamma\left(\frac{\gamma+1}{\gamma}\right), \tag{14}$$

and $\Gamma(\mu, t)$ is an incomplete gamma function.

Considering the standard constraints $\sum_{i=1}^{W} p_i = 1$ and

$$\langle O^{(r)} \rangle = \sum_{i=1}^{W} p_i O_i^{(r)} = O_\gamma^{(r)}, \tag{15}$$

where $\{O^{(r)}\}$ are observed quantities and applying the maximum entropy principle we determine the distribution function

$$p_i = exp\left(-\left[\Gamma\left(\frac{\gamma+1}{\gamma}\right) + \alpha + \sum_{r=1}^{R} \beta_r O_i^{(r)}\right]^\gamma\right), \tag{16}$$

where $i = 1, \dots, w$, and α and $\{\beta_r\}$ are the Lagrange multipliers determined from the constraints. Distribution (16) is characteristic of the systems with a skewed degree distribution. Note that distribution function (16) describes the degree distribution of the functional brain network precisely enough in case $r_c = 0.5$.

Using the constraints $\sum_i p_i = 1$, $c_q^{-1} \sum_k k p_k^q = \mu$, $c_q^{-1} \sum_k k^2 p_k^q = \rho^2$ and entropy (11) in the framework of the maximum entropy principle we obtain the degree distribution in the form (6) describing the degree distribution of the functional brain network in case $r_c = 0.6$, and where $c_q = \sum_k p_k^q$ [9].

In case of constraints $\sum_i p_i = 1$, $\sum_i p_i k_i = k$ and entropy (11) the maximum entropy principle gives a distribution function in the form

$$P_{st}(k) = Z^{-1}\left(1 - (1 - q)\frac{k}{\kappa_0}\right)^{\frac{1}{1-q}}, \tag{17}$$

where

Journal of Physics: Conference Series **597** (2015) 012038 doi:10.1088/1742-6596/597/1/012038

$$Z = \int_0^\infty \left(1 - (1-q)\frac{k}{\kappa_0}\right)^{\frac{1}{1-q}} dk, \quad (18)$$

describing the degree distribution of the functional brain network precisely enough in case $r_c = 0.7$.

3. Order parameter in functional brain networks

The exact solution of the Ising model has been found only for the one-dimensional and two-dimensional cases in the zero external field [12]. So for the description of spatio-temporal structures occurring in the functional brain networks we shall develop a theory of phase transitions of Landau – Ginzburg type [13]. It should be stressed that the theory of phase transitions in graphs is discussed in [14] and analysis of the systems with long-range space interactions and temporal memory is given in [15]. We shall proceed from the construction of a small-world network which is a regular graph with additional shortening links. Under such consideration a regular substructure of a small-world network could possess a symmetry of the discrete subgroup of the Galilean group.

There occurs a problem to determine transformation properties of the order parameter. To solve this problem we first have to determine a number of the components of the order parameter involved into the process under consideration. We proceed from the measured time series $x(t)$ and use the Takens method [16, 17]. Choosing from a set of experimental data N equidistant points, we obtain a set of discrete variables $x_0(t_i), \dots, x_{n-1}(t_i + (n-1)\tau)$, where $i = 1, \dots, n-1$ and τ is a temporal shift. It is necessary construct the points of the phase space and calculate a correlation function of the attractor.

$$C(r) = \frac{1}{N^2} \sum_{\substack{i,j=1 \\ i \neq j}}^{N} \theta\left(r - |x_i - x_j|\right), \quad (19)$$

where $\theta(x)$ is the Heaviside function and r is distance. Further we construct the dependence $\ln C(r)$ on $\ln r$ [16, 17]. If the value of the slope d depending on n reaches a plateau above a certain n then the system represented by the present temporal sequence must have an attractor. The value d which reached saturation should be considered as a dimension of the attractor and the value n above which the saturation is observed should be considered as the minimum number of variables necessary for modulation of the behavior corresponding to the current attractor [16, 17]. Based on the results of the analysis of the time series we can choose the n as a number of the order parameters.

If $D_0(\vec{k})$ is an irreducible representation of the discrete subgroup of the Galilean group of a regular subgraph and $\{\eta_{ij}\}$, $i = 1, \dots, m$, $j = 1, \dots, p$ are the order parameters which are transformed in accord with this irreducible representation where m is a number of wave vectors in the star $\{\vec{\boldsymbol{k}}^*\}$ of the irreducible representation $D_0(\vec{k})$ and p is dimension of a small representation, then the dimension $D_0(\vec{k})$ is $= m \times p$. Thus the choice of a particular irreducible representation could be made from the standard reference books of irreducible representations [18].

It should be stressed that if the wave vector under consideration is expressed irrationally in terms of vectors of the reciprocal lattice then only invariants of the rotation group have corresponding irreducible representations. If the wave vector is expressed rationally in terms of vectors of the reciprocal lattice then the irreducible representation under consideration admits anisotropic invariants.

The phenomenon of hysteresis is inherent in a human brain and thus an irreducible representation may not satisfy the Landau condition that is contain invariants of the third order.

As small-world systems are initially inhomogeneous a free energy functional must have the Lifshitz invariant. So the task is to consider irreducible representations corresponding to the internal points of the Brillouin zone. During the transition into modulated structures the point symmetry elements of the initial phase remain [19]. As the Landau and Lifshitz conditions are violated we deal with phase transitions of the first order.

4. Derivation of the equation of motion for the order parameter near the critical point in the functional brain network

To derive an equation of motion for the order parameter in the system with a small-world property we determine the free energy functional $F[\eta]$ in the form [13, 15]:

$$F[\eta] = F_0[\eta] + F_I[\eta], \quad (20)$$

Where

$$F_0[\eta] = \int_R drdt \int_R dr'dt' \left\{\frac{1}{2}\frac{\partial\eta(r,t)}{\partial t} g_0(r,t,r',t')\frac{\partial\eta(r',t')}{\partial t'} + \frac{1}{2}\frac{\partial\eta(r,t)}{\partial r} g_1(r,t,r',t')\frac{\partial\eta(r',t')}{\partial r'}\right\} \quad (21)$$

and

$$F_I[\eta] = -\int_R drdt \int_R dr'dt' V(\eta(r,t), u(r',t')). \quad (22)$$

We took into account the presence of random shortening links in the structure. Here r is a spatial coordinate, t is time and the functions $g_0(r,t,r',t')$ and $g_1(r,t,r',t')$ describe the influence of the small-world property on the critical properties of the system. Integration is performed over the region R in two-dimensional space R^2 to which (r,t) belong.

The equation of motion for the order parameter $\eta(r,t)$ is derived with the use of the Gateaux derivative of the functional $F[\eta(r,t)]$, determined as:

$$\delta F[\eta,h] = \left[\frac{d}{d\varepsilon}F[\eta+\varepsilon,h]\right]_{\varepsilon=0} = lim_{\varepsilon\to 0}\frac{F[\eta+\varepsilon,h]-F[\eta]}{\varepsilon}, \quad (23)$$

where $h(r) = \delta\eta(r)$ is a smooth integrable function.

Further it is suitable to introduce the functions:

$$K_0(r,t,r',t') = \frac{1}{2}\left(g_0(r,t,r',t') + g_0(r',t',r,t)\right), \quad (24)$$

$$K_1(r,t,r',t') = \frac{1}{2}\left(g_1(r,t,r',t') + g_1(r',t',r,t)\right) \quad (25)$$

and the symmetrical function

$$U(\eta(r,t),\eta(r',t')) = U(r,t)\delta(r-r')\delta(t-t'). \quad (26)$$

We note that such choice of $K_0(r,t,r',t')$ and $K_1(r,t,r',t')$ allows to separate spatio-temporal derivatives. The dynamic equation for the order parameter is determined from the principle of stationarity $\delta F[\eta,h] = 0$ and for the arbitrary function $h(r,t)$ has the form:

$$\int dt' \int dr' \frac{\partial K_0(r,t,r',t')}{\partial t}\cdot\frac{\partial\eta(r',t')}{\partial t'} + \int dt' \int dr' \frac{\partial K_1(r,t,r',t')}{\partial t}\cdot\frac{\partial\eta(r',t')}{\partial r'} + \frac{\partial U(\eta(r,t))}{\partial\eta(r,t)} = 0. \quad (27)$$

This is an integro-differential equation allowing to obtain an equation of motion for the order parameter for various kernels $K_0(r,t,r',t')_0$ and $K_1(r,t,r',t')$.

We suppose that $K_0(r,t,r',t') = \delta(r-r')G_0(t,t')$, $K_1(r,t,r',t') = \delta(t-t')G_1(r,r')$. In this case time and the spatial coordinate separate. In case of one component order parameter in accord with the symmetry arguments $U(\eta(r,t)) = \frac{\alpha}{2}\eta^2(r,t) + \frac{\beta}{4}\eta^4(r,t)$, then from (27) we obtain:

$$\int_0^t G_0(t-t')\frac{\partial\eta(r,t')}{\partial}dt' = -\int_0^r G_1(r-r')\frac{\partial\eta(r',t)}{\partial}dr' - \alpha\eta(r,t) - \beta\eta^3(r,t). \quad (28)$$

Here α is a control parameter of the system and β is a positive parameter. The form of the function $U(\eta(r))$ is determined from the condition of invariance. Hence $U(\eta(r))$ includes the integer basis of invariants of the irreducible representation of the space group of the high symmetry phase. The solutions of (28) are investigated in [20].

5. Systems described by two-component order parameter

In case of the system described by the two-component order parameter which is transformed according to the irreducible representation corresponding to the internal rational point of the Brillouin zone we have an integer basis consisting of two invariants. In the polar coordinate system these invariants have the form ρ^2 and $\rho^n \cos n\varphi$, where ρ and φ are an amplitude and a phase of the order parameter correspondingly and n is a parameter of anisotropy.

In this case a spatial dependence of the order parameter in accord with (27) takes the form:

$$\int dz\left[-\frac{k}{2}\rho^2\frac{\partial k_1(x,z)}{\partial x}\frac{\partial\varphi(z)}{\partial z}\right]+n\gamma\rho^n\cos n\varphi=0. \tag{29}$$

The integral expression in (29) could be considered as averaging over the distribution function $\frac{\partial k_1(x,z)}{\partial x}=P(x,z)$. The q-exponential distribution is characteristic of the topology of small-world systems and thus we could suppose that $P(r)=Z^{-1}\exp_q(-\varsigma r)^q$ where $\exp_q(x)=(1+(1-q)x)^{\frac{1}{1-q}}$. Asymptotic behavior of $P(r)=Z^{-1}\exp_q(-\varsigma r)^q$ at $q>1$ has the form $P(r)\sim r^{-\frac{q}{q-1}}$. In this case equation (29) takes the form:

$$\frac{d^v\phi(x)}{dx^v}+\omega_0^v\sin\phi(x)=0, \tag{30}$$

where $\phi(x)$ is a renormalized phase of the order parameter. When $v=2$ the spatial distribution of $\phi(x)$ is determined by the equation

$$\frac{d^2\phi(x)}{dx^2}+\omega_0^2\sin\phi(x)=0. \tag{31}$$

The first integral of this equation has the form

$$h=\frac{1}{2}\left(\frac{d\phi(x)}{dx}\right)^2-\omega_0^2\cos\phi(x). \tag{32}$$

When $h_0<\omega_0^2$ the solution is the periodic function

$$\phi(x)=B\sin x. \tag{33}$$

In case $h_0>\omega_0^2$, the trajectory $\phi(x)$ increases indefinitely.

$$\phi(x)=2am(x), \tag{34}$$

where $\mathrm{am}(x)$ is the Jacobi elliptic function.

The solution corresponding to $h_0=\omega_0^2$ determines a separatrices. In this case $\frac{d\phi(x)}{dx}=\pm2\omega_0\cos\frac{\phi(x)}{2}$, and for the initial condition $\phi(x=0)=0$ the solution of this equation has the form:

$$\phi(x)=4arctg(\pm\omega_0x)-\pi. \tag{35}$$

For finite motion the solution of fractional equation (30) is determined as

$$\phi(x)=-B(\omega_0x)^{\alpha-1}E_{\alpha,\alpha}(-(\omega_0x)^\alpha). \tag{36}$$

Here $E_{\alpha,\beta}(z)=\sum_{n=0}^{\infty}\frac{z^{\alpha n}}{\Gamma(\alpha n+\beta)}$ is the Mittag–Leffler function and the function $\Gamma(\alpha n+\beta)$ is the Euler gamma function.

When $\alpha=2$ we have $E_{2,2}(-z^2)=\frac{\sin z}{z}$ and the solution of the linear fractional equation has the form $\phi(x)=B\sin z$. When $1<\alpha<2$, we have a continual number of differential equations. The oscillating function with a decreasing amplitude is a solution of (30).

6. Conclusion

Order and disorder are inherent in small-world systems and consequently in functional brain networks. We propose the algorithm constructing a network equivalent to functional brain networks. We suppose that a regular substructure of a small-world network could be described by a discrete subgroup of the Galilean group. In such approach random connections generate the medium in which the order parameter moves.

We have presented the equations allowing to determine an entropy for fractal and multifractal small-world networks and applying the maximum entropy principle we have derived characteristic degree distribution functions. The obtained various distribution functions in the log-log scale have a rectilinear region. Using this property with the help of the principle of the least action we have derived an equation of motion for the order parameter in the form of a fractional differential equation.

We have presented the algorithm for determination of the transformation properties of the order parameter. The Rapp diagram is constructed from the analysis of the measured time series. The number of the components of the order parameter is determined from this diagram. On the other hand the number of the components of the order parameter coincides with the dimension of the irreducible representation equal to that of the irreducible representation of a small group multiplied by a number of wave vectors in the star corresponding to the wave vector under consideration. These data can be found in literature [13].

We show that in case of the functional brain networks only internal points of the Brillouin zone should be considered. Therefore we deal with the phase transitions without the loss of the point symmetry elements [14]. We show that in case of two-component order parameter the spatio distribution of the order parameter is determined by the one–dimensional sine-Gordon fractional differential equation. The change of the fractional dimension due to the fractal dimension of the structure changes the medium in which the order parameter moves. It could be shown that if the fractional dimension of the order parameter equation of motion changes from 2 to 1 the solution different from zero becomes identically zero. Thus our results discover the principles of a human brain functioning near the threshold of criticality.

References

[1] Eguiluz V M, Chialvo D R, Cecchi G A, Baliki M and Apkarian A V 2005 Scale-free brain functional networks *Physical Review Letters* **94** 018102

[2] Bullmore E and Sporns O 2009 Complex brain networks: graph theoretical analysis of structural and functional systems *Nature Reviews Neuroscience* **10** 186–198

[3] Gadjiev B 2013 Superstatistics for fractional systems *Preprint* arXiv:1305.0933

[4] Anteneodo C and Plastino A R 1999 Maximum entropy approach to stretched exponential probability distributions *Journal of Physics* A: *Mathematical and General* **32** 1089–1097

[5] Cosma R S 2007 Maximum likehood estimation for q-exponential distributions *Preprint* arXiv: math/0701854

[6] Fraiman D, Balenzuela P, Foss J and Chialvo D R 2009 Ising-like dynamics in large-scale functional brain networks *Physical Review* E **79** 061922

[7] Barrat A and Weigt M 2000 On the properties of small-world network models *European Physical Journal* B **13** 547–560

[8] Newman M E J 2010 *Networks: An Introduction* (Oxford: Oxford University Press)

[9] Hasegawa H 2006 Nonextensive aspects of small-world networks *Physica* A: *Statistical Mechanics and its Applications* **365** 383–401

[10] Shafee F 2007 Lambert function and a new nonextensive form of entropy *IMA J Applied Mathematics* **72** 785–800

[11] Tsallis C 2009 *Introduction to Nonextensive Statistical Mechanics: Approaching a Complex World* (New York: Springer)

[12] Dotsenko Vik S 1995 Critical phenomena and quenched disorder *Phys. Usp.* **38 (5)** 457–496

[13] Gadjiev B R 2011 Phase transition in generalized inhomogeneous 'cubic' systems *Journal of Physics: Conference Series* **284** 012026

[14] Dorogovtsev S N, Goltsev A V and Mendes J F F 2008 Critical phenomena in complex networks *Rev.*

Mod. Phys. **80** 1275–1335
[15] Zaslavsky G M and Tarasov V E 2007 Fractional dynamics of systems with long-range space interactions and temporal memory *Physica* A **383** 291–308
[16] Haken H 1996 *Principles of Brain Functioning: Synergetic Approach to Brain Activity, Behavior, and Cognition* (Berlin: Springer)
[17] Nicolis G and Prigogine I 1989 *Exploring Complexity: An Introduction* (New York: W H Freeman and Company)
[18] Kovalev O V 1965 *Irreducible Representations of the Space Groups* (New York: Gordon and Breach)
[19] Gadjiev B 2009 Features of incommensurate phases in multiferroics *Physica Status Solidi* C **6** 1321–1324
[20] Tarasov V E 2011 *Fractional Dynamics: Application of Fractional Calculus to Dynamics of Particles, Fields and Media* (Berlin: Springer, HEP)

Backlund transformation for integrable hierarchies: example - mKdV hierarchy

J F Gomes, A L Retore, N I Spano and A H Zimerman

Instituto de Física Teórica IFT-Unesp
São Paulo - Brasil

E-mail: jfg@ift.unesp.br, retore@ift.unesp.br, natyspano@ift.unesp.br, zimerman@ift.unesp.br

Abstract. In this note we present explicitly the construction of the mKdV hierarchy and show that it decomposes into positive and negative graded sub-hierarchies. We extend the construction of the Backlund transformation for the sinh-Gordon model to all other positive and negative odd graded equations of motion generated by the same affine algebraic structure. Some simple examples of solutions are explicitly verified to satisfy, in a universal manner, the Backlund transformations for the first few odd (positive and negative) sub-hierarchies.

1. Introduction

A subclass of non linear integrable models underlined by an affine $\hat{sl}(2)$ Lie algebra is well known to be connected to the mKdV equation. These are in fact higher flows obtained from the zero curvature representation [1] and by general algebraic arguments these flows are restricted to be related to certain positive odd grade generators. Another subclass of nonlinear integrable models containing for instance, the sinh-Gordon model may be formulated, also within the zero curvature representation, but now associated to negative grade generators. The relation between the mKdV and sine-Gordon models was already observed some time ago [2], [3] in terms of conservation laws and by more algebraic arguments it was generalized for the AKNS hierarchy in [4] and to other integrable hierarchies associated to mixed gradations allowing internal degrees of freedom in [5], [6] and [7]. The general construction and classification of the hierarchy maybe understood in terms of a decomposition of the affine Lie algebra into graded subspaces by a judicious choice of a grading operator. The hierarchy is further specified by choosing a constant grade one operator. These algebraic ingredients define a series of non-linear equations of motion, each corresponding to a different time evolution and hence to a hamiltonian structure (see for instance [8] and references therein).

Backlund transformation has recently been employed to extend the set of integrable models to incorporate defects. Such defects preserve the integrability when they are described by Backlund transformations connecting two distinct solutions of the same equation of motion at its location. This was firstly observed in [9] for the sine-Gordon and extended to affine Toda field theories with defects in [10] and to other non-relativistic models [11]. In this last reference the same space component of the Backlund transformation of the sine-Gordon models is employed to describe integrable defects within the mKdV model. It thus seem natural the extend the same space component of the Backlund transformation to other members of the integrable hierarchy.

In this note we start by reviewing, in section 2, the algebraic construction for general integrable hierarchies. We discuss explicitly the mKdV Hierarchy constructed out of the affine $\hat{sl}(2)$ Kac-Moody algebra and principal gradation. We show how the *positive and negative odd* sub-hierarchis naturally arises from the zero curvature representation. An important point to notice is that, while the time component for the construction of each model varies with the gradation, the form of the space component remains the same for all models within the hierarchy. Moreover, we discuss, in section 3, the construction of Backlund transformation in terms of gauge transformation that preserves the form of the space Lax operator. By general arguments, the space Backlund transformation is understood to be universal within the hierarchy. This fact can be seen in ref. [12] where the space Backlund transformation for the sinh-Gordon and mKdV were derived by canonical transformations and shown to agree. Also, by the same procedure the space Backlund transformation for the KdV and the Sawada-Kotera equations (belonging to the same hierarchy) obtained in [12] and [13] respectively agree. Its time components can be extended to the whole hierarchy by using the appropriated equations of motion. As an explicit example, we derive the space Backlund transformation for the sinh-Gordon model and construct its time components o for higher positive and negative grade time evolutions.

This, in fact maybe observed by the explicit space-time dependence of the soliton solutions. We explicit display the 1- and 2-soliton solutions for all (odd positive and negative grade) models within the hierarchy and show how they can be arranged in pairs in order to satisfy the Backlund relations. Some explicit examples are verified in section 4.

2. The mKdV Hierarchy

The main ingredient underlying the construction of integrable hierarchies (IH) is the Lax operator,

$$L_x = \partial_x + E^{(1)} + A_0 \tag{1}$$

where $E^{(1)}$ and A_0 are Lie algebra $\mathcal{G}$ valued elements and carry an affine structure which classifies the IH.

The systematic construction of IH and its Lax operators of the form (1) consists in the decomposition of an affine algebra $\hat{\mathcal{G}}$ into integer graded subspaces

$$\hat{\mathcal{G}} = \sum_a \mathcal{G}_a, \quad a \in Z \tag{2}$$

induced by a choice of a grading operator Q, such that

$$[Q, \mathcal{G}_a] = a\mathcal{G}_a, \qquad [\mathcal{G}_a, \mathcal{G}_b] \in \mathcal{G}_{a+b}, \qquad a, b \in Z. \tag{3}$$

Furthermore, the IH is determined by fixing the semi simple grade one operator $E^{(1)} \in \mathcal{G}_1$ such that it decomposes $\hat{\mathcal{G}} = \mathcal{K} \oplus \mathcal{M}$ where $\mathcal{K}$ is the Kernel of $E^{(1)}$ and $\mathcal{M}$ is its complement, i.e.,

$$\mathcal{K} = \{x \in \hat{\mathcal{G}}, [x, E^{(1)}] = 0\}, \tag{4}$$

such that

$$[\mathcal{K}, \mathcal{K}] \subset \mathcal{K}, \qquad [\mathcal{K}, \mathcal{M}] \subset \mathcal{M}, \qquad [\mathcal{M}, \mathcal{M}] \subset \mathcal{K}.$$

The equations of motion are determined by solving the zero curvature equation

$$[\partial_x + E^{(1)} + A_0, \partial_{t_N} + D^{(N)} + D^{(N-1)} + \cdots + D^{(0)}] = 0, \tag{5}$$

Journal of Physics: Conference Series **597** (2015) 012039 doi:10.1088/1742-6596/597/1/012039

The solution of eq. (5) may be systematically constructed by considering $D^{(j)} \in \mathcal{G}_j$ and $A_0 \in \mathcal{K}_0$, $\mathcal{K}_0 \in \mathcal{G}_0$ and can be decomposed according to the graded structure as

$$[E, D^{(N)}] = 0 \tag{6}$$

$$[E, D^{(N-1)}] + [A_0, D^{(N)}] + \partial_x D^{(N)} = 0$$

$$\vdots = \vdots$$

$$[A_0, D^{(0)}] + \partial_x D^{(0)} - \partial_{t_N} A_0 = 0. \tag{7}$$

The unknown $D^{(j)}$'s can be solved starting from the highest to the lowest grade projections as functionals of A_0 and its $x-$ derivatives. Notice that, in particular the highest grade equation, namely $[E, D^{(N)}] = 0$ implies $D^{(N)} \in \mathcal{K}$. If we consider the fields of the theory to parametrize $A_0 \in \mathcal{K}_0$, the equations of motion are obtained form the zero grade component (7).

We shall now work with an explicit example of the *mkdV hierarchy* based upon the $\hat{\mathcal{G}} = \hat{sl}(2)$ affine algebra,

$$[h^{(m)}, E_{\pm\alpha}^{(n)}] = \pm 2E_{\pm\alpha}^{(n)}, \qquad [E_{\alpha}^{(m)}, E_{-\alpha}^{(n)}] = h^{(m+n)} \tag{8}$$

The grading operator is $Q = 2\lambda\frac{d}{d\lambda} + \frac{1}{2}h$ and decomposes the affine algebra $\hat{\mathcal{G}}$ into even and odd graded subspaces

$$\mathcal{G}_{2m} = \{h^{(m)} = \lambda^m h\},$$

$$\mathcal{G}_{2m+1} = \{\lambda^m (E_\alpha + \lambda E_{-\alpha}),\ \lambda^m (E_\alpha - \lambda E_{-\alpha})\} \tag{9}$$

for $m = 0, \pm 1, \pm 2, \cdots$ and $[\mathcal{G}_a, \mathcal{G}_b] \subset \mathcal{G}_{a+b}$. The integrable hierarchy is then specified by a choice of a semi-simple element $E = E^{(1)}$, where

$$E^{(2n+1)} = \lambda^n (E_\alpha + \lambda E_{-\alpha}) \tag{10}$$

and $A_0 = v(x, t_n)h^{(0)}$. The Kernel of $E^{(1)}$ is therefore given by

$$\mathcal{K} = \mathcal{K}_{2n+1} = \{\lambda^n (E_\alpha + \lambda E_{-\alpha})\} \tag{11}$$

and has grade $2n+1$. It thus follows from (6) that the highest grade component of $D^{(N)}$ has grade $N = 2n+1$. The component within the $\mathcal{M}$ of the zero grade projection of (7) leads to the evolution equations according to time $t = t_{2n+1}$. Notice that $D^{(0)}$ lies within the Cartan subalgebra and hence $[A_0, D^{(0)}] = 0$. The equations of motion are then simplified to $\partial_{t_{2n+1}} A_0 = \partial_x D^{(0)}$, Examples are,

$$n = 1 \qquad 4\partial_{t_3} v = \partial_x \left(\partial_x^2 v - 2v^3\right) \qquad mKdV \tag{12}$$

$$n = 2 \qquad 16\partial_{t_5} v = \partial_x \left(\partial_x^4 v - 10v^2(\partial_x^2 v) - 10v(\partial_x v)^2 + 6v^5\right), \tag{13}$$

$$n = 3 \qquad 64\partial_{t_7} v = \partial_x \left(\partial_x^6 v - 70(\partial_x v)^2(\partial_x^2 v) - 42v(\partial_x^2 v)^2 - 56v(\partial_x v)(\partial_x^3 v)\right)$$
$$- \partial_x \left(14v^2\partial_x^4 v - 140v^3(\partial_x v)^2 - 70v^4(\partial_x^2 v) + 20v^7\right) \tag{14}$$

$\cdots etc$

For the negative mKdV sub-hierarchy let us propose the following form for the zero curvature representation

$$[\partial_x + E^{(1)} + A_0, \partial_{t_{-N}} + D^{(-N)} + D^{(-N+1)} + \cdots + D^{(-1)}] = 0. \tag{15}$$

Differently from the positive hierarchy case, the lowest grade projection now yields,

$$\partial_x D^{(-N)} + [A_0, D^{(-N)}] = 0,$$

a nonlocal equation for $D^{(-N)}$. Having solved for $D^{(-N)}$, the second lowest projection of grade $-N+1$, leads to

$$\partial_x D^{(-N+1)} + [A_0, D^{(-N+1)}] + [E^{(1)}, D^{(-N)}] = 0$$

which determines $D^{(-N+1)}$. The proccess follows recursively until we reach the zero grade projection

$$\partial_{t_{-N}} A_0 - [E^{(1)}, D^{(-1)}] = 0 \tag{16}$$

which yields the evolution equation for field A_0 according to time $t = t_{-N}$ Notice that in this case there is no condition upon N.

The simplest example is to take $N = 1$ when the zero curvature decomposes into

$$\begin{aligned} \partial_x D^{(-1)} + [A_0, D^{(-1)}] &= 0, \\ \partial_{t_{-1}} A_0 - [E^{(1)}, D^{(-1)}] &= 0. \end{aligned} \tag{17}$$

In order to solve the first equation, we define the zero grade group element $B = \exp(\mathcal{G}_0)$ and define

$$D^{(-1)} = BE^{(-1)}B^{-1}, \qquad A_0 = -\partial_x BB^{-1}, \tag{18}$$

Under such parametrization the second eqn. (18) becomes the well known (relativistic) Leznov-Saveliev equation,

$$\partial_{t_{-1}} \left(\partial_x BB^{-1}\right) + [E^{(1)}, BE^{(-1)}B^{-1}] \tag{19}$$

which for $\hat{sl}(2)$ with principal gradation $Q = 2\lambda\frac{d}{d\lambda} + \frac{1}{2}h$, yields the sinh-Gordon equation

$$\partial_{t_{-1}}\partial_x \phi = e^{2\phi} - e^{-2\phi}, \qquad B = e^{-\phi h}. \tag{20}$$

where $t_{-1} = z, x = \bar{z}, A_0 = vh \equiv \partial_x \phi h$.

For higher values of $N = 2, 3, 4, \cdots$ we found

$$\partial_{t_{-2}}\partial_x \phi = 4e^{-2\phi}d^{-1}e^{2\phi}4e^{2\phi}d^{-1}e^{-2\phi} \tag{21}$$

$$\partial_{t_{-3}}\partial_x \phi = 4e^{-2\phi}d^{-1}\left(e^{2\phi}d^{-1}(\sinh 2\phi)\right) + 4e^{2\phi}d^{-1}\left(e^{-2\phi}d^{-1}(\sinh 2\phi)\right) \tag{22}$$

$$\begin{aligned} \partial_{t_{-4}}\partial_x \phi &= 4e^{-2\phi}d^{-1}\left(e^{2\phi}d^{-1}(e^{-2\phi}d^{-1}e^{2\phi} + e^{2\phi}d^{-1}e^{-2\phi})\right) \\ &+ 4e^{2\phi}d^{-1}\left(e^{-2\phi}d^{-1}(e^{-2\phi}d^{-1}e^{2\phi} + e^{2\phi}d^{-1}e^{-2\phi})\right) \end{aligned} \tag{23}$$

$$\begin{aligned}\partial_{t_{-5}}\partial_x\phi &= 8e^{-2\phi}d^{-1}\left(e^{2\phi}d^{-1}\left(e^{-2\phi}d^{-1}\left(e^{2\phi}d^{-1}(\sinh 2\phi)\right)+e^{2\phi}d^{-1}\left(e^{-2\phi}d^{-1}(\sinh 2\phi)\right)\right)\right)\\ &+ 8e^{2\phi}d^{-1}\left(e^{-2\phi}d^{-1}\left(e^{-2\phi}d^{-1}\left(e^{2\phi}d^{-1}(\sinh 2\phi)\right)+e^{2\phi}d^{-1}\left(e^{-2\phi}d^{-1}(\sinh 2\phi)\right)\right)\right),\end{aligned} \tag{24}$$

where $d^{-1}f=\int^x f(y)dy$. For $N=2$ eqn. (21) was derived in [17] using recurssion operators.

We now consider soliton solutions for the entire hierarchy. The general algebraic structure of the zero curvature representation yields a general method for constructing soliton solutions based on the fact that $v=0$ (and/or $\phi=0$) is the vacuum solution for all positive and negative odd sub-hierarchies, i.e., eqns. (12)-(14) and (20), (22), (24), $\cdots$, etc. [1] The idea of the dressing method is to map the *trivial vacuum* into an nontrivial configuration by gauge transformation, i.e,

$$\partial_x + A_0 = (\Theta_\pm)^{-1}\left(\partial_x + E^{(1)} + A_0^{vac}\right)\Theta_\pm,$$

$$\partial_{t_k} + D^{(k)} + \cdots + D^{(0)} = (\Theta_\pm)^{-1}\left(\partial_{t_k} + D^{k)}_{vac} + \cdots + D^{(0)}_{vac}\right)\Theta_\pm$$

where $\Theta_\pm$ are group elements of the form

$$\Theta_-^{-1} = e^{p(-1)}e^{p(-2)}\cdots, \qquad \Theta_+^{-1} = e^{q(0)}e^{q(1)}e^{q(2)}\cdots,$$

$p^{(-i)}$ and $q^{(i)}$ are linear combinations of grade $(-i)$ and (i) generators respectively.

It thus follows that one and two soliton solutions for the mKdV hierarchy with $A_0^{vac}=0$ can be written as

$$\begin{aligned}\phi_{1-sol} &= \ln\left(\frac{1-a_1\rho_1}{1+a_1\rho_1}\right)\\ \phi_{2-sol} &= \ln\left(\frac{1-a_1\rho_1-a_2\rho_2+a_1a_2a_{12}\rho_1\rho_2}{1+a_1\rho_1+a_2\rho_2+a_1a_2a_{12}\rho_1\rho_2}\right)\\ \vdots &= \vdots\end{aligned} \tag{25}$$

$a_{12}=\left(\frac{\kappa_1-\kappa_2}{\kappa_1+\kappa_2}\right)^2$ and

$$\rho_i(\kappa_i) = \exp 2(k_i x + t_a (k_i)^a). \tag{26}$$

These are solutions for all equations of the positive and negative odd hierarchies, i.e., eqns. (12)-(14) and (20), (22), (24), $\cdots$, for $a=3,5,7,\cdots$ and $a=-1,-3,-5,\cdots$ respectively.

It is clear that $\phi_{vac}=\phi_0=0$ do not satisfy eqns. (21) or (23). It follows that the negative even cases, (21), (23), etc, do not admit zero vacuum solution, i.e. $v_{vac}=0$. is not a solution. The soliton solutions (25) have to be modified accordingly to $v_{vac}=v_0\neq 0$, (see [8]) and these cases shall be discussed elsewhere.

[1] For negative even sub-hierarchy, the vacuun solution is obtained for $v=v_0\neq 0$ and the dressing method works equally well generating general formulae for multisoliton solutions but with a deformation parameter v_0, see for instance [8]

3. Backlund Transformation

In this section we start by noticing that the zero curvature representation (5) or (15) of the form

$$[\partial_x + A_x, \partial_t + A_t] = 0 \tag{27}$$

are invariant under gauge transformations of the type

$$A_\mu(\phi, \partial_x\phi, \cdots) \to \tilde{A}_\mu = U^{-1}A_\mu U + U^{-1}\partial_\mu U \tag{28}$$

If we now choose $U(\phi_1, \phi_2)$ such that it maps one field configuration ϕ_1 into another field configuration ϕ_2 preserving the equations of motion (i.e., zero curvature (27)) see [18],

$$U A_\mu(\phi_1) = A_\mu(\phi_2)U + \partial_\mu U \tag{29}$$

If we now take $A_\mu = A_x = E^{(1)} + A_0$ which is *common to all members* of the hierarchy, we find that

$$U = \begin{bmatrix} 1 & -\frac{\beta}{2\lambda}e^{-(\phi_1+\phi_2)} \\ -\frac{\beta}{2}e^{(\phi_1+\phi_2)} & 1 \end{bmatrix} \tag{30}$$

satisfies (29) provided

$$\partial_x(\phi_1 - \phi_2) = -\beta \sinh(\phi_1 + \phi_2). \tag{31}$$

For the sinh-Gordon model, the equations of motion (20) are satisfied if we further introduce the time component of the Backlund transformation,

$$\partial_{t_{-1}}(\phi_1 + \phi_2) = \frac{1}{\beta}\sinh(\phi_2 - \phi_1). \tag{32}$$

where $\phi_a, a = 1, 2$ satisfy the sinh-Gordon eqn, $\partial_{t_1}\partial_x\phi_a = \frac{1}{2}\sinh\phi_a$. The gauge transformation (30) leads to the Backlund transformation for the negative odd sub-hierarchy. Consider first the $t = t_{-3}$ evolution equation (22) where

$$A_{t_{-3}} = D^{(-3)} + D^{(-2)} + D^{(-1)} \tag{33}$$

where

$$\begin{aligned} D^{(-3)}(\phi) &= -a\left(e^{-2\phi}E_{\alpha}^{(-1)} + e^{2\phi}E_{-\alpha}^{(-1)}\right), \qquad D^{(-2)}(\phi) = 2aI(\phi)h^{(-1)} \\ D^{(-1)}(\phi) &= -4a\left(e^{-2\phi}\int^x e^{2\phi}I(\phi)E_{\alpha}^{(-1)} + e^{2\phi}\int^x e^{-2\phi}I(\phi)E_{-\alpha}^{(0)}\right) \end{aligned} \tag{34}$$

where $I(\phi_i) = \int^x \sinh(2\phi_i)$, $i = 1, 2$. Inserting the gauge transformation U (30) into (29) for $A_\mu = A_{t_{-3}}$ given in (33) and (34) we find the Backlund transformation

$$\partial_{t_{-3}}(\phi_1 + \phi_2) = \frac{8}{\beta}e^{\phi_1-\phi_2}\int^x e^{2\phi_2}I(\phi_2) - \frac{8}{\beta}e^{-\phi_1+\phi_2}\int^x e^{2\phi_1}I(\phi_1), \tag{35}$$

together with the subsidiary conditions

$$\begin{aligned} I(\phi_2) - I(\phi_1) &= \beta e^{\phi_1-\phi_2}\int^x \left(e^{-2\phi_1}I(\phi_1) + e^{2\phi_2}I(\phi_2)\right) \\ I(\phi_2) + I(\phi_1) &= \frac{2}{\beta}\sinh(\phi_2 - \phi_1) \end{aligned} \tag{36}$$

The very same argument follows for $t = t_{-5}$ where $A_{t_{-5}} = D^{(-5)} + D^{(-4)} + D^{(-3)} + D^{(-2)} + D^{(-1)}$. Here (30) into (29) yields,

$$\partial_{t_{-5}}(\phi_1 + \phi_2) = \frac{16}{\beta}e^{\phi_1-\phi_2}\int^x e^{2\phi_2}W(\phi_2) - \frac{16}{\beta}e^{-\phi_1+\phi_2}\int^x e^{-2\phi_2}W(\phi_2) \quad (37)$$

where $W(\phi_i) = \int^x \left(e^{-2\phi_i}\int^y e^{2\phi_i}I(\phi_i)\right)dy + \int^x \left(e^{2\phi_i}\int^y e^{-2\phi_i}I(\phi_i)\right)dy$ together with the subsidiary conditions

$$\begin{aligned} W(\phi_2) - W(\phi_1) &= \beta e^{\phi_1-\phi_2}\left(\int^x e^{-2\phi_1}W(\phi_1) + e^{2\phi_2}W(\phi_2)\right) \\ I(\phi_2) - I(\phi_1) &= \beta e^{\phi_1-\phi_2}\left(\int^x e^{-2\phi_1}I(\phi_1) + e^{2\phi_2}I(\phi_2)\right) \\ I(\phi_2) + I(\phi_1) &= \frac{2}{\beta}\sinh(\phi_2 - \phi_1). \end{aligned} \quad (38)$$

Other subsidiary relations are obtained from (36) and (38) by replacing $\phi_i \to -\phi_i$.

The key observation that allows us to extend such Backlund transformation to other positive higher grade members of the hierarchy (12 - 14), etc is to notice that the zero grade component of equation (7) is trivially solved by parametrizing $A_0 = -\partial_x BB^{-1}$ and $D^{(0)} = d_0 h^{(0)} = -\partial_{t_{2n+1}}BB^{-1}$. On the other hand by solving grade by grade the zero curvature eqn. (6)-(7) we find explicit expressions (42)-(44) for $D^{(0)}$. We define then the multi-time evolution for the field $v(x,t_N) = \partial_x\phi(x,t_N)$ to be

$$n=1 \qquad \partial_{t_3}\phi(x,t) \equiv d_0 = \frac{1}{4}\partial_x^3\phi - \frac{1}{2}(\partial_x\phi)^3 \quad (39)$$

$$n=2 \qquad \partial_{t_5}\phi(x,t) \equiv d_0 = \frac{1}{16}\partial_x^5\phi - \frac{5}{8}(\partial_x\phi)^2\partial_x^3\phi - \frac{5}{8}\partial_x\phi(\partial_x^2\phi)^2 + \frac{3}{8}(\partial_x\phi)^3, \quad (40)$$

$$\begin{aligned} n=3 \qquad \partial_{t_7}\phi(x,t) \equiv d_0 &= \frac{1}{64}\partial_x^7\phi - \frac{35}{32}(\partial_x^2\phi)^2(\partial_x^3\phi) - \frac{21}{32}(\partial_x\phi)(\partial_x^3\phi)^2 - \frac{7}{8}(\partial_x\phi)(\partial_x^2\phi)(\partial_x^4\phi) \\ &- \frac{7}{32}(\partial_x\phi)^2(\partial_x^5\phi) + \frac{35}{16}(\partial_x\phi)^3(\partial_x^2\phi)^2 + \frac{35}{32}(\partial_x\phi)^4(\partial_x^3\phi) - \frac{5}{16}\partial_x^8\phi \end{aligned} \quad (41)$$

$\cdots etc$

It follows that the Backlund transformation for the time component t_3 may be derived from (40) by considering

$$4\partial_{t_3}(\phi_1 - \phi_2) = \partial_x^3\phi_1 - \partial_x^3\phi_2 - 2(\partial_x\phi_1)^3 + 2(\partial_x\phi_2)^3,$$

Eliminating $\partial_x\phi_2$ from $x-$ component of Backlund transf. for Sinh-Gordon, i.e., $\partial_x(\phi_1 - \phi_2) = -\beta\sinh(\phi_1 + \phi_2)$, we find

$$\begin{aligned} 4(\partial_{t_3}\phi_2 - \partial_{t_3}\phi_1) &= \beta(\partial_x^2\phi_1 + \partial_x^2\phi_2)\cosh(\phi_1+\phi_2) - \frac{\beta}{2}(\partial_x\phi_1 + \partial_x\phi_2)^2\sinh(\phi_1+\phi_2) \\ &- \frac{\beta^3}{2}\sinh^3(\phi_1+\phi_2). \end{aligned} \quad (42)$$

which is in agreement with [14].

Analogously the same follows from (41) and (42) for t_5 and t_7 respectively, yielding

$$\begin{aligned} 16\partial_{t_5}(\phi_2 - \phi_1) &= 2\beta\partial_x^4\phi_1\cosh(\phi_1+\phi_2) - 4\beta\partial_x\phi_1(\partial_x^3\phi_1)\sinh(\phi_1+\phi_2) \\ &- 12\beta(\partial_x\phi_1)^2\partial_x^2\phi_1\cosh(\phi_1+\phi_2) + 2\beta(\partial_x^2\phi_1)^2\sinh(\phi_1+\phi_2) \\ &+ 6\beta(\partial_x\phi_1)^4\sinh(\phi_1+\phi_2) - 4\beta^2(\partial_x\phi_1)^3 + 2\beta^2\partial_x^3\phi_1 \\ &- 2\beta^3(\partial_x\phi_1)^2\sinh(\phi_1+\phi_2) + 2\beta^3\partial_x^2\phi_1\cosh(\phi_1+\phi_2) \\ &+ 2\beta^4\partial_x\phi_1 + \beta^5\sinh(\phi_1+\phi_2) \end{aligned} \quad (43)$$

and

$$\begin{aligned}
64(\partial_{t_7}\phi_2 - \partial_{t_7}\phi_1) &= -20\beta(\partial_x\phi_1)^6\sinh(\phi_{12}) + 20\beta\partial_x(\phi_1)^2(\partial_x^2\phi_1)^2\sinh(\phi_{12}) \\
&- 20\beta(\partial_x^2\phi_1)^3\cosh(\phi_{12}) + 40\beta(\partial_x\phi_1)^3\partial_x^3\phi_1\sinh(\phi_{12}) \\
&- 80\beta\partial_x\phi_1\partial_x^2\phi_1\partial_x^3\phi_1\cosh(\phi_{12}) - 2\beta(\partial_x^3\phi_1)^2\sinh(\phi_{12}) \\
&- 20\beta(\partial_x\phi_1)^2\partial_x^4\phi_1\cosh(\phi_{12}) + 4\beta\partial_x^2\phi_1\partial_x^4\phi_1\sinh(\phi_{12}) \\
&+ 60\beta(\partial_x\phi_1)^4\partial_x^2\phi_1\cosh(\phi_{12}) - 4\beta\partial_x\phi_1\partial_x^5\phi_1\sinh(\phi_{12}) \\
&+ 2\beta\partial_x^6\phi_1\cosh(\phi_{12}) + 12\beta^2(\partial_x\phi_1)^5 - 20\beta^2\partial_x\phi_1(\partial_x^2\phi_1)^2 \\
&- 20\beta^2(\partial_x\phi_1)^2\partial_x^3\phi_1 + 2\beta^2\partial_x^5\phi_1 - 12\beta^3(\partial_x\phi_1)^4\sinh(\phi_{12}) \\
&- 12\beta^3(\partial_x\phi_1)^2\partial_x^2\phi_1\cosh(\phi_{12}) + 2\beta^3(\partial_x^2\phi_1)^2\sinh(\phi_{12}) \\
&- 4\beta^3\partial_x\phi_1\partial_x^3\phi_1\sinh(\phi_{12}) + 2\beta^3\partial_x^4\phi_1\cosh(\phi_{12}) \\
&- 4\beta^4(\partial_x\phi_1)^3 + 2\beta^4\partial_x^3\phi_1 - 2\beta^5(\partial_x\phi_1)^2\sinh(\phi_{12}) \\
&+ 2\beta^5\partial_x^2\phi_1\cosh(\phi_{12}) + 2\beta^6\partial_x\phi_1 + \beta^7\sinh(\phi_{12}) \qquad (44)
\end{aligned}$$

where $\phi_{12} \equiv \phi_1 + \phi_2$.

4. Examples

In this section we shall consider few solutions for the Backlund solutions for (12)-(14) and (20), (22) and (24).

4.1. Vacuum - 1-soliton

Let

$$\phi_1 = \phi_{vac} = 0, \qquad \phi_2 = \phi_{1-sol} = \ln\left(\frac{1+R\rho}{1-R\rho}\right), \qquad \rho = e^{2kx+2k^N t_N} \qquad (45)$$

and R is a constant. It becomes clear that the Backlund equations (42)-(44) are satisfied by (45) for $\beta = 2k$.

4.2. 1-soliton - 1-soliton

$$\phi_1 = \phi_{1-sol}(k_1) = \ln\left(\frac{1+R_1\rho_1}{1-R_1\rho_1}\right), \qquad \phi_2 = \phi_{1-sol}(k_2) = \ln\left(\frac{1+R_2\rho_2}{1-R_2\rho_2}\right), \qquad \rho_i = e^{2k_ix+2k_i^N t_N} \qquad (46)$$

and $R_i = R_i(k_i)$. Backlund solutions given in (46) satisfy the Backlund equations (42)-(44) for

$$k_1 = k_2 = k, \qquad R_2 = \left(\frac{2k+\beta}{2k-\beta}\right) R_1. \qquad (47)$$

respectively for $N = 3, 5$ and 7. It was verified that the same also satisfy eqns. (20), (22) and (24) with $N = -1, -3$ and -5 respectively.

4.3. 1-soliton - 2-soliton

$$\phi_1 = \phi_{1-sol} = \ln\left(\frac{1+\rho_1}{1-\rho_1}\right), \qquad \phi_2 = \phi_{2-sol} = \ln\left(\frac{1+\delta(\rho_1-\rho_2)-\rho_1\rho_2}{1-\delta(\rho_1-\rho_2)-\rho_1\rho_2}\right), \qquad (48)$$

where $\delta = \frac{k_1+k_2}{k_1-k_2}$.

We have verified that the Backlund equations for the higher members of the hierarchy, namely,(42)-(44) are satisfied by (48) for $N = 3, 5$ and 7. Also, the permutability theorem, which establishes the equality of the 2-soliton solution obtained from vacuum to 1-soliton solution for $\beta = 2k_1$ and subsequently this 1-soliton to 2-soliton solutions and $\beta = 2k_2$. On the other hand the very same 2-soliton solution is obtained from the vacuum to 1-soliton solution for $\beta = 2k_2$ and subsequently this 1-soliton to 2-soliton solutions and $\beta = 2k_1$. For technical reasons we were unable to verify this case for the negative odd equations (20), (22) and (24).

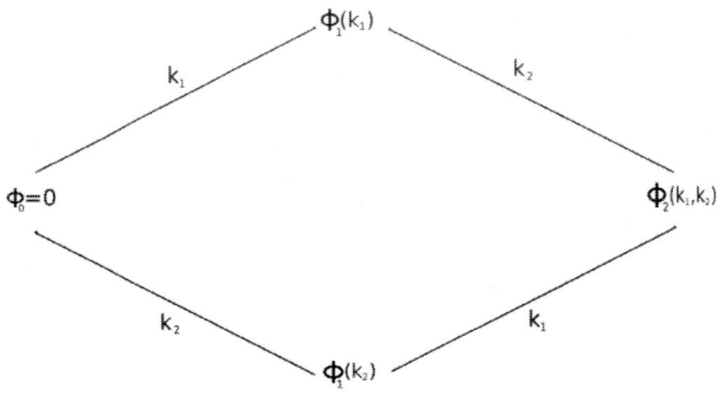

Figure 1. Permutability theorem for 2-solitons solution

5. Conclusions and Further Remarks

The construction of a set of non-linear integrable equations of motion were defined in terms of a zero curvature representation and an affine Lie algebra. Explicitly, we have considered the mKdV hierarchy for positive and negative odd graded time evolutions. A class of soliton solutions, labeled according to different graded time evolutions, were constructed in a universal manner to all members of the hierarchy. These were derived, by the dressing method from the trivial vacuum solution , i.e. $v_{vac} = 0$.

We have extended the Backlund transformation of the sinh-Gordon model to other higher and lower graded members of the mKdV hierarchy. We have shown that the spatial component of the Backlund transformation is common to all members of the hierarchy and is a direct consequence of the common Lax operator. These equations are verified to be satisfied by the same pair of solutions for the first few members of the hierarchy.

For the negative even graded equations the trivial vacuum is observed *not* to be solution of the equations of motion and a deformation of the dressing method need to be employed (see [8]).

We are extending the same arguments from the $N = 1$ super sinh-Gordon model to the supersymmetric mKdV hierarchy [15].

So far, our arguments were based upon known spatial component of the Backlund transformation of the Sinh-Gordon model. Such relation defines the so called type I Backlund transformation. The extension to type II Backlund transformation [16] , [18] is a subject for future investigation.

Acknowledgments

We would like to thank Capes, CNPq and Fapesp for support.

References

[1] Aratyn H, Gomes J F, Nissimov E, Pacheva S and Zimerman A H 2000 Symmetry flows, conservation laws and dressing approach to the integrable models *NATO Advanced Research Workshop on Integrable Hierarchies and Modern Physical Theories July 22-26 2000 Chicago* 243–275 (nlin/0012042)
[2] Chodos A 1980 *Phys. Rev.* D **21** 2818
[3] Tracy C A and Widom H 1976 *Commun. Math. Phys.* **179** 1
[4] Aratyn H, Ferreira L A, Gomes J F and Zimerman A H 2000 *J. Phys.* A **33** L331 (nlin/0007002)
[5] Gomes J F, Gueuvoghlanian E P, Sotkov G M and Zimerman A H 2001 *Nucl. Phys.* B **606** 441 (hep-th/0007169)
[6] Cabrera-Carnero I, Gomes J F, Sotkov G M and Zimerman A H 2002 *Nucl. Phys.* B **634** 433 (hep-th/0201047)
[7] Gomes J F, Sotkov G M and Zimerman A H 2005 *Nucl. Phys.* B **714** 179 (hep-th/0405182)
[8] Gomes J F, Starvaggi França G, de Melo G R and Zimerman A H 2009 *J. Phys.* A **42** 445204 (arXiv:0906.5579)
[9] Bowcock P, Corrigan E and Zambon C 2004 *Int. J. Mod. Phys.* A **19** S2 82 (hep-th/0305022)
[10] Bowcock P, Corrigan E and Zambon C 2004 *JHEP* 0401 056 (hep-th/0401020)
[11] Corrigan E and Zambon C 2006 *Nonlinearity* **19** 1447-1469 (nlin/0512038)
[12] Kodama Y and Wadati M 1976 *Progr. Theor. Phys.* **56** 1740
[13] Villani A and Zimerman A H 1977 *Rev. Bras. de Fisica* **7** 649
[14] Rogers C and Shadwick W F 1982 *Backlund Transformation and their Application* (New York: Academic Press)
[15] Gomes J F, Ymai L H and Zimerman A H 2006 *Phys. Lett.* A **359** 630 (arXiv:hep-th/0607107)
[16] Corrigan E and Zambon C 2009 *J. of Phys.* A **42** 475203 (arXiv:0908.3126)
[17] Qiao Z and Strampp W 2002 *Physica* A **313** 365
[18] Aguirre A R, Araujo T R, Gomes J F and Zimerman A H 2011 *J. High. En. Phys.* **12** 56 (arXiv:1110.1589)

Natural limits of Electroweak Model as contraction of its gauge group

N A Gromov

Department of Mathematics, Komi Science Center UrD RAS, Kommunisticheskaya st. 24, Syktyvkar 167982, Russia

E-mail: gromov@dm.komisc.ru

Abstract. The low and higher energy limits of the Electroweak Model are obtained from first principles of gauge theory. Both limits are given by the same contraction of the gauge group, but for the different consistent rescalings of the field space. Mathematical contraction parameter in both cases is interpreted as energy. The very weak neutrino-matter interactions is explained by zero tending contraction parameter, which depend on neutrino energy. The second consistent rescaling corresponds to the higher energy limit of the Electroweak Model. At the infinite energy all particles lose masses, electroweak interactions become long-range and are mediated by the neutral currents. The limit model represents the development of the early Universe.

1. Introduction

The modern theory of electroweak processes is the Electroweak Model, which is in good agreement with experimental dates, including the latest ones from LHC. This model is a gauge theory based on the gauge group $SU(2) \times U(1)$, which is the direct product of two simple groups. The operation of group contraction [1] transforms a simple or semisimple group to a nonsemisimple one. In particular the special unitary group $SU(2)$ is contracted to the group isomorphic to Euclid group $E(2)$ [2]. For better understanding of a complicated physical system it is useful to investigate its behaviour for limiting values of its physical parameters. In this paper we discuss at the level of classical gauge fields the modified Electroweak Model with the contracted gauge group $SU(2;j) \times U(1)$. It was shown [3]–[6] that at low energies the contraction parameter depends on the energy s in center-of-mass system, so the contracted gauge group corresponds to the zero energy limit of the Electroweak Model. The very weak neutrinos-matter interactions and the linear dependence of their cross-section on neutrino energy both are explained from first principles of the Electroweak Model as contraction of its gauge group. But for the same contraction of the gauge group there is another consistent rescaling of the representation space, which lead to the infinite energy limit of the Electroweak Model. In this paper we consider the second possibility and discuss some particle properties in early Universe, where similar higher energies can exist.

2. Standard Electroweak Model

We shall follow the books [7]–[9] in description of standard Electroweak Model. Its Lagrangian is the sum of boson, lepton and quark Lagrangians

$$L = L_B + L_L + L_Q. \tag{1}$$

Boson sector $L_B = L_A + L_\phi$ involve two parts: the gauge field Lagrangian

$$L_A = \frac{1}{8g^2}\mathrm{Tr}(F_{\mu\nu})^2 - \frac{1}{4}(B_{\mu\nu})^2 = -\frac{1}{4}[(F^1_{\mu\nu})^2 + (F^2_{\mu\nu})^2 + (F^3_{\mu\nu})^2] - \frac{1}{4}(B_{\mu\nu})^2 \tag{2}$$

and the matter field Lagrangian

$$L_\phi = \frac{1}{2}(D_\mu\phi)^\dagger D_\mu\phi - \frac{\lambda}{4}\left(\phi^\dagger\phi - v^2\right)^2, \tag{3}$$

where $\phi = \begin{pmatrix} \phi_1 \\ \phi_2 \end{pmatrix} \in C_2$ are the matter fields. The covariant derivatives are given by

$$D_\mu\phi = \partial_\mu\phi - ig\left(\sum_{k=1}^{3} T_k A^k_\mu\right)\phi - ig'YB_\mu\phi, \tag{4}$$

where $T_k = \frac{1}{2}\tau_k, k = 1,2,3$ are generators of $SU(2)$, $Y = \frac{1}{2}\mathbf{1}$ is generator of $U(1)$, g and g' are constants. The gauge fields

$$A_\mu(x) = -ig\sum_{k=1}^{3} T_k A^k_\mu(x), \quad B_\mu(x) = -ig'B_\mu(x) \tag{5}$$

take their values in Lie algebras $su(2)$, $u(1)$ respectively, and the stress tensors are as follows

$$F_{\mu\nu}(x) = \mathcal{F}_{\mu\nu}(x) + [A_\mu(x), A_\nu(x)], \quad B_{\mu\nu} = \partial_\mu B_\nu - \partial_\nu B_\mu.$$

To generate mass for the vector bosons the special mechanism of spontaneous symmetry breaking is used. One of L_B ground states

$$\phi^{vac} = \begin{pmatrix} 0 \\ v \end{pmatrix}, \quad A^k_\mu = B_\mu = 0$$

is taken as a vacuum state of the model, and small field excitations $v + \chi(x)$ with respect to this vacuum are regarded.

The fermion sector is represented by the lepton L_L and quark L_Q Lagrangians. The lepton Lagrangian is taken in the form

$$L_L = L_l^\dagger i\tilde{\tau}_\mu D_\mu L_l + e_r^\dagger i\tau_\mu D_\mu e_r - h_e[e_r^\dagger(\phi^\dagger L_l) + (L_l^\dagger\phi)e_r], \tag{6}$$

where $L_l = \begin{pmatrix} \nu_l \\ e_l \end{pmatrix}$ is the $SU(2)$-doublet, e_r is the $SU(2)$-singlet, h_e is constant, $\tau_0 = \tilde{\tau}_0 = \mathbf{1}$, $\tilde{\tau}_k = -\tau_k$, τ_μ are Pauli matrices and e_r, e_l, ν_l are two component Lorentzian spinors. Last terms with factor h_e represent electron mass.The covariant derivatives are given by the formulas:

$$D_\mu L_l = \partial_\mu L_l - i\frac{g}{\sqrt{2}}\left(W^+_\mu T_+ + W^-_\mu T_-\right)L_l -$$

$$-i\frac{g}{\cos\theta_w}Z_\mu\left(T_3 - Q\sin^2\theta_w\right)L_l - ieA_\mu QL_l,$$

$$D_\mu e_r = \partial_\mu e_r - ig'QA_\mu e_r\cos\theta_w + ig'QZ_\mu e_r\sin\theta_w, \tag{7}$$

Journal of Physics: Conference Series **597** (2015) 012040 doi:10.1088/1742-6596/597/1/012040

where $T_\pm = T_1 \pm iT_2$, $Q = Y + T_3$ is the electrical charge, $Y = \frac{1}{2}\mathbf{1}$ is the hypercharge, $e = gg'(g^2+g'^2)^{-\frac{1}{2}}$ is the electron charge and $\sin\theta_w = eg^{-1}$. The new gauge fields

$$Z_\mu = \frac{1}{\sqrt{g^2+g'^2}}\left(gA_\mu^3 - g'B_\mu\right), \quad A_\mu = \frac{1}{\sqrt{g^2+g'^2}}\left(g'A_\mu^3 + gB_\mu\right),$$

$$W_\mu^\pm = \frac{1}{\sqrt{2}}\left(A_\mu^1 \mp iA_\mu^2\right) \tag{8}$$

are introduced instead of (5).

The quark Lagrangian is given by

$$L_Q = Q_l^\dagger i\tilde{\tau}_\mu D_\mu Q_l + u_r^\dagger i\tau_\mu D_\mu u_r + d_r^\dagger i\tau_\mu D_\mu d_r -$$

$$-h_d[d_r^\dagger(\phi^\dagger Q_l) + (Q_l^\dagger\phi)d_r] - h_u[u_r^\dagger(\tilde{\phi}^\dagger Q_l) + (Q_l^\dagger\tilde{\phi})u_r], \tag{9}$$

where left quark fields form the $SU(2)$-doublet $Q_l = \begin{pmatrix} u_l \\ d_l \end{pmatrix}$, right quark fields u_r, d_r are the $SU(2)$-singlets, $\tilde{\phi}_i = \epsilon_{ik}\bar{\phi}_k, \epsilon_{00} = 1, \epsilon_{ii} = -1$ is the conjugate representation of $SU(2)$ group and h_u, h_d are constants. All fields u_l, d_l, u_r, d_r are two component Lorentzian spinors. Last four terms with factors h_d and h_u specify the d- and u-quark mass. The covariant derivatives are given by

$$D_\mu Q_l = \left(\partial_\mu - ig\sum_{k=1}^{3}\frac{\tau_k}{2}A_\mu^k - ig'\frac{1}{6}B_\mu\right)Q_l,$$

$$D_\mu u_r = \left(\partial_\mu - ig'\frac{2}{3}B_\mu\right)u_r, \quad D_\mu d_r = \left(\partial_\mu + ig'\frac{1}{3}B_\mu\right)d_r. \tag{10}$$

From the viewpoint of electroweak interactions all known leptons and quarks are divided on three generations. Next two lepton generation are introduced in a similar way to (6). They are left $SU(2)$-doublets

$$\begin{pmatrix} \nu_\mu \\ \mu \end{pmatrix}_l, \quad \begin{pmatrix} \nu_\tau \\ \tau \end{pmatrix}_l, \quad Y = -\frac{1}{2} \tag{11}$$

and right $SU(2)$-singlets: $\mu_r, \tau_r, Y = -1$. In addition to u- and d-quarks of the first generation there are (c, s) and (t, b) quarks of the next generations, which left fields

$$\begin{pmatrix} c_l \\ s_l \end{pmatrix}, \quad \begin{pmatrix} t_l \\ b_l \end{pmatrix}, \quad Y = \frac{1}{6}, \tag{12}$$

are described by the $SU(2)$-doublets and the right fields are $SU(2)$-singlets: $c_r, t_r, Y = \frac{2}{3}$; $s_r, b_r, Y = -\frac{1}{3}$.Their Lagrangians are introduced in a similar way to (9). Full lepton and quark Lagrangians are obtained by the summation over all generations. In what follows we shall regarded only first generations of leptons and quarks.

3. Low energy limit of the Electroweak Model

We consider a model where the contracted gauge group $SU(2;j)\times U(1)$ acts in the boson, lepton and quark sectors. The contracted group $SU(2;j)$ is obtained [10, 11]by the consistent rescaling of the fundamental representation of $SU(2)$ and the space C_2

$$z'(j) = \begin{pmatrix} jz_1' \\ z_2' \end{pmatrix} = \begin{pmatrix} \alpha & j\beta \\ -j\bar{\beta} & \bar{\alpha} \end{pmatrix}\begin{pmatrix} jz_1 \\ z_2 \end{pmatrix} = u(j)z(j),$$

$$\det u(j) = |\alpha|^2 + j^2|\beta|^2 = 1, \quad u(j)u^\dagger(j) = 1 \tag{13}$$

in such a way that the hermitian form

$$z^\dagger z(j) = j^2|z_1|^2 + |z_2|^2 \tag{14}$$

remains invariant, when contraction parameter tends to zero $j \to 0$ or is equal to the nilpotent unit $j = \iota$, $\iota^2 = 0$. The actions of the unitary group $U(1)$ and the electromagnetic subgroup $U(1)_{em}$ in the space $C_2(\iota)$ with the base $\{z_2\}$ and the fiber $\{z_1\}$ are given by the same matrices as on the space C_2.

The space $C_2(j)$ of the fundamental representation of $SU(2;j)$ group can be obtained from C_2 by substituting z_1 by jz_1. Substitution $z_1 \to jz_1$ induces another ones for Lie algebra generators $T_1 \to jT_1$, $T_2 \to jT_2$, $T_3 \to T_3$. As far as the gauge fields take their values in Lie algebra, we can substitute the gauge fields instead of transforming the generators, namely:

$$A_\mu^1 \to jA_\mu^1, \;\; A_\mu^2 \to jA_\mu^2, \;\; A_\mu^3 \to A_\mu^3, \;\; B_\mu \to B_\mu. \tag{15}$$

Indeed, due to commutativity and associativity of multiplication by j

$$su(2;j) \ni \left\{A_\mu^1(jT_1) + A_\mu^2(jT_2) + A_\mu^3 T_3\right\} = \\ = \left\{(jA_\mu^1)T_1 + (jA_\mu^2)T_2 + A_\mu^3 T_3\right\}. \tag{16}$$

For the gauge fields (8) these substitutions are as follows:

$$W_\mu^\pm \to jW_\mu^\pm, \;\; Z_\mu \to Z_\mu, \;\; A_\mu \to A_\mu. \tag{17}$$

The left lepton $L_l = \begin{pmatrix} \nu_l \\ e_l \end{pmatrix}$ and quark $Q_l = \begin{pmatrix} u_l \\ d_l \end{pmatrix}$ fields are $SU(2)$-doublets, so their components are transformed in the similar way as components of the vector z, namely:

$$\nu_l \to j\nu_l, \quad e_l \to e_l, \quad u_l \to ju_l, \quad d_l \to d_l. \tag{18}$$

The right lepton and quark fields are $SU(2)$-singlets and therefore are not changed.

The full Lagrangian of the modified model is obtained from the standard one [7]–[9] by the above substitutions and can be represented in the form

$$L(j) = L_b + j^2 L_f, \tag{19}$$

where L_b is Lagrangian of the base fields and L_f is Lagrangian of the fiber fields. The transformed boson Lagrangian was discussed in [10, 11] on the level of classical fields, where it was shown that masses of all particles involved in the Electroweak Model remain the same under contraction $j^2 \to 0$. In this limit the contribution $j^2 L_f$ of neutrino, W-boson and u-quark fields as well as their interactions with other fields to the Lagrangian (19) will be vanishingly small in comparison with contribution L_b of electron, d-quark and remaining boson fields. So Lagrangian (19) describes very weak interaction of neutrino fields with the matter. On the other hand, contribution of the neutrino part $j^2 L_f$ to the full Lagrangian is risen when the parameter j^2 is increased, that again corresponds to the experimental facts. So contraction parameter can be phenomenologically connected with neutrino energy.

To establish the precise physical meaning of the contraction parameter we need more fine consideration on the level of quantized fields. It is necessary to regard neutrino elastic scattering on electron and quarks [8]. We have found that the contraction parameter is connected with the neutrino energy $\sqrt{s}$ in center-of-mass system and is evaluated through the fundamental parameters of the Electroweak Model as follows

$$j^2(s) = \frac{g}{m_W}\sqrt{s}, \tag{20}$$

where m_W is W-boson mass and g is constant [4, 5]. Therefore, contraction $j \to 0$ result in low energy limit of the Electroweak Model.

4. High-Energy Lagrangian of Electroweak Model

We shown in previous section that contraction $j \to 0$ of the gauge group (13) of the Electroweak Model corresponds to its zero energy limit. In this limit the first components of the lepton and quark doublets become infinite small in comparison with their second components. On the contrary, when energy increase the first components of the doublets become greater then their second ones. In the infinite energy limit the second components of the lepton and quark doublets will be infinite small as compared with their first components. To describe this limit we introduce new contraction parameter ϵ and **new consistent rescaling** of the group $SU(2)$ and the space C_2 as follows [12]

$$z'(\epsilon) = \left(\begin{array}{c} z_1' \\ \epsilon z_2' \end{array} \right) = \left(\begin{array}{cc} \alpha & \epsilon\beta \\ -\epsilon\bar{\beta} & \bar{\alpha} \end{array} \right) \left(\begin{array}{c} z_1 \\ \epsilon z_2 \end{array} \right) = u(\epsilon)z(\epsilon),$$

$$\det u(\epsilon) = |\alpha|^2 + \epsilon^2|\beta|^2 = 1, \quad u(\epsilon)u^\dagger(\epsilon) = 1 \tag{21}$$

Both contracted groups $SU(2;j)$ (13) and $SU(2;\epsilon)$ (21) are the same and are isomorphic to Euclid group $E(2)$, but the space $C_2(\epsilon)$ is split in the limit $\epsilon \to 0$ on the one-dimension base $\{z_1\}$ and the one-dimension fiber $\{z_2\}$. From the mathematical point of view it is not important first or second Cartesian axis forms the base of fibering and in this sence constructions (13) and (21) are equivalent. But the doublet components are interpreted as a certain physical fields, therefore the fundamental representations (13) and (21) of the same contracted unitary group lead to the different limit cases of the Electroweak Model, namely, its zero energy and infinite energy limits.

In the second contraction scheme (21) all gauge bosons are transformed according to the rules (17) with the natural substitution of j by ϵ. Instead of (18) the lepton and quark fields are transformed now as follows

$$e_l \to \epsilon e_l, \quad d_l \to \epsilon d_l, \quad \nu_l \to \nu_l, \quad u_l \to u_l. \tag{22}$$

The next reason for inequality of the first and second doublet components is the special mechanism of spontaneous symmetry breaking, which is used to generate mass of vector bosons and other elementary particles of the model. In this mechanism one of Lagrangian ground states $\phi^{vac} = \left(\begin{array}{c} 0 \\ v \end{array} \right)$ is taken as vacuum of the model and then small field excitations $v + \chi(x)$ with respect to this vacuum are regarded. So Higgs boson field χ and constant v are multiplied by ϵ. As far as masses of all particles are proportionate to v we obtain the following transformation rule for the contraction (21)

$$\chi \to \epsilon\chi, \quad v \to \epsilon v, \quad m_p \to \epsilon m_p, \quad p = \chi, W, Z, e, u, d. \tag{23}$$

After transformations (17), (22)–(23) the boson Lagrangian of the Electroweak Model can be represented in the form

$$L_B(\epsilon) = -\frac{1}{4}\mathcal{Z}_{\mu\nu}^2 - \frac{1}{4}\mathcal{F}_{\mu\nu}^2 + \epsilon^2 L_{B,2} + \epsilon^3 g W_\mu^+ W_\mu^- \chi + \epsilon^4 L_{B,4},$$

$$L_{B,4} = m_W^2 W_\mu^+ W_\mu^- - \frac{1}{2}m_\chi^2\chi^2 - \lambda v\chi^3 - \frac{\lambda}{4}\chi^4 +$$

$$+\frac{g^2}{4}\left(W_\mu^+ W_\nu^- - W_\mu^- W_\nu^+\right)^2 + \frac{g^2}{4}W_\mu^+ W_\nu^- \chi^2,$$

$$L_{B,2} = \frac{1}{2}\left(\partial_\mu\chi\right)^2 + \frac{1}{2}m_Z^2\left(Z_\mu\right)^2 - \frac{1}{2}\mathcal{W}_{\mu\nu}^+\mathcal{W}_{\mu\nu}^- +$$

$$
\begin{aligned}
&+\frac{gm_z}{2\cos\theta_W}(Z_\mu)^2\chi+\frac{g^2}{8\cos^2\theta_W}(Z_\mu)^2\chi^2-\\
&-2ig\left(W_\mu^+W_\nu^- - W_\mu^-W_\nu^+\right)\Big(\mathcal{F}_{\mu\nu}\sin\theta_W+\mathcal{Z}_{\mu\nu}\cos\theta_W\Big)-\\
&-\frac{i}{2}e\left[A_\mu\left(\mathcal{W}_{\mu\nu}^+W_\nu^- - \mathcal{W}_{\mu\nu}^-W_\nu^+\right)+\frac{i}{2}eA_\nu\left(\mathcal{W}_{\mu\nu}^+W_\mu^- - \mathcal{W}_{\mu\nu}^-W_\mu^+\right)\right]-\\
&-\frac{i}{2}g\cos\theta_W\left[Z_\mu\left(\mathcal{W}_{\mu\nu}^+W_\nu^- - \mathcal{W}_{\mu\nu}^-W_\nu^+\right)-Z_\nu\left(\mathcal{W}_{\mu\nu}^+W_\mu^- - \mathcal{W}_{\mu\nu}^-W_\mu^+\right)\right]-\\
&-\frac{e^2}{4}\left\{\left[\left(W_\mu^+\right)^2+\left(W_\mu^-\right)^2\right](A_\nu)^2-\right.\\
&\left.-2\left(W_\mu^+W_\nu^+ + W_\mu^-W_\nu^-\right)A_\mu A_\nu+\left[\left(W_\nu^+\right)^2+\left(W_\nu^-\right)^2\right](A_\mu)^2\right\}-\\
&-\frac{g^2}{4}\cos\theta_W\left\{\left[\left(W_\mu^+\right)^2+\left(W_\mu^-\right)^2\right](Z_\nu)^2-\right.\\
&\left.-2\left(W_\mu^+W_\nu^+ + W_\mu^-W_\nu^-\right)Z_\mu Z_\nu+\left[\left(W_\nu^+\right)^2+\left(W_\nu^-\right)^2\right](Z_\mu)^2\right\}-\\
&-eg\cos\theta_W\left[W_\mu^+W_\mu^-A_\nu Z_\nu+W_\nu^+W_\nu^-A_\mu Z_\mu-\right.\\
&\left.-\frac{1}{2}\left(W_\mu^+W_\nu^- + W_\nu^+W_\mu^-\right)\left(A_\mu Z_\nu+A_\nu Z_\mu\right)\right].
\end{aligned}
\tag{24}
$$

The lepton Lagrangian in terms of electron and neutrino fields takes the form

$$
\begin{aligned}
&L_L(\epsilon)=L_{L,0}+\epsilon^2L_{L,2}=\\
&=\nu_l^\dagger i\tilde{\tau}_\mu\partial_\mu\nu_l+e_r^\dagger i\tau_\mu\partial_\mu e_r+g'\sin\theta_w e_r^\dagger\tau_\mu Z_\mu e_r-\\
&-g'\cos\theta_w e_r^\dagger\tau_\mu A_\mu e_r+\frac{g}{2\cos\theta_w}\nu_l^\dagger\tilde{\tau}_\mu Z_\mu\nu_l+\\
&+\epsilon^2\left\{e_l^\dagger i\tilde{\tau}_\mu\partial_\mu e_l-m_e(e_r^\dagger e_l+e_l^\dagger e_r)+\right.\\
&+\frac{g\cos 2\theta_w}{2\cos\theta_w}e_l^\dagger\tilde{\tau}_\mu Z_\mu e_l-ee_l^\dagger\tilde{\tau}_\mu A_\mu e_l+\\
&\left.+\frac{g}{\sqrt{2}}\Big(\nu_l^\dagger\tilde{\tau}_\mu W_\mu^+e_l+e_l^\dagger\tilde{\tau}_\mu W_\mu^-\nu_l\Big)\right\}.
\end{aligned}
\tag{25}
$$

The quark Lagrangian in terms of u- and d-quarks fields can be written as

$$
L_Q(\epsilon)=L_{Q,0}-\epsilon m_u(u_r^\dagger u_l+u_l^\dagger u_r)+\epsilon^2L_{Q,2},
$$

$$
\begin{aligned}
&L_{Q,0}=d_r^\dagger i\tau_\mu\partial_\mu d_r+u_l^\dagger i\tilde{\tau}_\mu\partial_\mu u_l+u_r^\dagger i\tau_\mu\partial_\mu u_r-\\
&-\frac{1}{3}g'\cos\theta_w d_r^\dagger\tau_\mu A_\mu d_r+\frac{1}{3}g'\sin\theta_w d_r^\dagger\tau_\mu Z_\mu d_r+\\
&+\frac{2e}{3}u_l^\dagger\tilde{\tau}_\mu A_\mu u_l+\frac{g}{\cos\theta_w}\left(\frac{1}{2}-\frac{2}{3}\sin^2\theta_w\right)u_l^\dagger\tilde{\tau}_\mu Z_\mu u_l+\\
&+\frac{2}{3}g'\cos\theta_w u_r^\dagger\tau_\mu A_\mu u_r-\frac{2}{3}g'\sin\theta_w u_r^\dagger\tau_\mu Z_\mu u_r,
\end{aligned}
$$

Journal of Physics: Conference Series **597** (2015) 012040 doi:10.1088/1742-6596/597/1/012040

$$L_{Q,2} = d_l^\dagger i\tilde{\tau}_\mu \partial_\mu d_l - m_d(d_r^\dagger d_l + d_l^\dagger d_r) -$$
$$-\frac{e}{3} d_l^\dagger \tilde{\tau}_\mu A_\mu d_l - \frac{g}{\cos\theta_w}\left(\frac{1}{2} - \frac{2}{3}\sin^2\theta_w\right) d_l^\dagger \tilde{\tau}_\mu Z_\mu d_l +$$
$$+\frac{g}{\sqrt{2}}\left[u_l^\dagger \tilde{\tau}_\mu W_\mu^+ d_l + d_l^\dagger \tilde{\tau}_\mu W_\mu^- u_l\right]. \quad (26)$$

The complete Lagrangian of the modified model is given by the sum $L(\epsilon) = L_B(\epsilon) + L_L(\epsilon) + L_Q(\epsilon)$ and can be written in the form

$$L(\epsilon) = L_\infty + \epsilon L_1 + \epsilon^2 L_2 + \epsilon^3 L_3 + \epsilon^4 L_4. \quad (27)$$

For the infinite energy ($\epsilon = 0$) limit Lagrangian is equal to

$$L_\infty = -\frac{1}{4}\mathcal{Z}_{\mu\nu}^2 - \frac{1}{4}\mathcal{F}_{\mu\nu}^2 + \nu_l^\dagger i\tilde{\tau}_\mu \partial_\mu \nu_l + u_l^\dagger i\tilde{\tau}_\mu \partial_\mu u_l +$$
$$+e_r^\dagger i\tau_\mu \partial_\mu e_r + d_r^\dagger i\tau_\mu \partial_\mu d_r + u_r^\dagger i\tau_\mu \partial_\mu u_r + L_\infty^{int}(A_\mu, Z_\mu), \quad (28)$$

where

$$L_\infty^{int}(A_\mu, Z_\mu) = \frac{g}{2\cos\theta_w}\nu_l^\dagger \tilde{\tau}_\mu Z_\mu \nu_l + \frac{2e}{3} u_l^\dagger \tilde{\tau}_\mu A_\mu u_l +$$
$$+\frac{g}{\cos\theta_w}\left(\frac{1}{2} - \frac{2}{3}\sin^2\theta_w\right) u_l^\dagger \tilde{\tau}_\mu Z_\mu u_l +$$
$$+g'\sin\theta_w e_r^\dagger \tau_\mu Z_\mu e_r - g'\cos\theta_w e_r^\dagger \tau_\mu A_\mu e_r -$$
$$-\frac{1}{3}g'\cos\theta_w d_r^\dagger \tau_\mu A_\mu d_r + \frac{1}{3}g'\sin\theta_w d_r^\dagger \tau_\mu Z_\mu d_r +$$
$$+\frac{2}{3}g'\cos\theta_w u_r^\dagger \tau_\mu A_\mu u_r - \frac{2}{3}g'\sin\theta_w u_r^\dagger \tau_\mu Z_\mu u_r. \quad (29)$$

The limit model includes only **massless particles**: photons A_μ and neutral bosons Z_μ, left quarks u_l and neutrinos ν_l, right electrons e_r and quarks u_r, d_r. The electroweak interactions become long-range because are mediated by the massless neutral Z-bosons and photons.

The infinite energies can exist only in the initial moment of creation when the Universe is point-like [13, 14] and it is not clear what means long-range electroweak interactions. However more interesting is the Universe evolution and the limit Lagrangian L_∞ can be considered as a good approximation near the Big Bang just as the nonrelativistic mechanics is a good approximation of the relativistic one at low velocities. From the explicit form of the interaction part $L_\infty^{int}(A_\mu, Z_\mu)$ it follows that there are no interactions between particles of different kind, for example neutrinos interact only with each other by neutral currents. It looks like some stratification of the Electroweak Model with only one sort of particles in each stratum.

The mass term of u-quark in the Lagrangian (27) is proportional to the first power $\epsilon\, m_u(u_r^\dagger u_l + u_l^\dagger u_r)$, whereas the mass terms of Z-boson, electron and d-quark are multiplied by the second power $\epsilon^2\left[\frac{1}{2}m_Z^2(Z_\mu)^2 + m_e(e_r^\dagger e_l + e_l^\dagger e_r) + m_d(d_r^\dagger d_l + d_l^\dagger d_r)\right]$, so u-quark first restores its mass in the evolution of the Universe. Higgs boson and charged W-boson, whose mass terms are multiplied by ϵ^4, last restore their masses after all other particles of the Electroweak Model.

5. Conclusion

We have investigated the low and higher energy limits of the Electroweak Model which are obtained from first principles of gauge theory as contraction of its gauge group. Above limits are given by the same contraction of the gauge group, but for the different consistent rescalings of the representation space. It was shown that mathematical contraction parameter in both cases is interpreted as typical energy.

The very weak neutrino-matter interactions especially at low energies can be explained by this model already at the level of classical (non-quantum) gauge fields. The zero tending contraction parameter is connected with neutrino energy and reproduce the linear energy dependence of the neutrino-matter cross-section.

The alternative rescaling of the gauge group and the field space corresponds to the infinite energy limit of the Electroweak Model, which goes in this limit through the five stages depending on the powers of the contraction parameters. At the infinite energy all particles are massless and electroweak interactions become long-range. But the infinite energies can exist only in the Big Bang, i.e. in the initial moment of creation when the Universe is point-like and it is not clear what means long-range. However more interesting is the Universe development and the limit Lagrangian L_∞ can be considered as a good approximation near the Big Bang just as the nonrelativistic mechanics is a good good approximation of the relativistic one at low velocities. Particularly we can conclude that according to the Electroweak Model u-quark first restores its mass among other particles in the evolution of the Universe.

References

[1] Inönü E and Wigner E P 1953 On the contraction of groups and their representations *Proc. Nat. Acad. Sci. U.S.A.* **39** 510
[2] Gromov N A and Man'ko V I 1990 The Jordan-Schwinger representations of Cayley-Klein groups. II. The unitary groups *J. Math. Phys.* **31** 1054
[3] Gromov N A 2012 Contraction of Electroweak Model can explain the interactions neutrinos with matter *Phys. Part. Nucl.* **43** 723
[4] Gromov N A 2012 Contraction of Electroweak Model and neutrino *Phys. Atom. Nucl.* **75** 1203
[5] Gromov N A 2013 Interpretation of neutrino-matter interactions at low energies as contraction of gauge group of Electroweak Model *Phys. Atom. Nucl.* **76** 1144
[6] Gromov N A 2012 *Contractions of Classical and Quantum Groups* (Moscow: Fizmatlit) [in Russian]
[7] Rubakov V A 1999 *Classical Gauge Fields* (Moscow: Editorial URSS) [in Russian]
[8] Okun' L B 2005 *Leptons and quarks* (Moscow: Editorial URSS) [in Russian]
[9] Peskin M E and Schroeder D V 1995 *An Introduction to Quantum Field Theory* (New York: Addison-Wesley)
[10] Gromov N A 2010 Analog of Electroweak Model for contracted gauge group *Phys. Atom. Nucl.* **73** 326
[11] Gromov N A 2011 Limiting case of modified Electroweak Model for contracted gauge group *Phys. Atom. Nucl.* **74** 908
[12] Gromov N A 2014 Higher and low energy limits of Electroweak Model *Proc. Komi SC UrB RAS* **1(17)** 4 [in Russian]
[13] Linde L D 1990 *Particle Physics and Inflationary Cosmology* (Moscow: Nauka) [in Russian].
[14] Gorbunov D S and Rubakov V A 2011 *Introduction to the Theory of the Early Universe: Hot Big Bang Theory* (Singapore: World Scientific)

The quantum formulation derived from assumptions of epistemic processes

Inge S Helland

Professor emeritus, Department of Mathemathics, University of Oslo

E-mail: ingeh@math.uio.no

Abstract. Motivated by Quantum Bayesianism I give background for a general epistemic approach to quantum mechanics, where complementarity and symmetry are the only essential features. A general definition of a symmetric epistemic setting is introduced, and for this setting the basic Hilbert space formalism is arrived at under certain technical assumptions. Other aspects of ordinary quantum mechanics will be developed from the same basis elsewhere.

1. Introduction

The ordinary textbook formulation of quantum mechanics is very abstract. Its starting point: 'The state of a physical system is a normalized vector in a separable Hilbert space' has lead to an extremely rich theory, a theory which has not been refuted by any experiment and whose predictions range over an extremely wide varity of situations. Nevertheless, it is still unclear how this state concept should be interpreted.

Many conferences on quantum foundation have been arranged in recent years, but this has only implied that the number of new interpretations have increased, and no one of the old have died out. In two of these conferences, a poll among the participants was carried out [1, 2]. The result was an astonishing disagreement on several simple and fundamental questions. One of these questions was whether quantum theory should be interpreted as an objective theory of the world (the ontological interpretation) or if it only expresses our knowledge of the world (the epistemic interpretation). According to Webster's Unabridged Dictionary, the adjective 'epistemic' means 'of or pertaining to knowledge, or the conditions for acquiring it'.

In Spekkens [3] a toy model was developed, a toy model which can be taken as a strong argument in favour of an epistemic interpretation of quantum mechanics. Partly based on this and partly on other basic premises, Fuchs [4, 5, 6] and others have argued for various versions of Quantum Bayesianism, a radical interpretation where the subjective observer plays an important part. See also the philosophical discussion by Timpson [7] and the popular account in [8].

The present paper takes as a point of departure that in some sense or other the epistemic interpretation should be important for the fundametal issues of the quantum world. The next question then arises: Can one find a new and more intuitive *foundation* of quantum theory, a foundation related to the epistemic interpretation? It is my view that such a foundation must have some relation to statistical inference theory, another scientific fundament, which gives tools for wide variety of empirical investigations, and which in its very essence is epistemic. It should also be a kind of decision theory, related to decisions taken in everyday life.

Every epistemic question together with an answer to this question imply at least two decisions: First a decision to focus in the selection of the question to ask. Then data are collected, and one ends up with a decision on which answer to the focused question one should give from these data. This last decision is covered by traditional decision theory, but one needs to pay more attention to the process of focusing.

A very tentative answer to the question on foundation given above, is then yes, but there is still more work to be done before such a foundation can be completely established. In [9] I will aim to carry out the arguments in more detail.

2. Background and definitions

The Quantum Bayesianism is founded on an observer's belief, quantified by a Bayesian probability. I want to relate my state concept to the notion of *certain* belief, which I call knowledge. The knowledge will be associated with an agent or with a group of communicating agents, and his/her/their knowledge will be knowledge about what I will call an e-variable.

An epistemic process is any process under which an agent or a group of communicating agents obtain knowledge about a physical system. In general there are many ways by which one can obtain knowledge about the world or about aspects of the world. In a given situation the observer has some background, in terms of his history, in terms of his physical environment, and in terms of the concepts that he is able to use in analyzing the situation. All this may limit his ability to obtain knowledge.

A *conceptual variable* is any variable related to the physical system, defined by an agent or by a group of communicating agents. The variable may be a scalar, a vector or belong to a larger space.

An *e-variable* or epistemic conceptual variable θ is a conceptual variable associated with an epistemic process: Before the process the agent (or agents) has (have) no knowledge about θ; after the process he/they has/have some knowledge, in the simplest case full knowledge: $\theta = u_k$. Here u_k is one of the possible values that θ can take. In this paper it is mostly assumed for simplicity that θ is discrete, which it will be in the elementary quantum setting below. For a continuous variable θ, knowledge on the e-variable will be taken to mean a statement to the effect that θ belongs to some given set, an interval if θ is a scalar.

The e-variable concept is a generalization of the parameter concept as used in statistical inference, introduced by Fisher [10], and today incorporated in nearly all applications of statistics. In statistics, a parameter θ is usually an index in the statistical model for the observations, and the purpose of an empirical investigation is to obtain statements about θ, in terms of point estimation, confidence interval estimation or conclusion from the testing of hypotheses. The parameter is often associated with a hypothetical infinite population. My e-variable will also be allowed to be associated with a finite physical system, a particle or a set of particles. But, in the same way as with a parameter, the purpose of any empirical investigation will be to try to conclude with some statement, a statement expressed in terms of an e-variable θ.

It is important that not all conceptual variables are e-variables. A conceptual variable ϕ is called *inaccessible* if there is no epistemic process by which one can get accurate knowledge about it. An example from the area of quantum mechanics is $\phi = (\xi, \pi)$, where ξ is the position of a particle, and π is the momentum. Also macroscopic examples abound, for instance connected to counterfactual situations [9]. Another physical example is $\phi = (\lambda^x, \lambda^y, \lambda^z)$, where λ^a is the component of an angular momentum or a spin for a particle or a system of particles, the component in direction a. Here each λ^a is an accessible conceptual variable, an e-variable, but the vector ϕ is inaccessible.

In such cases, where a vector of e-variables is inaccessible, it is equivalent to say that the components are complementary. Since introduced by Bohr, the concept of complementarity has played a fundamental and important role in quantum mechanics.

It is essential to stress that the inaccessible conceptual variables above are not hidden variables. Variables like ϕ are just mathematical variables, but variables upon which group actions may be defined. $\phi = (\xi, \pi)$ may be subject to Galilean transformations, time translations or changes of units, while $\phi = (\lambda^x, \lambda^y, \lambda^z)$ may be subject to rotations. This will of course also induce transformation of the components, the e-variables. Important transformations in the group of rotations of ϕ are: 1) Those leading to the change in the values of λ^x (or of any other fixed component); 2) Those leading to an exchange of λ^x and λ^y (or any other pair of components).

3. The maximal symmetrical epistemic setting

The purpose of this paper is to indicate how one can derive the essence of the quantum formalism from reasonable assumptions in a setting which generalizes the angular momentum setting described above.

Let in general ϕ be an inaccessible conceptual variable taking values in some measurable space $(\Phi, \mathcal{F})$, and let $\lambda^a = \lambda^a(\phi)$ be accessible functions for a belonging to some index set $\mathcal{A}$. I will repeat that a conceptual variable is accessible if it in the given context can be estimated with arbitrary accuracy by *some* experiment. Technically I will without further mention assume that all functions defined on Φ are measurable. To begin with, I will assume that the functions λ^a are maximal, and also that they can be transformed into each other by an automorphism in Φ.

Assumption 1 *a) Consider the partial ordering defined by $\alpha < \beta$ iff $\alpha = f(\beta)$ for some function f. Under this partial ordering each $\lambda^a(\phi)$ is maximally accessible.*

b) For $a \neq b$ there is an invertible transformation g_{ab} such that $\lambda^b(\phi) = \lambda^a(g_{ab}\phi)$.

Note that the partial ordering in a) is consistent with accessibility: If β is accessible and $\alpha = f(\beta)$, then α is accessible. Also, ϕ is an upper bound under this partial ordering. The existence of maximal accessible conceptual variables follows then from Zorn's lemma.

For the angular momentum case, the transformation g_{ab} means exchange of the component λ^a and the component λ^b. In this process, one can imagine that the vector ϕ is decomposed as $\phi = (\lambda^a, \lambda^b, \lambda^c)$, where the axis c is orthogonal to the (a, b)-plane, and g_{ab} then changes the sign of λ^c, moving along a great circle.

Below, I will often single out a particular index $0 \in \mathcal{A}$. Then a), given b), can be formally weakened to the assumption that $\lambda^0(\phi)$ is maximally accessible, and b) can be weakened to the existence for all a of an invertible transformation g_{0a} such that $\lambda^a(\phi) = \lambda^0(g_{0a}(\phi))$. Take $g_{ab} = g_{0a}^{-1} g_{0b}$.

Even though ϕ is inaccessible, it is possible to operate on ϕ with functions, in particular group actions. The group of automorphisms on Φ, all transformations of elements ϕ, always exists from a mathematical point of view, and one can imagine many subgroups of this group. Some of these will now be defined.

Definition 1 *For each a, let $\tilde{G}^a$ be the group of automorphisms on Λ^a, the space upon which λ^a varies. For $\tilde{g}^a \in \tilde{G}^a$ let g^a be any transformation on Φ for which $\tilde{g}^a\lambda^a(\phi) = \lambda^a(g^a\phi)$.*

It is easily verified that 1) For fixed $\tilde{g}^a$ the transformations g^a form a set which is invariant under decompositions and inversion. 2) For fixed a the transformations g^a form a group G^a.

Thus the group G^a is the group transforming values of λ^a into the same or other values of this e-variable, and the corresponding group $\tilde{G}^a$ is the group of all transformations of these values.

Obvious consequences of Definition 1 are that $\tilde{G}^a$ is transitive over Λ^a and that G^a is transitive over Φ.

Now single out a fixed index $0 \in \mathcal{A}$.

Definition 2 *Let G be the group of transformations generated by G^0 and the transformations g_{0a}, $a \in \mathcal{A}$.*

It is easily verified that $G^a = g_{0a}^{-1} G^0 g_{0a}$. Together with $g_{ab} = g_{0a}^{-1} g_{0b}$ this implies that G also is the group generated by G^a, $a \in \mathcal{A}$ and g_{ab}, $a, b \in \mathcal{A}$.

Now I want to introduce the further

Assumption 2 *a) The group G is a locally compact topological group, and satisfies weak assumptions such that an invariant measure ρ on Φ exists.*
b) The group generated by products of elements of $G^a, G^b, ...; a, b, ... \in \mathcal{A}$ is equal to G.

Assumption 2a) is a technical one, needed in the next section. Note that G is defined in terms of transformations upon Φ, so that the topology must be introduced in terms of these transformations. Technically this can for instance be achieved by assuming Φ to be a metric space with metric d, and letting $g_n \to g$ if $\sup_\phi d(g_n\phi, g\phi) \to 0$. Concerning Assumption 2b), it follows from $g^a g^b ... = g_{0a}^{-1} g^0 g_{0a} g_{0b}^{-1} g^{0'} g_{0b}, ...,$ where $g^a \in G^a, g^b \in G^b, ...$ and $g^0, g^{0'}, ... \in G^0$, that the group of products is contained in G. That it is equal to G, is an assumption on the richness of the set $\mathcal{A}$ or the richness of G^0.

Here is a proof that assuption 2b) holds in the angular momentum/ spin case: Let c be any direction perpendicular to the midline between a and b. Let g^c be the group element changing the sign of λ^c, the c-component of ϕ, moving along a great cicle. Then $g_{ab} = g^c$.

The setting described here, where Assumption 1 and Assumption 2 are satisfied, includes many quantum mechanical situations including angular momenta and spins. I will call it *the maximal symmetrical epistemic setting.*

An important special case is when each Λ^a is discrete. Then $\tilde{G}^a$ is the group of permutations of elements of Λ^a.

4. Quantum reconstruction

Fix $0 \in \mathcal{A}$, and let H be the Hilbert space

$$H = \{f \in L^2(\Phi, \rho) : \ f(\phi) = \tilde{f}(\lambda^0(\phi)) \text{ for some } \tilde{f}\}.$$

I will assume that the λ^a's are discrete. Then H is separable. If the λ^a's take d different values, H is d-dimensional. Since all separable Hilbert spaces are isomorphic, it is enough to arrive at the quantum formulation on this H.

Lemma 1 *The values u_k^a of λ^a can always be arranged such that $u_k^a = u_k$ is the same for each a $(k = 1, 2, ...)$.*

<u>Proof.</u> By Assumption 1

$$\{\phi : \lambda^b = u_k^b\} = \{\phi : \lambda^a(g_{ab}\phi) = u_k^b\} = g_{ba}(\{\phi : \lambda^a(\phi) = u_k^b\}).$$

The sets in brackets on the lefthand side here are disjoint with union Φ. But then the sets in brackets on the righthand side are disjoint with union $g_{ab}(\Phi) = \Phi$, and this implies that $\{u_k^b\}$ gives all possible values of λ^a.

Now I am able to formulate the main result of this paper:

Theorem 1 *a) For every a, u_k and associated with every indicator function $I(\lambda^a(\phi) = u_k)$ there is a vector $|a; k\rangle \in H$, which is unique in the sense that $|a; k\rangle \neq |b; j\rangle$ for all a, b, j, k except* *in the trivial case $a = b, j = k$. This inequality is interpreted to mean that there is no phase factor $e^{i\gamma}$ such that $|a;k\rangle = e^{i\gamma}|b;j\rangle$.*

b) For each a the vectors $|a;k\rangle$ form an orthonormal basis for H.

This gives us the possibility to interprete the vectors $|a;k\rangle$, corresponding to the indicators $I(\lambda^a(\phi) = u_k)$, as follows:

1) The question 'What is the value of λ^a?' has been focused on.

2) Through an epistemic process we have obtained the answer '$\lambda^a = u_k$'.

When a ket vector is defined, a corresponding bra vector can be defined. The operator corresponding to λ^a can be defined as

$$A^a = \sum_k u_k |a;k\rangle\langle a;k|.$$

In the maximal setting this has non-degenerate eigenvalues.

5. Proof under an extra assumption

Let U be the left regular representation of G on $L^2(\Phi, \rho)$: $U(g)f(\phi) = f(g^{-1}\phi)$. It is well known that this is a unitary representation. We will seek a corresponding representation of G on the smaller space H.

In the following, recall that upper indices as in g^a indicate variables related to a particular λ^a, here a group element of G^a. Also recall that 0 is a fixed index in $\mathcal{A}$. Lower indices as in g_{ab} has to do with the relation between two different λ^a and λ^b.

Proposition 1 *a) A (multivalued) representation V of G on the Hilbert space H can always be found.*

b) There is an extended group G' such that V is a univalued representation of G' on H.

c) There is a homomorphism $G' \to G^0$ such that $V(g') = U(g^0)$. If $g' \neq e'$ in G', then $g^0 \neq e$ in G^0.

Proof. a) For each a and for $g^a \in G^a$ define $V(g^a) = U(g_{0a})U(g^a)U(g_{a0})$. Then $V(g^a)$ is an operator on H, since it is equal to $U(g_{0a}g^a g_{a0})$, and $g_{0a}g^a g_{a0} \in G^0 = g_{0a}G^a g_{a0}$. For a product $g^a g^b g^c$ with $g^a \in G^a$, $g^b \in G^b$ and $g^c \in G^c$ we define $V(g^a g^b g^c) = V(g^a)V(g^b)V(g^c)$, and similarly for all elements of G that can be written as a finite product of elements from different subgroups.

Let now g and h be any two elements in G such that g can be written as a product of elements from G^a, G^b and G^c, and similarly h (the proof is similar for other cases.) It follows that $V(gh) = V(g)V(h)$ on these elements, since the last factor of g and the first factor of h either must belong to the same subgroup or to different subgroups; in both cases the product can be defined by the definition of the previous paragraph. In this way we see that V is a representation on the set of finite products, and since these generate G by Assumption 2b) , it is a representation of G.

Since different representations of g as a product may give different solutions, we have to include the possibility that V may be multivalued.

b) Assume as in a) that we have a multivalued representation V of G. Define a larger group G' as follows: If $g^a g^b g^c = g^d g^e g^f$, say, with $g^k \in G^k$ for all k, we define $g'_1 = g^a g^b g^c$ and $g'_2 = g^d g^e g^f$. Let G' be the collection of all such new elements that can be written as a formal product of elements $g^k \in G^k$. The product is defined in the natural way, and the inverse by for example $(g^a g^b g^c)^{-1} = (g^c)^{-1}(g^b)^{-1}(g^a)^{-1}$. By Assumption 2b), the group G' generated by this construction must be at least as large as G. It is clear from the proof of a) that V also is a representation of the larger group G' on H, now a one-valued representation.

c) Consider the case where $g' = g^a g^b g^c$ with $g^k \in G^k$. Then by the proof of a):

$$V(g') = U(g_{0a})U(g^a)U(g_{a0})U(g_{0b})U(g^b)U(g_{b0})U(g_{0c})U(g^c)U(g_{c0})$$
$$= U(g_{0a}g^a g_{a0}g_{0b}g^b g_{b0}g_{0c}g^c g_{c0}) = U(g^0),$$

where $g^0 \in G^0$. The group element g^0 is unique since the decomposition $g' = g^a g^b g^c$ is unique for $g' \in G'$. The proof is similar for other decompositions. By the construction, the mapping $g' \to g^0$ is a homomorphism.

Assume now that $g^0 = e$ and $g' \neq e'$. Since $U(g^0)\tilde{f}(\lambda^0(\phi)) = \tilde{f}(\lambda^0((g^0)^{-1}(\phi)))$, it follows from $g^0 = e$ that $U(g^0) = I$ on H. But then from what has been just proved, $V(g') = I$, and since V is a univariate representation, it follows that $g' = e'$, contrary to the assumption.

Now choose an orthonormal basis for H: $f_1, ..., f_d$ where $f_k(\phi) = \tilde{f}_k(\lambda^0(\phi))$, and where the interpretation of f_k is that $\lambda^0 = u_k$. Write $|0;k\rangle = f_k(\phi)$.

Introduce the assumption that the representation V really is multivalued, in a sense to be made precise below. Let g'_{0a1} and g'_{0a2} be two different elements of the group G', both corresponding to g_{0a} of G. Define $g'_a = (g'_{0a1})^{-1}g'_{0a2}$. Then $g'_a \neq e'$ in G'. By the homomorphism of Proposition 1c), let $g'_a \to g^0_a$. Then $g^0_a \neq e$ in G^0. Now define

$$|a;k\rangle = \tilde{f}_k(\lambda^0(g^0_a\phi)) = U((g^0_a)^{-1})|0;k\rangle.$$

Assumption 3 a) g'_{0a1} and g'_{0a2} can be chosen so that $\tilde{g}^0_a\lambda \neq \lambda$ for all $\lambda \in \Lambda^0$ when $a \neq 0$.
b) $g'_{0a1}, g'_{0a2}, g'_{0b1}$ and g'_{0b2} can be chosen so that $\tilde{g}^0_a\lambda \neq \tilde{g}^0_b\lambda$ for all $\lambda \in \Lambda^0$ when $a \neq b$.

Proof of Theorem 1 under Assumption 3. Let $j \neq k$. I will first prove that the basis functions $f_1, ..., f_d$ can be chosen so that $|a;k\rangle \neq |0;j\rangle$ for all a. Here and below, inequality of state vectors is interpreted to mean that they can not be made equal by introducing a phase factor. To this end, choose $\tilde{f}_j$ and $\tilde{f}_k$ in such a way that there exists an λ^j_0 such that $\tilde{f}_k(\lambda) \neq \tilde{f}_j(\lambda^j_0)$ for all $\lambda \in \Lambda^0$. A possible choice is the indicator function $\tilde{f}_j(\lambda) = I(\lambda = u_j)$, but there are other choices. Then for any fixed g, $\tilde{f}_{kg}$ defined by $\tilde{f}_{kg}(\lambda^0(\phi)) = \tilde{f}(\lambda^0(g\phi))$ is different from $\tilde{f}_j$, and $|a;k\rangle \neq |0;j\rangle$ for all a.

Next fix k and choose $\tilde{f}_k$ in such a way that there exists a λ^k_1 such that $\tilde{f}_k(\lambda) \neq \tilde{f}_k(\lambda^k_1)$ when $\lambda \neq \lambda^k_1$ and $\lambda \in \Lambda^0$. A possible choice is again $\tilde{f}_k(\lambda) = I(\lambda = u_k)$. For $|a;k\rangle \neq |0;k\rangle$ it is sufficient that $\tilde{f}_k(\lambda^0(g^0_a\phi)) \neq \tilde{f}_k(\lambda^0(\phi))$ for at least one ϕ, and this will hold if $\lambda^0(g^0_a\phi) \neq \lambda^0(\phi)$ for all ϕ. This is equivalent to Assumption 3a).

The proof that $|a;k\rangle \neq |b;j\rangle$ (except in the trivial case $a = b, k = j$) holds under assumption 3b), is similar.

The vectors $|0;k\rangle$ are chosen to be an orthonormal basis for H. Since $|a;k\rangle = U|0;k\rangle$ for some unitary U, it follows that the vectors $|a;k\rangle$ form an orthonormal basis.

Assumption 3 is not satisfied for the spin/ angular momentum case. Nevertheless it is shown in [16] that Theorem 1 holds also for this case.

6. The general symmetrical epistemic setting

Go back to the definition of the maximal symmetrical epistemic setting. Let again ϕ be the inaccessible conceptual variable and let $\lambda^a = \lambda^a(\phi)$ for $a \in \mathcal{A}$ be the maximal accessible conceptual variables satisfying Assumption 1. Let the corresponding induced groups G^a and G satisfy Assumption 2. Let us assume either the spin/ angular momentum case or that Assumption 3 holds. Finally, let t^a for each a be an arbitrary real function on the range of λ^a, and assume that we instead of focusing on λ^a, focus on $\theta^a = t^a(\lambda^a)$ for each $a \in \mathcal{A}$. I will call this the symmetrical epistemic setting; the e-variables θ^a are no longer maximal.

Consider first how to define the quantum states $|a;k\rangle$. We are no longer interested in the full information on λ^a, but keep first the Hilbert space as in Section 4. For definiteness look at the case where Assumption 3 holds. Then let $f_k^a(\phi) = I(t^a(\lambda^a) = t^a(u_k)) = I(\theta^a = u_k^a)$, where $u_k^a = t^a(u_k)$. We let again g'_{0a1} and g'_{0a2} be two distinct elements of G' such that $g'_{0ai} \rightarrow g_{0a}$, define $g'_a = (g'_{0a1})^{-1} g'_{0a2}$ and let $g'_a \rightarrow g_a^0$ under the homomorphism of Proposition 1c). Then define

$$|a;k\rangle = V((g'_a)^{-1})U(g_{0a})f_k^a(\phi) = U((g_a^0)^{-1})f_{ka}^0(\phi) = f_{ka}^0(g_a^0\phi)$$

with $f_{ka}^0(\phi) = I(t^a(\lambda^0(\phi)) = u_k^a)$.

Interpretation of the state vector $|a;k\rangle$**:** *1) The question: 'What is the value of θ^a?' has been posed. 2) We have obtained the answer $\theta^a = u_k^a$. Both the question and the answer are contained in the state vector.*

From this we may define the operator connected to the e-variable θ^a:

$$A^a = \sum_k u_k^a |a;k\rangle\langle a;k| = \sum_k t^a(u_k)|a;k\rangle\langle a;k|.$$

Then A^a is no longer necessarily an operator with distinct eigenvalues, but A^a is still Hermitian: $A^{a\dagger} = A^a$.

Interpretation of the operator A^a**:** *This gives all possible states and all possible values corresponding to the accessible e-variable θ^a.*

The projectors $|a;k\rangle\langle a;k|$ and hence the ket vectors $|a;k\rangle$ are no longer uniquely determined by A^a: They can be transformed arbitrarily by unitary transformations in each space corresponding to one eigenvalue. As long as the focus is only on θ^a, or A^a, I will redefine $|a;k\rangle$ by allowing it to be subject to such transformations. These transformed eigenvectors all still correspond to the same eigenvalue, that is, the same observed value of θ^a and they give the same operators A^a. In particular, in the maximal symmetric epistemic setting I will allow an arbitrary constant phase factor in the functions defining the $|a;k\rangle$'s.

As an important example of the general construction, assume that λ^a is a vector: $\lambda^a = (\theta^{a_1}, ..., \theta^{a_m})$. Then one can write a state vector corresponding to λ^a as

$$|a;k\rangle = |a_1;k_1\rangle \otimes ... \otimes |a_m;k_m\rangle$$

in an obvious notation, where $a = (a_1, ..., a_m)$ and $k = (k_1, ..., k_m)$. The different θ's may be connected to different subsystems. This construction can also be made, and is of great interest, in the spin/ angular momentum case.

So far I have kept the same groups G^a and G when going from λ^a to $\theta^a = t^a(\lambda^a)$, that is from the maximal symmetrical epistemic setting to the general symmetrical epistemic setting. This implies that the (large) Hilbert space will be the same. A special case occurs if t^a is a reduction to an orbit of G^a. Then the construction of this section can also be carried with a smaller group action acting just upon an orbit, resulting then in a smaller Hilbert space. In the example of the previous paragraph it may be relevant to consider one Hilbert space for each subsystem. The large Hilbert space is however the correct space to use when the whole system is considered.

These considerations are highly relevant when considering several observers. One single observer may have access to just a few subsystems. In addition he has his own context. From this context one can define what is his accessible and inaccessible conceptual variables. In the same way a group of several observers may through verbal communication arrive at a common context, and from this context one can define their accessible and inaccessible conceptual variables.

Journal of Physics: Conference Series **597** (2015) 012041 doi:10.1088/1742-6596/597/1/012041

Imagine that these observers together observe a particular physical system, and consider the corresponding Hilbert space.

Assumption 4. *For most physical system at some particular time one can either imagine an observer or a group of communicating observers for which the assumptions of the symmetrical epistemic setting are satisfied. In many cases all real and imagined observers agree on the physical observations, in which case this is considered part of the ontic world.*

It is important that for every physical system we can consider imagined observers. All real and imagined observers may agree on observations like charge or non-relativistic mass of a particle. For these observations we do not need the formalism of quantum mechanics. For other variables the construction of this section applies.

At any time we can also imagine non-communicating observers. For each observer then the general symmetrical setting may apply. Particular state vectors in each observer's Hilbert space might then be linear combinations of primitive state vectors of the form $|a_1; k_1\rangle \otimes ... \otimes |a_s; k_s\rangle$. As is well known, when linear combinations of primitive state vectors can not be reduced to a primitive form, they are called entangled state vectors. Entangled state vectors play an important role in many discussions of quantum mechanics.

When the physical system available to an observer A is subject to a non-maximal symmetrical setting resulting in an operator $A^a = \sum_k u_k^a |a; k\rangle\langle a; k|$, we can always imagine another observer B, communicating with A, with an operator $B^b = \sum_j v_j^b |b; j\rangle\langle b; j|$, so that the operator $A^a \otimes B^a = \sum_{k,j} u_k^a v_j^b (|a; k\rangle \otimes |b; j\rangle)(\langle a; k| \otimes \langle b; j|)$ corresponds to a maximal setting, i.e., has distinct eigenvalues.

Assumption 4 is assumed to hold for nearly all physical systems, and also for combinations of physical systems. Through the imagined observers the construction of Section 5 can be carried out, and for each case a Hilbert space can be constructed.

Connected to any general physical system, one may have several e-variables θ and corresponding operators A. In the ordinary quantum formalism there is well-known theorem saying that, in my formulation, $\theta^1, ..., \theta^n$ are compatible, that is, there exists an e-variable λ such that $\theta^i = t^i(\lambda)$ for some functions t^i if and only if the corresponding operators commute:

$$[A^i, A^j] \equiv A^i A^j - A^j A^i = 0 \text{ for all } i, j.$$

(See Holevo [13].) Compatible e-variables may in principle be estimated simultaneously with arbitrary accuracy.

7. Further developments

In [9] I discuss a number of further implications of a possible Hilbert space formulation of epistemic processes.

First, the density operators are defined by

$$\sigma^a = \pi_k^a \sum_k |a; k\rangle\langle a; k|.$$

Here the π_k^a are either prior probabilities, posterior probabilities, or derived from confidence distributions. Traditionally, statisticians are divided into Bayesians and frequentists; I want to include both cultures. Prior distributions and posterion distributions are Bayesian concepts for parameters. Confidence distributions is a recent corresponding frequentist concept for distribution of parameters [14].

For simple epistemic states, Born's formula can be written as

$$P(\theta^b = u_j^b | \theta^a = u_k^a) = |\langle a; k | b; j\rangle|^2.$$

Journal of Physics: Conference Series **597** (2015) 012041 doi:10.1088/1742-6596/597/1/012041

This formula is proved in [9] from 1) An extension of the likelihood principle from statistics; 2) An assumption of rationality (Dutch book principle); 3) Paul Busch's version of Gleason's theorem [15].

In statistics, the likelihood is the joint density of the observation, and the likelihood principle states that observations with the same likelihood produce the same experimental evidence. This principle can be proved from fairly obvious presuppositions. The Dutch book principle states: For a rational actor, no choice of payoffs in a series of bets shall lead to sure loss for the bettor. Thus in this approach, Born's formula, which is formulated as an independent axiom in most texbooks, can be seen to follow from quite intuitive assumptions.

In [9] also the quantum mechanics for position and momentum is discussed. The Schrödinger equation $i\hbar\frac{d}{dt}|\psi\rangle_t = H|\psi\rangle_t$ for one-dimensional position is shown to follow from 1) Position as an inaccessible stochastic process; 2) Conditioning both on past events and on future events; 3) Elements from Nelson's stochastic mechanics.

EPR and violation of Bell's inequality for simple epistemic states can be given interesting interpretations. The conditionality principle from statistics states: When doing a coin toss to choose an experiment, an observer should condition upon the value of the coin toss. Imagine that this principle is used both for Alice and for Bob. This implies that their observations are interpreted as some sort of experiment. That is, observation means inference: Decision on what has been observed. Under this condition the EPR paradox turns out to be no paradox any more, and the violation of Bell's inequality can be simply explained. The assumption of *realism* is violated by this version of quantum mechanics; the assumption of locality is not violated.

It is well known from the Quantum Bayesian literature that the so-called paradox of Wigner's friend is no paradox any more with an epistemic interpretation of the state vector.

8. Concluding remarks

Despite its enormous success, there is still no consensus among physicists about what quantum theory is saying about the nature of reality. In my view, its foundation should be related to the process of obtaining knowledge about the world. It seems to be possible to formalize this in a precise mathematical way.

In this theory something resembling the formal state definition of traditional quantum mechanics is derived from a simpler and more natural conceptual basis. This derivation is done under certain technical assumptions. The use of group theory seems to be essential.

It also seems to be important that the approach here and in [9] uses ideas both from the quantum mechanical tradition and from the statistical tradition. In the long run one might perhaps hope for some sort of synthesis in the foundation of the two sciences.

Since space is limited here, I prepare an extended version [16], where I will discuss the validity of Theorem 1 for the spin/angular momentum case, and where the relation to other approaches towards the foundation of quantum mechanics and to the interpretation of quantum mechanics will be discussed.

References

[1] Schlossbauer M, Koller J and Zellinger A 2013 A snapshot of fundamental attitudes toward quantum mechanics (*Preprint* quant-ph/1301.1069)

[2] Norsen T and Nelson S 2013 Yet another snapshot of fundamental attitudes toward quantum mechanics (*Preprint* quant-ph/1306.4646)

[3] Spekkens R W 2007 *Phys. Rev.* A **75** 032110

[4] Fuchs C A 2010 QBism, the Perimeter of Quantum Bayesianism (*Preprint* quant-ph/1003.5209)

[5] Fuchs C A and Schack R 2011 *Foundations of Physics* **41** 345

[6] Fuchs C A, Mermin N D and Schack R 2013 An introduction to QBism with application to the locality of quantum mechanics (*Preprint* quant-ph/1311.5253)

[7] Timpson C G 2008 *Studies in History and Philosophy of Modern Physics* **39** 579

[8] von Baeyer H C 2013 *Scientific American* **308** (6) 38
[9] Helland I S In preparation *Epistemic Processes. A Basis for Statistics and for Quantum Mechanics.* Book manuscript
[10] Fisher R A 1922 On the mathematical foundation of theoretical statistics. Reprinted in Fisher R A 1950 *Contribution to Mathematical Statistics* (New York: Wiley)
[11] Helland I S 2004 *International Statistical Review* **72** 409
[12] Helland I S 2010 *Steps Towards a Unified Basis for Scientific Models and Methods* (Singapore: World Scientific)
[13] Holevo A S 1982 *Probabilistic and Statistical Aspects of Quantum Theory* (Amsterdam: North Holland)
[14] Schweder T and Hjort N L 2002 *Scandinavian Journal of Statistics* **29** 309
[15] Busch P 2003 *Physical Review Letters* **91** (12) 120403
[16] Helland I S 2015 Approach towards the quantum formulation from assumptions of epistemic processes. Article in preparation

The orthogonal planes split of quaternions and its relation to quaternion geometry of rotations[1]

Eckhard Hitzer

Osawa 3-10-2, Mitaka 181-8585, International Christian University, Japan

E-mail: hitzer@icu.ac.jp

Abstract. Recently the general orthogonal planes split with respect to any two pure unit quaternions $f, g \in \mathbb{H}$, $f^2 = g^2 = -1$, including the case $f = g$, has proved extremely useful for the construction and geometric interpretation of general classes of double-kernel quaternion Fourier transformations (QFT) [7]. Applications include color image processing, where the orthogonal planes split with $f = g =$ the grayline, naturally splits a pure quaternionic three-dimensional color signal into luminance and chrominance components. Yet it is found independently in the quaternion geometry of rotations [3], that the pure quaternion units f, g and the analysis planes, which they define, play a key role in the geometry of rotations, and the geometrical interpretation of integrals related to the spherical Radon transform of probability density functions of unit quaternions, as relevant for texture analysis in crystallography. In our contribution we further investigate these connections.

1. Introduction to quaternions

Gauss, Rodrigues and Hamilton's four-dimensional (4D) quaternion algebra $\mathbb{H}$ is defined over $\mathbb{R}$ with three imaginary units:

$$\boldsymbol{ij} = -\boldsymbol{ji} = \boldsymbol{k},\ \boldsymbol{jk} = -\boldsymbol{kj} = \boldsymbol{i},\ \boldsymbol{ki} = -\boldsymbol{ik} = \boldsymbol{j},\ \boldsymbol{i}^2 = \boldsymbol{j}^2 = \boldsymbol{k}^2 = \boldsymbol{ijk} = -1. \tag{1}$$

The explicit form of a quaternion $q \in \mathbb{H}$ is $q = q_r + q_i\boldsymbol{i} + q_j\boldsymbol{j} + q_k\boldsymbol{k} \in \mathbb{H}$, $q_r, q_i, q_j, q_k \in \mathbb{R}$. We have the isomorphisms $Cl(3,0)^+ \cong Cl(0,2) \cong \mathbb{H}$, i.e. $\mathbb{H}$ is isomorphic to the algebra of rotation operators in $Cl(3,0)$. The quaternion conjugate (equivalent to Clifford conjugation in $Cl(3,0)^+$ and $Cl(0,2)$) is defined as $\widetilde{q} = q_r - q_i\boldsymbol{i} - q_j\boldsymbol{j} - q_k\boldsymbol{k}$, $\widetilde{pq} = \widetilde{q}\,\widetilde{p}$, which leaves the scalar part q_r unchanged. This leads to the norm of $q \in \mathbb{H}$ $|q| = \sqrt{q\widetilde{q}} = \sqrt{q_r^2 + q_i^2 + q_j^2 + q_k^2}$, $|pq| = |p|\,|q|$. The part $\boldsymbol{q} = V(q) = q - q_r = \frac{1}{2}(q - \widetilde{q}) = q_i\boldsymbol{i} + q_j\boldsymbol{j} + q_k\boldsymbol{k}$ is called a *pure* quaternion, it squares to the *negative* number $-(q_i^2 + q_j^2 + q_k^2)$. Every unit quaternion $\in \mathbb{S}^3$ (i.e. $|q| = 1$) can be written as: $q = q_r + q_i\boldsymbol{i} + q_j\boldsymbol{j} + q_k\boldsymbol{k} = q_r + \sqrt{q_i^2 + q_j^2 + q_k^2}\,\widehat{\boldsymbol{q}} = \cos\alpha + \widehat{\boldsymbol{q}}\sin\alpha = \exp(\alpha\widehat{\boldsymbol{q}})$, where $\cos\alpha = q_r$, $\sin\alpha = \sqrt{q_i^2 + q_j^2 + q_k^2}$, $\widehat{\boldsymbol{q}} = \boldsymbol{q}/|q| = (q_i\boldsymbol{i} + q_j\boldsymbol{j} + q_k\boldsymbol{k})/\sqrt{q_i^2 + q_j^2 + q_k^2}$, and $\widehat{\boldsymbol{q}}^2 = -1$, $\widehat{\boldsymbol{q}} \in \mathbb{S}^2$. The left and right *inverse* of a non-zero quaternion is $q^{-1} = \widetilde{q}/|q|^2 = \widetilde{q}/(q\widetilde{q})$. The *scalar part* of a quaternion is defined as $\mathrm{S}(q) = q_r = \frac{1}{2}(q + \widetilde{q})$, with *symmetries* $\forall p, q \in \mathbb{H}$: $\mathrm{S}(pq) = \mathrm{S}(qp) =$

[1] In memory of Hans Wondratschek, *07 Mar. 1925 in Bonn, †26 Oct. 2014 in Durlach.

$p_rq_r - p_iq_i - p_jq_j - p_kq_k$, $\mathrm{S}(q) = \mathrm{S}(\tilde{q})$, and *linearity* $\mathrm{S}(\alpha p + \beta q) = \alpha\,\mathrm{S}(p) + \beta\,\mathrm{S}(q) = \alpha p_r + \beta q_r$, $\forall p, q \in \mathbb{H}$, $\alpha, \beta \in \mathbb{R}$.

The scalar part and the quaternion conjugate allow the definition of the $\mathbb{R}^4$ inner product of two quaternions p, q as $p \cdot q = \mathrm{S}(p\tilde{q}) = p_rq_r + p_iq_i + p_jq_j + p_kq_k \in \mathbb{R}$. Accordingly we interpret in this paper the four quaternion coefficients as coordinates in $\mathbb{R}^4$. In this interpretation selecting any two-dimensional plane subspace[2] and its orthogonal complement two-dimensional subspace allows to split four-dimensional quaternions $\mathbb{H}$ into pairs of orthogonal two-dimensional planes (compare Theorem 3.5 of [7]). Dealing with rotations in this paper includes general rotations in $\mathbb{R}^4$.

Definition 1.1 (Orthogonality of quaternions). *Two quaternions $p, q \in \mathbb{H}$ are* orthogonal $p \perp q$, *if and only if* $\mathrm{S}(p\tilde{q}) = 0$.

2. Motivation for quaternion split

2.1. Splitting quaternions and knowing what it means

We deal with a split of quaternions, motivated by the consistent appearance of two terms in the *quaternion Fourier transform* $\mathcal{F}\{f\}(u,v) = \int_{\mathbb{R}^2} e^{-\boldsymbol{i}xu} f(x,y) e^{-\boldsymbol{j}yv} dxdy$ [4]. This observation[3] (note that in the following always $\boldsymbol{i}$ is on the left, and $\boldsymbol{j}$ is on the right) and that every quaternion can be rewritten as $q = q_r + q_i\boldsymbol{i} + q_j\boldsymbol{j} + q_k\boldsymbol{k} = q_r + q_i\boldsymbol{i} + q_j\boldsymbol{j} + q_k\boldsymbol{ij}$, motivated the quaternion *split*[4] with respect to the pair of orthonormal pure unit quaternions $\boldsymbol{i}, \boldsymbol{j}$

$$q = q_+ + q_-, \quad q_\pm = \frac{1}{2}(q \pm \boldsymbol{i}q\boldsymbol{j}). \tag{2}$$

Using (1), the detailed results of this split can be expanded in terms of real components $q_r, q_i, q_j, q_k \in \mathbb{R}$, as

$$q_\pm = \{q_r \pm q_k + \boldsymbol{i}(q_i \mp q_j)\}\frac{1 \pm \boldsymbol{k}}{2} = \frac{1 \pm \boldsymbol{k}}{2}\{q_r \pm q_k + \boldsymbol{j}(q_j \mp q_i)\}. \tag{3}$$

The analysis of these two components leads to the following Pythagorean *modulus identity* [5].

Lemma 2.1 (Modulus identity). *For $q \in \mathbb{H}$, $|q|^2 = |q_-|^2 + |q_+|^2$.*

Lemma 2.2 (Orthogonality of OPS split parts [5]). *Given any two quaternions $p, q \in \mathbb{H}$ and applying the OPS split of (2) the resulting two parts are* orthogonal, *i.e., $p_+ \perp q_-$ and $p_- \perp q_+$,*

$$\mathrm{S}(p_+\widetilde{q_-}) = 0, \qquad \mathrm{S}(p_-\widetilde{q_+}) = 0. \tag{4}$$

Next, we discuss the map $\boldsymbol{i}(\,)\boldsymbol{j}$, which will lead to an adapted orthogonal basis of $\mathbb{H}$. We observe, that $\boldsymbol{i}q\boldsymbol{j} = q_+ - q_-$, i.e. under the map $\boldsymbol{i}(\,)\boldsymbol{j}$ the q_+ part is *invariant*, but the q_- part *changes sign*. Both parts are *two-dimensional* (3), and by Lemma 2.2 they span *two completely orthogonal planes*, therefore also the name *orthogonal planes split (OPS)*. The q_+ plane has the

[2] The notion of two-dimensional plane employed here is thus *different* from a two-dimensional plane in $\mathbb{R}^3$. The latter can be characterized by a unit bivector area element of the plane, which corresponds via the isomorphism $Cl(3,0)^+ \cong \mathbb{H}$ to a pure unit quaternion. This difference in interpretation means also that despite of the isomorphism $Cl(3,0)^+ \cong \mathbb{H}$, the notion and expression of rotations in $\mathbb{R}^4$ cannot be automatically carried over to rotation operators of $Cl(3,0)$. Only in the case when rotations are restricted to the three-dimensional subset of pure quaternions, then Hamilton's original $\mathbb{R}^3$ interpretation of these rotations is obvious.

[3] Replacing e.g. $\boldsymbol{i} \to \boldsymbol{j}$, $\boldsymbol{j} \to \boldsymbol{k}$ throughout would merely change the notation, but not the fundamental observation.

[4] Also called *orthogonal planes split* (OPS) as explained below.

orthogonal quaternion basis $\{\boldsymbol{i}-\boldsymbol{j}=\boldsymbol{i}(1+\boldsymbol{ij}), 1+\boldsymbol{ij}=1+\boldsymbol{k}\}$, and the q_- plane has orthogonal basis $\{\boldsymbol{i}+\boldsymbol{j}=\boldsymbol{i}(1-\boldsymbol{ij}), 1-\boldsymbol{ij}=1-\boldsymbol{k}\}$. All four basis quaternions (if normed: $\{q_1, q_2, q_3, q_4\}$)

$$\{\boldsymbol{i}-\boldsymbol{j}, 1+\boldsymbol{ij},\ \boldsymbol{i}+\boldsymbol{j}, 1-\boldsymbol{ij}\}, \tag{5}$$

form an orthogonal basis of $\mathbb{H}$ interpreted as $\mathbb{R}^4$. Moreover, we obtain the following geometric picture on the left side of Fig. 1. The map $\boldsymbol{i}()\boldsymbol{j}$ *rotates* the q_- plane by 180° around the two-dimensional q_+ axis plane. This interpretation of the map $\boldsymbol{i}()\boldsymbol{j}$ is in perfect agreement with Coxeter's notion of *half-turn* [2]. In agreement with its geometric interpretation, the map $\boldsymbol{i}(\,)\boldsymbol{j}$ is an *involution*, because applying it twice leads to identity

$$\boldsymbol{i}(\boldsymbol{i}q\boldsymbol{j})\boldsymbol{j} = \boldsymbol{i}^2 q \boldsymbol{j}^2 = (-1)^2 q = q. \tag{6}$$

We have the important exponential factor identity

$$e^{\alpha \boldsymbol{i}} q_\pm e^{\beta \boldsymbol{j}} = q_\pm e^{(\beta \mp \alpha)\boldsymbol{j}} = e^{(\alpha \mp \beta)\boldsymbol{i}} q_\pm. \tag{7}$$

This equation should be compared with the kernel construction of the quaternion Fourier transform (QFT). The equation is also often used in our present context for values $\alpha = \pi/2$ or $\beta = \pi/2$.

Finally, we note the interpretation [7] of the QFT integrand $e^{-\boldsymbol{i}x_1\omega_1} h(\boldsymbol{x})\, e^{-\boldsymbol{j}x_2\omega_2}$ as a *local rotation* by phase angle $-(x_1\omega_1 + x_2\omega_2)$ of $h_-(\boldsymbol{x})$ in the two-dimensional q_- plane, spanned by $\{\boldsymbol{i}+\boldsymbol{j}, 1-\boldsymbol{ij}\}$, and a *local rotation* by phase angle $-(x_1\omega_1 - x_2\omega_2)$ of $h_+(\boldsymbol{x})$ in the two-dimensional q_+ plane, spanned by $\{\boldsymbol{i}-\boldsymbol{j}, 1+\boldsymbol{ij}\}$. This concludes the geometric picture of the OPS of $\mathbb{H}$ (interpreted as $\mathbb{R}^4$) with respect to two orthonormal pure quaternion units.

2.2. Even one pure unit quaternion can do a nice split

Let us now analyze the involution $\boldsymbol{i}(\,)\boldsymbol{i}$. The map $\boldsymbol{i}(\,)\boldsymbol{i}$ gives

$$\boldsymbol{i}q\boldsymbol{i} = \boldsymbol{i}(q_r + q_i\boldsymbol{i} + q_j\boldsymbol{j} + q_k\boldsymbol{k})\boldsymbol{i} = -q_r - q_i\boldsymbol{i} + q_j\boldsymbol{j} + q_k\boldsymbol{k}. \tag{8}$$

The following orthogonal planes split (OPS) with respect to the *single quaternion unit* $\boldsymbol{i}$ gives

$$q_\pm = \frac{1}{2}(q \pm \boldsymbol{i}q\boldsymbol{i}), \quad q_+ = q_j\boldsymbol{j} + q_k\boldsymbol{k} = (q_j + q_k\boldsymbol{i})\boldsymbol{j}, \quad q_- = q_r + q_i\boldsymbol{i}, \tag{9}$$

where the q_+ plane is two-dimensional and manifestly *orthogonal* to the two-dimensional q_- plane. The basis of the two planes are (if normed: $\{q_1, q_2\}$, $\{q_3, q_4\}$)

$$q_+\text{-basis: } \{\boldsymbol{j}, \boldsymbol{k}\}, \qquad q_-\text{-basis: } \{1, \boldsymbol{i}\}. \tag{10}$$

The geometric interpretation of $\boldsymbol{i}(\,)\boldsymbol{i}$ as Coxeter *half-turn* is perfectly analogous to the case $\boldsymbol{i}(\,)\boldsymbol{j}$. This form (9) of the OPS is identical to the quaternionic *simplex/perplex* split applied in quaternionic signal processing, which leads in color image processing to the *luminosity/chrominance* split [6].

3. General orthogonal two-dimensional planes split (OPS)

Assume in the following an arbitrary pair of pure unit quaternions f, g, $f^2 = g^2 = -1$. The *orthogonal 2D planes split (OPS)* is then defined with respect to *any two pure unit quaternions* f, g as

$$q_\pm = \frac{1}{2}(q \pm fqg) \qquad \Longrightarrow \qquad fqg = q_+ - q_-, \tag{11}$$

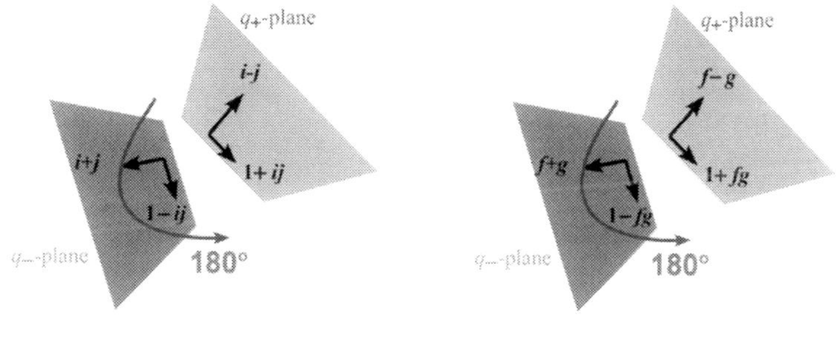

Figure 1. Geometric pictures of the involutions $\boldsymbol{i}(\,)\boldsymbol{j}$ and $f()g$ as half turns.

i.e. under the map $f()g$ the q_+ part is invariant, but the q_- part changes sign.

Both parts are two-dimensional, and span two completely orthogonal planes. For $f \neq \pm g$ the q_+ plane is spanned by two orthogonal quaternions $\{f-g, 1+fg = -f(f-g)\}$, the q_- plane is e.g. spanned by $\{f+g, 1-fg = -f(f+g)\}$. For $g = f$ a fully *orthonormal* four-dimensional basis of $\mathbb{H}$ is (R acts as rotation operator (rotor))

$$\{1, f, \boldsymbol{j}', \boldsymbol{k}'\} = R^{-1}\{1, \boldsymbol{i}, \boldsymbol{j}, \boldsymbol{k}\}R, \qquad R = \boldsymbol{i}(\boldsymbol{i}+f), \tag{12}$$

and the two orthogonal two-dimensional planes basis:

$$q_+\text{-basis: } \{\boldsymbol{j}', \boldsymbol{k}'\}, \qquad q_-\text{-basis: } \{1, f\}. \tag{13}$$

Note the notation for normed vectors in [3] $\{q_1, q_2, q_3, q_4\}$ for the resulting total *orthonormal basis of* $\mathbb{H}$.

Lemma 3.1 (Orthogonality of two OPS planes). *Given any two quaternions q, p and applying the OPS with respect to any two pure unit quaternions f, g we get zero for the scalar part of the mixed products*

$$Sc(p_+\widetilde{q}_-) = 0, \qquad Sc(p_-\widetilde{q}_+) = 0. \tag{14}$$

Note, that the two parts $x_\pm$ can be *represented* as

$$x_\pm = x_{+f}\,\frac{1 \pm fg}{2} + x_{-f}\,\frac{1 \mp fg}{2} = \frac{1 \pm fg}{2}\,x_{+g} + \frac{1 \mp fg}{2}\,x_{-g}, \tag{15}$$

with commuting and anticommuting parts $x_{\pm f}f = \pm f x_{\pm f}$, etc.

Next we mention the possibility to perform a split along any given set of two (two-dimensional) analysis planes. It has been found, that any two-dimensional plane in $\mathbb{R}^4$ determines in an elementary way an OPS split and vice versa, compare Theorem 3.5 of [7].

Let us turn to the geometric interpretation of the map $f()g$. It *rotates* the q_- plane by 180° around the q_+ axis plane. This is in perfect agreement with Coxeter's notion of *half-turn* [2], see the right side of Fig. 1. The following *identities* hold

$$e^{\alpha f} q_\pm e^{\beta g} = q_\pm e^{(\beta \mp \alpha)g} = e^{(\alpha \mp \beta)f} q_\pm. \tag{16}$$

This leads to a straightforward geometric interpretation of the integrands of the quaternion Fourier transform (OPS-QFT) with two pure quaternions f, g, and of the orthogonal 2D planes phase rotation Fourier transform [7].

We can further incorporate quaternion conjugation, which consequently provides a geometric interpretation of the QFT involving quaternion conjugation of the signal function. For $d =$

$e^{\alpha g}, t = e^{\beta f}$ the map $d\widetilde{()}t$ represents a *rotary-reflection* in four dimensions with pointwise *invariant line* $d + t$, a rotary-reflection *axis* $d - t$: $d\widetilde{(d-t)}t = -(d-t)$, and rotation *angle* $\Gamma = \pi - \arccos \mathrm{S}\big(\widetilde{dt}\big)$ in the plane $\perp \{d+t, d-t\}$. (The derivation of Γ will be shown later.) We obtain the following Lemma.

Lemma 3.2. *For OPS* $q_\pm = \frac{1}{2}(q \pm fqg)$, *and left and right exponential factors we have the identity*

$$e^{\alpha g}\,\widetilde{q_\pm} e^{\beta f} = \widetilde{q_\pm} e^{(\beta\mp\alpha)f} = e^{(\alpha\mp\beta)g}\,\widetilde{q_\pm}. \tag{17}$$

4. Coxeter on Quaternions and Reflections [2]

The four-dimensional *angle* Θ between two unit quaternions $p, q \in \mathbb{H}$, $|p| = |q| = 1$, is defined by

$$\cos\Theta = Sc(p\widetilde{q}). \tag{18}$$

The right and left *Clifford translations* are defined by Coxeter [2] as

$$q \to q' = qa, \quad q \to q'' = aq, \quad a = e^{\widehat{a}\Theta}, \quad \widehat{a}^2 = -1. \tag{19}$$

Both Clifford translations represent *turns by constant angles* $\Theta_{q,q'} = \Theta_{q,q''} = \Theta$. We analyze the following special cases, assuming the split $q_\pm$ w.r.t. $f = g = \widehat{a}$:

- For $\widehat{a} = \boldsymbol{i}$, $aq_- = q_- a = (q_- a)_-$, is a mathematically *positive* (anti clockwise) *rotation* in the q_- plane $\{1, \boldsymbol{i}\}$.
- Similarly, $aq_+ = (aq_+)_+$, is a mathematically *positive rotation* in the q_+ plane $\{\boldsymbol{j}, \boldsymbol{k}\}$.
- Finally, $q_+ a = \widetilde{a} q_+ = (q_+ a)_+$, is a mathematically *negative rotation* (clockwise) by Θ in the q_+ plane $\{\boldsymbol{j}, \boldsymbol{k}\}$.

Next, we compose Clifford translations, assuming the *split* $q_\pm$ w.r.t. $f = g = \widehat{a}$. For the unit quaternion $a = e^{\widehat{a}\Theta}$, $\widehat{a}^2 = -1$ we find that

$$q \to aqa = a^2 q_- + q_+ \tag{20}$$

is a *rotation* only in the q_- plane by the *angle* 2Θ, and

$$q \to aq\tilde{a} = q_- + a^2 q_+ \tag{21}$$

is a *rotation* only in the q_+ plane by the *angle* 2Θ.

Let us now revisit Coxeter's **Lemma 2.2** in [2]: For any two quaternions $a, b \in \mathbb{H}$, $|a| = |b|$, $a_r = b_r$, we can find a $y \in \mathbb{H}$ such that

$$ay = yb. \tag{22}$$

We now further ask for the set of *all* $y \in \mathbb{H}$ such that $ay = yb$? Based on the OPS, the answer is straightforward. For $a = |a|e^{\Theta\hat{a}}$, $b = |a|e^{\Phi\hat{b}}$, $\Phi = \pm\Theta$, $\hat{a}^2 = \hat{b}^2 = -1$ we use the split $q_\pm = \frac{1}{2}(q \pm \hat{a}q\hat{b})$ to obtain:

- For $\Theta = \Phi$: The set of all y spans the q_- plane. Moreover,

$$aq_+b = q_+, q_+b = \tilde{a}q_+, aq_+ = q_+\tilde{b}, aq_-b = a^2q_- = q_-b^2, aq_- = q_-b. \tag{23}$$

- For $\Theta = -\Phi$: The set of all y spans the q_+ plane. Moreover,

$$aq_-b = q_-, q_-b = \tilde{a}q_-, aq_- = q_-\tilde{b}, aq_+b = a^2q_+ = q_+b^2, aq_+ = q_+b. \tag{24}$$

Let us turn to a reflection in a hyperplane. **Theorem 5.1** in [2] says: The reflection in the hyperplane $\perp a \in \mathbb{H}$: $Sc(a\tilde{q}) = 0$, $|a|^2 = 1$, $a = |a|e^{\Theta\hat{a}}$, $\hat{a}^2 = -1$, is represented by

$$q \to -a\tilde{q}a. \tag{25}$$

We analyze the situation using the OPS. We define the split $q_\pm = \frac{1}{2}(q \pm \hat{a}q\hat{a})$ to obtain

$$q_+ \to -a\widetilde{q_+}a = q_+, \qquad a \to -a\tilde{a}a = -a. \tag{26}$$

and for $a' = ae^{-\frac{\pi}{2}\hat{a}}$

$$a' \to -a\widetilde{a'}a = a'. \tag{27}$$

We further consider a general rotation. **Theorem 5.2** in [2] states: The general rotation through 2Φ (about a plane) is $q \to aqb$, $a = |a|e^{\Phi\hat{a}}$, $b = |a|e^{\Theta\hat{b}}$, $\Phi = \pm\Theta$, $\hat{a}^2 = \hat{b}^2 = -1$.

We again apply the OPS. We define the split $q_\pm = \frac{1}{2}(q \pm \hat{a}q\hat{b})$ to obtain:

- For $\Theta = \Phi$: Rotation of q_- plane by 2Φ around q_+-plane.
- For $\Theta = -\Phi$: Rotation of q_+ plane by 2Φ around q_--plane.

Let us illustrate this with an *example*: $\hat{a} = \hat{b} = \boldsymbol{i}$, $\Phi = -\Theta$,

$$aq_-b = q_-. \tag{28}$$

For $q_+ = \boldsymbol{j}$:

$$aq_+b = a\boldsymbol{j}b = \boldsymbol{j}b^2 = \boldsymbol{j}e^{-2\theta\boldsymbol{i}} = \boldsymbol{j}\cos 2\Phi - \boldsymbol{k}\sin 2\Phi, \tag{29}$$

a rotation in the q_+-plane around the q_- plane. Note, that the detailed analysis of general $q \to aqb$, $q_\pm = \frac{1}{2}(q \pm \hat{a}q\hat{b})$, $|a| = |b| = 1$, can be found in [7].

As for the *rotary inversion*, we follow the discussion in [7], sec. 5.1, but add a simple formula for determining the rotation angle. The rotary inversion is given by, $d, t \in \mathbb{H}$, $|d| = |t| = 1$, $q \to d\tilde{q}t$. For $d \neq \pm t$, $[d,t] = dt - td$, we obtain two vectors in the rotation plane $v_{1,2} = [d,t](1 \pm \tilde{d}t)$, with $d\widetilde{v_{1,2}}t = -v_{1,2}\tilde{d}t$. The angle Γ of rotation can therefore be simply found from

$$\cos\Gamma = Sc(\frac{1}{|v_1|^2}\tilde{v}_1 d\tilde{v}_1 t) = Sc(-\tilde{d}t) = \cos(\pi - \gamma), \tag{30}$$

with γ the angle between d and t : $\cos\gamma = \tilde{d}t$.

5. Quaternion geometry of rotations [3] analyzed by 2D OPS

According to [3] the circle $C(q_1, q_2)$ of all unit quaternions[5], which rotate $g \to f$, $f \neq \pm g$ is given by

$$q(t) = \frac{1 - fg}{|1 - fg|}e^{\frac{t}{2}g} = q_1 e^{\frac{t}{2}g}, \quad t \in [0, 2\pi), \quad q(t)g\widetilde{q(t)} = f, \quad q_2 = \frac{f+g}{|f+g|}.$$

The two-dimensional OPS $q_\pm^{f,g} = \frac{1}{2}(q \pm fqg)$ tells us, that all $q(t), t \in [\,0, 2\pi\,)$ are elements of the q_- plane. And indeed, $fq_-g = -q_-$ for all $q_- \in \mathbb{H}$ leads to

$$f = q_- g q_-^{-1}, \tag{31}$$

[5] Note that apart from the shortest (and longest) rotation(s) along a geodesic connecting two end points of two unit vectors attached to the center of a unit sphere, any circle on the unit sphere, which includes the end points gives another trajectory of (two) rotation(s) between the two endpoints with rotation axis through the center of the unit sphere and the center of the circle on the unit sphere.

for all q_- in the q_--plane. Note, that this is valid for all $f, g \in \mathbb{H}$, $f^2 = g^2 = -1$, even for $f = \pm g$! We therefore get a *one line proof*, which at the same time generalizes from the unit circle to the whole plane.

Meister and Schaeben [3] state that for $q \in C(q_1, q_2)$: $fq, qg, fqg \in C(q_1, q_2)$. This can easily be generalized to the whole q_--plane, because

$$(fq_-)_- = fq_-, \quad (q_-g)_- = q_-g, \quad (fq_-g)_- = fq_-g. \tag{32}$$

We can use the exponential form, and show that the circle $C(q_1, q_2)$ parametrization of (34), (35) in [3] is a *specialization of the general relation*

$$e^{\frac{t}{2}f} q_- = q_- e^{\frac{t}{2}g} \tag{33}$$

which means that the two parametrizations are element wise identical.

Now we look at the quaternion circles for the rotations $g \to \pm f$. **Prop. 5** of [3] states: Two circles $C(q_1, q_2) = G(g, f)$ and $C(q_3, q_4) = G(g, -f) = G(-g, f)$, representing all rotations $g \to f$ and $g \to -f$, respectively, are orthonormal to each other. Here four orthogonal unit quaternions are defined as:

$$q_1 = \frac{1 - fg}{|1 - fg|}, \quad q_2 = \frac{f + g}{|f + g|}, \quad q_3 = \frac{1 + fg}{|1 + fg|}, \quad q_4 = \frac{f - g}{|f - g|}. \tag{34}$$

We provide a simple proof: We already know that all q_1, q_2 span the q_- plane of the split $q_\pm^{f,g} = \frac{1}{2}(q \pm fqg)$, and all q_3, q_4 span the q_+-plane. And that

$$fq_\pm g = \pm q_\pm \qquad \Leftrightarrow \qquad f = q_\pm(\mp g)q_\pm^{-1}. \tag{35}$$

QED. Note, the proof is again much faster than in [3]. We see that $G(g, f) = \{q_-^{f,g}/|q_-^{f,g}|, \forall q \in \mathbb{H}\}$, and $G(g, -f) = \{q_+^{f,g}/|q_+^{f,g}|, \forall q \in \mathbb{H}\}$.

For later use, we translate the notation of [3] (38),(39):

$$n_3 = -n_1 = \frac{[f, g]}{|[f, g]|}, \quad n_4 = q_4, \quad n_2 = n_4 n_1, \quad n_4 = n_1 n_2, \quad n_1 = n_2 n_4, \tag{36}$$

which shows that $\{n_1, n_2, n_4\}$ is a right handed set of three orthonormal pure quaternions, obtained by rotating $\{\boldsymbol{i}, \boldsymbol{j}, \boldsymbol{k}\}$.

The two circles $G(g, f), G(g', f)$ do not intersect for $g \neq g'$, see Cor. 1(i) of [3]. We provide a simple proof: Assume $\exists_1 q \in H : fqg = -q, fqg' = -q$ for $g \neq g'$. Then

$$fqg = fqg' \Leftrightarrow g = g' \Rightarrow G(g, f) \bigcap G(g', f) = \emptyset. \tag{37}$$

QED.

Cor. 1 (iii) of [3] further states that for every 3D rotation R and given g_0, $g_0^2 = -1$ we can always find f, $f^2 = -1$, such that R is represented by a (unit) quaternion q in $G(g_0, f)$. We can equivalently ask for f, such that q representing the rotation R is $\in q_-^{f,g_0}$-plane. We find

$$fqg_0 = -q \Leftrightarrow f = qg_0q^{-1}. \tag{38}$$

The left side of Fig. 2 shows two small circles $C(g, \rho), C(f, \rho) \subset \mathbb{S}^2$ [3]. We now analyze the mapping between pairs of small circles. A small circle with center g and radius ρ is defined as

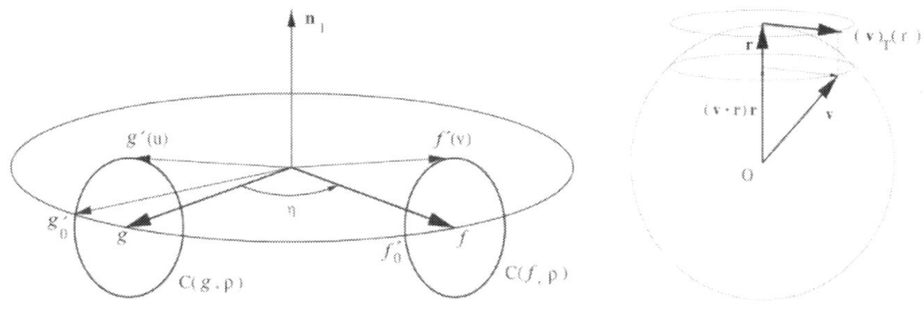

Figure 2. Small circles and tangential plane projection. Adapted from Figs. 2 and 3 of [3].

$C(g,\rho) = \{g' \in \mathbb{S}^2 : g \cdot g' = \cos\rho\}$, and all $q \in q_-^{f,g}$-plane map $C(g,\rho)$ to the small circle $C(f,\rho)$ of the same radius (a slight generalization of [3], Prop. 6), with the correspondence

$$q(t)g'(u)q(t)^{-1} = f'(u+2t),$$
$$q(t) = q_1 e^{tg}, \quad g'(u) = e^{\frac{u}{2}g} g'_0 e^{-\frac{u}{2}g}, \quad f'(u) = e^{\frac{u}{2}f} f'_0 e^{-\frac{u}{2}f} \tag{39}$$

starting with the corresponding circle points $q_1 g'_0 = f'_0 q_1$.

We provide the following direct proof: We repeatedly apply (16) to obtain

$$\begin{aligned} q_1 g'_0 = f'_0 q_1 &\Leftrightarrow e^{\frac{u}{2}f} q_1 g'_0 e^{-\frac{u}{2}g} = e^{\frac{u}{2}f} f'_0 q_1 e^{-\frac{u}{2}g} \\ &\Leftrightarrow q_1 e^{\frac{u}{2}g} g'_0 e^{-\frac{u}{2}g} = e^{\frac{u}{2}f} f'_0 e^{-\frac{u}{2}f} q_1 \\ &\Leftrightarrow e^{tf} q_1 e^{\frac{u}{2}g} g'_0 e^{-\frac{u}{2}g} = e^{tf} e^{\frac{u}{2}f} f'_0 e^{-\frac{u}{2}f} e^{-tf} e^{tf} q_1 \\ &\Leftrightarrow q_1 e^{tg}\, e^{\frac{u}{2}g} g'_0 e^{-\frac{u}{2}g} = e^{tf} e^{\frac{u}{2}f} f'_0 e^{-\frac{u}{2}f} e^{-tf}\, q_1 e^{tg}. \end{aligned} \tag{40}$$

QED. Note, that this proof is much shorter than in [3], and we do not need to use addition theorems.

We consider the *projection onto the tangential plane* of a $\mathbb{S}^2$ vector, see the right side of Fig. 2. Assume $v, r \in \mathbb{S}^2$. Note, that $(v)_T(r) = v - (v\cdot r)r$ of [3] can be simplified to $(v)_T(r) = V(vr)r^{-1}$, valid for all pure (non-unit) quaternions r.

Finally we consider the *torus theorem* for all maps $g \to$ small circle $C(f, 2\Theta)$. We slightly reformulate the theorem **Prop. 13** of [3]. We will use the two-dimensional OPS with respect to $f, g \in \mathbb{S}^2$, and the corresponding orthonormal basis $\{q_1, q_2, q_3, q_4\}$ of (34). The theorem says, that the circle $C(q_1, q_2) \in q_-$-plane: $q_-(s) = q_1 \exp(sg/2)$, $s \in [0, 2\pi)$, represents all rotations $g \to f$, while the orthogonal circle $C(q_3, q_4) \in q_+$-plane: $q_+(t) = q_3 \exp(-tg/2)$, $t \in [0, 2\pi)$, represents all rotations $g \to -f$. Then the *spherical torus* $T(q_-(s), q_+(t); \Theta)$ is defined as the set of quaternions

$$q(s,t;\Theta) = q_-(2s)\cos\Theta + q_+(2t)\sin\Theta, \quad s,t \in [0,2\pi), \quad \Theta[0,\pi/2], \tag{41}$$

and represents all rotations $g \to C(f, 2\Theta) \subset \mathbb{S}^2$.

In particular, the set $q(s, -s; \Theta)$ maps g for all $s \in [0, 2\pi)$ onto f'_0 in the f, g plane with $g \cdot f'_0 = \cos(\eta - 2\Theta)$, $g \cdot f = \cos\eta$,

$$q(s,-s;\Theta) g q(s,-s;\Theta)^{-1} = f'_0 \quad \forall s \in [0, 2\pi). \tag{42}$$

Moreover, for arbitrary $s_0 \in [0, 2\pi)$, the set $q(s_0, t - s_0; \Theta)$ (or equivalently $q(s_0 + t, s_0; \Theta)$) maps $g \to f' \in C(f, 2\Theta)$, which results from positive rotation (counter-clockwise) of f'_0 about f by the angle $t \in [0, 2\pi)$,

$$q(s_0, t - s_0; \Theta) g q(s_0, t - s_0; \Theta)^{-1} = e^{\frac{t}{2}f} f_0' e^{-\frac{t}{2}f} \quad \forall s_0 \in [0, 2\pi). \tag{43}$$

We state the following *direct proof* of the torus theorem.

$$\begin{aligned} q(s, t; \Theta) &= q_-(2s)\cos\Theta + q_+(2t)\sin\Theta = q_1 e^{sg}\cos\Theta + q_3 e^{-tg}\sin\Theta \\ &= (q_1\cos\Theta + q_3 e^{-(t+s)g}\sin\Theta)e^{sg} \\ &= (\cos\Theta + (q_3/q_1)e^{-(t+s)f}\sin\Theta)q_1 e^{sg} \\ &= (\cos\Theta + (-n_1)e^{-(t+s)f}\sin\Theta)q_1 e^{sg} \\ &= (\cos\Theta + e^{(t+s)f}(-n_1)\sin\Theta)e^{sf}q_1 \\ &= e^{-n_1'\Theta}e^{sf}q_1, \qquad n_1' = e^{(t+s)f}(-n_1), (n_1')^2 = -1. \end{aligned} \tag{44}$$

We observe, that n_1' is n_1 rotated around f by angle $s + t$. Application to g gives

$$q(s, t; \Theta)g\, q(s, t; \Theta)^{-1} = e^{-n_1'\Theta} f e^{n_1'\Theta}, \tag{45}$$

so geometrically g is rotated into $f = e^{sf}q_1 g\, q_1^{-1}e^{-sf}$, which in turn is rotated around n_1' on the circle $C(f, 2\Theta)$. For $t = -s$ obviously

$$q(s, -s; \Theta)gq(s, -s; \Theta)^{-1} = e^{-n_1\Theta} f e^{n_1\Theta} = f_0', \tag{46}$$

is a rotation in the f, g plane of g into f_0', with $g \cdot f_0' = \cos\eta - 2\Theta$. We further note, that for $s = s_0$, $t \to t - s_0$: $n_1' = e^{tf}(-n_1)$, such that

$$q(s_0, t - s_0; \Theta)gq(s_0, t - s_0; \Theta)^{-1} = e^{-n_1'\Theta} f e^{n_1'\Theta} = e^{tf} f_0' e^{-tf}, \tag{47}$$

describes the small circle $C(f, 2\Theta)$. QED.

Our proof is very *compact*, obtained by *direct* computation of *monomial* results, which in turn permit direct *geometric interpretation*.

6. Conclusions

We have exposed the geometric understanding of the *general OPS split of quaternions* [7] into two *orthogonal planes* ($\mathbb{R}^4$ interpretation). Moreover, we have consolidated the OPS with the geometric understanding by Altmann [1], Coxeter [2], and Meister and Schaeben [3].

Acknowledgments

Soli Deo gloria. R. Ablamowicz, F. Brackx, R. Bujack, D. Eelbode, S. Georgiev, J. Helmstetter, B. Mawardi, S. Sangwine, G. Scheuermann, J. Van der Jeugt, anonymous reviewer.

References

[1] Altmann S 1986 *Rotations, Quaternions, and Double Groups* (New York: Dover)
[2] Coxeter H S M 1946 Quaternions and Reflections *The Amer. Math. Monthly* **53**(3) 136
[3] Meister L and Schaeben H 2005 A concise quaternion geometry of rotations *Math. Meth. in the Appl. Sci.* **28** 101
[4] Hitzer E 2007 Quaternion Fourier Transform on Quaternion Fields and Generalizations *Adv. in App. Cliff. Alg.* **17** 497
[5] Hitzer E 2010 Directional Uncertainty Principle for Quaternion Fourier Transforms *Adv. in App. Cliff. Alg.* **20**(2) 271

[6] Ell T A and Sangwine S J 2007 Hypercomplex Fourier Transforms of Color Images *IEEE Trans. on Image Processing* **16**(1) 22

[7] Hitzer E and Sangwine S J 2013 The Orthogonal 2D Planes Split of Quaternions and Steerable Quaternion Fourier Transform *Quaternion and Clifford Fourier Transforms and Wavelets* TIM **27** eds Hitzer E and Sangwine S J (Heidelberg: Springer)

[8] Bülow T, Felsberg M and Sommer G 2001 Non-commutative Hypercomplex Fourier Transforms of Multidimensional Signals *Geometric Computing with Clifford Algebras: Theoretical Foundations and Applications in Computer Vision and Robotics* ed Sommer G (Berlin: Springer) 187

Journal of Physics: Conference Series **597** (2015) 012043 doi:10.1088/1742-6596/597/1/012043

Nonabelian T-duals of the flat background

L Hlavatý and I Petr

Czech Technical University in Prague, Faculty of Nuclear Sciences and Physical Engineering, Břehová 7, 115 19 Prague 1, Czech Republic

E-mail: hlavaty@fjfi.cvut.cz, ivo.petr@fjfi.cvut.cz

Abstract. Classification of nonabelian T-duals of the flat matric in D=4 with respect to the four-dimensional continuous subgroups of the Poincare group is given. After dualizing the flat background we identify majority of dual models as conformal sigma models in plane-parallel wave backgrounds, most of them having torsion. We give their form in Brinkmann coordinates. Besides the pp-waves we find several diagonalizable curved metrics with nontrivial scalar curvature and torsion.

Due to the nonabelian T-duality, we find general solution of classical field equations of all the sigma models in terms of d'Alembert solutions of wave equation.

1. Introduction

String theory in curved and/or time-dependent background can be formulated as a sigma model satisfying supplementary conditions. Finding solutions of equations of motion in these backgrounds is usually very complicated. That's why every solvable case attracts considerable attention. An example of such a model is a string theory in the homogeneous plane-parallel wave background, solved in Ref. [1] in terms of Bessel functions. Plane-parallel (pp-)wave backgrounds in string theory were repeatedly investigated in the past (see e.g. references in [2]). They give not only solvable models [3], but also allow to study the behavior of strings near spacetime singularity [4, 5]. Moreover, to extract information about string behavior in general curved background one can take the Penrose limit [6], extended for fields in string theory in [7], and study string behavior in the resulting plane-wave background.

Particular cases of four-dimensional pp-wave background obtained from gauged WZW models were given in [8],[9],[2] as

$$ds^2 = dudv + \frac{g_1(u')}{g_1(u')g_2(u)+q^2}\,dx_1^2 + \frac{g_2(u)}{g_1(u')g_2(u)+q^2}\,dx_2^2, \tag{1}$$

$$B_{12} = \frac{q}{g_1(u')g_2(u)+q^2},$$

where $u' = au + d$ $(a, d = const)$ and the functions g_i can take any pair of the following values

$$g(u) = 1\,,\quad u^2\,,\quad \tanh^2 u\,,\quad \tan^2 u\,,\quad u^{-2}\,,\quad \coth^2 u\,,\quad \cot^2 u. \tag{2}$$

It is mentioned in [8] that this background is dual to the flat space for $g_1 = 1$ $g_2 = u^2$. We shall show that several other cases of these backgrounds are dual to the flat space as well. Moreover,

Journal of Physics: Conference Series **597** (2015) 012043 doi:10.1088/1742-6596/597/1/012043

we shall use this fact for finding general solution of classical sigma model field equations in these pp-wave backgrounds. Beside pp-waves, we find several curved backgrounds with diagonalizable metrics and solve their field equations as well.

We understand the non-Abelian T-duality [10] as a special case of Poisson–Lie T-plurality [11] based on the structure of Drinfeld double. For technical reasons we shall restrict to four spacetime dimensions, but the discussion can be extended to higher dimensions using the spectator fields or subgroups of Poincare group in higher dimensions.

2. Non-Abelian T-duality

The sigma model on a manifold M is given by the classical action

$$S_F[X] = -\int d\sigma_+ d\sigma_- \left(\partial_- X^\mu F_{\mu\nu}(X)\partial_+ X^\nu\right) = \tag{3}$$
$$= \frac{1}{2}\int d\tau d\sigma \left[-\partial_\tau X^\mu G_{\mu\nu}(X)\partial_\tau X^\nu + \partial_\sigma X^\mu G_{\mu\nu}(X)\partial_\sigma X^\nu - 2\partial_\tau X^\mu B_{\mu\nu}(X)\partial_\sigma X^\nu\right],$$

where F is a second order tensor field on M, with the metric and the NS–NS 2-form (torsion potential) given by the symmetric and antisymmetric part of F

$$G_{\mu\nu} = \frac{1}{2}(F_{\mu\nu} + F_{\nu\mu}),\ B_{\mu\nu} = \frac{1}{2}(F_{\mu\nu} - F_{\nu\mu}).$$

The worldsheet coordinates are

$$\sigma_+ = \frac{1}{\sqrt{2}}(\tau + \sigma),\ \sigma_- = \frac{1}{\sqrt{2}}(\tau - \sigma).$$

The functions X^μ are determined by the composition $X^\mu(\tau,\sigma) = x^\mu(X(\tau,\sigma))$, where $X : \mathbb{R}^2 \ni (\tau,\sigma) \mapsto X(\tau,\sigma) \in M$ and $x^\mu : \mathbf{U}_p \to \mathbb{R}$ are components of a coordinate map on a neighborhood $\mathbf{U}_p$ of an element $X(\tau,\sigma) = p \in M$.

The non-Abelian T-duality [10] of sigma models is a special case of Poisson–Lie T-duality [11],[12] that can be formulated by virtue of the Drinfeld double – a connected Lie group whose Lie algebra $\mathfrak{d}$ can be decomposed into a pair of equally dimensional subalgebras $\mathfrak{g}, \tilde{\mathfrak{g}}$ that are maximally isotropic with respect to a symmetric ad-invariant non-degenerate bilinear form $< .,. >$ on $\mathfrak{d}$.

The Drinfeld double suitable for non-Abelian T-duality is the semidirect product $G \ltimes \widetilde{G}$, where the group G can be taken as a subgroup of the isometry group of the background, which, in our case, will be flat. The group $\widetilde{G}$ has to be chosen Abelian in order to satisfy the conditions of dualizability of the sigma model [11]. We shall focus on the case, when the isometry subgroup acts freely and transitively on the manifold, so that we can identify $G \approx M$. This is usually referred to as atomic duality. Let us summarize the main points of the construction of dual models.

Given the four-dimensional subgroup G of the isometry group generated by Killing vectors of the flat metric, the tensor F is symmetric and can be written as

$$F_{\mu\nu}(x) = G_{\mu\nu}(x) = e_\mu^a(g(x))(E_0)_{ab}e_\nu^b(g(x)), \tag{4}$$

where E_0 is a constant non-singular symmetric matrix, and $e_\mu^a(g(x))$ are components of right-invariant forms $(dg)g^{-1}$ expressed in coordinates $\{x^\mu\}$ on the group G and basis of its Lie algebra $\{T_a\}$.

Denoting the mutually dual bases of $\mathfrak{g}$ and Abelian $\tilde{\mathfrak{g}}$ as $\{T_i\}$, $\{\tilde{T}^j\}$, we construct subspaces $\varepsilon^+ = Span(T_i + E_{0,ij}\tilde{T}^j)$, $\varepsilon^- = Span(T_i - E_{0,ji}\tilde{T}^j)$ that are orthogonal w.r.t. $<,>$ and span the whole Lie algebra $\mathfrak{d}$. The field equations for the sigma model on the group G can be rewritten as equation

$$< (\partial_{\pm} l)l^{-1}, \varepsilon^{\pm} >= 0 \tag{5}$$

for mapping l from the worldsheet in $\mathbb{R}^2$ into the Drinfeld double D.

Due to Drinfeld, there exists a unique decomposition (at least in the vicinity of the unit element of D) of an arbitrary element l of D as a product of elements from G and $\tilde{G}$. Solutions of equation (5) and solution of the equations of motion for the sigma model $X^\mu(\tau,\sigma) = x^\mu(g(\tau,\sigma))$ are related by

$$l(\tau,\sigma) = g(\tau,\sigma)\tilde{h}(\tau,\sigma) \in D,$$

where $\tilde{h}(\tau,\sigma) \in \tilde{G}$ fulfills the equations

$$\partial_\tau \tilde{h}_j = -v_j^\lambda G_{\lambda\nu}\partial_\sigma X^\nu, \tag{6}$$

$$\partial_\sigma \tilde{h}_j = -v_j^\lambda G_{\lambda\nu}\partial_\tau X^\nu, \tag{7}$$

with v_j^λ representing components of the left-invariant fields v_j on G in the group coordinates x^μ.

The metric and the torsion potential of the non-Abelian T-dual model can be obtained from the tensor $\widetilde{F}$

$$\widetilde{F}_{\mu\nu}(\tilde{x}) = [E_0 + \widetilde{\Pi}(\tilde{g}(\tilde{x}))]^{-1}, \tag{8}$$

where the matrix $\widetilde{\Pi}$ is given by the adjoint representation of the Abelian subgroup $\widetilde{G}$ on the Lie algebra of the Drinfeld double in the mutually dual bases

$$Ad(\tilde{g})^T = \begin{pmatrix} \mathbf{1} & 0 \\ \widetilde{\Pi}(\tilde{g}) & \mathbf{1} \end{pmatrix}.$$

The relation between solution $X^\mu(\tau,\sigma)$ of the equations of motion of the sigma model given by F and solution $\tilde{X}^\mu(\tau,\sigma) := \tilde{x}^\mu(\tilde{g}(\tau,\sigma))$ of the sigma model given by $\widetilde{F}$ follows from two possible decompositions of elements l of the Drinfeld double

$$g(\tau,\sigma)\tilde{h}(\tau,\sigma) = \tilde{g}(\tau,\sigma)h(\tau,\sigma), \tag{9}$$

where $g, h \in G,\ \tilde{g}, \tilde{h} \in \tilde{G}$. The map $\tilde{h} : \mathbb{R}^2 \to \tilde{G}$ that we need for this transformation is obtained from the equations (6),(7).

The equation (9) defines the Poisson–Lie transformation between solutions of the equations of motion of the original sigma model and its dual. Its application may be rather complicated. To use it for finding solution of the dual model, the following three steps must be done:

- Step 1: One has to know the solution $X^\mu(\tau,\sigma)$ of the sigma model given by $F_{\mu\nu}(x)$.
- Step 2: Given $X^\mu(\tau,\sigma)$, one has to find $\tilde{h}(\tau,\sigma)$, i.e. solve the system of PDEs (6),(7).
- Step 3: Given $l(\tau,\sigma) = g(\tau,\sigma)\tilde{h}(\tau,\sigma) \in D$, one has to find the dual decomposition $l(\tau,\sigma) = \tilde{g}(\tau,\sigma)h(\tau,\sigma)$, where $\tilde{g}(\tau,\sigma) \in \tilde{G}$, $h(\tau,\sigma) \in G$. Functions $\tilde{X}^\mu(\tau,\sigma) := \tilde{x}^\mu(\tilde{g}(\tau,\sigma))$ then solve the field equations of the dual sigma model.

3. Strings in pp-wave background

We will be interested in special subclass of metrics called pp-waves. Their metric in the so called Brinkmann coordinates $(u, v, z_3, z_4, \ldots, z_D)$ can be written as

$$ds^2 = 2dudv - K(u,\vec{z})du^2 + d\vec{z}^{\,2}, \tag{10}$$

where $d\vec{z}^2$ is the Euclidean metric in the transversal space with coordinates $\vec{z} = (z_3, z_4, \ldots, z_D)$. We denote the number of transversal coordinates by d, such that $D = 2 + d$. The NS–NS 2-form of particular interest to us has the form

$$B = B_j(u, \vec{z}) du \wedge dz_j. \tag{11}$$

The metric (10) has covariantly constant null Killing vector ∂_v and particularly simple curvature properties, because the Ricci tensor has only one nonzero component

$$R_{uu} = \frac{1}{2}(\partial_3^2 K + \partial_4^2 K + \ldots + \partial_D^2 K),$$

and the scalar curvature vanishes. Special case is background resulting from the Penrose–Güven limit [6, 7], with

$$K(u, \vec{z}) = K_{ij}(u) z_i z_j, \tag{12}$$

and the torsion

$$H = H_{ij}(u) du \wedge dz_i \wedge dz_j \tag{13}$$

that follows from the NS–NS 2-form (11) if $B_j(u, \vec{z})$ is linear in z. The one-loop conformal invariance conditions then simplify to solvable differential equation for the dilaton $\Phi = \Phi(u)$

$$\Phi''(u) - K_{jj}(u) + \frac{1}{4} H_{ij}(u) H_{ij}(u) = 0. \tag{14}$$

We are going to show that the sigma models in pp-wave backgrounds with special forms of functions K_{ij} (12) and H_{ij} (13) can be obtained as non-Abelian T-duals of sigma models in flat background. Killing vectors of the flat metric $\eta = \mathrm{diag}(-1, 1, 1, 1)$ in coordinates (t, x, y, z) are

$$P_0 = \partial_t,\ P_j = \partial_j,\ L_j = -\varepsilon_{ijk} x^j \partial_k,\ M_j = -x^j \partial_t - t\partial_j, \tag{15}$$

and form ten-dimensional Poincare Lie algebra. To apply the (atomic) non-Abelian T-duality on sigma models in the flat background, we shall need four-dimensional subalgebras of Poincare Lie algebra, classified in [13].

4. Example

As an example we shall present dual model obtained from subalgebra

$$Span[\mathcal{K}_1 = L_3 + \epsilon\,(P_0 + P_3), \mathcal{K}_2 = P_1, \mathcal{K}_3 = P_2, \mathcal{K}_4 = P_0 - P_3],\ \epsilon = \pm 1$$

that produces dual model with torsion. The commutation relations of this subalgebra are

$$[\mathcal{K}_1, \mathcal{K}_2] = \mathcal{K}_3, \qquad [\mathcal{K}_1, \mathcal{K}_3] = -\mathcal{K}_2.$$

Transformation of coordinates in the flat background

$$t = x^1\epsilon + x^4, \quad x = x^2, \quad y = x^3, \quad z = x^1\epsilon - x^4, \tag{16}$$

yields components of the flat metric in the group coordinates as

$$F_{\mu\nu}(x) = \begin{pmatrix} 0 & 0 & 0 & -2\epsilon \\ 0 & 1 & 0 & 0 \\ 0 & 0 & 1 & 0 \\ -2\epsilon & 0 & 0 & 0 \end{pmatrix}.$$

Dual background in this case is

$$\widetilde{F}_{\mu\nu}(\tilde{x}) = \begin{pmatrix} 0 & 0 & 0 & -\frac{1}{2\epsilon} \\ 0 & 1 & 0 & \frac{\tilde{x}_3}{2\epsilon} \\ 0 & 0 & 1 & -\frac{\tilde{x}_2}{2\epsilon} \\ -\frac{1}{2\epsilon} & -\frac{\tilde{x}_3}{2\epsilon} & \frac{\tilde{x}_2}{2\epsilon} & -\frac{\tilde{x}_2^2+\tilde{x}_3^2}{4\epsilon^2} \end{pmatrix},$$

and transformation to Brinkmann coordinates

$$\tilde{x}_1 = -v, \qquad \tilde{x}_2 = z_3, \qquad \tilde{x}_3 = z_4, \qquad \tilde{x}_4 = 2\epsilon\, u, \tag{17}$$

brings the dual metric to the homogeneous and isotropic form

$$ds^2 = 2dudv - ({z_3}^2 + {z_4}^2)\, du^2 + d{z_3}^2 + d{z_4}^2. \tag{18}$$

Torsion in Brinkmann coordinates is constant

$$H = -2\, du \wedge dz_3 \wedge dz_4, \tag{19}$$

and the dilaton is

$$\Phi(u) = c_1 + c_2\, u.$$

To find general solution of field equations of the dual sigma model with torsion we have to express the coordinates $\tilde{x}_\mu$ in terms of x^ν and $\tilde{h}_k$. Solution of the equation (9) is

$$\tilde{x}_1 = \tilde{h}_1 + x^2\tilde{h}_3 - x^3\tilde{h}_2, \qquad \tilde{x}_2 = \tilde{h}_2 \cos x^1 - \tilde{h}_3 \sin x^1,$$
$$\tilde{x}_4 = \tilde{h}_4, \qquad \tilde{x}_3 = \tilde{h}_2 \sin x^1 + \tilde{h}_3 \cos x^1.$$

Combining this with (17) and (16), we find general solution of field equations of the sigma model with metric (18) and torsion (19) as

$$\begin{aligned} U(\tau,\sigma) &= \frac{\tilde{h}_4(\tau,\sigma)}{2\epsilon}, \\ V(\tau,\sigma) &= -\tilde{h}_1(\tau,\sigma) - \tilde{h}_3(\tau,\sigma)W^x(\tau,\sigma) + \tilde{h}_2(\tau,\sigma)W^y(\tau,\sigma), \\ Z_3(\tau,\sigma) &= \cos(\Omega(\tau,\sigma))\, \tilde{h}_2(\tau,\sigma) - \sin(\Omega(\tau,\sigma))\, \tilde{h}_3(\tau,\sigma), \\ Z_4(\tau,\sigma) &= \cos(\Omega(\tau,\sigma))\tilde{h}_3(\tau,\sigma) + \sin(\Omega(\tau,\sigma))\tilde{h}_2(\tau,\sigma), \end{aligned}$$

where $W^I(\tau,\sigma)$ are solutions of the wave equations,

$$\Omega(\tau,\sigma) = \frac{W^t + W^z}{2\epsilon},$$

and $\tilde{h}_\mu$ are solutions of the PDEs (6), (7),

$$\tilde{h}_1 = \gamma_1 + \int \left[\epsilon\,(\partial_\tau W^t - \partial_\tau W^z) + W^x\partial_\tau W^y - W^y\partial_\tau W^x\right] d\sigma,$$

$$\tilde{h}_2 = \gamma_2 - \int \partial_\tau W^x\, d\sigma, \qquad \tilde{h}_3 = \gamma_3 - \int \partial_\tau W^y\, d\sigma,$$

$$\tilde{h}_4 = \gamma_4 + \int \left(\partial_\tau W^t + \partial_\tau W^z\right)\, d\sigma.$$

Journal of Physics: Conference Series **597** (2015) 012043 doi:10.1088/1742-6596/597/1/012043

String-type solutions in the light-cone gauge are obtained if we choose

$$W^t(\tau,\sigma) + W^z(\tau,\sigma) = 2\,\epsilon\,\kappa\,\sigma,$$

$$W^x(\tau,\sigma) = \sum_{n=-\infty}^{\infty} e^{2in\sigma} \int Z_3^n(\tau)(\kappa\sin(\kappa\sigma) - 2i\,n\cos(\kappa\sigma)) \\ - Z_4^n(\tau)(\kappa\cos(\kappa\sigma) + 2i\,n\sin(\kappa\sigma))\,d\tau,$$

$$W^y(\tau,\sigma) = \sum_{n=-\infty}^{\infty} e^{2in\sigma} \int Z_3^n(\tau)(\kappa\cos(\kappa\sigma) + 2i\,n\sin(\kappa\sigma)) \\ + Z_4^n(\tau)(\kappa\sin(\kappa\sigma) - 2i\,n\cos(\kappa\sigma))\,d\tau.$$

where $Z_3^n(\tau)$ and $Z_4^n(\tau)$ solve the system of differential equations

$$Z_3^{n\prime\prime}(\tau) + \left(4\,n^2 + \kappa^2\right) Z_3^n(\tau) - 4i\,n\,\kappa Z_4^n(\tau) = 0,$$

$$Z_4^{n\prime\prime}(\tau) + \left(4\,n^2 + \kappa^2\right) Z_4^n(\tau) + 4i\,n\,\kappa Z_3^n(\tau) = 0.$$

5. Results for other subalgebras

The classification of subalgebras of the Poincare algebra in [13] was carried out up to group of inner automorphisms of the connected component of the Poincare group (proper orthochronous Poincare transformations). Up to this equivalence, there are 35 four–dimensional subalgebras of the Poincare algebra spanned by Killing vectors (15). Only 17 of them act transitively and freely on the flat manifold and can be used for nonabelian T–duality.

Eleven of the duals are plane-parallel wave backgrounds whose metrics can be brought to the Brinkmann form

$$ds^2 = 2dudv - [K_3(u){z_3}^2 + K_4(u){z_4}^2]du^2 + d{z_3}^2 + d{z_4}^2,$$

and the torsion is

$$H = dB = H(u)\,du \wedge dz_3 \wedge dz_4.$$

Functions $K_3(u), K_4(u), H(u)$ acquire one of the following forms

$$K_3(u) = K_4(u) = 1, \qquad H(u) = -2, \tag{20}$$

$$K_3(u) = \frac{3}{(u^2+1)^2},\quad K_4(u) = -\frac{(2u^2-1)}{(u^2+1)^2},\quad H(u) = \pm\tfrac{2}{u^2+1}, \tag{21}$$

$$K_3(u) = 2\,\mathrm{sech}^2(u),\quad K_4(u) = 2\,\delta\,\mathrm{sech}^2(u),\ \delta = 0,1, \qquad H(u) = 0, \tag{22}$$

$$K_3(u) = -2\,\mathrm{csch}^2(u),\quad K_4(u) = -2\,\delta\,\mathrm{csch}^2(u),\ \delta = 0,1, \qquad H(u) = 0, \tag{23}$$

$$K_3(u) = K_4(u) = \frac{(1 + 2\beta^2\mathrm{sech}^2(u))}{\beta^2}, \qquad H(u) = -\tfrac{2}{\beta}, \tag{24}$$

$$K_3(u) = K_4(u) = \frac{(1 - 2\beta^2\mathrm{csch}^2(u))}{\beta^2}, \qquad H(u) = -\tfrac{2}{\beta}, \tag{25}$$

$$\begin{aligned} K_3(u) &= -\frac{\operatorname{sech}^4(u)\left(2\left(\beta^2+1\right)\sinh^2(u)-\beta^2\right)}{\left(\tanh^2(u)+\beta^2\right)^2}, \\ K_4(u) &= \frac{\beta^2\operatorname{sech}^4(u)\left(2(\beta^2+1)\cosh^2(u)+1\right)}{\left(\tanh^2(u)+\beta^2\right)^2}, \\ H(u) &= \frac{2\beta}{\beta^2\cosh^2(u)+\sinh^2(u)}, \end{aligned} \tag{26}$$

$$\begin{aligned} K_3(u) &= \frac{\operatorname{csch}^4(u)\left(2(\beta^2+1)\cosh^2(u)+\beta^2\right)}{\left(\coth^2(u)+\beta^2\right)^2}, \\ K_4(u) &= -\frac{\beta^2\operatorname{csch}^4(u)\left(2(\beta^2+1)\sinh^2(u)-1\right)}{\left(\coth^2(u)+\beta^2\right)^2}, \\ H(u) &= -\frac{2\beta}{\cosh^2(u)+\beta^2\sinh^2(u)}, \end{aligned} \tag{27}$$

where $\beta \in \mathbb{R} \smallsetminus \{0\}$.

Even though the B-fields obtained by T-duality are usually not of the form $B = B_i(u)\, du \wedge dz_i$, they are gauge equivalent to

$$B' = H(u)\, du \wedge (z_3 dz_4 - z_4 dz_3),$$

and corresponding sigma models are exactly conformal [3].

Except for (24), (25), these pp-wave backgrounds can be transformed to the gauged WZW background forms (1) by the standard transformation from Brinkmann to Rosen coordinates. In most of the transformed backgrounds the functions g_1, g_2 acquire forms $g_1(u) = 1$ and g_2 any of the functions (2), but some other combinations of functions (g_1, g_2) also arise, namely $(u^{-2}, \tanh^2 u)$, $(u^{-2}, \coth^2 u)$, $(\tanh^2 u, \tanh^2 u)$ and $(\coth^2 u, \coth^2 u)$.

Consequently, the pp-waves of the form (1) are duals of the flat metric not only for $g_1(u) = 1$ and $g_2(u) = u^2$, as mentioned in the Introduction, but also for many other combinations of functions g_1, g_2 from the set (2).

Besides the pp-waves, we get dual metrics with non-vanishing scalar curvature

$$ds^2 = -dy_1{}^2 + dy_2{}^2 + \frac{y_1^2}{y_1^2+\alpha^2}\, dy_3{}^2 + \frac{1}{y_1^2+\alpha^2}\, dy_4{}^2, \tag{28}$$

$$H = -\frac{2y_1\alpha}{\left(y_1^2+\alpha^2\right)^2}\, dy_1 \wedge dy_3 \wedge dy_4,$$

$$ds^2 = \frac{1}{y_3^2-\alpha^2}\, dy_1{}^2 + \frac{y_3^2}{\alpha^2-y_3^2}\, dy_2{}^2 + dy_3{}^2 + dy_4{}^2, \tag{29}$$

$$H = \frac{2y_3\alpha}{\left(y_3^2-\alpha^2\right)^2}\, dy_1 \wedge dy_2 \wedge dy_3$$

$$ds^2 = -dy_1{}^2 + dy_2{}^2 + \frac{y_2^2}{y_2^2+\alpha^2}\, dy_3{}^2 + \frac{1}{y_2^2+\alpha^2}\, dy_4{}^2 \tag{30}$$

Journal of Physics: Conference Series **597** (2015) 012043 doi:10.1088/1742-6596/597/1/012043

$$H = \frac{2y_2\alpha}{\left(y_2^2 + \alpha^2\right)^2} dy_2 \wedge dy_3 \wedge dy_4$$

They are obtained as non-Abelian duals with respect to S_{11}, S_{18}, S_{19}. Note that isomorphic (but non-equivalent under proper ortochronous Poincare transformation) subalgebras S_{17} respectively S_{18}, S_{19} lead to backgrounds with vanishing and non vanishing curvature.

The metrics (28)–(30) remind black hole [15] and cosmological backgrounds [16] rewritten in [2] to diagonal forms depending again on particular functions g_1, g_2. The difference from (28) – (30) is in these functions.

References

[1] Papadopoulos G, Russo J G and Tseytlin A A 2003 Solvable model of strings in a time-dependent plane-wave background *Class. Quant. Grav.* **20** 969 (*Preprint* hep-th/0211289)
[2] Tseytlin A A 1995 Exact solutions of closed string theory *Class. Quant. Grav.* **12** 2365 (*Preprint* hep-th/9505052)
[3] Blau M, O'Loughlin M, Papadopoulos G and Tseytlin A A 2003 Solvable models of strings in homogeneous plane wave backgrounds *Nucl. Phys.* B **673** 57 (*Preprint* hep-th/0304198)
[4] Horowitz G T and Steif A R 1990 Strings in strong gravitational fields *Phys. Rev.* D 42 1950
[5] de Vega H J and Sanchez N 1992 Strings falling into space-time singularities *Phys. Rev.* D **45** 2783
[6] Penrose R 1976 in: *Differential Geometry and Relativity* (Dordrecht: Reidel)
[7] Gueven R 2000 Plane wave limits and T-duality *Physics Letters* B **482** 255 (*Preprint* hep-th/0005061)
[8] Sfetsos K and Tseytlin A A 1994 Four Dimensional Plane wave string solutions with coset CFT description *Nucl. Phys.* B **427** 245 (*Preprint* hep-th/9404063)
[9] Tseytlin A A 1994 Exact string solutions and duality (*Preprint* hep-th/9407063)
[10] de la Ossa X C and Quevedo F 1993 Duality symmetries from non-abelian isometries in string theories *Nucl. Phys.* B **403** 377 (*Preprint* hep-th/9210021)
[11] Klimčík C and Ševera P 1995 Dual non-Abelian duality and the Drinfeld double *Phys. Lett.* B **351** 455 (*Preprint* hep-th/9502122)
[12] Klimčík C 1996 Poisson–Lie T-duality *Nucl. Phys.* B (Proc. Suppl.) **46** 116 (*Preprint* hep-th/9509095)
[13] Patera J, Sharp R T, Winternitz P and Zassenhauss H 1976 Subgroups of the Poincar group and their invariants *J. Math. Phys.* **17** 977
[14] Hlavatý L and Petr I 2014 New solvable sigma models in plane-parallel wave background *Int. J. Mod. Phys.* A **29** 1450009 (*Preprint* arXiv:1308.0153)
[15] Hořava P 1992 Some exact solutions of string theory in four and five dimensions *Phys. Lett.* B **278** 101 (*Preprint* hep-th/9110067).
[16] Nappi C and Witten E 1992 A closed expanding universe in string theory *Phys. Lett.* B **293** 309 (*Preprint* hep-th/9206078)

Journal of Physics: Conference Series **597** (2015) 012044 doi:10.1088/1742-6596/597/1/012044

On some differential transformations of hypergeometric equations

M N Hounkonnou[1] **and A Ronveaux**[1,2]

[1]International Chair in Mathematical Physics and Applications (ICMPA-UNESCO Chair), University of Abomey-Calavi, 072 B.P. 50 Cotonou, Republic of Benin

[2]Département de Mathématiques, Catholic University of Louvain, 2, Chemin du Cyclotron, B-1348, Louvain-la-Neuve, Belgium

E-mail: norbert.hounkonnou@cipma.uac.bj, andre.michel.ronveaux@gmail.com

Abstract. Many algebraic transformations of the hypergeometric equation $\sigma(x)z''(x) + \tau(x)z'(x) + lz(x) = 0$, where σ, τ, l are polynomial functions of degrees 2 (at most), 1, 0, respectively, are well known. Some of them involve $x = x(t)$, a polynomial of degree r, in order to recover the Heun equation, extension of the hypergeometric equation by one more singularity. The case $r = 2$ was investigated by K. Kuiken (see 1979 SIAM J. Math. Anal. **10** (3) 655-657) and extended to $r = 3, 4, 5$ by R. S. Maier (see 2005 J. Differ. Equat. **213** 171 - 203). The transformations engendered by the function $y(x) = A(x)z(x)$, also very popular in mathematics and physics, are used to get from the hypergeometric equation, for instance, the Schroedinger equation with appropriate potentials, as well as Heun and confluent Heun equations. This work addresses a generalization of Kimura's approach proposed in 1971, based on differential transformations of the hypergeometric equations involving $y(x) = A(x)z(x) + B(x)z'(x)$. Appropriate choices of A(x) and B(x) permit to retrieve the Heun equations as well as equations for some exceptional polynomials. New relations are obtained for Laguerre and Hermite polynomials.

1. Introduction

In 1971, Kimura [1] investigated in detail all Fuchsian differential equations $F\{y(x)\} = 0$ reducible to hypergeometric equations $H\{z(x)\} = 0$ by a linear transformation $y(x) = P_0(x)z(x) + P_1(x)z'(x)$, with P_0 and P_1 rational functions of x. To compute $P_0(x)$ and $P_1(x)$, he assumed the same set of singularities and the same monodromy group for $H\{y(x)\} = 0$ and $F\{z(x)\} = 0$, giving information on the structures and properties of functions P_0 and P_1. The equation $F\{y\} = y'' + A_1(x)y' + A_2(x)y = 0$, $(y = y(x))$, being Fuchsian and written as a generalized Heun equation, contains regular singular points with many parameters, (not all free), in order to eliminate logarithmic situations and irreducibility (equation not factorizable).

Many theorems and properties are provided to compute $P_0(x)$ and $P_1(x)$ in particular situations, but without using the coupled differential equations satisfied by P_0 and P_1.

The aim of this work is to reverse this approach, making it possible to eliminate the Fuchsian constraints. We proceed as follows: inject an *arbitrary* linear transformation $y = A(x)z + B(x)z'$ into the hypergeometric equation and build a second order differential equation for y, (not always Fuchsian), depending on the two differentiable functions $A(x)$ and $B(x)$, polynomial or not.

Exceptional orthogonal polynomials (EOPs) like exceptional Jacobi, exceptional Laguerre, and so on, (see [2] and references therein), can be retrieved from the same linear transformation given a long time ago in the above mentioned seminal paper by Kimura.

Recall that the notion of the exceptional (X_l) orthogonal polynomials was introduced in 2008 by Gomez-Ullate, Kamran and Milson [3] in the framework of Sturm Liouville theory. These orthogonal polynomials are exceptional in the sense that they start at degree $l(l \geq 1)$ instead of degree 0 constant term, thus avoiding restrictions of Bochner's theorem and they satisfy second-order differential equations. These authors constructed the lowest examples, the X_1 Laguerre and X_1 Jacobi polynomials explicitly. A quantum mechanical reformulation with shape-invariant potentials was given by Quesne and collaborators [4, 5] within the framework of one-dimensional quantum mechanics. The merit of quantum mechanical reformulation resides in the fact that the orthogonality and completeness of the obtained eigenfunctions are guaranteed. Besides, the well established solution mechanism of shape invariance combined with the Crum's method [6], or the so-called factorization method [7], or the susy quantum mechanics [8] is available.

Then, two sets of infinitely many shape invariant potentials, the deformed radial oscillator potentials and the deformed trigonometric/hyperbolic Darboux-Pöschl-Teller (DPT) potentials, and the corresponding X_l Laguerre and Jacobi polynomials ($l = 1, 2, \ldots \infty$) were investigated by Sasaki *et al* [9, 10, 11] in 2009.

This paper is organised as follows. In section 2, we provide a quick overview on hypergeometric equations and Heun's equations. Section 3 deals with our generalization of Kimura's approach, based on a differential transformation of hypergeometric equations. Known special functions and orthogonal polynomials are retrieved from the presented general formalism. Concluding remarks are presented in section 4.

2. Hypergeometric differential equation and Heun's equations

The hypergeometric differential equation

$$\frac{d^2z}{dx^2} + \left(\frac{\gamma}{x} + \frac{\delta}{x-1}\right)\frac{dz}{dx} + \frac{\alpha\beta}{x(x-1)}z = 0 \tag{1}$$

is characterized as a Fuchsian differential equation of the second order which has three singularities at $x = 0, 1, \infty$ corresponding to the exponents $(0, 1-\gamma); (0, 1-\delta)$ and (α, β) respectively, satisfying the Fuchs relation

$$\alpha + \beta - \gamma - \delta + 1 = 0. \tag{2}$$

This equation has been studied in detail and extended in several directions by many mathematicians. As one of those extensions we are led in a natural way to a second order equation of Fuchsian type whose singularities and exponents are given by the following Riemannian scheme

$$\begin{matrix} 0 & 1 & a_1 & \cdots & a_k & \infty \\ 0 & 0 & 0 & \cdots & 0 & \alpha \\ 1-\gamma & 1-\delta & 1-\epsilon_1 & \cdots & 1-\epsilon_k & \beta, \end{matrix}$$

where the exponents are connected by Fuchs's relation

$$\alpha + \beta - \gamma - \delta - \epsilon_1 - \cdots - \epsilon_k + 1 = 0.$$

The equation defined by these data contains just k accessory parameters, denoted by $\rho_1, \cdots, \rho_k$, and can be written in the form:

$$\frac{d^2y}{dx^2} + \left(\frac{\gamma}{x} + \frac{\delta}{x-1} + \sum_{i=1}^{k} \frac{\epsilon_i}{x-a_i}\right)\frac{dy}{dx} + \frac{\alpha\beta x^k + \rho_1 x^{k-1} + \cdots + \rho_k}{x(x-1)\prod_{i=1}^{k}(x-a_i)}y = 0. \tag{3}$$

In the case when $k = 1$, the equation is often called Heun's equation that we write in the following form [12, 13]:

$$\frac{d^2\omega}{dz^2} + \left(\frac{\gamma}{z} + \frac{\delta}{z-1} + \frac{\epsilon}{z-a}\right)\frac{d\omega}{dz} + \frac{\alpha\beta z - q}{z(z-1)(z-a)}\omega = 0, \tag{4}$$

with the Fuchs relation:

$$\alpha + \beta + 1 = \gamma + \delta + \epsilon. \tag{5}$$

This equation has four regular singularities at $0, 1, a, \infty$ with corresponding exponents $\{0, 1-\gamma\}, \{0, 1-\delta\}, \{0, 1-\epsilon\}, \{\alpha, \beta\}$, respectively. All other homogeneous linear differential equations of the second order having four regular singularities in the extended complex plane $\mathbb{C} \cup \{\infty\}$, can be transformed into (4). The parameters play different roles:

- a is the singularity parameter
- $\alpha, \beta, \gamma, \delta, \epsilon$ are exponent parameters
- q is the accessory parameter.

The total number of free parameters is six.

Making the substitution $\omega(z) = z^{-\gamma/2}(z-1)^{-\delta/2}(z-a)^{-\epsilon/2}W(z)$, (4) can be transformed into the normal form:

$$\frac{d^2W}{dz^2} = \left(\frac{A}{z} + \frac{B}{z-1} + \frac{C}{z-a} + \frac{D}{z^2} + \frac{E}{(z-1)^2} + \frac{F}{(z-a)^2}\right)W, \tag{6}$$

$$A + B + C = 0,$$

$$A = -\frac{\gamma\delta}{2} - \frac{\gamma\epsilon}{2a} + \frac{q}{a}, \qquad B = \frac{\gamma\delta}{2} - \frac{\delta\epsilon}{2(a-1)} - \frac{q-\alpha\beta}{a-1},$$

$$C = \frac{\gamma\epsilon}{2a} + \frac{\delta\epsilon}{2(a-1)} - \frac{a\alpha\beta - q}{a(a-1)}, \qquad D = \frac{1}{2}\gamma\left(\frac{1}{2}\gamma - 1\right),$$

$$E = \frac{1}{2}\delta\left(\frac{1}{2}\delta - 1\right), \qquad F = \frac{1}{2}\epsilon\left(\frac{1}{2}\epsilon - 1\right).$$

Confluent forms of Heun's differential equation (4) arise when two or more of the regular singularities merge to form an irregular singularity. This is analogous to the derivation of the confluent hypergeometric equation. There are four standard forms, as follows (see [14] and references therein):

(i) Confluent Heun Equation

$$\frac{d^2y(x)}{dx^2} + \left(\alpha + \frac{\beta+1}{x} + \frac{\gamma+1}{x-1}\right)\frac{dy(x)}{dx} + \\ + \frac{(2\delta + \alpha(\beta+\gamma+2))x + 2\eta + \beta + (\gamma-\alpha)(\beta+1)}{2x(x-1)}y(x) = 0. \quad (7)$$

It has regular singularities at $x = 0$ and 1, and an irregular singularity of rank 1 at $x = \infty$. Mathieu functions, spheroidal wave functions and Coulomb spheroidal functions are special cases of solutions of the confluent Heun equation.

(ii) Doubly-confluent Heun equation

$$x^2\frac{d^2y(x)}{dx^2} + \alpha\left(x + \frac{1}{x}\right)\left(x\frac{dy(x)}{dx}\right) + \left[\left(\beta_1 + \frac{1}{2}\right)\alpha x + \left(\frac{\alpha^2}{2} - \gamma\right) \\ + \left(\beta_{-1} - \frac{1}{2}\right)\frac{\alpha}{x}\right]y(x) = 0 \quad (8)$$

has irregular singularities at $x = 0$ and ∞, each of rank 1.

(iii) Biconfluent Heun equation

$$\frac{d^2y(x)}{dx^2} + \left(\frac{\alpha+1}{x} - \beta - 2x\right)\frac{dy(x)}{dx} + \left(\gamma - \alpha - 2 - \frac{\delta + (\alpha+1)\beta}{2x}\right)y(x) = 0 \quad (9)$$

has a regular singularity at $x = 0$, and an irregular singularity at ∞ of rank 2.

(iv) Triconfluent Heun equation

$$\frac{d^2y(x)}{dx^2} - (\gamma + 3x^2)\frac{dy(x)}{dx} + [\alpha + (\beta - 3)x]y(x) = 0 \quad (10)$$

has one singularity, an irregular singularity of rank 3 at $x = \infty$.

3. Differential transformation of hypergeometric equations

In this section, we consider the following hypergeometric equation:

$$\sigma(x)z''(x) + \tau(x)z'(x) + \lambda z(x) = 0, \quad (11)$$

and transformation:

$$y(x) = A(x)z(x) + B(x)z'(x), \quad (12)$$

where $\sigma \equiv \sigma(x)$ is a polynomial of degree less or equal to 2, $\tau \equiv \tau(x)$ a polynomial of degree exactly equal to 1, λ is a constant; $A = A(x)$ and $B = B(x)$ are polynomials of degrees r and s, respectively. Then, we build a general linear second order ordinary differential equation $\mathcal{L}_2[y(x)] = 0$ satisfied by the function y, polynomial or not, and investigate a set of situations. For that, we proceed as follows:

- First derive (12) with respect to x, using (11), and get:

$$\sigma y' = \bar{A}z + \bar{B}z' \quad (13)$$

where $\bar{A} = \sigma A' - B\lambda$, $\bar{B} = \sigma A + \sigma B' - \tau B$.

- Derive (13) with respect to x and obtain

$$\sigma(\sigma y')' = \bar{\bar{A}} z + \bar{\bar{B}} z', \tag{14}$$

where $\bar{\bar{A}} = \sigma \bar{A}' - \bar{B}\lambda$, $\bar{\bar{B}} = \sigma \bar{A} + \sigma \bar{B}' - \tau \bar{B}$.

- Rewrite the coefficients $\bar{\bar{A}}$ and $\bar{\bar{B}}$ of (14) in terms of σ, τ, λ, A, B and their derivatives as follows:

$$\begin{aligned} \bar{\bar{A}} &= A''\sigma^2 + \sigma A'\sigma' - 2\lambda\sigma B' - \lambda A\sigma + \lambda B\tau & (15) \\ \bar{\bar{B}} &= 2A'\sigma^2 - \lambda\sigma B + \sigma A\sigma' + \sigma^2 B'' + \sigma B'\sigma' \\ & \quad -2\sigma B'\tau - \sigma B\tau' - \tau A\sigma + B\tau^2. & (16) \end{aligned}$$

- Then, using (13) and (14) we form the following determinantal equation giving the ordinary differential equation satisfied by y :

$$\mathcal{L}_2[y(x)] = \begin{vmatrix} y & A & B \\ \sigma y' & \bar{A} & \bar{B} \\ \sigma(\sigma y')' & \bar{\bar{A}} & \bar{\bar{B}} \end{vmatrix} = 0, \tag{17}$$

or, equivalently,

$$\hat{\mathcal{L}}_2[y(x)] = \begin{vmatrix} y & A & B \\ \sigma y' & \bar{A} & \bar{B} \\ \sigma^2 y'' & \hat{A} & \hat{B} \end{vmatrix} = 0 \tag{18}$$

with

$$\begin{aligned} \hat{A} &= \sigma \bar{A}' - \bar{B}\lambda - \sigma'\bar{A} & (19) \\ \hat{B} &= \sigma \bar{A} + \sigma \bar{B}' - \bar{B}(\tau + \sigma'). & (20) \end{aligned}$$

The following observations are in order:

(i) The coefficient of y'' is $\sigma^2(A\bar{B} - B\bar{A})$ which introduces new singular points depending on the degrees r and s of the polynomials A and B. In general the obtained ODE is no more Fuchsian.

(ii) This general formulation does not take into account the possible logarithmic solutions and apparent singularities for $\mathcal{L}_2[y(x)] = 0$ which strongly restrict the choice of the polynomials A and B. Such a freedom allows to cover a larger class of equations with $r+s+2$ parameters such as Heun, generalized Heun equations and their various confluencies and to propose some solutions.

(iii) An obvious way to remain in the Fuchsian's class is to choose $B(x) = \sigma(x)$ or more generally $B(x) = S(x)\sigma(x)$. Such a simplified choice has been already considered in the work by Kimura [1] and in recent literature on exceptional polynomials (see [2] and references therein).

Let us investigate some relevant particular cases in the sequel.

3.1. Basic choice: $A(x)$ is arbitrary and $B(x) = \sigma(x)$

For arbitrary $A(x)$ and $B(x) = \sigma(x)$, the equation (17) can be transformed into the following second order differential equation:

$$\mathcal{L}_2[y(x)] = \sigma P y'' + Q y' + R y = 0 \tag{21}$$

where

$$P \equiv P(x) = A^2 + A(\sigma' - \tau) + \sigma(\lambda - A') \tag{22}$$

$$\begin{aligned} Q \equiv Q(x) &= A''\sigma^2 + \sigma(A\tau' + \tau\lambda - \lambda\sigma' - A\sigma'' - 2AA') \\ &\quad + A\tau(\sigma' + A - \tau) \end{aligned} \tag{23}$$

$$\begin{aligned} R \equiv R(x) &= \sigma[A''(\tau - A - \sigma') + A'(2A' + \sigma'' - 3\lambda - \tau') + \lambda(\lambda + \tau' - \sigma'')] \\ &\quad + \sigma'[2\lambda A + \lambda\sigma' - \lambda\tau - \tau A'] + (A - \tau)(\lambda A - A'\tau). \end{aligned} \tag{24}$$

Note that by using appropriate choices of the coefficients in the polynomial A, we can reduce the above equation into hypergeometric equations and generate known or unknown contiguous relations between the solutions $y(x)$. For instance, setting:

$$\sigma(x) = B(x) = 1 - x^2,\ \tau(x) = \beta - \alpha - (\alpha + \beta + 2)x \quad \text{and} \quad \lambda = (n-1)(n + \alpha + \beta),\ n \in \mathbb{N}, \tag{25}$$

then, the equation (21), after simplification, coincides with the hypergeometric equation satisfied by the Jacobi polynomials $P_{n-1}^{(\alpha,\beta)}(x)$:

$$\begin{aligned} \mathcal{L}_2[y(x)] &= (1 - x^2)y''(x) + [\beta - \alpha - (\alpha + \beta + 2)x]y'(x) \\ &\quad + (n-1)(n + \alpha + \beta)y(x) = 0 \end{aligned} \tag{26}$$

giving $y(x) = P_{n-1}^{(\alpha,\beta)}(x)$, polynomial of degree $n - 1$, as solution. One can then use the well known contiguous relation [15]:

$$\begin{aligned} &(2n + \alpha + \beta)(1 - x^2)\frac{d}{dx}P_n^{(\alpha,\beta)}(x) - n[\alpha - \beta - (2n + \alpha + \beta)x]P_n^{(\alpha,\beta)}(x) \\ &= 2(n + \alpha)(n + \beta)P_{n-1}^{(\alpha,\beta)}(x) \end{aligned} \tag{27}$$

to find the polynomial coefficient $A(x)$ of $P_n^{(\alpha,\beta)}(x)$. Of course if λ is arbitrary, non polynomial solutions like the second kind solutions of (11) can be exploited to generate similar contiguous relations for the $z(x)$ as for the polynomial solutions $y(x)$.

3.2. Second choice: A and B are constant

Let us consider now the trivial case where A and B are constants. Then, the equation (17) can be reduced to a simpler second order differential equation as follows:

$$\mathcal{L}_2[y(x)] = \sigma P y'' + Q y' + R y = 0 \tag{28}$$

where

$$P \equiv P(x) = A(A\sigma - \tau B) + \lambda B^2 \tag{29}$$

$$Q \equiv Q(x) = \sigma(\tau A^2 + AB\tau') + (\lambda B^2\tau + \lambda\sigma' B^2 - AB\tau^2 - \sigma' AB\tau) \tag{30}$$

$$R \equiv R(x) = \lambda A^2\sigma + (\lambda^2 B^2 - \lambda AB\sigma' + \lambda B^2\tau' - \lambda BA\tau). \tag{31}$$

This equation is also not hypergeometric in general, but suitable choices of the quantities σ, τ, λ and A, B can also permit to deduce relevant contiguous relations. For instance, we retrieve the following particular equations:

- Laguerre equation
  (i) If $\sigma(x) = x$, $\tau(x) = 1 + \alpha - x$ and $\lambda = n$, the equation (28) yields the hypergeometric equation respected by Laguerre polynomials $L_n^{(\alpha)}(x)$.
  (ii) The choice $A = 1 = -B$ gives again a Laguerre polynomial for $y(x)$ confirming the well known relation [15]

$$L_n^{(\alpha)}(x) - [L_n^{(\alpha)}(x)]' = [L_{n+1}^{(\alpha)}(x)]' = L_n^{(\alpha+1)}(x). \tag{32}$$

- Hermite equation
  If $\sigma = 1$ and $\tau = -2x$, the equation (28) takes the form:

$$\begin{aligned}\mathcal{L}_2[y(x)] &= y''(A^2 + B^2\lambda + 2ABx) - y'[2AB + 2x(A^2 + B^2\lambda) + 4x^2AB] \\ &\quad + y\lambda[B^2(\lambda - 2) + A(A + 2xB)] = 0\end{aligned} \tag{33}$$

  which can be further simplified to give:
  (i) For the particular case of $B = 0$ with arbitrary A, the well known Hermite equation:

$$\mathcal{L}_2[y(x)] = y'' - 2xy' + \lambda y = 0. \tag{34}$$

  (ii) In the opposite case, when $A = 0$ and B is arbitrary, the equation (33) is transformed into a modified Hermite equation

$$\mathcal{L}_2[y(x)] = y'' - 2xy' + \bar{\lambda} y = 0 \text{ with } \bar{\lambda} = \lambda - 2. \tag{35}$$

3.3. Link with the Heun equations and exceptional Jacobi polynomials
Besides, Heun general equation, its confluent forms as well as exceptional Jacobi polynomials can also be retrieved from the general approach developed here.

3.3.1. General Heun's equation

- With the choice $A = ax + b$ and $\sigma = x^2 - x$, the degrees of P, Q, R in equation (21) are 2, 3 and 2, respectively. Then, the HEUN equation

$$\frac{d^2y}{dx^2}\left[\frac{\gamma}{x} + \frac{\delta}{x-1} + \frac{\epsilon}{x-\mu}\right]\frac{dy}{dx} + \frac{\alpha\beta x + \rho}{x(x-1)(x-\mu)}y = 0, \tag{36}$$

  is recovered when the coefficients a, b and τ are chosen so that $P(x)$ reduces to $x - \mu$, and $Q(x) = (x - \mu)Q_1(x)$ and $R(x) = (x - \mu)R_1(x)$, where $Q_1(x)$ and $R_1(x)$ are polynomials.

- Other appropriate choices also produce general Heun and confluent Heun equations [13] with explicit solutions $y(x) = A(x)z(x) + B(x)z'(x)$. Kimura gives [1] solutions in two relevant cases:
  (i) for $\epsilon = -1$ with $A(x)$ of degree 1 and $B(x) = x(x-1)$,
  and
  (ii) for $\epsilon = -2$ with $A(x)$ of degree 2 and $B(x)$ of degree 3.

  The Kimura method is very nice, but complicated and restrictive, the confluent equations being excluded. Its intrinsic complexity resides in the step $\epsilon = -1$ to -2 increasing the degree of polynomials $A(x)$ and $B(x)$. In our approach, the problem is entirely algebraic, although not excluding, of course, also some difficulties.

3.3.2. Confluent Heun equation

- Setting $\sigma = 2x$, $A = B = 1$, the relations (29)-(31) give:

$$\tau = x;\ P(x) = x - 1;\ Q(x) = x^2 - x - 2 \quad \text{and} \quad R(x) = -x + 2 \tag{37}$$

providing $\lambda = -1$ and the confluent Heun equation (7) with the coefficients

$$(\alpha, \beta, \gamma, \delta, \eta) = \left(\frac{1}{2}, 0, -2, -\frac{1}{2}, \frac{9}{4}\right).$$

3.3.3. Biconfluent Heun equation

- Taking $\tau = x$, $A = B = 2$, equations (29)-(31) yield:

$$\sigma = -\frac{1}{2};\ \lambda = \frac{1}{2};\ P(x) = -4x;\ Q(x) = -4x^2 - 2 \quad \text{and} \quad R(x) = -2x + 2 \tag{38}$$

leading to the biconfluent version of Heun equation (7) with the parameters

$$(\alpha, \beta, \delta, \gamma) = (-2, 0, -2, -1). \tag{39}$$

3.3.4. Exceptional χ_1-Jacobi polynomials Finally, let us mention that the exceptional χ_1−Jacobi polynomials investigated in [2] can be also easily retrieved from our method. Indeed, set, for $g, h \notin \{-1/2, -3/2, -5/2, \ldots\}$:

$$\zeta(\eta) = \frac{g-h}{2}\eta + \frac{g+h+1}{2}, \tilde{\zeta}(\eta) = \frac{g-h}{2}\eta + \frac{g+h+3}{2}$$

and consider the polynomials

$$A(\eta) = \frac{1}{k+h+\frac{1}{2}}\Big(h + \frac{1}{2}\Big)\tilde{\zeta}(\eta)$$

and

$$B(\eta) = \frac{1}{k+h+\frac{1}{2}}\Big(1+\eta\Big)\zeta(\eta)$$

of degrees 1 and 2, respectively. Let $z(\eta)$ be the Jacobi polynomial parametrized as [2]:

$$P_k(\eta) = \frac{(g+\frac{1}{2})_k}{k!}\sum_{j=0}^{k}\frac{(-k)_j(k+g+h+2)_j}{j!(g+\frac{1}{2})_j}\Big(\frac{1-\eta}{2}\Big)^j. \tag{40}$$

Then the transformation (12) gives the χ_1− Jacobi polynomials $y(\eta) \equiv \hat{P}_k(\eta)$ satisfying the following differential equation:

$$\begin{aligned}(1-\eta^2)y''(\eta) &+ \Big(h - g - (g+h+3)\eta - 2\frac{(1-\eta^2)\zeta'(\eta)}{\zeta(\eta)}\Big)y'(\eta) \\ &+\Big(-\frac{2(h+\frac{1}{2})(1-\eta)\tilde{\zeta}'(\eta)}{\zeta(\eta)} + k(k+g+h+2) + g - h\Big)y(\eta) = 0.\end{aligned} \tag{41}$$

Here also, transforming η into a new variable and assigning adequate relations between parameters lead to specific Heun's differential equations as shown, for instance, with the particular variable change $\eta = 1 - 2x$. For more details about such relations between Heun's equations and differential equation describing the exceptional Jacobi polynomials, see [2] and references therein.

4. Concluding remarks

This work addressed a generalization of Kimura's approach based on differential transformation of hypergeometric equations. This transformation involving $y(x) = A(x)z(x) + B(x)z'(x)$, where $z(x)$ is a solution of a hypergeometric equation, led to a general second order differential equation for y that encompasses various ODEs of mathematical physics as particular cases such as Heun equations and equations for exceptional polynomials. In particular, the equations (21) and (28), defined with the relations (22)-(24) and (29)-(31), permited us to generate new as well as known relations for Laguerre and Hermite polynomials.

Acknowledgments

The authors are grateful to anonymous referees for their useful comments and careful reading of the manuscript. MNH thanks Professor Tudor Ratiu, Chair of Geometric Analysis, and his collaborators for their hospitality during his stay as Visiting Professor at the Centre Interfacultaire Bernoulli (CIB) at the Ecole Polytechnique Fédérale de Lausanne (EPFL) where this paper was completed. This work is partially supported by the Abdus Salam International Centre for Theoretical Physics (ICTP, Trieste, Italy) through the Office of External Activities (OEA) - Prj-15. The ICMPA is in partnership with the Daniel Iagolnitzer Foundation (DIF), France.

References

[1] Kimura T 1971 *Funkcialaj Ekvacioj* **13** 213-232
[2] Takemura K 2012 *J. Phys. A: Math. Theor* **45** 085211
[3] Gómez-Ullate D, Kamran N and Milson R 2008 (*arXiv*:0805.3376 [math-ph]); Gómez-Ullate D, Kamran N and Milson R 2009 *J. Math. Anal. Appl.* **359** 352–367 (*arXiv*:0807.3939 [math-ph])
[4] Quesne C 2008 *J. Phys. A* **41** 392001
[5] Bagchi B, Quesne C and Roychoudhury R 2009 *Pramana J. Phys* **73** 337-347
[6] Crum M M 1955 *Quart. J. Math. Oxford Ser.*(2) **6** 121-127
[7] Infeld L and Hull T E 1951 *Rev. Mod. Phys.* **23** 21-68
[8] Cooper F, Khare A and Sukhatme U 1995 *Phys. Rep.* **251** 267-385
[9] Odake S, Sasaki R 2009 *Phys. Lett B.* **679**
[10] Ho C-L, Odake S and Sasaki R 2011 *SIGMA* **7** 107 (*arXiv*:0912.5447 [math-ph])
[11] Sasaki R, Tsujimoto S and Zhedanov A 2010 *J. Phys. A* **43** 315204
[12] Sleeman B D and Kuznetsov V B 2010 *Heun Functions in NIST Handbook of Mathematical Functions, Chapter 31*(Cambridge: Cambridge University Press)
[13] Ronveaux A 1995 *Heun's differential equations* (Oxford: Oxford University Press) (ed.)
[14] Hounkonnou M N and Ronveaux A 2009 *Appl. Math. Comput.* **209** 421-424
[15] Rainville E O 2002 *Special functions* (New York: Mcmillan)

Characteristic identities for Lie (super)algebras

Phillip S Isaac[1], **Jason L Werry and Mark D Gould**

School of Mathematics and Physics, The University of Queensland, St Lucia, QLD 4072, Australia

E-mail: [1]psi@maths.uq.edu.au

Abstract. We present an overview of characteristic identities for Lie algebras and superalgebras. We outline methods that employ these characteristic identities to deduce matrix elements of finite dimensional representations. To demonstrate the theory, we look at the examples of the general linear Lie algebras and Lie superalgebras.

1. Introduction

Characteristic identities have proven to be a useful tool for revealing important and explicit information about representations of Lie algebras and superalgebras. The aim of this article is to give an overview of how characteristic identities may be used for such purposes.

It was Dirac in 1936 [1] who first employed polynomial identities related to $sl(2)$ in the context of relativistically invariant wave equations. Later, Lehrer-Ilamed [2] noted more generally that n^2 elements of the enveloping algebra $U(g)$ of a Lie algebra g satisfy n^2 identities, which in some cases may be expressed as a single polynomial identity of degree n for an $n \times n$ matrix with entries from $U(g)$. There were subsequent works, for example, by Louck [3], Mukunda [4] and Louck and Galbraith [5] in which polynomial identities were encountered for a variety of Lie algebras. In [6–8], Bracken and Green established a general theory of characteristic identities for classical Lie algebras, which was soon followed by the work of O'Brien, Cant and Carey [9] in which the authors instituted a suitable algebraic formalism. In a series of papers [10–16], Gould made use of the characteristic identities to deduce information such as invariants, Wigner coefficients and matrix elements pertaining to the irreducible representations of the classical Lie algebras. It is also worth noting that these polynomial identities also extend to finite groups [17] and quantum groups [18, 19].

Following the work of Kac [20, 21], Jarvis and Green [22, 23] investigated characteristic identities and invariants related to the vector representation of the general linear, special linear and orthosymplectic Lie superalgebras. Other works soon followed that were related to other Lie superalgebras (e.g. [24]) and more general representations (e.g. [25]). It was not until the more recent work of the current authors [26, 27], however, in which the approach of Gould for Lie algebras was generalised to the case of Lie superalgebras, particularly for the general linear case.

2. Characteristic identities for Lie algebras

To demonstrate how the characteristic identities may be utilised, we exhibit the theory in the context of the Lie algebra $gl(n)$.

2.1. *Characteristic matrix*

We denote the generators of $gl(n)$ by a_{ij}, $i,j = 1,\ldots,n$ which satisfy the defining relations

$$[a_{ij}, a_{k\ell}] = \delta_{kj}a_{i\ell} - \delta_{i\ell}a_{kj}.$$

We may assemble the generators into a square $n \times n$ matrix $\mathcal{A}$ with (i,j) entry a_{ij}, i.e.

$$\mathcal{A} = \begin{pmatrix} a_{11} & a_{12} & \cdots & a_{1n} \\ a_{21} & a_{22} & \cdots & a_{2n} \\ \vdots & \vdots & & \vdots \\ a_{n1} & a_{n2} & \cdots & a_{nn} \end{pmatrix}$$

and define powers of $\mathcal{A}$ recursively by

$$\left(\mathcal{A}^{M+1}\right)_{ij} = \left(\mathcal{A}^{M}\right)_{ik} \mathcal{A}_{kj} = \mathcal{A}_{ik} \left(\mathcal{A}^{M}\right)_{kj}, \quad \left(\mathcal{A}^{0}\right)_{ij} = \delta_{ij}.$$

2.2. *Examples: $gl(1)$, $gl(2)$*

Casimir invariants are defined as $\sigma_1 = a_{ii}$, $\sigma_2 = a_{ij}a_{ji}$, $\sigma_3 = a_{ij}a_{jk}a_{ki}, \ldots, \sigma_M = \mathrm{tr}\left(\mathcal{A}^M\right)$. To assist the reader in understanding the notation used throughout, consider the following straightforward examples that were originally presented in [7]. For $gl(1)$, (trivial) direct calculation yields

$$\mathcal{A} - \sigma_1 = 0,$$

which is interpreted to mean $\mathcal{A}_{ij} - \sigma_1\delta_{ij} = 0$ for each i,j. For $gl(2)$, we may similarly calculate

$$\mathcal{A}^2 - (\sigma_1 + 1)\mathcal{A} + \frac{1}{2}\left(\sigma_1^2 + \sigma_1 - \sigma_2\right) = 0.$$

This gives two examples of polynomial identities satisfied by the matrix $\mathcal{A}$ with coefficients in the centre of the enveloping algebra of $gl(1)$ and $gl(2)$ respectively. It is worth noting that in these cases $\sigma_2 = \sigma_1^2$, $\sigma_3 = \sigma_1^3$ for $gl(1)$ and $\sigma_3 = \frac{3}{2}\sigma_1\sigma_2 - \frac{1}{2}\sigma_1^3 + \sigma_2 - \frac{1}{2}\sigma_1^2$ for $gl(2)$, demonstrating that the Casimir invariants so defined are not generally independent.

2.3. *The characteristic identity*

The characteristic identity of $\mathcal{A}$ for $gl(n)$ on a finite dimensional irreducible representation $V(\lambda)$ is constructed in the following way.

In the previous section we saw examples of how the matrix $\mathcal{A}$ satisfies polynomial identities with coefficients expressible in terms of the Casimir invariants. If we now look at such polynomial identities on a finite dimensional irreducible representation $V(\lambda)$, with highest weight highest weight $\lambda = (\lambda_1, \lambda_2, \ldots, \lambda_n)$, we then have from Schur's Lemma that the Casimir invariants take on constant values on $V(\lambda)$. For example, for $gl(n)$,

$$\sigma_1 = \sum_{j=1}^{n} \lambda_j, \qquad \sigma_2 = \sum_{j=1}^{n} \lambda_j(\lambda_j + n + 1 - 2j).$$

Bracken and Green [6, 7] showed that the identity

$$\prod_{j=1}^{n} (\mathcal{A} - \alpha_j) = 0$$

holds on a finite dimensional irreducible representation of $gl(n)$, where $\alpha_j = \lambda_j + n - r$ are referred to as the characteristic roots. We highlight the fact that the entries of $\mathcal{A}$ are now representation matrices.

2.4. The adjoint matrix

The matrix $\mathcal{A}$ is not the only such matrix one may define. Consider taking the negative transpose of $\mathcal{A}$ as

$$\overline{\mathcal{A}}_{ij} = -a_{ji}.$$

One may also show that

$$\prod_{j=1}^{n}(\overline{\mathcal{A}} - \overline{\alpha}_j) = 0$$

holds on a finite dimensional irreducible representation of $gl(n)$, where the characteristic roots are in this case given by $\overline{\alpha}_j = r - 1 - \lambda_j = n - 1 - \alpha_j$.

2.5. General characteristic matrix

The construction of these characteristic matrices at this stage may seem ad hoc, but there is in fact a general formalism which relies on the quadratic Casimir element [9, 16]. Such a matrix can then be constructed for an arbitrary semisimple Lie algebra.

Let Δ be the co-product of the enveloping algebra $U(g)$ of a semisimple Lie algebra g, and let π_μ denote any irreducible representation of g corresponding to module $V(\mu)$. We may define the matrix with algebraic entries

$$\mathcal{A}_\mu = -\frac{1}{2}\left[\ (\pi_\mu \otimes \mathrm{id})\Delta(\sigma_2) - \pi_\mu(\sigma_2)\otimes 1 - I \otimes \sigma_2\ \right].$$

Considering its action on an arbitrary finite dimensional irreducible representation π_λ gives

$$\mathcal{A}_\mu^\lambda = -\frac{1}{2}\left[\ (\pi_\mu \otimes \pi_\lambda)\Delta(\sigma_2) - \pi_\mu(\sigma_2)\otimes I - I \otimes \pi_\lambda(\sigma_2)\ \right].$$

In the case $V(\mu)$ is the vector representation of $g = gl(n)$ (i.e. highest weight $\mu = (1, 0, \ldots, 0)$) we obtain $\overline{\mathcal{A}}$, and for the case we take the dual vector representation of $g = gl(n)$, we obtain $\mathcal{A}$.

2.6. Characteristic roots

For the characteristic roots α_ν in the general case, first consider the Clebsch-Gordan decomposition

$$V(\mu)\otimes V(\lambda) = \bigoplus_\nu m_{\mu\lambda}^\nu V(\nu).$$

On each $V(\nu)$, $\mathcal{A}_\mu^\lambda \in$End$(V(\mu)\otimes V(\lambda))$ takes on the constant value

$$\alpha_\nu = -\frac{1}{2}\left[\ \chi_\nu(\sigma_2) - \chi_\mu(\sigma_2) - \chi_\lambda(\sigma_2)\ \right],$$

where $\chi_\nu(\sigma_2)$ denotes the eigenvalue of σ_2 on $V(\nu)$, $\chi_\lambda(\sigma_2)$ is the eigenvalue of σ_2 on $V(\lambda)$, and $\chi_\mu(\sigma_2)$ the eigenvalue of σ_2 on $V(\mu)$. This results in the characteristic identity

$$\prod_\nu(\mathcal{A}_\mu^\lambda - \alpha_\nu) = 0.$$

Journal of Physics: Conference Series **597** (2015) 012045 doi:10.1088/1742-6596/597/1/012045

2.7. Projection operators

We may use the characteristic identities to define projection operators, which, as we shall see, turn out to be a crucial ingredient for determining matrix elements.

We modify our notation slightly, so that $\mathcal{A}_n$ is understood to be the characteristic matrix associated with $gl(n)$. Explicitly, we have from the characteristic identity

$$P\begin{bmatrix} n \\ r \end{bmatrix} = \prod_{\ell \neq r} \left(\frac{\mathcal{A}_n - \alpha_{\ell,n}}{\alpha_{r,n} - \alpha_{\ell,n}} \right),$$

and

$$\overline{P}\begin{bmatrix} n \\ r \end{bmatrix} = \prod_{\ell \neq r} \left(\frac{\overline{\mathcal{A}}_n - \overline{\alpha}_{\ell,n}}{\overline{\alpha}_{r,n} - \overline{\alpha}_{\ell,n}} \right).$$

If we set V to be the vector representation of $gl(n)$, we have the decomposition

$$V \otimes V(\lambda) = \bigoplus_{k=1}^{n} V(\lambda + \Delta_k),$$

and so

$$\overline{P}\begin{bmatrix} n \\ r \end{bmatrix} : V \otimes V(\lambda) \longrightarrow V(\lambda + \Delta_r)$$

is a projection. Here, Δ_k is simply an n-tuple with a 1 in the kth entry and zeroes elsewhere.

By contrast, denoting by V^* the dual vector representation of $gl(n)$, we have the decomposition

$$V^* \otimes V(\lambda) = \bigoplus_{k=1}^{n} V(\lambda - \Delta_k),$$

which implies that

$$P\begin{bmatrix} n \\ r \end{bmatrix} : V^* \otimes V(\lambda) \longrightarrow V(\lambda - \Delta_r)$$

is a projection. These projections will be used in what follows.

3. Matrix elements of irreducible representations

We may utilise the characteristic identity to determine the matrix elements of finite dimensional irreducible representations. Here we demonstrate the procedure, at the same time highlighting the power of this approach. Of particular note is the fact that we may deduce, up to a phase factor, not only the matrix elements of the elementary generators of $gl(n)$, i.e. those of the form $a_{i\ i+1}$, but also the nonelementary generators.

3.1. Vector operators

Before presenting the details of the matrix element formulae, we first need to describe vector operators. Define a $gl(n)$ vector operator as a collection of n operators ψ_i satisfying

$$[a_{ij}, \psi_k] = \delta_{jk}\psi_i.$$

Related to the dual vector representation, we also define a $gl(n)$ contragredient vector operator as a collection of n operators ϕ_j satisfying

$$[a_{ij}, \phi_k] = -\delta_{ik}\phi_j.$$

From the work of Green [7], we have

$$\psi_j = \sum_{r=1}^{n} \psi \begin{bmatrix} n \\ r \end{bmatrix}_j,$$

where

$$\psi \begin{bmatrix} n \\ r \end{bmatrix}_j = \psi_i \overline{P} \begin{bmatrix} n \\ r \end{bmatrix}_{ij} = P \begin{bmatrix} n \\ r \end{bmatrix}_{ji} \psi_i$$

are linearly independent and change that value of the highest weight label $\lambda_{r,n}$ by one, leaving the other $\lambda_{k,n}$ unchanged, i.e.

$$\lambda_{k,n}\psi \begin{bmatrix} n \\ r \end{bmatrix}_j = \psi \begin{bmatrix} n \\ r \end{bmatrix}_j (\lambda_{k,n} + \delta_{kr}).$$

Also, in the case of the contragredient vector operator, we have

$$\phi_j = \sum_{r=1}^{n} \phi \begin{bmatrix} n \\ r \end{bmatrix}_j,$$

such that

$$\lambda_{k,n}\phi \begin{bmatrix} n \\ r \end{bmatrix}_j = \phi \begin{bmatrix} n \\ r \end{bmatrix}_j (\lambda_{k,n} - \delta_{kr}),$$

where

$$\phi \begin{bmatrix} n \\ r \end{bmatrix}_j = \phi_i P \begin{bmatrix} n \\ r \end{bmatrix}_{ij} = \overline{P} \begin{bmatrix} n \\ r \end{bmatrix}_{ji} \phi_i.$$

3.2. Branching rules

The $gl(n) \subset gl(n+1)$ branching rules give the decomposition of a $gl(n+1)$ irreducible representation into a direct sum of $gl(n)$ irreducible representations:

$$V(\lambda_{n+1}) = \bigoplus_{\lambda_n} V(\lambda_n),$$

where the direct sum is taken over an admissible set of dominant $gl(n)$ weights λ_n. This determines the branching rules. Specifically, the highest weights

$$\lambda_{n+1} = (\lambda_{1,n+1}, \lambda_{2,n+1}, \ldots, \lambda_{n+1,n+1}), \quad \lambda_n = (\lambda_{1,n}, \lambda_{2,n}, \ldots, \lambda_{n,n})$$

are known to satisfy

$$\lambda_{i,n+1} - \lambda_{i,n} \in \mathbb{Z}_+, \quad \lambda_{i,n} - \lambda_{i+1,n+1} \in \mathbb{Z}_+.$$

By recursion we may deduce the Gelfand-Tsetlin [28] patterns for the entire subalgebra chain

$$gl(1) \subset gl(2) \subset \cdots \subset gl(n) \subset gl(n+1),$$

which is in one-to-one correspondence with an orthonormal basis of $V(\lambda_{n+1})$.

Now consider the $gl(n+1)$ characteristic matrix:

$$\mathcal{A}_{n+1} = \left(\begin{array}{cccc|c} a_{11} & a_{12} & \cdots & a_{1n} & a_{1\ n+1} \\ a_{21} & a_{22} & \cdots & a_{2n} & a_{2\ n+1} \\ \vdots & \vdots & & \vdots & \vdots \\ a_{n1} & a_{n2} & \cdots & a_{nn} & a_{n\ n+1} \\ \hline a_{n+1\ 1} & a_{n+1\ 2} & \cdots & a_{n+1\ n} & a_{n+1\ n+1} \end{array}\right) = \left(\begin{array}{c|c} \mathcal{A}_n & \psi \\ \hline \phi & a_{n+1\ n+1} \end{array}\right),$$

where $\psi_j = a_{j\ n+1}$, $\phi_j = a_{n+1\ j} = \psi_j^\dagger$ are $gl(n)$ vector and contragredient vector operators respectively. The matrix $\mathcal{A}_{n+1}$ satisfies

$$\prod_{j=1}^{n+1}(\mathcal{A}_{n+1} - \alpha_{j,n+1}) = 0,$$

and the projections are given explicitly by

$$P\begin{bmatrix} n+1 \\ r \end{bmatrix} = \prod_{\ell \neq r}^{n+1}\left(\frac{\mathcal{A}_{n+1} - \alpha_{\ell,n+1}}{\alpha_{r,n+1} - \alpha_{\ell,n+1}}\right).$$

3.3. Construction of invariants

Following the work of Gould [13], we are now in a position to construct eigenvalues of invariants $C_{k,n+1}$ (defined below). Consider the $gl(n+1)$ identity expressed in the form

$$\mathcal{A}_{n+1}P\begin{bmatrix} n+1 \\ k \end{bmatrix} = \alpha_{k,n+1}P\begin{bmatrix} n+1 \\ k \end{bmatrix}.$$

Looking at the $(j, n+1)$ entry, $j = 1, 2, \ldots, n$ gives

$$\psi_j C_{k,n+1} = (\alpha_{k,n+1} - \mathcal{A}_n)_{j\ell}P\begin{bmatrix} n+1 \\ k \end{bmatrix}_{\ell\ n+1},$$

where

$$C_{k,n+1} = P\begin{bmatrix} n+1 \\ k \end{bmatrix}_{n+1\ n+1}$$

is a $gl(n)$ invariant. After some manipulation, we arrive at

$$P\begin{bmatrix} n+1 \\ k \end{bmatrix}_{j\ n+1} = \sum_{r=1}^{n}\psi\begin{bmatrix} n \\ r \end{bmatrix}_j(\alpha_{k,n+1} - \alpha_{r,n} - 1)^{-1}C_{k,n+1}.$$

It is known [8, 10] that for any polynomial $p(x)$,

$$\begin{aligned} p(\mathcal{A}_{n+1}) &= \sum_{k=1}^{n+1}p(\alpha_{k,n+1})P\begin{bmatrix} n+1 \\ k \end{bmatrix} \\ \Rightarrow \delta_{j\ n+1} &= \sum_{k=1}^{n+1}P\begin{bmatrix} n+1 \\ k \end{bmatrix}_{j\ n+1} = 0, \\ \Rightarrow &\sum_{r=1}^{n}\psi\begin{bmatrix} n \\ r \end{bmatrix}_j\left(\sum_{k=1}^{n+1}(\alpha_{k,n+1} - \alpha_{r,n} - 1)^{-1}C_{k,n+1}\right) = 0 \\ \Rightarrow &\sum_{k=1}^{n+1}(\alpha_{k,n+1} - \alpha_{r,n} - 1)^{-1}C_{k,n+1} = 0, \end{aligned}$$

since the $\psi\begin{bmatrix} n \\ r \end{bmatrix}_j$ are linearly independent. Also,

$$\sum_{k=1}^{n+1}C_{k,n+1} = 1,$$

since the $C_{k,n+1}$ are defined in terms of projections. This gives a set of linear equations that can be solved for $C_{k,n+1}$, giving

$$C_{k,n+1} = \prod_{p\neq k}^{n+1}(\alpha_{k,n+1} - \alpha_{p,n+1})^{-1}\prod_{\ell=1}^{n}(\alpha_{k,n+1} - \alpha_{\ell,n} - 1).$$

Similarly, for

$$\overline{C}_{k,n+1} = \overline{P}\begin{bmatrix} n+1 \\ k \end{bmatrix}_{n+1\ n+1},$$

we have

$$\overline{C}_{k,n+1} = \prod_{p\neq k}^{n+1}(\alpha_{k,n+1} - \alpha_{p,n+1})^{-1}\prod_{\ell=1}^{n}(\alpha_{k,n+1} - \alpha_{\ell,n}).$$

3.4. Matrix element formula

In [10], Gould showed that

$$\begin{aligned} \psi\begin{bmatrix} n \\ r \end{bmatrix}\psi^\dagger\begin{bmatrix} n \\ r \end{bmatrix} &= \overline{M}_{r,n}P\begin{bmatrix} n \\ r \end{bmatrix}, \\ \psi^\dagger\begin{bmatrix} n \\ r \end{bmatrix}\psi\begin{bmatrix} n \\ r \end{bmatrix} &= M_{r,n}\overline{P}\begin{bmatrix} n \\ r \end{bmatrix}, \end{aligned} \tag{1}$$

where

$$\overline{M}_{r,n} = (-1)^n\prod_{p=1}^{n+1}(\alpha_{p,n+1} - \alpha_{r,n})\prod_{\ell\neq r}^{n}(\alpha_{r,n} - \alpha_{\ell,n} - 1)^{-1},$$

$$M_{r,n} = (-1)^n\prod_{p=1}^{n+1}(\alpha_{p,n+1} - \alpha_{r,n} - 1)\prod_{\ell\neq r}^{n}(\alpha_{r,n} - \alpha_{\ell,n} + 1)^{-1}.$$

Taking the (n,n) entry of equation (1) above leads to the following expression for the action of the elementary generators $a_{n\ n+1}$:

$$a_{n\ n+1}\left|\lambda_{j,k}\right\rangle = \sum_{r=1}^{n}\psi\begin{bmatrix} n \\ r \end{bmatrix}\left|\lambda_{j,k}\right\rangle = \sum_{r=1}^{n}N_r^n\left|\lambda_{j,k} + \Delta_{r,n}\right\rangle,$$

where $|\lambda_{j,k}\rangle$ denotes a state corresponding to a Gelfand pattern, $\Delta_{r,n}$ denotes a shift in the rth highest weight label at the nth level, and

$$N_r^n = \langle\lambda_{j,k}|M_{r,n}\overline{C}_{r,n}\left|\lambda_{j,k}\right\rangle^{1/2}$$

$$\Rightarrow\quad N_r^n = \left(\frac{(-1)^n\prod_{p=1}^{n+1}(\lambda_{p,n+1} - \lambda_{r,n} + r - p)\prod_{\ell=1}^{n-1}(\lambda_{r,n} - \lambda_{\ell,n-1} + \ell - r + 1)}{\prod_{\ell\neq r}^{n}(\lambda_{r,n} - \lambda_{\ell,n} + \ell - r)(\lambda_{r,n} - \lambda_{\ell,n} + \ell - r - 1)}\right)^{1/2}.$$

Similar expressions may be obtained for $a_{n+1\ n}$.

Using similar techniques, we may also determine matrix element formulae for $a_{\ell\ n+1}$:

$$a_{\ell\ n+1}\left|\lambda_{j,k}\right\rangle = \sum_{i_n=1}^{n}\sum_{i_{n-1}=1}^{n-1}\cdots\sum_{i_\ell=1}^{\ell}N\begin{bmatrix} n & \cdots & \ell \\ i_n & \cdots & i_\ell \end{bmatrix}\left|\lambda_{j,k} + \Delta_{i_n,n} + \cdots + \Delta_{i_\ell,\ell}\right\rangle,$$

Journal of Physics: Conference Series **597** (2015) 012045 doi:10.1088/1742-6596/597/1/012045

where

$$N\begin{bmatrix} n & \cdots & \ell \\ i_n & \cdots & i_\ell \end{bmatrix} = \pm \prod_{r=\ell}^{n} N_{i_r}^r \prod_{r=\ell+1}^{n} \left[(\lambda_{i_r,r} - \lambda_{i_{r-1},r-1} + i_{r-1} - i_r)(\lambda_{i_r,r} - \lambda_{i_{r-1},r-1} + i_{r-1} - i_r + 1)\right]^{-1/2}. \tag{2}$$

Similarly for $a_{n+1\ \ell}$.

4. Generalisation to Lie superalgebras

For the Lie superalgebra $gl(m|n)$, we now have generators a_{pq}, $p,q = 1,2,\ldots,m+n$ satisfying

$$[a_{pq}, a_{rs}] \equiv a_{pq}a_{rs} - (-1)^{[(p)+(q)][(r)+(s)]}a_{rs}a_{pq} = \delta_{qr}a_{ps} - (-1)^{[(p)+(q)][(r)+(s)]}\delta_{ps}a_{rq}.$$

where

$$(p) = \begin{cases} 0, & p = 1,\ldots,m, \\ 1, & p = m+1,\ldots,m+n. \end{cases}$$

Analogous to the Lie algebra case, we define the characteristic matrices by setting

$$\mathcal{A}_{pq} = -(1)^{(p)}a_{pq}, \quad \overline{\mathcal{A}}_{pq} = -(-1)^{(p)(q)}a_{qp}.$$

On the irreducible representation $V(\Lambda)$, with $\Lambda = (\Lambda_1, \Lambda_2, \ldots, \Lambda_m|\Lambda_{m+1}, \ldots, \Lambda_{m+n})$, it is also possible to show [25] that the following characteristic identities hold

$$\prod_{p=1}^{m+n}(\mathcal{A} - \alpha_p) = 0, \quad \prod_{p=1}^{m+n}(\overline{\mathcal{A}} - \overline{\alpha}_p) = 0,$$

with associated characteristic roots

$$\alpha_p = (-1)^{(p)}(\Lambda_p + m - p) - n, \quad \overline{\alpha}_p = m - (-1)^{(p)}(\Lambda_p + m + 1 - p).$$

By analogy with the methods outlined for $gl(n)$, we may construct similar objects for $gl(m|n)$ as listed below.

1. Projection operators $P\begin{bmatrix} m+n \\ r \end{bmatrix}$;
2. Vector operators $\psi_r = (-1)^{(r)}a_{r\ m+n+1}$;
3. Branching rules;
4. Invariants ($C_{r,m+n}$, $\overline{C}_{r,m+n}$, $M_{r,m+n}$ and $\overline{M}_{r,m+n}$);
5. Matrix elements.

There is, however, one major complication in the context of Lie superalgebras. The representation theory is not as well-understood as that of Lie algebras, and in particular the branching rules for an arbitrary irreducible representation are not known. In fact, in general it is not even known if an irreducible representation is completely reducible, i.e. there is no analogue of Weyl's Theorem in general. To avoid this problem, we look at a special subset of the irreducible representations first considered by Scheunert, Nahm and Rittenberg [29].

4.1. On complete reducibility

The so-called star representations were first considered by Scheunert, Nahm and Rittenberg [29] in order to ensure complete reducibility. Let $\langle\ |\ \rangle$ be a well-defined, positive-definite, invariant, sesquilinear form defined on an irreducible highest weight module $V(\Lambda)$ of $gl(m|n)$:

$$\langle a_{pq}v|w\rangle = (-1)^{\epsilon[(p)+(q)]}\langle v|a_{qp}w\rangle, \quad \epsilon = 0,1.$$

For the case $\epsilon = 0$, $V(\Lambda)$ is said to be a type 1 star irrep. Otherwise, for $\epsilon = 1$, $V(\Lambda)$ is called a type 2 star irrep. We have complete reducibility in these cases, and so determine branching rules, which is the one complication in the above procedure.

Journal of Physics: Conference Series **597** (2015) 012045 doi:10.1088/1742-6596/597/1/012045

4.2. Classification of unitary (star) irreducible representations
The star representations, which we now refer to as "unitary" due to the existence of the positive-definite form (i.e. an inner product), were classified by Gould and Zhang [30,31]. Here we only focus on the type 1 case.

In particular, $V(\Lambda)$, with highest weight $\Lambda = (\lambda_1, \lambda_2, \ldots, \lambda_m|\lambda_{\bar{1}}, \ldots, \lambda_{\bar{n}})$ is a type 1 star irrep. iff

(i) $\lambda_m + \lambda_{\bar{n}} > n - 1$ (typical), or

(ii) there exists an odd index $\mu = \{\bar{1}, \ldots, \bar{n}\}$ such that $\lambda_m + \lambda_\mu + 1 - \mu = 0 = \lambda_{\bar{n}} - \lambda_\mu$ (atypical).

There are similar conditions for type 2 star irreps which may be found in [30], but we will not include the details here.

As indicated in the paper [27], we set

$$\varepsilon = (1, 1, \ldots, 1|0, 0, \ldots, 0), \quad \delta = (-1, -1, \ldots, -1|1, 1, \ldots, 1).$$

Then a type 1 star irrep. $V(\Lambda)$ of $gl(m|n)$ has highest weight expressible in the form

$$\Lambda = \Lambda_0 + \gamma\varepsilon + \omega\delta, \tag{3}$$

where Λ_0 corresponds to a covariant tensor representation, i.e. it occurs within a finite tensor product of the vector module $V(1, \dot{0}|\dot{0})$ with itself, $\gamma = 0, 1, 2, \ldots, n-1$ or $n - 1 < \gamma \in \mathbb{R}$, and $\omega \in \mathbb{R}$.

4.3. Form of the matrix elements
The main outcome of our research so far is that for type 1 star irreps., generators of $gl(m|n+1)$ have action of the form

$$a_{\ell\ p+1}\,|\Lambda_{q,s}\rangle = \sum_u N[u_p, u_{p-1}, \ldots u_{\ell+1}, u_\ell]\,\left|\Lambda_{q,s} + \Delta_{u_p,p} + \cdots + \Delta_{u_\ell,\ell}\right\rangle, \quad \ell < p+1,$$

and similarly for $a_{p+1\ \ell}$. The precise form of the coefficients $N[u_p, u_{p-1}, \ldots u_{\ell+1}, u_\ell]$ (i.e. matrix elements) will not be given here, but are similar in form to the matrix elements of equation (2) above which are related to the Lie algebra $gl(n)$. In fact, for the cases under consideration, we make the point that the procedure closely follows that already presented for $gl(n)$. For the fully explicit details of matrix element formulae and other features such as branching rules related to the type 1 unitary irreps of $gl(m|n)$, see [27].

Based on the convenient form of the highest weight given by equation (3), we see that our results agree with those presented in the work of Stoilova and van der Jeugt [32], Molev [33] and also the essentially typical representations considered by Palev [34,35].

5. Future work

Currently work is in preparation for providing the explicit details of the type 2 unitary irreps. It is also of interest to consider how to treat the cases of representations that do not admit an invariant inner product, and hence are not guaranteed to be completely reducible. One possibility is to investigate mixed tensor representations. That is, those representations that occur within the tensor product of a type 1 unitary irrep. with a type 2 unitary irrep.

Having made some progress for the classical Lie superalgebras, it is also of interest to determine the analogous matrix element formulae for the quantum group counterparts.

Finally, our procedure may be considered beyond the vector representation (a type 1 unitary irrep.) and the dual vector representation (a type 2 unitary irrep.), leading to a more general pattern calculus. This remains the topic of future work.

Acknowledgments
M.D.G. and P.S.I are supported by the Australian Research Council through Discovery Project DP140101492. J.L.W. acknowledges the support of an Australian Postgraduate Award.

References
[1] Dirac P A M 1936 *Proc. R. Soc. Lond. A* **155** 447
[2] Lehrer Y 1956 *Bull. Res. Counc. Israel* **5A** 197
[3] Louck J D 1965 *J. Math. Phys.* **6** 1786
[4] Mukunda N 1967 *J. Math. Phys.* **8** 1069
[5] Louck J D and Galbraith H W 1972 *Rev. Mod. Phys.* **44** 540
[6] Bracken A J and Green H S 1971 *J. Math. Phys.* **12** 2099
[7] Green H S 1971 *J. Math. Phys.* **12** 2106
[8] Green H S 1975 *J. Austral. Math. Soc. (Series B)* **19** 129
[9] O'Brien D M, Cant A and Carey A L 1977 *Ann. Inst. Henri Poincaré, Section A: Physique théorique* **26** 405
[10] Gould M D 1978 *J. Austral. Math. Soc. (Series B)* **20** 290
[11] Gould M D 1978 *J. Austral. Math. Soc. (Series B)* **20** 401
[12] Gould M D 1980 *J. Math. Phys.* **21** 444
[13] Gould M D 1981 *J. Math. Phys.* **22** 15
[14] Gould M D 1981 *J. Math. Phys.* **22** 2376
[15] Gould M D 1984 *J. Phys. A: Math. Gen.* **17** 1
[16] Gould M D 1985 *J. Austral. Math. Soc. (Series B)* **26** 257
[17] Gould M D 1987 *J. Phys. A: Math. Gen.* **20** 2657
[18] Gould M D, Zhang R B and Bracken A J 1991 *J. Math. Phys.* **32** 2298
[19] Gould M D, Links J and Bracken A J 1992 *J. Math. Phys.* **33** 1008
[20] Kac V G 1977 *Adv. in Math.* **26** 8
[21] Kac V G 1978 *Lecture Notes in Math.* **676** (Berlin: Springer) 597
[22] Jarvis P D and Green H S 1979 *J. Math. Phys.* **20** 2115
[23] Green H S and Jarvis P D 1983 *J. Math. Phys.* **24** 1681
[24] Jarvis P D and Murray M K 1983 *J. Math. Phys.* **24** 1705
[25] Gould M D 1987 *J. Austral. Math. Soc. (Series B)* **28** 310
[26] Gould M D, Isaac P S and Werry J L 2013 *J. Math. Phys.* **54** 013505
[27] Gould M D, Isaac P S and Werry J L 2014 *J. Math. Phys.* **55** 011703
[28] Gelfand I M and Tsetlin M L 1950 *Dokl. Akad. Nauk., SSSR* **71** 825 (Russian) English translation in: Gelfand I M 1988 "Collected Papers", Vol II (Berlin: Springer-Verlag) 653
[29] Scheunert M, Nahm W and Rittenberg V 1977 *J. Math. Phys.* **18** 146
[30] Gould M D and Zhang R B 1990 *J. Math. Phys.* **31** 2552
[31] Gould M D and Zhang R B 1990 *Lett. Math. Phys.* **20** 221
[32] Stoilova N I and Van der Jeugt J 2010 *J. Math. Phys.* **51** 093523
[33] Molev A I 2011 *Bull. Inst. Math. Acad. Sinica* **6** 415
[34] Palev T D 1987 *Funct. Anal. Appl.* **21** 245
[35] Palev T D 1989 *Funct. Anal. Appl.* **23** 141

Journal of Physics: Conference Series **597** (2015) 012046 doi:10.1088/1742-6596/597/1/012046

Irreducible decompositions of the elasticity tensor under the linear and orthogonal groups and their physical consequences

Yakov Itin[1] **and Friedrich W Hehl**[2]
[1] Inst. Mathematics, Hebrew Univ. of Jerusalem
and Jerusalem College of Technology, Jerusalem 91160, Israel
[2] Inst. Theor. Physics, Univ. of Cologne, 50923 Köln, Germany
and Dept. Physics & Astron., Univ. of Missouri, Columbia, MO 65211, USA

E-mail: itin@math.huji.ac.il, hehl@thp.uni-koeln.de

Abstract. We study properties of the fourth rank elasticity tensor C within *linear* elasticity theory. First C is irreducibly decomposed under the linear group into a "Cauchy piece" S (with 15 independent components) and a "non-Cauchy piece" A (with 6 independent components). Subsequently, we turn to the physically relevant orthogonal group, thereby using the metric. We find the finer decomposition of S into pieces with 9+5+1 and of A into those with 5+1 independent components. Some reducible decompositions, discussed earlier by numerous authors, are shown to be inconsistent. — Several physical consequences are discussed. The Cauchy relations are shown to correspond to $A = 0$. Longitudinal and transverse sound waves are basically related by S and A, respectively.

1. Introduction

The constitutive law in linear elasticity theory for an anisotropic homogeneous body, the generalized Hooke law, postulates a linear relation between the two second-rank tensor fields, the stress σ^{ij} and the strain ε_{kl} (see [17, 23, 18, 13]):

$$\sigma^{ij} = C^{ijkl}\,\varepsilon_{kl}\,. \tag{1}$$

In standard linear elasticity, the stress and the strain tensors are assumed to be symmetric. Consequently the elasticity tensor has the so called *minor symmetries*[1]

$$C^{[ij]kl} = C^{ij[kl]} = 0\,. \tag{2}$$

Another restriction for the elasticity tensor is based on the energy consideration. The energy density of a deformed material is expressed as $W = \frac{1}{2}\sigma^{ij}\varepsilon_{ij}$. When the Hooke law (2) is substituted here, this expression takes the form

$$W = \frac{1}{2}C^{ijkl}\varepsilon_{ij}\varepsilon_{kl}\,. \tag{3}$$

[1] The Bach parenthesis () and [] denote symmetrization and antisymmetrization, respectively: $(ij) := \{ij + ji\}/2$ and $[ij] := \{ij - ji\}/2$, see [21].

The right-hand side of (3) involves only those combinations of the elasticity tensor components which are symmetric under permutation of the first and the last pairs of indices. Consequently, the so-called *major symmetry*,

$$C^{ijkl} - C^{klij} = 0 \tag{4}$$

is assumed. In 3-dimensional space, a fourth rank tensor with the symmetries (2) and (4) has 21 independent components.

In the literature on elasticity, a special decomposition of C^{ijkl} into two tensorial parts is frequently used, see, for example, Cowin [5], Campanella & Tonton [4], Podio-Guidugli [20], Weiner [25], and Haussühl [10]. It is obtained by symmetrization and antisymmetrization of the elasticity tensor with respect to its *two middle indices:*

$$M^{ijkl} := C^{i(jk)l}\,, \quad N^{ijkl} := C^{i[jk]l}\,, \qquad \text{with} \qquad C^{ijkl} = M^{ijkl} + N^{ijkl}\,. \tag{5}$$

It is straightforward to show [15] that M and N fulfill the major symmetry (2) but *not* the minor symmetries (4). Moreover, the tensor M can be further decomposed. Accordingly, the reducible decomposition in $(5)_3$ does not correspond to a direct sum decomposition of the vector space defined by C. The vector spaces of M and C both turn out to be 21-dimensional, that of N is 6-dimensional. Thus, the tensors M and N are auxiliary quantities but, due to the lack of the minor symmetries, they they do *not* represent elasticity tensors, that is, they cannot be used to characterize a certain material elastically. These quantities are placeholders without direct physical interpretation. However, in the elasticity literature some physical interpretations of the tensors M and N are offered. Still, these results are inconsistent as we showed earlier in [15].

In this paper, we present various decompositions of the elasticity tensor based on group-theoretic arguments and discuss some physical applications of these decompositions.

The organization of the paper is as follows: In Sec.2 we discuss the relation between the decomposition of a tensor and the group of transformations acted on the basic vector space. Sec.3 is devoted to the $GL(3)$ decomposition of the elasticity tensor. It is based on the Young diagram technique. Similarly, in Sec.4 we treat the case of the $SL(3)$ and in Sec.5 that of the $SO(3)$. In Sec.6 we delineate some physical applications of the irreducible decomposition described.

2. Groups of transformations and corresponding decompositions of tensors

In a precise algebraic treatment, a tensor must be viewed as a multi-linear map from a vector space V, or the dual of it V^*, to the field of real numbers, rather than as a collection of real valued components.

A contravariant tensor of rank p is defined (see [7], p.59) as a multi-linear map from the Cartesian product of p copies of V^* into the field $\mathbb{R}$,

$$T : \underbrace{V^* \times V^* \times \cdots \times V^*}_{p} \to \mathbb{R}\,. \tag{6}$$

The set of all tensors T of the rank p compose a vector space by itself, say $\mathcal{T}$. The dimension of this *tensor space* is equal to n^p, that is, a 4th rank tensor in 3 dimensions (3d) provides for $\mathcal{T}$ $3^4 = 81$ dimensions. As a basis in $\mathcal{T}$, we can take the tensor products of the basis elements of V,

$$e_{i_1} \otimes e_{i_2} \otimes \cdots \otimes e_{i_p}\,. \tag{7}$$

Accordingly, an arbitrary contravariant tensor of rank p can be expressed as

$$T = T^{i_1 i_2 \cdots i_p}\, e_{i_1} \otimes e_{i_2} \otimes \cdots \otimes e_{i_p}\,. \tag{8}$$

Thus, the properties of the tensor T in (8), in particular its proper decomposition, is related to the group of general linear transformations $GL(3)$ acting on the basis of V. This group encompasses all non-singular 3×3 matrices.

The vector space V has a basis e_i and a dual cobasis ϑ^i. Under the $GL(3)$, the cobasis transforms as $\vartheta^{i'} = L_i{}^{i'}\vartheta^i$. We can attribute a volume element to this cobasis,

$$\epsilon := \frac{1}{3!}\epsilon_{ijk}\vartheta^i \wedge \vartheta^j \wedge \vartheta^k ; \qquad (9)$$

here ϵ_{ijk} is the totally antisymmetric Levi-Civita permutation symbol, which is $\epsilon_{ijk} = \pm 1$ for even or odd permutations of 123, respectively, and is $\epsilon_{ijk} = 0$ otherwise. The primed volume element ϵ' for the transformed cobasis $\vartheta^{i'}$ is defined analogously. If we substitute $\vartheta^{i'} = L_i{}^{i'}\vartheta^i$, we find

$$\epsilon' = \frac{1}{3!}\epsilon_{i'j'k'}L_i{}^{i'}L_j{}^{j'}L_k{}^{k'}\vartheta^i \wedge \vartheta^j \wedge \vartheta^k . \qquad (10)$$

Since $\epsilon_{i'j'k'} = \epsilon_{ijk} = 0, \pm 1$ is numerically invariant under linear basis transformations, the volume element transforms as a scalar *density*,

$$\epsilon' = (\det L_k{}^{k'})^{-1}\,\epsilon . \qquad (11)$$

If we require that $\det L_k{}^{k'} \stackrel{!}{=} +1$, then the volume element is an invariant (or scalar) and the $GL(3)$ reduces to the special linear group $SL(3)$.

Furthermore, we know from elasticity theory that the description of a deformation is based on the concept of the *distance* within a continuum. A deformation is meant to be the *change* of the distances between nearby points. However, the (Euclidean) distance $ds^2 = g := g_{ij}dx^i \otimes dx^j$ as such, with $g_{ij} = \mathrm{diag}(1,1,1)$, is, by definition, invariant under the Euclidean group $T(3) \rtimes SO(3)$, the semidirect product of the translations $T(3)$ with the rotations $SO(3)$. Accordingly, at one point in a continuum, the special orthogonal group $SO(3)$ is the one relevant for the representations of the tensors in elasticity theory.

The $SO(3)$ corresponds to all orthogonal matrices (reciprocal equals to the transpose) with determinant +1. If we allow also parity transformations, we arrive at the $O(3)$, the orthogonal group, and, if we drop the orthogonality requirement, eventually at the $GL(3)$. Alternatively, we can keep parity invariance in the first step and then arrive at the $GL(3)$ via the special linear group $SL(3)$, which has determinant +1:

$$GL(3) \longleftarrow \begin{Bmatrix} SL(3)\,, \det = +1 \\ O(3)\,,\ \det = \pm 1 \end{Bmatrix} \longleftarrow SO(3) . \qquad (12)$$

Incidentally, between the set of subgroups of $GL(3)$ we will use only those that do not preserve any spatial direction. For physical problems that involve some special direction, elastic material in exterior magnetic field, for instance, smaller subgroups such as $GL(2)$ are also relevant.

The process of going down from the $GL(3)$ to the $SO(3)$ is described in detail in Schouten [22], Chap.III. On the level of the $O(3)$, we can define a *scalar* volume element[2] according to $\eta := \sqrt{\det(g_{kl})}\,\epsilon$. The transformations of the $O(3)$ are volume preserving. The ϵ is a *pre*metric concept, whereas the η requires the existence of a metric g.

We can collect there results in a table:

space	volume element	line element	transformation	group
V	ϵ		$\det L \neq 0$	$GL(3)$
V	ϵ		$\det L = +1$	$SL(3)$
V	η	g	$\det L = \pm 1$	$O(3)$
V	η	g	$\det L = +1$	$SO(3)$

[2] This is also discussed in some detail, for example, in [12], Secs.A.1.9 and C.2.3, respectively.

Journal of Physics: Conference Series **597** (2015) 012046 doi:10.1088/1742-6596/597/1/012046

Table 1. The different groups involved in the irreducible decomposition of the elasticity tensor.

Although the group $SO(3)$ is highly relevant in elasticity theory, it is convenient to provide the decomposition of the elasticity tensor relative to the bigger groups and to arrive at the $SO(3)$ only at the last stage. In this case, we are able to identify the different origins of the irreducible parts and to derive the stratified hierarchy of the invariants. Such a method is also useful from the technical point of view because we need to use different algebraic methods for different groups of transformations. We begin with the decomposition relative to the biggest group $GL(3)$.

3. $GL(3)$-decomposition

3.1. Young decomposition

Let us recall some basic facts about 4th rank tensors and their Young decomposition:

1) Covariant tensor of 4th rank over the vector space V (with $\dim V = 3$) is a multi-linear map from the Cartesian product of 4 copies of the vector space V into the real numbers (see [7], p.58),
$$T : V \times V \times V \times V \to \mathbb{R}\,. \tag{13}$$
2) The set of the 4th rank tensors constitutes a new vector space $\mathcal{T}$ called a 4th rank tensor space. Its dimension is equal to $3^4 = 81$.
3) The tensor space $\mathcal{T}$ can be decomposed into the direct sum of its subspaces that are invariant under the action of the group $GL(3, \mathbb{R})$.
4) Due to the well-known Schur-Weyl duality, the irreducible decomposition of the space of the 4th rank tensors under $GL(3, \mathbb{R})$ corresponds to the irreducible decomposition of the permutation group S_4.
5) A known practical way to derive the irreducible decomposition of S_4 is by use of Young diagrams.
6) A generic 4th rank tensor over the 3d tensor space can be decomposed into a direct sum of four subspaces. This decomposition is depicted by the Young diagrams, see [3] or [8],

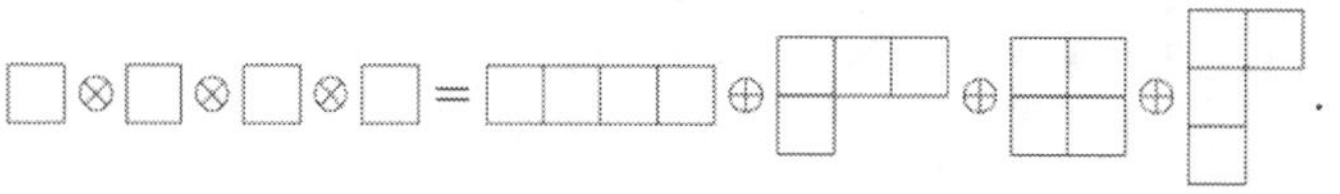

(14)

The left-hand side describes a generic 4th rank tensor. On the right-hand side, the diagrams represent the 4 subspaces.

7) The diagrams in (14) come with a weight factor f^λ, called the dimension of the irreducible representation λ of the permutation group S_p. For a diagram of the shape λ, the f^λ-factor is calculated by the use of the hook-length formula,
$$f^\lambda = p! \prod_{(\alpha,\beta)\in\lambda} \frac{1}{\text{hook}(\alpha,\beta)}\,. \tag{15}$$
The hook length $\text{hook}(\alpha, \beta)$ of the element (α, β) of the diagram is defined as the number of squares directly below or to the right of (α, β) (counting (α, β) once). Thus, for the diagrams depicted in (14),
$$f^{\lambda_1} = 1\,, \quad f^{\lambda_2} = 3\,, \quad f^{\lambda_3} = 2\,, \quad f^{\lambda_4} = 3\,, \quad f^{\lambda_5} = 1\,. \tag{16}$$

Journal of Physics: Conference Series **597** (2015) 012046 doi:10.1088/1742-6596/597/1/012046

8) The dimensions of the irreducible decomposition of $GL(n,\mathbb{R})$ are calculated by the Stanley hook-content formula

$$\dim V^{\lambda} = \prod_{(\alpha,\beta)\in\lambda} \frac{n+\alpha-\beta}{\mathrm{hook}(\alpha,\beta)} \,. \tag{17}$$

Consequently, for the diagrams depicted in (14) and for the dimension $n=3$ of the vector space, we have

$$\dim V^{\lambda_1} = 15, \quad \dim V^{\lambda_2} = 15, \quad \dim V^{\lambda_3} = 6, \quad \dim V^{\lambda_4} = 3\,. \tag{18}$$

9) Every diagram represents a subspace of the tensor space $\mathcal{T}$,

$$\mathcal{T} = {}^{(1)}\mathcal{T} \oplus {}^{(2)}\mathcal{T} \oplus {}^{(3)}\mathcal{T} \oplus {}^{(4)}\mathcal{T}\,. \tag{19}$$

These subspaces intersect only at zero. Moreover, they all are mutually non-isomorphic. Hence we have a direct sum decomposition of the tensor space. The dimension of the initial tensor space is separated into the dimension of the subspaces as follows:

$$\dim \mathcal{T} = \sum_i f^{\lambda_i} \times \dim V^{\lambda_i}\,. \tag{20}$$

The dimension of the tensor space is distributed between the subspaces according to

$$81 = 1\times 15 + 3\times 15 + 2\times 6 + 3\times 3\,. \tag{21}$$

10) The decomposition (21) is unique but reducible. In accordance with (16), the subspaces ${}^{(2)}\mathcal{T}, {}^{(3)}\mathcal{T}$, and ${}^{(4)}\mathcal{T}$ can be decomposed still further into sub-subspaces:

$${}^{(2)}\mathcal{T} = \left({}^{(2,1)}\mathcal{T} \oplus {}^{(2,2)}\mathcal{T} \oplus {}^{(2,3)}\mathcal{T}\right), \tag{22}$$

$${}^{(3)}\mathcal{T} = \left({}^{(3,1)}\mathcal{T} \oplus {}^{(3,2)}\mathcal{T}\right), \tag{23}$$

$${}^{(4)}\mathcal{T} = \left({}^{(4,1)}\mathcal{T} \oplus {}^{(4,2)}\mathcal{T} \oplus {}^{(4,3)}\mathcal{T}\right). \tag{24}$$

These decompositions are irreducible but not unique.

3.2. Irreducible decomposition of C^{ijkl}

The elasticity tensor is not a general 4th rank tensor; rather, it carries its minor and major symmetries. Accordingly, we are looking for a a decomposition of an invariant subspace of the tensor space $\mathcal{T}$ into a direct sum of its sub-subspaces,

$$\mathcal{C} = \alpha\,{}^{(1)}\mathcal{C} \oplus \beta\,{}^{(2)}\mathcal{C} \oplus \gamma\,{}^{(3)}\mathcal{C} \oplus \delta\,{}^{(4)}\mathcal{C}\,, \tag{25}$$

where

$$\alpha = 0,1\,; \qquad \beta = 0,1,2,3\,; \qquad \gamma = 0,1,2\,; \qquad \delta = 0,1,2,3\,. \tag{26}$$

Using the minor and major symmetries, we find $\alpha = 1$ and $\delta = 0$. Since $\dim \mathcal{C} = 21$, we find as a unique solution of (25),

$$\alpha = \gamma = 1\,; \qquad \beta = \delta = 0\,.$$

Thus, we proved

Proposition 1: The decomposition

$$C^{ijkl} = S^{ijkl} + A^{ijkl}\,, \tag{27}$$

with

$$S^{ijkl} := C^{(ijkl)} = \frac{1}{3}(C^{ijkl} + C^{iklj} + C^{iljk})\,, \quad \text{and} \quad A^{ijkl} := \frac{1}{3}(2C^{ijkl} - C^{ilkj} - C^{iklj})\,, \tag{28}$$

is irreducible and unique.

Proposition 2: The partial tensors satisfy the minor symmetries,

$$S^{[ij]kl} = S^{ij[kl]} = 0 \qquad \text{and} \qquad A^{[ij]kl} = A^{ij[kl]} = 0\,, \tag{29}$$

and the major symmetry,

$$S^{ijkl} = S^{klij} \qquad \text{and} \qquad A^{ijkl} = A^{klij}\,. \tag{30}$$

Proposition 3: The irreducible decomposition of C signifies the reduction of the tensor space $\mathcal{C}$ into a direct sum of two subspaces $\mathcal{S} \subset \mathcal{C}$ (for the tensor S) and $\mathcal{A} \subset \mathcal{C}$ (for the tensor A),

$$\mathcal{C} = \mathcal{S} \oplus \mathcal{A}\,. \tag{31}$$

In particular, we have

$$\dim C = 21\,, \qquad \dim S = 15\,, \qquad \dim A = 6\,. \tag{32}$$

Proposition 4: The irreducible piece A^{ijkl} of the elasticity tensor is a fourth rank tensor. Alternatively, it can be represented as a symmetric second rank tensor density

$$\Delta_{mn} := \frac{1}{4}\epsilon_{mil}\epsilon_{njk}A^{ijkl}\,, \tag{33}$$

where $\epsilon_{ijk} = 0, \pm 1$ denotes the Levi-Civita permutation symbol. The proof of this proposition is given in the next section.

4. $SL(3)$-decomposition

Since the elasticity tensor C^{ijkl} satisfies the symmetries (2) and (4), most of its contractions with ϵ_{ijk} vanish. Indeed, for the completely symmetric part the contraction in two indices is identically zero,

$$\epsilon_{mij}S^{ijkl} = 0\,. \tag{34}$$

The second part of the elasticity tensor yields a non-vanishing contraction,

$$K_m{}^{jk} := \frac{1}{2}\epsilon_{mil}A^{ijkl}\,. \tag{35}$$

This tensor $K_i{}^{jk}$ has vanishing traces and is antisymmetric in the upper indices,

$$K_i{}^{ik} = 0\,, \qquad K_k{}^{jk} = 0\,, \qquad K_i{}^{(jk)} = 0\,. \tag{36}$$

Thus, $K_i{}^{jk}$ has 6 independent components, exactly like the initial tensor A^{ijkl}.

Because of the antisymmetry of $K_i{}^{jk}$, we do not loose anything if we contract the upper indices with ϵ_{njk}:

$$\Delta_{mn} := \frac{1}{2}\epsilon_{njk}K_m{}^{jk} = \frac{1}{4}\epsilon_{mil}\epsilon_{njk}A^{ijkl}\,. \tag{37}$$

We can check that this tensor is symmetric

$$\Delta_{[mn]} = 0\,. \tag{38}$$

Thus, it has the same 6 independent components. This proves the Proposition 4.

Summing up: there are no additional $SL(3)$ invariants of the elasticity tensor, and this tensor cannot be decomposed further under the action of the $SL(3)$. Moreover, we derived a representation of A^{ijkl} in terms of Δ_{mn}. Under the action of the special linear group $SL(3)$, the latter quantity is an ordinary tensor.

5. $SO(3)$-decomposition

Since for the elasticity tensor the invariance of the volume element does not yield additional tensor decompositions, we skip the group $O(3)$ and pass directly to its subgroup $SO(3)$. Consider a vector space W endowed with a metric tensor g_{ij}. The norm and the scalar product of vectors are defined now in the conventional way by $(u,v) = g_{ij}u^i v^j$ and $|v| = (v,v)^{1/2}$. In order to preserve the scalar product, we must restrict ourselves to the orthogonal group $O(3)$. Invariance of the volume element η is guaranteed if we restrict ourselves to the group $SO(3)$. Now we can use the metric tensor g_{ij} and its inverse g^{kl}.

From the contraction of the metric with the totally symmetric Cauchy part S^{ijkl}, a unique symmetric second rank tensor and its scalar contraction can be constructed,

$$S^{ij} := g_{kl}S^{ijkl} \qquad \text{and} \qquad S := g_{ij}g_{kl}S^{ijkl}\,. \tag{39}$$

Define the traceless part of the tensor S^{ij} as

$$\not{S}^{ij} := S^{ij} - \frac{1}{3}Sg^{ij}\,, \qquad \text{with} \qquad g_{ij}\not{S}^{ij} = 0\,. \tag{40}$$

Now we turn to the decomposition of the tensor S^{ijkl}. We define the subtensors

$$^{(2)}S^{ijkl} := \alpha\,\not{S}^{(ij}g^{kl)} \qquad \text{and} \qquad ^{(3)}S^{ijkl} := \beta S g^{(ij}g^{kl)}\,. \tag{41}$$

We denote the remaining part as

$$^{(1)}S^{ijkl} := S^{ijkl} - {}^{(2)}S^{ijkl} - {}^{(3)}S^{ijkl}\,. \tag{42}$$

Now we require the tensor $^{(1)}S^{ijkl}$ to be traceless. This yields,

$$\alpha = \frac{6}{7}\,, \qquad \beta = \frac{1}{5}\,. \tag{43}$$

Hence we obtain the unique irreducible decomposition of the tensor S^{ijkl} into three pieces,

$$S^{ijkl} = {}^{(1)}S^{ijkl} + {}^{(2)}S^{ijkl} + {}^{(3)}S^{ijkl}\,. \tag{44}$$

These pieces are invariant under the action of the group $SO(3)$.

In order to decompose the non-Cauchy part A^{ijkl}, it is convenient to use its representation by the tensor density Δ_{ij}. The latter is irreducibly decomposed to a scalar and traceless parts.

$$\Delta_{ij} = \not{\Delta}_{ij} + \frac{1}{3}\Delta g_{ij}\,, \qquad \text{where} \qquad \Delta := g^{ij}\Delta_{ij}\,. \tag{45}$$

Consequently, we obtain the decomposition of A^{ijkl} into two independent parts

$$A^{ijkl} = {}^{(1)}A^{ijkl} + {}^{(2)}A^{ijkl}, \tag{46}$$

where the scalar and the traceless parts are given by

$${}^{(2)}A^{ijkl} := \frac{2}{3}\Delta\left(g^{ij}g^{kl} - g^{il}g^{jk}\right) \qquad \text{and} \qquad {}^{(1)}A^{ijkl} := A^{ijkl} - {}^{(2)}A^{ijkl}, \tag{47}$$

respectively. This way we derived a composition of the elasticity tensor into five irreducible parts

$$C^{ijkl} = \left({}^{(1)}S^{ijkl} + {}^{(2)}S^{ijkl} + {}^{(3)}S^{ijkl}\right) + \left({}^{(1)}A^{ijkl} + {}^{(2)}A^{ijkl}\right). \tag{48}$$

Between the first parentheses we collected the terms corresponding to the Cauchy part, the second parentheses enclose the non-Cauchy terms. The dimension of the tensor space of the elasticity tensor is decomposed into the sum of the corresponding dimensions of the subspaces,

$$21 = (9 + 5 + 1) + (5 + 1)\,. \tag{49}$$

Thus, C^{ijkl} can be represented by one totally traceless and totally symmetric 4th rank tensor ${}^{(1)}C^{ijkl}$ *plus* two symmetric traceless 2nd rank tensors $\not{S}^{ij}$, $\not{\Delta}_{ij}$ *plus* two scalars S, Δ. This decomposition is unique and irreducible under the action of the $SO(3,\mathbb{R})$.

It is quite remarkable that the same decomposition was derived Backus [1] (see also Baerheim [2]) in a rather different way. In our group-theoretical treatment, we obtained all irreducible pieces in a covariant form. Moreover, we derived a tree of the independent invariances and their relations to different transformation groups. We describe this stratified structure in the following diagram:

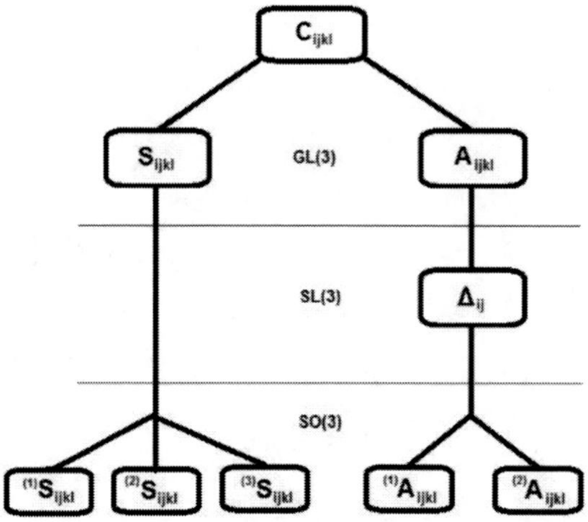

Figure 1. The decomposition tree of the elasticity tensor C_{ijkl}. Note that Δ_{ij} is a tensor (and *not* a density) under the $SL(3)$.

6. Applications

Some non-trivial applications were recently discussed by us in [11, 15] (for the analogous case in electrodynamics, see [14]):

6.1. *Cauchy relations*
The Cauchy relations are given by

$$N^{ijkl} = 0 \qquad \text{or} \qquad C^{ijkl} = C^{ikjl}\,, \tag{50}$$

for their history, see Todhunter [24]. The representation in (50) is widely used in the elasticity literature, see, for example, [9, 5, 6, 4, 20, 25]. In [11, 15], we presented an alternative form of these conditions, namely

$$A^{ijkl} = 0 \qquad \text{or} \qquad 2C^{ijkl} - C^{ilkj} - C^{iklj} = 0\,. \tag{51}$$

The last equation can be presented in a more economical form,

$$\Delta_{mn} = 0\,. \tag{52}$$

The basic difference between (50) on the one hand and (51) or (52) at the other hand is that our conditions (51) or (52) are represented in an irreducible form. This can have decisive consequences.

For most materials the Cauchy relations do not hold, even approximately. In fact, the elasticity of a generic anisotropic material is described by the whole set of the 21 independent components, it is not restricted to the set of 15 independent components obeying the Cauchy relations. This fact seems to nullify the importance of the Cauchy relations for modern solid state theory and leave them only as historical artifact.

However, a lattice-theoretical approach to the elastic constants shows, see Leibfried [16], that the Cauchy relations are valid provided (i) the interaction forces between the atoms or molecules of a crystal are central forces, as, for instance, in rock salt, (ii) each atom or molecule is a center of symmetry, and (iii) the interaction forces between the building blocks of a crystal can be well approximated by a harmonic potential, see Perrin [19]. Accordingly, a study of the *violations* of the Cauchy relations yields important information about the intermolecular forces of elastic bodies. Accordingly, one should look for the *deviation* A^{ijkl} of the elasticity tensor C^{ijkl} from its Cauchy part S^{ijkl}.

6.2. *Acoustic waves*
If u_i is the displacement field, the propagation of acoustic waves in anisotropic media is determined by the following equation (ρ = mass density):

$$\rho g^{il}\ddot{u}_l - C^{ijkl}\,\partial_j\partial_k u_l = 0\,. \tag{53}$$

With a plane wave ansatz $u_i = U_i e^{i\left(\zeta n_j x^j - \omega t\right)}$, we obtain a system of three homogeneous algebraic equations

$$\left(v^2 g^{il} - \Gamma^{il}\right) U_l = 0\,, \tag{54}$$

where the *Christoffel tensor* $\Gamma^{il} := (1/\rho)\,C^{ijkl} n_j n_k$ and of the *phase velocity* $v := \omega/\zeta$ are used. Substituting the irreducible *SA*-decomposition into the Christoffel tensor, we obtain

$$\Gamma^{il} = \mathcal{S}^{il} + \mathcal{A}^{il}\,, \tag{55}$$

where the Cauchy and non-Cauchy Christoffel tensors are defined by

$$\mathcal{S}^{il} := S^{ijkl} n_j n_k = \mathcal{S}^{li}\,, \qquad \text{and} \qquad \mathcal{A}^{il} := A^{ijkl} n_j n_k = \mathcal{A}^{li}\,. \tag{56}$$

Acoustic wave propagation in an elastic medium is an eigenvector problem, see (54), with the phase velocity v^2 as the eigenvalues. In general, three distinct real positive solutions correspond to three independent waves ${}^{(1)}U_l$, ${}^{(2)}U_l$, and ${}^{(3)}U_l$, called *acoustic polarizations*. For isotropic materials, there are three pure polarizations: one *longitudinal (or compression) wave* with $\vec{U}\times\vec{n}=0$ and two *transverse (or shear) waves* with $\vec{U}\cdot\vec{n}=0$.

In general, for anisotropic materials, three pure modes do not exist. The identification of the pure modes and the condition for their existence is an interesting problem. Let us show how the irreducible decomposition of the elasticity tensor, which we applied to the Christoffel tensor, can be used here.

Proposition 5: Let n^i denote an allowed direction for the propagation of a *compression* wave. Then the velocity $v_{\rm L}$ of this wave in the direction of n^i is determined only by the *Cauchy* part of the elasticity tensor:

$$v_{\rm L}=\sqrt{S^{ij}n_in_j}\,. \tag{57}$$

Moreover, for a medium with a given elasticity tensor, all three *purely polarized* waves (one longitudinal and two transverse) can propagate in the direction $\vec{n}$ if and only if

$$S^{ij}n_j=S^{kl}n_kn_ln^i\,. \tag{58}$$

Accordingly, for a given medium, the directions of the purely polarized waves depend on the Cauchy part of the elasticity tensor *alone*. In other words, two materials, with the same Cauchy parts S^{ijkl} of the elasticity tensor but different non-Cauchy parts A^{ijkl}, have the same pure wave propagation directions and the same longitudinal velocity.

Acknowledgments
This work was supported by the German-Israeli Foundation for Scientific Research and Development (GIF), Research Grant No. 1078-107.14/2009. Y.I. would like to thank Jan-Willem Van Holten for the invitation to the Group 30 Colloquium.

References

[1] Backus G 1970 Geometrical picture of anisotropic elastic tensors *Reviews Geophys. Space Phys.* **8** 633–671
[2] Baerheim R 1993 Harmonic decomposition of the anisotropic elasticity tensor *Quarterly J. Mech. Appl. Math.* **46** 391–418
[3] Boerner H 1970 *Representations of groups* (Amsterdam: North Holland Publishing Co.)
[4] Campanella A and Tonon M L 1994 A note on the Cauchy relations *Meccanica* **29** 105–108
[5] Cowin S C 1989 Properties of the anisotropic elasticity tensor *Quarterly J. Mech. Appl. Math.* **42** 249–266 Corrigenda 1993 ibid. **46** 541–542
[6] Cowin S C and Mehrabadi M M 1992 The structure of the linear anisotropic elastic symmetries *J. Mech. Phys. Solids* **40** 1459–1471
[7] Frankel T 2012 *The Geometry of Physics* 3rd edn. (Cambridge, UK: Cambridge University Press)
[8] Hamermesh M 1989 *Group Theory and its Application to Physical Problems* (New York: Dover)
[9] Haussühl S 1983 *Physics of Crystals (in German)* (Weinheim Germany: Physik-Verlag)
[10] Haussühl S 2007 *Physical Properties of Crystals: An Introduction* (Weinheim Germany: Wiley-VCH)
[11] Hehl F W and Itin Y 2002 The Cauchy relations in linear elasticity theory *J. Elasticity* **66** 185–192 [arXiv:cond-mat/0206175]
[12] Hehl F W and Obukhov Yu N 2003 *Foundations of Classical Electrodynamics: Charge, flux and metric* (Boston: Birkhäuser)
[13] Hetnarski R B and Ignaczak J 2011 *The Mathematical Theory of Elasticity* 2nd edn. (Boca Raton: CRC Press)
[14] Itin Y 2009 On light propagation in premetric electrodynamics. Covariant dispersion relation *J. Phys. A* **42** 475–402
[15] Itin Y and Hehl F W 2012 The constitutive tensor of linear elasticity: its decompositions, Cauchy relations, null Lagrangians, and wave propagation *J. Math. Phys.* **54** 042903 (2013) [arXiv:1208.1041]

[16] Leibfried G 1955 Gittertheorie der mechanischen und thermischen Eigenschaften der Kristalle *Handbuch der Physik/Encyclopedia of Physics (ed. S. Flügge), vol. VII/1, Kristallphysik I* 104–324 (Berlin: Springer)
[17] Love A E H 1927 *A Treatise on the Mathematical Theory of Elasticity* 4th edn. (Cambridge, UK: Cambridge University Press)
[18] Marsden J E and Hughes T J R 1983 *Mathematical Foundations of Elasticity* (Englewood Cliffs, NJ: Prentice-Hall)
[19] Perrin B 1979 Cauchy relations revisited *Phys. Stat. Sol. B* **91** K115–K120
[20] Podio-Guidugli P 2000 *A Primer in Elasticity* (Dordrecht: Kluwer)
[21] Schouten J A 1954 *Ricci-Calculus* 2nd edn. (Berlin: Springer)
[22] Schouten J A 1989 *Tensor Analysis for Physicists* reprinted 2nd edn. (New York: Dover)
[23] Sokolnikoff I S 1956 *Mathematical Theory of Elasticity* 2nd edn. (New York: McGraw-Hill)
[24] Todhunter I 1960 *A History of the Theory of Elasticity and the Strength of Materials, from Galilei to Lord Kelvin,* edited and completed by K. Pearson. Vol. I: Galilei to Saint-Venant 1639–1850 496–505 (New York: Dover)
[25] Weiner J H 2002 *Statistical Mechanics of Elasticity* (Mineola, NY: Dover)

The $\mathfrak{su}(2)$ Krawtchouk oscillator model under the $\mathcal{CP}$ deformed symmetry

E I Jafarov[1], **A M Jafarova**[2] **and J Van der Jeugt**[3]

[1]Institute of Physics, Azerbaijan National Academy of Sciences, Javid av. 33, AZ-1143 Baku, Azerbaijan

[2]Institute of Mathematics and Mechanics, Azerbaijan National Academy of Sciences, Baku, Azerbaijan

[3]Department of Applied Mathematics, Computer Science and Statistics, Ghent University, Krijgslaan 281-S9, B-9000 Gent, Belgium

E-mail: ejafarov@physics.ab.az; Joris.VanderJeugt@UGent.be

Abstract. We define a new algebra, which can formally be considered as a $\mathcal{CP}$ deformed $\mathfrak{su}(2)$ Lie algebra. Then, we present a one-dimensional quantum oscillator model, of which the wavefunctions of even and odd states are expressed by Krawtchouk polynomials with fixed $p = 1/2$, $K_{2n}(k; 1/2, 2j)$ and $K_{2n}(k-1; 1/2, 2j-2)$. The dynamical symmetry of the model is the newly introduced $\mathfrak{su}(2)_{\mathcal{CP}}$ algebra. The model itself gives rise to a finite and discrete spectrum for all physical operators (such as position and momentum). Among the set of finite oscillator models it is unique in the sense that any specific limit reducing it to a known oscillator models does not exist.

1. Introduction

Krawtchouk polynomials are the simplest finite-discrete polynomials of the Askey scheme of orthogonal polynomials. They play an essential role in the study of problems coming from different branches of modern science. Image analysis based on their moments [1], exactly solvable birth and death processes [2], perfect qubit transfer in a spin chain with nearest-neighbour interaction [3] are only a few of such problems solved by employing Krawtchouk polynomials. Also in the "discretisation" of the quantum-mechanical formulation of the harmonic oscillator problem they play an important role. The non-relativistic explicit formulation of this problem in the infinite continuous configuration space is well known within the canonical approach. The wavefunctions of stationary states are expressed in terms of the Hermite polynomials both in momentum and position representation [4]. Under the assumption that the quantum world is quite different from the classical one, various approaches and methods allow to construct a number of exactly solvable mathematical models of the quantum harmonic oscillator, where the wavefunctions are expressed in terms of other known orthogonal polynomials. Also the Krawtchouk polynomials allow one to construct a number of interesting quantum harmonic oscillator models in finite-discrete configuration space. At present, the known models using Krawtchouk polynomials are the $\mathfrak{su}(2)$ model [5–7] and the $\mathfrak{sl}(2|1)$ model [8] for the finite quantum harmonic oscillator. In the case of $\mathfrak{su}(2)$ dynamical symmetry, the wavefunctions of the model are defined through the special case of Krawtchouk polynomials with a fixed parameter $p = 1/2$. For the 'supersymmetric' model with $\mathfrak{sl}(2|1)$ dynamical symmetry, the wavefunctions are expressed by means of general Krawtchouk polynomials with parameter $0 < p < 1$. In the present work, we present a new oscillator model with wavefunctions of even and odd states also expressed by Krawtchouk polynomials with fixed $p = 1/2$, but in a very different way: the polynomials that appear are $K_{2n}(k; 1/2, 2j)$ and $K_{2n}(k-1; 1/2, 2j-2)$ respectively. The dynamical symmetry is formally called a $\mathcal{CP}$ deformation of the $\mathfrak{su}(2)$ Lie algebra.

The current paper is structured as follows; in section 2, the $\mathcal{CP}$ deformation of the $\mathfrak{su}(2)$ Lie algebra and its representations are introduced. Then, this algebra is used for the construction of a model of the finite-discrete quantum harmonic oscillator. In section 4, we discuss explicit expression of the wavefunctions of the constructed model both in momentum and position representations. Finally, section 5 covers a brief summary and discussion of the obtained results.

2. The algebra $\mathfrak{su}(2)$ under the $\mathcal{CP}$ deformed symmetry

Our starting point is the definition of the formal algebra $\mathfrak{su}(2)$ under the $\mathcal{CP}$ deformed symmetry by its basis elements J_0, J_+ and J_-.

Definition 1 *The algebra $\mathfrak{su}(2)_{\mathcal{CP}}$ is a unital algebra with basis elements J_0, J_+, J_-, $\mathcal{C}$ and $\mathcal{P}$ subject to the following relations:*

- *The operator $\mathcal{C}$ commutes with all basis elements.*
- *$\mathcal{P}$ is a parity operator satisfying $\mathcal{P}^2 = 1$ and*

$$[\mathcal{P}, J_0] = \mathcal{P}J_0 - J_0\mathcal{P} = 0, \qquad \{\mathcal{P}, J_\pm\} = \mathcal{P}J_\pm + J_\pm\mathcal{P} = 0. \tag{1}$$

- *The $\mathfrak{su}(2)$ commutation relations are $\mathcal{CP}$ deformed as follows:*

$$[J_0, J_\pm] = \pm J_\pm, \tag{2}$$
$$[J_+, J_-] = 2J_0\,(\mathcal{CP} - 1). \tag{3}$$

The next step is to describe representations of this algebra, by defining the actions of the basis elements J_0, J_+ and J_- in such a way that they are compatible with the commutation relations (2) and (3). It is easy to verify that an action exists on the usual standard basis vectors $|j, m\rangle$ (with j being an integer and $m = -j, -j+1, \ldots, +j$), which is reminiscent of but different from the common $\mathfrak{su}(2)$ algebra action. This is given by:

$$J_0|j,m\rangle = m\,|j,m\rangle, \tag{4}$$

$$J_+|j,m\rangle = \begin{cases} \sqrt{(j-m)(j-m-1)}\,|j,m+1\rangle, & \text{if } j+m \text{ is even;} \\ \sqrt{(j+m)(j+m+1)}\,|j,m+1\rangle, & \text{if } j+m \text{ is odd,} \end{cases} \tag{5}$$

$$J_-|j,m\rangle = \begin{cases} \sqrt{(j+m)(j+m-1)}\,|j,m-1\rangle, & \text{if } j+m \text{ is even;} \\ \sqrt{(j-m)(j-m+1)}\,|j,m-1\rangle, & \text{if } j+m \text{ is odd.} \end{cases} \tag{6}$$

Note that j must be integer (and not half-integer) in order to be a representation (one way to see it is that one should still have $J_+|j,j\rangle = 0$ and $J_-|j,-j\rangle = 0$). The algebra representation is completed by the action of the extra operators $\mathcal{C}$ and $\mathcal{P}$:

$$\mathcal{C}|j,m\rangle = 2j\,|j,m\rangle, \tag{7}$$
$$\mathcal{P}|j,m\rangle = (-1)^{j+m}\,|j,m\rangle. \tag{8}$$

Despite the fact that this algebra is at first sight very close to $\mathfrak{su}(2)$, its behaviour is quite different. For example, the representations matrices of $J_0, J_\pm$ in the $\mathfrak{su}(2)_{\mathcal{CP}}$ algebra never coincide with the representation matrices of these elements in the $\mathfrak{su}(2)$ algebra, except for the trivial representation $j = 0$. Note also that there is no "deformation parameter" that takes the $\mathfrak{su}(2)_{\mathcal{CP}}$ algebra to the $\mathfrak{su}(2)$ algebra for some limit.

Journal of Physics: Conference Series **597** (2015) 012047 doi:10.1088/1742-6596/597/1/012047

3. A one-dimensional oscillator model based on the algebra $\mathfrak{su}(2)_{CP}$

We now consider an oscillator model based on the algebra introduced, and choose the Hamiltonian, position and momentum operators in a way that is similar to the non-deformed $\mathfrak{su}(2)$ one-dimensional oscillator case [6, 7], i.e.

$$\hat{H} = J_0 + j + \frac{1}{2}, \qquad \hat{q} = \frac{1}{2}(J_+ + J_-), \qquad \hat{p} = \frac{\mathrm{i}}{2}(J_+ - J_-). \tag{9}$$

One can easily check that these three operators satisfy the Heisenberg equations:

$$\left[\hat{H}, \hat{q}\right] = -\mathrm{i}\hat{p}, \quad \left[\hat{H}, \hat{p}\right] = \mathrm{i}\hat{q}. \tag{10}$$

Then, one can also observe that

$$\hat{H}|j,m\rangle = (m + j + \frac{1}{2})|j,m\rangle. \tag{11}$$

So the spectrum of $\hat{H}$ is equidistant but finite:

$$n + \frac{1}{2} \qquad (n = 0, 1, \ldots, 2j). \tag{12}$$

From the actions (5) and (6) one obtains for the position operator $\hat{q}$ that

$$2\hat{q}\,|j,m\rangle = \sqrt{(j+m)\,(j+m-1)}\,|j,m-1\rangle + \sqrt{(j-m)\,(j-m-1)}\,|j,m+1\rangle\,, \tag{13}$$

when $j + m = 2n$ is even, and

$$2\hat{q}\,|j,m\rangle = \sqrt{(j-m)\,(j-m+1)}\,|j,m-1\rangle + \sqrt{(j+m)\,(j+m+1)}\,|j,m+1\rangle\,, \tag{14}$$

when $j + m = 2n + 1$ is odd. One can do the same computations for the operator $2\mathrm{i}\hat{p}$. In matrix form, both position and momentum operators $2\hat{q}$ and $2\mathrm{i}\hat{p}$ take the following form:

$$2\hat{q} = \begin{pmatrix} 0 & M_0 & 0 & \cdots & 0 \\ M_0 & 0 & M_1 & \cdots & 0 \\ 0 & M_1 & 0 & \ddots & \\ \vdots & \vdots & \ddots & \ddots & M_{2j-1} \\ 0 & 0 & & M_{2j-1} & 0 \end{pmatrix} \equiv M^q, \tag{15}$$

and

$$2\mathrm{i}\hat{p} = \begin{pmatrix} 0 & M_0 & 0 & \cdots & 0 \\ -M_0 & 0 & M_1 & \cdots & 0 \\ 0 & -M_1 & 0 & \ddots & \\ \vdots & \vdots & \ddots & \ddots & M_{2j-1} \\ 0 & 0 & & -M_{2j-1} & 0 \end{pmatrix} \equiv M^p, \tag{16}$$

with matrix elements

$$M_k = \begin{cases} \sqrt{k\,(k+1)}, & \text{if } k \text{ is odd;} \\ \sqrt{(2j-k)\,(2j-k-1)}, & \text{if } k \text{ is even.} \end{cases} \tag{17}$$

The purpose is now to study the eigenvalues of the position (and momentum) operator in this model. In other words: we need to find eigenvalues and eigenvectors of these matrices M^q and

M^p. It turns out that these eigenvectors can be constructed in terms of symmetric Krawtchouk polynomials. In general, Krawtchouk polynomials $K_n(x;p,N)$ of degree n $(n=0,1,\ldots,N)$ in the variable x, with parameter $0<p<1$, are defined in terms of the ${}_2F_1$ hypergeometric function of argument $\frac{1}{p}$ as follows [9]:

$$K_n(x;p,N) = {}_2F_1\left(\begin{matrix} -n, -x \\ -N \end{matrix}; \frac{1}{p}\right). \tag{18}$$

Their orthogonality holds for discrete values of x, and is given by:

$$\sum_{x=0}^{N} w(x,N)\, K_n(x;p,N)\, K_{n'}(x;p,N) = h(n,N)\,\delta_{n,n'}, \tag{19}$$

where,

$$w(x,N) = \binom{N}{x} p^x (1-p)^x \quad (x=0,1,\ldots,N), \tag{20}$$

$$h(n,N) = \frac{(-1)^n\, n!}{(-N)_n}\left(\frac{1-p}{p}\right)^n. \tag{21}$$

Here, $(a)_k = a(a+1)\cdots(a+k-1)$ are Pochhammer symbols. We can introduce the following orthonormal Krawtchouk functions:

$$\tilde{K}_n(x;p,N) = \sqrt{\frac{w(x,N)}{h(n,N)}}\, K_n(x;p,N). \tag{22}$$

Let us also mention that for $p=1/2$, the corresponding Krawtchouk polynomials are called "symmetric", since

$$K_n(x;\tfrac{1}{2},N) = (-1)^n K_n(N-x;\tfrac{1}{2},N). \tag{23}$$

In previous examples, the construction of a quantum oscillator model in term of orthogonal polynomials is based on the differential (or difference) equation, the solution of which is expressed in terms of that polynomial. In this sense, the $\mathfrak{su}(2)$ one-dimensional Krawtchouk oscillator model [6, 7] is based on the difference equation for Krawtchouk polynomials [9, (9.11.5)], whereas the $\mathfrak{sl}(2|1)$ one-dimensional Krawtchouk oscillator model [8] is based on the forward and backward shift operator equations for them [9, (9.11.6), (9.11.8)]. For our current construction, the essential point here is a set of new difference equations for symmetric Krawtchouk polynomials ($p=\frac{1}{2}$). These new relations involve a pair of symmetric Krawtchouk polynomials shifted in variables $2x$ or $2(x+1)$ and parameters $2(j-1)$ and $2j$.

Proposition 2 *The symmetric Krawtchouk polynomials satisfy the following difference equations:*

$$\begin{aligned} j(2j-1)\,K_{j+n}\left(2(x+1);\tfrac{1}{2},2j\right) &= -(x+1)(2x+1)\,K_{j+n-1}\left(2x;\tfrac{1}{2},2(j-1)\right) \\ &\quad + (j-x-1)(2j-2x-3)\,K_{j+n-1}\left(2(x+1);\tfrac{1}{2},2(j-1)\right); \end{aligned} \tag{24}$$

$$\begin{aligned} 2(j+n)(j-n)\,K_{j+n-1}\left(2x;\tfrac{1}{2},2(j-1)\right) &= j(2j-1)\,K_{j+n}\left(2x;\tfrac{1}{2},2j\right) \\ &\quad - j(2j-1)\,K_{j+n}\left(2(x+1);\tfrac{1}{2},2j\right). \end{aligned} \tag{25}$$

In these equations, the integer n is arbitrary but should be taken appropriately (e.g. for the first Krawtchouk polynomial, the degree $j+n$ should belong to $\{0,1,\ldots,2j\}$).

Proof. The validity of both equations is related to a higher level hypergeometric series than ${}_2F_1$, namely ${}_3F_2$. For this purpose, we can use the following relation [10, (39)]:

$$ {}_2F_1\left(\begin{matrix}-2x,-j-n\\ -2j\end{matrix};2\right)=(-1)^x\frac{\binom{j}{x}}{\binom{2j}{2x}}\,{}_3F_2\left(\begin{matrix}-n,n,-x\\ 1/2,-j\end{matrix};1\right). \tag{26}$$

Using this equation for the ${}_2F_1$ expressions in (24) leads to the following identity to verify:

$$ j\,{}_3F_2\left(\begin{matrix}-n,n,-x-1\\ 1/2,-j\end{matrix};1\right)=(x+1)\,{}_3F_2\left(\begin{matrix}-n,n,-x\\ 1/2,-j+1\end{matrix};1\right) \\ -(x-j+1)\,{}_3F_2\left(\begin{matrix}-n,n,-x-1\\ 1/2,-j+1\end{matrix};1\right). \tag{27}$$

Now, to prove the correctness of (27) it is sufficient to simplify the Pochhammer symbols in the expansion of the ${}_3F_2$'s in the right hand side, collect terms, and the left hand side appears automatically.

The proof of (25) is similar. In this case, using (26) leads to the following identity:

$$ \frac{j^2-n^2}{j}\,{}_3F_2\left(\begin{matrix}-n,n,-x\\ 1/2,-j+1\end{matrix};1\right)=(x+\frac{1}{2})\,{}_3F_2\left(\begin{matrix}-n,n,-x-1\\ 1/2,-j\end{matrix};1\right) \\ -(x-j+\frac{1}{2})\,{}_3F_2\left(\begin{matrix}-n,n,-x\\ 1/2,-j\end{matrix};1\right). \tag{28}$$

The verification of this again depends on elementary manipulations of Pochhammer symbols. □

Now, our aim is to construct eigenvalues and eigenvectors of M^q (15). This is rather technical, but is essentially based on (24) and (25). For this purpose, we construct the matrix U in the following proposition.

Proposition 3 *Let $M^q \equiv 2\hat{q}$ be the tridiagonal $(2j+1)\times(2j+1)$-matrix (15) and let $U=(U_{kl})_{0\le k,l\le 2j}$ be the $(2j+1)\times(2j+1)$-matrix with matrix elements as follows. For an even row index $2i$ $(i=0,1,\ldots,j)$, let*

$$ \begin{aligned} &U_{2i,j-k}=U_{2i,j+k}=(-1)^i 2^{k-j}\tilde{K}_{2i}(k;\tfrac{1}{2},2j),\ k\in\{1,\ldots,j-1\}; \\ &U_{2i,0}=U_{2i,2j}=(-1)^i 2^{-1/2}\tilde{K}_{2i}(j;\tfrac{1}{2},2j); \\ &U_{2i,j}=(-1)^i 2^{-j+1/2}\tilde{K}_{2i}(0;\tfrac{1}{2},2j). \end{aligned} \tag{29}$$

For an odd row index $2i+1$ $(i=0,1,\ldots,j-1)$, let

$$ \begin{aligned} &U_{2i+1,j-k}=-U_{2i+1,j+k}=(-1)^{i+1}2^{k-j}\tilde{K}_{2i}(k-1;\tfrac{1}{2},2j-2),\ k\in\{1,\ldots,j-1\} \\ &U_{2i+1,0}=-U_{2i+1,2j}=(-1)^{i+1}2^{-1/2}\tilde{K}_{2i}(j-1;\tfrac{1}{2},2j-2); \\ &U_{2i+1,j}=0. \end{aligned} \tag{30}$$

Then U is an orthogonal matrix:

$$ UU^T=U^TU=I. \tag{31}$$

Furthermore, the columns of U are the eigenvectors of M^q, i.e.

$$ M^qU=UD^q, \tag{32}$$

where $D^q=\mathrm{diag}(\epsilon_0,\epsilon_1,\ldots,\epsilon_{2j})$ is a diagonal matrix containing the eigenvalues ϵ_k of M^q:

$$ \epsilon_k=-2\sqrt{(j-k)(j+k)},\quad \epsilon_j=0,\quad \epsilon_{2j-k}=2\sqrt{(j-k)(j+k)},\quad (k=0,\ldots,j-1). \tag{33}$$

Proof. The proof of this proposition is quite similar to [10, Proposition 3]. The matrix relations (32) follow directly from the two difference equations (24) and (25). The orthogonality of the matrix U is slightly tricky, and does not follow (as one would expect) directly from the orthogonality of symmetric Krawtchouk polynomials. Instead, just as in the proof of Proposition 2 we need to go to higher level hypergeometric series. Reexpressing the ${}_2F_1$'s in terms of ${}_3F_2$'s by means of (26), the corresponding ${}_3F_2$'s can be related to Hahn polynomials [9] with parameters $\alpha = \beta = -1/2$. The orthogonality of these Hahn polynomials leads to:

$$\sum_{x=0}^{N} \frac{1}{2^{2N-1}} \binom{2N}{n} \binom{2N}{2x} K_n(2x; \tfrac{1}{2}, 2N) K_{n'}(2x; \tfrac{1}{2}, 2N) = \begin{cases} \delta_{n,n'}, & \text{if } n \neq N, \\ 2\delta_{n,n'}, & \text{if } n = N, \end{cases} \tag{34}$$

for $n, n' \in \{0, 1, \ldots, N\}$. The extra factor 2 for $n = n' = N$ appears because for that case one has to take a limit of the squared norm of the Hahn polynomial, so one gets another factor than in the other cases where one can just substitute $\alpha = \beta = -1/2$. Equation (34) is the underlying equation proving the orthogonality of the columns of U. The extra factor 2 is also responsible for the extra factors $2^{1/2}$ when $k = j$. □

This proposition gives us essentially the eigenvalues q of the position operator $\hat{q}$. There are $2j + 1$ distinct eigenvalues and they are given by:

$$-j, -\sqrt{j^2 - 1}, \ldots - \sqrt{j^2 - (j-1)^2}, 0, \sqrt{j^2 - (j-1)^2}, \ldots, \sqrt{j^2 - 1}, j, \tag{35}$$

or in general:

$$q_{\pm(j-k)} = \pm\sqrt{(j-k)(j+k)}, \quad k = 0, 1, 2, \ldots, j. \tag{36}$$

In Figure 1, we present a plot of the spectrum of the position operator for the $\mathfrak{su}(2)$, $\mathfrak{sl}(2|1)$ and $\mathfrak{su}(2)_{\mathcal{CP}}$ oscillator models. One observes that the spectrum of the position operator is quite different compared to the finite-discrete oscillator models already known. The main difference is the distribution of positions. For the $\mathfrak{su}(2)$ oscillator model, the positions are equidistant ($q_{\pm k}^{\mathfrak{su}(2)} = \pm k$). The distribution of positions for the $\mathfrak{sl}(2|1)$ model is not equidistant, due to $q_{\pm k}^{\mathfrak{sl}(2|1)} = \pm\sqrt{k}$. In that case, the interval covering the positions is smaller than the corresponding one for the $\mathfrak{su}(2)$ case. Comparing the $\mathfrak{su}(2)_{\mathcal{CP}}$ case with two others, we observe that some of its properties overlap with $\mathfrak{su}(2)$ and others with $\mathfrak{sl}(2|1)$. The interval covering the positions is the same for the $\mathfrak{su}(2)$ and the $\mathfrak{su}(2)_{\mathcal{CP}}$ model: $q_{\pm j}^{\mathfrak{su}(2)} = q_{\pm j}^{\mathfrak{su}(2)_{\mathcal{CP}}}$. Both the $\mathfrak{sl}(2|1)$ and $\mathfrak{su}(2)_{\mathcal{CP}}$ oscillator models have a non-equidistant distribution of the position eigenvalues. The main feature of the $\mathfrak{su}(2)_{\mathcal{CP}}$ oscillator model is that the position values are concentrated near the border of the inteval $[-j, +j]$, whereas near the middle of the interval the distribution is sparse.

4. Position and momentum wavefunctions

The position wavefunctions of the $\mathfrak{su}(2)_{\mathcal{CP}}$ finite-discrete oscillator model are defined in a similar way as in [10]. Their explicit expression is computed through the overlap between the $\hat{q}$-eigenvectors $|j, q_{j-k}\rangle$ and the $\hat{H}$-eigenvectors $|j, m\rangle$. Let us denote them by $\Psi_{j+m}(q)$, where $m = -j, -j+1, \ldots, +j$, and where q assumes one of the discrete values of q_k with $(k = -j, -j+1, \ldots, +j)$. Simple computations show that

$$\Psi_{j+m}(q_{j-k}) = \langle j, m | j, q_{j-k}\rangle = U_{j+m, j+k}, \tag{37}$$

where $U_{j+m,j+k}$ are the matrix elements of M^q. Then, one can find that for the even case $j + m = 2n$ and positive values of position, the wavefunctions have the following expression:

$$\Psi_{2n}(q_k) = (-1)^n 2^{k-j} \tilde{K}_{2n}(k; \tfrac{1}{2}, 2j), \quad n = 0, 1, \ldots, j, \quad k = 1, \ldots, j-1, \tag{38}$$

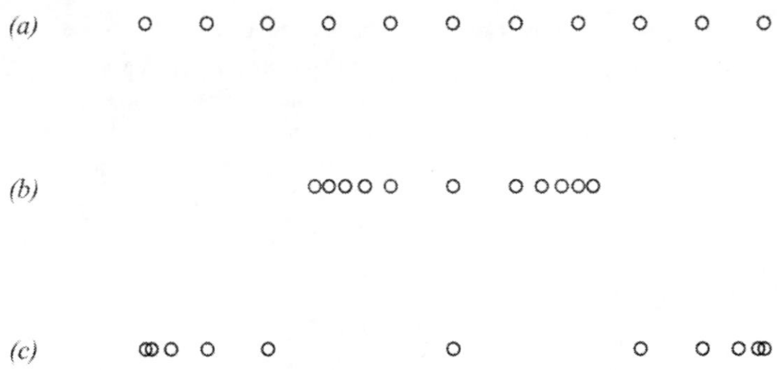

Figure 1. Spectrum of the position operator for $j = 5$, (a) in the case of the $\mathfrak{su}(2)$ model, (b) in the case of the $\mathfrak{sl}(2|1)$ model and (c) in the case of the $\mathfrak{su}(2)_{\mathcal{CP}}$ model.

whereas, for the odd case $j + m = 2n + 1$ and positive values of position, we have

$$\Psi_{2n+1}(q_k) = (-1)^n 2^{k-j} \tilde{K}_{2n}(k-1; \tfrac{1}{2}, 2j-2). \tag{39}$$

One can extend these computations and obtain similar expressions for zero and negative values of the positions.

In Figure 2 we plot the discrete wavefunctions (38) and (39) for some large j-value, and for some n-values. Due to the unique property of the position eigenvalues to be located near the border of the interval $[-j, +j]$, the behaviour of the wavefunctions also reflects this peculair distrubution.

We can define momentum wavefunctions $\Phi_{j+m}(p)$ in a similar way. Their explicit expression is computed through the overlap between the $\hat{p}$-eigenvectors $|j, p_{j-k}\rangle$ and the $\hat{H}$-eigenvectors $|j, m\rangle$. In this case, the simple relation between M^p and M^q shows that

$$\Phi_{j+m}(p_{j-k}) = \langle j, m | j, p_{j-k}\rangle = V_{j+m,j+k}, \quad p_{j-k} \equiv q_{j-k}, \quad V_{k,l} \equiv \mathrm{i}^{k+1} U_{k,l}. \tag{40}$$

Then, in accordance with equations (38) and (39), one finds that for the even case $j + m = 2n$ and for positive values of momentum, the wavefunctions have the following expression:

$$\Phi_{2n}(p_k) = \mathrm{i}\, 2^{k-j} \tilde{K}_{2n}(k; \tfrac{1}{2}, 2j), \quad n = 0, 1, \ldots, j, \quad k = 1, \ldots, j-1, \tag{41}$$

whereas for the odd case $j + m = 2n + 1$ and positive values of momentum, we have

$$\Phi_{2n+1}(p_k) = -2^{k-j} \tilde{K}_{2n}(k-1; \tfrac{1}{2}, 2j-2). \tag{42}$$

These computations can easily be extended for zero and negative values of the momentum. Both wavefunctions $\Psi(q)$ and $\Phi(p)$ are related by the so called $\mathcal{CP}$ symmetric Krawtchouk transform:

$$\Phi(p_l) = \sum_{k=-j}^{j} \mathcal{K}_{kl} \Psi(q_k), \tag{43}$$

where, $\mathcal{K} = U^T V$. The matrix elements $\mathcal{K}_{k,l}$ can be computed by using (34).

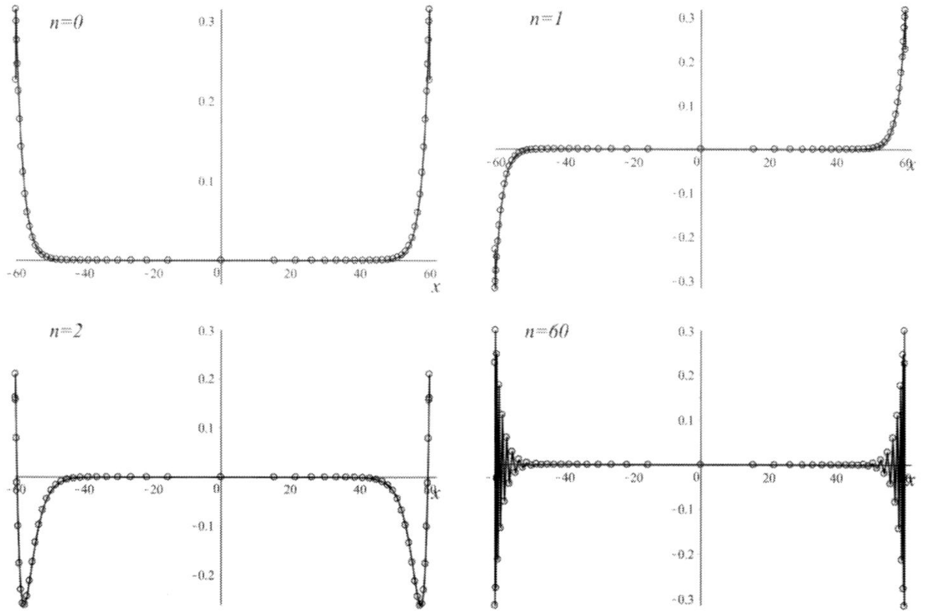

Figure 2. Plots of the discrete wavefunctions $\Psi_n(q)$ in the representation with $j = 30$. We plot the wavefunctions for $n = 0, 1, 2, 60$.

5. Discussion

Despite the fact that the non-relativistic one-dimensional quantum oscillator in the canonical approach has a nice explicit solution in terms of Hermite polynomials, there are still attempts to construct new oscillator models inspired by different starting points. One of these starting points is the assumption of discrete space, in which case one is looking for oscillator models with a discrete position spectrum (which could be finite of infinite discrete). The most popular finite discrete oscillator model [6,7] is based on the $\mathfrak{su}(2)$ algebra, and uses Krawtchouk polynomials.

In current paper, we have constructed a new model for the one-dimensional oscillator, with wavefunctions expressed in terms of the Krawtchouk polynomials with parameter $p = 1/2$ (the symmetric Krawtchouk polynomials). In order to construct a solvable oscillator model, we have first defined a new algebra, formally called the $\mathfrak{su}(2)$ algebra under the $\mathcal{CP}$ deformation. The representations of this algebra were given and are similar to representations of $\mathfrak{su}(2)$, but the action of the raising and lowering operators J_+ and J_- differs considerably. As a consequence, when considering the oscillator model related to this algebra, the matrix representations of position and momentum operators is also different. In the standard $|j, m\rangle$ bases, the position operator is tridiagonal, and we have shown that its eigenvalues and eigenvectors involve Krawtchouk polynomials. The matrix relation for the eigenvalue problem is equivalent to certain finite difference equations for Krawtchouk polynomials.

The resulting model is unique in the sense that it does not generalize any other existing oscillator model. Peculiar properties are the non-equidistant distribution of the position values as well as the behaviour of the wavefunctions around zero and near the border of the maximum and minimum eigenvalues.

Acknowledgements
EIJ kindly acknowledges support from Research Grant **EIF-2012-2(6)-39/08/1** of the Science Development Foundation under the President of the Republic of Azerbaijan.

References
[1] Yap P T, Paramesran R and Ong S H 2003 *IEEE Trans. Image Processing* **12** 1367
[2] Sasaki R 2009 *J. Math. Phys.* **50** 103509
[3] Albanese C, Christandl M, Datta N and Ekert A 2004 *Phys. Rev. Lett.* **93** 230502
[4] Landau L D and Lifshitz E M 1997 *Quantum Mechanics: Non-Relativistic Theory* (Oxford: Butterworth-Heinemann)
[5] Atakishiyev N M and Suslov S K 1990 *Theor. Math. Phys.* **85** 1055
[6] Atakishiyev N M, Pogosyan G S, Vicent L E and Wolf K B 2001 *J. Phys.* A **34** 9381
[7] Atakishiyev N M, Pogosyan G S and Wolf K B 2005 *Phys. Part. Nuclei* **36** 247
[8] Jafarov E I and Van der Jeugt J 2012 *J. Phys.* A **45** 275301
[9] Koekoek R, Lesky P A and Swarttouw R F 2010 *Hypergeometric orthogonal polynomials and their q-analogues* (Berlin: Springer-Verlag).
[10] Jafarov E I, Stoilova N and Van der Jeugt J 2011 *J. Phys.* A **44** 355205

Journal of Physics: Conference Series **597** (2015) 012048 doi:10.1088/1742-6596/597/1/012048

Projective representations from quantum enhanced graph symmetries

R M Kaufmann[1], **S Khlebnikov**[2] **and B Wehefritz–Kaufmann**[1,2]

[1]Department of Mathematics, Purdue University, West Lafayette, IN 47907
[2]Department of Physics, Purdue University, West Lafayette, IN 47907

E-mail: rkaufman@math.purdue.edu, skhleb@physics.purdue.edu, ebkaufma@math.purdue.edu

Abstract. We define re–gaugings and enhanced symmetries for graphs with group labels on their edges. These give rise to interesting projective representations of subgroups of the automorphism groups of the graphs. We furthermore embed this construction into several higher levels of generalization using category theory and show that they are natural in that language. These include projective representations of the re–gauging groupoid and a novel generalization to all symmetries of the graph.

Introduction

In [4], we developed a method of re–gaugings and actions of enhanced graphs symmetries for labelled graphs. The upshot were projective representations that are of interest in condensed matter physics. In those applications there is an underlying geometry at work, but the method itself is more general and can be generalized or reduced to a combinatorial group theoretic framework, which we will present. Presently, we will label the edges of a graph by elements of a group G and we give a presentation of the actions that is precise and concise. The precision is needed, since there are several actions (both left and right) at work which need to be disentangled. We present a new result on the action of general symmetries.

We will also recall the particular projective representations we found in [4] since their occurrence "in nature" as natural symmetries might be of interest to group theorists as well as physicists and give an example of the new result on the action of general symmetries. Lastly, we give a new presentation of our constructions in the language of categories and show that they become very natural.

This paper is at the same time more general and more specific than [4]. In *loc. cit.* the labels were invertible elements of a C^*–algebra, here they live in a general group. The geometry of [4] is then recovered by specializing the group to invertible functions on tori, or more generally invertible elements in a not necessarily commutative C^* algebra. We also give a more technically precise account of the actions. On the other hand, [4] deals with the possibility of an actual groupoid representation, that is invertible morphisms between different vector spaces, while here, we work in the situation where there is only one underlying vector space. The categorical interpretation is entirely new.

1. Combinatorics and graphs

In this section, we give the details of the following construction of representations: Fix a connected graph Γ with k vertices and a labelling lab of its directed edges by group elements of a fixed group G such that edges in opposite orientation are labelled by inverse group elements. We define a group of quantum automorphisms $Aut_q(\Gamma, lab)$ of such a Γ–labelled graph. This is a subgroup of the automorphisms of the graph, which preserves the labels up to re-gauging, defined below. We show that after fixing an ordered rooted spanning tree (ORST) τ of the graph, there is a natural way to attach a $k \times k$ matrix with coefficients in G to each quantum automorphism. In the case that G is Abelian these matrices give a projective representation of the group of quantum automorphisms of (Γ, lab), into the group of $k \times k$ matrix with coefficients in G. The co–cycle is explicitly given and defines a group extension $\widehat{Aut}_q(\Gamma, lab)$. Thus applying any representation to G, we get a projective representation of $Aut_q(\Gamma, lab)$ and a representation of $\widehat{Aut}_q(\Gamma, lab)$. A different choice of ORST gives rise to a projective representation.

1.1. Graphs, paths and spanning trees

In this paragraph, we fix the notations we will be using. This is necessary to be able to be precise later on. A graph is a collection of vertices V, flags or half–edges F, a boundary map $\partial : F \to V$ which attaches to each half–edge its vertex and a fixed point free involution $\imath : F \to F$ on F. An edge is then a pair of half–edges $\{f_1, f_2\}$ constituting an orbit of $\imath$: $f_1 = \imath(f_2)$. We denote the set of edges by E. Then ∂ associates to each edge the set of its endpoints. An orientation of an edge is the choice of ordering of its two flags. Each edge has two possible orientations, which we call opposite. For an oriented edge $\vec{e} = (f_1, f_2)$ we set $s(e) = \partial(f_1)$ and $t(e) = \partial(f_2)$ and call them source and target. Notice that for small loops, whose flags are incident to the same vertex, both orientations have the same sources and targets. This is why we chose to use the more elaborate way to present graphs above. We can also identify an orientation of an edge by the choice of the first half edge. In this way the set F is naturally the set of all oriented edges. The bijection is given by $f \leftrightarrow (f, \imath(f))$. Using this bijection, we define $s(f) = f$ and $t(f) = \imath(f)$ and from now on think of flags as oriented edges. In this notation $\imath$ flips the orientation if $f \leftrightarrow (f, \imath(f)) = \vec{e}$ then $\imath(f) \leftrightarrow (\imath(f), f) =: \imath(\vec{e}) =: \overleftarrow{e}$.

1.1.1. Spanning trees, orders and action of the permutation groups A spanning tree of a graph is a subgraph that is a tree (i.e. its realization is contractible) whose vertices are all the vertices of the tree. A rooted spanning tree (RST) is such a spanning tree together with the choice of a root. The edges in a rooted tree have a natural orientation by directing the edges *away* from the root; see Figure 1. In any rooted tree τ, any vertex v has a unique shortest path to and from the root vertex v_{rt}, which we will call $\gamma^{\tau}_{v_{rt}v}$ and $\gamma^{\tau}_{vv_{rt}} = (\gamma^{\tau}_{v_{rt}v})^{-1}$.

An order on a graph is a bijection $ord : V \to \{1, \ldots, |V|\}$. A compatible order for a rooted spanning tree has to have $ord(v_{rt}) = 1$. Denote $v = ord^{-1}$ and write $v(i) = v_i$ then $ord(v_i) = i$, and for a rooted spanning tree with a compatible order $v_{rt} = v_1$[1]. An ordered rooted spanning tree (ORST) is a spanning tree together with a compatible order. Notice that an order for a spanning tree gives an order for the graph. The permutation group $\mathbb{S}_k$, $k = |V|$, naturally acts on orders and inverse orders via $\sigma(v) = v \circ \sigma^{-1}$. Setting $v' = \sigma(v)$ this means that $v'_i = v_{\sigma^{-1}(i)}$. Furthermore, $\sigma \in \mathbb{S}_n$ induces a bijection $\sigma_V : V \to V$ on the set of vertices via

$$\sigma_V \circ v = v \circ \sigma^{-1} \tag{1.1}$$

that is $\sigma_V = v \circ \sigma^{-1} \circ ord$. In other words, $\sigma_V(v_i) = v'_i = v_{\sigma^{-1}(i)}$. And vice–versa given $\Sigma_V : V \to V$ a bijection, it determines an element σ of $\mathbb{S}_k$ via $\sigma = ord \circ \sigma_V^{-1} \circ v$.

[1] In [4] we used v_0 for the root, which would mean that $v_0 = v_1$.

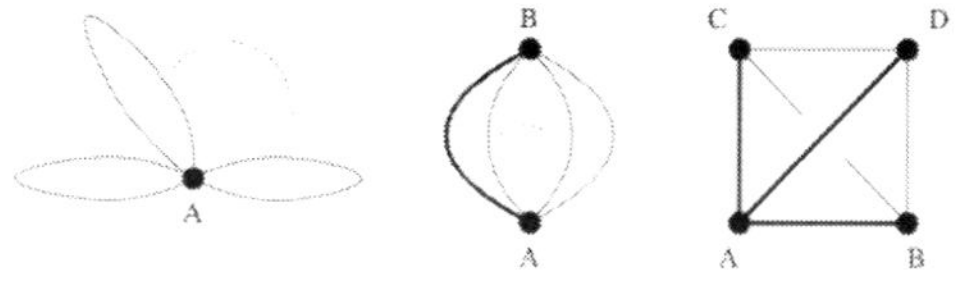

P_n D_n G

Figure 1. Graphs with spanning trees. The root is A. The petal graphs P_n, the graphs D_n and the graph G

Remark 1.1. Given any order on Γ, one can obtain a map $Aut(V) \to \mathbb{S}_k$. The way the action is set up is that composition in $\mathbb{S}_k$ corresponds to composition of σ_V. In particular, if $v' = \sigma(v)$ and $v'' = \sigma'(v')$ then $v''_i = v'_{(\sigma')^{-1}(i)} = v_{\sigma^{-1}(\sigma')^{-1}(i)} = v_{(\sigma'\sigma)^{-1}}(i)$. Notice that the order used at each step is the induced order.

1.1.2. π_1 of a graph An (edge)-path on a graph is a sequence of oriented edges, $\vec{e}_i$: $i = 1 \dots n$ together with an orientation of each e_i such that for any two consecutive $\vec{e}_i, \vec{e}_{i+1}$: $t(\vec{e}_i) = s(\vec{e}_{i+1})$. An edge path is called reduced, if no two consecutive oriented edges are opposites of each other. Concatenation of paths and inverting a path makes the set of all reduced paths into the so–called path groupoid $\pi_1(\Gamma)$. If in the concatenation of reduced paths there is a pair of adjacent opposite edges, they are simply deleted in the product of π_1. The reduced loops at a fixed vertex $\pi_1(\Gamma, v)$ form a group, the fundamental group at v.[2]

Contracting all of the edges of a spanning tree of Γ leaves a graph with one vertex whose realization is a bouquet of n S^1s, we call such a graph a petal graph P_n. It is well known that $\pi_1(\Gamma) \simeq \pi_1(P_n) = \mathbb{F}_n$, the free group in $n = 1 + \chi(\Gamma)$ generators, where $\chi(\Gamma) = |V| - |E|$ is the Euler characteristic. Each spanning tree τ gives a different isomorphism. The inverse map is given by associating the following path to a small loop l around a petal of P_n. Let $\vec{e}$ be the unique oriented edge of the preimage of l that is not in the spanning tree and let $s(e) = v, t(e) = w$. Then the inverse image of l is the path $\gamma^{\tau}_{v_{rt}w} \circ \vec{e} \circ \gamma^{\tau}_{vv_{rt}}$. Here and below, we use the concatenation of paths $\circ$ in the order of functions, that is $p \circ q$ is: first go through q then through p.

1.2. Labelled graphs, gauging and spanning tree gauge.
A G–labelling for a graph Γ is a morphism $lab : F \to G$ such that $lab(\imath(f)) = lab(f)^{-1}$. Or identifying flags with oriented edges in the notation of §1.1: $lab(\vec{e}) = lab(\overleftarrow{e})^{-1}$. A Γ–labelling lab extends to paths in the following way: let $p = (\vec{e}_1, \dots, \vec{e}_n)$ be a path in Γ, then we define $lab(p) := lab(e_n) \dots lab(e_1)$. Since going back and forth along an edge yields the identity in G, this descends to $lab : \pi(\Gamma) \to G$. A G–labelled graph is a pair of a graph and a G–labelling. A G–labelled graph with a RST τ is in spanning tree gauge, if $lab(\vec{e}) = 1$ for all oriented edges/flags $\vec{e}$ in τ.

Remark 1.2. Of course by composing a G–labelleing by a group homomorphism $G \to H$ one obtains an H–labelleing and all constructions below will push–forward. The most general

[2] The data of vertices and edges and ∂ define a 1–dimensional CW complex. If Γ is a graph we denote by $|\Gamma|$ its realization, given by gluing together intervals (one for each edges) at their end points if these correspond to shared vertices. An edge path gives an actual path on $|\Gamma|$ starting and ending at a vertex. A path is reduced if the realization does not go back and forth through an edge. It is a fact that $\pi(|G|) = \pi(\Gamma)$ and $\pi_1(|G|, v) = \pi_1(\Gamma, v)$ are the topological fundamental groupoid and group.

labelling one can have in a spanning tree gauge is to label the edges not in the spanning tree by the generators of a free group. This will then be the free group on $1-\chi(\Gamma)$ generators, where $\chi(\Gamma)$ is the Euler characteristic of Γ. Another useful example for physics is when $G=U(1)$. This is of course also the universal receptacle of unitary characters.

Definition 1.3. A gauge element ϕ is a map $\phi: V \to G$. The gauging of a G-labelling lab by a gauge element ϕ is defined to be the labelling lab^ϕ given by $lab^\phi(\vec{e}) = \phi(t(\vec{e}))lab(\vec{e})\phi(s(\vec{e}))^{-1}$, for any oriented edge $\vec{e}$. The gauge elements form a group, via point–wise multiplication, the gauge group of Γ. It is isomorphic to G^V and after the choice of an order to G^k where $k=|V|$.

Two labellings lab and lab' are called gauge–equivalent if they differ by a gauging ϕ, i.e. $\exists\phi : lab' = lab^\phi$.

Lemma 1.4. *Given a G–labelled graph (Γ, lab) and a rooted spanning tree (τ, v_{rt}), then there is a gauge element ϕ such that (Γ, lab^ϕ) is in spanning tree gauge w.r.t. τ. Explicitly, $\phi(v) = lab(p)$, where p is the shortest edge path from v to v_{rt} along τ, will be such an element and it is the unique such element satisfying he condition $\phi(v_{rt}) = 1 \in G$.*

Proof. If one sets $\phi(v'_{rt}) = 1 \in G$ then one can iteratively solve for ϕ of all the other vertices using the distance from the root as determined by the RST. The result is the above explicit gauge element. □

To make all dependencies clear, we will sometimes write $\phi = \phi^{lab}(\to \tau)$. Given lab, we will write $lab_\tau := lab^{\phi(\to\tau)}$ and write $\phi^{lab_\tau}(\tau \to \tau')$ for the re–gauging of lab_τ into spanning tree gauge w.r.t. τ'.

There is some ambiguity in the choice of ϕ above. One parameter is the initial condition $\phi(v'_{rt}) = 1 \in G$.

Lemma 1.5. *Given any lab in spanning tree gauge, its stabilizer subgroup under gauging will be the* constant functions *$\phi : V \to Z(H)$, where H is the subgroup of G generated by all the $lab(\vec{e})$ and $Z(H)$ is its centralizer in G. If G is Abelian this subgroup is isomorphic to G in the diagonal embedding into G^V and the different choices of ϕ of Lemma 1.4 are distinguished by their value on the root $\phi(v_{rt})$. For any lab its stabilizer subgroup is conjugate (inside G^V) to such a subgroup.*

Proof. Fix $\phi(v_{rt}) = g$ then along any edge of the spanning tree the label is 1, so that invariance of lab means that the function ϕ must be constant $\phi(v) \equiv g$. Furthermore for lab to be invariant, we need that for any non–spanning tree edge $\vec{e}$: $g\, lab(\vec{e})g^{-1} = lab(\vec{e})$, so that indeed $g \in Z(H)$. The other statements follow from this. □

Corollary 1.6. *If the image of lab generates G then $G^V/Z(G)$ acts as the effective gauge group. If G is Abelian, then for any lab its stabilizer group is G and the quotient G^V/G acts effectively.* □

Corollary 1.7. *If lab and lab′ are gauge equivalent, and lab is in spanning tree gauge w.r.t. the RST τ and lab′ is in spanning tree gauge w.r.t. τ', then there is a gauge element ϕ of the form given in Lemma 1.4 such that $lab' = lab^{\phi\circ\psi}$, where ψ is a constant regauging that is unique up to $Z(H)$. In particular if G is Abelian we can choose $\psi = id$.*

Proof. The assumption is that there exists a ϕ, such that $lab' = lab^\phi$. Any other gauging element ϕ' satisfying $lab \circ \sigma_E^{-1} = lab^{\phi'}$ would differ by an element in the stabilizer of lab which by the above lemma is a constant function to G. So, by using such a function, we can (a) normalize the given ϕ to the one of Lemma 1.4 and (b) obtain any other ϕ that solves the problem from this one via elements in the stabilizer of lab. □

Corollary 1.8. *Up to a constant regauging* ψ, $lab_\tau^{\phi^{lab_\tau}(\tau\to\tau')} = lab_{\tau'}^\psi$ *and if* G *is Abelian* $lab_\tau^{\phi^{lab_\tau}(\tau\to\tau')} = lab_{\tau'}$.

Proof. Obviously the two labellings are gauge equivalent, so since they are both in spanning tree gauge w.r.t. τ' they agree up to a constant regauging, which acts trivially in the Abelian case. □

1.2.1. Re–gauging matrices For any triple (lab, τ, τ'), with τ, τ' ORSTs and lab in spanning tree gauge with respect to τ, let $\sigma \in \mathbb{S}_k$ be the permutation, such that $ord' = \sigma \circ ord$ or, in other words, $v'_i = v_{\sigma^{-1}(i)}$. Let $\phi = \phi^{lab}(\tau \to \tau')$ be the gauge element of Lemma 1.4 that re–gauges lab for spanning tree gauge τ to spanning tree gauge τ' and set $\Phi = diag(\phi(v'_1), \dots, \phi(v'_k))$. Then define

$$M_{lab_\tau}(\tau \to \tau') := \Phi P_\sigma \tag{1.2}$$

here and sometimes below we added the redundant subscript τ as an indicator that lab is in spanning tree gauge w.r.t. τ to help keep track.

1.2.2. Cocycle and re–gauging groupoid action For any ordered pair of RST, (τ, τ') with roots v_{rt} and v'_{rt}, we define $p(\tau, \tau') := \gamma^\tau_{v'_{rt}v_{rt}}$ to be the shortest path from v_{rt} to v'_{rt} in τ' and for any triple of RSTs (τ, τ', τ'') with roots $v_{rt}, v'_{rt}, v''_{rt}$ set:

$$l(\tau, \tau', \tau'') := p^{-1}(\tau, \tau')p(\tau', \tau'')p(\tau, \tau'') = \gamma^\tau_{v_{rt}v'_{rt}}\gamma^{\tau'}_{v'_{rt}v''_{rt}}\gamma^\tau_{v''_{rt}v_{rt}} \tag{1.3}$$

which is a loop at v_{rt}.

Proposition 1.9. *[4] The matrices above satisfy*

$$M_{lab_{\tau'}}(\tau' \to \tau'')M_{lab_\tau}(\tau \to \tau') = C(\tau, \tau', \tau'')M_{lab_\tau}(\tau \to \tau') \tag{1.4}$$

where $C(\tau, \tau', \tau'') = lab_\tau(l(\tau, \tau', \tau''))$.

1.3. Quantum enhanced symmetries for labelled graphs
A symmetry σ_Γ of a graph Γ as above is a pair (σ_V, σ_F) of a bijection of the set of vertices $\sigma_V : V \to V$ and a bijection of set of flags σ_F which is compatible with ∂, and $\imath$ i.e. $\partial(\sigma_F(f)) = \sigma_F(\partial(f))$ and $\imath(\sigma_F(f)) = \sigma_V(\imath(f))$. These symmetries from the group $Aut(\Gamma)$. The map on flags induces a map on directed edges which we denote by σ_E.

Any element σ_Γ of $Aut(\Gamma)$ pushes forward RSTs. If τ is an RST, then $\tau' = \sigma_V(\tau)$, the image of τ, is again an RST, with root $\sigma_V(v_{rt})$. Symmetries of the graph act on a G–labelling as follows: $\sigma_\Gamma(lab) = lab \circ \sigma_E^{-1}$. On orders they act according to $\sigma_\Gamma(ord) = ord \circ \sigma_V^{-1}$, so that $ord'(v_i) = \sigma(i)$, and $v'_i = v_{\sigma^{-1}(i)}$ give the induced order on τ'.

Remark 1.10. Given an order ord there is a map $Aut(\Gamma) \to \mathbb{S}_k$. The composition of automorphisms on vertices corresponds to the composition of elements of symmetric group as in Remark 1.1. An element of $\mathbb{S}_k$ is liftable if it is in the image of this map. A symmetric graph is a graph where this map is bijective. An example is the graph G of Figure 1.1.2 or more generally the complete graph K_n.

A classical symmetry for a labelled graph (Γ, lab) is a symmetry of the graph that also satisfies $\sigma_\Gamma(lab) = lab \circ \sigma_E = lab$. These from the group $Aut(\Gamma, lab)$. If the labelling is constant this is the symmetry group of the graph. In general this will be a subgroup of it.

Definition 1.11. A *quantum enhanced symmetry* of a labelled graph is a symmetry σ_Γ of the underlying graph, such that there is a gauge group element ϕ for which $(lab \circ \sigma_E)^\phi = lab$. If lab is in spanning tree gauge w.r.t. τ, then a quantum enhanced symmetry is strict if the ϕ can be chosen to be $\phi^{lab\circ\sigma}(\sigma^{-1}(\tau) \to \tau)$.[3]

The quantum enhanced symmetry group of the labelled graph will be a subgroup $Aut_q(\Gamma, lab) \subset Aut(\Gamma)$ of the symmetry group of the underlying graph which contains the classical symmetry group of the labelled graph $Aut_q(\Gamma, lab) \supset Aut(\Gamma, lab)$. It usually is strictly bigger, see §3.[4] For a lab in spanning tree gauge w.r.t. the strict quantum enhanced symmetry group $Aut^0_q(\Gamma, lab)$ is a subgroup of $Aut_q(\Gamma, lab)$. Notice that if G is Abelian $Aut^0_q(\Gamma, lab) = Aut_q(\Gamma, lab)$.

Remark 1.12. Notice that if lab is in spanning tree gauge w.r.t. τ, then $\sigma_V(lab)$ is in spanning tree gauge w.r.t. $\sigma_V(\tau)$. Thus to check if σ_Γ is an enhanced quantum symmetry, for a lab in spanning tree gauge w.r.t. a tree τ, by Corollary 1.7, we only have to check if $lab \circ \sigma_E = lab^{\phi\circ\psi}$ for the preferred quantum enhanced symmetry ϕ given in Lemma 1.4 up to a constant ψ that transforms $lab \circ \sigma_V^{-1}$ into spanning tree gauge w.r.t. to τ. If G is Abelian or σ_V is strict, then we can choose $\psi \equiv 1$.

1.4. Projective representation for quantum enhanced symmetries

Fix a G–labelled graph (Γ, lab) and an ORST τ. Assume that lab is in spanning tree gauge with respect to τ. This is without loss of generality, since by Lemma 1.4, for any RST τ there is an invertible gauge element $\phi(\tau)$ such that $lab^{\phi(\tau)}$ is in spanning tree gauge w.r.t. τ. For $\sigma_\Gamma \in Aut_q(\Gamma, lab)$, set

$$\rho_{lab_\tau}(\sigma_\Gamma) := M_{lab_\tau}(\tau \to \sigma(\tau))$$

Theorem 1.13. *[4] Let lab_τ be in spanning tree gauge w.r.t. the ORST τ, then ρ_{lab_τ} is a projective representation of $Aut^0_q(\Gamma, lab)$ with cocycle $C(\sigma'_\Gamma, \sigma_\Gamma) = \rho(l(\sigma'_\Gamma \circ \sigma_\Gamma(\tau_0), \sigma_\Gamma(\tau_0), \tau_0)$.* □

Recall that if G is Abelian, then $Aut^0_q = Aut_q$.

Remark 1.14. The concrete form of the action depends on the choice of the preferred gauge elements. If one would consistently use the normalisation $\phi(v_{rt}) = g$ then one would get an equivalent representation. The fact that this is a projective representation is, however, due exactly to this choice of normalisation. When performing several of these gaugings monodromy can appear from the movement of the root of the spanning trees, this is what is captured by the cocyle.

1.4.1. The action as pre–gauging or re–gauging The relationship between the actions on lab and the spanning trees is as follows. Given the data $(lab, \tau, \sigma_\Gamma)$ with lab in spanning tree gauge w.r.t. τ, and σ_Γ strict, there are the two ways to check whether lab and $\sigma_\Gamma^{-1}(lab) = lab \circ \sigma$ are gauge equivalent, provided by Corollary 1.7.

The first is to act by σ^{-1} sending lab to $lab \circ \sigma$ which is in spanning tree gauge $\sigma^{-1}(\tau)$ and then to re–gauge to τ.

$$lab \overset{\text{act by } \sigma_\Gamma^{-1}}{\longrightarrow} lab \circ \sigma \overset{\text{re–gauge}}{\longrightarrow} (lab \circ \sigma)^{\phi^{lab\circ\sigma}(\sigma^{-1}(\tau)\to\tau)}$$

[3] Notice we switched to a right action since spanning trees will push–forward and we will get a representation of these groups in this fashion.

[4] These quantum enhanced symmetries naturally appear in quantum contexts, where the labelling by is $U(1)$ and the re–gauging is a choice of phase–shifts. The stabilizer is then an overall phase–shift, so that the effective phase shifts are in $U(1)^k/U(1)$.

The second pre–gauges lab from τ to $\sigma(\tau)$, so that then applying σ_Γ will put it back into spanning tree gauge w.r.t. τ:

$$lab \overset{\text{pre–gauge}}{\longrightarrow} lab^{\phi^{lab}(\tau\to\sigma(\tau))} \overset{\text{act by }\sigma_\Gamma^{-1}}{\longrightarrow} lab^{\phi^{lab}(\tau\to\sigma(\tau))}\circ\sigma$$

Lemma 1.15. *There is the following symmetry between pre–gauging and re–gauging:*

$$M_{lab_\tau}(\tau\to\sigma_\Gamma(\tau)) = M_{(lab\circ\sigma)_{\sigma_\Gamma^{-1}(\tau)}}(\sigma_\Gamma^{-1}(\tau)\to\tau) \tag{1.5}$$

Proof. The proof is a tedious but straightforward unraveling of definitions. □

This equation (1.5) states that the two matrices obtained from the two gaugings above coincide. We used the latter version above. In the situation where $\sigma_\Gamma \in Aut_q^0(\Gamma, lab)$, we furthermore know that in both cases the final labelling is again lab (*a priori* up to possibily a constant re–gauging) due to Corollary 1.7 and hence *a posteriori* agree. This is why there is an action. The cocycle stems from the fact that the normalisation of two repeated pre– or re–gaugings need not be 1 on the *original* root v_{rt}.

1.4.2. Liftings and extensions We can actually take any triple of an element $\sigma \in Aut(\Gamma)$, an ORST τ and a labelling lab and associate to it the matrix $\rho_{lab,\tau}(\sigma_\Gamma)$, if σ is not in $Aut_q(\Gamma, lab)$ then the re–gauged $(lab\circ\sigma)^{\phi^{lab\circ\sigma}(\sigma^{-1}(\tau)\to\tau)}$ will not equal lab and there is no obvious group action.

There is however an action of a certain groupoid and if the action of $Aut(\Gamma)$ is liftable there is a fixed point group that acts. These will be described in the next section. In this section, we do however define the relevant matrices and matrix products.

Definition 1.16. $\sigma \in Aut(\Gamma)$ is liftable for lab if there is an element $\Psi = \Psi(\sigma) \in Aut(G)$ such that

$$(lab\circ\sigma)_{\sigma^{-1}(\tau)}^{\phi^{lab\circ\sigma}(\sigma^{-1}(\tau)\to\tau)} = (lab\circ\sigma)_\tau = (\Psi\circ lab)_\tau \tag{1.6}$$

Consider the composition of two re–gaugings induced by $\sigma_\Gamma, \sigma'_\Gamma \in Aut(\Gamma)$.

$$\tau \to \sigma_\Gamma(\tau) \to \sigma\sigma'_\Gamma(\tau)$$

Theorem 1.17. *If σ is liftable for lab, then* $\Psi(\rho_{lab_\tau})(\sigma'_\Gamma)\rho_{lab_\tau}(\sigma_\Gamma) = C(\sigma',\sigma)\rho_{lab_\tau}(\sigma'_\Gamma\sigma_\Gamma)$.

Proof. There is the following chain of equalities

$$\begin{aligned}\Psi\circ(M_{lab_\tau}(\tau\to\sigma'(\tau))) &= M_{\Psi\circ lab_\tau}(\tau\to\sigma'(\tau))\\ &= M_{(lab\circ\sigma)_{\sigma^{-1}\tau}^{\phi^{lab\circ\sigma}(\sigma^{-1}(\tau)\to\tau)}}(\tau\to\sigma'(\tau))\\ &= M_{lab_{\sigma(\tau)}}(\sigma(\tau)\to\sigma'(\sigma(\tau))\end{aligned}$$

where the first equality is by definition, the second equality is due to (1.5) and the last equation is a straightforward, but tedious check using the definitions. The claim now follows from (1.4). □

1.5. Remark on Quivers

Instead of graphs, one can use quivers. A quiver or a directed graph $\vec{\Gamma}$ is a graph Γ in which all edges are oriented, this means that we have two maps $s, t : E \to V$. Given a quiver there is a natural forgetful map which forgets the orientation of the edges. We call two quivers groupoid equivalent, if they have the same underlying graph. We will use the notation $\vec{\Gamma}$ for quivers and Γ for the underlying graph. A (directed–edge) path on a quiver is a sequence of directed edges $e_i : i = 1 \dots n$, such that for two consecutive directed edges e_i, e_{i+1}: $t(e_i) = s(e_{i+1})$.

Fix a monoid G. A G–labelled quiver is a quiver $\vec{\Gamma}$ together with a map $lab : E \to G$. There is an obvious extension of ρ to all directed edge paths.

If G is a group, we can extend the definition of ρ to the full path groupoid as follows:. Given an edge path on the underlying graph $p = (e_1, \dots, e_n)$, we define $\epsilon_i = 1$ if the orientation of e_i in the path agrees with that of Γ and $\epsilon_i = -1$ if the orientations are opposite. The formula $\hat{lab}(p) = lab(e_n)^{\epsilon_n} \dots lab(e_1)^{\epsilon_1}$ defines a morphism of the path groupoid into G. We call two G–labelled quivers (Γ, lab) and (Γ', lab') groupoid equivalent, if the underlying graphs are equivalent and $lab(e) = lab'(e)$ whenever their orientations agree and $lab(e) = lab^{-1}(e)$ if the orientations differ in the two graphs. A Γ labelled graph is equivalence class of G–labelled quivers under groupoid equvialence.

2. Categorical formulation and extension of the actions and representations

The notions become totally natural in the language of categories and it is possible to generalize the actions by the individual enhanced quantum automorphisms groups of fixed labellings into an action of a bigger groupoid.

2.1. Categories and fundamental groups

A category $\mathcal{C}$ is a collection/class of objects and for each pair of objects X, Y a collection of morphisms $Hom_{\mathcal{C}}(X, Y)$, called morphisms from X to Y, together with an associative composition morphism $Hom_{\mathcal{C}}(X, Y) \times Hom_{\mathcal{C}}(Y, Z) \xrightarrow{\circ} Hom(_{\mathcal{C}}X, Y)$, and identities $id_x \in Hom(X, X)$. Writing $g \circ f := \circ(f, g)$, being an identity means that $id_X \circ f = f, g \circ id_X = g$ for f any morphism into X and g any morphism out of X. A category is small if the objects and all the $Hom_{\mathcal{C}}(X, Y)$ are sets. If f is a map from X to Y, then X is called the source of f and Y is called the target of f, and one writes $X \xrightarrow{f} Y$.[5]

Maps between categories are called functors. Functors from a category $\mathcal{C}$ to a category $\mathcal{D}$ again form a category $Fun(\mathcal{C}, \mathcal{D})$ whose morphisms are called natural transformations. To keep track for two functors $F, G : \mathcal{C} \to \mathcal{D}$ one writes $Nat(F, G)$ for $Hom_{Fun(\mathcal{C},\mathcal{D})}(F, G)$.

2.2. Groupoids

A groupoid is a category whose morphisms are all isomorphisms. A group correspondes to a groupoid with only one object and vice–versa via the following identification: $Hom(*, *) = G$, where $*$ is the unique object and with composition of morphisms given by group multiplication. For any object x in a groupoid, $Hom(x, x) = End(x) = Aut(x)$ is a group called the automorphisms group of x. A groupoid has a natural (contravariant) automorphism op which is the identity on objects and sends any morphism to its inverse.

In this language a group representation becomes a functor $\underline{G} \to Vect_k$ to the category of

[5] A slick way to give a small category is as a pair of sets, M, O (morphisms and objects) with two maps $M \rightrightarrows^s_t O$, the source and target maps, together with an associative composition morphism $\circ : M\, {}_s\!\times_t M \to M$ and a section $id : O \to M$ of both s and t such that $id_X = id(X)$ is an identity. Here $M\, {}_s\!\times_t M \to M \subset M \times M$ is the relative product consisting of elements (f, g) with $t(f) = s(g)$ and being an identity means the same thing as above.

k-vector spaces.[6] A functor from a groupoid to $\mathcal{V}ect_k$ is given by a collection of vector spaces and isomorphisms between them — one for every morphism in the groupoid. We consider not only functors to $\mathcal{V}ect_k$ but also functors of groupoids from a groupoid Γ to a fixed $\underline{G}$.

A set X determines a groupoid in several ways. One is the discrete groupoid, which just has id_x for $x \in X$ as its morphisms. There is another canonical groupoid, which we call $\mathcal{K}(X)$, the complete groupoid. It has as morphisms pairs of elements (x, y) which we write $x \to y$ with the composition $(y \to z) \circ (x \to y) = x \to z$. This makes $(x, x) = id_x$ and $y \to x = (x \to y)^{-1}$.[7]

A set X with an action of a group G determines a groupoid in the following fashion. The objects are the set X the morphisms are $X \times G$ with the source map s given by $s(x, g) = x$ and target map $t(x, g) = g(x)$. This becomes clear in the notation $x \xrightarrow{g}$ or $x \xrightarrow{g} g(x)$. Composition is defined for $x \xrightarrow{g}$ and $y \xrightarrow{h}$ if $g(x) = y$ then $(x \xrightarrow{g}) \circ (y \xrightarrow{h}) = x \xrightarrow{hg}$, this groupoid is called the action groupoid. It is also a semi–direct product and this is why we write $\underline{X \rtimes G}$ with the underline to stress that it is a groupoid. This groupoid coincides with $\mathcal{K}(X)$ if the action is regular, that is X is a G–torsor.

There is a canonical functor $st : \underline{X \rtimes G} \to \mathcal{K}(X)$ which is identity on objects and on morphisms is defined as $st(x \xrightarrow{g} g(x)) = x \to g(x)$ or in other notation $st(\phi) = (s(\phi), t(\phi))$. The functor st is bijective on objects, but does not need to be surjective on morphisms. It will usually partition X into components $X = X_1 \amalg \cdots \amalg X_n$, such that $st(\underline{X \rtimes G}) = \mathcal{K}(X_1) \amalg \cdots \amalg \mathcal{K}(X_n)$

2.3. The constructions above in category theoretical language

A quiver $\vec{\Gamma}$, determines a category $P(\vec{\Gamma})$, whose objects are vertices and whose morphisms are paths on the quiver. This is also the free category on the morphisms given by $\vec{\Gamma}$. A graph Γ determines a groupoid $\pi(\Gamma)$, its path–groupoid, in the following way. Choose any quiver whose underlying graph is the given graph and consider the category it defines. Now add additional morphisms which are inverses to the morphisms given by the directed edges. This is equivalent to saying that the morphisms are the reduced paths on Γ. A functor on such a groupoid is given by the values on the vertices and on the directed edges of any quiver representing the graph. A G–labelling of a quiver is a functor lab from the category $P(\vec{\Gamma})$ to the groupoid $\underline{G}$. Such a functor, by a universal property, factors through the groupoid $\pi(\Gamma)$. Groupoid equivalence of labellings then means that the induced functors from $\pi(\Gamma)$ to $\underline{G}$ agree.

In other workds, a G–labelling of Γ is equivalent to a functor $lab : \pi(\Gamma) \to \underline{G}$. lab is in spanning tree gauge with respect to τ if $lab|_{\pi(\tau)}$ is trivial, i.e. maps all morphisms to the id_*.

Regauging in this language is a natural transformation from one labelling functor to another. Such a natural transformation is given by a collection of maps: $\phi(v) \in Hom_{\underline{G}}(F(v), F'(v)) = G$, one for each vertex v, which for each directed edge $\vec{e}$ form v to w satisfy $\phi(w)^{-1} \circ F(\vec{e}) = F'(\vec{e}) \circ \phi(v)$. These natural transformations are invertible and yield isomorphisms of functors. In other words, lab and lab' are gauge equivalent if they are representatives of the same isomorphism class of functors $[lab] = [lab']$.

Proposition 2.1. *$Nat(lab, -) = Map(V, G)$ and each such map ϕ gives an invertible natural transformation $lab \to lab^\phi$. Thus, $Fun(\pi(\Gamma), \underline{G})$ is a groupoid and re–gaugings $\phi \in Map(V, G)$ act transitively on the elements of a given isomorphism class $[lab]$.*

In the case that the image of lab generates G, the group of automorphisms of lab is G and hence the group G^V/G acts simply transitively on the functors in a given equivalence class lab. □

[6] A general groupoid can have many components, but is is equivalent as a category to a disjoint union of group categories. In our case the groupoids have one component and hence all of them are equivalent to some $\underline{G}$.

[7] This is the same as using a morphism $X \times X \rightrightarrows^{s=p_1}_{t=p_2} X$ with composition given by the map $(X \times X)\ {}_s\times_t (X \times X) \to X \times X \times X \times X \xrightarrow{p_{14}} X \times X$ where p_{14} is the projection to the first and fourth component.

An automorphism of $\pi(\Gamma)$ is a functor from $\pi(\Gamma) \to \pi(\Gamma)$ that is invertible. Any graph symmetry $\sigma \in Aut(\Gamma)$ yields an automorphism on $\pi(\Gamma)$, which we also denote by σ. On objects it is given by $\sigma(v) = \sigma_V(v)$ and on morphisms as follows. Pick $\vec{e}$ with orientation $s(e) = v$ and $t(e) = w$ then define $\sigma(\vec{e}) = \sigma_F(\vec{e})$ which has the orientation given by the source $\sigma_V(v)$ and target $\sigma_V(w)$. This gives an action of $Aut(\Gamma)$ on $Fun(\pi(\Gamma), \underline{\Gamma})$ given by $lab \mapsto lab \circ \sigma^{-1}$. Note if lab is in spanning tree gauge w.r.t. τ then $\sigma(lab) = lab \circ \sigma^{-1}$ is in spanning tree gauge w.r.t. the push–forward $\sigma(\tau)$. To check that $[lab] = [lab \circ \sigma^{-1}]$ we can apply Corollary 1.7.

Lemma 2.2. *A quantum enhanced symmetry of G–labelled graph, that is a functor* $lab : \pi(\Gamma) \to \underline{G}$, *is an element* $\sigma \in Aut(\Gamma)$ *which satisfies that* $[lab] = [lab \circ \sigma]$, *i.e. there exists an equivalence* ϕ: $(lab \circ \sigma^{-1})^{\phi} \simeq lab$. *If lab is in spanning tree gauge w.r.t.* τ *then* $\sigma \in Aut(\Gamma)$ *is a strict quantum enhanced symmetry if* $lab = \sigma(lab)^{\phi}$ *in the notation above.*

The group of automorphisms Ψ of $\underline{G}$, which correspond to automorphisms of G, also acts on $Fun(\pi(\Gamma), \underline{G})$ via $lab \to \Psi \circ lab$.

Definition 2.3. An element of $\sigma \in Aut(\Gamma)$ is called transferable via lab if there exists a $\Psi \in Aut(\underline{G})$ such that $(lab \circ \sigma)^{\phi^{lab\circ\sigma}(\sigma^{-1}(\tau)\to\tau)} = \Psi \circ lab$.

There are usually many more automorphisms of $\underline{G}$ than those transferred from symmetries of the underlying graph. As an example lets look at the n–petal graph with a functor to the free group $\mathbb{F}_n$ which associates one generator to each (directed) edge. The automorphisms correspond to automorphism of the free group, while the graph symmetries only give the permutations of the generators.

2.4. Groupoid extensions and representations

With this language in place, we can also enlarge the setting of Theorem 1.13.

Fix a graph Γ and let X be the set of all ORSTs of Γ. The re–gauging groupoid is by definition $\mathcal{K} = \mathcal{K}(X)$. With this notation we can translate the results of

Theorem 2.4. *Let G be Abelian Fix a G–labelling lab and assume it is in spanning tree gauge with respect to some ORST* τ *then then the matrices* $M_{lab_\tau}(\tau \to \tau')$ *of equation (1.2) yield projective representation of* $\mathcal{K}$ *in* $M_{|V|}(G)$ *with cocycles* $lab(l)$ *with* l *defined in equation (1.3). Moreover this lifts to the groupoid extension of* $\mathcal{K}$ *by* $\pi(\Gamma)$ *given by* $\mathcal{L} = \amalg_{v \in V} \pi_1(\pi(\Gamma), v)$ *defined in Proposition 2.5 below.*

Moreover via st this yields a projective representation of $\underline{X \rtimes Aut(\Gamma)}$, *which induce the projective representations of 1.13.*

2.4.1. Groupoid extensions Even in the case that G is not Abelian, there is a Groupoid extension, which in the Abelian case lift the projective action to an honest action. There are functors $p^{op} : \mathcal{K}^{op} \to \pi(\Gamma)$ and $l^{op} : \mathcal{K}_1^{op}{}_t \times_s \mathcal{K}_1^{op} \to \mathcal{L}$

$$p^{op}(g^{op}) := \gamma^{\tau}_{v_{rt}v'_{rt}}, \quad l^{op}(g^{op}, h^{op}) := l(h, g) \tag{2.1}$$

Proposition 2.5. *[4] The pair* (p^{op}, l^{op}) *are an element of* $C^2_{\pi(\Gamma)}(\mathcal{K}, \mathcal{L})$ *that is a* $\pi(\Gamma)$*–crossed* $\mathcal{K}$ *2–cocycle with values in* $\mathcal{L}$. *By general theory, [8, 10, 11] the noncommutative cocycle* (p, l) *gives rise to a groupoid extension* (Σ, b) *over* $\pi(\Gamma)$

$$\Sigma : 1 \to \mathcal{L} \to \hat{\mathcal{K}} \to \mathcal{K} \to 1 \quad b : \hat{\mathcal{K}} \to \pi(\Gamma) \tag{2.2}$$

2.5. Choice of spanning tree gauges as functors
One can add one more layer of functoriality to the choices of spanning tree gauges to understand how the action of $Aut_q(\Gamma, lab)$ is induced as well as its generalization to $Aut(G)$.

Assume G is Abelian. Fix lab, then this defines a projective/weak functor $gauge : \mathcal{K} \to Fun(\pi(\Gamma), \underline{G})$. On objects it is defined by $gauge(\tau) = lab_\tau$ where $lab_\tau = lab^{\phi(\to\tau)}$ is the re–gauging of lab into spanning tree gauge w.r.t. τ. On morphisms we set $gauge(\tau \to \tau')$ to be the natural transformation given by re–gauing lab_τ to be in spanning tree gauge w.r.t. τ' using the re–gauging of Lemma 1.4. $gauge(\tau)$ fails to be a functor precisely because the compositions are not "on the nose" but only up to a constant re–gauging given by the co–cycle. The exact behaviour is the content of Theorem 1.13.

Even more is true, $gauge$ is also a projective functor $gauge : Fun(\pi(\Gamma), \underline{G}) \to Fun_{pr}(\mathcal{K}, Fun(\pi(\Gamma), \underline{G}))$, where Fun_{pr} are projective/weak functors. Given a natural transformation $\phi : lab \to lab^\phi$ on the left it is mapped to the identity transformation on the right.

Now $Aut(\pi(\Gamma))$ acts on $Fun(\pi(\Gamma), \underline{G})$ and this action gets transferred by $gauge$. The content of Theorem 1.13 is that if one restricts to strict quantum enhanced automorphisms, one again obtains an action by projective functors.

If G is not Abelian, for $gauge(lab)$, the re–gauging has to be followed if necessary by a constant re–gauging, which is a fixed choice and likewise for $gauge$ as a functor, one has to send a natural transformation ϕ to the choices of ψ, one for each τ according to Corollary 1.7. These modification are best described by 2–morphisms. Indeed, since we are looking at maps between categories, we should actually be looking at the 2–category of categories. This is well known in the theory of orbifolds and stacks. This higher categorical interpretation provides links to the theory of gerbes and their twists, which have recently entered into condensed matter physics as well [5].

2.6. Lifts
Finally to explain Theorem 1.17 we notice that there is also an action of $Aut(\underline{G})$ on $Fun(\pi(\Gamma), \underline{G})$. If one takes both actions together, one obtains the bi–groupoid $\rho, s, \lambda : Aut(\pi(\Gamma)) \ltimes Fun(\pi(\Gamma), \underline{G}) \rtimes Aut(\underline{G}) \to Fun(\pi(\Gamma), \underline{G})$.

By restricting to liftable enhanced symmetries and following the morphism $lab \xrightarrow{\sigma} lab \circ \sigma^{-1}$ by the morphism $lab \circ \sigma^{-1} \xrightarrow{\Psi(\sigma)} lab$ (or vice–versa, since the left and right actions commute), we get automorphisms of lab, which are sequences of σ's and Ψ's which via $gauge$ get transferred to projective functors.

3. Representations
We will now give the details for the example of the tetrahedral graph or the full square. It has symmetry group $\mathbb{S}_4$. It acts transitively on all ORST. The subgroup of $\mathbb{S}_3$ acts transitively on all orders for a fixed RST. Fix $G = \mathbb{F}_3$ the free group on three generators. We will denote inverses by $*$ to keep with the application in physics, where we apply a character to $U(1)$. We fix an initial rooted spanning tree and order as in the first picture of Figure 2 and fix the $G = \mathbb{F}_3$–labelling as indicated.

We will use the following graphical calculation technique:

1. Write down the graph together with the initial spanning tree and order.
2. Push–forward the spanning tree and the order. This is given by replacing i by $\sigma(i)$.
3. Write the re–gauging at the vertices. They are given by going along the new spanning tree to the root and multiplying the labels in order.
   (a) Read off the matrix Φ by putting the G label of a vertex onto the diagonal in the place indicated by the label $\sigma(i)$.

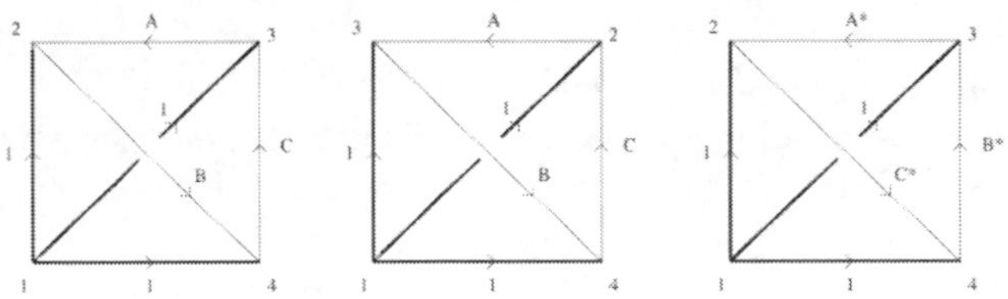

Figure 2. Calculation of the action of (23) on T^3. The original graph, the pushed forward order and the move into the old position to read off the morphism

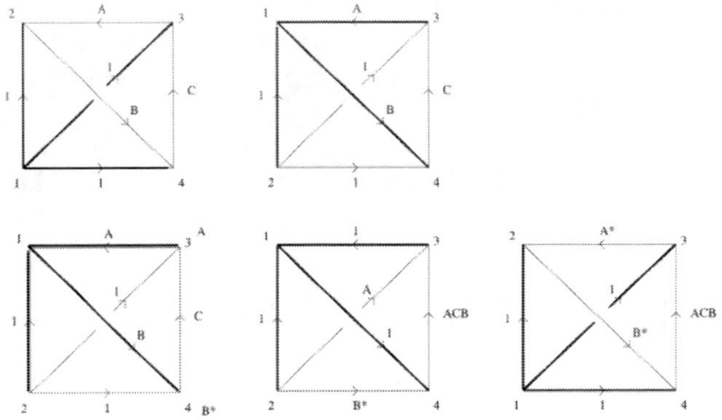

Figure 3. Calculation of the action of (12) on T^3

(b) Obtain $M = \Phi P_\sigma$

4. Perform the regauging, by multiplying the label of an ordered edge from the left by the label of its target and on the right by the inverse label of its source.
5. Rearrange the vertices with σ_V. Formally, this is done by writing v_i next to $\sigma(i)$, but one can just use the new numbering to move the graph back into its old position and read off the transformation Ψ.

3.1. Lifts of the action

The lift of the action of $\mathbb{S}_4$ on $\mathbb{F}_3$ is fixed once we know the action of the generators (12), (23) and (34).

The action of (23) is graphically calculated in Figure 2, from which one reads off $\Psi((23))(A, B, C) = (A^*, C^*, B^*)$. Here (A, B, C) is the notation for the initially chosen basis of $\mathbb{T}^3$.

A similar calculation shows that $\Psi((34))(A, B, C) = (B^*, A^*, C^*)$. A consequence is that the cycle $(234) = (23)(34)$ acts as $\Psi((234))(A, B, C) = \Psi((23))(B^*, A^*, C^*) = (B, C, A)$ and is the cyclic permutation.

The action of (12) is more complicated as the root is moved. For this we calculate graphically, see Figure 3, and read off Ψ as: $(A, B, C) \mapsto (A^*, B^*, ACB)$.

This allows us to compute fixed points and stabilizer groups if we take a character. This is equivalent to treating A, B, C as variables in $U(1)$. We will first concentrate on non-Abelian stabilizer groups. There are only two fixed points under the full $\mathbb{S}_4$ action and these are $(1, 1, 1)$ and $(-1, -1, -1)$. The group A_4, the subgroup of all even permutations, is the stabilizer group of the two points (i,i,i) and $(-i,-i,-i)$. One can readily check that these are the only non–Abelian stabilizer groups. The other possibility would be $\mathbb{S}_3$, but a short calculation shows that anything that is stabilized by any $\mathbb{S}_3$ subgroup is stabilized by all of $\mathbb{S}_4$.

3.1.1. Representations We collect together the matrices M needed for further calculation. Again, we fix our initial ordered rooted spanning tree as before.

Using short hand notation, the matrix for the re–gauging induced by the transpositions $(12),(13),(14)$ from the initial spanning tree to the pushed forward one are

$$\rho_{12}=\begin{pmatrix}0&1&&\\1&0&&\\&&A&\\&&&B^*\end{pmatrix},\rho_{13}=\begin{pmatrix}0&&1&\\&A^*&&\\1&&0&\\&&&C\end{pmatrix},\rho_{14}=\begin{pmatrix}0&&&1\\&B&0&\\&0&C^*&\\1&&&0\end{pmatrix}$$

The calculation for ρ_{12} can be read off from Figure 3. For this we read off the matrix Φ from the re–gauging parameter and the matrix M_σ is given by the permutation we are considering. The other calculations are similar. All other transpositions, viz. those not involving 1, simply yield permutation matrices as there is no re–gauging involved. It is convenient to also have the following matrices as a reference:

$$\rho_{(12)(34)}=\begin{pmatrix}0&1&&\\1&0&&\\&&0&B^*\\&&A&0\end{pmatrix},\quad \rho_{(14)(23)}=\begin{pmatrix}&&&1\\&&C^*&\\&B&&\\1&&&\end{pmatrix}$$

and finally

$$\rho_{(132)}=\begin{pmatrix}0&1&0&\\0&0&A&\\1&0&0&\\&&&B^*\end{pmatrix}$$

3.1.2. The point $(1,1,1)$. At $(1,1,1)$, the matrices $\rho_{12},\rho_{23},\rho_{34}$ give the usual representation of $\mathbb{S}_4$ on $\mathbb{C}^4$. As is well known this representation decomposes into the trivial representation and an irreducible 3–dim representation.

3.1.3. The point $(-1,-1,-1)$. In this case, the matrices $\rho_{12},\rho_{23},\rho_{34}$ only give a projective representation. As one can check $\rho_{12}\rho_{23}\rho_{12}=-\rho_{13}$ while $\rho_{23}\rho_{12}\rho_{23}=\rho_{13}$ for instance. Define the 1–cocycle λ by $\lambda(\sigma)=(-1)$ if 1 appears in a cycle of length >1 and 1 else. So that $\lambda((12))=\lambda((13))=\lambda((123))=-1$ while $\lambda((23))=\lambda((24))=\lambda((234))=1$. Then one calculates that $\tilde{\rho}:=\rho\circ\lambda$ has a trivial cocycle c and thus ρ is isomorphic to a true linear representation of $\mathbb{S}_4$. Checking the characters, one sees again that in this case the irreducible components of $\tilde{\rho}$ are again the one–dimensional trivial representation and the 3–dimensional standard representation. The trivial representation is spanned by $(-1,1,1,1)$.

Remark 3.1. We would like to remark that the choice of λ amounts to choosing a different gauge for the root vertex, namely -1 instead of 1.

3.1.4. The points (i,i,i) *and* $(-i,-i,-i)$. These points are similar to each other. We will treat the first one in detail. Again, we have only a projective representation of A_4 aka. the tetrahedral group T. Namely, $\rho_{(12)(34)}\rho_{(13)(24)}=-i\rho_{(14)(23)}$. Again we can scale by a 1–cocycle λ. This time $\lambda(id)=1$, $\lambda((ij)(kl))=i$, $\lambda(ijk)=1$ if $1\notin\{i,j,k\}$, and $\lambda((ijk))=i$ if $1\in\{i,j,k\}$. The

Representative	1	−1	s^3	t^2	s^2	t	s
Elts in Conj. Class	1	1	6	4	4	4	4
Order	1	2	3	3	4	4	6
χ_1	1	1	1	1	1	1	1
χ_2	1	1	1	ω	ω^2	ω^2	ω
χ_3	1	1	1	ω^2	ω	ω	ω^2
χ_4	2	−2	0	−1	−1	1	1
χ_5	2	−2	0	$-\omega$	$-\omega^2$	ω^2	ω
χ_6	2	−2	0	$-\omega^2$	$-\omega$	ω	ω^2
χ_7	3	3	−1	0	0	0	0

Table 1. Character table of $2 \cdot A_4$ [15], where $\omega = e^{\frac{2\pi i}{3}}$.

resulting representation $\tilde{\rho} = \rho \circ \lambda$ is then still a projective representation, but is it a representation of the unique non–trivial $\mathbb{Z}/2\mathbb{Z}$ extension of A_4, which goes by the names $2T, 2A_4, SL(2,3)$ or the binary tetrahedral group. This group is well known. It is presented by generators s and t with the relations $s^3 = t^3 = (st)^2$. In $SL(2,3)$ (that is the special linear group of 2×2 matrices over the field with three elements $\mathbb{F}_3$), one can choose $s = \begin{pmatrix} -1 & -1 \\ 0 & -1 \end{pmatrix}$ and $t = \begin{pmatrix} -1 & 0 \\ -1 & -1 \end{pmatrix}$.

For $2A_4$ using a set theoretic section $\wedge$ of the extension sequence

$$1 \longrightarrow \mathbb{Z}/2\mathbb{Z} \longrightarrow 2A_4 \underset{\wedge}{\rightleftarrows} A_4 \longrightarrow 1 \tag{3.1}$$

and z as a generator for $\mathbb{Z}/2\mathbb{Z}$, we can pick $s = z\widehat{(123)}, t = z\widehat{(234)}$ as generators. Now we can check the character table, Table 1, and find that the representation $\tilde{\rho}$ over the complex numbers decomposes as the sum of two irreducible two–dimensional representations $\chi_5 \oplus \chi_6$. In fact, these are the two representations into which the unique real irreducible 4–dimensional representation of complex type splits over $\mathbb{C}$.

The explicit computation for the representation

$$\begin{aligned} \tilde{\rho}(s) &= -\lambda((123))\rho_{(123)} = \begin{pmatrix} 0 & 0 & -i & 0 \\ -i & 0 & 0 & 0 \\ 0 & 1 & 0 & 0 \\ 0 & 0 & 0 & -1 \end{pmatrix} \\ \tilde{\rho}(t) &= -\lambda((234))\rho_{(234)} = -\begin{pmatrix} 1 & 0 & 0 & 0 \\ 0 & 0 & 0 & 1 \\ 0 & 1 & 0 & 0 \\ 0 & 0 & 1 & 0 \end{pmatrix} \end{aligned} \tag{3.2}$$

is as follows. Suppose the $\tilde{\rho} = \bigoplus_{i=1}^{7} a_i \rho_i$, where ρ_i is the irrep with character χ_i. Now $tr(id) = 4, tr(-1) = -4$, using the character table this implies that the coefficients $a_1 = a_2 = a_3 = a_7 = 0$ and furthermore $(*)\, a_4 + a_5 + a_6 = 2$. We furthermore have that $tr(s) = -1$ so that $a_4 + \omega a_5 + \omega^2 a_6 = -1$ which together with (*) implies that $a_4 = 0, a_5 = a_6 = 1$. This fixes the decomposition into irreps. As a double check one can verify that the rest of the equations are also satisfied.

The analysis of the complex conjugate point $(-\pi/2, -\pi/2, -\pi/2)$ is analogous.

We would briefly like to remark that these special points have a meaning in the study of real materials which are in the form of a Gyroid. In these materials, band sticking is forced by the presence of the symmetries and leads to special properties of the material [2].

a, b, c	Group	Iso class of of extension	type	Dim of Irreps
$(0,0,0)$	$\mathbb{S}_4$	$\mathbb{S}_4$	trivial	1,3
(π,π,π)	$\mathbb{S}_4$	$\mathbb{S}_4$	trivialisable	1,3
$(\frac{\pi}{2},\frac{\pi}{2},\frac{\pi}{2})$ $(\frac{3\pi}{2},\frac{3\pi}{2},\frac{3\pi}{2})$	A_4	$2A_4$	isomorphic extension	2,2

Table 2. Possible choices of parameters (a, b, c) leading to non–Abelian enhanced symmetry groups

3.2. Groupoid calculation

Here we calculate one example of Theorem 1.17:

$$\rho_{(12)(34)} = \begin{pmatrix} 0 & 1 & & \\ 1 & 0 & & \\ & & 0 & A \\ & & B^* & 0 \end{pmatrix}, \quad \rho_{(13)(24)} = \begin{pmatrix} & & 1 & 0 \\ & & 0 & A^* \\ 1 & 0 & & \\ 0 & C & & \end{pmatrix}$$

Using the substitution $\rho'_{(13)(24)} = \rho_{(13)(24)}(A \to B, C \to ABC)$, we indeed obtain

$$\rho_{(1234)}\rho'_{(13)(24)} = B^*\rho_{(14)(23)} = B^* \begin{pmatrix} & & & 1 \\ & & B & \\ & C^* & & \\ 1 & & & \end{pmatrix}$$

Acknowledgments

We would like to thank the organizers of the 30th International Colloquium on Group Theoretical Methods in Physics, Ghent (Belgium) for a very interesting conference. RK thankfully acknowledges support from NSF under the grants DMS-1007846 and DMS-0805881. BWK thankfully acknowledges support from the NSF under the grants PHY-0969689 and PHY-1255409.

Any opinions, findings and conclusions or recommendations expressed in this material are those of the authors and do not necessarily reflect the views of the National Science Foundation.

This work was partially supported by grants from the Simons Foundation (#267481 to Erika Birgit Kaufmann and #267555 and #207591 to Ralph Kaufmann). Both RK and BK thank the Simons Foundation for this support.

References

[1] Kaufmann R M, Khlebnikov S and Wehefritz–Kaufmann B 2012 The geometry of the double gyroid wire network: quantum and classical *Journal of Noncommutative Geometry* **6** 623-664

[2] Kaufmann R M, Khlebnikov S and Wehefritz–Kaufmann B 2012 Singularities, swallowtails and Dirac points. An analysis for families of Hamiltonians and applications to wire networks, especially the Gyroid *Annals of Physics* **327** 2865-2884

[3] Kaufmann R M, Khlebnikov S and Wehefritz–Kaufmann B 2012 The noncommutative geometry of wire networks from triply periodic surfaces *J. Phys.: Conf. Ser.* **343** 012054

[4] Kaufmann R M, Khlebnikov S and Wehefritz–Kaufmann B 2012 Re-gauging groupoid, symmetries and degeneracies for Graph Hamiltonians and applications to the Gyroid wire network *DESY preprint* 12-133 (*Preprint* arXiv:1208.3266)

[5] Freed D S and Moore G W 2013 Twisted Equivariant Matter *Ann. Henri Poincaré* **14** 1927-2023

[6] Bellissard J 1992 Gap labelling theorems for Schrödinger operators *From number theory to physics* (Berlin: Springer) 538–630

[7] Serre J P 1994 *Cohomologie galoisienne.* Fifth edition. Lecture Notes in Mathematics **5** (Berlin: Springer-Verlag)

[8] Dedecker P 1964 Les foncteurs Ext_Π, $\mathbf{H}^2_\Pi$ et H^2_Π non abéliens *C. R. Acad. Sci. Paris* **258** 4891–4894

[9] Kurosh A G. 1956 *The theory of groups. Vol. II.* Translated from the Russian and edited by K. A. Hirsch (New York: Chelsea Publishing Company) 308

[10] Giraud J 1971 *Cohomologie non-abelienne* Grundle. Math. Wiss. **179** (New York: Springer-Verlag)

[11] Moerdijk I 2003 Lie groupoids, gerbes, and non-abelian cohomology *K-Theory* **28** 207–258

[12] Schur I 1911 Über die Darstellung der symmetrischen und der alternierenden Gruppe durch gebrochene lineare Substitutionen *J. für Math* **139** 155-250

[13] Karpilovsky G 1985 *Projective Representations of Finite Groups* (New York: Dekker)

[14] Brown K S 1972 *Cohomology of Groups* Graduate Texts in Mathematics **87** (Berlin: Springer Verlag)

[15] Schur I 1907 Untersuchungen über die Darstellung der endlichen Gruppen durch gebrochene lineare Substitutionen *J. für Math.* **132** 85-137

Ternary Z_3-graded generalization of Heisenberg's algebra

Richard Kerner
LPTMC, Tour 23-13, 5-e, Boîte 121,
Université Pierre et Marie Curie - CNRS UMR 7600,
4 Place Jussieu, 75005 Paris, France

E-mail: `richard.kerner@upmc.fr`

Abstract. We investigate a ternary, Z_3-graded generalization of the Heisenberg algebra. It turns out that introducing a non-trivial cubic root of unity, $j = e^{\frac{2\pi i}{3}}$, one can define two types of creation operators instead of one, accompanying the usual annihilation operator. The two creation operators are non-hermitian, but they are mutually conjugate. Together, the three operators form a ternary algebra, and some of their cubic combinations generate the usual Heisenberg algebra.

A cubic analogue of Hamiltonian operator is constructed by analogy with the usual harmonic oscillator. A set of eigenstates in coordinate representation is constructed in terms of functions satisfying linear differential equation of third order.

1. Introduction

Our goal being a ternary generalization of Heisenberg's algebra, let us start with recalling basic facts about ternary algebras [1], [2] The usual definition of an algebra involves a linear space $\mathcal{A}$ (over real or complex numbers) endowed with a *binary* constitutive relations

$$\mathcal{A} \times \mathcal{A} \to \mathcal{A} \tag{1}$$

In a finite dimensional case, dim $\mathcal{A}$ = N, in a chosen basis $\mathbf{e}_1, \mathbf{e}_2, ..., \mathbf{e}_N$, the constitutive relations (1) can be encoded in *structure constants* f^k_{ij} as follows:

$$\mathbf{e}_i \mathbf{e}_j = f^k_{ij}\, \mathbf{e}_k. \tag{2}$$

With the help of these structure constants all essential properties of a given algebra can be expressed, e.g. they will define a *Lie algebra* if they are antisymmetric and satisfy the Jacobi identity:

$$f^k_{ij} = -f^k_{ji}, \quad f^k_{im} f^m_{jl} + f^k_{jm} f^m_{li} + f^k_{lm} f^m_{ij} = 0, \tag{3}$$

whereas an abelian algebra will have its structure constants symmetric, $f^k_{ij} = f^k_{ji}$.

Usually, when we speak of algebras, we mean *binary algebras*, understanding that they are defined via *quadratic* constitutive relations (2). On such algebras the notion of $Z_2 - grading$ can be naturally introduced. An algebra $\mathcal{A}$ is called a $Z_2 - graded$ algebra if it is a direct sum of two parts, with symmetric (abelian) and anti-symmetric product respectively,

$$\mathcal{A} = \mathcal{A}_0 \oplus \mathcal{A}_1, \tag{4}$$

with *grade* of an element being 0 if it belongs to $\mathcal{A}_0$, and 1 if it belongs to $\mathcal{A}_1$. Under the multiplication in a Z_2-graded algebra the grades add up reproducing the composition law of the Z_2 permutation group: if the grade of an element A is a, and that of the element B is b, then the grade of their product will be $a+b$ *modulo 2*:

$$grade(AB) = grade(A) + grade(B), \quad \text{so that} \quad AB = (-1)^{ab} BA. \tag{5}$$

It is worthwhile to notice at this point that the above relationship can be written in an alternative form, with all the expressions on the left side as follows:

$$AB - (-1)^{\alpha\beta} BA = 0, \text{ or } AB + (-1)^{(\alpha\beta+1)} BA = 0 \tag{6}$$

The equivalence between these two alternative definitions of commutation (anticommutation) relations inside a Z_2-graded algebra is no more possible if by analogy we want to impose *cubic* relations on algebras with Z_3-symmetry properties, in which the non-trivial cubic root of unity, $j = e^{\frac{2\pi i}{3}}$ plays the role similar to that of -1 in the binary relations displaying a Z_2-symmetry.

The Z_3 cyclic group is an abelian subgroup of the S_3 symmetry group of permutations of three objects. The S_3 groups contains *six* elements, including the group unit e (the identity permutation, leaving all objects in place: $(abc) \rightarrow (abc)$, the two cyclic permutations

$$(abc) \rightarrow (bca) \quad \text{and} \quad (abc) \rightarrow (cab),$$

and three odd permutations,

$$(abc) \rightarrow (cba), \quad (abc) \rightarrow (bac) \quad \text{and} \quad (abc) \rightarrow (acb).$$

There was a unique definition of *commutative* binary algebras given in two equivalent forms,

$$xy + (-1)yx = 0 \quad \text{or} \quad xy = yx. \tag{7}$$

In the case of cubic algebras [2] we have the following four generalizations of the notion of *commutative* algebras:

a) Generalizing the first form of the commutativity relation (7), which amounts to replacing the -1 generator of Z_2 by j-generator of Z_3 and binary products by products of three elements, we get

$$S: \; x^\mu x^\nu x^\lambda + j\, x^\nu x^\lambda x^\mu + j^2\, x^\lambda x^\mu x^\nu = 0, \tag{8}$$

where $j = e^{\frac{2\pi i}{3}}$ is a primitive third root of unity.

b) Another primitive third root, $j^2 = e^{\frac{4\pi i}{3}}$ can be used in place of the former one; this will define the conjugate algebra $\bar{S}$, satisfying the following cubic constitutive relations:

$$\bar{S}: \; x^\mu x^\nu x^\lambda + j^2\, x^\nu x^\lambda x^\mu + j\, x^\lambda x^\mu x^\nu = 0. \tag{9}$$

Clearly enough, both algebras are infinitely-dimensional and have the same structure. Each of them is a possible generalization of infinitely-dimensional algebra of usual commuting variables with a finite number of generators. In the usual Z_2-graded case such algebras are just polynomials in variables $x^1,\ x^2, ... x^N$; the algebras S and $\bar{S}$ defined above are also spanned by polynomials, but with different symmetry properties, and as a consequence, with different dimensions corresponding to a given power.

c) Then we can impose the following "weak" commmutation, valid only for cyclic permutations of factors:

$$S_1: \; x^\mu x^\nu x^\lambda = x^\nu x^\lambda x^\mu \neq x^\nu x^\mu x^\lambda, \tag{10}$$

d) Finally, we can impose the following "strong" commutation, valid for arbitrary (even or odd) permutations of three factors:

$$S_0: \quad x^\mu x^\nu x^\lambda = x^\nu x^\lambda x^\mu = x^\nu x^\mu x^\lambda \tag{11}$$

The four different associative algebras with cubic commutation relations can be represented in the following diagram, in which all arrows correspond to *surjective homomorphisms*. The commuting generators can be given the common grade 0.

$$\begin{array}{ccc} S & & \bar{S} \\ & \searrow \quad \swarrow & \\ & S_1 & \\ & \downarrow & \\ & S_0 & \end{array}$$

Let us turn now to the Z_3 generalization of anti-commuting generators, which in the usual homogeneous case with Z_2-grading define Grassmann algebras. Here, too, we have four different choices:

a) The "strong" cubic anti-commutation,

$$\Lambda_0: \quad \Sigma_{\pi \in S_3} \, \theta^{\pi(A)} \theta^{\pi(B)} \theta^{\pi(C)} = 0, \tag{12}$$

i.e. the sum of *all* permutations of three factors, even and odd ones, must vanish.

b) The somewhat weaker "cyclic" anti-commutation relation,

$$\Lambda_1: \quad \theta^A \theta^B \theta^C + \theta^B \theta^C \theta^A + \theta^C \theta^A \theta^B = 0, \tag{13}$$

i.e. the sum of *cyclic* permutations of three elements must vanish. The same independent relation for the odd combination $\theta^C \theta^B \theta^A$ holds separately.

c) The j-skew-symmetric algebra:

$$\Lambda : \theta^A \theta^B \theta^C = j \, \theta^B \theta^C \theta^A. \tag{14}$$

and its conjugate algebra $\bar{\Lambda}$, isomorphic with Λ, which we distinguish by putting a bar on the generators and using dotted indices:

d) The j^2-skew-symmetric algebra:

$$\bar{\Lambda}: \quad \bar{\theta}^{\dot{A}} \bar{\theta}^{\dot{B}} \bar{\theta}^{\dot{C}} = j^2 \bar{\theta}^{\dot{B}} \bar{\theta}^{\dot{C}} \bar{\theta}^{\dot{A}} \tag{15}$$

Both these algebras are finite dimensional. For j or j^2-skew-symmetric algebras with N generators the dimensions of their subspaces of given polynomial order are given by the following generating function:

$$H(t) = 1 + Nt + N^2 t^2 + \frac{N(N-1)(N+1)}{3} t^3, \tag{16}$$

where we include pure numbers (dimension 1), the N generators θ^A (or $\bar{\theta}^{\dot{B}}$), the N^2 independent quadratic combinations $\theta^A \theta^B$ and $N(N-1)(N+1)/3$ products of three generators $\theta^A \theta^B \theta^C$.

The above four cubic generalization of Grassmann algebra are represented in the following diagram, in which all the arrows are surjective homomorphisms.

$$\begin{array}{ccc} & \Lambda_0 & \\ & \downarrow & \\ & \Lambda_1 & \\ & \swarrow \quad \searrow & \\ \Lambda & & \bar{\Lambda} \end{array}$$

2. Examples of Z_3-graded ternary algebras

2.1. The Z_3-graded analog of Grassman algebra

Let us introduce N generators spanning a linear space over complex numbers, satisfying the following cubic relations [3], [4]:

$$\theta^A\theta^B\theta^C = j\,\theta^B\theta^C\theta^A = j^2\,\theta^C\theta^A\theta^B, \tag{17}$$

with $j = e^{2i\pi/3}$, the primitive root of 1. We have $1 + j + j^2 = 0$ and $\bar{j} = j^2$.

Let us denote the algebra spanned by the θ^A generators by $\mathcal{A}$ [3], [4].

We shall also introduce a similar set of *conjugate* generators, $\bar{\theta}^{\dot{A}}$, $\dot{A}, \dot{B}, ... = 1, 2, ..., N$, satisfying similar condition with j^2 replacing j:

$$\bar{\theta}^{\dot{A}}\bar{\theta}^{\dot{B}}\bar{\theta}^{\dot{C}} = j^2\,\bar{\theta}^{\dot{B}}\bar{\theta}^{\dot{C}}\bar{\theta}^{\dot{A}} = j\,\bar{\theta}^{\dot{C}}\bar{\theta}^{\dot{A}}\bar{\theta}^{\dot{B}}, \tag{18}$$

Let us denote this algebra by $\bar{\mathcal{A}}$.

We shall endow the algebra $\mathcal{A} \oplus \bar{\mathcal{A}}$ it with a natural Z_3 grading, considering the generators θ^A as grade 1 elements, their conjugates $\bar{\theta}^{\dot{A}}$ being of grade 2.

The grades add up modulo 3, so that the products $\theta^A\theta^B$ span a linear subspace of grade 2, and the cubic products $\theta^A\theta^B\theta^C$ being of grade 0. Similarly, all quadratic expressions in conjugate generators, $\bar{\theta}^{\dot{A}}\bar{\theta}^{\dot{B}}$ are of grade $2 + 2 = 4_{mod\,3} = 1$, whereas their cubic products are again of grade 0, like the cubic products od θ^A's. [7]

Combined with the associativity, these cubic relations impose finite dimension on the algebra generated by the Z_3 graded generators. As a matter of fact, cubic expressions are the highest order that does not vanish identically. The proof is immediate:

$$\theta^A\theta^B\theta^C\theta^D = j\,\theta^B\theta^C\theta^A\theta^D = j^2\,\theta^B\theta^A\theta^D\theta^C = j^3\,\theta^A\theta^D\theta^B\theta^C = j^4\,\theta^A\theta^B\theta^C\theta^D, \tag{19}$$

and because $j^4 = j \neq 1$, the only solution is $\theta^A\theta^B\theta^C\theta^D = 0$.

2.2. Ternary Clifford algebra

Let us introduce the following three 3×3 matrices:

$$Q_1 = \begin{pmatrix} 0 & 1 & 0 \\ 0 & 0 & j \\ j^2 & 0 & 0 \end{pmatrix}, \quad Q_2 = \begin{pmatrix} 0 & j & 0 \\ 0 & 0 & 1 \\ j^2 & 0 & 0 \end{pmatrix}, \quad Q_3 = \begin{pmatrix} 0 & 1 & 0 \\ 0 & 0 & 1 \\ 1 & 0 & 0 \end{pmatrix}, \tag{20}$$

and their hermitian conjugates

$$Q_1^\dagger = \begin{pmatrix} 0 & 0 & j \\ 1 & 0 & 0 \\ 0 & j^2 & 0 \end{pmatrix}, \quad Q_2^\dagger = \begin{pmatrix} 0 & 0 & j \\ j^2 & 0 & 0 \\ 0 & 1 & 0 \end{pmatrix}, \quad Q_3^\dagger = \begin{pmatrix} 0 & 0 & 1 \\ 1 & 0 & 0 \\ 0 & 1 & 0 \end{pmatrix}. \tag{21}$$

These matrices can be allowed natural Z_3 grading,

$$\text{grade}(Q_k) = 1, \quad \text{grade}(Q_k^\dagger) = 2, \tag{22}$$

The above matrices span a very interesting ternary algebra. Out of three independent Z_3-graded ternary combinations, only one leads to a non-vanishing result. One can check without much effort that both j and j^2 skew ternary commutators do vanish:

$$\{Q_1, Q_2, Q_3\}_j = Q_1Q_2Q_3 + jQ_2Q_3Q_1 + j^2Q_3Q_1Q_2 = 0,$$

$$\{Q_1, Q_2, Q_3\}_{j^2} = Q_1Q_2Q_3 + j^2Q_2Q_3Q_1 + jQ_3Q_1Q_2 = 0,$$

and similarly for the odd permutation, $Q_2Q_1Q_3$. On the contrary, the totally symmetric combination does not vanish; it is proportional to the 3×3 identity matrix $\mathbf{1}$:

$$Q_aQ_bQ_c + Q_bQ_cQ_a + Q_cQ_aQ_b = \eta_{abc}\,\mathbf{1}, \quad a, b, ... = 1, 2, 3. \tag{23}$$

with η_{abc} given by the following non-zero components:

$$\eta_{111} = \eta_{222} = \eta_{333} = 1, \quad \eta_{123} = \eta_{231} = \eta_{312} = 1, \quad \eta_{213} = \eta_{321} = \eta_{132} = j^2. \tag{24}$$

all other components vanishing. The relation 23) may serve as the definition of *ternary Clifford algebra*.

Another set of three matrices is formed by the hermitian conjugates of Q_a, which we shall endow with dotted indeces $\dot{a}, \dot{b}, ... = 1, 2, 3$: satisfying conjugate identities

$$Q_{\dot{a}}Q_{\dot{b}}Q_{\dot{c}} + Q_{\dot{b}}Q_{\dot{c}}Q_{\dot{a}} + Q_{\dot{c}}Q_{\dot{a}}Q_{\dot{b}} = \eta_{\dot{a}\dot{b}\dot{c}}\,\mathbf{1}, \quad \dot{a}, \dot{b}, ... = 1, 2, 3. \tag{25}$$

with $\eta_{\dot{a}\dot{b}\dot{c}} = \bar{\eta}_{abc}$.

It is obvious that any similarity transformation of the generators Q_a will keep the ternary anti-commutator (24) invariant. As a matter of fact, if we define $\tilde{Q}_b = P^{-1}Q_bP$, with P a non-singular 3×3 matrix, the new set of generators will satisfy the same ternary relations, because

$$\tilde{Q}_a\tilde{Q}_b\tilde{Q}_c = P^{-1}Q_aPP^{-1}Q_bPP^{-1}Q_cP = P^{-1}(Q_aQ_bQ_c)P,$$

and on the right-hand side we have the unit matrix which commutes with all other matrices, so that $P^{-1}\,\mathbf{1}\,P = \mathbf{1}$.

2.3. Ternary Z_3-graded commutator

In any associative algebra $\mathcal{A}$ one can introduce a new binary operation, the *commutator*, using the generator of the Z_2 group in form of multiplication by -1 :

$$X, Y \in A \to [X, Y] = XY + (-1)YX = XY - YX. \tag{26}$$

In the case of the Z_2 group the generator of its representation on complex numbers was equal to -1; note that $-1 + (-1)^2 = 0$ In the case of the Z_3 group, the generator of its complex representation can be chosen to be $j = e^{\frac{2\pi i}{3}}$, with $j + j^2 + j^3 = 0$

Consider the following cubic combination defined on an associative algebra $\mathcal{A}$:

$$X, Y, Z \in \mathcal{A}, \qquad \{X, Y, Z\} := XYZ + jYZX + j^2ZXY.$$

One obviously has:

$$\{X, Y, Z\} = j\{Y, Z, X\} = j^2\{Z, X, Y\}, \quad \text{and consequently} \quad \{X, X, X\} = 0.$$

In the case when $\mathcal{A}$ is a *unital* algebra, i.e. it contains a unit element $\mathbf{1}$ such that $\mathbf{1}X = X\mathbf{1} = X$ for any $X \in \mathcal{A}$, a unique Lie algebra is naturally generated by the cubic commutator:

$$\{X, 1, Y\} = X \cdot 1 \cdot Y + j\,1 \cdot Y \cdot X + j^2\,Y \cdot X \cdot 1 =$$

$$XY + j\,YX + j^2\,YX = XY + (j + j^2)YX = XY - YX = [X, Y].$$

The following simple example of a Z_3 cubic algebra can be constructed with 2×2 complex matrices. Consider the Lie algebra spanned by three Pauli's matrices:

$$\sigma_1 = \begin{pmatrix} 0 & 1 \\ 1 & 0 \end{pmatrix}, \quad \sigma_2 = \begin{pmatrix} 0 & -i \\ i & 0 \end{pmatrix}, \quad \sigma_3 = \begin{pmatrix} 1 & 0 \\ 0 & -1 \end{pmatrix}$$

satisfying the well known commutation relations:

$$[\sigma_i, \sigma_j] = C_{ij}^k \, \sigma_k, \quad i, j, k = 1, 2, 3, \quad C_{ij}^k = 2i\epsilon_{kij}.$$

The enveloping algebra $\mathcal{A}_\sigma$ contains the unit matrix $\mathbf{1}$:

$$\sigma_i \sigma_j = i \, \epsilon_{ijk} + \delta_{ij} \mathbf{1}$$

Let us define the *cubic j-commutator* on the algebra $\mathcal{A}_\sigma$:

$$\{\sigma_i, \sigma_j, \sigma_k\} = \sigma_i \sigma_j \sigma_k + j \, \sigma_j \sigma_k \sigma_i + j^2 \, \sigma_k \sigma_i \sigma_j, \; ; j = e^{\frac{2\pi i}{3}}.$$

This cubic algebra contains three cubic subalgebras generated by two σ-matrices out of three: for example

$$\{\sigma_1, \sigma_2, \sigma_1\} = -2\,\sigma_2, \quad \{\sigma_2, \sigma_1\,\sigma_2\} = -2\,\sigma_1,$$

and similarly for the couples σ_2, σ_3 and σ_3, σ_1. We have also $\{\sigma_1, \sigma_2, \sigma_3\} = 0$.

3. Ternary Heisenberg algebra

3.1. The Z_2 Heisenberg algebra

Let us first remind the original construction of the Heisenberg algebra used in the analysis of *quantum harmonic oscillator.* The Hamiltonian of classical harmonic oscillator, expressed in reduced dimensionless variables reads:

$$H = \frac{p^2}{2} + \frac{x^2}{2}$$

after first quantization which attributes to canonical variables (p, x) corresponding quantum operators acting on the Hilbert space $L^2(R)$ of square-integrable functions

$$p \to \hat{p} = -i\hbar \frac{d}{dx}, \quad x \to \hat{x}$$

The quantum version of he Hamiltonian is the hermitian operator

$$\hat{H} = -\frac{d^2}{dx^2} + x^2$$

The classical Heisenberg algebra [9]is generated by the following two operators:

$$a = \frac{1}{\sqrt{2}}\left(\bar{x} + i\bar{p}\right), \quad a^\dagger = \frac{1}{\sqrt{2}}\left(\bar{x} - i\bar{p}\right),$$

(in terms of dimensionless operators $\bar{x}$ and $\bar{p}$ defined as follows:

$$\bar{x} = \frac{x}{\lambda}, \quad \bar{p} = \frac{\lambda p}{\hbar}.$$

with λ some unit of length) satisfying the well known commutation relations

$$[\bar{x}, \, \bar{p}] = i. \quad \left[a, \, a^\dagger\right] = 1.$$

Now the quantum version of he Hamiltonian becomes a hermitian operator, which can be expressed by means of the operators a and $a^\dagger$:

$$\hat{H} = -\frac{d^2}{dx^2} + x^2 = aa^\dagger + a^\dagger a,$$

The eigenfunctions of this operator, corresponding to fixed values of energy E, are obtained by defining first the lowest energy state corresponding to the zero-eigenvalue state of the operator a, $a \mid 0 >= 0$. In coordinate representation the zero state is represented by a Gaussian function:

$$a \mid 0 >= 0 \quad \rightarrow [\frac{d}{dx} + x]\, f(x) = 0 \quad \rightarrow f(x) = e^{-\frac{x^2}{2}}. \tag{27}$$

Positive eigenvalue functions of the Hamiltonian $\hat{H}$ are then obtained by acting on $f(x)$ with consecutive powers of $a^\dagger$. [9]

3.2. The Z_3 ternary Heisenberg algebra

In the Z_2-graded case the two independent combinations of the operators $\frac{d}{dx}$ and x are given by the following non-singular matrix:

$$\begin{pmatrix} 1 & 1 \\ -1 & 1 \end{pmatrix}.$$

By analogy, in the Z_3-graded case, the following transformation matrix should be used, producing three independent combinations of the operators $\frac{d}{dx}$, x and $\mathbf{1}$:

$$\begin{pmatrix} 1 & 1 & 1 \\ j & j^2 & 1 \\ j^2 & j & 1 \end{pmatrix}$$

A natural ternary generalization of Heisenberg's algebra should be spanned by three generators [10]instead of two. By analogy with the Z_2 case, the Z_3-generalization will be generated by the following three operators:

$$c_1 = \lambda \frac{d}{dx} + jx + j^2 1, \qquad c_2 = \lambda \frac{d}{dx} + j^2 x + j1, \qquad c_3 = \lambda \frac{d}{dx} + x + 1,$$

which span the following *binary* Lie algebra:

$$[c_1, c_2] = \lambda(j^2 - j)\, 1, \qquad [c_2, c_3] = \lambda(1 - j^2)\, 1, \qquad [c_3, c_1] = \lambda(j - 1)\, 1,$$

The following linear relations hold:

$$\frac{1}{3}(c_1 + c_2 + c_3) = \lambda \frac{d}{dx}, \qquad \frac{1}{3}\left(j\, c_1 + j^2\, c_2 + c_3\right) = \mathbf{1}, \qquad \frac{1}{3}\left(j^2\, c_1 + j c_2 + c_3\right) = x,$$

Here are the independent 3-commutators between the generators of our algebra:

$$\{c_1, c_2, c_1\} = -3\lambda\, c_1, \quad \{c_2, c_1, c_2\} = 3\lambda\, c_2,$$

$$\{c_2, c_3, c_2\} = -3j\,\lambda\, c_2, \quad \{c_3, c_2, c_3\} = 3j\lambda\, c_3,$$
$$\{c_3, c_1, c_3\} = -3j^2\,\lambda\, c_3, \quad \{c_1, c_3, c_1\} = 3j^2\,\lambda\, c_1,$$
$$\{c_2, c_3, c_1\} = \lambda\left[(1 - j)c_1 + (j^2 - 1)c_2 + (j - j^2)c_3\right],$$
$$\{c_1, c_3, c_2\} = \lambda\left[(j^2 - j)c_1 + (j^2 - j)c_2 + (j^2 - j)c_3\right],$$

If we add the unit operator $\mathbf{1}$ to the three generators c_1, c_2 and c_3, the ordinary commutators between the generators can be interpreted as ternary commutators involving the unit operator 1 in the middle:

$$[c_1, c_2] = \{c_1, 1, c_2\} = \lambda(j^2 - j)\, 1,$$
$$[c_2, c_3] = \{c_2, 1, c_3\} = \lambda(1 - j^2)\, 1,$$
$$[c_3, c_1] = \{c_3, 1, c_1\} = \lambda(j - 1)\, 1,$$

$$(c_1c_3c_2 + c_3c_2c_1 + c_2c_1c_3) - (c_2c_3c_1 + c_3c_1c_2 + c_1c_2c_3) = 3\lambda(j - j^2).$$

after renormalization $c_k \to \frac{1}{\sqrt{3}} c_k$ and setting $\lambda = -i$, the right-hand side of this cubic commutation relation will become equal to $\hbar\mathbf{1}$. Normalized operators are as follows

$$c_1 = \frac{1}{\sqrt{3}}\left[\frac{\sqrt{\hbar}}{i}\frac{d}{dx} + jx + j^2 1\right], \quad c_2 = \frac{1}{\sqrt{3}}\left[\frac{\sqrt{\hbar}}{i}\frac{d}{dx} + j^2 x + j1\right], \quad c_3 = \frac{1}{\sqrt{3}}\left[\frac{\sqrt{\hbar}}{i}\frac{d}{dx} + x + 1\right],$$

3.3. Invariance group of Heisenberg's algebra

Let us consider a general linear transformation

$$a \to A = \alpha a + \beta a^\dagger, \ ; \quad a^\dagger \to B = \gamma a + \delta a^\dagger,$$

Imposing the same constitutive relations on new generators $\tilde{a}$ and $\tilde{a}^\dagger$,

$$AB - BA = 1,$$

leads to the following conditions on the coefficients α, β, γ, δ: $\alpha\delta - \gamma\beta = 1$, which defines the $SL(2, C)$ group.

Keeping invariant just the commutation relations does not preserve the particular structure of the Heisenberg algebra. If we want to make sure that the new generators define a Heisenberg algebra isomorphic with the original one, we should impose supplementary condition, making sure that $\tilde{a}^\dagger$ is indeed a hermitian conjugate of $\tilde{a}$:

$$a \to \tilde{a} = \alpha a + \beta a^\dagger, \ ; \quad a^\dagger \to \tilde{a}^\dagger = \gamma a + \delta a^\dagger,$$

Imposing the same constitutive relations on new generators $\tilde{a}$ and $\tilde{a}^\dagger$,

$$\tilde{a}\tilde{a}^\dagger - \tilde{a}^\dagger\tilde{a} = 1,$$

leads to the following conditions on the coefficients α, β, γ, δ: $\alpha\delta - \gamma\beta = 1$, $\quad \bar{\alpha} = \delta$, $\bar{\beta} = \gamma$. which defines the Bogolyubov group.

The Bogolyubov group, isomorphic with $SU(1, 1)$ can be represented by 2×2 matrices constructed as follows:

$$\begin{pmatrix} \cosh\psi\, e^{i\theta_1} & \sinh\psi\, e^{i\theta_2} \\ \sinh\psi\, e^{-i\theta_2} & \cosh\psi\, e^{-i\theta_1} \end{pmatrix}$$

It has three real parameters, ψ, θ_1 and θ_2.

Now we would like to determine the group of linear transformations preserving the algebra $\mathcal{A}_{3H}$ The 3×3 matrices $U^i_{j'}$ spanning such group transform the set of three generators c_i into a new set $\tilde{c}_{j'}$

$$c_i \to \tilde{c}_{j'} = U^i_{j'}\, c_i.$$

$$\{c_i, c_j, c_k\} = \rho^m_{ijk}\, c_m, \quad \{\tilde{c}_{i'}, \tilde{c}_{j'}, \tilde{c}_{k'}\} = \rho^{m'}_{i'j'k'}\, \tilde{c}_{m'},$$

$$U^i_{i'}\, \rho^{i'}{}_{j'k'm'} = \rho^i{}_{jkm}\, U^j_{j'} U^k_{k'} U^m_{m'}.$$

Journal of Physics: Conference Series **597** (2015) 012049 doi:10.1088/1742-6596/597/1/012049

with supplementary condition

$$\tilde{c}^{\dagger}_{1'} = \tilde{c}_{2'}, \quad \tilde{c}^{\dagger}_{2'} = \tilde{c}_{1'} \quad \tilde{c}^{\dagger}_{3'} = \tilde{c}_{3'}.$$

It is easy to see that there are three subgroups of this group of transformations, each of the isomorphic with $SL(2, C)$:

$$\begin{pmatrix} U^1_{1'} & U^1_{2'} & 0 \\ U^2_{1'} & U^2_{2'} & 0 \\ 0 & 0 & 1 \end{pmatrix}, \quad \begin{pmatrix} U^1_{1'} & 0 & U^1_{3'} \\ 0 & 1 & 0 \\ U^3_{1'} & 0 & U^3_{3'} \end{pmatrix} \quad \begin{pmatrix} 1 & 0 & 0 \\ 0 & U^2_{2'} & U^2_{3'} \\ 0 & U^3_{2'} & U^3_{3'} \end{pmatrix}.$$

Each of these matrices would belong to an $SL(2, C)$ group if we do not impose hermiticity conditions. But if we want to conserve the relations between the new generators as they were between the old ones, we shall restrain the three subgroups to corresponding Bogolyubov groups. The products of these matrices, belonging now to $SU(1, 1)$, span the analogue of the Bogolyubov group preserving ternary Z_3-graded Heisenberg algebra constructed above.

4. Ternary analogue of quantum oscillator

The Z_3-graded analogues of creation and annihilation operators a and $a^{\dagger}$ are the operators $c_3 = c_3^{\dagger}$, c_1 and $c_2 = c_1^{\dagger}$. Imposing hermiticity means that the factor λ in front of the derivation must be pure imaginary, so we shall set $\lambda = -i$, as in the usual quantization scheme.

The expression $aa^{\dagger} + a^{\dagger}a$ can be interpreted as a sum of all Z_2 permutations, or all S_2 permutations, because here all permutations are cyclic, which is not the case of Z_3 and S_3 groups. The sum of all cyclic permutations of three operators $c_1c_2c_3$, or even all (six) permutations does not lead to any simple expression. Moreover, in such combinations the complex representation of Z_3 does not appear.

However, the quadratic harmonic oscillator Hamiltonian $aa^{\dagger} + a^{\dagger}a$ can be represented in an alternative manner, in which the representation of Z_2 by multiplications by 1 and -1 does appear explicitly:

$$\hat{H} = \frac{1}{2} \left[(a + a^{\dagger})^2 + (-1)(a - a^{\dagger})^2 \right] = aa^{\dagger} + a^{\dagger}a. \tag{28}$$

Now the Z_3 generalization becomes obvious: we should form the sum of *cubes* of similar expressions with three generators each, multiplied by j and j^2 in all possible combinations so as to ensure the hermiticity of the resulting expression, like in (28) above. Two expressions of this type can be formed with the generators c_1, c_2 and c_3,

$$(c_3 + c_1 + c_2)^3 + j(c_3 + jc_1 + j^2c_2)^3 + j^2(c_3 + j^2c_1 + jc_2)^3,$$

$$\text{and} \quad (c_3 + c_1 + c_2)^3 + j^2(c_3 + jc_1 + j^2c_2)^3 + j(c_3 + j^2c_1 + jc_2)^3.$$

Neither of these two combinations is hermitian, but their sum $\hat{H}_{Z_3}$ is (if we set λ pure imaginary, e.g. $-i$):

$$\hat{H}_{Z_3} = 2\lambda^3 \frac{d^3}{dx^3} - x^3 - 1 = 2i \frac{d^3}{dx^3} - x^3 - 1. \tag{29}$$

This operator is the Z_3-graded generalization of quantum harmonic oscillator Hamiltonian. The expression $x^3 + 1$ appearing as the "potential" part factorizes as follows:

$$x^3 + 1 = (x + 1)(x + j)(x + j^2).$$

The eigenvalue equation

$$\hat{H}_{Z_3} f(x) = E f(x)$$

can be solved explicitly for the particular value of $E = -1$, which amounts to solve the following differential equation of third order:

$$2i\frac{d^3 f}{dx^3} - x^3\, f(x) = 0. \tag{30}$$

This equation can be solved by expanding $f(x)$ into a power series

$$f(x) = \sum_{k=0}^{\infty} c_k\, x^k.$$

The solution is given in terms of the *generalized hypergeometric function* $F([\ \ ];p,q;\xi)$. In the convention used in "Maple" computing programme, this symbol defines the following power series:

$$F([];p,q;\xi) = \sum_{k=0}^{\infty} c_k x^k, \tag{31}$$

with the coefficients c_k defined as follows using gamma-functions:

$$c_k = \frac{\Gamma(p)\Gamma(q)}{\Gamma(p+k)\Gamma(q+k)} \tag{32}$$

The eigenfunction satisfying (30) is a linear combination of three independent solutions:

$$F([\,];\frac{2}{3},\frac{5}{6};\frac{ix^6}{432}), \quad x\,F([\,];\frac{5}{6},\frac{7}{6};\frac{ix^6}{432}), \quad \text{and} \quad x^2\,F([\,];\frac{7}{3},\frac{4}{3};\frac{ix^6}{432}),$$

corresponding to three different initial conditions, as it should be for a third order differential equation. All these eigenfunctions are complex.

In the classical harmonic oscillator case the operators a and $a^\dagger$ were not hermitian; nevertheless their combination $aa^\dagger + a^\dagger a$ became hermitian. We might as well abandon the requirement of hermiticity, i.e. admit real value for the factor λ in the definition of our operators c_k, and still get real eigenvalues for the non-hermitian third order operator, in the spirit of C. Bender's work on the subject [11].

Acknowledgements

The author is greatly indebted to Karol Penson and Michel Dubois-Violette for numerous enlightening discussions and many useful remarks.

References

[1] Gelfand I M, Kapranov M M and Zelevinsky A V 1994 *Multidimensional Determinants, Discriminants and Resultants* (Birkhauser: Boston)
[2] Vainerman L and Kerner R 1996 *Journal of Mathematical Physics* **37** 2553
[3] Kerner R 1991 *Comptes Rendus Acad. Sci. Paris.* **10** 1237
[4] Kerner R 1992 *Journal of Mathematical Physics* **33** 403
[5] Abramov V, Kerner R and Le Roy B 1997 *Journal of Mathematical Physics* **38** 1650
[6] Kerner R 1997 *Classical and Quantum Gravity* **14** A203
[7] Kerner R 2001 *Proceedings of the 23-rd ICGTMP colloquium, Dubna 2000* (*Preprint* arXiv:math-ph/0011023)
[8] Kerner R and Suzuki O 2012 *Int. J. Geom. Methods Mod. Phys.* **09** 1261007
[9] Schrödinger E 1940 *Proceedings of the Royal Irish Academy* **46** 9
[10] Nambu Y 1973 *Physical Review D* **7** 2405
[11] Bender C M and Boettcher S 1998 *Phys. Rev. Lett.* **80** 5243

Journal of Physics: Conference Series **597** (2015) 012050 doi:10.1088/1742-6596/597/1/012050

Group theory for embedded random matrix ensembles

V K B Kota

Physical Research Laboratory, Ahmedabad 380 009, India

E-mail: `vkbkota@prl.res.in`

Abstract. Embedded random matrix ensembles are generic models for describing statistical properties of finite isolated quantum many-particle systems. For the simplest spinless fermion (or boson) systems with say m fermions (or bosons) in N single particle states and interacting with say k-body interactions, we have EGUE(k) [embedded GUE of k-body interactions) with GUE embedding and the embedding algebra is $U(N)$. In this paper, using EGUE(k) representation for a Hamiltonian that is k-body and an independent EGUE(t) representation for a transition operator that is t-body and employing the embedding $U(N)$ algebra, finite-N formulas for moments up to order four are derived, for the first time, for the transition strength densities (transition strengths multiplied by the density of states at the initial and final energies). In the asymptotic limit, these formulas reduce to those derived for the EGOE version and establish that in general bivariate transition strength densities take bivariate Gaussian form for isolated finite quantum systems. Extension of these results for other types of transition operators and EGUE ensembles with further symmetries are discussed.

1. Introduction

Wigner introduced random matrix theory (RMT) in physics in 1955 primarily to understand statistical properties of neutron resonances in heavy nuclei [1, 2]. Depending on the global symmetry properties of the Hamiltonian of a quantum system, namely rotational symmetry and time-reversal symmetry, we have Dyson's tripartite classification of random matrices giving the classical random matrix ensembles, the Gaussian orthogonal (GOE), unitary (GUE) and symplectic (GSE) ensembles. In the last three decades, RMT has found applications not only in all branches of quantum physics but also in many other disciplines such as Econophysics, Wireless communication, information theory, multivariate statistics, number theory, neural and biological networks and so on [3, 4, 5, 6]. However, in the context of isolated finite many-particle quantum systems, classical random matrix ensembles are too unspecific to account for important features of the physical system at hand. One refinement which retains the basic stochastic approach but allows for such features consists in the use of embedded random matrix ensembles.

Finite quantum systems such as nuclei, atoms, quantum dots, small metallic grains, interacting spin systems modeling quantum computing core and ultra-cold atoms, share one common property - their constituents predominantly interact via two-particle interactions. Therefore, it is more appropriate to represent an isolated finite interacting quantum system, say with m particles (fermions or bosons) in N single particle (sp) states by random matrix models generated by random k-body (note that $k < m$ and most often we have $k = 2$) interactions and propagate the information in the interaction to many particle spaces. Then we have random matrix ensembles in m-particle spaces - these ensembles are defined by representing the k-particle Hamiltonian (H) by GOE/GUE/GSE and then the m particle H matrix is generated by the m-particle Hilbert space geometry. The key element here is the recognition that there is a Lie algebra that transports the information in the two-particle spaces to many-particle spaces. As a GOE/GUE/GSE random matrix ensemble in two-particle spaces is embedded in the m-particle H matrix, these ensembles are more generically called embedded ensembles (EE).

Embedded ensembles are proved to be rich in their content and wide in their scope. A book giving detailed discussion of the various properties and applications of a wide variety of embedded matrix ensembles is now available [7]. Significantly, the study of embedded random matrix ensembles is still developing. Partly this is due to the fact that mathematical tractability of these ensembles is still a problem and this is the topic of the present paper. A general formulation for deriving analytical results is to use the Wigner-Racah algebra of the embedding Lie algebra [7]. The focus in the present paper is on transition strengths. Note that transition strengths probe the structure of the eigenfunctions of a quantum many-body system.

Given a transition operator $\mathcal{O}$ acting on the m particle eigenstates $|E\rangle$ of H will give transition matrix elements $|\langle E_f \mid \mathcal{O} \mid E_i\rangle|^2$ and the corresponding transition strength density (this will take into account degeneracies in the eigenvalues) is,

$$I_{\mathcal{O}}(E_i, E_f) = I(E_f)\, |\langle E_f \mid \mathcal{O} \mid E_i\rangle|^2\, I(E_i)\,. \tag{1}$$

In Eq. (1), $I(E)$ are state densities normalized to the dimension of the m particle spaces. Note that E_i and E_f belong to the same m particle system or different systems depending on the nature of the transition operator $\mathcal{O}$. In the discussion ahead, we will consider both situations. Random matrix theory has been used in the past to derive the form of $I(E)$, the state densities. In particular, exact (finite N) formulas for lower order moments $\langle H^p\rangle$, $p = 2$ and 4 of $I(E)$ are derived both for EGOE and EGUE ensembles using group theoretical methods directly [8] or indirectly [9]. Going beyond the eigenvalue densities, here we will apply group theoretical methods and derive for the first time some exact formulas for the lower order moments of the transition strength densities $I_{\mathcal{O}}(E_i, E_f)$. Using these, the general form of transition strength densities is deduced. We will restrict to fermion systems and at the end discuss the extension to boson systems and to other situations. Let us add that some results valid in the asymptotic ($N \to \infty$) limit for both the state densities and transition strength densities are available in literature [10, 11, 12, 13].

In general, the Hamiltonian may have many symmetries with the fermions (or bosons) carrying other degrees of freedom such as spin, orbital angular momentum, isospin and so on. Also we may have in the system different types of fermions (or bosons) and for example in atomic nuclei we have protons and neutrons. In addition, a transition operator may preserve particle number and other quantum numbers or it may change them. Off all these various situations, here we have considered three different systems: (i) a system of m spinless fermions and a transition operator that preserves the particle number; (ii) a transition operator that removes say k_0 number of particles from the m fermion system; (iii) a system with two types of spinless fermions with the transition operator changing k_0 number of particles of one type to k_0 number of other particles as in nuclear beta and double beta decay. In all these we will restrict to EGUE. Now we will give a preview.

Section 2 gives some basic results for EGUE(k) for spinless fermion systems as derived in [8] and some of their extensions. Using these results, formulas for the lower order bivariate moments of the transition strength densities for the situation (i) above are derived and they are presented in detail in Section 3. Using these, results in the asymptotic limit are derived and they are presented in Section 4. Some basic results for the situations (ii) and (iii) above are given in Section 5. Finally, in Section 6 briefly discussed are some open group theoretical problems in the embedded ensembles theory for transition strength densities.

2. Basic EGUE(k) results for a spinless fermion system

Let us consider m spinless fermions in N degenerate sp states with the Hamiltonian $\hat{H}$ a k-body operator,

$$\hat{H} = \sum_{i,j} V_{ij}(k)\, A_i^\dagger(k)\, A_j(k)\;, \quad V_{ij}(k) = \left\langle k, i \mid \hat{H} \mid k, j \right\rangle \;. \tag{2}$$

Here $A_i^\dagger(k)$ is a k particle (normalized) creation operator and $A_i(k)$ is the corresponding annihilation operator (a hermitian conjugate). Also, i and j are k-particle indices. Note that the k and m particle space dimensions are $\binom{N}{k}$ and $\binom{N}{m}$ respectively. We will consider $\hat{H}$ to be EGUE(k) in m-particle spaces. Then V_{ij} form a GUE with V matrix being Hermitian. The real and imaginary parts of V_{ij} are independent zero centered Gaussian random variables with variance satisfying,

$$\overline{V_{ab}(k)\, V_{cd}(k)} = V_H^2\, \delta_{ad}\delta_{bc} \;. \tag{3}$$

Here the 'over-line' indicates ensemble average. From now on we will drop the hat over H and denote when needed H by $H(k)$. Let us add that in physical systems, $k = 2$ is of great interest and in some systems such as atomic nuclei it is possible to have $k = 3$ and even $k = 4$ [14, 15].

The $U(N)$ algebra that generates the embedding, as shown in [8], gives formulas for the lower order moments of the one-point function, the eigenvalue density $I(E) = \overline{\langle\langle \delta(H - E)\rangle\rangle}$ and also for the two-point function in the eigenvalues. In particular, explicit formulas are given in [8, 7] for $\overline{\langle H^P \rangle^m}$, $P = 2, 4$ and $\overline{\langle H^P \rangle^m \, \langle H^Q \rangle^m}$, $P + Q = 2, 4$. Used here is the $U(N)$ tensorial decomposition of the $H(k)$ operator giving $\nu = 0, 1, \ldots, k$ irreducible parts $B^{\nu,\omega_\nu}(k)$ and then,

$$H(k) = \sum_{\nu=0;\omega_\nu\in\nu}^{k} W_{\nu,\omega_\nu}(k)\; B^{\nu,\omega_\nu}(k) \;. \tag{4}$$

With the GUE(k) representation for the $H(k)$ operator, the expansion coefficients W's will be independent zero centered Gaussian random variables with, by an extension of Eq. (3),

$$\overline{W_{\nu_1,\omega_{\nu_1}}(k)\; W_{\nu_2,\omega_{\nu_2}}(k)} = V_H^2\; \delta_{\nu_1,\nu_2}\delta_{\omega_{\nu_1}\omega_{\nu_2}} \;. \tag{5}$$

For deriving formulas for the various moments, the first step is to apply the Wigner-Eckart theorem for the matrix elements of $B^{\nu,\omega_\nu}(k)$. Given the m-fermion states $|f_m v_i\rangle$, we have with respect to the $U(N)$ algebra, $f_m = \{1^m\}$, the antisymmetric irreducible representation in Young tableaux notation and v_i are additional labels. Note that ν introduced above corresponds to the Young tableaux $\{2^\nu 1^{N-2\nu}\}$ and ω_ν are additional labels. Now, Wigner-Eckart theorem gives

$$\langle f_m v_f \mid B^{\nu,\omega_\nu}(k) \mid f_m v_i\rangle = \langle f_m \mid\mid B^\nu(k) \mid\mid f_m\rangle \; C^{\nu,\omega_\nu}_{f_m v_f\,,\, \overline{f_m v_i}} \;. \tag{6}$$

Here, $\langle - - \mid\mid - - \mid\mid - - \rangle$ is the reduced matrix element and $C^{---}_{--\,,\,--}$ is a $U(N)$ Clebsch-Gordan (C-G) coefficient [note that we are not making a distinction between $U(N)$ and $SU(N)$]. Also, $|\overline{f_m v_i}\rangle$ represent a m hole state (see [8] for details). In Young tableaux notation $\overline{f_m} = \{1^{N-m}\}$. Definition of $B^{\nu,\omega_\nu}(k)$ and the $U(N)$ Wigner-Racah algebra will give,

$$\begin{gathered} |\langle f_m \mid\mid B^\nu(k) \mid\mid f_m\rangle|^2 = \Lambda^\nu(N, m, m-k)\;, \\ \Lambda^\mu(N', m', r) = \binom{m'-\mu}{r}\binom{N'-m'+r-\mu}{r}\;. \end{gathered} \tag{7}$$

Note that $\Lambda^{\nu}(N,m,k)$ is nothing but, apart from a N and m dependent factor, a $U(N)$ Racah coefficient [8]. This and the various properties of the $U(N)$ Wigner and Racah coefficients give two formulas for the ensemble average of a product any two m particle matrix elements of H,

$$\begin{array}{l}\overline{\langle f_m v_1 \mid H(k) \mid f_m v_2\rangle \, \langle f_m v_3 \mid H(k) \mid f_m v_4\rangle} \\ = V_H^2 \displaystyle\sum_{\nu=0;\omega_\nu}^{k} \Lambda^{\nu}(N,m,m-k)\, C^{\nu,\omega_\nu}_{f_m v_1\,,\,\overline{f_m}\,\overline{v_2}}\, C^{\nu,\omega_\nu}_{f_m v_3\,,\,\overline{f_m}\,\overline{v_4}}\,,\end{array} \tag{8}$$

and also

$$\begin{array}{l}\overline{\langle f_m v_1 \mid H(k) \mid f_m v_2\rangle \, \langle f_m v_3 \mid H(k) \mid f_m v_4\rangle} \\ = V_H^2 \displaystyle\sum_{\nu=0;\omega_\nu}^{m-k} \Lambda^{\nu}(N,m,k)\, C^{\nu,\omega_\nu}_{f_m v_1\,,\,\overline{f_m}\,\overline{v_4}}\, C^{\nu,\omega_\nu}_{f_m v_3\,,\,\overline{f_m}\,\overline{v_2}}\end{array} \tag{9}$$

Eq. (9) follows by applying a Racah transform to the product of the two C-G coefficients appearing in Eq. (8). Let us mention two important properties of the $U(N)$ C-G coefficients that are quite useful,

$$\sum_{v_i} C^{\nu,\omega_\nu}_{f_m v_i\,,\,\overline{f_m}\,\overline{v_i}} = \sqrt{\binom{N}{m}}\,\delta_{\nu,0}\,, \quad C^{0,0}_{f_m v_i\,,\,\overline{f_m}\,\overline{v_j}} = \binom{N}{m}^{-1/2} \delta_{v_i\,,v_j}\,. \tag{10}$$

From now on we will use the symbol f_m only in the C-G coefficients, Racah coefficients and the reduced matrix elements. However, for the matrix elements of an operator we will use m implying totally antisymmetric state for fermions (symmetric state for bosons).

Starting with Eq. (4) and using Eqs. (5), (9) and (10) will immediately give the formula,

$$\overline{\langle [H(k)]^2 \rangle^m} = \binom{N}{m}^{-1} \sum_{v_i} \overline{\langle m v_i \mid [H(k)]^2 \mid m v_i\rangle} = V_H^2\, \Lambda^0(N,m,k)\,. \tag{11}$$

Similarly, for $\overline{\langle H^4\rangle^m}$ first the ensemble average is decomposed into 3 terms as,

$$\begin{array}{l}\overline{\langle\langle [H(k)]^4 \rangle\rangle^m} = \displaystyle\sum_{v_i} \overline{\langle m v_i \mid [H(k)]^4 \mid m v_i\rangle} \\[2ex] = \displaystyle\sum_{v_i,v_j,v_p,v_l} \Big[\overline{\langle m v_i \mid H(k) \mid m v_j\rangle \, \langle m v_j \mid H(k) \mid m v_p\rangle}\;\overline{\langle m v_p \mid H(k) \mid m v_l\rangle \, \langle m v_l \mid H(k) \mid m v_i\rangle} \\[2ex] + \overline{\langle m v_i \mid H(k) \mid m v_j\rangle \, \langle m v_l \mid H(k) \mid m v_i\rangle}\;\overline{\langle m v_j \mid H(k) \mid m v_p\rangle \, \langle m v_p \mid H(k) \mid m v_l\rangle} \\[2ex] + \overline{\langle m v_i \mid H(k) \mid m v_j\rangle \, \langle m v_p \mid H(k) \mid m v_l\rangle}\;\overline{\langle m v_j \mid H(k) \mid m v_p\rangle \, \langle m v_l \mid H(k) \mid m v_i\rangle}\Big]\,.\end{array} \tag{12}$$

Note that the trace $\langle\langle H^4\rangle\rangle^m = \binom{N}{m} \langle H^4\rangle^m$. It is easy to see that the first two terms simplify to give $2[\overline{\langle H^2\rangle^m}]^2$ and the third term is simplified by applying Eq. (8) to the first ensemble average and Eq. (9) to the second ensemble average. Then, the final result is

$$\overline{\langle [H(k)]^4 \rangle^m} = 2\left[\overline{\langle H^2\rangle^m}\right]^2 + V_H^4 \binom{N}{m}^{-1} \sum_{\nu=0}^{min(k,m-k)} \Lambda^{\nu}(N,m,k)\, \Lambda^{\nu}(N,m,m-k)\, d(N:\nu)\,; \tag{13}$$

$$d(N:\nu) = \binom{N}{\nu}^2 - \binom{N}{\nu-1}^2\,. \tag{14}$$

Also, a by-product of Eqs. (9) and (10) is

$$\sum_{v_j} \overline{\langle mv_i \mid H(k) \mid mv_j \rangle \, \langle mv_j \mid H(k) \mid mv_k \rangle} = \overline{\langle [H(k)]^2 \rangle^m} \, \delta_{v_i,v_k} \tag{15}$$

and we will use this in Section 3. Now we will discuss the results for the moments of the transition strength densities generated by a transition operator $\mathcal{O}$.

3. Lower-order moments of transition strength densities: H EGUE(k) and $\mathcal{O}$ an independent EGUE(t)

For a spinless fermion system, similar to the H operator, we will consider the transition operator $\mathcal{O}$ to be a t-body and represented by EGUE(t) in m-particle spaces. Then, the matrix of $\mathcal{O}$ in t-particle spaces will be a GUE with the matrix elements $\mathcal{O}_{ab}(t)$ being zero centered independent Gaussian variables with the variance satisfying,

$$\overline{\mathcal{O}_{ab}(t) \, \mathcal{O}_{cd}(t)} = V_{\mathcal{O}}^2 \, \delta_{ad}\delta_{bc} \; . \tag{16}$$

Further, we will assume that the GUE representing H in k particle space and the GUE(t) representing $\mathcal{O}$ in t particle space are independent (this is equivalent to the statement that $\mathcal{O}$ do not generate diagonal elements $\langle E_i \mid \mathcal{O} \mid E_i \rangle$; see [11]). It is important to mention that in the past there are attempts to study numerically transition strengths using the eigenvectors generated by a EGOE (with symmetries) for H for nuclear systems and a realistic transition operator with results in variance with those obtained from realistic interactions [16]. As described in the present paper, a proper random matrix theory for transition strengths has to employ ensemble representation for both the Hamiltonian and the transition operator.

With EGUE representation, $\mathcal{O}$ is Hermitian and hence $\mathcal{O}^\dagger = \mathcal{O}$. Moments of the transition strength densities $I_{\mathcal{O}}(E_i, E_f)$ are defined by

$$M_{PQ}(m) = \overline{\langle \mathcal{O}^\dagger H^Q \mathcal{O} H^P \rangle^m} = \overline{\langle \mathcal{O} H^Q \mathcal{O} H^P \rangle^m} \; . \tag{17}$$

Here the ensemble average is w.r.t. both EGUE(k) and EGUE(t). Now we will derive formulas for M_{PQ} with $P+Q=2$ and 4; the moments with $P+Q$ odd will vanish by definition.

Firstly, the unitary decomposition of $\mathcal{O}(t)$ gives,

$$\mathcal{O}(t) = \sum_{\nu=0;\omega_\nu \in \nu}^{t} U_{\nu,\omega_\nu}(t) \, B^{\nu,\omega_\nu}(t) \; . \tag{18}$$

The U's satisfy a relation similar to Eq. (5). Now, for $P=Q=0$ we have using Eq. (11),

$$\overline{\langle \mathcal{O}\mathcal{O} \rangle^m} = V_{\mathcal{O}}^2 \, \Lambda^0(N,m,t) \; . \tag{19}$$

Also, we have the relations

$$\overline{\langle \mathcal{O}\mathcal{O} H^P \rangle^m} = \overline{\langle \mathcal{O}\mathcal{O} \rangle^m} \; \overline{\langle H^P \rangle^m} = \overline{\langle \mathcal{O} H^P \mathcal{O} \rangle^m} \; . \tag{20}$$

For the proof consider $\overline{\langle \mathcal{O}\mathcal{O}H^P \rangle^m} = \binom{N}{m}^{-1} \sum_{v_i,v_j} \overline{\langle mv_i \mid \mathcal{O}\mathcal{O} \mid mv_j \rangle} \; \overline{\langle mv_j \mid H^P \mid mv_i \rangle}$. Now, applying Eq. (15) gives $\overline{\langle mv_i \mid \mathcal{O}(t)\mathcal{O}(t) \mid mv_j \rangle} = \overline{\langle \mathcal{O}\mathcal{O} \rangle^m} \, \delta_{v_i,v_j}$ and substituting this will give the first equality in Eq. (20). Finally, $\overline{\langle \mathcal{O}\mathcal{O}H^P \rangle^m} = \overline{\langle \mathcal{O}H^P\mathcal{O} \rangle^m}$ follows from the cyclic invariance of the m-particle average. Eq. (20) gives,

$$M_{20}(m) = M_{02}(m) = \overline{\langle \mathcal{O}\mathcal{O} \rangle^m} \; \overline{\langle H^2 \rangle^m} \, , \quad M_{40}(m) = M_{04}(m) = \overline{\langle \mathcal{O}\mathcal{O} \rangle^m} \; \overline{\langle H^4 \rangle^m} \; . \tag{21}$$

Formulas for $\overline{\langle \mathcal{O}\mathcal{O}\rangle^m}$, $\overline{\langle H^2\rangle^m}$ and $\overline{\langle H^4\rangle^m}$ follow from Eqs. (19), (11) and (13). Thus, the non-trivial moments M_{PQ} for $P+Q \leq 4$ are M_{11}, $M_{13} = M_{31}$ and M_{22}.

It is easy to recognize that the bivariate moment M_{11} has same structure as the third term in Eq. (12) for $\overline{\langle H^4\rangle^m}$ as the $\mathcal{O}$ and H ensembles are independent. Then, the formula for M_{11} follows directly from the second term in Eq. (13). This gives,

$$\begin{array}{l} M_{11}(m) = \overline{\langle \mathcal{O}(t)\, H(k)\, \mathcal{O}(t)\, H(k)\rangle^m} = \\ V_{\mathcal{O}}^2\, V_H^2\, \binom{N}{m}^{-1} \displaystyle\sum_{\nu=0}^{min(k,m-t)} \Lambda^\nu(N,m,t)\, \Lambda^\nu(N,m,m-k)\, d(N:\nu)\;. \end{array} \tag{22}$$

This equation has the correct $t \leftrightarrow k$ symmetry. Using Eqs. (22) and (11), we have for the bivariate correlation coefficient ξ, the formula

$$\xi(m) = \frac{M_{11}(m)}{\sqrt{M_{20}(m)\, M_{02}(m)}} = \frac{\displaystyle\sum_{\nu=0}^{min(t,m-k)} \Lambda^\nu(N,m,k)\, \Lambda^\nu(N,m,m-t)\, d(N:\nu)}{\binom{N}{m}\, \Lambda^0(N,m,t)\, \Lambda^0(N,m,k)}\;. \tag{23}$$

Turning to M_{PQ} with $P+Q=4$, the first trivial moment is $M_{13} = M_{31}$. For M_{31} we have,

$$\begin{array}{l} \binom{N}{m}\, M_{31}(m) = \overline{\langle\langle \mathcal{O}(t) H(k) \mathcal{O}(t) [H(k)]^3\rangle\rangle^m} \\ = \displaystyle\sum_{v_i,v_j,v_k,v_l} \overline{\langle m v_i \mid \mathcal{O}(t) \mid m v_j\rangle\, \langle m v_k \mid \mathcal{O}(t) \mid m v_l\rangle}\; \overline{\langle m v_j \mid H(k) \mid m v_k\rangle\, \langle m v_l \mid [H(k)]^3 \mid m v_i\rangle} \end{array} \tag{24}$$

and the last equality follows from the result that the EGUE's representing H and $\mathcal{O}$ are independent. The ensemble average of the product of two $\mathcal{O}$ matrix elements follows easily from Eq. (9) giving,

$$\overline{\langle m v_i \mid \mathcal{O}(t) \mid m v_j\rangle\, \langle m v_k \mid \mathcal{O}(t) \mid m v_l\rangle} = V_{\mathcal{O}}^2 \sum_{\nu=0}^{m-t} \Lambda^\nu(N,m,t) \sum_{\omega_\nu} C^{\nu,\omega_\nu}_{f_m v_i\,,\, \overline{f_m v_l}}\, C^{\nu,\omega_\nu}_{f_m v_k\,,\, \overline{f_m v_j}}\;. \tag{25}$$

The ensemble average of the product of a H matrix element and H^3 matrix element appearing in Eq. (24) is,

$$\begin{array}{l} \overline{\langle m v_j \mid H(k) \mid m v_k\rangle\, \langle m v_l \mid [H(k)]^3 \mid m v_i\rangle} = \\ \displaystyle\sum_{v_p,v_q} \Big[\overline{\langle m v_j \mid H(k) \mid m v_k\rangle\, \langle m v_l \mid H(k) \mid m v_p\rangle}\; \overline{\langle m v_p \mid H(k) \mid m v_q\rangle\, \langle m v_q \mid H(k) \mid m v_i\rangle} \\ +\overline{\langle m v_j \mid H(k) \mid m v_k\rangle\, \langle m v_q \mid H(k) \mid m v_i\rangle}\; \overline{\langle m v_l \mid H(k) \mid m v_p\rangle\, \langle m v_p \mid H(k) \mid m v_q\rangle} \\ +\, \overline{\langle m v_j \mid H(k) \mid m v_k\rangle\, \langle m v_p \mid H(k) \mid m v_q\rangle}\; \overline{\langle m v_l \mid H(k) \mid m v_p\rangle\, \langle m v_q \mid H(k) \mid m v_i\rangle} \Big]\;. \end{array} \tag{26}$$

The first two terms in Eq. (26) combined with Eq. (24) will give $2\overline{\langle [H(k)]^2\rangle^m}\, M_{11}(m)$. For the third term we can use Eqs. (8) and (9) giving

$$\begin{array}{l} \overline{\langle m v_j \mid H(k) \mid m v_k\rangle\, \langle m v_p \mid H(k) \mid m v_q\rangle}\; \overline{\langle m v_l \mid H(k) \mid m v_p\rangle\, \langle m v_q \mid H(k) \mid m v_i\rangle} \\ = V_H^4 \displaystyle\sum_{\nu_1=0;\omega_{\nu_1}}^{k}\; \sum_{\nu_2=0;\omega_{\nu_2}}^{m-k} \Lambda^{\nu_1}(N,m,m-k)\, \Lambda^{\nu_2}(N,m,k) \\ \times\, C^{\nu_1,\omega_{\nu_1}}_{f_m v_j\,,\, \overline{f_m v_k}}\, C^{\nu_1,\omega_{\nu_1}}_{f_m v_p\,,\, \overline{f_m v_q}}\, C^{\nu_2,\omega_{\nu_2}}_{f_m v_l\,,\, \overline{f_m v_i}}\, C^{\nu_2,\omega_{\nu_2}}_{f_m v_q\,,\, \overline{f_m v_p}}\;. \end{array} \tag{27}$$

Combining this with Eq. (25) and applying the orthonormal properties of the C-G coefficients will give the final formula for M_{31},

$$
\begin{array}{l}
M_{31}(m) = \overline{\langle \mathcal{O}(t)\, H(k)\, \mathcal{O}(t)\, [H(k)]^3 \rangle^m} = \\[2mm]
V_{\mathcal{O}}^2\, V_H^4 \binom{N}{m}^{-1} \left\{ 2\Lambda^0(N,m,k) \displaystyle\sum_{\nu=0}^{min(k,m-t)} \Lambda^\nu(N,m,t)\, \Lambda^\nu(N,m,m-k)\, d(N:\nu) \right. \\[2mm]
\left. + \displaystyle\sum_{\nu=0}^{min(k,m-k,m-t)} \Lambda^\nu(N,m,t)\, \Lambda^\nu(N,m,k)\, \Lambda^\nu(N,m,m-k)\, d(N:\nu) \right\} .
\end{array}
\tag{28}
$$

For deriving the formula for M_{22}, we will make use of the decompositions similar to those in Eqs. (24) and (26). Then it is easy to see that M_{22} will have three terms and simplifying these will give (details are not given here due to lack of space),

$$
\begin{array}{l}
M_{22}(m) = \overline{\langle O(t) O(t) \rangle^m} \left\{ \overline{\langle H(k) H(k) \rangle^m} \right\}^2 \\[2mm]
+ V_{\mathcal{O}}^2 V_H^4 \left\{ \binom{N}{m}^{-1} \displaystyle\sum_{\nu=0}^{min(t,m-k)} \Lambda^\nu(N,m,m-t)\; [\Lambda^\nu(N,m,k)]^2\; d(N:\nu) \right. \\[2mm]
\left. + \binom{N}{m}^{-1} \displaystyle\sum_{\nu=0}^{min(k+t,m-k)} \Lambda^\nu(N,m,k)\, d(N:\nu) \sum_{\nu_1=0}^{t} \sum_{\nu_2=0}^{k} \sum_{\rho} |\langle f_m \,||\, [B^{\nu_1}(t) B^{\nu_2}(k)]^{\nu,\rho} \,||\, f_m \rangle|^2 \right\}
\end{array}
\tag{29}
$$

Here ρ labels multiplicity of the irrep ν in the Kronecker product $\nu_1 \times \nu_2 \to \nu$. The third term above simplifies to,

$$
\begin{array}{l}
\binom{N}{m}^{-2} \displaystyle\sum_{\nu=0}^{min(k+t,m-k)} \sum_{\nu_1=0}^{t} \sum_{\nu_2=0}^{k} \Lambda^\nu(N,m,k)\, \Lambda^{\nu_1}(N,m,m-t)\, \Lambda^{\nu_2}(N,m,m-k) \\[2mm]
\times d(N:\nu_1) d(N:\nu_2) \displaystyle\sum_{\rho} [U(f_m\, \nu_1\, f_m\, \nu_2\, ;\, f_m\, \nu)_\rho]^2 \;.
\end{array}
$$

The U or Racah coefficient here is with respect to $U(N)$. Deriving formulas for this U-coefficient is an important open problem.

4. Asymptotic results for the bivariate moments for H EGUE(k) and $\mathcal{O}$ an independent EGUE(t)

Lowest order (sufficient for most purposes) shape parameters of the bivariate strength density are the bivariate reduced cumulants of order four, i.e. k_{rs}, $r+s=4$. The k_{rs} can be written in terms of the normalized central moments $\widetilde{M}_{PQ}$ where $\widetilde{M}_{PQ} = M_{PQ}/M_{00}$. Then, the scaled moments μ_{PQ} are

$$
\mu_{PQ} = \left\{ \left[\widetilde{M}_{20}\right]^{P/2} \left[\widetilde{M}_{02}\right]^{Q/2} \right\}^{-1} \widetilde{M}_{PQ} \,, \quad P+Q \geq 2 \,. \tag{30}
$$

Note that $\mu_{20} = \mu_{02} = 1$ and $\mu_{11} = \xi$. Now the fourth order cumulants are,

$$
k_{40} = \mu_{40} - 3\,, \;\; k_{04} = \mu_{04} - 3\,, \;\; k_{31} = \mu_{31} - 3\,\xi\,, \;\; k_{13} = \mu_{13} - 3\,\xi\,, \;\; k_{22} = \mu_{22} - 2\,\xi^2 - 1\,. \tag{31}
$$

The $|k_{rs}| \sim 0$ for $r+s \geq 3$ implies that the transition strength density is close to a bivariate Gaussian (note that in our EGUE applications, $k_{rs} = 0$ for $r+s$ odd by definition). We have verified in large number of numerical examples, obtained using the formulas in Section 3 for some typical values of N, m, k and t, that the cumulants $|k_{rs}|$ with $r+s=4$ are in general very small implying that for EGUE, transition strength densities approach bivariate Gaussian form. For a better understanding of this result, it is useful to derive expressions for μ_{PQ} and thereby for k_{PQ}, using Eq. (31), in the asymptotic limit defined by $N\to\infty$, $m\to\infty$, $m/N\to 0$ and k and t fixed. First we will consider $N\to\infty$ and m fixed with $k,t<<m$. We will use,

$$\binom{N-p}{r} \stackrel{p/N\to 0}{\longrightarrow} \frac{N^r}{r!}\;,\quad d(N:\nu) \stackrel{\nu/N\to 0}{\longrightarrow} \frac{N^{2\nu}}{(\nu!)^2}\;. \tag{32}$$

Note that $d(N:\nu)$ was defined by Eq. (14). Let us start with the formula for ξ given by Eq. (23). Applying Eq. (32) it is easily seen that only the term with $\nu=t$ in the sum in Eq. (23) will contribute in the asymptotic limit. Using this, applying Eqs. (32) and the formula for Λ^ν given by Eq. (7) will lead to,

$$\xi(m) \longrightarrow \frac{m!\,N^{2t}}{N^m\,(t!)^2}\frac{\binom{m-t}{k}\binom{N-m+k-t}{k}\binom{N-2t}{m-t}}{\binom{m}{k}\binom{N-m+k}{k}\binom{m}{t}\binom{N-m+t}{t}} \longrightarrow \binom{m}{k}^{-1}\binom{m-t}{k}\;. \tag{33}$$

Similarly, μ_{40} will be, using Eqs. (21), (13) and (11),

$$\mu_{40}(m) \longrightarrow 2+\frac{m!\,N^{2k}}{N^m\,(k!)^2}\frac{\binom{m-k}{k}\binom{N-m}{k}\binom{N-2k}{m-k}}{\binom{m}{k}\binom{N-m+k}{k}\binom{m}{k}\binom{N-m+k}{k}} \longrightarrow 2+\binom{m}{k}^{-1}\binom{m-k}{k}\;. \tag{34}$$

Turning to μ_{31}, it is easy to see from Eq. (28) that M_{31} has two terms. The first term is $(2\xi)\,\langle O(t)O(t)\rangle^m\,[\langle H(k)H(k)\rangle^m]^2$ and in the sum in the second term only $\nu=k$ will survive in the asymptotic limit. These then will give,

$$\begin{aligned}\mu_{31}(m) &= (2\xi)+\binom{N}{m}^{-1}\frac{\Lambda^k(N,m,t)\Lambda^k(N,m,k)\Lambda^k(N,m,m-k)\,d(N:k)}{\Lambda^0(N,m,t)[\Lambda^0(N,m,k)]^2}\\ &\longrightarrow \xi\left[2+\binom{m}{k}^{-1}\binom{m-k}{k}\right]=\xi\mu_{40}\;.\end{aligned} \tag{35}$$

Note that in the final simplifications we have used Eq. (32). Finally, let us consider μ_{22}. Firstly, M_{22} has three terms as seen from Eq. (29) and then, there will be corresponding three terms in μ_{22}. Let us call them $T1$, $T2$ and $T3$. It is seen that $T1=1$ and $T2$ is (in the corresponding sum in M_{22}, only $\nu=t$ term will contribute in the asymptotic limit),

$$T2 \longrightarrow \binom{N}{m}^{-1}\frac{\Lambda^t(N,m,m-t)\left[\Lambda^t(N,m,k)\right]^2\,d(N:t)}{\Lambda^0(N,m,t)[\Lambda^0(N,m,k)]^2} \longrightarrow \binom{m}{k}^{-2}\binom{m-t}{k}^2\;. \tag{36}$$

Similarly, T_3 in the asymptotic limit will be

$$T_3 \longrightarrow \frac{\Lambda^{t+k}(N,m,k)}{\Lambda^0(N,m,k)}\,[\xi(N\to\infty)] \to \binom{m}{k}^{-2}\binom{m-t-k}{k}\binom{m-t}{k}\;. \tag{37}$$

Here we have used Eq. (33) for ξ in the $N\to\infty$ limit. Now combining T_1, T_2 and T_3 we have

$$\mu_{22}(m) \longrightarrow 1+\binom{m}{k}^{-2}\binom{m-t}{k}^2+\binom{m}{k}^{-2}\binom{m-t-k}{k}\binom{m-t}{k}\;. \tag{38}$$

Comparing Eq. (37) with the formula for T_3, it is easy to see that $[U(f_m\,t\,f_m\,k\,;\,f_m\,t{+}k)]^2 \stackrel{asymp}{\longrightarrow} \binom{m}{k}^{-1}\binom{m-t}{k}$. The asymptotic formulas given by Eqs. (33), (34), (35) and (38) show that the finite N results derived in Section 3 reduce in the asymptotic limit exactly to those derived before using binary correlation approximation [11, 7]. More importantly, they show that the cumulants tend to zero in the dilute (or asymptotic) limit.

5. Lower-order moments of transition strength densities: results for particle transfer operator and beta decay type operator

In this Section we will consider transition operators that are non-hermitian and that are of great interest for example in nuclear physics. These are particle removal (or addition) operators and beta decay type operators. Let us begin with particle removal operator $\mathcal{O}$ and say it removes k_0 number of particles when acting on a m fermion state. Then the general form of $\mathcal{O}$ is,

$$\mathcal{O} = \sum_{\alpha_0} V_{\alpha_0}\, A_{\alpha_0}(k_0)\;. \tag{39}$$

Here, $A_{\alpha_0}(k_0)$ is a k_0 particle annihilation operator and α_0 are indices for a k_0 particle state. We will choose V_{α_0} to be zero centered independent Gaussian random variables with variance satisfying $\overline{V_\alpha V_\beta^\dagger} = V_{\mathcal{O}}^2\, \delta_{\alpha\beta}$. Then we have,

$$\overline{\langle \mathcal{O}^\dagger \mathcal{O}\rangle^m} = V_{\mathcal{O}}^2 \binom{m}{k_0}\,, \quad \overline{\langle \mathcal{O}\mathcal{O}^\dagger\rangle^m} = V_{\mathcal{O}}^2 \binom{N-m}{k_0}\;. \tag{40}$$

Eq. (15) also gives the important relations,

$$\overline{\langle \mathcal{O}^\dagger \mathcal{O} H^p\rangle^m} = \overline{\langle \mathcal{O}^\dagger \mathcal{O}\rangle^m}\;\overline{\langle H^p\rangle^m}\,, \quad \overline{\langle \mathcal{O}^\dagger H^p \mathcal{O}\rangle^m} = \overline{\langle \mathcal{O}^\dagger \mathcal{O}\rangle^m}\;\overline{\langle H^p\rangle^{m-k_0}}\;. \tag{41}$$

Another useful result that follows from Eq. (40) is,

$$\left\langle m \mid\mid A^\dagger(k_0) \mid\mid m-k_0 \right\rangle \langle m-k_0 \mid\mid A(k_0) \mid\mid m\rangle = \binom{N-k_0}{m-k_0}\;. \tag{42}$$

Following the procedure used in Section 3 it is possible to derive formulas for the lower order moments of the transition strength density generated by $\mathcal{O}$ given by Eq. (39). Eqs. (41) and (40) along with Eqs. (11) and (13) will give M_{20}, M_{02}, M_{40} and M_{04}. Then, the first non-trivial moment is M_{11}. Formula for $M_{11} = \overline{\langle \mathcal{O}^\dagger H \mathcal{O} H\rangle^m}$ is derived by starting with the definition of $\overline{\langle \mathcal{O}^\dagger H \mathcal{O} H\rangle^m}$,

$$\begin{aligned}\tbinom{N}{m}\,\overline{\langle \mathcal{O}^\dagger H \mathcal{O} H\rangle^m} = &\sum_{v_1,v_2,v_3,v_4} \overline{\langle m,v_1 \mid \mathcal{O}^\dagger \mid m-k_0,v_2\rangle \langle m-k_0,v_3 \mid \mathcal{O} \mid m,v_4\rangle} \\ &\times \overline{\langle m-k_0,v_2 \mid H \mid m-k_0,v_3\rangle \langle m,v_4 \mid H \mid m,v_1\rangle}\;.\end{aligned} \tag{43}$$

Applying Eqs. (4) - (7) and the Wigner-Eckart theorem along with Eq. (42) will give,

$$\begin{aligned}M_{11}(m) = \overline{\langle \mathcal{O}^\dagger H(k) \mathcal{O} H(k)\rangle^m} = V_{\mathcal{O}}^2 V_H^2 \tbinom{N}{m}^{-1} \tbinom{N-k_0}{m-k_0} \tbinom{N}{k_0}^{1/2} \times \sum_{\nu=0}^{k} \\ \left[d(N:\nu)\,\Lambda^\nu(N,m-k_0,m-k_0-k)\,\Lambda^\nu(N,m,m-k)\right]^{1/2} (-1)^\phi\, U(f_m\,\overline{f_{m-k_0}}\,f_m\,f_{m-k_0}\,;\,f_{k_0}\,\nu)\;.\end{aligned} \tag{44}$$

The phase factor ϕ depends on ν. Formula for the U-coefficient appearing in Eq. (44) was available in [17]. It is possible to proceed forward and derive formulas for M_{31}, M_{13} and M_{22} (these will be reported elsewhere). Now we will turn to beta decay type operators.

Let us consider a system with m_1 fermions in N_1 sp states and m_2 fermions in N_2 sp states with H preserving (m_1, m_2). Then, the H operator, assumed to be k-body, is given by,

$$\begin{aligned}H(k) &= \sum_{i+j=k}\sum_{\alpha,\beta\in i}\sum_{a,b\in j} V_{\alpha a:\beta b}(i,j)\, A_\alpha^\dagger(i)\, A_\beta(i)\, A_a^\dagger(j)\, A_b(j)\,, \\ V_{\alpha a:\beta b}(i,j) &= \langle i,\alpha : j,a \mid H \mid i,\beta : j,b\rangle\;.\end{aligned} \tag{45}$$

Journal of Physics: Conference Series **597** (2015) 012050 doi:10.1088/1742-6596/597/1/012050

Here we are using Greek labels $\alpha, \beta, \ldots$ to denote the many particle states generated by fermions occupying the orbit with N_1 sp states and the Roman labels $a, b, \ldots$ for the many particle states generated by the fermions occupying the orbit with N_2 sp states. For each (i, j) pair with $i + j = k$, we have a matrix $V(i, j)$ in the k-particle space and we assume that the $V(i, j)$ matrices are represented by independent GUE's with their matrix elements being zero centered with variance,

$$\overline{V_{\alpha a;\beta b}(i,j)\, V_{\alpha' a';\beta' b'}(i',j')} = V_H^2(i,j)\; \delta_{ii'}\delta_{jj'}\delta_{\alpha\beta'}\delta_{ab'}\delta_{\beta\alpha'}\delta_{ba'}\;. \tag{46}$$

It is important to note that the embedding algebra for the EGUE generated by the action of the $H(k)$ operator on $|m_1, v_\alpha : m_2, v_a\rangle$ states is the direct sum algebra $U(N_1) \oplus U(N_2)$. Thus we have EGUE(k)-$[U(N_1) \oplus U(N_2)]$ ensemble. For this system, we will consider a beta decay ($k_0 = 1$ below) type transition operator,

$$\mathcal{O} = \sum_{\alpha,a} O_{\alpha a} A^\dagger_\alpha(k_0)\, A_a(k_0)\;;\quad O_{\alpha a} = \langle k_0, \alpha \mid \mathcal{O} \mid k_0, a\rangle \tag{47}$$

and assume a GUE representation for the $\mathcal{O}$ matrix in the defining space giving $\overline{O^\dagger_{\alpha,a}O_{\beta,b}} = V^2_{\mathcal{O}}\;\delta_{\alpha\beta}\delta_{ab}$. Note that in general the $\mathcal{O}$ matrix is a rectangular matrix. Now, the ensemble averaged bivariate moments of the transition strength density are $M_{PQ}(m_1, m_2) = \overline{\langle \mathcal{O}^\dagger H^Q \mathcal{O} H^P\rangle^{(m_1,m_2)}}$. Note that $\mathcal{O}$ takes (m_1, m_2) to $(m_1 + k_0, m_2 - k_0)$.

Firstly, recognize that $A^\dagger_\alpha(k_0)$ and $A_\beta(k_0)$ transform as the $U(N_1)$ tensors $f_{k_0} = \{1^{k_0}\}$ and $\{\overline{f_{k_0}}\} = \{1^{N_1-k_0}\}$ and similarly $A^\dagger_a(k_0)$ and $A_b(k_0)$ with respect to $U(N_2)$. These will give easily the results,

$$\overline{\langle \mathcal{O}^\dagger\mathcal{O}\rangle^{(m_1,m_2)}} = V^2_{\mathcal{O}} \binom{N_1 - m_1}{k_0}\binom{m_2}{k_0}\,,\quad \overline{\langle \mathcal{O}\mathcal{O}^\dagger\rangle^{(m_1,m_2)}} = V^2_{\mathcal{O}} \binom{N_2 - m_2}{k_0}\binom{m_1}{k_0}\,. \tag{48}$$

For deriving formulas for the bivariate moments with P and/or $Q \neq 0$, unitary decomposition of H is carried out with respect to the $U(N_1) \oplus U(N_2)$ algebra. This gives,

$$\overline{\langle H^2\rangle^{(m_1,m_2)}} = \sum_{i+j=k} V_H^2(i,j)\; \Lambda^0(N_1, m_1, i)\, \Lambda^0(N_2, m_2, j)\;. \tag{49}$$

It is also easy to show that $\overline{\langle \mathcal{O}^\dagger\mathcal{O}H^P\rangle^{(m_1,m_2)}} = \overline{\langle \mathcal{O}^\dagger\mathcal{O}\rangle^{(m_1,m_2)}}\;\overline{\langle H^P\rangle^{(m_1,m_2)}}$ and $\overline{\langle \mathcal{O}^\dagger H^P\mathcal{O}\rangle^{(m_1,m_2)}} = \overline{\langle \mathcal{O}^\dagger\mathcal{O}\rangle^{(m_1,m_2)}}\;\overline{\langle H^P\rangle^{(m_1+k_0,m_2-k_0)}}$. Also, the first nontrivial bivariate moment $M_{11}(m_1, m_2) = \overline{\langle \mathcal{O}^\dagger H\mathcal{O}H\rangle^{(m_1,m_2)}}$ is given by,

$$\begin{aligned} M_{11}(m_1, m_2) &= V^2_{\mathcal{O}} \tbinom{N_1-k_0}{m_1}\tbinom{N_2-k_0}{m_2-k_0}\left[\tbinom{N_1}{m_1}\tbinom{N_2}{m_2}\right]^{-1} \sqrt{\binom{N_1}{k_0}\binom{N_2}{k_0}}\; \sum_{i+j=k} V_H^2(i,j) \\ &\times \sum_{\nu=0}^{i}\sum_{\nu'=0}^{j} \left[d(N_1:\nu)\, d(N_2:\nu')\, \Lambda^\nu(N_1, m_1, m_1 - i)\Lambda^\nu(N_1, m_1 + k_0, m_1 + k_0 - i)\right]^{1/2} \\ &\times \left[\Lambda^{\nu'}(N_2, m_2, m_2 - j)\Lambda^{\nu'}(N_2, m_2 - k_0, m_2 - k_0 - j)\right]^{1/2} \quad (-1)^{\phi_1(\nu)+\phi_2(\nu')} \\ &\times\; U(f_{m_1+k_0}\,\overline{f_{m_1}}\,f_{m_1+k_0}\,f_{m_1}\,;\,f_{k_0}\,\nu)\; U(f_{m_2}\,\overline{f_{m_2-k_0}}\,f_{m_2}\,f_{m_2-k_0}\,;\,f_{k_0}\,\nu')\;. \end{aligned} \tag{50}$$

Here, ϕ_1 and ϕ_2 are phase factors. Also, $\overline{f_r} = \{1^{N-r}\}$ and $\nu = \{2^\nu, 1^{N-2\nu}\}$ with $N = N_1$ or N_2 as appropriate. Formula for the U-coefficients in Eq. (50) is available in [17].

6. Conclusions and future outlook

In this paper we have presented for the first time exact (finite N) results for the moments of the transition strength densities using $U(N)$ Wigner-Racah algebra for EGUE random matrix ensembles. In particular, formulas for the moments up to fourth order are derived in detail for the Hamiltonian a EGUE(k) and the transition operator a EGUE(t) for spinless fermion systems. Numerical results on one hand and the asymptotic results derived from the exact results on the other, showed that the fourth order cumulants approach zero in the dilute limit implying that the strength densities approach bivariate Gaussian form. As discussed in Section 5, the formulation given in Sections 2 and 3 extends to transition operators that are particle removal (or particle addition) operators and also to beta decay (also neutrinoless double beta decay) type operators. Complete results for these (i.e. for M_{PQ} with $P+Q=4$) will be presented elsewhere. All these results are useful in nuclear spectroscopy [11, 7]. Another important extension is to EGUE with $U(\Omega) \times SU(r)$ embedding discussed in [18]. With applications in mesoscopic systems, it is important to derive formulas for the bivariate moments of transition strength densities for systems with $r = 2$. For fermions, this corresponds to spin degree of freedom and then we have spin scalar and spin vector transition operators. For this system, formulas for the Racah coefficients that appear, even for the fourth moment of the state densities, are not yet available [7] and an exploration of the asymptotic methods suggested in [19] could prove to be fruitful. Let us mention that the results presented in Sections 3 and 5 extend to boson systems by using the $N \to -N$ and $N \to m$ symmetries discussed in [8, 7]. In future, it is also important and useful to extend the present work to EGOE and EGSE ensembles. Finally, future in the subject of embedded random matrix ensembles in quantum physics is exciting with enormous scope for developing new group theoretical methods for their analysis and with the possibility of their applications in a variety of isolated finite quantum many-particle systems.

Acknowledgments

Thanks are due to Prof. H.A. Weidenmüller, for many useful discussions.

References

[1] Porter C E 1965 *Statistical Theories of Spectra: Fluctuations* (New York: Academic Press)
[2] Weidenmüller H A 2014 *J. Phys. G: Nucl. Part. Phys.* **41** 094010
[3] Mehta M L 2004 *Random Matrices* (Netherlands: Elsevier B.V.)
[4] Haake F 2010 *Quantum signatures of Chaos* (Heidelberg: Springer-Verlag)
[5] Akemann G, Baik J and Di Francesco P 2011 *The Oxford Handbook of Random Matrix Theory* (Oxford: Oxford University Press)
[6] Couillet R and Debbah M 2012 *Random Matrix Methods for Wireless Communications* (New York: Cambridge University Press)
[7] Kota V K B 2014 *Embedded Random Matrix Ensembles in Quantum Physics* (Heidelberg: Springer-Verlag).
[8] Kota V K B 2005 *J. Math. Phys.* **46** 033514
[9] Benet L, Rupp T and Weidenmüller H A 2001 *Ann. Phys. (N.Y.)* **292** 67
[10] Mon K K and French J B 1975 *Ann. Phys. (N.Y.)* **95** 90
[11] French J B, Kota V K B, Pandey A and Tomsovic S 1988 *Ann. Phys. (N.Y.)* **181** 235
[12] Vyas Manan and Kota V K B 2011 *Preprint* nucl-th/1106.0395
[13] Small R A and Müller S 2014 *Phys. Rev. E* **90** 010102(R)
[14] Zelevinsky V 2009 *Phys. Atomic Nuclei* **72** 1107
[15] Furnstahl R J and Hebeler K 2013 *Rep. Prog. Phys.* **76** 126301
[16] Horoi H, Brown B A and Zelevinsky V 2001 *Phys. Rev. Lett.* **87** 062501
[17] Hecht K T 1975 *Comm. Math. Phys.* **41** 135
[18] Vyas Manan and Kota V K B 2012 *J. Math. Phys.* **53** 123303
[19] French J B and Draayer J P 1979 *Symmetries and statistical behavior in Fermion systems*, in the proceedings of the International Colloquium on Group Theoretical Methods in Physics, eds. W. Beiglbock, A. Bohm and E. Takasugi (Berlin: Springer) 394

Journal of Physics: Conference Series **597** (2015) 012051 doi:10.1088/1742-6596/597/1/012051

Tetrahedron equation and quantum R matrices for q-oscillator representations

A Kuniba[1] **and M Okado**[2]

[1] Graduate School of Arts and Sciences, University of Tokyo, Komaba, 153-8902, Japan
[2] Department of Mathematics, Osaka City University, Osaka, 558-8585, Japan

E-mail: atsuo@gokutan.c.u-tokyo.ac.jp, okado@sci.osaka-cu.ac.jp

Abstract. We review and supplement the recent result by the authors on the reduction of the three dimensional R (3d R) satisfying the tetrahedron equation to the quantum R matrices for the q-oscillator representations of $U_q(D^{(2)}_{n+1})$, $U_q(A^{(2)}_{2n})$ and $U_q(C^{(1)}_n)$. A new formula for the 3d R and a quantum R matrix for $n = 1$ are presented and a proof of the irreducibility of the tensor product of the q-oscillator representations is detailed.

1. Introduction

This paper is a summary and supplement of the recent result [9] by the authors, which is motivated by the earlier works [13, 2, 11]. The tetrahedron equation (1) [14] is a three dimensional generalization of the Yang-Baxter equation [1]. In [11] a new prescription was proposed to reduce it to the Yang-Baxter equation $R_{1,2}R_{1,3}R_{2,3} = R_{2,3}R_{1,3}R_{1,2}$ by using the special boundary vectors defined by (3) and (10). Applied to a particular solution of the tetrahedron equation (3d L operator [2]), the reduction was shown [11] to give the quantum R matrices for the spin representations [12].

In [9] a similar reduction was studied for the distinguished solution of the tetrahedron equation which we call 3d R. The 3d R was obtained as the intertwiner of the quantum coordinate ring $A_q(sl_3)$ [6], (The original formula on p194 therein contains a misprint.) and was found later also in a different setting [2]. They were shown to coincide and to constitute the solution of the 3d reflection equation in [7]. See [9, App. A] for more detail. The main result of [9] was the identification of the reduction of the 3d R with the quantum R matrices for the quantum affine algebras $U_q = U_q(D^{(2)}_{n+1})$, $U_q(A^{(2)}_{2n})$ and $U_q(C^{(1)}_n)$. Their relevant representations turned out to be new infinite dimensional ones which we called the *q-oscillator representations.* There are two kinds of boundary vectors, which curiously correspond to the choices of the above three algebras. See Remark 5.

This paper contains a summary of these results and a few supplements. The formula (9) for the 3d R and (19) for the quantum R matrix for $n = s = t = 1$ case are new. Section 4 recollects a proof of the irreducibility of the tensor product of the q-oscillator representations whose detail was omitted in [9]. The result for $n = 1$ was reported earlier in [8]. More recently it has been shown that the q-oscillator representations [9] quoted in Prop. 1-3 here actually factor through a homomorphism from U_q to the n fold tensor product of the *q-oscillator algebra* [10].

Throughout the paper we assume that q is generic and use the following notations:

$$(z;q)_m = \prod_{k=1}^{m}(1 - zq^{k-1}), \quad (q)_m = (q;q)_m, \quad \binom{m}{k}_q = \frac{(q)_m}{(q)_k(q)_{m-k}}, \quad [m]_q = \frac{q^m - q^{-m}}{q - q^{-1}},$$

where the q-binomial is to be understood as zero unless $0 \le k \le m$. $[m]_{q^t}$ with $t = 1$ will simply be denoted by $[m]$.

2. Reducing the tetrahedron equation to the Yang-Baxter equation

2.1. General scheme using boundary vectors

Let F be a vector space and $R \in \mathrm{End}(F^{\otimes 3})$. Consider the tetrahedron equation:

$$R_{1,2,4}R_{1,3,5}R_{2,3,6}R_{4,5,6} = R_{4,5,6}R_{2,3,6}R_{1,3,5}R_{1,2,4} \in \mathrm{End}(F^{\otimes 6}) \tag{1}$$

where $R_{i,j,k}$ acts as R on the i, j, k th components from the left in $F^{\otimes 6}$.

We recall the prescription which produces an infinite family of solutions to the Yang-Baxter equation from a solution to the tetrahedron equation based on special boundary vectors [11]. First we regard (1) as a one-site relation, and extend it to the n-site version. Let $\overset{\alpha_i}{F}, \overset{\beta_i}{F}, \overset{\gamma_i}{F}$ be the copies of F, where α_i, β_i and γ_i $(i = 1, \ldots, n)$ are just labels. Renaming the spaces 1, 2, 3 by them gives $R_{\alpha_i,\beta_i,4}R_{\alpha_i,\gamma_i,5}R_{\beta_i,\gamma_i,6}R_{4,5,6} = R_{4,5,6}R_{\beta_i,\gamma_i,6}R_{\alpha_i,\gamma_i,5}R_{\alpha_i,\beta_i,4}$ for each i. Thus for any i one can carry $R_{4,5,6}$ through $R_{\alpha_i,\beta_i,4}R_{\alpha_i,\gamma_i,5}R_{\beta_i,\gamma_i,6}$ to the left reversing it into $R_{\beta_i,\gamma_i,6}R_{\alpha_i,\gamma_i,5}R_{\alpha_i,\beta_i,4}$. Applying this n times leads to

$$\begin{aligned}&\big(R_{\alpha_1,\beta_1,4}R_{\alpha_1,\gamma_1,5}R_{\beta_1,\gamma_1,6}\big)\cdots\big(R_{\alpha_n,\beta_n,4}R_{\alpha_n,\gamma_n,5}R_{\beta_n,\gamma_n,6}\big)R_{4,5,6}\\ &= R_{4,5,6}\big(R_{\beta_1,\gamma_1,6}R_{\alpha_1,\gamma_1,5}R_{\alpha_1,\beta_1,4}\big)\cdots\big(R_{\beta_n,\gamma_n,6}R_{\alpha_n,\gamma_n,5}R_{\alpha_n,\beta_n,4}\big).\end{aligned} \tag{2}$$

This is an equality in $\mathrm{End}(\overset{\boldsymbol{\alpha}}{F} \otimes \overset{\boldsymbol{\beta}}{F} \otimes \overset{\boldsymbol{\gamma}}{F} \otimes \overset{4}{F} \otimes \overset{5}{F} \otimes \overset{6}{F})$, where $\boldsymbol{\alpha} = (\alpha_1, \ldots, \alpha_n)$ is the array of labels and $\overset{\boldsymbol{\alpha}}{F} = \overset{\alpha_1}{F} \otimes \cdots \otimes \overset{\alpha_n}{F}\ (= F^{\otimes n})$. The notations $\overset{\boldsymbol{\beta}}{F}$ and $\overset{\boldsymbol{\gamma}}{F}$ should be understood similarly.

Next we introduce special boundary vectors. Suppose one has a vector $|\chi_s(x)\rangle \in F$ depending on a variable x such that its tensor product

$$|\chi_s(x, y)\rangle = |\chi_s(x)\rangle \otimes |\chi_s(xy)\rangle \otimes |\chi_s(y)\rangle \in F \otimes F \otimes F \tag{3}$$

satisfies the relation

$$R|\chi_s(x, y)\rangle = |\chi_s(x, y)\rangle. \tag{4}$$

The index s is put to distinguish possibly more than one such vectors. Suppose there exist vectors in the dual space

$$\langle\chi_s(x, y)| = \langle\chi_s(x)| \otimes \langle\chi_s(xy)| \otimes \langle\chi_s(y)| \in F^* \otimes F^* \otimes F^*$$

having the similar property

$$\langle\chi_s(x, y)|R = \langle\chi_s(x, y)|. \tag{5}$$

Then evaluating (2) between $\langle\chi_s(x, y)|$ and $|\chi_t(1, 1)\rangle$, one obtains

$$S_{\boldsymbol{\alpha},\boldsymbol{\beta}}(z) = \varrho^{s,t}(z)\langle\chi_s(z)|R_{\alpha_1,\beta_1,3}R_{\alpha_2,\beta_2,3}\cdots R_{\alpha_n,\beta_n,3}|\chi_t(1)\rangle \in \mathrm{End}(\overset{\boldsymbol{\alpha}}{F} \otimes \overset{\boldsymbol{\beta}}{F}), \tag{6}$$

where $\varrho^{s,t}(z)$ is inserted to control the normalization. The composition of R and matrix elements are taken for the space signified by 3. One may simply write it as $S(z) \in \mathrm{End}(F^{\otimes n} \otimes F^{\otimes n})$ dropping the dummy labels. The $S(z)$ depends on s and t although they have been temporarily suppressed. It follows from (2), (4) and (5) that $S(z)$ satisfies the Yang-Baxter equation:

$$S_{\boldsymbol{\alpha},\boldsymbol{\beta}}(x)S_{\boldsymbol{\alpha},\boldsymbol{\gamma}}(xy)S_{\boldsymbol{\beta},\boldsymbol{\gamma}}(y) = S_{\boldsymbol{\beta},\boldsymbol{\gamma}}(y)S_{\boldsymbol{\alpha},\boldsymbol{\gamma}}(xy)S_{\boldsymbol{\alpha},\boldsymbol{\beta}}(x) \in \mathrm{End}(\overset{\boldsymbol{\alpha}}{F} \otimes \overset{\boldsymbol{\beta}}{F} \otimes \overset{\boldsymbol{\gamma}}{F}). \tag{7}$$

2.2. A realization of the scheme

We focus on the solution R of the tetrahedron equation mentioned in the introduction. Take F to be an infinite dimensional space $F = \bigoplus_{m\geq 0}\mathbb{Q}(q)|m\rangle$ with the dual $F^* = \bigoplus_{m\geq 0}\mathbb{Q}(q)\langle m|$ having the bilinear pairing $\langle l|m\rangle = (q^2)_m\delta_{l,m}$. Then the 3d R is given by

$$R(|i\rangle \otimes |j\rangle \otimes |k\rangle) = \sum_{a,b,c\geq 0} R^{a,b,c}_{i,j,k}|a\rangle \otimes |b\rangle \otimes |c\rangle, \tag{8}$$

$$R^{a,b,c}_{i,j,k} = \delta^{a+b}_{i+j}\delta^{b+c}_{j+k}\sum_{\lambda+\mu=b}(-1)^\lambda q^{ik+b+\lambda(c-a)+\mu(\mu-i-k-1)}\binom{\lambda+a}{a}_{q^2}\binom{i}{\mu}_{q^2}, \tag{9}$$

where $\delta^i_j = 1$ if $i=j$ and 0 otherwise. The sum is over $\lambda,\mu \in \mathbb{Z}_{\geq 0}$ such that $\lambda+\mu = b$ with the further condition $\mu \leq i$. It satisfies $(q^2)_a(q^2)_b(q^2)_c\, R^{a,b,c}_{i,j,k} = (q^2)_i(q^2)_j(q^2)_k\, R^{i,j,k}_{a,b,c}$ [9, eq.(A.1)]. The formula (9) is simpler than [9, eq.(2.10)]. Its derivation will be given elsewhere.

The two boundary vectors satisfying (4) and (5) are known [11] and given by

$$\langle\chi_s(z)| = \sum_{m\geq 0}\frac{z^m}{(q^{s^2})_m}\langle sm|, \quad |\chi_s(z)\rangle = \sum_{m\geq 0}\frac{z^m}{(q^{s^2})_m}|sm\rangle \quad (s=1,2). \tag{10}$$

Given two boundary vectors, one can construct four families of solutions to the Yang-Baxter equation $S(z) = S^{s,t}(z) = S^{s,t}(z,q)$ $(s,t=1,2)$ by (6) by substituting (9) and (10). Each family consists of the solutions labeled with $n \in \mathbb{Z}_{\geq 1}$. They are the matrices acting on $F^{\otimes n}\otimes F^{\otimes n}$ whose elements read

$$S^{s,t}(z)\big(|\mathbf{i}\rangle \otimes |\mathbf{j}\rangle\big) = \sum_{\mathbf{a},\mathbf{b}} S^{s,t}(z)^{\mathbf{a},\mathbf{b}}_{\mathbf{i},\mathbf{j}}|\mathbf{a}\rangle \otimes |\mathbf{b}\rangle, \tag{11}$$

$$S^{s,t}(z)^{\mathbf{a},\mathbf{b}}_{\mathbf{i},\mathbf{j}} = \varrho^{s,t}(z)\sum_{c_0,\ldots,c_n\geq 0}\frac{z^{c_0}(q^2)_{sc_0}}{(q^{s^2})_{c_0}(q^{t^2})_{c_n}}R^{a_1,b_1,sc_0}_{i_1,j_1,c_1}R^{a_2,b_2,c_1}_{i_2,j_2,c_2}\cdots R^{a_{n-1},b_{n-1},c_{n-2}}_{i_{n-1},j_{n-1},c_{n-1}}R^{a_n,b_n,c_{n-1}}_{i_n,j_n,tc_n}, \tag{12}$$

where $|\mathbf{a}\rangle = |a_1\rangle\otimes\cdots\otimes|a_n\rangle \in F^{\otimes n}$ for $\mathbf{a} = (a_1,\ldots,a_n) \in (\mathbb{Z}_{\geq 0})^n$, etc. By Applying [9, eq.(A.1)] to (12) it is straightforward to show

$$S^{t,s}(z)^{\mathbf{a},\mathbf{b}}_{\mathbf{i},\mathbf{j}}/\varrho^{t,s}(z) = \left(\prod_{r=1}^{n}\frac{z^{\frac{1}{t}j_r}(q^2)_{i_r}(q^2)_{j_r}}{z^{\frac{1}{t}b_r}(q^2)_{a_r}(q^2)_{b_r}}\right)S^{s,t}(z^{\frac{s}{t}})^{\bar{\mathbf{i}},\bar{\mathbf{j}}}_{\bar{\mathbf{a}},\bar{\mathbf{b}}}/\varrho^{s,t}(z^{\frac{s}{t}}), \tag{13}$$

where $\bar{\mathbf{a}} = (a_n,\ldots,a_1)$ is the reverse array of $\mathbf{a} = (a_1,\ldots,a_n)$ and similarly for $\bar{\mathbf{b}},\bar{\mathbf{i}}$ and $\bar{\mathbf{j}}$. Henceforth we shall only consider $S^{1,1}(z), S^{1,2}(z)$ and $S^{2,2}(z)$ in the rest of the paper. The matrix elements $R^{a,b,c}_{i,j,k}$ (9) and $S^{s,t}(z)^{\mathbf{a},\mathbf{b}}_{\mathbf{i},\mathbf{j}}$ (12) are depicted as follows:

$R^{a,b,c}_{i,j,k} =$ 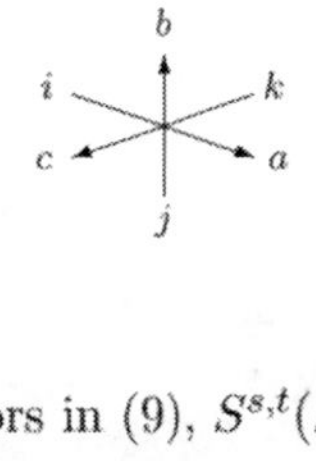

$S^{s,t}(z)^{\mathbf{a},\mathbf{b}}_{\mathbf{i},\mathbf{j}} =$ 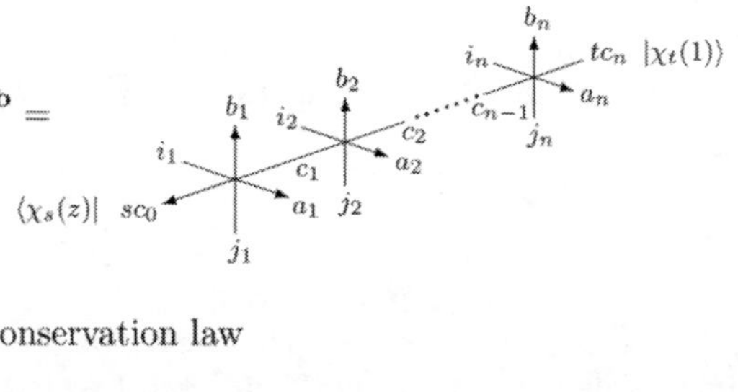

Due to δ factors in (9), $S^{s,t}(z)$ obeys the conservation law

$$S^{s,t}(z)^{\mathbf{a},\mathbf{b}}_{\mathbf{i},\mathbf{j}} = 0 \text{ unless } \mathbf{a}+\mathbf{b} = \mathbf{i}+\mathbf{j} \tag{14}$$

and the sum (12) is constrained by the n conditions $b_1 + sc_0 = j_1 + c_1, \ldots, b_n + c_{n-1} = j_n + tc_n$ leaving effectively a *single* sum. For $(s,t) = (2,2)$, they further enforce a parity constraint

$$S^{2,2}(z)^{\mathbf{a},\mathbf{b}}_{\mathbf{i},\mathbf{j}} = 0 \ \text{unless} \ |\mathbf{a}| \equiv |\mathbf{i}|, \ |\mathbf{b}| \equiv |\mathbf{j}| \mod 2, \tag{15}$$

where $|\mathbf{a}| = a_1 + \cdots + a_n$, etc. Thus we have a direct sum decomposition

$$S^{2,2}(z) = S^{+,+}(z) \oplus S^{+,-}(z) \oplus S^{-,+}(z) \oplus S^{-,-}(z), \tag{16}$$

$$S^{\epsilon_1,\epsilon_2}(z) \in \mathrm{End}\big((F^{\otimes n})_{\epsilon_1} \otimes (F^{\otimes n})_{\epsilon_2}\big), \qquad (F^{\otimes n})_{\pm} = \bigoplus_{\mathbf{a} \in (\mathbb{Z}_{\geq 0})^n, (-1)^{|\mathbf{a}|} = \pm 1} \mathbb{Q}(q)|\mathbf{a}\rangle. \tag{17}$$

We dare allow the coexistence of somewhat confusing notations $S^{s,t}(z)$ and $S^{\epsilon_1,\epsilon_2}(z)$ expecting that they can be properly distinguished from the context. (A similar warning applies to $\varrho^{s,t}(z)$ in the sequel.) We choose the normalization factors as

$$\varrho^{1,1}(z) = \frac{(z;q)_\infty}{(-zq;q)_\infty}, \quad \varrho^{1,2}(z) = \frac{(z^2;q^2)_\infty}{(-z^2q;q^2)_\infty}, \quad \varrho^{\epsilon_1,\epsilon_2}(z) = \left(\frac{(z;q^4)_\infty}{(zq^2;q^4)_\infty}\right)^{\epsilon_1\epsilon_2}. \tag{18}$$

Then the matrix elements of $S^{1,1}(z), S^{1,2}(z)$ and $S^{\epsilon_1,\epsilon_2}(z)$ are rational functions of q and z.

2.3. Example

Let us present an explicit form of the matrix element (12) for $n = 1$. It was worked out earlier in [8, Prop.2] by using a formula for $R^{a,b,c}_{i,j,k}$ different from (9). For simplicity we concentrate on the case $s = t = 1$ and write $S^{s,t}(z)^{\mathbf{a},\mathbf{b}}_{\mathbf{i},\mathbf{j}}$ as $S(z)^{a,b}_{i,j}$ with $a,b,i,j \in \mathbb{Z}_{\geq 0}$. A direct calculation using (9) and (18) leads to

$$\begin{aligned} &S(z)(|i\rangle \otimes |j\rangle) = \sum_{a,b \geq 0} S(z)^{a,b}_{i,j} |a\rangle \otimes |b\rangle, \qquad S(z)^{a,b}_{i,j} = z^{a-i} \frac{(q^2)_i (q^2)_j}{(q^2)_a (q^2)_b} S(z)^{i,j}_{a,b}, \\ &S(z)^{a,b}_{i,j} = \delta^{a+b}_{i+j} \sum_{\lambda,\mu} (-1)^\lambda q^{j(1-a)+\mu(\mu-1)} \binom{j}{\lambda}_{q^2} \binom{\lambda+i}{b}_{q^2} \frac{(-q;q)_{i-a}(z;q)_{a+\lambda-\mu}}{(-zq,q)_{i+\lambda-\mu}} \quad (0 \leq a \leq i). \end{aligned} \tag{19}$$

The last sum is over $\lambda, \mu \in \mathbb{Z}_{\geq 0}$ such that $\lambda + \mu = j$ and $\lambda + i \geq b$. Thus it is actually a single sum over $\max(0, b-i) \leq \lambda \leq j$. The formula (19) is simpler than [8, eq.(2.19)]. From our main Theorem 4 it follows that $S^{a,b}_{i,j}(z=1) = \delta^a_j \delta^b_i$, which is consistent with the above result.

3. Quantum R matrices for q-oscillator representations

The Drinfeld-Jimbo quantum affine algebras without derivation $U_q = U_q(D^{(2)}_{n+1})$, $U_q(A^{(2)}_{2n})$ and $U_q(C^{(1)}_n)$ are the Hopf algebras generated by $e_i, f_i, k_i^{\pm 1}$ $(0 \leq i \leq n)$ satisfying the relations [3, 4]:

$$k_i k_i^{-1} = k_i^{-1} k_i = 1, \ [k_i, k_j] = 0, \ k_i e_j k_i^{-1} = q_i^{a_{ij}} e_j, \ k_i f_j k_i^{-1} = q_i^{-a_{ij}} f_j, \ [e_i, f_j] = \delta_{ij} \frac{k_i - k_i^{-1}}{q_i - q_i^{-1}},$$

$$\sum_{\nu=0}^{1-a_{ij}} (-1)^\nu e_i^{(1-a_{ij}-\nu)} e_j e_i^{(\nu)} = 0, \qquad \sum_{\nu=0}^{1-a_{ij}} (-1)^\nu f_i^{(1-a_{ij}-\nu)} f_j f_i^{(\nu)} = 0 \ \ (i \neq j),$$

where $e_i^{(\nu)} = e_i^\nu / [\nu]_{q_i}!$, $f_i^{(\nu)} = f_i^\nu / [\nu]_{q_i}!$ with $[\nu]_q = [\nu]_q [\nu-1]_q \cdots [1]_q$. The Cartan matrix $(a_{ij})_{0 \leq i,j \leq n}$ [5] is given by $a_{i,j} = 2\delta_{i,j} - \max((\log q_j)/(\log q_i), 1)\delta_{|i-j|,1}$. The data q_i is specified above the corresponding vertex $i\,(0 \le i \le n)$ in the Dynkin diagrams:

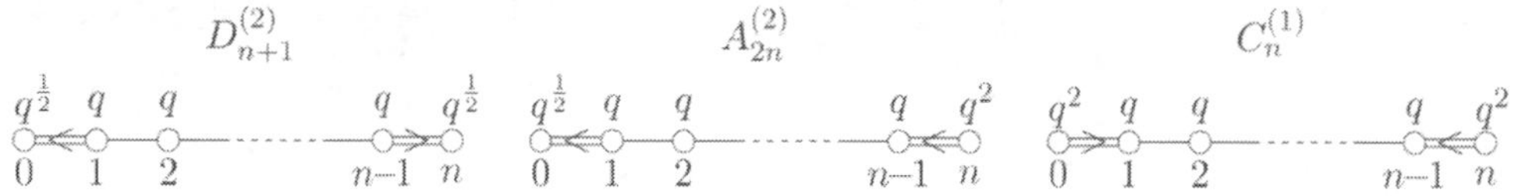

We employ the coproduct Δ of the form $\Delta(k_i^{\pm 1}) = k_i^{\pm 1} \otimes k_i^{\pm 1}$, $\Delta(e_i) = 1 \otimes e_i + e_i \otimes k_i$ and $\Delta(f_i) = f_i \otimes 1 + k_i^{-1} \otimes f_i$.

3.1. q-oscillator representations

We introduce representations of U_q on the tensor product of the Fock space $\hat{F}^{\otimes n}$ or $F^{\otimes n}$, where $\hat{F} = \bigoplus_{m \ge 0} \mathbb{C}(q^{\frac{1}{2}})|m\rangle$ is a slight extension of the coefficient field of F. They all factor through an algebra homomorphism from U_q to the q-oscillator algebra as shown in [10, Prop. 2.1]. As in the previous section we write the elements of $\hat{F}^{\otimes n}$ as $|\mathbf{m}\rangle = |m_1\rangle \otimes \cdots \otimes |m_n\rangle \in \hat{F}^{\otimes n}$ for $\mathbf{m} = (m_1, \ldots, m_n) \in (\mathbb{Z}_{\ge 0})^n$ and describe the changes in $\mathbf{m}$ by the vectors $\mathbf{e}_i = (0, \ldots, 0, \overset{i}{1}, 0, \ldots, 0) \in \mathbb{Z}^n$. In the following propositions $\kappa = \frac{q+1}{q-1}$ and x is a nonzero parameter.

Proposition 1. *The following defines an irreducible $U_q(D^{(2)}_{n+1})$ module structure on $\hat{F}^{\otimes n}$.*

$$
\begin{aligned}
e_0|\mathbf{m}\rangle &= x|\mathbf{m} + \mathbf{e}_1\rangle, \\
f_0|\mathbf{m}\rangle &= i\kappa[m_1]x^{-1}|\mathbf{m} - \mathbf{e}_1\rangle, \\
k_0|\mathbf{m}\rangle &= -iq^{m_1 + \frac{1}{2}}|\mathbf{m}\rangle, \\
e_j|\mathbf{m}\rangle &= [m_j]|\mathbf{m} - \mathbf{e}_j + \mathbf{e}_{j+1}\rangle \quad (1 \le j \le n-1), \\
f_j|\mathbf{m}\rangle &= [m_{j+1}]|\mathbf{m} + \mathbf{e}_j - \mathbf{e}_{j+1}\rangle \quad (1 \le j \le n-1), \\
k_j|\mathbf{m}\rangle &= q^{-m_j + m_{j+1}}|\mathbf{m}\rangle \quad (1 \le j \le n-1), \\
e_n|\mathbf{m}\rangle &= i\kappa[m_n]|\mathbf{m} - \mathbf{e}_n\rangle, \\
f_n|\mathbf{m}\rangle &= |\mathbf{m} + \mathbf{e}_n\rangle, \\
k_n|\mathbf{m}\rangle &= iq^{-m_n - \frac{1}{2}}|\mathbf{m}\rangle.
\end{aligned}
$$

Proposition 2. *The following defines an irreducible $U_q(A^{(2)}_{2n})$ module structure on $\hat{F}^{\otimes n}$.*

$$
\begin{aligned}
e_0|\mathbf{m}\rangle &= x|\mathbf{m} + \mathbf{e}_1\rangle, \\
f_0|\mathbf{m}\rangle &= i\kappa[m_1]x^{-1}|\mathbf{m} - \mathbf{e}_1\rangle, \\
k_0|\mathbf{m}\rangle &= -iq^{m_1 + \frac{1}{2}}|\mathbf{m}\rangle, \\
e_j|\mathbf{m}\rangle &= [m_j]|\mathbf{m} - \mathbf{e}_j + \mathbf{e}_{j+1}\rangle \quad (1 \le j \le n-1), \\
f_j|\mathbf{m}\rangle &= [m_{j+1}]|\mathbf{m} + \mathbf{e}_j - \mathbf{e}_{j+1}\rangle \quad (1 \le j \le n-1), \\
k_j|\mathbf{m}\rangle &= q^{-m_j + m_{j+1}}|\mathbf{m}\rangle \quad (1 \le j \le n-1), \\
e_n|\mathbf{m}\rangle &= \frac{[m_n][m_n - 1]}{[2]^2}|\mathbf{m} - 2\mathbf{e}_n\rangle, \\
f_n|\mathbf{m}\rangle &= |\mathbf{m} + 2\mathbf{e}_n\rangle, \\
k_n|\mathbf{m}\rangle &= -q^{-2m_n - 1}|\mathbf{m}\rangle.
\end{aligned}
$$

Proposition 3. *The following defines an irreducible $U_q(C_n^{(1)})$ module structure on $(F^{\otimes n})_+$ and $(F^{\otimes n})_-$ defined in (17).*

$$
\begin{aligned}
e_0|\mathbf{m}\rangle &= x|\mathbf{m}+2\mathbf{e}_1\rangle,\\
f_0|\mathbf{m}\rangle &= \frac{[m_1][m_1-1]}{[2]^2}x^{-1}|\mathbf{m}-2\mathbf{e}_1\rangle,\\
k_0|\mathbf{m}\rangle &= -q^{2m_1+1}|\mathbf{m}\rangle,\\
e_j|\mathbf{m}\rangle &= [m_j]|\mathbf{m}-\mathbf{e}_j+\mathbf{e}_{j+1}\rangle \quad (1\le j\le n-1),\\
f_j|\mathbf{m}\rangle &= [m_{j+1}]|\mathbf{m}+\mathbf{e}_j-\mathbf{e}_{j+1}\rangle \quad (1\le j\le n-1),\\
k_j|\mathbf{m}\rangle &= q^{-m_j+m_{j+1}}|\mathbf{m}\rangle \quad (1\le j\le n-1),\\
e_n|\mathbf{m}\rangle &= \frac{[m_n][m_n-1]}{[2]^2}|\mathbf{m}-2\mathbf{e}_n\rangle,\\
f_n|\mathbf{m}\rangle &= |\mathbf{m}+2\mathbf{e}_n\rangle,\\
k_n|\mathbf{m}\rangle &= -q^{-2m_n-1}|\mathbf{m}\rangle.
\end{aligned}
$$

We call these irreducible representations the *q-oscillator representations* of U_q. For the twisted case $U_q(D_{n+1}^{(2)})$ and $U_q(A_{2n}^{(2)})$, they are *singular* at $q=1$ because of the factor κ.

3.2. Quantum R matrices

Let $V=\hat{F}^{\otimes n}$ for $U_q(D_{n+1}^{(2)}), U_q(A_{2n}^{(2)})$ and $V=F^{\otimes n}$ for $U_q(C_n^{(1)})$. First we consider $U_q(D_{n+1}^{(2)})$ and $U_q(A_{2n}^{(2)})$. Let $V_x=\hat{F}^{\otimes n}[x,x^{-1}]$ be the representation space of U_q in Propositions 1 and 2. By the existence of the universal R matrix [3] there exists an element $R\in\mathrm{End}(V_x\otimes V_y)$ such that

$$\Delta'(g)R = R\Delta(g) \quad \forall g\in U_q \tag{20}$$

up to an overall scalar. Here Δ' is the opposite coproduct defined by $\Delta' = P\circ\Delta$, where $P(u\otimes v)=v\otimes u$ is the exchange of the components. A little inspection of our representations shows that R depends on x and y only through the ratio $z=x/y$. Moreover $V_x\otimes V_y$ is irreducible ([9, Prop. 12] and Sec. 4 of this paper) hence R is determined only by postulating (20) for $g=k_r, e_r$ and f_r with $0\le r\le n$. Thus denoting the R by $R(z)$, we may claim [4] that it is determined by the conditions

$$
\begin{aligned}
(k_r\otimes k_r)R(z) &= R(z)(k_r\otimes k_r), &(21)\\
(e_r\otimes 1 + k_r\otimes e_r)R(z) &= R(z)(1\otimes e_r + e_r\otimes k_r), &(22)\\
(1\otimes f_r + f_r\otimes k_r^{-1})R(z) &= R(z)(f_r\otimes 1 + k_r^{-1}\otimes f_r) &(23)
\end{aligned}
$$

for $0\le r\le n$ up to an overall scalar. We fix the normalization of $R(z)$ by

$$R(z)(|\mathbf{0}\rangle\otimes|\mathbf{0}\rangle) = |\mathbf{0}\rangle\otimes|\mathbf{0}\rangle, \tag{24}$$

where $|\mathbf{0}\rangle\in\hat{F}^{\otimes n}$ is defined in the beginning of Section 3.1 with $\mathbf{0}=(0,\dots,0)$. We call the intertwiner $R(z)$ the *quantum R matrix* for q-oscillator representation. It satisfies the Yang-Baxter equation

$$R_{12}(x)R_{13}(xy)R_{23}(y) = R_{23}(y)R_{13}(xy)R_{12}(x). \tag{25}$$

Next we consider $U_q(C_n^{(1)})$. Denote by $V_x^{\pm} = (F^{\otimes n})_{\pm}[x, x^{-1}]$ the representation spaces in Proposition 3 and set $V_x = V_x^+ \oplus V_x^- = F^{\otimes n}[x, x^{-1}]$. See (17) for the definition of $(F^{\otimes n})_{\pm}$. We define the quantum R matrix $R(z)$ to be the direct sum

$$R(z) = R^{+,+}(z) \oplus R^{+,-}(z) \oplus R^{-,+}(z) \oplus R^{-,-}(z), \tag{26}$$

where each $R^{\epsilon_1,\epsilon_2}(z) \in \mathrm{End}(V_x^{\epsilon_1} \otimes V_y^{\epsilon_2})$ is the quantum R matrix with the normalization condition

$$\begin{aligned} &R^{+,+}(z)(|\mathbf{0}\rangle \otimes |\mathbf{0}\rangle) = |\mathbf{0}\rangle \otimes |\mathbf{0}\rangle, && R^{+,-}(z)(|\mathbf{0}\rangle \otimes |\mathbf{e}_1\rangle) = \frac{-iq^{1/2}}{1-z}|\mathbf{0}\rangle \otimes |\mathbf{e}_1\rangle, \\ &R^{-,+}(z)(|\mathbf{e}_1\rangle \otimes |\mathbf{0}\rangle) = \frac{-iq^{1/2}}{1-z}|\mathbf{e}_1\rangle \otimes |\mathbf{0}\rangle, && R^{-,-}(z)(|\mathbf{e}_1\rangle \otimes |\mathbf{e}_1\rangle) = \frac{z-q^2}{1-zq^2}|\mathbf{e}_1\rangle \otimes |\mathbf{e}_1\rangle. \end{aligned} \tag{27}$$

The R matrix $R(z)$ satisfies the Yang-Baxter equation (25). In fact it is decomposed into the finer equalities ($\epsilon_1, \epsilon_2, \epsilon_3 = \pm$)

$$R_{12}^{\epsilon_1,\epsilon_2}(x) R_{13}^{\epsilon_1,\epsilon_3}(xy) R_{23}^{\epsilon_2,\epsilon_3}(y) = R_{23}^{\epsilon_2,\epsilon_3}(y) R_{13}^{\epsilon_1,\epsilon_3}(xy) R_{12}^{\epsilon_1,\epsilon_2}(x).$$

3.3. Main theorem

Define the operator K acting on $\hat{F}^{\otimes n}$ by $K|\mathbf{m}\rangle = (-iq^{\frac{1}{2}})^{m_1+\cdots+m_n}|\mathbf{m}\rangle$. Introduce the gauge transformed quantum R matrix by

$$\tilde{R}(z) = (K^{-1} \otimes 1) R(z) (1 \otimes K). \tag{28}$$

It is easy to see that $\tilde{R}(z)$ also satisfies the Yang-Baxter equation (25).

In Section 2.2 the solutions $S^{s,t}(z)$ of the Yang-Baxter equation have been constructed from the 3d R in (11), (12) and (18). In Section 3.2 the quantum R matrices for q-oscillator representations of $U_q(D_{n+1}^{(2)})$, $U_q(A_{2n}^{(2)})$ and $U_q(C_n^{(1)})$ have been defined. The next theorem, which is the main result of [9], states the precise relation between them. (See (13) for $S^{2,1}(z)$.)

Theorem 4. *Denote by $\tilde{R}_{\mathfrak{g}}(z)$ the gauge transformed quantum R matrix (28) for $U_q(\mathfrak{g})$. Then the following equalities hold:*

$$S^{1,1}(z) = \tilde{R}_{D_{n+1}^{(2)}}(z), \quad S^{1,2}(z) = \tilde{R}_{A_{2n}^{(2)}}(z), \quad S^{2,2}(z) = \tilde{R}_{C_n^{(1)}}(z),$$

where the last one means $S^{\epsilon_1,\epsilon_2}(z) = \tilde{R}^{\epsilon_1,\epsilon_2}(z)$ between (16) and (26) with the gauge transformation (28).

Remark 5. Theorem 4 suggests the following correspondence between the boundary vectors (10) with the end shape of the Dynkin diagrams:

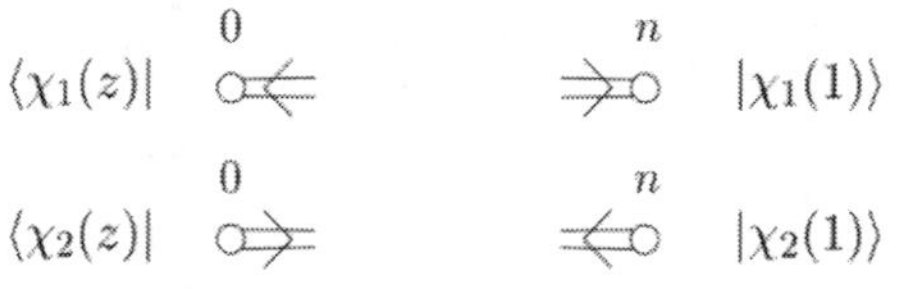

Consistently with Remark 5, $S^{2,1}(z)$, which is reducible to $S^{1,2}(z^{1/2})$ by (13), is identified [10] with the quantum R matrix for q-oscillator representation of another $U_q(A_{2n}^{(2)})$ realized as the affinization of the classical part $U_q(B_n)$. (Proposition 2 corresponds to taking the classical part to be $U_q(C_n)$.) As far as $\langle\chi_1(z)|$ and $|\chi_1(1)\rangle$ are concerned, the above correspondence agrees with the observation made in [11, Remark 7.2] on the similar result concerning a 3d L operator. With regard to $\langle\chi_2(z)|$ and $|\chi_2(1)\rangle$, the relevant affine Lie algebras $A_{2n}^{(2)}$ and $C_n^{(1)}$ in this paper are the subalgebras of $B_{n+1}^{(1)}$ and $D_{n+2}^{(1)}$ in [11, Theorem 7.1] obtained by folding their Dynkin diagrams.

4. Proof of the irreducibility of the tensor product

In [9] we gave a proof of the following proposition.

Proposition 6 (Prop. 12 of [9]). *As a $U_q(D_{n+1}^{(2)})$ or $U_q(A_{2n}^{(2)})$ module $V_x \otimes V_y$ is irreducible. As a $U_q(C_n^{(1)})$ module each $V_x^{\epsilon_1} \otimes V_y^{\epsilon_2}$ ($\epsilon_1, \epsilon_2 = \pm$) is irreducible.*

Since the explanation there was not sufficient, we give the detailed proof here. Let $\mathfrak{g} = D_{n+1}^{(2)}, A_{2n}^{(2)}$ or $C_n^{(1)}$, $I = \{0, 1, \ldots, n\}$, and for a subset J of I let $U_q(\mathfrak{g}_J)$ be the subalgebra of $U_q(\mathfrak{g})$ generated by $\{e_j, f_j, k_j^{\pm 1} \mid j \in J\}$. Recall the vector v_l [9, Prop. 4] for $\mathfrak{g} = D_{n+1}^{(2)}$ and v_l^ϵ [9, Prop. 5] for $A_{2n}^{(2)}, C_n^{(1)}$.

Proposition 7. *For $\mathfrak{g} = D_{n+1}^{(2)}$*

$$V^{\otimes 2} = \sum_{l=0}^{\infty} U_q(\mathfrak{g}_{I \setminus \{0\}}) v_l,$$

and for $\mathfrak{g} = A_{2n}^{(2)}, C_n^{(1)}$

$$V^{\epsilon_1} \otimes V^{\epsilon_2} = \sum_{\substack{l=0 \\ (-1)^l = \epsilon_1 \epsilon_2}}^{\infty} U_q(\mathfrak{g}_{I \setminus \{0\}}) v_l^{\epsilon_1}.$$

This is an immediate consequence of the following two lemmas. Set $w_{l,k} = |k\mathbf{e}_{n-1}\rangle \otimes |(l-k)\mathbf{e}_n\rangle \in V^{\otimes 2}$ ($l \geq 0, 0 \leq k \leq l$).

Lemma 8. *For $\mathfrak{g} = D_{n+1}^{(2)}$*

$$w_{l,k} \in \sum_{j=0}^{l} U_q(\mathfrak{g}_{\{n-1,n\}}) v_j,$$

and for $\mathfrak{g} = A_{2n}^{(2)}, C_n^{(1)}$

$$w_{l,k} \in \sum_{\substack{j=0 \\ j \equiv l \pmod 2}}^{l} U_q(\mathfrak{g}_{\{n-1,n\}}) v_j^\epsilon \quad \textit{where } \epsilon = (-1)^k.$$

Proof. We treat the $\mathfrak{g} = D_{n+1}^{(2)}$ case first. Note that the set of vectors $B = \{v_l, f_n v_{l-1}, \ldots, f_n^l v_0\}$ is linearly independent in the vector subspace spanned by $\{|k\mathbf{e}_n\rangle \otimes |(l-k)\mathbf{e}_n\rangle \mid 0 \leq k \leq l\}$. Hence, B is also a basis and

$$w_{l,0} = |\mathbf{0}\rangle \otimes |l\mathbf{e}_n\rangle \in \sum_{j=0}^{l} U_q(\mathfrak{g}_{\{n\}}) v_j. \tag{29}$$

Next note that

$$w_{l,k} = (f_{n-1} f_n - q^{-1} f_n f_{n-1}) w_{l-1,k-1} + i \frac{q^{l-k+1/2}}{[l-k+1]} f_{n-1} w_{l,k-1} \quad (1 \leq k \leq l).$$

This relation together with (29) shows the result.

Suppose now $\mathfrak{g} = A_{2n}^{(2)}, C_n^{(1)}$. We compare $B' = \{v_l^\epsilon, f_n v_{l-1}^\epsilon, \ldots, f_n^{l'} v_{l-2l'}^\epsilon\}$ ($l' = \lfloor l/2 \rfloor$ ($\epsilon = +$), $= \lfloor (l-1)2 \rfloor$ ($\epsilon = -$)) and the subspace spanned by $\{|k\mathbf{e}_n\rangle \otimes |(l-k)\mathbf{e}_n\rangle \mid 0 \leq k \leq l, (-1)^k = \epsilon\}$.

We have

$$w_{l,0} = |\mathbf{0}\rangle \otimes |l\mathbf{e}_n\rangle \in \sum_{\substack{j=0 \\ j\equiv l \pmod 2}}^{l} U_q(\mathfrak{g}_{\{n\}})v_j^+,$$

$$|\mathbf{e}_n\rangle \otimes |(l-1)\mathbf{e}_n\rangle \in \sum_{\substack{j=0 \\ j\equiv l \pmod 2}}^{l} U_q(\mathfrak{g}_{\{n\}})v_j^- \quad (l \geq 1).$$

From

$$w_{1,1} = f_{n-1}v_1^-, \quad w_{2,1} = \frac{q^{-1}}{[2]}v_0^- + \frac{1}{[2]^2}f_{n-1}v_2^-,$$

$$w_{l,1} = -\frac{[l-1]}{[l]}f_n w_{l-2,1} + \frac{q^{-1}[l-1]}{[2][l]}(f_{n-1}f_n - q^{-2}f_n f_{n-1})|\mathbf{e}_n\rangle \otimes |(l-3)\mathbf{e}_n\rangle$$
$$+ \frac{q^{l-1}}{[l]}f_{n-1}|\mathbf{e_n}\rangle \otimes |(l-1)\mathbf{e}_n\rangle \quad (l \geq 3),$$

$$w_{l,k} = (\frac{1}{[2]}f_{n-1}^2 f_n - q^{-1}f_{n-1}f_n f_{n-1} + q^{-2}f_n f_{n-1}^2)w_{l-2,k-2} + \frac{q^{2l-2k+1}}{[l-k+1][l-k+2]}f_{n-1}^2 w_{l,k-2}$$
$$(2 \leq k \leq l),$$

we obtain the result. □

Lemma 9. *Let W_l be the vector subspace of $V^{\otimes 2}$ spanned by $|\sum_{j=1}^n k_j\mathbf{e}_j\rangle \otimes |\sum_{j=1}^n k'_j\mathbf{e}_j\rangle$ such that $\sum_{j=1}^n (k_j + k'_j) = l$. Then we have*

$$W_l = \sum_{0\leq k\leq l} U_q(\mathfrak{g}_{I\setminus\{0,n\}})w_{l,k}.$$

Proof. As a $U_q(\mathfrak{g}_{I\setminus\{0,n\}})(= U_q(A_{n-1}))$-module W_l is isomorphic to $\bigoplus_{k=0}^l L(k\mathbf{e}_n) \otimes L((l-k)\mathbf{e}_n)$, where $L(\lambda)$ stands for the irreducible highest weight $U_q(A_{n-1})$-module with highest weight λ. By representation theory of $U_q(A_{n-1})$, $L(k\mathbf{e}_n) \otimes L((l-k)\mathbf{e}_n)$ is generated by the highest weight vectors of weight of the form $j\mathbf{e}_{n-1} + (l-j)\mathbf{e}_n$ for some $0 \leq j \leq \min(k, l-k)$. Hence, it is enough to show that any vector in W_l of weight of the form $(l-j)\mathbf{e}_n + j\mathbf{e}_{n-1}$ is generated by $w_{l,j}$ over $U_q(\mathfrak{g}_{\{n-1\}})(= U_q(sl_2))$. But it is a well-known fact from representation theory of $U_q(sl_2)$, namely, $L(a\mathbf{e}_n) \otimes L(b\mathbf{e}_n)$ is generated by $|a\mathbf{e}_{n-1}\rangle \otimes |b\mathbf{e}_n\rangle$, where $|a\mathbf{e}_{n-1}\rangle$ (resp. $|b\mathbf{e}_n\rangle$) is the lowest (resp. highest) weight vector. □

Proof of Prop. 6. Suppose $\mathfrak{g} = D_{n+1}^{(2)}$ and let W be a nonzero submodule of $V^{\otimes 2}$. In the proof of Prop. 12 of [9], we have shown that W contains v_l for any $l \geq 0$. Similarly, for $\mathfrak{g} = A_{2n}^{(2)}$ (resp. $C_n^{(1)}$), using Lemma 8 (resp. 10) of [9], we can show a nonzero submodule W of $V^{\otimes 2}$ (resp. $V^{\epsilon_1} \otimes V^{\epsilon_2}$) contains v_l^ϵ for any $l \geq 0, \epsilon = \pm$ (resp. $v_l^{\epsilon_1}$ for any l such that $(-1)^l = \epsilon_1\epsilon_2$). The claim now follows from Prop. 7. □

Acknowledgments
The authors thank S. Sergeev for collaboration in their previous work. This work is supported by Grants-in-Aid for Scientific Research No. 23340007 and No. 24540203 from JSPS. A.K. thanks the organizers of the 30th International Colloquium on Group Theoretical Methods in Physics at Ghent University during 14-18 July 2014 for hospitality.

References

[1] Baxter R J 2007 *Exactly solved models in statistical mechanics* (New York: Dover Publications)
[2] Bazhanov V V and Sergeev S M 2006 *J. Phys. A: Math. Theor.* **39** 3295
[3] Drinfeld V G 1987 *Proc. International Congress of Mathematicians* (American Mathematical Society) p798
[4] Jimbo M 1985 *Lett. Math. Phys.* **10** 63
[5] Kac V G 1990 *Infinite dimensional Lie algebras* (Cambridge: Cambridge University Press)
[6] Kapranov M M and Voevodsky V A 1994 *Proc. Symposia in Pure Math.* **56** 177
[7] Kuniba A and Okado M 2012 *J. Phys. A: Math. Theor.* **45** 465206
[8] Kuniba A and Okado M 2013 *J. Phys. A: Math. Theor.* **46** 485203
[9] Kuniba A and Okado M 2013 Tetrahedron equation and quantum R matrices for q-oscillator representations of $U_q(A^{(2)}_{2n}), U_q(C^{(1)}_n)$ and $U_q(D^{(2)}_{n+1})$ (*Preprint* arXiv:1311.4258v3)
[10] Kuniba A, Okado M and Sergeev S 2014 Tetrahedron equation and quantum R matrices for modular double of $U_q(D^{(2)}_{n+1}), U_q(A^{(2)}_{2n})$ and $U_q(C^{(1)}_n)$ (*Preprint* arXiv:1409.1986)
[11] Kuniba A and Sergeev S 2013 *Commun. Math. Phys.* **324** 695
[12] Okado M 1990 *Commun. Math. Phys.* **134** 467
[13] Sergeev S M 1997 *Modern Phys. Lett. A* **12** 1393
[14] Zamolodchikov A B 1980 *Soviet Phys. JETP* **79** 641

Phase amplitude conformal symmetry in Fourier transforms

S Kuwata

Department of Frontier Sciences, Graduate School of Information Sciences, Hiroshima City University, Asaminami-ku, Hiroshima 731-3194, Japan

E-mail: kuwata@hiroshima-cu.ac.jp

Abstract. For the Fourier transform $\mathfrak{F} : L^2(\mathbb{R}) \to L^2(\mathbb{R})$ of a complex-valued even or odd function φ, it is found that the amplitude invariance $|\mathfrak{F}\varphi| = |\varphi|$ leads to a phase invariance or inversion as $\arg(\mathfrak{F}\varphi) = \pm \arg\varphi + \theta$ (θ = constant). The converse holds unless $\arg\varphi$ = constant. The condition $|\varphi| = |\mathfrak{F}\varphi|$ is required in dealing with, for example, the minimum uncertainty relation between position and momentum. Without the evenness or oddness of φ, $|\mathfrak{F}\varphi| = |\varphi|$ does not necessarily imply $\arg(\mathfrak{F}\varphi) = \pm \arg\varphi + \theta$, nor is the converse.

1. Introduction

The relation between the phase and amplitude in the Fourier transform of a function $\varphi : \mathbb{R}^n \to \mathbb{C}$ has been studied when we deal with, for example, the problem of whether the function φ can be obtained from the amplitude (or magnitude) of its Fourier transform $\mathfrak{F}\varphi$ by retrieving its phase $\arg(\mathfrak{F}\varphi)$. This kind of phase retrieval problem [1, 2] has arisen in many fields, such as optics, electron microscopy, X-ray crystallography. However, due to the lack of uniqueness of $\arg(\mathfrak{F}\varphi)$, some conditions should be imposed on $\mathfrak{F}\varphi$ so as to reconstruct $\arg(\mathfrak{F}\varphi)$ uniquely.

One assumption, which is adopted in many authors [3, 4], is that $\mathfrak{F}\varphi$ is band limited, that is, φ has a compact support $\Omega \subset \mathbb{R}^n$. In this case, $\mathfrak{F}\varphi$ turns out to be an entire function of a finite order by Paley-Wiener theorem [5]. However, this assumption on φ does not imply the uniqueness of $\arg(\mathfrak{F}\varphi)$ (except a trivial factor). So some further restriction is necessary for the phase retrieval. As such a condition, the irreducibility of $\mathfrak{F}\varphi$ has been known to date [6].

For $n = 1$, on the one hand, the irreducibility of $\mathfrak{F}\varphi$ may be too strong, because the entire function of one variable with a finite order can always be factorized using its complex zeros by Hadamard factorization theorem [7]. So the irreducibility indicates that the function $\mathfrak{F}\varphi(z)$ is restricted to such that has only one simple zero, as in the form $(z - z_0)\,\mathrm{e}^{p(z)}$, where $z_0 \in \mathbb{C}$ and $p(z)$ represents a polynomial. For $n \geq 2$, on the other hand, the irreducibility of $\mathfrak{F}\varphi$ is not so strong a condition. This is because the entire function $\mathfrak{F}\varphi(z_1, \ldots, z_n)$ for $n \geq 2$ cannot always be factorized using their zeros, so that the $\mathfrak{F}\varphi(z_1, \ldots, z_n)$ in itself is likely to be irreducible.

It is to be mentioned that the analogous characterization of the uniqueness problem of the phase retrieval holds for discrete Fourier transform for $\varphi : \mathbb{Z}^n \to \mathbb{C}$ and $\mathfrak{F}\varphi : T^n \to \mathbb{C}$ (where $\mathbb{Z}$ and T^n represent the set of integers and the n-dimensional torus, respectively); more generally, for Fourier transform on groups as $\varphi : G \to \mathbb{C}$ and $\mathfrak{F}\varphi : \Gamma \to \mathbb{C}$ [8] (where G and Γ represent a (locally) compact abelian group and its dual group, respectively). As in the case of continuous Fourier transform, $\arg(\mathfrak{F}\varphi)$ for discrete Fourier transform can be determined from the irreducibility of $\mathfrak{F}\varphi$ [9, 10] up to a trivial factor.

In this paper, however, we do not assume the irreducibility nor the band limitedness of $\mathfrak{F}\varphi$. Instead, we will deal with a symmetric relation under the exchange of $|\varphi| \leftrightarrow |\mathfrak{F}\varphi|$. To be more precise, let $\mathcal{M} : L^2(\mathbb{R}) \times L^2(\mathbb{R}) \to \mathbb{R}$ be a functional of $|\varphi|$ and $|\mathfrak{F}\varphi|$, and consider the relation of $\mathcal{M}[|\varphi|, |\mathfrak{F}\varphi|] = 0$. Such a relation is found, for example, in the position-momentum uncertainty relation (whether it is based on the variance [11, 12] or on the Shannon entropy [13]). In the case of minimum uncertainty, it is required further that the functional form of $|\mathfrak{F}\varphi|$ should be the same as $|\varphi|$ (after an appropriate dilation of argument). Although the condition of $|\mathfrak{F}\varphi| = |\varphi|$ in itself does not necessarily lead to the uniqueness of $\arg(\mathfrak{F}\varphi)$ (hence that of $\arg\varphi$), we can relate $\arg(\mathfrak{F}\varphi)$ to $\arg\varphi$ under a certain condition.

The aim of the present paper is to examine how the amplitude invariance under Fourier transform reflects on the phase invariance (or inversion), and vice versa. In Sect. 2, we give some preliminaries of the Fourier transform $\mathfrak{F} : L^2(\mathbb{R}) \to L^2(\mathbb{R})$ of a function $\varphi : \mathbb{R} \to \mathbb{C}$. From a physical point of view, we assume that φ is continuous on the real axis, so that φ can be expanded in terms of the Hermite polynomials. In Sect. 3, we show that for φ being an even or odd function, the condition of $|\mathfrak{F}\varphi| = |\varphi|$ leads to $\arg(\mathfrak{F}\varphi) = \pm\arg\varphi + \text{constant}$. In Sect. 4, we show the converse holds in an analogous way. In Sec. 5, we make a simple generalization of the phase amplitude relation obtained in Sects. 3 and 4. Conclusion is given in Sec. 6.

2. Preliminaries

Let $\mathfrak{F} : L^2(\mathbb{R}) \to L^2(\mathbb{R})$ be the Fourier transform of a function $\varphi : \mathbb{R} \to \mathbb{C}$ by

$$(\mathfrak{F}\varphi)(x) = \frac{1}{\sqrt{2\pi}} \int_{\mathbb{R}} \varphi(k)\, \mathrm{e}^{\mathrm{i}kx}\, \mathrm{d}k,$$

from which

$$\mathfrak{F}^2 = \mathfrak{R}, \quad \mathfrak{F}^3 = \mathfrak{F}^{-1} = \mathfrak{C}\mathfrak{F}\mathfrak{C}, \tag{1}$$

where $\mathfrak{R} : \varphi(x) \mapsto \varphi(-x)$ and $\mathfrak{C} : \varphi(x) \mapsto \varphi^*(x)$ represent the space inversion and complex conjugate, respectively. For later convenience, we introduce some following sets:

$$\begin{aligned} A &:= \{\varphi(x) \in Q \,|\, |\mathfrak{F}\varphi| = |\varphi|\}, \\ P_\pm &:= \{\varphi(x) \in Q \,|\, \arg(\mathfrak{F}\varphi) = \pm\arg\varphi + \theta \quad (\theta = \text{constant})\}, \\ E_\pm &:= \{\varphi(x) \in Q \,|\, \mathfrak{F}\varphi = \lambda\mathfrak{C}^{(1\pm1)/2}\varphi \quad (\lambda \in \mathbb{C})\}, \end{aligned}$$

where $Q := \{\varphi(x) \in L^2(\mathbb{R}) \,|\, \varphi(x) \neq 0 \text{ (almost everywhere)}\}$. From (1), it is found that $|\lambda| = 1$ for $\varphi \in E_+$, and $\lambda \in \{1, -1, \mathrm{i}, -\mathrm{i}\}$ for $\varphi \in E_-$, so that $A \cap P_\pm = E_\mp$. Hence, we get

$$A \cap P = E, \tag{2}$$

where $P = P_+ \cup P_-$ and $E = E_+ \cup E_-$.

We assume that $\varphi \in L^2(\mathbb{R})$ is continuous on the real axis. This implies that $\varphi(x)$ can be expanded using the Hermite polynomial $H_n(x)$ as

$$\varphi = \sum_{n=0}^{\infty} \alpha_n h_n \quad (\alpha_n \in \mathbb{C}), \qquad h_n(x) = H_n(x)\, \mathrm{e}^{-x^2/2}.$$

Noticing that h_n is an eigenfunction for $\mathfrak{F}$ (with its eigenvalue being i^n):

$$(\mathfrak{F}h_n)(x) = \mathrm{i}^n h_n(x),$$

Journal of Physics: Conference Series **597** (2015) 012052 doi:10.1088/1742-6596/597/1/012052

we find that $\varphi \in A, P$ can be rewritten as

$$\begin{cases} \varphi \in A \iff \sum_{n,m\in\mathbb{N}} \beta_{nm} H_n(x) H_m(x) \equiv 0, \\ \varphi \in P \iff \sum_{n,m\in\mathbb{N}} \gamma_{nm} H_n(x) H_m(x) \equiv 0, \end{cases} \tag{3}$$

where $\beta_{nm} = [\alpha_n^* \alpha_m - (\mathrm{i}^n \alpha_n)^* \mathrm{i}^m \alpha_m] + (m \leftrightarrow n)$ and

$$\gamma_{nm} = \begin{cases} \left[\mathrm{e}^{\mathrm{i}\theta} (\mathrm{i}^n \alpha_n)^* \alpha_m - (\text{c.c.})\right] + (m \leftrightarrow n) & (\text{for } \varphi \in P_+), \\ \left[\mathrm{e}^{\mathrm{i}\theta} (\mathrm{i}^n \alpha_n)^* \alpha_m^* - (\text{c.c.})\right] + (m \leftrightarrow n) & (\text{for } \varphi \in P_-), \end{cases}$$

with $\mathbb{R} \ni \theta = \arg(\mathfrak{F}\varphi) \mp \arg \varphi$ for $\varphi \in P_\pm$, and (c.c.) representing the complex conjugate. Here we have symmetrized β_{nm} and γ_{nm} under $m \leftrightarrow n$, due to the invariance of the sum over n and m in (3) under $n \leftrightarrow m$.

The followings are basic properties concerning β_{nm}. By its definition, it is easily found that

$$\beta_{nm} \equiv 0 \quad (\text{for } n - m \in 4\mathbb{N}). \tag{4}$$

In addition, it is found that for $|\,n - m\,|, |\,n - m'\,| \notin 4\mathbb{N}$,

$$(\beta_{nm} = \beta_{nm'} = 0,\ \alpha_n \neq 0) \implies \beta_{mm'} = 0. \tag{5}$$

To show (5), it is sufficient to deal with the case $\alpha_m, \alpha_{m'} \neq 0$; otherwise, $\beta_{mm'}$ would vanish by its definition. For $|\,n - m\,| \notin 4\mathbb{N}$ (with $\alpha_n, \alpha_m \neq 0$), the condition of $\beta_{nm} = 0$ indicates the relative direction between $\alpha_n \in \mathbb{C}$ and $\alpha_m \in \mathbb{C}$ is determined in the complex plane. Thus the condition of $\beta_{nm} = \beta_{nm'} = 0$ implies that the relative direction between α_m and $\alpha_{m'}$ is determined, in which the relation of $\beta_{mm'} = 0$ is satisfied.

Moreover, we obtain

$$\begin{cases} \beta_{nm} = 0 \quad (\forall n, m \in \mathbb{N}) \iff \varphi \in E, \\ \gamma_{nm} = 0 \quad (\forall n, m \in \mathbb{N}) \iff \varphi \in E \cup C, \end{cases} \tag{6}$$

where C represents the set such that

$$C := \{\varphi \in Q \mid \arg(\mathfrak{F}\varphi),\ \arg \varphi = \text{constant}\}.$$

To show (6), it may be useful to classify the set $\Phi = \{n \in \mathbb{N} \mid \alpha_n \neq 0\}$ into three following cases (although in solving $\beta_{nm} = 0$, the case (b) is not necessary):

$$\Phi \subset \begin{cases} 4\mathbb{N} + k \quad (k = 0, 1, 2, 3), & \text{(a)} \\ 2\mathbb{N} + k \quad (k = 0, 1), & \text{(b)} \\ \text{otherwise.} & \text{(c)} \end{cases}$$

The case where $\varphi \in C$ is realized for the cases (a) and (b). In the case of (b) but not (a), $\varphi \in C$ does not necessarily implies $\varphi \in A$, as is found in the following example:

$$\varphi(x) = \big(\alpha_0 + \alpha_2 H_2(x)\big)\, \mathrm{e}^{-x^2/2} \qquad (\alpha_0, \alpha_2 \in \mathbb{R} \setminus \{0\}), \tag{7}$$

from which $(\mathcal{F}\varphi)(x) = \big(\alpha_0 - \alpha_2 H_2(x)\big)\, \mathrm{e}^{-x^2/2}$. Thus the condition $\gamma_{nm} = 0$ (for all $n, m \in \mathbb{N}$) does not necessarily implies that $\varphi \in A$.

Table 1. Examples of $\varphi \in A\setminus E$, where $N = \deg_x p(x)+\deg_x q(x)$, and p_1, p_2 represent the real, imaginary parts of p, respectively. For the second row for $N=7$, $x_0 = q^{1/3} - q^{-1/3}$ $(q = 2+\sqrt{3})$, $c = \frac{1}{2}x_0^3(5+x_0^2)$, $d = -\frac{3}{4}x_0(-12+9x_0^2+x_0^4)$, $b_2 = 1+\mathrm{i}$, $b_1 = \frac{1}{2}x_0[4+\mathrm{i}(4+5x_0^2+4x_0^4)]$, and $b_0 = \frac{1}{2}[(4-7x_0^2-3x_0^4)+\mathrm{i}(4+5x_0^2+x_0^4)]$. From the property $\varphi' \in A\setminus E \Longleftrightarrow \varphi \in A\setminus E$ under (10), and the linear independence of p and p^* and that of q and q^*, it is required that $|\deg_x p_2(x) - \deg_x p_1(x)| \in 4\mathbb{N}\setminus\{0\}$, and $\deg_x q(x) \geq 2$. Thus $N \geq 6$ is necessary for $\varphi \in A\setminus E$.

N	$p_1(x)$	$p_2(x)$	$q(x)$	rank β_{nm}
6	1	$x(x^3-7x+4)$	$\mathrm{i}H_2(x) - 2(1-\mathrm{i})H_1(x) - 4H_0(x)$	4
7	1	$x(x^3-9x+4\sqrt{3})$	$(1+\mathrm{i})H_3(x) + 4\sqrt{3}\mathrm{i}H_2(x) - 12(1-\mathrm{i})H_1(x) - 8\sqrt{3}H_0(x)$	4
	$x-x_0$	$x(x^3-9x+c)(x-x_0)+d$	$b_2H_2(x)+b_1H_1(x)+b_0H_0(x)$	4

3. Amplitude to phase

In this section, we obtain a function $\varphi \in L^2(\mathbb{R})$ such that $\varphi \in A$ implies that $\varphi \in P$. Before proceeding further, it should be remarked that the relation of $\varphi \in A$ does not necessarily implies that $\varphi \in P$. The element of $\varphi \in A$ can be rewritten using polynomials $p, q : \mathbb{R} \to \mathbb{C}$ such that (Hadamard decomposition)

$$\varphi(x) = p(x)q(x)\,\mathrm{e}^{-x^2/2}, \quad (\mathfrak{F}\varphi)(x) = p(x)q^*(x)\,\mathrm{e}^{-x^2/2}. \tag{8}$$

In this case, the condition of $\varphi \notin P$ can be rephrased by the linear independence of $p(x)$ and $p^*(x)$ and that of $q(x)$ and $q^*(x)$. For $\deg_x p(x) + \deg_x q(x) \geq 6$ (where "deg" represents the degree), such linear independence can be realized, as summarized in Table 1, where p_1 and p_2 represent the real and imaginary parts of p, respectively.

It should be noticed that from (1), there exist two following transformations $T : \varphi \mapsto \varphi'$ such that

$$\varphi \in A\setminus E \Longrightarrow \varphi' \in A\setminus E. \tag{9}$$

(i) $T = \mathfrak{R}$, where $\mathfrak{R} : \varphi(x) \mapsto \varphi'(x) = \varphi(-x)$ represents the space inversion. Due to the commutativity of $[\mathfrak{R},\mathfrak{F}] = [\mathfrak{R},\mathfrak{C}] = 0$ and the isomorphism $\mathfrak{R}(\varphi_1\varphi_2) = (\mathfrak{R}\varphi_1)(\mathfrak{R}\varphi_2)$, the statement of (9) holds.

(ii) $T = \mathfrak{L}$, where $\mathfrak{L}$ is defined by

$$\mathfrak{L} : \varphi \mapsto \varphi' = \begin{pmatrix}1 & \mathrm{i}\end{pmatrix} \mathsf{M} \begin{pmatrix}\mathfrak{f}_1\\ \mathfrak{f}_2\end{pmatrix}\varphi, \qquad (\mathsf{M} \in \mathrm{GL}_2(\mathbb{R})),$$

where $\mathfrak{f}_1 = \frac{1}{2}(1+\mathfrak{C}\mathfrak{F})$ and $\mathfrak{f}_2 = \frac{1}{2\mathrm{i}}(1-\mathfrak{C}\mathfrak{F})$. For $\varphi \in A$, where (8) is realized, $\mathfrak{L}$ is reduced to

$$\mathfrak{L} : \begin{pmatrix}p_1\\ p_2\\ q\end{pmatrix} \mapsto \begin{pmatrix}p_1'\\ p_2'\\ q'\end{pmatrix} = (\mathsf{M}\oplus 1)\begin{pmatrix}p_1\\ p_2\\ q\end{pmatrix}. \tag{10}$$

Considering the relation

$$\langle \mathfrak{f}_1\varphi, \mathrm{i}\mathfrak{f}_2\varphi\rangle = 0 \qquad (\text{for } \varphi \in A),$$

where $\langle z_1, z_2\rangle = \frac{1}{2}(z_1^* z_2 + (\text{c.c.}))$ represents the Euclidean inner product, we find that $\varphi \in A \Longrightarrow \varphi' \in A$. The linear independence of p and p^* is conserved due to $\det \mathsf{M} \neq 0$. Hence (9) is satisfied for $T = \mathfrak{L}$.

In both transformations $T = \mathfrak{R}, \mathfrak{L}$, the rank of β_{nm} remains invariant:

$$\operatorname{rank} \beta'_{nm} = \operatorname{rank} \beta_{nm}.$$

This is because β_{nm} is transformed as

$$\beta_{nm} \to \beta'_{nm} = \begin{cases} (-1)^{n+m} \beta_{nm} & (\text{for } T = \mathfrak{R}), \\ \det \mathsf{M} \cdot \beta_{nm} & (\text{for } T = \mathfrak{L}). \end{cases}$$

For $N = 6, 7$, there is no other solution for $\varphi \in A \setminus E$ than those listed in Table 1, up to the solutions obtained under the transformations $T = \mathfrak{R}, \mathfrak{L}$, so that it can be concluded that $\operatorname{rank} \beta_{nm} = 4$ for all $\varphi \in A \setminus E$. From this finding, it is conjectured that

$$\operatorname{rank} \beta_{nm} = 4 \qquad (\text{for all } \varphi \in A \setminus E), \tag{11}$$

despite the value of N (≥ 6).

Now we go back to obtain a function φ such that $\varphi \in A$ implies that $\varphi \in P$ (so that by (2), $\varphi \in P$ can be replaced by $\varphi \in E$). This corresponds to obtaining a set X whose element $\chi \subset \mathbb{N}$ is given by

$$\forall \chi \in X, \quad \left\{ (\beta_{nm})_{n,m \in \chi} \,\middle|\, \sum_{n,m \in \chi \subset \mathbb{N}} \beta_{nm} H_n(x) H_m(x) \equiv 0 \right\} = \{(0_{nm})_{n,m \in \chi}\},$$

where (0_{nm}) represents a zero matrix. Although it seems to be somewhat challenging to obtain all the elements in X, the followings are the examples of those:

$$S_5, \ \mathbb{T}_0 \cup \mathbb{T}_2, \ \mathbb{T}_1 \cup \mathbb{T}_3 \in X,$$

where $S_n := \{0, 1, \ldots, n\}$ and $\mathbb{T}_k := 4\mathbb{N} + k$. The relation of $\mathbb{T}_k \cup \mathbb{T}_{k+2} \in X$ (for $k = 0, 1$) indicates that for φ an even or odd function, $\varphi \in A$ implies $\varphi \in P$. The above elements in X are slightly generalized to such that

$$S_6 \setminus \{n\}, \ \mathbb{T}_k \cup \mathbb{T}_\ell \cup \mathbb{T}_m \in X \tag{12}$$

for all $n \in S_6$ and $k, \ell, m \in S_3$. It should be recalled that

$$S_6 \notin X, \tag{13}$$

as is found from the example for $N = 6$ in Table 1.

To show (12), the subsequent identity may be useful.

$$\sum_{n,m \in \mathbb{N}} \beta_{nm} H_n(x) H_m(x) = \sum_{k \in \mathbb{N}} \hat{\beta}_k H_k(x), \tag{14}$$

where the coefficients $\hat{\beta}_k$'s are given by multiplying both sides of (14) with $H_k(x)\, \mathrm{e}^{-x^2/2}$ and integrating over $x \in \mathbb{R}$, with the result that (for the integral formula, see, for example [14])

$$\hat{\beta}_k = \sum_{\substack{n,m \in \mathbb{N} \\ k+m+n \in 2\mathbb{N}}} \frac{2^{s-k}\, n!\, m!}{(s-k)!\,(s-m)!\,(s-n)!} \beta_{nm},$$

with $s = \frac{k+m+n}{2}$. Notice that the lengths k, m, n form a triangle, due to $s-k, s-m, s-n \geq 0$. For the proof of $S_5 \in X$, see Appendix A. In a similar way, we can prove $S_6 \setminus \{n\} \in X$ for all $n \in S_6$.

Moreover, we can show $\mathbb{T}_k \cup \mathbb{T}_\ell \cup \mathbb{T}_m \in X$ in an analogus way. In this case, we first restrict $\varphi = \sum_{n \in \mathbb{T}_k \cup \mathbb{T}_\ell \cup \mathbb{T}_m} \alpha_n h_n$ to such that the sum over n is bounded as $n \leq N$. Solving $\hat{\beta}_k = 0$ for $k = 2N$ to 0 in a descending order, where the repeated application of (5) is made, we obtain $\beta_{nm} = 0$ inductively for all $n, m \leq N$. Considering that N can be chosen as arbitrary in $\mathbb{N}$, we eventually find that $\mathbb{T}_k \cup \mathbb{T}_\ell \cup \mathbb{T}_m \in X$.

To summarize, we obtain

$$A\big|_{I_P} \subset P\big|_{I_P}, \tag{15}$$

where $\cdot\big|_{I_P}$ represents $\cdot \cap I_P$, with $I_P := \left\{ \sum_{n\in\chi} \alpha_n h_n(x) \,\middle|\, \alpha_n \in \mathbb{C}, \chi \in X \right\}$.

4. Phase to amplitude

In the previous section, we have found that for $\varphi \in I_P$, the relation of $\varphi \in A$ implies $\varphi \in P$. In this section, we deal with the converse. For this aim, we introduce a set Y, which corresponds to X, such that

$$\forall v \in Y, \quad \left\{ (\gamma_{nm})_{n,m\in v} \,\middle|\, \sum_{n,m\in v\subset\mathbb{N}} \gamma_{nm} H_n(x) H_m(x) \equiv 0 \right\} = \{(0_{nm})_{n,m\in v}\}.$$

As in (12) and (13), we can show that

$$S_6 \setminus \{n\},\ \mathbb{T}_k \cup \mathbb{T}_\ell \cup \mathbb{T}_m \in Y; \quad S_6 \notin Y,$$

for all $n \in S_6$ and $k, \ell, m \in S_3$. This suggests that the set Y might be identical to X:

$$Y = X, \tag{16}$$

although a rigorous proof has not been made.

Now we are ready to examine whether or not the converse of (15) holds. From the second relation of (6), it is found that $\varphi \in P \cap I_P$ [assuming (16); if not, we should simply replace X in the definition of I_P by $X \cap Y$] implies either $\varphi \in E$ or $\varphi \in C$. In the case of $\varphi \in E$, the relation of $\varphi \in A$ is indeed satisfied. However, in the case of $\varphi \in C$, the relation of $\varphi \in A$ does not necessarily hold, as in an example of (7). If we eliminate C from I_P, we obtain $P\big|_{I'_P} \subset A\big|_{I'_P}$, where $I'_P = I_P \setminus C$. This, together with $A\big|_{I'_P} \subset A\big|_{I_P}$, leads to

$$P\big|_{I'_P} \subset A\big|_{I_P}. \tag{17}$$

From (15) and (17), it is found that

$$A\big|_{I_P} \neq P\big|_{I_P}, \quad A\big|_{I'_P} = P\big|_{I'_P}.$$

The first inequality comes from $P\big|_{I'_P} \subsetneq P\big|_{I_P}$.

At the end of this section, we make a remark about $\operatorname{rank} \gamma_{nm}$. From an explicit solution of $P \setminus (E \cup C) \ni \varphi = \sum_{n=0}^{N} \alpha_n h_n$ for $N = 6, 7, \ldots$, it is conjectured that

$$\operatorname{rank} \gamma_{nm} = 4 \qquad (\text{for all } \varphi \in P \setminus (E \cup C)),$$

which is analogous to (11).

5. Simple generalization

We have found that for $\varphi \in I_P$, the condition of $\varphi \in A$ implies the relation of $\varphi \in P$. In this section, we relax the condition of $\varphi \in A$ in such a way that the relation between $\arg(\mathcal{F}\varphi)$ and $\arg\varphi$ still holds in an analogous way. Such a generalization can be given by a (one parameter) modification which is commutative with $\mathfrak{F}$ and with $\mathfrak{C}$.

Let $\varphi_\epsilon(x)$ denote $\varphi(x,\epsilon)$ (where $x \in \mathbb{R}, \epsilon \in I := [-\Lambda, \Lambda] \subset \mathbb{R}$). Suppose that φ_ϵ is continuous with respect to ϵ on I, and assume the uniform convergence of $\mathfrak{F}\varphi_\epsilon$ with respect to ϵ. Then the modification $\tau_\delta : \varphi_\epsilon \mapsto \varphi_{\epsilon+\delta}$ is commutative with $\mathfrak{F}$ and $\mathfrak{C}$ (for all $\epsilon, \epsilon+\delta \in I$). Denote by $A^{(\Lambda)}$ and $P^{(\Lambda)}$ the sets such that

$$A^{(\Lambda)} := \{\varphi_\epsilon \in Q \,|\, \forall\epsilon \in I,\ |\,\mathfrak{F}\varphi_\epsilon\,| = |\,\varphi_{-\epsilon}\,|\},$$
$$P^{(\Lambda)} := \{\varphi_\epsilon \in Q \,|\, \forall\epsilon \in I,\ \arg(\mathfrak{F}\varphi_\epsilon) = \pm\arg\varphi_{-\epsilon} + \theta \quad (\theta = \text{constant})\}.$$

It is apparent that $A = A^{(0)}$ and $P = P^{(0)}$. In what follows, we first show that

$$A^{(\Lambda)}\big|_{I_P} \subset P^{(\Lambda)}\big|_{I_P}. \tag{18}$$

In deriving (18), notice that there is no loss of generality that φ_ϵ and $\varphi_{-\epsilon}$ (for $\epsilon \neq 0$) are linearly independent of each other. Otherwise, the condition of $|\,\mathfrak{F}\varphi_\epsilon\,| = |\,\varphi_{-\epsilon}\,|$ would turn out to be $|\,\mathfrak{F}\varphi_\epsilon\,| = |\,\varphi_\epsilon\,|$, the same problem dealt with in Sect. 3.

To begin with, it may be convenient to decompose $\varphi_{\pm\epsilon} \in A^{(\Lambda)} \cap I_P$ into

$$\varphi_{\pm\epsilon} = \varphi^+ \pm \varphi^-,$$

where $\varphi^\pm = \frac{1}{2}(\varphi_\epsilon \pm \varphi_{-\epsilon})$ (double-sign corresponds). Then the condition of $\varphi_{\pm\epsilon} \in A^{(\Lambda)}$ can be rewritten as

$$|\,\mathfrak{F}\varphi^+\,|^2 - |\,\varphi^+\,|^2 = |\,\varphi^-\,|^2 - |\,\mathfrak{F}\varphi^-\,|^2, \tag{19}$$
$$\langle \mathfrak{F}\varphi^+, \mathfrak{F}\varphi^- \rangle = -\langle \varphi^+, \varphi^- \rangle. \tag{20}$$

Considering that the left-hand side and right-hand side of (19) are even and odd function of ϵ, respectively, and that the relation of (19) holds for all $\epsilon \in I$, we find that the both sides of (19) should be constant c with respect to ϵ:

$$c = |\,\mathfrak{F}\varphi^+\,|^2 - |\,\varphi^+\,|^2.$$

By taking the above relation in the limit of $\epsilon \to 0$, which is commutative with $\mathfrak{C}$ and $\mathfrak{F}$, and by recalling that $\lim_{\epsilon\to 0}\varphi_\epsilon = \varphi_0 \in A$, it is found that $c = 0$, so that

$$\varphi^\pm \in A. \tag{21}$$

Recall also the condition of $\varphi_{\pm\epsilon} \in I_P$, namely, $\varphi^\pm \in I_P$. This, together with (21), leads to

$$\varphi^\pm \in E \tag{22}$$

by $A\big|_{I_P} \subset P\big|_{I_P}$.

The remaining thing is to derive $\varphi_\epsilon \in P^{(\Lambda)}$ from (22). Depending upon the case where φ^+, φ^- belong to E_+ or E_-, we deal with the four following cases:

(i) For $\varphi^{\pm} \in E_{+}$, we can write $\begin{pmatrix} \mathfrak{F}\varphi^{+} \\ \mathfrak{F}\varphi^{-} \end{pmatrix} = \begin{pmatrix} \lambda_{+}\varphi^{+} \\ \lambda_{-}\varphi^{-} \end{pmatrix}$, where $(\lambda_{\pm})^{4} = 1$. Substituting this relation into (20), we obtain

$$\langle \lambda_{+}\lambda_{-}^{*} + 1,\ (\varphi^{+})^{*}\varphi^{-} \rangle = 0, \tag{23}$$

where use has been made of $\langle z_1,\ z_2 z_3 \rangle = \langle z_1 z_2^{*},\ z_3 \rangle$ for all $z_1, z_2, z_3 \in \mathbb{C}$. Recall that φ_{ϵ} and $\varphi_{-\epsilon}$ can be chosen as linearly independent, in which $\arg\varphi^{+} - \arg\varphi^{-}$ is not constant. In this case, the relation of (23) indicates that $\lambda_{+}\lambda_{-}^{*} + 1 = 0$, or equivalently, $\lambda_{+} + \lambda_{-} = 0$ by $|\lambda_{-}| = 1$. As a consequence, we obtain

$$\begin{aligned} \mathfrak{F}\varphi_{\epsilon} &= \mathfrak{F}(\varphi^{+} + \varphi^{-}) \\ &= \lambda_{+}\varphi^{+} + \lambda_{-}\varphi^{-} \\ &= \lambda_{+}\varphi_{-\epsilon}, \end{aligned}$$

so that $\varphi_{\epsilon} \in P^{(\Lambda)}$.

(ii) For $\varphi_{\pm} \in E_{-}$, we can write $\begin{pmatrix} \mathfrak{F}\varphi^{+} \\ \mathfrak{F}\varphi^{-} \end{pmatrix} = \begin{pmatrix} e^{i\eta_{+}}(\varphi^{+})^{*} \\ e^{i\eta_{-}}(\varphi^{-})^{*} \end{pmatrix}$, where $\eta_{\pm} \in \mathbb{R}$. In this case, we can obtain $\mathfrak{F}\varphi_{\epsilon} = e^{i\eta_{+}}(\varphi_{-\epsilon})^{*}$ in an analogous way to the case (i), so that $\varphi_{\epsilon} \in P^{(\Lambda)}$.

(iii) For $\varphi^{\pm} \in E_{\pm}$, we can write $\begin{pmatrix} \mathfrak{F}\varphi^{+} \\ \mathfrak{F}\varphi^{-} \end{pmatrix} = \begin{pmatrix} \lambda_{+}\varphi^{+} \\ e^{i\eta_{-}}(\varphi^{-})^{*} \end{pmatrix}$. Substituting this relation into (20), we obtain either $\arg\varphi^{+}$ or $\arg\varphi^{-}$ is constant, which corresponds to a special case of (ii) or (i), respectively. Thus in either case, it follows that $\varphi_{\epsilon} \in P^{(\Lambda)}$.

(iv) For $\varphi^{\pm} \in E_{\mp}$, we also obtain $\varphi_{\epsilon} \in P^{(\Lambda)}$ in quite an analogous way to the case (iii).

In all cases (i)–(iv), it is found that $\varphi_{\epsilon} \in P^{(\Lambda)}$. This ends the proof of (18).

The converse of (18) does not necessarily hold, as in the case of $\Lambda = 0$. However, if we eliminate the case of $\varphi_{\epsilon} \in C$, the converse holds as in the form

$$P^{(\Lambda)}|_{I_P'} \subset A^{(\Lambda)}|_{I_P}, \tag{24}$$

which is reduced to (17) in the limit of $\Lambda \to 0$.

6. Conclusion

We have shown that under the Fourier transform $\mathfrak{F} : L^2(\mathbb{R}) \to L^2(\mathbb{R})$ of an even or odd complex-valued function φ, the invariance of the amplitude ($\varphi \in A$) leads to either the invariance or inversion of the phase up to an additive constant ($\varphi \in P$). We have also shown that the converse holds in an analogous way, unless $\arg\varphi$ is constant. Without the evenness or oddness of φ, this phase amplitude relation does not hold in general; such an example is found in Table 1. It does not seem to be so simple a matter to obtain a necessary and sufficient condition for $A \subset P$ and that for $P \subset A$; we have just presented a sufficient condition, as in the form of (12).

The generalization of this phase amplitude relation can be easily made if the modification τ of φ is commutative with $\mathfrak{F}$ and with the complex conjugate $\mathfrak{C}$, as is summarized in (18) and (24). To make further application, we should relax the commutativity of τ with $\mathfrak{F}$ and $\mathfrak{C}$.

Acknowledgments

The author is indebted to K. Terada for stimulating discussions, and M. Hirokado for critically reading the manuscript. This work was supported in part by Grant-in-Aid specified by HCU.

Journal of Physics: Conference Series **597** (2015) 012052 doi:10.1088/1742-6596/597/1/012052

Appendix A. Proof of $S_5 \in X$

To prove $S_5 \in X$, we show that the condition $\sum_{n,m\in S_5} \beta_{nm} H_n(x) H_m(x) \equiv 0$ leads to $\beta_{nm} = 0$ for all $n, m \in S_5$. In this Appendix, we deal with the case where

$$\alpha_5 \neq 0. \tag{A.1}$$

The case of $\alpha_5 = 0$ can be dealt with in an analogous way. To begin with, the relation of $\sum_{n,m\in S_5} \beta_{nm} H_n(x) H_m(x) \equiv 0$ can be rewritten using (14) as

$$\begin{aligned}
0 &= \beta_{45} && \text{(A.2)}\\
&= \beta_{35} && \text{(A.3)}\\
&= \beta_{25} + \beta_{34} && \text{(A.4)}\\
&= \beta_{24} && \text{(A.5)}\\
&= \beta_{05} + \beta_{14} + \beta_{23} + 20\,\beta_{25} + 24\,\beta_{34} && \text{(A.6)}\\
&= \beta_{13} && \text{(A.7)}\\
&= \beta_{03} + \beta_{12} + 8\,\beta_{14} + 12\,\beta_{23} + 80\,\beta_{25} + 144\,\beta_{34} && \text{(A.8)}\\
&= \beta_{02} && \text{(A.9)}\\
&= \beta_{01} + 4\,\beta_{12} + 24\,\beta_{23} + 192\,\beta_{34}, && \text{(A.10)}
\end{aligned}$$

each of which corresponds to $\hat{\beta}_k = 0$ for $k = 9, 8, \ldots, 1$, respectively, where use has been made of the first relation of (4). For the above relations corresponding to $\hat{\beta}_\ell = 0$ (for $\ell = 7, 6, \ldots, 1$), some relations of $\hat{\beta}_m = 0$ (for $m = \ell + 2, \ell + 4, \ldots, 8$ or 9) have been taken into account so as to simplify the relation of $\hat{\beta}_\ell = 0$. Under the condition of (A.1), we obtain from (A.2) and (A.3)

$$\beta_{34} = 0, \tag{A.11}$$

which, together with (A.4), leads to

$$\beta_{25} = 0. \tag{A.12}$$

In deriving (A.11), use has been made of (5). Using (5) again, we obtain from (A.3) and (A.12)

$$\beta_{23} = 0. \tag{A.13}$$

At this stage, the relations of (A.6), (A.8), and (A.10) are reduced to

$$\begin{cases} \beta_{05} + \beta_{14} = 0, \\ \beta_{03} + \beta_{12} + 8\,\beta_{14} = 0, \\ \beta_{01} + 4\,\beta_{12} = 0. \end{cases} \tag{A.14}$$

The rest thing is to show that each of the terms in the left-hand side of (A.14) is vanishing. Here we classify the case where (α_2, α_3) is equal to $(0, 0)$ or not.

(i) For $\alpha_2 \neq 0$, it is found from (A.9), (A.12), and (A.13) that $\beta_{05} = \beta_{03} = 0$ by using (5). The remaining β_{nm}'s are all vanishing by (A.14).

(ii) For $\alpha_3 \neq 0$, we similarly obtain $\beta_{14} = \beta_{12} = 0$ from (A.7), (A.11), and (A.13), so that the remaining β_{nm}'s are all vanishing by (A.14).

(iii) For $\alpha_2 = \alpha_3 = 0$, it follows that $\beta_{03} = \beta_{12} = 0$ by definition, so that the remaining β_{nm}'s are all vanishing by (A.14).

In either case, we obtain $\beta_{nm} = 0$ for all $m, m \in S_5$.

References

[1] Wolf E 1962 *Proc. Phys. Soc.* **80** 1269
[2] Roman P and Marathay A S 1963 *Nuovo Cimento* **6** 1452
[3] Walther A 1963 *Opt. Act.* **10** 41
[4] Hofstetter E M 1964 *IEEE Trans. Inform. Theory* **IT-10** 119
[5] Paley R E A C and Wiener N 1934 *Fourier Transforms in the Complex Domain* (*Am. Math. Soc. Colloquium Publ.* vol 19) (Providence RI: Am. Math. Soc.)
[6] Sanz J L C and Huang T S 1984 *J. Math. Anal. Appl.* **104** 302
[7] Levin B Y 1996 *Lectures on Entire Functions* (*Translations of Mathematical Monographs* vol 150) (Providence RI: Am. Math. Soc.)
[8] Rosenblatt J 1984 *Commun. Math. Phys.* **95** 317
[9] Bruck Y M and Sodin L G 1979 *Opt. Commun.* **30** 304
[10] Hayes M H 1982 *IEEE Trans. Acoust., Speech, Signal Proc.* **ASSP-30** 140
[11] Kennard E H 1927 *Z. Phys.* **44** 326
[12] Robertson H P 1929 *Phys. Rev.* **34** 163
[13] Bialynicki-Birula I and Mycielski J 1975 *Commun. Math. Phys.* **44** 129
[14] Gradshteyn I S and Ryzhik I M 2007 *Table of Integrals, Series, and Products* ed Jeffrey A and Zwillinger D (Burlington, MA: Academic) p 804 7th ed

Journal of Physics: Conference Series **597** (2015) 012053 doi:10.1088/1742-6596/597/1/012053

Supersymmetry and conformal Galilei algebras

Zhanna Kuznetsova

Centre of Mathematics, Computing and Cognition, Federal University of ABC, av. dos Estados 5001, Santo Andre, SP, Brazil

E-mail: zhanna.kuznetsova@ufabc.edu.br

Abstract. Nonrelativistic l - conformal Galilei algebras and their supersymmetric extensions are considered. The construction presented in this article is different from usual based on supersymmetrization of constructed l-conformal Galilei algebras. We start instead from N-extended supersymmetric quantum mechanics, require in addition the conformal symmetry and construct the l-Galilei algebras.

The superconformal algebras obtained from the irreducible representations of one-dimensional SUSY are simple superalgebras given in the Kac classification. With enlarging the number N of SUSY extensions different restrictions on conformal weight of fields appear.

1. Introduction

In recent years (super)conformal nonrelativistic algebras have been intensively studied. The interest is due to an appearance of this kind of symmetries in very different areas of physics and mathematics. The applications go from anisotropic critical systems [1] to conformal invariance of unitary Fermi gas [2], AdS/cold atom correspondence [3] and AdS/CFT duality in nonrelativistic systems [4].

A *kinematical* algebra is a Lie algebra spanned by the generators P_i of spatial displacements, $P_t = H$ of time translations, M_{ij} of rotations and B_i of inertial transformations. Space and time reversals are automorphisms. Kinematical algebras associated with d dimensional space have been classified by Bacry and Levy-Leblond in 1968 [5]. According to the classification they are divided in 2 classes, relativistic and nonrelativistic and in 3 subclasses each one. In the nonrelativistic class there are 2 types of algebras such that time translations and inertial transformations do not commute, the Newton-Hooke algebras and the Galilei algebras. Any nonrelativistic algebra admits only one nontrivial central extension.

A nonrelativistic *conformal* algebra contains a nonrelativistic kinematical algebra as a proper subalgebra. There is a countable class of finite-dimensional nonrelativistic conformal algebras [6] associated with the pair of numbers (d, l), where d is a dimension of space and l is a positive integer or half-integer number. For each pair (d, l) there is only one inequivalent nonrelativistic conformal algebra which is called in the literature **l-conformal Galilei** or **nonrelativistic Galilei**. The Newton-Hooke algebra is an example of $l = \frac{1}{2}$-conformal Galilei and the "Galilei" from Bacry and Levi-Leblond classification is $l = 1$-conformal Galilei in this sense.

The number l measures the "non-relativiticity" of spacetime in the following way. $(\Delta t)^{2l}$ and Δx transform with the same conformal factor. The inverse $z = \frac{1}{l}$ is called the *dynamical exponent* of the nonrelativistic l-conformal algebra.

The l-conformal algebras for $l = \frac{1}{2}$ and $l = 1$ are largely studied together with their applications. The algebras with bigger numbers of l are under active investigations in recent years, see, for example, [7] and [8].

If we consider l-conformal symmetries together with supersymmetry, we arrive at a class of superalgebras characterized by 3 numbers (d, l, N), where N is the number of supersymmetries. For some particular choices of l, d and N clear results are obtained. It is important to mention the Duval and Horvathy paper [9], where $l = \frac{1}{2}$-superconformal algebras are constructed for any N and generalized for pseudo-supersymmetry governed by a pair (N_+, N_-). In the paper by Fedoruk and Lukierski [10] supersymmetric $N = 4$ finite extensions are obtained for $d = 1, 2, 3, 4, 5$ and $l = \frac{3}{2}$ (this can be generalized for bigger number of d). For $N = 2$, any d and l, the supersymmetric extensions and their dynamics are investigated in [12], [13] and [14].

In this article I describe the construction of D-module representations of nonrelativistic superconformal algebras from the representations of 1-dimensional N-extended supersymmetry. The initial dimension is related with time coordinate. The d space dimensions can be added in a different way for nonrelativistic systems.

The plan of the paper is the following. In the 2nd Section a classification of 1-dimensional supersymmetric algebras is presented. The 3rd Section is devoted to superconformal models obtained from the multiplets of the N-extended supersymmetries in Section 2. In the 4th Section I present the extension of superconformal algebras to l-superconformal using the results of the previous Section, considering the cases $N = 1, 2, 4$.

2. N-extended supersymmetry

Let us consider representations of N-extended 1-dimensional supersymmetry algebra

$$\{Q_i, Q_j\} = \delta_{ij} H, \quad [H, Q_i] = 0. \quad (i, j = 1, \ldots, N) \tag{1}$$

The anti-commutational relations of (1) are similar to the relations for Clifford algebra generators Γ_i

$$\{\Gamma_i, \Gamma_j\} = 2\eta_{ij}, \tag{2}$$

where $\eta_{ij} = diag(+1, +1, \ldots, +1, -1, -1, \ldots, -1)$ is a diagonal matrix with p positive and q negative entries and $i, j = 1, 2, \ldots, (p+q)$. The difference is in the fact that the coefficient in the r.h.s. of (1) is only positive and that the constant 2 in (2) should be replaced by the operator H in (1).

Therefore, for systematical construction of N-extended SUSY representations for any N, the matrix representations of Clifford algebras were used. These representations can be found in [15], [16]. We wll use the term *Gamma matrix* for the generators of matrix representations of Clifford algebras.

From supersymmetry requirements (the supersymmetry generator should exchange bosons with fermions) the Gamma matrices used for the construction are of the form

$$\Gamma_i = \begin{pmatrix} 0 & \gamma_i \\ \pm\gamma_i & 0 \end{pmatrix} \quad or \quad \begin{pmatrix} 0 & \mathbf{1} \\ \mathbf{1} & 0 \end{pmatrix}, \tag{3}$$

where $\mathbf{1}$ is $n \times n$ identity matrix, 0 is $n \times n$ zero matrix and γ_i are the Gamma matrices from $(p+q-2)$ representation.

In the representation of (1) $H = \mathbf{1}_{(2n)}\partial_t$ and

$$Q_i = \begin{pmatrix} 0 & \gamma_i \\ \pm\gamma_i\partial_t & 0 \end{pmatrix}. \tag{4}$$

This representation acts on a multiplet with n bosonic and n fermionic fields. This multiplet is denoted as (n,n) root multiplet. The mass dimensions of the fields in the multiplet are λ and $\lambda+\frac{1}{2}$. The number of fields n for each number N of supersymmetries is presented in the table

$N=1$	$n=1$	$N=9$	16	$N=17$	256	$N=25$	4096
$N=2$	$n=2$	$N=10$	32	$N=18$	512	$N=26$	8192
$N=3$	4	$N=11$	64	$N=19$	1024	$N=27$	16384
$N=4$	4	$N=12$	64	$N=20$	1024	$N=28$	16384
$N=5$	8	$N=13$	128	$N=21$	2048	$N=29$	32768
$N=6$	8	$N=14$	128	$N=22$	2048	$N=30$	32768
$N=7$	8	$N=15$	128	$N=23$	2048	$N=31$	32768
$N=8$	8	$N=16$	128	$N=24$	2048	$N=32$	32768

The SUSY algebra representation (4) which acts on (n,n) root multiplets can be changed via dressing transformations $A \mapsto A' = T^{-1}AT$. Here $A = \{H, Q_i\}$ are any generators of (4) and $T = diag(\frac{d}{dt},\ldots,\frac{d}{dt},1,\ldots 1)$ with some number m of derivatives $\frac{d}{dt}$ in the diagonal. The new representation acts on a new multiplet with $(k,n,n-k)$ fields $(m=n-k)$ with mass dimensions λ, $\lambda+\frac{1}{2}$ and $\lambda+1$. The string of numbers of fields with different mass dimensions is called *field content* of the multiplet and the number of mass dimensions is called the *length* of the multiplet. For example, (n,n) is a length-2 multiplet and $(k,n,n-k)$ is a length-3 multiplet.

The procedure can be continued with $m = n,\ldots,2n$ derivatives in T dressing operator and even with second derivatives in T, but should be stopped if a singular operator in a new algebra $\{A'\}$ appears. So all length-3 multiplets $(k,n,n-k)$, $k=1,2,\ldots,n-1$ can be constructed for any N, while length-4 and length-5 are exceptional cases. The full classification and examples up to $N=10$ are given in [15]. The results for $N \leq 8$ are given in the table below

	length-2	length-3	length-4
N=1	(1,1)		
N=2	(2,2)	(1,2,1)	
N=3	(4,4)	(3,4,1), (2,4,2), (1,4,3)	(1,3,3,1)
N=4	(4,4)	(3,4,1), (2,4,2), (1,4,3)	
N=5	(8,8)	(7,8,1), (6,8,2), (5,8,3), (4,8,4) (3,8,5), (2,8,6), (1,8,7)	(1,5,7,3), (3,7,5,1) (1,6,7,2), (2,7,6,1), (2,6,6,2) (1,7,7,1)
N=6	(8,8)	(7,8,1), (6,8,2), ..., (1,8,7)	(1,6,7,2), (2,7,6,1), (2,6,6,2) (1,7,7,1)
N=7	(8,8)	(7,8,1), (6,8,2), ..., (1,8,7)	(1,7,7,1)
N=8	(8,8)	(7,8,1), (6,8,2), ..., (1,8,7)	

(5)

Representations acting on different multiplets are non-equivalent in physical applications. We will call them, for example, as $N=4$ $(1,4,3)$ representation, to stress that it acts on the $(1,4,3)$ multiplet.

3. Superconformal algebras in one dimension

In this section the representations of superconformal algebras are obtained from N-SUSY representations adding conformal transformations and closing the algebras.

A superconformal algebra in one dimension contains $2N$ generators in the odd sector and the $sl(2) \oplus R$ structure of generators in the even sector. R is called "R-symmetry".

The generators of $sl(2)$ act on a component field with mass dimension λ as

$$H = \frac{\partial}{\partial t}, \tag{6}$$

$$D = -(t\frac{\partial}{\partial t} + \lambda),$$
$$K = -(t^2\frac{\partial}{\partial t} + 2\lambda t).$$

They form diagonal matrix operators acting on a multiplet of fields.

Let us remember simple superconformal algebras as abstract algebras.

- The **N = 2** super-conformal algebra is $A(1,0) = sl(2|1)$. Its R-symmetry is $u(1)$.
- The **N = 4** superconformal algebras are
  - **i)** the $A(1,1) = sl(2|2)/\mathbf{Z}$ superalgebra with 6 even generators ($sl(2) \oplus sl(2)$);
  - **ii)** the $sl(2|2)$ superalgebra, with bosonic sector given by $sl(2) \oplus sl(2) \oplus u(1)$;
  - **iii)** the exceptional superalgebras $D(2,1;\alpha)$, parametrized by a complex $\alpha \neq 0, -1$ with 9 even generators in $sl(2) \oplus sl(2) \oplus sl(2)$. In the limits $\alpha \to 0$, $\alpha \to -1$ and $\alpha \to \infty$ the superalgebra becomes $A(1,1)$.
    The superalgebras related with α, $\frac{1}{\alpha}$, $-(1+\alpha)$, $-\frac{1}{1+\alpha}$, $-\frac{1+\alpha}{\alpha}$ and $-\frac{\alpha}{1+\alpha}$ are isomorphic.
    For $\alpha = -2, -\frac{1}{2}, 1$ the $D(2,1;\alpha)$ becomes $D(2,1)$ superalgebra.
- The **N = 8** superalgebras with 16 odd generators are:
  - **i)** the $A(3,1) = sl(4|2)$ superalgebra with 19 even generators organized as $sl(2) \oplus sl(4) \oplus u(1)$;
  - **ii)** the $D(4,1) = osp(8|2)$ superalgebra with even sector given by $sl(2) \oplus so(8)$ (31 even generators);
  - **iii)** the $D(2,2) = osp(4|4)$ superalgebra with 16 even generators in $sl(2) \oplus so(3) \oplus sp(4)$;
  - **iv)** the $F(4)$ exceptional superalgebra with 24 even generators in $sl(2) \oplus so(7)$.

We use the N-extended supersymmetry representations associated with the irreducible multiplets from (5) for constructing superconformal algebras.

The **N = 2** supersymmetric multiplets are $(2,2) = (x_1, x_2; \psi_1, \psi_2)$ and $(1,2,1) = (x; \psi_1, \psi_2; g)$. The corresponding superalgebra $sl(2|2)$ representations for $(2,2)$ and $(1,2,1)$ can be found in [17].

The **N = 4** irreps are the root multiplet $(4,4)$ and $(1,4,3)$, $(2,4,2)$, $(3,4,1)$. Scaling dimensions of the component fields are $\lambda, \lambda + \frac{1}{2}, \lambda + 1$. Here we present the results from [17] for different multiplets.

The *(4,4) root multiplet* for $\lambda = 0$ together with the generators (6) give D-module representation of $A(1,1)$ superalgebra.

For $\lambda \neq 0$ the SUSY representation corresponding to $(4,4)$ gives $D(2,1;\alpha)$ representation with α and scaling dimension λ related as

$$\alpha = -2\lambda$$

Other results for the $N = 4$ superconformal representations are presented in

N	multiplet	superalgebra	λ
N=2	(2,2)	$sl(2\|1)$	any λ
	(1,2,1)	$sl(2\|1)$	any λ
N=4	(4,4),(1,4,3),(2,4,2), (3,4,1)	$A(1,1)$	$\lambda = 0$
	(4,4)	$D(2,1;\alpha)$	$\lambda = -\frac{\alpha}{2}$
	(1,4,3)	$D(2,1;\alpha)$	$\lambda = \alpha$
	(2,4,2)	$sl(2\|2)$	any $\lambda \neq 0$
	(3,4,1)	$D(2,1;\alpha)$	$\lambda = -\alpha$

(7)

For **N = 8** the irreducible multiplets are the $(8,8)$ root multiplet and $(1,8,7)$, $(2,8,6)$, $(3,8,5)$, $(4,8,4)$, $(5,8,3)$, $(6,8,2)$, $(7,8,1)$. For each multiplet physically nonequivalent D-module representations of supersymmetry are constructed in [15]. As in $N<8$ cases, the (6) $sl(2)$ generators have been added with the requirement to close new $N=8$ superconformal algebras with *finite* number of generators. The $N=8$ superconformal algebras are finite only for particular values of the conformal weight [18]. All irreps can be written as $(k,8.8-k)$ multiplets with $(8,8)\equiv(0,8,8)$ or $(8,8,0)$. The conformal weight λ for all *finite* representations is restricted by

$$\lambda = \frac{1}{k-4}. \tag{8}$$

Therefore, a finite superconformal representation for the $(4,8,4)$ supermultiplet can not be obtained by this procedure.

The results are presented in the table below.

N	multiplet	superalgebra	λ
N=8	(8,8,0)	$osp(8\|2)$	$\lambda=1/4$
	(7,8,1)	$F(4)$	$\lambda=1/3$
	(6,8,2)	$sl(4\|2)$	$\lambda=1/2$
	(5,8,3)	$osp(4\|4)$	$\lambda=1$
	(3,8,5)	$osp(4\|4)$	$\lambda=-1$
	(2,8,6)	$sl(4\|2)$	$\lambda=-1/2$
	(1,8,7)	$F(4)$	$\lambda=-1/3$
	(0,8,8)	$osp(8\|2)$	$\lambda=-1/4$
N=7	(1,7,7,1)	$G(3)$	$\lambda=-1/4$

(9)

Therefore, N-extended supersymmetry representations with different irreps induce superconformal simple superalgebra representations with $2N$ odd generators.

4. l - superconformal Galilei

In this Section I consider superconformal nonrelativistic Galilei class of algebras associated with the triple (d,l,N). The representations for $N=2$ and $N=4$ are presented for any l and d.

4.1. General considerations

The (d,l,N)-superconformal Galilei algebra generators are:

(i) $sl(2)$ generators H,D,K;

(ii) $so(d)$ subalgebra of space rotations generators M_{ij} (the $sl(2)$ generators commute with $so(d)$);

(iii) N-extended supersymmetry algebra (1) with odd generators Q_a $(a=1,\dots,N)$ and their superconformal partners S_a $(a=1,\dots,N)$;

(iv) R-symmetry bosonic generators R_α;

(v) Abelian subalgebra with generators $P_i^{(n)}$ $(i=1,\dots,d,\ n=0,\dots 2l)$;

(vi) additional odd generators $X_{a,i}^{(n)}$ from the commutators $[Q_a,P_i^{(n)}]$;

(vii) additional even generators $J_i^{(n)}$ from $[R,P_i^{(n)}]$.

The algebras for $N=2$ are presented in [14]. These algebras are constructed for any d and l.

As in superconformal case, the representations acting on different N-SUSY multiplets give rise to different superconformal Galilei algebras. For example, in $N=2$ [14], different nonrelativistic Galilei algebras are constructed from the $(1,2,1)$ and $(2,2)$ superconformal representations.

The (d,l,N)-symmetry generators are acting on multiplets of bosonic and fermionic fields which depend on $t, x_1, \ldots, x_d$. The dependence on space coordinates are introduced by replacing

$$\lambda \to \hat{\lambda} = \lambda + l\frac{\partial}{\partial x_i} \tag{10}$$

in D and K generators of $sl(2)$.

The Abelian subalgebra generators are diagonal multiple of identity with entries $t^n\frac{\partial}{\partial x_i}$ for $P_i^{(n)}$. The $so(d)$ rotational symmetry is also presented alsy by diagonal matrices with differential operators $x_i\frac{\partial}{\partial x_j} - x_j\frac{\partial}{\partial x_i}$ in the entries.

4.2. $N=2$

The $N=2$ SUSY multiplets are $(2,2)$ and $(1,2,1)$.

The $(1,2,1)$ superconformal Galilei algebra $\mathcal{G}_{(1,2,1)}$ coincides with the superalgebra realized on superspace by Masterov in [11].

The $(2,2)$ algebra $\mathcal{G}_{(2,2)}$ has different (from the previous one) number of $J_i^{(n)}$ and $X_{i,a}^{(n)}$ generators.

The D-module representation of $\mathcal{G}_{(2,2)}$ in terms of (2×2) matrix blocks is given by

$$\begin{aligned}
H &= \begin{pmatrix} \partial_t & 0 \\ 0 & \partial_t \end{pmatrix}, \quad D = \begin{pmatrix} -(t\partial_t + lx_i\partial_i + \lambda) & 0 \\ 0 & -(t\partial_t + lx_i\partial_i + \lambda + \frac{1}{2}) \end{pmatrix} \\
K &= \begin{pmatrix} -(t^2\partial_t + 2ltx_i\partial_i + 2\lambda t) & 0 \\ 0 & -(t^2\partial_t + 2ltx_i\partial_i + (2\lambda+1)t) \end{pmatrix} \\
R &= \begin{pmatrix} -2\sigma_A(\partial_t + lx_i\partial_i + \lambda) & 0 \\ 0 & [-2\sigma_A(\partial_t + lx_i\partial_i + \lambda) + 1] \end{pmatrix} \\
Q_1 &= \begin{pmatrix} 0 & 1 \\ \partial_t & 0 \end{pmatrix}, \quad Q_2 = \begin{pmatrix} 0 & \sigma_A \\ -\sigma_A\partial_t & 0 \end{pmatrix} \\
S_1 &= \begin{pmatrix} 0 & t \\ t\partial_t + 2lx_i\partial_i + 2\lambda & 0 \end{pmatrix}, \quad S_2 = \begin{pmatrix} 0 & t\sigma_A \\ -\sigma_A(t\partial_t + 2lx_i\partial_i + 2\lambda) & 0 \end{pmatrix} \\
P_i^{(n)} &= \begin{pmatrix} t^n\partial_i & 0 \\ 0 & t^n\partial_i \end{pmatrix}, \quad J_i^{(n)} = \begin{pmatrix} \sigma_A t^n\partial_i & 0 \\ 0 & \sigma_A t^n\partial_i \end{pmatrix} \\
X_{1,i}^{(n)} &= \begin{pmatrix} 0 & 0 \\ t^n\partial_i & 0 \end{pmatrix}, \quad X_{1,i}^{(n)} = \begin{pmatrix} 0 & 0 \\ -\sigma_A t^n\partial_i & 0 \end{pmatrix} \\
M_{ij} &= \begin{pmatrix} x_i\partial_j - x_j\partial_i & 0 \\ 0 & x_i\partial_j - x_j\partial_i \end{pmatrix},
\end{aligned} \tag{11}$$

where the 0 block is understood as (2×2) zero matrix, $\sigma_A = \begin{pmatrix} 0 & 1 \\ -1 & 0 \end{pmatrix}$ and all operators are multiplied by the (2×2) identity matrix. For example, $\partial_t \equiv \frac{\partial}{\partial t}\begin{pmatrix} 1 & 0 \\ 0 & 1 \end{pmatrix}$. Also $\partial_i \equiv \frac{\partial}{\partial x_i}$ and the sum over repeating indexes is understood.

Let us present the (anti)commutation relations of $\mathcal{G}_{(2,2)}$.

The $sl(2|1)$ subalgebra reads as follows

$$\begin{aligned}
[D,H] &= H, \quad [D,K]=-K, \quad [H,K]=2D, \\
\{Q_a,Q_b\} &= 2\delta_{ab}H, \quad \{S_a,S_b\}=-2\delta_{ab}K, \quad \{Q_a,S_b\}=-2\delta_{ab}D+\sigma_A R \\
[D,Q_a] &= \frac{1}{2}Q_a, \quad [K,Q_a]=S_a, \quad [R,Q_a]=-\sigma_A Q_b \\
[D,S_a] &= -\frac{1}{2}S_a, \quad [H,S_a]=A_a, \quad [R,S_a]=-\sigma_A S_b
\end{aligned} \tag{12}$$

The commutators involving the Abelian subalgebra generators $P_i^{(n)}$ are

$$\begin{aligned}
[H,P_i^{(n)}] &= nP_i^{(n-1)}, \quad [D,P_i^{(n)}]=-(n-l)P_i^{(n)}, \quad [K,P_i^{(n)}]=-(n-2l)P_i^{(n+1)} \\
[Q_a,P_i^{(n)}] &= nX_{a,i}^{(n-1)}, \quad [S_a,P_i^{(n)}]=(n-2l)X_{a,i}^{(n)}, \quad [R,P_i^{(n)}]=2lJ_i^{(n)}.
\end{aligned} \tag{13}$$

The odd generators $X_{a,i}^{(n)}$ (anti)commutators are

$$\begin{aligned}
[H,X_{a,i}^{(n)}] &= nX_{a,i}^{(n-1)}, \quad [D,X_{a,i}^{(n)}]=-(n-l+\frac{1}{2})X_{a,i}^{(n)} \\
[K,X_{a,i}^{(n)}] &= -(n-2l+1)X_{a,i}^{(n+1)}, \\
[R,X_{1,i}^{(n)}] &= -(2l+1)X_{2,i}^{(n)}, \quad [R,X_{2,i}^{(n)}]=(2l+q)X_{2,i}^{(n)} \\
\{Q_a,X_{b,i}^{(n)}\} &= \delta_{ab}P_i^{(n)}-\epsilon_{ab}J_i^{(n)}, \quad \{S_a,X_{a,i}^{(n)}\}=\delta_{ab}P_i^{(n+1)}-\epsilon_{ab}J_i^{(n+1)}.
\end{aligned} \tag{14}$$

The even generators $J_i^{(n)}$ relations are

$$\begin{aligned}
[H,J_i^{(n)}] &= nJ_i^{(n-1)}, \quad [D,J_i^{(n)}]=-(n-l)J_i^{(n)}, \quad [K,J_i^{(n)}]=-(n-2l)J_i^{(n+1)} \\
[Q_a,J_i^{(n)}] &= -n\epsilon_{ab}X_{b,i}^{(n-1)}, \quad [S_a,J_i^{(n)}]=-(n-2l)\epsilon_{ab}X_{a,i}^{(n)}, \quad [R,J_i^{(n)}]=-2lP_i^{(n)}.
\end{aligned} \tag{15}$$

The $so(d)$ commutators are

$$\begin{aligned}
[M_{ij},M_{kl}] &= -\delta_{ik}M_{jl}-\delta_{jl}M_{ik}+\delta_{il}M_{jk}+\delta_{jk}M_{il} \\
[P_k^{(n)},M_{ij}] &= \delta_{ik}P_j^{(n)}-\delta_{jk}P_i^{(n)} \\
[J_k^{(n)},M_{ij}] &= \delta_{ik}J_j^{(n)}-\delta_{jk}J_i^{(n)} \\
[X_{a,k},M_{ij}] &= \delta_{ik}X_{a,j}^{(n)}-\delta_{jk}X_{a,i}^{(n)}.
\end{aligned} \tag{16}$$

From the two $N=2$ supersimmetric representations, two inequivalent $(d,l,2)$-Galilei algebras are obtained.

4.3. $N=4$

Here we consider only the representation associated to the $(4,4)$ SUSY multiplet. The representations associated with the other $N=4$ multiplets have similar properties.

For $N=4$ supersymmetric irrep $(4,4)$ the D-module representation of $\mathcal{G}_{(4,4)}$ is constructed similarly with (11). In $\mathcal{G}_{(4,4)}$ the (2×2) 0-matrix blocks are substituted by (4×4) 0-matrix blocks, all operators are multiplied by (4×4) identity matrix, etc. The supersymmetry generators are

$$Q_a = \begin{pmatrix} 0 & \gamma_a \\ -\gamma_a\partial_t & 0 \end{pmatrix}, \quad Q_4=\begin{pmatrix} 0 & 1 \\ \partial_t & 0 \end{pmatrix}, \tag{17}$$

where $a = 1, 2, 3$ and $\gamma_a\gamma_b = -\delta_{ab}\mathbf{1}_4 + \epsilon_{abc}\gamma_c$ are representation of $su(2)$ by (4×4) real matrices.

The superconformal partners are presented by

$$S_a = tQ_a + 2(\lambda - lx_i\partial_i), \quad S_4 = tQ_4 - 2(\lambda - lx_i\partial_i)\begin{pmatrix} 0 & 0 \\ \gamma_a & 0 \end{pmatrix}. \tag{18}$$

The R-symmetry generators are

$$R_a = \begin{pmatrix} -2\gamma_a(\lambda_l x_i\partial_i) & 0 \\ 0 & \gamma_a(1-2\lambda+2lx_i\partial_i) \end{pmatrix}, \quad \bar{R}_a = \begin{pmatrix} \gamma_a & 0 \\ 0 & 0 \end{pmatrix}. \tag{19}$$

The Abelian subalgebra generators and the $so(d)$ generators have the same form as in $\mathcal{G}_{(2,2)}$. The odd generators $X^{(n)}_{a,i}$ are of the form

$$X^{(n)}_{a,i} = \begin{pmatrix} 0 & 0 \\ -\gamma_a & 0 \end{pmatrix} t^n\partial_i, \quad X_{4,i} = \begin{pmatrix} 0 & 0 \\ 1 & 0 \end{pmatrix} t^n\partial_i. \tag{20}$$

The even generators induced by the Abelian subalgebra are of two types

$$J^{(n)}_a = t^n\partial_i\begin{pmatrix} \gamma_a & 0 \\ 0 & \gamma_a \end{pmatrix}, \quad W^{(n)}_a = t^n\partial_i\begin{pmatrix} -\gamma_a & 0 \\ 0 & \gamma_a \end{pmatrix} \tag{21}$$

These generators are analogous to the $\mathcal{G}_{(2,2)}$ superalgebra generators. However, the $\mathcal{G}_{(4,4)}$ algebra does not close as a finite Lie algebra.

The non-vanishing (anti)commutators of the generators presented above close a non-linear W-algebra together with an L generator $L = \begin{pmatrix} \lambda + lx_i\partial_i & 0 \\ 0 & \lambda + lx_i\partial_i \end{pmatrix}$. This W-algebra is at most quadratic in the generators. The nonvanishing relations are

$$\begin{aligned}
[D,H] &= H, \quad [D,K] = -K, \quad [H,K] = 2D, \\
[D,Q_A] &= \frac{1}{2}Q_A, \quad [K,Q_A] = S_A, \quad [H,S_A] = Q_A, \quad [D,S_A] = -\frac{1}{2}S_A, \\
\{Q_A,Q_B\} &= 2\delta_{AB}H, \quad \{S_A,S_B\} = -2\delta_{AB}K, \\
\{Q_4,S_4\} &= -2D, \quad \{Q_4,S_a\} = R_a, \quad \{Q_a,S_4\} = -R_a, \quad \{Q_a,S_b\} = -2\delta_{ab}D - \epsilon_{abc}(R_c + 4L\overline{R}_c), \\
[R_a,Q_4] &= -Q_a, \quad [R_a,Q_b] = \delta_{ab}Q_4 + \epsilon_{abc}(1-4L)Q_c, \\
[R_a,S_4] &= -S_a, \quad [R_a,S_b] = \delta_{ab}S_4 + \epsilon_{abc}(1-4L)S_c, \\
[\overline{R}_a,Q_4] &= Q_a, \quad [\overline{R}_a,Q_b] = -\delta_{ab}Q_4 + \epsilon_{abc}Q_c, \\
[\overline{R}_a,S_4] &= S_a, \quad [\overline{R}_a,S_b] = -\delta_{ab}S_4 + \epsilon_{abc}S_c, \\
[R_a,R_b] &= \epsilon_{abc}[2(1-2L)R_c + 4L\overline{R}_c], \quad [R_a,\overline{R}_b] = -\epsilon_{abc}4L\overline{R}_c, \quad [\overline{R}_a,\overline{R}_b] = 2\epsilon_{abc}\overline{R}_c.
\end{aligned} \tag{22}$$

These (anti)commutators correspond to the exceptional algebra $D(2,1;\alpha)$ for a specific relation connecting α and L. Let us consider for simplicity $d = 1$. One should note that the generator L commutes with all the generators of $D(2,1;\alpha)$. Therefore L entering (22) can be taken as a c-number, making this subalgebra a linear Lie superalgebra.

The remaining non-vanishing (anti)commutators are given by

$$\begin{aligned}
[H,P^{(n)}] &= nP^{(n-1)}, \quad [D,P^{(n)}] = -(n-\ell)P^{(n)}, \quad [K,P^{(n)}] = -(n-2\ell)P^{(n+1)}, \\
[Q_A,P^{(n)}] &= nP^{(n-1)}X_A, \quad [S_A,P^{(n)}] = (n-2\ell)P^{(n)}X_A, \\
[R_a,P^{(n)}] &= 2\ell P^{(n)}J_a,
\end{aligned}$$

$$\begin{aligned}
[D, X_A] &= -\frac{1}{2}X_A, \quad [K, X_A] = -TX_A, \\
\{Q_4, X_4\} &= J, \quad \{Q_4, X_a\} = W_a, \quad \{Q_a, X_4\} = -W_a, \quad \{Q_a, X_b\} = \delta_{ab}J + \epsilon_{abc}W_c, \\
\{S_4, X_4\} &= TJ, \quad \{S_4, X_a\} = TW_a, \quad \{S_a, X_4\} = -TW_a, \quad \{S_a, X_b\} = \delta_{ab}TJ - \epsilon_{abc}TJ_c, \\
[R_a, X_4] &= -X_a, \quad [R_a, X_b] = \delta_{ab}X_4 + \epsilon_{abc}(1-4L)X_c, \\
[\overline{R}_a, X_4] &= X_a, \quad [\overline{R}_a, X_b] = -\delta_{ab}X_4 + \epsilon_{abc}X_c, \\
[Q_a, J_b] &= 2\epsilon_{abc}Q_c, \quad [S_a, J_b] = 2\epsilon_{ab}S_c, \quad [R_a, J_b] = 2\epsilon_{abc}R_c, \quad [\overline{R}_a, J_b] = 2\epsilon_{abc}\overline{R}_c, \\
[X_a, J_b] &= 2\epsilon_{abc}X_c, \quad [J_a, J_b] = 2\epsilon_{abc}J_c, \\
[Q_A, J] &= 2Q_A - 4Y_A, \quad [S_A, J] = 2S_A - 4TY_A, \quad [X_A, J] = 2X_A
\end{aligned} \tag{23}$$

and

$$\begin{aligned}
[D, Y_A] &= \frac{1}{2}Y_A, \quad [K, Y_A] = TY_A, \\
\{Q_4, Y_4\} &= H, \quad \{Q_a, Y_4\} = -HJ_a, \quad \{S_4, Y_4\} = -D + L + \frac{1}{4}(J - C), \\
\{S_a, Y_4\} &= [D - L - \frac{1}{4}(J - C)]J_a, \\
[R_a, Y_4] &= -Y_a, \quad [\overline{R}_a, Y_4] = Y_a, \\
\{X_4, Y_4\} &= C, \quad \{Q_4, Y_a\} = HJ_a, \quad \{Q_a, Y_b\} = \delta_{ab}H - \epsilon_{abc}W_cH, \\
\{S_0, Y_a\} &= [-D + L + \frac{1}{4}(J - C)]J_a, \\
\{S_a, Y_b\} &= [-D + L + \frac{1}{4}(J - C)](\delta_{ab} - \epsilon_{abc}J_c), \\
[R_a, Y_b] &= \delta_{ab}Y_0 - \epsilon_{abc}Y_c, \quad [\overline{R}_a, Y_b] = -\delta_{ab}Y_0 + \epsilon_{abc}Y_c, \\
\{X_4, Y_a\} &= J_a, \quad \{X_a, Y_b\} = \delta_{ab}C - \epsilon_{abc}J_c, \quad [J_a, Y_b] = 2\epsilon_{abc}Y_c, \\
[H, T] &= C, \quad [D, T] = -T, \quad [K, T] = -T^2, \\
[Q_A, T] &= X_A, \quad [S_A, T] = TX_A, \quad [P^{(n)}, L] = \ell P^{(n)},
\end{aligned} \tag{24}$$

where $W_a = J_a - 2\overline{R}_a$, $A = 1, 2, 3, 4$, $Y_4 = \begin{pmatrix} 0 & 1 \\ 0 & 0 \end{pmatrix}$, $Y_a = \begin{pmatrix} 0 & \gamma_a \\ 0 & 0 \end{pmatrix}$, $C = \mathbf{1}_8$, $T = t \cdot \mathbf{1}_8$. All other (anti)commutators are vanishing.

It seems that, for generic values of d and l, the exists an obstruction in obtaining finitely generated N-extended superconformal Galilei algebras of Lie type for $N > 2$. The possibility to get finitely generated Lie superalgebras for special values of d and l, beyond the cases of [9] and [10], is currently investigated.

Acknowledgments

This research was supported by the FAPESP under Grant 2014/08397-0.

References

[1] Henkel M 1994 *J. Statist. Phys.* **75** 1023; Mehen T, Stewart I W and Wise M B 2000 *Phys. Lett.* B **474** 145
[2] Son D T and Wingate M 2006 *Annals Phys.* **321** 197
[3] Son D T 2008 *Phys. Rev.* D **78** 046003
[4] Balasubramanian K and McGreevy J 2008 *Phys. Rev. Lett.* **101** 061601; Goldberger W D 2009 *JHEP* **0903** 069; Barbon J L B and Fuertes C A 2008 *JHEP* **0809** 030
[5] Bacry H and Lévy-Leblond J 1968 *J. Math. Phys.* **9** 1605
[6] Negro J, del Olmo M A and Rogriguez-Marco A 1997 *J. Math. Phys.* **38** 3786

Journal of Physics: Conference Series **597** (2015) 012053 doi:10.1088/1742-6596/597/1/012053

[7] Aizawa N, Chandrashekar R and Segar J (arXiv:1408.4842); Galajinsky A and Masterov I 2011 *Phys. Lett.* B **702** 265; Galajinsky A and Masterov I 2013 *Nucl. Phys.* B **866** 212; Andrzejewski K, Gonera J and Maslanka P 2012 *Phys. Rev.* D **86** 065009

[8] Aizawa N, Kimura Y and Segar J 2013 *J. Phys.* A **46** 405204; Lu R, Mazorchuk V and Zhao K 2014 *J. Pure Appl. Algebra* **218** 1885; Duval C and Horvathy P 2011 *Journal of Physics* A **44** 335203

[9] On Schrdinger superalgebras Duval C and Horvathy P A 1994 *J. Math. Phys.* **35** 2516

[10] Fedoruk S and Lukierski J 2011 *Phys. Rev.* D **84** 065002

[11] Masterov I 2012 *J. Math. Phys.* **53** 072904

[12] Galajinsky A and Masterov I 2009 *Phys. Lett.* B **675** 116; Masterov I (arXiv:1410.5335)

[13] Aizawa N 2012 *J. Phys.* A **45** 475203

[14] Aizawa N, Kuznetsova Z and Toppan F 2013 *J. Math. Phys.* **54** 093506

[15] Kuznetsova Z, Rojas M and Toppan F 2006 *JHEP* **0603** 098

[16] Kuznetsova Z 2008 *ROMP* **61** 295

[17] Kuznetsova Z and Toppan F 2012 *J. Math. Phys.* **53** 043513

[18] Khodaee S and Toppan F 2012 *J. Math. Phys.* **53** 103518

Journal of Physics: Conference Series **597** (2015) 012054 doi:10.1088/1742-6596/597/1/012054

Emergent symmetries in atomic nuclei from first principles

K D Launey, A C Dreyfuss, R B Baker, J P Draayer and T Dytrych

Department of Physics and Astronomy, Louisiana State University, Baton Rouge, LA 70803, USA

E-mail: kristina@baton.phys.lsu.edu

Abstract. An innovative symmetry-guided approach and its applications to light and intermediate-mass nuclei is discussed. This approach, with $\mathrm{Sp}(3,\mathbb{R})$ the underpinning group, is based on our recent remarkable finding, namely, we have identified the symplectic Sp(3,R) as an approximate symmetry for low-energy nuclear dynamics. This study presents the results of two complementary studies, one that utilizes realistic nucleon-nucleon interactions and unveils symmetries inherent to nuclear dynamics from first principles (or *ab initio*), and another study, which selects important components of the nuclear interaction to explain the primary physics responsible for emergent phenomena, such as enhanced collectivity and alpha clusters. In particular, within this symmetry-guided framework, *ab initio* applications of the theory to light nuclei reveal the emergence of a simple orderly pattern from first principles. This provides a strategy for determining the nature of bound states of nuclei in terms of a relatively small fraction of the complete shell-model space, which, in turn, can be used to explore ultra-large model spaces for a description of alpha-cluster and highly deformed structures together with associated rotations. We find that by using only a fraction of the model space extended far beyond current no-core shell-model limits and a long-range interaction that respects the symmetries in play, the outcome reproduces characteristic features of the low-lying 0^+ states in ^{12}C (including the elusive Hoyle state of importance to astrophysics) and agrees with *ab initio* results in smaller spaces. For these states, we offer a novel perspective emerging out of no-core shell-model considerations, including a discussion of associated nuclear deformation, matter radii, and density distribution. The framework we find is also extensible beyond ^{12}C, namely, to the low-lying 0^+ states of ^{8}Be as well as the ground-state rotational band of Ne, Mg, and Si isotopes.

1. Introduction

Approximate symmetries in atomic nuclei that favor large deformation along with monopole and quadrupole excitations thereof, associated with the SU(3) group and the symplectic $\mathrm{Sp}(3,\mathbb{R})$ group, have been long recognized in selected cases [1–10]. But only recently, we have confirmed the existence of these symmetries, within the *ab initio* (or "from first principles") symmetry-adapted no-core shell model (SA-NCSM) framework [11]. This study unveiled the orderly patterns associated with such symmetries in nuclear wavefunctions from first principles, without *a priori* symmetry constraints. This suggests a novel theoretical framework for nuclear structure modeling that, in addition to the exact conventional symmetries, such as parity and rotational invariance, can take advantage of these approximate symmetries. The latter are utilized and further understood in the framework of the microscopic no-core symplectic model (NCSpM) [12], which combines a long-range many-nucleon interaction that respects the symmetries in play together with a symmetry-mixing spin-orbit term. The findings of the present study point to a new insight, namely, understanding the mechanism and the primary physics responsible for the emergence of simple structures in complex nuclei from a no-core shell-model perspective.

One of the most successful particle-driven models is the *ab initio* no-core shell model (NCSM) [13], which can accommodate any type of inter-nucleon interactions, including modern two- and three-nucleon realistic interactions. Specifically, for a general problem, the NCSM adopts the intrinsic non-relativistic nuclear plus Coulomb interaction Hamiltonian defined as follows:

$$H = T_{\rm rel} + V_{NN} + V_{NNN} + \ldots + V_{\rm Coulomb}, \qquad (1)$$

where the V_{NN} nucleon-nucleon interaction is included along with the $V_{\rm Coulomb}$ Coulomb interaction between the protons. The Hamiltonian may include a V_{NNN} 3-nucleon interaction and additional terms such as multi-nucleon interactions among more than three nucleons simultaneously and higher-order electromagnetic interactions such as magnetic dipole-dipole terms. It adopts the harmonic oscillator (HO) single-particle basis characterized by the $\hbar\Omega$ oscillator strength (or equivalently, the oscillator length $b = \sqrt{\frac{\hbar}{m\Omega}}$ for a nucleon mass m) and retains many-body basis states of a fixed parity, consistent with the Pauli principle, and limited by a cutoff $N_{\rm max}$. The $N_{\rm max}$ cutoff is defined as the maximum number of HO quanta allowed in a many-body basis state above the minimum for a given nucleus. It divides the space in "horizontal" HO shells and is dictated by particle-hole excitations (this is complementary to the NCSpM, which divides the space in vertical slices selected by collectivity-driven rules). It seeks to obtain the lowest few eigenvalues and eigenfunctions of the Hamiltonian (1). The NCSM has achieved remarkable descriptions of low-lying states from the lightest nuclei up through ^{12}C, ^{16}O, and ^{14}F [13, 14], and is further augmented by several techniques, such as NCSM/RGM [15], Importance Truncation NCSM [16] and Monte Carlo NCSM [17]. This supports and complements results of other first-principle approaches, such as Green's function Monte Carlo (GFMC) [18], Coupled-cluster (CC) method [19], In-Medium SRG [20], and Lattice Effective Field Theory (EFT) [21].

However, the established NCSM methods are not able to reach the physics regime necessary for a description of largely deformed nuclear states, such as the ^{12}C Hoyle state that was predicted based on observed abundances of heavy elements in the universe [22], and which has attracted much recent attention both in theory (e.g., see [21, 23, 24]) and experiment ([25–32]). In this study, we address this problem, within a no-core shell-model framework, by utilizing a small subset of symplectic Sp(3, $\mathbb{R}$) basis states [6, 7] (with the complete set yielding results equivalent to those of the NCSM), an Sp(3, $\mathbb{R}$)-preserving part of the long-range inter-nucleon interaction [12], and an important symmetry-breaking spin-orbit term.

2. Discovery of a highly structured pattern within the first-principle SA-NCSM framework

The *ab initio* symmetry-adapted no-core shell model (SA-NCSM) [11] combines the first-principle concept of the NCSM with symmetry-guided considerations. For the first time, we show the emergence – in the framework of the SA-NCSM from first principles – of orderly patterns that favor large deformation/low spin in nuclear wavefunctions [11]. These patterns are linked to the SU(3) group and the symplectic Sp(3, $\mathbb{R}$) group, where Sp(3, $\mathbb{R}$)$\supset$SU(3). SU(3) describes the 3-dimensional harmonic oscillator and its irreducible representations (irreps) are specified by $(\lambda_\omega\, \mu_\omega)$ quantum numbers that can be related to the intrinsic quadrupole deformation [33–35] – e.g., $(0\,0)$, $(\lambda_\omega\, 0)$, and $(0\, \mu_\omega)$ describe spherical, prolate, and oblate shapes, respectively. The Sp(3, $\mathbb{R}$) group is described in detail in Section 3.

In particular, the *ab initio* $N_{\rm max} = 8$ SA-NCSM results with the bare N^{3}LO [36] (similarly, for JISP16 [37]) realistic interaction for the ^{8}Be 0^+ ground state (*gs*) and its rotational band

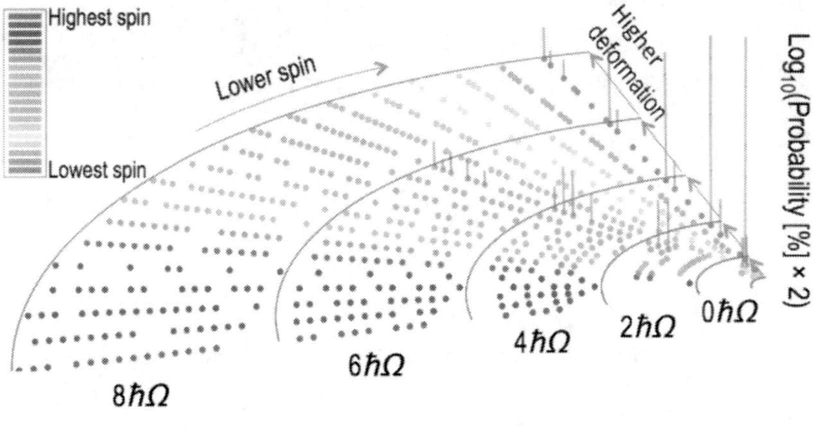

Figure 1. Probability distribution across Pauli-allowed intrinsic spin and deformation configurations for the *ab initio* SA-NCSM 0^+ *gs* of ^{8}Be for $N_{\max} = 8$ and $\hbar\Omega = 25$ MeV with the N^{3}LO interaction. The concentration of strengths to the far right demonstrates the dominance of collectivity in the calculated eigenstates. This novel feature enables the SA-NCSM – by using symmetry-dictated subspaces – to reach new domains currently inaccessible by *ab initio* calculations, such as isotopes of Ne, Mg, and Si.

reveal the dominance of the $0\hbar\Omega$ component with the foremost contribution coming from the leading $(\lambda_\omega\,\mu_\omega) = (4\,0)$ SU(3) irrep (Fig. 1). Furthermore, we find that important SU(3) configurations are then organized into structures according to the $\mathrm{Sp}(3,\mathbb{R})$ symplectic group, that is, the $(\lambda_\sigma\,\mu_\sigma) = (4\,0)$ $S=0$ symplectic irrep contains the $(\lambda_\omega\,\mu_\omega) = (4\,0)$ configuration in the $0\hbar\Omega$ subspace, $(\lambda_\omega\,\mu_\omega) = (6\,0)$, $(4\,1)$, and $(2\,2)$ configurations in the $2\hbar\Omega$ subspace, and so on. Those configurations, all part of a single symplectic irrep, the $S=0$ $(4\,0)$, indeed realize the major components of the wavefunction in the respective subspaces. Similar results are observed for other *p*-shell nuclei, such as ^{6}Li, ^{6}He, and ^{12}C. This further reveals the significance of the symplectic symmetry to nuclear dynamics. Moreover, the outcome supports a symmetry-guided concept [11], based on the dominance of only a few configurations, as evident by the concentration of probability to the far right in Fig. 1. That the relevant model space can be systematically selected, using a quantified cutoff, starting from the largest deformation and associated symplectic excitations thereof, and including ever smaller deformation until convergence of results is achieved, is a key feature of the SA-NCSM.

3. Symplectic $\mathrm{Sp}(3,\mathbb{R})$ group

The significance of the symplectic $\mathrm{Sp}(3,\mathbb{R})$ group for a microscopic description of a quantum many-body system of interacting particles has been recognized by Rowe and Rosensteel [6, 7]. Indeed, the 21 symplectic generators are directly related to the particle momentum ($p_{s\alpha}$) and coordinate ($r_{s\alpha}$) operators, with $\alpha = x$, y, and z for the 3 spatial directions and s labeling an individual nucleon, and realize important observables. Namely, $\mathrm{Sp}(3,\mathbb{R})$-preserving operators include: the many-particle kinetic energy $\sum_{s,\alpha} p_{s\alpha}^2/2m$, the HO potential, $\sum_{s,\alpha} m\Omega^2 r_{s\alpha}^2/2$, the mass quadrupole moment $Q_{(2\mathcal{M})} = \sum_s q_{(2\mathcal{M})s} = \sum_s \sqrt{16\pi/5}r_s^2 Y_{(2\mathcal{M})}(\hat{\mathbf{r}}_s)$, and angular momentum L operators, together with multi-shell collective vibrations and vorticity degrees of freedom for a description from irrotational to rigid rotor flows. Indeed, an important $\mathrm{Sp}(3,\mathbb{R})$-preserving interaction is $\frac{1}{2}Q\cdot Q = \frac{1}{2}\sum_s q_s \cdot (\sum_t q_t)$, as this realizes the physically relevant interaction of each particle with the total quadrupole moment of the nuclear system.

The symplectic Sp(3, $\mathbb{R}$) symmetry underpins the symplectic shell model, which provides a microscopic formulation of the Bohr-Mottelson collective model [1] and is a multiple oscillator shell generalization of the successful Elliott SU(3) model [2]. The symplectic model with Sp(3, $\mathbb{R}$)-preserving interactions has achieved a remarkable reproduction of rotational bands and transition rates without the need for introducing effective charges, while only a single Sp(3, $\mathbb{R}$) configuration is used [7, 9]. A shell-model study in a symplectic basis that allows for mixing of Sp(3, $\mathbb{R}$) configurations due to pairing and non-degenerate single-particle energies above a ^{16}O core [8] has found that using only seven Sp(3, $\mathbb{R}$) configurations is sufficient to achieve a remarkable reproduction of the ^{20}Ne energy spectrum as well as of $E2$ transition rates without effective charges.

3.1. The $\mathfrak{sp}(3,\mathbb{R})$ algebra and generators

As described in [38], the translationally invariant (intrinsic) symplectic generators can be written as SU(3) tensor operators in terms of the HO raising, $b^{\dagger(1\,0)}_{i\alpha} = \frac{1}{\sqrt{2}}(X_{i\alpha} - iP_{i\alpha})$, and lowering $b^{(0\,1)}_{i\alpha}$ dimensionless operators (with $\mathbf{X}$ and $\mathbf{P}$ the lab-frame position and momentum coordinates and $\alpha = 1, 2, 3$ for the three spatial directions) [7],

$$A^{(2\,0)}_{\mathcal{L}M} = \frac{1}{\sqrt{2}}\sum_{i=1}^{A}\left[b^\dagger_i \times b^\dagger_i\right]^{(2\,0)}_{\mathcal{L}M} - \frac{1}{\sqrt{2}A}\sum_{s,t=1}^{A}\left[b^\dagger_s \times b^\dagger_t\right]^{(2\,0)}_{\mathcal{L}M} \tag{2}$$

$$C^{(1\,1)}_{\mathcal{L}M} = \sqrt{2}\sum_{i=1}^{A}\left[b^\dagger_i \times b_i\right]^{(1\,1)}_{\mathcal{L}M} - \frac{\sqrt{2}}{A}\sum_{s,t=1}^{A}\left[b^\dagger_s \times b_t\right]^{(1\,1)}_{\mathcal{L}M}, \tag{3}$$

together with $B^{(0\,2)}_{\mathcal{L}M} = (-)^{\mathcal{L}-M}(A^{(2\,0)}_{\mathcal{L}-M})^\dagger$ ($\mathcal{L} = 0, 2$) and $H^{(00)}_{00} = \sqrt{3}\sum_i\left[b^\dagger_i \times b_i\right]^{(00)}_{00} - \frac{\sqrt{3}}{A}\sum_{s,t}\left[b^\dagger_s \times b_t\right]^{(0\,0)}_{00} + \frac{3}{2}(A-1)$, where the sums run over all A particles of the system. The eight generators $C^{(1\,1)}_{\mathcal{L}, M}$ ($\mathcal{L} = 1, 2$) close the $\mathfrak{su}(3)$ subalgebra of $\mathfrak{sp}(3,\mathbb{R})$. They realize the angular momentum operator:

$$L_{1M} = C^{(1\,1)}_{1M},\ M = 0, \pm 1, \tag{4}$$

and the Elliott algebraic quadrupole moment tensor $\mathcal{Q}^a_{2M} = \sqrt{3}C^{(1\,1)}_{2M}$, $M = 0, \pm1, \pm2$. The mass quadrupole moment can be constructed in terms of the symplectic generators as,

$$Q_{2M} = \sqrt{3}(A^{(2\,0)}_{2M} + C^{(1\,1)}_{2M} + B^{(0\,2)}_{2M}). \tag{5}$$

Equivalently, the symplectic generators, being one-body-plus-two-body operators can be expressed in terms of the creation operators $a^\dagger_{(\eta\,0)} = a^\dagger_\eta$, which create a nucleon in the HO shell $\eta = 0, 1, 2, \ldots$, together with its SU(3)-conjugate annihilation operator, $\tilde{a}_{(0\,\eta)}$. This is achieved by using the known matrix elements of the position ($\mathbf{X}_i$) and momentum ($\mathbf{P}_i$) operators in a HO basis, and hence, e.g., the first sum of $A^{(2\,0)}_{\mathcal{L}M}$ in Eq. (2) becomes, $\sum_\eta \sqrt{\frac{(\eta+1)(\eta+2)(\eta+3)(\eta+4)}{12}}\left[a^\dagger_{(\eta+2\,0)} \times \tilde{a}_{(0\,\eta)}\right]^{(2\,0)}_{\mathcal{L}M}$ [39]. Note that this operator describes excitations of a nucleon from the η shell to the $\eta+2$ shell, which corresponds to creating two single-particle HO excitation quanta, as manifested in the first term of Eq. (2).

Finally, the translationally invariant generators can be realized in terms of intrinsic coordinates, that is, particle position ($\mathbf{r}_i$) and momentum ($\mathbf{p}_i$) coordinates relative to the CM ($\sum_i \mathbf{r}_i = 0$ and $\sum_i \mathbf{p}_i = 0$), and, e.g., Eq. (2) becomes, $A^{(2\,0)}_{\mathcal{L}M} = \frac{1}{\sqrt{2}}\sum_i\left[(b^\dagger_i - \frac{1}{A}\sum_s b^\dagger_s) \times (b^\dagger_i - \frac{1}{A}\sum_t b^\dagger_t)\right]^{(2\,0)}_{\mathcal{L}M} \equiv \frac{1}{\sqrt{2}}\sum_{i=1}^{A}\left[\mathfrak{b}^\dagger_i \times \mathfrak{b}^\dagger_i\right]^{(2\,0)}_{\mathcal{L}M}$, with the dimensionless intrinsic operators, $\mathfrak{b}^{\dagger(1\,0)}_{i\alpha} = \frac{1}{\sqrt{2}}(r_{i\alpha} - ip_{i\alpha})$.

3.2. Irreducible representations and many-body basis states

A many-body basis state of a symplectic irrep (detailed in [7]) is labeled according to the group chain,

$$\begin{array}{ccccccc} \mathrm{Sp}(3,\mathbb{R}) & \supset & U(3) & \supset & \mathrm{SO}(3) & \supset & SO(2) \\ \sigma & n\rho & \omega & \kappa & L & & M \end{array} \tag{6}$$

and constructed by acting with symmetrically coupled polynomials in the symplectic raising operators, $A^{(2\,0)}$, on a unique symplectic bandhead configuration, $|\sigma\rangle$,

$$|\sigma n\rho\omega\kappa LM\rangle = \left[\left[A^{(2\,0)} \times A^{(2\,0)} \ldots \times A^{(2\,0)}\right]^n \times |\sigma\rangle\right]^{\rho\omega}_{\kappa LM}, \tag{7}$$

where $\sigma \equiv N_\sigma\,(\lambda_\sigma\,\mu_\sigma)$ labels the Sp(3, $\mathbb{R}$) irrep, $n \equiv N_n\,(\lambda_n\,\mu_n)$, $\omega \equiv N_\omega\,(\lambda_\omega\,\mu_\omega)$, and $N_\omega = N_\sigma + N_n$ is the total number of HO quanta (ρ and κ are multiplicity labels). This can be generalized to include spin, $|\sigma n\rho\omega\kappa(LS_\sigma)JM_J\rangle = \sum_{MM_S}\langle LM; S_\sigma M_S|\,JM_J\rangle|\sigma n\rho\omega\kappa LMS_\sigma M_S\rangle$, and also isospin. States within a symplectic irrep have the same spin (isospin) value, which is given by the spin S_σ (isospin T_σ) of the bandhead $|\sigma; S_\sigma(T_\sigma)\rangle$ [40]. Symplectic basis states span the entire shell-mode space. A complete set of labels includes additional quantum numbers $|\{\alpha\}\,\sigma\rangle$ that distinguish different bandheads with the same $N_\sigma\,(\lambda_\sigma\,\mu_\sigma)$. Sp(3, $\mathbb{R}$)-preserving Hamiltonians render energy spectra degenerate with respect to $\{\alpha\}$. However, for all present calculations for *gs* rotational bands and associated observables, $\{\alpha\}$ is unique (an only set).

The symplectic structure accommodates relevant particle-hole (p-h) configurations in a natural way (see also Fig. 1 of Ref. [12]). According to Eq. (7), the basis states of an Sp(3, $\mathbb{R}$) irrep are built over a bandhead $|\sigma\rangle$ by $2\hbar\Omega$ 1p-1h (one particle raised by two shells) monopole ($\mathfrak{L} = 0$) or quadrupole ($\mathfrak{L} = 2$) excitations, realized by the first term in $A^{(2\,0)}_{\mathfrak{L}M}$ of Eq. (2), together with a smaller $2\hbar\Omega$ 2p-2h correction for eliminating the spurious center-of-mass motion, realized by the second term in $A^{(2\,0)}_{\mathfrak{L}M}$. The symplectic bandhead $|\sigma\rangle$ is the lowest-weight Sp(3, $\mathbb{R}$) state, which is defined by the usual requirement that the symplectic lowering operators annihilate it. The bandhead, $|\sigma; \kappa_\sigma L_\sigma M_\sigma\rangle$, is an SU(3)-coupled many-body state with a given nucleon distribution over the HO shells and while not utilized here, can be obtained in terms of the above-mentioned creation operators $a^\dagger_{(\eta\,0)} = a^\dagger_\eta$. E.g., for a $0\hbar\Omega$ bandhead, the nucleon distribution is a single configuration,

$$\left[a^\dagger_{(\eta_1\,0)} \times a^\dagger_{(\eta_2\,0)} \times \ldots \times a^\dagger_{(\eta_A\,0)}\right]^{(\lambda_\sigma\,\mu_\sigma)}_{\kappa_\sigma L_\sigma M_\sigma} |0\rangle \tag{8}$$

with $N_\sigma = \eta_1 + \eta_2 + \ldots + \eta_A + \frac{3}{2}(A-1)$, such that $N_\sigma\hbar\Omega$ includes the HO zero-point energy. Note that $3/2$ is subtracted from N_σ to ensure a proper treatment of the CM – in addition to this, the NCSpM uses translationally invariant symplectic generators that can be expressed in $\mathbf{r}_i$ and $\mathbf{p}_i$ relative to the CM; these symplectic generators are used to build the basis, the interaction, the many-particle kinetic energy operator, as well as to evaluate observables.

An example for the basis states within a symplectic irrep follows for ^{24}Mg. Its lowest HO-energy configuration is given by $N_\sigma = 62.5$ or $0\hbar\Omega$ (no HO excitation quanta), while the $4\hbar\Omega$ (20 0) symplectic irrep includes:

(i) A bandhead ($N_n = 0$) with $N_\sigma = 66.5$ (or $4\hbar\Omega$) and $(\lambda_\sigma\,\mu_\sigma) = (20\,0)$;

(ii) $N_n = 2$ states with N_ω=68.5 and $(\lambda_\omega\,\mu_\omega) = (22\,0)$, $(20\,1)$, and $(18\,2)$;

(iii) and so forth for higher N_n.

For each $(\lambda_\omega\,\mu_\omega)$, the quantum numbers κ, L and M are given by Elliott [2, 4]. E.g., for (22 0), $\kappa = 0$, $L = 0, 2, 4, \ldots, 22$, and $M = -L, -L+1, \ldots, L$.

3.3. Reduced matrix elements of the symplectic generators

The SU(3)-reduced matrix elements of the Sp(3, ℝ) generators are analytically known [7,41–44]. The steps to compute $\langle \sigma n_f \rho_f \omega_f \| A^{(20)} \| \sigma n_i \rho_i \omega_i \rangle$, similarly for $B^{(02)}_{\mathcal{L}M} = (-)^{\mathcal{L}-M}(A^{(20)}_{\mathcal{L}-M})^\dagger$, are outlined in what follows:

(i) Calculations of non-normalized matrix elements $(n_f \| \mathcal{A}^{(20)} \| n_i)$ (with $A^{(20)} \to \mathcal{A}^{(20)}$ in the contraction u(3) [Weyl] limit) using Eq. (4.51) of Ref. [7] (see also, [41]) with $n_1 = \frac{N_n+2\lambda_n+\mu_n}{3}$, $n_2 = \frac{N_n-\lambda_n+\mu_n}{3}$, and $n_3 = \frac{N_n-\lambda_n-2\mu_n}{3}$ associated with $n_i = N_{n,i}\,(\lambda_{n,i}\,\mu_{n,i})$ and n_f (note that Ref. [7] uses '$a^\dagger$' instead of '$\mathcal{A}^{(20)}$', not to be confused with the creation fermion operator referenced above);

(ii) Calculations of non-normalized $(\sigma n_f \rho_f \omega_f \| \mathcal{A}^{(20)} \| \sigma n_i \rho_i \omega_i)$ from $(n_f \| \mathcal{A}^{(20)} \| n_i)$ using Eq. (4.50) of Ref. [7];

(iii) Calculations of $\langle \sigma n_f \rho_f \omega_f \| A^{(20)} \| \sigma n_i \rho_i \omega_i \rangle$ from the non-normalized reduced matrix elements (step 2) using the $\mathcal{K}$-matrix approach [42,43]. The present calculations utilize the full $\mathcal{K}$ matrix (exact calculations). However, in the multiplicity-free case ($\rho_i^{\max} = \rho_f^{\max} = 1$) or in the limit of large σ [45], the normalization matrix reduces to normalization coefficients (a diagonal $\mathcal{K}$ matrix) given by Eq. (17) of Ref. [45].

For the $C^{(11)}_{\mathfrak{L}M}$ SU(3)-reduced matrix elements, see, e.g., Eq. (19) of Ref. [44]. Using the reduced matrix elements of the Sp(3, ℝ) generators and the relation (5), the analytical formula for the $Q \cdot Q$ matrix elements has been derived in Ref. [44].

For example, Sp(3, ℝ)-preserving Hamiltonians can include the many-particle kinetic energy:

$$\frac{T}{\hbar\Omega} = \frac{1}{\hbar\Omega}\sum_i \frac{\mathbf{p}_i^2}{2m} = \frac{1}{2}H^{(00)}_{00} - \sqrt{\frac{3}{8}}(A^{(20)}_{00} + B^{(02)}_{00}), \tag{9}$$

the HO potential:

$$\frac{V_{HO}}{\hbar\Omega} = \frac{1}{\hbar\Omega}\sum_i \frac{m\Omega^2\mathbf{r}_i^2}{2} = \frac{1}{2}H^{(00)}_{00} + \sqrt{\frac{3}{8}}(A^{(20)}_{00} + B^{(02)}_{00}), \tag{10}$$

as well as terms dependent on L, see Eq. (4), and Q, see Eq. (5). These interactions have analytical matrix elements in the symplectic basis (7) and act within a symplectic irrep ($\sigma_f = \sigma_i \equiv \sigma$). For example, for the dimensionless many-particle kinetic energy, $\frac{T}{\hbar\Omega}$, the matrix elements are given as:

$$\begin{aligned}
&\left\langle \sigma n_f \rho_f \omega_f \kappa_f L_f M_f \middle| \frac{T}{\hbar\Omega} \middle| \sigma n_i \rho_i \omega_i \kappa_i L_i M_i \right\rangle = \\
&\frac{1}{2}\left\langle \sigma n_f \rho_f \omega_f \kappa_f L_f M_f \middle| H^{(00)}_{00} \middle| \sigma n_i \rho_i \omega_i \kappa_i L_i M_i \right\rangle \\
&-\sqrt{\frac{3}{8}}\left\langle \sigma n_f \rho_f \omega_f \kappa_f L_f M_f \middle| A^{(20)}_{00} + B^{(02)}_{00} \middle| \sigma n_i \rho_i \omega_i \kappa_i L_i M_i \right\rangle = \\
&\frac{1}{2}N_\omega \delta_{f,i} - \sqrt{\frac{3}{8}}\Big(\langle \omega_i \kappa_i L_i M_i; (2\,0)00 | \,\omega_f \kappa_f L_f M_f \rangle \langle \sigma n_f \rho_f \omega_f \| A^{(20)} \| \sigma n_i \rho_i \omega_i \rangle + \text{conjugate} \Big),
\end{aligned} \tag{11}$$

where $\langle \omega_i \kappa_i L_i M_i; (2\,0)00 | \,\omega_f \kappa_f L_f M_f \rangle$ is the SU(3) Clebsch-Gordan coefficient and the reduced matrix elements of the symplectic generators are calculated according to the steps outlined above.

4. Utilizing approximate symmetries in nuclear modeling

We employ the no-core symplectic model (NCSpM), outlined in Ref. [9], with a novel interaction that is effectively realized by an exponential dependence on the quadrupole-quadrupole ($Q.Q$) two-body interaction [12,38]. This introduces simple but important many-body interactions that enter in a prescribed hierarchical way given in powers of a small parameter, the only adjustable parameter in the model. The model offers a microscopic no-core shell-model description of nuclei in terms of mixed deformed configurations and allows the inclusion of higher-energy particle excitations [12] that are currently inaccessible by *ab initio* shell models. It reduces to the successful Elliott model [2,4] in the limit of a single valence shell and a zero model parameter.

The underlying groups of the NCSpM are the symplectic $\mathrm{Sp}(3,\mathbb{R})$ group [6,7] and its SU(3) subgroup [2,4]. The symplectic basis, described above and utilized in NCSpM, is related, via a unitary transformation (see the review [40]), to the three-dimensional HO (m-scheme) many-body basis used in the NCSM [13]. Indeed, the NCSpM employed within a complete model space up through $N_{\rm max}$, will coincide with the NCSM for the same $N_{\rm max}$ cutoff. Important features of the NCSpM model are (1) the $\mathrm{Sp}(3,\mathbb{R})$ irreps divide the space into 'vertical cones' that are comprised of basis states of good $(\lambda\,\mu)$ quantum numbers, and (2) its ability to down-select to the most relevant configurations, which are chosen among all possible $\mathrm{Sp}(3,\mathbb{R})$ irreps within an $N_{\rm max}$ model space.

4.1. Schematic many-nucleon interaction

As discussed in Ref. [12], we employ a microscopic many-body interaction, which allows for large $N_{\rm max}$ no-core shell-model applications. This interaction utilizes two central components: a single-particle piece, consisting of the harmonic oscillator potential and a spin-orbit term, together with a collective piece, which enters through the use of the quadrupole-quadrupole interaction. Specifically, we utilize an elementary form tied to a long-range expansion of the nucleon-nucleon central force $V(|\mathbf{r}_i-\mathbf{r}_j|)$ [46], simplified by considering the most relevant degrees of freedom necessary to describe deformed spatial configurations,

$$H_\gamma = \sum_{i=1}^{A}\left(\frac{\mathbf{p}_i^2}{2m}+\frac{m\Omega^2\mathbf{r}_i^2}{2}-\kappa l_i\cdot s_i\right) + \frac{\chi}{2}\frac{\left(e^{-\gamma(Q\cdot Q-\langle Q\cdot Q\rangle_{N_n})}-1\right)}{\gamma}, \tag{12}$$

where $\hbar\Omega$, κ, and χ are parameters, for which we use empirical estimates, while $\gamma\geq 0$ is the only adjustable parameter in the model (as discussed below). H_γ is given in terms of particle momentum and position coordinates relative to the center of mass. The average contribution $\langle Q\cdot Q\rangle_{N_n}$ of $Q\cdot Q$ for given N_n HO excitations [47] introduces a considerable renormalization of the HO shell structure and hence, is removed in multi-shell studies [35].

We use $\chi=\hbar\Omega/(4\sqrt{N_{\omega(f)}N_{\omega(i)}})$ for a $\langle f|H_\gamma|i\rangle$ matrix element for final (f) and initial (i) many-body states. The decrease of χ with N_ω, to leading order in λ/N_ω, has been shown by Rowe [48] based on self-consistent arguments. We also use the empirical estimates $\hbar\Omega\approx 41/A^{1/3}=18$ MeV and $\kappa\approx 20/A^{2/3}=3.8$ MeV (e.g., see [1]). The only adjustable parameter of the NCSpM model is γ, which controls the presence of the many-body interactions and thus cannot be informed by existing two-body and three-body interactions. The effective interaction (12) introduces hierarchical many-body interactions in a prescribed way, e.g.,

$$\frac{\chi}{2\gamma}(e^{-\gamma Q\cdot Q}-1)=-\frac{1}{2}[\chi(\sum_{k=0}^{\infty}\frac{(-\gamma)^k(Q\cdot Q)^k}{(k+1)!})]Q\cdot Q, \tag{13}$$

such that they become quickly negligible for a reasonably small $\gamma\ll 1$. E.g., we find that for ^{12}C, besides $Q\cdot Q$, only one term is sufficient for the ground-state band ($k=1$), while three

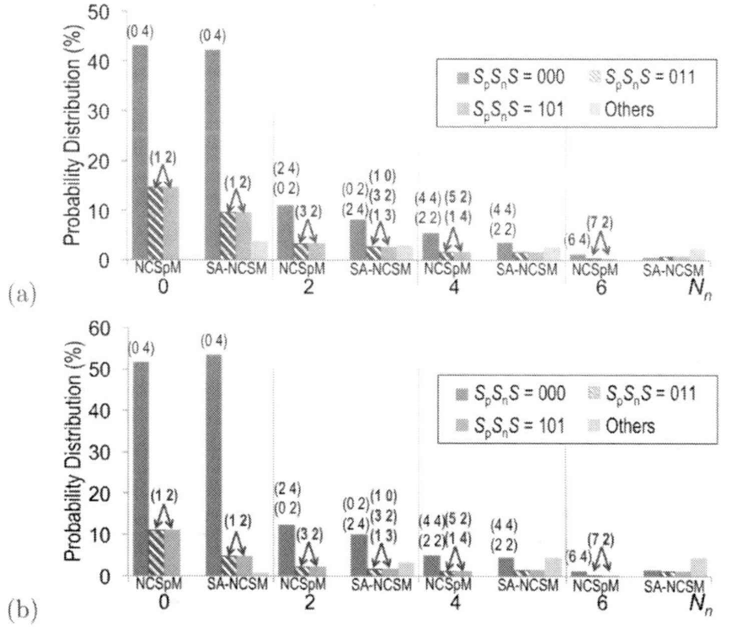

Figure 2. Probability distribution for ^{12}C as a function of the N_n total excitations of (a) the lowest 0^+ state and (b) the lowest 4^+ state as calculated by the NCSpM with H_γ (left set of bars within each N_n value) and the *ab initio* SA-NCSM with the bare JISP16 NN interaction (right set of bars within each N_n value). Both models are limited to an $N_{\max} = 6$ model space for comparison. The dominant shape deformations, specified by $(\lambda\,\mu)$, are shown. Very similar results are obtained for the lowest 2^+ state.

terms ($k = 3$) are sufficient for the Hoyle-state band [12]. This ties directly to the interaction used in Ref. [49], which is given as a polynomial in Q, namely, $Q \cdot Q$, $(Q \times Q) \cdot Q$, and $(Q \cdot Q)^2$, and applied to the ^{24}Mg *gs* rotational band.

4.2. Comparison to ab initio *no-core shell model*

A comparison of the present NCSpM results for ^{12}C to *ab initio* outcomes is possible in smaller model spaces, for example, for the *gs* rotational band in $N_{\max} = 6$. This space appears to be reasonable for these states for both models. In particular, we compare to wavefunctions obtained in the SA-NCSM [50] with the bare JISP16 realistic interaction [37]. Consistent with the outcome of Refs. [50] and [11] (see, e.g., Fig. 1 in Ref. [11] for ^{6}Li and ^{8}Be wavefunctions in $N_{\max} = 8-10$, as well as Fig. 1 above), the *ab initio* $N_{\max} = 6$ SA-NCSM results for the 0^+ *gs*, first 2^+ and first 4^+ states of ^{12}C reveal the dominance of the $0\hbar\Omega$ component with the foremost contribution coming from the leading $(0\,4)$ $S = 0$ irrep (see Fig. 2 for the *gs* and the 4^+_1 state). Furthermore, we find that important SU(3) configurations are then organized into structures according to the Sp$(3, \mathbb{R})$ symplectic group, that is, the $(0\,4)$ symplectic irrep gives rise to dominant $(0\,2)$ and $(2\,4)$ configurations in the $2\hbar\Omega$ subspace and so on (see Fig. 2), similarly to the discussion above for the ^{8}Be ground state. The next important configuration is spin-1 $(1\,2)$ at $0\hbar\Omega$ with the associated symplectic excitations (Fig. 2). Therefore, among all possible configurations present in the SA-NCSM, only the states of the $(0\,4)$ and then $(1\,2)$ symplectic irreps appear dominant.

In addition, we find a close similarity to *ab initio* results for the NCSpM wavefunctions of the gs rotational band, calculated with H_γ (12) for $\gamma = 0.74 \times 10^{-4}$ and symplectic irreps with bandheads (0 4) and (1 2) (Fig. 2). NCSpM and SA-NCSM calculations are performed for $\hbar\Omega = 18$ and $N_{\max} = 6$ model spaces. The close agreement of the probability distribution across the $N_n\hbar\Omega$ subspaces and of the SU(3) content of the wavefunctions for the two models indicates that the two $0\hbar\Omega$ bandheads of the NCSpM model space account for the part of the complete $N_{\max} = 6$ model space that is most relevant to the physics of the ^{12}C gs rotational band. If the SA-NCSM model space is reduced to only the spin components used in this study, $S_pS_nS = 000$, 011, and 101, NCSpM observables such as gs matter rms radius and $Q_{2_1^+}$ reproduce their *ab initio* counterparts as much as 80-90% and 70-90%, respectively, for the same $\hbar\Omega$ and $N_{\max} = 2$, 4 and 6. This suggests that the interaction used in NCSpM has effectively captured a major portion of the underlying physics of the realistic interaction important to the low-lying nuclear states [12].

4.3. Clustering and collectivity in ^{12}C

The NCSpM outcome reveals a quite remarkable agreement with the experiment (Fig. 3a). The results are shown for $N_{\max} = 20$, which we found sufficient to yield convergence. This $N_{\max}$ model space is further reduced by selecting the most relevant symplectic irreps, namely, five symplectic bandheads and all symplectic multiples thereof up through $N_{\max} = 20$ of total dimensionality of 6.6×10^3. The four bandheads are the spin-zero ($S = 0$) $0\hbar\Omega$ 0p-0h (0 4), $2\hbar\Omega$ 2p-2h (6 2), and $4\hbar\Omega$ 4p-4h (12 0) symplectic bandheads together with the $S = 1$ $0\hbar\Omega$ 0p-0h (1 2). In comparison to experiment, the outcome (Fig. 3a, dark colors) reveals that the lowest 0^+, 2^+, and 4^+ states of the 0p-0h symplectic irreps (Fig. 3a, blue) calculated with $\gamma = -1.71 \times 10^{-4}$ closely reproduce the gs rotational band, indicating clear oblate shapes (Fig. 3b, top). In addition, the calculated lowest states of the $4\hbar\Omega$ 4p-4h (12 0) irrep (Fig. 3a, red) are found to lie close to the Hoyle-state rotational band, revealing alpha-clustering and prolate shapes (Fig. 3b, bottom). Furthermore, the lowest 0^+ of the $2\hbar\Omega$ 2p-2h (6 2) irrep (Fig. 3a, green) is found to lie around the 10-MeV 0^+ resonance (third 0^+ state).

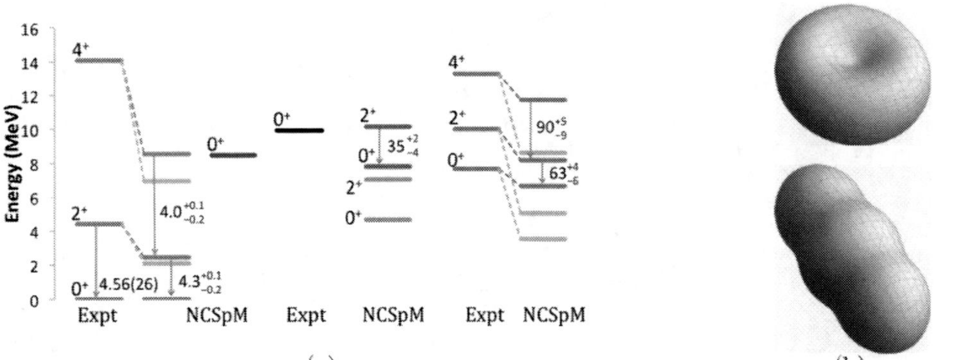

(a) (b)

Figure 3. (a) Energy spectrum for ^{12}C calculated using NCSpM in the $N_{\max} = 20$ model space of Sp(3, $\mathbb{R}$) irreps with $N_\sigma = 24.5$ (blue, left), $N_\sigma = 26.5$ (green, middle), and $N_\sigma = 28.5$ (red, right), and $(\lambda_\sigma\, \mu_\sigma) = (0\,4)$, $(6\,2)$, and $(12\,0)$ (light colors) and with the addition of the (1 2) Sp(3, $\mathbb{R}$) irrep (dark colors). Experimental data is from [51], except the latest results for 0_3^+ [28] and the states above the Hoyle state, 2^+ [31] and 4^+ [52]. $B(E2)$ transition rates are in W.u. with theoretical uncertainties estimated for a $\pm 60\%$ deviation of the Hoyle state energy. (b) NCSpM matter-density profile (with respect to the intrinsic frame) of the ^{12}C ground state (top) and the Hoyle state (bottom) showing the formation of three clusters in the Hoyle state within a no-core shell-model framework.

As compared to the model space consisting of the spin-0 irreps only (Fig. 3a, light colors), the addition of only one spin-1 irrep yields a NCSpM energy spectrum that is further improved and found to lie reasonably close to the experimental data. The Sp(3, $\mathbb{R}$)-nonpreserving spin-orbit term mixes the spin-0 (0 4) and spin-1 (1 2) irreps for all $J^\pi = 0^+$, 2_1^+, and 4_1^+, which results in a more realistic energy spacing between the excited states. Specifically, we see the *gs* separating from the higher-lying 0^+ states, and a slight stretching in the *gs* rotational band. This is similar to the findings of early cluster models, which remedy this by allowing for alpha-cluster dissociation due to a spin-orbit force as discussed in [24].

The outcome of the present analysis is not limited to ^{12}C. The model we find is also applicable to the low-lying states of other *p*-shell nuclei, such as ^{8}Be, as well as *sd*-shell nuclei without any adjustable parameters [38, 53]. In particular, using the same $\gamma = -1.71 \times 10^{-4}$ as determined for ^{12}C, we describe selected low-lying states in ^{8}Be in an $N_{\max} = 24$ model space with only 3 spin-zero $0\hbar\Omega$ (4 0), $2\hbar\Omega$ (6 0), and $4\hbar\Omega$ (8 0) symplectic irreps [53]. Furthermore, we have successfully applied the NCSpM without any adjustable parameters to the *gs* rotational band of heavier nuclei, such as ^{20}O, ^{20,22,24}Ne, ^{20,22}Mg, and ^{24}Si [38]. This suggests that the fully microscopic NCSpM model has indeed captured an important part of the physics that governs the low-energy nuclear dynamics and informs key features of the interaction and nuclear structure primarily responsible for the formation of such simple patterns.

In short, by utilizing approximate symmetries that are found to underpin nuclear dynamics, we offer a novel framework, based on the no-core shell model, to further understand highly deformed states, exemplified by the Hoyle state and its rotational band.

Acknowledgments

We thank D. Rowe and G. Rosensteel, as well as the PetaApps Collaboration, in particular, J. P. Vary and P. Maris, for useful discussions. This work was supported by the U.S. NSF (OCI-0904874), the U.S. DOE (DE-SC0005248), and SURA. ACD acknowledges support by the U.S. NSF (grant 1004822) through the REU Site in Dept. of Physics & Astronomy at LSU. We also acknowledge DOE/NERSC, LSU/LONI, and NSF & U. of Illinois/Blue Waters for providing HPC resources.

References

[1] Bohr A and Mottelson B R 1969 *Nuclear Structure* vol 1 (New York: Benjamin)
[2] Elliott J P 1958 *Proc. Roy. Soc. A* **245** 128
[3] Elliott J P 1958 *Proc. Roy. Soc. A* **245** 562
[4] Elliott J P and Harvey M 1962 *Proc. Roy. Soc. A* **272** 557
[5] Hecht K T 1971 *Nucl. Phys. A* **170** 34
[6] Rosensteel G and Rowe D J 1977 *Phys. Rev. Lett.* **38** 10
[7] Rowe D J 1985 *Reports on Progr. in Phys.* **48** 1419
[8] Draayer J, Weeks K and Rosensteel G 1984 *Nucl. Phys.* **A419** 1
[9] Bahri C and Rowe D J 2000 *Nucl. Phys. A* **662** 125
[10] Vargas C, Hirsch J and Draayer J 2001 *Nucl. Phys. A* **690** 409
[11] Dytrych T, Launey K D, Draayer J P, Maris P, Vary J P, Saule E, Catalyurek U, Sosonkina M, Langr D and Caprio M A 2013 *Phys. Rev. Lett.* **111** 252501
[12] Dreyfuss A C, Launey K D, Dytrych T, Draayer J P and Bahri C 2013 *Phys. Lett. B* **727** 511
[13] Navrátil P, Vary J P and Barrett B R 2000 *Phys. Rev. Lett.* **84** 5728
[14] Maris P, Shirokov A M and Vary J P 2010 *Phys. Rev. C* **81** 021301(R)
[15] Navrátil P, Quaglioni S, Stetcu I and Barrett B R 2009 *J. Phys. G: Nucl. Part.* **36** 083101
[16] Roth R and Navrátil P 2007 *Phys. Rev. Lett.* **99** 092501
[17] Abe T, Maris P, Otsuka T, Shimizu N, Utsuno Y and Vary J 2012 *Phys. Rev. C* **86** 054301
[18] Pieper S C, Varga K and Wiringa R B 2002 *Phys. Rev. C* **66** 044310
[19] Wloch M, Dean D J, Gour J R, Hjorth-Jensen M, Kowalski K, Papenbrock T and Piecuch P 2005 *Phys. Rev. Lett.* **94** 212501

[20] Tsukiyama K, Bogner S K and Schwenk A 2011 *Phys. Rev. Lett.* **106** 222502
[21] Epelbaum E, Krebs H, Lee D and Meissner U G 2011 *Phys. Rev. Lett.* **106** 192501
[22] Hoyle F 1954 *Astrophys. J. Suppl. Ser.* **1** 121
[23] Chernykh M, Feldmeier H, Neff T, von Neumann-Cosel P and Richter A 2007 *Phys. Rev. Lett.* **98** 032501
[24] Khoa D T, Cuonga D C and Kanada-En'yo Y 2011 *Phys. Letts. B* **695** 469
[25] Fynbo H O U *et al.* 2005 *Nature* **433** 136
[26] Freer M *et al.* 2009 *Phys. Rev. C* **80** 041303
[27] Hyldegaard S *et al.* 2010 *Phys. Rev. C* **81** 024303
[28] Itoh M *et al.* 2011 *Phys. Rev. C* **84** 054308
[29] Zimmerman W R, Destefano N E, Freer M, Gai M and Smit F D 2011 *Phys. Rev. C* **84** 027304
[30] Raduta A R *et al.* 2011 *Phys. Letts. B* **705** 65
[31] Zimmerman W R *et al.* 2013 *Phys. Rev. Lett.* **110** 152502
[32] Marin-Lambarri D J, Bijker R, Freer M, Gai M, Kokalova T, Parker D and Wheldon C 2014 *Phys. Rev. Lett.* **113** 012502
[33] Rosensteel G and Rowe D J 1977 *Ann. Phys. N.Y.* **104** 134
[34] Leschber Y and Draayer J P 1987 *Phys. Letts. B* **190** 1
[35] Castaños O, Draayer J P and Leschber Y 1988 *Z. Phys. A* **329** 33
[36] Entem D R and Machleidt R 2003 *Phys. Rev. C* **68** 041001
[37] Shirokov A, Vary J, Mazur A and Weber T 2007 *Phys. Lett. B* **644** 33
[38] Tobin G K, Ferriss M C, Launey K D, Dytrych T, Draayer J P and Bahri C 2014 *Phys. Rev. C* **89** 034312
[39] Escher J and Draayer J P 1999 *Phys. Rev. Lett.* **82** 5221
[40] Dytrych T, Sviratcheva K D, Draayer J P, Bahri C and Vary J P 2008 *J. Phys. G: Nucl. Part. Phys.* **35** 123101
[41] Rosensteel G and Rowe D J 1983 *J. Math. Phys.* **24** 2461
[42] Rowe D 1984 *J. Math. Phys.* **25** 2662
[43] Hecht K 1985 *J. Phys. A* **18** L1003
[44] Rosensteel G 1990 *Phys. Rev. C* **42** 2463
[45] Rowe D, Rosensteel G and Carr R 1984 *J. Phys. A: Math. Gen.* **17** L399
[46] Harvey M 1968 *Adv. Nucl. Phys.* **1** 67
[47] Rosensteel G and Draayer J P 1985 *Nucl. Phys. A* **436** 445
[48] Rowe D J 1967 *Phys. Rev.* **162** 866
[49] Peterson D and Hecht K 1980 *Nucl. Phys. A* **344** 361
[50] Dytrych T, Maris P, Launey K D, Draayer J P, Vary J P, Caprio M, Langr D, Catalyurek U and Sosonkina M 2014 *to be submitted to Phys. Rev. C*
[51] Ajzenberg-Selove F and Kelley J 1990 *Nucl. Phys. A* **506** 1
[52] Freer M *et al.* 2011 *Phys. Rev. C* **83** 034314
[53] Launey K D, Dytrych T, Draayer J P, Tobin G K, Ferriss M C, Langr D, Dreyfuss A C, Maris P, Vary J P and Bahri C 2013 Symmetry-adapted No-core Shell Model for Light Nuclei *Proceedings of the 5th International Conference on Fission and properties of neutron-rich nuclei, ICFN5, November 4 - 10, 2012, Sanibel Island, Florida, edited by J. H. Hamilton and A. V. Ramayya* (Singapore: World Scientific) p 29

Rotations of the most asymmetric molecules via 4-step and 1-step ladder operators

E Ley-Koo[1]
Instituto de Física, Universidad Nacional Autónoma de México
Apartado Postal 20-364, CP 01000 México, D.F., México

E-mail: eleykoo@fisica.unam.mx

Abstract. The symmetries of the free rotations of asymmetric molecules - at the levels of their respective Hamiltonians, spheroconal harmonic eigenfuntions, and eigenenergies - are reviewed in general. The situation of the most asymmetric molecules is analyzed in particular; their eigenstates with vanishing asymmetry-distribution eigenenergies are identified together with the Valdés and Piña 4-step ladder operators connecting them. Three sets of ladder operators identified by Méndez-Fragoso and Ley-Koo, connecting specific eigenstates of molecules with any symmetry, are also reviewed briefly. The successive applications of the 4-step ladder operators, to the lower eigenstates 1, y, xz, xyz, of the most asymmetric molecules, lead to their respective angular momentum $\ell = 4N, 4N+1, 4N+2, 4N+3$, for $N = 0,1,2...$, companions of the same species with vanishing asymmetry-distribution energy; in turn, the successive application of the cartesian components of the angular momentum L_x, L_y, L_z operators to the above states, for each value of ℓ, lead to the respective companion states with the same angular momentum and companion species [xy, xz, yz,1], [z, xyz, x, y], [yz, 1, xy, xz], [x, y, z, xyz] and complementary excitations, thus completing each set of $(2\ell+1)$ eigenstates, with symmetrically distributed energy levels.

1. Introduction

Kramers and Ittmann pioneered the study of the quantum rotations of asymmetric molecules using Lamé spheroconal harmonic polynomials [1], and Patera and Winternitz identified the corresponding representations of the rotational group [2]. This contribution is focused on the successive use of the 4-step ladder operators identified in [3], and of the angular momentum operators as one of the sets identified in [4], on the appropriate eigenstates of the most asymmetric molecules, exhibiting the respective connections among successive pairs of such states.

The article reviews in Sect. 2 the symmetries of the free rotations of asymmetric molecules, including the most asymmetric ones, at the levels of their Hamiltonians, eigenfunctions and eigenenergies, referring the readers to [5-8] for the details. Sections 3 and 4 describe the actions of the respective 4-step and 1-step ladder operators on the individual pairs of eigenstates. Section 5 provides the overall picture of the successive and combined actions of both types of operators, starting from the lower eigenstates with $\ell = 0, 1, 2, 3$ and vanishing asymmetry-distribution energy. Section 6 contains a discussion of the specific results of this investigation and its relevance for developing the theory of angular momentum in bases of Lamé spheroconal harmonics.

[1] To whom any correspondence should be addressed

2. Symmetries in the rotations of asymmetric molecules

The section reviews briefly and successively the symmetries of the Hamiltonians, their eigenfunctions and energy spectra for asymmetric molecules, including the most asymmetric ones.

2.1. Symmetries in the Hamiltonians

The starting point is the Hamiltonian for the free rotations of asymmetric molecules with moments of inertia $I_1 < I_2 < I_3$ in the body fixed reference frame with principal axes:

$$H = \frac{1}{2}\left(\frac{L_x^2}{I_1} + \frac{L_y^2}{I_2} + \frac{L_z^2}{I_3}\right)$$

in terms of the squares of the cartesian components of the angular momentum operators.

Piña introduced the reparametrization [5]:

$$\frac{1}{I_i} = Q + Pe_i$$

in terms of the average of the trace of the matrix of inverses of moments of inertia,

$$Q = \frac{1}{3}\left(\frac{1}{I_1} + \frac{1}{I_2} + \frac{1}{I_3}\right) \quad ;$$

the parameters of asymmetry-distribution, $e_1 > e_2 > e_3$, subject to the conditions of not contributing to the trace, and on their magnitudes

$$e_1 + e_2 + e_3 = 0, \quad e_1^2 + e_2^2 + e_3^2 = \tfrac{3}{2};$$

and the magnitude of the asymmetry P, such that

$$P^2 = \frac{2}{9}\left[\left(\frac{1}{I_2} - \frac{1}{I_3}\right)^2 + \left(\frac{1}{I_1} - \frac{1}{I_3}\right)^2 + \left(\frac{1}{I_1} - \frac{1}{I_2}\right)^2\right] .$$

In this way, the Hamiltonian $H = H^Q + PH^*$ is separated into its spherical rotor and asymmetry-distribution contributions

$$H^Q = \frac{1}{2}QL^2, \quad H^* = \frac{1}{2}\left(e_1L_x^2 + e_2L_y^2 + e_3L_z^2\right) .$$

The three pairs of Hamiltonians commute with each other and consequently share common eigenfunctions. Their invariance under the individual parity transformations $x \to -x$, $y \to -y$, $z \to -z$ lead also to eigenfunctions with the respective well-defined parities.

The two conditions on the asymmetry-distribution parameters indicate that only one of them can be chosen independently. An alternative parametrization with a single angle, $0 < \sigma < 60^\circ$, can also be used:

$$e_1 = \cos\sigma,\ e_2 = \cos(\sigma - 120^\circ),\ \ e_3 = \cos(\sigma + 120^\circ).$$

The particular cases of $(\sigma = 0 : e_1 = 1,\ e_2 = e_3 = -0.5)$ and $(\sigma = 60^\circ : e_1 = e_2 = 0.5,\ e_3 = -1)$ correspond to prolate and oblate symmetric molecules with the respective x and z axes of rotational symmetry. The most asymmetric molecules are defined by ($\sigma = 30^\circ$: $e_1 = \sqrt{3}/2, e_2 = 0,\ e_3 = -\sqrt{3}/2$), for which H^* loses its dependence on L_y, and becomes proportional to $L_x^2 - L_z^2$. Also, for molecules with values of σ and $60^\circ - \sigma$, it is easy to establish the relationships $e_1(60^\circ - \sigma) = -e_3(\sigma)$, $e_2(60^\circ - \sigma) = -e_2(\sigma)$ and $e_3(60^\circ - \sigma) = -e_1(\sigma)$, with the corresponding changes of sign between $H^*(\sigma)$ and $H^*(60^\circ - \sigma)$, complemented with the exchange of L_x^2 and L_z^2.

2.2. *Symmetries in the eigenfunctions in spheroconal coordinates*

The spheroconal coordinates are expressed in terms of their transformation equations to cartersian coordinates: $x = r\mathrm{dn}(\chi_1 \mid k_1^2)\mathrm{sn}(\chi_2 \mid k_2^2)$, $y = r\mathrm{cn}(\chi_1 \mid k_1^2)\mathrm{cn}(\chi_2 \mid k_2^2)$, $z = r\mathrm{sn}(\chi_1 \mid k_1^2)\mathrm{dn}(\chi_2 \mid k_2^2)$ using the Jacobi elliptical integrals. Their parameters k_i^2 are subject to the condition that their sum is one, ensuring that the sum of the squares of the cartesian coordinates is equal to the square of the spherical radial coordinate. Fixed values of the angular coordinates $\chi_1 = \chi_{10}$ and $\chi_2 = \chi_{20}$ correspond to elliptical cones with axis along the *z*-axis and the *x*-axis, respectively. Notice that the change of the indices 1 and 2 in the coordinates and parameters is equivalent to the exchange of *x* and *z*, while *y* remains the same.

The scale factors for these coordinates are readily obtained [3-8]:

$$h_r = 1 \;\; h_{\chi_1} = h_{\chi_2} = r\sqrt{1 - k_1^2\mathrm{sn}^2(\chi_1 \mid k_1^2) - k_2^2\mathrm{sn}^2(\chi_2 \mid k_2^2)}.$$

leading to the spheroconal coordinate representation of the operators of the square of the angular momentum and the asymmetry-distribution Hamiltonian,

$$L^2 = -\frac{\hbar^2 r^2}{h_\chi^2}\left[\frac{\partial^2}{\partial\chi_1^2} + \frac{\partial^2}{\partial\chi_2^2}\right]$$

$$H^* = -\frac{\hbar^2 r^2}{2h_\chi^2}\left\{[e_1 - (e_1 - e_2)\mathrm{sn}^2(\chi_2 \mid k_2^2)]\frac{\partial^2}{\partial\chi_1^2} + [e_3 - (e_2 - e_3)\mathrm{sn}^2(\chi_1 \mid k_1^2)]\frac{\partial^2}{\partial\chi_2^2}\right\}.$$

The respective eigenvalue equations with eigenvalues $\hbar^2\ell(\ell+1)$ and E^*, respectively, admit common factorizable eigenfuntions $\Lambda_1(\chi_1)\Lambda_2(\chi_2)$. Both equations can be simultaneously separated into ordinary Lamé differential equations, provided the geometric parameters and the asymmetry-distribution parameters are chosen to be connected by

$$k_1^2 = \frac{e_2 - e_3}{e_1 - e_3}, \quad k_2^2 = \frac{e_1 - e_2}{e_2 - e_3}.$$

In such a case, the symmetric prolate and oblate molecules are characterized by ($k_1^2 = 0$, $k_2^2 = 1$) and ($k_1^2 = 1$, $k_2^2 = 0$), respectively, and the most asymmetric by $k_1^2 = k_2^2 = 0.5$.
In fact, the separated equations for $i = 1, 2$ have the common form

$$\left[-\frac{d^2}{d\chi_i^2} + \ell(\ell+1)k_i^2\mathrm{sn}^2(\chi_i \mid k_i^2)\right]\Lambda_i(\chi_i) = h_i\Lambda_i(\chi_i),$$

with separation constants

$$h_1 = -\frac{\ell(\ell+1)e_3}{e_1 - e_3} + \frac{2E^*}{\hbar^2(e_1 - e_3)}, \quad h_2 = \frac{\ell(\ell+1)e_1}{e_1 - e_3} - \frac{2E^*}{\hbar^2(e_1 - e_3)}.$$

After the Lamé equations are solved, their respective eigenvalues can be combined, so that

$$h_1 + h_2 = \ell(\ell+1), \;\; e_1h_1 + e_3h_2 = 2E^*/\hbar^2.$$

The derivatives of the Jacobi elliptical functions sn, cn and dn equal the products of the other two: cn dn, -dn sn, -k^2sn cn, respectively. Consequently, the Lamé equation has removable singularities of eight different species *A*: [1, d, c, s, dc, ds, cs, dcs] and *B*: [1, s, c, d, sc, sd, cd, scd] using only the initial of the respective functions, for the variables χ_1 and χ_2. Their corresponding products *AB*:[1, *x*, *y*, *z*, *xy*, *xz*, *yz*, *xyz*] allow the identifications of the respective *x*, *y* and *z* parities for the successive eigenfunctions, as well as their number and type of cartesian nodal planes: $n^{AB} = 0, 1, 2$ or 3.

The general form of the Lamé functions includes the product of the singularity removing factor and a polynomial in even powers of $\mathrm{sn}(\chi_i \mid k_i^2)$: $\Lambda^A(\chi_i, h_n^A) = A(\chi_i)\sum_{s=0}^{N_{max}} a_s(h_{n_i}^A)\mathrm{sn}^{2s}(\chi_i \mid k_i^2)$.

Substitution in the Lamé differential equation leads to three term recurrence relations for the expansion coefficients, equivalent to a tridiagonal matrix eigenvalue problem; diagonalization of the latter yields the eigenvalues $h_{n_i}^A$, and the associated coefficients $a_s(h_{n_i}^A)$. The subindices in the eigenvalues count the number of nodal elliptical cones in the eigenfunction which come in symmetrical pairs. The values of N_{max} are: $(\ell+2)/2$ for ℓ even and $A=1$, and $\ell/2$ for ℓ even and $A=$ dc, ds, cs; and $(\ell+1)/2$ for ℓ odd, and $A=$ d, c, s, and $(\ell-1)/2$ for ℓ odd and $A=$ dcs. In both cases the total number of states is $(2\ell+1)$.

The matching of the eigenvalues, $h_{n_1}^A$ and $h_{n_2}^B$ is guided by their species and parities, and also by the condition that their sum is $\ell(\ell+1)$. Additionally, the number of nodes for the product eigenstates with a given value of ℓ, using the notation $\ell[AB]n_1n_2$, is such that: $n^{AB}+n_1+n_2=\ell$. Correspondingly, the combination $e_1h_{n_1}^A + e_3h_{n_2}^B$ yields the asymmetry-distribution energy $E_{n_1n_2}^{*\ell AB}$ for each eigenstate.

Figures 2 in [6] and 2 in [7] illustrate 3-dimensional polar graphs for the eigenfunctions with $\ell=4$ and 5, respectively, for $\sigma = 0^\circ, 15^\circ, 30^\circ, 45^\circ, 60^\circ$. The readers may examine them in order to recognize some of their properties as described in the previous paragraphs. Additionally, notice the same shapes in the first and fifth columns, second and fourth, as well as in the upper and lower parts of the third one, differing only in orientations. The latter follow from the symmetries in the last paragraph in Sect. 2.1 and in the first paragraph of this section.

2.3 Symmetries in the asymmetry-distribution energies

Figure 1 in [7] illustrates the asymmetry-distribution energy spectra $E_{n_1n_2}^{*AB}(\sigma)$ for the values of $\ell = 0,1,2,...,10,20,30$, as functions of the asymmetry parameter in the interval $[0,60^\circ]$. The doubly degenerate energy levels for the prolate and oblate symmetric molecules at the left and right of the individual figures, with energies such that $E^*(\sigma=0) = -E^*(\sigma=60^\circ)$ can be recognized. The reader may interpolate or extrapolate, between or from them, following the energy lines from the left and the right with equal steps in σ accompanied by equal vertical displacements towards or away from the horizontal line with $E^*=0$. The reasons mentioned in the closing paragraph of the previous section also explain the general connections $E_{n_1n_2}^{*\ell AB}(\sigma) = -E_{n_2n_1}^{*\ell A'B'}(60^\circ-\sigma)$ for species $[AB]$ and $[A'B']$ in which x and z are exchanged, accompanied by the exchange of excitations in χ_1 and χ_2. The figures under consideration contain lines that are antisymmetric relative to the point $(\sigma=30^\circ, E^*=0)$; alternatively, a rotation of each figure around such a point by 180° is self-mapping.

2.4 Symmetries in the most asymmetric molecules

The most asymmetric molecules defined by the parameters $\sigma=30^\circ$, $k_1^2=k_2^2=0.5$, are described by the Hamiltonian $H^*=(\sqrt{3}/4)[L_x^2-L_z^2]$, which is antisymmetric under the exchange of L_x and L_z. This allows us to expect that it may have eigenstates with vanishing asymmetry-distribution energy, $E^*=0$.

The species of such eigenstates should be invariant under the exchange of x and z. The favoured options are $1, y=$ cc, $xz=$ ds sd, and $xyz=$ dcs scd, with the common $[A]=[B]$. The vanishing of the energy $E^{*\ell[Ab]}_{n_1 n_2}$, according with Sect. 2.2 requires that $h^A_{n_1} = h^A_{n_2}$. Then the excitation of both degrees of freedom must also be the same $n_1 = n_2$. From the distribution of nodes for a given ℓ, the states of our interest belong in the following categories: $|\ell = 4N[1]2N, 2N\rangle$, $|\ell = 4N+1[y]2N, 2N\rangle$, $|\ell = 4N+2[xz]2N, 2N\rangle$, and $|\ell = 4N+3[xyz]2N, 2N\rangle$.

3. Angular Momentum 4-step ladder operators for the most asymmetric molecules

Valdés and Piña investigated the rotational spectra of the most asymmetric molecules, identifying their eigenstates with vanishing energies as well as the 4-step ladder operators connecting them [3]. Here we simply borrow their equations describing the pertinent states and ladder operators.

In their notation, the Lamé polynomials $\Lambda_{\ell E'}$ become the products of the singularity removing factors and Jacobi polynomials of degree n with labels α, γ in the variable $w = cn^2(\chi_i \mid k_i^2 = 0.5)$ squared:

$$\Lambda_{4n,0} = f_n(1/4, 3/4, w^2)$$

$$\Lambda_{4n+1,0} = \sqrt{w} f_n(3/4, 5/4, w^2)$$

$$\Lambda_{4n+2,0} = \sqrt{1-w^2}\, f_n(5/4, 3/4, w^2)$$

$$\Lambda_{4n+3,0} = \sqrt{w(1-w^2)}\, f_n(7/4, 5/4, w^2)$$

The 4-step ladder operators, connecting eigenstates of the same species, do it as follows:

$$\left[2w\left(1-w^2\right)\frac{d}{dw} - (\ell+1)\left(w^2 - \frac{\ell+3}{2\ell+5}\right)\right]\Lambda_{\ell,0} = \frac{\sqrt{(\ell+1)(\ell+2)(\ell+3)(\ell+4)}}{2\ell+5}\Lambda_{\ell+4,0}$$

$$\left[2w\left(1-w^2\right)\frac{d}{dw} + (\ell+4)\left(w^2 - \frac{\ell+2}{2\ell+5}\right)\right]\Lambda_{\ell+4,0} = \frac{\sqrt{(\ell+1)(\ell+2)(\ell+3)(\ell+4)}}{2\ell+5}\Lambda_{\ell,0}$$

in their up and down versions, respectively.

4. Three sets of ladder operators for molecules of any asymmetry

The ladder operators for the Lamé spheroconal harmonic polynomials, identified in [4], come in three sets: 1) For eigenstates with the same angular momentum ℓ, same species *AB*, and different and complementary angular excitations $n_1 + n_2 = n'_1 + n'_2 = \ell - n^{AB}$, $h^A_{n_1} + h^B_{n_2} = h^A_{n'_1} + h^B_{n'_2} = \ell(\ell+1)$ according to Sect. 2.2. 2) For eigenstates with the same angular momentum ℓ and different species, corresponding to the cartesian components of the angular momentum operators L_x, L_y, L_z. 3) For eigenstates with neigbouring values of angular momentum $\ell' = \ell \pm 1$, different parities and different species, which correspond to the cartesian components of the linear momentum operators p_x, p_y, p_z.

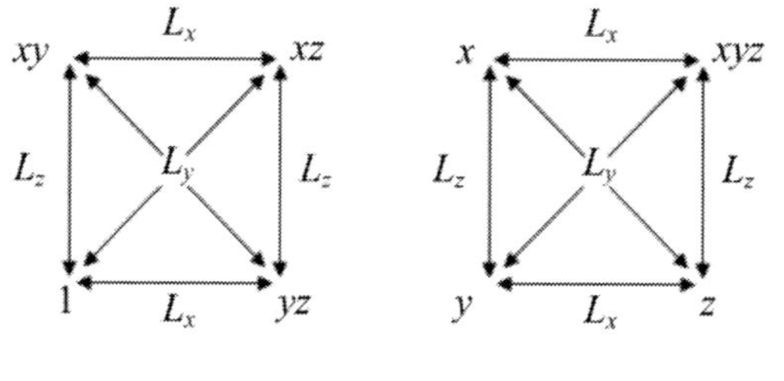

Figure 1. Actions of angular momentum operators between pairs of eigenstates of companion species for a) even and b) odd values of ℓ.

Notice that both figures contain the operators L_x, L_y and L_z in the same positions. The species of the eigenstates on the lower left and upper right corners belong to the species symmetric under the exchange between x and z, while the pairs in the upper left and lower right corners change one into the other under the same exchange.

5. Connections of successive rotational eigenstates of the most asymmetric molecules using the 4-step ladder and L_x, L_y, L_z operators

This section exhibits graphically, via figures 2 and 3, the overall picture of the connections between the eigenstates with vanishing asymmetry-distribution energy and $\ell = 4N, 4N+1, 4N+2, 4N+3$ for $N = 0, 1, 2, 3$, provided by the 4-step ladder operators described in Sect. 3; as well as their companion eigenstates for each value of ℓ of different species and angular excitations, using the angular momentum operators as described in Sect. 4.

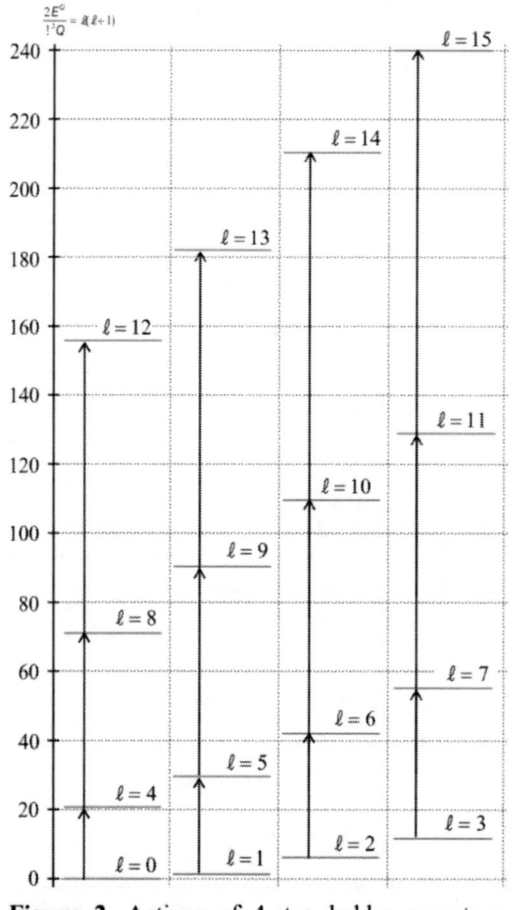

Figure 2. Actions of 4-step ladder operators between eigenstates of the same species with $E^* = 0$ for $\ell = 0,1,2,3$ modulo 4.

The actions depicted in figure 2 combine the raising actions for both degrees of freedom described in Sect. 4, for $n_1=n_2=0$, 2, 4, 6. The lowest eigenstates are the monomials 1, y, xz, xyz in their cartesian representation.

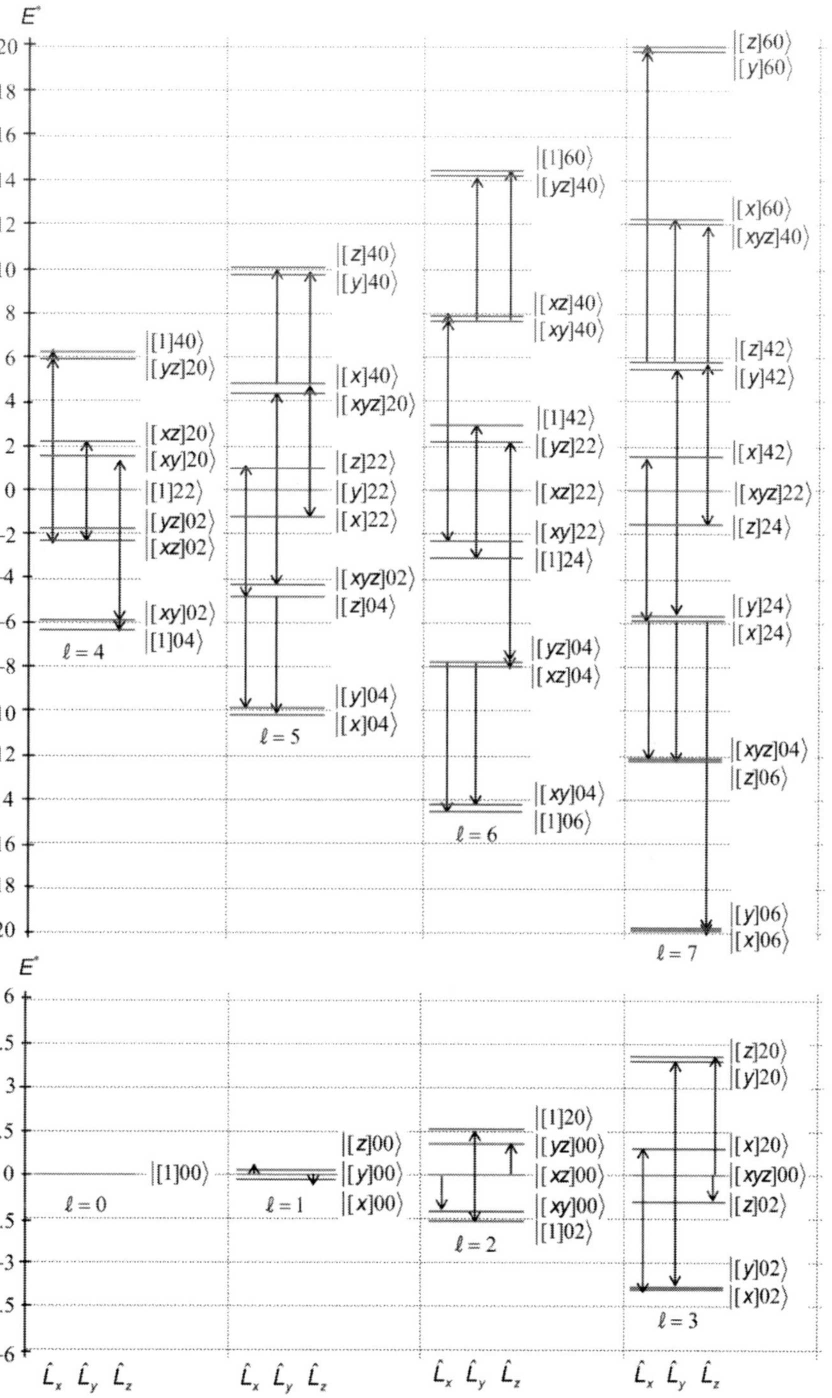

Figure 3a. Actions of angular momentum operators L_x, L_y, L_z between eigenstates with a given value of $\ell = 4N$, $4N+1$, $4N+2$, $4N+3$ and companion species for N=0, 1.

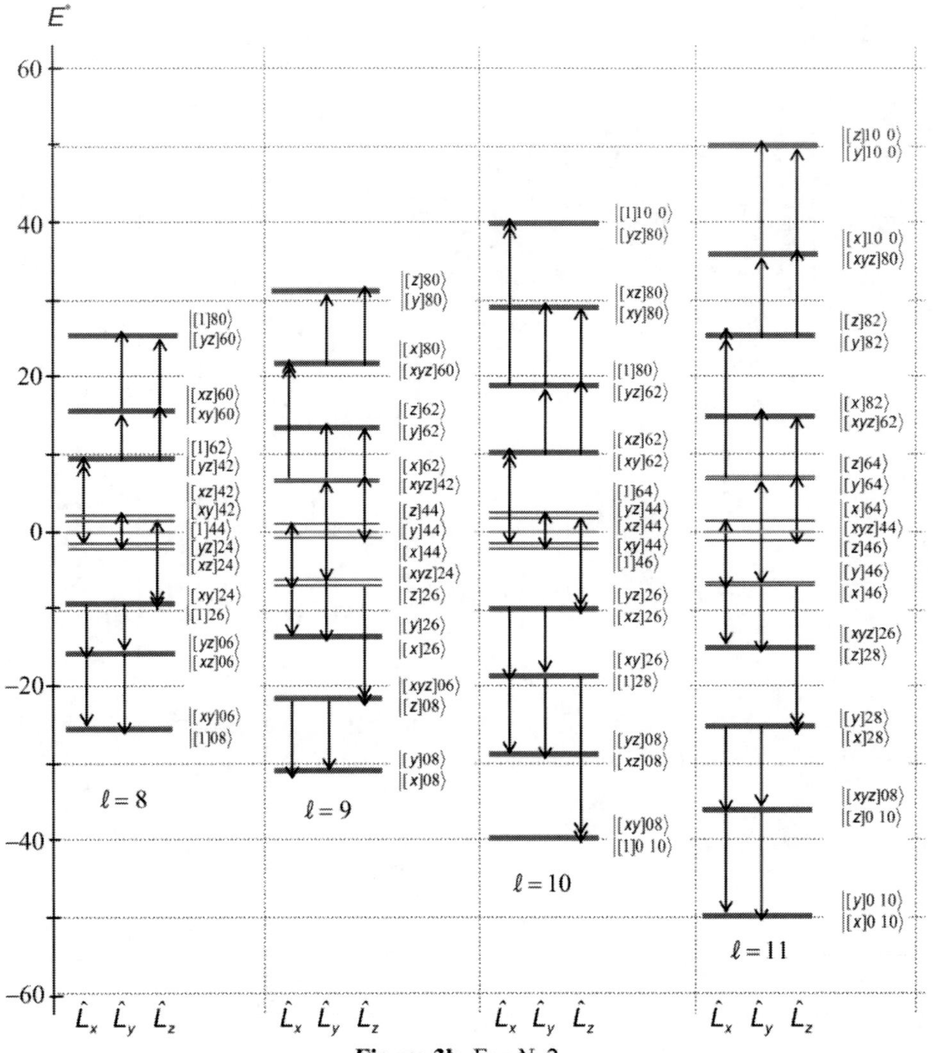

Figure 3b. For *N*=2.

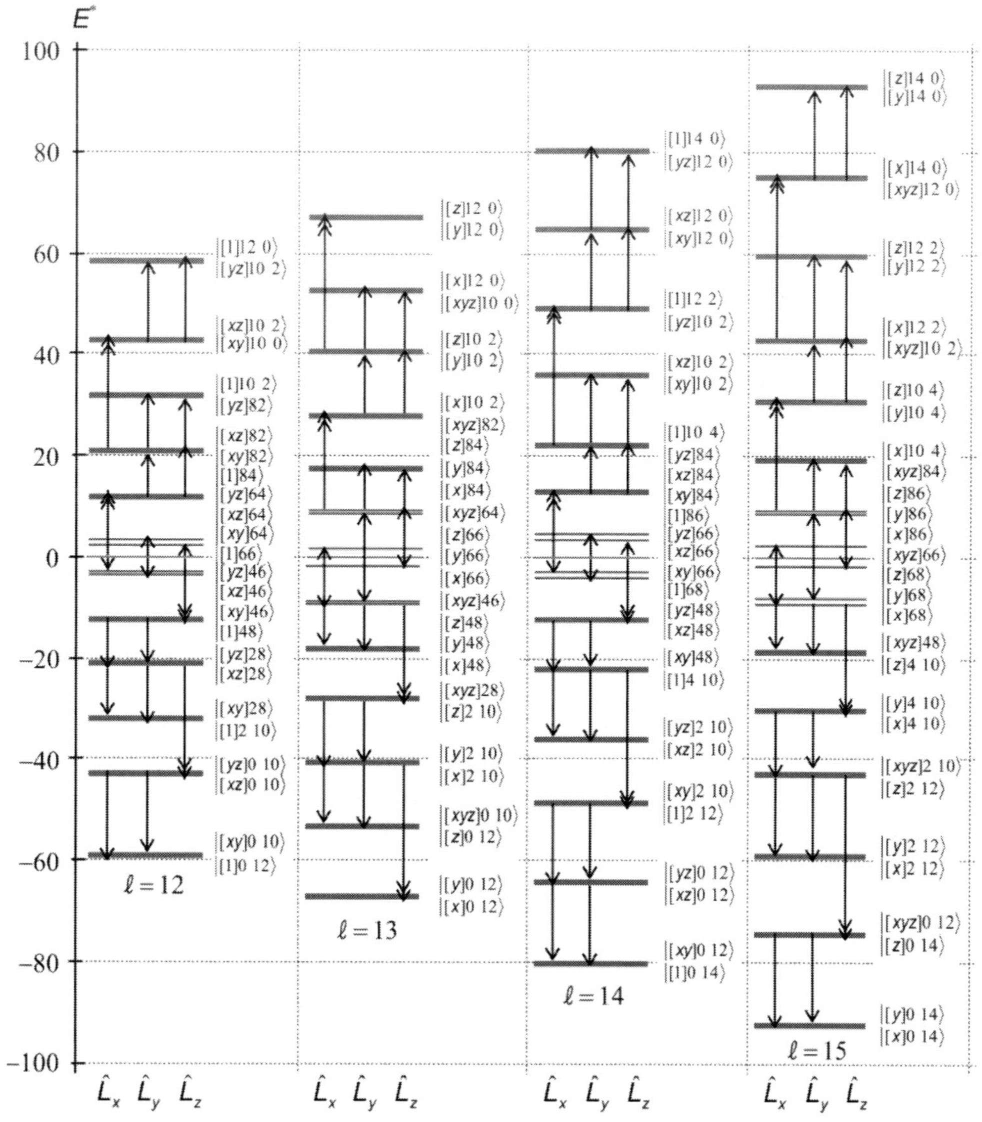

Figure 3c. For N=3.

The actions depicted in figure 3 follow from the successive applications of the operators of figure 1, starting from the eigenstates with $E^{*} = 0$ of figure 2. In figure 3a the lower eigenstates : 1 has zero eigenvalues for any component of the angular momentum; y becomes z and x when L_x and L_z act on it, with positive and negative energies because $e_1 > 0$ and $e_3 < 0$, respectively; xz, under the action of the same operators become yz and xy with the corresponding positive and negative energies, respectively, and under L_y leads to the two states of species [1] with positive energy for n_1=2, n_2=0 and negative for n_1=0, n_2=2; and xyz under the actions of L_x, L_y and L_z becomes the companions of species x, y, z with n_1=2, n_2=0 above, and with n_1=0, n_2=2 in the order z, y, x symmetrically below. The extensions for N=1, 2, 3 can be identified in the upper part of figure 3a, and figures 3b and 3c, with the successive applications of the same operators. In the last two figures, notice that the energy levels far above and below $E^* = 0$ come in pairs with almost the same energy, including the connections of some of those pairs represented only by the tip of the respective arrows. The readers may also ascertain, for each value of ℓ, the numbers of states for the respective species, as described in Sect. 2.2, as well as the matching and the types of their nodal surfaces when going from the lower to the higher energy levels, following the changes in the χ_1 and χ_2 excitations.

6. Discussion

The symmetries of the rotations of asymmetric molecules reviewed in Sect. 2 allow the identification of the commuting set of operators H, H^Q, H^* and the individual *x*, *y* and *z* parity operators; with the corresponding labeling of their common eigenstates $\ell[AB]n_1n_2$ characterizing their angular momentum, species and excitations, and also of their respective eigenenergies, especially E^*. The 4-step angular momentum ladder operators in Sect. 3 provide the connections between the eigenstates of the most asymmetric molecules with vanishing asymmetry-distribution energy $E^* = 0$, $\ell = 0,\ 1,\ 2,\ 3$ modulo 4 as graphically illustrated in Sect. 5. The cartesian components of the angular momentum operators in Sect. 4 applied successively to each of the above eigenstates lead to their sets of 2ℓ companion eigenstates with the common value of ℓ, companion species, and complementary excitations and, with energies $\pm E^*$ of the same magnitude above and below the level $E^* = 0$, as illustrated also in Sect 5.

In addition to this specific investigation for the most asymmetric molecules, a few remarks about molecules of any asymmetry can be added to the discussion. First, we recall the connection between the Hamiltonians $H^*(\sigma)$ and $H^*(60° - \sigma)$ involving the change of sign in their respective asymmetry-distribution parameters as well as the exchange of L_x and L_z; and also the consequent antisymmetry of the lines in the (σ, E^*) planes in figure 1 of [7] relative to the point $(\sigma = 30°, E^* = 0)$. The labeling of the eigenstates of the most asymmetric molecules is also valid for the molecules with the other asymmetries, $0 < \sigma < 60°$; for any of them with a fixed value of σ and their respective energies $E^*(\sigma)$, the connections illustrated in figure 3 also hold, with the corresponding changes in the energy positions. While the present investigation involves only vertical displacements, the linear momentum operators in Sect. 4 produce horizontal displacements connecting states in different columns in figure 3, for molecules of any asymmetry; for the most asymmetric molecules, four successive applications should take us from ℓ to $\ell + 4$ making contact with Sect. 3. Our concluding remark is that these investigations contribute to the development of the theory of angular momentum in bases of Lamé spheroconal harmonic polynomials.

References

[1] Kramers N A and Ittmann G P 1929 *Zeitschrift Phys.* **52** 553
[2] Patera J and Winternitz P 1973 *J. Math. Phys.* **14** 1130
[3] Valdés M T and Piña E 2006 *Rev. Mex. Fis.* **52** 220
[4] Méndez-Fragoso R and Ley-Koo E 2012 *SIGMA* **8** 074 16 pages
[5] Piña E 1999 *J. Mol. Struct. (THEOCHEM)* **493** 159
[6] Ley-Koo E and Méndez-Fragoso R 2008 *Rev. Mex. Fis.* **54** 69
[7] Ley-Koo and Méndez-Fragoso 2008 *Rev. Mex. Fis.* **54** 162
[8] Méndez-Fragoso R and Ley-Koo E 2011 *Adv. Quant. Chem.* **62** 137

The second order pole over split quaternions

Matvei Libine

Department of Mathematics, Indiana University, Rawles Hall, 831 East 3rd St, Bloomington, IN 47405, USA

E-mail: mlibine@indiana.edu

Abstract. This is an addition to a series of papers [1, 2, 3, 4], where we develop quaternionic analysis from the point of view of representation theory of the conformal Lie group and its Lie algebra. In this paper we develop split quaternionic analogues of certain results from [4]. Thus we introduce a space of functions $\mathcal{D}^h \oplus \mathcal{D}^a$ with a natural action of the Lie algebra $\mathfrak{gl}(2,\mathbb{H}_\mathbb{C}) \simeq \mathfrak{sl}(4,\mathbb{C})$, decompose $\mathcal{D}^h \oplus \mathcal{D}^a$ into irreducible components and find the $\mathfrak{gl}(2,\mathbb{H}_\mathbb{C})$-equivariant projectors onto each of these irreducible components.

1. Introduction

This is an addition to a series of papers [1, 2, 3, 4], where we develop quaternionic analysis from the point of view of representation theory of the conformal Lie group and its Lie algebra. Since the paper [4] serves as a starting point for this article, we highlight the most relevant results. Let $\mathbb{H}$ denote the algebra of quaternions, $\mathbb{H}_\mathbb{C} = \mathbb{H} \otimes_\mathbb{R} \mathbb{C}$ and

$$\mathbb{H}_\mathbb{C}^\times = \{Z \in \mathbb{H}_\mathbb{C};\ Z \text{ is invertible}\},$$

then let

$$\mathcal{K} = \mathbb{C}[z_{11}, z_{12}, z_{21}, z_{22}, N(Z)^{-1}]$$

be the space of polynomial functions on $\mathbb{H}_\mathbb{C}^\times$. We can think of $\mathcal{K}$ as a quaternionic analogue of Laurent polynomials $\mathbb{C}[z, z^{-1}]$. There is a natural action of the Lie algebra $\mathfrak{gl}(2,\mathbb{H}_\mathbb{C}) \simeq \mathfrak{sl}(4,\mathbb{C})$ on $\mathcal{K}$ denoted by ρ_1 (see Lemma 1). According to Theorem 7 in [4] (restated here as Theorem 3), $(\rho_1, \mathcal{K})$ decomposes into three irreducible components:

$$\mathcal{K} = \mathcal{K}^- \oplus \mathcal{K}^0 \oplus \mathcal{K}^+.$$

This decomposition is followed by a derivation of projectors of $\mathcal{K}$ onto its irreducible components $\mathcal{K}^-$, $\mathcal{K}^0$ and $\mathcal{K}^+$. These projectors are regarded as quaternionic analogues of the Cauchy's formula for the second order pole

$$f'(z_0) = \frac{1}{2\pi i} \oint \frac{f(z)\,dz}{(z-z_0)^2}.$$

The cases of $\mathcal{K}^-$ and $\mathcal{K}^+$ are fairly straightforward. On the other hand, the case of $\mathcal{K}^0$ is more subtle and requires a certain regularization of infinities which is well known in four-dimensional quantum field theory as "scalar vacuum polarization". This vacuum polarization is usually

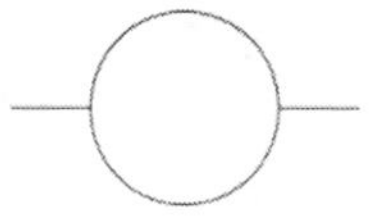

Figure 1. Scalar vacuum polarization diagram.

encoded by a Feynman diagram shown in Figure 1 and plays a key role in renormalization theory (see, for example, [5]).

Corollary 6 in [4] (restated here as Proposition 2) asserts that

$$\mathcal{H} = \mathbb{C}\text{-span of } \{t^l_{n\,\underline{m}}(Z) \cdot N(Z)^k;\ k \in \mathbb{Z}\},$$

where $t^l_{n\,\underline{m}}(Z)$'s are the matrix coefficients of $SU(2)$. In this article we replace the matrix coefficients of $SU(2)$ with the matrix coefficients of the discrete series representations of $SU(1,1)$ – denoted by $\tau^l_{n\,\underline{m}}(Z)$ – and consider a space

$$\mathcal{D}^h \oplus \mathcal{D}^a = \mathbb{C}\text{-span of } \{\tau^l_{n\,\underline{m}}(Z) \cdot N(Z)^k;\ k \in \mathbb{Z}\}. \tag{1}$$

(We denote by $\mathcal{D}^h$ and $\mathcal{D}^a$ the subspaces of (1) spanned by $\tau^l_{n\,\underline{m}}(Z) \cdot N(Z)^k$, where $\tau^l_{n\,\underline{m}}(Z)$'s are the matrix coefficients of holomorphic and antiholomorphic discrete series respectively.) We regard the space (1) as a split quaternionic analogue of $\mathcal{H}$. Our first result is Theorem 5, which shows that this space is invariant under the ρ_1-action of $\mathfrak{gl}(2,\mathbb{H}_\mathbb{C})$ and decomposes into six irreducible components that we denote by $\mathcal{D}^h_<$, $\mathcal{D}^{--}$, $\mathcal{D}^h_>$, $\mathcal{D}^a_<$, $\mathcal{D}^{++}$ and $\mathcal{D}^a_>$. Then in Proposition 8 we find the reproducing kernels for each of these irreducible components. Finally, in Theorem 9 we obtain the $\mathfrak{gl}(2,\mathbb{H}_\mathbb{C})$-equivariant projectors onto the irreducible components of the space (1). This result can be regarded as a split analogue of Corollary 14 and Theorem 15 in [4]. Perhaps surprisingly, unlike the case of $\mathcal{H}^0$, there are absolutely no convergence issues for the reproducing kernels or the integral operators realizing the projectors onto the irreducible components.

2. Preliminaries

Let $\mathbb{H}_\mathbb{C}$ denote the space of complexified quaternions: $\mathbb{H}_\mathbb{C} = \mathbb{H} \otimes \mathbb{C}$, it can be identified with the algebra of 2×2 complex matrices:

$$\mathbb{H}_\mathbb{C} = \mathbb{H} \otimes \mathbb{C} \simeq \left\{ Z = \begin{pmatrix} z_{11} & z_{12} \\ z_{21} & z_{22} \end{pmatrix};\ z_{ij} \in \mathbb{C} \right\} = \left\{ Z = \begin{pmatrix} z^0 - iz^3 & -iz^1 - z^2 \\ -iz^1 + z^2 & z^0 + iz^3 \end{pmatrix};\ z^k \in \mathbb{C} \right\}.$$

For $Z \in \mathbb{H}_\mathbb{C}$, we write

$$N(Z) = \det \begin{pmatrix} z_{11} & z_{12} \\ z_{21} & z_{22} \end{pmatrix} = z_{11}z_{22} - z_{12}z_{21} = (z^0)^2 + (z^1)^2 + (z^2)^2 + (z^3)^2$$

and think of it as the norm of Z. Let

$$\mathbb{H}_\mathbb{C}^\times = \{Z \in \mathbb{H}_\mathbb{C};\ N(Z) \neq 0\}$$

be the group of invertible complexified quaternions, $\mathbb{H}_\mathbb{C}^\times \simeq GL(2,\mathbb{C})$.

Let $\widetilde{\mathcal{H}}$ denote the space of $\mathbb{C}$-valued functions on $\mathbb{H}_{\mathbb{C}}$ (possibly with singularities) which are holomorphic with respect to the complex variables $z_{11}, z_{12}, z_{21}, z_{22}$. We recall the action of $GL(2,\mathbb{H}_{\mathbb{C}})$ on $\widetilde{\mathcal{H}}$ given by equation (49) in [1]:

$$\rho_1(h) : f(Z) \quad \mapsto \quad \bigl(\rho_1(h)f\bigr)(Z) = \frac{f\bigl((aZ+b)(cZ+d)^{-1}\bigr)}{N(cZ+d)\cdot N(a'-Zc')},$$
$$h = \begin{pmatrix} a' & b' \\ c' & d' \end{pmatrix},\ h^{-1} = \begin{pmatrix} a & b \\ c & d \end{pmatrix} \in GL(2,\mathbb{H}_{\mathbb{C}}).$$

Differentiating this action, we obtain an action of the Lie algebra $\mathfrak{gl}(2,\mathbb{H}_{\mathbb{C}}) \simeq \mathfrak{gl}(4,\mathbb{C})$, which we still denote by ρ_1.

Lemma 1 (Lemma 68 from [1]). *Let $\partial = \begin{pmatrix} \partial_{11} & \partial_{21} \\ \partial_{12} & \partial_{22} \end{pmatrix}$, where $\partial_{ij} = \frac{\partial}{\partial z_{ij}}$. The Lie algebra action ρ_1 of $\mathfrak{gl}(2,\mathbb{H}_{\mathbb{C}})$ on $\widetilde{\mathcal{H}}$ is given by*

$$\rho_1\begin{pmatrix} A & 0 \\ 0 & 0 \end{pmatrix} : f \mapsto \mathrm{Tr}\bigl(A\cdot(-Z\cdot\partial f - f)\bigr)$$
$$\rho_1\begin{pmatrix} 0 & B \\ 0 & 0 \end{pmatrix} : f \mapsto \mathrm{Tr}\bigl(B\cdot(-\partial f)\bigr)$$
$$\rho_1\begin{pmatrix} 0 & 0 \\ C & 0 \end{pmatrix} : f \mapsto \mathrm{Tr}\Bigl(C\cdot\bigl(Z\cdot(\partial f)\cdot Z + 2Zf\bigr)\Bigr) = \mathrm{Tr}\Bigl(C\cdot\bigl(Z\cdot\partial(Zf)\bigr)\Bigr)$$
$$\rho_1\begin{pmatrix} 0 & 0 \\ 0 & D \end{pmatrix} : f \mapsto \mathrm{Tr}\Bigl(D\cdot\bigl((\partial f)\cdot Z + f\bigr)\Bigr) = \mathrm{Tr}\Bigl(D\cdot\bigl(\partial(Zf) - f\bigr)\Bigr).$$

This lemma implies that $\mathfrak{gl}(2,\mathbb{H}_{\mathbb{C}})$ preserves the spaces

$$\mathcal{H}^+ = \{\text{polynomial functions on } \mathbb{H}_{\mathbb{C}}\} = \mathbb{C}[z_{11}, z_{12}, z_{21}, z_{22}] \qquad \text{and}$$
$$\mathcal{H} = \{\text{polynomial functions on } \mathbb{H}_{\mathbb{C}}^{\times}\} = \mathbb{C}[z_{11}, z_{12}, z_{21}, z_{22}, N(Z)^{-1}].$$

Recall that $t^l_{n\,\underline{m}}(Z)$'s denote the matrix coefficients of $SU(2)$ described by equation (27) of [1] (cf. [6]). These are polynomial functions on $\mathbb{H}_{\mathbb{C}}$. The parameters range as follows:

$$l = 0, \frac{1}{2}, 1, \frac{3}{2}, \dots, \qquad m, n = -l, -l+1, -l+2, \dots, l.$$

We recall some results from [4].

Proposition 2 (Corollary 6 in [4]). *The functions*

$$t^l_{n\,\underline{m}}(Z)\cdot N(Z)^k, \qquad l = 0, \frac{1}{2}, 1, \frac{3}{2}, \dots, \quad m, n = -l, -l+1, \dots, l, \quad k \in \mathbb{Z},$$

form a vector space basis of $\mathcal{H} = \mathbb{C}[z_{11}, z_{12}, z_{21}, z_{22}, N(Z)^{-1}]$.

Theorem 3 (Theorem 7 in [4]). *The representation $(\rho_1, \mathcal{H})$ of $\mathfrak{gl}(2,\mathbb{H}_{\mathbb{C}})$ has the following decomposition into irreducible components:*

$$(\rho_1, \mathcal{H}) = (\rho_1, \mathcal{H}^-) \oplus (\rho_1, \mathcal{H}^0) \oplus (\rho_1, \mathcal{H}^+),$$

where

$$\mathcal{H}^+ = \mathbb{C}\text{-span of } \bigl\{t^l_{n\,\underline{m}}(Z)\cdot N(Z)^k;\ k \ge 0\bigr\},$$
$$\mathcal{H}^- = \mathbb{C}\text{-span of } \bigl\{t^l_{n\,\underline{m}}(Z)\cdot N(Z)^k;\ k \le -(2l+2)\bigr\},$$
$$\mathcal{H}^0 = \mathbb{C}\text{-span of } \bigl\{t^l_{n\,\underline{m}}(Z)\cdot N(Z)^k;\ -(2l+1) \le k \le -1\bigr\}$$

(see Figure 2).

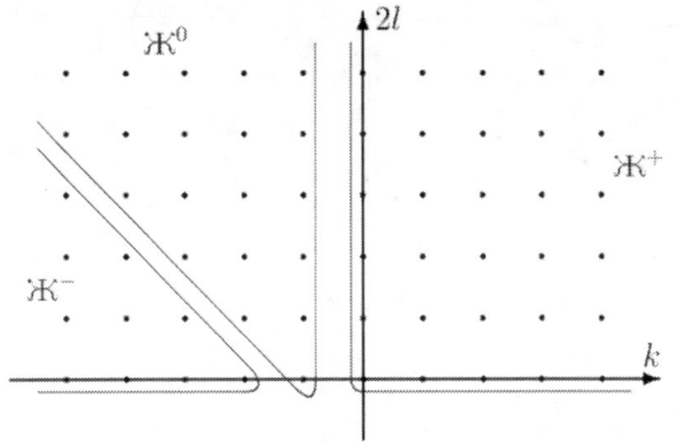

Figure 2. Decomposition of $(\rho_1, Ж)$ into irreducible components.

3. Split Quaternions

In this section we essentially replace the classical quaternions $\mathbb{H}$ with split quaternions $\mathbb{H}_\mathbb{R}$ by replacing the matrix coefficients $t^l_{m\,\underline{n}}(Z)$ of $SU(2)$ with the matrix coefficients of $SU(1,1)$, which we denote by $\tau^l_{m\,\underline{n}}(Z)$ in order to avoid conflict of notations.

We are primarily interested in a real form of $\mathbb{H}_\mathbb{C}$ called split quaternions:

$$\mathbb{H}_\mathbb{R} = \left\{ Z = \begin{pmatrix} z_{11} & z_{12} \\ z_{21} & z_{22} \end{pmatrix} \in \mathbb{H}_\mathbb{C};\ z_{22} = \overline{z_{11}},\ z_{21} = \overline{z_{12}} \right\}.$$

This algebra is studied in detail in [2]. The "unit sphere" in $\mathbb{H}_\mathbb{R}$ is a hyperboloid

$$SU(1,1) = \{X \in \mathbb{H}_\mathbb{R};\ N(X) = 1\},$$

which is a Lie group isomorphic to $SL(2,\mathbb{R})$. Another important group sitting inside $\mathbb{H}_\mathbb{C}$ is

$$U(1,1) = \{Z = e^{i\theta} \cdot X \in \mathbb{H}_\mathbb{C};\ X \in SU(1,1) \subset \mathbb{H}_\mathbb{R},\ \theta \in \mathbb{R}\}.$$

We orient $U(1,1)$ so that $\{\tilde{e}_0, \tilde{e}_1, \tilde{e}_2, e_3\}$ is a positive basis of the tangent space at $1 \in U(1,1)$ (same orientation as in [2]), where

$$\tilde{e}_0 = \begin{pmatrix} -i & 0 \\ 0 & -i \end{pmatrix}, \quad \tilde{e}_1 = \begin{pmatrix} 0 & 1 \\ 1 & 0 \end{pmatrix}, \quad \tilde{e}_2 = \begin{pmatrix} 0 & i \\ -i & 0 \end{pmatrix}, \quad e_3 = \begin{pmatrix} -i & 0 \\ 0 & i \end{pmatrix}.$$

We regard groups $U(1,1) \times U(1,1)$ and $SU(1,1) \times SU(1,1)$ as subgroups of $GL(2,\mathbb{H}_\mathbb{C})$:

$$U(1,1) \times U(1,1) = \left\{ \begin{pmatrix} a & 0 \\ 0 & d \end{pmatrix} \in GL(2,\mathbb{H}_\mathbb{C});\ a, d \in U(1,1) \subset \mathbb{H}_\mathbb{C} \right\}. \tag{2}$$

$$SU(1,1) \times SU(1,1) = \left\{ \begin{pmatrix} a & 0 \\ 0 & d \end{pmatrix} \in GL(2,\mathbb{H}_\mathbb{C});\ a, d \in \mathbb{H}_\mathbb{R},\ N(a) = N(d) = 1 \right\}. \tag{3}$$

The matrix coefficients of the discrete series representations of $SU(1,1)$ are rational functions on $\mathbb{H}_\mathbb{C}$ described by equation (13) of [2] (cf. [6]):

$$\tau^l_{n\,\underline{m}}(Z) = \frac{1}{2\pi i} \oint (sz_{11} + z_{21})^{l-m} (sz_{12} + z_{22})^{l+m} s^{-l+n} \frac{ds}{s}, \qquad Z = \begin{pmatrix} z_{11} & z_{12} \\ z_{21} & z_{22} \end{pmatrix}, \tag{4}$$

where the integral is taken over the unit circle $\{s \in \mathbb{C};\ |s| = 1\}$ traversed once in the counterclockwise direction. The parameters l, m and n range over

$$l = -1, -\frac{3}{2}, -2, -\frac{5}{2}, \dots, \qquad m, n \in \mathbb{Z} + l, \qquad m, n \le l \text{ or } m, n \ge -l.$$

When $m, n \ge -l$, we get the matrix coefficients of the holomorphic discrete series representations. And when $m, n \le l$, we get the matrix coefficients of the antiholomorphic discrete series representations. While the integral expressions for $t^l_{n\,\underline{m}}(Z)$'s and $\tau^l_{n\,\underline{m}}(Z)$'s are exactly the same, the indices l, m and n have different ranges, and we insist on using different notations, since the former are matrix coefficients of $SU(2)$ and the latter are matrix coefficients of $SU(1,1)$.

We define split analogues of the space $\mathcal{K}$:

$$\mathcal{D}^h = \mathbb{C}\text{-span of } \{\tau^l_{n\,\underline{m}}(Z) \cdot N(Z)^k;\ m, n \ge -l,\ k \in \mathbb{Z}\},$$
$$\mathcal{D}^a = \mathbb{C}\text{-span of } \{\tau^l_{n\,\underline{m}}(Z) \cdot N(Z)^k;\ m, n \le l,\ k \in \mathbb{Z}\}$$

(h stands for "holomorphic" and a stands for "antiholomorphic").

As in Subsection 3.6 in [2], we define a $\mathfrak{gl}(2, \mathbb{H}_\mathbb{C})$-invariant symmetric bilinear pairing for functions on $\mathcal{D}^h \oplus \mathcal{D}^a$

$$\langle f_1, f_2 \rangle_1 = \frac{i}{2\pi^3} \int_{U(1,1)} f_1(Z) \cdot f_2(Z)\, dV, \tag{5}$$

where dV is a holomorphic 4-form on $\mathbb{H}_\mathbb{C}$

$$dV = dz^0 \wedge dz^1 \wedge dz^2 \wedge dz^3 = \frac{1}{4} dz_{11} \wedge dz_{12} \wedge dz_{21} \wedge dz_{22}.$$

Proposition 4 (Proposition 56 in [2]). *We have the following orthogonality relations:*

$$\langle \tau^{l'}_{n'\,\underline{m'}}(Z) \cdot N(Z)^{k'}, \tau^l_{m\underline{n}}(Z^{-1}) \cdot N(Z)^{-k-2} \rangle_1 = -\frac{1}{2l+1} \delta_{kk'} \delta_{ll'} \delta_{mm'} \delta_{nn'}, \tag{6}$$

where the indices k, l, m, n are $l = -1, -\frac{3}{2}, -2, \dots$, $m, n \in \mathbb{Z} + l$, $m, n \ge -l$ or $m, n \le l$, $k \in \mathbb{Z}$ and similarly for k', l', m', n'. In particular, the functions $\tau^l_{n\,\underline{m}}(Z) \cdot N(Z)^k$'s are linearly independent.

Since $\tau^l_{m\underline{n}}(Z^{-1}) = \tau^l_{m\underline{n}}(Z^+) \cdot N(Z)^{-2l}$ is proportional to $\tau^l_{-n\,\underline{-m}}(Z) \cdot N(Z)^{-2l}$, the functions $\tau^l_{m\underline{n}}(Z^{-1}) \cdot N(Z)^{-k-2}$ indeed lie in $\mathcal{D}^h \oplus \mathcal{D}^a$. Hence the spaces $\mathcal{D}^h$ and $\mathcal{D}^a$ are dual to each other under the pairing (5).

4. Decomposition of $\mathcal{D}^h$ and $\mathcal{D}^a$ into Irreducible Components

In this section we discuss a split analogue of Theorem 3:

Theorem 5. *The spaces $\mathcal{D}^h$ and $\mathcal{D}^a$ are invariant under the ρ_1-action of $\mathfrak{gl}(2, \mathbb{H}_\mathbb{C})$. We have the following decompositions into irreducible components:*

$$(\rho_1, \mathcal{D}^h) = (\rho_1, \mathcal{D}^h_<) \oplus (\rho_1, \mathcal{D}^{--}) \oplus (\rho_1, \mathcal{D}^h_>) \quad \text{and} \quad (\rho_1, \mathcal{D}^a) = (\rho_1, \mathcal{D}^a_<) \oplus (\rho_1, \mathcal{D}^{++}) \oplus (\rho_1, \mathcal{D}^a_>),$$

where

$$\mathcal{D}^h_< = \mathbb{C}\text{-span of } \{\tau^l_{n\,\underline{m}}(Z) \cdot N(Z)^k;\ m, n \ge -l,\ k \le -1\},$$
$$\mathcal{D}^{--} = \mathbb{C}\text{-span of } \{\tau^l_{n\,\underline{m}}(Z) \cdot N(Z)^k;\ m, n \ge -l,\ 0 \le k \le -(2l+2)\},$$
$$\mathcal{D}^h_> = \mathbb{C}\text{-span of } \{\tau^l_{n\,\underline{m}}(Z) \cdot N(Z)^k;\ m, n \ge -l,\ k \ge -(2l+1)\},$$
$$\mathcal{D}^a_< = \mathbb{C}\text{-span of } \{\tau^l_{n\,\underline{m}}(Z) \cdot N(Z)^k;\ m, n \le l,\ k \le -1\},$$
$$\mathcal{D}^{++} = \mathbb{C}\text{-span of } \{\tau^l_{n\,\underline{m}}(Z) \cdot N(Z)^k;\ m, n \le l,\ 0 \le k \le -(2l+2)\},$$
$$\mathcal{D}^a_> = \mathbb{C}\text{-span of } \{\tau^l_{n\,\underline{m}}(Z) \cdot N(Z)^k;\ m, n \le l,\ k \ge -(2l+1)\}$$

(see Figure 3).

Journal of Physics: Conference Series **597** (2015) 012056 doi:10.1088/1742-6596/597/1/012056

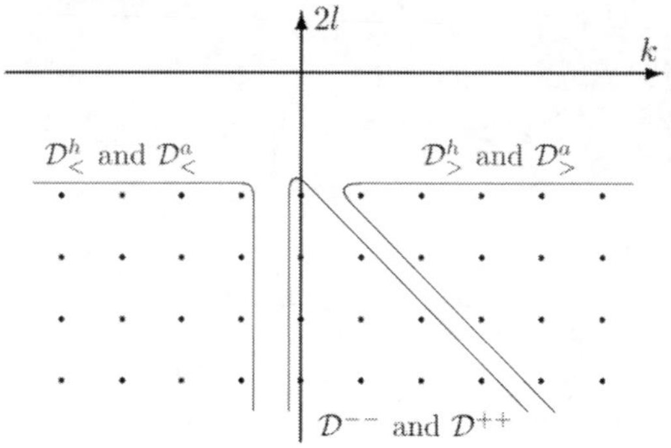

Figure 3. Decompositions of $(\rho_1, \mathcal{D}^h)$ and $(\rho_1, \mathcal{D}^a)$ into irreducible components.

Figure 4.
Action of $\rho_1\left(\begin{smallmatrix} A & 0 \\ 0 & 0 \end{smallmatrix}\right)$.

Figure 5.
Action of $\rho_1\left(\begin{smallmatrix} 0 & B \\ 0 & 0 \end{smallmatrix}\right)$.

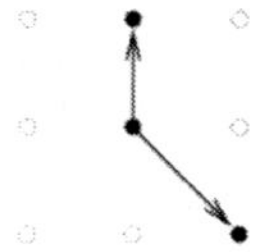

Figure 6.
Action of $\rho_1\left(\begin{smallmatrix} 0 & 0 \\ C & 0 \end{smallmatrix}\right)$.

Figure 7.
Action of $\rho_1\left(\begin{smallmatrix} 0 & 0 \\ 0 & D \end{smallmatrix}\right)$.

Proof. Since the proof is essentially the same as that of Theorem 7 in [4], we only give an outline. Note that the basis elements of $\mathcal{D}^h$ and $\mathcal{D}^a$ consist of functions of the kind

$$f_l(Z) \cdot N(Z)^k, \qquad \Box f_l(Z) = 0, \quad l = -1, -\frac{3}{2}, -2, -\frac{5}{2}, \dots, \quad k \in \mathbb{Z},$$

where the functions $f_l(Z)$ range over a basis of harmonic functions which are homogeneous of degree $2l$. Recall that we consider $U(1,1) \times U(1,1)$ and $SU(1,1) \times SU(1,1)$ as subgroups of $GL(2, \mathbb{H}_\mathbb{C})$ via (2) and (3). For k and l fixed, these functions span an irreducible representation of the Lie algebra $\mathfrak{u}(1,1) \times \mathfrak{u}(1,1)$, which remains irreducible when restricted to $\mathfrak{su}(1,1) \times \mathfrak{su}(1,1)$, belongs to the holomorphic discrete series when $m, n \geq -l$ and the antiholomorphic discrete series when $m, n \leq l$.

By Lemma 1, matrices of the kind $\left(\begin{smallmatrix} 0 & B \\ 0 & 0 \end{smallmatrix}\right) \in \mathfrak{gl}(2, \mathbb{H}_\mathbb{C})$ with $B \in \mathbb{H}_\mathbb{C}$ act by

$$f \mapsto \operatorname{Tr}\bigl(B \cdot (-\partial f)\bigr),$$

and, by equation (27) in [4],

$$\partial\bigl(f_l(Z) \cdot N(Z)^k\bigr) = \frac{2l+k+1}{2l+1} \partial f_l \cdot N(Z)^k + \frac{k}{2l+1}\bigl(Z^+ \cdot (\partial^+ f_l) \cdot Z^+ + Z^+ f_l\bigr) \cdot N(Z)^{k-1}$$

with ∂f_l and $Z^+ \cdot (\partial^+ f_l) \cdot Z^+ + Z^+ f_l$ being harmonic and having degrees $2l-1$ and $2l+1$ respectively.

Using Lemma 1 again, matrices of the kind $\left(\begin{smallmatrix} 0 & 0 \\ C & 0 \end{smallmatrix}\right) \in \mathfrak{gl}(2, \mathbb{H}_\mathbb{C})$ with $C \in \mathbb{H}_\mathbb{C}$ act by

$$f \mapsto \operatorname{Tr}\bigl(C \cdot (Z \cdot (\partial f) \cdot Z + 2Zf)\bigr) = \operatorname{Tr}\bigl(C \cdot (Z \cdot \partial(Zf))\bigr),$$

and, by equation (28) in [4],

$$Z \cdot \partial(f_l \cdot N(Z)^k) \cdot Z + 2Zf_l \cdot N(Z)^k$$
$$= \frac{2l+k+2}{2l+1}\big(Z\cdot(\partial f_l)\cdot Z + Zf_l\big)\cdot N(Z)^k + \frac{k+1}{2l+1}\partial^+ f_l \cdot N(Z)^{k+1}$$

with $Z\cdot(\partial f_l)\cdot Z + Zf_l$ and $\partial^+ f_l$ being harmonic and having degrees $2l+1$ and $2l-1$ respectively.

Since

$$\partial(f(Z^{-1})) = -Z^{-1}\cdot(\partial f)\big|_{Z^{-1}}\cdot Z^{-1},$$

it follows that

$$Z\cdot(\partial f_l)\cdot Z + Zf_l = -\operatorname{Inv}\circ\partial\circ\operatorname{Inv},$$

where

$$\operatorname{Inv}: \quad f \mapsto \frac{1}{N(Z)}\cdot f(Z^{-1}).$$

Hence $Z\cdot(\partial f_l)\cdot Z + Zf_l$ and its conjugate $Z^+\cdot(\partial^+ f_l)\cdot Z^+ + Z^+ f_l$ become zero if and only if $l=-1$.

The actions of $\left(\begin{smallmatrix} A & 0\\ 0 & 0\end{smallmatrix}\right)$, $\left(\begin{smallmatrix} 0 & B\\ 0 & 0\end{smallmatrix}\right)$, $\left(\begin{smallmatrix} 0 & 0\\ C & 0\end{smallmatrix}\right)$ and $\left(\begin{smallmatrix} 0 & 0\\ 0 & D\end{smallmatrix}\right)$ are illustrated in Figures 4-7. In Figure 5 the vertical arrow disappears when $2l+k+1=0$ and the diagonal arrow disappears when $k=0$ or $l=-1$. Similarly, in Figure 6 the vertical arrow disappears when $l=-1$ or $2l+k+2=0$ and the diagonal arrow disappears when $k=-1$. This proves that the spaces $\mathcal{D}^h$, $\mathcal{D}^h_<$, $\mathcal{D}^{--}$, $\mathcal{D}^h_>$, $\mathcal{D}^a$, $\mathcal{D}^a_<$, $\mathcal{D}^{++}$ and $\mathcal{D}^a_>$ are $\mathfrak{gl}(2,\mathbb{H}_\mathbb{C})$-invariant. This also proves that $\mathcal{D}^h_<$, $\mathcal{D}^{--}$ and $\mathcal{D}^h_>$ are irreducible $\mathfrak{gl}(2,\mathbb{H}_\mathbb{C})$-invariant subspaces of $\mathcal{D}^h$ and that $\mathcal{D}^a_<$, $\mathcal{D}^{++}$ and $\mathcal{D}^a_>$ are irreducible $\mathfrak{gl}(2,\mathbb{H}_\mathbb{C})$-invariant subspaces of $\mathcal{D}^a$. □

Remark 6. *Theorem 53 in [2] gives explicit isomorphisms of* $\mathfrak{gl}(2,\mathbb{H}_\mathbb{C})$*-modules*

$$(\rho_1, \mathcal{D}^{--}) \simeq (\rho_1, \mathcal{H}^-) \qquad \textit{and} \qquad (\rho_1, \mathcal{D}^{++}) \simeq (\rho_1, \mathcal{H}^+).$$

One immediately notices the similarity of Figures 2 and 3 illustrating the decompositions of Theorems 3 and 5 respectively. The figures look similar because the matrix coefficients $t^l_{n\,\underline{m}}(Z)$*'s of* $SU(2)$ *and* $\tau^l_{n\,\underline{m}}(Z)$*'s of* $SU(1,1)$ *span irreducible representations of the diagonal subalgebras for* l *fixed, are homogeneous of degree* $2l$*, except* $l\geq 0$ *in the case of* $SU(2)$*-coefficients and* $l\leq -1$ *in the case of* $SU(1,1)$*-coefficients. The "border conditions" separating irreducible components* $k\geq 0$ *or* $k\leq -1$ *and* $k\geq -(2l+1)$ *or* $k\leq -(2l+2)$ *are formally the same in the cases of* $\mathcal{H}$*,* $\mathcal{D}^h$ *and* $\mathcal{D}^a$*.*

5. Reproducing Kernels

In this section we obtain the reproducing kernels for the irreducible components of $\mathcal{D}^h$ and $\mathcal{D}^a$. Recall that Γ^- and Γ^+ are certain open Ol'shanskii semigroups lying inside $GL(2,\mathbb{C})\subset\mathbb{H}_\mathbb{C}$ which were described in Subsection 3.3 in [2]. For our purposes they can be defined as

$$\Gamma^- = \left\{ g\cdot\begin{pmatrix}\lambda_1 & 0\\ 0 & \lambda_2\end{pmatrix}\cdot g^{-1}\in GL(2,\mathbb{C});\ g\in SU(1,1),\ \lambda_1,\lambda_2\in\mathbb{C},\ |\lambda_1|>1>|\lambda_2|>0\right\},$$

$$\Gamma^+ = \left\{ g\cdot\begin{pmatrix}\lambda_1 & 0\\ 0 & \lambda_2\end{pmatrix}\cdot g^{-1}\in GL(2,\mathbb{C});\ g\in SU(1,1),\ \lambda_1,\lambda_2\in\mathbb{C},\ |\lambda_2|>1>|\lambda_1|>0\right\}.$$

We already have the reproducing kernels for $\mathcal{D}^{--}$ and $\mathcal{D}^{++}$:

Proposition 7 (Proposition 54 in [2]). *We have the following matrix coefficient expansions for the reproducing kernels of $\mathcal{D}^{--}$ and $\mathcal{D}^{++}$ respectively:*

$$\frac{1}{N(Z-W)^2} = \sum_{\substack{k,l,m,n \\ m,n \geq -l \geq 1 \\ 0 \leq k \leq -2l-2}} -(2l+1)\tau^l_{n\underline{m}}(W) \cdot N(W)^k \cdot \tau^l_{m\underline{n}}(Z^{-1}) \cdot N(Z)^{-k-2}$$

which converges pointwise absolutely whenever $WZ^{-1} \in \Gamma^-$ and

$$\frac{1}{N(Z-W)^2} = \sum_{\substack{k,l,m,n \\ m,n \leq l \leq -1 \\ 0 \leq k \leq -2l-2}} -(2l+1)\tau^l_{n\underline{m}}(W) \cdot N(W)^k \cdot \tau^l_{m\underline{n}}(Z^{-1}) \cdot N(Z)^{-k-2}$$

which converges pointwise absolutely whenever $WZ^{-1} \in \Gamma^+$. The sums are taken first over all m, n, then over $k = 0, 1, 2, \dots, -(2l+2)$, and then over $l = -1, -\frac{3}{2}, -2, -\frac{5}{2}, \dots$.

It remains to find the reproducing kernels for the other four irreducible components.

Proposition 8. *Suppose that WZ^{-1} is $SU(1,1)$-conjugate to a diagonal matrix $\begin{pmatrix} \lambda_1 & 0 \\ 0 & \lambda_2 \end{pmatrix}$. We have the following matrix coefficient expansions for the reproducing kernels of $\mathcal{D}^h_<$, $\mathcal{D}^h_>$, $\mathcal{D}^a_<$ and $\mathcal{D}^a_>$ respectively:*

(i) When $WZ^{-1} \in \Gamma^-$ and $|N(WZ^{-1})| > 1$ we have a pointwise absolutely convergent series

$$\sum_{\substack{k,l,m,n \\ m,n \geq -l \geq 1 \\ k \leq -1}} -(2l+1)\tau^l_{n\underline{m}}(W) \cdot N(W)^k \cdot \tau^l_{m\underline{n}}(Z^{-1}) \cdot N(Z)^{-k-2} = \frac{-N(Z)^{-2} \cdot \lambda_1}{(\lambda_1 - \lambda_2)(1-\lambda_1\lambda_2)(1-\lambda_1)^2};$$

(ii) When $WZ^{-1} \in \Gamma^-$ and $|N(WZ^{-1})| < 1$ we have a pointwise absolutely convergent series

$$\sum_{\substack{k,l,m,n \\ m,n \geq -l \geq 1 \\ k \geq -(2l+1)}} -(2l+1)\tau^l_{n\underline{m}}(W) \cdot N(W)^k \cdot \tau^l_{m\underline{n}}(Z^{-1}) \cdot N(Z)^{-k-2} = \frac{-N(Z)^{-2} \cdot \lambda_2}{(\lambda_1 - \lambda_2)(1-\lambda_1\lambda_2)(1-\lambda_2)^2};$$

(iii) When $WZ^{-1} \in \Gamma^+$ and $|N(WZ^{-1})| > 1$ we have a pointwise absolutely convergent series

$$\sum_{\substack{k,l,m,n \\ m,n \leq l \leq -1 \\ k \leq -1}} -(2l+1)\tau^l_{n\underline{m}}(W) \cdot N(W)^k \cdot \tau^l_{m\underline{n}}(Z^{-1}) \cdot N(Z)^{-k-2} = \frac{N(Z)^{-2} \cdot \lambda_2}{(\lambda_1 - \lambda_2)(1-\lambda_1\lambda_2)(1-\lambda_2)^2};$$

(iv) When $WZ^{-1} \in \Gamma^+$ and $|N(WZ^{-1})| < 1$ we have a pointwise absolutely convergent series

$$\sum_{\substack{k,l,m,n \\ m,n \leq l \leq -1 \\ k \geq -(2l+1)}} -(2l+1)\tau^l_{n\underline{m}}(W) \cdot N(W)^k \cdot \tau^l_{m\underline{n}}(Z^{-1}) \cdot N(Z)^{-k-2} = \frac{N(Z)^{-2} \cdot \lambda_1}{(\lambda_1 - \lambda_2)(1-\lambda_1\lambda_2)(1-\lambda_1)^2}.$$

The sums are taken first over all m, n, then over k, and then over $l = -1, -\frac{3}{2}, -2, -\frac{5}{2}, \dots$.

Journal of Physics: Conference Series **597** (2015) 012056 doi:10.1088/1742-6596/597/1/012056

Proof. We will prove part *(i)* only, the other parts can be proved in the same way. The proof is similar to that of Proposition 54 in [2]. Since $WZ^{-1} \in \Gamma^-$, WZ^{-1} is $SU(1,1)$-conjugate to a diagonal matrix $\left(\begin{smallmatrix} \lambda_1 & 0 \\ 0 & \lambda_2 \end{smallmatrix}\right)$ for some $\lambda_1, \lambda_2 \in \mathbb{C}$ with $|\lambda_1| > 1 > |\lambda_2| > 0$. Using equations (33) and (34) from [2] and letting the indices k, l, m, n run over

$$m, n = -l, -l+1, -l+2, \dots, \qquad k = -1, -2, -3, \dots, \qquad l = -1, -\frac{3}{2}, -2, -\frac{5}{2}, \dots,$$

we obtain:

$$\begin{aligned}
&\sum_{k,l,m,n} (2l+1)\tau^l_{n\underline{m}}(W) \cdot N(W)^k \cdot \tau^l_{m\underline{n}}(Z^{-1}) \cdot N(Z)^{-k-2} \\
&\quad = \sum_{k,l,n} \frac{2l+1}{N(Z)^2} \cdot \tau^l_{n\underline{n}}(WZ^{-1}) \cdot N(WZ^{-1})^k = \sum_{k,l} \frac{2l+1}{N(Z)^2} \cdot \tilde{\Theta}^-_l(WZ^{-1}) \cdot N(WZ^{-1})^k \\
&\qquad = \sum_{k,l} \frac{2l+1}{N(Z)^2} \cdot \frac{\lambda_1^{2l+1} \cdot (\lambda_1\lambda_2)^k}{\lambda_1 - \lambda_2} = \sum_{l \le -1} \frac{(2l+1)N(Z)^{-2} \cdot \lambda_1^{2l+1}}{(\lambda_1 - \lambda_2)(\lambda_1\lambda_2 - 1)} \\
&\qquad\qquad = \frac{N(Z)^{-2} \cdot \lambda_1}{(\lambda_1 - \lambda_2)(1 - \lambda_1\lambda_2)(1 - \lambda_1)^2}.
\end{aligned}$$

□

From these expansions of reproducing kernels and orthogonality relations (6) we immediately obtain the following extension of Theorem 57 from [2], which can be regarded as a split analogue of Corollary 14 and Theorem 15 in [4].

Theorem 9. *Let $R > 0$. The integral operators*

$$f \mapsto \frac{i}{2\pi^3} \int_{Z \in R \cdot U(1,1)} k(Z,W) \cdot f(Z)\, dV, \qquad f \in \mathcal{D}^h \oplus \mathcal{D}^a,$$

provide the $\mathfrak{gl}(2, \mathbb{H}_\mathbb{C})$-equivariant projectors onto the irreducible components of $\mathcal{D}^h \oplus \mathcal{D}^a$ as follows:

(i) If $k(Z,W) = N(Z-W)^{-2}$ and $W \in R \cdot \Gamma^-$, this operator provides the projector onto $\mathcal{D}^{--}$;

(ii) If $k(Z,W) = N(Z-W)^{-2}$ and $W \in R \cdot \Gamma^+$, this operator provides the projector onto $\mathcal{D}^{++}$;

(iii) If $k(Z,W) = \frac{-N(Z)^{-2} \cdot \lambda_1}{(\lambda_1-\lambda_2)(1-\lambda_1\lambda_2)(1-\lambda_1)^2}$ and $W \in R \cdot \Gamma^-$, $|N(W)| > 1$, this operator provides the projector onto $\mathcal{D}^h_<$;

(iv) If $k(Z,W) = \frac{-N(Z)^{-2} \cdot \lambda_2}{(\lambda_1-\lambda_2)(1-\lambda_1\lambda_2)(1-\lambda_2)^2}$ and $W \in R \cdot \Gamma^-$, $|N(W)| < 1$, this operator provides the projector onto $\mathcal{D}^h_>$;

(v) If $k(Z,W) = \frac{N(Z)^{-2} \cdot \lambda_2}{(\lambda_1-\lambda_2)(1-\lambda_1\lambda_2)(1-\lambda_2)^2}$ and $W \in R \cdot \Gamma^+$, $|N(W)| > 1$, this operator provides the projector onto $\mathcal{D}^a_<$;

(vi) If $k(Z,W) = \frac{N(Z)^{-2} \cdot \lambda_1}{(\lambda_1-\lambda_2)(1-\lambda_1\lambda_2)(1-\lambda_1)^2}$ and $W \in R \cdot \Gamma^+$, $|N(W)| < 1$, this operator provides the projector onto $\mathcal{D}^a_>$;

where WZ^{-1} is $SU(1,1)$-conjugate to $\left(\begin{smallmatrix} \lambda_1 & 0 \\ 0 & \lambda_2 \end{smallmatrix}\right)$.

References

[1] Frenkel I and Libine M 2008 Quaternionic analysis, representation theory and physics *Adv. in Math.* **218** 1806-1877 (also arXiv:0711.2699)

[2] Frenkel I and Libine M 2011 Split quaternionic analysis and the separation of the series for $SL(2,\mathbb{R})$ and $SL(2,\mathbb{C})/SL(2,\mathbb{R})$ *Adv. in Math.* **228** 678-763 (also arXiv:1009.2532)

[3] Frenkel I and Libine M 2012 Quaternionic analysis and the Schrödinger model for the minimal representation of $O(3,3)$ *Int. Math. Res. Notices* **2012** 4904-4923 (also arXiv:1110.2106)

[4] Frenkel I and Libine M 2014 Anti De Sitter deformation of quaternionic analysis and the second-order pole *Int. Math. Res. Notices* to appear (doi: 10.1093/imrn/rnu083; also arXiv:1404.7098)

[5] Smirnov V A 2004 Evaluating Feynman integrals *Springer Tracts in Modern Physics* **211** (Berlin: Springer-Verlag)

[6] Vilenkin N Ja 1968 Special functions and the theory of group representations (translated from the Russian by V. N. Singh) *Translations of Mathematical Monographs* **22** (Providence: American Mathematical Society)

New families of superintegrable systems from k-step rational extensions, polynomial algebras and degeneracies

Ian Marquette

School of Mathematics and Physics, The University of Queensland, Brisbane, QLD 4072, Australia

E-mail: i.marquette@uq.edu.au

Abstract. Four new families of two-dimensional quantum superintegrable systems are constructed from k-step extension of the harmonic oscillator and the radial oscillator. Their wavefunctions are related with Hermite and Laguerre exceptional orthogonal polynomials (EOP) of type III. We show that ladder operators obtained from alternative construction based on combinations of supercharges in the Krein-Adler and Darboux Crum (or state deleting and creating) approaches can be used to generate a set of integrals of motion and a corresponding polynomial algebra that provides an algebraic derivation of the full spectrum and total number of degeneracies. Such derivation is based on finite dimensional unitary representations (unirreps) and doesn't work for integrals build from standard ladder operators in supersymmetric quantum mechanics (SUSYQM) as they contain singlets isolated from excited states. In this paper, we also rely on a novel approach to obtain the finite dimensional unirreps based on the action of the integrals of motion on the wavefunctions given in terms of these EOP. We compare the results with those obtained from the Daskaloyannis approach and the realizations in terms of deformed oscillator algebras for one of the new families in the case of 1-step extension. This communication is a review of recent works.

1. Introduction

Superintegrable systems in classical and quantum mechanics have many properties that make them very interesting models from both mathematics and physics point of view [1]. Among these properties, the algebra of integrals of motion is non-Abelian and usually it is a finitely generated polynomial algebra. Moreover, this polynomial algebra in quantum mechanics provides energy spectra and information on wave functions and leads to an additional degeneracy of energy levels. This is however a quite complicated task for systems with higher order integrals to obtain the polynomial algebra generated by the integrals even in two-dimensions and in the case of cubic algebras [2]. Moreover, the analysis of the representations in order to obtain an algebraic derivation of the energy spectrum and total number of degeneracies is also a difficult problem. A method was introduced by Dakaloyannis [3] to study representations of quadratic algebras based on Casimir operators and realizations as deformed oscillator algebras and was applied on Smorodinsky-Winternitz systems [4]. This approach was generalized to cubic algebras [2] and even to a class of polynomial algebras of arbitrary order with three generators [5].

It has been pointed out that not all the spectrum and the degeneracies were recovered from the finite dimensional unitary representations (unirreps) [2] for many of the Gravel systems [6]. More recently, we showed that these systems are first members of families of superintegrable systems involving 1-step extension of harmonic oscillator [7] and related to Hermite exceptionnal orthogonal polynomials (EOP) of type III (X_m). Such orthogonal polynomials were introduced quite recently in context of Sturm-Liouville problems and supersymmetric quantum mechanics (SUSYQM) [8, 9, 10, 10, 12, 13, 14, 15] but the connection with superintegrability was much more unexplored and only one example related to Jacobi EOP was studied [16]. It was pointed out that the incomplete algebraic solutions were due to the construction of integrals from standard ladder operators in supersymmetric quantum mechanics and that these ladder operators admit isolated singlets from excited states [7]. However, all the energy levels and degeneracies were obtained algebraically for the 1-step extension of radial oscillator related Laguerre EOP of type I and II, but not for the type III. It was demonstrated in another paper that new type of ladder operators with $m+1$ infinite chains of levels can be created in the case of 1-step extension of the harmonic oscillator from which a new set of integrals and a corresponding polynomial algebra provide via finite dimensional unirreps the whole spectrum and total number of degeneracies [17]. We also pointed out that another type of ladder operators with doublet states and many infinite chains of levels [18] for 2-step extension of harmonic oscillator exists, which showed that ladder operators for a given systems are far than unique. However, these new ladder operators for these 1D exactly solvable systems related to type III Hermite EOP (X_{m_1,m_2}) do not allow to generate integrals from which all the spectrum can be obtained for 2D superintegrable generalisations. From these results, we presented a way to construct new ladder from deleting and creating approach of supersymmetric quantum mechanics for k-step extension of the harmonic oscillator and radial oscillator both related to type III EOP (Hermite and Laguerre). We constructed new superintegrable models and showed how from a new set of integrals and its corresponding polynomial algebra a complete algebraic derivation of the whole spectrum and total number of degeneracies can be performed using a novel approach based on the action of the integrals [19].
The purpose of this communication is to present a review of some of these results obtained in collaboration with C.Quesne [7, 17, 18, 19] and point out also the more complex patterns of the energy levels and unirreps for these superintegrable Hamiltonians.
The paper is organized as follow. In Section 2, we present four families of 2D superintegrable systems and their degeneracies. In Section 3, we present the polynomial algebras of these systems. In Section 4, we present an algebraic derivation of the degeneracies of the complete spectrum using a direct approach for the four cases involving k-step extension. In Section 5, we compare these results obtained in Section 4 in the case of 1-step for one of the families using Daskaloyannis approach to obtain finite dimensional unirreps.

2. New superintegrable systems and degeneracies

In previous paper we introduce new families of quantum superintegrable systems from k-step extension of the harmonic oscillator

$$\begin{aligned} H_a = H_x^{(ko)} + H_y^{(o)} = -\frac{d^2}{dx^2} - \frac{d^2}{dy^2} + x^2 + y^2 - 2k \\ - 2\frac{d^2}{dx^2} \log \mathcal{W}(\mathcal{H}_{m_1}, \mathcal{H}_{m_2}, \ldots, \mathcal{H}_{m_k}), \end{aligned} \tag{1}$$

$$\begin{aligned} H_b = H_x^{(kro)} + H_y^{(o)} = -\frac{d^2}{dx^2} - \frac{d^2}{dy^2} + \frac{1}{4}x^2 + \frac{l(l+1)}{x^2} + y^2 - k \\ - 2\frac{d^2}{dx^2} \log \tilde{\mathcal{W}}(L_{m_1}^{(-\alpha-k)}(-z), L_{m_2}^{(-\alpha-k)}(-z), \ldots, L_{m_k}^{(-\alpha-k)}(-z)), \end{aligned} \tag{2}$$

$$H_c = H_x^{(ko)} + H_y^{(ro)} = -\frac{d^2}{dx^2} - \frac{d^2}{dy^2} + x^2 + \frac{1}{4}y^2 + \frac{l(l+1)}{y^2} - 2k \\ - 2\frac{d^2}{dx^2}\log \mathcal{W}(\mathcal{H}_{m_1}, \mathcal{H}_{m_2}, \ldots, \mathcal{H}_{m_k}), \tag{3}$$

$$H_d = H_x^{(kro)} + H_y^{(ro)} = -\frac{d^2}{dx^2} - \frac{d^2}{dy^2} + \frac{1}{4}x^2 + \frac{l(l+1)}{x^2} + \frac{1}{4}y^2 + \frac{l(l+1)}{y^2} - k \\ - 2\frac{d^2}{dx^2}\log \tilde{\mathcal{W}}(L_{m_1}^{(-\alpha-k)}(-z), L_{m_2}^{(-\alpha-k)}(-z), \ldots, L_{m_k}^{(-\alpha-k)}(-z)), \tag{4}$$

where the 1D components $H^{(o)}$, $H^{(ro)}$, $H^{(ko)}$ and $H^{(kro)}$ in Eqs. (1), (2), (3) and (4) are respectively the harmonic oscillator, the radial oscillator, the k-step extension of the harmonic oscillator and the k-step extension of the radial oscillator [19]. We have the condition $\alpha + k > m_k + 1$ and $\mathcal{H}_m(x) = (-\mathrm{i})^m H_m(\mathrm{i}x)$ is a mth-degree pseudo-Hermite polynomial and $L_\nu^{(\alpha)}(z)$ denotes a νth-degree Laguerre polynomial. $m_1 < m_2 < \cdots < m_k$ with m_i even (resp. odd) for i odd (resp. even).

For convenience, the parameter $\hbar$ have been set to 1 in all the quantum Hamiltonians, supercharges, ladder and integrals [19], however the classical limit $\hbar \to 0$ of systems $H^{(ko)}$ and $H^{(kro)}$ will be only the harmonic oscillator and the radial oscillator. This is a consequence that these k-step extensions and their ladder operators are build from supercharges in supersymmetric quantum mechanics which is by essence a quantum formalism. However, ladder operators and related deformed Heisenberg algebras can still be defined in classical mechanics by replacing the commutator by the Poisson bracket and obtained by other methods [20, 21] as SUSYQM can not be used to generate such operators. However, as for the case of superintegrable systems, the search and classification of 1D classical systems with analog of ladder operators needs to be performed separately than the quantum case, but such 1D systems can also be used to generate 2D classical superintegrable systems [21].

The wave functions can be calculated using Darboux-Crum or Krein-Adler formalism and are related to Hermite and Laguerre EOP of type III. These four families of systems generalise the Smorodinsky-Winternitz Hamiltonians [4]. Moreover, it can be shown using the separation of variables that the energy spectrum $E_{i,N}$ of the corresponding Hamiltonian H_i $(i = a, b, c, d)$ is

$$E_{i,N} = 2N + \gamma_i, \quad N = \nu_x + \nu_y + 1, \tag{5}$$

where the constants $\gamma_a = 0$, $\gamma_b = \alpha + k$, $\gamma_c = \alpha$, $\gamma_d = 2\alpha + k$, and the quantum numbers ν_x and ν_y can take the following values

$$\nu_x = -m_k - 1, \ldots, -m_1 - 1, 0, 1, 2, \ldots, \qquad \nu_y = 0, 1, 2 \ldots. \tag{6}$$

A direct calculation of the degeneracies of the spectrum given by Eq. (5) by an enumeration process leads to the following pattern

$$\deg(E_N) = \begin{cases} k - j + 1 & \text{if } N = -m_j, -m_j + 1, \ldots, -m_{j-1} - 1, \\ & \text{for } j = 2, 3, \ldots, k, \\ k & \text{if } N = -m_1, -m_1 + 1, \ldots, 0, \\ N + k & \text{if } N = 1, 2, 3, \ldots. \end{cases} \tag{7}$$

This is one of the distinctive feature of these superintegrable Hamiltonians to exhibit more complicate patterns for the degeneracies given by Eq. (7) than what was previously observed in other models [1]. Let us emphasis that in other superintegrable Hamiltonians obtained even if the finitely generated polynomial algebra contains higher order terms and even of arbitrary order as for the 2D Tremblay-Turbiner-Winternitz (TTW) systems [22, 1, 23, 24], the degeneracies are similar to the 2D anisotropic harmonic oscillator. We thus need to provide an algebraic explanation for unusual degeneracies in Eq.(7) of these four quantum systems H_a, H_b, H_c and H_d i.e. we need to obtain the whole energy spectrum and total number of degeneracies from the unirreps of the polynomial algebra and the Section 3, 4 and 5 of this papers are devoted to this task.

3. Construction of the integrals and the polynomial algebras

These four families of systems presented in previous section and given by Eqs. (1), (2), (3) and (4) belong to the following class of Hamiltonian that consists in sum of two 1D Hamiltonians and allowing separation of variables in Cartesian coordinates

$$H = H_x + H_y = -\frac{d^2}{dx^2} - \frac{d^2}{dy^2} + V_x(x) + V_y(y), \tag{8}$$

and allowing ladder operators of order l_x and l_y that satisfy polynomial Heisenberg algebras (PHA's) $(i = x, y)$

$$[H_i, a_i^\dagger] = \lambda_i a_i^\dagger, \quad [H_i, a_i] = -\lambda_i a_i, \tag{9}$$

$$a_i a_i^\dagger = Q_i(H_i + \lambda_i), \quad a_i^\dagger a_i = Q_i(H_i), \tag{10}$$

where λ_x and λ_y are constants. Q_x and Q_y are polynomials in H_x and H_y. The second order integrals K consists only in the difference of the 1D parts and the higher order intgrals I_- and I_+ (with $n_1\lambda_x = n_2\lambda_y = \lambda$, n_1,$n_2 \in \mathbb{Z}^*$) take a very convenient factorized form

$$K = \frac{1}{2\lambda}(H_x - H_y), \quad I_- = a_x^{n_1} a_y^{\dagger n_2}, \quad I_+ = a_x^{\dagger n_1} a_y^{n_2}. \tag{11}$$

The integrals are respectively of order 2, $l_x n_1 + l_y n_2$ and $l_x n_1 + l_y n_2$ and generate a polynomial algebra with three generators $PA(3)$.

$$[K, I_\pm] = \pm I_\pm, \quad [I_-, I_+] = F_{n_1,n_2}(K+1, H) - F_{n_1,n_2}(K, H), \tag{12}$$

$$F(K) = \prod_{i=1}^{n_1} Q_x\left(\frac{H}{2} + \lambda K - (n_1 - i)\lambda_x\right) \prod_{j=1}^{n_2} Q_y\left(\frac{H}{2} - \lambda K + j\lambda_y\right), \tag{13}$$

where $F(K)$ is a polynomial of degree $n_1 l_x + n_2 l_y$ in the generator K and the polynomial algebra is of order $n_1 l_x + n_2 l_y - 1$. As it is expected for a 2D systems it involved the central element H.

3.1. Polynomial algebra of the four families

Explicit form of the ladder operators (c_o,$c_o^\dagger$), (c_{ro},$c_{ro}^\dagger$), (c_{ko},$c_{ko}^\dagger$) and (c_{kro},$c_{kro}^\dagger$) that are respectively ladder operators of the harmonic oscillator $H^{(o)}$, radial oscillator $H^{(ro)}$, k-step extension of the harmonic oscillator $H^{(ko)}$ and k-step extension of the radial oscillator $H^{(kro)}$ have been given and their corresponding PHA [19]. The operators c_{ko} and c_{kro} were build using deleting and creating approaches of SUSYQM.

Let us summarize by Table 1, the parameters of the PHA's and their order for each of the four 1D components involved in the four families of superintegrable systems.

These ladder operators allow to generate integrals of motion factorized forms as given by Eq. (11) and the polynomial algebra from Eq. (12) and (13) for the four families of superintegrable systems. We present in Table 2, the order of these integrals I_- and I_+, the

Table 1. The values $|a_x|$, $|a_y|$, $|Q_x|$ and $|Q_y|$ denote the order of the ladder operators and the order of the polynomials Q_x and Q_y. The parameters λ_x and λ_y of the polynomial Heisenberg algebras (PHA's) are also given.

Families	λ_x	λ_y	a_x	a_y	$\|a_x\|$	$\|a_y\|$	Q_x	Q_y	$\|Q_x\|$	$\|Q_y\|$
H_a	$2m_k+2$	2	c_{ko}	c_o	m_k+1	1	Q_{ko}	Q_o	m_k+1	1
H_b	$2m_k+2$	2	c_{kro}	c_o	$2m_k+2$	1	Q_{kro}	Q_o	$2m_k+1$	1
H_c	$2m_k+2$	2	c_{ko}	c_{ro}	m_k+1	2	Q_{ko}	Q_{ro}	m_k+1	2
H_d	$2m_k+2$	2	c_{kro}	c_{ro}	$2m_k+2$	2	Q_{kro}	Q_{ro}	$2m_k+1$	2

Table 2. The parameters n_1, n_2 and λ for integrals of the four families of systems H_a, H_b, H_c and H_d are given. $|I_+|$, $|I_-|$, $|F_{n_1,n_2}|$ and $|PA(3)|$ denote respectively the order of the integrals of motion I_+ and I_-, the order of the polynomial F_{n_1,n_2} and the order of the related polynomial algebra.

Families	n_1	n_2	λ	$\|I_+\|$	$\|I_-\|$	$\|F_{n_1,n_2}\|$	$\|PA(3)\|$
H_a	1	m_k+1	$2m_k+2$	$2m_k+2$	$2m_k+2$	$2m_k+2$	$2m_k+1$
H_b	1	m_k+1	$2m_k+2$	$3m_k+3$	$3m_k+3$	$3m_k+2$	$3m_k+1$
H_c	1	m_k+1	$2m_k+2$	$3m_k+3$	$3m_k+3$	$3m_k+3$	$3m_k+2$
H_d	1	m_k+1	$2m_k+2$	$4m_k+4$	$4m_k+4$	$4m_k+3$	$4m_k+2$

order of the polynomial $|F_{n_1,n_2}|$ and the order of the polynomial aLgebra $|PA(3)|$. The integrals K is of order 2 in all cases. Let us also present explicitly the polynomial algebra by provinding $F_{n_1,n_2}(K)$ for the Hamiltonians H_a, H_b, H_c and H_d :

$$F_a(K) = \prod_{n=1}^{k}(\frac{H}{2} + (2m_k+2)K + 2m_n + 1) \tag{14}$$

$$\prod_{r=1,r\neq T}^{m_k}(\frac{H}{2} + (2m_k+2)K - 2r - 1)\prod_{j=1}^{m_k+1}(\frac{H}{2} - (2m_k+2)K + 2j),$$

$$F_b(K) = \prod_{n=1}^{k}(\frac{H}{2} + (2m_k+2)K - \alpha + 2m_n - k + 1) \tag{15}$$

$$\prod_{r=0}^{m_k}(\frac{H}{2} + (2m_k+2)K + \alpha - 2r + k - 1)$$

$$\prod_{s=1,r\neq T}^{m_k}(\frac{H}{2} + (2m_k+2)K - \alpha - 2s - k - 1)\prod_{j=1}^{m_k+1}(\frac{H}{2} - (2m_k+2)K + 2j),$$

$$F_c(K) = \prod_{n=1}^{k}(\frac{H}{2} + (2m_k+2)K + 2m_n + 1)\prod_{r=1,r\neq T}^{m_k}(\frac{H}{2} + (2m_k+2)K - 2r - 1) \tag{16}$$

$$\prod_{j=1}^{m_k+1}\frac{1}{4}(\frac{H}{2} - (2m_k+2)K + 2j - \frac{3}{2} - l)(\frac{H}{2} - (2m_k+2)K + 2j - \frac{1}{2} + l),$$

$$F_d(K) = \prod_{n=1}^{k}(\frac{H}{2} + (2m_k+2)K - \alpha + 2m_n - k + 1) \tag{17}$$

$$\prod_{r=0}^{m_k}(\frac{H}{2} + (2m_k+2)K + \alpha - 2r + k - 1) \prod_{s=1, r\neq T}^{m_k}(\frac{H}{2} + (2m_k+2)K - \alpha - 2s - k - 1)$$

$$\prod_{j=1}^{m_k+1}\frac{1}{4}(\frac{H}{2} - (2m_k+2)K + 2j - \frac{3}{2} - l)(\frac{H}{2} - (2m_k+2)K + 2j - \frac{1}{2} + l),$$

where $T = \{m_k - m_{k-1}, \ldots, m_k - m_1\}$.

4. Direct approach of constructing unirreps

The polynomial algebras given by Eq. (12) and (13) with Eqs. (14), (15), (16) and (17) are deformed $u(2)$ Lie algebras and the finite-dimensional unirreps may be characterized by their basis states by $|N, \tau, s, \sigma\rangle$ where $\sigma = -s, -s+1, \ldots, s$ and τ distinguishes between repeated representations specified by the same s (integer or half-integer). Moreover, σ denotes the eigenvalue of $I_0 = K + c$, where c is some representation-dependent constant and I_+, I_- are such that $I_+|N, \tau, s, s\rangle = I_-|N, \tau, s, -s\rangle = 0$. From these constraints, we obtain the value of s for every N value. In order to implement the method and obtain the unirreps, we need to denote the states differently. These states can be represented in unified way by the two quantum numbers ν_x and ν_y as $|\nu_x\rangle_1|\nu_y\rangle_2$ but also using the quantum numbers ν_x and N , where $N = \nu_x + \nu_y + 1$ as $|N, \nu_x\rangle = |\nu_x\rangle_1|N - \nu_x - 1\rangle_2$. The action of the integrals I_- and I_+ in this new basis can be obtained from the one of the lowering and raising operators as I_- and I_+ are in fact polynomials of the ladders operators. The action of the operators c_{ko} or c_{kro} in the x axis was shown to be [19] :

$$c\psi_\nu^{(2)} = 0, \qquad \nu = -m_k - 1, \ldots, -m_1 - 1, 1, 2, \ldots, m_k - m_{k-1} - 1, \\ m_k - m_{k-1} + 1, \ldots, m_k - m_1 - 1, m_k - m_1 + 1, \ldots, m_k, \tag{18}$$

$$c\psi_0^{(2)} \propto \psi_{-m_k-1}^{(2)}, \quad c\psi_{m_k - m_i}^{(2)} \propto \psi_{-m_i-1}^{(2)}, \qquad i = 1, 2, \ldots, k-1, \tag{19}$$

$$c\psi_\nu^{(2)} \propto \psi_{\nu - m_k - 1}^{(2)}, \qquad \nu = m_k + 1, m_k + 2, \ldots. \tag{20}$$

and for c_o and c_{ro} in the y axis is given by

$$c\psi_0^{(2)} = 0, \quad c\psi_\nu^{(2)} \propto \psi_{\nu+1}^{(2)}, \qquad \nu = 1, 2, 3, \ldots \quad . \tag{21}$$

The key element here consists in the fact the action of the ladder operators of the harmonic oscillator and radial oscillator on their respective eigenstates is similar even if their order and their PHA are different. The same observation can be made for their k-step extension as the action of the k-step extension of the harmonic oscillator and the k-step extension of the radial oscillator on their respective eigenfunctions in terms of EOP are similar even if their order as differential operators differ greatly and their corresponding PHA. Thus the action of integrals in the $|N, \nu_x\rangle$ basis is obtained in unified way for the four superintegrable systems with $K|N, \nu_x\rangle \propto |N, \nu_x\rangle$ and

$$I_+|N, \nu_x\rangle \propto |N, \nu_x + m_k + 1\rangle, \quad I_-|N, \nu_x\rangle \propto |N, \nu_x - m_k - 1\rangle. \tag{22}$$

The zero modes of these operators are enumerated in term of the values of ν_x and N by considering

$$I_-|N, \nu_x\rangle = 0, \quad I_+|N, \nu_x\rangle = 0. \tag{23}$$

Tables of these states were presented in the paper [19]. Starting from states annihilated by I_- and acting with I_+ until we reach a state annihilated by I_+ we identify the values of s where $2s = p$ is the number of states in the sequences. Furthermore, the σ is associated with each state forming a given sequence. Using notation $N = \lambda n_1 n_2 + \mu$ with appropriate values of λ and μ, we enumerate all the states and obtain the values of τ. This construction is based on the action on physical states and thus there is no problem of nonphysical states as for the Daskaloyannis approach. All the states are classified in sequences of different length that correspond to finite dimensional unirreps and given in Table 3. There are 11 cases and 21 types of sequences.

Table 3. Set of $p = 2s$ values with their number of occurrences, number $\mathcal{N}$ of unirreps per level, and total level degeneracy for the polynomial algebra $PA(3)$ given by Eq. (12) and (13) with Eqs. (14), (15), (16) and (17) related to Hamiltonians H_a, H_b, H_c, or H_d. In this table, j runs over $2, 3, \ldots, k-1$.

Case	λ	μ	$2s = p$	$\mathcal{N}$	$\deg(E_N)$
1	-1	$1, \ldots, m_k - m_{k-1}$	0	1	1
2	-1	$m_k - m_j + 1, \ldots, m_k - m_{j-1}$	0^{k-j+1}	$k - j + 1$	$k - j + 1$
3	-1	$m_k - m_1 + 1, \ldots, m_k$	0^k	k	k
4	0	0	0^k	k	k
5	0	$1, \ldots, m_k - m_{k-1}$	1 $0^{\mu+k-2}$	$\mu + k - 1$	$N + k$
6	0	$m_k - m_j + 1, \ldots, m_k - m_{j-1}$	1^{k-j+1} $0^{\mu-k+2j-2}$	$\mu + j - 1$	$N + k$
7	0	$m_k - m_1 + 1, \ldots, m_k$	1^k $0^{\mu-k}$	μ	$N + k$
8	$1, 2, \ldots$	0	λ^k $(\lambda - 1)^{m_k-k+1}$	$m_k + 1$	$N + k$
9	$1, 2, \ldots$	$1, \ldots, m_k - m_{k-1}$	$\lambda + 1$ $\lambda^{\mu+k-2}$ $(\lambda - 1)^{m_k-\mu-k+2}$	$m_k + 1$	$N + k$
10	$1, 2, \ldots$	$m_k - m_j + 1, \ldots, m_k - m_{j-1}$	$(\lambda + 1)^{k-j+1}$ $\lambda^{\mu-k+2j-2}$ $(\lambda - 1)^{m_k-\mu-j+2}$	$m_k + 1$	$N + k$
11	$1, 2, \ldots$	$m_k - m_1 + 1, \ldots, m_k$	$(\lambda + 1)^k$ $\lambda^{\mu-k}$ $(\lambda - 1)^{m_k-\mu+1}$	$m_k + 1$	$N + k$

5. Daskaloyannis approach of constructing unirreps

Alternatively, the polynomial algebras given by Eq. (12) and (13) with Eqs. (14), (15), (16) and (17) belong to the class of polynomial algebras of arbitrary order studied [5] and for which constraints from Jacobi identity, the Casimir operators and the realizations as deformed oscillator algebras were obtained. A formula can be obtained for the structure function at any given order.

In general the construction of unirreps involve the realization of the Casimir operators in terms of central elements i.e. H for 2D Hamiltonians. In our case, we have a very specific structure and the integrals are build from ladder operators generated by various supercharges and these ladder operators generate PHA's. Due to these underlying structures the realization of the polynomial algebra with three generators as deformed oscillator algebra can be obtained in a straightforward manner by taking $b^\dagger = I_+$, $b = I_-$ and $N = K - u$ (where u is a representation dependent parameter that is determined using further constraints) :

$$[N, b^\dagger] = b^\dagger, \quad [N, b] = -b, \tag{24}$$

$$b^\dagger b = \Phi(N), \quad bb^\dagger = \Phi(N+1), \quad \Phi(H, u, N) = F_{n_1,n_2}(K, H). \tag{25}$$

In general the Fock basis is $|c, n>$ with $C|c, n>= c|c, n>$ and $N|c, n>= n|c, n>$ (where C is the Casimir operator) and using realization of the Casimir operators in term of H we can in fact use a basis $|E, n>$ such that $H|E, n>= E|E, n>$ and $N|E, n>= n|E, n>$. For our 4 systems, we can simply take a such basis $|E, n>$. Furthermore, we have $b|E, 0>= 0$. and the action of operators b and $b^\dagger$ are given by $b^\dagger|n>\propto |E, n+1>$ and $b|n>\propto |E, n-1>$. We have the existence of finite dimensional unitary representations if we impose the following three constraints : $\Phi(E, p+1, u) = 0$, $\Phi(E, 0, u) = 0$ and $\Phi(E, n, u) > 0$, $\forall\, n > 0$. The energy E and the constant u can be obtained from this set of constraints that are algebraic equations as

$$E_i = E_i(p), \quad \Phi_i(x) = x(p+1-x)\prod_j^L(\alpha_i^{(j)}p + \beta_i^{(j)}x + \gamma_i^{(j)}), \quad u_i = u_i(p).$$

For most cases, the solutions will be valid for $p \in \mathbb{N}$ and p correspond to a degeneracy of $p+1$. However, truncated solutions that are valid only for $p \leq p_{max}$ were observed [2, 7] and such phenomenon is associated with isolated multiplets of the ladder operators that are used to generate the integrals. However, even if the solutions are not truncated, $p+1$ is not necessarily the total number of degeneracies as for a given energy level $E(p)$ more than one solutions $E_i(p)$ can produce the same energy E with a different values of p. The total number of degeneracies will be given by unions of finite dimensional unirreps. Let us point out the fact that the structure function is a polynomial allows the existence of many solutions and the possibility to obtain algebraically these more complicated patterns. Let us present a calculation for the Case H_a and with one 1-step. There are two families of unirreps with $p \in \mathbb{N}$, $k \in \{1, 2, \ldots, m+1\}$, $l \in \{1, 2, \ldots, m\}$,

$$\Phi_1 = x\prod_{i=1}^{m}[(m+1)x - m - 1 - i]\prod_{j=1}^{m+1}[(m+1)(p+1-x) - m + j - k], \tag{26}$$

$$\Phi_2 = [(m+1)x + m + 1 + l]\prod_{i=1}^{m}[(m+1)x + l - i]\prod_{j=1}^{m+1}[(m+1)(p+1-x) + j - k]. \tag{27}$$

The associated energy spectrum is given by

$$E_1(p) = 2[(m+1)p + 1 - k], \quad 2(m+1)u_1 = -\frac{E}{2} - 2m - 1, \tag{28}$$

$$E_2(p) = 2[(m+1)(p+1) + l - k + 1], \quad 2(m+1)u_2 = -\frac{E}{2} + 2l + 1, \tag{29}$$

where $p \in \mathbb{N}$. We set $\nu_x = (m+1)n_x + a_1$, $\nu_y = (m+1)n_y + a_2$, with $n_x, n_y \in \mathbb{N}$, $a_1 \in \{-m-1, 1, 2, \ldots, m\}$, and $a_2 \in \{0, 1, \ldots, m\}$. The energy E obtained from separation of variables and wavefunctions involving Hermite EOP of type III can be rewritten as $E = 2[(m+1)(n_x+n_y)+a_1+a_2+1]$. E_1 and E_2, correspond to E with $n_x+n_y = p \in \mathbb{N}$, $a_2 = m+1-k \in \{0,1,\ldots,m\}$, and $a_1 = -m-1$ or $a_1 = l \in \{1,2,\ldots,m\}$. The total number of degeneracies is not obtained directly as these $(m+1)^2$ solutions can provide the same energy for different value of p and one needs to enumerate the states from these expressions (i.e. $E_1(p)$ and $E_2(p)$) and taking into account that an energy with a given p represent $p+1$ states.There are only 5 different cases and 9 types of sequences [17] that are contained in the more general results for k-step extensions and presented in Table 3.

6. Conclusion

We presented four new families of systems involving k-step extension of the harmonic oscillator and radial oscillator related to Hermite and Laguerre EOP of type III. Let us denote, that families that are superintegrable and involve k-step extension in both axis were also obtained and studied [19]. These four superintegrable models are caracterised by interesting degeneracies patterns that are related with non trivial finite dimensional unirreps given by Table 3. These results demonstrate how the symmetry algebra and its representations explain the degeneracies of these systems as observed in other superintegrable Hamiltonians with less complicated pattern. These results also showed that superintegrable systems with spectrum beyond isotropic and anisotropic harmonic oscillator type of degeneracies exist and can be build by using these 1D components related to type III EOP.

Furthermore, we showed that the Daskaloyannis approach can be used for these systems with higher order integrals, however appropriate set of integrals generated from ladder operators without multiplets isolated from excited states need to be used. We pointed out that a direct approach can be taken to study in an unified way the four quantum models by using the action of the integrals. Moreover, this novel approach relying on action of the integrals also reveals the underlying structure of the unirreps in a more explicit way in terms of states than the Daskaloyannis approach that is very convenient when no factorized form in terms of ladder have been identified for the integrals. An open problem is the generalisation of these approaches for higher dimensional superintegrable quantum systems and their related finitely generated higher rank polynomial algebras which possess more complicated commutation relations [25].

Acknowledgments

The research of I.M. was supported by a Discovery Early Career Researcher Award DE130101067.

References

[1] Miller Jr W, Post S and Winternitz P 2013 *J. Phys.* A **46** 423001 (*Preprint* arXiv:1309.2694)
[2] Marquette I 2009 *J. Math. Phys.* **50** 012101 (*Preprint* arXiv:0807.2858)
[3] Daskaloyannis C 2001 *J. Math. Phys.* **42** 1100 (*Preprint* arXiv:math-ph/0003017)
[4] Winternitz P, Smorodinsky Ya A, Uhlir U and Fris I 1967 *Sov. J. Nucl. Phys.* **4** 444
[5] Isaac P S and Marquette I 2014 *J. Phys. A: Math. Theor.* **47** 205203 (*Preprint* arXiv:1402.3361)
[6] Gravel S 2004 *J. Math. Phys.* **45** 1003 (*Preprint* arXiv:math-ph/0302028)
[7] Marquette I and Quesne C 2013 *J. Math. Phys.* **54** 042102 (*Preprint* arXiv:1211.2957)
[8] Gómez-Ullate D, Kamran N and Milson R 2010 *J. Approx. Theory* **162** 987 (*Preprint* arXiv:0805.3376)
[9] Gómez-Ullate D, Kamran N and Milson R 2009 *J. Math. Anal. Appl.* **359** 352 (*Preprint* arXiv:0807.3939)
[10] Gómez-Ullate D, Grandati Y and Milson R 2014 *J. Phys. A: Math. Theor.* **47** 015203 (*Preprint* arXiv:1306.5143)
[11] Fellows J M and Smith R A 2009 *J. Phys. A: Math. Theor.* **42** 335303 (*Preprint* math-ph/0604008)
[12] Quesne C 2008 *J. Phys. A: Math. Theor.* **41** 392001 (*Preprint* arXiv:0807.4087)
[13] Quesne C 2011 *Mod. Phys. Lett.* A **26** 1843 (*Preprint* arXiv:1106.1990)
[14] Odake S and Sasaki R 2010 *Phys. Lett.* B **684** 173 (*Preprint* arXiv:0911.3442)
[15] Odake S and Sasaki R 2011 *Phys. Lett.* B **702** 164 (*Preprint* arXiv:1105.0508)
[16] Post S, Tsujimoto S and Vinet L 2012 *J. Phys. A: Math. Theor.* **45** 405202 (*Preprint* arXiv:1206.0480)

[17] Marquette I and Quesne C 2013 *J. Math. Phys.* **54** 102102 (*Preprint* arXiv:1303.7150)
[18] Marquette I and Quesne C 2013 *J. Phys. A: Math. Theor.* **46** 155201 (*Preprint* arXiv:1212.3474)
[19] Marquette I and Quesne C 2014 *J. Math. Phys.* **55** 112103 (*Preprint* arXiv:1402.6380)
[20] Kuru S and Negro J 2008 *Ann. Phys.* **323** 413 (*Preprint* arXiv:0709.4649)
[21] Marquette I 2012 *J. Math. Phys.* **53** 012901 (*Preprint* arXiv:0709.4649)
[22] Tremblay F, Turbiner A V and Winternitz P 2009 *J. Phys. A: Math. Theor.* **42** 242001 (*Preprint* arXiv:0904.0738)
[23] Kalnins E G, Kress J M and Miller Jr W 2011 *SIGMA* **7** 031 (*Preprint* arXiv:1011.6548)
[24] Kalnins E G, Kress J M and Miller Jr W 2012 *J. Phys.: Conf. Ser.* **343** 012075
[25] Tanoudis Y and Daskaloyannis C 2011 *SIGMA* **7** 054 (*Preprint* arXiv:1102.0397)

Second order symmetries of the conformal laplacian and R-separation

Jean-Philippe Michel[1], **Fabian Radoux**[2] **and Josef Silhan**[3]

[1] Institut de recherche en mathématique et physique (IRMP), Université Catholique de Louvain (UCL), Chemin du Cyclotron 2, 1348 Louvain-la-Neuve, Belgium
[2] Department of Mathematics, University of Liège, Grande Traverse 12, 4000 Liège, Belgium
[3] Department of Algebra and Geometry of the Masaryk University in Brno, Janàčkovo nàm. 2a, 662 95 Brno, Czech Republic

E-mail: jean-philippe.michel@ulg.ac.be, fabian.radoux@ulg.ac.be, silhan@math.muni.cz

Abstract. Let (M, g) be an arbitrary pseudo-Riemannian manifold of dimension at least 3, let $\Delta := \nabla_a \mathrm{g}^{ab} \nabla_b$ be the Laplace-Beltrami operator and let Δ_Y be the conformal Laplacian. In some references, Kalnins and Miller provide an intrinsic characterization for R-separation of the Laplace equation $\Delta\Psi = 0$ in terms of second order conformal symmetries of Δ. The main goal of this paper is to generalize this result and to explain how the (resp. conformal) symmetries of $\Delta_Y + V$ (where V is an arbitrary potential) can be used to characterize the R-separation of the Schrödinger equation $(\Delta_Y + V)\Psi = E\Psi$ (resp. the Schrödinger equation at zero energy $(\Delta_Y + V)\Psi = 0$). Using a result exposed in our previous paper, we obtain characterizations of the R-separation of the equations $\Delta_Y \Psi = 0$ and $\Delta_Y \Psi = E\Psi$ uniquely in terms of (conformal) Killing tensors pertaining to (conformal) Killing-Stäckel algebras.

1. Introduction

We work over a pseudo-Riemannian manifold (M, g) of dimension $n \geq 3$, with Levi-Civita connection ∇ and scalar curvature Sc. The main result of [1] was the classification of all the second order differential operators D_1 such that the relation

$$\Delta_Y D_1 = D_2 \Delta_Y \tag{1}$$

holds for some differential operator D_2, where $\Delta_Y := \nabla_a \mathrm{g}^{ab} \nabla_b - \frac{n-2}{4(n-1)}\mathrm{Sc}$ is the conformal Laplacian. Such operators D_1 are called conformal symmetries of order 2 of Δ_Y. They preserve the kernel of Δ_Y, i.e. the solution space of the equation $\Delta_Y \psi = 0$, $\psi \in \mathcal{C}^\infty(M)$.

The main goal of this paper is to explain how the second-order (resp. conformal) symmetries of $\Delta_Y + V$ can be used to characterize the existence of R-separating coordinates systems for the Schrödinger equation $(\Delta_Y + V)\Psi = E\Psi$ (resp. the Schrödinger equation at zero energy $(\Delta_Y + V)\Psi = 0$), extending in this way the results in [2] and [3], where the authors characterized the R-separation of the Laplace equation $\Delta\Psi = 0$ and the Helmholtz equation $\Delta\Psi = E\Psi$ using (conformal) symmetries of the Laplace-Beltrami operator Δ.

The paper is organized as follows.

In the first section, in a first step, we recall briefly the basic notions that are necessary to understand the paper: the conformal Laplacian Δ_Y, (conformal) symmetries of Δ_Y, (conformal)

Killing tensors. In a second step, we recall the characterization of the existence of (conformal) symmetries of Δ_Y obtained in [1]. This characterization uses an obstruction operator defined in [1] from curvature tensors. Then, we recall the structure of (conformal) symmetries of Δ_Y given in [1] thanks to the natural and conformally invariant quantization.

In the second section, we recall first the definition of the R-separability of the equations $(\Delta_Y + V)\Psi = E\Psi$ and $(\Delta_Y + V)\Psi = 0$. Next, we give the definitions of (conformal) Killing-Stäckel algebras which can be used to characterize the additive separation of the Hamilton-Jacobi equation. We give then the characterizations of the R-separation of the equations $(\Delta_Y+V)\Psi = E\Psi$ and $(\Delta_Y+V)\Psi = 0$ in terms of these (conformal) Killing-Stäckel algebras and of (conformal) symmetries of $\Delta_Y + V$. Using a result in [1], we obtain then characterizations of the R-separation of the equations $\Delta_Y\Psi = 0$ and $\Delta_Y\Psi = E\Psi$ uniquely in terms of (conformal) Killing tensors belonging to (conformal) Killing-Stäckel algebras. Finally, we give a sufficient (but not necessary) condition which ensures the R-separation of the equations $\Delta_Y\Psi = E\Psi$ and $\Delta_Y\Psi = 0$. This condition is expressed uniquely in terms of components of curvature tensors and is similar to the Robertson condition which ensures the separation of the Schrödinger equation $\Delta\Psi = E\Psi$.

In the appendices, we give the proof of the main theorem giving the characterization of the R-separation of the equation $(\Delta_Y+V)\Psi = 0$ in terms of conformal symmetries of Δ_Y+V (Theorem 5). The proof is divided in two steps: first, we characterize the R-separation of $(\Delta_Y + V)\Psi = 0$ in terms of the notions of conformal Stäckel metric and pseudo-Stäckel multiplier. We can notice that this characterization was already obtained by the authors of [4] by means of a definition of R-separation different from ours (but equivalent). In a second step, we prove Theorem 5 by using the previous characterization and by adapting the proof given in [2] of the characterization of the R-separation of the equation $\Delta\psi = 0$.

2. Second order symmetries of the conformal Laplacian

We are going to recall the fundamental notions that are necessary to understand the paper. The reader who would to know more details about them is invited to consult [1].

2.1. *The conformal Laplacian*

Starting from a pseudo-Riemannian manifold (M, g) of dimension n, the conformal Laplacian Δ_Y is defined in the following way:

$$\Delta_Y := \nabla_a \mathrm{g}^{ab} \nabla_b - \frac{n-2}{4(n-1)}\mathrm{Sc},$$

where ∇ denotes the Levi-Civita connection of g and Sc the scalar curvature.

2.2. *The algebra of symmetries of the conformal Laplacian*

The symmetries of Δ_Y are defined as differential operators that commute with Δ_Y. Hence, they preserve the eigenspaces of Δ_Y.

More generally, conformal symmetries D_1 are defined by the weaker algebraic condition

$$\Delta_Y \circ D_1 = D_2 \circ \Delta_Y, \qquad (2)$$

for some differential operator D_2, so that they only preserve the kernel of Δ_Y.

2.3. *Killing tensors and conformal Killing tensors*

Definition 1. *Let (M, g) be a pseudo-Riemannian manifold and $H = \mathrm{g}^{ij}p_i p_j$ the function on T^*M associated with* g*, where (x^i, p_i) denotes the canonical coordinates on T^*M. If we denote* *by $\mathcal{S}_0$ the algebra of polynomial functions on T^*M, the algebra of symmetries of the null geodesic flow of* g, *denoted by* $\mathcal{K}$, *is given by the following subalgebra of* $\mathcal{S}_0$:

$$\mathcal{K} = \{K \in \mathcal{S}_0 \colon \{H, K\} \in (H)\}.$$

A function in $\mathcal{K}$ can be viewed as a conformal Killing tensor, when we view it as a symmetric contravariant tensor.

Definition 2. *The algebra of symmetries of the geodesic flow of* g *is given by the set of functions in* $\mathcal{S}_0$ *that Poisson commute with* H.

An element of this algebra can be viewed as a Killing tensor, when we view it as a symmetric contravariant tensor.

2.4. Structure of the second order symmetries of Δ_Y

In the sequel, $\mathcal{S}^k_\delta$ denotes the space of polynomial functions of degree k on T^*M with values in the space of δ-densities whereas $\mathcal{D}^k_{\lambda,\mu}$ denotes the space of differential operators of order k acting between λ and μ-densities, with $\mu - \lambda = \delta$.

Let us recall the formula giving the natural and conformally equivariant quantization at the order two (see [5]):

Theorem 1. *[5] Let $\delta \notin \{\frac{2}{n}, \frac{n+2}{2n}, 1, \frac{n+1}{n}, \frac{n+2}{n}\}$. A natural and conformally invariant quantization $\mathcal{Q}_{\lambda,\mu} : \mathcal{S}^{\leq 2}_\delta \to \mathcal{D}^2_{\lambda,\mu}$ is provided, on a pseudo-Riemannian manifold* (M, g) *of dimension n, by the formulas*

$$\begin{aligned}
\mathcal{Q}_{\lambda,\mu}(f) &= f \\
\mathcal{Q}_{\lambda,\mu}(X) &= X^a\nabla_a + \frac{\lambda}{1-\delta}(\nabla_a X^a) \\
\mathcal{Q}_{\lambda,\mu}(S) &= S^{ab}\nabla_a\nabla_b + \beta_1(\nabla_a S^{ab})\nabla_b + \beta_2 \mathrm{g}^{ab}(\nabla_a \mathrm{Tr} S)\nabla_b \\
&+ \beta_3(\nabla_a\nabla_b S^{ab}) + \beta_4\, \mathrm{g}^{ab}\nabla_a\nabla_b(\mathrm{Tr} S), + \beta_5\, \mathrm{Ric}_{ab} S^{ab} + \beta_6\, \mathrm{Sc}\,(\mathrm{Tr}\, S)
\end{aligned} \tag{3}$$

where f, X, S are symbols of degrees 0, 1, 2 respectively and $\mathrm{Tr}\, S = \mathrm{g}_{ab}S^{ab}$. *The coefficients β_i entering the last formula are given e.g. in [1] (see Theorem 2.4).*

The principal symbol of a (conformal) symmetry of Δ_Y has to be a (conformal) Killing tensor. On an arbitrary pseudo-Riemannian manifold, a (conformal) Killing tensor has to satisfy some condition to be the principal symbol of a (conformal) symmetry. This condition can be expressed by means of a natural and conformally invariant operator which is denoted here by **Obs** and which is defined below. In the following definition, C and A denote respectively the Weyl tensor and the Cotton-York tensor.

Definition 3. *The operator* **Obs** *is defined as follows:*

$$\mathbf{Obs} : \mathcal{S}^2_0 \to \mathcal{S}^1_{2/n} : S \mapsto \frac{2(n-2)}{3(n+1)}\mathbf{F}(S),$$

where $(\mathbf{F}(S))^a = \mathrm{C}^r{}_{st}{}^a\nabla_r S^{st} - 3\mathrm{A}_{rs}{}^a S^{rs}$.

In the following statement, $\mathcal{Q}_{\lambda_0,\lambda_0}$ denotes the natural and conformally invariant quantization introduced in Theorem 1. If we denote by $F_{\frac{2}{n}}$ the fiber bundle of $\frac{2}{n}$-densities, the isomorphism $\Gamma(TM \otimes F_{\frac{2}{n}}) \cong \Gamma(T^*M)$ provided by the metric is denoted by $^\flat$ (see [1] for more details).

Theorem 2. *The second order (conformal) symmetries of Δ_Y are exactly the operators*

$$\mathcal{Q}_{\lambda_0,\lambda_0}(K + X) + f,$$

where X is a (conformal) Killing vector field, K is a (conformal) Killing 2-tensor such that $\mathbf{Obs}(K)^\flat$ *is an exact one-form and $f \in \mathcal{C}^\infty(M)$ is defined up to a constant by* $\mathbf{Obs}(K)^\flat = -2\mathrm{d}f$.

3. Application to the R-separation of the Schrödinger equations

Following [2, 3], we provide an intrinsic characterization for R-separation of the Schrödinger equation and of the Schrödinger equation at fixed energy in terms of second order (conformal) symmetries of the conformal Laplacian. Resorting to our previous results, this leads to a new criterion for having R-separation of these equations. Proofs are deferred to appendices.

3.1. Definition of R-separation

The Schrödinger equation, with fixed potential $V \in \mathcal{C}^\infty(M)$, reads as

$$(\Delta_Y + V)\psi = E\psi, \tag{4}$$

where $\psi \in \mathcal{C}^\infty(M)$ is the unknown and $E \in \mathbb{R}$ is called the energy. Solving the Schrödinger equation means to determine the solutions for all $E \in \mathbb{R}$.

We consider also the Schrödinger equation at fixed energy, i.e., up to changing V,

$$(\Delta_Y + V)\psi = 0. \tag{5}$$

Up to modifying V by the curvature term $\frac{(n-2)}{4(n-1)}\mathrm{Sc}$, one can replace the conformal Laplacian by the Laplace-Beltrami one, as done usually. Multiplicative R-separation for one of these equations is usually understood [4, 6] as the search for a coordinate system (x^i) and a family of solutions of the form

$$\psi(x) = R(x)\prod_{i=1}^{n}\phi_i(x^i, c_\alpha), \tag{6}$$

parametrized by $c_\alpha \in \mathbb{R}$ with $\alpha = 1, \ldots, 2n-1$, and satisfying the completeness condition

$$\mathrm{rank}\left[\frac{\partial}{\partial c_\alpha}\left(\frac{\phi_i'}{\phi}\right)\Bigg|\frac{\partial}{\partial c_\alpha}\left(\frac{\phi_i''}{\phi}\right)\right] = 2n-1, \qquad \text{with } \alpha = 1, \ldots, 2n-1 \text{ and } i = 1, \ldots, n.$$

We restrict ourself to separation along orthogonal coordinates (x^i), i.e. such that $\mathrm{g}_{ij} = 0$ if $i \neq j$. We use an alternative working definition which is equivalent, see e.g. Remark 1.

Definition 4. *[2, 3] Equation* (5) *is R-separable in an orthogonal coordinate system* (x^i) *if there exist* $n+1$ *functions* $R, h_i \in \mathcal{C}^\infty(M)$ *and* n *differential operators* $L_i := \partial_i^2 + l_i(x^i)\partial_i + m_i(x^i)$ *such that*

$$R^{-1}(\Delta_Y + V)R = \sum_{i=1}^{n} h_i L_i.$$

The Schrödinger equation (4) *is also R-separable in the coordinate system* (x^i) *if, for all* $E \in \mathbb{R}$, *there exist* $n+1$ *functions* $R, h_i \in \mathcal{C}^\infty(M)$ *and* n *differential operators* $L_i := \partial_i^2 + l_i(x^i)\partial_i + m_i(x^i)$ *such that*

$$R^{-1}(\Delta_Y + V)R - E = \sum_{i=1}^{n} h_i L_i.$$

In the R-separation for the Schrödinger equation, it is clear that only the zero order coefficients of the operators L_i depend on E. More precisely, the functions m_i are then affine functions in E. Looking at the principal symbol of the Schrödinger equation or of the equation (5), we deduce that the functions h_i are equal to g^{ii} for all $i = 1, \ldots, n$. One can show that ψ, defined in (6), is a solution of the Schrödinger equation or of the equation (5) if and only if $L_i\phi_i = 0$ for all i. One has the following easy but crucial fact:

Proposition 1. *[6] The R-separability of the equation* $\Delta_Y\psi = 0$ *is a conformally invariant property.*

3.2. Intrinsic characterizations for R-separation

Let (M, g) be a pseudo-Riemannian manifold, (x^i, p_i) a canonical coordinate system on T^*M and $H = g^{ij}p_ip_j \in \mathcal{S}^2$. Remark that a quadratic symmetric tensor $K \in \mathcal{S}^2$ identifies via the metric as a symmetric endomorphism of TM. In that way, H identifies with the identity. The geodesic Hamilton-Jacobi equation is the coordinate independent equation

$$g^{ij}(\partial_i W)(\partial_j W) = E,$$

where $W \in \mathcal{C}^\infty(M)$ is the unknown and $E \in \mathbb{R}$ is the energy. If $E = 0$, this is the null geodesic Hamilton-Jacobi equation.

We need the following notions, borrowed from [7, 8] and [9, 10] respectively.

Definition 5. *A* Killing-Stäckel algebra *is an n-dimensional linear space $\mathcal{I}$ of Killing 2-tensors that satisfy the three following properties:*

(i) they commute as linear operators,

(ii) they are diagonalizable as linear operators,

(iii) they are in involution: $\{K_1, K_2\} = 0$ for all $K_1, K_2 \in \mathcal{K}$.

Note that the metric Hamiltonian $H = g^{ij}p_ip_j$ belongs to any Killing-Stäckel algebra (for more details on this point, see [8], Theorem 7.8).

Definition 6. *A* conformal Killing-Stäckel algebra *is an n-dimensional linear space $\mathcal{I}$ of conformal Killing 2-tensors that satisfy the following conditions:*

(i) they commute as linear operators,

(ii) they are diagonalizable as linear operators,

(iii) $\{K_1, K_2\} \in (H)$ for all K_1 and K_2 in $\mathcal{I}$.

As pointed out in [10], each conformal Killing-Stäckel space contains a tensor of the kind fH, for some function f.

In the Riemannian case, the hypothesis *(ii)* in Definitions 5 and 6 is not necessary: indeed, in this situation, the 2-tensors in the (conformal) Killing-Stäckel algebra are automatically diagonalizable as linear operators because they are symmetric with respect to a Riemannian metric.

Kalnins and Miller provide in [7] (resp. [9]) an intrinsic characterization for additive separation of the (resp. null) geodesic Hamilton-Jacobi equation, in terms of (resp. conformal) Killing-Stäckel algebra. These classical results, completing those of Stäckel [11] and Eisenhart [12] can be summarized up as follows, see [7, 9] or [8, 10].

Theorem 3. *If $\mathcal{I}$ is a (conformal) Killing-Stäckel algebra and if $K \in \mathcal{I}$, then K, viewed as an endomorphism, admits n principal directions which integrate in a local coordinate system (x^i) on M. In this coordinate system, the tensors in $\mathcal{I}$ are diagonal.*

Theorem 4. *The (resp. null) geodesic Hamilton-Jacobi equation admits additive separation in an orthogonal coordinate system if and only if there exists a (resp. conformal) Killing-Stäckel algebra.*

Actually, it turns out that the coordinate system (x^i) in Theorem 3 is a coordinate system in which the (null) geodesic Hamilton-Jacobi equation is separable.

In [2, 3], Kalnins and Miller provide an analogue of the latter theorem for R-separation of the equations $\Delta\psi = 0$ and $\Delta\psi = E\psi$. In the presence of potentials, their results extend easily, leading to the two following theorems. We provide a proof for the first one in Appendix B, it relies on [2] and on the material in Appendix A. The proof of the second one is similar and is left to the reader.

Journal of Physics: Conference Series **597** (2015) 012058 doi:10.1088/1742-6596/597/1/012058

Theorem 5. *Equation (5), $(\Delta_Y + V)\psi = 0$, R-separates in an orthogonal coordinate system if and only if:*

(a) there exists a conformal Killing-Stäckel algebra $\mathcal{I}$,

(b) for all $K \in \mathcal{I}$, there exists $D \in \mathcal{D}^2(M)$ with principal symbol $\sigma_2(D) = K$ such that $[\Delta_Y + V, D] = A \circ (\Delta_Y + V)$, for some $A \in \mathcal{D}(M)$.

Theorem 6. *The Schrödinger equation, $(\Delta_Y + V)\psi = E\psi$, R-separates in an orthogonal coordinate system if and only if:*

(a) there exists a Killing-Stäckel algebra $\mathcal{I}$,

(b) for all $K \in \mathcal{I}$, there exists $D \in \mathcal{D}^2(M)$ with principal symbol $\sigma_2(D) = K$ such that $[\Delta_Y + V, D] = 0$.

Remark that, compared to the analogous theorems in [2, 3], we relax the hypothesis of commutation of the symmetries of $\Delta_Y + V$ between themselves. From our results on the second order (resp. conformal) symmetries of the conformal Laplacian we deduce:

Corollary 1. *The equation $\Delta_Y\psi = 0$ separates in orthogonal coordinates if and only if there exists a conformal Killing-Stäckel algebra $\mathcal{I}$ such that $\mathbf{Obs}(K)^\flat$ is exact for all $K \in \mathcal{I}$.*

Corollary 2. *The Schrödinger equation separates in orthogonal coordinates if and only if there exists a Killing-Stäckel algebra $\mathcal{I}$ such that $\mathbf{Obs}(K)^\flat$ is exact for all $K \in \mathcal{I}$.*

There exists a convenient and geometrical condition which ensures separation for the usual Schrödinger equation $\Delta\psi = E\psi$, where $\Delta = \nabla_i \mathrm{g}^{ij}\nabla_j$ is the Laplace-Beltrami Laplacian. Namely, if the geodesic Hamilton-Jacobi equation separates in a coordinate system (x^i), then the Schrödinger equation separates in the same coordinate system if and only if the Ricci tensor satisfies the Robertson condition: $\mathrm{Ric}_{ij} = 0$ if $i \neq j$ [12]. In particular, this condition implies the R-separation for the Schrödinger equation in the coordinate system (x^i). From the preceding corollaries and the explicit form of the operator **Obs**, we get an analogous but distinct condition for R-separation.

Proposition 2. *Let (M, g) be a pseudo-Riemannian manifold of dimension at least 4. If there exists a (resp. conformal) Killing-Stäckel algebra such that $C_{ijjk} = 0$ in the corresponding coordinate system, then the Schrödinger equation $\Delta_Y\psi = E\psi$ (resp. the equation $\Delta_Y\psi = 0$) admits R-separation in the same coordinate system. If M is of dimension 3, the condition $C_{ijjk} = 0$ should be replaced by $A_{jji} = 0$.*

Proof. Indeed, on one hand, the tensors belonging to the (conformal) Killing-Stäckel algebra ensuring the separation of the Hamilton-Jacobi equation are diagonal in the coordinate system (x^i) (see Theorem 3). On the other hand, recall that the operator **Obs** is equal to

$$\frac{2(n-2)}{3(n+1)}\mathbf{g}^{im}p_i\partial_{p_j}\partial_{p_l}\left(\mathrm{C}^k{}_{jlm}\nabla_k - 3\mathrm{A}_{jlm}\right).$$

If (M, g) has a dimension greater than or equal to 4, then the fact that $C_{ijjk} = 0$ implies that $A_{jji} = 0$ thanks to the relation

$$(n-3)\mathrm{A}_{abc} = \nabla_r \mathrm{C}_{bc}{}^r{}_a.$$

We can then conclude that $\mathbf{Obs}(K) = 0$ for all $K \in \mathcal{I}$ using the fact that the tensors in $\mathcal{I}$ are diagonal.

If M is of dimension 3, then C vanishes and the condition $A_{jji} = 0$ is then sufficient to conclude that $\mathbf{Obs}(K) = 0$ for all $K \in \mathcal{I}$.

Eventually, Corollaries 1 and 2 yield the result. □

However, neither the Robertson's condition nor the one in Proposition 2 is necessary to get R-separation of the Schrödinger equation. This is illustrated thanks to the DiPirro metrics given in [1] (section 5, page 23). Let us recall that the Hamiltonians associated with these metrics are given by

$$H = \frac{1}{2(\gamma(x_1,x_2)+c(x_3))}\left(a(x_1,x_2)p_1^2 + b(x_1,x_2)p_2^2 + p_3^2\right) \tag{7}$$

where a, b, c and γ are arbitrary functions and (x^i, p_i) are canonical coordinates on $T^*\mathbb{R}^3$. The DiPirro metrics admit diagonal Killing tensors K given by

$$K = \frac{1}{\gamma(x_1,x_2)+c(x_3)}\left(c(x_3)a(x_1,x_2)p_1^2 + c(x_3)b(x_1,x_2)p_2^2 - \gamma(x_1,x_2)p_3^2\right).$$

We know thanks to Proposition 5.1 in [1] that there exists a symmetry of Δ_Y given by the Killing tensor K defined above. If $c(x^3) = 1$, DiPirro metrics admit a third Killing tensor $K' = p_3^2$, with a corresponding symmetry operator given by $D' = \frac{\partial^2}{\partial(x^3)^2}$. Thus, the 3-dimensional linear space generated by H, K and K' constitutes a Killing-Stäckel algebra whose elements satisfy the condition (b) in Theorem 6 because this condition is verified for each generator of this algebra. Then, the Schrödinger equation R-separates in the coordinate system (x^1, x^2, x^3). However, a direct computation, performed e.g. with Mathematica, leads to $\mathrm{Ric}_{12} \neq 0$ and $A_{112} \neq 0$.

Let (x^i) be a coordinate system which separates the (resp. null) geodesic Hamilton-Jacobi equation. This is an open question to find a Robertson's like condition, which exactly determines if the coordinate system (x^i) R-separates the Schrödinger equation (resp. the equation $\Delta_Y\psi = 0$). Such a condition should be ideally written only in terms of the curvature tensors, in the coordinate system (x^i).

Acknowledgments
This research has been partially funded by the Interuniversity Attraction Poles Program initiated by the Belgian Science Policy Office. J. Šilhan was supported by the grant agency of the Czech republic under the grant P201/12/G028.

Appendix A. Stäckel metrics

Appendix A.1. Stäckel metrics and Stäckel multipliers

Let (x^i) be an orthogonal coordinate system for the metric g. We introduce two coordinates dependent notions. First, the Stäckel operators are given by

$$\mathrm{St}_{ij} = \partial_i\partial_j - (\partial_i \log|\mathrm{g}^{jj}|)\partial_j - (\partial_j \log|\mathrm{g}^{ii}|)\partial_i, \quad i \neq j, \tag{A.1}$$

with no summation on the repeated indices. Second, a Stäckel matrix is an invertible matrix of functions $(S_{ij}(x^i))$, whose i^{th}-row depends on the coordinate x^i only.

Proposition 3. *A metric* g *is called a Stäckel metric if there exist orthogonal coordinates* (x^i), *called Stäckel coordinates, such that one of the four following equivalent conditions is satisfied:*

(i) $\mathrm{g}^{jj} = \frac{S^{j1}}{S}$, *where* S^{j1} *is the cofactor in* $(j, 1)$ *of a Stäckel matrix* (S_{ij}) *and* S *its determinant,*
(ii) $\mathrm{St}_{ij}(\mathrm{g}^{kk}) = 0$, *for all* $i, j, k = 1, \ldots, n$ *and* $i \neq j$,
(iii) there exists a Killing-Stäckel algebra with diagonal elements in the coordinates (x^i),
(iv) the geodesic Hamilton-Jacobi equation separates in the coordinates (x^i).

The notion of Stäckel metric goes back to Stäckel himself [11] and he proves the equivalence between (i) and (iv). The remaining equivalences are proved in [12]. In particular, (i) and (iii) are linked through the fact that $\mathcal{I}$ is generated by the tensors $K_i = \frac{S^{ji}}{S}p_j^2$, for $i = 1, \ldots, n$. Remark that a given Stäckel metric can admit several inequivalent Stäckel coordinate systems. They have been classified in the flat case in low dimensions, see e.g. [13] for the 3-dimensional case.

Proposition 4. *A metric* g *is called a conformal Stäckel metric if there exist orthogonal coordinates* (x^i), *called conformal Stäckel coordinates, such that one of the four following equivalent conditions is satisfied:*

(i) *there exists a Stäckel metric* $\hat{g}$, *with Stäckel coordinates* (x^i), *conformally related to* g,

(ii) $\frac{\mathrm{St}_{ij}(g^{kk})}{g^{kk}} = \frac{\mathrm{St}_{ij}(g^{ll})}{g^{ll}}$, *for all* $i, j, k, l = 1, \dots, n$ *and* $i \neq j$,

(iii) *there exists a conformal Killing-Stäckel algebra with diagonal elements in the coordinates* (x^i),

(iv) *the null geodesic Hamilton-Jacobi equation separates in the coordinates* (x^i).

The equivalence between (i) and (iv) goes back to Stäckel [11]. The proof of the equivalence between the remaining statements can be found in [9]. We introduce now Stäckel multipliers. They relate different Stäckel metrics in a conformal class, which admit common Stäckel coordinates.

Proposition 5. *Let* g *be a Stäckel metric with Stäckel coordinates* (x^i). *A Stäckel multiplier is a function* Q *satisfying one of the three following equivalent conditions:*

(i) *the metric* Qg *is a Stäckel metric, with Stäckel coordinates* (x^i),

(ii) $\mathrm{St}_{ij}(Q) = 0$, *for all* $i, j = 1, \dots, n$ *and* $i \neq j$,

(iii) *there exist functions* $k_j(x^j)$ *such that* $Q(x) = \sum_{j=1}^{n} g^{jj} k_j(x^j)$.

The equivalence of (i), (ii) and (iii) is established in [14]. The latter notion can be straightforwardly extended to the case of conformal Stäckel metrics.

Proposition 6. *Let* g *be a conformal Stäckel metric with conformal Stäckel coordinates* (x^i). *A pseudo-Stäckel multiplier is a function* Q *satisfying one of the three following equivalent conditions:*

(i) *the metric* Qg *is a Stäckel metric, with Stäckel coordinates* (x^i),

(ii) $\frac{\mathrm{St}_{ij}(Q)}{Q} = \frac{\mathrm{St}_{ij}(g^{kk})}{g^{kk}}$, *for all* $i, j, k = 1, \dots, n$ *and* $i \neq j$,

(iii) *there exist functions* $k_j(x^j)$ *such that* $Q(x) = \sum_{j=1}^{n} g^{jj} k_j(x^j)$.

Remark that, by the very definition of a conformal Stäckel metric, there always exist pseudo-Stäckel multipliers.

Appendix A.2. Necessary and sufficient conditions for R-separation

It is well-known that multiplicative R-separation for the Schrödinger equation (4) implies additive separation for the geodesic Hamilton-Jacobi equation, and that multiplicative R-separation of the equation (5) implies additive separation of the null geodesic Hamilton-Jacobi equation. The converse statement is not true and suppose an extra condition involving the potential and the metric. The following theorem is a straightforward generalization of a result in [2], see also [4].

Theorem 7. *Equation* (5), $(\Delta_Y + V)\psi = 0$, *R-separates in the orthogonal coordinate system* (x^i) *if and only if the two following conditions hold:*

(i) g *is a conformal Stäckel metric with conformal Stäckel coordinates* (x^i),

(ii) $V + \frac{\Delta_Y R}{R}$ *is a pseudo-Stäckel multiplier.*

Moreover, the function R *is then any function of the form* $\frac{S}{Q(\det g)^{1/2}} \prod_j a_j$, *where* $\hat{g} = Q$g *is a Stäckel metric,* S *is the determinant of a Stäckel matrix* (S_{ij}) *of* $\hat{g}$ *and where the functions* a_j *are such that their derivatives* $\partial_j a_j$ *depend only on the coordinate* x^j.

Proof. The R-separation of Equation (5) occurs in coordinates (x^i) if and only if there exists a function $R \in \mathcal{C}^\infty(M)$ such that

$$R^{-1}(\Delta_Y + V)R = \sum_{j=1}^{n} \mathrm{g}^{jj} L_j, \tag{A.2}$$

where $L_j = \partial_j^2 + l_j\partial_j + m_j$, with the functions l_j and m_j which depend only on the coordinate x^j. Since g is diagonal in the coordinates (x^i), we have the formula $\nabla_i \mathrm{g}^{ij}\nabla_j = \mathrm{g}^{jj}\partial_j^2 + \Gamma^j\partial_j$, where $\Gamma^j = (\det \mathrm{g})^{-1/2}\partial_j(\mathrm{g}^{jj}(\det \mathrm{g})^{1/2})$ is the trace of the Christoffel symbol of ∇. A direct computation shows that

$$R^{-1}\Delta_Y R = \mathrm{g}^{jj}\partial_j^2 + 2\mathrm{g}^{jj}(\partial_j \log R)\partial_j + \Gamma^j\partial_j + \frac{\Delta_Y R}{R}.$$

In consequence, Equation (A.2) holds if and only if:

$$\begin{cases} \partial_j(\log \mathrm{g}^{jj}) = l_j - \partial_j(2\log R + \log(\det \mathrm{g})^{1/2}), \quad \forall j = 1, \dots, n, \\ V + \frac{\Delta_Y R}{R} = \sum_i \mathrm{g}^{ii} m_i. \end{cases} \tag{A.3}$$

These two equations hold if and only if the metric g is a conformal Stäckel metric and $V + \frac{\Delta_Y R}{R}$ is a pseudo-Stäckel multiplier. We determine now the form of the function R. Since g is a conformal Stäckel metric, there exists a pseudo-Stäckel multiplier Q and a Stäckel metric $\hat{\mathrm{g}} = Q\mathrm{g}$, with Stäckel matrix (S_{ij}). Denoting by S its determinant, one sets $f = \log \frac{Q(\det \mathrm{g})^{1/2}}{S}$. By definition of a Stäckel matrix we have $\partial_j S^{j1} = 0$, and the equality $\mathrm{g}^{jj}\partial_j f = (\det \mathrm{g})^{-1/2}\partial_j(\mathrm{g}^{jj}(\det \mathrm{g})^{1/2})$ follows. In consequence, we deduce that

$$2\partial_j \log R = l_j - \partial_j f, \quad \forall j = 1, \dots, n,$$

whose solutions are $R = \left(\frac{S}{Q(\det \mathrm{g})^{1/2}}\right)^{1/2} \prod_j a_j$ with $\partial_j a_j = -\frac{1}{2} l_j$, and the functions l_j are free. □

Remark 1. *Theorem 7 is proved in [4, Proposition 23], but resorting to the definition of R-separation of variables provided by Equation* (6) *rather than the one given in Definition 4. This shows that these two definitions of R-separation of variables are indeed equivalent.*

The following theorem is easily deduced from the preceding one.

Theorem 8. *The Schrödinger equation* (4) *R-separates in the orthogonal coordinate system* (x^i) *if and only if the two following conditions hold:*

(i) g *is a Stäckel metric with Stäckel coordinates* (x^i),

(ii) $V + \frac{\Delta_Y R}{R}$ *is a Stäckel multiplier.*

Proof. If the R-separation condition (A.3) holds for all $E \in \mathbb{R}$, we deduce that constants are pseudo-Stäckel multipliers. Hence, the metric g is Stäckel and the pseudo-Stäckel multiplier $V + \frac{\Delta_Y R}{R}$ is then a Stäckel multiplier. □

Appendix B. Proof of Theorem 5

First, we suppose that (x^i) is an R-separable coordinate system for Equation (5) and prove conditions (a) and (b). By the very definition of R-separation, there exists a function $R \in \mathcal{C}^\infty(M)$ such that

$$R^{-1}(\Delta_Y + V)R = \sum_{j=1}^{n} \mathrm{g}^{jj} L_j, \tag{B.1}$$

where $L_j = \partial_j^2 + l_j\partial_j + m_j$, with the functions l_j and m_j which depend only on the coordinate x^j. Moreover, Theorem 7 applies and g is then a conformal Stäckel metric. In consequence, there exists a pseudo-Stäckel multiplier Q and a Stäckel metric $\hat{\mathrm{g}} = Q\mathrm{g}$.

We denote by (S_{ij}) a Stäckel matrix attached to the metric $\hat{\mathrm{g}}$ and by (S^{ij}) the matrix of cofactors, so that $S^{ij}S_{jk} = S\delta^i_k$ with S the determinant of (S_{ij}). Hence, we obtain n independent diagonal Killing tensors $K_i = \frac{S^{ji}}{S}p_j^2$ for $\hat{\mathrm{g}}$, with $K_1 = \hat{\mathrm{g}}^{ij}p_ip_j$. They are automatically conformal Killing tensors for g and they generate a conformal Killing-Stäckel algebra $\mathcal{I}$, so the condition (a) is satisfied. Following [2], we set

$$A_i = \sum_j \frac{S^{ji}}{S}\left(\partial_j^2 + (\partial_j f)\partial_j + a_j\right),$$

where $f = \log\frac{Q(\det \mathrm{g})^{1/2}}{S}$ and $a_j = m_j + \frac{1}{2}\partial_j(\partial_j f - l_j) + \frac{1}{4}((\partial_j f)^2 - l_j^2)$. The principal symbol of A_i is then K_i. Setting $\rho^j_{(i)} = \hat{\mathrm{g}}_{jj}\frac{S^{ji}}{S}$, the Poisson commutation relations, $\{K_1, K_i\} = 0$, translate as $\partial_k(\rho^j_{(i)}\hat{\mathrm{g}}^{jj}) = \rho^k_{(i)}(\partial_k\hat{\mathrm{g}}^{jj})$ for all indices i, j, k. Then a direct computation shows that (see [2]),

$$[A_1, A_i] = 0, \quad \forall i = 1, \ldots, n. \tag{B.2}$$

The next step is to prove that $QA_1 = \Delta_Y + V$. By definition of a Stäckel matrix we have $\mathrm{g}^{jj} = Q\frac{S^{j1}}{S}$ and $\partial_j S^{j1} = 0$. The equality $\mathrm{g}^{jj}\partial_j f = (\det \mathrm{g})^{-1/2}\partial_j(\mathrm{g}^{jj}(\det \mathrm{g})^{1/2})$ follows. In particular, one has $\Delta := \nabla_i \mathrm{g}^{ij}\nabla_j = \mathrm{g}^{jj}(\partial_j^2 + (\partial_j f)\partial_j)$, which leads to

$$QA_1 = \Delta + \sum_j \mathrm{g}^{jj}a_j.$$

According to the proof of Theorem 7, the equation (B.1) is equivalent to the system (A.3), which reads now as

$$\begin{cases} 2\partial_j(\log R) = l_j - \partial_j f, & \forall j = 1, \ldots, n, \\ V + \frac{\Delta_Y R}{R} = \sum_i \mathrm{g}^{ii}m_i. \end{cases}$$

Combining these two equations with the expression of a_j and the equality $\frac{\Delta R}{R} = \mathrm{g}^{jj}(R^{-1}\partial_j^2(R) + (\partial_j f)(\partial_j \log R))$, we get $QA_1 = \Delta_Y + V$. Therefore, Equation (B.2) implies that

$$[\Delta_Y + V, A_i] = [Q, A_i] \circ A_1$$

for $i = 1, \ldots, n$. Since $\sigma_2(A_i) = K_i$ for $i = 1, \ldots, n$ and since the 2-tensors K_i $(i = 1, \ldots, n)$ generate $\mathcal{I}$, the condition (b) is satisfied.

We prove now that the conditions (a) and (b) imply that g is a conformal Stäkel metric and $V + \frac{\Delta_Y R}{R}$ is a pseudo-Stäckel multiplier. This leads to the conclusion thanks to Theorem 7. We straightforwardly get that g is a conformal Stäckel metric by Proposition 4. Its conformally related Stäckel metric $\hat{\mathrm{g}} = Q\mathrm{g}$ admits a Stäckel matrix (S_{ij}) so that the tensors $K_i = \frac{S^{ji}}{S}p_j^2$

$(i = 1, \ldots, n)$ are generators of the conformal Killing-Stäckel algebra $\mathcal{I}$. Hence, we can introduce $f = \log \frac{Q(\det g)^{1/2}}{S}$ as above and we set $\rho^j_{(i)} = \hat{g}_{jj} \frac{S^{ji}}{S}$. Setting $V' = V - \frac{n-2}{4(n-1)}\mathrm{Sc}$, we get that

$$\Delta_Y + V = g^{jj}(\partial_j^2 + (\partial_j f)\partial_j) + V'.$$

Let $D_1 = Q^{-1}(\Delta_Y + V)$ and D_i, $(i = 2, \ldots, n)$ be the conformal symmetries corresponding to the tensors K_i $(i = 1, \ldots, n)$ defined above. If we set $\tilde{D}_i = e^{f/2} D_i e^{-f/2}$, the operators $\tilde{D}_i$ read as

$$\sum_j \rho^j_{(i)} \hat{g}^{jj} \partial_j^2 + \xi^j_{(i)} \partial_j + \mu_{(i)},$$

where $\rho^j_{(1)} = 1$, $\xi^j_1 = 0$ and $\mu_{(1)} = Q^{-1}(V + \frac{\Delta_Y R}{R})$ where $2 \log R = -f$. According to [2], the equations

$$[\tilde{D}_1, \tilde{D}_j] \in (\tilde{D}_1), \quad \forall j = 2, \ldots, n,$$

imply that $\mu_{(1)} = V + \frac{\Delta_Y R}{R}$ is a pseudo-Stäckel multiplier. Moreover, $R = e^{-f/2}$ is one of the admissible functions R for R-separation and Theorem 5 is then proved.

References

[1] Michel J P, Radoux F and Šilhan J 2014 *SIGMA* **10** 26
[2] Kalnins E G and Miller Jr W 1982 *J. Phys. A* **15** 2699–2709
[3] Kalnins E G and Miller Jr W 1984 *Adv. in Math.* **51** 91–106
[4] Chanu C and Rastelli G 2006 *Int. J. Geom. Methods Mod. Phys.* **3** 489–508
[5] Duval C and Ovsienko V Y 2001 *Selecta Math. (N.S.)* **7** 291–320 (*Preprint* math.DG/9801122)
[6] Chanachowicz M, Chanu C M and McLenaghan R G 2009 *J. Geom. Phys.* **59** 876 – 884
[7] Kalnins E and Miller Jr W 1980 *SIAM Journal on Mathematical Analysis* **11** 1011–1026
[8] Benenti S, Chanu C and Rastelli G 2002 *J. Math. Phys.* **43** 5183–5222
[9] Kalnins E and Miller Jr W 1983 *SIAM Journal on Mathematical Analysis* **14** 126–137
[10] Benenti S, Chanu C and Rastelli G 2005 *J. Math. Phys.* **46** 042901, 29
[11] Stckel P 1891 *Uber die Integration der Hamilton-Jacobischen differential Gleichung mittelst Separation der Variabelen* Habilitationsschrift Halle
[12] Eisenhart L P 1934 *Ann. of Math. (2)* **35** 284–305
[13] Boyer C P, Kalnins E G and Miller Jr W 1976 *Nagoya Math. J.* **60** 35–80
[14] Kalnins E and Miller Jr W 1981 *SIAM Journal on Mathematical Analysis* **12** 617–629

Journal of Physics: Conference Series **597** (2015) 012059 doi:10.1088/1742-6596/597/1/012059

Wilson polynomials/functions and intertwining operators for the generic quantum superintegrable system on the 2-sphere

W Miller Jr and Q Li

School of Mathematics, University of Minnesota, Minneapolis, Minnesota, 55455, U.S.A.

E-mail: `miller@ima.umn.edu`

Abstract. The Wilson and Racah polynomials can be characterized as basis functions for irreducible representations of the quadratic symmetry algebra of the quantum superintegrable system on the 2-sphere, $H\Psi = E\Psi$, with generic 3-parameter potential. Clearly, the polynomials are expansion coefficients for one eigenbasis of a symmetry operator L_2 of H in terms of an eigenbasis of another symmetry operator L_1, but the exact relationship appears not to have been made explicit. We work out the details of the expansion to show, explicitly, how the polynomials arise and how the principal properties of these functions: the measure, 3-term recurrence relation, 2nd order difference equation, duality of these relations, permutation symmetry, intertwining operators and an alternate derivation of Wilson functions – follow from the symmetry of this quantum system. This paper is an exercise to show that quantum mechancal concepts and recurrence relations for Gausian hypergeometrc functions alone suffice to explain these properties; we make no assumptions about the structure of Wilson polynomial/functions, but derive them from quantum principles. There is active interest in the relation between multivariable Wilson polynomials and the quantum superintegrable system on the n-sphere with generic potential, and these results should aid in the generalization. Contracting function space realizations of irreducible representations of this quadratic algebra to the other superintegrable systems one can obtain the full Askey scheme of orthogonal hypergeometric polynomials. All of these contractions of superintegrable systems with potential are uniquely induced by Wigner Lie algebra contractions of $so(3, C)$ and $e(2, C)$. All of the polynomials produced are interpretable as quantum expansion coefficients. It is important to extend this process to higher dimensions.

1. Introduction

We define a quantum superintegrable system as an integrable Hamiltonian system on an n-dimensional pseudo-Riemannian manifold with potential: $H = \Delta_n + V$ that admits $2n-1$ algebraically independent partial differential operators commuting with H, the maximum possible,[1]. Thus $[H, L_j] = 0, \quad n = 1, 2, \cdots, 2n-1$ where Δ_n is the Laplace-Beltrami operator on the manifold and we choose the generators L_j such that the sum of their orders is a small as possible. Superintegrability captures the properties of quantum Hamiltonian systems that allow the Schrödinger eigenvalue problem $H\Psi = E\Psi$ to be solved exactly, analytically and algebraically. Typically, the basis symmetries L_j generate an algebra under commutation, not usually a Lie algebra that closes at finite order. It is this algebra that is responsible for the solvability of the quantum system.

The generic superintegrable system on the 2-sphere, system $S9$ in our listing [2], is

$$H = J_1^2 + J_2^2 + J_3^2 + \frac{a_1}{s_1^2} + \frac{a_2}{s_2^2} + \frac{a_3}{s_3^2}, \qquad a_j = \frac{1}{4} - k_j^2,$$

where $J_3 = s_1\partial_{s_2} - s_2\partial_{s_1}$ and J_2, J_3 are obtained by cyclic permutations of indices. Here, $s_1^2 + s_2^2 + s_3^2 = 1$. The operators J_k preserve the order of homogeneity when acting on functions in $\mathcal{R}^3$ so they act on C^∞ functions on the sphere. The basis symmetries can be chosen as

$$L_1 = J_1^2 + \frac{a_3 s_2^2}{s_3^2} + \frac{a_2 s_3^2}{s_2^2}, \; L_2 = J_2^2 + \frac{a_1 s_3^2}{s_1^2} + \frac{a_3 s_1^2}{s_3^2}, \; L_3 = J_3^2 + \frac{a_2 s_1^2}{s_2^2} + \frac{a_1 s_2^2}{s_1^2},$$

where J_1, J_2, J_3 are rotation generators: $J_3 = s_2\partial_{s_1} - s_1\partial_{s_2}, \cdots$. Note the discrete symmetry of $H = L_1 + L_2 + L_3 + a_1 + a_2 + a_3$ with respect to permutations of the indices $1, 2, 3$. With the commutator $R = [L_1, L_2] = [L_2, L_3] = [L_3, L_1]$, the algebra generated by these symmetries has the structure

$$[L_i, R] = 4\{L_i, L_k\} - 4\{L_i, L_j\} - (8 + 16a_j)L_j + (8 + 16a_k)L_k + 8(a_j - a_k), \qquad (1)$$

where $i, j, k = 1, 2, 3$, pairwise distinct. Here, $\{A, B\} = AB + BA$, and $\{A, B, C\}$ is the symmetrizer of 3 operators. R^2 is contained in the algebra and expressed by

$$R^2 = \frac{8}{3}\{L_1, L_2, L_3\} - (16a_1 + 12)L_1^2 - (16a_2 + 12)L_2^2 - (16a_3 + 12)L_3^2 + \qquad (2)$$

$$\frac{52}{3}(\{L_1, L_2\} + \{L_2, L_3\} + \{L_3, L_1\}) + \frac{1}{3}(16 + 176a_1)L_1 + \frac{1}{3}(16 + 176a_2)L_2 + \frac{1}{3}(16 + 176a_3)L_3$$

$$+ \frac{32}{3}(a_1 + a_2 + a_3) + 48(a_1a_2 + a_2a_3 + a_3a_1) + 64a_1a_2a_3.$$

This algebra can be matched with $QR(3)$, the structure algebra of the Racah and Wilson polynomials, [3, 4, 5, 6, 7, 8]. The significance of these special functions is that they are the expansion coefficients of a basis of eigenfunctions of one of the spherical coordinate generators L_i in terms of an eigenbasis of L_j, $i \neq j$. Long before, Dunkl in the remarkable paper [9], (see also [10, 11] and references contained therein) had computed these coefficients as expansions of polynomial bases (not eigenbases) and shown them to be Racah/Wilson polynomials. However, to our knowledge no one has worked out the details of this relationship between the functions and the expansion coefficients to understand how important properties of Wilson polynomials can be interpreted from a quantum mechanical viewpoint. Here we show the critical role of intertwining operators for the quantum system in determining parameter changing recurrences for Wilson/Racah polynomials. In particular, the ${}_4F_3$ expressions for Wilson polynomials/functions follow from intertwining operators alone.

We recall the definition of ${}_4F_3$ hypergeometric functions:

$${}_4F_3\left(\begin{matrix} a_1, & a_2, & a_3, & a_4 \\ b_1, & b_2, & b_3 & \end{matrix}; x\right) = \sum_{k=0}^{\infty} \frac{(a_1)_k(a_2)_k(a_3)_k(a_4)_k}{(b_1)_k(b_2)_k(b_3)_k k!} x^k$$

where $(a)_0 = 1$, $\quad (a)_k = a(a+1)(a+2)\cdots(a+k-1)$ if $k \geq 1$. If $a_1 = -n$ for n a nonnegative integer then the sum is finite with $n+1$ terms. The Wilson polynomials of order n in t^2 are

$$\Phi_n(\alpha, \beta, \gamma, \delta; t) = {}_4F_3\left(\begin{matrix} -n, & \alpha+\beta+\gamma+\delta+n-1, & \alpha - t, & \alpha + t \\ \alpha+\beta, & \alpha+\gamma, & \alpha+\delta & \end{matrix}; 1\right).$$

If $\alpha + \beta = -m$, m a nonnegative integer we have the finite set $\Phi_0, \cdots, \Phi_m$ of Racah polynomials.

2. Structure algebra and interbasis expansion coefficients

The basis simultaneous eigenfunctions of L_1 and H: take the form

$$\Psi_{N-n,n} = (s_1^2+s_2^2)^{\frac{1}{2}(2n+k_1+k_2+1)}(1-s_1^2-s_2^2)^{\frac{1}{2}(k_3+\frac{1}{2})}(\frac{s_2^2}{s_1^2+s_2^2})^{\frac{1}{2}(k_2+\frac{1}{2})}(\frac{s_1^2}{s_1^2+s_2^2})^{\frac{1}{2}(k_1+\frac{1}{2})}$$

$$\times P_n^{(k_2,k_1)}(\frac{s_1^2-s_2^2}{s_1^2+s_2^2})P_{N-n}^{(2n+k_1+k_2+1,k_3)}(1-2s_1^2-2s_2^2),$$

$$\frac{1}{4}(L_1+2k_1k_2+2k_1+2k_2+\frac{3}{2})\Psi_{N-n,n} = -n(n+k_1+k_2+1)\Psi_{N-n,n}, \quad n=0,1,\cdots,N,$$

$$H\Psi_{N-n,n} = E_N\Psi_{N-n,n}, \quad E_N = -(2N+k_1+k_2+k_3+2)^2+\frac{1}{4}, \quad N=0,1,\cdots,$$

separable in spherical coordinates, $x = \cos(2\varphi), y = \cos(2\theta)$ where $s_1 = \sin\theta\cos\varphi, s_2 = \sin\theta\sin\varphi, s_3 = \cos\varphi$, orthogonal with respect to area measure on the 1st octant of the 2-sphere. The dimension of eigenspace E_N is $N+1$.

$$P_n^{(\alpha,\beta)}(y) = \left(\begin{array}{c} n+\alpha \\ n \end{array}\right) {}_2F_1\left(\begin{array}{cc} -n & \alpha+\beta+n+1 \\ \alpha+1 & \end{array}; \frac{1-y}{2}\right), \quad \text{Jacobi polynomials}$$

Define the functions $\Lambda_{N-n,n}$ by the permutation $1 \leftrightarrow 3$, in $\Psi_{N-n,n}$. Thes are eigenfunctions of L_2, H:

$$\frac{1}{4}(L_2+2k_2k_3+2k_2+2k_3+\frac{3}{2})\Lambda_{N-q,q} = -q(q+k_2+k_3+1)\Lambda_{N-q,q}, \quad q=0,1,\cdots,N,$$

$$H\Lambda_{N-q,q} = E_N\Lambda_{N-q,q}, \quad E_N = -(2N+k_1+k_2+k_3+2)^2+\frac{1}{4}, \quad N=0,1,\cdots.$$

They are separable in a different set of spherical coordinates X, Y expressible in terms of x, y by $X = \frac{1+x+3y-xy}{xy-x+y+3}, Y = \frac{x-y-1-xy}{2}$, and orthogonal with respect to area measure on the 1st octant of the 2-sphere. Due to the S_3 permutation symmetry there is also a basis of eigenfunctions of L_3 and H, but we will not consider it here. Using the method described in [12] we can derive the structure algebra of $S9$, just from the 1st order Gaussian differential recurrences for

$${}_2F_1\left(\begin{array}{cc} a, & b \\ c & \end{array}; z\right) \to {}_2F_1\left(\begin{array}{cc} a\pm 1, & b \\ c & \end{array}; z\right), \quad {}_2F_1\left(\begin{array}{cc} a, & b \\ c & \end{array}; z\right) \to {}_2F_1\left(\begin{array}{cc} a\pm 1, & b\mp 1 \\ c\mp 1 & \end{array}; z\right)$$

and a limiting process. One consequence from Section 5.1 of that paper, is:

$$L_2\Psi_{m,n} = A_n\Psi_{m-1,n+1}+B_n\Psi_{m,n}+C_n\Psi_{m+1,n-1} =$$

$$-\frac{4(N+k_3-n)(N+n+k_1+k_2+k_3+2)(n+1)(n+k_1+k_2+1)}{(2n+k_1+k_2+2)(2n+k_1+k_2+1)}\Psi_{m-1,n+1}$$

$$[-\frac{(k_1^2-k_2^2)\left(k_3^2-(2N+k_1+k_2+k_3+2)^2\right)}{2(2n+k_1+k_2+2)(2n+k_1+k_2)}+\frac{1}{2}(2n+k_1+k_2+1)^2+\frac{1}{4}-k_1^2-\frac{1}{2}(2N+2+k_1+k_2+k_3)^2$$

$$+\frac{1}{2}k_3^2]\Psi_{m,n}-\frac{4(N-n+1)(N+n+k_1+k_2+1)(n+k_1)(n+k_2)}{(2n+k_1+k_2)(2n+k_1+k_2+1)}\Psi_{m+1,n-1},$$

The action of L_1 on the L_2 eigenbasis follows immediately from permutation symmetry. Now we expand the L_2 eigenbasis in terms of the L_1 eigenbasis:

$$\Lambda_{N-q,q}^{(k_1,k_2,k_3)} = \sum_{n=0}^{N} R_q^n(k_1,k_2,k_3)\,\Psi_{N-n,n}^{(k_1,k_2,k_3)}, \quad q=0,\cdots.N$$

Using the self-adjoint properties of L_1, L_2 we can find recurrences to compute the norm:

$$||\Psi_{N-n,n}||^2 = \frac{1}{4n!\Gamma(N-n+1)}\frac{\Gamma(n+k_1+1)\Gamma(n+k_2+1)\Gamma(N-n+k_3+1)\Gamma(N+n+k_1+k_2+2)}{(2N+k_1+k_2+k_3+2)(2n+k_1+k_2+1)\Gamma(n+k_1+k_2+1)\Gamma(N+n+k_1+k_2+k_3+2)}.$$

The squared norm of $\Lambda_{N-n,n}$ follows by applying the permutation $1 \leftrightarrow 3$ to the above expression. To better exploit the symmetry of our system we rescale the bases and expansion:

$$\Psi'^{(k_1,k_2,k_3)}_{N-n,n} = \frac{(-1)^n n!\Gamma(N-n+1)}{\Gamma(N-n+k_3+1)\Gamma(n+k_2+1)}\Psi^{(k_1,k_2,k_3)}_{N-n,n},\quad \Lambda'^{(k_1,k_2,k_3)}_{N-q,q} = \sum_{n=0}^{N} R'^n_q(k_1,k_2,k_3)\,\Psi'^{(k_1,k_2,k_3)}_{N-n,n},\quad q = 0,\cdots,N$$

The two sets of $N+1$ basis vectors $\{\frac{\Psi'_{m,n}}{||\Psi'_{m,n}||}\}$, $\{\frac{\Lambda'_{p,q}}{||\Lambda'_{p,q}||}\}$ are each orthonormal, implying that the $(N+1)\times(N+1)$ matrix $\left(\frac{||\Psi'_{m,n}||R'^n_q}{||\Lambda'_{p,q}||}\right)$, $\quad 0 \le n, q \le N$, is orthogonal. We have identities

$$\sum_{\ell=0}^{N}\frac{R'^{n_1}_{\ell}R'^{n_2}_{\ell}}{||\Lambda'_{N-\ell,\ell}||^2} = \frac{\delta_{n_1,n_2}}{||\Psi'_{N-n_1,n_1}||^2},\quad \sum_{\ell=0}^{N}\frac{R'^{\ell}_{q_1}R'^{\ell}_{q_2}}{||\Lambda_{N-q_1,q_1}||^2} = \frac{\delta_{q_1,q_2}}{||\Psi'_{N-\ell,\ell}||^2}.$$

By permutation symmetry: $\Lambda'^{(k_1,k_2,k_3)}_{p,q} = \Psi'^{(k_3,k_2,k_1)}_{p,q}$. We set

$$R'^n_q(k_1,k_2,k_3)_N \cdot ||\Psi'^{(k_1,k_2,k_3)}_{N-n,n}||^2 \equiv \Xi'\begin{pmatrix} k_1 & k_2 & k_3 \\ n & N & q\end{pmatrix}.$$

Note that Ξ' is invariant under the transposition of its 1st and 3rd columns, and satisfies

$$\sum_{q=0}^{N}\frac{\Xi'\begin{pmatrix} k_1 & k_2 & k_3 \\ n_1 & N & q\end{pmatrix}\Xi'\begin{pmatrix} k_1 & k_2 & k_3 \\ n_2 & N & q\end{pmatrix}}{||\Lambda^{(k_1,k_2,k_3)}_{N-q,q}||^2} = ||\Psi'^{(k_1,k_2,k_3)}_{N-n_1,n_1}||^2\delta_{n_1,n_2},$$

$$\sum_{\ell=0}^{N}\frac{\Xi'\begin{pmatrix} k_1 & k_2 & k_3 \\ \ell & N & q_1\end{pmatrix}\Xi'\begin{pmatrix} k_1 & k_2 & k_3 \\ \ell & N & q_2\end{pmatrix}}{||\Psi'^{(k_1,k_2,k_3)}_{N-\ell,\ell}||^2} = ||\Lambda'^{(k_1,k_2,k_3)}_{N-q_1,q_1}||^2\delta_{q_1,q_2}.$$

Applying L_1, L_2 to both sides of the expansion we have $L_1\Lambda'_{N-q,q} = \sum_{n=0}^{N}\mu_n R'^n_q \Psi'_{N-n,n}$ and

$$L_2\Lambda'_{q,p} = \sum_{n=0}^{N}\lambda_q R'^n_q\Psi'_{N-n,n} = \sum_{n=0}^{N}\left(A'_{n-1}R'^{n-1}_q + B'_n R'^n_q + C'_{n+1}R'^{n+1}_q\right)\Psi'_{N-n,n},$$

$$\mu_n = -(2n+1)^2 - 2(2n+1)(k_1+k_2) + 2k_1k_2 - \frac{1}{2},\quad \lambda_q = -(2q+1)^2 - 2(2q+1)(k_2+k_3) + 2k_2k_3 - \frac{1}{2}.$$

By equating coefficients of $\Psi'_{N-n,n}$, we find a 3-term recurrence formula for R'^n_q, hence for Ξ'. If we make the identifications

$$k_1 = \delta+\beta-1,\ k_2 = \alpha+\gamma-1,\ k_3 = \alpha-\gamma,\ N = -\alpha-\beta, t = q+\frac{k_2+k_3+1}{2},$$

this formula for Ξ' agrees exactly with the 3-term recurrence formula for the Racah polynomials, so, by symmetry, $\Xi'\begin{pmatrix} k_1 & k_2 & k_3 \\ n & N & q\end{pmatrix} = K(N,k_1,k_2,k_3)R_n(k_1k_2k_3,q)$ where R_n is proportional to a Racah polynomial in t^2 of order n and there is permutaion symmetry $\begin{pmatrix} k_1 \\ n \end{pmatrix} \leftrightarrow \begin{pmatrix} k_3 \\ q \end{pmatrix}$. Moreover,

$$||\Lambda'^{(k_1,k_2,k_3)}_{N-q,q}||^2 \sim \frac{\Gamma(t-\alpha+1)\Gamma(t-\beta+1)\Gamma(t-\gamma+1)\Gamma(t-\delta+1)\Gamma(t)}{\Gamma(t+\alpha)\Gamma(t+\beta)\Gamma(t+\gamma)\Gamma(t+\delta)\Gamma(t+1)},$$

so the measure defined by $||\Lambda'^{(k_1,k_2,k_3)}_{N-q,q}||^{-2}$ is a scalar times a function of t that is symmetric with respect to all permutations of $\alpha, \beta, \gamma, \delta$. This implies that the family of orthogonal polynomials determined by this measure must admit this symmetry up to a multiplicative factor. Further, the left hand side of the orthogonality relation is proportional to

$$\sum_{q=0}^{N} \frac{(2\alpha)_q(\alpha+1)_q(\alpha+\beta)_q(\alpha+\gamma)_q(\alpha+\delta)_q}{(\alpha)_q(\alpha-\beta+1)_q(\alpha-\gamma+1)_q(\alpha-\delta+1)_q q!} \Xi' \begin{pmatrix} k_1 & k_2 & k_3 \\ n_1 & N & q \end{pmatrix} \Xi' \begin{pmatrix} k_1 & k_2 & k_3 \\ n_2 & N & q \end{pmatrix},$$

precisely the measure for orthogonality of the Racah polynomials $\Phi_n^{(\alpha,\beta\gamma,\delta)}(t^2)$ where $t = q + \alpha$.

Duality: By making the transpositions $k_1 \leftrightarrow k_3$, $n \leftrightarrow q$ we obtain the result of applying L_1 to the expansion of $\Psi'_{N-n,n}$ in an L_2 eigenbasis. This gives a 3-term recurrence relation for Ξ, defining a family of orthogonal polynomials $p'_q(n)$ in the variable n. It is a 2nd order difference equation in q, hence t, for the Racah polynomials as eigenfunctions. This action induces a model of an irreducible representation of the structure algebra of $S9$ in which the basis functions are Racah polynomials in t^2 and the symmetry operators map to difference operators.

3. Intertwining operators

Let $\mathcal{W}_{k_1,k_2,k_3}$ be the space of functions on the first octant of the 2-sphere and with Hamiltonian $H^{(k_1,k_2,k_3)}$, symmetry operators $L_j^{(k_1,k_2,k_3)}$ and inner product $< \Phi, \Psi >_{k_1,k_2,k_3}$. Let $\mathcal{W}_{k'_1,k'_2,k'_3}$ be another such space. An *intertwining operator* is a mapping $X^{(k_1,k_2,k_3)} : \mathcal{W}_{k_1,k_2,k_3} \to \mathcal{W}_{k'_1,k'_2,k'_3}$ such that

$$X^{(k_1,k_2,k_3)} H^{(k_1,k_2,k_3)} = H^{(k'_1,k'_2,k'_3)} X^{(k_1,k_2,k_3)}.$$

Note that $X^{(k_1,k_2,k_3)}$ maps eigenfunctions of $H^{(k_1,k_2,k_3)}$ to eigenfunctions of $H^{(k'_1,k'_2,k'_3)}$ and its adjoint $X^{*(k_1,k_2,k_3)}$ reverses the action.

Such energy shifting transformations are induced by the basic differential recurrence relations obeyed by Gaussian hypergeometric functions. For example the standard recurrence

$$\left[z(1-z)\frac{d}{dz} - (b+a-1)z + c - 1\right] {}_2F_1\left(\begin{matrix} a & b \\ c & \end{matrix} ; z \right) = (c-1) {}_2F_1\left(\begin{matrix} a-1 & b-1 \\ c-1 & \end{matrix} ; z \right),$$

induces a 1st order differential operator $T^{(k_1,k_2,k_3)} : \mathcal{W}_{k_1,k_2,k_3} \to \mathcal{W}_{k_1-1,k_2-1,k_3}$ such that, in terms of the x, y variables,

$$T^{(k_1,k_2,k_3)} = \sqrt{1-x^2}\,\partial_x - \frac{1}{2}(k_2 - \frac{1}{2})\sqrt{\frac{1+x}{1-x}} + \frac{1}{2}(k_1 - \frac{1}{2})\sqrt{\frac{1-x}{1+x}},\quad T^{(k_1,k_2,k_3)}\Psi_{m,n}^{(k_1,k_2,k_3)} = -(n+1)\Psi_{m,n+1}^{(k_1-1,k_2-1,k_3)}.$$

The adjoint is induced by $\frac{d}{dz} {}_2F_1\left(\begin{matrix} a & b \\ c & \end{matrix} ; z \right) = \frac{ab}{c} {}_2F_1\left(\begin{matrix} a+1 & b+1 \\ c+1 & \end{matrix} ; z \right)$,

$$T^{*(k_1,k_2,k_3)} : \mathcal{W}_{k_1,k_2,k_3} \to \mathcal{W}_{k_1+1,k_2+1,k_3},\quad \Psi_{m,n}^{(k_1,k_2,k_3)} \to -(k_1+k_2+n+1)\Psi_{m,n-1}^{(k_1+1,k_2+1,k_3)},$$

$$T^{*(k_1,k_2,k_3)} = -\sqrt{1-x^2}\,\partial_x - \frac{1}{2}(k_2 + \frac{1}{2})\sqrt{\frac{1+x}{1-x}} + \frac{1}{2}(k_1 + \frac{1}{2})\sqrt{\frac{1-x}{1+x}},$$

Note that these intertwining operators are defined independent of basis. The action of T and T^* on the Λ-basis can again be computed from 1st order relations obeyed by Gaussian hypergeometric functions. To find these we transform to X, Y coordinates and again make use of first order hypergeometric differential recurrences. We obtain

$$T^{(k_1,k_2,k_3)}\Lambda'^{(k_1,k_2,k_3)}_{p,q} = -\frac{1}{2q+k_2+k_3+1}\Lambda'^{(k_1-1,k_2-1,k_3)}_{p+1,q} + \frac{1}{2q+k_2+k_3+1}\Lambda'^{(k_1-1,k_2-1,k_3)}_{p,q+1},$$

$$\tau^{(\alpha-\frac{1}{2},\beta-\frac{1}{2},\gamma-\frac{1}{2},\delta-\frac{1}{2})}\,\Xi'\begin{pmatrix} k_1-1 & k_2-1 & k_3 \\ n & N+1 & q \end{pmatrix} = n(k_1+k_2+n-1)\,\Xi'\begin{pmatrix} k_1 & k_2 & k_3 \\ n-1 & N & q \end{pmatrix},$$

$$\tau^{(\alpha-\frac{1}{2},\beta-\frac{1}{2},\gamma-\frac{1}{2},\delta-\frac{1}{2})}f(t) = \frac{1}{2t}\left[(f(t+\frac{1}{2}) - f(t-\frac{1}{2})\right], \quad t = q + \frac{k_2+k_3+1}{2}.$$

$$\tau^{*(\alpha+\frac{1}{2},\beta+\frac{1}{2},\gamma+\frac{1}{2},\delta+\frac{1}{2})}\,\Xi'\begin{pmatrix} k_1+1 & k_2+1 & k_3 \\ n & N-1 & q \end{pmatrix} = \Xi'\begin{pmatrix} k_1 & k_2 & k_3 \\ n+1 & N & q \end{pmatrix},$$

$$\tau^{*(\alpha+\frac{1}{2},\beta+\frac{1}{2},\gamma+\frac{1}{2},\delta+\frac{1}{2})}f(t) = \frac{1}{2t}\left[(\alpha+t)(\beta+t)(\gamma+t)(\delta+t)f(t+\frac{1}{2}) - (\alpha-t)(\beta-t)(\gamma-t)(\delta-t)f(t-\frac{1}{2})\right].$$

Note that (with $t = q + \frac{k_2+k_3+1}{2}$),

$$\tau^{*(\alpha+\frac{1}{2},\beta+\frac{1}{2},\gamma+\frac{1}{2},\delta+\frac{1}{2})}\,\tau^{(\alpha,\beta,\gamma,\delta)}\,\Xi'\begin{pmatrix} k_1 & k_2 & k_3 \\ n & N & q \end{pmatrix} = n(k_1+k_2+n+1)\,\Xi'\begin{pmatrix} k_1 & k_2 & k_3 \\ n & N & q \end{pmatrix}$$

a 2nd order difference equation for Ξ' as a polynomial in t^2. We will solve this equation.

3.1. Calculation of Racah polynomials

Note that $\Xi'_n(t)$ can be written as $\Xi' = G(n, N, k_1, k_2, k_3)\,\Phi_n(t)$ where $\Phi_n(t)$ is a polynomial in t^2 such that $\Phi_n(0) = 1$. Thus we can write $\Phi_n(t) = \sum_{k=0}^{n} w_k P_k(\alpha, t)$, where $P_k(\alpha, t) = (\alpha+t)_k(\alpha-t)_k$, $w_0 = 1$. Applying τ to $P_k(\alpha, t)$, we get,

$$\tau P_k(\alpha, t) = -kP_{k-1}(\alpha + \frac{1}{2}, t) \tag{3}$$

Applying τ^* to the shifted basis, we get

$$\tau^* P_k(\alpha + \frac{1}{2}, t) = -(\alpha+\beta+\gamma+\delta+k)P_{k+1}(\alpha, t) + (\alpha+\beta+k)(\alpha+\gamma+k)(\alpha+\delta+k)P_k(\alpha, t)_k. \tag{4}$$

Thus,

$$\tau^*\tau P_k(\alpha, t) = k(\alpha+\beta+\gamma+\delta+k-1)P_k(\alpha, t) - k(\alpha+\beta+k-1)(\alpha+\gamma+k-1)(\alpha+\delta+k-1)P_{k-1}(\alpha, t),$$

a 2-term recurrence relation, which implies $w_{k+1} = \frac{(-n+k)(n+\alpha+\beta+\gamma+\delta+k-1)}{(k+1)(\alpha+\beta+k)(\alpha+\gamma+k)(\alpha+\delta+k)} w_k$, $w_0 = 1$. It is easy to solve this recurrence to obtain

$$w_k = \frac{(-n)_k(n+\alpha+\beta+\gamma+\delta-1)_k}{k!(\alpha+\beta)_k(\alpha+\gamma)_k(\alpha+\delta)_k}, \; k = 0, 1, \cdots. \tag{5}$$

Hence, unique up to a scalar multiple,

$$\Phi_n(t) = {}_4F_3\left(\begin{matrix} -n & n+\alpha+\beta+\gamma+\delta-1 & \alpha+t & \alpha-t \\ \alpha+\beta & \alpha+\gamma & \alpha+\delta & \end{matrix} ;1\right). \tag{6}$$

3.2. More permutation symmetry:

Since the measure for the Racah polynomials is invariant under all permutations of $\alpha, \beta, \gamma, \delta$, and the Racah polynomials can be obtained from the measure by the Gram-Schmidt process. Each such permutation must take a Racah polynomial to a scalar multiple of itself. We find $(\alpha+\beta)_n(\alpha+\gamma)_n(\alpha+\delta)_n\,\Phi_n^{(\alpha,\beta,\gamma,\delta)}(t)$ is invariant under all permutations.

3.3. More intertwining operators

The recurrence

$$\left(z\frac{d}{dz}+c-1\right){}_2F_1\left(\begin{matrix} a & & b \\ & c & \end{matrix};z\right)=(c-1){}_2F_1\left(\begin{matrix} a & & b \\ & c-1 & \end{matrix};z\right)$$

leads to an intertwining operator $U^{(k_1,k_2,k_3)}_{(-,+,-,+)}:\ \mathcal{W}_{k_1,k_2,k_3}\longrightarrow\mathcal{W}_{k_1+1,k_2-1,k_3}$,

$$\mu^{(\beta,\delta)}\,\Phi_n^{(\alpha-\frac12,\beta+\frac12,\gamma-\frac12,\delta+\frac12)}(t)=\frac{(n+\beta+\delta)(n+\alpha+\gamma-1)}{(\alpha+\gamma-1)}\,\Phi_n^{(\alpha,\beta,\gamma,\delta)}(t)$$

Its action on the Λ' basis induces the recurrence

$$\mu^{(\beta,\delta)}f(t)=\frac{1}{2t}\left[(\beta+t)(\delta+t)f(t+\frac12)-(\beta-t)(\delta-t)f(t-\frac12)\right],$$

$$\mu^{(\beta,\delta)}\,\Xi'\left(\begin{matrix} k_1+1 & k_2-1 & k_3\\ n & N & q\end{matrix}\right)=(n+k_1+1)(n+k_2)\,\Xi'\left(\begin{matrix} k_1 & k_2 & k_3\\ n & N & q\end{matrix}\right).$$

The permutation invariance of the Racah polynomials leads to a family of recurrences in the μ such that any pair of $\alpha,\beta,\gamma,\delta$ can be raised by $\frac12$ and the other pair lowered by $\frac12$. These also follow from intertwining operators induced by Gaussian hypergeometric differential recurrences. In particular, the hypergeometric recurrence

$$\left(z\frac{d}{dz}+a\right){}_2F_1\left(\begin{matrix} a & & b \\ & c & \end{matrix};z\right)=a\,{}_2F_1\left(\begin{matrix} a+1 & & b \\ & c & \end{matrix};z\right)$$

induces the operator

$$U^{(k_1,k_2,k_3)}_{(+,+,-,-)}=\sqrt{\frac{1+y}{2}}\left[-(1-y)\partial_y-N-\frac{k_1}{2}-\frac{k_2}{2}-\frac12+\frac12(k_3+\frac12)\left(\frac{1-y}{1+y}\right)\right],\qquad(7)$$

$$U^{(k_1,k_2,k_3)}_{(+,+,-,-)}\Psi^{(k_1,k_2,k_3)}_{m,n}=-(n+N+k_1+k_2+1)\Psi^{(k_1,k_2,k_3+1)}_{m-1,n},\quad m\ge1,n\ge0.\qquad(8)$$

The action on the L_2 eigenbasis is

$$\mu^{(\alpha,\beta)}\,\Xi'\left(\begin{matrix} k_1 & k_2 & k_3+1\\ n & N-1 & q\end{matrix}\right)=-\frac{2N+k_1+k_2+k_3+2}{2N+k_1+k_2+k_3+1}\,\Xi'\left(\begin{matrix} k_1 & k_2 & k_3\\ n & N & q\end{matrix}\right),$$

$$\mu^{(\alpha,\beta)}\Phi_n^{(\alpha+\frac12,\beta+\frac12,\gamma-\frac12,\delta-\frac12)}=(\alpha+\beta)\Phi_n^{(\alpha,\beta,\gamma,\delta)},\quad \alpha+\beta=-N.\qquad(9)$$

The recurrence $\left(z\frac{d}{dz}+b\right){}_2F_1\left(\begin{matrix} a & & b \\ & c & \end{matrix};z\right)=b\,{}_2F_1\left(\begin{matrix} a & & b+1 \\ & c & \end{matrix};z\right)$, induces the operator

$$U^{(k_1,k_2,k_3)}_{(+,-,-,+)}=\sqrt{\frac{1+y}{2}}\left[(y-1)\partial_y+N+\frac{k_1}{2}+\frac{k_2}{2}+k_3+\frac32+\frac12(k_3+\frac12)\left(\frac{y-1}{y+1}\right)\right],\qquad(10)$$

$$U^{(k_1,k_2,k_3)}_{(+,-,-,+)}\Psi^{(k_1,k_2,k_3)}_{m,n}=(n+N+k_1+k_2+k_3+2)\Psi^{(k_1,k_2,k_3+1)}_{m,n},\quad m\ge0,n\ge0.\qquad(11)$$

In terms of Ξ' and Φ_n the action is

$$\mu^{(\alpha,\delta)}\,\Xi'\left(\begin{matrix} k_1 & k_2 & k_3+1\\ n & N & q\end{matrix}\right)=\frac{2N+k_1+k_2+k_3+2}{2N+k_1+k_2+k_3+3}\,\Xi'\left(\begin{matrix} k_1 & k_2 & k_3\\ n & N & q\end{matrix}\right),$$

$$\mu^{(\alpha,\delta)}\Phi_n^{(\alpha+\frac12,\beta-\frac12,\gamma-\frac12,\delta+\frac12)}=(\alpha+\delta)\Phi_n^{(\alpha,\beta,\gamma,\delta)},\qquad \alpha+\delta=N+k_1+k_2+k_3+2.$$

Note that the operators (7,10) are not basis independent, since they depend on N. However, the intertwining operator $V^{(k_1,k_2,k_3)}:\ \mathcal{W}_{k_1,k_2,k_3}\longrightarrow\mathcal{W}_{k_1,k_2,k_3+1}$,

$$V^{(k_1,k_2,k_3)}=U^{(k_1,k_2,k_3)}_{(+,-,-,+)}+U^{(k_1,k_2,k_3)}_{(+,+,-,-)}=\sqrt{\frac{1+y}{2}}\left[2(y-1)\partial_y+k_3+1\right].\qquad(12)$$

is basis independent, so defined intrinsically, as is its adjoint.

3.4. The expansion coefficients
Solving all of these recurrences for Ξ', we find

$$R'^{n}_{q}(k_1,k_2,k_3)_N \cdot ||\Psi'_{N-n,n}(k_1,k_2,k_3)||^2 = \Xi'\begin{pmatrix} k_1 & k_2 & k_3 \\ n & N & q \end{pmatrix} =$$

$$\frac{4c\ (2N+k_1+k_2+k_3+2)\Gamma(N+1)}{\Gamma(N+k_1+k_2+k_3+2)\Gamma(k_2+1)}\Phi_n^{(\alpha,\beta,\gamma,\delta)}(t^2), \quad t = q+\frac{k_2+k_3+1}{2},$$

$$\alpha = \frac{k_2+k_3+1}{2},\ \beta = -N-\frac{k_2+k_3+1}{2},\ \gamma = \frac{k_2-k_3+1}{2},\ \delta = N+k_1+\frac{k_2+k_3+3}{2},$$

$$\Phi_n = {}_4F_3\left(\begin{matrix} -n, & k_1+k_2+n+1, & -q, & k_3+k_2+q+1 \\ -N, & k_2+1, & N+k_1+k_2+k_3+2 & \end{matrix};1\right)$$

The overall scaling factor $c(k_1,k_2,k_3)$ can be determined by evaluating the double integral for Ξ' in the simplest case $n=q=N=0$ where it factors into a product of beta integrals.

4. Extension to Wilson polynomials

Racah polynomials are $S9$ expansion coefficients for finite dimensional representations on the real 2-sphere. Wilson polynomials are expansion coefficients related to infinite dimensional representations for the Schrödinger eigenvalue equation of the generic potential on the upper sheet of the 2d hyperboloid. (Much earlier, Koornwinder [10, 11] pointed out the connection between expansion coefficients on higher dimensional hyperboloids but not specifically to the 2d case.) A Hilbert space structure is imposed on the eigenspace corresponding to a single continuous spectrum eigenvalue, where N is a negative real number, not an integer.

We expand the L_2 basis vectors in terms of the L_1 basis:

$$\Lambda_q = \sum_{n=0}^{\infty} R_q^n \Psi_n, \quad \sum_{\ell=0}^{\infty} \frac{||\Psi_{n_1}||^2}{||\Lambda_\ell||^2} R_\ell^{n_1} R_\ell^{n_2} = \delta_{n_1,n_2},$$

Applying L_2 to both sides of the expansion we can show that the R_q^n satisfy a three term recurrence relation and a difference equation as before, and the orthogonality relation can be rewritten in the form

$$\sum_{q=0}^{\infty} \frac{(2\alpha)_q(\alpha+1)_q(\alpha+\beta)_q(\alpha+\gamma)_q(\alpha+\delta)_q}{(\alpha)_q(\alpha-\beta+1)_q(\alpha-\gamma+1)_q(\alpha-\delta+1)_q\, q!} R'^{n_1}_{q} R'^{n_2}_{q} = \delta_{n_1,n_2} h_{n_1},$$

where $R'^{n}_{q} \sim \Phi_n(t)$ is a Wilson polynomial. This is equivalent to a ${}_5F_4$ identity and can all be made rigorous. Wilson recast the orthogonality into the form of a contour integral which greatly extended its domain of validity. All of the Racah-intertwining operators extend to this case.

The quantum problem on the 2d hyperboloid has mixed spectrum, both bound states and continuous spectra, [13]. An interesting task for future research is to work out the interbasis expansion associated with the spectral decomposition in this case and to relate it explicitly to partially discrete, partially continuous orthogonality relations for Wilson polynomials.

5. Extension to Wilson functions

If n is not an integer a formal calculation using (3) and (4) that ignores series convergence still gives (6) as a solution of the eigenvalue equation $\tau^*\tau\,\Phi_n = n(n+\alpha+\beta+\gamma+\delta-1)\Phi_n$. However, a careful calculation gives

$$[\tau^*\tau - n(n+\alpha+\beta+\gamma+\delta)]\sum_{k=0}^{K}\omega_k\, P(t,\alpha)_k = \frac{(-n)_{K+1}(n+\alpha+\beta+\gamma+\delta-1)_{K+1}}{K!\ (\alpha+\beta)_K(\alpha+\gamma)_K(\alpha+\delta)_K}(\alpha+t)_K(\alpha-t)_K.$$

Journal of Physics: Conference Series **597** (2015) 012059 doi:10.1088/1742-6596/597/1/012059

Taking the limit as $K \to +\infty$, and making use of the Stirling formula, we obtain

$$[\tau^*\tau - n(n+\alpha+\beta+\gamma+\delta)]\,\Phi_n(t) = \frac{\Gamma(\alpha+\beta)\Gamma(\alpha+\gamma)\Gamma(\alpha+\delta)}{\Gamma(-n)\Gamma(n+\alpha+\beta+\gamma+\delta-1)\Gamma(\alpha+t)\Gamma(\alpha-t)}. \quad (13)$$

Since the Γ function has a pole at the negative integers, we see that $\Phi_n(t)$ satisfies the eigenvalue equation for n a nonnegative integer, but otherwise it does not, except for isolated choices of the parameters. Now consider the functions with relations

$$Q(t,\alpha,\beta)_k = \frac{\Gamma(1-\beta+t)\Gamma(1-\beta-t)}{\Gamma(\alpha+t)\Gamma(\alpha-t)}(1-\beta+t)_k(1-\beta-t)_k,\ \tau\, Q(t,\alpha,\beta)_k = (\alpha+\beta-k-1)\, Q(t,\alpha+\frac{1}{2},\beta+\frac{1}{2})_k,$$

$$\tau^*\, Q(t,\alpha+\frac{1}{2},\beta+\frac{1}{2})_k = -(\gamma+\delta+k)\, Q(t,\alpha,\beta)_k + k(k-\beta+\delta)(k-\beta+\gamma)\, Q(t,\alpha,\beta)_{k-1}.$$

Then computing formally without regard to series convergence, we obtain the nonpolynomial solution of the $\tau^*\tau$ eigenvalue equation:

$$\Psi_n(t) = \frac{\Gamma(1-\beta+t)\Gamma(1-\beta-t)}{\Gamma(\alpha+t)\Gamma(\alpha-t)}\,{}_4F_3\left(\begin{matrix}1-n-\alpha-\beta & n+\gamma+\delta & 1-\beta+t & 1-\beta-t\\ 2-\alpha-\beta & 1-\beta+\gamma & 1-\beta+\delta & \end{matrix};1\right). \quad (14)$$

However, a careful computation, exactly analogous to that for $\Phi_n(t)$, yields the result

$$(\tau^*\tau - n(n+\alpha+\beta+\gamma+\delta))\,\Psi_n(t) = \frac{\Gamma(2-\alpha-\beta)\Gamma(1-\beta+\gamma)\Gamma(1-\beta+\delta)}{\Gamma(1-n-\alpha-\beta)\Gamma(n+\gamma+\delta)\Gamma(\alpha+t)\Gamma(\alpha-t)}, \quad (15)$$

so $\Psi_n(t)$ doesn't satisfy the eigenvalue equation. Now, comparing (13),(15), we see that functions

$$\tilde{\Phi}_n^{(\alpha,\beta\gamma,\delta)}(t) = \Phi_n(t) - \frac{\Gamma(\alpha+\beta)\Gamma(\alpha+\gamma)\Gamma(\alpha+\delta)\Gamma(1-n-\alpha-\beta)\Gamma(n+\gamma+\delta)}{\Gamma(-n)\Gamma(n+\alpha+\beta+\gamma+\delta-1)\Gamma(2-\alpha-\beta)\Gamma(1-\beta+\gamma)\Gamma(1-\beta+\delta)}\Psi_n(t) \quad (16)$$

do satisfy the eigenvalue equation for general $n = n' + c$ where n' runs over the integers and c is a fixed noninteger. Furthermore it is straightforward to verify that $\tilde{\Phi}_n^{(\alpha,\beta\gamma,\delta)}(t)$ satisfies all of the recurrence formulas induced by the intertwining operators τ, μ, τ^*, μ^* for general n, such as (9), that are satisfied by Φ_n for nonnegative integer values. The functions $\tilde{\Phi}_n(t)$ have the duality property, hence since they satisfy the 2nd order difference eigenvalue equation for general n they must also satisfy the 3-term recurrence formula. Thus for fixed c these basis functions define an infinite-dimensional irreducible representation of the quadratic algebra of $S9$ in which the eigenvalues of L_1 are indexed by an arbitrary integer n'. In the original quantum problem this representation is easy to construct: all of the recurrence relations that we have derived using hypergeometric functions remain valid for nonpolynomial hypergeometric functions. These nonpolynomial bases are no longer normalizable, but the recurrence relations remain valid. Not clear was how such representations could be realized in terms of difference operators. The solution is that the analytic continuation of the basis functions is (16), the Wilson functions which lead to associated Wilson polynomials [14, 15], and to the Wilson transform [16], which corresponds to infinite dimensional irreducible representations of the quadratic algebra of $S9$ in which the spectrum of L_1 is (partly) continuous. In the original quantum mechanical system the analytic continuation is evident, but the integral over the sphere, i.e. in x, y, giving the interbasis expansion coefficients is deformed into a Pochhammer contour integral on a Riemann surface over the x-plane with branch points at ± 1 and a similar surface over the y-plane. All of the intertwining operator recurrences can be verified by integration by parts.

6. Discussion and Conclusions

- We showed explicitly how Racah and Wilson polynomials, and the Wilson functions, arise as expansion coefficients for the generic superintegrable system on the complex 2-sphere, relating two different sets of spherical coordinate bases. We employed only techniques from quantun theory and facts about the differential recurrence relations of ordinary ${}_2F_1$ hypergeometric series. We did not use results from the established theory of Wilson polynomials and functions, although we did identify instances where our results correspond to known facts about the functions.
- We showed how the principal properties of these functions: the measure, 3-term recurrence relation, the orthogonality measure, 2nd order difference equation, duality, permutation symmetry, and intertwining operators – follow from the symmetry of this quantum system.
- The parameter changing difference relations for the polynomials follow from intertwining operators for the quantum system. All of the properties of this system are induced by the fundamental differential recurrence relations of the Gaussian hypergeometric functions.
- There is active interest in the relation between multivariable Wilson polynomials and the quantum superintegrable system on the n-sphere with generic potential, e.g. [17], and these results should aid in the generalization.
- By contracting function space realizations of irreducible representations of the $S9$ quadratic algebra to the other superintegrable systems one obtains the full Askey scheme of orthogonal hypergeometric polynomials, uniquely induced by Lie algebra contractions of $so(3, C)$ and $e(2, C)$, [18, 19]. This work should be extended to multivariable orthogonal polynomials.

Acknowledgments

W. M. benefited greatly from conversations with Charles Dunkl, Mourad Ismail, Eric Koelink and Tom Koornwinder. We thank the refferee for several important suggestions. This work was partially supported by a grant from the Simons Foundation (# 208754 to Willard Miller, Jr.).

References

[1] Miller W Jr, Post S and Winternitz P 2013 Classical and Quantum Superintegrability with Applications *J. Phys. A: Math. Theor.* **46** 423001
[2] Kalnins E G, Kress J M, Miller W Jr and Pogosyan G S 2001 Completeness of superintegrability in two dimensional constant curvature spaces *J. Phys. A Math Gen.* **34** 4705
[3] Kalnins E G, Miller W Jr and Post S 2007 Wilson polynomials and the generic superintegrable system on the 2-sphere *J. Phys. A: Math. Theor.* **40** 11525-11538
[4] Zhedanov A S 1991 Hidden symmetry of Askey-Wilson polynomials *Theoret. and Math. Phys.* **89** 1146-1157
[5] Granovskii Y I, Zhedanov A S and Lutsenko I M 1992 Quadratic algebras and dynamics in curved spaces. I. Oscillator *Theor. Math. Phys.* **91** 474-480
[6] Granovskii Y I, Zhedanov A S and Lutsenko I M 1992 Quadratic algebras and dynamics in curved spaces. II. The Kepler problem *Theor. Math. Phys.* **91** 604-612
[7] Granovskii Y I, Lutsenko I M and Zhedanov A S 1992 Mutual integrability, quadratic algebras, and dynamical symmetry *Ann. Phys.* **217** 1-20 (section 3)
[8] Genest V, Vinet L and Zhedanov A 2014 Superintegrability in two dimensions and the Racah-Wilson algebra *Lett. Math. Phys.* **104** 8 931-952
[9] Dunkl C F 1984 Orthogonal polynomials with symmetry of order three *Canad. J. Math.* **36** 685-717
[10] Koornwinder T H 1986 A group theoretic interpretation of Wilson polynomials *Complex analysis and applications '85* (Publishing House of the Bulgarian Academy of Sciences Sofia) 335-355
[11] Koornwinder T H 1988 Group theoretic interpretations of Askey's scheme of hypergeometric orthogonal polynomials *Orthogonal polynomials and their applications* Lecture Notes in Math 1329 (Berlin: Springer-Verlag)
[12] Kalnins E G, Kress J M and Miller W Jr 2011 A recurrence relation approach to higher order quantum superintegrability *SIGMA* **7** 031
[13] Frank A and Wolf K B 1985 Lie algebras for systems with mixed spectra. I. The scattering Pöschl-Teller potential *J. Math. Phys.* **26** 977-983

[14] Ismail M E H, Lettessier J, Valent G and Wimp J 1990 Two families of associated Wilson polynomials *Can. J. Math.* **42** 659-695

[15] Masson D R 1991 Associated Wilson polynomials *Constr. Approx.* **7** 521-534

[16] Groenevelt W 2003 The Wilson function transform *Int. Math. Res. Not.* **52** 2779-2817

[17] Kalnins E G, Miller W Jr and Post S 2011 Two-variable Wilson polynomials and the generic superintegrable system on the 3-sphere *SIGMA* **7** 051

[18] Kalnins E G, Miller W Jr and Post S 2013 Contractions of 2D 2nd order quantum superintegrable systems and the Askey scheme for hypergeometric orthogonal polynomials *SIGMA* **9** 057

[19] Kalnins E G and Miller W Jr 2014 Quadratic algebra contractions and 2nd order superintegrable systems *Analysis and Application* **12** 583-612

Journal of Physics: Conference Series **597** (2015) 012060 doi:10.1088/1742-6596/597/1/012060

Neutral bions in the $\mathbb{C}P^{N-1}$ model for resurgence

Tatsuhiro Misumi, Muneto Nitta and Norisuke Sakai

Department of Physics, and Research and Education Center for Natural Sciences, Keio University, Hiyoshi 4-1-1, Yokohama, Kanagawa 223-8521, Japan

E-mail: misumi@phys-h.keio.ac.jp, nitta@phys-h.keio.ac.jp, norisuke.sakai@gmail.com

Abstract. Classical configurations in the $\mathbb{C}P^{N-1}$ model on $\mathbb{R}^1 \times S^1$ is studied with twisted boundary conditions. Starting from fractional instantons with the $\mathbb{Z}_N$ twisted boundary conditions, we review briefly the relevance of our study to resurgence phenomenon in field theory. We consider primarily configurations composed of multiple fractional instantons, called "neutral bions", which are identified as "perturbative infrared renormalons". We construct an explicit ansatz corresponding to topologically trivial configurations containing one fractional instanton ($\nu = 1/N$) and one fractional anti-instanton ($\nu = -1/N$), which is guaranteed to become a solution of field equations asymptotically for large separations. The attractive interactions between the instanton constituents from small to large separations are found to be consistent with the standard separated-instanton calculus. Our results suggest that the ansatz enables us to study bions and the related physics for a wide range of separations. This talk is mainly based on our recent work published in JHEP **1406**, 164 (2014) [arXiv:1404.7225 [hep-th]].

1. Introduction

Recently a progress has been brought about to understand the confinement mechanism in QCD-like theories by compactifing the theories in one spatial dimension with the period L. Because of the asymptotic freedom, the small L theory is in the weak-coupling regime ($L \ll 1/\Lambda_{\rm QCD}$) in QCD(adj.) on $\mathbb{R}^3 \times S^1$. Instanton is a particle-like soliton of codimension D in D-dimensional spacetime. If $\mathbb{Z}_N$ twisted boundary conditions are imposed, we can obtain fractional instantons which have fractional amount of instanton number. A molecule of fractional instantons and anti-instantons are generally called bions. For compactified space with small L, the perturbative analysis is reliable and shows that the Wilson holonomy stabilizes the $\mathbb{Z}_N$ twisted boundary condition. In this case, one can demonstrate the condensation of classical field configurations composed of fractional instantons and anti-instantons with magnetic charge, which are called magnetic bions. Condensation of magnetic bions implies the confinement in the compactified theory [1, 2, 3, 4, 5, 6]. It is anticipated that this phenomenon continues to hold for large L, suggesting the confinement even for uncompactified theory.

It has also been pointed out that neutral bions play an important role in explaining the resurgence phenomenon [7, 8, 9, 10, 11, 12, 13, 14]. Neutral bions are composite of fractional instantons and anti-instantons with vanishing topological charge (instanton number) and zero magnetic charge. Although pertubation series in field thoeries are divergent, the Borel transformation is useful to obtain the resummation of such divergent series. The Borel transform often exhibits singularities on the positive real axis producing an imaginary ambiguities in the Borel-resummed results. It has been found that these ambiguities in qunatum mechanics are cancelled by contributions from molecule of instanton and anti-instantons [15, 16, 17, 18]. However, asymptotically free field theories like QCD and the $\mathbb{C}P^{N-1}$ nonlinear sigma models exhibits similar but more serious singularities near the origin, which are called infrared renormalons [19, 20]. Recently it has been found that the nonperturbative contributions around the neutral saddle points also have imaginary ambiguities. These imaginary ambiguities of bion amplitudes are found to cancel [7, 8] precisely the imaginary ambiguities associated with the infrared renormalons, resulting in a more rigorous foundation of field thoery. This phenomenon is called the resurgence. It is expected that full semi-classical expansion including perturbative and non-perturbative sectors, which is called resurgent expansion [21], leads to unambiguous and self-consistent definition of field theories in the spirit of quantum mechanical examples [15, 16, 17]. However, it is not straightforward to verify these arguments in gauge theories directly, since it is difficult to find an explicit ansatz of bion configurations.

At this stage, it is quite interesting and useful to study simpler models in lower dimensions [8, 9] instead of gauge theories in four spacetime dimensions. In particular, models in two spacetime dimensions are more tractable compared to gauge theories in four dimensions, but have many features which are common to gauge theories in four dimensions. Among them, the $\mathbb{C}P^{N-1}$ model in two spacetime dimensions has been studied as a toy model [22] of the Yang-Mills theory in four spacetime dimensions, because of their interesting features such as the dynamical mass gap, the asymptotic freedom and the existence of instantons as Bogomol'nyi-Prasad-Sommerfield (BPS) solutions [23]. The $\mathbb{C}P^{N-1}$ model on $\mathbb{R}^1 \times S^1$ with twisted boundary conditions admits BPS fractional instantons as configurations with the minimal topological charge [24, 25, 26, 27] (see also Refs. [28, 29, 30]). In Ref. [8], generic arguments on bion configurations were given in the $\mathbb{C}P^{N-1}$ model on $\mathbb{R}^1 \times S^1$ with $\mathbb{Z}_N$ twisted boundary conditions, based on the independent instanton description taking account of interactions between far-separated fractional instantons and anti-instantons. More recently, exact non-BPS solutions were found in the $\mathbb{C}P^{N-1}$ model on $\mathbb{R}^1 \times S^1$ with the $\mathbb{Z}_N$ twisted boundary condition [9]. The simplest non-BPS solution that they found is a four-instanton configuration composed of two fractional instantons (instanton charge $\nu = 1/N$) and two fractional anti-instantons ($\nu = -1/N$) for $N \geq 3$ placed at particular relative positions. Their study suggests that it is likely that bion configurations needed to understand the resurgence phenomenon may not be solutions of field equations. It has been known that field configurations other than the solution of the equations of motion may give significant contributions in the functional integral. Recently we constructed Ansatz of bion configurations that are guaranteed to become solutions of the field equations asymptotically as separations between constituent fractional instantons get larger, and studied interactions of constituent fractional instantons in bion configuration [31]. More recently, we have also studied bions in the Grassmann sigma models which include the $\mathbb{C}P^{N-1}$ model as a subclass [32].

2. $\mathbb{C}P^{N-1}$ model

Since bions in nonlinear sigma models in two spacetime dimensions are interesting and more tractable, let us consider the $\mathbb{C}P^{N-1}$ model. To describe the target space $\mathbb{C}P^{N-1} = \frac{SU(N)}{SU(N-1)\times U(1)}$, we denote an N-component vector of complex scalar fields as $\omega(x)$, and a normalized complex N-component vector composed from ω as $n(x)$

$$n(x) \equiv \omega(x)/|\omega(x)|, \quad |\omega| = \sqrt{\omega^\dagger \omega}. \tag{1}$$

The Lagrangian in Euclidean two dimensions for the nonlinear sigma model with $\mathbb{C}P^{N-1}$ as the target space is given by

$$\mathcal{L} = \frac{1}{2\pi g^2}(D_\mu n)^\dagger (D_\mu n), \tag{2}$$

and the topological charge representing $\pi_2(\mathbb{C}P^{N-1}) \simeq \mathbb{Z}$ is defined by

$$Q = \frac{1}{2\pi}\int d^2x\; i\epsilon_{\mu\nu}(D_\mu n)^\dagger(D_\nu n) = \frac{1}{2\pi}\int d^2x \epsilon_{\mu\nu}\partial_\mu A_\nu\,, \tag{3}$$

where $d^2x \equiv dx_1dx_2$ and $\mu,\nu = 1,2$. Our convention of the covariant derivative is $D_\mu = \partial_\mu - iA_\mu$ with a composite gauge field

$$A_\mu(x) \equiv -in^\dagger\partial_\mu n. \tag{4}$$

We consider $\mathbb{R}^1 \times S^1$ as the geometry of base manifold, and configurations on it satisfying periodicity in the x_2 direction with the period L. The Lagrangian $\mathcal{L}$ and topological charge Q can be expressed in terms of the projection operator $\mathbf{P} \equiv nn^\dagger = \frac{\omega\omega^\dagger}{\omega^\dagger\omega}$ and using the complex coordinate $z \equiv x_1 + ix_2$,

$$\mathcal{L} = \frac{2}{g^2}\mathrm{Tr}\left[\partial_z\mathbf{P}\partial_{\bar{z}}\mathbf{P}\right], \tag{5}$$

$$Q = 2\int d^2x\mathrm{Tr}\left[\mathbf{P}(\partial_{\bar{z}}\mathbf{P}\partial_z\mathbf{P} - \partial_z\mathbf{P}\partial_{\bar{z}}\mathbf{P})\right]. \tag{6}$$

We define the euclidean energy density (Lagrangian) $s(x_1)$, and the topological charge density $q(x_1)$ as functions of x_1 after the integration over x_2:

$$s(x_1) = \frac{1}{g^2\pi}\int dx_2\mathrm{Tr}\left[\partial_z\mathbf{P}\partial_{\bar{z}}\mathbf{P}\right], \tag{7}$$

$$q(x_1) = \frac{1}{\pi}\int dx_2\mathrm{Tr}\left[\mathbf{P}(\partial_{\bar{z}}\mathbf{P}\partial_z\mathbf{P} - \partial_z\mathbf{P}\partial_{\bar{z}}\mathbf{P})\right]. \tag{8}$$

We can perform the Bogomol'nyi completion to obtain the Bogomol'nyi bound [33] for the total energy S

$$S = \frac{1}{2\pi g^2}\int d^2x\left(\frac{1}{2}\left|D_\mu n \mp i\epsilon_{\mu\nu}D_\nu n\right|^2 \pm i\epsilon_{\mu\nu}(D_\nu n)^\dagger(D_\mu n)\right) \geq \pm\frac{Q}{g^2}. \tag{9}$$

The Bogomol'nyi bound with the upper sign is saturated when the following BPS equation is satisfied

$$D_1 n - iD_2 n = \frac{1}{2}D_{\bar{z}}n = 0. \tag{10}$$

The solution of the BPS equation is precisely holomorphic $\omega(z)$. The Bogomol'nyi bound with the lower sign is saturated when the following anti-BPS equation is satisfied

$$D_1 n + iD_2 n = \frac{1}{2}D_{\bar{z}}n = 0. \tag{11}$$

The solution of the anti-BPS equation is anti-holomorphic $\omega(\bar{z})$.

In the following we omit the coupling $1/g^2$ for simplicity unless stated otherwise.

3. Borel resummation

Let us give a brief review of resurgence and the importance of bion contributions. It has been known that the number of Feynman diagrams in quantum field theory grows factorially. This factorially divergent perturbation series of quantum field theory is usually discussed by means

Journal of Physics: Conference Series **597** (2015) 012060 doi:10.1088/1742-6596/597/1/012060

of the Borel transform. The Borel transform method is applicable to the following class of divergent series (called Gevrey-1)

$$P(g^2) = \sum_{q=0}^{\infty} a_q (g^2)^q, \quad |a_q| \le Cq! \left(\frac{1}{A}\right)^q, \tag{12}$$

where C, A are constants. The Borel transform $BP(t)$ is defined as

$$BP(t) = \sum_{q=0}^{\infty} \frac{a_q}{q!} t^q, \tag{13}$$

and the Borel resummation $\mathbb{B}(g^2)$ is defined as

$$\mathbb{B}(g^2) = \int_0^{\infty} \frac{dt}{g^2} e^{-t/g^2} BP(t). \tag{14}$$

One can easily see that the Borel resummation $\mathbb{B}(g^2)$ reproduces the original sum $P(g^2)$ correctly whenever one can exchange the integral and the sum. Otherwise, we need to define the sum in terms of the Borel resummation.

As a simplified toy model, let us consider a factorially divergent series of the following one with alternating signs

$$P(g^2) = C \sum_{q=0}^{\infty} q! \left(\frac{-g^2}{A}\right)^q. \tag{15}$$

Then the Borel transform becomes an analytic function without singularities on the positive real axis

$$BP(t) = C \sum_{q=0}^{\infty} \left(\frac{-t}{A}\right)^q = \frac{CA}{A+t}. \tag{16}$$

Therefore the Borel resummation is well-defined as an integral along the positive real axis

$$\mathbb{B}(g^2) = \int_0^{\infty} \frac{dt}{g^2} e^{-t/g^2} \frac{CA}{A+t}. \tag{17}$$

This altenating factorially divergent series is a typical example of Borel summable divergent series.

On the other hand, if perturbation series is not alternating, the factorially divergent series gives the Borel transform with singularities on positive real axis and the Borel resummmation has imaginary ambiguities. For instance, suppose that the perturbation series $P_{\text{pert}}(g^2)$ gives non-alternating factorially divergent series like

$$P_{\text{pert}}(g^2) = C \sum_{q=0}^{\infty} q! \left(\frac{g^2}{A}\right)^q. \tag{18}$$

The Borel transform has a singularity on positive real axis

$$BP_{\text{pert}}(t) = C \sum_{q=0}^{\infty} \left(\frac{t}{A}\right)^q = \frac{CA}{A-t}, \tag{19}$$

$$\mathbb{B}_{\text{pert}}(g^2) = \int_0^{\infty} \frac{dt}{g^2} e^{-t/g^2} \frac{CA}{A-t}. \tag{20}$$

Therefore the Borel resummation has imaginary ambiguities depending on the choice of integration contours to avoid the singularity (equivalently the analytic continuation from negative real axis)

$$\mathbb{B}_{\rm pert}(g^2 \pm \epsilon) = {\rm Re}\mathbb{B}_{\rm pert}(|g^2|) \pm i{\rm Im}\mathbb{B}_{\rm pert}(|g^2|), \quad {\rm Im}\mathbb{B}_{\rm pert}(|g^2|) \sim -\pi e^{-A/g^2}. \tag{21}$$

This ambiguous imaginary part has to be cancelled by contributions from other saddle points in the path-integral. In quantum mechanics or scalar field theories, the position of the singularity in the Borel plane has been found to correspond to molecules of instanton and anti-instanton [15, 16, 17]. However, the asymptotically free theories like QCD and the $\mathbb{C}P^{N-1}$ model give additional singularities due to IR renormalons which are much closer to the origin and hence give much larger contributions. Recent study [7, 8] showed that the bion amplitudes give the nonperturbative contribution needed to cancell the imaginary ambiguity due to the IR renormalons.

4. Fractional instantons and neutral-bion in $\mathbb{Z}_N$ twisted boundary conditions

When one space direction is compactified to S^1 with the circumference L, there is a possibility to impose non-periodic boundary condition. Moreover, asymptotic free theory becomes weakly coupled, and the stable configurations of Wilson holonomies of $U(1)^{N-1}$ subgroup of $SU(N)$ can be analyzed perturbatively [8, 9]. It turns out that the following Z_N twisted boundary condition is most favorable under certain conditions

$$\omega(x_1, x_2 + L) = \Omega\, \omega(x_1, x_2), \qquad \Omega = {\rm diag.}\left[1, e^{2\pi i/N}, e^{4\pi i/N}, \cdots, e^{2(N-1)\pi i/N}\right]. \tag{22}$$

This $\mathbb{Z}_N$ twisted boundary condition corresponds to the vacuum with the symmetry breaking $SU(N) \to U(1)^{N-1}$. The Wilson-loop holonomy in the compactified direction is given by

$$\langle A_2 \rangle = (0, 2\pi/N, \cdots, 2(N-1)\pi/N), \tag{23}$$

where the gauge fields A_2 is the composite gauge field defined in Eq. (4) of the $\mathbb{C}P^{N-1}$ model.

The simplest solution of the BPS equation (10) satisfying the Z_N twisted boundary condition in Eq. (22) is given by

$$\omega_L = \left(0, \cdots, 0, 1, \lambda e^{+2\pi z/N}, 0, \cdots\right)^T, \quad \omega_R = \left(0, \cdots, 0, 1, \lambda e^{-2\pi \bar{z}/N}, 0, \cdots, 0\right)^T. \tag{24}$$

One can easily see that these BPS solutions give the total action S and the topological charge Q as

$$S = 1/N, \quad Q = \pm 1/N. \tag{25}$$

Therefore we call this BPS solution as a fractional instanton. The fractional instanton is located at $x_1 = \frac{N}{2\pi}\log\frac{1}{\lambda}$.

A neutral bion is a molecule of fractional instantons and anti-instatons and is unstable under the annihilation process. Then we anticipate that generic neutral bion configuration is not a solution of field equations. Therefore we wish to construct a field configuration that reduces to a solution at least for far-separated fractional instantons and anti-instantons asymptotically.

From the BPS solution in Eq.(24) and their complex conjugates, we are naturally led to consider the following ansatz for the $\mathbb{C}P^1$ model satisfying the $\mathbb{Z}_N$ twisted boundary condition in Eq. (22)

$$\omega = \left(0, \cdots, 0, 1 + \lambda_2 e^{2\pi(z+\bar{z})/N},\ \lambda_1 e^{2\pi z/N}, 0, \cdots, 0\right)^T. \tag{26}$$

A fractional instanton is located at $x_1 = \frac{N}{2\pi}\log\frac{1}{\lambda_1}$, and a fractional anti-instanton is at $x_1 = \frac{N}{2\pi}\log\frac{\lambda_1}{\lambda_2}$, respectively. This configuration becomes a solution of field equations asymptotically as the separation $\frac{N}{2\pi}\log\frac{\lambda_1^2}{\lambda_2}$ goes to infinity. For $\lambda_1^2 \gg \lambda_2$, this configuration corresponds to a $1/N$ instanton ($S = 1/N$, $Q = 1/N$) and a $1/N$ anti-instanton ($S = 1/N$, $Q = -1/N$) at large separations. The total action and the net topological charge in the large-separation limit are given by

$$S = 2/N, \quad Q = 0\,. \tag{27}$$

The topological charge is unchanged as the separation between the fractional instanton and the anti-instanton changes. However, the total action monotonically decreases as the instantons get closer, which shows an attractive force.

5. Interactions between fractional instanton and anti-instanton

We have evaluated analytically the total action for various values of separation between the fractional instanton and anti-instanton.

In Fig. 1, we show the total action S as a function of the separation $\tau = (1/\pi)\log\lambda_1^2/\lambda_2$ for $N = 2$. We also show the attractive force defined by $F = -\frac{dS}{d\tau}$ as a function of the separation τ. For negative values, the moduli parameter τ loses the physical meaning as the separation between the fractional instanton and anti-instanton, and merely describes a deformation of the field configuration.

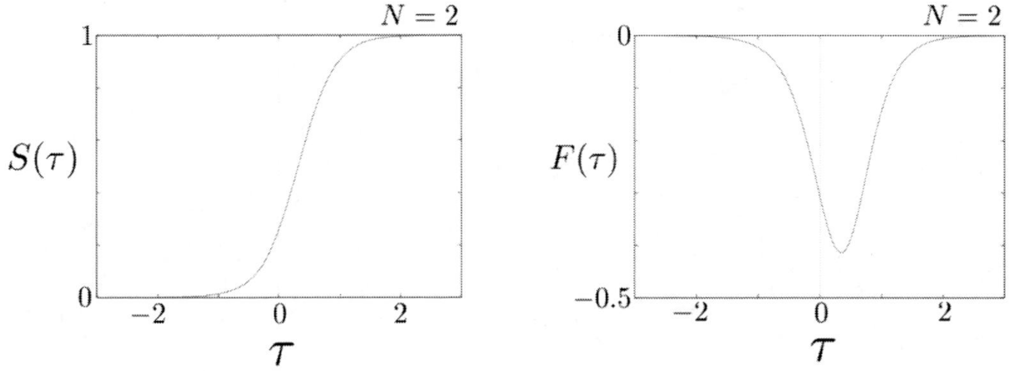

Figure 1. The $\tau = (1/\pi)\log\lambda_1^2/\lambda_2$ dependence of the total action S and the force $F = -\frac{dS}{d\tau}$ for (26) with $N = 2$. For $\tau \geq 0$, we can interpret τ as separation between the instanton constituents. The configuration is changed from $S = 1$ to $S = 0$, due to the attractive force. The configuration for $\tau \geq 1$ corresponds to neutral bions.

To compare our concrete ansatz (26) to the far-separated instanton argument in Ref. [8], we analyze the interaction part of the action for our configuration. The interaction part of the action density is written as the difference of the action density $s(x_1)$ compared to the sum of the one fractional-instanton density and one fractional-anti-instanton density $s_{\nu=1/N}(x_1)+s_{\nu=-1/N}(x_1)$,

$$s_{\rm int}(x_1) = s(x_1) - (s_{\nu=1/N}(x_1) + s_{\nu=-1/N}(x_1))\,. \tag{28}$$

The total interaction action is given by integrating over the space

$$S_{\rm int}(N,\tau) = \frac{1}{\pi}\int dx\, s_{\rm int}(x_1)\,. \tag{29}$$

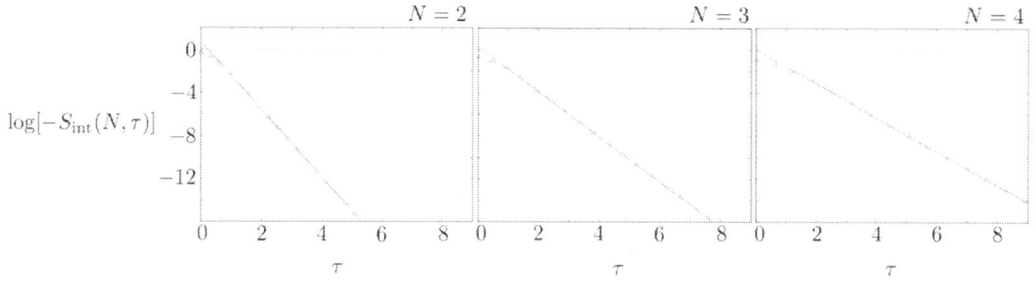

Figure 2. Plot of $\log(-S_{\rm int}(N,\tau))$ as a function of τ for $N=2$ (left), $N=3$ (center) and $N=4$ (right) for (26) (red curves with triangle points). For $\tau > 1$, the curve is almost equivalent to $-(2\pi/N)\tau + C(N)$ (blue curves).

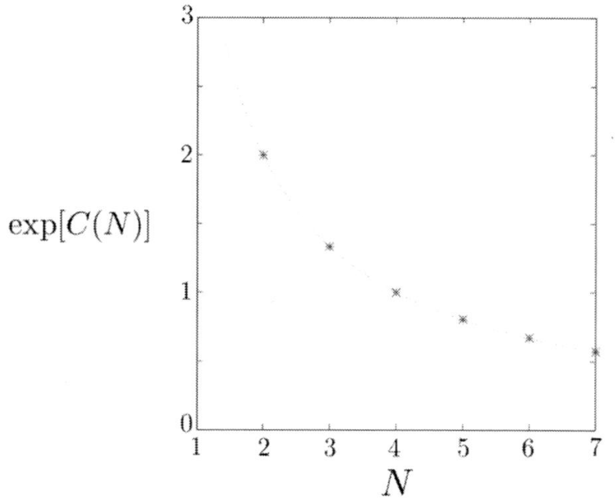

Figure 3. The coefficient of the interaction action $\exp[C(N)]$ in Eq.(32) as a function of N for $N = 2,3,4,5,6,7$ for the Ansatz (26) (blue points). The coefficient can be approximated by $4/N$ (a red curve).

In Fig. 2, we plot the logarithm of the total interaction action $S_{\rm int}(N,\tau)$ as a function of τ for $N = 2,3,4$. In the $\tau \geq 1$ region, $\log(-S_{\rm int}(N,\tau))$ can be well approximated by analytic lines,

$$\log\left[-S_{\rm int}(N,\tau)\right] \;\sim\; -\xi(N)\,\tau \;+\; C(N)\,, \qquad (\tau \geq 1)\,, \tag{30}$$

where $\xi(N)$ is a slope and $C(N)$ is a y-intercept. In Fig. 2 we simultaneously depict these analytic lines for the three cases. The slopes ξ of the approximate lines read $\xi \sim \pi$ for $N=2$, $\xi \sim 2\pi/3$ for $N=3$ and $\xi \sim \pi/2$ for $N=4$, which indicates that the slope ξ can be generally expressed as

$$\xi(N) \;\sim\; \frac{2\pi}{N}\,. \tag{31}$$

Therefore we observe that the interaction action can be written as the following form for $\tau \geq 1$ region,

$$S_{\rm int}(N,\tau) \sim -e^{C}e^{-\xi\tau}, \qquad \xi = \frac{2\pi}{N}, \qquad (\tau \geq 1). \tag{32}$$

This ξ is equivalent to the (dimensionless) lowest Kaluza-Klein spectrum Lm_{LKK}, which is given as $Lm_{LKK} = |q_i - q_j| = 2\pi/N$, where q_i and q_j are two nonzero components of Wilson-loop holonomies in (23). Restoring the coupling constant g^2, we finally obtain that the fractional instanton and anti-instanton exert attractive interaction from large to small separations as

$$S_{\rm int}(\tau) = -\frac{8\xi}{g^2}e^{-\xi\tau}, \qquad \xi \equiv \frac{2\pi}{N}. \tag{33}$$

This is precisely in accord with the approximate result for far-separated case [8].

6. Bion amplitudes

Now we are in a position to sketch briefly how our results fit into the resurgence phenomenon.

According to the study, the imaginary ambiguity arising in non-Borel-summable perturbative series is compensated by the contributions of neutral bions. This phenomenon, which is called "resurgence", works as follows [8]: The effective interaction energy by bosonic exchange between one fractional instanton and one fractional anti-instanton is given in Eq. (33). The total bion amplitude including the fermion zero-mode exchange contribution is given by

$$\mathcal{B} \propto -e^{-2S_I/N}\int_0^\infty d\tau\, e^{-V^{ij}_{\rm eff}(\tau)}, \tag{34}$$

with $V_{\rm eff}(\tau) = S_{\rm int}(\tau) + 2N_f\xi\tau$ and S_I being the instanton action. N_f stands for fermion flavors. For neutral bion, semiclassical description of independent fractional instantons breaks down since the interaction is attractive and instantons are merged in the end. Here, the BZJ-prescription, replacing $g^2 \to -g^2$, works to extract meaningful information from this amplitude. The prescription turns the interaction (spuriously) into a repulsive one and the amplitude becomes well-defined as

$$\mathcal{B}(g^2,N_f) \;\to\; \tilde{\mathcal{B}}(-g^2,N_f) \;\propto\; (-g^2N/8\pi)^{2N_f}\Gamma(2N_f)e^{-2S_I/N}. \tag{35}$$

By using the analytic continuation in the g^2 complex plane, we can continue back to the original g^2. For $N_f = 0$ case, we then encounter the following imaginary ambiguity in the amplitude as

$$\tilde{\mathcal{B}}(g^2,0) \;\propto\; \left(\log(g^2N/8\pi) - \gamma \pm i\pi\right)e^{-2S_I/N}. \tag{36}$$

We can rephrase this situation as follows: unstable negative modes of bions give rise to imaginary ambiguities of the amplitude. The imaginary ambiguity has the same magnitude with an opposite sign as the leading-order ambiguity ($\sim \mp i\pi e^{-2S_I/N}$) arising from the non-Borel-summable series expanded around the perturbative vacuum. The ambiguities at higher orders ($\mp i\pi e^{-4S_I/N}$, $\mp i\pi e^{-6S_I/N}$,...) are cancelled by amplitudes of bion molecules (2-bion, 3-bion,...), and the full trans-series expansion around the perturbative and non-perturbative vacua results in unambiguous definition of field theories.

Finally we would like to mention the results of our more recent study of bions in the Grassmann sigma model [32]. The Grassmann sigma model with the Grassmann manifold $Gr_{N_{\rm F},N_{\rm C}}$ as a target space include the $\mathbb{C}P^{N_{\rm F}-1}$ model as a subclass with $N_{\rm C} = 1$ (and $N_{\rm F} - N_{\rm C} = 1$). We have found that charged bions do not exist in the $\mathbb{C}P^{N_{\rm F}-1}$ model, whereas the genuine Grassmann sigma model with $N_{\rm C} \geq 2, N_{\rm F} - N_{\rm C} \geq 2$ admits charged bions. The Grassmann sigma model admits BPS fractional instantons [25, 26, 27] with instanton number greater than unity (of order N_C), which cannot be reduced to composite of instantons and fractional instantons. Interaction between fractional instanton and anti-instanton in the Grassmann model is obtained to form neutral bions. We have also obtained exact non-BPS solutions of field equations for charged bions. To obtain these results, D-brane configurations are found to be a valuable tool to analyze fractoinal instantons and bions.

Acknowledgments
We are grateful to Mithat Ünsal and Gerald Dunne for their interest and valuable comments and correspondences on their related work during the entire course of our study. T. M. and N. S. thank Philip Argyres, Alexei Cherman, Falk Bruckmann, and Tin Sulejmanpasic and other participants of CERN theory institute 2014, "Resurgence and Transseries in quantum, gauge and string theories" for the fruitful discussion and useful correspondence. T. M. is in part supported by the Japan Society for the Promotion of Science (JSPS) Grants Number 26800147. The work of M. N. is supported in part by Grant-in-Aid for Scientific Research (No. 25400268) and by the "Topological Quantum Phenomena" Grant-in-Aid for Scientific Research on Innovative Areas (No. 25103720) from the Ministry of Education, Culture, Sports, Science and Technology (MEXT) of Japan. N. S. is supported by Grant-in Aid for Scientific Research No. 25400241 from the Ministry of Education, Culture, Sports, Science and Technology (MEXT) of Japan.

References

[1] Unsal M 2008 Abelian duality, confinement, and chiral symmetry breaking in QCD(adj) *Phys. Rev. Lett.* **100** 032005 (arXiv:0708.1772 [hep-th])
[2] Unsal M 2009 Magnetic bion condensation: A New mechanism of confinement and mass gap in four dimensions *Phys. Rev.* D **80** 065001 (arXiv:0709.3269 [hep-th])
[3] Shifman M and Unsal M 2008 QCD-like Theories on R(3) x S(1): A Smooth Journey from Small to Large r(S(1)) with Double-Trace Deformations *Phys. Rev.* D **78** 065004 (arXiv:0802.1232 [hep-th])
[4] Poppitz E and Unsal M 2009 Conformality or confinement: (IR)relevance of topological excitations *JHEP* **0909** 050 (arXiv:0906.5156 [hep-th])
[5] Anber M M and Poppitz E 2011 Microscopic Structure of Magnetic Bions *JHEP* **1106** 136 (arXiv:1105.0940 [hep-th])
[6] Poppitz E, Schaefer T and Unsal M 2012 Continuity, Deconfinement, and (Super) Yang-Mills Theory *JHEP* **1210** 115 (arXiv:1205.0290 [hep-th])
[7] Argyres P and Unsal M 2012 A semiclassical realization of infrared renormalons *Phys. Rev. Lett.* **109** 121601 (arXiv:1204.1661 [hep-th])
[8] Dunne G V and Unsal M 2012 Resurgence and Trans-series in Quantum Field Theory: The CP(N-1) Model *JHEP* **1211** 170 (arXiv:1210.2423 [hep-th])
[9] Dabrowski R and Dunne G V 2013 fractionalized Non-Self-Dual Solutions in the CP(N-1) Model *Phys. Rev.* D **88** 2 025020 (arXiv:1306.0921 [hep-th])
[10] Dunne G V and Unsal M 2014 Generating Non-perturbative Physics from Perturbation Theory *Phys. Rev.* D **89** 041701 (arXiv:1306.4405 [hep-th])
[11] Cherman A, Dorigoni D, Dunne G V and Unsal M 2014 Resurgence in QFT: Unitons, Fractons and Renormalons in the Principal Chiral Model *Phys. Rev. Lett.* **112** 021601 (arXiv:1308.0127 [hep-th])
[12] Basar G, Dunne G V and Unsal M 2013 Resurgence theory, ghost-instantons, and analytic continuation of path integrals *JHEP* **1310** 041 (arXiv:1308.1108 [hep-th])
[13] Dunne G V and Unsal M 2014 Uniform WKB, Multi-instantons, and Resurgent Trans-Series *Phys. Rev.* D **89** 105009 (arXiv:1401.5202 [hep-th])
[14] Cherman A, Dorigoni D and Unsal M 2013 Decoding perturbation theory using resurgence: Stokes phenomena, new saddle points and Lefschetz thimbles (arXiv:1403.1277 [hep-th])
[15] Bogomolny E B 1980 Calculation Of Instanton - Anti-instanton Contributions In Quantum Mechanics *Phys. Lett.* B **91** 431
[16] Zinn-Justin J 1981 Multi - Instanton Contributions in Quantum Mechanics *Nucl. Phys.* B **192** 125
[17] Zinn-Justin J and Jentschura U D 2004 Multi-instantons and exact results I: Conjectures, WKB expansions, and instanton interactions *Annals Phys.* **313** 197 (quant-ph/0501136)

[18] Zinn-Justin J 1981 Perturbation Series at Large Orders in Quantum Mechanics and Field Theories: Application to the Problem of Resummation *Phys. Rept.* **70** 109
[19] 't Hooft G 1979 Can We Make Sense Out of Quantum Chromodynamics? *Subnucl. Ser.* **15** 943
[20] Beneke M 1999 Renormalons *Phys. Rept.* **317** 1 (hep-ph/9807443)
[21] Ecalle J 1981 *Les Fonctions Resurgentes* Vol. I - III (Orsay: Publ. Math.)
[22] Polyakov A M 1989 *Gauge Fields and Strings* (Contemporary Concepts in Physics) (London: Harwood Academic Publishers)
[23] Polyakov A M and Belavin A A 1975 Metastable States of Two-Dimensional Isotropic Ferromagnets *JETP Lett.* **22** 245 [Pisma Zh. Eksp. Teor. Fiz. **22** 503 (1975)]
[24] Eto M, Isozumi Y, Nitta M, Ohashi K and Sakai N 2005 Instantons in the Higgs phase *Phys. Rev.* D **72** 025011 (hep-th/0412048)
[25] Eto M, Fujimori T, Isozumi Y, Nitta M, Ohashi K, Ohta K and Sakai N 2006 Non-Abelian vortices on cylinder: Duality between vortices and walls *Phys. Rev.* D **73** 085008 (hep-th/0601181)
[26] Eto M, Fujimori T, Nitta M, Ohashi K, Ohta K and Sakai N 2008 Statistical mechanics of vortices from D-branes and T-duality *Nucl. Phys.* B **788** 120 (hep-th/0703197)
[27] Eto M, Isozumi Y, Nitta M, Ohashi K and Sakai N 2006 Solitons in the Higgs phase: The Moduli matrix approach *J. Phys.* A **39** R315 (hep-th/0602170)
[28] Bruckmann F 2008 Instanton constituents in the O(3) model at finite temperature *Phys. Rev. Lett.* **100** 051602 (arXiv:0707.0775 [hep-th])
[29] Brendel W, Bruckmann F, Janssen L, Wipf A and Wozar C 2009 Instanton constituents and fermionic zero modes in twisted CP^n models *Phys. Lett.* B **676** 116 (arXiv:0902.2328 [hep-th])
[30] Harland D 2009 Kinks, chains, and loop groups in the CP^n sigma models *J. Math. Phys.* **50** 122902 (arXiv:0902.2303 [hep-th])
[31] Misumi T, Nitta R and Sakai N 2014 Neutral bions in the $\mathbb{C}P^{N-1}$ model *JHEP* **1406** 164 (arXiv:1404.7225 [hep-th])
[32] Misumi T, Nitta M and Sakai N 2014 Classifying bions in Grassmann sigma models and non-Abelian gauge theories by D-branes (arXiv:1409.3444 [hep-th])
[33] Bogomolny E B 1976 Stability of Classical Solutions *Sov. J. Nucl. Phys.* **24** 449 [Yad. Fiz. **24** 861 (1976)]

Journal of Physics: Conference Series **597** (2015) 012061 doi:10.1088/1742-6596/597/1/012061

Localization and the connection between $U_q(\mathfrak{so}(3))$ and $U_{\tilde{q}}(\mathfrak{osp}(1|2))$

Patrick Moylan

Department of Physics, Pennsylvania State University, The Abington College, Abington, PA 19001 USA

E-mail: pjm11@psu.edu

Abstract. In [1] we constructed homomorphisms from the quantized enveloping algebras of the Euclidean and Euclidean super Lie algebras, $U_q(\mathfrak{iso}(2))$ and $U_{\tilde{q}}(\widetilde{\mathfrak{iso}}(2))$, onto their images in localizations of $U_q(\mathfrak{sl}(2))$ and $U_{\tilde{q}}(\mathfrak{osp}(1|2))$, respectively, and, conversely, we described homomorphisms of $U_q(\mathfrak{sl}(2))$ and $U_{\tilde{q}}(\mathfrak{osp}(1|2))$ into localizations of $U_q(\mathfrak{iso}(2))$ and $U_{\tilde{q}}(\widetilde{\mathfrak{iso}}(2))$. Herewith we use these homomorphisms to obtain a new relationship between $U_q(\mathfrak{osp}(1|2))$ and $U_q(\mathfrak{so}\,(3))$.

1. Introduction

Localization, or formation of quotients, is a powerful tool in mathematics with many applications. It can be used to relate different algebraic structures which share some common similarities. An important example of this is the Gelfand-Kirillov conjecture [2] which asserts that, for the universal enveloping algebra of a Lie algebra over an algebraically closed field, its quotient field is isomorphic to some skew field extension of a Weyl algebra over a transcendental extension of the base field.

Another related example, which is of importance to physics, is given in [3], where they demonstrate a $*$-isomorphism between commutative algebraic extensions of the Lie fields of $SO(1,4)$ and the Poincaré group. In [4] we used this isomorphism of Lie fields to construct representations of the Poincaré Lie algebra out of unitary representations of $SO(1,4)$. Analogous result for SO(2,3) and the Poincaré Lie algebra are established in [5] except that the isomorphism of Lie fields is no longer a *-isomorphism. From this result and a similar study of the other real forms of $SO(5)$ and the much easier to analyze lower dimensional cases, it is clear that similar such isomorphisms of Lie fields of $G = SO(p,q)$ and associated semidirect products of the form $SO(p-1,q) \times_s V$ and $SO(p,q-1) \times_s V$ exists for all $SO(p,q)$ groups with $p+q \leq 5$.

In [1] we generalized the ideas of the previous paragraph to quantized Lie algebras and to quantized Lie superalgebras. Specifically, we considered $U_q(\mathfrak{sl}(2))$ and $U_{\tilde{q}}(\mathfrak{osp}(1,2))$, and their relationship to $U_q(\mathfrak{iso}(2))$ and $U_{\tilde{q}}(\widetilde{\mathfrak{iso}}(2))$, respectively. Here $U_q(\mathfrak{iso}(2))$ is the q deformed universal enveloping algebra of the Euclidean group in the plane which, if we neglect its Hopf algebra structure, is isomorphic to $U(\mathfrak{iso}(2))$, the universal enveloping algebra of the Euclidean group [6]. Similarly, $U_{\tilde{q}}(\widetilde{\mathfrak{iso}}(2))$ is the $\tilde{q}$ deformed universal enveloping algebra of the Euclidean Lie superalgebra, which is likewise isomorphic to $U(\widetilde{\mathfrak{iso}}(2))$, again, only as an isomorphism of enveloping algebras, not as a Hopf algebra isomorphism. $U_q(\mathfrak{sl}(2))$ and $U_{\tilde{q}}(\mathfrak{osp}(1,2))$ are the standard Drinfeld-Jimbo q deformations of the universal enveloping algebras of $\mathfrak{sl}(2)$ and its supersymmetric counterpart, $\mathfrak{osp}(1,2)$, respectively. In [1] we constructed homomorphisms from $U_q(\mathfrak{sl}(2))$ and $U_q(\mathfrak{iso}(2))$ onto their images into extensions of localizations of $U_q(\mathfrak{iso}(2))$ and $U_q(\mathfrak{sl}(2))$, respectively, and we used these homomorphism to obtain results on the relationship between representations of the respective algebras. There we also constructed homomorphisms from $U_{\tilde{q}}(\mathfrak{osp}(1|2))$ and $U_{\tilde{q}}(\widetilde{\mathfrak{iso}}(2))$ onto their images into extensions of localizations of $U_{\tilde{q}}(\widetilde{\mathfrak{iso}}(2))$ and $U_{\tilde{q}}(\mathfrak{osp}(1|2))$, respectively, and we used these homomorphism to obtain new representations of $U_{\tilde{q}}(\mathfrak{osp}(1|2))$.

Here we use results about $U_{\tilde{q}}(\mathfrak{osp}(1|2))$ and $U_q(\mathfrak{sl}(2))(\simeq U_q(\mathfrak{so}(3)))$, mentioned in the previous paragraph with an aim towards obtaining a better understanding of the results in [7] for $U_{\tilde{q}}(\mathfrak{osp}(1|2n))$ and $U_q(\mathfrak{so}(2n+1))$, something which we we have succeeded in demonstrating completely and quite clearly in this paper for the $n=1$ case. For an alternative approach to understanding the relationship between $U_{\tilde{q}}(\mathfrak{osp}(1|2n))$ and $U_q(\mathfrak{so}(2n+1))$ see [8].

The existence of a one-to-one correspondence between the finite dimensional representations of $\mathfrak{osp}(1|2n)$ and $\mathfrak{so}(2n+1)$, except in the case of what they call *spinorial representations*, has been known for quite some time [9]. For quantum algebras, this is the result of [7], namely, a connection between $U_{\tilde{q}}(\mathfrak{osp}(1|2n))$ and $U_{-\tilde{q}}(\mathfrak{so}(2n+1))$, which only holds for non-spinorial representations. Work of Aizawa et. al in [10] on $U_{\tilde{q}}(\mathfrak{osp}(1,2))$ shows that the even-dimensional representations, for which the said isomorphism is not applicable, still possess properties reflecting this isomorphism. In this paper we specialize to the $n=1$ case and construct a mapping from $U_{\tilde{q}}(\mathfrak{osp}(1|2))$ into a localization of $U_q(\mathfrak{so}(3))$ which can be used to obtain a homomorphism of a quotient of $U_{\tilde{q}}(\mathfrak{osp}(1|2))$ into its image in a localization of $U_q(\mathfrak{so}(3))$. Since we work at the level of abstract (quantized enveloping) algebras and their localizations our results for $n=1$ are more general that those in [7]. In particular, they are applicable not only to finite dimensional representations but also to infinite dimensional ones. Furthermore, it is possible that our results could help to explain the phenomenon in [10] to which we just alluded.

Remarks on notation and definitions: We follow the conventions about notation established in [1]. Except for elements of the Cartan subalgebras for which we always use plain faced letters, quantities made out of elements of $U_q(\mathfrak{iso}(2))$ and $U_{\tilde{q}}\left(\widetilde{\mathfrak{iso}}(2)\right)$ and of their localizations are usually denoted with bold faced letters and we use plain faced letters to denote elements of $U_q(\mathfrak{so}(2n+1))$ and $U_{\tilde{q}}(\mathfrak{osp}(1|2n)$ and their localizations. Elements of Lie superalgebras are always denoted with tildes placed over the letters.

Our definitions of $U_q(\mathfrak{so}(2n+1))$ and $U_{\tilde{q}}(\mathfrak{osp}(1|2n)$ differ slightly from those given in [7]. For the Drinfeld-Jimbo $U_q(\mathfrak{so}(2n+1))$ we take the standard definition found in [11], which uses $q_i = q^{d_i}$ where $d_i := (\alpha_i, \alpha_i)/2$ with α_i being the ith simple root. Presumably the two definitions can be related to one another by a simple rescaling of some of the generators.

2. $U_q(\mathfrak{so}(2n+1))$ and $U_{\tilde{q}}(\mathfrak{osp}(1|2n))$

We recall the q-deformation $U_q(so(2n+1)) = U_q(so(2n+1,\mathbb{C})$ is defined as follows. Let (,) be the standard inner product on $\mathbb{R}^n$. The root system of $U_q(so(2n+1))$ is defined in terms of the standard orthonormal unit vectors, $\epsilon_1,\dots,\epsilon_n$, which form a basis of $\mathbb{R}^n$ and satisfy $(\epsilon_i,\epsilon_j)=\delta_{ij}$, $i,j=1,\dots,n$. For the simple root system we have:

$$\Pi = \{\alpha_1,\dots,\alpha_n\}$$

$$\alpha_j = \epsilon_j - \epsilon_{j+1}\ ,\ j=1,\dots,n-1;\ \alpha_n = \epsilon_n.$$

The Cartan matrix of $U_q(\mathfrak{so}(2n+1))$ is the $n \times n$ matrix $(a_{jk}) = 2(\alpha_j, \alpha_k)/(\alpha_k, \alpha_k)$. Explicitly we have:

$$A = (a_{ij}) = \begin{pmatrix} 2 & -1 & 0 & \dots & 0 & 0 & 0 \\ -1 & 2 & -1 & \dots & 0 & 0 & 0 \\ \dots & \dots & \dots & \dots & \dots & \dots & \dots \\ 0 & 0 & 0 & \dots & 2 & -1 & 0 \\ 0 & 0 & 0 & \dots & -1 & 2 & -2 \\ 0 & 0 & 0 & \dots & 0 & -1 & 2 \end{pmatrix}$$

We introduce the Cartan-Weyl basis with H_i, $i = 1, \dots, n$, being the basis of the Cartan subalgebra $\mathcal{H}$ associated with the simple roots and $X_i^\pm$, $i = 1, \dots, n$, being the simple root vectors. Let $q \in \mathbb{C}$ and set $K_i = q_i^{H_i}$ with $q_i = (q)^{d_i}$ where $d_i := (\alpha_i, \alpha_i)/2$ and q_i such that $q_i^2 \neq 1$ for $i = 1, 2, \dots, n$. Further let $[x]_{q_i} = \frac{q_i^x - q_i^{-x}}{q_i - q_i^{-1}}$. We then define $U_q(\mathfrak{so}(2n+1))$ as the unital associative complex algebra with generators K_i, K_i^{-1} and $X_j^\pm$ $(1 \le i \le n)$ and defining relations [11]:

$$K_iK_j = K_jK_i, K_iK_i^{-1} = K_i^{-1}K_i = 1, K_iX_j^\pm K_i^{-1} = q_i^{\pm a_{ij}} X_j^\pm \tag{2.1a}$$

$$[X_i^+, X_j^-] = \delta_{ij} \frac{K_i - K_i^{-1}}{q_i - q_i^{-1}} \tag{2.1b}$$

$$\sum_{r=0}^{1-a_{ji}} (-1)^r \frac{[1-a_{ji}]_{q_i}!}{[r]_{q_i}![1-a_{ji}-r]_{q_i}!} (X_i^\pm)^{1-a_{ji}-r} X_j^\pm (X_i^\pm)^r = 0, (i \neq j). \tag{2.1c}$$

Expressing the first two of the above equations in terms of the H_i and explicitly writing out the last set of equations, we have that Eqs. (2.1a), (2.1b) and (2.1c) are (formally) equivalent to

$$[H_i, H_j] = 0, [H_i, X_j^\pm] = \pm a_{ij} X_j^\pm, \tag{2.1a'}$$

$$[X_i^+, X_j^-] = \delta_{ij}[H_i]_{q_i}, \tag{2.1b'}$$

$$\left.\begin{array}{c} (X_{i\pm1}^\pm)^2 X_i^\pm - [2]_q X_{i\pm1}^\pm X_i^\pm X_{i\pm1}^\pm + X_i^\pm (X_{i\pm1}^\pm)^2 = 0 \text{ for } i \pm 1 \neq n \\ (X_n^\pm)^3 X_{n-1}^\pm - [3]_{q^{1/2}} (X_n^\pm)^2 X_{n-1}^\pm X_n^\pm + [3]_{q^{1/2}} X_n^\pm X_{n-1}^\pm (X_n^\pm)^2 - X_{n-1}^\pm (X_n^\pm)^3 = 0 \\ X_i^\pm X_j^\pm - X_j^\pm X_i^\pm = 0 \text{ for } j \neq i \pm 1. \end{array}\right\} \tag{2.1c'}$$

We note that the $\pm$ in the superscripts are not correlated with the $\pm$ in the subscripts. We do not bother to write down the definitions of the Cartan-Weyl generators corresponding to the non-simple roots, since we do not need to make use of them in this paper.

We now state the definition of $U_{\bar{q}}(\mathfrak{osp}(1|2n))$. We again introduce $\epsilon_1, \dots, \epsilon_n$, $(\epsilon_i, \epsilon_j) = \delta_{ij}$, $i, j = 1, \dots, n$ and the root system of $U_{\bar{q}}(\mathfrak{osp}(1|2n))$ is given by

$$\Delta_{\bar{0}} = \{+\epsilon_i \pm \epsilon_j, -\epsilon_i \pm \epsilon_j, 1 \le i < j \le n, \pm 2\epsilon_i, 1 \le i \le n\}, \quad \text{(even roots)}$$

$$\Delta_{\bar{1}} = \{\pm\epsilon_i, 1 \le i \le n\}, \quad \text{(odd roots)}.$$

For the simple root system we have:

$$\Pi = \{\alpha_1, \dots, \alpha_n\}$$

with

$$\alpha_j = \epsilon_j - \epsilon_{j+1}, j = 1, \ldots, n-1; \alpha_n = \epsilon_n.$$

The α_i for $i < n$ are even and the root α_n is odd. We note that the even root system $\Delta_{\bar{0}}$ is the root system of $U_{\tilde{q}}(\mathfrak{sp}(2n))$ i.e. that of $U_q(C_n)$ [12]. (Note that $\mathfrak{sp}(2n)$ is also denoted by $\mathfrak{sp}(n)$ in mathematical literature.) We introduce the Cartan-Weyl basis with $\tilde{H}_i$, $i = 1, \ldots, n$, being the basis of the Cartan subalgebra $\tilde{\mathcal{H}}$ associated with the simple roots and $\tilde{X}_i^{\pm}$, $i = 1, \ldots, n$ being the simple root vectors. We let

$$A = (a_{ij}) = \begin{pmatrix} 2 & -1 & 0 & \ldots & 0 & 0 & 0 \\ -1 & 2 & -1 & \ldots & 0 & 0 & 0 \\ \ldots & \ldots & \ldots & \ldots & \ldots & \ldots & \ldots \\ 0 & 0 & 0 & \ldots & 2 & -1 & 0 \\ 0 & 0 & 0 & \ldots & -1 & 2 & -1 \\ 0 & 0 & 0 & \ldots & 0 & -2 & 2 \end{pmatrix}$$

denote the Cartan matrix of $U_{\tilde{q}}(\mathfrak{sp}(2n))$ and consider its decomposition $A = A^d A^s$ with $A^d = \mathrm{diag}(1, \ldots, 1, 2)$ and $A^s = (a^s_{ij})$. Except for its last row, the matrix for A^s is the same as the matrix for A. The last row of A^s is obtained from the matrix for A by dividing all terms in the last row by $\frac{1}{2}$.

Let $\tilde{K}_i = \tilde{q}_i^{\tilde{H}_i}$ with $\tilde{q}_i = (\tilde{q})^{d_i}$ where, as before, $d_i := (\alpha_i, \alpha_i)/2$ and $\tilde{q}_i$ such that $\tilde{q}_i^2 \neq 1$ for $i = 1, 2, \ldots, n$. We can now define $U_{\tilde{q}}(\mathfrak{osp}(1|2n))$ as the unital associative complex algebra with generators $\tilde{K}_i, \tilde{K}_i^{-1}$ and $\tilde{X}_j^{\pm}$ $(1 \leq i \leq n)$ and relations [13] [7]:

$$\tilde{K}_i\tilde{K}_j = \tilde{K}_j\tilde{K}_i, \tilde{K}_i\tilde{K}_i^{-1} = \tilde{K}_i^{-1}\tilde{K}_i = 1, K_i\tilde{X}_j^{\pm}\tilde{K}_i^{-1} = \tilde{q}_i^{\pm a^s_{ij}}\tilde{X}_j^{\pm} \tag{2.2a}$$

$$[\tilde{X}_i^{+}, \tilde{X}_j^{-}] = \delta_{ij}\frac{\tilde{K}_i - \tilde{K}_i^{-1}}{\tilde{q}_i - \tilde{q}_i^{-1}} = \delta_{ij}[\tilde{H}_i]_{\tilde{q}_i}, \tag{2.2b'}$$

$$\left.\begin{aligned} (\tilde{X}^{\pm}_{i\pm1})^2\tilde{X}^{\pm}_i - [2]_{\tilde{q}}\tilde{X}^{\pm}_{i\pm1}\tilde{X}^{\pm}_i\tilde{X}^{\pm}_{i\pm1} + \tilde{X}^{\pm}_i(\tilde{X}^{\pm}_{i\pm1})^2 &= 0 \text{ for } i\pm1\neq n \\ (\tilde{X}^{\pm}_n)^3\tilde{X}^{\pm}_{n-1} + [3]_{(-\tilde{q})^{\frac{1}{2}}}(\tilde{X}^{\pm}_n)^2\tilde{X}^{\pm}_{n-1}\tilde{X}^{\pm}_n + [3]_{(-\tilde{q})^{\frac{1}{2}}}\tilde{X}^{\pm}_n\tilde{X}^{\pm}_{n-1}(\tilde{X}^{\pm}_n)^2 + \tilde{X}^{\pm}_{n-1}(\tilde{X}^{\pm}_n)^3 &= 0 \\ \tilde{X}^{\pm}_i\tilde{X}^{\pm}_j - \tilde{X}^{\pm}_j\tilde{X}^{\pm}_i = 0 \text{ for } j \neq i\pm1. & \end{aligned}\right\} \tag{2.2c'}$$

Just as in the $U_q(\mathfrak{so}(2n+1))$ case considered above, the $\pm$ in the superscripts are not correlated with the $\pm$ in the subscripts. Furthermore, Eq. (2.2a) is, as for $U_q(\mathfrak{so}(2n+1))$, formally equivalent to the following:

$$[\tilde{H}_i, \tilde{H}_j] = 0 \ , \quad [\tilde{H}_i, \tilde{X}_j^{\pm}] = \pm a^s_{ij}\tilde{X}_j^{\pm}. \tag{2.2a'}$$

The $\mathbb{Z}_2$ grading on $U_{\tilde{q}}(\mathfrak{osp}(1|2n))$ is that induced from the following grading on the generators: $\deg\tilde{X}_n^{\pm} = 1$ and $\deg a = 0$ for a being any one of the other generators of $U_{\tilde{q}}(\mathfrak{osp}(1|2n))$.

Journal of Physics: Conference Series **597** (2015) 012061 doi:10.1088/1742-6596/597/1/012061

3. The Algebras $U_q(\mathfrak{iso}(2))$ and $U_{\tilde{q}}(\widetilde{\mathfrak{iso}}(2))$

The quantized universal enveloping algebra $U_q(\mathfrak{iso}(2))$ is the complex, unital associative algebra with generators A, B, K, K^{-1} satisfying the relations $KK^{-1} = K^{-1}K = 1$ and

$$KBK^{-1} = q^2B, \quad KCK^{-1} = q^{-2}C \tag{3.1}$$

$$BC = CB. \tag{3.2}$$

We have equivalently, with $K = q^H$, generators H, $\mathbf{P}^+ = B$ and $\mathbf{P}^- = C$ and relations

$$[H, \mathbf{P}^\pm] = \pm 2\mathbf{P}^\pm , \tag{3.1'}$$

$$[\mathbf{P}^+, \mathbf{P}^-] = 0 . \tag{3.2'}$$

The center $\mathfrak{Z}(U_q(\mathfrak{iso}(2)))$ of $U_q(\mathfrak{iso}(2))$ is generated by

$$Y^2 = \mathbf{P}^+\mathbf{P}^- = \mathbf{P}^-\mathbf{P}^+. \tag{3.3}$$

Note that algebra defined by Eqs. (3.1′) and (3.2′) is the same as the enveloping algebra of $\mathfrak{iso}(2)$, the classical Euclidean Lie algebra of two dimensional Euclidean space; however, as a Hopf algebra it is different [6]. $U_q(\mathfrak{iso}(2))$ can be obtained from $U_q(\mathfrak{so}(3))$ by a contraction process analogous to that defined by Segal [14] and Inönü and Wigner [15] for Lie algebras, namely, replace K, E, F by K, ρE, ρF and let $\rho \downarrow 0$ [16].

The quantized universal enveloping algebra of the Euclidean Lie superalgebra, $U_{\tilde{q}}(\widetilde{\mathfrak{iso}}(2))$, is defined exactly as in the previous paragraph except with anticommutator instead of commutator in Eq. (3.2′) and tildes placed over the $\mathbf{P}^\pm$, K and H. Thus, in terms of the generators $\tilde{H}$ and $\tilde{\mathbf{P}}^\pm$, we have the relations

$$[\tilde{H}, \tilde{\mathbf{P}}^\pm] = \pm 2\tilde{\mathbf{P}}^\pm , \tag{3.4}$$

$$\{\tilde{\mathbf{P}}^+, \tilde{\mathbf{P}}^-\} = 0 . \tag{3.5}$$

Let $\tilde{Y}^2$ be the element of $\mathfrak{Z}(U_{\tilde{q}}(\widetilde{\mathfrak{iso}}(2)))$ given by

$$\tilde{Y}^2 = -i(-1)^{\tilde{H}}\, \tilde{\mathbf{P}}^+\tilde{\mathbf{P}}^- = i(-1)^{\tilde{H}}\tilde{\mathbf{P}}^-\tilde{\mathbf{P}}^+. \tag{3.6}$$

In order to give a precise meaning to expressions like $(-1)^{\tilde{H}} = e^{i\pi\tilde{H}}$ which occurs in this equation and other similar expressions which occur frequently throughout the rest of the paper, we refer the reader to [17]. There they describe how to construct extensions of enveloping algebras which accommodate such quantities as $e^{i\pi\tilde{H}}$ defined by infinite series (cf. [17]). To keep things simple we do not make a distinction between enveloping algebras and necessary such extensions for incorporating such formal series expansions.

4. Localizations of $U_q(\mathfrak{so}(3))$ and $U_q(\mathfrak{iso}(2))$

In this section and the next we recall our results in [1] on localizations of $U_q(\mathfrak{so}(3))$ and $U_{\tilde{q}}(\mathfrak{osp}(1|2)$ and associated homomorphisms of $U_q(\mathfrak{so}(3))$ and $U_{\tilde{q}}(\mathfrak{osp}(1|2)$ into various localizations. In the next remaining sections, we make the substitutions $q \to q^2$ and $\tilde{q} \to \tilde{q}^2$, i.e. replace q in section 2 everywhere by q^2 etc. We also use a slightly different definition for q numbers, namely, $[x]_q = \frac{(q^{x/2}-q^{-x/2})}{(q^{1/2}-q^{-1/2})}$, in order to conform with that used in [1]. (The change amounts simply to the following: $[x]_{q^{\frac{1}{2}}}$ in section 2 = $[x]_q$ in this section and the next two, and similarly with $\tilde{q}$ replacing q.) Let $X^\pm$ and H denote the generators of $U_q(\mathfrak{so}(3))$, the relations for which are given by Eqs. $(2.1a')$ and $(2.1b')$ for $n = 1$ with subscripts on the generators omitted, i.e.

$$[H, X^{\pm}] = \pm 2X^{\pm} \qquad (4.1a)$$

$$[X^{+}, X^{-}] = [H]_{q^2} \qquad (4.1b)$$

For q not a root of unity, $\mathfrak{Z}(U_q(\mathfrak{so}(3)))$ is generated by

$$\Delta_q = X^+X^- + \left(\left[\frac{H-1}{2}\right]_{q^2}\right)^2 - \frac{1}{4}\cdot I = X^-X^+ + \left(\left[\frac{H+1}{2}\right]_{q^2}\right)^2 - \frac{1}{4}\cdot I \ . \qquad (4.2)$$

For a description of $\mathfrak{Z}(U_q(\mathfrak{so}(3)))$ when q is a root of unity we refer the reader to [11].

Now define

$$\mathbf{X}^{\pm} = \left\{\pm\frac{1}{Y}\left[\frac{H\mp 1}{2}\right]_{q^2} + I\right\}\mathbf{P}^{\pm} = \mathbf{P}^{\pm}\left\{\pm\frac{1}{Y}\left[\frac{H\pm 1}{2}\right]_{q^2} + I\right\} \qquad (4.3\pm)$$

where Y is a solution of the algebraic equation $Y^2 - \mathbf{P}^+\mathbf{P}^- = 0$ in $U_q(\mathfrak{iso}(2))$ and is assumed to commute with all elements of $U_q(\mathfrak{iso}(2))$.[1] Let $\tau(X^{\pm}) = \mathbf{X}^{\pm}$ and $\tau(H) = H$. This defines a mapping τ from $U_q(\mathfrak{so}(3))$ into a commutative algebraic extension of the localization of $U_q(\mathfrak{iso}(2))$ with denominators consisting of powers of Y.

Proposition 1 The map τ defines a homomorphism from $U_q(\mathfrak{sl}(2))$ onto its image. In particular the $\mathbf{X}^{\pm}$ defined by Eqs. (4.3±) together with H satisfy the relations, Eqs. (4.1a) and (4.1b), of the generators of $U_q(\mathfrak{so}(3))$. Furthermore, let Δ_q be defined by Eq. (4.2) but with $\mathbf{X}^{\pm}$ replacing $X^{\pm}$. Then $\Delta_q = Y^2 - \frac{1}{4}\cdot I$.

For a proof of this result we refer the reader to [1].

Now let Y be such that it commutes with all elements of $U_q(\mathfrak{sl}(2))$ and satisfies the equation

$$Y^2 = \Delta_q + \frac{1}{4}\cdot I \ . \qquad (4.4)$$

Let $P^{\pm} = (D_L^{\pm})^{-1}X^{\pm} = X^{\pm}(D_R^{\pm})^{-1}$ with $D_L^{\pm} = \left(\pm\frac{1}{Y}\left[\frac{H\mp1}{2}\right]_{q^2} + I\right)$ and $D_R^{\pm} = \left(\pm\frac{1}{Y}\left[\frac{H\pm1}{2}\right]_{q^2} + I\right)$ and define τ' by $\tau'(\mathbf{P}^{\pm}) = P^{\pm}$ and $\tau'(H) = H$.

Proposition 2 τ' extends by linearity to a homomorphism from $U_q(\mathfrak{iso}(2))$ into a localization of $U_q(\mathfrak{sl}(2))$. In particular, $\tau'(\mathbf{P}^{\pm})$ and $\tau'(H)$ satisfy the commutation relations, Eqs. (3.1′) and (3.2′), of $U_q(\mathfrak{iso}(2))$ and, furthermore, $P^+P^- = Y^2$.

For the proof of this proposition we again refer to [1].

5. Localizations of $U_{\tilde{q}}(\mathfrak{osp}(1|2))$ and $U_{\tilde{q}}(\widetilde{\mathfrak{iso}}(2))$

The definition of the $\tilde{q}$-deformation $U_{\tilde{q}}(\mathfrak{osp}(1|2))$ of the enveloping algebra of $\mathfrak{osp}(1|2)$ is obtained from Eqs. (2.2a′), $(2.2b')$ by specializing to $n = 1$. Just as for the $U_q(\mathfrak{so}(3))$ treated in the previous section, we omit the subscripts on the generators, and for $U_{\tilde{q}}(\mathfrak{osp}(1|2))$ we have generators $\tilde{H}$, $\tilde{X}^{\pm}$ with relations ($\{\cdot,\cdot\}$ denote anticommutator):

$$[\tilde{H}, \tilde{X}^{\pm}] = \pm\tilde{X}^{\pm} \ , \qquad (5.1a)$$

$$\{\tilde{X}^+, \tilde{X}^-\} = [\tilde{H}]_{\tilde{q}^2} \ . \qquad (5.1b)$$

[1] In other words we consider a commutative algebraic extension $\hat{U}_q(\mathfrak{iso}(2))$ of $U_q(\mathfrak{iso}(2))$ defined as follows: $\hat{U}_q(\mathfrak{iso}(2)) = \{a + bY | a, b \in U_q(\mathfrak{iso}(2)) \ , \ [Y, a] = 0 \ \forall \ a \in U_q(\mathfrak{iso}(2), \ Y^2 - \mathbf{P}^+\mathbf{P}^- = 0. \ \}$

The Casimir operator of $U_{\tilde q}(\mathfrak{osp}(1|2))$ is $\tilde\Delta_{\tilde q} = \tilde S_{\tilde q}^{\,2} + 2\cdot I$ with [1]

$$\tilde S_{\tilde q} = \frac{\tilde q^{1/2}k - \tilde q^{-1/2}k^{-1}}{\tilde q - \tilde q^{-1}} - (\tilde q^{1/2} + \tilde q^{-1/2})\tilde X^- \tilde X^+ = [\tilde H + \frac{1}{2}]_{\tilde q^2} - [2]_{\tilde q}\tilde X^- \tilde X^+ =$$

$$= -\frac{\tilde q^{-1/2}k - \tilde q^{1/2}k^{-1}}{(\tilde q - \tilde q^{-1})} + (\tilde q^{1/2} + \tilde q^{-1/2})\tilde X^+ \tilde X^- = -[\tilde H - \frac{1}{2}]_{\tilde q^2} + [2]_{\tilde q}\tilde X^+ \tilde X^- . \qquad (5.2)$$

We now describe homomorphisms associated with $U_{\tilde q}(\widetilde{\mathfrak{iso}}(2))$ and $U_q(\mathfrak{iso}(2))$. These will be used in the next section in order to relate $U_q(\mathfrak{so}(2n+1))$ and $U_{\tilde q}(\mathfrak{osp}(1|2n))$ for $n=1$. Define τ_0 by $\tau_0(H) = -2\tilde H$, $\tau_0(\mathbf{P}^+) = -\tilde{\mathbf{P}}^-$, $\tau_0(\mathbf{P}^-) = e^{-i\pi\tilde H}\tilde{\mathbf{P}}^+$ and $\tilde\tau_0$ by $\tilde\tau_0(\tilde H) = -\frac{1}{2}H$, $\tilde\tau_0(\tilde{\mathbf{P}}^-) = -\mathbf{P}^+$, $\tilde\tau_0(\tilde{\mathbf{P}}^+) = e^{-\frac{i\pi}{2}H}\mathbf{P}^-$.

Proposition 3 τ_0 and $\tilde\tau_0$ define Lie algebra and Lie superalgebra homomorphisms from $U_q(\mathfrak{iso}(2))$ and $U_{\tilde q}(\widetilde{\mathfrak{iso}}(2))$ onto their images in $U_{\tilde q}(\widetilde{\mathfrak{iso}}(2))$ and $U_q(\mathfrak{iso}(2))$, respectively.

In order to prove this result we extend τ_0 and $\tilde\tau_0$ by linearity to $\mathfrak{iso}(2)$ and $\widetilde{\mathfrak{iso}}(2)$, respectively. We verify the respective commutation relations and then lift the mappings to the enveloping algebras.

Next we describe homomorphisms of $U_{\tilde q}(\mathfrak{osp}(1|2))$ and $U_{\tilde q}(\widetilde{\mathfrak{iso}}(2))$ into extensions of localizations of $U_{\tilde q}(\widetilde{\mathfrak{iso}}(2))$ and $U_{\tilde q}(\mathfrak{osp}(1|2))$, respectively. Let

$$\tilde{\mathbf{X}}^\pm = \left(\frac{1}{\tilde Y}\sqrt{e^{i\pi\tilde H}\left[\tilde H \mp \frac{1}{2}\right]_{\tilde q^2}} + \sqrt{\pm I}\right)\frac{\tilde{\mathbf{P}}^\pm}{\sqrt{[2]_{\tilde q}}} = \frac{\tilde{\mathbf{P}}^\pm}{\sqrt{[2]_{\tilde q}}}\left(\frac{\pm i}{\tilde Y}\sqrt{e^{i\pi\tilde H}\left[\tilde H \pm \frac{1}{2}\right]_{\tilde q^2}} + \sqrt{\pm I}\right) \qquad (5.3\pm)$$

where I is the identity in $U_{\tilde q}(\widetilde{\mathfrak{iso}}(2))$.

Proposition 4 If $\tilde Y$ is such that it commutes with all elements of $U_{\tilde q}(\widetilde{\mathfrak{iso}}(2))$ and satisfies Eq. (3.6), then Eqs. (5.3±) define a homomorphism $\tilde\tau_{\tilde q}$ from $U_{\tilde q}(\mathfrak{osp}(1|2))$ into a localization of $U_{\tilde q}(\widetilde{\mathfrak{iso}}(2))$, with $\tilde\tau_{\tilde q}(\tilde X^\pm) = \tilde{\mathbf{X}}^\pm$ and $\tilde\tau_{\tilde q}(\tilde H) = \tilde H$. Furthermore, let $\tilde{\mathbf{S}}_{\tilde q}$ be defined by Eqn. (5.2) but with $\tilde{\mathbf{X}}^\pm$ replacing $\tilde X^\pm$, then $\tilde{\mathbf{S}}_{\tilde q} = -e^{-i\pi\tilde H}\tilde Y^2$.

Now let $\tilde P^\pm = (\tilde D_L^\pm)^{-1}\tilde X^\pm = \tilde X^\pm(\tilde D_R^\pm)^{-1}$ with $\sqrt{[2]_{\tilde q}}\tilde D_L^\pm = \frac{1}{\tilde Y}\sqrt{e^{i\pi\tilde H}\left[\tilde H \mp \frac{1}{2}\right]_{\tilde q^2}} + \sqrt{\pm I}$ and $\sqrt{[2]_{\tilde q}}\tilde D_R^\pm = \frac{\pm i}{\tilde Y}\sqrt{e^{i\pi\tilde H}\left[\tilde H \pm \frac{1}{2}\right]_{\tilde q^2}} + \sqrt{\pm I}$ where now I is the identity in $U_{\tilde q}(\mathfrak{osp}(1|2))$.

Proposition 5 Let $\tilde\tau'_{\tilde q}(\tilde{\mathbf{P}}^\pm) = \tilde P^\pm$ and $\tilde\tau'_{\tilde q}(\tilde H) = \tilde H$. If $\tilde Y^2$ is such that it commutes with all elements of $U_{\tilde q}(\mathfrak{osp}(1|2))$ and satisfies

$$\tilde Y^2 + e^{i\pi\tilde H}\tilde S_{\tilde q} = 0, \qquad (5.4)$$

then $\tilde\tau'_{\tilde q}$ extends to a homomorphism of $\widetilde{\mathfrak{iso}}(2)$ onto its image in a localization of $U_{\tilde q}(\mathfrak{osp}(1|2))$. In particular $\tilde P^\pm$ and $\tilde H$ satisfy the commutation relations (3.4) and (3.5) of $U(\widetilde{\text{iso}}(2))$ and, furthermore, $\tilde P^+\tilde P^- = ie^{-i\pi\tilde H}\tilde Y^2$.

We refer the reader to [1] for proofs of these propositions.

6. A mapping of $U_{\tilde q}(\mathfrak{osp}(1|2))$ into a localization of $U_q(\mathfrak{so}(3))$ and the connection between $U_q(\mathfrak{so}(2n+1))$ and $U_{\tilde q}(\mathfrak{osp}(1|2n))$

We now make use of the homomorphisms described in the above propositions to construct a mapping from $U_{\tilde q}(\mathfrak{osp}(1|2))$ into a localization of $U_q(\mathfrak{so}(3))$ and determine conditions for this mapping to be a homomorphism. Having established this we can construct representations of $U_{\tilde{q}}(\mathfrak{osp}(1|2))$ out of representations of $U_q(\mathfrak{so}(3))$, provided the representation of $U_q(\mathfrak{so}(3))$ lifts to a representation of its localization into which $U_{\tilde{q}}(\mathfrak{osp}(1|2))$ is embedded. Necessary conditions for $U_q(\mathfrak{so}(3))$ representations to lift to representations of their localizations are stated in Lemma 1 of [1], and it is easy to see that there are nontrivial such representations of $U_q(\mathfrak{so}(3))$ for which the conditions of the Lemma are satisfied, for example, extensions of infinitesimally unitarizable principle series representations of $U_q(so(2,1))$ to its complexification, $U_q(so(3))$ (cf. [18]).

We define the mapping π from $U_{\tilde{q}}(\mathfrak{osp}(1|2))$ into a localization of $U_q(\mathfrak{so}(3))$ using the composition of homomorphisms $\tau' \circ \tilde{\tau}_0 \circ \tilde{\tau}$. (For simplicity, we set $\tilde{\tau} = \tilde{\tau}_{\tilde{q}}$, where $\tilde{\tau}_{\tilde{q}}$ is defined in Proposition 4 .) We have:

$$\pi(\tilde{H}) = (\tau' \circ \tilde{\tau}_0 \circ \tilde{\tau})(H) , \tag{6.1}$$

$$\pi(\tilde{X}^{\pm}) = (\tau' \circ \tilde{\tau}_0 \circ \tilde{\tau})(X^{\pm}). \tag{6.2$\pm$}$$

Using the definitions of $\tilde{\tau}$, $\tilde{\tau}_0$ and τ' given in the previous two sections, we are led to the following:

$$\pi(\tilde{H}) = -\frac{1}{2}H , \tag{6.3}$$

$$\begin{aligned} \pi(\tilde{X}^{\pm}) &= \pm\frac{1}{\sqrt{[2]_{\tilde{q}}}}(-1)^{-\frac{H}{4}\mp\frac{H}{4}}\left(\frac{\left\{\frac{1}{\tilde{Y}}\sqrt{-(-1)^{-\frac{H}{2}}[\frac{H\pm1}{2}]_{\tilde{q}^2}}+\sqrt{\pm I}\right\}}{\{\mp\frac{1}{Y}[\frac{H\pm1}{2}]_{q^2}+I\}}\right)X^{\mp} \\ &= -\frac{1}{\sqrt{[2]_{\tilde{q}}}}X^{\mp}(-1)^{-\frac{H}{4}\mp\frac{H}{4}}\left(\frac{\left\{\pm\frac{i}{\tilde{Y}}\sqrt{-(-1)^{-\frac{H}{2}}[\frac{H\mp1}{2}]_{\tilde{q}^2}}+\sqrt{\pm I}\right\}}{\{\mp\frac{1}{Y}[\frac{H\mp1}{2}]_{q^2}+I\}}\right) \end{aligned} \tag{6.4$\pm$}$$

and

$$\tilde{Y}^2 = i(-1)^{-H}Y^2. \tag{6.5}$$

From calculations using these equations and with the help of some identities for q numbers given in [19] we obtain

$$[\pi(\tilde{H}), \pi(\tilde{X}^{\pm})] = \pm\pi(\tilde{X}^{\pm}) \tag{6.6$\pm$}$$

$$\{\pi(\tilde{X}^{+}), \pi(\tilde{X}^{-})\} = (-1)^{H}[\pi(\tilde{H})]_{\tilde{q}^2}. \tag{6.7}$$

In order for π to define a homomorphism we clearly must have

$$(-1)^{H} = I . \tag{6.8}$$

In other words, we must consider a quotient algebra for which this additional condition, along with the defining relations (5.1a) and (5.2b) of $U_{\tilde{q}}(\mathfrak{osp}(1|2))$ hold true. We have thus recovered, for the special case of $n = 1$, the main criterion established in [7] in order that $U_{\tilde{q}}(\mathfrak{osp}(1|2n))$ be related to $U_q(\mathfrak{so}(2n+1))$. The homomorphism defined by Eqs. (6.3) and (6.4±) permits us to relate representations of the two algebras, $U_{\tilde{q}}(\mathfrak{osp}(1|2n))$ and $U_q(\mathfrak{so}(2n+1))$, in which representations the actions of $\pi(\tilde{X}^{\pm})$ and $\pi(\tilde{H})$ are well-defined. We have already mentioned in the introduction to this section a class of representations for which this is the case. In addition, Eqs. (6.4±) are also applicable for studying the relationship between highest weight representations of $U_{\tilde{q}}(\mathfrak{osp}(1|2))$ and $U_q(\mathfrak{so}(3))$ even though the actions of $\{\mp\frac{1}{Y}[\frac{H\mp1}{2}]_{q^2}+I\}^{-1}$ and $\left\{\pm\frac{i}{\tilde{Y}}\sqrt{-(-1)^{-\frac{H}{2}}[\frac{H\mp1}{2}]_{\tilde{q}^2}}+\sqrt{\pm I}\right\}^{-1}$ in such representations will not be well-defined as operators on the representation spaces (cf. [18]); nevertheless, the operators representing their ratios will still be well-defined for appropriate choices of q and $\tilde{q}$. We leave it to the reader to work out the straightforward details, in order to convince himself of this.

Notice that we did not need to impose any relationship between q and $\tilde{q}$ in order to establish the above homomorphism. In the general case (n arbitrary) studied in [7], it turns out that $q^2 = -\tilde{q}^2$ (in notation adopted in sections 4 and 5). The reason for this can be understood, at least partially, by comparing the q-Serre relations, Eqs. (2.1c′) and (2.2c′), which for the $n = 1$ case are vacuous.

Finally, it would be interesting to try to to generalize our results to arbitrary n by combining them with a variant of the mapping from $U_q(\mathfrak{so}(2n+1))$ to $U_{\tilde{q}}(\mathfrak{osp}(1|2n))$ which is given in [7]. A variant of it is necessary in order to account for the above noted difference in the definitions given in [7] with the standard definitions of the q deformed Drinfeld-Jimbo enveloping algebras. This might provide us with even more insight into understanding the connection between $U_q(\mathfrak{so}(2n+1))$ and $U_{\tilde{q}}(\mathfrak{osp}(1|2n))$ for arbitrary n.

References

[1] Moylan P 2014 *J. Phys. Conf. Ser.* **512** 012026 9
[2] Gelfand I M and Kirillov A A 1966 *Dokl. Akad. Nauk SSSR* **167** 503 (*Sov. Math.* **7** 403 (1966))
[3] Božek P Havlíček M and Navrátil O 1985 A new relationship between the Lie algebras of Poincaré and de Sitter groups *Preprint Universitas Carolina Pragensis NCITF/1*
[4] Havlíček M and Moylan P 1993 *J. Math. Phys.* **34** 11 5320–5332
[5] Moylan P 2004 Representations of the Poincaré group from positive energy representations of SO(2,3) *Proc. 3rd Int. Sympos. Quantum Theory and Symmetries* ed Argyres P C et al (New Jersey: World Scientific) 105–110
[6] Schupp P Watts P and Zumino B 1992 *Lett. Math. Phys.* **24** 141–145
[7] Zhang R B 1992 *Lett. Math. Phys.* **25** 317–325
[8] Lanzman E 2002 *Algebras and Representation Theory* **5** 3 235-238
[9] Rittenberg V and Scheunert M 1982 *Comm. Math. Phys.* **83** 1
[10] Aizawa N, Chakrabarti R, Mohammed S S N and Naina S S 2007 *J. Phys. A: Math. Theor.* **40** 14985-15000
[11] Klimyk A and Schmüdgen K 1997 *Quantum Groups and Their Representations* (Berlin: Springer Verlag)
[12] Dobrev V K and Zhang R B 2005 *Yadrenda Fizika* **68** 10 1724–1732 (*Physics of Atomic Nuclei* **68** 10 1660–1669)
[13] Palev T D and Van der Jeugt J 1995 *J. Phys. A: Math. Gen.* **28** 2605–2616
[14] Segal I E 1951 *Duke Math. Jour.* **18** 221
[15] Inönü E and Wigner E P 1953 *Proc. Nat. Acad. Sci. USA* **39** 510
[16] Koelink H T 1994 *Duke Math. Jour.* **76** 2 483–508
[17] Jacobson N 1979 *Lie Algebras* (New York: Dover) 171; Zhelobenko D P 1994 *Representations of Reductive Lie Algebras* (Moscow: Nauka) Ch. 3
[18] Moylan P 2014 Localizations of $U_q(\mathfrak{sl}(2))$ and $U_q(\mathfrak{osp}(1|2))$ associated with Euclidean and super Euclidean algebras *Lie Theory and Its Applications in Physics* Springer Proceedings in Mathematics and Statistics **111** ed Dobrev V K (Tokyo: Springer-Verlag)
[19] Kachurik I I 1998 *Ukrainian Math. Jour.* **50** 8 1201-1211

Journal of Physics: Conference Series **597** (2015) 012062 doi:10.1088/1742-6596/597/1/012062

Šapovalov elements and the Jantzen filtration for contragredient Lie superalgebras: a survey

Ian M Musson

Department of Mathematical Sciences, University of Wisconsin-Milwaukee, USA.

E-mail: musson@uwm.edu

Abstract. This is a survey of some recent results on Šapovalov elements and the Jantzen filtration for contragredient Lie superalgebras. The topics covered include the existence and uniqueness of the Šapovalov elements, bounds on the degrees of their coefficients and the behavior of Šapovalov elements when the Borel subalgebra is changed. There is always a unique term whose coefficient has larger degree than any other term. This allows us to define some new highest weight modules. If X is a set of orthogonal isotropic roots and $\lambda \in \mathfrak{h}^*$ is such that $\lambda + \rho$ is orthogonal to all roots in X, we construct highest weight modules with character $\epsilon^\lambda p_X$. Here p_X is a partition function that counts partitions not involving roots in X. When $|X| = 1$, these modules are used to give a Jantzen sum formula for Verma modules in which all terms are characters of modules in the category $\mathcal{O}$ with positive coefficients.

1. Introduction

Throughout this paper we work over an algebraically closed field $\mathtt{k}$ of characteristic zero. If $\mathfrak{g}$ is a semisimple Lie algebra, necessary and sufficient conditions for the existence of a non-zero homomorphism between Verma modules can be obtained by combining work of Verma [15] with work of Bernstein, Gelfand and Gelfand [1], [2]. Such maps can be described explicitly in terms of certain elements introduced by Šapovalov in [14]. Necessary and sufficient conditions for a simple highest weight module to be a composition factor of a Verma module were also obtained in [1], [2]. A more elementary proof of the latter result can be given using the Jantzen filtration and sum formula [8]. [1] However neither Šapovalov elements nor the Jantzen filtration have received much attention for classical simple Lie superalgebras. In this paper we review some recent results on Šapovalov elements, the Jantzen filtration and sum formula in the super case. New phenomena arise due to the presence of isotropic roots. Proofs not found here are given in [11] Chapter 9 and 10, [12] or [13].

Let $\mathfrak{g} = \mathfrak{g}(A, \tau)$ be a finite dimensional contragredient Lie superalgebra with Cartan subalgebra $\mathfrak{h}$, and set of simple roots Π, see [11] Chapter 5. Let Δ^+ be the set of positive roots containing Π, and

$$\mathfrak{g} = \mathfrak{n}^- \oplus \mathfrak{h} \oplus \mathfrak{n}^+$$

[1] We remark that the multiplicities of the composition factors in Verma modules for semisimple Lie algebras are given by the Kazdhan-Lusztig conjecture [10]. For type A Lie superalgebras, they are given by Brundan's analog of the Kazdhan-Lusztig conjecture, proven in [5], see also [4] Theorem A and [3] Theorem 3.6.

the corresponding triangular decomposition of $\mathfrak{g}$. We use the Borel subalgebras $\mathfrak{b} = \mathfrak{h} \oplus \mathfrak{n}^+$ and $\mathfrak{b}^- = \mathfrak{n}^- \oplus \mathfrak{h}$. The Verma module $M(\lambda)$ with highest weight $\lambda \in \mathfrak{h}^*$, and highest weight vector v_λ is induced from $\mathfrak{b}$. Let ρ_0 (resp. ρ_1) be the half-sum of the positive even (resp. odd) positive roots and $\rho = \rho_0 - \rho_1$.

Fix a non-degenerate invariant symmetric bilinear form (,) on $\mathfrak{h}^*$ as in [11] Theorem 5.4.1, and for all $\alpha \in \mathfrak{h}^*$, let $h_\alpha \in \mathfrak{h}$ be the unique element such that $(\alpha, \beta) = \beta(h_\alpha)$ for all $\beta \in \mathfrak{h}^*$. Then for all $\alpha \in \Delta^+$, choose elements $e_{\pm\alpha} \in \mathfrak{g}^{\pm\alpha}$ such that

$$[e_\alpha, e_{-\alpha}] = h_\alpha.$$

We give bounds on the degrees of the coefficients of H_π of the Šapovalov elements in Equation (1) below. The exact form of the coefficients depends on the way the positive roots are ordered. However there is always a unique coefficient of highest degree, and we determine the leading term of this coefficient up to a scalar multiple. These results appear to be new even for simple Lie algebras. The existence of a unique coefficient of highest degree is used to construct some new highest weight modules $M^\gamma(\lambda)$, where γ is an isotropic root and $(\lambda+\rho, \gamma) = 0$, [13]. There is a version of the Jantzen sum formula in [11] Theorem 10.3.1, see also [6], which involves infinite sums of Verma modules with alternating sign coefficients. The module $M^\gamma(\lambda)$ has character $e^\lambda p_\gamma$ see (8) for notation. The infinite sums can be replaced by a finite sum of characters of modules of this form leading to a formula where both sides are sums of characters in the category $\mathcal{O}$. In type A there are explicit expressions for Šapovalov elements, see [12] Section 7.

2. Preliminaries

To simplify the exposition, we assume in this survey that any non-isotropic root of $\mathfrak{g}$ is even. This should give the reader the flavor of the results in general, but the proofs and some of the statements should be modified if this is not the case.

If $\eta \in Q^+ = \sum_{\alpha\in\Pi} \mathbb{N}\alpha$, a *partition* of η is a map $\pi : \Delta^+ \longrightarrow \mathbb{N}$ such that $\pi(\alpha) = 0$ or 1 for all isotropic roots α, and

$$\sum_{\alpha\in\Delta^+} \pi(\alpha)\alpha = \eta.$$

Let $\mathbf{P}(\eta)$ be the set of partitions of η. The *degree* of a partition π is defined to be $|\pi| = \sum_{\alpha\in\Delta^+} \pi(\alpha)$. Partitions are useful because they can be used to index a basis for $U(\mathfrak{n}^-)$. Fix an ordering on the set Δ^+, and for π a partition, set

$$e_{-\pi} = \prod_{\alpha\in\Delta^+} e_{-\alpha}^{\pi(\alpha)},$$

the product being taken with respect to this order. Then the elements $e_{-\pi}$, with $\pi \in \mathbf{P}(\eta)$ form a basis of $U(\mathfrak{n}^-)^{-\eta}$.

Suppose γ is a positive root, m is a positive integer and set

$$\mathcal{H}_{\gamma,m} = \{\lambda \in \mathfrak{h}^* | (\lambda + \rho, \gamma) = m(\gamma,\gamma)/2\},$$

and let $\mathcal{I}(\mathcal{H}_{\gamma,m})$ be the ideal of $S(\mathfrak{h})$ consisting of functions vanishing on $\mathcal{H}_{\gamma,m}$. The *Šapovalov element* $\theta_{\gamma,m}$ corresponding to the pair (γ, m) has the form

$$\theta_{\gamma,m} = \sum_{\pi\in\mathbf{P}(m\gamma)} e_{-\pi} H_\pi, \tag{1}$$

where $H_\pi \in U(\mathfrak{h})$, and satisfies

$$e_\alpha \theta_{\gamma,m} \in U(\mathfrak{g})(h_\gamma + \rho(h_\gamma) - m(\gamma,\gamma)/2) + U(\mathfrak{g})\mathfrak{n}^+, \text{ for all } \alpha \in \Delta^+. \tag{2}$$

This means that if $\lambda \in \mathcal{H}_{\gamma,m}$, then $\theta_{\gamma,m}v_\lambda$ is a highest weight vector in $M(\lambda)$. If γ is simple, then $\theta_{\gamma,m} = e^m_{-\gamma}$, so now assume that γ is not simple. Let $\pi^0 \in \mathbf{P}(m\gamma)$ be the unique partition of $m\gamma$ such that $\pi^0(\alpha) = 0$ if $\alpha \in \Delta^+\backslash\Pi$. The partition $m\pi^\gamma$ of $m\gamma$ is given by $m\pi^\gamma(\gamma) = m$, and $m\pi^\gamma(\alpha) = 0$ for all positive roots α different from γ. We normalize $\theta_{\gamma,m}$ so that the coefficient H_{π^0} is equal to 1. This guarantees that $\theta_{\gamma,m}v_\lambda$ is never zero. For a semisimple Lie algebra, the existence of such elements was shown by Šapovalov, [14] Lemma 1.

For a non-isotropic root α, we set $\alpha^\vee = 2\alpha/(\alpha,\alpha)$, and write s_α for the reflection corresponding to α. As usual the Weyl group W is the subgroup of $GL(\mathfrak{h}^*)$ generated by all such reflections. For $u \in W$ set

$$N(u) = \{\alpha \in \Delta_0^+ | u\alpha < 0\}, \qquad \ell(u) = |N(u)|.$$

Suppose Π_{even} is the set of even simple roots, and let W_{even} be the subgroup of W generated by the reflections s_α, where $\alpha \in \Pi_{\text{even}}$. Note that W_{even} can be a proper subgroup of W. For example this happens when $\mathfrak{g} = \mathfrak{osp}(2m, 2n)$. If $\Pi = \{\alpha_i | i = 1, \ldots, t\}$ is the set of simple roots, and γ is a positive root such that $\gamma = \sum_{i=1}^t a_i\alpha_i$, then the *height* $\text{ht}\gamma$ of γ is $\text{ht}\gamma = \sum_{i=1}^t a_i$.

Theorem 2.1. *Let $\mathfrak{g}$ be semisimple Lie algebra or a contragredient Lie superalgebra, and γ a positive root. If γ is isotropic assume that $m = 1$. Suppose $\gamma = w\beta$ for a simple root β and $w \in W_{\text{even}}$, and for $\alpha \in N(w^{-1})$, set $q(w,\alpha) = (w\beta, \alpha^\vee)$. Then*

(a) *there exists a Šapovalov element $\theta_{\gamma,m} \in U(\mathfrak{b}^-)^{-m\gamma}$, which is unique modulo the left ideal $U(\mathfrak{b}^-)\mathcal{I}(\mathcal{H}_{\gamma,m})$.*

(b) *the coefficients of $\theta_{\gamma,m}$ satisfy*

$$|\pi| + \deg H_\pi \leq m\text{ht}\gamma, \tag{3}$$

and

$$H_{m\pi^\gamma} \text{ has leading term } \prod_{\alpha \in N(w^{-1})} h_\alpha^{mq(w,\alpha)}. \tag{4}$$

Let $d_{m\gamma}$ be the degree of $H_{m\pi^\gamma}$.

Corollary 2.2. *In Theorem 2.1 $H_{m\pi^\gamma}$ is the unique term of degree $d_{m\gamma}$ in $\theta_{\gamma,m}$.*

Proof. This follows easily from (3) and (4). □

3. Šapovalov Elements and their Coefficients

A finite dimensional contragredient Lie superalgebra $\mathfrak{g}$ has, in general several conjugacy classes of Borel subalgebras, and this both complicates and enriches the representation theory of $\mathfrak{g}$. The complications are partially resolved by at first fixing a Borel subalgebra $\mathfrak{b}$ (or equivalently a basis of simple roots for $\mathfrak{g}$) with special properties. The effect of changing the Borel subalgebra is studied in detail in [13], see also the next Section.

In [9] Table VI, Kac gave a particular diagram in each case that we will call *distinguished*. The corresponding set of simple roots and Borel subalgebra are also called distinguished.

The distinguished Borel subalgebra contains at most one simple isotropic root vector. Unless $\mathfrak{g} = \mathfrak{osp}(1, 2n)$, or $\mathfrak{g} = \mathfrak{osp}(2, 2n)$ there is exactly one other Borel subalgebra with this property up to conjugacy in Aut $\mathfrak{g}$. A representative of this class (and its set of simple roots) will be called *anti-distinguished*. We assume that $\mathfrak{b}$ is either distinguished or anti-distinguished.

Theorem 2.1 is proved by looking at the proofs given in [7] or [11] and keeping track of the coefficients. Given $\lambda \in \mathfrak{h}^*$ and $x = \sum_i a_i \otimes b_i \in U(\mathfrak{b}^-) = U(\mathfrak{n}^-) \otimes S(\mathfrak{h})$, set $x(\lambda) = \sum_i a_i b_i(\lambda) \in U(\mathfrak{n}^-)$. Let (γ, m) be as in the statement of the Theorem and set $\mathcal{H} = \mathcal{H}_{\gamma,m}$. The idea of the proof is to construct elements $\theta^\lambda \in U(\mathfrak{n}^-)^{-m\gamma}$ for all λ in a dense subset of $\mathcal{H}$ such that $\theta^\lambda v_\lambda$ is a highest weight vector in $M(\lambda)^{\lambda - m\gamma}$, and that

$$\theta^\lambda = \sum_{\pi \in \mathbf{P}(m\gamma)} a_{\pi,\lambda} e_{-\pi}. \tag{5}$$

where $a_{\pi,\lambda}$ is a polynomial function of $\lambda \in \Lambda$ satisfying suitable conditions. For $\pi \in \mathbf{P}(m\gamma)$, the assignment $\lambda \to a_{\pi,\lambda}$ for $\lambda \in \Lambda$ determines a polynomial map from $\mathcal{H}$ to $U(\mathfrak{n}^-)^{-m\gamma}$, so there exists an element $H_\pi \in U(\mathfrak{h})$ uniquely determined modulo $\mathcal{I}(\mathcal{H})$ such that $H_\pi(\lambda) = a_{\pi,\lambda}$ for all $\lambda \in \Lambda$. We define the element $\theta \in U(\mathfrak{b}^-)$ by setting

$$\theta = \sum_{\pi \in \mathbf{P}(m\gamma)} e_{-\pi} H_\pi.$$

Note that $\theta(\lambda) = \theta^\lambda$. The Šapovalov element in Theorem 2.1 is constructed inductively using the next Lemma, see for example [7] Section 4.13 or [11] Theorem 9.4.3.

Lemma 3.1. *Let γ be a positive root, and m a positive integer which is equal to 1 if γ is isotropic. Suppose that $\alpha \in \Pi_{\text{even}}$, and set*

$$\mu = s_\alpha \cdot \lambda,\ \gamma' = s_\alpha \gamma,\ p = (\mu + \rho, \alpha^\vee),\ q = (\gamma, \alpha^\vee).$$

Assume that $p, q \in \mathbb{N} \backslash \{0\}$ and $\theta' \in U(\mathfrak{n}^-)^{-m\gamma'}$ is such that $v = \theta' v_\mu \in M(\mu)$ is a highest weight vector. Then there is a unique $\theta \in U(\mathfrak{n}^-)^{-m\gamma}$ such that

$$e_{-\alpha}^{p+mq} \theta' = \theta e_{-\alpha}^p. \tag{6}$$

Using Lemma 3.1 it follows that the coefficients of the Šapovalov elements $\theta_{\gamma,m}$ are obtained by taking $\mathbb{k}$-linear combinations of products of coefficients of the $\theta_{\gamma',m}$, with binomial coefficients.

4. Changing the Borel subalgebra

Using adjacent Borel subalgebras (equivalently odd reflections), it is possible to give an alternative construction of Šapovalov elements corresponding to an isotropic root γ, provided γ is a simple root for some Borel subalgebra. This condition always holds in type A, but for other types, it is quite restrictive: if $\mathfrak{g} = \mathfrak{osp}(2m, 2n + 1)$, (an algebra that does not satisfy the restrictions imposed at the start of section 2) the assumption only holds for roots of the form $\pm(\epsilon_i - \delta_j)$, while if $\mathfrak{g} = \mathfrak{osp}(2m, 2n)$ it holds only for these roots and the root $\epsilon_m + \delta_n$.

Suppose that $\mathfrak{b}$ is the distinguished Borel subalgebra, and let $\mathfrak{b}'$ be another Borel subalgebra with the same even part as $\mathfrak{b}$. Consider a sequence of Borel subalgebras

$$\mathfrak{b} = \mathfrak{b}^{(0)}, \mathfrak{b}^{(1)}, \ldots, \mathfrak{b}^{(r)}.$$

Assume there are isotropic roots α_i such that $\mathfrak{g}^{\alpha_i} \subset \mathfrak{b}^{(i-1)}, \mathfrak{g}^{-\alpha_i} \subset \mathfrak{b}^{(i)}$ for $1 \le i \le r$, and $\alpha_1, \ldots, \alpha_r$ are distinct positive roots of $\mathfrak{b}$.

Theorem 4.1. *Set $F(\gamma) = \{i | 1 \le i \le r$ and $(\gamma, \alpha_i) = 0\}$. There is a nonzero $c \in \mathtt{k}$ such that for all $\lambda \in \mathcal{H}_\gamma$,*

$$e_{\alpha_1} \dots e_{\alpha_r} e_{-\gamma} e_{-\alpha_r} \dots e_{-\alpha_1} v_\lambda = c \prod_{i \in F(\gamma)} (\lambda + \rho, \alpha_i) \theta_\gamma v_\lambda.$$

5. The Square of a Šapovalov Element

When γ is an isotropic root we write $\mathcal{H}_\gamma$ and θ_γ in place of $\mathcal{H}_{\gamma,1}$ and $\theta_{\gamma,1}$ respectively. Here we record an elementary but important property of the Šapovalov element θ_γ corresponding to such a root.

Theorem 5.1. *If $\lambda \in \mathcal{H}_\gamma$, then $\theta_\gamma^2 v_\lambda = 0$. Equivalently, $\theta_\gamma(\lambda - \gamma)\theta_\gamma(\lambda) = 0$.*

Proof. We assume $\gamma = w\beta$ for $\beta \in \Pi$ where β is isotropic, and $w \in W_{\rm even}$. For a suitable Zariski dense subset Λ of $\mathcal{H}_\beta$, suppose $\lambda \in w \cdot \Lambda$. The proof is by induction on $\ell(w)$. We can assume that $w \neq 1$. Suppose $w = s_\alpha u$ with $\ell(u) = \ell(w) - 1$, and set $\gamma' = s_\alpha \gamma$. Replace μ with $\mu - \gamma'$ and $\lambda = s_\alpha \cdot \mu$ with $s_\alpha \cdot (\mu - \gamma') = \lambda - \gamma$ in Equation (6). Then p is replaced by $p + q$ and adopting the notation of (5),

$$e_{-\alpha}^{p+2q} \theta_{\gamma'}^{\mu-\gamma'} = \theta_\gamma^{\lambda-\gamma} e_{-\alpha}^{p+q}.$$

Combining this with Equation (6) and using induction we have

$$0 = e_{-\alpha}^{p+2q} \theta_{\gamma'}^{\mu-\gamma'} \theta_{\gamma'}^{\mu} = \theta_\gamma^{\lambda-\gamma} \theta_\gamma^{\lambda} e_{-\alpha}^{p}.$$

The result follows since $e_{-\alpha}$ is not a zero divisor in $U(\mathfrak{n}^-)$. □

6. Modules with Prescribed Characters

We introduce some new highest weight modules whose characters are given by generating functions for certain kinds of partitions. If X is a set of pairwise orthogonal isotropic positive roots, set

$$\mathbf{P}_X(\eta) = \{\pi \in \mathbf{P}(\eta) | \pi(\alpha) = 0 \text{ for all } \alpha \in X\}.$$

and $\mathbf{p}_X(\eta) = |\mathbf{P}_X(\eta)|$. Set $p_X = \sum \mathbf{p}_X(\eta) \epsilon^{-\eta}$. We have

$$p_X = \prod_{\alpha \in \Delta_1^+ \setminus X} (1 + \epsilon^{-\alpha}) / \prod_{\alpha \in \Delta_0^+} (1 - \epsilon^{-\alpha}). \tag{7}$$

If X is empty, set $p = p_X$, and if $X = \{\alpha\}$ is a singleton write

$$\mathbf{P}_\alpha(\eta), \quad \mathbf{p}_\alpha(\eta), \quad \text{and } p_\alpha \tag{8}$$

instead of $\mathbf{P}_X(\eta), \mathbf{p}_X(\eta)$, and p_X. For a module M in the BGG category $\mathcal{O}$, the character of M is defined by ch $M = \sum_{\eta \in \mathfrak{h}^*} \dim_{\mathtt{k}} M^\eta \epsilon^\eta$. Recall that the Verma module $M(\lambda)$ has character $\epsilon^\lambda p$.

Theorem 6.1. *Suppose that X is an isotropic set of positive roots and $\lambda \in \mathcal{H}_\gamma$ for all $\gamma \in X$. Then there exists a factor module $M^X(\lambda)$ of $M(\lambda)$ such that*

$$\text{ch } M^X(\lambda) = \epsilon^\lambda p_X.$$

Journal of Physics: Conference Series **597** (2015) 012062 doi:10.1088/1742-6596/597/1/012062

If $X = \{\gamma\}$ we write $M^\gamma(\lambda)$ in place of $M^X(\lambda)$. The construction of the modules $M^X(\lambda)$ involves a process of deformation and specialization. First we extend scalars to $A = \mathtt{k}[T]$ and $B = \mathtt{k}(T)$. If R is either of these algebras we set $U(\mathfrak{g})_R = U(\mathfrak{g}) \otimes R$. Choose $\xi \in \mathfrak{h}^*$ such that $(\xi, \gamma) = 0$ for all $\gamma \in X$, and $(\xi, \alpha^\vee) \notin \mathbb{Z}$ for all even roots α. Next consider the $U(\mathfrak{g})_B$-module $M(\widetilde{\lambda})_B$ with highest weight $\widetilde{\lambda} = \lambda + T\xi$, and form the factor module of $M(\widetilde{\lambda})_B$ obtained by setting $\theta_\gamma v_{\widetilde{\lambda}}$ equal to zero for $\gamma \in X$. Then take a suitable $U(\mathfrak{g})_A$-submodule of this factor module and reduce mod T to obtain the module $M^X(\lambda)$. In more detail, we set

$$M^X(\widetilde{\lambda})_B = M(\widetilde{\lambda})_B / \sum_{\gamma \in X} U(\mathfrak{g})_B \theta_\gamma v_{\widetilde{\lambda}}.$$

Then $M^X(\widetilde{\lambda})_B$ is a $U(\mathfrak{g})_B$-module generated by a highest weight vector $v^X_{\widetilde{\lambda}}$ (the image of $v_{\widetilde{\lambda}}$) with weight $\widetilde{\lambda}$. Set $M^X(\widetilde{\lambda})_A = U(\mathfrak{g})_A v^X_{\widetilde{\lambda}} \subset M^X(\widetilde{\lambda})_B$, and

$$M^X(\lambda) = M^X(\widetilde{\lambda})_A / T M^X(\widetilde{\lambda})_A. \tag{9}$$

Then

$$M^X(\widetilde{\lambda})_A \otimes_A B = M^X(\widetilde{\lambda})_B.$$

Based on Corollary 2.2 we can show

Lemma 6.2. *Let $M = U(\mathfrak{g})_B v$ be a module with highest weight $\widetilde{\lambda}$ and highest weight vector v. Suppose that $\theta_\gamma v = 0$ for all $\gamma \in X$. Then*

(a) *for all η the weight space $M^{\widetilde{\lambda}-\eta}$ is spanned over B by all $e_{-\pi} v$ where $\pi \in \mathbf{P}_X(\eta)$.*

(b) $\dim_B M^{\widetilde{\lambda}-\eta} \leq \mathbf{p}_X(\eta)$.

Proof of Theorem 6.1 when $X = \{\gamma\}$. We set

$$v = v_{\widetilde{\lambda}} \in M_B = M(\widetilde{\lambda})_B, \quad u = \theta_\gamma v, \quad N_B = U(\mathfrak{g})_B u.$$

For $R = A$ or B we write $M^\gamma(\widetilde{\lambda})_R$ in place of $M^X(\widetilde{\lambda})_R$. Then the module $M^\gamma(\lambda)$ defined by (9) is generated by the image $\overline{v}$ of v which is a highest weight vector of weight λ. Also by Theorem 5.1, $\theta_\gamma u = 0$, so we can apply Lemma 6.2 to both N_B and M_B/N_B. This gives

$$\dim(M_B/N_B)^{\widetilde{\lambda}-\eta} \leq \mathbf{p}_\gamma(\eta), \quad \dim N_B^{\widetilde{\lambda}-\eta} \leq \mathbf{p}_\gamma(\eta - \gamma). \tag{10}$$

Since

$$\mathbf{p}(\eta) = \dim {M_B}^{\widetilde{\lambda}-\eta} = \dim(M_B/N_B)^{\widetilde{\lambda}-\eta} + \dim N_B^{\widetilde{\lambda}-\eta},$$

and $\mathbf{p}_\gamma(\eta-\gamma)+\mathbf{p}_\gamma(\eta) = \mathbf{p}(\eta)$, it follows that equality holds in (10). Now we obtain the result from the following considerations applied to the weight spaces of the modules $M^\gamma(\widetilde{\lambda})_R$ for $R = A, B$. If K is an A-submodule of a finite dimensional B-module L such that $K_A \otimes_A B = L$, then $\dim_{\mathtt{k}} K/TK = \dim_B L$. □

Remark 6.3. If M' is the kernel of the natural map $M(\lambda) \longrightarrow M^\gamma(\lambda)$, then $U(\mathfrak{g})\theta_\gamma v_\lambda \subseteq M'$, but the inclusion can be strict. Indeed this happens when $\mathfrak{g} = \mathfrak{sl}(2,1)$ and $\lambda = -\rho$ [13], see also [11] Exercise 10.5.4.

Journal of Physics: Conference Series **597** (2015) 012062 doi:10.1088/1742-6596/597/1/012062

7. The Jantzen Sum Formula

The Jantzen sum formula for a semisimple Lie algebra expresses the sum of the characters of the terms in the Jantzen filtration as a sum of characters of Verma modules. There is a version of the formula for contragredient Lie superalgebras in [11] Theorem 10.3.1, but it contains some terms that are not characters of Verma modules. Here we see that these extra terms are actually characters of the modules $M^\gamma(\lambda)$ introduced in Theorem 6.1.

For $\lambda \in \mathfrak{h}^*$ define

$$A(\lambda) = \{\alpha \in \Delta_0^+ | (\lambda + \rho, \alpha^\vee) \in \mathbb{N}\backslash\{0\}\},$$
$$B(\lambda) = \{\alpha \in \Delta_1^+ | (\lambda + \rho, \alpha) = 0\}.$$

Now we state our improved version of the Jantzen sum formula. At the same time, rather than using characters as in [11], it is useful to rewrite the result using the Grothendieck group $K(\mathcal{O})$ of the category $\mathcal{O}$. We define $K(\mathcal{O})$ to be the free abelian group generated by the symbols $[L(\lambda)]$ for $\lambda \in \mathfrak{h}^*$. If $M \in \mathcal{O}$, the class of M in $K(\mathcal{O})$ is defined as $[M] = \sum_{\lambda\in\mathfrak{h}^*} |M : L(\lambda)|[L(\lambda)]$, where $|M : L(\lambda)|$ is the multiplicity of the composition factor $L(\lambda)$ in M.

Theorem 7.1. *For all* $\lambda \in \mathfrak{h}^*$

$$\sum_{i>0}[M_i(\lambda)] = \sum_{\alpha\in A(\lambda)} [M(s_\alpha \cdot \lambda)] + \sum_{\gamma\in B(\lambda)} [M^\gamma(\lambda - \gamma)]. \tag{11}$$

Proof. Combine Theorem 6.1 with the result from [11] Theorem 10.3.1. □

The advantage of using this version of the formula is that $K(\mathcal{O})$ has a natural partial order. For $A, B \in \mathcal{O}$ we write $A \geq B$ if $[A] - [B]$ is a linear combination of classes of simple modules with non-negative integer coefficients. Clearly if B is a subquotient of A we have $[A] \geq [B]$.

8. Orthogonal Isotropic Roots

We consider the structure of $M(\lambda)$ when $B(\lambda)$ consists of two orthogonal roots γ, γ'. We say that $\lambda \in \mathcal{H}_\gamma \cap \mathcal{H}_{\gamma'}$ is *weakly generic* if $A(\lambda) = \emptyset$, and $B(\lambda) = \{\gamma, \gamma'\}$, and *generic* if μ is weakly generic for all $\mu \in \lambda + \mathbb{Z}\gamma + \mathbb{Z}\gamma'$. Choose $\xi \in \mathfrak{h}^*$ such that $(\xi, \gamma) = 0$ for all $\gamma \in X$, and $(\xi, \alpha) \neq 0$ for all even roots α.

Lemma 8.1. *For* $\lambda \in \mathcal{H}_\gamma \cap \mathcal{H}_{\gamma'}$ *there are only finitely many* $c \in \mathbf{k}$ *such that*

$$\theta_\gamma(\lambda + c\xi - \gamma)\theta_{\gamma'}(\lambda + c\xi) = 0 \tag{12}$$

or

$$\theta_{\gamma'}(\lambda + c\xi - \gamma)\theta_\gamma(\lambda + c\xi) = 0. \tag{13}$$

Proof. Set $\widetilde{\lambda} = \lambda + T\xi$. It follows from (4) and Corollary 2.2 that when $\theta_\gamma(\widetilde{\lambda} - \gamma)\theta_{\gamma'}(\widetilde{\lambda})v_{\widetilde{\lambda}}$ is written as a A-linear combination of terms $e_{-\pi}v_{\widetilde{\lambda}}$, the coefficient of $e_{-\gamma}e_{\gamma'}v_{\widetilde{\lambda}}$ is a polynomial in T of degree $d_\gamma + d_{\gamma'}$. Hence there are only finitely many c such that Equation (12) holds, and a similar argument applies to Equation (13). □

It follows from Lemma 8.1 that the set

$$\Lambda = \{\lambda \text{ generic in } \mathcal{H}_\gamma \cap \mathcal{H}_{\gamma'} | \theta_{\gamma'}(\lambda - \gamma)\theta_\gamma(\lambda) \neq 0 \neq \theta_\gamma(\lambda - \gamma')\theta_{\gamma'}(\lambda)\}$$

is Zariski dense in $\mathcal{H}_{\gamma'} \cap \mathcal{H}_\gamma$.

For $\lambda \in \Lambda$, the Jantzen sum formula (11) reads

$$\sum_{i>0}[M_i(\lambda)] = [M^\gamma(\lambda-\gamma)] + [M^{\gamma'}(\lambda-\gamma')]. \tag{14}$$

Next we quote a result from [13] without proof.

Lemma 8.2. *For* $\lambda \in \Lambda$,

$$|M(\lambda) : L(\lambda-\gamma-\gamma')| = 1.$$

Now based on equation (14) we can show

Lemma 8.3. *For* $\lambda \in \Lambda$ *we have*

(a) *If* $|M(\lambda) : L(\mu)| > 0$, *then* $\mu \in \lambda - \mathbb{N}\gamma - \mathbb{N}\gamma'$.

(b) $|M^\gamma(\lambda) : L(\lambda)| = 1$.

(c) $|M^{\gamma'}(\lambda-\gamma') : L(\lambda - r\gamma)| = 0$, *for* $r > 0$.

(d) $|M^\gamma(\lambda) : L(\lambda-\gamma)| = 0$.

(e) $\sum_{i>0} |M_i(\lambda) : L(\lambda-\gamma)| = 1$.

(f) $|M(\lambda) : L(\lambda - r\gamma)| = 0$ *for* $r \geq 2$.

(g) $|M(\lambda) : L(\lambda-\gamma-\gamma')| = |M_2(\lambda) : L(\lambda-\gamma-\gamma')| = 1$.

(h) $|M^\gamma(\lambda) : L(\lambda-\gamma-\gamma')| = 0$.

(i) $|M(\lambda) : L(\lambda - r\gamma - s\gamma')| = 0$ *if* $r+s \geq 3$.

Proof. (a) Follows from Equation (14) and induction on $\lambda-\mu$, while (b) holds since $M^\gamma(\lambda)$ is an image of $M(\lambda)$. We note that (c) follows from (a). To prove (d) set $L = L(\lambda-\gamma)$, and suppose for a contradiction, that $a = |M^\gamma(\lambda) : L| > 0$. Then

$$a \leq |M(\lambda) : L| \leq |M^\gamma(\lambda-\gamma) : L| + |M^{\gamma'}(\lambda-\gamma') : L| = 1,$$

by (b) and (c). Hence $a = |M(\lambda) : L| = 1$, but then

$$|M(\lambda) : L| = |M^\gamma(\lambda) : L| + |M^\gamma(\lambda-\gamma) : L| = a+1,$$

since in the Grothendieck group $K(\mathcal{O})$, we have

$$[M(\lambda)] = [M^\gamma(\lambda)] + [M^\gamma(\lambda-\gamma)]. \tag{15}$$

Now (e) follows from (b), (c) and (14), while (f) holds because for $r \geq 2$,

$$|M(\lambda) : L(\lambda-r\gamma)| \leq |M^\gamma(\lambda-\gamma) : L(\lambda-r\gamma)| + |M^{\gamma'}(\lambda-\gamma') : L(\lambda-r\gamma)| = 0,$$

by (c), (d) and induction. Next set $L = L(\lambda - \gamma - \gamma')$. To prove (g), it is enough, by Lemma 8.2 to show that $\sum_{i>0} |M_i(\lambda) : L| = 2$. First note that $|M^{\gamma}(\lambda - 2\gamma) : L| = 0$ by (c), so $|M^{\gamma}(\lambda - \gamma) : L| = |M(\lambda - \gamma) : L|$ and likewise $|M^{\gamma'}(\lambda - \gamma') : L| = |M(\lambda - \gamma') : L|$. So by (e),

$$\sum_{i>0} |M_i(\lambda) : L| = |M^{\gamma}(\lambda - \gamma) : L| + |M^{\gamma'}(\lambda - \gamma') : L| = |M(\lambda - \gamma) : L| + |M(\lambda - \gamma') : L| = 2.$$

By the proof of (g) $|M^{\gamma}(\lambda - \gamma) : L| = 1$, so (h) follows from (15). Finally (i) follows from (14) and induction on $r + s$. For example

$$|M(\lambda) : L(\lambda - \gamma - 2\gamma')| \leq |M^{\gamma}(\lambda - \gamma) : L(\lambda - \gamma - 2\gamma')| + |M^{\gamma'}(\lambda - \gamma') : L(\lambda - \gamma - 2\gamma')| = 0,$$

using (f) and (g). □

Theorem 8.4. *Suppose γ and γ' are orthogonal isotropic roots, and $\lambda \in \Lambda$. Define*

$$V_1 = U(\mathfrak{g})\theta_{\gamma}v_{\lambda}, \quad V_2 = U(\mathfrak{g})\theta_{\gamma'}v_{\lambda}.$$

Then the Jantzen filtration $\{M_i = M_i(\lambda)\}_{i>0}$ on $M(\lambda)$ is given by

$$M_3 = 0, \quad M_2 = V_1 \cap V_2 \cong L(\lambda - \gamma - \gamma'), \quad M_1 = V_1 + V_2.$$

Moreover M_1/M_2 is the direct sum of $L(\lambda - \gamma)$ and $L(\lambda - \gamma')$.

Proof. By Lemma 8.3 the only composition factors of M_1 are $L(\lambda-\gamma)$, $L(\lambda-\gamma')$, and $L(\lambda-\gamma-\gamma')$ each with multiplicity one. By (e) and (g) in the Lemma $L(\lambda - \gamma - \gamma') \subseteq M_2$ and

$$\sum_{i>0}[M_i] = [L(\lambda - \gamma)] + [L(\lambda - \gamma')] + 2[L(\lambda - \gamma - \gamma')].$$

Note that V_1 and V_2 are generated by highest weight vectors of weights $\lambda - \gamma$ and $\lambda - \gamma'$. The result follows since for $\lambda \in \Lambda$, $\theta_{\gamma'}\theta_{\gamma}v_{\lambda}$ and $\theta_{\gamma}\theta_{\gamma'}v_{\lambda}$ are highest weight vectors with weight $\lambda - \gamma - \gamma'$. □

Corollary 8.5. *There is a rational function p of $\lambda \in \mathcal{H}_{\gamma} \cap \mathcal{H}_{\gamma'}$ such that*

$$\theta_{\gamma'}(\lambda - \gamma)\theta_{\gamma}(\lambda) = p(\lambda)\theta_{\gamma}(\lambda - \gamma')\theta_{\gamma'}(\lambda). \tag{16}$$

Proof. The highest weight vectors referred to in the last sentence of the proof are necessarily proportional. Hence (16) holds since the coefficients of $e_{-\gamma'}e_{-\gamma}$ in these highest weight vectors (when written as linear combinations of the $e_{-\pi}$ with $\pi \in \mathbf{P}(\gamma + \gamma')$) are polynomials in λ. □

Corollary 8.5 is applied in the proof of Theorem 6.1 in the case $|X| \geq 2$. We sketch the main new ingredient when $X = \{\gamma, \gamma'\}$ as in the Corollary. In this case $M(\widetilde{\lambda})_B$ has a series of submodules

$$M(\widetilde{\lambda})_B = W_0 \supset W_1 \supset W_2 \supset W_3 \supset W_4 = 0, \tag{17}$$

where

$$W_1 = U(\mathfrak{g})_B\theta_{\gamma}v_{\widetilde{\lambda}} + U(\mathfrak{g})_B\theta_{\gamma'}v_{\widetilde{\lambda}}, \quad W_2 = U(\mathfrak{g})_B\theta_{\gamma}v_{\widetilde{\lambda}}, \quad W_3 = U(\mathfrak{g})_B\theta_{\gamma}\theta_{\gamma'}v_{\widetilde{\lambda}}.$$

From the Corollary and Theorem 5.1 we deduce that $\theta_{\gamma}\theta_{\gamma'}v_{\widetilde{\lambda}} \in W_2$, $\theta_{\gamma'}\theta_{\gamma}v_{\widetilde{\lambda}} \in W_3$ and $\theta_{\gamma'}\theta_{\gamma}\theta_{\gamma'}v_{\widetilde{\lambda}} = 0$. Thus from Theorem 5.1 and Lemma 6.2 we obtain bounds on the dimensions of the weight spaces of the factors in the series (17) which are analogous to (10). The statement about characters follows as before.

9. Non-Orthogonal Isotropic Roots

If γ, γ' are non-orthogonal isotropic roots, then $\gamma' = s_\alpha \gamma$ for some even root α. In this situation we relate the Šapovalov elements for γ, γ' and α. We set $\theta_{\alpha,0} = 1$ and $Q_0^+ = \sum_{\alpha \in \Pi_{\text{even}}} \mathbb{N}\alpha$.

Theorem 9.1. *Let γ be a positive isotropic root and α a non-isotropic root contained in Q_0^+. Let v_λ be a highest weight vector in a Verma module with highest weight λ, and set $\gamma' = s_\alpha\gamma$. Suppose $p = (\lambda + \rho, \alpha^\vee) \in \mathbb{N}\backslash\{0\}$. Then*

(a) *If $(\lambda + \rho, \gamma') = 0$ we have*

$$\theta_\gamma \theta_{\alpha,p} v_\lambda = \theta_{\alpha,p+1} \theta_{\gamma'} v_\lambda. \tag{18}$$

(b) *If $(\lambda + \rho, \gamma) = 0$, and $p - 1 \geq 0$, we have*

$$\theta_{\gamma'} \theta_{\alpha,p} v_\lambda = \theta_{\alpha,p-1} \theta_\gamma v_\lambda. \tag{19}$$

Proof. It suffices to prove (a) for all λ in the Zariski dense subset Λ of $\mathcal{H}_{\gamma'} \cap \mathcal{H}_{\alpha,p}$ given by

$$\Lambda = \{\lambda \in \mathcal{H}_{\gamma'} \cap \mathcal{H}_{\alpha,p} | A(\lambda) = \{\alpha\},\ B(\lambda) = \{\gamma'\}\}.$$

However for $\lambda \in \Lambda$, $M(\lambda)$ contains a unique highest weight vector of weight $s_\alpha \cdot \lambda - \gamma$. Thus, $\theta_\gamma \theta_{\alpha,p} v_\lambda$ and $\theta_{\alpha,p+1} \theta_{\gamma'} v_\lambda$ are equal up to a scalar multiple. If π^0 is the partition of $p\alpha + \gamma$ with $\pi^0(\sigma) = 0$ for all non-simple roots σ, it follows easily from the definition of Šapovalov elements, that $e_{-\pi^0} v_\lambda$ occurs with coefficient equal to one in both $\theta_\gamma \theta_{\alpha,p} v_\lambda$ and $\theta_{\alpha,p+1} \theta_{\gamma'} v_\lambda$, and this gives the desired result. The proof of (b) is similar. □

Remark 9.2. The most interesting case of Equation (18) arises when $p = 0$, since then we have an inclusion between submodules of a Verma module obtained by multiplying the highest weight vector v_λ by θ_γ and $\theta_{\gamma'}$. Similarly the most interesting case of Equation (19) is when $p = 1$.

Acknowledgments

I thank Kevin Coulembier for helpful comments. The research that led to this paper was partly supported by NSA Grant H98230-12-1-0249 and Simons Foundation grant 318264.

References

[1] Bernstein I N, Gelfand I M and Gelfand S I 1971 Structure of representations that are generated by vectors of higher weight *Funkcional Anal. i Priložen.* **5** 1–9
[2] Bernstein I N, Gelfand I M and Gelfand S I 1975 Differential operators on the base affine space and a study of $\mathfrak{g}$-modules *Lie groups and their representations* (*Proc. Summer School Bolyai János Math. Soc. Budapest 1971*) (Halsted, New York) 21–64
[3] Brundan J 2014 Representations of the general linear Lie superalgebra in the BGG category $\mathcal{O}$ *Developments and Retrospectives in Lie Theory* ed G Mason, I Penkov and J A Wolf (Berlin: Springer) 71–98
[4] Brundan J, Losev I and Webster B Tensor product categorifications and the super Kazhdan-Lusztig conjecture (arXiv:math/1310.0349)
[5] Cheng S-J, Wang W and Lam N 2008 Brundan-Kazhdan-Lusztig and super duality conjectures *Publ. Res. Inst. Math. Sci.* **44** 1219–1272
[6] Gorelik M 2004 The Kac construction of the centre of $U(\mathfrak{g})$ for Lie superalgebras *J. Nonlinear Math. Phys.* **11** 325–349
[7] Humphreys J E 2008 *Representations of Semisimple Lie Algebras in the BGG Category $\mathcal{O}$* (Providence: American Mathematical Society)
[8] Jantzen J C 1979 *Moduln mit einem höchsten Gewicht* (Berlin: Springer)
[9] Kac V G 1977 Lie superalgebras *Advances in Math.* **26** 8–96
[10] Kazhdan D and Lusztig G 1979 Representations of Coxeter groups and Hecke algebras *Invent. Math.* **53**
[11] Musson I M 2012 *Lie Superalgebras and Enveloping Algebras* (Providence: American Mathematical Society)

[12] Musson I M Coefficients of Šapovalov elements for simple Lie algebras and contragredient Lie superalgebras (arXiv:math/1311.0570)

[13] Musson I M The Jantzen filtration and sum formula for basic classical Lie superalgebras (in preparation)

[14] Šapovalov N N 1972 A certain bilinear form on the universal enveloping algebra of a complex semisimple Lie algebra *Funkcional. Anal. i Priložen.* **6** 65–70

[15] Verma D-N 1968 Structure of certain induced representations of complex semisimple Lie algebras *Bull. Amer. Math. Soc.* **74** 160–166

Meromorphic solutions to the q-Painlevé equations around the origin

Y Ohyama

Department of Mathematics, Osaka University 1-1 Machikaneyama-machi, Toyonaka, 560-0043, Japan

E-mail: ohyama@ist.osaka-u.ac.jp

Abstract. We study meromorphic solutions to the q-Painlevé equations with type VI, V, III around the origin. Since the origin of q-$P_{\rm VI}$, q-$P_{\rm V}$ and q-$P_{\rm III}$ are singular points of the q-Briot-Bouquet type, they have holomorphic solutions for generic parameters. The q-$P_{\rm VI}$, q-$P_{\rm V}$ and q-$P_{\rm III}$ have four, three and two meromorphic solutions around the origin respectively. We can solve linear connection problems for holomorphic solutions.

1. Introduction

A q-analogue of the Painlevé equations has been studied for a quarter of a century. Many discrete Painlevé equations are found in 1980s, but Grammaticos, Ramani, Papageorgiou and Hietarinta gave a systematic study on discrete Painlevé equations in 1991 [5], [14]. They discovered many discrete equations with singularity confinement property, which is a discrete analogue of the Painlevé property (no movable branch points). There are also many studies on the q-Painlevé equations, such as bilinear forms, the Bäcklund transformations, determinant formula for hypergeometric solutions and so on. Sakai has describes a degeneration diagram of discrete Painlevé equations from the viewpoint of spaces of initial conditions [16]. His theory elegantly explains the reason why many different discrete analogues of Painlevé equations exist for each differential Painlevé equations. Different types of discrete Painlevé equations can live on the same space of initial conditions. They correspond to directions of translation on the affine Weyl group symmetries, which act on the spaces of initial conditions. The reader may consult two good survey papers on discrete Painlevé equations [4] [11].

But there exist few studies on transcendental solutions to q-Painlevé equations, although it is believed that generic solutions for q-Painlevé equations are transcendental. Here "transcendental" solutions mean solutions neither of the hypergeometric type nor algebraic.

We study some examples of special solutions to q-Painlevé equations, which are holomorphic around the fixed singular points. These solutions are q-analogue of meromorphic solutions around the fix points to differential Painlevé equations [8]. In [8], we have studied meromorphic solutions around the origin to the third, fifth and sixth Painlevé differential equations. These equations have a singular point of the Briot-Bouquet type at the origin. In general the Painlevé transcendents have a wild singularity around the fixed singular points. But such tempered solutions play an important role in mathematical physics. Moreover, we can easily determine monodromy or the Stokes data of the linearized equations (the Lax pair) of the Painlevé equations. In this paper, we classify all of meromorphic solutions around the origin to q-$P_{\rm VI}$,

q-$P_{\rm V}$, q-$P_{\rm III}$ and give connection matrices of the linearized equations. The case of q-$P_{\rm VI}$ has been studied in [12].

Isomonodromic methods gives a powerful tool in the analytic study of the Painlevé equations,. For q-Painlevé equations, we consider *connection preserving deformations* instead of monodromy preserving deformations. We take a linear q-difference equation $Y(qx) = A(x)Y(x)$, where $A(x)$ is a matrix valued polynomial. We choose local solutions $Y_0(x)$ around $x = 0$ and $Y_\infty(x)$ around $x = \infty$. The connection matrix $P(x)$ is given by $Y_\infty(x) = Y_0(x)P(x)$. We introduce an extra parameter t and study $Y(qx) = A(x,t)Y(x)$. When the connection matrix $P(x,t)$ is a pseudo-constant in t, i.e. $P(x,t) = P(x,qt)$, we say that the equation $Y(qx) = A(x,t)Y(x)$ has a connection preserving deformation on t. This is a discrete analogue of monodromy preserving deformations. Jimbo and Sakai found a q-analog of the sixth Painlevé equation using a connection preserving deformation [6]. Murata constructed Lax forms of the q-Painlevé equations for the type I to V [10].

We can take a limit $t \to 0$ in the Lax form for our holomorphic solutions of q-Painlevé equations around $t = 0$. Since the Lax form goes to q-hypergeometric equations by taking the limit, we can determine the connection matrix exactly. Since our holomorphic solutions exist for generic parameters of the q-Painlevé equations, our solutions are different from the hypergeometric type solutions (particular solutions) shown in [7], which exist only on special parameters of the q-Painlevé equations.

1.1. q-Painlevé equations

We list up three q-Painlevé equations: q-$P_{\rm VI}$, q-$P_{\rm V}$, q-$P_{\rm III}$. We denote $y = y(t), z = z(t)$, $\bar{y} = y(qt), \bar{z} = z(qt)$. We take $a_1, a_2, a_3, a_4, b_1, b_2, b_3, b_4$ as parameters of equations.

$$q\text{-}P_{\rm VI}: \frac{y\bar{y}}{a_3a_4} = \frac{(\bar{z}-b_1t)(\bar{z}-b_2t)}{(\bar{z}-b_3)(\bar{z}-b_4)}, \quad \frac{z\bar{z}}{b_3b_4} = \frac{(y-a_1t)(y-a_2t)}{(y-a_3)(y-a_4)} \quad \left(\frac{b_1b_2}{b_3b_4} = q\frac{a_1a_2}{a_3a_4}\right).$$

$$q\text{-}P_{\rm V}: \frac{y\bar{y}}{a_3a_4} = -\frac{(\bar{z}-b_1t)(\bar{z}-b_2t)}{\bar{z}-b_3}, \quad \frac{z\bar{z}}{b_3} = -\frac{(y-a_1t)(y-a_2t)}{a_4(y-a_3)} \quad \left(\frac{b_1b_2}{b_3} = q\frac{a_1a_2}{a_3a_4}\right).$$

$$q\text{-}P_{\rm III}: \frac{y\bar{y}}{a_3a_4} = -\frac{\bar{z}(\bar{z}-b_2t)}{\bar{z}-b_3}, \quad \frac{z\bar{z}}{b_3} = -\frac{y(y-a_1t)}{a_4(y-a_3)}.$$

Taking a limit $q \to 1$, q-Painlevé equations go to differential Painlevé equations of the given type.

We have few applications of the q-Painlevé equations to mathematical sciences, but our new solutions would be useful in other fields in the future.

1.2. Basic notations

We set a q-shifted factorial

$$(a_1,\ldots,a_r;q)_n = \prod_{i=1}^{n}(a_i;q)_n, \quad (a;q)_n = (1-a)(1-qa)\cdots(1-q^{n-1}a).$$

A basic hyper geometric series [3] is defined by

$${}_r\varphi_s\left(a_1,\ldots,a_r;b_1,\ldots,b_s;q,z\right) = \sum_{n=0}^{\infty}\frac{(a_1,\ldots,a_r;q)_n}{(b_1,\ldots,b_s;q)_n(q;q)_n}\left[(-1)^n q^{n(n-1)/2}\right]^{1+s-r} z^n.$$

We use Jacobi's theta function as follows:

$$\theta_q(x) := \sum_{n=-\infty}^{\infty} q^{n(n-1)/2} x^n = (q, -x, -q/x; q)_\infty.$$

In this notation, $\theta_q(x)$ satisfies the equations $x\theta_q(xq) = \theta_q(x)$ and $\theta(1/x) = \frac{1}{x}\theta(x)$.

We set

$$e_c(x) := \frac{\theta(x)}{\theta(cx)},$$

for $c \in \mathbb{C}^\times$. We have $e_c(xq) = ce_c(x)$.

2. q-linear equations and connections

We review a q-connection and the q-Stokes phenomenon. Adams gave a pioneer work on irregular singular points of linear q-difference equations [1]. Modern study on irregular singular points of linear q-difference equations are explained in [15].

We consider a q-difference equation of size r:

$$Y(qx) = A(x)Y(x), \quad A(x) = A_0 + xA_1 + \cdots + x^N A_N. \tag{1}$$

Here A_j $(j = 0, 1, ..., N)$ and $Y(x)$ are $r \times r$ matrices.

2.1. Connection matrix

When all of the eigenvalues of A_0 are not zero, $x = 0$ in (1) is called *a regular singular point.* When all of the eigenvalues of A_N are not zero, $x = \infty$ in (1) is called *a regular singular point.* In the following, we set the eigenvalues of A_0 is $\rho_1, ..., \rho_r$ and

$$A_N = \mathrm{diag}(\kappa_1, ...\kappa_r).$$

We assume that $\rho_j/\rho_k \notin q^{\mathbb{Z}}$ and $\kappa_j/\kappa_k \notin q^{\mathbb{Z}}$ $(j \neq k)$.

Under these assumptions, (1) has local solutions around $x = 0$ and $x = \infty$ in the following forms [2]

$$Y_0(x) = L(x)\ \mathrm{diag}(e_{\rho_1}, ..., e_{\rho_r}),$$
$$Y_\infty(x) = \theta(x)^{-N} R(x)\ \mathrm{diag}(e_{\kappa_1}, ..., e_{\kappa_r}).$$

Here

$$L(x) = \sum_{n=0}^{\infty} L_n x^n, \qquad R(x) = \sum_{n=0}^{\infty} R_n x^{-n}$$

We assume that $R_0 = I_r$ and $\det L_0 = 1$. Moreover $L_0^{-1} A_0 L_0 = \mathrm{diag}(\rho_1, ..., \rho_r)$. Such solutions are unique up to the action of diagonal matrices D_0, D_∞.

Definition. We define the connection matrix $P(x)$ as

$$Y_\infty(x) = Y_0(x)P(x).$$

Matrix elements of $P(x)$ are elliptic functions, not constants.

2.2. Basic hypergeometric series

Heine's hypergeometric series ${}_2\varphi_1(a,b;c;x)$ satisfies the equation

$$(c-abqx)u(xq^2)-[c+q-(a+b)qx]u(qx)+q(1-x)u(x)=0. \tag{2}$$

Local solutions around $x=0$ are

$$u_1={}_2\varphi_1\left(a,b;c;x\right),\quad u_2=e_{q/c}(x){}_2\varphi_1\left(qa/c,qb/c;q^2/c;x\right).$$

Local solutions around $x=\infty$ are

$$v_1=\frac{1}{e_a(-x)}{}_2\varphi_1\left(a,aq/c;aq/b;cq/abx\right),\ v_2=(a\leftrightarrow b).$$

The connection matrix of (2) is determined by Watson [17]. Since our local solutions are different from Watson's local solutions, the connection matrix is also different.

Theorem 1 *The connection formula between (u_1,u_2) around $x=0$ and (v_1,v_2) around $x=\infty$ is given by*

$$u_1=\frac{(b,c/a;q)_\infty}{(c,b/a;q)_\infty}v_1+\frac{(a,c/b;q)_\infty}{(c,a/b;q)_\infty}v_2,$$
$$u_2=\frac{(qb/c,q/a;q)_\infty}{(q^2/c,b/a;q)_\infty}\frac{e_a(-x)e_{q/c}(x)}{e_{qa/c}(-x)}v_1+\frac{(qa/c,q/b;q)_\infty}{(q^2/c,a/b;q)_\infty}\frac{e_b(-x)e_{q/c}(x)}{e_{qb/c}(-x)}v_2.$$

2.3. q-confluent hypergeometric equation

When some exponents ρ_j (or κ_i) are zero, $x=0$ (or $x=\infty$) is irregular singular. In cases of confluent functions of ${}_{r+1}\varphi_r$, we may determine the Stokes coefficients.

The q-confluent hypergeometric equation is given by

$$(1-abqx)u(q^2x)-\{1-(a+b)qx\}\,u(qx)-qxu(x)=0. \tag{3}$$

Local solutions around $x=0$ are

$$u_1(x)={}_2\varphi_0(a,b;-;q,x),\quad u_2(x)=\frac{(abx;q)_\infty}{\theta(-qx)}{}_2\varphi_1\left(\frac{q}{a},\frac{q}{b};0;q,abx\right)$$

Local solutions around $x=\infty$ are

$$v_1(\lambda,x)=\frac{\theta\left(\frac{qax}{\lambda}\right)}{\theta\left(\frac{qx}{\lambda}\right)}{}_2\varphi_1\left(a,0;\frac{aq}{b};q,\frac{q}{abx}\right),\quad v_2(\lambda,x)=(a\leftrightarrow b)$$

For (3), $x=0$ is an irregular singular point and u_1 is a divergent series. Since $x=\infty$ is a regular singular point, the power series in $v_1(\lambda,x)$ and $v_2(\lambda,x)$ are convergent.

We explain the q-Borel-Laplace method to obtain a resummation of u_1 following Zhang [18]. For a power series $f(x)=\sum_{n\geq0}a_n$, the q-Borel transformation is given by

$$\left(\mathcal{B}_q^+f\right)(\xi):=\sum_{n\geq0}a_nq^{\frac{n(n-1)}{2}}\xi^n.$$

We fix $\lambda \in \mathbb{C}^*$. The q-Laplace transformation of $\varphi(\xi)$ is given by

$$\left(\mathcal{L}^+_{q,\lambda}\varphi\right)(x) := \sum_{n\in\mathbb{Z}} \frac{\varphi(\lambda q^n)}{\theta_q\left(\frac{\lambda q^n}{x}\right)}.$$

If $f(x)$ is a convergent series, $f = (\mathcal{L}^+_{q,\lambda} \circ \mathcal{B}^+_q)f$. But we may apply the q-Borel-Laplace transformation to divergent series. We set $\tilde{u}_1(\lambda, x) = (\mathcal{L}^+_{q,\lambda} \circ \mathcal{B}^+_q)u_1$. The resummation $\tilde{u}_1(\lambda, x)$ on is holomorphic on $\mathbb{C}^* \setminus (-\lambda)q^Z$ and $u_1(x)$ is an asymptotic series of $\tilde{u}_1(\lambda, x)$ [18]. The Stokes region is an open dense subset $(-\lambda)q^Z$ in $\mathbb{C}^*$. We remark that the series in $u_2(x)$ is convergent even though $x = 0$ is an irregular singular point. The reader may consult [15] and its references on resummation process of divergent q-series.

In irregular case, the connection matrix depends on the resummation parameter λ:

Theorem 2 *The connection formula of* (3) *is as follows:*

$$\tilde{u}_1(\lambda, x) = \frac{(b;q)_\infty}{\left(\frac{b}{a};q\right)_\infty}\frac{\theta(a\lambda)}{\theta(\lambda)}v_1(\lambda, x) + \frac{(a;q)_\infty}{\left(\frac{a}{b};q\right)_\infty}\frac{\theta(b\lambda)}{\theta(\lambda)}v_2(\lambda, x),$$

$$u_2(x) = \frac{(q/a;q)_\infty}{(b/a;q)_\infty}v_1(-1, x) + \frac{(q/b;q)_\infty}{(a/b;q)_\infty}v_2(-1, x).$$

Remark. The first formula is given by Zhang [18], and the second is given by Morita [9].

3. Meromorphic solutions to q-P_{VI}, q-P_{V} and q-P_{III}

For differential Painlevé equations P_{VI}, P_{V} and q-P_{III}, generic solutions are multi-valued around the fixed singular point $x = 0$ and have essential singularities. But there exist finite number of single-valued meromorphic solutions around $x = 0$. Such solutions are transcendental in general. We have shown that we can determine exactly monodromy or the Stokes data of linearized equations of the Painlevé equations for such meromorphic solutions of P_{VI}, P_{V} and P_{III} around the origin [8].

In this section we list up all of meromorphic solutions q-P_{VI}, q-P_{V} and q-P_{III} around the origin:

Theorem 3 *For generic parameters, all of meromorphic solutions to* q-P_{VI}, q-P_{V} *and* q-P_{III} *around the origin are holomorphic.*

(1) For generic parameters, there exist four holomorphic solutions to q-P_{VI} *around the origin:*

$$\text{I)}\quad y(t) = \frac{a_3b_3 - a_4b_4}{b_3 - b_4} + O(t), \qquad z(t) = \frac{a_3b_3 - a_4b_4}{a_3 - a_4} + O(t),$$

$$\text{II)}\quad y(t) = \frac{a_4b_3 - a_3b_4}{b_3 - b_4} + O(t), \qquad z(t) = \frac{a_3b_4 - a_4b_3}{a_3 - a_4} + O(t),$$

$$\text{III)}\quad y(t) = \frac{a_1a_2(b_1 - b_2)}{a_2b_1 - a_1b_2}t + O(t^2), \qquad z(t) = -\frac{b_1b_2(a_1 - a_2)}{(a_2b_1 - a_1b_2)q}t + O(t^2),$$

$$\text{IV)}\quad y(t) = \frac{a_1a_2(b_1 - b_2)}{a_1b_1 - a_2b_2}t + O(t^2), \qquad z(t) = \frac{b_1b_2(a_1 - a_2)}{(a_1b_1 - a_2b_2)q}t + O(t^2).$$

(2) There exist three holomorphic solutions to q-$P_{\rm V}$ around the origin

$$\text{I)}\quad y(t) = (a_3 - a_4 b_3) + O(t), \qquad z(t) = \left(b_3 - \frac{a_3}{a_4}\right) + O(t),$$

$$\text{II)}\quad y(t) = \frac{a_1 a_2 (b_1 - b_2)}{a_2 b_1 - a_1 b_2} t + O(t^2), \quad z(t) = \frac{b_1 b_2 (a_1 - a_2)}{(a_1 b_2 - a_2 b_1) q} t + O(t^2),$$

$$\text{III)}\quad y(t) = \frac{a_1 a_2 (b_1 - b_2)}{a_1 b_1 - a_2 b_2} t + O(t^2), \quad z(t) = \frac{b_1 b_2 (a_1 - a_2)}{(a_1 b_1 - a_2 b_2) q} t + O(t^2).$$

(3) There exist two holomorphic solutions to q-$P_{\rm III}$ around the origin:

$$\text{I)}\quad y(t) = (a_3 - a_4 b_3) + O(t), \qquad z(t) = \left(b_3 - \frac{a_3}{a_4}\right) + O(t),$$

$$\text{II)}\quad y(t) = \frac{a_1 a_3 a_4 b_2^2}{a_3 a_4 b_2^2 - a_1^2 b_3 q} t + O(t^2), \quad z(t) = \frac{a_1^2 b_2 b_3}{-a_3 a_4 b_2^2 + a_1^2 b_3 q} t + O(t^2).$$

*P*roof. 1) By direct calculations, we can easily checked that q-$P_{\rm J}$ (J=III, V, VI) has formal solutions shown in the Theorem.

2) We can show the formal solutions are convergent by a q-analogue of the Briot-Bouquet theorem shown by Poincaré [13]:

Lemma 4 (Poincaré) *For $f_j(t;y) = f_j(t, y_1, y_2, ..., y_n)$ $(j = 1, 2, ..., n)$, we assume that*
a) f_j is holomorphic around $(t, y) = (0, 0)$ and $f_j(0; 0) = 0$
b) the eigenvalues of matrix

$$\left(\frac{\partial f_j}{\partial y_k}(0;0)\right)_{1\le j,k\le n}$$

are not q^n $(n = 1, 2, 3, ...)$.

Then q-difference equation

$$y_j(qt) = f_j(t;y) \qquad (j = 1, 2, ..., n) \tag{4}$$

has a unique convergent solution of the type

$$y_j = \sum_{k=1}^{\infty} a_{jk} t^k \qquad (j = 1, 2, ..., n).$$

We call a system (4) *with the properties (a) and (b) 'q-Briot-Bouquet type'.*

Since we can easily modify $P_{\rm VI}$, $P_{\rm V}$ and $P_{\rm III}$ to q-Briot-Bouquet type, the solutions in the theorem are convergent. □

Remark. In differential cases, $P_{\rm VI}$, $P_{\rm V}$ and $P_{\rm III}$ have the same number of meromorphic solutions around the origin as in Theorem 3 (2,3,4).

4. Connection Preserving Deformation

A q-analogue of the sixth Painlevé equation is obtained as a connection preserving deformation by Jimbo and Sakai [6]. Connection preserving deformations of other types of q-analogue of the Painlevé equations are shown by Murata [10].

A connection preserving deformation is a compatibility condition of the following two difference equations:

$$\begin{aligned} Y(qx,t) &= A(x,t)Y(x,t), \\ Y(x,qt) &= B(x,t)Y(x,t). \end{aligned} \tag{5}$$

In the cases of q-Painlevé equations, $A(x,t)$ and $B(x,t)$ are rank two matrices and

$$A(x,t) = A_0(t) + xA_1(t) + x^2A_2.$$

Moreover

$$\begin{aligned} B(x,t) &= \frac{x}{(x-a_1qt)(x-a_2qt)}(xI + B_0(t)), \quad q\text{-}P_{\mathrm{VI}}, q\text{-}P_{\mathrm{V}}, \\ &= \frac{x}{x-a_1qt}(xI + B_0(t)), \qquad q\text{-}P_{\mathrm{III}}. \end{aligned}$$

We call (5) a *linearized equation* of the q-Painlevé equation.

We tale variables $y = y(t)$, $z_i = z_i(t)$ $(i = 1, 2)$ such that

$$A_{12}(y,t) = 0, \quad A_{11}(y,t) = \kappa_1 z_1, \quad A_{22}(y,t) = z_2, \tag{6}$$

We set a variable z as

$$z_2 = \kappa_1\kappa_2 qz(y - a_3). \tag{7}$$

(1) Connection preserving deformation of q-P_{VI}

We take a normalization of A_2: $A_2 = \mathrm{diag}(\kappa_1, \kappa_2)$. We assume that eigenvalues of $A_0(t)$ are $\rho_1 = \theta_1 t, \rho_2 = \theta_2 t$, and $\det A(x,t) = \kappa_1\kappa_2(x-a_1t)(x-a_2t)(x-a_3)(x-a_4)$. We have

$$\kappa_1\kappa_2 a_1a_2a_3a_4 = \theta_1\theta_2.$$

$$A(x,t) = \begin{pmatrix} \kappa_1((x-y)(x-\alpha)+z_1) & \kappa_2 w(x-y) \\ \kappa_1 w^{-1}(\gamma x + \delta) & \kappa_2((x-y)(x-\beta)+z_2) \end{pmatrix}.$$

Here

$$\begin{aligned} \alpha &= \frac{[y^{-1}((\rho_1+\rho_2)t - \kappa_1 z_1 - \kappa_2 z_2) - \kappa_2((a_1+a_2)t + a_3 + a_4 - 2y)]}{\kappa_1 - \kappa_2}, \\ \beta &= \frac{[-y^{-1}((\rho_1+\rho_2)t - \kappa_1 z_1 - \kappa_2 z_2) + \kappa_1((a_1+a_2)t + a_3 + a_4 - 2y)]}{\kappa_1 - \kappa_2}, \\ \gamma &= z_1 + z_2 + (y+\alpha)(y+\beta) + (\alpha+\beta)y - a_1a_2t^2 - (a_1+a_2)(a_3+a_4)t - a_3a_4, \\ \delta &= y^{-1}(a_1a_2a_3a_4t^2 - (\alpha y + z_1)(\beta y + z_2)), \end{aligned}$$

$$b_1 = \frac{a_1a_2}{\rho_1}, \quad b_2 = \frac{a_1a_2}{\rho_2}, \quad b_3 = \frac{1}{\kappa_1 q}, \quad b_4 = \frac{1}{\kappa_2}. \tag{8}$$

(2) Connection preserving deformation of q-P_{V}

We take a normalization of A_2: $A_2 = \mathrm{diag}(\kappa_1, 0)$. We assume that eigenvalues of $A_0(t)$ are $\rho_1 = \theta_1 t, \rho_2 = \theta_2 t$, and that $\det A(x,t) = \kappa_1\kappa_2(x-a_1t)(x-a_2t)(x-a_3)$. We have

$$-\kappa_1\kappa_2 a_1a_2a_3 = \theta_1\theta_2.$$

$$z_1 z_2 = \kappa_2(y - a_1 t)(y - a_2 t)(y - a_3). \tag{9}$$

Therefore z_1 is represented by z:

$$z_1 = \frac{(y - a_1 t)(y - a_2 t)}{\kappa_1 q z} \tag{10}$$

$$A(x,t) = \begin{pmatrix} \kappa_1((x-y)(x-\alpha)+z_1) & w(x-y) \\ \kappa_1 w^{-1}(\gamma x + \delta) & \kappa_2(x-y)+z_2 \end{pmatrix}. \tag{11}$$

Here

$$\begin{aligned} \alpha &= \frac{1}{\kappa_1}[y^{-1}((\theta_1+\theta_2)t - \kappa_1 z_1 - z_2) + \kappa_2], \\ \gamma &= z_2 - \kappa_2((2y+\alpha) - (a_1+a_2)t - a_3), \\ \delta &= y^{-1}(-\kappa_2 a_1 a_2 a_3 t^2 - (\alpha y + z_1)(-\kappa_2 y + z_2)). \end{aligned}$$

(3) Connection preserving deformation of q-P_{III}

We take a normalization of A_2: $A_2 = \mathrm{diag}(\kappa_1, 0)$, We assume that eigenvalues of $A_0(t)$ are $\theta_1 t, 0$ and that $\det A(x,t) = \kappa_1\kappa_2 x(x-a_1 t)(x-a_3)$. We have

$$-\kappa_1\kappa_2 a_1 a_2 a_3 = \theta_1\theta_2.$$

$$z_1 z_2 = \kappa_2(y - a_1 t)(y - a_2 t). \tag{12}$$

Therefore z_1 is represented by z:

$$z_1 = \frac{y(y - a_1 t)}{\kappa_1 q z}. \tag{13}$$

$$A(x,t) = \begin{pmatrix} \kappa_1((x-y)(x-\alpha)+z_1) & w(x-y) \\ \kappa_1 w^{-1}(\gamma x + \delta) & \kappa_2(x-y)+z_2 \end{pmatrix}. \tag{14}$$

Here

$$\begin{aligned} \alpha &= \frac{1}{\kappa_1}[y^{-1}(\theta_1 t - \kappa_1 z_1 - z_2) + \kappa_2], \\ \gamma &= z_2 - \kappa_2(2y + \alpha - t(a_1+a_2)), \\ \delta &= -y^{-1}(\alpha y + z_1)(-\kappa_2 y + z_2). \end{aligned}$$

For detail, see [6] and [10].

5. Connection formula for holomorphic solutions around the origin

We show that connection matrix of linearized equations of q-P_{VI}, q-P_{V} and q-P_{III} by taking the limit $t \to 0$.

Theorem 5 *For solutions (I-IV) to* q-P_{VI} *and the solution (I) of* q-P_V, $A(x,0)$ *reduces to basic hypergeometric function* ${}_2\varphi_1(t)$.

For solutions (I, II) of q-P_{III} *and for solutions (II),(III) of* q-P_V , $A(x,0)$ *reduces to basic hypergeometric function* ${}_1\varphi_1(t)$.

Journal of Physics: Conference Series **597** (2015) 012063 doi:10.1088/1742-6596/597/1/012063

In this sense, we can determine the connection problem of linearized q-Painlevé equations because the connection formula of ${}_2\varphi_1(t)$ is shown Watson and connection formula of ${}_1\varphi_1(t)$ is shown by Zhang and Morita as explained in section two. Theorem 5 is just a q-analogue of results in [8].

The case of q-P_{VI} has been shown in [12].
Since we obtain other cases by similar ways, we explain the cases of q-P_{V} (I) and (II)

[I] The case of the solution (I) to q-P_{V} :
For the linearized equation

$$Y(qx) = A(x,t)Y(x),$$

we set $x = \xi t$. Then

$$Y(q\xi t) = (A_0 + A_1\xi t + t^2 A_2\xi^2)Y(\xi t).$$

We set $Y(\xi t) = \xi^{\mathrm{lq}\, t} Z(\xi)$. (Here $\mathrm{lq}\, t = \log t/\log q$)

$$Z(q\xi) = (A_0/t + A_1\xi + tA_2\xi^2)Z(\xi)$$

Taking the limit $t \to 0$, we have (Eigenvalues of $A_0(t)$ are $\theta_1 t, \theta_2 t$)

$$Z(q\xi) = (A_0/t + A_1\xi)Z(\xi)$$

$$A_0/t = \begin{pmatrix} \frac{a_1a_2(b_1+b_2)}{b_1b_2} & a_3\left(\frac{a_4{}^2b_1b_2}{a_1a_2q} - 1\right)w'(0) \\ \frac{a_1{}^3a_2{}^3q}{a_3b_1b_2(a_1a_2q-a_4^2b_1b_2)w'(0)} & 0 \end{pmatrix}$$

$$A_1 = \begin{pmatrix} -\frac{a_1a_2}{a_4b_1b_2} & 0 \\ \frac{a_1^2a_2^2(a_1+a_2-a_4b_1-a_4b_2)q}{a_3b_1b_2(a_1a_2q-a_4^2b_1b_2)w'(0)} & -a_4 \end{pmatrix}$$

Therefore it reduces to hypergeometric equations ${}_2\varphi_1\left(\frac{a_4b_2}{a_2}, \frac{a_1q}{a_4b_1}; \frac{b_2q}{b_1}; \frac{\xi}{a_1}\right)$.

[II] The case of the solution (II) to q-P_{V} :

We can take a limit $t \to 0$ directly:

$$A(x,0) = \tilde{A}_1x + \tilde{A}_2x^2.$$

Here is

$$\tilde{A}_1 = \begin{pmatrix} \frac{a_3b_1b_2+a_4b_1b_2-a_2b_1-a_1b_2}{b_1b_2(-b_3q)} & w(0) \\ \frac{(a_1-a_4b_1)(a_4b_2-a_2)(a_4b_1-a_1b_3q)(a_4b_2-a_2b_3q)}{a_4^2b_1^2b_2^2(1-b_3q)^2w(0)} & \frac{a_3a_4b_2^2+a_2^2b_3q-a_2a_3b_2-a_2a_4b_2}{a_2b_2(1-b_3q)} \end{pmatrix}$$

$$\tilde{A}_2 = A_2 = \mathrm{diag}\,(\kappa_1, 0)$$

Solution to $Y(qx) = A(x,0)Y(x)$ around $x = 0$ are given by

$$Y^{(0)}(x) = \frac{q^{u(u-1)/2}}{(x/a_3)_\infty}\begin{pmatrix} y_{11}^{(0)} & y_{12}^{(0)} \\ y_{21}^{(0)} & y_{22}^{(0)} \end{pmatrix} x^{D_0}$$

Here each elements are represented by ${}_1\varphi_1$ as follows:

$$y_{11}^{(0)} = C_{11} \cdot {}_1\varphi_1 \left(\frac{a_4 b_2}{a_2 b_3}, \frac{a_1 b_2}{a_2 b_1} q; \frac{x}{a_4} \right), \quad y_{12}^{(0)} = C_{12} \cdot {}_1\varphi_1 \left(\frac{a_4 b_1}{a_1 b_3}, \frac{a_2 b_1}{a_1 b_2} q; \frac{x}{a_4} \right),$$
$$y_{21}^{(0)} = C_{21} \cdot {}_1\varphi_1 \left(\frac{a_1 b_4}{a_3 b_1}, \frac{a_1 b_2}{a_2 b_1} q; \frac{x}{a_4} \right), \quad y_{22}^{(0)} = C_{22} \cdot {}_1\varphi_1 \left(\frac{a_2 b_4}{a_3 b_2}, \frac{a_2 b_1}{a_1 b_2} q; \frac{x}{a_4} \right),$$

where

$$C_{11} = b_2(a_1 a_2 - a_3 a_4 b_1 b_2), \qquad C_{12} = b_1(a_1 a_2^2 - a_3 a_4 b_1 b_2) w(0),$$
$$C_{21} = a_1(a_2 - a_3 b_2)(a_2 - a_4 b_2)/w(0), \quad C_{22} = a_2(a_1 - a_3 b_1)(a_1 - a_4 b_1).$$

which reduced to the q-confluent hypergeometric equations.

6. Summary and future problems

As the same as differential Painlevé equations, the origin $t = 0$ of q-analogue of q-P_{VI}, q-P_{V}, q-P_{III} is a singular point of q-Briot-Bouquet type. For such q-Painlevé equations, there exist a finite number of holomorphic solutions around singular points. we can determine connection/Stokes coefficients of linear q-difference equations for holomorphic solutions of q-Painlevé equations at the fixed singular point.

We will study other q-Painlevé equations. There are no results on higher order cases. We should study nonlinear q-Stokes phenomenon around singular points of non q-Briot-Bouquet type. Such singularity appears at the infinity of q-P_{J} except for J=VI.

References

[1] Adams C R 1929 On the linear ordinary q-difference equation *Ann. Math* **30** 195–205
[2] Birkhoff G D 1914 The generalized Riemann problem for linear differential equations and the allied problems for linear difference and q-difference equations *Proc. Am. Acad. Arts and Sciences* **49** 521–568
[3] Gasper G and Rahman R 2004 *Basic Hypergeometric Series, second edition* (Cambridge: Cambridge University Press)
[4] Grammaticos B and Ramani A 2004 Discrete Painlevé equations: a review, in: *Discrete Integrable Systems*, Lecture Notes in Physics **644** 245–321
[5] Grammaticos B, Ramani A and Papageorgiou V 1991 Do integrable mappings have the Painlevé property? *Phys. Rev. Lett.* **67** 1825–1828
[6] Jimbo M and Sakai H 1996 A q-analog of the sixth Painlevé equation *Lett. Math. Phys.* **38** 145–154
[7] Kajiwara K, Masuda T, Noumi M, Ohta Y and Yamada Y 2005 Construction of hypergeometric solutions to the q-Painlevé equations *Int. Math. Res. Notices* 1439–1463
[8] Kaneko K and Ohyama Y 2013 Meromorphic Painlevé transcendents at a fixed singularity *Math. Nachr.* **286** 861–875
[9] Morita T 2013 A connection formula for the q-confluent hypergeometric function *SIGMA* **9** 050, 13 pages
[10] Murata M 2009 Lax forms of the q-Painlevé equations *J. Phys. A: Math. Theor.* **42** 115201
[11] Noumi M 2007 Special functions arising from discrete Painlevé equations: A survey *J. Comput. Appl. Math.* **202** 48–55
[12] Ohyama Y 2009 Analytic solutions to the q-Painlevé equations around the origin *RIMS Kokyuroku Bessatsu* **B13** 45–52
[13] Poincaré H 1880 Sur une classe nouvelle de transcendantes uniformes *J. Math. Pures et Appl.* **6** 313–365
[14] Ramani A, Grammaticos B and Hietarinta J 1991 Discrete version of the Painlevé equations *Phys. Rev. Lett.* **67** 1829–1832
[15] Ramis J-P, Sauloy J and Zhang C 2013 Local Analytic classification of q-difference equations *Astérisque* **355**
[16] Sakai H 2001 Rational surfaces associated with affine root systems and geometry of the Painlevé equations *Comm. Math. Phys.* **220** 165–229
[17] Watson G N 1910 The continuation of functions defined by generalized hypergeometric series *Trans. Cambridge Phil. Soc.* **21** 281–299
[18] Zhang C 2002 Une sommation discrete pour des equations aux q-differences lineaires et a coefficients analytiques: theorie generale et exemples *Differential equations and the Stokes phenomenon* (Singapore: World Sci. Publ.) 309–329

30th International Colloquium on Group Theoretical Methods in Physics (Group30) IOP Publishing
Journal of Physics: Conference Series **597** (2015) 012064 doi:10.1088/1742-6596/597/1/012064

Ladder operators for solvable potentials connected with exceptional orthogonal polynomials

C Quesne

Physique Nucléaire Théorique et Physique Mathématique, Université Libre de Bruxelles, Campus de la Plaine CP229, Boulevard du Triomphe, B-1050 Brussels, Belgium

E-mail: `cquesne@ulb.ac.be`

Abstract. Exceptional orthogonal polynomials constitute the main part of the bound-state wavefunctions of some solvable quantum potentials, which are rational extensions of well-known shape-invariant ones. The former potentials are most easily built from the latter by using higher-order supersymmetric quantum mechanics (SUSYQM) or Darboux method. They may in general belong to three different types (or a mixture of them): types I and II, which are strictly isospectral, and type III, for which k extra bound states are created below the starting potential spectrum. A well-known SUSYQM method enables one to construct ladder operators for the extended potentials by combining the supercharges with the ladder operators of the starting potential. The resulting ladder operators close a polynomial Heisenberg algebra (PHA) with the corresponding Hamiltonian. In the special case of type III extended potentials, for this PHA the k extra bound states form k singlets isolated from the higher excited states. Some alternative constructions of ladder operators are reviewed. Among them, there is one that combines the state-adding and state-deleting approaches to type III extended potentials (or so-called Darboux-Crum and Krein-Adler transformations) and mixes the k extra bound states with the higher excited states. This novel approach can be used for building integrals of motion for two-dimensional superintegrable systems constructed from rationally-extended potentials.

1. Introduction

In 2008, the introduction of exceptional orthogonal polynomials (EOP) by Gómez-Ullate, Kamran, and Milson [1, 2] has constituted a big advance in both mathematics and physics. They indeed form complete and orthogonal polynomial systems generalizing the classical orthogonal polynomials (COP) of Hermite, Laguerre, and Jacobi. In contrast with the latter, there are indeed some gaps in the sequence of their degrees, which does not impede them to form complete systems. Furthermore, they occur in bound-state wavefunctions of some supersymmetric partners of shape-invariant potentials, which are rational extensions of the latter [3].

Since then, a lot of work has been carried out in this area (see, e.g., [4, 5, 6, 7, 8, 9, 10, 11, 12, 13, 14, 15, 16, 17, 18, 19, 20, 21, 22, 23, 24, 25, 26, 27, 28, 29, 30]). It turned out that the easiest approach to construct solvable potentials connected with EOP is nth-order supersymmetric quantum mechanics (SUSYQM) or, equivalently, some extensions of the Darboux method [31]. In such an approach, one starts from n different seed solutions $\varphi_1, \varphi_2, \ldots, \varphi_n$ of a starting Hamiltonian $H^{(1)}$ and one gets the potential of the partner as $V^{(2)}(x) = V^{(1)}(x) - 2\frac{d^2}{dx^2}\mathcal{W}(\varphi_1, \varphi_2, \ldots, \varphi_n)$, provided the Wronskian $\mathcal{W}$ is nonsingular.

In general, there may occur three types of EOP (I, II, and III). For I and II, $H^{(2)}$ is strictly isospectral to $H^{(1)}$ and shape invariant. For III, there are some additional levels below the spectrum of $H^{(1)}$ and $H^{(2)}$ is not shape invariant. Note that for the harmonic oscillator, only type III EOP occur.

The purpose of the present communication is to review some recent works, made in collaboration with I Marquette, on the construction of ladder operators for the potentials connected with EOP [17, 32, 33, 34].

2. Ladder operators in SUSYQM

In nth-order SUSYQM, let us consider a set of two partner Hamiltonians

$$H^{(i)} = -\frac{d^2}{dx^2} + V^{(i)}(x), \qquad i = 1, 2, \tag{1}$$

and let us assume that $H^{(1)}$ has some ladder operators a and $a^\dagger$ of order k. These generate with $H^{(1)}$ a polynomial Heisenberg algebra (PHA) of order $k-1$ [35, 36, 37], whose commutation relations are given by

$$\begin{gathered}[H^{(1)}, a^\dagger] = \lambda a^\dagger, \qquad [H^{(1)}, a] = -\lambda a, \\ [a, a^\dagger] = P^{(1)}(H^{(1)} + \lambda) - P^{(1)}(H^{(1)}),\end{gathered} \tag{2}$$

where λ is some constant and $P^{(1)}(H^{(1)})$ is some kth-degree polynomial in $H^{(1)}$, $P^{(1)}(H^{(1)}) = \prod_{i=1}^k (H^{(1)} - \epsilon_i)$.

Since a and/or $a^\dagger$ may have zero modes, this PHA may have infinite-dimensional unitary irreducible representations (unirreps), as well as finite-dimensional ones (singlet, doublet, and more generally multiplet).

Well-known examples are those of the harmonic oscillator and of the radial harmonic oscillator. In the former case, a and $a^\dagger$ may be written as

$$a = \frac{d}{dx} + x, \qquad a^\dagger = -\frac{d}{dx} + x, \tag{3}$$

and we have $k = 1$, $\lambda = 2$ and $P^{(1)}(H^{(1)}) = H^{(1)} - 1$. The corresponding PHA then reduces to the Heisenberg algebra. In the latter case, we may write

$$\begin{aligned} a &= \frac{1}{4}\left(2\frac{d^2}{dx^2} + 2x\frac{d}{dx} + \frac{1}{2}x^2 - \frac{2\ell(\ell+1)}{x^2} + 1\right), \\ a^\dagger &= \frac{1}{4}\left(2\frac{d^2}{dx^2} - 2x\frac{d}{dx} + \frac{1}{2}x^2 - \frac{2\ell(\ell+1)}{x^2} - 1\right), \end{aligned} \tag{4}$$

where ℓ denotes the angular momentum quantum number. Now $k = 2$, $\lambda = 2$, $P^{(1)}(H^{(1)}) = \frac{1}{16}(2H^{(1)} - 3 - 2\ell)(2H^{(1)} - 1 + 2\ell)$ and the resulting PHA is the su(1,1) Lie algebra. In more general cases, the PHA turns out to be nonlinear.

In nth-order SUSYQM, the two Hamiltonians $H^{(1)}$ and $H^{(2)}$ intertwine with nth-order differential operators $\mathcal{A}$ and $\mathcal{A}^\dagger$ as $\mathcal{A}H^{(1)} = H^{(2)}\mathcal{A}$, $\mathcal{A}^\dagger H^{(2)} = H^{(1)}\mathcal{A}^\dagger$. The operators $\mathcal{A}$ and $\mathcal{A}^\dagger$, which may be written in terms of the n seed solutions $\varphi_1, \varphi_2, \ldots, \varphi_n$ of $H^{(1)}$, enable one to relate the energy spectra, the wavefunctions and the ladder operators of $H^{(1)}$ and $H^{(2)}$ [31, 36].

In particular, from the ladder operators a and $a^\dagger$ of $H^{(1)}$, one gets ladder operators b and $b^\dagger$ of $H^{(2)}$, which are $(2n+k)$)th-order differential operators given by $b = \mathcal{A}a\mathcal{A}^\dagger$ and $b^\dagger = \mathcal{A}a^\dagger\mathcal{A}^\dagger$, respectively [35, 36]. The new operators $H^{(2)}$, $b^\dagger$ and b satisfy another PHA, which is entirely determined by the first one and the SUSYQM transformation. Their commutation relations are indeed given by

$$\begin{aligned} &[H^{(2)}, b^\dagger] = \lambda b^\dagger, \qquad [H^{(2)}, b] = -\lambda b, \\ &[b, b^\dagger] = P^{(2)}(H^{(2)} + \lambda) - P^{(2)}(H^{(2)}), \end{aligned} \tag{5}$$

where $P^{(2)}(H^{(2)})$ is the $(2n+k)$th-degree polynomial in $H^{(2)}$ defined by

$$\begin{aligned} &P^{(2)}(H^{(2)}) = P^{(1)}(H^{(2)}) f(H^{(2)} - \lambda) f(H^{(2)}), \\ &f(H^{(2)}) = \mathcal{A}\mathcal{A}^\dagger. \end{aligned} \tag{6}$$

3. Ladder operators for EOP-related problems

In Ref. [32], we used the method presented in Sect. 2 to construct ladder operators for potentials related to (type III) Hermite and (type I, II, or III) Laguerre EOP.

For type I or II, corresponding to isospectrality, the properties of $(H^{(1)}, a, a^\dagger)$ are transferred to $(H^{(2)}, b, b^\dagger)$ and b, $b^\dagger$ act on all the wavefunctions of $H^{(2)}$. In contrast, for type III, although the properties of $(H^{(1)}, a, a^\dagger)$ are still transferred to $(H^{(2)}, b, b^\dagger)$, the operators b, $b^\dagger$ do not see the states that have been added below the spectrum of $H^{(1)}$. Hence, apart from some unirreps similar to those present for $H^{(1)}$, there are also n singlets corresponding to the added states.

When one constructs higher-dimensional superintegrable systems whose constituent Hamiltonians are connected with EOP families, ladder operators of these one-dimensional Hamiltonians are very useful to build integrals of the motion of order higher than two by using a well-known prescription [38]. However, in order to be able to algebraically derive the energy spectrum of the superintegrable systems, together with the degeneracies of their levels, from the representations of the polynomial algebra generated by the integrals of motion, it is necessary that such ladder operators act on all eigenstates of the constituent Hamiltonians. In Ref. [32], this condition was only satisfied for the type I or II case, but not for type III.

In a first attempt to deal with this problem [17], we proposed some new ladder operators, different from b and $b^\dagger$, for rational extensions of the harmonic oscillator, associated with type III Hermite EOP and constructed in second-order SUSYQM. It turned out, however, that the two added states below the spectrum of the oscillator formed a doublet and were not connected to higher excited states.

In a second attempt, we succeeded in solving the problem in the cases of rational extensions of the harmonic oscillator and of the radial harmonic oscillator, constructed in nth-order SUSYQM [33, 34]. The procedure used will be reviewed in the next section.

4. State adding versus state deleting in nth-order SUSYQM

In general, there are several possibilities for choosing the n different seed solutions $\varphi_1, \varphi_2, \ldots, \varphi_n$ of the starting Hamiltonian $H^{(1)}$ and these go back to the Darboux intertwining method [39].

Broadly speaking, there are two (equivalent) possibilities: that of adding a number of states below the spectrum of the starting Hamiltonian, which is the method alluded to above, and that of deleting a number of bound states in this spectrum. The former is referred to as the Darboux-Crum method and takes its origin in the Crum extension [40] of the Darboux method. In such a case, the seed functions are unphysical solutions of the Schrödinger equation for $H^{(1)}$ with energy less than the ground-state, which are converted into physical solutions of $H^{(2)}$. The latter is called the Krein-Adler method and comes from a generalization of the Crum method made independently by Krein [41] and Adler [42]. The seed functions are then chains of bound-state wavefunctions that may be lacunary with some even gaps. Note that in general the two n values are different in both approaches.

To distinguish both possibilities, we shall use different notations, as follows:

1) $(H^{(1)}, H^{(2)})$ with $H^{(i)} = -\frac{d^2}{dx^2} + V^{(i)}(x)$, $i = 1, 2$, in the case of Darboux-Crum method;

2) $(\bar{H}^{(1)}, \bar{H}^{(2)})$ with $\bar{H}^{(i)} = -\frac{d^2}{dx^2} + \bar{V}^{(i)}(x)$, $i = 1, 2$, in the case of Krein-Adler method.

We plan to show that in the cases of the harmonic oscillator and of the radial harmonic oscillator, it is possible to arrive at $\bar{V}^{(2)}(x) = V^{(2)}(x) +$ constant, thereby allowing the construction of ladder operators in a novel way [34].

5. Ladder operators for rational extensions of the harmonic oscillator

In this case, we start from the same harmonic oscillator Hamiltonian, $\bar{H}^{(1)} = H^{(1)}$, which means that $\bar{V}^{(1)}(x) = V^{(1)}(x) = x^2$, $-\infty < x < \infty$.

In the state-adding case, we take $n \to k$ and

$$\varphi_i(x) \to \phi_{m_i}(x) = \mathcal{H}_{m_i}(x)e^{x^2/2}, \qquad \mathcal{H}_{m_i}(x) = (-\mathrm{i})^{m_i} H_{m_i}(\mathrm{i}x), \qquad i = 1, 2, \ldots, k, \tag{7}$$

where $H_m(x)$ denotes a Hermite polynomial. The $\phi_{m_i}(x)$'s are eigenfunctions of $H^{(1)}$, corresponding to eigenvalues $E_{m_i} = -2m_i - 1$ below the ground state. For even m values, they are nodeless on the whole real line, while for odd m ones, they have a single zero at $x = 0$. With the choice $m_1 < m_2 < \cdots < m_k$ with m_i even (resp. odd) for i odd (resp. even), the partner $V^{(2)}(x)$ turns out to be nonsingular [36]. On using standard properties of Wronskians [43], one can then show that $V^{(2)}(x)$ becomes

$$V^{(2)}(x) = x^2 - 2k - 2\frac{d^2}{dx^2} \log \mathcal{W}(\mathcal{H}_{m_1}, \mathcal{H}_{m_2}, \ldots, \mathcal{H}_{m_k}) \tag{8}$$

with corresponding spectrum

$$E^{(2)}_\nu = 2\nu + 1, \qquad \nu = -m_k - 1, \ldots, -m_2 - 1, -m_1 - 1, 0, 1, 2, \ldots. \tag{9}$$

On the other hand, in the state-deleting case, we take $n \to m_k - k + 1$ and

$$(\varphi_1, \varphi_2, \ldots, \varphi_n) \to (\psi_1, \psi_2, \ldots, \check{\psi}_{m_k - m_{k-1}}, \ldots, \check{\psi}_{m_k - m_2}, \ldots, \check{\psi}_{m_k - m_1}, \ldots, \psi_{m_k}), \tag{10}$$

where ψ_ν denotes a bound-state wavefunction and $\check{\psi}_\nu$ means that ψ_ν is excluded from the list. The wavefunctions in (10) are then suppressed from the spectrum. Standard properties of Wronskians lead again to

$$\bar{V}^{(2)}(x) = x^2 + 2(m_k + 1 - k) - 2\frac{d^2}{dx^2} \log \mathcal{W}(H_1, H_2, \ldots, \check{H}_{m_k - m_{k-1}}, \ldots, \check{H}_{m_k - m_1}, \ldots, H_{m_k}), \tag{11}$$

which is nonsingular provided the gaps between the surviving levels with $\nu = 0, m_k - m_{k-1}, \ldots, m_k - m_2, m_k - m_1, m_k + 1, m_k + 2, \ldots$, correspond to even numbers of consecutive levels [41, 42]. This leads to the same conditions on m_1, m_2, $\ldots$, m_k as in the state-adding case. After redefining ν, one gets the spectrum

$$\bar{E}^{(2)}_\nu = 2m_k + 2\nu + 3, \qquad \nu = -m_k - 1, \ldots, -m_2 - 1, -m_1 - 1, 0, 1, 2, \ldots. \tag{12}$$

As it has been proved that the two Wronskians in (8) and (11) only differ by some multiplicative constant [22], the two obtained potentials $V^{(2)}(x)$ and $\bar{V}^{(2)}(x)$ are the same up to some additive constant,

$$\bar{V}^{(2)}(x) = V^{(2)}(x) + 2m_k + 2. \tag{13}$$

This result agrees with the corresponding energies (9) and (12).

SUSYQM intertwining relations show that one can go from $H^{(2)}$ to $H^{(2)}+2m_k+2$ along the following path

$$H^{(2)} \xrightarrow{A^\dagger} H^{(1)} = \bar{H}^{(1)} \xrightarrow{\bar{A}} \bar{H}^{(2)} = H^{(2)} + 2m_k + 2 \quad (\text{with } H^{(2)} \xrightarrow{c} \bar{H}^{(2)}) \tag{14}$$

The (m_k+1)th-order differential operator

$$c = \bar{A}A^\dagger \tag{15}$$

that performs such a transformation is therefore a lowering operator for $H^{(2)}$.

Together with its Hermitian conjugate and $H^{(2)}$, it satisfies a PHA of m_kth order, defined by the commutation relations

$$\begin{aligned}
&[H^{(2)}, c^\dagger] = (2m_k+2)c^\dagger, \qquad [H^{(2)}, c] = -(2m_k+2)c, \\
&[c, c^\dagger] = Q(H^{(2)}+2m_k+2) - Q(H^{(2)}), \\
&Q(H^{(2)}) = \left(\prod_{i=1}^{k}(H^{(2)}+2m_i+1)\right)\left(\prod_{\substack{j=1 \\ j\neq m_k-m_{k-1},\ldots,m_k-m_1}}^{m_k}(H^{(2)}-2j-1)\right).
\end{aligned} \tag{16}$$

As explained in [34], the explicit form of $Q(H^{(2)})$, given in (16), can be obtained from standard properties of SUSYQM.

From the action of $Q(H^{(2)}) = c^\dagger c$ on $\psi_\nu^{(2)}(x)$ obtained by replacing $H^{(2)}$ by $E_\nu^{(2)}$, that of c and $c^\dagger$ on the same can be easily derived. By choosing the normalization constants of the $\psi_\nu^{(2)}$'s in such a way that all matrix elements of the ladder operators $c^\dagger$, c are nonnegative, one can write the results for c as

$$\begin{aligned}
c\psi_\nu^{(2)} = 0, \qquad & \nu = -m_k-1,\ldots,-m_1-1,1,2,\ldots,m_k-m_{k-1}-1, \\
& m_k-m_{k-1}+1,\ldots,m_k-m_1-1,m_k-m_1+1,\ldots,m_k,
\end{aligned} \tag{17}$$

$$c\psi_0^{(2)} = \left[2^{m_k+1}(m_k+1)!\left(\prod_{i=1}^{k-1}\frac{m_i+1}{m_k-m_i}\right)\right]^{1/2}\psi_{-m_k-1}^{(2)}, \tag{18}$$

$$\begin{aligned}
c\psi_{m_k-m_i}^{(2)} = \Bigg[&2^{m_k+1}(m_k+1)(2m_k-m_i+1)(m_k-m_i-1)!m_i! \\
&\times\left(\prod_{j=1}^{i-1}\frac{m_k+m_j-m_i+1}{m_i-m_j}\right)\left(\prod_{l=i+1}^{k-1}\frac{m_k+m_l-m_i+1}{m_l-m_i}\right)\Bigg]^{1/2}\psi_{-m_i-1}^{(2)}, \\
& i = 1,2,\ldots,k-1,
\end{aligned} \tag{19}$$

$$\begin{aligned}
c\psi_\nu^{(2)} = &\left[2^{m_k+1}(\nu+m_k+1)\frac{(\nu-1)!}{(\nu-m_k-1)!}\left(\prod_{i=1}^{k-1}\frac{\nu+m_i+1}{\nu+m_i-m_k}\right)\right]^{1/2}\psi_{\nu-m_k-1}^{(2)}, \\
&\nu = m_k+1, m_k+2,\ldots,
\end{aligned} \tag{20}$$

those for $c^\dagger$ being deduced by using Hermitian conjugation. This proves that the above PHA has m_k+1 infinite-dimensional unirreps, whose lowest-weight states are $\psi_i^{(2)}(x)$ with $i=-m_k-1$, $\ldots$, $-m_1-1$, 1, 2, $\ldots$, $m_k-m_{k-1}-1$, $m_k-m_{k-1}+1$, $\ldots$, m_k-m_1-1, m_k-m_1+1, $\ldots$, m_k, respectively. Hence, the k lowest states are mixed with the higher ones.

Journal of Physics: Conference Series **597** (2015) 012064 doi:10.1088/1742-6596/597/1/012064

6. Ladder operators for rational extensions of the radial harmonic oscillator

The radial harmonic oscillator case is more complicated because SUSYQM changes the ℓ value in the potential $V_\ell(x) = \frac{1}{4}x^2 + \frac{\ell(\ell+1)}{x^2}$, $0 < x < \infty$. We sketch it below and refer the reader to [34] for more details.

In the state-adding case, we take $n \to k$, $V^{(1)}(x) = V_{\ell+k}(x)$, and

$$\varphi_i(x) \to \phi_{m_i}^{(\ell+k)}(x) \propto x^{-\ell-k} e^{\frac{1}{4}x^2} L_{m_i}^{(-\ell-k-\frac{1}{2})}\left(-\frac{1}{2}x^2\right), \qquad i = 1, 2, \ldots, k, \tag{21}$$

where $L_m^{(-\alpha-k)}(-z)$, with $z = \frac{1}{2}x^2$ and $\alpha = \ell + \frac{1}{2}$, denotes a Laguerre polynomial, $m_1 < m_2 < \cdots < m_k$, m_i is even (resp. odd) for i odd (resp. even), and $\alpha + k > m_k$. Then it turns out that $V^{(2)}(x)$ is a rationally-extended $V_\ell(x)$ potential,

$$V^{(2)}(x) = V_\ell(x) - k - 2\frac{d^2}{dx^2} \log \tilde{\mathcal{W}}\big(L_{m_1}^{(-\alpha-k)}(-z), L_{m_2}^{(-\alpha-k)}(-z), \ldots, L_{m_k}^{(-\alpha-k)}(-z)\big), \tag{22}$$

where $\tilde{\mathcal{W}}\big(f_1(z), f_2(z), \ldots, f_k(z)\big)$ denotes the Wronskian of the functions $f_1(z), f_2(z), \ldots, f_k(z)$ with respect to z, and that

$$E_{\ell,\nu}^{(2)} = 2\nu + \ell + k + \frac{3}{2}, \qquad \nu = -m_k - 1, \ldots, -m_2 - 1, -m_1 - 1, 0, 1, 2, \ldots. \tag{23}$$

In the state-deleting case, we take $n \to m_k + 1 - k$, $\bar{V}^{(1)}(x) = V_{\ell+k-m_k-1}(x)$, where $\alpha + k$ is assumed greater than $m_k + 1$. Then, with the choice

$$\begin{aligned}(\varphi_1, \varphi_2, \ldots, \varphi_n) \to \big(&\psi_1^{(\ell+k-m_k-1)}, \psi_2^{(\ell+k-m_k-1)}, \ldots, \check{\psi}_{m_k-m_{k-1}}^{(\ell+k-m_k-1)}, \ldots, \check{\psi}_{m_k-m_2}^{(\ell+k-m_k-1)}, \ldots, \\ &\check{\psi}_{m_k-m_1}^{(\ell+k-m_k-1)}, \ldots, \psi_{m_k}^{(\ell+k-m_k-1)}\big),\end{aligned} \tag{24}$$

we obtain that $\bar{V}^{(2)}(x)$ is a rationally-extended $V_\ell(x)$,

$$\begin{aligned}\bar{V}^{(2)}(x) = V_\ell(x) + m_k + 1 - k - 2\frac{d^2}{dx^2} \log \tilde{\mathcal{W}}\big(&L_1^{(\alpha+k-m_k-1)}(z), L_2^{(\alpha+k-m_k-1)}(z), \ldots, \\ &\check{L}_{m_k-m_{k-1}}^{(\alpha+k-m_k-1)}(z), \ldots, \check{L}_{m_k-m_1}^{(\alpha+k-m_k-1)}(z), \ldots, L_{m_k}^{(\alpha+k-m_k-1)}(z)\big),\end{aligned} \tag{25}$$

and

$$\bar{E}_{\ell,\nu}^{(2)} = 2\nu + \ell + k + m_k + \frac{5}{2}, \qquad \nu = -m_k - 1, \ldots, -m_2 - 1, -m_1 - 1, 0, 1, 2, \ldots. \tag{26}$$

Here, $V^{(2)}(x)$ and $\bar{V}^{(2)}(x)$ differ by an additive constant,

$$\bar{V}^{(2)}(x) = V^{(2)}(x) + m_k + 1, \tag{27}$$

again.

Since $\bar{H}^{(1)} \neq H^{(1)}$, we need a third SUSYQM transformation relating them to be able to construct ladder operators. As shown in Ref. [34], such a transformation is of $(m_k + 1)$th order and the corresponding supercharge operators $\tilde{\mathcal{A}}$ and $\tilde{\mathcal{A}}^\dagger$ can be written in terms of $m_k + 1$ seed solutions of class II, i.e., $n \to m_k + 1$ and

$$\varphi_i(x) \to \tilde{\phi}_{i-1}^{(\ell+k)}(x), \quad \tilde{\phi}_i^{(\ell+k)}(x) \propto x^{-\ell-k} e^{-\frac{1}{4}x^2} L_i^{(-\ell-k-\frac{1}{2})}\left(\frac{1}{2}x^2\right). \tag{28}$$

The corresponding partner potentials now read

$$\begin{aligned}\tilde{V}^{(1)}(x) &= V^{(1)}(x) = V_{\ell+k}(x),\\ \tilde{V}^{(2)}(x) &= \tilde{V}^{(1)}(x) - 2\frac{d^2}{dx^2}\log \mathcal{W}\big(\tilde{\phi}_0^{(\ell+k)}(x), \tilde{\phi}_1^{(\ell+k)}(x), \ldots, \tilde{\phi}_{m_k}^{(\ell+k)}(x)\big),\end{aligned} \tag{29}$$

where the latter is nonsingular provided $\ell + k + 1/2 > m_k$. The relation $\mathcal{W}\big(\tilde{\phi}_0^{(\ell+k)}, \tilde{\phi}_1^{(\ell+k)}, \ldots, \tilde{\phi}_{m_k}^{(\ell+k)}\big) \propto \left(x^{-\ell-k}e^{-x^2/4}\right)^{m_k+1} x^{m_k(m_k+1)/2}$, resulting from standard properties of Wronskians [43], leads to

$$\tilde{V}^{(2)}(x) = V_{\ell+k-m_k-1}(x) + m_k + 1 = \bar{V}^{(1)}(x) + m_k + 1, \tag{30}$$

which establishes the desired relation between $V^{(1)}(x)$ and $\bar{V}^{(1)}(x)$. Since type II seed functions lead to isospectral transformations, the corresponding energies satisfy the relation

$$\tilde{E}^{(2)}_{\ell+k-m_k-1,\nu} = \tilde{E}^{(1)}_{\ell+k,\nu} = E^{(1)}_{\ell+k,\nu} = 2\nu + \ell + k + \frac{3}{2}, \qquad \nu = 0, 1, 2, \ldots. \tag{31}$$

We conclude that this third transformation allows us to go from $\tilde{H}^{(1)} = H^{(1)}$ to $\tilde{H}^{(2)} = \bar{H}^{(1)} + m_k + 1$.

We can now go from $H^{(2)}$ to $H^{(2)} + 2m_k + 2$ along the following path:

$$H^{(2)} \xrightarrow{A^\dagger} H^{(1)} = \tilde{H}^{(1)} \xrightarrow{\tilde{A}} \tilde{H}^{(2)} = \bar{H}^{(1)} + m_k + 1 \xrightarrow{\bar{A}} \bar{H}^{(2)} + m_k + 1 = H^{(2)} + 2m_k + 2 \tag{32}$$

(with an arrow labelled c going directly from $H^{(2)}$ to $\bar{H}^{(2)} + m_k + 1$)

which shows that the $(2m_k + 2)$th-order operator

$$c = \bar{A}\tilde{A}A^\dagger \tag{33}$$

is a lowering operator for $H^{(2)}$.

Together with its Hermitian conjugate and $H^{(2)}$, it satisfies a PHA of $(2m_k + 1)$th order, similar to (16) except that $Q(H^{(2)})$ is now given by

$$\begin{aligned}Q(H^{(2)}) = &\left(\prod_{i=1}^{k}(H^{(2)} - \alpha + 2m_i - k + 1)\right)\left(\prod_{j=0}^{m_k}(H^{(2)} + \alpha - 2j + k - 1)\right)\\ &\times\left(\prod_{\substack{n=1\\ n\neq m_k-m_{k-1},\ldots,m_k-m_1}}^{m_k}(H^{(2)} - \alpha - 2n - k - 1)\right).\end{aligned} \tag{34}$$

The action of c and $c^\dagger$ on the wavefunctions of $H^{(2)}$ is not substantially different from that obtained in the harmonic oscillator case, so that the present PHA has m_k+1 infinite-dimensional unirreps too. Hence, the k lowest states are mixed with the higher ones again.

7. Conclusion

In the present contribution, it has been shown that for the multi-step extensions of the harmonic oscillator and of the radial harmonic oscillator connected with type III EOP, it is possible to build new ladder operators c, $c^\dagger$ that satisfy a PHA with only infinite-dimensional unirreps. These new ladder operators differ from the usual ones b, $b^\dagger$, well known in SUSYQM. This has been made possible by combining the state-adding (or Darboux-Crum) and state-deleting (or Krein-Adler) approaches to the construction of these extensions. Such new ladder operators are very useful to construct integrals of motion for superintegrable systems based on one-dimensional multi-step extensions connected with type III EOP.

An interesting open question is whether such a construction is possible for other types of potentials than the harmonic oscillator and the radial harmonic oscillator, such as potentials connected with Jacobi type III EOP.

References

[1] Gómez-Ullate D, Kamran N and Milson R 2010 *J. Approx. Theor.* **162** 987 (*Preprint* arXiv:0805.3376)
[2] Gómez-Ullate D, Kamran N and Milson R 2009 *J. Math. Anal. Appl.* **359** 352 (*Preprint* arXiv:0807.3939)
[3] Quesne C 2008 *J. Phys. A: Math. Theor.* **41** 392001 (*Preprint* arXiv:0807.4087)
[4] Gómez-Ullate D, Kamran N and Milson R 2010 *J. Phys. A: Math. Theor.* **43** 434016 (*Preprint* arXiv:1002.2666)
[5] Gómez-Ullate D, Kamran N and Milson R 2012 *J. Math. Anal. Appl.* **387** 410 (*Preprint* arXiv:1103.5724)
[6] Gómez-Ullate D, Kamran N and Milson R 2012 *Contemp. Math.* **563** 51 (*Preprint* arXiv:1101.5584)
[7] Gómez-Ullate D, Kamran N and Milson R 2013 *Found. Comput. Math.* **13** 615 (*Preprint* arXiv:1203.6857)
[8] Gómez-Ullate D, Grandati Y and Milson R 2014 *J. Phys. A: Math. Theor.* **47** 015203 (*Preprint* arXiv:1306.5143)
[9] Gómez-Ullate D, Grandati Y and Milson R 2014 *J. Math. Phys.* **55** 043510 (*Preprint* arXiv:1309.3756)
[10] Fellows J M and Smith R A 2009 *J. Phys. A: Math. Theor.* **42** 335303
[11] Bagchi B, Quesne C and Roychoudhury R 2009 *Pramana J. Phys.* **73** 337 (*Preprint* arXiv:0812.1488)
[12] Quesne C 2009 *SIGMA* **5** 084 (*Preprint* arXiv:0906.2331)
[13] Quesne C 2011 *Mod. Phys. Lett.* A **26** 1843 (*Preprint* arXiv:1106.1990)
[14] Quesne C 2011 *Int. J. Mod. Phys.* A **26** 5337 (*Preprint* arXiv:1110.3958)
[15] Quesne C 2012 *Int. J. Mod. Phys.* A **27** 1250073 (*Preprint* arXiv:1203.1812)
[16] Quesne C 2012 *SIGMA* **8** 080 (*Preprint* arXiv:1208.6165)
[17] Marquette I and Quesne C 2013 *J. Phys. A: Math. Theor.* **46** 155201 (*Preprint* arXiv:1212.3474)
[18] Odake S and Sasaki R 2009 *Phys. Lett.* B **679** 414 (*Preprint* arXiv:0906.0142)
[19] Odake S and Sasaki R 2010 *Phys. Lett.* B **684** 173 (*Preprint* arXiv:0911.3442)
[20] Sasaki R, Tsujimoto S and Zhedanov A 2010 *J. Phys. A: Math. Theor.* **43** 315204 (*Preprint* arXiv:1004.4711)
[21] Odake S and Sasaki R 2011 *Phys. Lett.* B **702** 164 (*Preprint* arXiv:1105.0508)
[22] Odake S and Sasaki R 2013 *J. Phys. A: Math. Theor.* **46** 245201 (*Preprint* arXiv:1212.6595)
[23] Odake S and Sasaki R 2013 *J. Phys. A: Math. Theor.* **46** 235205 (*Preprint* arXiv:1301.3980)
[24] Grandati Y 2011 *Ann. Phys. (NY)* **326** 2074 (*Preprint* arXiv:1101.0055)
[25] Grandati Y 2011 *J. Math. Phys.* **52** 103502 (*Preprint* arXiv:1103.5023)
[26] Grandati Y 2012 *Ann. Phys. (NY)* **327** 2411 (*Preprint* arXiv:1108.4503)
[27] Grandati Y 2012 *Phys. Lett.* A **376** 2866 (*Preprint* arXiv:1203.4149)
[28] Grandati Y and Quesne C 2013 *J. Math. Phys.* **54** 073512 (*Preprint* arXiv:1211.5308)
[29] Ho C-L 2011 *Prog. Theor. Phys.* **126** 185 (*Preprint* arXiv:1104.3511)
[30] Ho C-L 2011 *J. Math. Phys.* **52** 122107 (*Preprint* arXiv:1105.3670)
[31] Andrianov A A and Ioffe M V 2012 *J. Phys. A: Math. Theor.* **45** 503001 (*Preprint* arXiv:1207.6799)
[32] Marquette I and Quesne C 2013 *J. Math. Phys.* **54** 042102 (*Preprint* arXiv:1211.2957)
[33] Marquette I and Quesne C 2013 *J. Math. Phys.* **54** 102102 (*Preprint* arXiv:1303.7150)
[34] Marquette I and Quesne C 2014 *J. Math. Phys.* **55** 112103 (*Preprint* arXiv:1402.6380)
[35] Fernández C D J and Hussin V 1999 *J. Phys. A: Math. Gen.* **32** 3603
[36] Fernández C D J and Fernández-García N 2005 *AIP Conf. Proc.* vol 744 (Melville, NY: Amer. Inst. Phys.) p 236 (*Preprint* quant-ph/0502098)
[37] Carballo J M, Fernández C D J, Negro J and Nieto L M 2004 *J. Phys. A: Math. Gen.* **37** 10349
[38] Marquette I 2010 *J. Phys. A: Math. Theor.* **43** 135203 (*Preprint* arXiv:0908.4432)
[39] Darboux G 1882 *Compt. Rend. Acad. Sci.* **94** 1456
[40] Crum M M 1955 *Q. J. Math. Oxford Ser. 2* **6** 121
[41] Krein M G 1957 *Dokl. Akad. Nauk SSSR* **113** 970
[42] Adler V É 1994 *Theor. Math. Phys.* **101** 1381
[43] Muir T 1960 *A Treatise on the Theory of Determinants* (New York: Dover) (revised and enlarged by Metzler W H)

Journal of Physics: Conference Series **597** (2015) 012065 doi:10.1088/1742-6596/597/1/012065

The Verlinde formula in logarithmic CFT

David Ridout[1,2] **and Simon Wood**[1,2]

[1] Department of Theoretical Physics, Research School of Physics and Engineering, Australian National University, Acton, ACT 2600, Australia

[2] Mathematical Sciences Institute, Australian National University, Acton, ACT 2600, Australia

E-mail: `david.ridout@anu.edu.au`, `simon.wood@anu.edu.au`

Abstract. In rational conformal field theory, the Verlinde formula computes the fusion coefficients from the modular S-transformations of the characters of the chiral algebra's representations. Generalising this formula to logarithmic models has proven rather difficult for a variety of reasons. Here, a recently proposed formalism [1] for the modular properties of certain classes of logarithmic theories is reviewed, and refined, using simple examples. A formalism addressing fusion rules in simple current extensions is also reviewed as a means to tackle logarithmic theories to which the proposed modular formalism does not directly apply.

1. Introduction

Logarithmic conformal field theory typically refers to a two-dimensional conformally invariant quantum field theory for which there exist correlation functions exhibiting logarithmic singularities. The first study of a logarithmic conformal field theory appears to be that of Rozansky and Saleur [2] where the model under consideration possesses symmetries that include the affine Kac-Moody superalgebra $\widehat{\mathfrak{gl}}(1|1)$ of central charge $c = 0$. Shortly thereafter, Gurarie noted [3] the appearance of logarithmic singularities in the $c = -2$ fermionic ghost system and showed that a sufficient condition for their presence was the non-diagonalisability of the hamiltonian $\mathbf{H} = L_0 + \overline{L}_0$. (He also seems to be responsible for the appellation "logarithmic conformal field theory".) We refer to [4] for a recent collection of reviews on this topic.

As with rational conformal field theory, much of the analysis of logarithmic theories begins with identifying the physically relevant spectrum of representations of the chiral symmetry algebra (vertex operator algebra). Unlike the rational case, where the representations are all completely reducible, the spectrum of a logarithmic theory necessarily includes representations which are reducible, yet indecomposable.[1] In particular, the eigenvectors of $\mathbf{H}$ always span a subrepresentation which will be proper when $\mathbf{H}$ has non-trivial Jordan blocks. Whilst the failure of complete reducibility does not by itself guarantee Jordan blocks, hence logarithmic singularities, it seems that all the known conformal field theories whose spectrum includes reducible but indecomposable representations are logarithmic. Moreover, being logarithmic seems to be a generic property for conformal field theories, rather than a pathological one.

It is important to note, however, that logarithmic conformal field theories are necessarily non-unitary. This is not a concern for applications to statistical mechanics where the unitarity

[1] We recall that a representation is said to be indecomposable if it cannot be written as the direct sum of two non-trivial representations.

of a given lattice model is of little importance to the study of its phase transitions. Examples of lattice models whose scaling limits are known, or are believed, to be described by logarithmic conformal field theories include:

- Critical dense polymers [5], where the chiral symmetry is Virasoro (or symplectic fermions [6]) with $c = -2$.
- Critical percolation [7, 8, 9, 10], where the symmetry is Virasoro with $c = 0$.
- Abelian sandpiles [11, 12], where the symmetry is again Virasoro with $c = -2$, but the theory appears to differ from critical dense polymers.
- The $\mathfrak{gl}(1|1)$ spin chain [7, 13, 14], where the symmetry is symplectic fermions with $c = -2$.

Non-unitarity is, of course, a concern for string-theoretic applications, but here we note that it is not the underlying conformal field theory, but only what survives the BRST quantisation (gauge-fixing) that needs to be unitary. Logarithmic theories relevant to string theory include:

- Fermionic (bc) ghosts or symplectic fermions [15, 16], where the chiral symmetry algebra is the affine Kac-Moody superalgebra $\widehat{\mathfrak{psl}}(1|1)$ with $c = -2$.
- Bosonic ($\beta\gamma$) ghosts of central charge $c = -1$ [17, 18] and $c = 2$ [19].
- Fractional level Wess-Zumino-Witten models with $\widehat{\mathfrak{sl}}(2)$ [20, 17, 21, 18] and $\widehat{\mathfrak{sl}}(3)$ [22] symmetry.
- Wess-Zumino-Witten models on Lie supergroups [2, 23, 24, 25, 26].

There are also logarithmic conformal field theories that have been extensively studied for their own sake, in particular the $(1, p)$ triplet models [27, 28, 29, 30, 31, 32, 33, 34] (see [35] for a proposed lattice-theoretic realisation) and their (p', p) generalisations [36, 37, 38, 39, 40, 41, 42, 43], also known as the (W-extended) logarithmic minimal models. Recently, these ideas have been extended to $\widehat{\mathfrak{sl}}(2)$ triplets [44].

The mathematical formalism of rational conformal field theory is very rich and it is natural to try to extend the key insights to the logarithmic case. In particular, the determination of the fusion rules in rational theories is often significantly simplified through the use of the Verlinde formula [45]. This relies on computing the modular S-transformations of the characters of the irreducible representations. Unfortunately, it has been known for quite some time now [46] that the irreducible characters of a logarithmic conformal field theory need not carry a representation of the modular group, an observation which appears to completely invalidate the application of any Verlinde-type formula.

An example illustrating this dilemma is the $(1, 2)$ triplet model [27, 28] whose central charge is $c = -2$. In this model, there are four irreducible representations, $\mathbb{L}_0$, $\mathbb{L}_1$, $\mathbb{L}_{-1/8}$ and $\mathbb{L}_{3/8}$, labelled by the conformal dimensions of their highest weight states. The corresponding characters are

$$\begin{aligned} \operatorname{ch}[\mathbb{L}_0](\tau) &= \frac{1}{2}\left(\frac{\vartheta_{1,2}(0;\tau)}{\eta(\tau)} + \eta(\tau)^2\right), & \operatorname{ch}[\mathbb{L}_{-1/8}](\tau) &= \frac{\vartheta_{0,2}(0;\tau)}{\eta(\tau)}, \\ \operatorname{ch}[\mathbb{L}_1](\tau) &= \frac{1}{2}\left(\frac{\vartheta_{1,2}(0;\tau)}{\eta(\tau)} - \eta(\tau)^2\right), & \operatorname{ch}[\mathbb{L}_{3/8}](\tau) &= \frac{\vartheta_{2,2}(0;\tau)}{\eta(\tau)}. \end{aligned} \tag{1.1}$$

One quickly checks that it is the factors of $\eta(\tau)^2$ that lead to τ-dependent coefficients and spoil the modularity. For example,

$$\operatorname{ch}[\mathbb{L}_0](-1/\tau) = -\frac{\mathrm{i}\tau}{2}\left(\operatorname{ch}[\mathbb{L}_0](\tau) - \operatorname{ch}[\mathbb{L}_1](\tau)\right) + \frac{1}{4}\left(\operatorname{ch}[\mathbb{L}_{-1/8}](\tau) - \operatorname{ch}[\mathbb{L}_{3/8}](\tau)\right). \tag{1.2}$$

This can be alleviated by extending the space of characters to include a new object

$$\operatorname{tor}(\tau) = -\mathrm{i}\tau\left(\operatorname{ch}[\mathbb{L}_0](\tau) - \operatorname{ch}[\mathbb{L}_1](\tau)\right) \tag{1.3}$$

and checking that the modular S- and T-transformations preserve the space spanned by the irreducible characters and tor.

Unsurprisingly, there is a fair bit of literature addressing the issue of modularity for logarithmic conformal field theories. In particular, Miyamoto showed [47] that one is only guaranteed to have an action of the modular group on the space of torus amplitudes (one-point functions on the torus). Characters are examples of these amplitudes, as is the function tor of (1.3), so Miyamoto's result meshes nicely with the extended character space of the $(1,2)$ triplet model. However, the proof requires generalising the notion of torus amplitude (and hence character) to insert arbitrary elements of the chiral symmetry algebra into the trace. While no additional elements were needed in the case of the $(1,2)$ triplet model, it is not clear whether this will be true more generally. If not, then evaluating these refined characters and torus amplitudes leads to severe computational difficulties in all but the simplest examples. It also leaves one wondering how such a result helps in determining modular invariant bulk partition functions (which are, after all, sesquilinear combinations of characters).

Finding modular invariants therefore seems to require new insight [48], but one can instead ask a different question. Given a logarithmic conformal field theory, is there a variant of the Verlinde formula which can be used to compute the fusion rules efficiently? The technical answer to this question is "no", as long as we assume that the input to the Verlinde formula is the S-transformation of characters, because characters cannot distinguish between a representation and the direct sum of its irreducible constituents (composition factors). Instead, the best that we can hope for is a Verlinde-type formula that determines the *Grothendieck* fusion rules that are obtained from the genuine fusion rules by identifying representations with the same character.[2]

Is there a Verlinde formula that computes the Grothendieck fusion rules? For the $(1,p)$ triplet models, the answer is "yes" [49]. Here, one computes the modular S-matrix S on the basis of irreducible characters directly, so that the entries are τ-dependent in general. But then, the action is twisted by a carefully chosen τ-dependent automorphy factor A such that the product $\mathsf{S}' = \mathsf{AS}$ is both independent of τ and, with T, defines a representation of the modular group on character space. Remarkably, this matrix S' successfully block-diagonalises the Grothendieck fusion rules in a Verlinde-type formula given in [49, p717]. Most importantly, the entries of the block-diagonal matrix are also expressed in terms of the entries of S'.

Admittedly, this formula is rather more complicated than the Verlinde formula of rational conformal field theory. However, this is to be expected from the non-diagonalisability of the fusion matrices in the logarithmic case. But, there seems to be little understanding of why twisting by this automorphy factor A is required, whether there are other possibilities for A, and whether, for more general models, the existence of such a factor is always guaranteed. As far as we are aware, this type of Verlinde formula has not been generalised beyond the $(1,p)$ triplet models. Moreover, there are now several other proposals [50, 30, 51, 52] addressing seemingly different Verlinde formulae for these theories.

Rather than try to compare and contrast these proposals, which would be a worthy goal in itself, we shall instead turn to a rather different formalism which has recently been introduced [1] for modular properties and Verlinde formulae in a more general class of logarithmic conformal field theories. One difference between this proposal and those mentioned above is that it does not directly apply to the triplet models. However, this is not a serious restriction because the triplet models have close relatives to which the new formalism does apply. We shall explain how this allows us to compute Grothendieck fusion rules for the triplet models in Section 3.

[2] There is a subtle mathematical question that we are overlooking here of whether the notion of Grothendieck fusion is well-defined. See [38] for an example where it is not.

2. Modularity for logarithmic CFT

The formalism discussed in this section arises from the observation [25, 1] that the majority of the logarithmic conformal field theories that have been adequately analysed to date all share certain key features. Moreover, the known exceptions may all be understood as simple current extensions of theories with the desired features. We will therefore describe these features before explaining, with the aid of examples, how they allow one to straightforwardly compute modular transformations and apply the Verlinde formula to obtain Grothendieck fusion coefficients.

The formalism centres [1] about the identification of a collection of representations of the chiral algebra (vertex operator algebra) that we call the *standard modules*. These standard modules and their characters should have the following properties:

(1) The characters of the standard modules are parametrised by a measurable space (M, μ).

(2) The irreducible standard modules are called *typical*. The complement of the subset of M corresponding to the typical characters has measure zero with respect to μ.

(3) The only indecomposable module in the spectrum having a typical module (irreducible standard) as a composition factor (irreducible subquotient) is that typical module.[3]

(4) The characters of the standard modules form a (topological) basis for the space of all characters. In particular, the standard module characters are linearly independent.

(5) The characters of the standard modules span a representation of the modular group $\mathsf{SL}(2;\mathbb{Z})$. In the standard basis, the S-transformation defines a symmetric, unitary operator whose square may be identified with conjugation.

(6) Combining the S-transformation with the standard Verlinde formula leads to the Grothendieck fusion rules (with non-negative integer coefficients).

Any module in the spectrum that is not a direct sum of typical modules is called *atypical*, the nomenclature here being inspired by that of Kac for Lie superalgebra representations. We remark that there will usually be more than one inequivalent standard module for each atypical parameter $m \in M$, but each of these atypical standard modules will have the same character.

We will not give any general means for identifying the standard modules in a given theory beyond the obvious need to satisfy the above requirements; in examples, this identification is often clear. In this vein, we mention that verifying (3) rigorously is almost always too difficult at present, so we usually have to content ourselves with conjecturing that it holds. We also remark that (4) is generally seen to be satisfied by the existence of resolutions for the irreducible atypical modules in terms of atypical standard modules. The Euler-Poincaré principle then expresses the character of each irreducible as an (infinite) alternating sum of standard characters.

The main observation that we want to emphasise is that the typical modules behave, for all intents and purposes, as if they were in a rational conformal field theory. Indeed, rational theories fit into the above formalism as the special case in which all the standard modules are typical. The logarithmic nature of the theory is therefore concentrated in the atypical modules and these are required to form a set of measure zero in character-space. This means that, heuristically at least, any potential problems that the logarithmic structure would create are measure-theoretically void. While there is an open question as to how one identifies the physically relevant measure in general, we will see that in examples, the natural choice works.

Speaking of which, the simplest (non-rational) example illustrating this formalism is the (uncompactified) free boson. However, as this model is not logarithmic, we will refer the reader to [1, Sec. 1.2] and instead turn to a closely related logarithmic conformal field theory.

[3] Another way of stating this would be to say that typical modules are both projective and injective in the physically relevant category of modules. However, this is extremely difficult to check in practice, partly because it is not clear how one actually identifies the physically relevant category.

Example 1 (The singlet model [1, 53]). The fermionic bc ghost system at $c = -2$ contains the symplectic fermions algebra as the subalgebra generated by b and ∂c. This subalgebra in turn contains a subalgebra corresponding to the states of zero ghost number. This is the $(1, 2)$ singlet algebra and it is generated by the energy-momentum tensor and a single Virasoro primary W of conformal dimension 3.

(1) The singlet algebra possesses a continuous family of highest weight modules $\mathcal{F}_\lambda$, with $\lambda \in \mathbb{R}$. These are the standard modules. Their characters count the eigenvalues of L_0, but these do not distinguish conjugate modules. We therefore augment them by counting the eigenvalues of a shifted ghost number, the zero mode of $G = \,:bc:\, - \frac{1}{2}$, and the central element $\mathbf{1}$:

$$\mathrm{ch}\big[\mathcal{F}_\lambda\big](y;z;q) = \mathrm{tr}_{\mathcal{F}_\lambda} y^{\mathbf{1}} z^{G_0} q^{L_0 + 1/12} = \frac{y z^{\lambda - 1/2} q^{(\lambda - 1/2)^2/2}}{\eta(q)}. \tag{2.1}$$

(2) The standard modules are irreducible if and only if $\lambda \notin \mathbb{Z}$. We therefore equip $\mathbb{R}$ with Lebesgue measure μ, noting that $\mu(\mathbb{Z}) = 0$.

(3) We cannot prove that the typicals are projective and injective, but it seems to be true.

(4) The standard characters (2.1) are clearly linearly independent. The atypical standard modules $\mathcal{F}_\lambda$, with $\lambda \in \mathbb{Z}$, are built from two composition factors (irreducible subquotients). The structure is given by the following short exact sequences:[4]

$$0 \longrightarrow \mathcal{L}_\lambda \longrightarrow \mathcal{F}_\lambda \longrightarrow \mathcal{L}_{\lambda-1} \longrightarrow 0. \tag{2.2}$$

We then splice (see [1, App. A.2] for example) these exact sequences to obtain resolutions:

$$\cdots \longrightarrow \mathcal{F}_{\lambda+3} \longrightarrow \mathcal{F}_{\lambda+2} \longrightarrow \mathcal{F}_{\lambda+1} \longrightarrow \mathcal{L}_\lambda \longrightarrow 0. \tag{2.3}$$

This is also an exact sequence, so the Euler-Poincaré principle leads to the character formula

$$\mathrm{ch}\big[\mathcal{L}_\lambda\big] = \sum_{n=1}^{\infty} (-1)^{n-1} \mathrm{ch}\big[\mathcal{F}_{\lambda+n}\big] \qquad (\lambda \in \mathbb{Z}), \tag{2.4}$$

verifying that the standard characters form a topological basis of the space of all characters.

(5) The standard singlet characters span a representation of the modular group whose dimension is uncountably-infinite. The S-transformation is essentially a Fourier transform:

$$\begin{gathered} \mathrm{ch}\big[\mathcal{F}_\lambda\big]\big(\theta - \frac{\zeta^2}{2\tau}; \frac{\zeta}{\tau}; -\frac{1}{\tau}\big) = \int_{\mathbb{R}} \mathsf{S}\big[\mathcal{F}_\lambda \to \mathcal{F}_{\lambda'}\big] \mathrm{ch}\big[\mathcal{F}_{\lambda'}\big](\theta;\zeta;\tau)\, \mathrm{d}\lambda', \\ \mathsf{S}\big[\mathcal{F}_\lambda \to \mathcal{F}_{\lambda'}\big] = \mathsf{e}^{-2\pi \mathsf{i}(\lambda - 1/2)(\lambda' - 1/2)}. \end{gathered} \tag{2.5}$$

This is symmetric and unitary and it squares to the conjugation permutation $\lambda \leftrightarrow 1 - \lambda$:

$$\mathsf{S}\big[\mathcal{F}_\lambda \to \mathcal{F}_{\lambda'}\big] = \mathsf{S}\big[\mathcal{F}_{\lambda'} \to \mathcal{F}_\lambda\big], \tag{2.6a}$$

$$\int_{\mathbb{R}} \mathsf{S}\big[\mathcal{F}_\lambda \to \mathcal{F}_{\lambda'}\big] \mathsf{S}\big[\mathcal{F}_{\lambda''} \to \mathcal{F}_{\lambda'}\big]^* \,\mathrm{d}\lambda' = \delta\big(\lambda - \lambda''\big), \tag{2.6b}$$

$$\int_{\mathbb{R}} \mathsf{S}\big[\mathcal{F}_\lambda \to \mathcal{F}_{\lambda'}\big] \mathsf{S}\big[\mathcal{F}_{\lambda'} \to \mathcal{F}_{\lambda''}\big] \,\mathrm{d}\lambda' = \delta\big(\lambda + \lambda'' - 1\big). \tag{2.6c}$$

The atypical S-transformations are now obtained from the character formula (2.4):

$$\mathsf{S}\big[\mathcal{L}_\lambda \to \mathcal{F}_{\lambda'}\big] = \sum_{n=1}^{\infty} (-1)^{n-1} \mathsf{S}\big[\mathcal{F}_{\lambda+n} \to \mathcal{F}_{\lambda'}\big] = \frac{\mathsf{e}^{-2\pi \mathsf{i}\lambda(\lambda' - 1/2)}}{2\cos\big(\pi(\lambda' - 1/2)\big)}. \tag{2.7}$$

We note that the these elements *diverge* precisely at the atypical parameters $\lambda' \in \mathbb{Z}$.

[4] Exactness means that at each position $\cdots \xrightarrow{f} M \xrightarrow{g} \cdots$ in the sequence, the image of f coincides with the kernel of g. If $0 \to L \to M \to N \to 0$ is exact, then L is a submodule of M and M/L is isomorphic to N.

Journal of Physics: Conference Series **597** (2015) 012065 doi:10.1088/1742-6596/597/1/012065

(6) The (Grothendieck) fusion rules are given by the following version of the Verlinde formula:

$$\begin{bmatrix} \mathcal{F}_{\lambda''} \\ \mathcal{F}_\lambda \ \ \mathcal{F}_{\lambda'} \end{bmatrix} = \int_{\mathbb{R}} \frac{\mathsf{S}\big[\mathcal{F}_\lambda \to \mathcal{F}_{\lambda'''}\big]\mathsf{S}\big[\mathcal{F}_{\lambda'} \to \mathcal{F}_{\lambda'''}\big]\mathsf{S}\big[\mathcal{F}_{\lambda''} \to \mathcal{F}_{\lambda'''}\big]^*}{\mathsf{S}\big[\mathcal{L}_0 \to \mathcal{F}_{\lambda'''}\big]} \,\mathrm{d}\lambda''',$$
$$\mathrm{ch}\big[\mathcal{F}_\lambda\big] \times \mathrm{ch}\big[\mathcal{F}_{\lambda'}\big] = \int_{\mathbb{R}} \begin{bmatrix} \mathcal{F}_{\lambda''} \\ \mathcal{F}_\lambda \ \ \mathcal{F}_{\lambda'} \end{bmatrix} \mathrm{ch}\big[\mathcal{F}_{\lambda''}\big] \,\mathrm{d}\lambda''. \tag{2.8}$$

Note that $\mathcal{L}_0$ is the vacuum module. There are atypical analogues obtained by replacing $\mathcal{F}_\lambda$ or $\mathcal{F}_{\lambda'}$ (but never $\mathcal{F}_{\lambda''}$) by atypical modules. We thereby obtain the Grothendieck fusion rules

$$\mathrm{ch}\big[\mathcal{L}_\lambda\big] \times \mathrm{ch}\big[\mathcal{L}_{\lambda'}\big] = \mathrm{ch}\big[\mathcal{L}_{\lambda+\lambda'}\big], \qquad \mathrm{ch}\big[\mathcal{L}_\lambda\big] \times \mathrm{ch}\big[\mathcal{F}_{\lambda'}\big] = \mathrm{ch}\big[\mathcal{F}_{\lambda+\lambda'}\big],$$
$$\mathrm{ch}\big[\mathcal{F}_\lambda\big] \times \mathrm{ch}\big[\mathcal{F}_{\lambda'}\big] = \mathrm{ch}\big[\mathcal{F}_{\lambda+\lambda'}\big] + \mathrm{ch}\big[\mathcal{F}_{\lambda+\lambda'-1}\big], \tag{2.9}$$

confirming that the Grothendieck fusion coefficients are all non-negative integers. The atypical-atypical rule lifts to a genuine fusion rule because the result is an irreducible character that determines the module uniquely. The same is true for the atypical-typical rule because $\lambda + \lambda' \notin \mathbb{Z}$. The typical-typical rule, however, can only be lifted to a direct sum when $\lambda + \lambda'$ is typical. When $\lambda + \lambda' \in \mathbb{Z}$, we expect that the fusion product is a staggered module in the sense of [54, 1].[5]

The observed divergence (2.7) of the atypical S-transformation kernel at atypical parameters seems to be a general feature of logarithmic theories. For this reason, it is absolutely vital that the atypical standards correspond to a set of measure zero in parameter space. We will develop a method to deal with the case when the atypical parameters have positive measure in Section 3. Before turning to this, however, we detail another, slightly more involved, logarithmic example.

Example 2 (Bosonic ghosts [19])**.** The $\beta\gamma$ ghost system with $c = 2$ possesses a unique highest weight representation, the vacuum module $\mathcal{V}$. This module is irreducible. However, the ghost algebra admits a family of spectral flow automorphisms σ^ℓ which yields a family of mutually inequivalent representations $\sigma^\ell(\mathcal{V})$, $\ell \in \mathbb{Z}$. These are *not* the standard modules of the theory because their characters do *not* span a representation of the modular group.

(1) The ghost algebra also admits a continuous family of *parabolic* Verma modules $\mathcal{W}_j$, with $0 < j < 1$. Moreover, there exist two additional inequivalent indecomposable modules, $\mathcal{W}_0^+$ and $\mathcal{W}_0^-$, characterised by the short exact sequences

$$0 \longrightarrow \mathcal{V} \longrightarrow \mathcal{W}_0^+ \longrightarrow \sigma^{-1}(\mathcal{V}) \longrightarrow 0, \qquad 0 \longrightarrow \sigma^{-1}(\mathcal{V}) \longrightarrow \mathcal{W}_0^- \longrightarrow \mathcal{V} \longrightarrow 0. \tag{2.10}$$

The spectral flow images $\sigma^\ell(\mathcal{W}_j)$, together with the $\sigma^\ell(\mathcal{W}_0^\pm)$, are the standard modules. The standard characters again involve the ghost number, obtained from the zero mode of $J = \,:\beta\gamma:$, as well as the central element $\mathbf{1}$:

$$\mathrm{ch}\big[\sigma^\ell(\mathcal{W}_j)\big](y;z;q) = \mathrm{tr}_{\sigma^\ell(\mathcal{W}_j)}\, y^{\mathbf{1}} z^{J_0} q^{L_0-1/12} = \frac{y z^j q^{\ell j + \ell(\ell-1)/2}}{\eta(q)^2} \sum_{n\in\mathbb{Z}} z^n q^{n\ell}. \tag{2.11}$$

The common character of $\sigma^\ell(\mathcal{W}_0^+)$ and $\sigma^\ell(\mathcal{W}_0^-)$ is also given by this formula (with j set to 0). The standard characters are thus naturally parametrised by $j \in \mathbb{R}/\mathbb{Z} \cong \mathrm{S}^1$ and $\ell \in \mathbb{Z}$. We equip this space with the product μ of the standard Haar measure on S^1 and counting measure on $\mathbb{Z}$.

[5] Unfortunately, nobody seems to have computed fusion products for the $(1,2)$ singlet algebra. However, this expectation is consistent with the known fusion rules for the $(1,2)$ *triplet* algebra [27] as we shall see.

(2) The $\sigma^\ell(\mathcal{W}_j)$, with $j \neq 0$, are irreducible. These are the typical modules: $\mu(\{0\} \times \mathbb{Z}) = 0$.
(3) Again, we content ourselves with conjecturing that the typicals are projective and injective.
(4) Splicing (2.10) to obtain resolutions, we arrive at the atypical character formula

$$\operatorname{ch}[\sigma^\ell(\mathcal{V})] = \sum_{n=1}^{\infty} (-1)^{n-1} \operatorname{ch}[\sigma^{\ell+n}(\mathcal{W}_0^+)], \tag{2.12}$$

which confirms that the standard characters form a topological basis.
(5) The modular transformations take a rather more complicated form for this model:

$$\begin{aligned} \mathsf{S}\colon (\theta;\zeta;\tau) &\longrightarrow \left(\theta + \frac{\zeta^2}{2\tau} - \frac{\zeta}{2\tau} + \frac{\zeta}{2} + \frac{1}{2\pi}\left(\arg\tau - \frac{\pi}{2}\right); \frac{\zeta}{\tau}; -\frac{1}{\tau}\right), \\ \mathsf{T}\colon (\theta;\zeta;\tau) &\longrightarrow \left(\theta + \frac{1}{12}; \zeta; \tau+1\right). \end{aligned} \tag{2.13}$$

Nevertheless, they determine a representation of $\mathsf{SL}(2;\mathbb{Z})$ on the span of the standard characters with S-transformation

$$\begin{gathered} \operatorname{ch}[\sigma^\ell(\mathcal{W}_j)](\mathsf{S}(\theta;\zeta;\tau)) = \sum_{\ell'\in\mathbb{Z}} \int_{\mathbb{R}/\mathbb{Z}} \mathsf{S}[\sigma^\ell(\mathcal{W}_j) \to \sigma^{\ell'}(\mathcal{W}_{j'})] \operatorname{ch}[\sigma^{\ell'}(\mathcal{W}_{j'})](\theta;\zeta;\tau)\, \mathrm{d}j', \\ \mathsf{S}[\sigma^\ell(\mathcal{W}_j) \to \sigma^{\ell'}(\mathcal{W}_{j'})] = (-1)^{\ell+\ell'} \mathsf{e}^{-2\pi\mathsf{i}(\ell j' + \ell' j)}. \end{gathered} \tag{2.14}$$

It is easy to check that S is symmetric, unitary and squares to conjugation. The atypical S-transformation kernel

$$\mathsf{S}[\sigma^\ell(\mathcal{V}) \to \sigma^{\ell'}(\mathcal{W}_{j'})] = (-1)^{\ell+\ell'+1} \frac{\mathsf{e}^{-2\pi\mathsf{i}(\ell+1/2)j'}}{\mathsf{e}^{\mathsf{i}\pi j'} - \mathsf{e}^{-\mathsf{i}\pi j'}}, \tag{2.15}$$

again diverges at the atypical point $j' = 0$.
(6) The obvious Verlinde formula (with sums over $\ell \in \mathbb{Z}$ and integrals over $j \in \mathbb{R}/\mathbb{Z}$) gives the following Grothendieck fusion rules:

$$\operatorname{ch}[\sigma^\ell(\mathcal{V})] \times \operatorname{ch}[\sigma^{\ell'}(\mathcal{V})] = \operatorname{ch}[\sigma^{\ell+\ell'}(\mathcal{V})], \tag{2.16a}$$

$$\operatorname{ch}[\sigma^\ell(\mathcal{V})] \times \operatorname{ch}[\sigma^{\ell'}(\mathcal{W}_{j'})] = \operatorname{ch}[\sigma^{\ell+\ell'}(\mathcal{W}_{j'})], \tag{2.16b}$$

$$\operatorname{ch}[\sigma^\ell(\mathcal{W}_j)] \times \operatorname{ch}[\sigma^{\ell'}(\mathcal{W}_{j'})] = \operatorname{ch}[\sigma^{\ell+\ell'}(\mathcal{W}_{j+j'})] + \operatorname{ch}[\sigma^{\ell+\ell'-1}(\mathcal{W}_{j+j'})]. \tag{2.16c}$$

The coefficients are clearly non-negative in each case.

We close this section by noting that this formalism has been extensively tested on other logarithmic conformal field theories including the fractional level $\widehat{\mathfrak{sl}}(2)$ models [55, 56], the $\widehat{\mathfrak{gl}}(1|1)$ model [25, 57] and its Takiff extension [58], the logarithmic minimal models [59] and their $N = 1$ extensions [60], and the (p', p) singlet theories [53]. The Takiff model is of interest because it demonstrates that the formalism can naturally accommodate different degrees of atypicality. The singlet theories with $p, p' \geq 2$ are interesting because it is known [38] that their fusion product does not induce a well-defined product on the Grothendieck group because of the poor behaviour (non-rigidity) of the minimal model representations. The formalism deals with this naturally, and rather effectively, by regarding the minimal model characters as being zero.[6]

[6] One is therefore led to wonder whether the minimal model representations should be excluded from the physically relevant category. They do not seem to appear as direct summands in the bulk, boundary or defect sectors. However, excluding them would amount to taking the physical category to be non-abelian.

Journal of Physics: Conference Series **597** (2015) 012065 doi:10.1088/1742-6596/597/1/012065

3. Fusion Rules for Simple Current Extensions

We recall that a simple current may be generally defined as an indecomposable representation for which fusing with the simple current defines a permutation on the set of all indecomposables. For the purposes of this article, we shall simplify this definition by replacing "indecomposable" by "irreducible", as in [61, 62]. Examples of such simple currents then include the $\mathcal{L}_\lambda$, with $\lambda \in \mathbb{Z}$, for the $(1,2)$ singlet model (Example 1) and the $\sigma^\ell(\mathcal{V})$, with $\ell \in \mathbb{Z}$, for the $c = 2$ bosonic ghost system (Example 2). In both cases, the simple currents are atypical irreducibles.

If the vacuum module is also irreducible, as is the case for both examples considered in Section 2, then for every simple current $\mathcal{J}$, there exists a simple current $\mathcal{J}^{-1}$ such that their fusion gives the vacuum module, that is, $\mathcal{J}$ has a fusion inverse. Any given simple current (more generally, any set of simple currents) then generates a group G of simple currents under the fusion operation. This group then generates an extended symmetry algebra in the sense that the representation $\bigoplus_{\mathcal{J}\in\mathsf{G}} \mathcal{J}$ carries the structure of a vertex operator algebra.[7] Moreover, every representation $\mathcal{M}$ of the original symmetry algebra determines[8] a representation of the extended symmetry algebra on $\bigoplus_{\mathcal{J}\in\mathsf{G}} (\mathcal{J} \times \mathcal{M})$.

Mathematically, this may be formalised using the notions of restriction and induction [66, 67]: Let A denote the original vertex operator algebra, B its simple current extension by G, and $\otimes_{\mathsf{A}}$ and $\otimes_{\mathsf{B}}$ their fusion products. Then, restricting B (interpreted as its vacuum module) to an A-module and inducing A (also interpreted as its vacuum module) to a B-module amounts to

$$\mathsf{B}{\downarrow} \cong \bigoplus_{\mathcal{J}\in\mathsf{G}} \mathcal{J}, \qquad \mathsf{A}{\uparrow} = \mathsf{B} \otimes_{\mathsf{A}} \mathsf{A} \cong \mathsf{B}. \tag{3.1}$$

Moreover, we may induce an A-module $\mathcal{M}$ to a B-module $\mathcal{M}{\uparrow}$, obtaining

$$\mathcal{M}{\uparrow} = \mathsf{B} \otimes_{\mathsf{A}} \mathcal{M} \qquad \Rightarrow \qquad (\mathcal{M}{\uparrow}){\downarrow} \cong \bigoplus_{\mathcal{J}\in\mathsf{G}} (\mathcal{J} \otimes_{\mathsf{A}} \mathcal{M}), \tag{3.2}$$

as before. This formalisation has a practical benefit as well. The associativity of fusion products allows one to compute the fusion rules for the simple current extension B from those of A:

$$\begin{aligned} \mathcal{M}{\uparrow} \otimes_{\mathsf{B}} \mathcal{N}{\uparrow} &= (\mathsf{B} \otimes_{\mathsf{A}} \mathcal{M}) \otimes_{\mathsf{B}} (\mathsf{B} \otimes_{\mathsf{A}} \mathcal{N}) \cong ((\mathsf{B} \otimes_{\mathsf{A}} \mathcal{M}) \otimes_{\mathsf{B}} \mathsf{B}) \otimes_{\mathsf{A}} \mathcal{N} \\ &\cong (\mathsf{B} \otimes_{\mathsf{A}} \mathcal{M}) \otimes_{\mathsf{A}} \mathcal{N} \cong \mathsf{B} \otimes_{\mathsf{A}} (\mathcal{M} \otimes_{\mathsf{A}} \mathcal{N}) = (\mathcal{M} \otimes_{\mathsf{A}} \mathcal{N}){\uparrow}. \end{aligned} \tag{3.3}$$

Here, we remark that inducing an irreducible A-module $\mathcal{M}$ results in an irreducible B-module $\mathcal{M}{\uparrow}$. This is a special property of (invertible) simple current extensions that follows from (3.2) because each direct summand $\mathcal{J} \otimes_{\mathsf{A}} \mathcal{M}$ is an irreducible that is obtained from any other direct summand by acting with one of the simple currents in G, thus with an element of B.

Of course, one needs to justify why the fusion product of a vertex operator algebra may be regarded as a tensor product that is permeable to the algebra action. That fusion constitutes an abstract tensor product is, at least for C_2-cofinite vertex operator algebras, one of the great results of Lepowsky's group (see [68] for a review). While the logarithmic theories we study here are not C_2-cofinite in general, it is very reasonable to assume that fusion still has the properties of a tensor product on the physical category of representations. The permeability, on the other hand, reflects the locality requirement for fields corresponding to the algebra and its modules. More precisely, if $a(z)$ is a field of A and $\phi(z)$ and $\psi(z)$ are fields corresponding to elements of the A-modules $\mathcal{M}$ and $\mathcal{N}$, respectively, then locality and associativity give

$$\begin{aligned} (a(z)\,\phi(x))\,\psi(w) &= (\phi(x)\,a(z))\,\psi(w) = \phi(x)\,(a(z)\,\psi(w)) \\ \Longleftrightarrow \qquad a \cdot \phi \otimes_{\mathsf{A}} \psi &= \phi \cdot a \otimes_{\mathsf{A}} \psi = \phi \otimes_{\mathsf{A}} a \cdot \psi. \end{aligned} \tag{3.4}$$

[7] There are some subtleties to this statement, however; see [64, 65, 21] for example.

[8] For simplicity, and because our examples do not require it, we ignore the issue of *fixed points* here, see [63].

We conclude by computing the Grothendieck fusion rules of the $(1,2)$ triplet model as an application of this simple current formalism. Recall from Equation (1.2) that the irreducible characters do not span a representation of the modular group. From the point of view of the modular formalism of Section 2, one identifies candidates for the standard modules: The typical irreducibles $\mathbb{L}_{-1/8}$ and $\mathbb{L}_{3/8}$ and the atypical indecomposables $\mathbb{A}_0$ and $\mathbb{A}_1$ are characterised by

$$0 \longrightarrow \mathbb{L}_0 \longrightarrow \mathbb{A}_0 \longrightarrow \mathbb{L}_1 \longrightarrow 0, \qquad 0 \longrightarrow \mathbb{L}_1 \longrightarrow \mathbb{A}_1 \longrightarrow \mathbb{L}_0 \longrightarrow 0. \tag{3.5}$$

Unfortunately, the standard characters do not span a representation of the modular group either. We remark that as $\mathrm{ch}[\mathbb{A}_0] = \mathrm{ch}[\mathbb{A}_1]$, the atypical characters have measure $\frac{1}{3}$ with respect to the natural measure (counting measure) on the parametrisation space $M = \{-\frac{1}{8}, 0, \frac{3}{8}\}$.

As mentioned above, (Grothendieck) fusion rules for the $(1,2)$ triplet model are nevertheless easily derived by regarding this theory as the simple current extension of the $(1,2)$ singlet model. The group G of simple currents is generated by the irreducible singlet representation $\mathcal{J} = \mathcal{L}_2$ (or $\mathcal{L}_{-2}$) [1] whose highest weight state has conformal dimension 3. We can therefore apply the above simple current formalism with A and B being the singlet and triplet vertex operator algebras, respectively. It is straightforward to identify the triplet irreducibles as induced singlet modules (see Example 1):

$$\begin{aligned} \mathbb{L}_0 &= \mathcal{L}_0\!\uparrow, & \mathbb{L}_1 &= \mathcal{L}_1\!\uparrow, & \mathbb{L}_{-1/8} &= \mathcal{F}_{1/2}\!\uparrow, & \mathbb{L}_{3/8} &= \mathcal{F}_{-1/2}\!\uparrow, \\ \mathbb{L}_0\!\downarrow &= \bigoplus_{n\in\mathbb{Z}} \mathcal{L}_{2n}, & \mathbb{L}_1\!\downarrow &= \bigoplus_{n\in\mathbb{Z}} \mathcal{L}_{2n+1}, & \mathbb{L}_{-1/8}\!\downarrow &= \bigoplus_{n\in\mathbb{Z}} \mathcal{F}_{2n+1/2}, & \mathbb{L}_{3/8}\!\downarrow &= \bigoplus_{n\in\mathbb{Z}} \mathcal{F}_{2n-1/2}. \end{aligned} \tag{3.6}$$

The remaining singlet typicals $\mathcal{F}_\lambda$ can likewise be induced, but they turn out to define *twisted* representations on which the triplet fields act with non-trivial monodromy.

The singlet fusion rules derived in Example 1 may now be lifted to the triplet using (3.3). For example,

$$\mathbb{L}_1 \otimes_{\mathsf{B}} \mathbb{L}_1 = \mathcal{L}_1\!\uparrow \otimes_{\mathsf{B}} \mathcal{L}_1\!\uparrow = (\mathcal{L}_1 \otimes_{\mathsf{A}} \mathcal{L}_1)\!\uparrow = \mathcal{L}_2\!\uparrow = \mathbb{L}_0. \tag{3.7a}$$

In the same manner, we compute

$$\mathbb{L}_1 \otimes_{\mathsf{B}} \mathbb{L}_{-1/8} = \mathbb{L}_{3/8}, \qquad \mathbb{L}_1 \otimes_{\mathsf{B}} \mathbb{L}_{3/8} = \mathbb{L}_{-1/8}. \tag{3.7b}$$

These results agree of course with direct, though significantly more difficult, computations [27]. For the fusion rules of $\mathbb{L}_{-1/8}$ and $\mathbb{L}_{3/8}$, the corresponding singlet rules were only derived at the Grothendieck level, so we may only conclude that

$$\mathrm{ch}[\mathbb{L}_h] \otimes_{\mathsf{B}} \mathrm{ch}[\mathbb{L}_{h'}] = 2\,\mathrm{ch}[\mathbb{L}_0] + 2\,\mathrm{ch}[\mathbb{L}_1], \qquad (h, h' \in \{-\tfrac{1}{8}, \tfrac{3}{8}\}). \tag{3.8}$$

This result is also consistent with the known fusion rules [27] where the fusion products are known to be indecomposable staggered modules formed by glueing two copies of $\mathbb{L}_0$ and two copies of $\mathbb{L}_1$ together.

This example illustrates how one can combine the standard module formalism of Section 2 with the simple current formalism of Section 3 to rather effortlessly deduce the (Grothendieck) fusion rules of logarithmic conformal field theories. The results have been generalised to all the (p', p) triplet models [53] and are in complete agreement with the fusion computations that have been computed, as well as those that have only been conjectured, for these theories. We conclude by remarking that we are unaware of theories to which this combined formalism cannot be applied. It would be very interesting to study an example of such a theory.

Acknowledgments
We thank Jürgen Fuchs and Aliosha Semikhatov for discussions on the Verlinde formula for the $(1, p)$ triplet models, Matthias Gaberdiel for helpful correspondence concerning the modular properties of the triplet characters, and Antun Milas for pointing out a mistake in an earlier version. DR's research is supported by the Australian Research Council Discovery Project DP1093910. SW's work is supported by the Australian Research Council Discovery Early Career Researcher Award DE140101825.

References
[1] Creutzig T and Ridout D 2013 *J. Phys.* **A46** 494006 (arXiv:1303.0847 [hep-th])
[2] Rozansky L and Saleur H 1992 *Nucl. Phys.* **B376** 461–509
[3] Gurarie V 1993 *Nucl. Phys.* **B410** 535–549 (arXiv:hep-th/9303160)
[4] Gainutdinov A, Ridout D and Runkel I (Editors) 2013 *J. Phys.* **A46** 490301
[5] Pearce P and Rasmussen J 2007 *J. Stat. Mech.* **0702** P02015 (arXiv:hep-th/0610273)
[6] Morin-Duchesne A, Pearce P and Rasmussen J 2013 *Nucl. Phys.* **B874** 312–357 (arXiv:1303.4895 [hep-th])
[7] Read N and Saleur H 2007 *Nucl. Phys.* **B777** 316–351 (arXiv:hep-th/0701117)
[8] Rasmussen J and Pearce P 2007 *J. Stat. Mech.* **0709** P09002 (arXiv:0706.2716 [hep-th])
[9] Mathieu P and Ridout D 2007 *Phys. Lett.* **B657** 120–129 (arXiv:0708.0802 [hep-th])
[10] Ridout D 2009 *Nucl. Phys.* **B810** 503–526 (arXiv:0808.3530 [hep-th])
[11] Jeng M, Piroux G and Ruelle P 2006 *J. Stat. Mech.* **0610** P10015 (arXiv:cond-mat/0609284)
[12] Ruelle P 2013 *J. Phys.* **A46** 494014 (arXiv:1303.4310 [hep-th])
[13] Gainutdinov A, Read N and Saleur H 2013 *Nucl. Phys.* **B871** 245–288 (arXiv:1112.3403 [hep-th])
[14] Gainutdinov A, Jacobsen J, Read N, Saleur H and Vasseur R 2013 *J. Phys.* **A46** 494012 (arXiv:1303.2082 [hep-th])
[15] Kausch H Curiosities at $c = -2$ (arXiv:hep-th/9510149)
[16] Kausch H 2000 *Nucl. Phys.* **B583** 513–541 (arXiv:hep-th/0003029)
[17] Lesage F, Mathieu P, Rasmussen J and Saleur H 2004 *Nucl. Phys.* **B686** 313–346 (arXiv:hep-th/0311039)
[18] Ridout D 2011 *Nucl. Phys.* **B848** 216–250 (arXiv:1012.2905 [hep-th])
[19] Ridout D and Wood S 2014 Bosonic Ghosts at $c = 2$ as a Logarithmic CFT (arXiv:1408.4185 [hep-th])
[20] Gaberdiel M 2001 *Nucl. Phys.* **B618** 407–436 (arXiv:hep-th/0105046)
[21] Ridout D 2009 *Nucl. Phys.* **B814** 485–521 (arXiv:0810.3532 [hep-th])
[22] Adamović D 2014 A Realization of Certain Modules for the $N = 4$ Superconformal Algebra and the Affine Lie Algebra $A_2^{(1)}$ (arXiv:1407.1527 [math.QA])
[23] Saleur H and Schomerus V 2006 *Nucl. Phys.* **B734** 221–245 (arXiv:hep-th/0510032)
[24] Gotz G, Quella T and Schomerus V 2007 *JHEP* **0703** 003 (arXiv:hep-th/0610070)
[25] Creutzig T and Ridout D 2013 *Nucl. Phys.* **B872** 348–391 (arXiv:1107.2135 [hep-th])
[26] Quella T and Schomerus V 2013 *J. Phys.* **A46** 494010 (arXiv:1307.7724 [hep-th])
[27] Gaberdiel M and Kausch H 1996 *Phys. Lett.* **B386** 131–137 (arXiv:hep-th/9606050)
[28] Gaberdiel M and Kausch H 1999 *Nucl. Phys.* **B538** 631–658 (arXiv:hep-th/9807091)
[29] Feigin B, Gainutdinov A, Semikhatov A and Tipunin I Y 2006 *Comm. Math. Phys.* **265** 47–93 (arXiv:hep-th/0504093)
[30] Gaberdiel M and Runkel I 2008 *J. Phys.* **A41** 075402 (arXiv:0707.0388 [hep-th])
[31] Adamović D and Milas A 2008 *Adv. Math.* **217** 2664–2699 (arXiv:0707.1857 [math.QA])
[32] Pearce P, Rasmussen J and Ruelle P 2008 *J. Phys.* **A41** 295201 (arXiv:0803.0785 [hep-th])
[33] Nagatomo K and Tsuchiya A 2011 *Adv. Stud. Pure Math.* **61** 1–49 (arXiv:0902.4607 [math.QA])
[34] Tsuchiya A and Wood S 2013 *J. Phys.* **A46** 445203 (arXiv:1201.0419 [hep-th])
[35] Gainutdinov A, Saleur H and Tipunin I Y 2012 Lattice W-algebras and logarithmic CFTs (arXiv:1212.1378 [hep-th])
[36] Feigin B, Gainutdinov A, Semikhatov A and Tipunin I Y 2006 *Nucl. Phys.* **B757** 303–343 (arXiv:hep-th/0606196)
[37] Rasmussen J 2009 *Nucl. Phys.* **B807** 495–533 (arXiv:0805.2991 [hep-th])
[38] Gaberdiel M, Runkel I and Wood S 2009 *J. Phys.* **A42** 325403 (arXiv:0905.0916 [hep-th])
[39] Wood S 2010 *J. Phys.* **A43** 045212 (arXiv:0907.4421 [hep-th])
[40] Adamović D and Milas A 2010 *Int. Math. Res. Not.* **2010** 3896–3934 (arXiv:0908.4053 [math.QA])
[41] Gaberdiel M, Runkel I and Wood S 2011 *J. Phys.* **A44** 015204 (arXiv:1008.0082 [hep-th])
[42] Adamović D and Milas A 2012 *J. Math. Phys.* **53** 073511 (arXiv:1202.6667 [math.QA])
[43] Tsuchiya A and Wood S 2014 *Int. Math. Res. Not.* (to appear) (arXiv:1302.6435 [hep-th])

[44] Semikhatov A and Tipunin I Y 2013 *J. Phys.* **A46** 494011 (arXiv:1301.2235 [math.QA])
[45] Verlinde E 1988 *Nucl. Phys.* **B300** 360–376
[46] Flohr M 1996 *Int. J. Mod. Phys.* **A11** 4147–4172 (arXiv:hep-th/9509166)
[47] Miyamoto M 2004 *Duke Math. J.* **122** 51–91 (arXiv:math/0209101 [math.QA])
[48] Fuchs J, Schweigert C and Stigner C 2013 *J. Phys.* **A46** 494012 (arXiv:1302.4683 [hep-th])
[49] Fuchs J, Hwang S, Semikhatov A and Tipunin I Y 2004 *Comm. Math. Phys.* **247** 713–742 (arXiv:hep-th/0306274)
[50] Flohr M and Knuth H 2007 On Verlinde-Like Formulas in $c(p,1)$ Logarithmic Conformal Field Theories (arXiv:0705.0545 [math-ph])
[51] Gainutdinov A and Tipunin I Y 2009 *J. Phys.* **A42** 315207 (arXiv:0711.3430 [hep-th])
[52] Pearce P, Rasmussen J and Ruelle P 2010 *J. Phys.* **A43** 045211 (arXiv:0907.0134 [hep-th])
[53] Ridout D and Wood S 2014 *Nucl. Phys.* **B880** 175–202 (arXiv:1310.6479 [hep-th])
[54] Kytölä K and Ridout D 2009 *J. Math. Phys.* **50** 123503 (arXiv:0905.0108 [math-ph])
[55] Creutzig T and Ridout D 2012 *Nucl. Phys.* **B865** 83–114 (arXiv:1205.6513 [hep-th])
[56] Creutzig T and Ridout D 2013 *Nucl. Phys.* **B875** 423–458 (arXiv:1306.4388 [hep-th])
[57] Creutzig T and Ridout D 2011 *Springer Proc. Math. Stat.* **36** 349–368 (arXiv:1111.5049 [hep-th])
[58] Babichenko A and Ridout D 2013 *J. Phys.* **A46** 125204 (arXiv:1210.7094 [math-ph])
[59] Morin-Duchesne A, Rasmussen J and Ridout D 2015 Kac Modules for the Logarithmic Minimal Models (in preparation)
[60] Canagasabey M, Rasmussen J and Ridout D 2015 Fusion Rules and Kac Modules for the $N = 1$ Logarithmic Minimal Models (in preparation)
[61] Schellekens A and Yankielowicz S 1990 *Int. J. Mod. Phys.* **A5** 2903–2952
[62] Dong C, Li H and Mason G 1996 *Comm. Math. Phys.* **180** 671–707 (arXiv:q-alg/9504008)
[63] Fuchs J, Schellekens A and Schweigert C 1996 *Nucl. Phys.* **B473** 323–366 (arXiv:hep-th/9601078)
[64] Mathieu P and Ridout D 2007 *Nucl. Phys.* **B765** 201–239 (arXiv:hep-th/0609226)
[65] Mathieu P and Ridout D 2007 *Nucl. Phys.* **B776** 365–404 (arXiv:hep-th/0701250)
[66] Miyamoto M 1998 *J. Alg.* **201** 115–150 (arXiv:hep-th/9612032)
[67] Lam C 2001 *J. Math. Soc. Japan* **53** 541–557
[68] Huang Y Z and Lepowsky J 2013 *J. Phys.* **A46** 494009 (arXiv:1304.7556 [hep-th])

Complex nonlinear Fourier transform and its inverse

Pavle Saksida[1,2]

[1]Faculty of Mathematics and Physics, University of Ljubljana
[2]Institute of Mathematics, Physics and Mechanics, Ljubljana

E-mail: pavle.saksida@fmf.uni-lj.si

Abstract. We study the nonlinear Fourier transform associated to the integrable systems of AKNS-ZS type. Two versions of this transform appear in connection with the AKNS-ZS systems. These two versions can be considered as two real forms of a single complex transform $\mathcal{F}^c$. We construct an explicit algorithm for the calculation of the inverse transform $(\mathcal{F}^c)^{-1}(h)$ for an arbitrary argument h. The result is given in the form of a convergent series of functions in the domain space and the terms of this series can be computed explicitly by means of finitely many integrations.

1. Introduction

A partial differential equation in temporal and one spatial dimensions is integrable, if it can be rewritten in the form of the so called zero-curvature condition. Let

$$E(q(x,t)) = 0 \tag{1}$$

be a partial differential equation for the unknown function $q(x,t)$. If the equation E is integrable, then there exists a pair of matrix-valued functions

$$L_q(x,t;z),\, A_q(x,t;z) : [a,b] \times \mathbb{R}_+ \times \mathbb{C} \longrightarrow \mathrm{Mat}^{n\times n}$$

such that

$$(L_q)_t - (A_q)_x + [L_q, A_q] = E(q) \cdot M(z),$$

where $M(z)$ is a matrix which is constant or it depends only on the parameter z, called the spectral parameter. This means that for every z the connection $d + L_q\mathrm{d}t + A_q\mathrm{d}x$ on the trivial $\mathbb{C}^n$-bundle over $[a,b] \times \mathbb{R}_+$ is flat if and only if $q(x,t)$ is a solution of equation (1).

Fix t_0 and let $\Phi(x,t_0;z)$ be the solution of the linear initial value problem

$$\Phi_x = L_q(x,t_0;z) \cdot \Phi, \quad \Phi(a,t_0;z) = Id. \tag{2}$$

Denote

$$\mathrm{Hol}_{q(x,t_0)}(z) = \Phi(b,t_0,z).$$

The fact that holonomies of flat connections along contractible loops are trivial implies that

$$\mathrm{Hol}_{q(x,t)}(z) = P(b,t;z) \cdot \mathrm{Hol}_{q(x,0)}(z) \cdot P(a,t;z)^{-1}, \tag{3}$$

where $P(x_0, t; z)$ is the solution of

$$P_t = A_q(x_0, t; z) \cdot P, \quad P(x_0, 0; z) = Id.$$

If $[a, b] = \mathbb{R} = (\infty, \infty)$, and if $q(x, 0)$ has some suitable asymptotic behaviour at both infinities, then often the matrix $P(-\infty, t; z) = P(\infty, t; z)$ is independent of $q(x, t)$ and one can explicitly compute $\mathrm{Hol}_{q(x,t)}(z)$ from $\mathrm{Hol}_{q(x,0)}$ by means of equation (3).

Let $f(x)$ be a function of one (spacial) variable, belonging to some suitable function space. Solution of the linear initial value problem (2) whose coefficient matrix is L_f induces the map

$$\mathcal{F} : f(x) \longmapsto \mathrm{Hol}_f(z)$$

from our function space into a suitable space of matrix-valued functions. Suppose we are able to compute the inverse map $\mathcal{F}^{-1}$. Then the above discussion can be summarized into the following scheme:

$$q(x, 0) \overset{\mathcal{F}}{\longmapsto} \mathrm{Hol}_{q(x,0)}(z) \overset{\text{time evolution}}{\longmapsto} \mathrm{Hol}_{q(x,t_0)} \overset{\mathcal{F}^{-1}}{\longmapsto} q(x, t_0).$$

If we were able to compute all the above arrows, this scheme would yield the solution of the Cauchy problem for our equation (1). This procedure is called the *inverse scattering transform* method. It is reminiscent of the algorithm for solving the Cauchy problems for certain linear partial differential equations by means of the Fourier analysis. This can be summarized in the diagram

$$q(x, 0) \overset{F}{\longmapsto} F(q(x, 0))(z) \overset{\text{time evolution}}{\longmapsto} F(q(x, t_0)) \overset{F^{-1}}{\longmapsto} q(x, t_0).$$

This similarity was first exposed by Ablowitz, Kaup, Newell and Segur in [1] and [2]. The operator $\mathcal{F}$ is called the inverse scattering transform or, often, the nonlinear Fourier transform. Important results on the connection of the nonlinear Fourier transform and integrable systems were presented e.g. in [6], [7], [8]. Deep analytical issues concerning the nonlinear Fourier transform were studied in [3], [4], [5], [9], [10], [11]. In certain cases the derivative $\mathcal{F}$ is (essentially) the linear Fourier transform F. This fact was exploited in [12], [13] to shed some light on certain Poisson geometric aspects of nonlinear integrable systems and their linearizations. The integrable systems are only one of the fields in which the nonlinear Fourier transform plays an important role. Other such fields are orthogonal polynomials, random matrices and others. An introduction and overview to these topics is given in [15].

One of the essential differences between the usual, linear Fourier transform F and the nonlinear transform $\mathcal{F}$ lies in the fact that while the calculation of F and F^{-1} are problems of similar difficulty, the calculation of $\mathcal{F}^{-1}$ is considerably more difficult than the calculation of $\mathcal{F}$. Finding $\mathcal{F}^{-1}$ is a difficult inverse problem. It can be rephrased in the form of a suitable Riemann-Hilbert problem. But this Riemann-Hilbert problem is almost never explicitly soluble. Therefore, it yields an explicit expression of

$$\mathcal{F}^{-1}[\text{time evolution}(\mathcal{F}(q(x, 0)))] = q(x, t_0)$$

only for some exceptional initial conditions $q(x, 0)$. In this paper, our goal will be to find the preimage $\mathcal{F}^{-1}(h(z))$ of an arbitrary element $h(z)$ for the AKNS-ZS class of integrable equations. Since there is no hope of finding a general closed analytic expression, we will give the solution in the form of a series which will be convergent in a suitable topology. This convergent series will provide a tool by means of which one can calculate the approximations of $\mathcal{F}^{-1}(h(z))$ to any desired accuracy.

The AKNS-ZS systems are a relatively large and a very important class of integrable partial differential equations. Some of the most salient integrable equations belong to this class. Notable examples are

(i) Nonlinear Schrödinger equations: $iq_t = -q_{xx} + 2\kappa|q|q, \quad \kappa \pm 1.$
(ii) modified KdV equation: $q_t = 6q^2q_x - q_{xxx}$.
(iii) sine-Gordon equation: $q_{tx} = -\sin q$.

The common feature of the AKNS-ZS systems is the form of the L_q-matrices associated to them. These L_q-matrices are of the form

$$L_q = \begin{pmatrix} \frac{iz}{2} & q(x,t) \\ \kappa\, \overline{q(x,t)} & -\frac{iz}{2} \end{pmatrix},$$

where κ is equal to $+1$ or to -1. We shall construct a complexified version $\mathcal{F}^c$ of the nonlinear Fourier transform associated to the AKNS-ZS systems. The nolinear transforms $\mathcal{F}_1$ and $\mathcal{F}_2$ for the cases $\kappa = +1$ and $\kappa = -1$ will turn out to be two different real forms of $\mathcal{F}^c$. In addition to that, $\mathcal{F}^c$ is important in its own right. Namely, it is the transform associated to the vector nonlinear Schrödinger system

$$iq_t = q_{xx} - 2rq^2, \quad ir_t = r_{xx} - 2qr^2. \tag{4}$$

We will construct the iterative scheme by means of which one can calculate arbitrarily good approximations of $(\mathcal{F}^c)^{-1}$ for an arbitrary choice of the argument. From this scheme the procedures for calculation of the real forms $\mathcal{F}_1$ and $\mathcal{F}_2$ can be derived immediately.

In section 2 we provide the precise definition of the complex nonlinear Fourier transform $\mathcal{F}^c$ and of its real forms. We present our main result in theorem 3 of section 3.

2. The complex nonlinear Fourier transform and its real forms

We begin by giving a precise definition of the complex nonlinear transform $\mathcal{F}^c$. The L-matrix of the Lax pair for the vector nonlinear Schrödinger system (4) is

$$L_{(q,r)} = \begin{pmatrix} \frac{iz}{2} & q(x) \\ r(x) & -\frac{iz}{2} \end{pmatrix}, \tag{5}$$

therefore, $\mathcal{F}^c$ will be given by the holonomy of a matrix of the above form. It turns out that the expressions become more manageable if one changes the gauge by the gauge transform

$$G(x;z) = \begin{pmatrix} e^{-\frac{izx}{2}} & 0 \\ 0 & e^{\frac{izx}{2}} \end{pmatrix}.$$

The transformed matrix

$$L^g_{(q,r)} = G_x G^{-1} + G \cdot L_{(q,r)} \cdot G^{-1}$$

is

$$L^g_{(q,r)}(x;z) = \begin{pmatrix} 0 & e^{-izx}q(x) \\ e^{izx}r(x) & 0 \end{pmatrix}.$$

In this paper, we shall confine ourselves to the case where the operator $\mathcal{F}^c$ will be defined on a suitable L^2-space over a finite interval. Let the space $L^2_c[0,2\pi]$ be defined by

$$L^2_c[0,2\pi] = \{\big(f(x), g(x)\big); f(x), g(x) \in L^2[0,2\pi]\}.$$

This space is equipped with the obvious L^2-norm and with the complex structure J, given by

$$J(f,g) = (-g,f). \tag{6}$$

Journal of Physics: Conference Series **597** (2015) 012066 doi:10.1088/1742-6596/597/1/012066

The holonomy of $L^g_{(q,r)}$ can be defined by the following Dyson's series

$$\mathrm{Hol}_{(q,r)}(z) = I + \sum_{k=1}^{\infty} \int_{\triangle_k} L^g_{(q,r)}(x_1; z) L^g_{(q,r)}(x_2; z) \cdots L^g_{(q,r)}(x_k; z)\, \mathrm{d}\vec{x} = \begin{pmatrix} h^d(z) & h(z) \\ k(z) & k^d(z) \end{pmatrix}, \tag{7}$$

where

$$\triangle_k = \{(x_1, x_2, \ldots, x_k); 2\pi \geq x_1 \geq x_2 \geq \ldots \geq x_k \geq 0\}$$

is the so-called ordered simplex of dimension k. Whenever the functions $q(x), r(x)$ are continuous, the above holonomy can be obtained by solving the initial value problem

$$\Phi_x(x; z) = L^g_{(q,r)} \cdot \Phi(x; z), \quad \Phi(0; z) = Id$$

and then evaluating the solution at $x = 2\pi$. By contrast, the expression (7) is defined for every $(q, r) \in L^2_c[0, 2\pi]$.

Finally, we define the target space of the operator $\mathcal{F}^c$. Let $l^2_{\mathbb{Z}}$ denote the space of square-integrable bi-infinite complex sequences. Its complexification is the space

$$l^2_{c,\mathbb{Z}} = \{\Big(\{a(z)\}_{z\in\mathbb{Z}}, \{b(z)\}_{z\in\mathbb{Z}}\Big), \ \{a(z)\}_{z\in\mathbb{Z}}, \{b(z)\}_{z\in\mathbb{Z}} \in l^2_{\mathbb{Z}}\}.$$

This space is also equipped with the natural complex structure

$$K\Big(\{a(z)\}_{z\in\mathbb{Z}}, \{b(z)\}_{z\in\mathbb{Z}}\Big) = \Big(-\{b(z)\}_{z\in\mathbb{Z}}, \{a(z)\}_{z\in\mathbb{Z}}\Big). \tag{8}$$

Definition 1 *The complex nonlinear Fourier transform, associated to the AKNS-ZS systems is the map*

$$\mathcal{F}^c : L^2_c[0, 2\pi] \longrightarrow l^2_{c,\mathbb{Z}},$$

given by the formula

$$\mathcal{F}^c(q, r) = \Big(\{h(z)\}_{z\in\mathbb{Z}}, \{k(z)\}_{\mathbb{Z}}\Big),$$

where the values $(h(z), k(z))$ are defined by (7). More explicitly, (7) gives

$$(\mathcal{F}^c(q, r))(z) = \begin{pmatrix} \sum_{k=1}^{\infty} \int_{\triangle_{2k-1}} e^{-iz(x_1 - x_2 + \ldots + x_{2k-1})} q(x_1) r(x_2) \cdots q(x_{2k-1})\, \mathrm{d}\vec{x} \\ \sum_{k=1}^{\infty} \int_{\triangle_{2k-1}} e^{+iz(x_1 - x_2 + \ldots + x_{2k-1})} r(x_1) q(x_2) \cdots r(x_{2k-1})\, \mathrm{d}\vec{x} \end{pmatrix}. \tag{9}$$

The fact that the target space of $\mathcal{F}^c$ is indeed the space of pairs of *square-integrable* sequences $l^2_{c,\mathbb{Z}}$ is proved in [14]. We shall see below that for (q, r) close enough to the origin the entire holonomy $\mathrm{Hol}_{(q,r)}$ is indeed completely determined by its off-diagonal part.

The essential theorem about $\mathcal{F}^c$, which is also proved in [14], is the following.

Theorem 1 *The map*

$$\mathcal{F}^c : L^2_c[0, 2\pi] \longrightarrow l^2_{c,\mathbb{Z}}$$

is holomorphic with respect to the complex structures J and K, given by (6) and (8).

It should be mentioned that analyticity of $\mathcal{F}^c$ or of any of its real forms should not be taken for granted. It is true that $\mathcal{F}^c$ remains holomorphic if we replace the spaces $L^2_c[0, 2\pi]$ and $l^2_{c,\mathbb{Z}}$ by the appropriate Schwartz spaces. But it is not analytic if we replace them by L^2-spaces on the entire real line. In this case $\mathcal{F}^c$ is not even $\mathcal{C}^3$. This follows from a result, obtained by Muscalu, Tao and Thiele in [11].

Another essential property of $\mathcal{F}^c$ is the fact that its derivative at the origin is given by the usual linear Fourier transform. Indeed, let $(t(x), u(x)) \in L_c^2[0, 2\pi]$ be an arbitrary element. Then

$$D_0\mathcal{F}^c(t, u) = \frac{\mathrm{d}}{\mathrm{d}s}|_{s=0}\mathcal{F}^c(st, su).$$

From (7) we immediately get

$$(D_0\mathcal{F}^c(t, u))(z) = \Big(\int_0^{2\pi} t(x)e^{-izx}\,\mathrm{d}x, \int_0^{2\pi} e^{izx}u(x)\,\mathrm{d}x\Big) = \Big(F(t)(z), F(u)(-z)\Big),$$

where F denotes the linear Fourier transform. The above calculation gives the weak derivative, but since $\mathcal{F}^c$ is holomorphic, this is also the Fréchet derivative.

The complex spaces $L_c^2[0, 2\pi]$ and $l^2_{c,\mathbb{Z}}$ can be equipped with two real structures which yield important real forms of $\mathcal{F}^c$. The first is given by

$$\tau_1(q, r) = (-\bar{r}, -\bar{q}) \text{ on } L_c^2[0, 2\pi] \quad \text{and} \quad \sigma_1(h, k) = (-\bar{k}, -\bar{h}) \text{ on } l^2_{c,\mathbb{Z}}$$

and the second by

$$\tau_2(q, r) = (\bar{r}, \bar{q}) \text{ on } L_c^2[0, 2\pi] \quad \text{and} \quad \sigma_2(h, k) = (\bar{k}, \bar{h}) \text{ on } l^2_{c,\mathbb{Z}}.$$

These real structures have natural lifting on L-matrices and their holonomies. Note that matrices $L_{(q,r)}$ and $L^g_{(q,r)}$ take values in the Lie algebra $\mathfrak{sl}(2;\mathbb{C})$. The two natural real forms of $\mathfrak{sl}(2;\mathbb{C})$ are the real Lie algebras $\mathfrak{su}(2)$ and $\mathfrak{su}(1,1)$. The corresponding real structures are given by

$$\hat{\tau}_1(\alpha) = -\alpha^*, \quad \text{and} \quad \hat{\tau}_2(\alpha) = -J\cdot\alpha^*\cdot J,$$

where $J = \mathrm{diag}(1, -1)$ and α^* is the conjugate transpose of the matrix $\alpha \in \mathfrak{sl}(2;\mathbb{C})$. These two real structures induce the real structures on the corresponding Lie group $SL(2, \mathbb{C})$, given by

$$\hat{\sigma}_1(g) = (g^{-1})^*, \quad \text{and} \quad \hat{\sigma}_2(g) = J\cdot(g^{-1})^*\cdot J$$

and the associated real forms are of course the groups $SU(2)$ and $SU(1,1)$.

Let us consider the restriction of $\mathcal{F}^c$ on the real forms $R_1 = \{(q(x), -\overline{q(x)})\}$ and $R_2 = \{(q(x), \overline{q(x)})\}$ of the real structures τ_1 and τ_2, respectively. The corresponding L-matrices are $\mathfrak{su}(2)$ and $\mathfrak{su}(1,1)$ valued functions

$$L_q^{g,1} = \begin{pmatrix} 0 & e^{-izx}q(x) \\ -e^{izx}\overline{q(x)} & 0 \end{pmatrix} \quad \text{and} \quad L_q^{g,2} = \begin{pmatrix} 0 & e^{-izx}q(x) \\ e^{izx}\overline{q(x)} & 0 \end{pmatrix},$$

and their respective holonomies are of the form

$$\mathrm{Hol}_q^1(z) = \begin{pmatrix} h_1^d(z) & h_1(z) \\ -\overline{h_1(z)} & \overline{h_1^d(z)} \end{pmatrix}, \quad \text{and} \quad \mathrm{Hol}_q^2(z) = \begin{pmatrix} h_2^d(z) & h_2(z) \\ \overline{h_2(z)} & \overline{h_2^d(z)} \end{pmatrix}.$$

They take values in the groups $SU(2)$ and $SU(1,1)$. In both cases the anti-diagonal of the matrix is determined by the upper right term. Therefore, we can define the two real forms of the complex nonlinear transform $\mathcal{F}^c$ by the formulae

$$\mathcal{F}_1(q) = \{h_1(z)\}_{z\in\mathbb{Z}} \quad \text{and} \quad \mathcal{F}_2(q) = \{h_2(z)\}_{z\in\mathbb{Z}}.$$

From formula (9) we immediately get the expressions for the real nonlinear Fourier transforms

$$\mathcal{F}_1(q)(z) = \sum_{k=1}^{\infty}(-1)^{k-1}\int_{\Delta_{2k-1}} e^{-iz(x_1-x_2+\ldots+x_{2k-1})}q(x_1)\overline{q(x_2)}\cdots q(x_{2k-1})\,\mathrm{d}\vec{x} \qquad (10)$$

$$\mathcal{F}_2(q)(z) = \sum_{k=1}^{\infty}\int_{\Delta_{2k-1}} e^{-iz(x_1-x_2+\ldots+x_{2k-1})}q(x_1)\overline{q(x_2)}\cdots q(x_{2k-1})\,\mathrm{d}\vec{x}. \qquad (11)$$

3. Inverse of the complex nonlinear Fourier transform

In the previous section we have seen that $\mathcal{F}^c$ is a holomorphic map and that its derivative at the origin is given by the linear Fourier transform. The norms on $L^2_c[0,2\pi]$ and $l^2_{c,\mathbb{Z}}$ are obtained by means of Pythagoras' theorem. Thus for every $(t,u) \in L^2_c[0,2\pi]$, we have

$$\|D_0\mathcal{F}^c(t,u)\|^2 = \|\{F(t)(z)\}_{z\in\mathbb{Z}}\|^2 + \|\{F(u)(-z)\}_{z\in\mathbb{Z}}\|^2.$$

By Plancharel theorem, this gives

$$\|D_0\mathcal{F}^c(t,u)\|^2 = \|t\|^2 + \|u\|^2 = \|(u,t)\|^2.$$

In other words, $D_0\mathcal{F}^c$ is an isometry. Theorem 1 and the analytic version of the inverse mapping theorem for Hilbert spaces provide the proof of the following theorem.

Theorem 2 *There exist neighbourhoods $V \subset l^2_{c,\mathbb{Z}}$ and $U \subset L^2_c[0,2\pi]$ of the respective origins and the map*

$$\mathcal{G}^c : V \longrightarrow U$$

which is the inverse of the complex nonlinear Fourier transform $\mathcal{F}^c$. The map $\mathcal{G}^c$ is holomorphic with respect to the complex structures K and J.

Our next task will be to construct an iterative procedure by means of which one can compute $\mathcal{G}^c(h(z),k(z)) = (\mathcal{F}^c)^{-1}(h(z),k(z))$ for an arbitrary element $(h(z),k(z)) \in V$ to any desired degree of accuracy.

Let $w \mapsto (q(x;w), r(x;w))$ be a holomorphic curve in $L^2_c[0,2\pi]$ defined on a disc $\Delta_\alpha \subset \mathbb{C}$ whose center is the origin, and let $(q(x;0), r(x;0)) = (0,0)$. Suppose we can expand our curve into a power series around 0 on the entire Δ_α. We have

$$(q(x;w), r(x;w)) = \sum_{n=1}^{\infty} \frac{w^n}{n!}\,(q_n(x), r_n(x)). \tag{12}$$

Since $\mathcal{F}^c$ is holomorphic, so is the composition $w \mapsto \mathcal{F}^c(q(x;w), r(x;w))$. This holomorphic curve in $l^2_{c,\mathbb{Z}}$ can be represented in the form of the convergent power series

$$\mathcal{F}^c(q(x;w), r(x;w)) = \sum_{n=1}^{\infty} \frac{w^n}{n!}\,(h_n(z), k_n(z)).$$

We shall give the formulae which connect the terms (q_n, r_n) and (h_n, k_n) of the above expansions. Let

$$L^N_{(q(w),r(w))}(x;z) = \begin{pmatrix} 0 & L_1 & L_2 & \cdots & \frac{1}{N!}L_N \\ 0 & 0 & L_1 & \cdots & \frac{1}{(N-1)!}L_{N-1} \\ 0 & 0 & 0 & \cdots & \frac{1}{(N-2)}L_{N-2} \\ \vdots & \vdots & \vdots & \vdots & \vdots \\ 0 & 0 & 0 & \cdots & L_1 \\ 0 & 0 & 0 & \cdots & 0 \end{pmatrix},$$

where

$$L_k = \begin{pmatrix} 0 & e^{-izx}q_k(x) \\ e^{izx}r_k(x) & 0 \end{pmatrix},$$

and (q_k, r_k) are the terms of the expansion (12). Because the matrix L^N is strictly upper triangular, any product of more than N matrices of this form is equal to zero. In particular

$$L^N_{(q,r)}(x_1;z)\cdot L^N_{(q,r)}(x_2,z)\cdots L^N_{(q,r)}(x_m;z) = 0, \quad \text{if } m \geq N+1.$$

Therefore, the Dyson series

$$\mathrm{Hol}^N_{(q(w),r(w))}(z) = I + \sum_{k=1}^{N} \int_{\Delta_k} L^N_{(q(w),r(w))}(x_1; z) \cdot L^N_{(q(w),r(w))}(x_2; z) \cdots L^N_{(q(w),r(w))}(x; z)\, \mathrm{d}\vec{x} \quad (13)$$

for the holonomy of $L^N_{(q(w),r(w))}$ is actually a *finite sum* and is therefore explicitly computable in the form of a finite sum of integrals. This holonomy is again an upper-triangular block-Toeplitz matrix of the form

$$\mathrm{Hol}^N_{(q(w),r(w))}(z) = \begin{pmatrix} I & H_1(z) & H_2(z) & \cdots & \frac{1}{N!}H_N(z) \\ 0 & I & H_1(z) & \cdots & \frac{1}{(N-1)!}H_{N-1}(z) \\ 0 & 0 & I & \cdots & \frac{1}{(N-2)}H_{N-2}(z) \\ \vdots & \vdots & \vdots & \vdots & \vdots \\ 0 & 0 & 0 & \cdots & H_1(z) \\ 0 & 0 & 0 & \cdots & I \end{pmatrix}.$$

Let us now consider the usual 2×2 holonomy $\mathrm{Hol}_{(q(w),r(w))}(z)$ as a function of $w \in \Delta_\alpha \subset \mathbb{C}$. The composition $w \mapsto \mathcal{F}^c(q(w), r(w))$ is a holomorphic curve in $l^2_{c,\mathbb{Z}}$, which means that the anti-diagonal of $\mathrm{Hol}_{(q(w),r(w))}$ is holomorphic. In a completely analogous way one can prove that also the diagonal is holomorphic, so $w \mapsto \mathrm{Hol}_{(q(w),r(w))}$ is a matrix-valued holomorphic curve, defined on the disc Δ_α. (For the details, see [14].) Therefore, we can expand it into the Taylor series around the origin in Δ_α,

$$\mathrm{Hol}_{(q(w),r(w))}(z) = I + \sum_{n=1}^{\infty} \frac{w^n}{n!}\, \hat{H}_n(z) = I + \sum_{n=1}^{\infty} \frac{w^n}{n!} \begin{pmatrix} h^d_n(z) & h_n(z) \\ k_n(z) & k^d_n(z) \end{pmatrix}.$$

It is easy to see that the blocks of $\mathrm{Hol}^N_{(q(w),r(w))}$ and the terms of the above expansion are equal,

$$\hat{H}_n(z) = H_n(z), \quad \text{for every } n \in \mathbb{N}. \quad (14)$$

On the one hand, we have $\mathrm{Hol}^N_{(q(w),r(w))}(z) = \Psi^N(2\pi; z)$, where $\Psi^N(x; z)$ is the solution of the initial value problem

$$\Psi^N_x(x; z) = L^N_{(q(w),r(w))} \cdot \Psi^N(x; z), \quad \Psi^N(0; z) = Id$$

and this is equivalent to the system

$$(\psi_n)_x = \sum_{k=1}^{n} \binom{n}{k} L_k \cdot \psi_{n-k}, \quad \psi_n(0; z) = 0 \text{ for } n > 0 \text{ and } \psi_0(x; z) = Id,$$

where ψ_n are the 2×2 blocks of Ψ^N. On the other hand, expanding the 2×2 initial value problem

$$\Phi_x(x; z; w) = L_{(q(w),r(w))} \cdot \Phi(x; z; w), \quad \Phi(0; z; w) = Id$$

with respect to the parameter w yields precisely the same system for the terms $\phi_n(x; z)$ of the expansion $\Phi(x; z; w) = I + \sum_{n=1}^{\infty} (w^n/n!)\ \phi_n(x; z)$. The uniqueness part of the fundamental existence theorem for the ordinary differential equations gives $\phi_n(x; z) = \psi_n(x; z)$ and upon evaluation at $x = 2\pi$ this yields (14).

Equation (14) enables us to use the finite expression (13) for the calculation of the components $h_n(z)$ and $k_n(z)$ of the complex nonlinear Fourier transform

$$\mathcal{F}^c(q(w), r(w)) = \sum_{n=1}^{\infty} \frac{w^n}{n!}(h_n(z), k_n(z))$$

of the holomorphic curve $(q(w), r(w))$. A straightforward calculation gives

$$h_n(z) = \sum_{k=1}^{\hat{n}} \sum_{|J|=n} \alpha^J \int_{\triangle_{2k-1}} e^{-iz(x_1-x_2+\ldots+x_{j_{2k-1}})} q_{j_1}(x_1) r_{j_2}(x_2) \cdots q_{j_{2k-1}}(x_{2k-1})\, \mathrm{d}\vec{x} \quad (15)$$

$$k_n(z) = \sum_{k=1}^{\hat{n}} \sum_{|J|=n} \alpha^J \int_{\triangle_{2k-1}} e^{iz(x_1-x_2+\ldots+x_{2k-1})} r_{j_1}(x_1) q_{j_2}(x_2) \cdots r_{j_{2k-1}}(x_{2k-1})\, \mathrm{d}\vec{x}, \quad (16)$$

where $J = (j_1, j_2, \ldots, j_{2k-1})$ are multi-indices with $j_r \geq 1$, and

$$\alpha^J = \binom{n}{j_1, j_2, \ldots, j_{2k-1}}$$

are the multinomial coefficients. The summation limit $\hat{n}$ is equal to $n/2$ when n is even and to $(n+1)/2$ when n is odd. Rewriting formulae (15) and (16) in a more compact way yields the following proposition.

Proposition 1 *For every positive integer n, we have*

$$h_n(z) = \int_0^{2\pi} q_n(x) e^{-izx}\, \mathrm{d}x + \mathcal{H}_n(q_1, \ldots, q_{n-2}, r_1, \ldots, r_{n-2}) \quad (17)$$

$$k_n(z) = \int_0^{2\pi} r(x) e^{izx}\, \mathrm{d}x + \mathcal{K}_n(q_1, \ldots, q_{n-2}, r_1, \ldots, r_{n-2}), \quad (18)$$

where for $n > 1$,

$$\mathcal{H}_n(z) = \sum_{k=2}^{\hat{n}} \sum_{|J|=n} \alpha^J \int_{\triangle_{2k-1}} e^{-iz(x_1-x_2+\ldots+x_{j_{2k-1}})} q_{j_1}(x_1) r_{j_2}(x_2) \cdots q_{j_{2k-1}}(x_{2k-1})\, \mathrm{d}\vec{x} \quad (19)$$

$$\mathcal{K}_n(z) = \sum_{k=2}^{\hat{n}} \sum_{|J|=n} \alpha^J \int_{\triangle_{2k-1}} e^{iz(x_1-x_2+\ldots+x_{2k-1})} r_{j_1}(x_1) q_{j_2}(x_2) \cdots r_{j_{2k-1}}(x_{2k-1})\, \mathrm{d}\vec{x}, \quad (20)$$

and $\mathcal{H}_1(z) = \mathcal{K}_1(z) = 0$.

Proof: All that remains to be checked is the fact that the expressions $\mathcal{H}_n$ and $\mathcal{K}_n$ depend only on $q_1, \ldots q_{n-2}, r_1, \ldots, r_{n-2}$. Recall that all components j_r of J are strictly positive and that $|J| = n$. Therefore the terms of $\mathcal{H}_n$ which contain the functions q_j and r_k with the highest values of indices j and k are integrals over $\triangle_3$ of the products of the form $q_{\sigma(1)}(x_1) r_{\sigma(1)}(x_2) q_{\sigma(n-2)}(x_3)$, where σ is an arbitrary permutation of the symbols $\{1, 1, n-2\}$. Such terms of $\mathcal{K}_n$ are the integrals over $\triangle_3$ of all the products $r_{\sigma(1)}(x_1) q_{\sigma(1)}(x_2) r_{\sigma(n-2)}(x_3)$, which proves our claim. □

The first summands on the right-hand sides of formulae (17) and (18) are the linear Fourier transforms of the functions q_n and r_n. Therefore, equations (17) and (18) can be solved for q_n and r_n. In our case, the inverse linear Fourier transform G is given by

$$G(l(z))(x) = \sum_{z=-\infty}^{\infty} l(z) e^{izx}.$$

Proposition 2 *Let $w \mapsto (h(z;w), k(z;w))$ be a holomorphic curve in $l^2_{c,\mathbb{Z}}$ which starts at the origin and lies entirely in the neighbourhood $V \subset l^2_{c,\mathbb{Z}}$, given in theorem 2. Let its Taylor expansion be*

$$(h(z;w), k(z;w)) = \sum_{n=1}^{\infty} \frac{w^n}{n!}\ (h_n(z), k_n(z)).$$

Then the inverse image of $(\mathcal{F}^c)^{-1}(h(z;w), k(z;w))$ is the convergent series

$$(\mathcal{F}^c)^{-1}(h(z;w), k(z;w))(x) = \sum_{n=1}^{\infty} \frac{w^n}{n!}\ (q_n(x), r_n(x))$$

of elements in $L^2_c[0, 2\pi]$ whose components are given by

$$q_n(x) = G(h_n)(x) - G(\mathcal{H}_n(q_1, \ldots q_{n-2}, r_1, \ldots, r_{n-2}))(x) \tag{21}$$
$$r_n(x) = G(k_n)(-x) - G(\mathcal{K}_n(q_1, \ldots, q_{n-2}, r_1, \ldots, r_{n-2}))(-x). \tag{22}$$

The terms $\mathcal{H}_n$ and $\mathcal{K}_n$ are given by (19) and (20).

□

Let now $(h(z), k(z))$ be an arbitrary element in $V \subset l^2_{c,\mathbb{Z}}$. If we assume V to be convex, the line $w \mapsto (wh(z), wk(z))$, defined for $|w| \leq 1$, is a holomorphic curve in V. If we apply the above proposition to such a line and evaluate at $w = 1$, we get the following theorem.

Theorem 3 *Let $(h(z), k(z))$ be an arbitrary element of $V \subset l^2_{c,\mathbb{Z}}$. The inverse image $(\mathcal{F}^c)^{-1}(h(z), k(z))$ is given by a convergent series*

$$(\mathcal{F}^c)^{-1}(h(z), k(z)) = \sum_{n=1}^{\infty} \frac{1}{n!}(q_n(x), r_n(x))$$

whose terms are given by the formulae

$$\begin{aligned} q_1(x) &= G(h_1)(x) \\ q_{2n}(x) &= 0 \\ q_{2n+1}(x) &= -G(\mathcal{H}_{2n+1}(q_1, \ldots, q_{2n-1}, r_1, \ldots, r_{2n-1}))(x) \end{aligned}$$

and

$$\begin{aligned} r_1(x) &= G(k_1)(-x) \\ r_{2n}(x) &= 0 \\ r_{2n+1}(x) &= -G(\mathcal{K}_{2n+1}(q_1, \ldots, q_{2n-1}, r_1, \ldots, r_{2n-1}))(-x), \end{aligned}$$

where

$$\begin{aligned} \mathcal{H}_{2n+1} &= \sum_{k=2}^{n+1} \sum_{|J|=2n+1} \alpha^J \int_{\triangle_{2k-1}} e^{-iz(x_1-x_2+\ldots+x_{2k-1})} q_{j_1}(x_1) r_{j_2}(x_2) \cdots q_{j_{2k-1}}(x_{2k-1}) \\ \mathcal{K}_{2n+1} &= \sum_{k=2}^{n+1} \sum_{|J|=2n+1} \alpha^J \int_{\triangle_{2k-1}} e^{iz(x_1-x_2+\ldots+x_{2k-1})} r_{j_1}(x_1) q_{j_2}(x_2) \cdots r_{j_{2k-1}}(x_{2k-1}). \end{aligned}$$

All the components j_k of the multi-indices $J = (j_1, j_2, \ldots j_k)$ are odd positive integers and α^J is the multinomial coefficient.

Proof: The expressions have been derived above and the convergence is a consequence of the holomorphicity of $\mathcal{F}^c$, proved in [14]. It only remains to prove that all the even terms (q_{2n}, r_{2n}) are equal to zero. This is equivalent to the claim that all the components of the multi-indices J are odd. This can be proved by induction. Suppose that for some n the term q_{2n} is non-zero. Then by formula (19), there should exist a multi-index $J = (j_1, \ldots, j_{2k-1})$ such that $j_1 + \ldots + j_{2k-1} = 2n$. But by the induction hypothesis this would mean that the sum of an odd number of odd summands would have to be even, which is impossible. In the same way we can show that $r_{2n} = 0$, for every $n \in \mathbb{N}$. □

From the above result and from formulae (10), (11) of section 2, we immediately get the expressions for the inverses $\mathcal{F}_i^{-1}$ of the real forms $\mathcal{F}_i$, $i = 1, 2$. For $i = 1, 2$, we have

$$\mathcal{F}_i^{-1}(h(z))(x) = \sum_{n=1}^{\infty} \frac{1}{n!}\, q_{n,i}(x),$$

where

$$\begin{aligned} q_{1,i}(x) &= G(h)(x) \\ q_{2n,i}(x) &= 0 \\ q_{2n+1,i}(x) &= -G(\mathcal{H}_{2n+1,i}(q_1, \ldots, q_{2n-1}))(x) \end{aligned}$$

and

$$\begin{aligned} \mathcal{H}_{2n+1,1(z)} &= \sum_{k=2}^{\infty}(-1)^{k-1} \sum_{|J|=2n+1} \alpha^J \int_{\triangle_{2k-1}} e^{-iz(x_1-x_2+\ldots+x_{2k-1})} q_{j_1}(x_1)\overline{q_{j_2}(x_2)} \cdots q_{j_{2k-1}}(x_{2k-1}), \\ \mathcal{H}_{2n+1,2}(z) &= \sum_{k=2}^{\infty} \sum_{|J|=2n+1} \alpha^J \int_{\triangle_{2k-1}} e^{-iz(x_1-x_2+\ldots+x_{2k-1})} q_{j_1}(x_1)\overline{q_{j_2}(x_2)} \cdots q_{j_{2k-1}}(x_{2k-1}). \end{aligned}$$

References

[1] Ablowitz M J, Kaup D J, Newell A C and Segur H 1973 Nonlinear-evolution equations of physical significance *Phys. Rev. Lett.* **31** 125–27
[2] Ablowitz M J, Kaup D J, Newell A C and Segur H 1974 The inverse scattering transform - Fourier analysis for nonlinear problems *Stud. Appl. Math.* **53** 249-315
[3] Christ M and Kiselev A 1998 Absolutely continuous spectrum for one-dimensional Schrödinger operators with slowly decaying potentials: some optimal results *J. Amer. Math. Soc.* **11** 771–97
[4] Christ M and Kiselev A 2001 Maximal functions associated to filtrations *J. Funct. Anal.* **179** 409–25
[5] Frayer C, Hryniv R O, Mykytyuk Ya V and Perry P A 2009 Inverse scattering for Schrödinger operators with Miura potentials: I. Unique Riccati representatives and ZS-AKNS systems *Inverse Problems* **25** 11500–32
[6] Fokas A S 2002 Integrable nonlinear evolution equations on the half-line *Commun. Math. Phys.* **230** 1–39
[7] Fokas A S and Gelfand I M 1994 Integrability of linear and lonlinear evolution equations and the associated nonlinear Fourier transform *Lett. Math. Phys.* **32** 189–210
[8] Fokas A S and Sung L Y 2005 Generalized Fourier transforms, their nonlinearization and the imaging of the Brain *Notices Amer. Math. Soc.* **52** 1178–92
[9] Hryniv R O, Mykytyuk Ya V and Perry P A 2010 Inverse scattering for Schrödinger operators with Miura potentials: II. Different Riccati representatives *Comm. Partial Differential Equations* **36** 1587–623
[10] Muscalu C, Tao T and Thiele C 2003 A Carleson type theorem for a Cantor group model of the scattering transform *Nonlinearity* **16** 219–46
[11] Muscalu C, Tao T and Thiele C 2003 A counterexample to a multilinear end-point question by Christ and Kiselev *Math. Res. Lett.* **10** 237–46
[12] Saksida P 2011 On zero-curvature condition and Fourier analysis *J. Phys. A: Math. Gen.* **44** 85203–85222
[13] Saksida P 2012 Lax pairs and Fourier analysis: The case of sine-Gordon and Klein-Gordon equations *J. Phys.: Conf. Ser.* **343** 012109 (10pp)
[14] Saksida P 2013 On the nonlinear Fourier transform associated with periodic AKNS-ZS system and itsinverse *J. Phys. A: Math. Theor.* **46** 465204 (22pp)
[15] Tao T and Thiele C 2012 Nonlinear Fourier analysis (arXiv:1201.5129)

Position and momentum bases on the sphere for the monochromatic Maxwell fish-eye

Cristina Salto-Alegre and Kurt Bernardo Wolf

Instituto de Ciencias Físicas, Universidad Nacional Autónoma de México, Av. Universidad s/n, Cuernavaca, Morelos 62251, México

E-mail: bwolf@fis.unam.mx

Abstract. The sphere is well understood manifold, as is the basis of spherical harmonics for functions thereof. A stereographic projection of the sphere realizes the Maxwell fish-eye, an optical medium whose refractive index distribution is such that light rays travel in circles. Definite angular momentum on the sphere corresponds with monochromatic light in the fish-eye. A contraction of these manifolds (as the radius of the sphere grows without bound) results in a plane medium of wave functions subject to the Helmholtz equation. In the latter, there is the continuous basis of 'momenta' (plane waves) and a denumerably infinite basis of 'positions' and 'normal derivatives' (Bessel $\sim J_0(x)$ and $\sim J_1(x)/x$ functions) for the Hilbert space of these wavefields. We present the pre-contraction of these bases to the finite-dimensional systems of fish-eye medium and the sphere. The 'momentum' basis is a subset of Sherman-Volobuyev functions; in this paper we propose new 'position' and 'normal derivative' bases for this finite system. The bases are not orthogonal, so their measure is non-local, and here we find their dual bases.

1. The harmonic basis on the sphere

Phase space is a concept created in classical mechanics that applies as well in geometric optics [1]. In quantum mechanics and in wave optics, there are several approaches to phase space through the definition of the Wigner [2] and Wigner-like [3] quasidistribution functions. Those that follow Wigner's original formulation hinge on the Heisenberg-Weyl algebra of noncommuting operators of position and momentum; other formulations also rely on a supporting Lie algebra [4].

In this work we start on a different path, based on the monochromatic Maxwell fish-eye wave-optical system, presenting what appear to be the natural wavefields of definite momentum and position. Since this system is a stereographic projection of the sphere on a plane or higher-dimensional flat manifold [5, 6, 7, 8, 9, 10, 11], [12, Sect. 6.4], the analysis remits us to the sphere, treated as in angular momentum theory through the well-known spherical harmonics, whose relevant properties are recalled in Sect. 2, while the link to the fish-eye model is summarized in Sect. 3. In our case we have not a continuous, but a *finite* system, i.e., the Hilbert space of wavefields for which we provide the bases is finite-dimensional.

The basis of functions of 'most definite momentum' are known as the Sherman-Volobuyev functions, of which we need only a finite number for fixed angular momentum in Sect. 4, because they correspond to monochromatic wavefields in the optical model. The basis of 'most definite position' presented in Sect. 5 actually involves two sub-bases, of positions and of normal

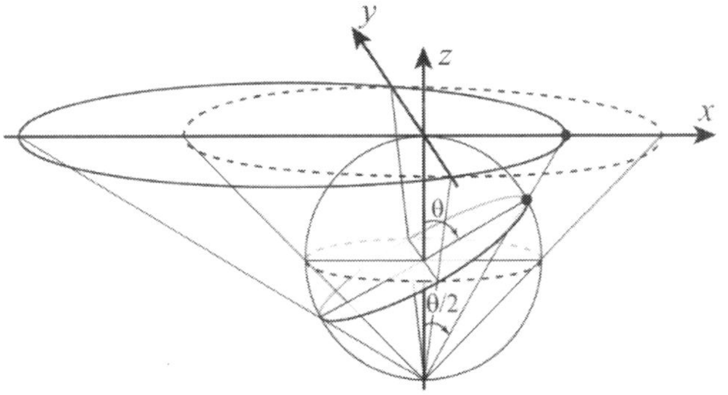

Figure 1. The stereographic map from the bottom pole of the sphere $\mathcal{S}^2$ of radius ρ to the $\mathcal{R}^2$ plane tangent to the top pole. This plane is the two-dimensional optical plane of a Maxwell fish-eye. Maximal circles in the former map onto circles in the latter (continuous lines). The equator of the sphere maps onto the centered 'minimal circle' of the fish-eye plane of radius 2ρ (dashed lines). Time evolution lets the sphere rotate freely around any axis, mapping as the motion of a 'light point' in the fish-eye.

derivatives, which were proposed in Ref. [13]. Here we also examine their dual bases in Sect. 6. From that reference, in Sect. 7 we summarize the contraction of the algebra, sphere and optical fish-eye system, to a homogeneous medium subject to the Helmholtz equation, where we find that the limit wavefields match those of a related development addressed some time ago [14], where the inner product of the latter is non-local.

As we underline in the concluding Sect. 8, we are building a *finite* system which is a deformed (i.e. pre-contracted) model of a wave-optical system that has had important applications in waveguide and radar theory, with the intent of setting up an appropriate phase space grounded on finite function bases, rather than on a Lie algebra of operators.

2. The harmonic basis on the sphere

The quantum (or classical) angular momentum operators $\vec{L} = -\mathrm{i}\,\vec{s} \times \vec{\nabla}_{\vec{s}}$ (or Poisson operators $\{\vec{L}, \circ\}$), $\vec{s} \equiv (x, y, z) \in \mathcal{R}^3$ generate the Lie algebra $\mathsf{so}(3)$ of rotations, and obey the commutation (or Poisson bracket) relations $[L_i, L_j] = \mathrm{i}L_k$, with i, j, k a cyclic permutation of x, y, z. Free rotation by ω about a unit axis $\hat{n}$ is generated by $\exp(-\mathrm{i}\omega t\, \hat{n} \cdot \vec{L})$; these rotations commute with the monochromatic wave equation $L^2 F(\vec{s}) = \ell(\ell+1)\, F(\vec{s})$, where $L^2 = \sum_i L_i^2$ is the invariant Casimir operator of $\mathsf{so}(3)$, reduced to the sphere $|\vec{s}| = \rho$, for fixed nonnegative ℓ. Spherical harmonics $Y_{\ell,m}(\theta, \phi)$ are the eigenfunctions of L^2 and L_z; their explicit expression is too well known to be reproduced here [15, Eqs. (3.141–153)]; the index m can take the $2\ell + 1$ integer-spaced values $m|_{-\ell}^{\ell}$. In particular, we shall use specifically the *extreme* ($m = \ell$) and the *middle* ($m = 0$) spherical harmonics:

$$Y_{\ell,\ell}(\theta, \phi) = \frac{(-1)^\ell}{2^\ell\, \ell!} \sqrt{\frac{(2\ell+1)!}{4\pi}}\, (\sin\theta\, e^{\mathrm{i}\phi})^\ell, \quad Y_{\ell,0}(\theta, \phi) = \sqrt{\frac{2\ell+1}{4\pi}}\, P_\ell(\cos\theta). \tag{1}$$

The spherical harmonics form an orthonormal and complete basis for the Hilbert space $\mathcal{L}^2(\mathcal{S}^2)$, whose inner product integrates over the sphere. We keep this inner product, but to be 'monochromatic' we fix ℓ, so $(Y_{\ell,m}, Y_{\ell,m'})_{\mathcal{S}^2} = \delta_{m,m'}$ and $\{Y_{\ell,m}\}_{m=-\ell}^{\ell}$ will be an orthonormal basis for a $(2\ell+1)$-dimensional space that we indicate by $\mathcal{M}_\ell$. Azimuthal and polar rotations yield the phases and linear combinations

$$\exp(-\mathrm{i}\alpha L_z)\,Y_{\ell,m}(\theta,\phi) \;=\; e^{-\mathrm{i}m\alpha}Y_{\ell,m}(\theta,\phi) = Y_{\ell,m}(\theta,\phi{-}\alpha), \tag{2}$$

$$\exp(-\mathrm{i}\beta L_y)\,Y_{\ell,m}(\theta,\phi) \;=\; \sum_{m'=-\ell}^{\ell} Y_{\ell,m'}(\theta,\phi)\,d^{\ell}_{m',m}(\beta), \tag{3}$$

where $d^{\ell}_{m',m}(\beta)$ are the Wigner *little-d* functions [15, Eq. (3.72)], whose explicit form we also need not reproduce here in full. For later use we record the following relations [15, Eq. (3.141)]

$$Y_{\ell,m}(\theta,\phi) = \sqrt{\frac{2\ell+1}{4\pi}}\,d^{\ell}_{m,0}(\theta)\,e^{\mathrm{i}m\phi}, \quad d^{\ell}_{m,\ell}(\tfrac{1}{2}\pi) = \frac{1}{2^{\ell}}\sqrt{\binom{2\ell}{\ell+m}}. \tag{4}$$

3. Stereographic projection on the fish-eye

The stereographic projection from the bottom pole of the sphere to a plane tangent to the top pole, is shown in geometric terms in Fig. 1. The map between polar coordinates and measures (θ,ϕ) on the sphere (of radius ρ) and polar coordinates on the plane (r,ϕ), is

$$\tan\tfrac{1}{2}\theta = r/2\rho, \qquad \mathrm{d}\cos\theta\,\mathrm{d}\phi = \Big(\frac{4\rho}{r^2+4\rho^2}\Big)^2 r\,\mathrm{d}r\,\mathrm{d}\phi. \tag{5}$$

Luneburg [5] investigated the transformation (5) and Buchdahl [7] recognized its symmetry of the fish-eye under rotations in one higher dimension. The stereographic map has been extended to a *canonical* map between the phase spaces of the sphere and plane in Ref. [9]. Since this map is point-to-point in the position spaces of Fig. 1, the map between momentum spaces will be *comatic*, i.e., if $(\mathbf{n},\sigma)$ is the projection of the unit vector $\hat{n}\in\mathcal{S}^2$ in the σ-hemisphere ($\sigma=\pm1$) on the equatorial x–y plane, and $\mathbf{r}\in\mathcal{R}^2$ is on the projected plane, then as detailed in [12, Sect. 6.4], when $U_\sigma(|\mathbf{n}|)$ is any function on the sphere, then there is a canonical transformation to the projected plane,

$$\mathbf{r} = U_\sigma(|\mathbf{n}|)\mathbf{n}, \quad \mathbf{p} = \frac{1}{U_\sigma(|\mathbf{n}|)}\mathbf{v} + W_\sigma(|\mathbf{n}|)\mathbf{n}\mathbf{n}\cdot\mathbf{v}, \tag{6}$$

where $\mathbf{p}$ is the conjugate momentum in the projected plane, $\mathbf{v}$ is the projection of momentum from the sphere on its x-y equatorial plane, and $W_\sigma(|\mathbf{n}|)$ is found in terms of $U_\sigma(|\mathbf{n}|)$ and its derivative $U'_\sigma(|\mathbf{n}|)$. Since this map is canonical, their Poisson brackets satisfy $\{n_i,v_j\}=\delta_{i,j} \Leftrightarrow \{r_i,p_j\}=\delta_{i,j}$. We know that classical and quantum mechanics follow each other under comatic maps such as (6), from Poisson brackets to commutators, so that canonicity becomes *unitarity* in wave-optical models. Since these transformations are invertible, they conserve the information contained in their corresponding states.

Note that two antipodal points on the sphere, (θ,ϕ) and $(\pi{-}\theta,\phi{+}\pi)$ map onto *conjugate* points in the fish-eye plane, (r,ϕ) and $(\bar{r},\phi{+}\pi)$, with $r\bar{r}=(2\rho)^2$. The sphere rotates by π between these points along any of the maximal circles that join the pair; if the sphere rotates freely, on the fish-eye plane thus, all rays departing from one point will meet at the conjugate point in the same time, so the Maxwell fish-eye is a perfect focusing system in space and also in time. This optical system is referred to Maxwell [16] because he found that a refractive index

$$n(r) = 4n_o\rho^2/(r^2+4\rho^2), \qquad n_o := n(0), \tag{7}$$

would make light go in circles. Actually, only *part* of the fish-eye can be realized optically, because $n(r)\geq 1$ is a physical constraint while n_o cannot be larger than 2.5 or so. But it is the *half*-fish-eye that receives technological application in ‘bubble-top’ radar installations using

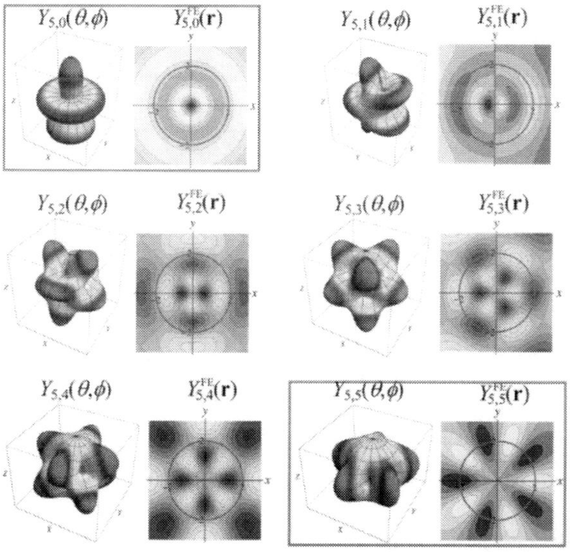

Figure 2. Real parts of the spherical harmonics (10) $Y_{5,m}(\theta,\phi)$, $m|_0^5$, on the sphere (left of every pair), and $Y^{\rm FE}_{5,m}(\mathbf{r})$ on the fish-eye plane (right of every pair); on the latter, the minimal circle projected from the equator of the sphere is shown. The proposed 'position' $Y_{5,0}$ and 'momentum' $Y_{5,5}$ functions are framed.

microwaves, where rubber foams serve to impress a refractive index in the concavity of a half-sphere antenna. A pulse at the back point of the antenna, seen as the source of circular arcs, will find the face of the half-sphere perpendicular to all of them, producing the same pulse in a beam of parallel lines. And conversely for detection of the target echo.

Finally, note that the measure in (5) is an *obliquity* factor for the map, and is proportional to the square of the refractive index (7). Hence, when functions F on the sphere $\Omega \equiv (\theta, \phi)$ are corresponded with functions f on the plane $\mathbf{r}$ through

$$f(\mathbf{r}) = \frac{4\rho}{r^2 + 4\rho^2} F(\Omega(\mathbf{r})), \quad F(\Omega) = \rho \sec^2 \tfrac{1}{2}\theta \; f(\mathbf{r}(\Omega)). \tag{8}$$

The inner products in the two spaces will relate directly as

$$(F, G)_{\mathcal{S}^2} := \int_{\mathcal{S}^2} \mathrm{d}^2\Omega \, F(\Omega)^* \, G(\Omega) = \int_{\mathsf{R}^2} \mathrm{d}^2\mathbf{r} \, f(\mathbf{r})^* \, g(\mathbf{r}) =: (f, g)_{\mathcal{R}^2}. \tag{9}$$

In particular, the basis of spherical harmonics of the first Section will map on a basis of wavefields on the fish-eye (FE) as

$$Y^{\rm FE}_{\ell,m}(\mathbf{r}) = \frac{4\rho}{r^2 + 4\rho^2} Y_{\ell,m}(\Omega(\mathbf{r})). \tag{10}$$

The set of functions $\{Y^{\rm FE}_{\ell,m}(\mathbf{r})\}_{m=-\ell}^{\ell}$ will be orthonormal and complete in $\mathcal{M}_\ell$ under the inner product $(f,g)_{\mathcal{R}^2}$ in (9). Figure 2 shows spherical harmonics on the sphere and their corresponding projections on the fish-eye medium.

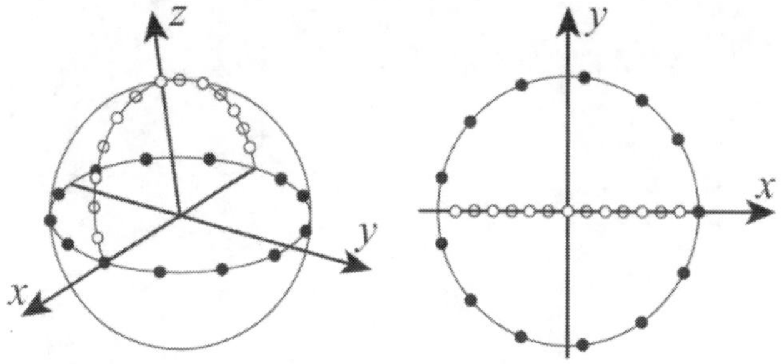

Figure 3. *Left*: The rotated $+z$ axes of the nShV basis of 'momentum' functions $V_{\ell,m}$ over the sphere (marked •) lie around the equator; those of the Legendre 'position' and 'normal derivative' bases $\Lambda_{\ell,m}$ (○) and $\Lambda'_{\ell,m}$ (⊖), are in the upper half-meridian in the x–z plane. *Right*: Their stereographic projections on the fish-eye plane for the nShV basis are points on the minimal circle, while for the two Legendre sub-bases are alternating points on the x-axis inside the minimal circle .

4. The momentum basis

We singled out the extreme and middle spherical harmonics $Y_{\ell,\ell}$ and $Y_{\ell,0}$ in (1) because we intend these to be the 'typical' momentum and position functions to be found on the sphere: the extreme $Y_{\ell,\ell}(\theta,\phi) \propto e^{i\ell\phi}$ exhibits an equatorial belt of oscillations which, on a sphere of radius ρ, will accommodate ℓ wavelengths $\lambda = 2\pi\rho/\ell$; the middle $Y_{\ell,0}(\theta,\phi) \propto P_\ell(\cos\theta)$ shows a maximum at the top pole $\theta = 0$, and a maximum (or minimum) at $\theta = \pi$ for ℓ even (or odd), and has the least second moment (with $\cos^2\theta$) among all $2\ell{+}1$ spherical harmonics of angular momentum ℓ in $\mathcal{M}_\ell$.

We propose to use these to generate bases of 'momentum' and of 'position' for wavefunctions on the sphere, supported by their appearance in the contraction limit $\rho, \ell \to \infty$, to be seen in Sect. 6, that mantains the ratio $\rho/\ell = \lambda/2\pi = 1/k$ constant, so that the functions on the sphere and fish-eye plane come to be solutions of the Helmholtz equation of wavenumber k, plane waves and localized Bessel J_0's, whose properties in this regard have been studied in [14] and [13].

By means of $2\ell{+}1$ distinct rotations of the sphere by $R_m, R_{m'} \in \mathsf{SO}(3)$, and as long as the rotated functions $R_m : Y_{\ell,\ell}$ and $R_{m'} : Y_{\ell,0}$ are linearly independent, they can be used to expand any function $F_\ell(\theta,\phi) \in \mathcal{M}_\ell$ of the same angular momentum. To obtain analytic results we should choose carefully these rotations $R(\alpha_m, \beta_m, \gamma_m)$ that we characterize by the axis $\hat{n}(\alpha_m, \beta_m)$ to which they send the top pole of the sphere.

For the momentum basis we choose $2\ell{+}1$ equidistant points α_m on the azimuthal circle $\alpha_m := 2\pi\, m/(2\ell+1)$, $m|_{-\ell}^{\ell}$, to span $\mathcal{M}_\ell$ with the extreme harmonics, defining the momentum functions as

$$\begin{aligned} V_{\ell,m}(\theta,\phi) &:= \exp(-\mathrm{i}\alpha_m L_z)\exp(-\mathrm{i}\tfrac{1}{2}\pi L_y) Y_{\ell,\ell}(\theta,\phi) \\ &= \sum_{m'=-\ell}^{\ell} \exp\Big(\frac{-2\pi\mathrm{i}mm'}{2\ell+1}\Big) d^{\ell}_{m',\ell}(\tfrac{1}{2}\pi)\, Y_{\ell,m'}(\theta,\phi) \\ &= V_{\ell,0}(\theta, \phi{-}\alpha_m), \quad m|_{-\ell}^{\ell}. \end{aligned} \tag{11}$$

These we call the *normalized* Sherman-Volobuyev (nShV) functions over the sphere, since they have unit norm under the $\mathcal{L}^2(\mathcal{S}^2)$ inner product, and whose ensemble $\ell|_0^\infty$ has been studied by Sherman and Volobuyev [17, 18] as 'Fourier analysis' for arbitrary functions or distributions over the sphere.

The $V_{\ell,m}(\theta,\phi)$'s chosen in (11) exhibit their oscillation belts on meridian circles equally distributed around the azimuthal circle α of directions, as shown in Fig. 3. These nShV functions

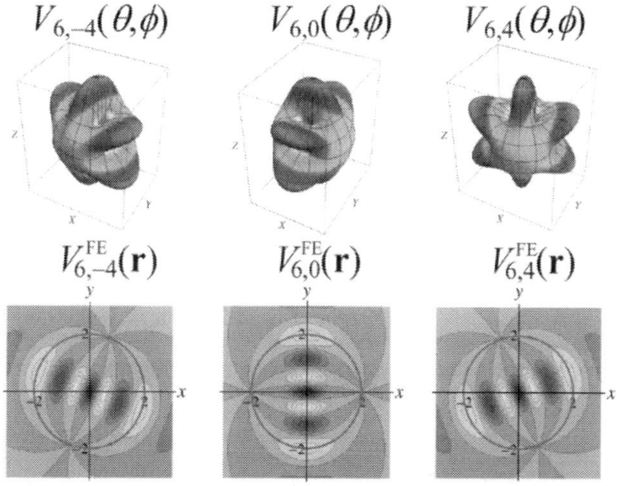

Figure 4. Real part of the normalized Sherman-Volobuyev functions $V_{6,m}(\theta,\phi)$, $m=-4,\,0,\,4$ in (11) on the sphere (top), and their projections on the fish-eye medium (bottom) with the minimal circle. The full basis is obtained through rotations around the z-axis by $\alpha_m = 2\pi\, m/(2\ell+1)$, $m|_{-\ell}^{\ell}$.

are not orthogonal, but

$$(V_{\ell,m}, V_{\ell,m'})_{\mathcal{S}^2} = d^{\ell}_{\ell,\ell}(\alpha_{m'}-\alpha_m) = \Big(\cos\frac{\pi(m'-m)}{2\ell+1}\Big)^{2\ell}. \tag{12}$$

No pair of nShV functions is orthogonal because $\frac{1}{2}(\alpha_{m'}-\alpha_m)=\frac{1}{2}\pi$ would imply $m'-m=\ell+\frac{1}{2}$, but all m's are integers. Also, one can verify that $\det|d^{\ell}_{\ell,\ell}(\alpha_{m'}-\alpha_m)|>0$, so the set of $2\ell+1$ nShV vectors form a complete basis for $\mathcal{M}_\ell$.

Any monochromatic function $F(\theta,\phi)\in\mathcal{L}^2(\mathcal{S}^2)$ can be thus expanded in the nShV basis,

$$F(\theta,\phi) = \sum_{m=-\ell}^{\ell} \widetilde{F}_m\, V_{\ell,m}(\theta,\phi). \tag{13}$$

Since the basis is not orthogonal, to recover the 'momentum' coefficients $\widetilde{F}_m$, $m|_{-\ell}^{\ell}$ we need the inner product in this basis, which is thus non-local:

$$(F,G)_{\mathcal{S}^2} = \sum_{m,m'=-\ell}^{\ell} \widetilde{F}^*_m \Big(\cos\frac{\pi(m-m')}{2\ell+1}\Big)^{2\ell} \widetilde{G}_{m'} =: (F,G)_{\mathcal{M}_\ell} \tag{14}$$

In Fig. 4 we show some of these functions over the sphere and their projection with the obliquity factor (8) which, following the convention used in (10) for bases on the fish-eye medium we denote by $V^{\rm FE}_{\ell,m}(\mathbf{r})$, but whose explicit expression we do not presently need.

5. The position and normal-derivative basis

For the proposed position basis we can use the $Y_{\ell,0}(\theta,\phi)$'s, rotated so that their axes point in $2\ell+1$ directions distributed on any azimuth circle parallel to the equator, or randomly over the sphere; these would provide acceptable bases provided they are linearly independent. For

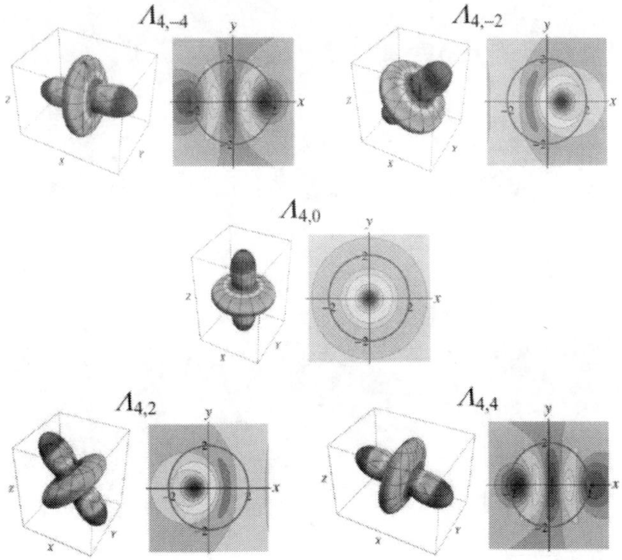

Figure 5. Legendre 'position' functions over the sphere (left of each pair) $\Lambda_{4,m}(\theta,\phi)$, $m = -4, -2, 0, 2, 4$ in (15) which are even in azimuth ϕ, and their projections $\Lambda^{\rm FE}_{4,m}(\mathbf{r})$ on the fish-eye medium (right of each pair), and the minimal circle.

reasons that will become clear after the contraction to the plane, we choose them to be on the upper half-meridian in the x-z plane of the sphere, as was shown in Fig. 3. We thus define the *Legendre harmonics* as given through rotations of $Y_{\ell,0}$ by angles $\beta_m = \pi m/(2\ell+1)$ around the y-axis,

$$\begin{aligned}
\Lambda_{\ell,m}(\theta,\phi) \;&:=\; e^{-i\beta_m L_y} Y_{\ell,0}(\theta,\phi) = \sum_{m'=-\ell}^{\ell} d^{\ell}_{m',0}(\beta_m)\, Y_{\ell,m'}(\theta,\phi) \\
&= \sum_{n=0}^{\ell} d^{\ell}_{n,0}(\beta_m)\, \tfrac{1}{2}(2-\delta_{n,0})\Big(Y_{\ell,n}(\theta,\phi) + Y_{\ell,-n}(\theta,\phi)^*\Big) \\
&= \sqrt{\frac{2\ell+1}{4\pi}}\left(d^{\ell}_{0,0}(\beta_m)\, d^{\ell}_{0,0}(\theta) + 2\sum_{n=1}^{\ell} d^{\ell}_{n,0}(\beta_m)\, d^{\ell}_{n,0}(\theta)\cos n\phi\right) \\
&= \Lambda_{\ell,m}(\theta,-\phi) = (-1)^{\ell}\Lambda_{\ell,m}(\theta\pm\pi,\,\phi).
\end{aligned} \tag{15}$$

where in the second line we used the symmetries $d^{\ell}_{m',0}(\beta) = (-1)^{m'} d^{\ell}_{-m',0}(\beta)$ and $Y_{\ell,-m'} = (-1)^{m'} Y^*_{\ell,m'}$, reducing the sum to $n|_0^{\ell}$. We show some Legendre functions $\Lambda_{\ell,m}(\theta,-\phi)$ in Fig. 5, together with their projections with obliquity on the fish-eye plane, that following our convention we write $\Lambda^{\rm FE}_{\ell,m}(\mathbf{r})$. On the sphere they are rotated versions of each other although on the fish-eye plane their appearance is different. The range $m|_{-\ell}^{\ell}$ for $\Lambda_{\ell,m}(\theta,\phi)$ is not warranted here, because all these functions are *even* under reflection across the x–z plane, or in azimuth, $\phi \leftrightarrow -\phi$; Hence, only $\ell+1$ of these functions can be linearly independent.

We thus need a second set of ℓ functions with axes on the remaining points in the half-meridian that be odd under $\phi \leftrightarrow -\phi$. To propose them, we note that from angular momentum theory,

$$L_x Y_{\ell,0}(\theta,\phi) = -i\sqrt{\frac{2\ell+1}{4\pi}}\, P^1_{\ell}(\cos\theta)\sin\phi, \tag{16}$$

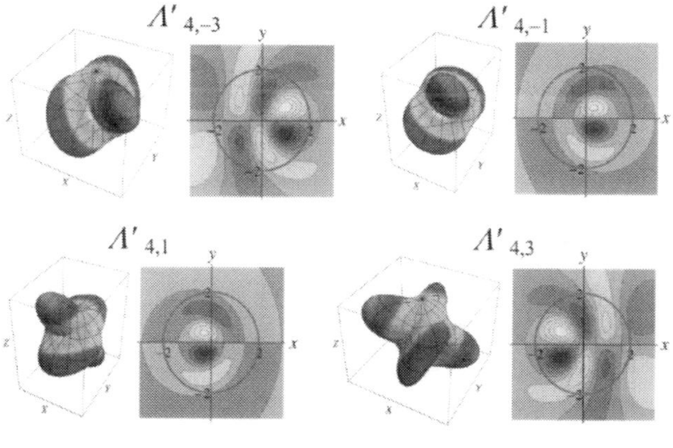

Figure 6. Legendre 'normal derivative' functions over the sphere (left of each pair) $\Lambda'_{4,m}(\theta,\phi)$, $m = -3, -1, 1, 3$ in (17) that are odd in azimuth ϕ, and their projections $\Lambda'^{\rm FE}_{4,m}(\theta,\phi)$ on the fish-eye medium (right of each pair) with the minimal circle.

so we rotate these by $\beta_m = \pi m/(2\ell{+}1)$, to define the 'normal derivative' Legendre functions, using the same symmetries as in (15),

$$\begin{aligned} \Lambda'_{\ell,m}(\theta,\phi) &:= \sqrt{\frac{2}{\ell(\ell{+}1)}}\, e^{-{\rm i}\beta_m L_y}\,(-{\rm i})\,L_x\,\Lambda_{\ell,0}(\theta,\phi) \\ &= \sum_{n=1}^{\ell}\Big(d^{\ell}_{n,1}(\beta_m)+d^{\ell}_{n,-1}(\beta_m)\Big)\,\tfrac{1}{2{\rm i}}\Big(Y_{\ell,n}(\theta,\phi)-Y_{\ell,-n}(\theta,\phi)^*\Big) \\ &= \sqrt{\frac{2\ell{+}1}{2\pi}}\sum_{n=1}^{\ell}\Big(d^{\ell}_{n,1}(\beta_m)+d^{\ell}_{n,-1}(\beta_m)\Big)d^{\ell}_{n,0}(\theta)\sin n\phi \\ &= -\Lambda'_{\ell,m}(\theta,-\phi) = (-1)^{\ell}\Lambda'_{\ell,m}(\theta{\pm}\pi,\,\phi), \end{aligned} \tag{17}$$

which are odd under $\phi \leftrightarrow -\phi$ and shown in Fig. 6, together with their projections on the fish-eye plane $\Lambda'^{\rm FE}_{\ell,m}(\mathbf{r})$.

We choose the ranges of m for the angles $\beta_m = \pi m/(2\ell{+}1)$ as follows:

$$\begin{aligned} &\text{for } \Lambda_{\ell,m}, \quad m \in \mathcal{E}_\ell := \{-\ell, -\ell{-}2, \ldots, 0, \ldots, \ell{-}2, \ell\}, \\ &\text{for } \Lambda'_{\ell,m}, \quad m \in \mathcal{E}'_\ell := \{-\ell{+}1, -\ell{+}3, \ldots, \ell{-}1\}. \end{aligned} \tag{18}$$

Since $|\beta_m| < \frac{1}{2}\pi$ the maximum of all these functions projects on the fish-eye plane *inside* the minimal circle $r = 2\rho$; their antipodes project to their conjugate points outside. To complement the set of Legendre functions, in Fig. 5 we show the normal-derivative Legendre functions with the m's that were 'absent' in Fig. 6.

Although all $\Lambda_{\ell,m}$'s are orthogonal to all $\Lambda'_{\ell,m'}$'s and each subset has been checked to be linearly independent, they are not orthogonal, but

$$(\Lambda_{\ell,m},\,\Lambda_{\ell,m'})_{\mathcal{S}^2} = d^{\ell}_{0,0}(\beta_{m'}{-}\beta_m) = P_\ell(\cos(\beta_{m'}{-}\beta_m)), \tag{19}$$

where the Legendre polynomials $P_\ell(\cos\beta)$ contain powers of $\cos^\nu\beta$, or equivalently of $\cos\nu\beta$, for $\nu \in \{\ell,\,\ell{-}2,\,\ldots,\,1 \text{ or } 0\}$. The inner products between 'normal derivative' Legendre functions are

$$(\Lambda'_{\ell,m},\,\Lambda'_{\ell,m'})_{\mathcal{S}^2} = \frac{1}{\ell}\,P^{(1,1)}_{\ell-1}(\cos(\beta_{m'}{-}\beta_m)) =: Q_{\ell-1}(\cos(\beta_{m'}{-}\beta_m)), \tag{20}$$

Journal of Physics: Conference Series **597** (2015) 012067 doi:10.1088/1742-6596/597/1/012067

where the Jacobi polynomial $P^{(1,1)}_{\ell-1} \equiv Q_{\ell-1}$ is of degree ν in $\cos^\nu\beta$ or equivalently $\cos\nu\beta$, for $\nu \in \{\ell-1,\ \ell-3,\ \ldots,\ 0 \text{ or } 1\}$, and they were checked to be also linearly independent.

Functions on the sphere of angular momentum ℓ are in $\mathcal{M}_\ell$ and can be written as linear combinations of both sets of Legendre functions,

$$F(\theta,\phi) = \sum_{m\in\mathcal{E}_\ell} F_m \Lambda_{\ell,m}(\theta,\phi) + \sum_{m\in\mathcal{E}'_\ell} F'_m \Lambda'_{\ell,m}(\theta,\phi). \tag{21}$$

Since non-orthogonal bases imply non-local measures, as in (14), we now have

$$\begin{aligned}(F,G)_{\mathcal{S}^2} &= \sum_{m,m'\in\mathcal{E}_\ell} F^*_m\, P_\ell\Big(\cos\frac{\pi(m-m')}{2\ell+1}\Big)\, G_{m'} \\ &\quad + \sum_{m,m'\in\mathcal{E}'_\ell} F'^*_m\, Q_{\ell-1}\Big(\cos\frac{\pi(m-m')}{2\ell+1}\Big)\, G'_{m'} \\ &=: (F,G)_{\mathcal{M}_\ell}.\end{aligned} \tag{22}$$

6. The dual bases

We saw in the previous Section that neither the set of nShV nor Legendre functions on the sphere are orthogonal on $\mathcal{M}_\ell$, although they have unit norm. It is natural to inquire on their duals, which will represent different wavefield bases on the sphere and fish-eye plane.

6.1. Dual of the nShV basis

The nShV functions $V_{\ell,m}(\theta,\phi)$ in (11), for the $2\ell{+}1$ equally-spaced angles $\alpha_m = 2\pi m/(2\ell{+}1)$, $m|^{\ell}_{-\ell}$, are complete in $\mathcal{M}_\ell$. If we denote the column vector of spherical harmonics by $\vec{Y}_\ell \equiv \{Y_{\ell,m}(\theta,\phi)\}^{\ell}_{m=-\ell}$ and the nShV column by $\vec{V}_\ell \equiv \{V_{\ell,m}(\theta,\phi)\}^{\ell}_{m=-\ell}$, then (11) can be written with a matrix $\mathbf{S}^\ell \equiv \{S^\ell_{m,m'}\}^{\ell}_{m,m'=-\ell}$ as

$$\vec{V}_\ell = \mathbf{S}^\ell\,\vec{Y}_\ell, \quad S^\ell_{m,m'} := \exp\Big(\frac{-2\pi\mathrm{i}m\,m'}{2\ell+1}\Big) d^\ell_{m,\ell}(\tfrac12\pi). \tag{23}$$

The dual basis functions can be then written as the components of a vector $\vec{\overline{V}}_\ell \equiv \{\overline{V}_{\ell,m}(\theta,\phi)\}^{\ell}_{m=-\ell}$, and related to the orthonormal spherical harmonic vector by a matrix $\overline{\mathbf{S}}^\ell \equiv \{\overline{S}^\ell_{m,m'}\}^{\ell}_{m,m'=-\ell}$, so that

$$(\overline{V}_{\ell,m}, V_{\ell,m'})_{\mathcal{S}^2} = \sum_{\overline{m}=-\ell}^{\ell} \overline{S}^{\ell\,*}_{\overline{m},m} S^\ell_{\overline{m},m'} = (\overline{\mathbf{S}}^{\ell\,\dagger}\mathbf{S}^\ell)_{m,m'} = \delta_{m,m'}, \tag{24}$$

where † indicates adjunction. We must thus have $\overline{\mathbf{S}}^\ell = (\mathbf{S}^\ell)^{\dagger\,-1}$. Knowing finite Fourier analysis, the matrix elements of $\overline{\mathbf{S}}^\ell$ can be found by inspection to be

$$\overline{S}^\ell_{m,m'} = \exp\Big(\frac{2\pi\mathrm{i}mm'}{2\ell+1}\Big)\Big/\,(2\ell{+}1)\, d^\ell_{m,\ell}(\tfrac12\pi). \tag{25}$$

Thus we have the dual nShV basis $\vec{\overline{V}}_\ell = \overline{\mathbf{S}}^\ell\,\vec{Y}_\ell$, whose normalized functions over the sphere are,

$$\overline{V}_{\ell,m}(\theta,\phi) = \frac{1}{2\ell{+}1}\sum_{m'=-\ell}^{\ell} \frac{\exp\big(-\frac{2\pi\mathrm{i}mm'}{2\ell+1}\big)}{d^\ell_{m',\ell}(\frac12\pi)} Y_{\ell,m'}(\theta,\phi) = \overline{V}_{\ell,0}(\theta,\phi{-}\alpha_m). \tag{26}$$

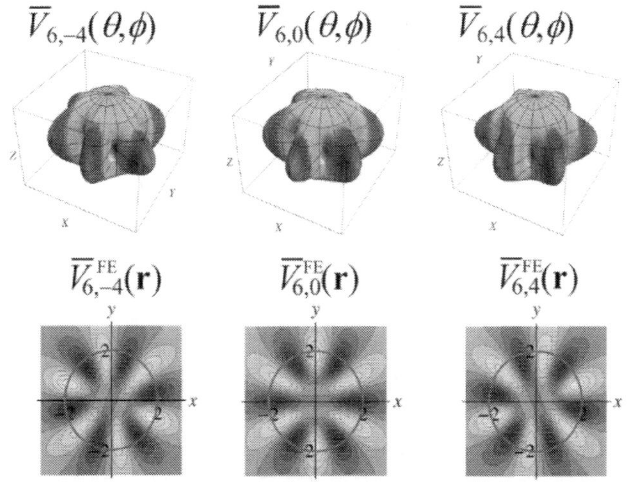

Figure 7. Real part of the normalized dual Sherman-Volobuyev functions $\overline{V}_{6,m}(\theta,\phi)$, $m = -4, 0, 4$ in (26) on the sphere (top), and their projections $\overline{V}^{\rm FE}_{6,m}(\mathbf{r})$ on the fish-eye medium (bottom) with the minimal circle. The full dual basis is obtained through rotations around the z-axis by $\alpha_m = 2\pi\, m/(2\ell+1)$, $m|_{-\ell}^{\ell}$.

We remark that it is unusual to see the Wigner little-d function in a denominator, but this is never zero. In Fig. 7 we show some of the dual functions $\overline{V}_{\ell,m}(\theta,\phi)$ over the sphere and $\overline{V}^{\rm FE}_{\ell,m}(\mathbf{r})$ on the fish-eye plane. Clearly, the dual functions on the sphere are rotated versions of each other; on the fish-eye plane they represent circular trajectories, all orthogonal to the line trajectories through the center that represent plane waves. They differ only by $2\ell+1$ equidistant phases.

6.2. *Dual of the Legendre bases*

The duals of the Legendre bases of positions and normal derivatives in (15) and (17) can be found as in (24), but now we should be careful to handle separately the symmetric and antisymmetric sub-bases, of dimensions $\ell+1$ and ℓ respectively. We accommodate them into column vectors related by matrices appropriate to their dimensions,

$$\vec{\Lambda}_\ell \equiv \{\Lambda_{\ell,\,2n-\ell}\}_{n=0}^{\ell}, \qquad \vec{Y}_\ell^{\rm S} \equiv \{\,{\rm Re}\,Y_{\ell,n}\}_{n=0}^{\ell}, \qquad \vec{\Lambda}_\ell = \mathbf{T}^\ell \vec{Y}_\ell^{\rm S}, \tag{27}$$

$$\vec{\Lambda}'_\ell \equiv \{\Lambda'_{\ell,\,2n-\ell-1}\}_{n=1}^{\ell}, \qquad \vec{Y}_\ell^{\rm A} \equiv \{\,{\rm Im}\,Y_{\ell,n}\}_{n=1}^{\ell}, \qquad \vec{\Lambda}'_\ell = \mathbf{T}'^\ell \vec{Y}_\ell^{\rm A}. \tag{28}$$

The $(\ell+1)\times(\ell+1)$ matrix $\mathbf{T}^\ell \equiv \{T^\ell_{n,n'}\}_{n,n'=0}^{\ell}$, and the $\ell\times\ell$ matrix $\mathbf{T}'^\ell \equiv \{T'^\ell_{n,n'}\}_{n,n'=1}^{\ell}$ have their elements given by the second lines in (15) and (17),

$$T^\ell_{n,n'} = (2-\delta_{n,0})\, d^\ell_{n,0}(\beta_{2n'-\ell}), \tag{29}$$

$$T'^\ell_{n,n'} = d^\ell_{n,1}(\beta_{2n'-\ell-1}) + d^\ell_{n,-1}(\beta_{2n'-\ell-1}). \tag{30}$$

As in the previous subsection, we must find the inverse adjoint matrices

$$\overline{\mathbf{T}}^\ell := (\mathbf{T}^\ell)^{\dagger\,-1}, \qquad \overline{\mathbf{T}}'^\ell := (\mathbf{T}'^\ell)^{\dagger\,-1}. \tag{31}$$

to build the two dual sub-bases,

$$\overline{\Lambda}_{\ell,\,2n-\ell}(\theta,\phi) = \sum_{n'=0}^{\ell} \overline{T}^\ell_{n,n'}\, Y^{\rm S}_{\ell,n'}(\theta,\phi), \quad n|_0^\ell, \tag{32}$$

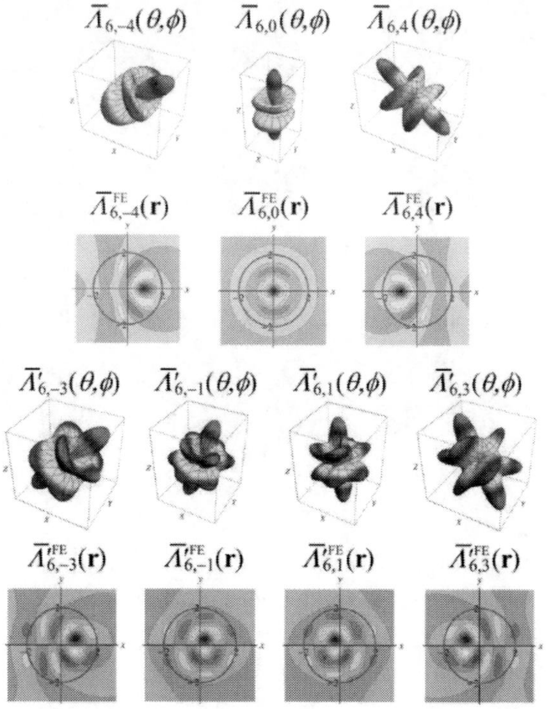

Figure 8. Real part of the duals of the Legendre position and normal-derivative functions $\overline{\Lambda}_{6,m}(\theta,\phi)$, $m = -4, 0, 4$ from (32) and $\overline{\Lambda}'_{6,m}(\theta,\phi)$, $m = -3, -1, 1, 3$ from (33) on the sphere (top), and their fish-eye projections $\overline{\Lambda}^{\rm FE}_{6,m}(\mathbf{r})$ and $\overline{\Lambda}'^{\rm FE}_{6,m}(\mathbf{r})$ on the fish-eye medium (bottom) with the minimal circle.

$$\overline{\Lambda}'_{\ell,\,2n-\ell-1}(\theta,\phi) \;=\; \sum_{n'=1}^{\ell} \overline{T}'^{\ell}_{n,n'}\, Y^{\rm A}_{\ell,n'}(\theta,\phi), \quad n|_1^{\ell}. \tag{33}$$

We cannot provide a closed expression for the matrix elements of $\overline{\mathbf{T}}^{\ell}$ and $\overline{\mathbf{T}}'^{\ell}$, so we had to resort to numerical computation of the matrices in order to show in Fig. 8 some of the dual basis functions (32)–(33), together with their fish-eye projections, $\overline{\Lambda}^{\rm FE}_{\ell,m}(\mathbf{r})$ and $\overline{\Lambda}'^{\rm FE}_{\ell,m}(\mathbf{r})$. We checked their duality properties by numerical integration over the sphere for several low values of ℓ. The dual vectors have the same parity as their original counterparts, and are also rotated copies of each other as (15) and (17). Again, on the fish-eye plane they superficially appear to be rotated versions of each other, but are slightly different from the original basis and from each other.

7. Contraction to the Helmholtz medium

By contraction of $\mathsf{so}(3)$ to the Euclidean Lie algebra $\mathsf{iso}(2)$ we shall see that the assignments of momentum and position to the nShV and Legendre bases are consistent with what we know of bases for solutions of the Helmholtz equation, and thus confirm their roles. We note the following limits of spherical harmonics and Wigner d-functions to Bessel functions:

$$\lim_{\substack{\rho\to\infty\\ \ell=k\rho}} Y_{\ell,m}\Big(\frac{r}{\rho},\phi\Big) = (-1)^m\sqrt{\frac{k\rho}{2\pi}}\, J_m(kr)\, e^{\mathrm{i}m\phi}, \quad \lim_{\substack{\rho\to\infty\\ \ell=k\rho}} d^{\ell}_{m,m'}\Big(\frac{r}{\rho}\Big) = J_{m-m'}(kr). \tag{34}$$

As we noted in Sect. 4, the limit $\rho \to \infty$ of the radius of the sphere in Fig. 1 should mantain a constant wavelength λ where the sphere is tangent to the fish-eye plane, so we must consider increasing $\ell \propto \rho$, related by the wavenumber as,

$$k := 2\pi/\lambda = \ell/\rho, \quad \ell = k\rho. \tag{35}$$

Also the polar angle θ in (34), to map onto points at a finite radius r on the fish-eye plane (5), must be taken in the limit $\rho \to \infty$ as $\theta \approx r/\rho \to 0$. In this limit the refractive index of the medium (7) becomes the constant n_o; finally, the inner product (9) and the obliquity factor (10) require renormalizations of functions to keep them finite.

The spherical harmonic basis $Y^{\rm FE}_{\ell,m}(\mathbf{r})$ in (10) is given by (34). We thus obtain the *multipole* basis for non-exponential solutions of the Helmholtz medium, that we indicate by lower-case letters,

$$y_{k,m}(\mathbf{r}) := \lim_{\rho\to\infty}^{\ell=k\rho} \frac{1}{\sqrt{\rho}} Y^{\rm FE}_{\ell,m}(\mathbf{r}) = (-1)^m \sqrt{\frac{k}{2\pi}} J_m(kr)\, e^{im\phi}, \quad m|^{\infty}_{-\infty}, \tag{36}$$

where m now ranges over all integers. The contraction of the Lie algebra generators yields

$$\begin{array}{lcl} L_x = -\mathrm{i}(y\partial_z - z\partial_y) & \to & \mathrm{i}\rho\partial_y, \\ L_y = -\mathrm{i}(z\partial_x - x\partial_z) & \to & -\mathrm{i}\rho\partial_x. \end{array} \tag{37}$$

while L_z remains invariant. The Casimir equation $L^2 = \ell(\ell{+}1)\mathit{1}$ after normalization becomes the Helmholtz equation for the wavefields $(\partial_x^2 + \partial_y^2)\, f(\mathbf{r}) = k^2 f(\mathbf{r})$; and $e^{-\mathrm{i}\beta_m L_y} \to e^{-x_m\partial_x}$ for $x_m = \rho\beta_m$ finite will apply to the angles of the Legendre basis.

7.1. Contraction of the nShV basis

The azimuth angles $\alpha_m = 2\pi\, m/(2\ell{+}1)$ of the nShV basis become dense on the α-circle, as $\Delta\alpha = 2\pi/(2\ell{+}1) \to 0$. We write α for α_m's on the circle and no longer count m as a label. Using (34), the Bessel generating function, other limits for the functions involved and renormalization, we obtain

$$\begin{aligned} v_{k,\alpha}(r,\phi) &:= \lim_{\rho\to\infty}^{\ell=k\rho} \Big(\frac{k\pi}{\rho}\Big)^{1/4} V_{\ell,m}(r/\rho,\, \phi) \\ &= \sqrt{\frac{k}{2\pi}} \exp(\mathrm{i}kr\sin(\phi-\alpha)). \\ &= \sqrt{\frac{k}{2\pi}} \sum_{m'=-\infty}^{\infty} e^{\mathrm{i}m'(\phi-\alpha+\pi)}\, J_{m'}(kr). \end{aligned} \tag{38}$$

These are plane waves of wavenumber k with wavefronts in the $\phi = \alpha$ direction, also expressed in terms of the multipole wavefields in (36).

The inner product on the sphere in (14) was non-local in the coefficients of the nShV basis. In the $\rho, \ell \to \infty$ limit, this kernel becomes a Dirac δ,

$$\lim_{\rho\to\infty}^{\ell=k\rho} \sqrt{\rho}\Big(\cos\frac{\pi(m-m')}{2\ell+1}\Big)^{2\ell} = \frac{1}{\sqrt{\pi k}}\delta(\alpha-\alpha'), \tag{39}$$

so the inner product for Helmholtz wavefields f, g in the plane-wave basis becomes local:

$$(f,g)_\circ := \int_{-\pi}^{\pi} d\alpha\, \widetilde{f}(\alpha)^* \widetilde{g}(\alpha), \tag{40}$$

Journal of Physics: Conference Series **597** (2015) 012067 doi:10.1088/1742-6596/597/1/012067

where [14]

$$f(\mathbf{r}) = \int_{-\pi}^{\pi} d\alpha \tilde{f}(\alpha)\, v_{k,\alpha}(\mathbf{r}) = \sqrt{\frac{k}{2\pi}} \int_{-\pi}^{\pi} d\alpha\, \tilde{f}(\alpha)\, e^{ikr\sin(\phi-\alpha)}, \tag{41}$$

$$\tilde{f}(\alpha) = \frac{\sigma}{2}\sqrt{\frac{k}{2\pi}} \int_{-\infty}^{\infty} dx \Big(f(x,0)\cos\alpha + \frac{1}{ik} \frac{\partial f(x,y)}{\partial y}\Big|_{y=0} \Big) e^{-ikx\sin\alpha}, \tag{42}$$

with $\sigma = \operatorname{sign}\cos\alpha$, and similarly for $g(\mathbf{r})$. This expression, termed the *wave* transform only requires us to know the values of the fields and normal derivatives on the x-axis. As understood by Sherman and Volobuyev [17, 18], we have a natural momentum basis on the sphere, which is reduced to a single wavenumber in the monochromatic Maxwell fish-eye.

The contraction limit of the dual basis $\{\overline{V}_{\ell,m}(\theta,\phi)\}_{m=-\ell}^{\ell}$ is the null vector; this is due to the factor $2\ell+1$ in the denominator of (26), and is also evident in (38) where the growth of $V_{\ell,m}(r/\rho,\,\phi)$ by $\sim \rho^{1/4}$ entails that its dual $\overline{V}_{\ell,m}(r/\rho,\,\phi)$ must vanish as $\sim \rho^{-1/4}$ to mantain the inner products $(\overline{V}_{\ell,m}, V_{\ell,m}) = 1$.

7.2. Contraction of the Legendre bases

The Legendre harmonics $\Lambda_{\ell,m}(\theta,\phi)$ in (15) point their 'fingers' at angles $\beta_m = \pi m/(2\ell+1)$ on an upper half-meridian, and which project on points x_m in the x-axis of the fish-eye plane. We saw that for $m \in \mathcal{E}_\ell$, these are a sub-basis in $\mathcal{M}_\ell$ for even functions of azimuth $\phi \leftrightarrow -\phi$, and of $y \leftrightarrow -y$ on the fish-eye plane. Similarly, the sub-basis of normal derivatives $\Lambda'_{\ell,m}(\theta,\phi)$ in (17) project the maximum of their derivatives on points x_m for $m \in \mathcal{E}_\ell$, and form the complementary sub-basis in $\mathcal{M}_\ell$ for functions that are odd in $\phi \leftrightarrow -\phi$ and $y \leftrightarrow -y$.

In the $\ell, \rho \to \infty$ limit (35), the set of points x_m, will become the infinite set of equally-spaced points

$$x_m = 2\rho\tan\tfrac{1}{2}\beta_m \quad \to \quad x_m = \rho\beta_m = \tfrac{1}{4}\lambda m, \quad m \in \mathcal{E}_\ell \text{ or } m \in \mathcal{E}'_\ell. \tag{43}$$

The two sets of functions of the Legendre bases, being rotated versions of each other, contract into *translated* copies, each set on points x_m separated by $\frac{1}{2}\lambda$; the even position functions we place on the points x_m with *even* $m \in \mathcal{E}$, and the odd normal derivative functions on x_m with *odd* $m \in \mathcal{E}'$. We use (34) for finite $r = \rho\theta$, to give their limits on the Helmholtz plane and renormalized by $1/\sqrt{\rho}$. The limit of the Legendre bases functions are

$$\lambda_{k,m}(\mathbf{r}) := \lim_{\substack{\ell=k\rho \\ \rho\to\infty}} \frac{1}{\sqrt{\rho}} \Lambda_{\ell,m}(r/\rho,\,\phi) = \sqrt{\frac{k}{2\pi}}\, J_0(k|\mathbf{r}-\mathbf{x}_m|), \tag{44}$$

$$\lambda'_{k,m}(\mathbf{r}) := \lim_{\substack{\ell=k\rho \\ \rho\to\infty}} \frac{-i}{\sqrt{\rho}} \Lambda'_{\ell,m}(\theta,\,\phi) = \sqrt{\frac{k}{\pi}}\, \frac{J_1(k|\mathbf{r}-\mathbf{x}_m|)}{|\mathbf{r}-\mathbf{x}_m|} y, \tag{45}$$

where $|\mathbf{r}-\mathbf{x}_m| = \sqrt{[(x-x_m)^2 + y^2]}$. The J_0's in (44) are the *narrowest* Helmholtz wavefields [14], and $J_1(x)/x$ have the largest normal derivatives, so we confirm that (44) plays the role of a *position* and *normal derivative* sub-bases for Helmholtz wavefields. This identification we can extend to the finite basis on the pre-contracted sphere. Again, the dual Legendre bases vanish: the growth $\sim \rho^{1/2}$ of the original $\Lambda_{\ell,m}$'s and $\Lambda'_{\ell,m}$'s that required renomalization in (44) and (45), implies that their duals decrease as $\sim \rho^{-1/2}$.

Expansions of the Legendre bases (44) and (45) in terms of the standard multipole basis $y_{k,m}(\mathbf{r})$ in (36) can be found from (15) and (17) also using (34),

$$\lambda_{k,m}(\mathbf{r}) = \sum_{m'\in\mathcal{E}} J_{m'}(\tfrac{1}{2}\pi m)\, y_{k,m'}(\mathbf{r}), \qquad m \in \mathcal{E} \text{ (even)}, \tag{46}$$

$$\lambda'_{k,m}(\mathbf{r}) = \sum_{m'\in\mathcal{E}'} \frac{m'\, J_{m'}(\frac{1}{2}\pi m)}{\frac{1}{2}\pi m}\, y_{k,m'}(\mathbf{r}), \quad m \in \mathcal{E}' \text{ (odd)}. \tag{47}$$

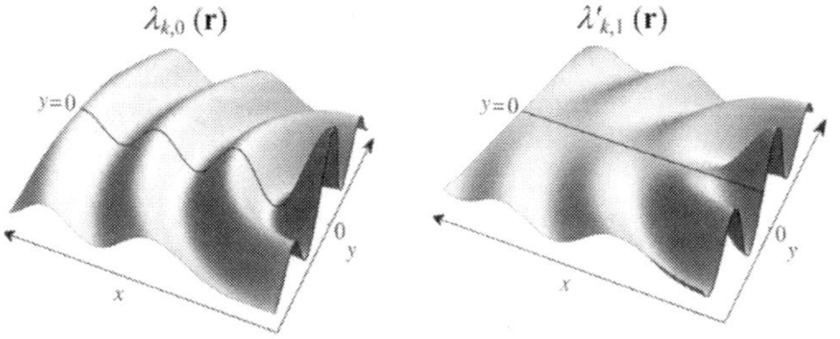

Figure 9. *Left*: The limit of the Legendre 'position' functions are the J_0-Bessel functions $\lambda_{k,m}(\mathbf{r})$ in (44) centered on x_m, $m \in \mathcal{E}$ even integers. *Right*: The limit of the 'normal derivative' Legendre functions $\lambda'_{k,m}(\mathbf{r})$ in (45) centered on x_m, $m \in \mathcal{E}'$ odd integers.

A generic non-exponential Helmholtz wavefield can thus be written in terms of the position and normal derivative bases (44) and (45) as

$$f(\mathbf{r}) = \sum_{m\in\mathcal{E}} f_m\, \lambda_{k,m}(\mathbf{r}) + \sum_{m\in\mathcal{E}'} f'_m\, \lambda'_{k,m}(\mathbf{r}). \tag{48}$$

In terms of these coefficients the inner product between two Helmholtz fields f, g, is the limit of the non-local inner product (22). We find that the limit inner product remains non-local in the discrete positions x_m,

$$(f,g)_{\mathcal{Z}} := \sum_{m,m'\in\mathcal{E}} f^*_m\, J_0(\tfrac{1}{2}\pi(m'-m))\, g_{m'} + \sum_{m,m'\in\mathcal{E}'} f'^*_m\, \frac{J_1(\frac{1}{2}\pi(m'-m))}{\frac{1}{2}\pi(m'-m)}\, g'_{m'}. \tag{49}$$

This result is known from Refs. [19, 14, 20].

8. Conclusions

Phase space is a well-defined object in classical mechanical and optical systems; in quantum systems, the Wigner quasidistribution function provides a proper and working representation of states both in position and momentum, even though they stem from noncommuting operators in the Heisenberg-Weyl Lie algebra. A variety of other systems have been endowed with phase space descriptions, as has the classical sphere, which is a symplectic manifold [1].

Our approach here has been to start with functions on the manifold of the sphere with definite angular momentum, and *propose* two function bases —rotated $Y_{\ell,\ell}$ and $Y_{\ell,0}$'s— to be (not *eigenfunctions* of any two operators, but) the closest analogues of quantum momentum and position wavefunctions. The use of Sherman-Volbuyev functions as 'Fourier' bases for functions on the sphere is known, but those we called Legendre functions of position and normal derivative appear to be new in this context. Their contraction limit to the Helmholtz medium binds these bases to previously known ones. Then, the stereographic projection of the sphere, the fish-eye medium, provides a physical model where the corresponding wavefields acquire a degree of physicality, which may be used later to extend this concept to other related Luneburg systems in wave optics.

The use of fish-eye optics for waveguides and radar antennas by itself suggests that our inquiry is not vacuous. Their perfect focusing properties in space and in time open other questions, such as bases for a sphere whose lower hemisphere is screened, corresponding to the fish-eye region bounded by the minimal circle where a waveguide is bounded by vacuum. Finally, the *half* fish-eye used in microwave radar antennas, coupled to a homogeneous, Euclidean-invariant medium may require some interesting group-theoretical techniques.

Acknowledgments

We thank Guillermo Krötzsch for his assistance in the preparation of the figures. Support for this research has been provided by the *Óptica Matemática* projects by the Universidad Nacional Autónoma de México PAPIIT IN101011, and by the Consejo Nacional de Ciencia y Tecnología project SEP-CONACYT CB220692.

References

[1] McDuff D and Salamon C 1998 *Introduction to Symplectic Topology* (Oxford: Oxford Univ. Press)
[2] Wigner E P 1932 On the quantum correction for thermodynamic equilibrium *Phys. Rev.* **40** 749-759
[3] Wolf K B, Alonso M A and Forbes G W 1999 Wigner functions for Helmholtz wavefields *J. Opt. Soc. Am. A* **16** 2476–2487
[4] Ali S T, Atakishiyev N M, Chumakov S M and Wolf K B 2000 The Wigner function for general Lie groups and the wavelet transform *Ann. Henri Poincaré* **1** 685–714
[5] Luneburg R K 1964 *Mathematical Theory of Optics* (Berkeley: Univ. of California Press)
[6] Demkov Yu N and Ostrovsky V N 1971 Internal symmetry of the Maxwell "fish-eye" problem and the Fock group for the hydrogen atom *Sov. Phys. JETP* **33** 1083–1087
[7] Buchdahl H A 1979 Kepler problem and Maxwell fish-eye *Am. J. Phys.* **46** 840–843
[8] Chen A C 1979 The Coulomb problem and the Maxwell fish-eye problem *Am. J. Phys.* **46** 214–217
[9] Frank A, Leyvraz F, and Wolf K B 1990 Hidden symmetry and potential group of the Maxwell fish-eye *J. Math. Phys.* **31** 2757–2768
[10] Roşu H C *et al* 1996 Supersymmetric features of the Maxwell fish-eye lens *SPIE Proc.* **2730** 436–439
[11] Makowski A J and Górska K J 2009 Quantization of the Maxwell fish-eye problem and the quantum-classical correspondence *Phys. Rev. A* **79** art. 052116
[12] Wolf K B 2004 *Geometric Optics on Phase Space* (Heidelberg: Springer-Verlag)
[13] Salto-Alegre C and Wolf K B 2014 Position and momentum bases for the monochromatic Maxwell fish-eye and the sphere (submitted)
[14] González-Casanova P and Wolf K B 1995 Interpolation of solutions to the Helmholtz equation *Numer. Methods Partial Differential Eq.* **11** 77–91
[15] Biedenharn L C and Louck J D 1981 *Angular Momentum in Quantum Physics, Theory and Application* Encyclopedia of Mathematics and Its Applications Vol. 8, Ed. by G.-C. Rota (Reading Mass.: Addison-Wesley)
[16] Maxwell J C 1854 Solution of a problem *Cambridge & Dublin Math. J.* **8** 188-193
[17] Sherman T O 1975 Fourier analysis on the sphere *Trans. Amer. Math. Soc.* **209** 1–31
[18] Volobuyev I P 1980 Plane waves on the sphere and some of their applications *Teor. Mat. Fiz.* **45** 421–426
[19] Steinberg S and Wolf K B 1981 Invariant inner products on spaces of solutions of the Klein–Gordon and Helmholtz equations *J. Math. Phys.* **22** 1660–1663
[20] Wolf K B 2013 Continuous versus discrete free Hamiltonian systems *J. Phys. A* **46** art. 335202

Richardson–Gaudin form of Bethe Ansatz solutions for an atomic-molecular Bose-Einstein condensate model

Yibing Shen and Jon Links[1]
School of Mathematics and Physics, The University of Queensland, 4072, Australia
E-mail: [1]jrl@maths.uq.edu.au

Abstract. An integrable model describing the interconversion between Bose-Einstein condensates of atomic and molecular degrees of freedom is studied. An exact solution for this model has been previously derived via the algebraic Bethe Ansatz formalism. Here we derive an alternative solution of the Richardson–Gaudin form, and highlight some of the implications of the result.

1. Introduction

In 2001 Vardi, Yurovsky, and Anglin [1] presented one of the earliest and simplest models for an atomic-molecular Bose–Einstein condensate. It was subsequently reported in [2] that this model admits an exact solution via the algebraic Bethe Ansatz, which was used to investigate ground-state and dynamical properties. A feature of the exact solution is that it is obtained in the *quasi-classical limit*. The connection between Yang–Baxter integrability, which underlies the algebraic Bethe Ansatz approach, and Richardson-Gaudin systems via the quasi-classical limit, was observed by Sklyanin in 1989 [3]. Strictly speaking Skylanin's work made connection to Gaudin's algebraic approach [4], which itself referenced Richardson's earlier work [5,6]. It is now commonplace to refer to a broad class of models as Richardson–Gaudin systems. Specifically, we say that a Hamiltonian admits a Bethe Ansatz solution of Richardson–Gaudin form if the eigenvalues of the Hamiltonian are expressible as a function of the roots of the system of coupled non-linear equations

$$f(v_j) = \sum_{k \neq j}^{M} \frac{2}{v_k - v_j}, \qquad j = 1, ..., M, \tag{1}$$

where $f(v)$ is a rational function that is model dependent. See [7] for examples. We mention that the form (1) also accommodates *trigonometric* solutions through an appropriate variable change. See [4, 8] for a discussion of this point.

A later study established that there exists a more general integrable atomic-molecular Bose–Einstein condensate model, which contains that of [1] as a limiting case, involving atom-atom, atom-molecule, and molecule-molecule scattering interactions. This is discussed in [9] and details the derivation of the algebraic Bethe Ansatz solution, not of the form (1), based on use of the Yang–Baxter equation. Subsequently there have been several studies of the extended model described in [9] including the calculation of quantum dynamics [10], molecular conversion efficiency [11], quantum phase transitions [12,13], and the effects of particle losses [14]. However it seems that no studies have been made which exploit the existence of an exact solution. The objective of the present work is to aid progress in this direction by deriving an exact solution of the extended model of [9] which retains the Richardson-Gaudin form of exact solution (1). The motivation for achieving this is that it opens up an avenue to make exact quantitative calculations following the methods described in [15], which were guided by the qualitative observations of [16].

The approach used to derive this exact solution in the Richardson–Gaudin form does not rely on a solution of the Yang-Baxter equation [3], nor the use of Gaudin's algebra [4]. We instead exploit a correspondence between Richardson–Gaudin systems and ordinary differential equations. This correspondence has received significant attention as a tool for numerically solving the Bethe Ansatz equations associated with Richardson–Gaudin systems [2,17–23]. Here we invert this correspondence to *derive* the Bethe Ansatz equations for the Hamiltonian under consideration. The method, which provides a simple and direct derivation, is based on the following observation. A polynomial

$$Q(x) = \prod_{k=1}^{M} (x - v_k),$$

where the roots v_k are distinct, satisfies the equations

$$-\frac{Q''(v_j)}{Q'(v_j)} = \sum_{k \neq j}^{M} \frac{2}{v_k - v_j}, \qquad j = 1, ..., M, \tag{2}$$

where the prime denotes differentiation. Eq. (2) has the same generic form as Eq. (1).

2. The Hamiltonian

The model for condensed, interacting, atomic and molecular bosons is given by the Hamiltonian [9–14]

$$H = U_{aa}N_a^2 + U_{bb}N_b^2 + U_{ab}N_aN_b + \mu_a N_a + \mu_b N_b + \Omega\left(a^\dagger a^\dagger b + b^\dagger aa\right) \tag{3}$$

where a, $a^\dagger$ are the annihilation and creation operators for an atomic mode, and b, $b^\dagger$ are the annihilation and creation operators for a diatomic, homonuclear molecular mode. These satisfy the canonical commutation relations

$$\left[a,\, a^\dagger\right] = \left[b,\, b^\dagger\right] = I$$

where I denotes the identity operator. Moreover

$$\left[a,\, b\right] = \left[a,\, b^\dagger\right] = \left[a^\dagger, b\right] = \left[a^\dagger,\, b^\dagger\right] = 0.$$

The parameters μ_i are chemical potentials for species i and Ω is the amplitude for the interconversion of atoms and molecules. The parameters U_j are scattering couplings, taking into account atom-atom (U_{aa}), atom-molecule (U_{ab}), and molecule-molecule (U_{bb}) interactions. The Hamiltonian commutes with the total atom number $N = N_a + 2N_b$ where $N_a = a^\dagger a$ and $N_b = b^\dagger b$. The limiting case $U_{aa} = U_{ab} = U_{bb} = 0$ is the model introduced in [1].

The Bethe Ansatz solution as derived in [9] provides energy eigenvalues given by

$$\begin{aligned} E = {} & \sigma + \delta(M+\kappa) + \gamma(M+\kappa)^2 \\ & + 2\eta^{-1}\kappa\Omega\left[(1-\eta\left(\alpha(M+\kappa)+\beta\right))\prod_{i=1}^{M}\frac{u_i-\eta}{u_i} - \prod_{i=1}^{M}\frac{u_i+\eta}{u_i}\right], \end{aligned}$$

where the parameters $\{u_i\}$ satisfy the Bethe Ansatz equations

$$[1 - \eta u_i - \eta(\alpha(M+\kappa)+\beta)]\left(\frac{u_i+\eta\kappa}{u_i-\eta\kappa}\right) = \prod_{j\neq i}^{M} \frac{u_i - u_j - \eta}{u_i - u_j + \eta}. \tag{4}$$

The parameters above are related to the coupling parameters of the Hamiltonian through

$$\begin{aligned}
\eta &= \frac{4U_{aa} + U_{bb} - 2U_{ab}}{2\Omega},\\
\alpha &= \frac{U_{bb} - 4U_{aa}}{2\Omega},\\
\beta &= \frac{2\mu_b - 4\mu_a + 4U_{aa} - U_{ab}}{4\Omega},\\
\sigma &= \frac{U_{aa} - 2\mu_a}{4},\\
\delta &= \frac{2\mu_b - U_{ab}}{2},\\
\gamma &= U_{bb}.
\end{aligned}$$

The allowed values for κ are $\kappa = 1/4, 3/4$, through which the total particle number is given by $N = 2M + 2\kappa - 1/2$.

Setting $U_{aa} = U_{ab} = U_{bb} = 0$ imposes $\eta = 0$, with which (4) becomes an identity. Furthermore setting $\mu_b = 0$ and expanding (4) in η yields the Bethe Ansatz solution in the quasi-classical limit:

$$E = 2\mu_a(M + \kappa - 1/4) - 2\Omega \sum_{i=1}^{M} u_i, \tag{5}$$

where the parameters $\{u_i\}$ satisfy the Bethe Ansatz equations

$$\frac{\mu_a}{\Omega} - u_i + \frac{2\kappa}{u_i} = \sum_{j\neq i}^{M} \frac{2}{u_j - u_i}. \tag{6}$$

The above solution has the Richardson–Gaudin form (1). Taking into account different conventions used, this is equivalent to the solution provided in [2].

3. The Bethe Ansatz solution of the full Hamiltonian in Richardson–Gaudin form

Our objective now is to provide an alternative derivation of Bethe Ansatz solution for (3) which has the Richardson–Gaudin form (1). Let $|\text{vac}\rangle$ denote the vacuum state. Since N is conserved, for each subspace of fixed N we have the basis states

$$|j\rangle = \left(a^\dagger\right)^{k+2j}\left(b^\dagger\right)^{M-j}|\text{vac}\rangle \tag{7}$$

for $j = 0, ..., M$, $M = (N-k)/2$ where $k = 1$ for odd N and $k = 0$ for even N. This implies the identity

$$k(k-1) = 0 \quad \Leftrightarrow \quad k^2 = k$$

which will be assumed below. Let

$$U = U_{aa}N_a^2 + U_{ab}N_aN_b + U_{bb}N_b^2 + \mu_a N_a + \mu_b N_b.$$

Then

$$U|j\rangle = \mathcal{U}_j|j\rangle$$

where

$$\begin{aligned}\mathcal{U}_j &= U_{aa}(k+2j)^2 + U_{ab}(k+2j)(M-j) + U_{bb}(M-j)^2 + \mu_a(k+2j) + \mu_b(M-j) \\ &= (4U_{aa} - 2U_{ab} + U_{bb})j(j-1) + (4(k+1)U_{aa} + (2M-k-2)U_{ab} + (1-2M)U_{bb} + 2\mu_a - \mu_b)j \\ &\quad + k^2U_{aa} + kMU_{ab} + M^2U_{bb} + k\mu_a + M\mu_b.\end{aligned}$$

We set

$$\begin{aligned}A &= 4U_{aa} - 2U_{ab} + U_{bb}, \\ B &= 4(k+1)U_{aa} + (2M-k-2)U_{ab} + (1-2M)U_{bb} + 2\mu_a - \mu_b, \\ C &= k^2U_{aa} + kMU_{ab} + M^2U_{bb} + k\mu_a + M\mu_b.\end{aligned}$$

We also have

$$\begin{aligned}b^\dagger aa|j\rangle &= a\left(aa^\dagger\right)\left(a^\dagger\right)^{k+2j-1}\left(b^\dagger\right)^{M-j+1}|\text{vac}\rangle \\ &= a\left(I+N_a\right)\left(a^\dagger\right)^{k+2j-1}\left(b^\dagger\right)^{M-j+1}|\text{vac}\rangle \\ &= (4j(j-1) + (4k+2)j)|j-1\rangle, \\ a^\dagger a^\dagger b|j\rangle &= \left(a^\dagger\right)^{k+2(j+1)} bb^\dagger\left(b^\dagger\right)^{M-j-1}|\text{vac}\rangle \\ &= \left(a^\dagger\right)^{k+2(j+1)}(I+N_b)\left(b^\dagger\right)^{M-j-1}|\text{vac}\rangle \\ &= (M-j)|j+1\rangle.\end{aligned}$$

It is now seen that the action of (3) on the basis elements (7) is equivalent to the action of the differential operator

$$\mathbf{H} = \left(Ax^2 + 4\Omega x\right)\frac{d^2}{dx^2} + \left(Bx + \Omega(4k+2-x^2)\right)\frac{d}{dx} + (C + \Omega Mx)$$

on the basis of monomials through the identification $x^j \equiv |j\rangle$. Consider

$$\mathbf{H}Q(x) = EQ(x),$$

or equivalently

$$\left(Ax^2 + 4\Omega x\right)Q''(x) + \left(Bx + \Omega(4k+2-x^2)\right)Q'(x) + (C + \Omega Mx)\,Q(x) = EQ(x), \tag{8}$$

where

$$Q(x) = \prod_{j=1}^{M}(x - v_j).$$

Then

$$\begin{aligned}\frac{Bv_j + \Omega(4k+2-v_j^2)}{Av_j^2 + 4\Omega v_j} &= -\frac{Q''(v_j)}{Q'(v_j)} \\ &= \sum_{k\neq j}^{M}\frac{2}{v_k - v_j}.\end{aligned} \tag{9}$$

Finally, by equating the Mth-order terms in (8), with

$$Q(x) \sim x^M - x^{M-1} \sum_{j=1}^{M} v_j,$$

we obtain

$$E = AM(M-1) + BM + C - \Omega \sum_{j=1}^{M} v_j. \tag{10}$$

Eqs. (9,10) provide the Bethe Ansatz solution of (3) in Richardson–Gaudin form. It can be verified that setting $U_{aa} = U_{ab} = U_{bb} = \mu_b = 0$ leads to (9) and (10) reducing to (6) and (5) respectively, with the change of variables $v_j = 2u_j$ and $k = 2\kappa - 1/2$.

An alternative formulation is to instead adopt the basis states

$$|l\rangle = \left(a^\dagger\right)^{N-2l} \left(b^\dagger\right)^l |\text{vac}\rangle \tag{11}$$

for $l = 0, ..., M$, $M = (N-k)/2$, and again $k = 1$ for odd N and $k = 0$ for even N. We set

$$\begin{aligned}
A &= 4U_{aa} - 2U_{ab} + U_{bb}, \\
\tilde{B} &= 4(N-1)U_{aa} - (N-2)U_{ab} - U_{bb} + 2\mu_a - \mu_b, \\
\tilde{C} &= N^2 U_{aa} + N\mu_a.
\end{aligned}$$

Now it can be shown that the action of (3) on the basis elements (11) is equivalent to the action of the differential operator

$$\tilde{\mathbf{H}} = \left(Ax^2 + 4\Omega x^3\right) \frac{d^2}{dx^2} + \left(-\tilde{B}x + \Omega(1 + (6-4N)x^2)\right) \frac{d}{dx} + \left(\tilde{C} + \Omega N(N-1)x\right) \tag{12}$$

on the basis of monomials through $x^l \equiv |l\rangle$. Consider

$$\tilde{\mathbf{H}} Q(x) = EQ(x)$$

where

$$Q(x) = \prod_{j=1}^{M} (x - w_j).$$

Then by the same considerations as above we find

$$\begin{aligned}
\frac{-\tilde{B}w_j + \Omega(1 + (6-4N)w_j^2)}{Aw_j^2 + 4\Omega w_j^3} &= -\frac{Q''(w_j)}{Q'(w_j)} \\
&= \sum_{k \neq j}^{M} \frac{2}{w_k - w_j}
\end{aligned} \tag{13}$$

and

$$E = AM(M-1) - \tilde{B}M + \tilde{C} - (4k+2)\Omega \sum_{j=1}^{M} w_j. \tag{14}$$

Set $w_j = v_j^{-1}$. Through standard algebraic manipulations it can be shown that (9) and (13) are equivalent by noting that

$$B = \tilde{B} - A(N-k-2),$$
$$C = AM(M-1) - \tilde{B}M + \tilde{C}.$$

Using (13) we may write

$$-\tilde{B} + \Omega(w_j^{-1} + (6-4N)w_j) = (Aw_j + 4\Omega w_j^2)\sum_{k\neq j}^{M} \frac{2}{w_k - w_j},$$

from which it follows that

$$-\tilde{B}M + \Omega\sum_{j=1}^{M} v_j - (4k+2)\Omega\sum_{j=1}^{M} w_j = -AM(M-1).$$

This leads to the conclusion that (10) and (14) are equivalent.

Eqs. (9) and (10), or equivalently Eqs. (13) and (14), provide the Bethe Ansatz solution of the Hamiltonian (3) in Richardson–Gaudin form. As mentioned earlier, the motivation for obtaining this form is that it opens up an avenue to make exact quantitative calculations for the ground-state energy and particular expectation values. This can be achieved following the methods described in [15] used for similar models. We look forward to reporting on developments in this direction in future work.

Acknowledgments

This work was supported by the Australian Research Council through Discovery Project DP110101414.

References

[1] Vardi A, Yurovsky V A and Anglin J R 2001 *Phys. Rev. A* **64** 063611
[2] Zhou H-Q, Links J and McKenzie R H 2003 *Int. J. Mod. Phys. B* **17** 5819
[3] Sklyanin E K 1989 *J. Sov. Math.* **47** 2473
[4] Gaudin M 1976 *J. Phys. (Paris)* **37** 1087
[5] Richardson R W and Sherman N 1964 *Nucl. Phys.* **52** 221
[6] Richardson R W 1965 *J. Math. Phys.* **6** 1034
[7] Ortiz G, Somma R, Dukelsky J and Rombouts S 2005 *Nucl. Phys. B* **707** 421
[8] Lukyanenko I, Isaac P S and Links J 2014 *Nucl. Phys. B* **886** 364
[9] Zhou H-Q, Links J, Gould M D and McKenzie R H 2003 *J. Math. Phys.* **44** 4690
[10] Santos G, Tonel A, Foerster A and Links J 2006 *Phys. Rev. A* **73** 023609
[11] Li J, Ye D-F, Ma C, Fu L-B and Liu J 2009 *Phys. Rev. A* **79** 025602
[12] Santos G, Foerster A, Links J, Mattei E and Dahmen S R 2010 *Phys. Rev. A* **81** 063621
[13] Li S-C and Fu L-B 2011 *Phys. Rev. A* **84** 023605
[14] Cui B, Wang L C and Yi X X 2012 *Phys. Rev. A* **85** 013618
[15] Links J and Marquette I 2015 *J. Phys. A: Math. Theor.* **48** 045204
[16] Rubeni D, Foerster A, Mattei E and Roditi I 2012 *Nucl. Phys. B* **856** 698
[17] Pan F, Bao L, Zhai L, Cui X and Draayer J P 2011 *J. Phys. A: Math. Theor.* **44** 39
[18] Faribault A, El Araby O, Strater C and Gritsev V 2011 *Phys. Rev. B* **83** 235124
[19] El Araby O, Gritsev V and Faribault A 2012 *Phys. Rev. B* **85** 115130
[20] Guan X, Launey K D, Xie M, Bao L, Pan F and Draayer J P 2012 *Phys. Rev. C* **86** 024313
[21] Marquette I and Links J 2012 *J. Stat. Mech.* P08019
[22] Lerma H S and Dukelsky J 2013 *Nucl. Phys. B* **870** 421
[23] Van Raemdonck M, De Baerdemacker S and Van Neck D 2014 *Phys. Rev. B* **89** 155136

Journal of Physics: Conference Series **597** (2015) 012069 doi:10.1088/1742-6596/597/1/012069

Momentum entanglement in relativistic quantum mechanics

W Smilga

Isardamm 135 d, 82538 Geretsried, Germany

E-mail: wsmilga@compuserve.com

Abstract. I present a new group-theoretical approach to the interaction mechanism of elementary particle physics. Within an irreducible unitary two-particle representation of the Poincaré group, the commutation relations of the Poincaré group require that the two-particle states be momentum entangled. As in gauge theories, momentum entanglement defines a correlation between two particles that can be described as an interaction provided by the exchange of virtual (gauge) quanta. The coupling constant of this interaction is uniquely determined by the structure of the irreducible two-particle state space. For two massive spin one-half particles, the coupling constant matches the empirical value of the electromagnetic coupling constant.

1. Introduction

It is generally accepted that the fundamental interactions of elementary particle physics are the result of symmetries. In the Standard Model (of particle physics), the symmetry with respect to gauge transformations is made responsible for the electromagnetic, weak, and strong interactions. It has been suggested that also the gravitational interaction can be based on gauge invariance, although to date no such theory exists. So far, attempts to derive gauge invariance from other, more basic principles have failed. Therefore, within the Standard Model, gauge invariance is understood as a first principle.

The Standard Model implements gauge invariance by coupling (charged) fermions to gauge fields. Within the perturbation algorithm, this coupling is described by the exchange of gauge bosons between two fermions. Looking at the structure of Feynman graphs, for example of electron scattering, it becomes apparent that it is not the gauge symmetry itself, but rather the momentum entangled structure generated by the exchange of virtual gauge bosons that defines the interaction, independently of what may have caused the entanglement. Hence, a more appropriate question than that of why Nature is gauge invariant, is the question: why are two-fermion states momentum entangled?

In the following, I will describe a fundamental mechanism that is inherent to relativistic quantum theories. It enforces, under very general conditions and independently of gauge invariance, momentum entangled two-particle states. I will show that this entanglement is responsible for at least two interactions: the electromagnetic and a gravitational interaction.

2. Short review of the Poincaré group

As a reminder, here is an excerpt from [1]: The Poincaré group PO(3,1) is a ten-parameter continuous group. The four generators for the infinitesimal translations are (represented by) Hermitian operators p_μ. The six Hermitian operators $m_{\mu\nu}$ generate the rotations in the $x^\mu - x^\nu$ plane. The rotations form the (homogeneous) Lorentz group SO(3,1). The commutation rules of these operators are

$$[p_\mu, p_\nu] = 0 \tag{1}$$
$$[m_{\mu\nu}, m_{\rho\sigma}] = -i(g_{\mu\rho}m_{\nu\sigma} - g_{\nu\rho}m_{\mu\sigma} + g_{\mu\sigma}m_{\rho\nu} - g_{\nu\sigma}m_{\rho\mu}) \tag{2}$$
$$[m_{\mu\nu}, p_\sigma] = i(g_{\nu\sigma}p_\mu - g_{\mu\sigma}p_\nu). \tag{3}$$

The Hermitian operators p_μ and $m_{\mu\nu}$ are identified with the observables of 4-momentum and angular momentum, respectively. The following scalar operators (Casimir operators)

$$P = p^\mu p_\mu \tag{4}$$

and

$$W = -w^\mu w_\mu, \quad \text{with} \quad w_\sigma = \frac{1}{2}\epsilon_{\sigma\mu\nu\lambda}m^{\mu\nu}p^\lambda, \tag{5}$$

commute with all the infinitesimal generators. Hence, for any irreducible representation of the Poincaré group, they are multiples of the identity, and their eigenvalues can be used to classify the irreducible representations. Within an irreducible representation, the states can be labelled by the eigenvalues of a complete set of commuting operators. For a complete set, the three spatial components of the momentum operator p_μ and one of the components of w_μ, say w_0, can be chosen. In vector notation (bold letters stand for 3-vectors):

$$w^0 = \mathbf{p}\cdot\mathbf{m}, \tag{6}$$
$$\mathbf{w} = p_0\,\mathbf{m} - \mathbf{p}\times\mathbf{n}, \tag{7}$$

where

$$\mathbf{m} = (m_{32}, m_{13}, m_{21}), \tag{8}$$
$$\mathbf{n} = (m_{01}, m_{02}, m_{03}). \tag{9}$$

Given an eigenstate of momentum, w_0 is proportional to the generator of the rotation in the plane perpendicular to the 3-momentum $\mathbf{p}$. These rotations are the only rotations that leave $\mathbf{p}$ invariant. For $\mathbf{p} = 0$, the rotational axis is arbitrary.

3. Uncorrelated versus correlated particles

A system of two independent particles with individual 4-momenta p_1 and p_2 is described by a product representation of the Poincaré group. The momenta p_1 and p_2 refer to two independent reference frames. The two-particle system is subject to changes of the frames of reference by the operations of SO(3,1) ⊗ SO(3,1). As long as the frames of reference are not correlated, I will call such a two-particle system *uncorrelated.*

In contrast, if the two particles are considered in the same space–time domain, their momenta refer to one and the same frame of reference. Whereas in an uncorrelated system the sum of the particle momenta does not make any sense, now the total momentum p can be defined by forming the sum $p = p_1 + p_2$. Furthermore, from the momenta p_1 and p_2 and the particle positions x_1 and x_2, an orbital angular momentum, described by the operator

$$m^{\mu\nu} = x^\mu q^\nu - x^\nu q^\mu, \text{ where } q^\mu = p_2^\mu - p_1^\mu \text{ and } x^\mu = x_2^\mu - x_1^\mu, \tag{10}$$

can be constructed. Two particles with a common frame of reference will be referred to as *correlated.*

The total momentum of a correlated two-particle system is subject to changes of the frame of reference by operations of a single SO(3,1) group. In addition, there is an internal rotational degree of freedom, similar to the spin degree of freedom of single particles. This degree of freedom is defined by rotations in the plane perpendicular to the 3-momentum $\mathbf{p}$, which leave the total momentum invariant but, in general, change the individual particle momenta.

For correlated two-particle systems, simultaneous eigenstates of the total momentum and of w_0 can be prepared. Now, w_0 refers to the internal rotational degree of freedom, possibly in combination with the spin of the particles. Under the action of SO(3,1), such an eigenstate can generate a complete basis for the state space of an irreducible two-particle representation of PO(3,1). Such a representation is characterized by the eigenvalues of the Casimir operators P and W.

An immediate consequence of the commutation relations (3) between the linear and the angular momentum is that the individual particle momenta in the operator of the orbital angular momentum (10) do not commute with the angular momentum that they create:

$$[p_k^\nu, m^{\mu\nu}] \neq 0 \neq [p_k^\mu, m^{\mu\nu}], \quad k = 1, 2. \tag{11}$$

This means that a simultaneous eigenstate of the total momentum and of the orbital angular momentum cannot, at the same time, be an eigenstate of $\mathbf{p}_1$ and $\mathbf{p}_2$, except when the 3-momenta $\mathbf{p}_1$ and $\mathbf{p}_2$, and therefore also $\mathbf{p}$, are parallel. This eigenstate must, therefore, have the structure of a momentum entangled superposition of product states. More precisely: since an eigenstate of a component of the angular momentum is invariant under the rotations generated by this component, the state is a rotationally symmetric superposition of product states with the same total momentum. In consequence, measurements of the individual particle momenta will yield non-deterministic results. Nevertheless, if the state is an eigenstate of the total momentum, the particle momenta will always add up to the (conserved) value of the total momentum. Hence an unbiased experimenter will conclude that the particles exchange momentum.

4. A complementary dynamical law

The result of the last section can be extended to all states of an irreducible two-particle representation of the Poincaré group: Since a complete set of momentum entangled eigenstates of total momentum and angular momentum form a basis, the entangled structure is passed on to all states of the same irreducible representation. If it were possible to construct from this basis a pure product state, then, because of Poincaré covariance, it would be possible to construct a complete set of such product states. These states would form a basis of a subspace of the product representation that is restricted only by the constancy of the Casimir operator P, but not of W, in contradiction to the assumed irreducibility.

Therefore, a general rule holds:

The states of an irreducible two-particle representation of the Poincaré group have the form of a momentum entangled superposition of product states.

Another formulation of this rule, in the pictorial language of relativistic perturbation theories, e.g., quantum electrodynamics (QED), is this:

In an irreducible two-particle representation of the Poincaré group, the particles exchange virtual quanta of momentum.

An exception to this rule is the representation where the 3-momenta $\mathbf{p}_1$ and $\mathbf{p}_2$, and therefore also $\mathbf{p}$, are parallel. In this representation, there is no entanglement and hence no exchange of virtual quanta.

In conjunction with the conservation laws for total momentum and angular momentum, this rule has the character of a *complementary dynamical law*, dictated by Poincaré symmetry. It defines a mechanism that forces two particles to exchange quanta of momentum in a controlled way. I call this rule 'a law' (in the following: 'the Dynamical Law') for two reasons: Firstly, it derives directly from the commutation relations of the Poincaré group and is, therefore, fundamental for relativistic quantum mechanics, as fundamental as the conservation laws of energy, momentum, and angular momentum. Secondly, it has, just as the mentioned conservation laws, far-reaching implications for the physics of elementary particles.

Actually, the Standard Model uses a concept of momentum exchange to describe three of the fundamental forces between elementary particles. Feynman, when formulating the perturbation theory of quantum electrodynamics [2–4], spoke of 'virtual quanta'. Based on the postulate of gauge invariance, the exchanged quanta became 'gauge bosons'. Integration over the exchanged momenta is an integral part of the Feynman rules in momentum space (cf., e.g., [5]). Together with the conservation of momentum at each vertex, this integration generates the kind of momentum entanglement that is demanded by the Dynamical Law.

As just shown, there is no need to explicitly postulate the presence of gauge bosons: they are there not for reasons of gauge invariance, but because the structure of the Poincaré group requires their existence. However, in contrast to the common understanding of QED, these gauge boson are not the quanta of an independent field. They rather stand for a field that describes the intrinsic structure of many-particle states as determined by Poincaré symmetry. Therefore, this field may be referred to as an *intrinsic gauge field*.

5. Irreducible two-particle states

The parameter space of an irreducible two-particle representation derives from the two-particle mass-shell relation

$$(p_1 + p_2)^2 = M^2, \tag{12}$$

which defines a seven-dimensional surface in an eight-dimensional space–time continuum. After inserting the mass-shell relations for the individual particles, this parameter space is described by a five-dimensional space Ω of independent momentum parameters. Ω is embedded into the six-dimensional momentum space $\mathbb{R}^3 \times \mathbb{R}^3$. It has spherical properties with respect to four dimensions and is unbounded in the radial direction. Accordingly, Ω has a spherical infinitesimal volume element $d\omega(\mathbf{p}_1, \mathbf{p}_2)$. On the other hand, the embedding into $\mathbb{R}^3 \times \mathbb{R}^3$ induces a Cartesian infinitesimal volume element d^5p on Ω.

Since the two-particle states of an irreducible representation are momentum entangled, a two-particle state is represented by an integral

$$|\Phi\rangle = \int d\omega(\mathbf{p_1}, \mathbf{p_2})\; c_{\mathbf{p_1 p_2}}\; |\mathbf{p_1}, \mathbf{p_2}\rangle \tag{13}$$

over pure product states $|\mathbf{p_1}, \mathbf{p_2}\rangle$, with state-specific coeffients $c_{\mathbf{p_1 p_2}}$. The range of integration is a finite (bounded) subspace of the full parameter space Ω. Since the product states are normalized to $\mathbb{R}^3 \times \mathbb{R}^3$, a normalization factor that maintains the normalization of the two-particle state is required. It equals the square root of the volume of the integration area. This area is essentially the spherical subspace of Ω. It makes sense to split the normalization factor into a state-independent factor ω, directly related to the square root of the volume of the spherical subspace, and a state-specific factor. In state (13), the first factor is assumed to be included in the volume element $d\omega$, the second one in the coefficients $c_{\mathbf{p_1 p_2}}$. The spherical volume element can then be replaced by the Cartesian volume element d^5p

$$d\omega(\mathbf{p}_1, \mathbf{p}_2) \;\Rightarrow\; \omega\, d^5p. \tag{14}$$

Using (14), a two-particle state can be written as

$$|\Phi\rangle = \int_\Omega \omega\, d^5p\; c_{\mathbf{p_1 p_2}}\; |\mathbf{p_1}, \mathbf{p_2}\rangle\,. \tag{15}$$

6. Scattering experiments

The experimental setup of a typical scattering experiment (Fig. 1) will help to reveal the physical meaning of ω. An incoming plane wave (lower left) with momentum $\mathbf{p}$ passes an aperture and hits a target (centre). Parts of the scattered wave pass the second aperture and are registered by a detector (right). Between the apertures, an incoming particle together with a particle of the target form an intermediate, correlated two-particle system. By positioning the first aperture at a distance $\mathbf{d}$ from the target, an orbital angular momentum $\mathbf{d} \times \mathbf{p}$ is selected. The observed transition is represented by a hyperbolic trajectory.

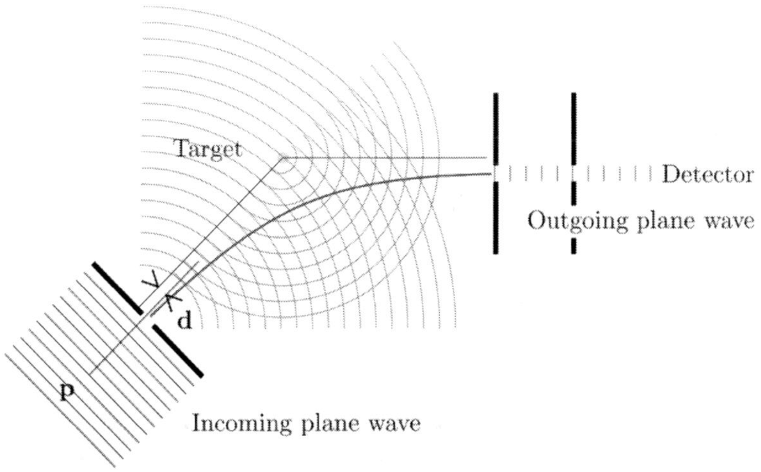

Fig. 1 Selection of orbital angular momentum $\mathbf{d} \times \mathbf{p}$

The transition amplitude from an incoming two-particle state, described by a pure product state $|\mathbf{p}_1, \mathbf{p}_2\rangle$, through an intermediate state with well-defined total momentum $\mathbf{p}$ and angular momentum m, to the outgoing product state $|\mathbf{p}_1 - \mathbf{k}, \mathbf{p}_2 + \mathbf{k}\rangle$, is then given by

$$S(\mathbf{k}) = \omega^2 \langle \mathbf{p}_1, \mathbf{p}_2 | \mathbf{p}, m\rangle \langle \mathbf{p}, m | \mathbf{p}_1 - \mathbf{k}, \mathbf{p}_2 + \mathbf{k}\rangle . \qquad (16)$$

Here, I have placed the state-independent normalization factor ω^2 of $|\mathbf{p}, m\rangle\langle \mathbf{p}, m|$ in front of the amplitude, which gives ω^2 the clear meaning of a *coupling constant*: ω determines, as a state-independent factor, the transition amplitude from an incoming product state $\langle \mathbf{p}_1, \mathbf{p}_2|$ to the same product state within $|\mathbf{p}, m\rangle$, and as well the transition amplitude from a product state $\langle \mathbf{p}_1 - \mathbf{k}, \mathbf{p}_2 + \mathbf{k}|$ within $\langle \mathbf{p}, m|$ to the same outgoing product state. Because the intermediate eigenstate of total and orbital angular momentum is momentum entangled, there will be non-zero transition amplitudes for $\mathbf{k} \neq \mathbf{0}$.

Generally speaking, any restriction on the relative momenta and positions of two correlated particles means a restriction of the available intermediate states. It is this restriction that makes the momentum exchange visible.

7. Coupling constant

As indicated above, ω^2 is determined by the geometry of the two-particle mass-shell relation (12), which is a little bit tricky, but, nevertheless, well defined. This allows calculating ω^2 from the volumes of certain homogeneous domains [6] D^5, Q^5 and S^4, which have been used in [7] to describe the structure of the two-particle mass-shell. An outline of the calculation can be found

Journal of Physics: Conference Series **597** (2015) 012069 doi:10.1088/1742-6596/597/1/012069

in the Appendix. The result (for massive leptons) is given by the expression

$$8\pi\, V(D^5)^{\frac{1}{4}} \,/\, (V(S^4)\, V(Q^5)) = \frac{9}{8\pi^4}\left(\frac{\pi^5}{2^4\, 5!}\right)^{1/4} = 1/137.03608245. \tag{17}$$

This expression is identical to the well-known semi-empirical Wyler formula [8]. Its numerical value closely matches the electromagnetic fine-structure constant α, whose empirical value is $1/137.035999084(51)$ [9].

This close match is strong evidence that the electromagnetic interaction is in fact the manifestation of the intrinsic gauge field (cf. Section 4) required by the Dynamical Law.

8. The Dynamical Law in particle theories

Since the Dynamical Law derives directly from Poincaré invariance, it is binding on every relativistic quantum theory of particle physics, including the existing gauge theories, as well as any future theory of quantum gravity. This, plus the fact that the concept of Poincaré invariance is experimentally extraordinarily well supported, makes the Dynamical Law a powerful instrument for the evaluation of existing theories and for setting up new theories.

In the following two subsections, I will analyse how two well-known relativistic theories comply with the Dynamical Law.

8.1. Quantum electrodynamics

Quantum electrodynamics is presently implemented with a perturbation algorithm. This algorithm has inconsistencies, in the form of divergent integrals (cf., e.g., [10]). The technique of renormalization was developed to remove these divergences. Since renormalization is more a recipe than a mathematical procedure, it makes the algorithm lack transparency. Nevertheless, the numerical results obtained in this way match the experimental data extremely well.

In contrast, the Dynamical Law offers a non-perturbative way to obtain the electromagnetic interaction. This makes it well suited to shed some light on the perturbation algorithm.

The perturbation algorithm of QED constructs states with a momentum entangled structure, in agreement with the structure of the states of irreducible two-particle representations. It also correctly normalizes the intermediate states by inserting the empirical value of the fine-structure constant α as the coupling constant. However, instead by restricting the product state space to an irreducible state space, the perturbation algorithm implements the entangled structure 'on top' of the product state space. An immediate consequence of this approach is that the integrations are carried out over the full six-dimensional momentum space of the product representation, instead of over the five-dimensional parameter space of an irreducible representation. As a result, we have logarithmically divergent integrals. It can be easily shown that reducing the number of integration variables by one turns the logarithmically divergent integrals of QED into finite integrals.

For practical calculations, the perturbation approach of QED remains the method of choice. This should not be taken as a license to overload the perturbation algorithm with questionable 'physical' interpretations. There are no indications, neither theoretical nor experimental, that 'virtual particles' are anything other than elements of a pictorial description of momentum entangled structures. This applies to virtual gauge bosons as well as to virtual electron–positron pairs, which, within the perturbation algorithm, are 'created' by gauge bosons.

Independently of the partial failure of the perturbation algorithm, it can be stated that the Poincaré group not only determines the structure of the electromagnetic interaction, it also explains, by means of the Dynamical Law, why there is an electromagnetic interaction, and it determines, in contrast to the principle of gauge invariance, the correct value of the electromagnetic coupling constant.

8.2. Gravitation

This subsection focuses on momentum entanglement in the limit of large quantum numbers. In contrast to the commutation relations between position and momentum, the commutation relations between momentum and angular momentum (3) do not contain Planck's constant h. Therefore, they remain valid when $h \to 0$. This indicates that the Dynamical Law, derived from these commutation relations, also applies to the classical limit.

According to Newton's laws, 'exchange of momentum' is another wording for 'curved trajectories'. From General Relativity it is known that curved trajectories can be described by an intrinsic curvature of space–time, such that the trajectories become geodesics with respect to the metric of space–time. Differential geometry describes this curvature by the Riemann curvature tensor $R_{\lambda\mu\nu\kappa}$. The Riemann tensor can be written as

$$\begin{aligned} R_{\lambda\mu\nu\kappa} &= C_{\lambda\mu\nu\kappa} - \frac{1}{6}R\left[g_{\lambda\nu}g_{\mu\kappa} - g_{\lambda\kappa}g_{\mu\nu}\right] \\ &+ \frac{1}{2}\left[g_{\lambda\nu}R_{\mu\kappa} - g_{\lambda\kappa}R_{\mu\nu} - g_{\mu\nu}R_{\lambda\kappa} + g_{\mu\kappa}R_{\lambda\nu}\right], \end{aligned} \tag{18}$$

where $C_{\lambda\mu\nu\kappa}$ is the traceless Weyl tensor, $R_{\mu\nu} = R^{\lambda}{}_{\mu\lambda\nu}$ is the Ricci tensor, and $R = R^{\alpha}{}_{\alpha}$ is the Ricci scalar [11].

In General Relativity (Einstein–Newton gravity), it is postulated that a certain combination of the Ricci tensor and the Ricci scalar be proportional to the energy–momentum tensor. In contrast, the curvature that results from the Dynamic Law is based on the exchange or flux of momentum between the particles. Unlike Einstein–Newton gravity, this is not a postulate, but rather a direct consequence of the momentum entangled structure of two-particle states. The flux of momentum associated with a particle at a point x is described by the off-diagonal elements $T^{0i}(x)$ and $T^{ik}(x)$ of the energy–momentum tensor $T^{\mu\nu}(x)$. There is no direct contribution from the diagonal elements $T^{ii}(x)$. Therefore, the curvature discussed here is generated by the traceless part of the energy–momentum tensor. In consequence, the generated curvature tensor must be traceless too, which means that the curvature must be described by the traceless Weyl tensor.

As shown in [11], this requirement lead to the field equations

$$W^{\mu\nu} = \alpha_g\left(T^{\mu\nu} - \frac{1}{2}g^{\mu\nu}T^{\lambda}{}_{\lambda}\right). \tag{19}$$

The tensor on the right is the energy–momentum tensor minus its trace. The tensor $W^{\mu\nu}$ on the left has the form

$$\begin{aligned} W^{\mu\nu} &= \frac{1}{2}g^{\mu\nu}(R^{\alpha}{}_{\alpha})^{;\beta}{}_{;\beta} + R^{\mu\nu;\beta}{}_{\beta} + R^{\mu\beta;\nu}{}_{\beta}R^{\mu\beta;\mu}{}_{\beta} \\ &- 2R^{\mu\beta}R^{\nu}{}_{\beta} + \frac{1}{2}g^{\mu\nu}R_{\alpha\beta}R^{\alpha\beta} - \frac{2}{3}g^{\mu\nu}(R^{\alpha}{}_{\alpha})^{;\beta}{}_{;\beta} \\ &+ \frac{2}{3}(R^{\alpha}{}_{\alpha})^{;\mu;\nu} + \frac{2}{3}R^{\alpha}{}_{\alpha}R^{\mu\nu} - \frac{1}{6}g^{\mu\nu}(R^{\alpha}{}_{\alpha})^2. \end{aligned} \tag{20}$$

The constant α_g acts as a coupling constant.

Equations (19) are nothing other than the field equations of conformal gravity. Conformal gravity is known to yield the same results as Einstein–Newton gravity as long as systems with the symmetry of the solar system are considered [12], see also [13]. On the scales of galaxies, the theories differ insofar as conformal gravity is able to describe the kinematics of galaxies solely on the basis of the distribution of visible matter [12], whereas in Einstein–Newton gravity large amounts of properly placed 'dark matter' are required to match the observational data.

What can be said about the value of the coupling constant α_g? A first calculation for a two-particle system of spinless particles, in analogy to the calculation of the electromagnetic coupling constant, reproduces this very same coupling constant. However, in contrast to the electromagnetic interaction, which can be neutralized, because there are positive and negative charges, the 'gravitational charge' always has the same sign, which makes it impossible to shield gravity. Therefore, to determine an effective coupling constant, in principle all particles of the universe must be taken into account. The number of protons derived from an estimated total mass of the visible universe of 10^{53} kg (cf., e.g., [14] and references cited therein) is 10^{80}. To exchange momentum with another particle, a given particle must, at first, find a second particle, before it can form a two-particle state with this partner. There are 10^{80} different potential partners. The quantum mechanical transition amplitude for a given particle to form a two-particle state with a specific second particle is therefore given by 10^{-40}, the square root of 10^{-80}. Hence the strength (or, better, the weakness) of the gravitational interaction is, first of all, determined by the factor 10^{-40}. This factor agrees quite well with the empirical ratio between the strengths of the gravitational and the electromagnetic interactions. It has no relation to the Planck mass, but rather to the number of particles in the universe.

Although the field equations (19) define a classical theory of gravitation, they describe, in the limit of large quantum numbers, nothing other than a basic property of relativistic quantum mechanics, namely the momentum entanglement of two-particle states. (Note that in momentum entangled two-particle states, there are none of those 'ghosts' that in the past have been associated with conformal gravity.) Conformal gravity is therefore compatible with relativistic quantum mechanics and especially with the Dynamical Law.

9. Conclusions

The Dynamical Law, which requires two-particle states to be momentum entangled, is mathematically derivable from the structure of the Poincaré group in combination with the basic principles of quantum mechanics. Therefore, as long as neither Poincaré invariance nor quantum mechanics is called into question, this law must be taken into account on the same footing as the conservation laws of energy, momentum, and angular momentum.

The Dynamical Law explains the success of gauge theories and at the same time it helps to understand their deficiencies. It is more restrictive than the principle of gauge invariance since it determines the coupling constants. On the other hand, it covers a wider range than that principle, since it does not require the gauge field to be massless. Therefore, in applying this law to the weak interaction, there is no need for the Higgs mechanism, but such a mechanism may well be needed in order to understand the different masses of leptons.

There is strong evidence that the interaction mechanism determined by the Dynamical Law is responsible for both the electromagnetic and the gravitational interaction. Therefore, quantum mechanics and gravitation can no longer be seen as incompatible. The analysis of how and to what extent quantum electrodynamics and Einstein–Newton gravity comply with the Dynamical Law identifies deficiencies in both theories, deficiencies that apparently contribute to the well-known problems of the divergences in QED and the failure of Einstein's general relativity to describe the kinematics of galaxies solely on the basis of visible matter.

The Standard Model describes two more interactions: the strong interaction and the weak interaction, which, similarly to QED, are described by the exchange of virtual gauge particles and result in momentum entangled structures. In quantum chromodynamics, momentum entanglement exists between the constituents of compound particles, whereas in the weak interaction the exchanged virtual quanta have a mass. In order to trace back these interactions explicitly to a conserved angular momentum, a more detailed theoretical understanding of the constituents of baryonic particles on the one hand and of the mechanism that is responsible for the different lepton masses on the other hand seems to be required.

Appendix: Calculation of the coupling constant

The following outlines the calculation of the factor ω^2. A more detailed description can be found in [7].

First consider the 5-dimensional unit ball B^5, given by

$$x_1^2 + x_2^2 + x_3^2 + x_4^2 + x_5^2 \leq 1. \tag{A.1}$$

Its boundary ∂B^5 is the 4-dimensional unit sphere S^4

$$x_1^2 + x_2^2 + x_3^2 + x_4^2 + x_5^2 = 1. \tag{A.2}$$

The volume of B^5 can be expressed by the integral

$$V(B^5) = \int_0^1 dr\, r^4 \int_{\partial B^5} dx_1\, dx_2\, dx_3\, dx_4. \tag{A.3}$$

This integral will be used to study the conversion of a spherical volume element into an Cartesian one (cf. Section 5, replacement (14)). At a given point b on ∂B^5, the infinitesimal spherical volume element $dr\, dx_1\, dx_2\, dx_3\, dx_4$ has a rectangular form with four edges of the same (infinitesimal) length and a fifth, along r, with a different length. Now replace the second integral in (A.3) by a 4-dimensional cube in the tangent space at b, such that the value of the integral is unchanged ('quadrature of the circle'). Then each edge of the cube will have length $s = V(B^5)^{\frac{1}{4}}$. A 5-dimensional cuboid with volume $V(B^5)$ can be formed in combination with the interval $[0, 1]$ on the r-axis. By subdividing this cuboid into N intervals in each direction, one obtains a cover of the cuboid by N^5 volume elements. Multiplying the length of the edges in the radial direction by s and letting $N \to \infty$ results in a cubical (Cartesian) infinitesimal volume element d^5x. The replacement of dr by $s\, dx_5$ introduces a factor of $V(B^5)^{\frac{1}{4}}$.

As mentioned in Section 5, a two-particle state, represented by an integral over product states, requires a normalization factor related to the volume of the integration area. This volume is defined by the spatial angle covered by the integral (13). The normalization factor is contained implicitly in $d\omega(\mathbf{p}_1, \mathbf{p}_2)$ but is made explicit in $\omega\, d^5p$ (cf. replacement (14)). By taking into account such normalization, the factor $V(\partial B^5)^{-1}$ is introduced into d^5x.

To complicate matters, restrict ∂B^5 by the two equations

$$x_1^2 + x_2^2 + x_3^2 = r^2 \quad \text{and} \quad x_4^2 + x_5^2 = 1 - r^2, \quad r \leq 1. \tag{A.4}$$

Equation (A.2) still holds, but whereas ∂B^5 has five orthogonal rotational symmetries, the space (A.4) is determined by rotations around $3 + 1$ orthogonal rotational axes. The addition of only one symmetry axis, for example, the rotational axis perpendicular to the x_3–x_4 plane, would convert the restricted space (A.4) into (A.2). The integral (A.3), calculated for the restricted space, is therefore smaller than the original one by a factor determined from the volume of the homogeneous space $\mathrm{SO}(5)/\mathrm{SO}(4) \cong S^4$. If, however, the restriction (A.4) is ignored and d^5x is maintained as a 5-dimensional volume element, for example because the integral is part of an algorithm that is based on a 5-dimensional Euclidean parameter space, then at least the result needs to be corrected ('renormalized') by the factor $V(S^4)^{-1}$.

These three factors combine to give the expression

$$V(B^5)^{\frac{1}{4}} \,/\, (V(\partial B^5)\, V(S^4)), \tag{A.5}$$

which, up to the factor 8π, exhibits the basic structure of Wyler's formula (17). (Although the derivation of (A.5) has been based on the unit ball, the result does not depend on the radius of the ball: A scaling of the radius is covered by the scaling property of d^5x.)

The 'formula' (A.5) can now be applied to the parameter space of two particles with equal mass m that are described by an irreducible two-particle representation of the Poincaré group. The total momentum $p = p_1 + p_2$ (p_1 and p_2 are 4-vectors) satisfies the mass-shell relation

$$p^2 = M^2. \tag{A.6}$$

For the relative momentum $q = p_1 - p_2$, one has

$$q^2 = 4m^2 - M^2 \leq 0. \tag{A.7}$$

Furthermore, the relations

$$2p_1p_2 = M^2 - 2m^2 \quad \text{and} \quad 2p_1\,p = 2p_2\,p = M^2 \tag{A.8}$$

let the particle momenta maintain a certain angle to the total momentum. This restricts the 4-vector q to a spacelike plane, which reduces q to a 2-component vector in this plane, respectively, to a 2+1-component vector in space–time.

Equations (A.6) and (A.7) can be combined to

$$p^2 + q^2 = 4m^2, \tag{A.9}$$

defining a parameter space with an internal structure comparable to the space defined by (A.4). It can be embedded into the 5+2-dimensional homogeneous space SO(5,2) / SO(5)×SO(2), similar to the embedding of space (A.4) into space (A.2). By replacing in (A.5) the volumes $V(B^5)$ and $V(\partial B^5)$ by the corresponding volumes $V(D^5)$ and $V(Q^5)$, where

$$V(D^5) = \frac{\pi^5}{2^4\,5!}, \quad V(Q^5) = \frac{8\pi^3}{3}, \quad V(S^4) = \frac{8\pi^2}{3}, \tag{A.10}$$

Wyler's formula (17), up to the factor of 8π, is obtained. The volumes of D^5 and Q^5 have been calculated in [6] from a bounded realization of SO(5,2) / SO(5)×SO(2) on the complex unit ball.

Two more factors, leading to the contribution of 8π in Wyler's formula, still have to be added: Firstly, a factor of 2 as part of the Jacobian that relates p and q to p_1 and p_2. Secondly, a factor of 4π, because in the momentum representation (the p-representation), the effective coupling constant in QED, to which must be compared the factor ω^2 as part of the transition amplitude (16), is not e but $e/4\pi$. Together with the convention $\alpha = e^2/4\pi$, this results in the factor 4π.

The normalization factor ω of a two-particle state is, as usual, obtained as the square root of the volume factor ω^2.

References

[1] Schweber S 1962 *An Introduction to Relativistic Quantum Field Theory*, 2nd printing (New York: Harper and Row) 38–46
[2] Feynman R P 1949 *Phys. Rev.* **76** 749–59
[3] Feynman R P 1949 *Phys. Rev.* **76** 769–89
[4] Feynman R P 1950 *Phys. Rev.* **80** 440–57
[5] Schweber S 1962 *An Introduction to Relativistic Quantum Field Theory*, 2nd printing (New York: Harper and Row) 483
[6] Hua L K 1963 *Harmonic Analysis of Functions of Several Complex Variables in the Classical Domains* (Translations of Mathematical Monographs Vol. 6) (Providence, RI: American Mathematical Society)
[7] Smilga W 2013 *J. Mod. Phys.* **4** 561–71 (*Preprint* arXiv:1004.0820v7)
[8] Wyler A 1971 *C. R. Acad. Sc. Paris* A **271** 186–8
[9] Hanneke D, Fogwell S and Gabrielse G 2008 *Phys. Rev. Lett.* **100** 120801
[10] Schweber S 1962 *An Introduction to Relativistic Quantum Field Theory*, 2nd printing (New York: Harper and Row) 584–645
[11] Mannheim P D 2012 *Found. Phys.* **42** 388–420 (*Preprint* arXiv:1101.2186)
[12] Mannheim P D 2006 *Prog. Part. Nucl. Phys.* **56** 340–445 (*Preprint* arXiv:astro-ph/0505266)
[13] Maldacena J 2011 *Preprint* arXiv:1105.5632
[14] Valev D 2014 *Phys. Int.* **5** 15–20 (*Preprint* arXiv:1004.1035)

Hermite polynomials and representations of the unitary group

A Strasburger, D Dziewa-Dawidczyk

Department of Aplied Mathematics, Warsaw University of Life Sciences (SGGW), Nowoursynowska 159, 02-787 Warsaw, Poland

E-mail: `aleksander_strasburger@sggw.pl`, `diana_dziewa_dawidczyk@sggw.pl`

Abstract. Spaces of homogeneous complex polynomials in D variables form carrier spaces for representations of the unitary group $U(D)$. These representations are well understood and their connections with certain families of classical orthogonal polynomials (Gegenbauer, Jacobi, and other) are widely studied. However, there is another realization for the action of the unitary group $U(D)$ on polynomials, not necessarily homogeneous, in which Hermite polynomials in D variables play an important role. This action is related to the metaplectic (oscillator) representation, and was studied some time ago by one of the present authors (A. S.) and, independently, by A. Wünsche for $D = 2$. In this note we want to concentrate on the latter realization and describe its properties in a more comprehensible way.

1. Introduction

This note pursues the common lines of investigations started some time ago by one of the present authors, [1, 2], and about the same time approached from a rather different perspective by A. Wünsche, [4, 5].

2. Hermite polynomials in one and several variables

2.1. The one variable case

Due to their exceptionally wide applicability Hermite polynomials in one variable belong to the toolkit of practically every theoretical physicist. They may be approached from a number of different directions — eigenvalue problem of the quantum mechanical oscillator Hamiltonian, recurrence formulas or as an instance of a family of orthogonal polynomials, to mention just a few most popular viewpoints. For the sake of brevity, we quote here in addition to the basic explicit representation,

$$H_n(x) = \sum_{k=0}^{[n/2]} (-1)^k \frac{n!}{k!(n-2k)!}(2x)^{n-2k}.$$

just two expressions, which are perhaps not so often remembered. An expression via a differential polynomial

$$H_n(x) = 2^n \left(\sum_{k=0}^{[n/2]} (-1)^k \frac{1}{2^{2k} k!} \frac{d^{2k}}{dx^{2k}} \right) x^n = \exp\left[-\frac{d^2}{2^2 dx^2} \right] x^n \qquad (2.1)$$

and via the conjugation with the Gaussian

$$H_n(x) = (-1)^n e^{x^2} \left(\frac{d}{dx}\right)^n (e^{-x^2}) \tag{2.2}$$

This later one is basic for deriving the extension to several variables.

2.2. The case of several variables

We work in the space $\mathbb{R}^d$ of real d dimension whose elements are denoted $x = (x_1, \ldots, x_d)$, and use the standard multi-index notation. Given a d-tuple of non-negative integers $\alpha = (\alpha_1, \ldots, \alpha_d) \in \mathbb{Z}^d_+$, to be called a multi-index, we write

$$x^\alpha = x_1^{\alpha_1} \cdots x_d^{\alpha_d}, \qquad \text{and} \qquad \partial^\alpha = \partial_1^{\alpha_1} \cdots \partial_d^{\alpha_d}.$$

The sum $|\alpha| = \sum_{j=1}^d \alpha_j$ is called the height of the multi-index α. For any polynomial $P(x) = \sum p_\alpha x^\alpha$ on $\mathbb{R}^d$ we shall denote by $P(\partial) = \sum p_\alpha \partial^\alpha$ the corresponding differential polynomial. In particular, if $r^2(x) = |x|^2 = \sum_{j=1}^d x_j^2$ is the square of the euclidean length, the corresponding differential polynomial $r^2(\partial) = \sum_{j=1}^d \partial_j^2 = \Delta$ is the usual Laplacian. A polynomial $P(x) = \sum p_\alpha x^\alpha$ is said to be a homogeneous of degree l if all its coefficients p_α are equal 0 unless $|\alpha| = l$. In this case the corresponding differential operator $P(\partial)$ is called homogeneous of degree l.

Finally, the algebra of complex valued polynomials in d variables is denoted by $\mathcal{P} = \mathcal{P}(\mathbb{R}^d)$ and its subspace of homogeneous of degree k elements by $\mathcal{P}^k$. Obviously each polynomial can be written as a sum of homogeneous polynomials in a unique way, hence

$$\mathcal{P} = \bigoplus_{k=0}^{\infty} \mathcal{P}^k.$$

Since the polynomial algebra $\mathcal{P}(\mathbb{R}^d)$ can be naturally identified with the algebra of symmetric tensors, it carries a natural representation of the full linear group $\mathbf{GL}(d, \mathbb{C})$. The subspaces $\mathcal{P}^k$ are invariant and irreducible with respect to this action. Formally, it is induced by the natural action of $\mathbf{GL}(d, \mathbb{C})$ on the generators of the algebra, i.e. the first degree polynomial functions (coordinates) x_j, $j = 1, \ldots, d$. It should be stressed, however, that it does not straightforwardly come from an action of $\mathbf{GL}(d, \mathbb{C})$ on the domain of polynomials, which in this case is $\mathbb{R}^d$.

The following property of polynomials in several variables is crucial in what follows. Recall that a function $f(x)$ (a polynomial) is called harmonic if it is annihilated by the Laplacian, $\Delta f = 0$. Now, any homogeneous polynomial $P \in \mathcal{P}^l(\mathbb{R}^d)$ can be uniquely decomposed as a sum of products of powers r^{2k} with homogeneous harmonic polynomials of degree $l - 2k$,

$$P = \sum_{k=0}^{[l/2]} r^{2k} h_{l-2k}(P), \qquad \text{where} \qquad \Delta(h_{l-2k}(P)) = 0. \tag{2.3}$$

This is called a canonical decomposition of P, cf. [14, 7].

In the common approach to Hermite polynomials in several variables, cf. [6], one views them as polynomial eigenfunctions of the operator

$$\Delta - 2E, \qquad \text{where } E \text{ is the Euler operator} \qquad E = \sum_{j=0}^{d} x_j \partial_j$$

By separation of variables it is immediately clear that they are linear combinations of products of Hermite polynomials in each coordinate separately,

$$H_\alpha(x) = H_{\alpha_1}(x_1)\cdots H_{\alpha_d}(x_d), \qquad \text{where} \quad \alpha = (\alpha_1, \ldots, \alpha_d), \quad x = (x_1, \ldots, x_d).$$

However it is not easy to make apparent properties of symmetry of so defined polynomials, e.g. for cases involving orthogonal or unitary symmetry. Thus another approach is desirable, which is better suited for expressing the action of symmetry groups. For that purposes the so called *Hermite map*, see [9, 7]), which is a suitable extension of the formula (2.2) is more appropriate.

Definition 1 (The Hermite map) Given a homogeneous polynomial $P(x)$ of degree l we assign it the polynomial H_P, which will be called Hermite polynomial associated with P, by means of the formula

$$H_P(x) = (-1)^l 2^{-l} e^{r^2} P(\partial) e^{-r^2}. \tag{2.4}$$

The Hermite map $P \mapsto H_P$ is a linear automorphism of the polynomial algebra. However, it does not preserve the grading with respect to the degree, nor the multiplication. In fact, using the Hobson formula one can derive the following formula (cf. [3])

$$H_P(x) = \sum_{k=0}^{[l/2]} (-1)^k \frac{1}{2^{2k} k!} \Delta^k P(x), \tag{2.5}$$

hence H_P is not necessarily homogeneous. We point out explicitly two special cases of the above:

For $\quad P(x) = x^\alpha = x_1^{\alpha_1}\cdots x_d^{\alpha_d} \quad$ there is $\quad H_P(x) = H_\alpha(x) = H_{\alpha_1}(x_1)\cdots H_{\alpha_d}(x_d)$;

For $\quad P(x) = r^{2l}(x) \quad$ there is $\quad H_P(x) = c_{l,d} L_l^{(d-2)/2}(r^2)$,

where $\quad L_l^{(d-2)/2}(r^2) = \frac{1}{l!} e^{r^2} r^{2-d} \frac{d}{d(r^2)} e^{-r^2} r^{d-2+2l}$

is the classical Laguerre polynomial of degree l and index $(d-2)/2$.

2.3. *Harmonic decomposition*

In general there seem to be no simple and direct way to describe the elements of the space

$$\operatorname{span}\{H_{\alpha_1}(x_1)\cdots H_{\alpha_d}(x_d) \mid \alpha_1 + \ldots + \alpha_d = l\} = \{\, H_P \mid P \in \mathcal{P}^l \,\},$$

avoiding the use of factorization. However, one can accomplish it by relation to harmonicity.

Proposition 1 *For a homogeneous polynomial $P \in \mathcal{P}^l$ with the harmonic decomposition* (2.3) *of the form*

$$P = \sum_{k=0}^{[l/2]} r^{2k} h_{l-2k}(P),$$

the Hermite polynomial H_P associated to P has a decomposition

$$H_P = \sum_{k=0}^{[l/2]} (-1)^k k! L_k^{(\alpha+l-2k)}(r^2) h_{l-2k}(P). \tag{2.6}$$

In particular, the Hermite map equals the identity on the space of harmonic polynomials.

Journal of Physics: Conference Series **597** (2015) 012070 doi:10.1088/1742-6596/597/1/012070

3. Defining an action of the linear group $\mathbf{GL}(d, \mathbb{C})$ on Hermite polynomials

3.1. Creation and annihilation operators

Let us briefly recall the formalism of creation and annihilation operators. We use a_j^+, a_j to denote the standard creation and annihilation operators

$$a_j^+ = \frac{1}{\sqrt{2}}(x_j - \partial_j), \quad a_j = \frac{1}{\sqrt{2}}(x_j + \partial_j), \qquad j = 1, \ldots, d$$

The associative algebra $\mathcal{W}$ they generate is called the *Weyl algebra* and is characterized by the Canonical Commutation Relations, thereafter referred to as CCR,

$$[a_j, a_k^+] = \delta_{jk}\,\mathbf{1}, \quad [a_j, a_k] = 0 = [a_j^+, a_k^+], \quad j, k = 1, \ldots, d,$$

where δ_{jk} denotes the Kronecker delta. Moreover, a_j^+ and a_j are conjugate to each other with respect to the usual inner product in $L^2(\mathbb{R}^d)$, the space of functions in $\mathbb{R}^d$, which are square-integrable with respect to the Lebesgue measure. Explicitly, $\mathcal{W}$ can be viewed as a algebra of partial differential operators with polynomial coefficients, cf. [11, 12]

3.2. The action of $\mathbf{GL}(d, \mathbb{C})$

Let $\mathcal{M}_{\mathbb{C}}$ be the span over $\mathbb{C}$ of $\{a_j^+, a_j \mid j = 1, \ldots, d\}$. The CCR impose on it a symplectic form given by

$$[X, Y] = iB(X, Y)\,\mathbf{1}, \quad X, Y \in \mathcal{M}_{\mathbb{C}}.$$

By virtue of general principles, every symplectic automorphism of $\mathcal{M}_{\mathbb{C}}$ can be uniquely extended to an automorphism of the Weyl algebra $\mathcal{W}$. Within the Weyl algebra we may form the symmetrized products of creation and annihilation operators,

$$q_{jk} = \frac{1}{2}(a_j a_k^+ + a_k^+ a_j) = a_k^+ a_j + \frac{1}{2}\delta_{jk}, \qquad j, k = 1, \ldots, d$$

By straightforward computation we see that

$$\begin{aligned} \operatorname{ad} q_{jk}(a_l^+) &= [q_{jk}, a_l^+] = \delta_{jl} a_k^+ \\ \operatorname{ad} q_{jk}(a_l) &= [q_{jk}, a_l] = -\delta_{kl} a_j. \end{aligned}$$

Thus we may view the elements q_{jk} as giving rise via the adjoint representation to the action of the general linear group $\mathbf{GL}(d, \mathbb{C})$ on $\mathcal{M}_{\mathbb{C}}$.

Proceeding more formally, we let $Q = (q_{jk})$ denote the $d \times d$ matrix with entries q_{jk}.

Proposition 2 *The elements q_{jk}, $j, k = 1, \ldots, d$, are linearly independent and span over $\mathbb{C}$ a Lie subalgebra $\mathfrak{g}$ of $\mathcal{W}$ isomorphic to the Lie algebra $\mathfrak{gl}(d,\mathbb{C})$ via the map*

$$\eta : \mathfrak{gl}(d,\mathbb{C}) \ni P \mapsto \operatorname{tr}(PQ) = \sum_{j,k=1}^{d} p_{jk} q_{kj} \in \mathfrak{g}, \quad P = (p_{jk}).$$

It is worth noting that the Hermite operator — the (anti-hermitian) Hamiltonian of the d-dimensional quantum mechanical harmonic oscillator given by

$$K = \frac{i}{2}(r^2 - \Delta) = \frac{i}{2}\sum_{j=1}^{d}(a_j a_j^+ + a_j^+ a_j),$$

is a generator of the center of $\mathfrak{g}$.

Journal of Physics: Conference Series **597** (2015) 012070 doi:10.1088/1742-6596/597/1/012070

To get the global, integrated form of the above action we proceed in the following way. Given a matrix $g = (g_{jk}) \in \mathbf{GL}(d, \mathbb{C})$ we denote its inverse by $g^{-1} = (G_{jk})$ and define $\rho(g) : \mathcal{M}_\mathbb{C} \to \mathcal{M}_\mathbb{C}$ by

$$\rho(g)a_j^+ = \sum_{k=1}^{d} g_{kj}a_k^+, \qquad \rho(g)a_j = \sum_{k=1}^{d} G_{jk}a_k, \qquad j = 1, \ldots, d.$$

This is a symplectic automorphism of $\mathcal{M}_\mathbb{C}$, hence it extends to an automorphism of $\mathcal{W}$, which we also denote by $\rho(g)$. Thus $g \mapsto \rho(g)$ is a representation of the $\mathbf{GL}(d, \mathbb{C})$ by automorphisms of $\mathcal{W}$. Moreover, the mapping $P \mapsto \mathrm{ad}(\eta(P))$ coincides with the differential $d\rho$ of the representation ρ defined above.

The representation ρ of $\mathbf{GL}(d, \mathbb{C})$ on $\mathcal{W}$ constructed above leaves invariant (separately) the subspaces $\mathcal{P}[a_1^+, \ldots, a_d^+]$ and $\mathcal{P}[a_1, \ldots, a_d]$ of polynomials in $\{a_j^+ \mid j = 1, \ldots, d\}$ and, resp. in $\{a_j \mid j = 1, \ldots, d\}$ and acts by the standard tensorial (fully symmetrized) representation on $\mathcal{P}[a_1^+, \ldots, a_d^+]$:

$$\rho(g)P(a_1^+, \ldots, a_d^+) = P(\rho(g^{-1})a_1^+, \ldots, \rho(g^{-1})a_d^+),$$

and its contragredient on $\mathcal{P}[a_1, \ldots, a_d]$.

Remark 1 In the papers [4, 5] Wünsche attempted to define an action of the group $\mathbf{GL}(2, \mathbb{C})$ on polynomials and differential operators on $\mathbb{R}^2$ by acting formally on them with matrices $U \in \mathbf{GL}(2, \mathbb{C})$, disregarding the fact that there is no *bona-fide* representation of $\mathbf{GL}(2, \mathbb{C})$ in the space of real vectors $\mathbb{R}^2$. The above definition from that point of view is more satisfactory, and allows to reproduce the formal computation from his papers in a more comprehensive way.

3.3. The conjugation with the Gaussian

Taking the clue form the construction of the Hermite map (2.4) we investigate the role played by the Gaussian. Since it is a smooth, rapidly vanishing function, it may be viewed as a multiplication operator on the space $L^2(\mathbb{R}^d)$ and thus it makes sense to compose it with other operators acting in that space.

Proposition 3 *Let us denote by G the map of $\mathcal{W}$ into $\mathcal{W}$ obtained by conjugation with the Gaussian, i.e.*

$$G : \mathcal{W} \ni D \mapsto G(D) \in \mathcal{W}, \quad G(D) = e^{r^2/2} \circ D \circ e^{-r^2/2}. \tag{3.1}$$

This map is an automorphism of the Weyl algebra such that for each homogeneous polynomial $P \in \mathcal{P}^l \subset \mathcal{W}$ the following holds

$$G(P(\partial)) = (-1)^l 2^{l/2} P(a^+), \qquad G^{-1}(P(\partial)) = 2^{l/2} P(a).$$

Clearly, this follows from

$$\partial_j(e^{-r^2/2} f(x)) = e^{-r^2/2}(x_j - \partial_j) f(x) = e^{-r^2/2} 2^{1/2} a_j^+ f(x).$$

by taking into account that conjugation is an automorphism of the Weyl algebra. Similar reasoning holds for the annihilation operators a_j.

3.4. The decomposition of $L^2(\mathbb{R}^d)$ into eigenspaces of the harmonic oscillator Hamiltonian

Let us denote by $\phi_0 = e^{-r^2/2}$ the (non normalized) ground state of the d-dimensional harmonic oscillator and for any non-negative integer l denote by $\mathcal{K}^l$ the space generated from ϕ_0 by the polynomials of degree l in creation operators,

$$\mathcal{K}^l = \mathcal{P}^l[a_1^+, \ldots, a_d^+]\phi_0 \subset L^(\mathbb{R}^d).$$

By the above constructions, the elements of that space are products of Hermite polynomials H_P with the Gaussian,

$$P(a^+)\phi_0 = 2^{l/2} H_P(x)\phi_0(x),$$

and more explicitly, in virtue of (2.6), they may be written in the form

$$\sum_{k=0}^{[l/2]} L_k^{(\alpha+l-2k)}(r^2) Y_{l-2k}(x)$$

where $Y_{l-2k}(x)$ is a homogeneous harmonic polynomial of degree $l-2k$. They are called Hermite–Weber functions — in physical terms they are eigenfunctions of the harmonic oscillator Hamiltonian K corresponding to the eigenvalue $l+d/2$. In fact, up to a factor $2^{-l/2}$

$$\frac{1}{2}(r^2-\Delta)H_P(x)\phi_0(x) = (-i)KP(a^+)\phi_0 = (l+d/2)H_P(x)\phi_0(x).$$

Since $d\rho(A)$ for $A \in \mathfrak{gl}(d,\mathbb{C})$ acts by differentiations,

$$(d\rho(A)P)(a^+)\phi_0 + P(a^+)\eta(A)\phi_0 = \eta(A)P(a^+)\phi_0,$$

and in virtue of the fact that the ground state ϕ_0 is annihilated by all $\eta(A) \in \mathfrak{sl}(d,\mathbb{C})$, the map

$$\mathcal{P}^l \ni P \mapsto \Psi(P) = P(a^+)\phi_0 \in \mathcal{S}(\mathbb{R}^d)$$

satisfies

$$\Psi(d\rho(A)P) = \eta(A)\Psi(P), \qquad A \in \mathfrak{sl}(d,\mathbb{C}),\ P \in \mathcal{P}^l.$$

i.e. it intertwines the actions of $\mathfrak{sl}(d,\mathbb{C})$ — the natural action on polynomials $d\rho$ with the restriction of the *metaplectic* representation $\eta(A)$.

This implies that each eigenspace of the Hamiltonian is invariant under the action of the unitary group — a somewhat mysterious fact apparently observed for the first time by Jauch and Hill in [13].

3.5. Connection with Wünsche's approach

The role of the factorized Hermite polynomials $H_\alpha(x)$ is a consequence of the following

Corollary 1 *The Hermite–Weber functions*

$$H_\alpha(x)e^{-r^2/2} = H_{\alpha_1}(x_1)\cdots H_{\alpha_d}(x_d)e^{-r^2/2}, \qquad |\alpha| = l$$

corresponding to the monomials $x^\alpha = x_1^{\alpha_1}\cdots x_d^{\alpha_d}$ *constitute the set of weight vectors of* $\mathcal{K}^l$ *with respect to the standard maximal torus of* $\boldsymbol{SU}(d)$ *consisting of diagonal matrices.*

Through some explicit computations for $d=2$ Wünsche was able to express the transforms of $H_{\alpha_1}(x_1)H_{\alpha_2}(x_2)$ with the help of Jacobi polynomials. His results may be viewed in the perspective of deriving matrix elements for representations of the unitary group $\mathbf{SU}(d)$ and will be examined at another place.

References

[1] Strasburger A 1998 Weyl algebra and a realization of the unitary symmetry *Rend. del Circ. Matem. di Palermo* serie II **Suppl. 54** 101–111

[2] Strasburger A 1999 On algebras of creation and annihilation operators, in: *Algebraic Methods and q-special functions* (Vinet L and van Diejen J F, eds.) (Providence, Rhode Island: Amer. Math. Soc.) 267–276

[3] Strasburger A 1991 Differential operators of gradient type associated with spherical harmonics *Annal. Polon. Math.* **53** 2 161–183

[4] Wünsche A 2000 General Hermite and Laguerre two-dimensional polynomials *J. Phys. A: Math. Gen.* **33** 1603–1629

[5] Wünsche A 2003 The role of Jacobi polynomials in the theory of Hermite and Laguerre $2D$ polynomials *J. Comp. Appl. Math.* **153** 521–529

[6] Ch F Dunkl and Y Xu 2001 *Orthogonal polynomials of several variables* (Cambridge: Cambridge University Press)

[7] Faraut J 1987 Analyse harmonique et fonctions spéciales in: Faraut J and Harzallah K *Deux Cours d'Analyse Harmonique* Ecole d'Été d'Analyse Harmonique de Tunis 1984 (Basel: Birkhäuser Verlag)

[8] Folland G B 1989 *Harmonic analysis in the phase space* (Princeton: Princeton University Press)

[9] Herz C 1965 *Analyse harmonique en plusieurs variables* (Orsay: Lecture notes)

[10] Hobson E W 1955 *The theory of spherical and elipsoidal harmonics* (Second ed.) (Cambridge: Cambridge University Press)

[11] Howe R 1980 Quantum mechanics and partial differential equations *J. Funct. Anal.* **38** 188–254

[12] Howe R 1985 Dual Pairs in Physics: harmonic oscillators, photons, electrons and singletons 179–207 in: *Applications of Group Theory in Physics and Mathematical Physics* Flato M et al. ed., *Lectures in Applied Mathematics* **21** (Providence, Rhode Island: Amer. Math. Soc.)

[13] Jauch J M and Hill E L 1940 On the Problem of Degeneracy in Quantum Mechanics *Phys. Rev.* **57**, 641–645

[14] Stein E M and Weiss G 1971 *Introduction to harmonic analysis on euclidean spaces* (Princeton: Princeton University Press)

[15] Wawrzyńczyk A 1984 *Group Representations and Special Functions* (Dordrecht–Warszawa: Reidel D and PWN)

Symmetries of the Schrödinger equation and algebra/superalgebra duality

Francesco Toppan

CBPF, Rua Dr. Xavier Sigaud 150, cep 22290-180, Rio de Janeiro (RJ), Brazil

E-mail: toppan@cbpf.br

Abstract. Some key features of the symmetries of the Schrödinger equation that are common to a much broader class of dynamical systems (some under construction) are illustrated. I discuss the algebra/superalgebra duality involving first and second-order differential operators. It provides different viewpoints for the spectrum-generating subalgebras. The representation-dependent notion of on-shell symmetry is introduced. The difference in associating the time-derivative symmetry operator with either a root or a Cartan generator of the $sl(2)$ subalgebra is discussed. In application to one-dimensional Lagrangian superconformal sigma-models it implies superconformal actions which are either supersymmetric or non-supersymmetric.

1. Introduction

Focusing on simple examples, I illustrate some general features that apply to a vast class of theories including non-relativistic Schrödinger equations in $1+d$ dimensions, the more general invariant equations associated with ℓ-conformal Galilei algebras, the $D=1$ Lagrangian (super)conformal models (one-dimensional sigma-models), together with several extended supersymmetric versions of these theories.

This broad class of dynamical equations share the properties that are discussed here. For simplicity I review these features in application to the Schrödinger equation in $1+1$-dimension with the three choices of the potential (constant, linear and quadratic) which induce the Schrödinger algebra [1] as the symmetry algebra of first-order differential operators. On the sigma-model side, I illustrate the simplest $osp(1|2)$-invariant case [2] with one bosonic and one fermionic time-dependent field.

The key issues that I am pointing out are the following. We have that 2 of the 6 first-order Schrödinger operators, the ones generating the Heisenberg-Lie algebra, have a natural half-integer grading. By taking their anticommutators one can construct 3 second-order differential operators. The total number of 9 operators so constructed define a finite closed structure, which can be either Lie-algebraic (by taking their commutators) or super-Lie algebraic, in terms of the (anti)-commutators which respect the $\mathbf{Z}_2$ grading. The fact that two compatible structures can be defined for the same set of operators is referred to as "algebra/superalgebra" duality (it is worth pointing out that the use of higher-order differential operators entering a universal enveloping algebra, together with their relations with higher spin theories, was also advocated in [3]; however, an important feature is that finite Lie or super-Lie algebras are recovered when we include only second-order differential operators of special type).

The operator Ω which defines the Schrödinger dynamics, being a second-order differential operator, does not belong to the Schrödinger algebra, but to the enlarged algebra (either its Lie-algebraic or its super-Lie algebraic version). The operator Ω does not commute with the first-order Schrödinger operators. On the other hand, *in the given representation*, the *on-shell* closure of the algebra is guaranteed once taking into account the $\Omega\Psi = 0$ dynamical equation (see Section **3**).

The algebra/superalgebra duality, applied to the spectrum generating subalgebras, is a reformulation of the celebrated result by Wigner in [4]. Essentially, for the harmonic oscillator, the Lie algebra side is based on the Heisenberg subalgebra and the construction of the eigenstates starts with the Fock's vacuum condition. In the dual super-Lie algebraic point of view, the same conditions are obtained from a highest weight representation of the $osp(1|2)$ spectrum-generating superalgebra.

For the free particle case, the operator Ω and the time-derivative operator belong to the grading 1 sector of the algebra (the time-derivative operator is a root of the $sl(2)$ subalgebra). In the harmonic oscillator case Ω and the time-derivative operator belong to the grading 0 sector of the algebra (the time-derivative operator is the Cartan of the $sl(2)$ subalgebra). This is the key difference which implies the continuum spectrum for the free particle and the discrete spectrum of the harmonic oscillator. In [2] a detailed analysis of one-dimensional (super)conformal models based on parabolic D-module reps (the time-derivative operator being associated with the positive $sl(2)$ root) versus hyperbolic/trigonometric D-module reps (the time-derivative operator being associated with the $sl(2)$ Cartan generator) was given. In the hyperbolic/trigonometric case extra potential terms are allowed. On the other hand, when invariance under superconformal algebras are considered, the difference is even more meaningful. In the parabolic case the dynamical system is both superconformal and supersymmetric. In the hyperbolic/trigonometric case the dynamical system is superconformal but not supersymmetric. This point is illustrated in Section **5**.

In the Conclusions I point out how the features here discussed enter more general dynamical systems. These features can be used to both identify and solve the invariant dynamics of this larger class of theories.

2. Schrödinger's equations in $d = 1$

We consider the differential operator Ω in $1+1$ dimensions:

$$\Omega = \partial_t + a\partial_x^2 - aV(x). \tag{1}$$

$V(x)$ is a potential term. If a is imaginary, the operator Ω defines the dynamics of the $1+1$ Schrödinger equation, written as

$$\Omega\Psi(x,t) = 0. \tag{2}$$

If a is real the above equation is the heat equation in the presence of the potential $V(x)$.

With standard methods, see [1], we can prove that, for three special cases of the potential, the invariance algebra of the equation (2), in terms of first-order differential operators, is given by the Schrödinger algebra ($l = \frac{1}{2}$ conformal Galilei algebra) in the presence of a central charge.

Indeed, the invariant condition $\Omega\delta\Psi(x,t) = 0$,
for $\delta\Psi(x,t) = f(x,t)\Psi_t(x,t) + g(x,t)\Psi_x(x,t) + h(x,t)\Psi(x,t)$,
leads to the set of equations

$$\begin{aligned} f_x &= 0, \\ f_t - 2g_x + af_{xx} &= 0, \end{aligned}$$

$$\begin{aligned} g_t + a\,(2aVf_x + 2h_x + g_{xx}) &= 0, \\ agV_x + h_t + 2a^2V_xf_x + aV(f_t + af_{xx}) + ah_{xx} &= 0. \end{aligned} \tag{3}$$

The three special cases correspond to the solutions with maximal number of generators:
i) the constant potential $V(x) = 0$ (free particle case),
ii) the linear potential $V(x) = \omega x$ and
iii) the quadratic potential $V(x) = \nu^2x^2$ (harmonic oscillator case).

In all the above cases, without loss of generality, the potential can be shifted by a constant u, $V(x) \to V(x) + u$, via a similarity transformation $\Omega \to e^{aut}\Omega e^{-aut}$.

In the quadratic case a linear term in the potential can always be eliminated by shifting the space coordinate, so that $x \to x + b$.

For all the other potentials, the invariance algebra of eq. (2) is smaller than the Schrödinger algebra.

The quadratic and constant realizations can be mutually recovered, see [5], from similarity transformations coupled with change of space and time coordinates. The existence of this set of transformations, however, is not essential for the following discussion.

A compatible assignment of the dimensions is

$$[t] = -1, \quad [x] = -\tfrac{1}{2}, \quad [\omega] = \tfrac{3}{2}, \quad [\nu] = 1, \quad ([\Omega] = 1). \tag{4}$$

The Schrödinger algebra is given by the 6 generators $z_{\pm 1}, z_0, w_{\pm}, c$. Their dimensions are

$$[z_{\pm 1}] = \pm 1, \quad [w_\pm] = \pm\tfrac{1}{2}, \quad [z_0] = [c] = 0. \tag{5}$$

The generators $z_{\pm 1}, z_0$ close an $sl(2)$ subalgebra with z_0 as the Cartan element. The generator c is the central charge.

The three explicit D-module reps are given by
i) $V(x) = 0$, constant potential case,

$$\begin{aligned} z_{+1} &= \partial_t, \\ z_0 &= t\partial_t + \frac{1}{2}x\partial_x + \frac{1}{4}, \\ z_{-1} &= t^2\partial_t + tx\partial_x - \frac{x^2}{4a} + \frac{1}{2}t, \\ w_+ &= \partial_x, \\ w_- &= t\partial_x - \frac{x}{2a}, \\ c &= 1; \end{aligned} \tag{6}$$

ii) $V(x) = \omega x$, linear potential case,

$$\begin{aligned} z_{+1} &= t^2\partial_t + (a^2\omega t^3 + tx)\partial_x + (\frac{t}{2} - \frac{1}{4}a^3\omega^2t^4 - \frac{3}{2}a\omega t^2x - \frac{x^2}{4a}), \\ z_0 &= -t\partial_t - (\frac{3}{2}a^2\omega t^2 + \frac{x}{2})\partial_x + (\frac{1}{2}a^3\omega^3t^3 + \frac{3}{2}a\omega tx - \frac{1}{4}), \\ z_{-1} &= \partial_t + 2a^2\omega t\partial_x - a^3\omega^2t^2 - a\omega x, \\ w_+ &= -t\partial_x + \frac{x}{2a} + \frac{1}{2}a\omega t^2, \\ w_- &= \partial_x - a\omega t, \\ c &= 1; \end{aligned} \tag{7}$$

iii) $V(x) = \nu^2 x^2$, quadratic potential case

$$\begin{aligned}
z_{+1} &= e^{4a\nu t}\left(\partial_t + 2a\nu x\partial_x + a\nu - 2a\nu^2x^2\right),\\
z_0 &= \partial_t,\\
z_{-1} &= e^{-4a\nu t}\left(\partial_t - 2a\nu x\partial_x - a\nu - 2a\nu^2x^2\right),\\
w_+ &= e^{2a\nu t}\left(\partial_x - \nu x\right),\\
w_- &= e^{-2a\nu t}\left(\partial_x + \nu x\right),\\
c &= 1.
\end{aligned} \tag{8}$$

In the quadratic case the non-vanishing commutation relations are given by

$$\begin{aligned}
[z_1, z_{-1}] &= -8a\nu z_0,\\
[z_0, z_{\pm1}] &= \pm 4a\nu z_{\pm1},\\
[z_{\pm1}, w_{\mp}] &= \mp 4a\nu w_{\pm},\\
[z_0, w_{\pm}] &= \pm 2a\nu w_{\pm},\\
[w_+, w_-] &= 2\nu c.
\end{aligned} \tag{9}$$

In the constant and linear cases the commutation relations are obtained from the above formulas with the substitution $\nu = -\frac{1}{4a}$.

The above equations give the structure constants of the one-dimensional, centrally extended, Schrödinger algebra.

The generator z_0 defines the grading corresponding to the (5) dimensions.

One should note that the Hamiltonian (i.e., the time-derivative operator), corresponds to a grading 1 generator (a root generator of the $sl(2)$ subalgebra) in the free particle case and to a grading 0 generator (the Cartan generator of the $sl(2)$ subalgebra) for the harmonic oscillator case.

The generators $w_\pm$ have half-integer grading with respect to the grading defined by z_0.

3. The algebra/superalgebra symmetry with higher differential operators

The second-order differential operators w_1, w_0, w_{-1}, obtained by taking the anticommutators of $w_\pm$, can be constructed:

$$\begin{aligned}
w_{+1} &= \{w_+, w_+\},\\
w_0 &= \{w_+, w_-\},\\
w_{-1} &= \{w_-, w_-\}.
\end{aligned} \tag{10}$$

Their explicit form, in the three respective cases above, is given by

i) the constant case,

$$\begin{aligned}
w_{+1} &= 2{\partial_x}^2,\\
w_0 &= 2t{\partial_x}^2 - \frac{x}{a}\partial_x - \frac{1}{2a},\\
w_{-1} &= 2t^2{\partial_x}^2 - \frac{2tx}{a}\partial_x + \frac{x^2}{2a^2} - \frac{t}{a};
\end{aligned} \tag{11}$$

ii) the linear case,

$$\begin{aligned}
w_1 &= \frac{1}{2a^2}\left(4a^2t^2{\partial_x}^2 - 4at(a^2t^2\omega + x)\partial_x + (a^4t^4\omega^2 - 2at + 2a^2t^2\omega x + x^2)\right),\\
w_0 &= \frac{1}{2a}\left(4at{\partial_x}^2 - (6a^2t^2\omega + 2x)\partial_x + (2a^3t^3\omega^2 + 2at\omega x - 1)\right),\\
w_{-1} &= 2{\partial_x}^2 - 4at\omega\partial_x + 2a^2t^2\omega^2;
\end{aligned} \tag{12}$$

Journal of Physics: Conference Series **597** (2015) 012071 doi:10.1088/1742-6596/597/1/012071

iii) the quadratic case,

$$\begin{aligned} w_{+1} &= e^{4at\nu}\left(2\partial_x{}^2 - 4\nu x\partial_x - 2\nu + 2\nu^2x^2\right), \\ w_0 &= 2\partial_x{}^2 - 2\nu^2x^2, \\ w_{-1} &= e^{-4at\nu}\left(2\partial_x{}^2 + 4\nu x\partial_x + 2\nu + 2\nu^2x^2\right). \end{aligned} \tag{13}$$

In all these cases we have two consistent closed structures which can be defined on the same set of differential operators, namely
1) the non-simple Lie algebra $eSch$ (the enlarged Schrödinger algebra), presented by the 9 generators $\{z_{\pm 1}, z_0, w_\pm, w_{\pm 1}, w_0, c\}$ and
2) the Lie superalgebra $sSch$ (the enlarged Schrödinger superalgebra) $sSch = S_0 \oplus S_1$, with 7 even generators ($z_{\pm 1}, z_0, w_{\pm 1}, w_0, c \in S_0$) and 2 odd generators ($w_\pm \in S_1$).

In all three cases (in the quadratic case for $\nu = -\frac{1}{4a}$), the extra non-vanishing structure constants besides (9) are given, for the $eSch$ algebra, by

$$\begin{aligned} [z_0, w_{\pm 1}] &= \mp w_{\pm 1}, \\ [z_{\pm 1}, w_0] &= \pm w_{\pm 1}, \\ [z_{\pm 1}, w_{\mp 1}] &= \pm 2w_0, \\ [w_\pm, w_0] &= \mp\frac{1}{a}w_\pm, \\ [w_\pm, w_{\mp 1}] &= \mp\frac{2}{a}w_\mp, \\ [w_0, w_{\pm 1}] &= \pm\frac{2}{a}w_{\pm 1}, \\ [w_1, w_{-1}] &= -\frac{4}{a}w_0. \end{aligned} \tag{14}$$

For the $sSch$ superalgebra we have the anti-commutators

$$\{w_+, w_+\} = w_{+1}, \quad \{w_+, w_-\} = w_0, \quad \{w_-, w_-\} = w_{-1} \tag{15}$$

(one should note that the Heisenberg-Lie algebra $[w_+, w_-] = 2\nu c$ is not a subalgebra of the $sSch$ superalgebra).

In all three cases (constant, linear and quadratic), the second-order differential operator Ω is a generator belonging to the enlarged Schrödinger algebra (either $eSch$ or $sSch$). We have the following identifications:
i) constant case,

$$\Omega = z_{+1} + \frac{1}{2}aw_{+1}, \tag{16}$$

ii) linear case,

$$\Omega = z_{-1} + \frac{a}{2}w_{-1}, \tag{17}$$

iii) quadratic case,

$$\Omega = z_0 + \frac{a}{2}w_0. \tag{18}$$

In each case, either the $eSch$ algebra or the $sSch$ superalgebra, is the *on-shell* symmetry algebra of the evolution equation determined by Ω. We have indeed that
i) in the constant case, all commutators involving Ω are vanishing, apart from

$$\begin{aligned} [z_0, \Omega] &= -z_{+1} - \frac{a}{2}w_{+1}, \\ [z_{-1}, \Omega] &= -2z_0 - aw_0. \end{aligned} \tag{19}$$

In the given representation, on the other hand, the above commutators are identified with the representation-dependent formulas

$$\begin{aligned} [z_{+1}, \Omega] &= -2t\Omega, \\ [z_0, \Omega] &= -\Omega. \end{aligned} \tag{20}$$

ii) in the linear case all commutators involving Ω are vanishing, apart from

$$\begin{aligned} [z_{+1}, \Omega] &= 2z_0 + aw_0, \\ [z_0, \Omega] &= z_{-1} + \frac{a}{2}w_{-1}. \end{aligned} \tag{21}$$

In the given linear representation, on the other hand, we have

$$\begin{aligned} [z_{+1}, \Omega] &= 2t\Omega, \\ [z_0, \Omega] &= \Omega. \end{aligned} \tag{22}$$

iii) In the quadratic case all commutators involving Ω are vanishing, apart from

$$\begin{aligned} [z_{+1}, \Omega] &= z_{+1} + \frac{1}{2}w_{+1}, \\ [z_{-1}, \Omega] &= -z_{-1} - \frac{1}{2}aw_{-1}. \end{aligned} \tag{23}$$

In the given quadratic representation, on the other hand, we have

$$\begin{aligned} [z_{+1}, \Omega] &= e^{-t}\Omega, \\ [z_{-1}, \Omega] &= -e^{t}\Omega. \end{aligned} \tag{24}$$

Comment: the fact that we obtain the representation-dependent commutators

$$[g, \Omega] = f_g \cdot \Omega, \tag{25}$$

for any generator g of either the $eSch$ or the $sSch$ algebra, with f_g a given function, tells us that $eSch$ or $sSch$ is the on-shell symmetry (super)algebra for the $\Omega\Psi(t, x) = 0$ equation.

4. Duality for spectrum generating algebras/superalgebras

The famous Wigner's analysis in [4], which allows in particular to solve the harmonic oscillator without using the canonical commutation relations, can be understood from the results discussed in the previous Section. In particular from the notion of algebra/superalgebra duality for the first and second order differential operators closing the on-shell symmetry (super)algebra of the one-dimensional oscillator.

We recall that the given set of 9 differential operators close, on the Lie algebra side, the $eSch$ enlarged Schrödinger algebra, while on the super-Lie algebra side they induce the $sSch$ superalgebra.

On the Lie algebra side the spectrum generating algebra allowing to reconstruct the eigenfunctions and eigenvalues of the harmonic oscillator is the Heisenberg-Lie algebra generated by $w_\pm = e^{\pm 2a\nu t}(\partial_x \pm \nu x)$.

The (unnormalized) vacuum solution $\Psi_{vac}(x,t)$ of the $\Omega\Psi(x,t) = \Psi_t + a\Psi_{xx} - a\nu^2 x^2 \Psi = 0$ equation, on the Lie algebra side, satisfies the Fock's vacuum condition

$$w_+ \Psi_{vac}(x,t) = 0, \tag{26}$$

together with the equations

$$\begin{aligned} z_0 \Psi_{vac}(x,t) &= -a\nu\Psi_{vac}(x,t), \\ w_0 \Psi_{vac}(x,t) &= a\nu\Psi_{vac}(x,t). \end{aligned} \tag{27}$$

The explicit solution is given by $\Psi_{vac}(x,t) = Ce^{-a\nu t}e^{-\frac{\nu}{2}x^2}$.

The eigenstates $\Psi_n(x,t)$, corresponding to higher energy eigenvalues of the harmonic oscillator, are constructed through the positions

$$\Psi_n(x,t) = (w_-)^n \Psi_{vac}(x,t). \tag{28}$$

By construction, they satisfy the $\Omega\Psi_n(x,t) = 0$ equation.

From the dual, superalgebraic, point of view, we have a spectrum-generating superalgebra given by the simple Lie superalgebra $osp(1|2) \subset sSch$. Its generators are $w_0, w_{\pm 1}$ and $w_\pm$.

In the superalgebra picture the same conditions to reconstruct eigenstates and eigenvalues of the harmonic oscillator are read differently. The Equation (26) and the second equation in (27) define a highest weight representation of $osp(1|2)$, with $\Psi_{vac}(x,t)$ being its highest weight vector.

5. Other cases: supersymmetric versus non-supersymmetric superconformal mechanics

The symmetry operator of the Schrödinger equation expressed via the time derivative corresponds, in the free particle case, to the Cartan generator of $sl(2)$ and, in the oscillatorial case, to a root generator of the $s(2)$-invariant subalgebra.

This feature is also present in other different contexts. In particular, in the case of (super)conformal mechanics in $0+1$ dimensions, realized in the Lagrangian setting.

Based on some results in [6], it was shown in [2] that the D-module reps of the (super)conformal algebras admit parabolic as well as hyperbolic/trigonometric realizations. These transformations define superconformally invariant actions. In the hyperbolic/trigonometric case, extra potentials, not allowed in the parabolic case, are present. This one, on the other hand, is not the only difference concerning the various types of realizations. In the parabolic case, the time-derivative operator (i.e., the "Hamiltonian") is associated with a positive root of the $sl(2)$-invariant subalgebra, while in the trigonometric/hyperbolic case it is associated with the Cartan element. As a consequence, we obtain different classes of superconformally-invariant models. In the parabolic case, the Hamiltonian, being associated to a bosonic root, is the square of the fermionic symmetry operators related to the simple fermionic roots. The resulting theory, besides being superconformal, is also supersymmetric in the ordinary sense of the word "supersymmetry". A different picture emerges in the hyperbolic/trigonometric case. The Hamiltonian is still a symmetry operator. On the other hand, it cannot be expressed as a square of fermionic symmetry operators. The resulting theory is superconformally-invariant, but not supersymmetric. Alternatively (following [7], which proposed this term in a different context), we can introduce the notion of "weak supersymmetry" to refer to this feature.

Indeed, the $\mathcal{N}$-extended ordinary supersymmetry requires, for a given $\mathcal{N}$, that a set of $\mathcal{N}$ fermionic symmetry generators Q_i closes the supersymmetry algebra $\{Q_i, Q_j\} = 2\delta_{ij}H$, $[H, Q_i] = 0$ $(i, j = 1, \ldots, \mathcal{N})$, where H is the time-derivative operator (the "Hamiltonian").

In the hyperbolic/trigonometric cases, $\mathcal{N}$ fermionic symmetry generators can be found. They are the square roots of a symmetry generator (let's call it Z), which does not coincide with the Hamiltonian H. As a matter of fact, in the hyperbolic/trigonometric cases, two independent symmetry subalgebras $\{Q_i^{\pm}, Q_j^{\pm}\} = 2\delta_{ij}Z^{\pm}$, $[Z^{\pm}, Q_i^{\pm}] = 0$ (with $Z^+ \neq H$ and $Z^- \neq H$) are encountered. In the parabolic cases two independent symmetry subalgebras are also encountered and one of them can be identified with the ordinary supersymmetry ($Z^- = H$, $Z^+ \neq H$).

In the hyperbolic/trigonometric cases the Hamiltonian H continues to be a symmetry operator. It belongs, however, to the 0-grading sector of the superconformal algebra and is not the square of any fermionic symmetry operator (contrary to the operators $Z^{\pm}$, which belong to the ± 1 grading sectors, respectively).

These points are conveniently illustrated with the simplest example, the $\mathcal{N} = 1$ theory based on the $(1, 1)$ supermultiplet (a single bosonic field φ and a single fermionic field ψ) admitting constant kinetic term and $osp(1|2)$ invariance. The action can be written as

$$\mathcal{S} \;=\; \int dt(\dot{\varphi}^2 - \psi\dot{\psi} + \epsilon\varphi^2). \tag{29}$$

The potential term is absent ($\epsilon = 0$) in the parabolic realization of the superconformal invariance. It is present in the hyperbolic ($\epsilon = 1$) and in the trigonometric ($\epsilon = -1$) realizations. In the hyperbolic case the five invariant operators (closing the $osp(1|2)$ algebra) are given by

$$\begin{aligned} Q^{\pm}\varphi &= e^{\pm t}\psi, & Q^{\pm}\psi &= e^{\pm t}(\dot{\varphi} \mp \varphi), \\ Z^{\pm}\varphi &= e^{\pm 2t}(\dot{\varphi} \mp \varphi), & Z^{\pm}\psi &= e^{\pm 2t}\dot{\psi}, \\ H\varphi &= \dot{\varphi}, & H\psi &= \dot{\psi}. \end{aligned} \tag{30}$$

One should note that $Z^{\pm} = (Q^{\pm})^2$.

No change of time variable $t \mapsto \tau(t)$ allows to represent either Z^+ or Z^- as a time-derivative operator with respect to the new time τ.

6. Conclusions

The several key features discussed in this paper can be extended to investigate the dynamics of more complicated systems. The algebra/superalgebra duality involving a finite number of first-order and second-order differential operators can be constructed not only only for Schrödinger equations in $1 + d$-dimensions, but also from ℓ-extended conformal Galilei algebras (a discussion of these first-order differential operator algebras can be found in [8], [9] and [10]), with half-integer ℓ ($\ell = \frac{1}{2}$ corresponds to the Schrödinger algebra). The generators associated with the half-integer grading can be naturally identified with the odd generators in the superalgebra picture. The representation-dependent notion of on-shell symmetry (as discussed in Section **3**) is applicable to construct the invariant dynamics associated with these conformal Galilei algebras. Invariant operators exist both at grading 1 (the generalization of the Schrödinger free case) and grading 0 (the generalization of the Schrödinger oscillatorial case). Unlike the Schrödinger case, where the invariant equation is given and the symmetry operators are derived with standard techniques, an inverse problem is defined. The algebra is now given; it is the invariant operator induced by the given representation that has to be computed. As a result we can identify new solvable differential equations. A joint paper with N. Aizawa and Z. Kuznetsova concerning this construction is currently under finalization.

Another feature which deserves to be noticed is that, if starting from a supersymmetric system, the construction leading to the algebra/superalgebra duality is replaced by a

Journal of Physics: Conference Series **597** (2015) 012071 doi:10.1088/1742-6596/597/1/012071

construction leading to superalgebra/ ($\mathbf{Z}_2 \times \mathbf{Z}_2$)-graded algebra duality, where the notion of $\mathbf{Z}_2 \times \mathbf{Z}_2$-graded superalgebra can be found, e.g., in [11]. Another paper about this construction is currently under finalization.

On the side of supersymmetric and non-supersymmetric superconformal one-dimensional sigma models, at present the quantization of these classical systems is under investigation.

Acknowledgments

This paper received support from CNPq through the PQ grant 306333/2013-9 .
I profited of useful discussions with N. Aizawa, N. L. Holanda, Z. Kuznetsova and M. Valenzuela.

References

[1] Olver P J 1998 *Applications of Lie Groups to Differential Equations* (Berlin: Springer)
[2] Holanda N L and Toppan F 2014 *J. Math. Phys.* **55** 061703
[3] Valenzuela M 2009 (*Preprint* arXiv:0912.0789)
[4] Wigner E P 1950 *Phys. Rev.* **77** 711
[5] Niederer U 1973 *Helv. Phys. Acta* **45** 802
[6] Papadopoulos G 2013 *Class. Quant. Grav.* **30** 075018
[7] Smilga A V 2004 *Phys. Lett.* **B 585** 173
[8] Galajinski A and Masterov I 2013 *Nucl. Phys.* **B 866** 212
[9] Andrzejewski K, Gonera J, Kosiński P and Maślanka P 2013 *Nucl. Phys.* **B 876** 309
[10] Aizawa N, Kimura Y and Segar J 2013 *J. Phys.* **A 46** 405204
[11] Rittenberg V and Wyler D 1978 *J. Math. Phys.* **19** 2193

Journal of Physics: Conference Series **597** (2015) 012072 doi:10.1088/1742-6596/597/1/012072

Gauge-covariant extensions of Killing tensors and conservation laws

Jan-Willem van Holten
Nikhef, Science Park 105, Amsterdam NL
and University of Leiden, Niels Bohrweg 2, Leiden NL
E-mail: v.holten@nikhef.nl

Abstract. In classical and quantum mechanical systems on manifolds with gauge-field fluxes, constants of motion are constructed from gauge-covariant extensions of Killing vectors and tensors. This construction can be carried out using a manifestly covariant procedure, in terms of covariant phase space with a covariant generalization of the Poisson brackets, c.q. quantum commutators. Some examples of this construction are presented.

1. Noether's theorem

This paper discusses symmetries and conservation laws in the context of hamiltonian dynamics. The discussion is framed predominantly in the language of classical dynamics, but the use of Poisson brackets and their correspondence with quantum commutators, guarantees that many results also apply to the operator formulation of quantum dynamics. The main difference is the operator ordering to be implemented in quantum theory, the technicalities of which are not relevant to the issues I focus on.

The connection between continuous symmetries and conservation laws is established by Noether's theorem [1]. I briefly review the theorem by considering infinitesimal transformations on phase-space variables (x, p) obtained from a generating function $G(x, p)$ through the Poisson brackets

$$\delta x = \{x, G\} = \frac{\partial G}{\partial p}, \qquad \delta G = \{p, G\} = -\frac{\partial G}{\partial x}. \tag{1}$$

Observe, that these variations are defined such that G itself is an invariant:

$$\delta G = \delta x \frac{\partial G}{\partial x} + \delta p \frac{\partial G}{\partial p} = \{G, G\} = 0. \tag{2}$$

Under such transformations the hamiltonian of the system changes by

$$\delta H = \{H, G\} = -\frac{dG}{dt}, \tag{3}$$

the change of G along the phase-space trajectory $(x(t), p(t))$ generated by the hamiltonian H. It follows immediately, that G is a constant of motion if the hamiltonian is invariant under the transformations (1).

It is also of some interest to consider the variation of the action

$$S = \int_1^2 dt \left(p \frac{dx}{dt} - H(x,p) \right). \tag{4}$$

Applying the variations (1)

$$\delta S = \int_1^2 dt \left[\frac{d}{dt} \left(p \frac{\partial G}{\partial p} - G \right) - \{H, G\} \right] = \left[p \frac{\partial G}{\partial p} - G \right]_1^2. \tag{5}$$

Thus variations under which the hamiltonian is invariant, leave the action invariant modulo boundary terms. This is sufficient for G to be a constant of motion.

2. Isometries of manifolds

On a manifold with (local) co-ordinates x^μ and metric $g_{\mu\nu}(x)$ the geodesics can be obtained as the trajectories of test-particles with proper-time hamiltonian

$$H = \frac{1}{2} g^{\mu\nu} p_\mu p_\nu. \tag{6}$$

Indeed, using the overdot notation for proper-time derivatives, the hamilton equations take the form

$$\dot{x}^\mu = g^{\mu\nu} p_\nu, \qquad \dot{p}_\mu = -\frac{1}{2} \frac{\partial g^{\nu\lambda}}{\partial x^\mu} p_\nu p_\lambda = \frac{1}{2} \frac{\partial g_{\nu\lambda}}{\partial x^\mu} \dot{x}^\nu \dot{x}^\lambda. \tag{7}$$

The last expression is equivalent to the geodesic equation

$$\ddot{x}^\mu + \Gamma_{\lambda\nu}^{\;\;\mu} \dot{x}^\lambda \dot{x}^\nu = 0. \tag{8}$$

In this language, isometries of the manifold are found as constants of motion which are linear in the momentum:

$$J(x,p) = J^\mu(x) p_\mu, \qquad \{J, H\} = (\nabla_\mu J_\nu) p^\mu p^\nu = 0, \tag{9}$$

where (in a somewhat hybrid notation) the contravariant components of the momentum are $p^\mu = \dot{x}^\mu$. Hence the covariant coefficient functions J^μ form a Killing vector, a solution of the Killing equation

$$\nabla_\mu J_\nu + \nabla_\nu J_\mu = 0 \quad \Leftrightarrow \quad J^\lambda \frac{\partial g_{\mu\nu}}{\partial x^\lambda} + \frac{\partial J^\lambda}{\partial x^\mu} g_{\lambda\nu} + \frac{\partial J^\lambda}{\partial x^\nu} g_{\mu\lambda} = 0. \tag{10}$$

The second (contravariant) form of the equation states that the Lie-derivative of the metric w.r.t. the vector J^μ vanishes, which is the usual definition of an isometry. Also note, that the constants of motion defined by Killing vectors are precisely those, for which

$$p \frac{\partial G}{\partial p} = G, \tag{11}$$

and therefore generates transformations under which the action is strictly invariant. This is to be expected for an isometry which by construction leaves the line element $ds^2 = g_{\mu\nu} dx^\mu dx^\nu$ invariant.

Although the coordinate transformations δx^μ generated by Killing vectors thus have an elegant interpretation, this does not hold for the corresponding transformations of the canonical momenta p_μ:

$$\delta p_\mu = A_\mu^{\;\nu}(x) p_\nu, \qquad A_\mu^{\;\nu} = -\frac{\partial J^\nu}{\partial x^\mu}. \tag{12}$$

This transformation rule is not general covariant; as $A_\mu^{\;\nu}(x)$ is point-dependent, covariance requires δp to be corrected for the parallel displacement generated by the translation δx^μ. Thus a set of covariant transformations in phase space is defined by

$$\Delta x^\mu = \delta x^\mu, \qquad \Delta p_\mu = \delta p_\mu - \delta x^\lambda \Gamma_{\lambda\mu}^{\;\;\;\nu} p_\nu. \tag{13}$$

In order for these transformations to respect the Poisson brackets (2) and (3), it is then necessary to introduce a covariant derivative

$$\mathcal{D}_\mu G = \frac{\partial G}{\partial x^\mu} + \Gamma_{\mu\nu}^{\;\;\;\lambda} p_\lambda \frac{\partial G}{\partial p_\nu} = -\Delta p_\mu, \tag{14}$$

such that

$$\Delta G = \Delta x^\mu \mathcal{D}_\mu G + \Delta p_\mu \frac{\partial G}{\partial p_\mu} = \{G, G\} = 0, \tag{15}$$

and

$$\Delta H = \Delta x^\mu \mathcal{D}_\mu H + \Delta p_\mu \frac{\partial H}{\partial p_\mu} = \{H, G\} = -\frac{dG}{d\tau}. \tag{16}$$

Thus we have constructed a covariant expression for the Poisson bracket of two arbitrary scalar phase-space functions:

$$\{G, K\} = \mathcal{D}_\mu G \frac{\partial K}{\partial p_\mu} - \frac{\partial G}{\partial p_\mu} \mathcal{D}_\mu K. \tag{17}$$

3. Killing tensors

The metric postulate guarantees that the geodesic hamiltonian (6) is covariantly constant:

$$\nabla_\lambda g_{\mu\nu} = 0 \quad \Leftrightarrow \quad \mathcal{D}H = 0. \tag{18}$$

The condition for a constant of geodesic motion then takes the simple form

$$\{G, H\} = p^\mu \mathcal{D}_\mu G = 0. \tag{19}$$

Applying this to a general expression of homogeneous rank n in the momenta

$$G(x,p) = G^{\mu_1 \dots \mu_n}(x) p_{\mu_1} \dots p_{\mu_n} \quad \Rightarrow \quad p^\mu \mathcal{D}_\mu G = \left(\nabla_{\mu_{n+1}} G_{\mu_1 \dots \mu_n} \right) p^{\mu_1} \dots p^{\mu_{n+1}}, \tag{20}$$

the condition for a constant of motion becomes a generalization of the Killing equation (10):

$$\nabla_{(\mu_{n+1}} G_{\mu_1 \dots \mu_n)} = 0. \tag{21}$$

The solutions of these equations are therefore known as *Killing tensors* [3, 4].

The geometrical interpretation of the transformations generated by constants of motion constructed from Killing tensors of rank 2 or higher, is more complicated than for Killing vectors. They do not generate transformations in the manifold, but in the tangent bundle, the physical phase-space:

$$\begin{aligned} \Delta x^\mu &= \frac{\partial G}{\partial p_\mu} = n G^{\mu\mu_1 \dots \mu_{n-1}} p_{\mu_1} \dots p_{\mu_{n-1}}, \\ \Delta p_\mu &= -\mathcal{D}_\mu G = -\left(\nabla_\mu G^{\mu_1 \dots \mu_n} \right) p_{\mu_1} \dots p_{\mu_n}. \end{aligned} \tag{22}$$

Note, that for transformations with $n \geq 2$ the action actually transforms by boundary terms:

$$p_\mu \frac{\partial G}{\partial p_\mu} = nG \quad \Rightarrow \quad \Delta S = (n-1) \left[G(x_2, p_2) - G(x_1, p_1) \right]. \tag{23}$$

Obviously, as G is a constant of motion the action is still invariant along geodesics.

Example: Kerr geometry [2, 3]
The Kerr geometry is a Ricci-flat geometry in 4-dimensional space-time: $R_{\mu\nu} = 0$, implying it is a solution of the Einstein equations of General Relativity in empty space-time. This solutions describes a rotating black hole with constant mass M and angular momentum J. The geodesic hamiltonian reads

$$H = \frac{1}{2\rho^2}\left[\Delta^2 p_r^2 + p_\theta^2 + \left(a\sin\theta\, p_t + \frac{p_\varphi}{\sin\theta}\right)^2 - \frac{1}{\Delta^2}\left((r^2+a^2)p_t + ap_\varphi\right)^2\right], \qquad (24)$$

where $a = J/M$ is the angular momentum per unit of mass, and the notation follows standard conventions:

$$\Delta^2 = r^2 - 2Mr + a^2, \qquad \rho^2 = r^2 + a^2\cos^2\theta.$$

This space-time geometry admits a rank-2 Killing tensor, coding the well-known Carter constant of motion:

$$\begin{aligned} K &= \frac{1}{2\rho^2}\left[-\Delta^2 a^2\cos^2\theta\, p_r^2 + r^2 p_\theta^2 + r^2\left(ap_t + \frac{p_\varphi}{\sin^2\theta}\right)^2\right. \\ &\qquad \left. + \frac{a^2\cos^2\theta}{\Delta^2}\left((r^2+a^2)p_t + ap_\varphi\right)^2\right]. \end{aligned} \qquad (25)$$

This constant of motion can be generalized to neutral and charged, spinless and spinning particles in Kerr-Newman space-time of charged black holes [3, 5, 6, 7].

4. Bracket algebra

The generators of hamiltonian symmetry transformations, defining constants of motion, define a Lie-algebra by their bracket relations (17); this follows from the Jacobi identity

$$\{\{G,K\},J\} + \{\{K,J\},G\} + \{\{J,G\},K\} = 0. \qquad (26)$$

Applying the identity to the special case where one of the functions is the hamiltonian: $J = H$, implies that constants of motion (G, K) satisfy

$$\{\{G,K\},H\} = 0. \qquad (27)$$

Hence the bracket of two constants of motion produces another constant of motion, and the set of such constants is closed under the bracket operation. Here we summarize some properties of this algebra [4].

The generators $J = J^\mu p_\mu$ linear in momentum define a Lie subalgebra of the full algebra:

$$\{J_1, J_2\} = J_3, \qquad J_3^\mu = J_2^\nu\,\nabla_\nu J_1^\mu - J_1^\nu\,\nabla_\nu J_2^\mu. \qquad (28)$$

The constants of motion of rank $n \geq 2$ then define representations of this subalgebra[1], characterized by the transformation rule

$$\{G_1, J\} = G_2, \qquad G_2^{\mu_1\ldots\mu_n} = J^\lambda\nabla_\lambda G^{\mu_1\ldots\mu_n} - nG^{\lambda(\mu_1\ldots\mu_{n-1}}\nabla_\lambda J^{\mu_n)}. \qquad (29)$$

The Jacobi identity

$$\{\{G,J_1\},J_2\} - \{\{G,J_2\},J_1\} = \{G,\{J_1,J_2\}\}, \qquad (30)$$

[1] The parenthesis around the indices $(\mu_1\ldots\mu_n)$ denote complete symmetrization with unit weight.

then guarantees that the transformations generated by the linear $J(x,p)$ on the rank-n tensors $G(x,p)$ obey the composition rules of the subalgebra spanned by the linear $J(x,p)$ themselves.

As for the brackets of a generator $G^{(n)}$ of rank n, and a generator $G^{(m)}$ of rank m, with both $(n,m) \geq 2$, it is easily recognized that any non-vanishing brackets of such quantities must be a generator $G^{(n+m-1)}$, of rank $n+m-1$ exceeding both n and m:

$$\left\{ G^{(n)}, G^{(m)} \right\} \sim G^{(n+m-1)}. \tag{31}$$

Therefore non-vanishing brackets of higher-rank generators will potentially lead to an infinite set of generators of arbitrary high rank. Well-known examples of such algebras are the Virasoro and Kac-Moody algebras of 2-D conformal field theories.

5. Abelian gauge interactions

Geodesic motion on a manifold applies to the motion of a pure mass point subject only to geometrical forces. However, Noether's theorem is very general and can be applied also in presence of external force fields [8, 9]. In this section I consider abelian gauge interactions transmitted by a vector field $A_\mu(x)$, acting on a point mass m with charge q. For such a particle the canonical hamiltonian reads

$$H = \frac{1}{2m}\, g^{\mu\nu} \left(p_\mu - qA_\mu \right) \left(p_\nu - qA_\nu \right). \tag{32}$$

The hamiltonian equations of motion (7), (8) are generalized to

$$p_\mu = m g_{\mu\nu} \dot{x}^\nu + qA_\mu, \qquad g_{\mu\nu} \left(\ddot{x}^\nu + \Gamma_{\kappa\lambda}^{\;\;\nu} \dot{x}^\kappa \dot{x}^\lambda \right) = \frac{q}{m}\, F_{\mu\nu}\, \dot{x}^\nu, \tag{33}$$

which is the Lorentz force law on curved manifolds.

A drawback of this hamiltonian formulation is, that the canonical momentum is not gauge invariant; under gauge transformations

$$A_\mu' = A_\mu + \nabla_\mu \Lambda, \qquad p_\mu' = p_\mu + q \nabla_\mu \Lambda. \tag{34}$$

Therefore it is preferable to work with the covariant momentum [7, 10]

$$\pi_\mu = p_\mu - qA_\mu = m g_{\mu\nu} \dot{x}^\nu, \tag{35}$$

in terms of which the hamiltonian takes the simple form

$$H = \frac{1}{2m}\, g^{\mu\nu} \pi_\mu \pi_\nu. \tag{36}$$

If the covariant derivative $\mathcal{D}$ now is generalized to

$$\mathcal{D}_\mu G = \frac{\partial G}{\partial x^\mu} + \Gamma_{\mu\nu}^{\;\;\lambda}\, \pi_\lambda\, \frac{\partial G}{\partial \pi_\nu}, \tag{37}$$

then as before the metric postulate can be used to show that

$$\mathcal{D}_\mu H = 0. \tag{38}$$

Moreover, the correct dynamics is reproduced as usual by

$$\frac{dG}{d\tau} = \left\{ G, H \right\}, \tag{39}$$

when supplemented by the covariant brackets

$$\{G, K\} = \mathcal{D}_\mu G \, \frac{\partial K}{\partial \pi_\mu} - \frac{\partial G}{\partial \pi_\mu} \, \mathcal{D}_\mu K + qF_{\mu\nu} \, \frac{\partial G}{\partial \pi_\mu} \frac{\partial K}{\partial \pi_\nu}. \tag{40}$$

In particular, this bracket reproduces the Ricci identity in the form

$$\{\pi_\mu, \pi_\nu\} = qF_{\mu\nu}. \tag{41}$$

In view of the identity (38) and the anti-symmetry of the field strength tensor $F_{\mu\nu}$, the condition for a scalar quantity $G(x, \pi)$ to be constant of motion becomes [10]

$$\{G, H\} = 0 \quad \Rightarrow \quad \pi^\mu \mathcal{D}_\mu G = q\pi^\mu F_{\mu\nu} \, \frac{\partial G}{\partial \pi_\nu}. \tag{42}$$

In particular, if G can be expressed as a power series

$$G(x, \pi) = \sum_n \frac{1}{n!} \, G^{(n)\, \mu_1 \ldots \mu_n}(x) \, \pi_{\mu_1} \ldots \pi_{\mu_n}, \tag{43}$$

then eq. (42) takes the form of a first-order p.d.e.

$$\nabla_{(\mu_1} G^{(n)}_{\mu_2 \ldots \mu_{n+1})} = qF_{(\mu_1}^{\;\;\;\nu} \, G^{(n+1)}_{\mu_2 \ldots \mu_{n+1})\nu}. \tag{44}$$

This is a generalization of equation (21) for Killing tensors, forming a hierarchy of equation connecting tensors of different rank. Nevertheless, the existence of a Killing tensor of rank n still provides a constant of motion: it allows the series expansion (43) to be truncated, with all higher-rank components vanishing:

$$\nabla_{(\mu_1} G^{(n)}_{\mu_2 \ldots \mu_{n+1})} = 0 \quad \Rightarrow \quad G^{(n+k)}_{\mu_1 \ldots \mu_{n+k}} = 0, \quad \forall k \geq 0, \tag{45}$$

while all lower-rank components are obtained by solving eq. (44) with the already known higher-rank ones as inhomogenous source terms on the right-hand side [10].

Finally note, that it is straightforward to include a scalar potential $\Phi(x)$ as well:

$$H = \frac{1}{2m} \, g^{\mu\nu} \pi_\mu \pi_\nu + \Phi, \quad \Rightarrow \quad \mathcal{D}_\mu H = \frac{\partial \Phi}{\partial x^\mu}. \tag{46}$$

This modifies the generalization of the Killing equation to include a gradient term [11]

$$\pi^\mu \mathcal{D}_\mu G = q\pi^\mu F_{\mu\nu} \frac{\partial G}{\partial \pi_\nu} + \frac{\partial \Phi}{\partial x^\mu} \frac{\partial G}{\partial \pi_\mu}. \tag{47}$$

Example: a quantum-dot model
A model of a quantum-dot, consisting of 2 electrons moving as a non-relativistic bound pair with Coulomb interaction in a confining harmonic potential and a magnetic field, was studied in ref. [12]. Ignoring the center-of-mass motion, the pair can be described in axial co-ordinates (ρ, z, φ) as a single particle with hamiltonian

$$H_{CM} = \frac{1}{2} \, g^{ij} \pi_i \pi_j + \Phi, \tag{48}$$

where

$$g_{ij} = \mathrm{diag}(1, 1, \rho^2), \qquad \Phi = \frac{1}{2}\left(\omega_0^2\rho^2 + \omega_z^2 z^2\right) - \frac{\kappa}{\sqrt{\rho^2 + z^2}}. \tag{49}$$

The definition of the dynamics is completed by the brackets

$$\{G, K\} = \mathcal{D}_i G \frac{\partial K}{\partial \pi_i} - \frac{\partial G}{\partial \pi_i}\mathcal{D}_i K - 2\omega_L \rho \left(\frac{\partial G}{\partial \pi_\rho}\frac{\partial K}{\partial \pi_\varphi} - \frac{\partial G}{\partial \pi_\varphi}\frac{\partial K}{\partial \pi_\rho}\right), \tag{50}$$

where $\omega_L = eB/2$ is the Larmor frequency.

To guarantee that a Killing tensor can be completed to a full constant of motion, solving eq. (47), it may be necessary to tune the parameters in the scalar potential. In the case of the quantum dot, there is a rank-4 Killing tensor which can be completed to a full constant of motion provided the magnetic field is tuned to a value such that the Larmor frequency is a fixed combination of the harmonic frequencies:

$$\omega_L^2 + \omega_0^2 = 4\omega_z^2. \tag{51}$$

Then the following expression is a solution of eq. (47) for this system:

$$\begin{aligned}
G = {} & \rho^2\pi_z^4 - 2\rho z \pi_\rho \pi_z^3 + z^2\pi_\rho^2\pi_z^2 + \frac{1}{\rho^2}\pi_\varphi^4 + \pi_\rho^2\pi_\varphi^2 + \left(2 + \frac{z^2}{\rho^2}\right)\pi_z^2\pi_\varphi^2 \\
& + 2\omega_L \pi_\varphi \left(\rho^2\pi_\rho^2 + (2\rho^2 + z^2)\pi_z^2\right) + \left[2\omega_z^2 z^3\rho + \frac{2\kappa z \rho}{\sqrt{\rho^2 + z^2}}\right]\pi_\rho\pi_z \\
& + \left[(2\omega_z^2 - \omega_0^2) z^2\rho^2 + 2\omega_L^2\rho^4 - \frac{2\kappa\rho^2}{\sqrt{\rho^2 + z^2}}\right]\pi_z^2 + \omega_L^2\,\rho^4\pi_\rho^2 \\
& + \left[2\omega_z^2 z^2 + (\omega_0^2 - 5\omega_L^2)\rho^2 - \frac{2\kappa}{\sqrt{\rho^2 + z^2}}\right]\pi_\varphi^2 \\
& - 2\omega_L\pi_\varphi\left[(3\omega_L^2 - \omega_0^2)\rho^4 - 2\omega_z^2\rho^2 z^2 + \frac{2\kappa\rho^2}{\sqrt{\rho^2 + z^2}}\right] + \omega_z^4\rho^2 z^4 + 2\omega_z^2\omega_L^2\rho^4 z^2 \\
& - \omega_L^2(3\omega_L^2 - 4\omega_z^2)\rho^6 + \frac{2\kappa}{\sqrt{\rho^2 + z^2}}\left(\omega_z^2\rho^2 z^2 - \omega_L^2\rho^4\right) + \frac{\kappa^2}{2}\frac{\rho^2 - z^2}{\rho^2 + z^2}.
\end{aligned} \tag{52}$$

6. Non-abelian gauge interactions

The dynamics of a particle with non-abelian gauge interactions is described by Wong's generalization of the Lorentz force law [13]

$$g_{\mu\nu}\left(\ddot{x}^\nu + \Gamma_{\kappa\lambda}^{\;\;\nu}\dot{x}^\kappa\dot{x}^\lambda\right) = \frac{g}{m}\, t_a F_{\mu\nu}^a \dot{x}^\nu, \qquad \dot{t}_a + g f_{ab}^{\;\;c} t_c A_\mu^b \dot{x}^\mu. \tag{53}$$

Here g is the coupling constant, $f_{ab}^{\;\;c}$ are the structure constants of the Lie algebra of gauge charges t_a, and the non-abelian field-strength tensor is

$$F_{\mu\nu}^a = \nabla_\mu A_\nu^a - \nabla_\nu A_\mu^a + g f_{bc}^{\;\;a} A_\mu^b A_\nu^c. \tag{54}$$

The theory can be cast in hamiltonian form [10] by introducing covariant momenta π_μ and a canonical hamiltonian

$$H = \frac{1}{2m}\, g^{\mu\nu} \pi_\mu \pi_\nu, \tag{55}$$

supplemented with brackets

$$\begin{aligned} \{G, K\} &= \mathcal{D}_\mu G \frac{\partial K}{\partial \pi_\mu} - \frac{\partial G}{\partial \pi_\mu} \mathcal{D}_\mu K + g t_a F_{\mu\nu}^a \frac{\partial G}{\partial \pi_\mu} \frac{\partial K}{\partial \pi_\nu} + f_{ab}^{\;\;\;c} t_c \frac{\partial G}{\partial t_a} \frac{\partial K}{\partial t_b}, \\ \mathcal{D}_\mu G &\equiv \frac{\partial G}{\partial x^\mu} + \Gamma_{\mu\nu}^{\;\;\;\lambda} \pi_\lambda \frac{\partial G}{\partial \pi_\nu} + g f_{ab}^{\;\;\;c}\, t_c\, A_\mu^a \frac{\partial G}{\partial t_b}. \end{aligned} \tag{56}$$

In particular these brackets guarantee the Ricci identity and the Lie-algebra of gauge charges:

$$\{\pi_\mu, \pi_\nu\} = g t_a\, F_{\mu\nu}^a, \qquad \{t_a, t_b\} = f_{ab}^{\;\;\;c} t_c. \tag{57}$$

It is now straightforward to derive the condition for the existence of constants of motion:

$$\{G, H\} = 0 \quad \Rightarrow \quad \pi^\mu \mathcal{D}_\mu G = g t_a F_{\mu\nu}^a\, \pi^\mu \frac{\partial G}{\partial \pi_\nu}. \tag{58}$$

In components this gives us a hierarchy of equations similar to (44) for the abelian case:

$$\begin{aligned} &\mathcal{D}_\mu G^{(0)} = g t_a F_\mu^{a\,\nu} G_\nu^{(1)}, \\ &\mathcal{D}_\mu G_\nu^{(1)} + \mathcal{D}_\nu G_\mu^{(1)} = g t_a \left(F_\mu^{a\,\lambda} G_{\lambda\nu}^{(2)} + F_\nu^{a\,\lambda} G_{\lambda\mu}^{(2)} \right), \\ &\mathcal{D}_\mu G_{\nu\lambda}^{(2)} + \mathcal{D}_\nu G_{\lambda\mu}^{(2)} + \mathcal{D}_\lambda G_{\mu\nu}^{(2)} = g t_a \left(F_\mu^{a\,\kappa} G_{\kappa\lambda\nu}^{(3)} + F_\nu^{a\,\kappa} G_{\kappa\lambda\mu}^{(2)} + F_\lambda^{a\,\kappa} G_{\kappa\mu\nu}^{(3)} \right), \\ &\ldots \end{aligned} \tag{59}$$

Example: 2-D $SU(2)$ Yang-Mills point charge

As a simple example consider the dynamics of a non-abelian point charge in a static magnetic $SU(2)$-field in the euclidean plane. In the 2-dimensional plane such a magnetic field can be factorised as

$$F_{ij}^a = \varepsilon_{ij} B^a, \tag{60}$$

where a represents an adjoint vector component in 3-dimensional internal $SU(2)$-space. The free Yang-Mills equation then implies

$$\nabla_i B^a + g \varepsilon^{abc} A_i^b B^c = 0 \quad \Rightarrow \quad \nabla_i B^{a\,2} = 0, \tag{61}$$

and the modulus of B^a is constant. Now the direction of B^a can be rotated point-wise in the plane by a local gauge transformation, and this freedom can be used to make B^a constant. Such a constant B^a-field is derived from a vector potential

$$A_i^a = -\frac{1}{2}\, \varepsilon_{ij} x^j B^a. \tag{62}$$

Now the 2-dimensional euclidean plane is invariant under translations and rotations, guaranteeing conservation of momentum and angular momentum. Our construction implies, that the angular momentum will obtain a field dependent addition:

$$J = \varepsilon_{ij} x_i \pi_j + \frac{g}{2} t_a B^a x_i^2. \tag{63}$$

In addition, there is also a pair of rank-2 Killing tensors, generating a conserved Runge-Lenz vector:

$$K_i = x_i \pi_j^2 - \pi_i \pi_j x_j + g B^a t_a \left(\frac{1}{2} \varepsilon_{ij} \pi_j x_k^2 + x_i x_j \varepsilon_{jk} \pi_k \right) + \frac{1}{2} \left(g B^a t_a \right)^2 x_i x_j^2. \tag{64}$$

Acknowledgments
The work described in this paper is supported by the Foundation for Fundamental Research of Matter (FOM), as part of the research programme *Theoretical Particle Physics in the Era of the LHC*.

References
[1] Noether E 1918 Invariante Variationsprobleme *Nachr. D. Knig. Gesellsch. D. Wiss. Zu Gttingen* Math-phys. Klasse 1918: 235
[2] Carter B 1968 Global Structure of the Kerr Family of Gravitational Fields *Phys. Rev.* **174** 1559
[3] Carter B 1977 Killing tensor quantum numbers and conserved currents in curved space *Phys. Rev.* D **16** 3395
[4] van Holten J W and Rietdijk R H 1993 Symmetries and motions in manifolds *J. Geom. and Physics* **11** 559 (arXiv:hep-th/9205074)
[5] Penrose R 1973 *Ann. N.Y. Acad. Sci.* **224** 125
[6] Floyd R 1973 *The Dynamics of Kerr Fields* PhD. Thesis (London)
[7] Gibbons G W, Rietdijk R H and van Holten J W 1993 SUSY in the sky *Nucl. Phys.* B **404** 42 (arXiv:hep-th/9303112)
[8] Jackiw R and Manton N S 1980 Symmetries and conservation laws in gauge theories *Ann. Phys.* (NY) **127** 257
[9] Duval C and Horvathy P 1982 Particles with internal structure: the geometry of classical motions and conservation laws *Ann. Phys.* (NY) **142** 10
[10] van Holten J W 2007 Covariant hamiltonian dynamics *Phys. Rev.* D **75** 025027 (arXiv:hep-th/0612216)
[11] Visinescu M 2010 Higher order first integrals of motion in a gauge covariant Hamiltonian framework *Mod. Phys. Lett.* A **25** 341 (arXiv:0910.3474v2)
[12] Cariglia M, Gibbons G W, van Holten J W, Horvathy P A, Kosínski P and Zhang P-M 2014 Killing tensors and canonical geometry in classical dynamics *Class. Quantum Grav.* **31** 125001 (arXiv:1401.8195)
[13] Wong S K 1970 Field and particle equations for the classical Yang-Mills field and particles with isotopic spin *Nuov. Cim.* A **65** 689

Non-additive probabilities and quantum logic in finite quantum systems

A Vourdas
Department of Computing, University of Bradford, Bradford BD7 1DP, United Kingdom
E-mail: A.Vourdas@bradford.ac.uk

Abstract. A quantum system $\Sigma(d)$ with variables in $\mathbb{Z}(d)$ and with Hilbert space $H(d)$, is considered. It is shown that the additivity relation of Kolmogorov probabilities, is not valid in the Birkhoff-von Neumann orthocomplemented modular lattice of subspaces $\mathcal{L}(d)$. A second lattice $\Lambda(d)$ which is distributive and contains the subsystems of $\Sigma(d)$ is also considered. It is shown that in this case also, the additivity relation of Kolmogorov probabilities is not valid. This suggests that a more general (than Kolmogorov) probability theory is needed, and here we adopt the Dempster-Shafer probability theory. In both of these lattices, there are sublattices which are Boolean algebras, and within these 'islands' quantum probabilities are additive.

1. Introduction

Quantum mechanics is a probabilistic theory, but quantum probabilities are more general than the standard (Kolmogorov) probabilities. Many authors have discussed various aspects of quantum probabilities[1, 2, 3]. Operational and convex geometry methods have been studied in [4, 5, 6, 7]. Test spaces have been studied in [8, 9], and fuzzy phase spaces in [10, 11]. Category theory methods have been studied in [12, 13]. We are interested in quantum probabilities as non-additive probabilities, and in particular in their link to logic. Non-additive probabilities describe aggregations, where 'the whole is greater than the sum of its parts'. In this general area, the Dempster-Shafer theory has been used extensively in Artificial Intelligence, Operations Research, Economics, etc[14, 15, 16, 17], and here we use it for quantum probabilities and link it to quantum logic. Quantum logic is based on the Birkhoff-von Neumann orthomodular lattice of closed subspaces of a Hilbert space and has been studied extensively in the literature[18, 19, 20, 21, 22].

Kolmogorov's probability theory is intimately related to set theory and Boolean algebra. It is a map from the powerset 2^{Ω} of some set Ω, to $[0, 1]$. The powerset 2^{Ω} is a Boolean algebra with subset ($\subseteq$), intersection ($\cap$) union ($\cup$), complement ($\overline{A} = \Omega - A$), as the logical connectives partial order ($\prec$), conjuction ($\wedge$), disjunction ($\vee$), negation ($\neg$), correspondingly. The additivity property of Kolmogorov probabilities is

$$\delta(A, B) = 0; \qquad \delta(A, B) = p(A \vee B) - p(A) - p(B) + p(A \wedge B), \tag{1}$$

where $A, B \subseteq \Omega$, and $p(A)$ are probabilities. From this follows that

$$p(A) + p(\overline{A}) = 1. \tag{2}$$

In recent work we have shown that the analogues of these additivity relations, are not valid in general, for quantum probabilities.

We consider a quantum system $\Sigma(d)$, with variables in $\mathbb{Z}(d)$ (the integers modulo d), described with the Hilbert space $H(d)$. We then consider two different lattices. The first lattice $\mathcal{L}(d)$ [23] is the Birkhoff-von Neumann orthomodular lattice of closed subspaces of $H(d)$, which in the finite case considered here, is actually the modular orthocomplemented lattice of subspaces. The second lattice $\Lambda(d)$ [24] is the distributive lattice (Heyting algebra) of subsystems of $\Sigma(d)$. A subsystem is a smaller system with variables in a subgroup of $\mathbb{Z}(d)$. The subgroups of $\mathbb{Z}(d)$ are $\mathbb{Z}(e)$ where $e|d$ (e is divisor of d), and therefore the subsystems of $\Sigma(d)$ are $\Sigma(e)$ with $e|d$. We will show explicitly how the states of a subsystem are embedded into the larger system. $\mathcal{L}(d)$ is a different lattice from $\Lambda(d)$. $\mathcal{L}(d)$ has an infinite number of elements, while $\Lambda(d)$ has a finite number of elements (corresponding to the divisors of d). Furthermore, the logical connectives in the two lattices are different.

In both of these lattices we show that the analogue of the additivity property $\delta(A, B) = 0$ is not valid in general. It is valid only within sublattices of $\mathcal{L}(d)$, and also sublattices of $\Lambda(d)$, which are Boolean algebras. As soon as we work with structures which are not Boolean algebras, the additivity property $\delta(A, B) = 0$ is not valid.

The Birkhoff-von Neumann lattice $\mathcal{L}(d)$ is not distributive, and this is a big deviation from Boolean algebras. Somebody might conjecture that the lack of distributivity is the only factor responsible for the non-additivity of quantum probabilities. For this reason we considered the second lattice which is distributive. But even in this case, quantum probabilities are non-additive.

There is a long history of the study of non-additive probabilities, and a practical scheme is the Dempster-Shafer theory[14, 15, 16, 17]. We adopt this scheme and interpret quantum probabilities as Dempster-Shafer probabilities. In the present paper we review and expand the work in [23, 24]. The emphasis is on the physical implications, and the interplay between quantum physics and logic (the formal proof of some statements, involves lattice theory techniques and has been given in [23, 24]). We also compare and contrast the two lattices, because there is different reason for the non-additivity of probability, in these two cases. In $\mathcal{L}(d)$ is non-commutativity, and in the Heyting algebra $\Lambda(d)$ the non-validity of the 'law of the exclusive middle' ($\neg a \vee a \prec \mathcal{I}$) and the fact that two negations of a statement are not equivalent to the statement ($a \prec \neg\neg a$).

2. The Dempster-Shafer theory: lower and upper probabilities

The Dempster-Shafer theory assigns two probabilities to a subset A of Ω. The lower probability (or belief) $\ell(A)$ and the upper probability (or plausibility) $u(A) = 1 - \ell(\overline{A})$. In Kolmogorov probability theory these two quantities are the same, due to Eq.(2). Kolmogorov probability theory is based on Boolean logic, according to which an element of Ω, belongs to either A or to $\overline{A}$. In Dempster-Shafer theory there are three categories: 'belongs to A', 'belongs to $\overline{A}$' and 'don't know'. In this case

$$\ell(A) + \ell(\overline{A}) \leq 1, \tag{3}$$

and the $1 - \ell(A) - \ell(\overline{A})$), corresponds to the 'don't know' category. The upper probability combines the 'belong to A' and the 'don't know'.

In his original work Dempster[14] considered Kolmogorov probabilities associated to subsets of a space X. He then considered a multivalued map from X to another space Ω. There is uncertainty related to probabilities in the space X, but there is an extra level of uncertainty due to the multivaluedness in the space Ω, which is sometimes called ambiguity in order to distinguish it from the original uncertainty. In a quantum context, the Dempster multivaluedness is the fact that a product of two classical quantities becomes the product of two operators which can be ordered in various ways.

An example similar to one in ref[25], is a company which does not know the exact age of its employees. It knows that n_1 employees are under 30 years old, n_2 employees are between 30 and 50, and n_3 are over 50. It wants to find the probability that a random employee is under 40 years old. The set A contains employees whose age is under 40. In this example, n_1 employees belong to A, n_2 employees belong to $\overline{A}$, and n_3 employees belong to Dempster's 'don't know' category. The lower probability is $n_1/(n_1+n_2+n_3)$ ('safe choice') and the upper probability $(n_1+n_2)/(n_1+n_2+n_3)$.

The lower and upper probabilities obey the inequalities

$$\begin{aligned} \ell(A\cup B)-\ell(A)-\ell(B)+\ell(A\cap B)\geq 0 \\ u(A\cup B)-u(A)-u(B)+u(A\cap B)\leq 0 \end{aligned} \tag{4}$$

3. Non-commutativity in $\mathcal{L}(d)$ and the non-additivity operator $\mathfrak{D}(H_1, H_2)$

Let H_1, H_2 be subspaces of $H(d)$. We define the conjunction and disjunction as

$$H_1 \wedge H_2 = H_1 \cap H_2; \quad H_1 \vee H_2 = \mathrm{span}(H_1 \cup H_2). \tag{5}$$

We also use the notation $\mathcal{O} = H(0)$ (the space which has only the zero vector) and $\mathcal{I} = H(d)$, for the smallest and greatest elements in the lattice. We denote as $\Pi(H_1)$ the projector to the subspace H_1.

In the lattice $\mathcal{L}(d)$, we define the following 'non-additivity operator':

$$\mathfrak{D}(H_1, H_2) = \Pi(H_1 \vee H_2) + \Pi(H_1 \wedge H_2) - \Pi(H_1) - \Pi(H_2). \tag{6}$$

The $\mathrm{Tr}[\rho\mathfrak{D}(H_1, H_2)]$ with a density matrix ρ, is analogous to $\delta(A, B)$ and it indicates deviations from additivity by quantum probabilities. The disjunctions describe 'aggregations' of subsystems, and the non-additivity shows that an aggregation (or clustering or coalition) is different than a naive sum of its parts. The commutator $[\Pi(H_1), \Pi(H_2)]$ is related to $\mathfrak{D}(H_1, H_2)$, through the relation[23]:

$$[\Pi(H_1), \Pi(H_2)] = \mathfrak{D}(H_1, H_2)[\Pi(H_1) - \Pi(H_2)]. \tag{7}$$

This shows a strong link between non-commutativity and non-additivity of quantum probabilities. Within the lattice $\mathcal{L}(d)$, there are sublattices which are Boolean algebras, and there the projectors commute, the $\mathfrak{D}(H_1, H_2) = 0$ and quantum probabilities obey the additivity relation $\mathrm{Tr}[\rho\mathfrak{D}(H_1, H_2)] = 0$ which is analogous to $\delta(A, B) = 0$ for Kolmogorov probabilities. But in the full lattice both $[\Pi(H_1), \Pi(H_2)]$ and $\mathfrak{D}(H_1, H_2)$ are in general non-zero and and therefore $\mathrm{Tr}[\rho\mathfrak{D}(H_1, H_2)]$ is in general non-zero. Taking into account Eq.(4), we interpret a pair $\mathrm{Tr}[\Pi(H_1)\rho]$ and $\mathrm{Tr}[\Pi(H_2)\rho]$ of quantum probabilities, as upper or lower probabilities according to whether $\mathrm{Tr}[\rho\mathfrak{D}(H_1, H_2)]$ is negative or positive correspondingly.

3.1. Example

We consider the 3-dimensional space $H(3)$ and its subspaces

$$H_1 = \{a(1,0,0)\}; \quad H_2 = \{a(1,1,0)\}. \tag{8}$$

Here we give the general vector in these subspaces. In this case

$$H_1 \vee H_2 = \{(a,b,0)\}; \quad H_1 \wedge H_2 = \mathcal{O}. \tag{9}$$

The corresponding projectors are

$$\Pi(H_1) = \begin{pmatrix} 1 & 0 & 0 \\ 0 & 0 & 0 \\ 0 & 0 & 0 \end{pmatrix}; \quad \Pi(H_2) = \frac{1}{2}\begin{pmatrix} 1 & 1 & 0 \\ 1 & 1 & 0 \\ 0 & 0 & 0 \end{pmatrix}; \quad \Pi(H_1 \vee H_2) = \begin{pmatrix} 1 & 0 & 0 \\ 0 & 1 & 0 \\ 0 & 0 & 0 \end{pmatrix} \tag{10}$$

Therefore

$$\mathfrak{D}(H_1, H_2) = \frac{1}{2}\begin{pmatrix} -1 & -1 & 0 \\ -1 & 1 & 0 \\ 0 & 0 & 0 \end{pmatrix} \tag{11}$$

We also consider the density matrix

$$\rho = \begin{pmatrix} a & 0 & 0 \\ 0 & 1-a & 0 \\ 0 & 0 & 0 \end{pmatrix}; \quad 0 \le a \le 1 \tag{12}$$

In this case $\mathrm{Tr}[\rho\mathfrak{D}(H_1, H_2)] = (1/2) - a$ and the $\mathrm{Tr}[\Pi(H_1)\rho]$ and $\mathrm{Tr}[\Pi(H_2)\rho]$ are upper or lower probabilities in the cases that $a > (1/2)$ or $a < (1/2)$, correspondingly.

4. The Heyting algebra $\Lambda(d)$

In this section we study the lattice $\Lambda(d)$ of subsystems of $\Sigma(d)$. This is a distributive lattice, which is a Heyting algebra (every finite distributive lattice is a Heyting algebra). We show that the analogue of Eq.(1) is not valid, and we interpret the corresponding quantum probabilities as Dempster-Shafer probabilities. There are sublattices of $\Lambda(d)$ which are Boolean algebras (all chains) and in these 'islands' the probabilities obey Eq.(1), and they are additive (Kolmogorov) probabilities. But in the full lattice they cannot be interpreted as Kolmogorov probabilities. This shows that the distributivity property in a lattice, is not enough to make the corresponding probabilities to be Kolmogorov probabilities. We need to have a Boolean algebra. Probabilities in structures which are not Boolean algebras, are not Kolmogorov probabilities, and in this paper we interpret them as Dempster-Shafer probabilities.

For simplicity, we use the same symbols for the logical operations in the two lattices $\mathcal{L}(d)$ and $\Lambda(d)$, although they are different.

4.1. The lattice $\Lambda_1(d)$ of divisors of d

We consider the set of all divisors of d. This set with

$$a \wedge b = \mathrm{GCD}(a, b); \qquad a \vee b = \mathrm{LCM}(a, b), \tag{13}$$

where GCD and LCM are the greatest common divisor and least common multiplier correspondingly, is a lattice which we denote as $\Lambda_1(d)$. The negation $\neg a$ is the largest divisor of d, which is coprime with a. The partial order in this lattice is divisibility ($e \prec d$ means $e|d$), and $\mathcal{I} = d$ and $\mathcal{O} = 1$ are the greatest and smallest elements. This lattice is a Heyting algebra and it is isomorphic to the lattice of subsystems below.

4.2. Subsystems of $\Sigma(d)$

A subsystem of $\Sigma(d)$, is a smaller system with variables in a subgroup of $\mathbb{Z}(d)$. The subgroups of $\mathbb{Z}(d)$, are the $\mathbb{Z}(e)$ with $e|d$. Therefore if $e|d$, the $\Sigma(e)$ is a subsystem of $\Sigma(d)$.

Let $|X_d; r\rangle$ where $r \in \mathbb{Z}(d)$, be an orthonormal basis in $H(d)$ which we call 'basis of position states'. The X_d in this notation is not a variable, but it indicates that they are position states in $H(d)$. $\Sigma(e)$ can be embedded into $\Sigma(d)$ as follows:

$$\sum_{r=0}^{e-1} a_r|X_e; r\rangle \quad \rightarrow \quad \sum_{r=0}^{d-1} a_r|X_d; \frac{rd}{e}\rangle; \quad e|d. \tag{14}$$

The set of subsystems of $\Sigma(d)$ with the logical operations

$$\begin{aligned} &\Sigma(e) \wedge \Sigma(f) = \Sigma(e \wedge f) \\ &\Sigma(e) \vee \Sigma(f) = \Sigma(e \vee f) \\ &\neg\Sigma(e) = \Sigma(\neg e) \end{aligned} \tag{15}$$

is a Heyting algebra with $\mathcal{O} = \Sigma(1)$ and $\mathcal{I} = \Sigma(n)$. It is isomorphic to $\Lambda_1(d)$ and we denote it as $\Lambda(d)$. The physical meaning of the logical operations is discussed in ref[26]).

For $e \prec f \prec d$, we define the projector from the subsystem $\Sigma(f)$ to a smaller subsystem $\Sigma(e)$, as

$$\varpi(f \rightarrow e) = \sum_{r=0}^{e-1} |X_f; \frac{fr}{e}\rangle\langle X_f; \frac{fr}{e}|; \quad e \prec f \prec d. \tag{16}$$

We also denote the $\varpi(d \rightarrow e)$ simply as $\varpi(e)$. Taking into account Eq.(14), we find the following compatibility condition:

$$\varpi(f \rightarrow e)\varpi(f) = \varpi(e); \quad e \prec f \prec d. \tag{17}$$

All these projectors commute with each other. Logical operations analogous to Eq.(15) can also be defined for the corresponding projectors. For example, $\varpi(e_1) \vee \varpi(e_2) = \varpi(e_1 \vee e_2)$ and $\varpi(e_1) \wedge \varpi(e_2) = \varpi(e_1 \wedge e_2)$. Also $\neg\varpi(e) = \varpi(\neg e)$ is projector to the space of the system $\neg\Sigma(e)$.

4.3. The non-additivity operator $\varpi_S(e_1, e_2)$

In [26] we have proved that

$$H(e_1 \vee e_2) = \text{span}[H(e_1) \cup H(e_2)] \oplus S(e_1, e_2). \tag{18}$$

where $S(e_1, e_2)$ is a s-dimensional space (where $s = e_1 \vee e_2 - e_1 - e_2 + e_1 \wedge e_2$), which is orthogonal to span$[H(e_1) \cup H(e_2)]$. Comparison of Eqs(5),(18) shows that the disjunction in the two lattices $\mathcal{L}(d)$ and $\Lambda(d)$ is different, and it proves again that the two lattices are different. The projector to the space $S(e_1, e_2)$ is

$$\varpi_S(e_1, e_2) = \varpi(e_1 \vee e_2) - \varpi(e_1) - \varpi(e_2) + \varpi(e_1 \wedge e_2). \tag{19}$$

Taking the trace of these projectors with a density matrix ρ we get probabilities:

$$\ell(e_1 \vee e_2) - \ell(e_1) - \ell(e_2) + \ell(e_1 \wedge e_2) = \text{Tr}[\rho\varpi_S(e_1, e_2)]. \tag{20}$$

This shows that these probabilities are not Kolmogorov probabilities. They are lower probabilities (the Tr$[\rho\varpi_S(e_1, e_2)]$ is non-negative) and the letter ℓ in the notation, indicates this. In the special case that $e_1 \prec e_2$ then $\varpi_S(e_1, e_2) = 0$. Therefore in chains within the lattice $\Lambda(d)$, the probabilities obey Eq.(1) and they can be interpreted as Kolmogorov probabilities.

We next insert $e_2 = \neg e_1$ in Eq.(19) and we get

$$\varpi_S(e, \neg e) = \varpi(e \vee \neg e) - \varpi(e) - \varpi(\neg e) + \varpi(1). \tag{21}$$

We use the notation $\neg\widetilde{\Sigma}(e)$ for the system $\neg\Sigma(e) = \Sigma(\neg e)$ without its lowest state $|X;0\rangle$ (the corresponding projector is $\varpi(\neg e) - \varpi(1)$). In set theory (which is a Boolean algebra) 'do not belong to $\overline{A}$' is equivalent to 'belong to A'. Also $A \cup \overline{A} = \Omega$ and this leads to Eq.(2). In the present formalism (which is a Heyting algebra) 'does not belong in $\neg\widetilde{\Sigma}(e)$' is not equivalent to 'belongs in $\Sigma(e)$'. Firstly, $e \vee \neg e \neq \mathcal{I}$ in general (here $\mathcal{I} = d$), and secondly $\varpi_S(e, \neg e)$ is in general non-zero.

4.4. Example:

We consider the $\Lambda(12)$ which comprises of the subsystems of $\Sigma(12)$. The projectors to these subsystems are

$$\begin{aligned}
&\varpi(2) = |X_{12};0\rangle\langle X_{12};0| + |X_{12};6\rangle\langle X_{12};6| \\
&\varpi(3) = |X_{12};0\rangle\langle X_{12};0| + |X_{12};4\rangle\langle X_{12};4| + |X_{12};8\rangle\langle X_{12};8| \\
&\varpi(2) \vee \varpi(3) = \varpi(6) = \sum_{\nu=0}^{5} |X_{12};2\nu\rangle\langle X_{12};2\nu| \\
&\varpi(2) \wedge \varpi(3) = \varpi(1) = |X_{12};0\rangle\langle X_{12};0|
\end{aligned} \tag{22}$$

In this case

$$\varpi_S(2,3) = \varpi(2 \vee 3) - \varpi(2) - \varpi(3) + \varpi(2 \wedge 3) = |X_{12};2\rangle\langle X_{12};2| + |X_{12};10\rangle\langle X_{12};10|. \tag{23}$$

The heart of Boolean algebras is the 'law of the excluded middle' $\neg a \vee a = \mathcal{I}$, i.e., something is either true or false. This is not valid in Heyting algebras, where $\neg a \vee a \prec \mathcal{I}$. Equivalently, in Boolean algebras $\neg\neg a = a$, i.e., two negations are equivalent to the positive statement. In Heyting algebras this is not true $(a \prec \neg\neg a)$.

We next exemplify the fact that 'does not belong in $\neg\widetilde{\Sigma}(e)$' is not equivalent to 'belongs in $\Sigma(e)$'. In $\Lambda(12)$, the spaces of subsystems $\Sigma(2)$ and $\neg\widetilde{\Sigma}(2) = \widetilde{\Sigma}(3)$ are span$\{|X_{12};0\rangle, |X_{12};6\rangle\}$ and span$\{|X_{12};4\rangle, |X_{12};8\rangle\}$, correspondingly. The state $|X_{12};5\rangle$ does not belong to either $\neg\widetilde{\Sigma}(2)$ or $\Sigma(2)$.

5. Discussion

Kolmogorov probabilities are intimatelly connected to Boolean logic, and their additivity property is that $\delta(A, B) = 0$ in Eq.(1). Quantum logic is based on the orthomodular lattice $\mathcal{L}(d)$ of Birkhoff and von Neumann. We have introduced the non-additivity operator $\mathfrak{D}(H_1, H_2)$ in Eq.(6), which quantifies the difference between an aggregation and the sum of its parts, and we have shown in Eq.(7) that it is related to the commutator of two projectors. Therefore non-commutativity is reponsible for the non-additivity of quantum probabilities. Within $\mathcal{L}(d)$ there are sublattices which are Boolean algebras and there the projectors commute and the corresponding probabilities are Kolmogorov probabilities. But in the full lattice $\mathcal{L}(d)$, we have interpreted quantum probabilities as Dempster-Shafer probabilities.

The lattice $\mathcal{L}(d)$ is not distributive (it is modular which is a weak version of distributivity). Lack of distributivity is a strong deviation from Boolean algebras. Therefore the question arises, if the lack of distributivity alone, is what makes quantum probabilities to be non-additive. In order to answer this question, we have considered the lattice $\Lambda(d)$ of subsystems, which is distributive (it is a Heyting algebra), and which is associated to commuting projectors. In this case also the quantum probabilities are non-additive, and they can be interpreted as lower probabilities within the Dempster-Shafer theory. This is because 'does not belong in $\neg\widetilde{\Sigma}(e)$' is not equivalent to 'belongs in $\Sigma(e)$' (in Heyting algebras two negations of a statement is not the same as the original statement). In chains within $\Lambda(d)$ quantum probabilities are additive, i.e., they are Kolmogorov probabilities.

The work shows the non-additive nature of quantum probabilities, and their interpretation in the context of the Dempster-Shafer theory. We note that without additivity the concept of integration needs revision. Choquet integrals [27] is the appropriate concept in this case, and further work is needed in this direction.

References

[1] Lahti P and Maczynski J 1992 *J. Math. Phys.* **33** 4133
[2] Caticha A 1998 *Phys. Rev. A* **57** 1572
[3] Schack R, Brun T A and Caves M C 2001 *Phys. Rev. A* **64** 014305
[4] Ludwig G 1967 *Comm. Math. Phys.* **4** 331
[5] Davies E B and Lewis J T 1970 *Commun. Math. Phys.* **17** 239
[6] Mielnik B 1974 *Comm. Math. Phys.* **37** 221
[7] Busch P, Grabowski M and Lahti P J 1995 *Operational Quantum Physics* (Berlin: Springer)
[8] Foulis D J and Randall C H 1972 *J. Math. Phys.* **13** 1667
[9] Randall C H and Foulis D J 1973 *J. Math. Phys.* **14** 1472
[10] Twareque Ali S and Emch G G 1974 *J. Math. Phys.* **15** 176
[11] Twareque Ali S and Doebner H D 1976 *J. Math. Phys.* **17** 1105
[12] Abramsky S and Coecke B 2004 *Proc. 19th Annual IEEE Symposium on Logic in Computer Science* 415
[13] Coecke B and Ross D 2011 *New J. Phys.* **13** 043016
[14] Dempster A P 1967 *Annal. Math. Stat.* **38** 325
[15] Shafer G 1976 *A Mathematical Theory of Evidence* (Princeton: Princeton Univ. Press)
[16] Halpern J Y and Fagin R 1992 *Artificial Intelligence* **54** 275
[17] Yager R and Liu L (Ed.) 2008 *Classic works of the Dempster-Shafer theory of belief functions* (Berlin: Springer)
[18] Birkhoff G and von Neumann J 1936 *Ann. Math.* **37** 823
[19] Piron C 1976 *Foundations of quantum physics* (New York: Benjamin)
[20] Mackey G W 1963 *Mathematical foundations of quantum mechanics* (New York: Benjamin)
[21] Jauch J 1968 *Foundations of quantum mechanics* (Reading: Addison-Wesley)
[22] Beltrametti E and Cassinelli G 1981 *The Logic of Quantum Mechanics* (Reading: Addison-Wesley)
[23] Vourdas A 2014 *J. Math. Phys.* **55** 082107
[24] Vourdas A 2014 *J. Phys. A: Math. Theor.* **47** 345203
[25] Zadeh L A 1986 *Artificial Intelligence Magazine* **7** 85
[26] Vourdas A 2013 *J. Math. Phys* **54** 082105
[27] Choquet G 1953 *Ann. Inst. Fourier* **5** 131

Sigma-models and homotopy algebras

Anton M Zeitlin

Department of Mathematics, Columbia University, 2990 Broadway, New York, NY 10027, USA IPME RAS, V.O. Bolshoj pr., 61, 199178, St. Petersburg
http://math.columbia.edu/~zeitlin

E-mail: zeitlin@math.columbia.edu

Abstract. We review the relation between homotopy algebras of conformal field theory and geometric structures arising in sigma models. In particular we formulate conformal invariance conditions, which in the quasi-classical limit are Einstein equations with extra fields, as generalized Maurer-Cartan equations.

1. Introduction: target space symmetries of sigma-models revisited

Sigma-models are among the most fruitful and puzzling objects in string theory. They can be treated as two dimensional conformal field theories CFT, describing movement of a string in the curved background with extra fields. The corresponding conformal invariance conditions have an interpretation of nonlinear equations of motion, like Einstein equations with extra fields [2], [3]. In this article we will review some of the results of manuscripts [20], [21],[22], which describe symmetries and conformal invariance conditions (nonlinear equations for the background fields) from the point of view of intrinsic homotopy Gerstenhaber algebras of CFT. To do this we have to reformulate the standard classical second order sigma-model using the special first order action [13] and describe standard symmetries in this new language. This reformulation is given in the first section. Then, in section 2, the homotopy algebras of chiral CFTs are described and finally, in section 3, we formulate conjectures concerning the homotopical description of Einstein equations with B-field and dilaton in this framework.

Let us consider a complex Riemann surface Σ, a complex manifold M of dimension d, and a map $\rho : \Sigma \to M$. Then one can write the following first order action functional:

$$S_0 = \frac{1}{2\pi i h}\int_{\Sigma}(\langle p \wedge \bar{\partial} X\rangle - \langle \bar{p} \wedge \partial X\rangle), \tag{1}$$

where p and $\bar{p}$ belong to $\rho^*(\Omega^{(1,0)}(M))\otimes\Omega^{(1,0)}(\Sigma)$ and $\rho^*(\Omega^{(0,1)}(M))\otimes\Omega^{(0,1)}(\Sigma)$ correspondingly, $X^i, X^{\bar{j}}$ stand for the pull-backs of the coordinate functions on M with respect to ρ, and $\langle\cdot,\cdot\rangle$ stands for standard pairing. This action has the following symmetries (we write them in components in the infinitesimal form):

$$X^i \to X^i - v^i(X), \quad p_i \to p_i + \partial_i v^k p_k; \quad X^{\bar{i}} \to X^{\bar{i}} - v^{\bar{i}}(\bar{X}), \quad p_{\bar{i}} \to p_{\bar{i}} + \partial_{\bar{i}} v^{\bar{k}} p_{\bar{k}}. \tag{2}$$

Here, the generators of the infinitesimal transformations $v, \bar{v}$ are the elements of $\Gamma(\mathcal{O}(T^{(1,0)}M))$ and $\Gamma(\mathcal{O}(T^{(0,1)}M))$ correspondingly, i.e. v^i ($v^{\bar{i}}$) are (anti)holomorphic. These symmetries illustrate invariance under the holomorphic coordinate transformations. There is another set of symmetries, induced by the (anti)holomorphic 1-forms. Let $\omega \in \mathcal{O}(T^{*(1,0)}M)$ and $\bar{\omega} \in \bar{\mathcal{O}}(T^{*(0,1)}M)$. Then the action (1) is invariant under the transformation of $p, \bar{p}$:

$$p_i \to p_i - \partial X^k(\partial_k \omega_i - \partial_i \omega_k), \quad p_{\bar{i}} \to p_{\bar{i}} - \bar{\partial} X^{\bar{k}}(\partial_{\bar{k}} \omega_{\bar{i}} - \partial_{\bar{i}} \omega_{\bar{k}}). \tag{3}$$

We want to generalize the action (1) so that it would be invariant under the diffeomorphism transformations and nonholomorphic generalizations of (3). In order to do that, one has to introduce extra (perturbation) terms to the action (1), so that resulting action is as follows:

$$\tilde{S} = \frac{1}{2\pi i h}\int_\Sigma (\langle p \wedge \bar{\partial} X\rangle - \langle \bar{p} \wedge \partial X\rangle - \langle \mu, p \wedge \bar{\partial} X\rangle - \langle \bar{\mu}, \partial X \wedge \bar{p}\rangle - \langle b, \partial X \wedge \bar{\partial} X\rangle),$$

where $b \in \Gamma(T^{*(1,0)}M \otimes T^{*(0,1)}M)$. The resulting symmetry transformations generated by $(v, \bar{v})$ can be written as follows:

$$\begin{aligned}
\mu^i_{\bar{j}} &\to \mu^i_{\bar{j}} - \partial_{\bar{j}} v^i + v^k \partial_k \mu^i_{\bar{j}} + v^{\bar{k}} \partial_{\bar{k}} \mu^i_{\bar{j}} + \mu^i_{\bar{k}} \partial_{\bar{j}} v^{\bar{k}} - \mu^k_{\bar{j}} \partial_k v^i + \mu^i_{\bar{l}} \mu^k_{\bar{j}} \partial_k v^{\bar{l}}, \\
b_{i\bar{j}} &\to b_{i\bar{j}} + v^k \partial_k b_{i\bar{j}} + v^{\bar{k}} \partial_{\bar{k}} b_{i\bar{j}} + b_{i\bar{k}} \partial_{\bar{j}} v^{\bar{k}} + b_{l\bar{j}} \partial_i v^l + b_{i\bar{k}} \mu^k_{\bar{j}} \partial_k v^{\bar{k}} + b_{l\bar{j}} \bar{\mu}^{\bar{k}}_i \partial_{\bar{k}} v^l,
\end{aligned} \tag{4}$$

and the formula for the transformation of $\bar{\mu}$ can be obtained from the one of μ by formal complex conjugation. This leads to the symmetry of the action $\tilde{S}$ if

$$\begin{aligned}
X^i &\to X^i - v^i(X, \bar{X}), \quad p_i \to p_i + p_k \partial_i v^k - p_k \mu^k_{\bar{l}} \partial_i v^{\bar{l}} - b_{j\bar{k}} \partial_i v^{\bar{k}} \partial X^j, \\
X^{\bar{i}} &\to X^{\bar{i}} - v^{\bar{i}}(X, \bar{X}), \quad \bar{p}_{\bar{i}} \to \bar{p}_{\bar{i}} + \bar{p}_{\bar{k}} \partial_{\bar{i}} v^{\bar{k}} - \bar{p}_{\bar{k}} \bar{\mu}^{\bar{k}}_l \partial_{\bar{i}} v^l - b_{\bar{j}k} \partial_{\bar{i}} v^k \bar{\partial} X^{\bar{j}}.
\end{aligned} \tag{5}$$

Therefore, the resulting action is invariant under the action of the infinitesimal diffeomorphism group. The component formulas (4) were first discovered in [7]. Similarly, we obtain that the transformations

$$b_{i\bar{j}} \to b_{i\bar{j}} + \partial_{\bar{j}} \omega_i - \partial_i \omega_{\bar{j}} + \mu^i_{\bar{j}}(\partial_i \omega_k - \partial_k \omega_i) + \bar{\mu}^{\bar{s}}_i(\partial_{\bar{j}} \omega_{\bar{s}} - \partial_{\bar{s}} \omega_{\bar{j}}) + \bar{\mu}^{\bar{i}}_j \mu^s_{\bar{k}}(\partial_s \omega_{\bar{i}} - \partial_{\bar{i}} \omega_s) \tag{6}$$

accompanied with $p_i \to p_i - \partial X^k(\partial_k \omega_i - \partial_i \omega_k) - \partial_{\bar{r}} \omega_i \partial X^{\bar{r}} - \bar{\mu}^{\bar{s}}_k \partial_i \omega_{\bar{s}} \partial X^k$, $p_{\bar{i}} \to p_{\bar{i}} - \bar{\partial} X^{\bar{k}}(\partial_{\bar{k}} \omega_{\bar{i}} - \partial_{\bar{i}} \omega_{\bar{k}}) - \partial_r \omega_{\bar{i}} \bar{\partial} X^r - \mu^s_{\bar{k}} \partial_i \omega_s \bar{\partial} X^{\bar{k}}$ leave $\tilde{S}$ invariant. Hereinafter, it is useful to consider $\mu, \bar{\mu}, b$ as matrix elements of $\tilde{\mathbb{M}} \in \Gamma((T^{(1,0)}M \oplus T^{*(1,0)}M) \otimes (T^{(0,1)}M \oplus T^{*(0,1)}M)$, i.e. $\tilde{\mathbb{M}} = \begin{pmatrix} 0 & \mu \\ \bar{\mu} & b \end{pmatrix}$. For simplicity of notation let us define $E = TM \oplus T^*M$, also $\mathcal{E} = T^{(1,0)}M \oplus T^{*(1,0)}M$ and $\bar{\mathcal{E}} = T^{(0,1)}M \oplus T^{*(0,1)}M$, so that $E = \mathcal{E} \oplus \bar{\mathcal{E}}$.

Let $\alpha \in \Gamma(E)$, i.e. $\alpha = (v, \bar{v}, \omega, \bar{\omega})$, where $v, \bar{v}$ are the elements of $\Gamma(T^{(1,0)}M)$ and $\Gamma(T^{(0,1)}M)$ correspondingly and $\omega \in \Omega^{(1,0)}(M)$, $\bar{\omega} \in \Omega^{(0,1)}(M)$. Next, we introduce an operator $D : \Gamma(E) \to \Gamma(\mathcal{E} \otimes \bar{\mathcal{E}})$, such that $D\alpha = \begin{pmatrix} 0 & \bar{\partial} v \\ \partial \bar{v} & \partial \bar{\omega} - \bar{\partial} \omega \end{pmatrix}$. Then the transformation of $\tilde{\mathbb{M}}$ under (4), (6) can be expressed by the following formula:

$$\tilde{\mathbb{M}} \to \tilde{\mathbb{M}} - D\alpha + \phi_1(\alpha, \tilde{\mathbb{M}}) + \phi_2(\alpha, \tilde{\mathbb{M}}, \tilde{\mathbb{M}}). \tag{7}$$

The second operation $\phi_1(\alpha, \tilde{\mathbb{M}})$ can be described as follows. Let us consider

$$\xi \in J^\infty(\mathcal{O}_M) \otimes J^\infty(\bar{\mathcal{O}}(\bar{\mathcal{E}})) \oplus J^\infty(\mathcal{O}(\mathcal{E})) \otimes J^\infty(\bar{\mathcal{O}}_M), \quad \mathbb{L} \in J^\infty(\mathcal{O}(\mathcal{E})) \otimes J^\infty(\bar{\mathcal{O}}(\bar{\mathcal{E}})), \tag{8}$$

where $J^\infty(E)$ for any bundle E over M stands for the corresponding ∞-jet bundle of E. In other words, let $\xi = \sum_J f^J \otimes \bar{b}^J + \sum_K b^K \otimes \bar{f}^K$, $\mathbb{L} = \sum_I a^I \otimes \bar{a}^I$, where $a^I, b^J \in J^\infty(\mathcal{O}(\mathcal{E}))$,

$f^I \in J^\infty(\mathcal{O}_M)$ and $\bar{a}^I, \bar{b}^J \in J^\infty(\bar{\mathcal{O}}(\bar{\mathcal{E}}))$, $\bar{f}^I \in J^\infty(\bar{\mathcal{O}}_M)$. Then we can introduce an operation $\phi_1(\xi, \mathbb{L})$ as follows:

$$\phi_1(\xi, \mathbb{L}) = \sum_{I,J} [b^J, a^I]_D \otimes \bar{f}^J \bar{a}^I + \sum_{I,K} f^K a^I \otimes [\bar{b}^K, \bar{a}^I]_D, \tag{9}$$

where $[\cdot,\cdot]_D$ is a Dorfman bracket (see the definition in the next section). Completing the tensor products in (8), we find that the operation ϕ_1 can be induced on $\alpha \in \Gamma(E)$ and $\tilde{\mathbb{M}} \in \Gamma(\mathcal{E} \otimes \bar{\mathcal{E}})$. One can explicitly check that (9) leads to the part of (4) and (6), linear in α and $\tilde{\mathbb{M}}$. The last part, bilinear in $\tilde{\mathbb{M}}$, also has an algebraic meaning of a similar kind: returning back to the jet notation, we find that on the jet counterparts of $\alpha, \tilde{\mathbb{M}}$, i.e. on $\xi, \mathbb{L}$ the expression for ϕ_2 is:

$$\phi_2(\xi, \mathbb{L}, \mathbb{L}) = \frac{1}{2} \sum_{I,J,K} \langle b^I, a^K \rangle a^J \otimes \bar{a}^J(\bar{f}^I)\bar{a}^K + \frac{1}{2} \sum_{I,J,K} a^J(f^I) a^K \otimes \langle \bar{b}^I, \bar{a}^K \rangle \bar{a}^J, \tag{10}$$

where $\bar{a}^J(\bar{f}^I)$, $a^J(f^I)$ correspond to the action of the differential operator, associated to the vector field, on a function ($\bar{a}^J(\bar{f}^I)$, $a^J(f^I)$ are set to be zero if $\bar{a}^J$, a^J are 1-forms). At the same time, the operation ϕ_2 has the following simple description: $\phi_2(\alpha, \tilde{\mathbb{M}}, \tilde{\mathbb{M}}) = \tilde{\mathbb{M}} \cdot D\alpha \cdot \tilde{\mathbb{M}}$ if we consider $\tilde{\mathbb{M}}$ as an element of $End(\Gamma(E))$.

Let us notice that we could generalize $\tilde{\mathbb{M}}$ in the following way: in the matrix expression for $\tilde{\mathbb{M}}$ let us fill in the empty spot, i.e. let us add extra element $g \in \Gamma(T^{(1,0)}M \otimes T^{(0,1)}M)$. Then the modified $\tilde{\mathbb{M}}$, i.e. $\mathbb{M} \in \Gamma(\mathcal{E} \otimes \bar{\mathcal{E}})$ can be expressed as follows: $\mathbb{M} = \begin{pmatrix} g & \mu \\ \bar{\mu} & b \end{pmatrix}$. The corresponding first order action functional is:

$$S_{fo} = \tag{11}$$
$$\frac{1}{2\pi i h} \int_\Sigma (\langle p \wedge \bar{\partial} X \rangle - \langle \bar{p} \wedge \partial X \rangle - \langle g, p \wedge \bar{p} \rangle - \langle \mu, p \wedge \bar{\partial} X \rangle - \langle \bar{\mu}, \bar{p} \wedge \partial X \rangle - \langle b, \partial X \wedge \bar{\partial} X \rangle).$$

It turns out that the symmetries of this action functional can be described by the same formula (7), where the algebraic meaning of operations on the jet level is given by the same formulas (9), (10). One can easily reproduce explicit component formulas for the infinitesimal symmetries of the action S_{fo}. The reason for introducing the g-term in the action functional is as follows. If the matrix $\{g^{i\bar{j}}\}$ is invertible, then using elementary variational calculus, one can find that the critical points for S_{fo} are the same as for the second-order action functional:

$$S_{so} = \frac{1}{4\pi h} \int_\Sigma d^2 z (G_{\mu\nu} + B_{\mu\nu}) \partial X^\mu \bar{\partial} X^\nu, \tag{12}$$

where G is a symmetric tensor and B is antisymmetric one, indices μ, ν run through the set $\{i, \bar{j}\}$. The expression for G and B via $\mathbb{M}$ is given by:

$$\begin{aligned}
G_{s\bar{k}} &= g_{i\bar{j}} \bar{\mu}^{\bar{i}}_s \mu^{\bar{j}}_{\bar{k}} + g_{s\bar{k}} - b_{s\bar{k}}, \quad B_{s\bar{k}} = g_{i\bar{j}} \bar{\mu}^{\bar{i}}_s \mu^{\bar{j}}_{\bar{k}} - g_{s\bar{k}} - b_{s\bar{k}}, \\
G_{si} &= -g_{i\bar{j}} \bar{\mu}^{\bar{j}}_s - g_{s\bar{j}} \bar{\mu}^{\bar{j}}_i, \quad G_{\bar{s}\bar{i}} = -g_{\bar{s}j} \mu^j_{\bar{i}} - g_{\bar{i}j} \mu^j_{\bar{s}}, \\
B_{si} &= g_{s\bar{j}} \bar{\mu}^{\bar{j}}_i - g_{i\bar{j}} \bar{\mu}^{\bar{j}}_s, \quad B_{\bar{s}\bar{i}} = g_{\bar{i}j} \mu^j_{\bar{s}} - g_{\bar{s}j} \mu^j_{\bar{i}},
\end{aligned} \tag{13}$$

where $\{g_{i\bar{j}}\}$ stands for the inverse matrix of $\{g^{i\bar{j}}\}$. Such parametrization of the second-order action in the case when M is a Riemann surface was first introduced in [15], [23]. The symmetries of the action functional S_{fo} transform into infinitesimal diffeomorphism transformations and the 2-form B symmetry $G \to G - L_{\mathbf{v}}G$, $B \to B - L_{\mathbf{v}}B$; $B \to B - 2d\boldsymbol{\omega}$, if $\alpha = (\mathbf{v}, \boldsymbol{\omega})$, so that $\mathbf{v} \in \Gamma(TM)$, $\boldsymbol{\omega} \in \Omega^1(M)$, i.e. the symmetries of S_{so}.

Let us formulate this as a theorem.

Theorem 1.1. *Let $\mathbb{M} \in \Gamma(\mathcal{E} \otimes \bar{\mathcal{E}})$, parametrized as above, so that its $\Gamma(T^{(1,0)}M \otimes T^{(0,1)}M)$ part is given by $\{g^{i\bar{j}}\}$, which is invertible, then the infinitesimal diffeomorphism transformations of the resulting symmetric and antisymmetric tensors G and B (see (13)), as well as the B-tensor shift by exact 2-forms are encoded in the formula $\mathbb{M} \to \mathbb{M} - D\alpha + \phi_1(\alpha, \mathbb{M}) + \phi_2(\alpha, \mathbb{M}, \mathbb{M})$, where $\alpha \in \Gamma(E)$ and operations ϕ_1, ϕ_2 are defined above.*

Note that if $\{G_{\mu\nu}\}$ is invertible and real, it gives rise to the metric tensor. Therefore, since $\mathbb{M}$ parametrizes both G and B, and transforms according to (7) under diffeomorphisms, it is analogous to Beltrami differential on the Riemann surface. So, from now on we will call the elements of $\Gamma(\mathcal{E} \otimes \bar{\mathcal{E}})$ as Beltrami-Courant differentials, since, as we see in the following sections, they are described by means of the Courant algebroid [12] structure on $\mathcal{E}, \bar{\mathcal{E}}$.

2. Vertex algebroids, G_∞-algebra and quasiclassical limit

In this section, we describe the constructions of the article [21] with some modifications and refer the reader to this article for some of the details. Namely, we discuss homotopy algebras of G_∞- and BV_∞- type (see e.g. [16]), emerging from the vertex algebras associated with semi-infinite complex (see e.g. [11]).

Each of the terms in the classical action S_0 from which we started the previous section, leads to the quantum theory which is well described locally on open neighborhoods of M by means of vertex algebra generated by operator products $X^i(z)p_j(w) \sim \frac{h\delta^i_j}{z-w}$, $X^{\bar{i}}(\bar{z})p_{\bar{j}}(\bar{w}) \sim \frac{h\delta^{\bar{i}}_{\bar{j}}}{\bar{z}-\bar{w}}$ and globally by means of gerbes of chiral differential operators on M [14]. Each of the corresponding vertex algebras, which provide the local description, form a $\mathbb{Z}_+$-graded vector space $V = \sum_{n=0}^{+\infty} V_n$, so that it is determined (see [14]) by means of a vertex algebroid. In our case, the vertex algebroid is described by means of the sheaf $\mathcal{V} = \mathcal{O}(\mathcal{E}) \otimes \mathbb{C}[h] \equiv \mathcal{O}(\mathcal{E})^h$ (resp. $\bar{\mathcal{O}}(\bar{\mathcal{E}})^h$), of vector spaces V_1, as well as the sheaf of V_0 spaces, which coincides with the structure sheaf $\mathcal{O}_M \otimes \mathbb{C}[h] = \mathcal{O}^h_M$ (resp. $\bar{\mathcal{O}}^h_M$), with certain algebraic operations between them.

Let us define a vertex algebroid (see e.g. [14], [1]) and then study our concrete case in detail.

A *vertex $\mathcal{O}_M$-algebroid* is a sheaf of $\mathbb{C}$-vector spaces $\mathcal{V}$ with a pairing $\mathcal{O}_M \otimes_{\mathbb{C}[h]} \mathcal{V} \to \mathcal{V}$, i.e. $f \otimes v \mapsto f * v$ such that $1 * v = v$, equipped with a structure of a Leibniz $\mathbb{C}[h]$-algebra $[\ ,\ ] : \mathcal{V} \otimes_{\mathbb{C}[h]} \mathcal{V} \to \mathcal{V}$, a $\mathbb{C}[h]$-linear map of Leibniz algebras $\pi : \mathcal{V} \to \Gamma(TM)$, which usually is referred to as an anchor, a symmetric $\mathbb{C}[h]$-bilinear pairing $\langle\ ,\ \rangle : \mathcal{V} \otimes_{\mathbb{C}[h]} \mathcal{V} \to \mathcal{O}^h_M$ a $\mathbb{C}$-linear map $\partial : \mathcal{O}_M \to \mathcal{V}$ such that $\pi \circ \partial = 0$, which satisfy certain relations identical to the ones satisfied between V_1 and V_0 subspaces of the nonnegatively graded vertex algebra V.

Let us concentrate on the case when $\mathcal{V} = \mathcal{O}(\mathcal{E})^h$. Explicitly, if $f \in \mathcal{O}_M$, $v, v_1, v_2 \in \mathcal{O}(T^{(1,0)}M)$, $\omega, \omega_1, \omega_2 \in \mathcal{O}(T^{*(1,0)}M)$, then locally in the neighborhood with the coordinates $\{X^i\}$

$$\begin{aligned} &\partial f = df, \quad \pi(v)f = -hv(f), \quad \pi(\omega) = 0, \quad f * v = fv + hdX^i\partial_i\partial_j f v^j, \quad f * \omega = f\omega, \\ &[v_1, v_2] = -h[v_1, v_2]_D - h^2 dX^i \partial_i \partial_k v_1^s \partial_s v_2^k, \quad [v, \omega] = -h[v, \omega]_D, \quad [\omega, v] = -h[\omega, v]_D, \\ &[\omega_1, \omega_2] = 0, \quad \langle v, \omega\rangle = -h\langle v, \omega\rangle^s, \quad \langle v_1, v_2\rangle = -h^2 \partial_i v_1^j \partial_j v_2^i, \quad \langle \omega_1, \omega_2\rangle = 0, \end{aligned} \tag{14}$$

where $\langle \cdot, \cdot\rangle^s$ is a standard pairing on $\mathcal{E}$ and $[\cdot, \cdot]_D$ is the Dorfman bracket:

$$[v_1, v_2]_D = [v_1, v_2]^{Lie}, \quad [v, \omega]_D = L_v\omega, \quad [\omega, v]_D = -i_v d\omega, \quad [\omega_1, \omega_2]_D = 0. \tag{15}$$

In [21], it was shown that given a holomorphic volume form on the open neighborhood U of M, one can associate a homotopy Gerstenhaber algebra to the vertex algebroid on U (although the main emphasis of [21] was on C_∞ part of it). This was done by considering semi-infinite complex associated to the vertex algebra [4]: due to the results of[11], [10], [9], [17], there is a structure of G_∞ algebra attached to it if the central charge of the corresponding Virasoro algebra is 26. Using this fact and considering the subcomplex corresponding to the elements of total conformal weight zero, we find out that the central charge condition can be dropped. The resulting complex $(\mathcal{F}^{\cdot}, Q)$ appears to be much shorter than original semi-infinite one:

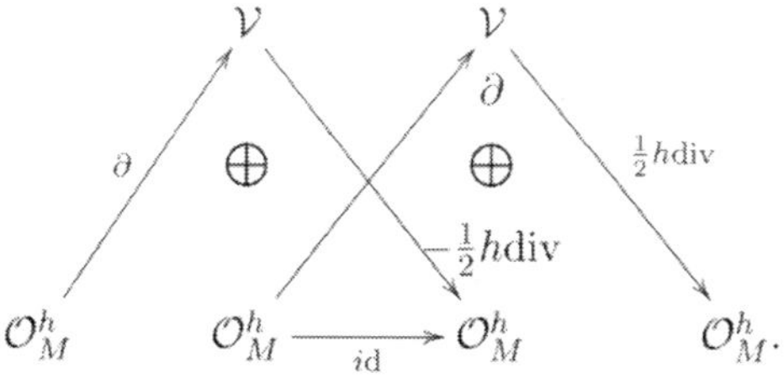

(16)

Here $\mathcal{F}^0_h \cong \mathcal{O}^h_M \cong \mathcal{F}^3_h$, $\mathcal{F}^1_h \cong \mathcal{O}^h_M \oplus \mathcal{V} \cong \mathcal{F}^3_h$, and div stands for divergence operator with respect to the nonvanishing volume form applied to sections of $\Gamma(U, T^{(1,0)}(M))$. Appropriate analogue of operator div in the case of general vertex algebroid is called *Calabi – Yau structure* on vertex algebroid [14] (since e.g. in our case to be defined globally M should possess a nonvanishing holomorphic volume form). According to [21], this complex has a bilinear operation $(\cdot,\cdot)_h$, which satisfies the Leibniz identity with respect to Q, it is also homotopy commutative and associative, and can be described via the operations from vertex algebroid.

We note that there is an operator $\mathbf{b}$ of degree -1 on $(\mathcal{F}_h^{\cdot}, Q)$ which anticommutes with Q:

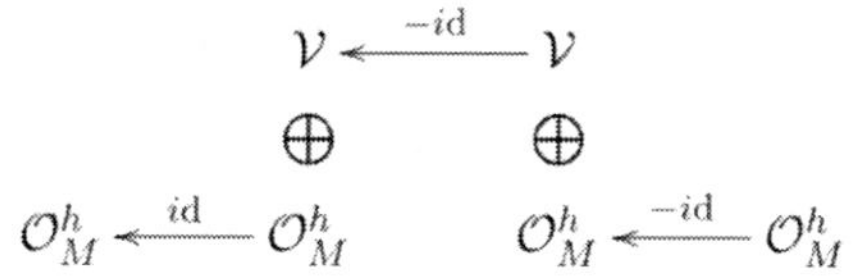

(17)

This operator gives rise to the bracket operation

$$(-1)^{|a_1|}\{a_1, a_2\}_h = \mathbf{b}(a_1, a_2)_h - (\mathbf{b}a_1, a_2)_h - (-1)^{|a_1|}(a_1, \mathbf{b}a_2)_h, \quad (18)$$

satisfying quadratic relations together with $(\cdot,\cdot)_h$ and Q, which follow from the properties of vertex algebra [11]. On the cohomology of Q these relations turn into the defining properties of Gerstenhaber algebra. Namely, the following Proposition holds.

Proposition 2.1.[21] *Symmetrized versions of operations $(\cdot,\cdot)_h$ together with (18) satisfy the* *relations of the homotopy Gerstenhaber algebra, which follow from these relations:*

$$\begin{aligned}
&Q(a_1,a_2)_h=(Qa_1,a_2)_h+(-1)^{|a_1|}(a_1,Qa_2)_h,\\
&(a_1,a_2)_h-(-1)^{|a_1||a_2|}(a_2,a_1)_h=Qm_h(a_1,a_2)+m_h(Qa_1,a_2)+(-1)^{|a_1|}m_h(a_1,Qa_2),\\
&Q(a_1,a_2,a_3)_h+(Qa_1,a_2,a_3)_h+(-1)^{|a_1|}(a_1,Qa_2,a_3)_h+\\
&(-1)^{|a_1|+|a_2|}(a_1,a_2,Qa_3)_h=((a_1,a_2)_h,a_3)_h-(a_1,(a_2,a_3)_h)_h,\\
&\{a_1,a_2\}+(-1)^{(|a_1|-1)(|a_2|-1)}\{a_2,a_1\}=\\
&(-1)^{|a_1|-1}(Qm'_h(a_1,a_2)-m'_h(Qa_1,a_2)-(-1)^{|a_2|}m'_h(a_1,Qa_2)),\\
&\{a_1,(a_2,a_3)_h\}_h=(\{a_1,a_2\}_h,a_3)_h+(-1)^{(|a_1|-1)|a_2|}(a_2,\{a_1,a_3\}_h)_h,\\
&\{(a_1,a_2)_h,a_3\}_h-(a_1,\{a_2,a_3\}_h)_h-(-1)^{(|a_3|-1)|a_2|}(\{a_1,a_3\}_h,a_2)_h=\\
&(-1)^{|a_1|+|a_2|-1}(Qn'_h(a_1,a_2,a_3)-n'_h(Qa_1,a_2,a_3)-\\
&(-1)^{|a_1|}n'_h(a_1,Qa_2,a_3)-(-1)^{|a_1|+|a_2|}n'_h(a_1,a_2,Qa_3),\\
&\{\{a_1,a_2\}_h,a_3\}_h-\{a_1,\{a_2,a_3\}_h\}_h+(-1)^{(|a_1|-1)(|a_2|-1)}\{a_2,\{a_1,a_3\}_h\}_h=0,
\end{aligned}\tag{19}$$

where m_h, m'_h are some bilinear operations of degrees -1, -2 correspondingly and n_h, n'_h are trilinear operations of degree -1, -2 correspondingly. There exist higher homotopies which turn this homotopy Gerstenhaber algebra into G_∞ algebra.

The last part of Proposition 2.1 follows from the results of [9], [17], [10] where it was show that the symmetrized versions of $(\cdot,\cdot)_h$, $\{,\}_h$ can be continued to the G_∞ algebra [16]. One should also consult [5], [6] for a different proof in the case of positively graded topological vertex algebras.

One of the central observations of [21] was that this G_∞ algebra has *quasiclassical* limit, which can be constructed as follows. Let $\mathcal{V}|_{h=0}=\mathcal{V}^0$ (in our example $\mathcal{V}^0=\mathcal{O}(\mathcal{E})$), then consider the subcomplex of $(\mathcal{F}^{\cdot}_h, Q)$, i.e. $(\mathcal{F}^{\cdot}, Q)\cong(\mathcal{F}^{\cdot}_1, Q)$, which is:

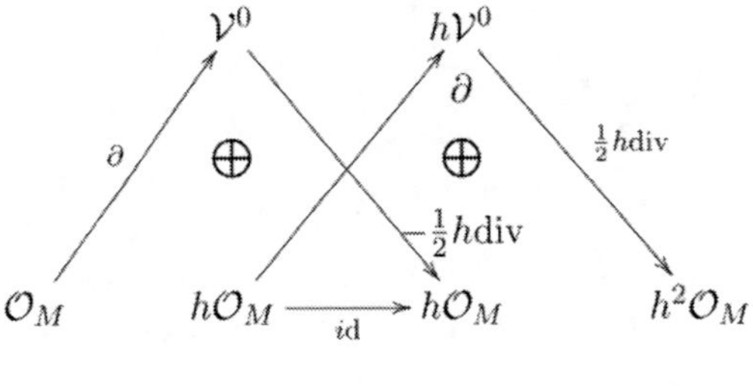

(20)

It is easy to see that

$$(\cdot,\cdot)_h:\mathcal{F}^i\otimes\mathcal{F}^j\to\mathcal{F}^{i+j}[h],\quad \{\cdot,\cdot\}_h:\mathcal{F}^i\otimes\mathcal{F}^j\to h\mathcal{F}^{i+j-1}[h],\quad \mathbf{b}:\mathcal{F}^i\to h\mathcal{F}^{i-1}[h],\tag{21}$$

so that $(\cdot,\cdot)_0=\lim_{h\to 0}(\cdot,\cdot)_h$, $\{\cdot,\cdot\}_0=\lim_{h\to 0}h^{-1}\{\cdot,\cdot\}_h$, $\mathbf{b}_0=\lim_{h\to 0}h^{-1}\mathbf{b}$ are well defined. The corresponding homotopy Gerstenhaber algebra is much less complicated: the corresponding L_∞ and C_∞ parts are only L_3 and C_3-algebras. Let us have a look in detail. On the level of the vertex algebroid of $\mathcal{O}(\mathcal{E})^h$, let us denote $\lim_{h\to 0}h^{-1}[v_1,v_2]=[v_1,v_2]_0$, $\lim_{h\to 0}h^{-1}\pi=\pi_0$, $\lim_{h\to 0}h^{-1}\langle\cdot,\cdot\rangle=\langle\cdot,\cdot\rangle_0$. Therefore, we can express the bilinear operations $(\cdot,\cdot)_0$ and $\{\cdot,\cdot\}_0$

on the complex

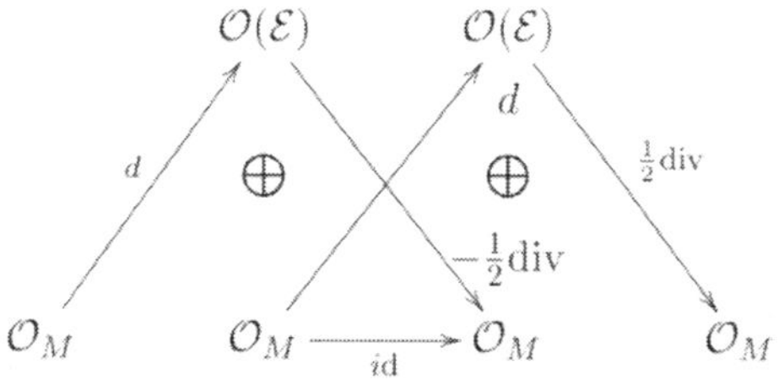

(22)

completely in terms of the operations of pure covariant parts of (14) (explicitly you can find it in [21]). Let us summarize results about the quasi-classical limit via proposition.

Proposition 2.2.[21] *The operations $(\cdot,\cdot)_0$, $\{\cdot,\cdot\}_0$ satisfy the relations (19) so that their symmetrized versions satisfy the relations of G_∞ algebra which is the quasiclassical limit of G_∞ algebra considered in Proposition 2.1. The resulting C_∞ and L_∞ algebras are reduced to C_3 and L_3 algebras.*

The classical limits for the corresponding homotopies $m_h = m_0 + O(h)$ and $n_h = n_0 + O(h)$ are as follows. The commutativity homotopy m_0 is nonzero iff its both arguments belong to $\mathcal{F}_1$: $m_0 = -\langle A_1, A_2\rangle_0$. The associativity homotopy n_0 is nonzero only when all three elements belong to $\mathcal{F}_1$ or one of the first two belongs to $\mathcal{F}_2$ and the other belongs to $\mathcal{F}_1$:

$$n_0(A_1, A_2, A_3) = A_2\langle A_1, A_3\rangle_0 - A_1\langle A_2, A_3\rangle_0, \quad n_0(A_1, \tilde{v}, A_2) = n_0(\tilde{v}, A_1, A_2) = -\tilde{v}\langle A_1, A_2\rangle_0.$$

As we stated before in the quasiclassical limit we get rid of all noncovariant terms in the expression for the product and the bracket: this is very close to the classical limit procedure for vertex algebroid. Namely, using our limit procedure, one can obtain Courant algebroid from vertex algebroid.

The definition of Courant algebroid is as follows (see e.g. [12], [1]). A Courant $\mathcal{O}_M$-algebroid is an $\mathcal{O}_M$-module $\mathcal{Q}$ equipped with the structure of a Leibniz $\mathbb{C}$-algebra $[,]_0 : \mathcal{Q}\otimes_{\mathbb{C}}\mathcal{Q}\to\mathcal{Q}$, an $\mathcal{O}_M$-linear map of Leibniz algebras (the anchor map) $\pi_0 : \mathcal{Q}\to\Gamma(TM)$, a symmetric $\mathcal{O}_M$-bilinear pairing $\langle\cdot,\cdot\rangle : \mathcal{Q}\otimes_{\mathcal{O}_M}\mathcal{Q}\to\mathcal{O}_M$, a derivation $\partial : \mathcal{O}_M\to\mathcal{Q}$, which satisfy

$$\begin{aligned}&\pi\circ\partial = 0, \quad [q_1, fq_2]_0 = f[q_1,q_2] + \pi_0(q_1)(f)q_2, \quad \langle[q,q_1],q_2\rangle + \langle q_1,[q,q_2]\rangle = \pi_0(q)(\langle q_1,q_2\rangle_0),\\ &[q,\partial(f)]_0 = \partial(\pi_0(q)(f)), \quad \langle q,\partial(f)\rangle = \pi_0(q)(f) \quad [q_1,q_2]_0 + [q_2,q_1]_0 = \partial(\langle q_1,q_2\rangle_0),\end{aligned} \tag{23}$$

where $f\in\mathcal{O}_M$ and $q, q_1, q_2\in\mathcal{Q}$. In our case $\mathcal{Q}\cong\mathcal{O}(\mathcal{E})$, π_0 is just a projection on $\mathcal{O}(TM)$ and $[q_1,q_2]_0 = -[q_1,q_2]_D$, $\langle q_1,q_2\rangle_0 = -\langle q_1,q_2\rangle^s$, $\partial = d$.

As we indicated earlier, both C_∞ and L_∞ parts of G_∞ algebra appear to be short. We expect this to happen with all the homotopies, i.e. it is natural to suggest that G_∞- algebra of Proposition 2.2 has only bilinear and trilinear operations, i.e. it is a G_3 algebra. In the following, since we are interested only in the quasiclassical algebra on the complex $(\mathcal{F}^\cdot, Q)$, we will neglect the 0 subscript for all multilinear operations of this algebra.

3. Homotopy Gerstenhaber algebra and Einstein equations

3.1. Nontrivial example. The homotopy Gerstenhaber algebra we studied in the previous section, has a subalgebra based on the following complex $(\mathcal{F}^{\cdot}_{sm}, \mathcal{Q})$.

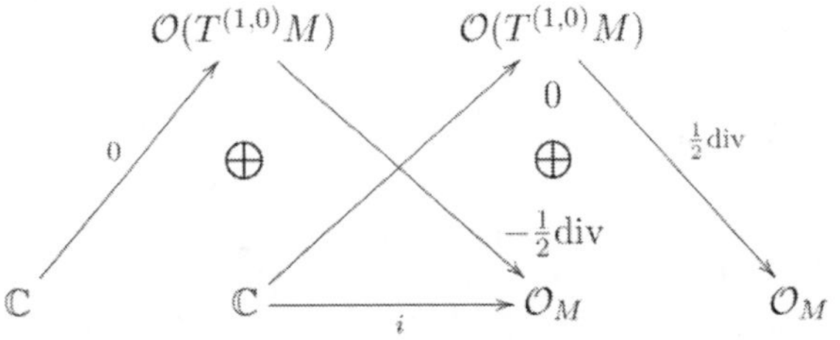

(24)

It is just a Gerstenhaber algebra (with no higher homotopies), moreover it is a BV algebra [16], since $\mathbf{b}$ operator also preserves $(\mathcal{F}^{\cdot}_{sm}, Q)$. Therefore, we have the following proposition.

Proposition 3.1. *Bilinear operations* $(\cdot,\cdot)$, $\{\cdot,\cdot\}$ *together with operator* $\mathbf{b}$ *generate the structure of BV algebra on* $(\mathcal{F}^{\cdot}_{sm}, Q)$.

Let us consider the ∞-jet version of the complex $(\mathcal{F}^{\cdot}_{sm}, Q)$: we substitute $\mathcal{O}_M$, $\mathcal{O}(T^{(0,1)}(M))$ by $J^\infty(\mathcal{O}_M)$, $J^\infty(\mathcal{O}(T^{(0,1)}(M))$. We denote the resulting complex as $(\mathcal{F}^{\cdot}_{sm,\infty}, Q)$. Then let us introduce a completed tensor product $\mathbf{F}^{\cdot}_{sm,\infty} = \mathcal{F}^{\cdot}_{sm,\infty}\hat{\otimes}\bar{\mathcal{F}}^{\cdot}_{sm,\infty}$, where $(\bar{\mathcal{F}}^{\cdot}_{sm,\infty}, \bar{Q})$ is the complex obtained from $(\mathcal{F}^{\cdot}_{sm,\infty}, Q)$ by complex conjugation. Complex $(\mathbf{F}^{\cdot}_{sm,\infty}, \mathcal{Q})$, where $\mathcal{Q} = Q + \bar{Q}$, is the jet version of the complex $(\mathbf{F}^{\cdot}_{sm}, \mathcal{Q})$, such that e.g. $\mathbf{F}^2_{sm} = \Gamma(T^{(1,0)}M \otimes T^{(0,1)}M) \oplus \bar{\mathcal{O}}(T^{(0,1)}M) \oplus \mathcal{O}(T^{(1,0)}M) \oplus \mathcal{O}_M \oplus \bar{\mathcal{O}}_M \oplus \mathbb{C}$. Clearly, the complex $(\mathbf{F}^{\cdot}_{sm}, \mathcal{Q})$ carries the structure of BV algebra inherited from $(\mathcal{F}^{\cdot}_{sm,\infty}, Q)$ and its complex conjugation, so that

$$(-1)^{|a_1|}\{a_1, a_2\} = \mathbf{b}^-(a_1, a_2) - (\mathbf{b}^-a_1, a_2) - (-1)^{|a_1|}(a_1\mathbf{b}^-a_2),$$

where $\mathbf{b}^- = \mathbf{b} - \bar{\mathbf{b}}$. Note, that the elements closed under $\mathbf{b}^-$ form a subalgebra in the differential graded Lie algebra (DGLA), generated by $\mathcal{Q}, \{\cdot,\cdot\}$. It turns out that the Maurer-Cartan equations of this DGLA and their symmetries have a very interesting meaning. To describe them, let us define some extra algebraic operations for convenience.

Let $g, h \in \Gamma(T^{(1,0)}M \otimes T^{(0,1)}M)$ so that their components are $g^{i\bar{j}}\partial_i \otimes \partial_{\bar{j}}$, $h^{i\bar{j}}\partial_i \otimes \partial_{\bar{j}}$. Then one can define the symmetric bilinear operation [13], [18]:

$$[[,]] : \Gamma(T^{(1,0)}M \otimes T^{(0,1)}M) \otimes \Gamma(T^{(1,0)}M \otimes T^{(0,1)}M) \to \Gamma(T^{(1,0)}M \otimes T^{(0,1)}M) \qquad (25)$$

written in components as follows: $[[g,h]]^{k\bar{l}} \equiv (g^{i\bar{j}}\partial_i\partial_{\bar{j}}h^{k\bar{l}} + h^{i\bar{j}}\partial_i\partial_{\bar{j}}g^{k\bar{l}} - \partial_i g^{k\bar{j}}\partial_{\bar{j}}h^{i\bar{l}} - \partial_i h^{k\bar{j}}\partial_{\bar{j}}g^{i\bar{l}})$ and looks much less complicated in the jet notation (see section 2). Namely, if $\tilde{\xi}, \tilde{\eta} \in J^\infty(\mathcal{O}(T^{(1,0)}M) \otimes J^\infty(\mathcal{O}(T^{(0,1)}M)$, so that $\xi = \sum_I v^I \otimes \bar{v}^I$, $\eta = \sum_J w^J \otimes \bar{w}^J$, where $v^I, w^J \in J^\infty(\mathcal{O}(T^{(1,0)}M)$, $\bar{v}^I, \bar{w}^J \in J^\infty(\mathcal{O}(T^{(0,1)}M)$, then $[[\xi,\eta]] = \sum_{I,J}[v^I, w^J] \otimes [\bar{v}^I, \bar{w}^J]$.

As noted in [13],[18], if bilinear tensor g is such that one can associate a Kähler metrics to it, then the Ricci tensor $R^{i\bar{j}}$ associated with such metric tensor is proportional to $[[g,g]]$, more precisely $R^{i\bar{j}}(g) = \frac{1}{2}[[g,g]]^{i\bar{j}}$.

If the complex manifold M has a volume form Ω, such that in local coordinates $\Omega = e^f dX^1 \cdots \wedge dX^n \wedge dX^{\bar{1}} \wedge \ldots dX^{\bar{n}}$. Let us denote the volume form which determines the differential $\mathcal{Q}$ as Ω', so that $f = -2\Phi'_0$, then Φ'_0 has to be locally a sum of holomorphic and antiholomorphic functions, i.e. it satisfies equation $\partial_i\partial_{\bar{j}}\Phi'_0 = 0$.

We will refer to the vector field $div_\Omega g$ such that $(div_\Omega g)^{\bar{j}} = \partial_i g^{i\bar{j}} + \partial_i f g^{i\bar{j}}$, $(div_\Omega g)^i = \partial_{\bar{j}} g^{i\bar{j}} + \partial_{\bar{j}} f g^{i\bar{j}}$ as the divergence of bivector field g with respect to the volume form Ω.

Now let the Maurer-Cartan element, closed under $\mathbf{b}^-$, namely the element of $\Gamma(T^{(1,0)}(M)\otimes T^{(0,1)}(M))\oplus\mathcal{O}(T^{(0,1)}(M)\oplus\mathcal{O}(T^{(1,0)}(M)\oplus\mathcal{O}_M\oplus\bar{\mathcal{O}}_M$ be defined by its components in the direct sum, i.e. as $(g,\bar{v},v,\phi,\bar{\phi})$.

Then the following Theorem holds, which can be proven by direct calculation.

Theorem 3.1a. *The Maurer-Cartan equation for the differential graded Lie algebra on* $\mathbf{F}^{\cdot}_{sm}|_{\mathbf{b}^-=0}$ *generated by* $\mathcal{Q}$ *and* $\{\cdot,\cdot\}$ *imposes the following system of equations on* $g,\phi,\bar{\phi}$ ($\bar{v},v$ *turn out to be auxiliary variables):*

1). *Vector field* $div_\Omega g$, *where* $\Omega=\Omega' e^{-2\phi+2\bar{\phi}}$ *is determined by* $f\equiv-2\Phi_0=-2(\Phi_0'+\phi-\bar{\phi})$ *and* $\partial_i\partial_{\bar{j}}\Phi_0=0$, *is such that its* $\Gamma(T^{(1,0)}M)$, $\Gamma(T^{(0,1)}M)$ *components are correspondingly holomorphic and antiholomorphic.*

2). *Bivector field* $g\in\Gamma(T'M\otimes T''M)$ *obeys the following equation:* $[[g,g]]+\mathcal{L}_{div_\Omega(g)}g=0$, *where* $\mathcal{L}_{div_\Omega(g)}$ *is a Lie derivative with respect to the corresponding vector fields.*

3). $div_\Omega div_\Omega(g)=0$.

The infinitesimal symmetries of the Maurer-Cartan equation coincide with the holomorphic coordinate transformations of the volume form and tensor $\{g^{i\bar{j}}\}$.

The constraints 1), 2), 3) coincide with the equations studied in [18], where it was shown that they are equivalent to Einstein equations, i.e. the following statement is valid.

Theorem 3.1b. *If tensor* $\{g^{i\bar{j}}\}$ *parametrises Hermitian metric, then the conditions 1), 2), 3) on* g *and* Φ_0 *from Theorem 3.1a are equivalent to Einstein equations*

$$\begin{aligned}
&R^{\mu\nu}=\frac{1}{4}H^{\mu\lambda\rho}H^{\nu}_{\lambda\rho}-2\nabla^\mu\nabla^\nu\Phi, \\
&\nabla_\mu H^{\mu\nu\rho}-2(\nabla_\lambda\Phi)H^{\lambda\nu\rho}=0, \\
&4(\nabla_\mu\Phi)^2-4\nabla_\mu\nabla^\mu\Phi+R+\frac{1}{12}H_{\mu\nu\rho}H^{\mu\nu\rho}=0,
\end{aligned} \tag{26}$$

where $H=dB$ *is a 3-form, so that metric* G, *2-form* B *and the dilaton field* $\Phi\in\mathcal{C}(M)$ *are expressed as follows:* $G_{i\bar{k}}=g_{i\bar{k}}$, $B_{i\bar{k}}=-g_{i\bar{k}}$, $\Phi=\log\sqrt{g}+\Phi_0$, $G_{ik}=G_{\bar{i}\bar{k}}=G_{ik}=G_{\bar{i}\bar{k}}=0$, *where by* g *under the square root we denote the determinant of* $\{g_{i\bar{j}}\}$.

Physically, the appearance of extra $\log(g)$ part in the dilaton corresponds to the fact that passing from the first order action involves integration over $p_i,p_{\bar{j}}$-variables, which leads to extra contribution to the dilatonic term.

3.2. Main conjecture. Following the ideas of section 3.1, we want to repeat the construction in the case of the complex $(\mathcal{F}^{\cdot},\mathcal{Q})$. Namely, we consider its jet version $(\mathcal{F}^{\cdot}_\infty,\mathcal{Q})$ and its complex conjugate $(\bar{\mathcal{F}}^{\cdot}_\infty,\mathcal{Q})$, so that $\mathbf{F}^{\cdot}_\infty=\mathcal{F}^{\cdot}_\infty\hat{\otimes}\bar{\mathcal{F}}^{\cdot}_\infty$.

It is the jet version of the complex $(\mathbf{F}^{\cdot},\mathcal{Q})$, such that e.g. the subspace of degree 1 is as follows: $\mathbf{F}^1\cong\Gamma(E)\oplus\mathcal{C}(M)\oplus\mathcal{C}(M)$. As in section 3.1, the divergence operator which determines $\mathcal{Q}$-operator, is based on the volume form, given in the local coordinates as $e^{-2\Phi_0'(X)}dX^1\ldots dX^n\wedge dX^{\bar{1}}\ldots dX^{\bar{n}}$, so that $\partial_i\partial_{\bar{j}}\Phi_0'=0$.

We can give $\mathbf{F}^{\cdot}$ the structure of the homotopy Leibniz bracket by means of the formula which is the same as in subsection 3.1, however now we have higher homotopies. We also note that $\mathbf{F}^{\cdot}|_{\mathbf{b}^-=0}\equiv\mathbf{F}^{\cdot}_-$ is invariant under $\{\cdot,\cdot\}$. Let us formulate the first part of the main conjecture.

Conjecture 3.1a. *The structure of homotopy Gerstenhaber algebra on* $\mathbf{F}^{\cdot}$ *can be extended to* G_∞*-algebra, so that the subcomplex* $\mathbf{F}^{\cdot}_-$ *is invariant under* L_∞ *operations.*

Let us focus on the subcomplex $(\mathbf{F}^{\cdot}_{-}, Q)$. The space of Maurer-Cartan elements, i.e. the subspace of the elements of degree 2 is: $\mathbf{F}^2_- \cong \Gamma(\mathcal{E}\otimes\bar{\mathcal{E}})\oplus\Gamma(E)\oplus\mathcal{C}(M)\oplus\mathcal{C}(M)$. The elements of this space are defined by means of the components from the direct sum above, i.e. $\Psi = (\mathbb{M}, \eta, \phi, \bar{\phi})$. We will denote the difference $\phi - \bar{\phi} \equiv \Phi''_0$ and $\Phi_0 \equiv \Phi'_0 + \Phi''_0$. Let us formulate the second part of the main conjecture.

Conjecture 3.1b. *Let $\Psi = (\mathbb{M}, \eta, \phi, \bar{\phi})$ be the solution of the generalized Maurer-Cartan (GMC) equation for L_∞-algebra on $\mathbf{F}^{\cdot}_{-}$, so that $\mathbb{M} = \begin{pmatrix} g & \mu \\ \bar{\mu} & b \end{pmatrix}$. Then the η-component is auxiliary and is expressed via $\mathbb{M}$ and $\phi, \bar{\phi}$. If $\{g^{i\bar{j}}\}$ is invertible, then G, B obtained from $\mathbb{M}$ via (13) together with $\Phi = \Phi_0 + \sqrt{g}$, where g is the determinant of $\{g_{i\bar{j}}\}$, satisfy the Einstein equations (26).*

The space of infinitesimal symmetry generators of GMC equation, i.e. $\mathbf{F}^1$ is given by $\mathbf{F}^1_- \cong \Gamma(E)\oplus\mathcal{C}(M)$, so that any element can be written in components as $\Lambda = (\xi, f)$.

The third part of the conjecture concerns the question how ξ, f are related to $\alpha \in \Gamma(E)$ in the transformation formula $\mathbb{M} \to \mathbb{M} - D\alpha + \phi_1(\alpha, \mathbb{M}) + \phi_2(\alpha, \mathbb{M}, \mathbb{M})$ from section 1. First, to justify the statement of Conjecture 3.1b, one can show the following via direct computation.

Proposition 3.2. *Let $\Lambda = (\xi, f) \in \mathbf{F}^1_-$ be the generator of the infinitesimal transformation of the solution of GMC equation. Then after the substitution $\xi = \alpha + \frac{1}{2}\mathbb{M}\cdot\alpha$ (where $\mathbb{M}$ is considered as an element of $End(\Gamma(E))$) the transformation of $\mathbb{M}$-component of the solution coincides with (7) up to the second order in $\mathbb{M}$.*

We note, that the symmetry generated by f-part of $\mathbf{F}^1$ element does not affect metric B-field or dilaton. It is easy to check that on the level of 0th order in $\mathbb{M}$: the symmetry transformation corresponds to the shift of ϕ and $\bar{\phi}$ by f. One can check, similar to Proposition 3.2, that this symmetry remains redundant for the first and second order. We claim that these statements are exact, namely the following Conjecture is true.

Conjecture 3.1c. *Let $\Lambda = (\xi, f) \in \mathbb{F}^1_-$, be the generator of the infinitesimal symmetries of the GMC equation. The corresponding transformation of $\mathbb{M}$-component of the solution of GMC coincide with the diffeomorphism and B-field transformation if $\xi = \alpha + \frac{1}{2}\mathbb{M}\cdot\alpha$. Under conditions of Conjecture 3.1b these transformations reproduce infinitesimal diffeomorphism transformations and shifts of B-field by exact 2-form, which are the symmetries of equations (26).*

Acknowledgments
I am grateful to A.N. Fedorova for careful reading of the manuscript and to the referee for useful remarks.

References
[1] Bressler P 2007 The first Pontryagin class *Compositio Math.* **143** 1127–1163
[2] Callan C G, Friedan D, Martinec E J and Perry M J 1985 *Nucl. Phys.* **B262** 593–609; Callan C G, Klebanov I R and Perry M J 1986 *Nucl. Phys.* **B278** 78–90
[3] Fradkin E S and Tseytlin A A 1985 Quantum String Theory Effective Action *Nucl. Phys.* **B261** 1–27; Non-linear electrodynamics from quantized strings *Phys. Lett.* **B163** (1985) 123–130
[4] Frenkel I B, Garland H and Zuckerman G J 1986 Semi-infinite cohomology and string theory *Proc. Nat. Acad. Sci.* **83** 8442–8446
[5] Galves I, Gorbounov V and Tonks A 2006 Homotopy Gerstenhaber Structures and Vertex Algebras (arXiv:math/0611231)
[6] Galvez-Carrillo I, Tonks A and Vallette B 2009 Homotopy Batalin-Vilkovisky algebras (arXiv:0907.2246)

[7] Gamayun O, Losev A S and A Marshakov 2009 First-order String Theory and the Kodaira-Spencer Equations. I *JHEP* 0909:028 (arXiv:0906.4746)
[8] Gorbounov V, Malikov F and Schechtman V 2004 Gerbes of chiral differential operators. II *Invent. Math.* **155** no. 3 605–680
[9] Huang Y Z and Zhao W 2000 Semi-infinite forms and topological vertex operator algebras *Comm. Contemp. Math.* **2** 191–241 (math.QA/9903014)
[10] Kimura T, Voronov T A and Zuckerman G J 1996 Homotopy Gerstenhaber algebras and topological field theory (arXiv:q-alg/9602009)
[11] Lian B and Zuckerman G 1993 New Perspectives on the BRST-algebraic structure of String Theory *Commun. Math. Phys.* **154** 613–646
[12] Liu Z J, Weinstein A and Xu P 1997 Manin triples for Lie Bialgebroids *J. Diff. Geom.* **45** 547–574
[13] Losev A S, Marshakov A and Zeitlin A M 2006 On the first-order Formalism in String Theory *Phys. Lett.* **B633** 375–381 (hep-th/0510065)
[14] Gorbounov V, Malikov F and Schechtman V 2004 Gerbes of chiral differential operators. II. Vertex algebroids *Invent. Math.* **155** 605–680
[15] Popov V N and Zeitlin A M 1985 Quasiconformal Instantons *Phys. Lett.* B **163** 185–188
[16] Tamarkin D and Tsygan B 2000 Noncommutative differential calculus, homotopy BV algebras and formality conjectures *Methods Funct. Anal. Topology* **6** No. 2 85–97
[17] Voronov A 2000 Homotopy Gerstenhaber algebras *Conference Moshe Flato 1999* (G. Dito and D. Sternheimer, eds.) vol. 2. (Amsterdam: Kluwer) 307–331 (arXiv:math/9908040)
[18] Zeitlin A M 2008 Perturbed Beta-Gamma Systems and Complex Geometry *Nucl. Phys.* **B794** 381 (arXiv:0708.0682)
[19] Zeitlin A M 2007 Formal Maurer-Cartan Structures: from CFT to Classical Field Equations JHEP12(2007)098 (arXiv:0708.0955) BRST, Generalized Maurer-Cartan Equations and CFT *Nucl. Phys.* **B759** (2006) 370–398 (hep-th/0610208)
[20] Zeitlin A M 2009 Beta-gamma systems and the deformations of the BRST operator *J. Phys.* **A42** 355401 (arXiv:0904.2234)
[21] Zeitlin A M 2011 Quasiclassical Lian-Zuckerman Homotopy Algebras, Courant Algebroid and Gauge Theory *Comm. Math. Physics* **303** 331–359 (arXiv:0910.3652)
[22] Zeitlin A M 2014 Beltrami-Courant Differentials and G_∞-algebras (arXiv:1404.3069) IHES: M-14-19
[23] Zeitlin M G 1988 Quasiconformal Instantons *Differential Geometry, Lie Groups, Mechanics* Proc. Scientific Seminars of Steklov Mathematical Institute 191–198 1984 (in Russian); Quasiconformal instantons *J. Sov. Math. Springer* **40** 149–155 (English translation)

AUTHOR INDEX

AUTHOR INDEX

Institute of Physics
Dirac House, Temple Back
Bristol BS1 6BE UK

CURRAN ASSOCIATES INC
proceedings.com